**DATE DUE FOR RETURN**

# Power Quality in Power Systems and Electrical Machines

# Power Quality in Power Systems and Electrical Machines

Ewald F. Fuchs

Mohammad A. S. Masoum

AMSTERDAM • BOSTON • HEIDELBERG • LONDON • NEW YORK
OXFORD • PARIS • SAN DIEGO • SAN FRANCISCO • SINGAPORE
SYDNEY • TOKYO

Academic Press is an imprint of Elsevier

Elsevier Academic Press
30 Corporate Drive, Suite 400, Burlington, MA 01803, USA
525 B Street, Suite 1900, San Diego, California 92101-4495, USA
84 Theobald's Road, London WC1X 8RR, UK

 This book is printed on acid-free paper.

**Library of Congress Cataloging-in-Publication Data**
Application submitted

**British Library Cataloguing in Publication Data**
A catalogue record for this book is available from the British Library

ISBN:  978-0-12-369536-9

For all information on all Elsevier Academic Press publications visit our Web site
at www.books.elsevier.com

Printed in The United States of America
08   09   10   11   12   9   8   7   6   5   4   3   2   1

# Preface

The increased use of power electronic components within the distribution system and the reliance on renewable energy sources which have converters as interface between the source and the power system lead to power quality problems for the operation of machines, transformers, capacitors and power systems. The subject of power quality is very broad by nature. It covers all aspects of power system engineering from transmission and distribution level analyses to end-user problems. Therefore, electric power quality has become the concern of utilities, end users, architects and civil engineers as well as manufacturers. The book is intended for undergraduate or graduate students in electrical and other engineering disciplines as well as for professionals in related fields. It is assumed that the reader has already completed electrical circuit analysis courses covering basic concepts such as Ohm's, Kirchhoff's, Ampere's and Faraday's laws as well as Norton and Thevenin equivalent circuits and Fourier analysis. In addition, knowledge of diodes and transistors and an introductory course on energy conversion (covering energy sources, transformers, simple control circuits, rudimentary power electronics, transformers, single- and three-phase systems as well as various rotating machine concepts such as brushless DC machines, induction and synchronous machines) is desirable.

This book has evolved from the content of courses given by the authors at the University of Colorado at Boulder, the Iran University of Science and Technology at Tehran and the Curtin University of Technology at Perth, Australia. The book is suitable for both electrical and non-electrical engineering students and has been particularly written for students or practicing engineers who want to teach themselves through the inclusion of about 150 application examples with solutions. More than 700 references are given in this book: mostly journal and conference papers as well as national and international standards and guidelines. The International System (SI) of units has been used throughout with some reference to the American/English system of units.

Power quality of power systems affects all connected electrical and electronic equipment, and is a measure of deviations in voltage, current, frequency, temperature, force, and torque of particular supply systems and their components. In recent years there has been considerable increase in nonlinear loads, in particular distributed loads such as computers, TV monitors and lighting. These draw harmonic currents which have detrimental effects including communication interference, loss of reliability, increased operating costs, equipment overheating, machine, transformer and capacitor failures, and inaccurate power metering. This subject is pertinent to engineers involved with power systems, electrical machines, electronic equipment, computers and manufacturing equipment. This book helps readers to understand the causes and effects of power quality problems such as nonsinusoidal wave shapes, voltage outages, harmonic losses, origins of single-time events such as voltage dips, voltage reductions, and outages, along with techniques to mitigate these problems. Analytical as well as measuring techniques are applied to power quality problems as they occur in existing systems based on central power stations and distributed generation mainly relying on renewable energy sources.

It is important for each power engineering student and professional who is active in the area of distribution systems and renewable energy that he/she knows solutions to power quality problems of electrical machines and power systems: this requires detailed knowledge of modeling, simulation and measuring techniques for transformers, machines, capacitors and power systems, in particular fundamental and harmonic power flow, relaying, reliability and redundancy, load shedding and emergency operation, islanding of power system and its voltage and frequency control, passive and active filtering methods, and energy storage combined with renewable energy sources. An intimate knowledge of guidelines and standards as well as industry regulations and practices is indispensable for solving power quality problems in a cost-effective manner. These aspects are addressed in this book which can be used either as a teaching tool or as a reference book.

## Key features:

- Provides theoretical and practical insight into power quality problems of machines and systems.
- 125 practical application (example) problems with solutions.
- Problems at the end of each chapter dealing with practical applications.
- Appendix with application examples, some are implemented in SPICE, Mathematica, and MATLAB.

## ACKNOWLEDGMENTS

The authors wish to express their appreciation to their families in particular to wives Wendy and Roshanak, sons Franz, Amir and Ali, daughters Heidi and Maryam for their help in shaping and proofreading the manuscript. In particular, the encouragement and support of Dipl.-Ing. Dietrich J. Roesler, formerly with the US Department of Energy, Washington DC, who was one of the first professionals coining the concept of power quality more than 25 years ago, is greatly appreciated. Lastly, the work initiated by the late Professor Edward A. Erdelyi is reflected in part of this book.

Ewald F. Fuchs, Professor
University of Colorado
Boulder, CO, USA
Mohammad A.S. Masoum, Associate Professor
Curtin University of Technology
Perth, WA, Australia

March 2008

# Contents

CHAPTER 3
**Modeling and Analysis of Induction Machines**
109

CHAPTER 4

**Modeling and Analysis of Synchronous Machines**
155

CHAPTER 7
**Power System Modeling under Nonsinusoidal
Operating Conditions**
261

CHAPTER 8

**Impact of Poor Power Quality on Reliability, Relaying, and Security**
301

CHAPTER 9

**The Roles of Filters in Power Systems**
359

CHAPTER 10

**Optimal Placement and Sizing of Shunt Capacitor
Banks in the Presence of Harmonics**
397

CHAPTER 11

**Unified Power Quality Conditioner (UPQC)**
443

# Introduction to Power Quality

The subject of power quality is very broad by nature. It covers all aspects of power system engineering, from transmission and distribution level analyses to end-user problems. Therefore, electric power quality has become the concern of utilities, end users, architects, and civil engineers as well as manufacturers. These professionals must work together in developing solutions to power quality problems:

- Electric utility managers and designers must build and operate systems that take into account the interaction between customer facilities and power system. Electric utilities must understand the sensitivity of the end-use equipment to the quality of voltage.
- Customers must learn to respect the rights of their neighbors and control the quality of their nonlinear loads. Studies show that the best and the most efficient solution to power quality problems is to control them at their source. Customers can perform this by careful selection and control of their nonlinear loads and by taking appropriate actions to control and mitigate single-time disturbances and harmonics before connecting their loads to the power system.
- Architects and civil engineers must design buildings to minimize the susceptibility and vulnerability of electrical components to power quality problems.
- Manufacturers and equipment engineers must design devices that are compatible with the power system. This might mean a lower level of harmonic generation or less sensitivity to voltage distortions.
- Engineers must be able to devise ride-through capabilities of distributed generators (e.g., wind and solar generating plants).

This chapter introduces the subject of electric power quality. After a brief definition of power quality and its causes, detailed classification of the subject is presented. The formulations and measures used for power quality are explained and the impacts of poor power quality on power system and end-use devices such as appliances are mentioned. A section is presented addressing the most important IEEE [1] and IEC [2] standards referring to power quality. The remainder of this chapter introduces issues that will be covered in the following chapters, including modeling and mitigation techniques for power quality phenomena in electric machines and power systems. This chapter contains nine application examples and ends with a summary.

## 1.1 DEFINITION OF POWER QUALITY

Electric power quality has become an important part of power systems and electric machines. The subject has attracted the attention of many universities and industries, and a number of books have been published in this exciting and relatively new field [3–12].

Despite important papers, articles, and books published in the area of electric power quality, its definition has not been universally agreed upon. However, nearly everybody accepts that it is a very important aspect of power systems and electric machinery with direct impacts on efficiency, security, and reliability. Various sources use the term "power quality" with different meaning. It is used synonymously with "supply reliability," "service quality," "voltage quality," "current quality," "quality of supply," and "quality of consumption."

Judging by the different definitions, power quality is generally meant to express the quality of voltage and/or the quality of current and can be defined as: the measure, analysis, and improvement of the bus voltage to maintain a sinusoidal waveform at rated voltage and frequency. This definition includes all momentary and steady-state phenomena.

## 1.2 CAUSES OF DISTURBANCES IN POWER SYSTEMS

Although a significant literature on power quality is now available, most engineers, facility managers, and consumers remain unclear as to what constitutes a power quality problem. Furthermore, due to the power system impedance, any current (or voltage) harmonic will result in the generation and propaga-

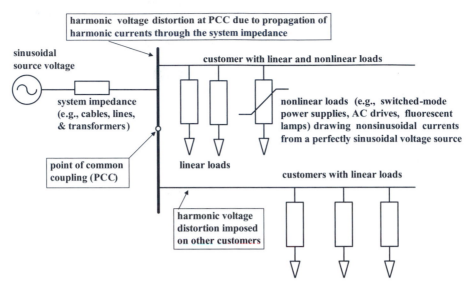

harmonic voltage distortion at PCC due to propagation of
harmonic currents through the system impedance

sinusoidal
source voltage

customer with linear and nonlinear loads

system impedance
(e.g., cables, lines,
& transformers)

nonlinear loads (e.g., switched-mode
power supplies, AC drives, fluorescent
lamps) drawing nonsinusoidal currents
from a perfectly sinusoidal voltage source

point of common
coupling (PCC)

linear loads

customers with linear loads

harmonic voltage
distortion imposed
on other customers

**FIGURE 1.1** Propagation of harmonics (generated by a nonlinear load) in power systems.

tion of voltage (or current) harmonics and affects the entire power system. Figure 1.1 illustrates the impact of current harmonics generated by a nonlinear load on a typical power system with linear loads.

What are the origins of the power quality problem? Some references [9] divide the distortion sources into three categories: small and predictable (e.g., residential consumers generating harmonics), large and random (e.g., arc furnaces producing voltage fluctuations and flicker), and large and predictable (e.g., static converters of smelters and high-voltage DC transmission causing characteristic and uncharacteristic harmonics as well as harmonic instability). However, the likely answers to the question are these: unpredictable events, the electric utility, the customer, and the manufacturer.

**Unpredictable Events.** Both electric utilities and end users agree that more than 60% of power quality problems are generated by natural and unpredictable events [6]. Some of these include faults, lightning surge propagation, resonance, ferroresonance, and geomagnetically induced currents (GICs) due to solar flares [13]. These events are considered to be utility related problems.

**The Electric Utility.** There are three main sources of poor power quality related to utilities:

• *The point of supply generation.* Although synchronous machines generate nearly perfect sinusoidal voltages (harmonic content less than 3%), there

are power quality problems originating at generating plants which are mainly due to maintenance activity, planning, capacity and expansion constraints, scheduling, events leading to forced outages, and load transferring from one substation to another.

• *The transmission system.* Relatively few power quality problems originate in the transmission system. Typical power quality problems originating in the transmission system are galloping (under high-wind conditions resulting in supply interruptions and/or random voltage variations), lightning (resulting in a spike or transient overvoltage), insulator flashover, voltage dips (due to faults), interruptions (due to planned outages by utility), transient overvoltages (generated by capacitor and/or inductor switching, and lightning), transformer energizing (resulting in inrush currents that are rich in harmonic components), improper operation of voltage regulation devices (which can lead to long-duration voltage variations), slow voltage variations (due to a long-term variation of the load caused by the continuous switching of devices and load), flexible AC transmission system (FACTS) devices [14] and high-voltage DC (HVDC) systems [15], corona [16], power line carrier signals [17], broadband power line (BPL) communications [18], and electromagnetic fields (EMFs) [19].

• *The distribution system.* Typical power quality problems originating in the distribution system are voltage dips, spikes, and interruptions, transient overvoltages, transformer energizing, improper operation of voltage regulation devices, slow voltage variations, power line carrier signals, BPL, and EMFs.

**The Customer.** Customer loads generate a considerable portion of power quality problems in today's power systems. Some end-user related problems are harmonics (generated by nonlinear loads such as power electronic devices and equipment, renewable energy sources, FACTS devices, adjustable-speed drives, uninterruptible power supplies (UPS), fax machines, laser printers, computers, and fluorescent lights), poor power factor (due to highly inductive loads such as induction motors and air-conditioning units), flicker (generated by arc furnaces [20]), transients (mostly generated inside a facility due to device switching, electrostatic discharge, and arcing), improper grounding (causing most reported customer problems), frequency variations (when secondary and backup power sources, such as diesel engine and turbine generators, are used), misapplication of technology, wiring regulations, and other relevant standards.

**Manufacturing Regulations.** There are two main sources of poor power quality related to manufacturing regulations:

- *Standards.* The lack of standards for testing, certification, sale, purchase, installation, and use of electronic equipment and appliances is a major cause of power quality problems.
- *Equipment sensitivity.* The proliferation of "sensitive" electronic equipment and appliances is one of the main reasons for the increase of power quality problems. The design characteristics of these devices, including computer-based equipment, have increased the incompatibility of a wide variety of these devices with the electrical environment [21].

Power quality therefore must necessarily be tackled from three fronts, namely:

- The utility must design, maintain, and operate the power system while minimizing power quality problems;
- The end user must employ proper wiring, system grounding practices, and state-of-the-art electronic devices; and
- The manufacturer must design electronic devices that keep electrical environmental disturbances to a minimum and that are immune to anomalies of the power supply line.

## 1.3 CLASSIFICATION OF POWER QUALITY ISSUES

To solve power quality problems it is necessary to understand and classify this relatively complicated subject. This section is based on the power quality classification and information from references [6] and [9].

There are different classifications for power quality issues, each using a specific property to categorize the problem. Some of them classify the events as "steady-state" and "non-steady-state" phenomena. In some regulations (e.g., ANSI C84.1 [22]) the most important factor is the duration of the event. Other guidelines (e.g., IEEE-519) use the wave shape (duration and magnitude) of each event to classify power quality problems. Other standards (e.g., IEC) use the frequency range of the event for the classification.

For example, IEC 61000-2-5 uses the frequency range and divides the problems into three main categories: low frequency (<9 kHz), high frequency (>9 kHz), and electrostatic discharge phenomena. In addition, each frequency range is divided into "radiated" and "conducted" disturbances. Table 1.1 shows

**TABLE 1.1** Main Phenomena Causing Electromagnetic and Power Quality Disturbances [6, 9]

| Conducted low-frequency phenomena |
| --- |
| Harmonics, interharmonics |
| Signaling voltage |
| Voltage fluctuations |
| Voltage dips |
| Voltage imbalance |
| Power frequency variations |
| Induced low-frequency voltages |
| DC components in AC networks |
| **Radiated low-frequency phenomena** |
| Magnetic fields |
| Electric fields |
| **Conducted high-frequency phenomena** |
| Induced continuous wave (CW) voltages or currents |
| Unidirectional transients |
| Oscillatory transients |
| **Radiated high-frequency phenomena** |
| Magnetic fields |
| Electric fields |
| Electromagnetic field |
| Steady-state waves |
| Transients |
| **Electrostatic discharge phenomena (ESD)** |
| **Nuclear electromagnetic pulse (NEMP)** |

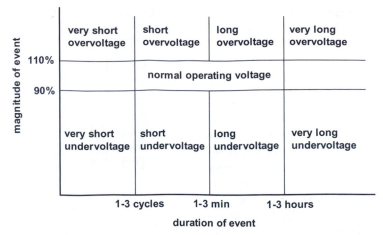

**FIGURE 1.2** Magnitude–duration plot for classification of power quality events [11].

the principal phenomena causing electromagnetic disturbances according to IEC classifications [9]. All these phenomena are considered to be power quality issues; however, the two conducted categories are more frequently addressed by the industry.

The magnitude and duration of events can be used to classify power quality events, as shown in Fig. 1.2. In the magnitude–duration plot, there are nine different parts [11]. Various standards give different names to events in these parts. The voltage magnitude is split into three regions:

- interruption: voltage magnitude is zero,
- undervoltage: voltage magnitude is below its nominal value, and
- overvoltage: voltage magnitude is above its nominal value.

The duration of these events is split into four regions: very short, short, long, and very long. The borders in this plot are somewhat arbitrary and the user can set them according to the standard that is used.

IEEE standards use several additional terms (as compared with IEC terminology) to classify power quality events. Table 1.2 provides information about categories and characteristics of electromagnetic phenomena defined by IEEE-1159 [23]. These categories are briefly introduced in the remaining parts of this section.

### 1.3.1 Transients

Power system transients are undesirable, fast- and short-duration events that produce distortions. Their characteristics and waveforms depend on the mechanism of generation and the network parameters

(e.g., resistance, inductance, and capacitance) at the point of interest. "Surge" is often considered synonymous with transient.

Transients can be classified with their many characteristic components such as amplitude, duration, rise time, frequency of ringing polarity, energy delivery capability, amplitude spectral density, and frequency of occurrence. Transients are usually classified into two categories: impulsive and oscillatory (Table 1.2).

An impulsive transient is a sudden frequency change in the steady-state condition of voltage, current, or both that is unidirectional in polarity (Fig. 1.3). The most common cause of impulsive transients is a lightning current surge. Impulsive transients can excite the natural frequency of the system.

An oscillatory transient is a sudden frequency change in the steady-state condition of voltage, current, or both that includes both positive and negative polarity values. Oscillatory transients occur for different reasons in power systems such as appliance switching, capacitor bank switching (Fig. 1.4), fast-acting overcurrent protective devices, and ferroresonance (Fig. 1.5).

### 1.3.2 Short-Duration Voltage Variations

This category encompasses the IEC category of "voltage dips" and "short interruptions." According to the IEEE-1159 classification, there are three different types of short-duration events (Table 1.2): instantaneous, momentary, and temporary. Each category is divided into interruption, sag, and swell. Principal cases of short-duration voltage variations are fault conditions, large load energization, and loose connections.

**TABLE 1.2** Categories and Characteristics of Electromagnetic Phenomena in Power Systems as Defined by IEEE-1159 [6, 9]

| Categories | | Typical spectral content | Typical duration | Typical voltage magnitude |
|---|---|---|---|---|
| 1. Transient | 1.1. Impulsive | | | |
| | • nanosecond | 5 ns rise | <50 ns | |
| | • microsecond | 1 µs rise | 50 ns–1 ms | |
| | • millisecond | 0.1 ms rise | >1 ms | |
| | 1.2. Oscillatory | | | |
| | • low frequency | <5 kHz | 0.3–50 ms | 0–4 pu |
| | • medium frequency | 5–500 kHz | 20 µs | 0–8 pu |
| | • high frequency | 0.5–5 MHz | 5 µs | 0–4 pu |
| 2. Short-duration variation | 2.1. Instantaneous | | | |
| | • interruption | | 0.5–30 cycles | <0.1 pu |
| | • sag | | 0.5–30 cycles | 0.1–0.9 pu |
| | • swell | | 0.5–30 cycles | 1.1–1.8 pu |
| | 2.2. Momentary | | | |
| | • interruption | | 0.5 cycle–3 s | <0.1 pu |
| | • sag | | 30 cycles–3 s | 0.1–0.9 pu |
| | • swell | | 30 cycles–3 s | 1.1–1.4 pu |
| | 2.3. Temporary | | | |
| | • interruption | | 3 s–1 min | <0.1 pu |
| | • sag | | 3 s–1 min | 0.1–0.9 pu |
| | • swell | | 3 s–1 min | 1.1–1.2 pu |
| 3. Long-duration variation | 3.1. Sustained interruption | | >1 min | 0.0 pu |
| | 3.2. Undervoltage | | >1 min | 0.8–0.9 pu |
| | 3.3. Overvoltage | | >1 min | 1.1–1.2 pu |
| 4. Voltage imbalance | | | steady state | 0.5–2% |
| 5. Waveform distortion | 5.1. DC offset | | steady state | 0–0.1% |
| | 5.2. Harmonics | 0–100th | steady state | 0–20% |
| | 5.3. Interharmonics | 0–6 kHz | steady state | 0–2% |
| | 5.4. Notching | | steady state | |
| | 5.5. Noise | Broadband | steady state | 0–1% |
| 6. Voltage fluctuation | | <25 Hz | intermittent | 0.1–7% |
| 7. Power frequency variations | | | <10 s | |

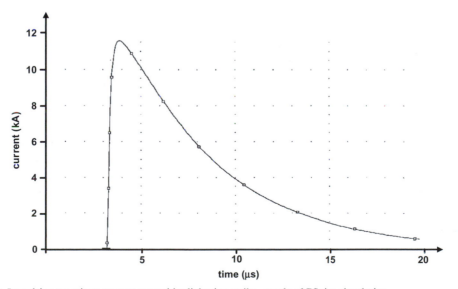

**FIGURE 1.3** Impulsive transient current caused by lightning strike, result of PSpice simulation.

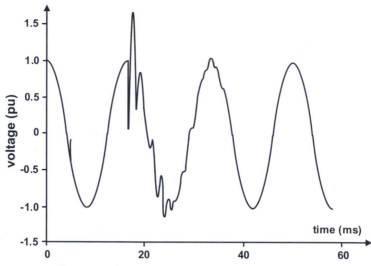

**FIGURE 1.4** Low-frequency oscillatory transient caused by capacitor bank energization.

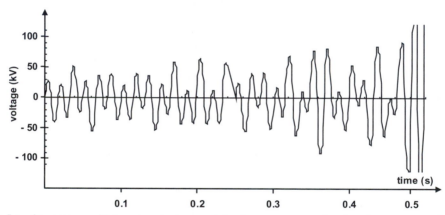

**FIGURE 1.5** Low-frequency oscillatory transient caused by ferroresonance of a transformer at no load, result of Mathematica simulation.

**Interruption.** Interruption occurs when the supply voltage (or load current) decreases to less than 0.1 pu for less than 1 minute, as shown by Fig. 1.6. Some causes of interruption are equipment failures, control malfunction, and blown fuse or breaker opening.

The difference between long (or sustained) interruption and interruption is that in the former the supply is restored manually, but during the latter the supply is restored automatically. Interruption is usually measured by its duration. For example, according to the European standard EN-50160 [24]:

- A short interruption is up to 3 minutes; and
- A long interruption is longer than 3 minutes.

However, based on the standard IEEE-1250 [25]:

- An instantaneous interruption is between 0.5 and 30 cycles;

- A momentary interruption is between 30 cycles and 2 seconds;
- A temporary interruption is between 2 seconds and 2 minutes; and
- A sustained interruption is longer than 2 minutes.

**Sags (Dips).** Sags are short-duration reductions in the rms voltage between 0.1 and 0.9 pu, as shown by Fig. 1.7. There is no clear definition for the duration of sag, but it is usually between 0.5 cycles and 1 minute. Voltage sags are usually caused by

- energization of heavy loads (e.g., arc furnace),
- starting of large induction motors,
- single line-to-ground faults, and
- load transferring from one power source to another.

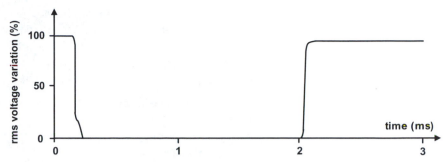

**FIGURE 1.6** Momentary interruptions due to a fault.

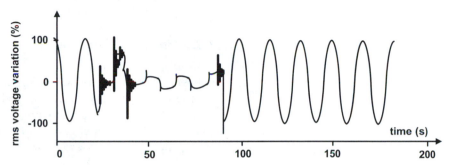

**FIGURE 1.7** Voltage sag caused by a single line-to-ground (SLG) fault.

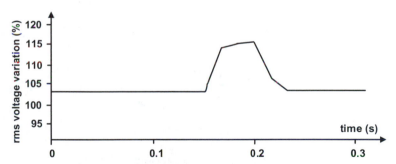

**FIGURE 1.8** Instantaneous voltage swell caused by a single line-to-ground fault.

Each of these cases may cause a sag with a special (magnitude and duration) characteristic. For example, if a device is sensitive to voltage sag of 25%, it will be affected by induction motor starting [11]. Sags are main reasons for malfunctions of electrical low-voltage devices. Uninterruptible power supply (UPS) or power conditioners are mostly used to prevent voltage sags.

**Swells.** The increase of voltage magnitude between 1.1 and 1.8 pu is called swell, as shown by Fig. 1.8. The most accepted duration of a swell is from 0.5 cycles to 1 minute [7]. Swells are not as common as sags and their main causes are

- switching off of a large load,
- energizing a capacitor bank, or
- voltage increase of the unfaulted phases during a single line-to-ground fault [10].

In some textbooks the term "momentary overvoltage" is used as a synonym for the term swell. As in the case of sags, UPS or power conditioners are typical solutions to limit the effect of swell [10].

### 1.3.3 Long-Duration Voltage Variations

According to standards (e.g., IEEE-1159, ANSI-C84.1), the deviation of the rms value of voltage from the nominal value for longer than 1 minute is called long-duration voltage variation. The main causes of long-duration voltage variations are load variations and system switching operations. IEEE-1159 divides these events into three categories (Table 1.2): sustained interruption, undervoltage, and overvoltage.

**Sustained Interruption.** Sustained (or long) interruption is the most severe and the oldest power quality event at which voltage drops to zero and does not return automatically. According to the IEC definition, the duration of sustained interruption is more than 3 minutes; but based on the IEEE definition the duration is more than 1 minute. The number and duration of long interruptions are very important characteristics in measuring the ability of a power system to deliver service to customers. The most important causes of sustained interruptions are

- fault occurrence in a part of power systems with no redundancy or with the redundant part out of operation,
- an incorrect intervention of a protective relay leading to a component outage, or
- scheduled (or planned) interruption in a low-voltage network with no redundancy.

**Undervoltage.** The undervoltage condition occurs when the rms voltage decreases to 0.8–0.9 pu for more than 1 minute.

**Overvoltage.** Overvoltage is defined as an increase in the rms voltage to 1.1–1.2 pu for more than 1 minute. There are three types of overvoltages:

- overvoltages generated by an insulation fault, ferroresonance, faults with the alternator regulator, tap changer transformer, or overcompensation;
- lightning overvoltages; and
- switching overvoltages produced by rapid modifications in the network structure such as opening of protective devices or the switching on of capacitive circuits.

### 1.3.4 Voltage Imbalance

When voltages of a three-phase system are not identical in magnitude and/or the phase differences between them are not exactly 120 degrees, voltage imbalance occurs [10]. There are two ways to calculate the degree of imbalance:

- divide the maximum deviation from the average of three-phase voltages by the average of three-phase voltages, or

- compute the ratio of the negative- (or zero-) sequence component to the positive-sequence component [7].

The main causes of voltage imbalance in power systems are

- unbalanced single-phase loading in a three-phase system,
- overhead transmission lines that are not transposed,
- blown fuses in one phase of a three-phase capacitor bank, and
- severe voltage imbalance (e.g., >5%), which can result from single phasing conditions.

### 1.3.5 Waveform Distortion

A steady-state deviation from a sine wave of power frequency is called waveform distortion [7]. There are five primary types of waveform distortions: DC offset, harmonics, interharmonics, notching, and electric noise. A Fourier series is usually used to analyze the nonsinusoidal waveform.

**DC Offset.** The presence of a DC current and/or voltage component in an AC system is called DC offset [7]. Main causes of DC offset in power systems are

- employment of rectifiers and other electronic switching devices, and
- geomagnetic disturbances [6, 7, 13] causing GICs.

The main detrimental effects of DC offset in alternating networks are

- half-cycle saturation of transformer core [26–28],
- generation of even harmonics [26] in addition to odd harmonics [29, 30],
- additional heating in appliances leading to a decrease of the lifetime of transformers [31–36], rotating machines, and electromagnetic devices, and
- electrolytic erosion of grounding electrodes and other connectors.

Figure 1.9a shows strong half-cycle saturation in a transformer due to DC magnetization and the influence of the tank, and Fig. 1.9b exhibits less half-cycle saturation due to DC magnetization and the absence of any tank. One concludes that to suppress DC currents due to rectifiers and geomagnetically induced currents, three-limb transformers with a relatively large air gap between core and tank should be used.

**Harmonics.** Harmonics are sinusoidal voltages or currents with frequencies that are integer multiples of the power system (fundamental) frequency (usually, f = 50 or 60 Hz). For example, the frequency

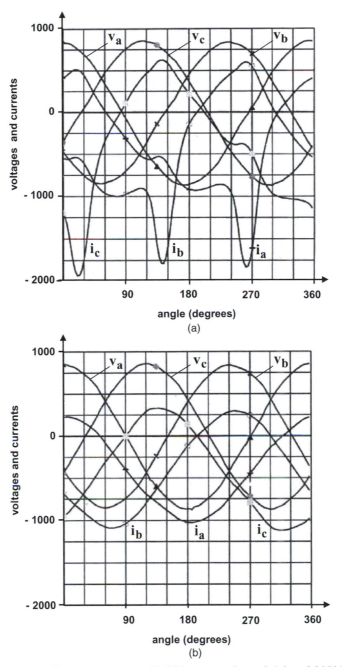

**FIGURE 1.9** Measured voltages and currents at balanced DC bias current $I_{DC} = -2$ A for a 2.3 kVA three-limb transformer (a) at full load with tank (note the strong half-cycle saturation) and (b) at full load without tank (note the reduced half-cycle saturation) [27]. Dividing the ordinate values by 2.36 and 203 the voltages in volts and the currents in amperes are obtained, respectively.

of the hth harmonic is (hf). Periodic nonsinusoidal waveforms can be subjected to Fourier series and can be decomposed into the sum of fundamental component and harmonics. Main sources of harmonics in power systems are

- industrial nonlinear loads (Fig. 1.10) such as power electronic equipment, for example, drives (Fig. 1.10a), rectifiers (Fig. 1.10b,c), inverters, or loads generating electric arcs, for example, arc furnaces, welding machines, and lighting, and

- residential loads with switch-mode power supplies such as television sets, computers (Fig. 1.11), and fluorescent and energy-saving lamps.

Some detrimental effects of harmonics are

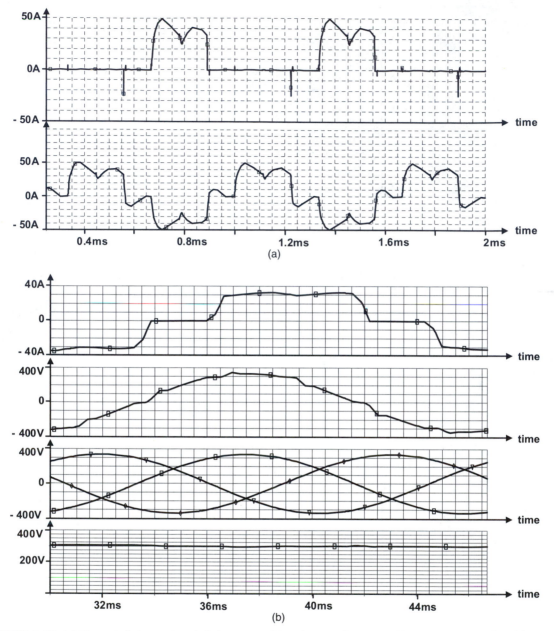

**FIGURE 1.10** (a) Computed electronic switch (upper graph) and motor (lower graph) currents of an adjustable-speed brushless DC motor drive for a phase angle of $\Theta = 0°$ [29]. (b) Voltage notching caused by a three-phase rectifier for a firing angle of $\alpha = 50°$, result of PSpice simulation. Top: phase current; second from top: line-to-line voltage of rectifier; third from top: line-to-line voltages of infinite bus; bottom: DC output voltage of rectifier.

- maloperation of control devices,
- additional losses in capacitors, transformers, and rotating machines,
- additional noise from motors and other apparatus,
- telephone interference, and
- causing parallel and series resonance frequencies (due to the power factor correction capacitor and cable capacitance), resulting in voltage amplification even at a remote location from the distorting load.

Recommended solutions to reduce and control harmonics are applications of high-pulse rectification, passive, active, and hybrid filters, and custom power devices such as active-power line conditioners (APLCs) and unified power quality conditioners (UPQCs).

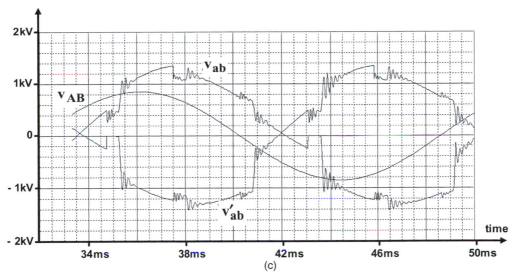

**FIGURE 1.10 (continued)** (c) Voltage notching caused by a three-phase rectifier with interphase reactor for a firing angle of $\alpha = 0°$, result of PSpice simulation. Waveshapes with notches: line-to-line voltages of rectifier, $V_{ab}$ and $V'_{ab}$ being the line-to-line voltages of the two voltage systems; sinusoidal waveshape: line-to-line voltage of infinite bus.

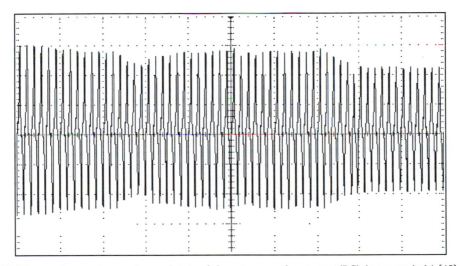

**FIGURE 1.11** Measured current wave shape of state-of-the-art personal computer (PC) (many periods) [45].

**Interharmonics.** Interharmonics are discussed in Section 1.4.1. Their frequencies are not integer multiples of the fundamental frequency.

**Notching.** A periodic voltage disturbance caused by line-commutated thyristor circuits is called notching. The notching appears in the line voltage waveform during normal operation of power electronic devices when the current commutates from one phase to another. During this notching period, there exists a momentary short-circuit between the two commutating phases, reducing the line voltage; the voltage reduction is limited only by the system impedance.

Notching is repetitive and can be characterized by its frequency spectrum (Figs. 1.10b,c). The frequency of this spectrum is quite high. Usually it is not possible to measure it with equipment normally used for harmonic analysis. Notches can impose extra stress on the insulation of transformers, generators, and sensitive measuring equipment.

Notching can be characterized by the following properties:

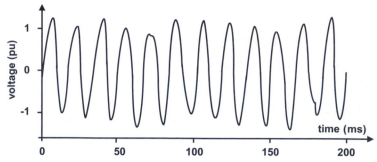

**FIGURE 1.12** Voltage flicker caused by arc furnace operation.

- *Notch depth:* average depth of the line voltage notch from the sinusoidal waveform at the fundamental frequency;
- *Notch width:* the duration of the commutation process;
- *Notch area:* the product of notch depth and width; and
- *Notch position:* where the notch occurs on the sinusoidal waveform.

Some standards (e.g., IEEE-519) set limits for notch depth and duration (with respect to the system impedance and load current) in terms of the notch depth, the total harmonic distortion THD$_v$ of supply voltage, and the notch area for different supply systems.

**Electric Noise.** Electric noise is defined as unwanted electrical signals with broadband spectral content lower than 200 kHz [37] superimposed on the power system voltage or current in phase conductors, or found on neutral conductors or signal lines. Electric noise may result from faulty connections in transmission or distribution systems, arc furnaces, electrical furnaces, power electronic devices, control circuits, welding equipment, loads with solid-state rectifiers, improper grounding, turning off capacitor banks, adjustable-speed drives, corona, and broadband power line (BPL) communication circuits. The problem can be mitigated by using filters, line conditioners, and dedicated lines or transformers. Electric noise impacts electronic devices such as microcomputers and programmable controllers.

### 1.3.6 Voltage Fluctuation and Flicker

Voltage fluctuations are systemic variations of the voltage envelope or random voltage changes, the magnitude of which does not normally exceed specified voltage ranges (e.g., 0.9 to 1.1 pu as defined by ANSI C84.1-1982) [38]. Voltage fluctuations are divided into two categories:

- step-voltage changes, regular or irregular in time, and
- cyclic or random voltage changes produced by variations in the load impedances.

Voltage fluctuations degrade the performance of the equipment and cause instability of the internal voltages and currents of electronic equipment. However, voltage fluctuations less than 10% do not affect electronic equipment. The main causes of voltage fluctuation are pulsed-power output, resistance welders, start-up of drives, arc furnaces, drives with rapidly changing loads, and rolling mills.

**Flicker.** Flicker (Fig. 1.12) has been described as "continuous and rapid variations in the load current magnitude which causes voltage variations." The term flicker is derived from the impact of the voltage fluctuation on lamps such that they are perceived to flicker by the human eye. This may be caused by an arc furnace, one of the most common causes of the voltage fluctuations in utility transmission and distribution systems.

### 1.3.7 Power–Frequency Variations

The deviation of the power system fundamental frequency from its specified nominal value (e.g., 50 or 60 Hz) is defined as power frequency variation [39]. If the balance between generation and demand (load) is not maintained, the frequency of the power system will deviate because of changes in the rotational speed of electromechanical generators. The amount of deviation and its duration of the frequency depend on the load characteristics and response of the generation control system to load changes. Faults of the power transmission system can also cause frequency variations outside of the accepted range for normal steady-state operation of the power system.

## 1.4 FORMULATIONS AND MEASURES USED FOR POWER QUALITY

This section briefly introduces some of the most commonly used formulations and measures of electric power quality as used in this book and as defined in standard documents. Main sources for power quality terminologies are IEEE Std 100 [40], IEC Std 61000-1-1, and CENELEC Std EN 50160 [41]. Appendix C of reference [11] presents a fine survey of power quality definitions.

### 1.4.1 Harmonics

Nonsinusoidal current and voltage waveforms (Figs. 1.13 to 1.20) occur in today's power systems due to equipment with nonlinear characteristics such as transformers, rotating electric machines, FACTS devices, power electronics components (e.g., rectifiers, triacs, thyristors, and diodes with capacitor smoothing, which are used extensively in PCs, audio, and video equipment), switch-mode power supplies, compact fluorescent lamps, induction furnaces, adjustable AC and DC drives, arc furnaces, welding tools, renewable energy sources, and HVDC networks. The main effects of harmonics are malopera-tion of control devices, telephone interferences, additional line losses (at fundamental and harmonic frequencies), and decreased lifetime and increased losses in utility equipment (e.g., transformers, rotating machines, and capacitor banks) and customer devices.

The periodic nonsinusoidal waveforms can be formulated in terms of Fourier series. Each term in the Fourier series is called the harmonic component of the distorted waveform. The frequency of harmonics are integer multiples of the fundamental frequency. Therefore, nonsinusoidal voltage and current waveforms can be defined as

$$v(t) = V_{DC} + \sum_{h=1}^{n} V_{rms}^{(h)} \cos(h\omega_o t + \alpha_h)$$
$$= V_{DC} + v^{(1)}(t) + v^{(2)}(t) + v^{(3)}(t) + v^{(4)}(t) + \dots\dots ,$$

(1-1a)

$$i(t) = I_{DC} + \sum_{h=1}^{n} I_{rms}^{(h)} \cos(h\omega_o t + \beta_h)$$
$$= I_{DC} + i^{(1)}(t) + i^{(2)}(t) + i^{(3)}(t) + i^{(4)}(t) + \dots\dots ,$$

(1-1b)

where $\omega_o$ is the fundamental frequency, $h$ is the harmonic order, and $V^{(h)}$, $I^{(h)}$, $\alpha_h$, and $\beta_h$ are the rms amplitude values and phase shifts of voltage and current for the hth harmonic.

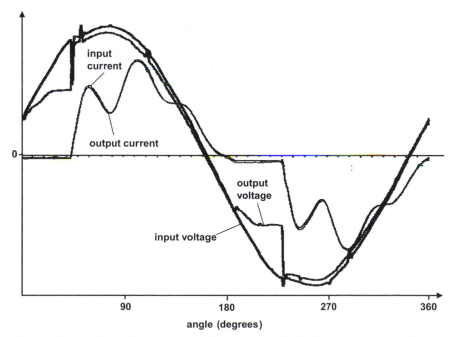

**FIGURE 1.13** Measured wave shapes of single-phase induction motor fed by thyristor/triac controller at rated operation [42].

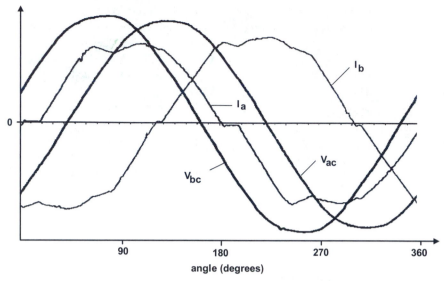

**FIGURE 1.14** Measured wave shapes of three-phase induction motor fed by thyristor/triac controller at rated operation [42].

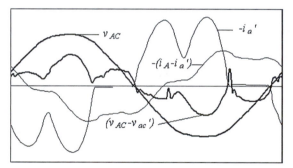

**FIGURE 1.15** Measured wave shapes of 4.5 kVA three-phase transformer feeding full-wave rectifier [43].

Even and odd harmonics of a nonsinusoidal function correspond to even (e.g., 2, 4, 6, 8, . . .) and odd (e.g., 3, 5, 7, 9, . . .) components of its Fourier series. Harmonics of order 1 and 0 are assigned to the fundamental frequency and the DC component of the waveform, respectively. When both positive and negative half-cycles of the waveform have identical shapes, the wave shape has half-wave symmetry and the Fourier series contains only odd harmonics. This is the usual case with voltages and currents of power systems. The presence of even harmonics is often a clue that there is something wrong (e.g., imperfect gating of electronic switches [42]), either with the load equipment or with the transducer used to make the measurement. There are notable exceptions to this such as half-wave rectifiers, arc furnaces (with random arcs), and the presence of GICs in power systems [27].

**Triplen Harmonics.** Triplen harmonics (Fig. 1.21) are the odd multiples of the third harmonic ($h = 3$, 9, 15, 21, . . .). These harmonic orders become an important issue for grounded-wye systems with current flowing in the neutral line of a wye configuration. Two typical problems are overloading of the neutral conductor and telephone interference.

For a system of perfectly balanced three-phase nonsinusoidal loads, fundamental current components in the neutral are zero. The third harmonic neutral currents are three times the third-harmonic phase currents because they coincide in phase or time.

Transformer winding connections have a significant impact on the flow of triplen harmonic currents caused by three-phase nonlinear loads. For the grounded wye-delta transformer, the triplen harmonic currents enter the wye side and since they are in phase, they add in the neutral. The delta winding provides ampere-turn balance so that they can flow in the delta, but they remain trapped in the delta and are absent in the line currents of the delta side of the transformer. This type of transformer connection is the most commonly employed in utility distribution substations with the delta winding connected to the transmission feeder. Using grounded-wye windings on both sides of the transformer allows balanced triplen harmonics to flow unimpeded from the low-voltage system to the high-voltage system. They will be present in equal proportion on both sides of a transformer.

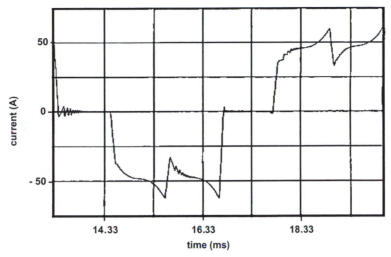

**FIGURE 1.16**  Calculated current of brushless DC motor in full-on mode at rated operation [29].

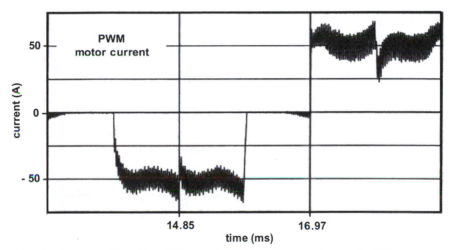

**FIGURE 1.17**  Calculated current of brushless DC motor in PWM mode at rated operation [29].

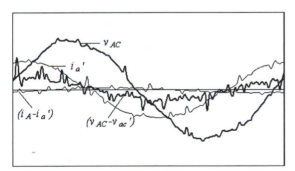

**FIGURE 1.18**  Measured wave shapes of 15 kVA three-phase transformer feeding resonant rectifier [43].

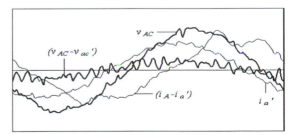

**FIGURE 1.19**  Measured wave shapes of 15 kVA three-phase transformer fed by PWM inverter [43].

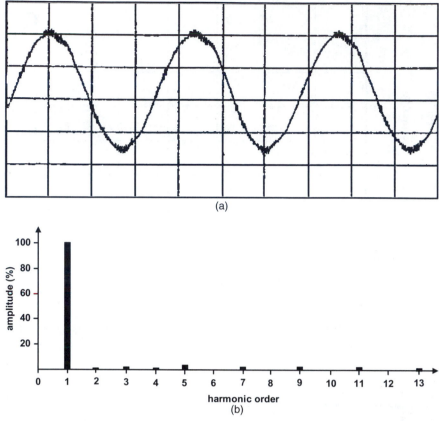

(a)

(b)

FIGURE 1.20 (a) Measured current and (b) measured current spectrum of 20 kW/25 kVA wind-power plant supplying power via inverter into the 240 V three-phase distribution system at rated load [44].

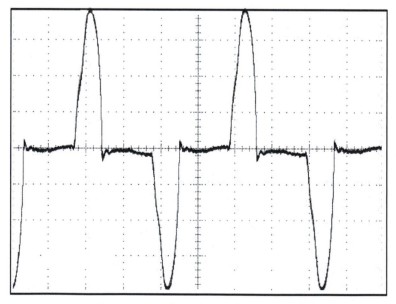

FIGURE 1.21 Input current to personal computer with dominant third harmonic [45].

**Subharmonics.** Subharmonics have frequencies below the fundamental frequency. There are rarely subharmonics in power systems. However, due to the fast control of electronic power supplies of computers, inter- and subharmonics are generated in the input current (Fig. 1.11) [45]. Resonance between the harmonic currents or voltages with the power system (series) capacitance and inductance may cause subharmonics, called subsynchronous resonance [46]. They may be generated when a system is highly inductive (such as an arc furnace during start-up) or when the power system contains large capacitor banks for power factor correction or filtering.

**Interharmonics.** The frequency of interharmonics are not integer multiples of the fundamental frequency. Interharmonics appear as discrete frequencies or as a band spectrum. Main sources of interharmonic waveforms are static frequency converters, cycloconverters, induction motors, arcing devices, and computers. Interharmonics cause flicker, low-frequency torques [32], additional temperature rise in induction machines [33, 34], and malfunctioning of protective (under-frequency) relays [35]. Interharmonics have been included in a number of guidelines such as the IEC 61000-4-7 [36] and the IEEE-519. However, many important related issues, such as the range of frequencies, should be addressed in revised guidelines.

**Characteristic and Uncharacteristic Harmonics.** The harmonics of orders $12k + 1$ (positive sequence) and $12k - 1$ (negative sequence) are called characteristic and uncharacteristic harmonics, respectively. The amplitudes of these harmonics are inversely proportional to the harmonic order. Filters are used to reduce characteristic harmonics of large power converters. When the AC system is weak [47] and the operation is not perfectly symmetrical, uncharacteristic harmonics appear. It is not economical to reduce uncharacteristic harmonics with filters; therefore, even a small injection of these harmonic currents can, via parallel resonant conditions, produce very large voltage distortion levels.

**Positive-, Negative-, and Zero-Sequence Harmonics** [48]. Assuming a positive-phase (abc) sequence balanced three-phase power system, the expressions for the fundamental currents are

$$i_a(t) = I_a^{(1)} \cos(\omega_o t)$$
$$i_b(t) = I_b^{(1)} \cos(\omega_o t - 120°) \qquad (1\text{-}2)$$
$$i_c(t) = I_c^{(1)} \cos(\omega_o t - 240°).$$

The negative displacement angles indicate that the fundamental phasors rotate clockwise in the space–time plane.

For the third harmonic (zero-sequence) currents,

$$i_a^{(3)}(t) = I_a^{(3)} \cos(3\omega_o t)$$
$$i_b^{(3)}(t) = I_b^{(3)} \cos 3(\omega_o t - 120°)$$
$$\qquad = I_b^{(3)} \cos(3\omega_o t - 360°) = I_b^{(3)} \cos(3\omega_o t) \qquad (1\text{-}3)$$
$$i_c^{(3)}(t) = I_c^{(3)} \cos 3(\omega_o t - 240°)$$
$$\qquad = I_c^{(3)} \cos(3\omega_o t - 720°) = I_c^{(3)} \cos(3\omega_o t).$$

This equation shows that the third harmonic phasors are in phase and have zero displacement angles between them. The third harmonic currents are known as zero-sequence harmonics.

The expressions for the fifth harmonic currents are

$$i_a^{(5)}(t) = I_a^{(5)} \cos(5\omega_o t)$$
$$i_b^{(5)}(t) = I_b^{(5)} \cos 5(\omega_o t - 120°) = I_b^{(5)} \cos(5\omega_o t - 600°)$$
$$\qquad = I_b^{(5)} \cos(5\omega_o t - 240°) = I_b^{(5)} \cos(5\omega_o t + 120°)$$
$$i_c^{(5)}(t) = I_c^{(5)} \cos 5(\omega_o t - 240°) = I_c^{(5)} \cos(5\omega_o t - 1200°)$$
$$\qquad = I_c^{(5)} \cos(5\omega_o t - 120°) = I_c^{(5)} \cos(5\omega_o t + 240°).$$
$$\qquad\qquad (1\text{-}4)$$

Note that displacement angles are positive; therefore, the phase sequence of this harmonic is counterclockwise and opposite to that of the fundamental. The fifth harmonic currents are known as negative-sequence harmonics.

Similar relationships exist for other harmonic orders. Table 1.3 categorizes power system harmonics in terms of their respective frequencies and sources.

Note that although the harmonic phase-shift angle has the effect of altering the shape of the composite waveform (e.g., adding a third harmonic component with 0 degree phase shift to the fundamental results in a composite waveform with maximum peak-to-peak value whereas a 180 degree phase shift will result in a composite waveform with minimum peak-to-peak value), the phase–sequence order of the harmonics is not affected. Not all voltage and current systems can be decomposed into positive-, negative-, and zero-sequence systems [49].

**Time and Spatial (Space) Harmonics.** Time harmonics are the harmonics in the voltage and current waveforms of electric machines and power systems due to magnetic core saturation, presence of nonlinear loads, and irregular system conditions (e.g., faults and imbalance). Spatial (space) harmonics are referred to the harmonics in the flux linkage of rotating electromagnetic devices such as induction and

**TABLE 1.3** Types and Sources of Power System Harmonics

| Type | Frequency | Source |
|------|-----------|--------|
| DC | 0 | Electronic switching devices, half-wave rectifiers, arc furnaces (with random arcs), geomagnetic induced currents (GICs) |
| Odd harmonics | $h \cdot f_1$ ($h$ = odd) | Nonlinear loads and devices |
| Even harmonics | $h \cdot f_1$ ($h$ = even) | Half-wave rectifiers, geomagnetic induced currents (GICs) |
| Triplen harmonics | $3h \cdot f_1$ ($h$ = 1, 2, 3, 4, . . .) | Unbalanced three-phase load, electronic switching devices |
| Positive-sequence harmonics | $h \cdot f_1$ ($h$ = 1, 4, 7, 10, . . .) | Operation of power system with nonlinear loads |
| Negative-sequence harmonics | $h \cdot f_1$ ($h$ = 2, 5, 8, 11, . . .) | Operation of power system with nonlinear loads |
| Zero-sequence harmonics | $h \cdot f_1$ ($h$ = 3, 6, 9, 12, . . .) (same as triplen harmonics) | Unbalanced operation of power system |
| Time harmonics | $h \cdot f_1$ ($h$ = an integer) | Voltage and current source inverters, pulse-width-modulated rectifiers, switch-mode rectifiers and inverters |
| Spatial harmonics | $h \cdot f_1$ ($h$ = an integer) | Induction machines |
| Interharmonic | $h \cdot f_1$ ($h$ = not an integer multiple of $f_1$) | Static frequency converters, cycloconverters, induction machines, arcing devices, computers |
| Subharmonic | $h \cdot f_1$ ($h < 1$ and not an integer multiple of $f_1$, e.g., $h$ = 15 Hz, 30 Hz) | Fast control of power supplies, subsynchronous resonances, large capacitor banks in highly inductive systems, induction machines |
| Characteristic harmonic | $(12k + 1) \cdot f_1$ ($k$ = integer) | Rectifiers, inverters |
| Uncharacteristic harmonic | $(12k - 1) \cdot f_1$ ($k$ = integer) | Weak and unsymmetrical AC systems |

synchronous machines. The main cause of spatial harmonics is the unsymmetrical physical structure of stator and rotor magnetic circuits (e.g., selection of number of slots and rotor eccentricity). Spatial harmonics of flux linkages will induce time harmonic voltages in the rotor and stator circuits that generate time harmonic currents.

### 1.4.2 The Average Value of a Nonsinusoidal Waveform

The average value of a sinusoidal waveform is defined as

$$I_{ave} = \frac{1}{T}\int_0^T i(t)\,dt. \qquad (1\text{-}5)$$

For the nonsinusoidal current of Eq. 1-1,

$$I_{ave} = \frac{1}{T}\int_0^T i(t)\,dt$$
$$= \frac{1}{T}\int_0^T [i^{(1)}(t) + i^{(2)}(t) + i^{(3)}(t) + i^{(4)}(t) + .....]\,dt.$$
$$(1\text{-}6)$$

Since all harmonics are sinusoids, the average value of a nonsinusoidal function is equal to its DC value:

$$I_{ave} = I_{DC}. \qquad (1\text{-}7)$$

### 1.4.3 The rms Value of a Nonsinusoidal Waveform

The rms value of a sinusoidal waveform is defined as

$$I_{rms} = \sqrt{\frac{1}{T}\int_0^T i^2(t)\,dt} = \sqrt{\frac{1}{2\pi}\int_0^T I_{max}^2 \cos^2(\omega t)\,dt}$$
$$= \sqrt{\frac{1}{2}I_{max}^2} = \frac{I_{max}}{\sqrt{2}}. \qquad (1\text{-}8)$$

For the nonsinusoidal current of Eq. 1-1,

$$I_{rms} = \sqrt{\frac{1}{T}\int_0^T i^2(t)\,dt}$$
$$= \left\{ \frac{1}{T}\int_0^T [I_{max}^{(1)}\cos(\omega t + \beta_1) + I_{max}^{(2)}\cos(2\omega t + \beta_2) \right.$$
$$\left. + I_{max}^{(3)}\cos(3\omega t + \beta_3) + ...]^2\,dt \right\}^{1/2}$$
$$= \left\{ \frac{1}{T}\int_0^T [(I_{max}^{(1)})^2 \cos^2(\omega t + \beta_1) \right.$$
$$+ (I_{max}^{(2)})^2 \cos^2(2\omega t + \beta_2)$$
$$+ (I_{max}^{(3)})^2 \cos^2(3\omega t + \beta_3) + ... + 2I_{max}^{(1)} \cdot$$
$$I_{max}^{(2)}\cos(\omega t + \beta_1)\cos(2\omega t + \beta_2) + ...]\,dt \right\}^{1/2}.$$
$$(1\text{-}9)$$

This equation contains two parts:

- The first part is the sum of the squares of harmonics:

$$\sum_{p=1}^{H} (I_{\max}^{(p)})^2 \cos^2(p\omega t + \beta_p). \quad (1\text{-}10)$$

- The second part is the sum of the products of harmonics:

$$\sum_{p=1}^{H}\sum_{q=1}^{H} I_{\max}^{(p)} I_{\max}^{(q)} \cos(p\omega t + \beta_p)\cos(q\omega t + \beta_q), \quad p \neq q.$$
$$(1\text{-}11)$$

After some simplifications it can be shown that the average of the second part is zero, and the first part becomes

$$I_{rms} = \int_0^{2\pi} \left[\sum_{p=1}^{H} (I_{\max}^{(p)})^2 \cos^2(p\omega t + \beta_p)\right] d(2\omega t)$$
$$= \int_0^{2\pi} \left[(I_{\max}^{(1)})^2 \cos^2(\omega t + \beta_1)\right] d(\omega t)$$
$$+ \left[\int_0^{2\pi} (I_{\max}^{(2)})^2 \cos^2(2\omega t + \beta_2)\right] d(2\omega t) + \dots. \quad (1\text{-}12)$$

Therefore, the rms value of a nonsinusoidal waveform is

$$I_{rms} = [(I_{rms}^{(1)})^2 + (I_{rms}^{(2)})^2 + (I_{rms}^{(3)})^2 + \dots + (I_{rms}^{(H)})^2]^{1/2}. \quad (1\text{-}13)$$

If the nonsinusoidal waveform contains DC values, then

$$I_{rms} = [I_{DC} + (I_{rms}^{(1)})^2 + (I_{rms}^{(2)})^2 + (I_{rms}^{(3)})^2 + \dots + (I_{rms}^{(H)})^2]^{1/2}. \quad (1\text{-}14)$$

### 1.4.4 Form Factor (FF)

The form factor (FF) is a measure of the shape of the waveform and is defined as

$$FF = \frac{I_{rms}}{I_{ave}}. \quad (1\text{-}15)$$

Since the average value of a sinusoid is zero, its average over one half-cycle is used in the above equation. As the harmonic content of the waveform increases, its FF will also increase.

### 1.4.5 Ripple Factor (RF)

Ripple factor (RF) is a measure of the ripple content of the waveform and is defined as

$$RF = \frac{I_{AC}}{I_{DC}}, \quad (1\text{-}16)$$

where $I_{AC} = \sqrt{(I_{rms})^2 - (I_{DC})^2}$. It is easy to show that

$$RF = \frac{\sqrt{(I_{rms})^2 - (I_{DC})^2}}{I_{DC}} = \sqrt{\frac{(I_{rms})^2}{(I_{DC})^2} - \frac{(I_{DC})^2}{(I_{DC})^2}}$$
$$= \sqrt{FF^2 - 1}. \quad (1\text{-}17)$$

### 1.4.6 Harmonic Factor (HF)

The harmonic factor (HF) of the hth harmonic, which is a measure of the individual harmonic contribution, is defined as

$$HF_h = \frac{I_{rms}^{(h)}}{I_{rms}^{(1)}}. \quad (1\text{-}18)$$

Some references [8] call HF the individual harmonic distortion (IHD).

### 1.4.7 Lowest Order Harmonic (LOH)

The lowest order harmonic (LOH) is that harmonic component whose frequency is closest to that of the fundamental and its amplitude is greater than or equal to 3% of the fundamental component.

### 1.4.8 Total Harmonic Distortion (THD)

The most common harmonic index used to indicate the harmonic content of a distorted waveform with a single number is the total harmonic distortion (THD). It is a measure of the effective value of the harmonic components of a distorted waveform, which is defined as the rms of the harmonics expressed in percentage of the fundamental (e.g., current) component:

$$THD_i = \frac{\sqrt{\sum_{h=2}^{\infty} (I^{(h)})^2}}{I^{(1)}}. \quad (1\text{-}19)$$

A commonly cited value of 5% is often used as a dividing line between a high and low distortion level. The ANSI standard recommends truncation of THD series at 5 kHz, but most practical commercially available instruments are limited to about 1.6 kHz (due to the limited bandwidth of potential and current transformers and the word length of the digital hardware [5]).

Main advantages of THD are

- It is commonly used for a quick measure of distortion; and
- It can be easily calculated.

Some disadvantages of THD are

- It does not provide amplitude information; and
- The detailed information of the spectrum is lost.

$THD_i$ is related to the rms value of the current waveform as follows [6]:

$$I_{rms} = \sqrt{\sum_{h=2}^{\infty}(I^{(h)})^2} = I^{(1)}\sqrt{1+THD_i^2}. \quad (1\text{-}20)$$

$THD$ can be weighted to indicate the amplitude stress on various system devices. The weighted distortion factor adapted to inductance is an approximate measure for the additional thermal stress of inductances of coils and induction motors [9, Table 2.4]:

THD adapted to inductance $= THD_{ind}$

$$= \frac{\sqrt{\sum_{h=2}^{50}\left(\frac{(V^{(h)})^2}{h^{\alpha}}\right)}}{V^{(1)}}, \quad (1\text{-}21)$$

where $\alpha = 1 \ldots 2$. On the other hand, the weighted $THD$ adapted to capacitors is an approximate measure for the additional thermal stress of capacitors directly connected to the system without series inductance [9, Table 2.4]:

THD adapted to capacitor $= THD_{cap}$

$$= \frac{\sqrt{\sum_{h=2}^{50}\left(h\times(V^{(h)})^2\right)}}{V^{(1)}}. \quad (1\text{-}22)$$

Because voltage distortions are maintained small, the voltage $THD_v$ nearly always assumes values which are not a threat to the power system. This is not the case for current; a small current may have a high $THD_i$ but may not be a significant threat to the system.

### 1.4.9 Total Interharmonic Distortion (TIHD)

This factor is equivalent to the (e.g., current) $THD_i$, but is defined for interharmonics as [9]

$$TIHD = \frac{\sqrt{\sum_{k=1}^{n}(I^{(k)})^2}}{I^{(1)}}, \quad (1\text{-}23)$$

where $k$ is the total number of interharmonics and $n$ is the total number of frequency bins present including subharmonics (e.g., interharmonic frequencies that are less than the fundamental frequency).

### 1.4.10 Total Subharmonic Distortion (TSHD)

This factor is equivalent to the (e.g., current) $THD_i$, but defined for subharmonics [9]:

$$TSHD = \frac{\sqrt{\sum_{s=1}^{S}(I^{(s)})^2}}{I^{(1)}}, \quad (1\text{-}24)$$

where s is the total number of frequency bins present below the fundamental frequency.

### 1.4.11 Total Demand Distortion (TDD)

Due to the mentioned disadvantages of THD, some standards (e.g., IEEE-519) have defined the total demand distortion factor. This term is similar to THD except that the distortion is expressed as a percentage of some rated or maximum value (e.g., load current magnitude), rather than as a percentage of the fundamental current:

$$TDD = \frac{\sqrt{\sum_{h=2}^{50}(I^{(h)})^2}}{I_{rated}}. \quad (1\text{-}25)$$

### 1.4.12 Telephone Influence Factor (TIF)

The telephone influence factor (TIF), which was jointly proposed by Bell Telephone Systems (BTS) and the Edison Electric Institute (EEI) and is widely used in the United States and Canada, determines the influence of power systems harmonics on telecommunication systems. It is a variation of THD in which the root of the sum of the squares is weighted using factors (weights) that reflect the response of the human ear [5]:

$$TIF = \frac{\sqrt{\sum_{i=1}^{\infty}(w_i V^{(i)})^2}}{\sqrt{\sum_{i=1}^{\infty}(V^{(i)})^2}}, \quad (1\text{-}26)$$

where $w_i$ are the TIF weighting factors obtained by physiological and audio tests, as listed in Table 1.4. They also incorporate the way current in a power circuit induces voltage in an adjacent communication system.

### 1.4.13 C-Message Weights

The C-message weighted index is very similar to the TIF except that the weights $c_i$ are used in place of $w_i$ [5]:

**TABLE 1.4** Telephone Interface ($w_i$) and C-Message ($c_i$) Weighting Factors [5]

| Harmonic order ($h, f_1 = 60$ Hz) | TIF weights ($w_i$) | C weights ($c_i$) | Harmonic order ($h, f_1 = 60$ Hz) | TIF weights ($w_i$) | C weights ($c_i$) |
|---|---|---|---|---|---|
| 1 | 0.5 | 0.0017 | 29 | 7320 | 0.841 |
| 2 | 10.0 | 0.0167 | 30 | 7570 | 0.841 |
| 3 | 30.0 | 0.0333 | 31 | 7820 | 0.841 |
| 4 | 105 | 0.0875 | 32 | 8070 | 0.841 |
| 5 | 225 | 0.1500 | 33 | 8330 | 0.841 |
| 6 | 400 | 0.222 | 34 | 8580 | 0.841 |
| 7 | 650 | 0.310 | 35 | 8830 | 0.841 |
| 8 | 950 | 0.396 | 36 | 9080 | 0.841 |
| 9 | 1320 | 0.489 | 37 | 9330 | 0.841 |
| 10 | 1790 | 0.597 | 38 | 9590 | 0.841 |
| 11 | 2260 | 0.685 | 39 | 9840 | 0.841 |
| 12 | 2760 | 0.767 | 40 | 10090 | 0.841 |
| 13 | 3360 | 0.862 | 41 | 10340 | 0.841 |
| 14 | 3830 | 0.912 | 42 | 10480 | 0.832 |
| 15 | 4350 | 0.967 | 43 | 10600 | 0.822 |
| 16 | 4690 | 0.977 | 44 | 10610 | 0.804 |
| 17 | 5100 | 1.000 | 45 | 10480 | 0.776 |
| 18 | 5400 | 1.000 | 46 | 10350 | 0.750 |
| 19 | 5630 | 0.988 | 47 | 10210 | 0.724 |
| 20 | 5860 | 0.977 | 48 | 9960 | 0.692 |
| 21 | 6050 | 0.960 | 49 | 9820 | 0.668 |
| 22 | 6230 | 0.944 | 50 | 9670 | 0.645 |
| 23 | 6370 | 0.923 | 55 | 8090 | 0.490 |
| 24 | 6650 | 0.924 | 60 | 6460 | 0.359 |
| 25 | 6680 | 0.891 | 65 | 4400 | 0.226 |
| 26 | 6790 | 0.871 | 70 | 3000 | 0.143 |
| 27 | 6970 | 0.860 | 75 | 1830 | 0.0812 |
| 28 | 7060 | 0.840 | | | |

$$C = \frac{\sqrt{\sum_{i=1}^{\infty}(c_i I^{(i)})^2}}{\sqrt{\sum_{i=1}^{\infty}(I^{(i)})^2}} = \frac{\sqrt{\sum_{i=1}^{\infty}(c_i I^{(i)})^2}}{I_{rms}}, \quad (1\text{-}27)$$

where $c_i$ are the C-message weighting factors (Table 1.4) that are related to the TIF weights by $w_i = 5(i)(f_0)c_i$. The C-message could also be applied to the bus voltage.

### 1.4.14  $V \cdot T$ and $I \cdot T$ Products

The THD index does not provide information about the amplitude of voltage (or current); therefore, BTS or the EEI use $I \cdot T$ and $V \cdot T$ products. The $I \cdot T$ and $V \cdot T$ products are alternative indices to the THD incorporating voltage or current amplitudes:

$$V \cdot T = \sqrt{\sum_{i=1}^{\infty}(w_i V^{(i)})^2}, \quad (1\text{-}28)$$

$$I \cdot T = \sqrt{\sum_{i=1}^{\infty}(w_i I^{(i)})^2}, \quad (1\text{-}29)$$

where the weights $w_i$ are listed in Table 1.4.

### 1.4.15  Telephone Form Factor (TFF)

Two weighting systems widely used by industry for interference on telecommunication system are [9]

- the sophomoric weighting system proposed by the International Consultation Commission on Telephone and Telegraph System (CCITT) used in Europe, and
- the C-message weighting system proposed jointly by Bell Telephone Systems (BTS) and the Edison Electric Institute (EEI), used in the United States and Canada.

These concepts acknowledge that the harmonic effect is not uniform over the audio-frequency range and use measured weighting factors to account for this nonuniformity. They take into account the type of telephone equipment and the sensitivity of the human ear to provide a reasonable indication of the interference from each harmonic.

The BTS and EEI systems describe the level of harmonic interference in terms of the telephone influence factor (Eq. 1-26) or the C-message (Eq. 1-27), whereas the CCITT system uses the telephone form factor (TFF):

$$TFF = \frac{1}{V^{(1)}} \sqrt{\sum_{h=1}^{\infty} K_h P_h (V^{(h)})^2}, \qquad (1\text{-}30)$$

where $K_h = h/800$ is a coupling factor and $P_h$ is the harmonic weight [9 (Fig. 2.5)] divided by 1000.

### 1.4.16 Distortion Index (DIN)

The distortion index (DIN) is commonly used in standards and specifications outside North America. It is also used in Canada and is defined as [5]

$$DIN = \frac{\sqrt{\sum_{i=2}^{\infty} (V^{(i)})^2}}{\sqrt{\sum_{i=1}^{\infty} (V^{(i)})^2}} = \frac{THD}{\sqrt{THD^2 + 1}}. \qquad (1\text{-}31)$$

For low levels of harmonics, a Taylor series expansion can be applied to show

$$DIN \approx THD \left(1 - \frac{1}{2} THD\right). \qquad (1\text{-}32)$$

### 1.4.17 Distortion Power (D)

Harmonic distortion complicates the computation of power and power factors because voltage and current equations (and their products) contain harmonic components. Under sinusoidal conditions, there are four standard quantities associated with power:

- Fundamental apparent power ($S_1$) is the product of the rms fundamental voltage and current;
- Fundamental active power ($P_1$) is the average rate of delivery of energy;
- Fundamental reactive power ($Q_1$) is the portion of the apparent power that is oscillatory; and
- Power factor at fundamental frequency (or displacement factor) $\cos \theta_1 = P_1/S_1$.

The relationship between these quantities is defined by the power triangle:

$$(S_1)^2 = (P_1)^2 + (Q_1)^2. \qquad (1\text{-}33)$$

If voltage and current waveforms are nonsinusoidal (Eq. 1-1), the above equation does not hold because S contains cross terms in the products of the Fourier series that correspond to voltages and currents of different frequencies, whereas $P$ and $Q$ correspond to voltages and currents of the same frequency. It has been suggested to account for these cross terms as follows [5, 50, 51]:

$$S^2 = P^2 + Q^2 + D^2, \qquad (1\text{-}34)$$

where

$$\begin{aligned} \text{Apparent power} &= S = V_{rms} I_{rms} \\ &= \sqrt{\sum_{h=0,1,2,3,\dots}^{H} (V_{rms}^{(h)})^2} \sqrt{\sum_{h=0,1,2,3,\dots}^{H} (I_{rms}^{(h)})^2}, \end{aligned}$$
$$(1\text{-}35)$$

$$\text{Total real power} = P = \sum_{h=0,1,2,3,\dots}^{H} V_{rms}^{(h)} I_{rms}^{(h)} \cos(\theta_h),$$
$$\text{where} \quad \theta_h = \alpha_h - \beta_h,$$
$$(1\text{-}36)$$

$$\text{Total reactive power} = Q = \sum_{h=0,1,2,3,\dots}^{H} V_{rms}^{(h)} I_{rms}^{(h)} \sin(\theta_h),$$
$$(1\text{-}37)$$

$$\begin{aligned} \text{Distortion power} = D = \sum_{m=0}^{H-1} \sum_{n=m+1}^{H} [(V_{rms}^{(m)})^2 (I_{rms}^{(n)})^2 \\ + V_{rms}^{(n)})^2 (I_{rms}^{(m)})^2 - 2V_{rms}^{(m)} I_{rms}^{(n)} \cos(\theta_m - \theta_n)]. \end{aligned}$$
$$(1\text{-}38)$$

Also, the fundamental power factor (displacement factor) in the case of sinusoidal voltage and nonsinusoidal currents is defined as [8]

$$\cos \theta_1 = \frac{P_1}{\sqrt{(P_1)^2 + (Q_1)^2}}, \qquad (1\text{-}39)$$

and the harmonic displacement factor is defined as [8]

$$\lambda = \frac{P_1}{\sqrt{(P_1)^2 + (Q_1)^2 + D^2}}. \qquad (1\text{-}40)$$

The power and displacement factor quantities are shown in addition to the power quantities in Fig. 1.22. A detailed comparison of various definitions of the distortion power $D$ is given in reference [51].

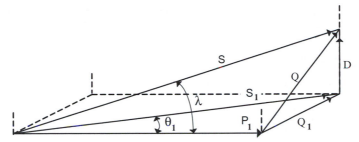

**FIGURE 1.22** Phasor diagram of different parameters of electric power under nonsinusoidal conditions.

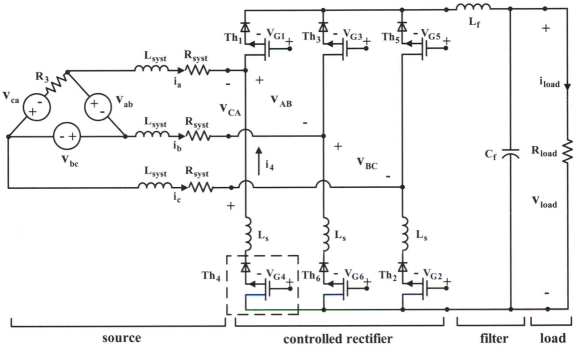

**FIGURE E1.1.1** Controlled three-phase, full-wave thyristor rectifier.

### 1.4.18 Application Example 1.1: Calculation of Input/Output Currents and Voltages of a Three-Phase Thyristor Rectifier

The circuit of Fig. E1.1.1 represents a phase-controlled, three-phase thyristor rectifier. The balanced input line-to-line voltages are $v_{ab} = \sqrt{2}\,240$ $\sin \omega t$, $v_{bc} = \sqrt{2}\,240 \sin(\omega t - 120°)$, and $v_{ca} = \sqrt{2}\,240$ $\sin(\omega t - 240°)$, where $\omega = 2\pi f$ and $f = 60$ Hz. Each of the six thyristors can be modeled by a self-commutated electronic switch and a diode in series, as is illustrated in Fig. E1.1.2. Use the following PSpice models for the MOSFET and the diode:

- Model for self-commutated electronic switch (MOSFET):
  .model SMM NMOS(Level = 3 Gamma = 0
  Delta = 0 Eta = 0 Theta = 0 Kappa = 0 Vmax = 0

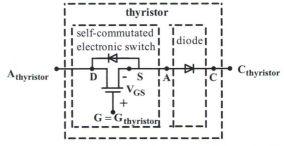

**FIGURE E1.1.2** Model for thyristor consisting of self-commuted switch and diode.

+ XJ = 0 TOX = 100 N UO = 600 PHI = 0.6 RS = 42.69 m KP = 20.87 u L = 2 u
+ W = 2.9 VTO = 3.487 RD = 0.19 CBD = 200 N PB = 0.8 MJ = 0.5 CGSO = 3.5 n
+ CGDO = 100 p RG = 1.2 IS = 10 F)

- Model for diode:
  .model D1N4001 D(IS = 10$^{-12}$)

The parameters of the circuit are as follows:

- System resistance and inductance $L_{syst} = 300 \, \mu H$, $R_{syst} = 0.05 \, \Omega$;
- Load resistance $R_{load} = 10 \, \Omega$;
- Filter capacitance and inductance $C_f = 500 \, \mu F$, $L_f = 1 \, mH$;
- Snubber inductance $L_s = 5 \, nH$;

Note that $R_3$ must be nonzero because PSpice cannot accept three voltage sources connected within a loop.

Perform a PSpice analysis plotting input line-to-line voltages $v_{ab}$, $v_{bc}$, $v_{ca}$, $v_{AB}$, $v_{BC}$, $v_{CA}$, input currents $i_a$, $i_b$, $i_c$, and the rectified output voltage $v_{load}$ and output current $i_{load}$ for $\alpha = 0°$ during the time interval $0 \le t \le 60$ ms. Print the input program. Repeat the computation for $\alpha = 50°$ and $\alpha = 150°$.

### 1.4.19 Application Example 1.2: Calculation of Input/Output Currents and Voltages of a Three-Phase Rectifier with One Self-Commutated Electronic Switch

An inexpensive and popular rectifier is illustrated in Fig. E1.2.1. It consists of four diodes and one self-commutated electronic switch operated at, for example, $f_{switch} = 600$ Hz. The balanced input line-to-

line voltages are $v_{ab} = \sqrt{2} \, 240 \sin \omega t$, $v_{bc} = \sqrt{2} \, 240 \sin(\omega t - 120°)$, and $v_{ca} = \sqrt{2} \, 240 \sin(\omega t - 240°)$, where $\omega = 2\pi f$ and $f = 60$ Hz. Perform a PSpice analysis. Use the following PSpice models for the MOSFET and the diodes:

- Model for self-commutated electronic switch (MOSFET):
  MODEL SMM NMOS(LEVEL = 3 GAMMA = 0 DELTA = 0 ETA = 0 THETA = 0
  + KAPPA = 0 VMAX = 0 XJ = 0 TOX = 100 N
  UO = 600 PHI = 0.6 RS = 42.69 M KP = 20.87 U
  + L = 2 U W = 2.9 VTO = 3.487 RD = 0.19
  CBD = 200 N PB = 40.8 MJ = 0.5 CGSO = 3.5 N
  + CGDO = 100 P RG = 1.2 IS = 10 F)
- Model for diode:
  .MODEL D1N4001 D(IS = 10$^{-12}$)

The parameters of the circuit are as follows:

- System resistance and inductance $L_{syst} = 300 \, \mu H$, $R_{syst} = 0.05 \, \Omega$;
- Load resistance $R_{load} = 10 \, \Omega$;
- Filter capacitance and inductance $C_f = 500 \, \mu F$, $L_f = 5 \, mH$; and
- Snubber inductance $L_s = 5 \, nH$.

Note that $R_3$ must be nonzero because PSpice cannot accept three voltage sources connected within a loop. The freewheeling diode is required to reduce the voltage stress on the self-commutated switch.

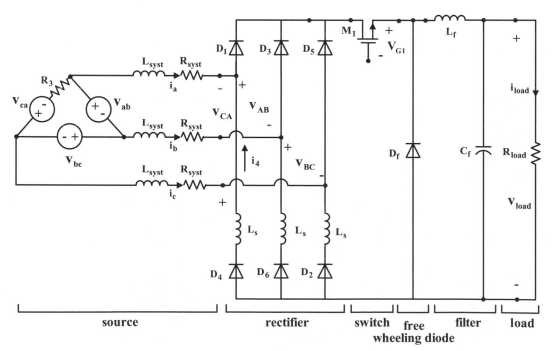

**FIGURE E1.2.1** Three-phase, full-wave rectifier with one self-commutated switch.

Print the PSpice input program. Perform a PSpice analysis plotting input line-to-line voltages $v_{ab}$, $v_{bc}$, $v_{ca}$, $v_{AB}$, $v_{BC}$, $v_{CA}$, input currents $i_a$, $i_b$, $i_c$, and the rectified output voltage $v_{load}$ and output current $i_{load}$ for a duty ratio of $\delta = 50\%$ during the time interval $0 \le t \le 60$ ms.

### 1.4.20 Application Example 1.3: Calculation of Input Currents of a Brushless DC Motor in Full-on Mode (Three-Phase Permanent-Magnet Motor Fed by a Six-Step Inverter)

In the drive circuit of Fig. E1.3.1 the DC input voltage is $V_{DC} = 300$ V. The inverter is a six-pulse or six-step or full-on inverter consisting of six self-commutated (e.g., MOSFET) switches. The electric machine is a three-phase permanent-magnet motor represented by induced voltages ($e_A$, $e_B$, $e_C$), resistances, and leakage inductances (with respect to stator phase windings) for all three phases. The induced voltage of the stator winding (phase A) of the permanent-magnet motor is

$$e_A = 160 \sin(\omega t + \theta) \qquad \text{[V]},$$

where $\omega = 2\pi f_1$ and $f_1 = 1500$ Hz. Correspondingly,

$$e_B = 160 \sin(\omega t + 240° + \theta) \qquad \text{[V]},$$
$$e_C = 160 \sin(\omega t + 120° + \theta) \qquad \text{[V]}.$$

The resistance $R_1$ and the leakage inductance $L_{1\ell}$ of one of the phases are $0.5\ \Omega$ and $50\ \mu$H, respectively.

The magnitude of the gating voltages of the six MOSFETs is $V_{Gmax} = 15$ V. The gating signals with their phase sequence are shown in Fig. E1.3.2. Note that the phase sequence of the induced voltages ($e_A$, $e_B$, $e_C$) and that of the gating signals (see Fig. E1.3.2) must be the same. If these phase sequences are not the same, then no periodic solution for the machine currents ($i_{MA}$, $i_{MB}$, $i_{MC}$) can be obtained.

The models of the enhancement metal-oxide semiconductor field-effect transistors and those of the (external) freewheeling diodes are as follows:

- Model for self-commutated electronic switch (MOSFET):
  .MODEL SMM NMOS(LEVEL = 3 GAMMA = 0 DELTA = 0 ETA = 0 THETA = 0
  + KAPPA = 0 VMAX = 0 XJ = 0 TOX = 100 N UO = 600 PHI = 0.6 RS = 42.69 M KP = 20.87 U
  + L = 2 U W = 2.9 VTO = 3.487 RD = 0.19 CBD = 200 N PB = 0.8 MJ = 0.5 CGSO = 3.5 N
  + CGDO = 100 P RG = 1.2 IS = 10 F)
- Model for diode:
  .MODEL D1N4001 D(IS = $10^{-12}$)

a) Using PSpice, compute and plot the current of MOSFET $Q_{Au}$ (e.g., $i_{QAU}$) and the motor current of phase A (e.g., $i_{MA}$) for the phase angles of the induced voltages $\theta = 0°$, $\theta = +30°$, $\theta = +60°$, $\theta = -30°$, and $\theta = -60°$. Note that the gating signal frequency of the MOSFETs corresponds to the frequency $f_1$, that is, full-on mode operation exists. For switching sequence see Fig. E1.3.2.

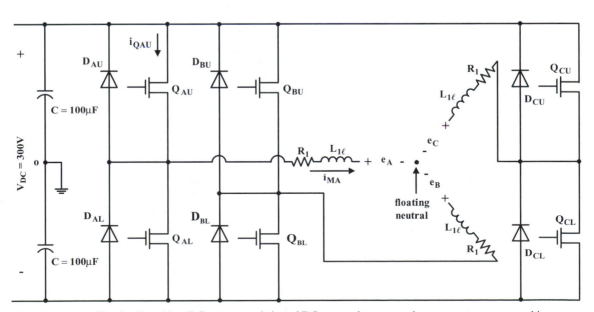

**FIGURE E1.3.1** Circuit of brushless DC motor consisting of DC source, inverter, and permanent-magnet machine.

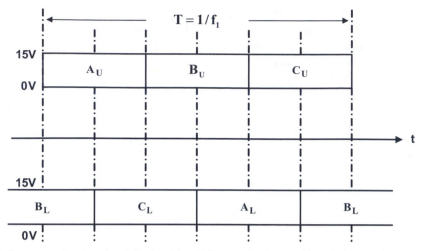

**FIGURE E1.3.2** Sequence of gating signals for brushless DC motor in six-step (six-pulse) operation.

b) Repeat part a for $\theta = +30°$ with reversed phase sequence.

Note the following:

- The step size for the numerical solution should be in the neighborhood of $\Delta t = 0.05\ \mu s$; and
- To eliminate computational transients due to inconsistent initial conditions compute at least three periods of all quantities and plot the last (third) period of $i_{QAU}$ and $i_{MA}$ for all five cases, where $\theta$ assumes the values given above.

### 1.4.21 Application Example 1.4: Calculation of the Efficiency of a Polymer Electrolyte Membrane (PEM) Fuel Cell Used as Energy Source for a Variable-Speed Drive

a) Calculate the power efficiency of a PEM fuel cell.

b) Find the specific power density of this PEM fuel cell expressed in W/kg.

c) How does this specific power density compare with that of a lead–acid battery [66]?

Hints:

- The nominal energy density of hydrogen is 28 kWh/kg, which is significantly larger than that of gasoline (12.3 kWh/kg). This makes hydrogen a desirable fuel for automobiles.
- The mass density of hydrogen is $\gamma = 0.0899$ g/liter.
- The oxygen atom has 8 electrons, 8 protons, and 8 neutrons.

A PEM fuel cell as specified by [65] has the following parameters:

| Performance: | Output power: $P_{rat} = 1200$ W[a] |
|---|---|
| | Output current: $I_{rat} = 46$ A[a] |
| | DC voltage range: $V_{rat} = 22$ to 50 V |
| | Operating lifetime: $T_{life} = 1500$ h[b] |
| Fuel: | Composition: $C = 99.99\%$ dry gaseous hydrogen |
| | Supply pressure: $p = 10$ to 250 PSIG |
| | Consumption: $V = 18.5$ SLPM[c] |
| Operating environment: | Ambient temperature: $t_{amb} = 3$ to 30°C |
| | Relative humidity: $RH = 0$ to 95% |
| | Location: Indoors and outdoors[d] |
| Physical: | Length × width × height: (56)(25)(33) cm |
| | Mass: $W = 13$ kg |
| Emissions: | Liquid water: $H_2O = 0.87$ liters maximum per hour. |

[a]Beginning of life, sea level, rated temperature range.
[b]CO within the air (which provides the oxygen) destroys the proton exchange membrane.
[c]At rated power output, SLPM ≡ standard liters per minute (standard flow).
[d]Unit must be protected from inclement weather, sand, and dust.

### 1.4.22 Application Example 1.5: Calculation of the Currents of a Wind-Power Plant PWM Inverter Feeding Power into the Power System

The circuit diagram of PWM inverter feeding power into the 240 $V_{L-L}$ three-phase utility system is shown in Fig. E1.5.1 consisting of DC source, inverter, filter, and power system. The associated control circuit is given by a block diagram in Fig. E1.5.2.

Use the PSpice program *windpower.cir* as listed below:

```
*windpower.cir;  Is=40Arms,VDC=450V,
phi=30
```

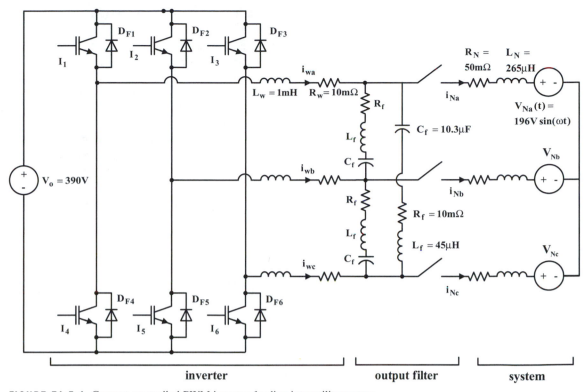

**FIGURE E1.5.1** Current-controlled PWM inverter feeding into utility system.

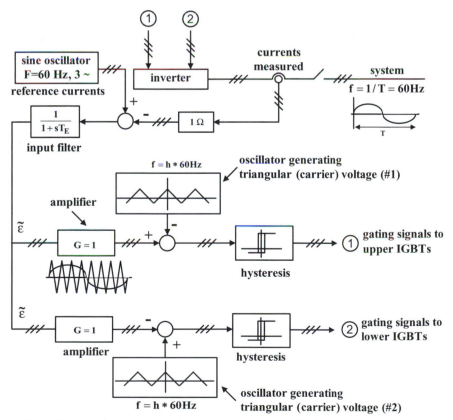

**FIGURE E1.5.2** Block diagram of control circuit for current-controlled PWM inverter.

```
VDC supply 2   0  450
***switches
msw1   2   11   10   10   Mosfet
dsw1   10   2   diode
msw2   2   21   20   20   Mosfet
dsw1   20   2   diode
msw3   2   31   30   30   Mosfet
dsw3   30   2   diode
msw4   10   41   0   0   Mosfet
dsw4   0   10   diode
msw5   20   51   0   0   Mosfet
dsw5   0   20   diode
msw6   30   61   0   0   Mosfet
dsw6   0   30   diode
*** inductors
L_W1   10   15   1m
L_W2   20   25   1m
L_W3   30   35   1m
*** resistors or voltage sources for
measuring current
R_W1   15   16   10m
R_W2   25   26   10m
R_W3   35   36   10m
***voltages serve as reference
currents
vref1   12   0   sin(0   56.6   60   0   0
0)
vref2   22   0   sin(0   56.6   60   0   0
-120)
vref3   32   0   sin(0   56.6   60   0   0
-240)
*** voltages derived from load
currents (measured with shunts)
eout1   13   0   15   16   100
eout2   23   0   25   26   100
eout1   33   0   35   36   100
*** error signals derived from the
difference between vref and eout
rdiff1   12   13a   1k
rdiff2   22   23a   1k
rdiff3   32   33a   1k
cdiff1 12 13a 1u
cdiff2 22 23a 1u
cdiff3 32 33a 1u
rdiff4   13a   13   1k
rdiff5   23a   23   1k
rdiff6   33a   33   1k
ecin1   14   0   12   13a   2
ecin2   24   0   22   23a   2
ecin3   34   0   32   33a   2
vtriangular   5   0   pulse(-10   10   0
86.5u   86.5u   0.6u   173.6u)

*** gating signals for upper
switches (mosfets) as a result
*** of comparison between triangular
voltage and error signals
xgs1 14   5   11   10 comp
xgs2 24   5   21   20 comp
xgs3 34   5   31   30 comp
*** gating of lower switches
egs4   41 0 poly(1)(11,10) 50 -1
egs5   51 0 poly(1)(21,20) 50 -1
egs6   61 0 poly(1)(31,30) 50 -1
*** filter removed because of node
limit of 64 for PSpice, not required
for Spice
lfi1   16   15b   45u
lfi2   26   25b   45u
lfi3   36   35b   45u
rfi1   15b   15c   0.01
rfi2   25b   25c   0.01
rfi3   35b   35c   0.01
cfi1   15c   26   10.3u
cfi2   25c   36   10.3u
cfi3   35c   16   10.3u
*** representation of utility system
RM1   16   18   50m
LM1   18   19   265u
Vout1   19   123   sin( 0   196   60   0
0   -30)
RM2   26   28   50m
LM2   28   29   265u
Vout2   29   123   sin( 0   196   60   0
0   -150)
RM3   36   38   50m
LM3   38   39   265u
Vout2   39   123   sin( 0   196   60   0
0   -270)
*** comparator: v1-v2, vgs
.subckt   comp 1   2   9   10
rin   1   3   2.8k
r1   3   2   20meg
e2   4   2   3   2   50
r2   4   5   1k
d1   5   6   zenerdiode1
d2   2   6   zenerdiode2
e3   7   2   5   2   1
r3   7   8   10
c3   8   2   10n
r4   3   8   100k
e4   9   10   8   2   1
.model zenerdiode1 D (Is=1p BV=0.1)
.model zenerdiode2 D (Is=1p BV=50)
.ends comp
```

```
*** models
.model Mosfet nmos(level=3 gamma=0
kappa=0 tox=100n rs=42.69m kp=20.87u
l=2u
+ w=2.9 delta=0 eta=0 theta=0 vmax=0
xj=0 uo=600 phi=0.6
+ vto=3.487 rd=0.19 cbd=200n pb=0.8
mj=0.5 cgso=3.5n cgdo=100p rg=1.2
is=10f)
.model diode d(is=1p)
***options
.options abstol=0.01m chgtol= 0.01m
reltol=50m vntol=1m itl5=0 itl4=200
***analysis request
.tran 5u 350m 16.67m 5u
***prepare for plotting
.probe
***final statement
.end
```

a) Use "reverse" engineering and identify the nodes of Figs. E1.5.1 and E1.5.2, as used in the PSpice program. It may be advisable that you draw your own detailed circuit.

b) Study the PSpice program *wr.cir*. In particular it is important that you understand the *poly* statements and the *subcircuit* for the comparator. You may ignore the statements for the filter between switch and inverter (see Fig. E1.5.1) if the node number exceeds the maximum number of 64 (note the student version of the PSpice program is limited to a maximum of 64 nodes).

c) Run this program with inverter inductance values of $L_w = 1$ mH for a DC voltage of $V_{DC} = 450$ V.

d) Plot the current supplied by the inverter to the power system and the phase power system's voltage.

## 1.5 EFFECTS OF POOR POWER QUALITY ON POWER SYSTEM DEVICES

Poor electric power quality has many harmful effects on power system devices and end users. What makes this phenomenon so insidious is that its effects are often not known until failure occurs. Therefore, insight into how disturbances are generated and interact within a power system and how they affect components is important for preventing failures. Even if failures do not occur, poor power quality and harmonics increase losses and decrease the lifetime of power system components and end-use devices. Some of the main detrimental effects of poor power quality include the following:

- Harmonics add to the rms and peak value of the waveform. This means equipment could receive a damagingly high peak voltage and may be susceptible to failure. High voltage may also force power system components to operate in the saturation regions of their characteristics, producing additional harmonics and disturbances. The waveform distortion and its effects are very dependent on the harmonic-phase angles. The rms value can be the same but depending on the harmonic-phase angles, the peak value of a certain dependent quantity can be large [52].

- There are adverse effects from heating, noise, and reduced life on capacitors, surge suppressors, rotating machines, cables and transformers, fuses, and customers' equipment (ranging from small clocks to large industrial loads).

- Utility companies are particularly concerned that distribution transformers may need to be derated to avoid premature failure due to overheating (caused by harmonics).

- Additional losses of transmission lines, cables, generators, AC motors, and transformers may occur due to harmonics (e.g., inter- and subharmonics) [53].

- Failure of power system components and customer loads may occur due to unpredicted disturbances such as voltage and/or current magnifications due to parallel resonance and ferroresonance.

- Malfunction of controllers and protective devices such as fuses and relays is possible [35].

- Interharmonics may occur which can perturb ripple control signals and can cause flicker at subharmonic levels.

- Harmonic instability [9] may be caused by large and unpredicted harmonic sources such as arc furnaces.

- Harmonic, subharmonic, and interharmonic torques may arise [32].

The effects of poor power quality on power systems and their components as well on end-use devices will be discussed in detail in subsequent chapters.

## 1.6 STANDARDS AND GUIDELINES REFERRING TO POWER QUALITY

Many documents for control of power quality have been generated by different organizations and institutes. These documents come in three levels of applicability and validity: guidelines, recommendations, and standards [5]:

- *Power quality guidelines* are illustrations and exemplary procedures that contain typical parameters and representative solutions to commonly encountered power quality problems;
- *Power quality recommended practices* recognize that there are many solutions to power quality problems and recommend certain solutions over others. Any operating limits that are indicated by recommendations are not required but should be targets for designs; and
- *Power quality standards* are formal agreements between industry, users, and the government as to the proper procedure to generate, test, measure, manufacture, and consume electric power. In all jurisdictions, violation of standards could be used as evidence in courts of law for litigation purposes.

Usually the first passage of a power quality document is done in the form of the guidelines that are often based on an early document from an industry or government group. Guides are prepared and edited by different working groups. A recommended practice is usually an upgrade of guidelines, and a standard is usually an upgrade of a recommended practice.

The main reasons for setting guidelines, recommendations, and standards in power systems with nonsinusoidal voltages or currents are to keep disturbances to user equipment within permissible limits, to provide uniform terminology and test procedures for power quality problems, and to provide a common basis on which a wide range of engineering is referenced.

There are many standards and related documents that deal with power quality issues. A frequently updated list of available documents on power quality issues will simplify the search for appropriate information. Table 1.5 includes some of the commonly used guides, recommendations, and standards on electric power quality issues. The mostly adopted documents are these:

- The North American Standards adopted by many countries of North and South America:
  - a) Institute of Electrical and Electronic Engineering (IEEE).
  - b) American National Standards Institute (ANSI).
  - c) Military Specifications (MIL-Specs) published by the U.S. Department of Defense and Canadian Electric Association (CEA).
- British Standards (BS).
- European (Standards) Norms (EN).

- International Electrotechnical Commission (IEC).
- Computer Business Equipment Manufacturers Association (CBEMA) curves.
- Information Technology Industry Council (ITIC) curves [6 (Fig. 2.13), 9 (Fig. 5.9)].
- VDE (Verein Deutscher Elektrotechniker) [8, page 1] of the German association of individuals and groups concerned with electrotechnics.
- NEMA [9, page 20] of the U.S. National Electric Manufacturers Association.

### 1.6.1 IEC 61000 Series of Standards for Power Quality

The IEC 61000 (or EN 61000) series [54], one of the most commonly used references for power quality in Europe, contains six parts, each with standards and technical reports [9]:

- Part 1 (General). Two sections cover application and interpretation aspects of EMC (electromagnetic compatibility).
- Part 2 (Environment). Twelve sections give classification of the electromagnetic environment and compatibility levels for different environments. Some aspects of this document include harmonic compatibility levels of residential LV (low voltage) systems (IEC 61000-2-2), industrial plants (IEC 61000-24), and residential MV (medium voltage) systems (IEC 61000-2-12).
- Part 3 (Limits). Eleven sections cover emission limits for harmonics and other disturbances. Some aspects of this document include harmonic current emission limits for equipment connected at LV with low (less than 16 A per phase) current (IEC 61000-3-2), flicker (IEC 61000-3-3), harmonic current emission limits for equipment connected at LV with high (more than 16 A per phase) current (IEC 61000-3-4), and assessment of emission limits for distorting loads in MV and HV (high voltage) power systems (IEC 61000-3-6).
- Part 4 (Testing and Measurement Techniques). Thirty-one sections describe standard methods for testing equipment of emission and immunity to different disturbances. Some aspects of this document include harmonic and interharmonic measurements and instrumentation (IEC 61000-4-7), dips and interruptions (EN 61000-4-11), interharmonics (EN 61000-4-13), and power quality measurement methods (IEC 61000-4-30).
- Part 5 (Installation and Mitigation Guidelines). Seven sections cover earthing (grounding), cabling,

**TABLE 1.5** Some Guides, Recommendations, and Standards on Electric Power Quality

| Source | Coverage |
|---|---|
| | **IEEE and ANSI Documents** |
| IEEE 4: 1995 | Standard techniques for high-voltage testing. |
| IEEE 100: 1992 | Standard dictionary of electrical and electronic terms. |
| IEEE 120: 1989 | Master test guide for electrical measurements in power circuits. |
| IEEE 141: 1993 | Recommended practice for electric power distribution for industrial plants. Effect of voltage disturbances on equipment within an industrial area. |
| IEEE 142: 1993 (The Green Book) | Recommended practice for grounding of industrial and commercial power systems. |
| IEEE 213: 1993 | Standard procedure for measuring conducted emissions in the range of 300 kHz to 25 MHz from television and FM broadcast receivers to power lines. |
| IEEE 241: 1990 (The Gray Book) | Recommended practice for electric power systems in commercial buildings. |
| IEEE 281: 1994 | Standard service conditions for power system communication equipment. |
| IEEE 299: 1991 | Standard methods of measuring the effectiveness of electromagnetic shielding enclosures. |
| IEEE 367: 1996 | Recommended practice for determining the electric power station ground potential rise and induced voltage from a power fault. |
| IEEE 376: 1993 | Standard for the measurement of impulse strength and impulse bandwidth. |
| IEEE 430: 1991 | Standard procedures for the measurement of radio noise from overhead power lines and substations. |
| IEEE 446: 1987 (The Orange Book) | Recommended practice for emergency and standby systems for industrial and commercial applications (e.g., power acceptability curve [5, Fig. 2-26], CBEMA curve). |
| IEEE 449: 1990 | Standard for ferroresonance voltage regulators. |
| IEEE 465 | Test specifications for surge protective devices. |
| IEEE 472 | Event recorders. |
| IEEE 473: 1991 | Recommended practice for an electromagnetic site survey (10 kHz to 10 GHz). |
| IEEE 493: 1997 (The Gold Book) | Recommended practice for the design of reliable industrial and commercial power systems. |
| IEEE 519: 1993 | Recommended practice for harmonic control and reactive compensation of static power converters. |
| IEEE 539: 1990 | Standard definitions of terms relating to corona and field effects of overhead power lines. |
| IEEE 859: 1987 | Standard terms for reporting and analyzing outage occurrences and outage states of electrical transmission facilities. |
| IEEE 944: 1986 | Application and testing of uninterruptible power supplies for power generating stations. |
| IEEE 998: 1996 | Guides for direct lightning strike shielding of substations. |
| IEEE 1048: 1990 | Guides for protective grounding of power lines. |
| IEEE 1057: 1994 | Standards for digitizing waveform recorders. |
| IEEE P1100: 1992 (The Emerald Book) | Recommended practice for powering and grounding sensitive electronic equipment in commercial and industrial power systems. |
| IEEE 1159: 1995 | Recommended practice on monitoring electric power quality. Categories of power system electromagnetic phenomena. |
| IEEE 1250: 1995 | Guides for service to equipment sensitive to momentary voltage disturbances. |
| IEEE 1346: 1998 | Recommended practice for evaluating electric power system compatibility with electronics process equipment. |
| IEEE P-1453 | Flicker. |
| IEEE/ANSI 18: 1980 | Standards for shunt power capacitors. |
| IEEE/ANSI C37 | Guides for surge withstand capability (SWC) tests. |
| IEEE/ANSI C50: 1982 | Harmonics and noise from synchronous machines. |
| IEEE/ANSI C57.110: 1986 | Recommended practice for establishing transformer capability when supplying nonsinusoidal load currents. |
| IEEE/ANSI C57.117: 1986 | Guides for reporting failure data for power transformers and shunt reactors on electric utility power systems. |
| IEEE/ANSI C62.45: 1992 (IEEE 587) | Recommended practice on surge voltage in low-voltage AC power circuits, including guides for lightning arresters applications. |
| IEEE/ANSI C62.48: 1995 | Guides on interactions between power system disturbances and surge protective devices. |
| ANSI C84.1: 1982 | American national standard for electric power systems and equipment voltage ratings (60 Hz). |

**TABLE 1.5** Some Guides, Recommendations, and Standards on Electric Power Quality (continued)

| Source | Coverage |
|---|---|
| ANSI 70 | National electric code. |
| ANSI 368 | Telephone influence factor. |
| ANSI 377 | Spurious radio frequency emission from mobile communication equipment. |
| **International Electrotechnical Commission (IEC) Documents** | |
| IEC 38: 1983 | Standard voltages. |
| IEC 816: 1984 | Guides on methods of measurement of short-duration transients on low-voltage power and signal lines. Equipment susceptible to transients. |
| IEC 868: 1986 | Flicker meter. Functional and design specifications. |
| IEC 868-0: 1991 | Flicker meter. Evaluation of flicker severity. Evaluates the severity of voltage fluctuation on the light flicker. |
| IEC 1000-3-2:1994 | Electromagnetic compatibility Part 3: Limits Section 2: Limits for harmonic current emissions (equipment absorbed current ≤16 A per phase). |
| IEC 1000-3-6: 1996 | Electromagnetic compatibility Part 3: Limits Section 6: Emission limits evaluation for perturbing loads connected to MV and HV networks. |
| IEC 1000-4: 1991 | Electromagnetic compatibility Part 4: Sampling and metering techniques. |
| EN 50160: 1994 | Voltage characteristics of electricity supplied by public distribution systems. |
| IEC/EN 60868-0 | Flicker meter implementation. |
| IEC 61000 standards on EMC | Electromagnetic compatibility (EMC). |
| **British Standards (BS) and European Norm Documents** | |
| BS5406 (based on IEC 555 part 2) | Control harmonic emissions from small domestic equipment. |
| **Other Documents** | |
| ER G5/3 | Basis of standards in some other (mostly commonwealth) countries, but it does not include notching and burst harmonics. |
| G5/4: 2001 | Limiting harmonic voltage distortion levels on public networks at the time of connection of new nonlinear loads to ensure compatibility of all connected equipment. |
| UIE-DWG-2-92-D | Produced by the Distribution Working Group (DWG) of Union Internationale Electrothermie (UIE). Includes guides for measurements of voltage dips and short-circuit interruptions occurring in industrial installations. |
| UIE-DWG-3-92-G | UIE guides for quality of electrical supply for industrial installations, including types of disturbances and relevant standards. |
| CBEMA Curves: 1983 | Produced by the Computer Business Equipment Manufacturers Association for the design of the power supply for computers and electronic equipment. |
| ITI Curves (new CBEMA curves) | Information Technology Industry Council (the new name for CBEMA) application. |

mitigation, and degrees of protection against EM (electromagnetic) disturbances.
- Part 6 (Generic Standards). Five sections cover immunity and emission standards for residential, commercial, industrial, and power station environments.

EN 61000-3-2 [2] introduces power quality limits (Table 1.6) for four classes of equipment:

- Class A: Balanced three-phase equipment and all other equipment, except those listed in other classes.
- Class B: Portable tools.

- Class C: Lighting equipment, including dimming devices.
- Class D: Equipment with a "special wave shape" and an input power of 75 to 600 W.

### 1.6.2 IEEE-519 Standard

The United States (ANSI and IEEE) do not have such a comprehensive and complete set of power quality standards as the IEC. However, their standards are more practical and provide theoretical background on the phenomena. This has made them very useful reference documents, even outside of the United States. IEEE-Std 519 [1] is the IEEE recommended practices and requirements for harmonic

**TABLE 1.6** Harmonic Limits Defined by the EN 61000 Standards for Different Classes of Equipment

| Harmonic order (h) | Class A (A) | Class B (A) | Class C (% of fundamental) | Class D (% of fundamental) |
|---|---|---|---|---|
| 2 | 1.08 | 1.62 | 2 | |
| 3 | 2.30 | 3.45 | $30 \times \lambda*$ | 3.4 |
| 4 | 0.43 | 0.65 | | |
| 5 | 1.44 | 2.16 | 10 | 1.9 |
| 6 | 0.30 | 0.45 | | |
| 7 | 0.77 | 1.12 | 7 | 1 |
| 8 | 0.23 | 0.35 | | |
| 9 | 0.40 | 0.60 | 5 | 0.5 |
| 10 | 0.18 | 0.28 | | |
| 11 | 0.33 | 0.50 | 3 | 0.35 |
| 12 | 0.15 | 0.23 | | |
| 13 | 0.21 | 0.32 | 3 | 0.296 |
| 14–40 (even) | 1.84/h | 2.76/h | | |
| 15–39 (odd) | 2.25/h | 3.338/h | 3 | 3.85/h |

$*\lambda$ is the circuit power factor.

control in electric power systems. It is one of the well-known documents for power quality limits. IEEE-519 is more comprehensive than IEC 61000-3-2 [2], but it is not a product standard. The first official version of this document was published in 1981. Product testing standards for the United States are now considered within TC77A/WG1 (TF5b) but are also discussed in IEEE. The current direction of the TC-77 working group is toward a global IEC standard for both 50/60 Hz and 115/230 V.

IEEE-519 contains thirteen sections, each with standards and technical reports [11]:

- Section 1 (Introduction and Scope). Includes application of the standards.
- Section 2 (Definition and Letter Symbols).
- Section 3 (References). Includes standard references.
- Section 4 (Converter Theory and Harmonic Generation). Includes documents for converters, arc furnaces, static VAr compensators, inverters for dispersed generation, electronic control, transformers, and generators.
- Section 5 (System Response Characteristics). Includes resonance conditions, effect of system loading, and typical characteristics of industrial, distribution, and transmission systems.
- Section 6 (Effect of Harmonics). Detrimental effects of harmonics on motors, generators, transformers, capacitors, electronic equipments, meters, relaying, communication systems, and converters.
- Section 7 (Reactive Power Compensation and Harmonic Control). Discusses converter power factor, reactive power compensation, and control of harmonics.

- Section 8 (Calculation Methods). Includes calculations of harmonic currents, telephone interference, line notching, distortion factor, and power factor.
- Section 9 (Measurements). For line notching, harmonic voltage and current, telephone interface, flicker, power factor improvement, instrumentation, and statistical characteristics of harmonics.
- Section 10 (Recommended Practices for Individual Consumers). Addresses standard impedance, customer voltage distortion limits, customer application of capacitors and filters, effect of multiple sources at a single customer, and line notching calculations.
- Section 11 (Recommended Harmonic Limits on the System). Recommends voltage distortion limits on various voltage levels, TIF limits versus voltage level, and IT products.
- Section 12 (Recommended Methodology for Evaluation of New Harmonic Sources).
- Section 13 (Bibliography). Includes books and general discussions.

IEEE-519 sets limits on the voltage and current harmonics distortion at the point of common coupling (PCC, usually the secondary of the supply transformer). The total harmonic distortion at the PCC is dependent on the percentage of harmonic distortion from each nonlinear device with respect to the total capacity of the transformer and the relative load of the system. There are two criteria that are used in IEEE-519 to evaluate harmonics distortion:

- limitation of the harmonic current that a user can transmit/inject into utility system ($THD_i$), and

**TABLE 1.7** IEEE-519 Harmonic Current Limits [1, 64] for Nonlinear Loads at the Point of Common Coupling (PCC) with Other Loads at Voltages of 2.4 to 69 kV

| | Maximum harmonic current distortion at PCC (% of fundamental) | | | | | |
|---|---|---|---|---|---|---|
| | Harmonic order (odd harmonics)[a] | | | | | |
| $I_{sc}/I_L$ | $h < 11$ | $11 \leq h < 17$ | $17 \leq h < 23$ | $23 \leq h < 35$ | $h \geq 35$ | $THD_i$ |
| $<20^b$ | 4.0 | 2.0 | 1.5 | 0.6 | 0.3 | 5.0 |
| 20–50 | 7.0 | 3.5 | 2.5 | 1.0 | 0.5 | 8.0 |
| 50–100 | 10.0 | 4.5 | 4.0 | 1.5 | 0.7 | 12.0 |
| 100–1000 | 12.0 | 5.5 | 5.0 | 2.0 | 1.0 | 15.0 |
| >1000 | 15.0 | 7.0 | 6.0 | 2.5 | 1.4 | 20.0 |

[a]Even harmonics are limited to 25% of the odd harmonic limits above.
[b]All power generation equipment is limited to these values of current distortion, regardless of the actual $I_{sc}/I_L$.
Here $I_{sc}$ = maximum short circuit current at PCC,
$I_L$ = maximum load current (fundamental frequency) at PCC.
For PCCs from 69 to 138 kV, the limits are 50% of the limits above. A case-by-case evaluation is required for PCCs of 138 kV and above.

**TABLE 1.8** IEEE-519 Harmonic Voltage Limits [1, 64] for Power Producers (Public Utilities or Cogenerators)

| | Harmonic voltage distortion (% at PCC) | | |
|---|---|---|---|
| | 2.3 to 69 kV | 69 to 138 kV | >138 kV |
| Maximum for individual harmonics | 3.0 | 1.5 | 1.0 |
| Total harmonic distortion ($THD_v$) | 5.0 | 2.5 | 1.5 |

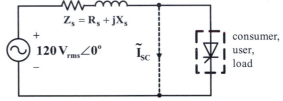

**FIGURE 1.23** Equivalent circuit of power system and nonlinear load. $Z_S$ is small (or $\tilde{I}_{SC}$ is large) for strong systems, and $Z_S$ is large (or $\tilde{I}_{SC}$ is small) for weak systems.

- limitation of the voltage distortion that the utility must furnish the user ($THD_v$).

The interrelationship of these two criteria shows that the harmonic problem is a system problem and not tied just to the individual load that generates the harmonic current.

Tables 1.7 and 1.8 list the harmonic current and voltage limits based on the size of the user with respect to the size of the power system to which the user is connected [1, 64].

The short-circuit current ratio ($R_{SC}$) is defined as the ratio of the short-circuit current (available at the point of common coupling) to the nominal fundamental load current (Fig. 1.23):

$$R_{SC} = \frac{|\tilde{I}_{SC}|}{\tilde{I}_L}. \qquad (1\text{-}41)$$

Thus the size of the permissible nonlinear user load increases with the size of the system; that is, the stronger the system, the larger the percentage of harmonic current the user is allowed to inject into the utility system.

Table 1.8 lists the amount of voltage distortion [1, 64] specified by IEEE-519 that is acceptable for a user as provided by a utility. To meet the power quality values of Tables 1.7 and 1.8, cooperation among all users and the utility is needed to ensure that no one user deteriorates the power quality beyond these limits. The values in Table 1.8 are low enough to ensure that equipment will operate correctly.

## 1.7 HARMONIC MODELING PHILOSOPHIES

For the simulation and modeling of power systems, the dynamic operation is normally subdivided into well-defined quasi steady-state regions [5]. Differential equations representing system dynamics in each region are transformed into algebraic relations, and the circuit is solved at the fundamental frequency (50 or 60 Hz) in terms of voltage and current phasors.

Modern power systems have many nonlinear components and loads that produce voltage and current harmonics. By definition, harmonics result from periodic steady-state conditions, and therefore their simulation should also be formulated in terms of harmonic phasors. Considering the complicated nature of many nonlinear loads (sources) and their couplings with the harmonic power flow, sophisticated modeling techniques are required for accurate simulation. Three techniques are usually used for harmonic analysis of power systems in the presence of nonlinear loads and/or components: time-domain simulation, frequency (harmonic)-domain modeling, and iterative procedures. The more recent approaches may use time-domain, frequency-domain or some combination of time- and frequency-domain techniques to achieve a more accurate solution (e.g., the main structure of many harmonic power flow algorithms are based on a frequency-domain technique, while nonlinear loads are modeled in a time-domain simulation).

### 1.7.1 Time-Domain Simulation

Dynamic characteristics of power systems are represented in terms of nonlinear sets of differential equations that are normally solved by numerical integration [5]. There are two commonly used time-domain techniques:

- state-variable approach, which is extensively used for the simulation of electronic circuits (SPICE [55]), and
- nodal analysis, which is commonly used for electromagnetic transient simulation of power system (EMTP [56]).

Two main limitations attached to the time-domain methods for harmonic studies are

- They usually require considerable computing time (even for small systems) for the calculation of harmonic information. This involves solving for the steady-state condition and then applying a fast Fourier transform (FFT); and
- There are some difficulties in time-domain modeling of power system components with distributed or frequency-dependent parameters.

The Electromagnetic Transient Program (EMTP) and PSpice are two of the well-known time-domain programs that are widely used for transient and harmonic analyses. Most examples of this book are solved using the PSpice software package.

### 1.7.2 Harmonic-Domain Simulation

The most commonly used model in the frequency domain assumes a balanced three-phase system (at fundamental and harmonic frequencies) and uses single-phase analysis, a single harmonic source, and a direct solution [5]. The injected harmonic currents

by nonlinear power sources are modeled as constant-current sources to make a direct solution possible. In the absence of any other nonlinear loads, the effect of a given harmonic source is often assessed with the help of equivalent harmonic impedances. The single-source concept is still used for harmonic filter design. Power systems are usually asymmetric. This justifies the need for multiphase harmonic models and power flow that considerably complicates the simulation procedures.

For more realistic cases, if more than one harmonic source is present in the power system, the single-source concept can still be used, provided that the interaction between them can be ignored. In these cases, the principle of superposition is relied on to compute the total harmonic distortion throughout the network.

### 1.7.3 Iterative Simulation Techniques

In many modern networks, due to the increased power ratings of nonlinear elements (e.g., HVDC systems, FACTS devices, renewable energy sources, and industrial and residential nonlinear loads) as compared to the system short-circuit power, application of superposition (as applied by harmonic-domain techniques) is not justified and will provide inaccurate results. In addition, due to the propagation of harmonic voltages and currents, the injected harmonics of each nonlinear load is a function of those of other sources. For such systems, accurate results can be obtained by iteratively solving nonlinear equations describing system steady-state conditions. At each iteration, the harmonic-domain simulation techniques can be applied, with all nonlinear interactions included. Two important aspects of the iterative harmonic-domain simulation techniques are:

- Derivation of system nonlinear equations [5]. The system is partitioned into linear regions and nonlinear devices (described by isolated equations). The system solution then consists predominantly of the solution for given boundary conditions as applied to each nonlinear device. Many techniques have been proposed for device modeling including time-domain simulation, steady-state analysis, analytical time-domain expressions [references 11, 13 of [5]], waveform sampling and FFT [reference 14 of [5]], and harmonic phasor analytical expressions [reference 15 of [5]].
- Solution of nonlinear equations [5]. Early methods used the fixed point iteration procedure of Gauss-Seidel that frequently diverges. Some techniques

replace the nonlinear devices at each iteration by a linear Norton equivalent (which might be updated at the next iteration). More recent methods make use of Newton-type solutions and completely decouple device modeling and system solution. They use a variety of numerical analysis improvement techniques to accelerate the solution procedure.

Detailed analyses of iterative simulation techniques for harmonic power (load) flow are presented in Chapter 7.

### 1.7.4 Modeling Harmonic Sources

As mentioned above, an iterative harmonic power flow algorithm is used for the simulation of the power system with nonlinear elements. At each iteration, harmonic sources need to be accurately included and their model must be updated at the next iteration.

For most harmonic power flow studies it is suitable to treat harmonic sources as (variable) harmonic currents. At each iteration of the power flow algorithm, the magnitudes and phase angles of these harmonic currents need to be updated. This is performed based on the harmonic couplings of the nonlinear load. Different techniques have been proposed to compute and update the values of injected harmonic currents, including:

- Thevenin or Norton harmonic equivalent circuits,
- simple decoupled harmonic models for the estimation of nonlinear loads (e.g., $I_h = 1/h$, where $h$ is the harmonic order),
- approximate modeling of nonlinear loads (e.g., using decoupled constant voltage or current harmonic sources) based on measured voltage and current characteristics or published data, and
- iterative nonlinear (time- and/or frequency-based) models for detailed and accurate simulation of harmonic-producing loads.

### 1.8 POWER QUALITY IMPROVEMENT TECHNIQUES

Nonlinear loads produce harmonic currents that can propagate to other locations in the power system and eventually return back to the source. Therefore, harmonic current propagation produces harmonic voltages throughout the power systems. Many mitigation techniques have been proposed and implemented to maintain the harmonic voltages and currents within recommended levels:

- high power quality equipment design,
- harmonic cancellation,
- dedicated line or transformer,
- optimal placement and sizing of capacitor banks,
- derating of power system devices, and
- harmonic filters (passive, active, hybrid) and custom power devices such as active power line conditioners (APLCs) and unified or universal power quality conditioners (UPQCs).

The practice is that if at PCC harmonic currents are not within the permissible limits, the consumer with the nonlinear load must take some measures to comply with standards. However, if harmonic voltages are above recommended levels – and the harmonic currents injected comply with standards – the utility will have to take appropriate actions to improve the power quality.

Detailed analyses of improvement techniques for power quality are presented in Chapters 8 to 10.

### 1.8.1 High Power Quality Equipment Design

The use of nonlinear and electronic-based devices is steadily increasing and it is estimated that they will constitute more than 70% of power system loading by year 2010 [10]. Therefore, demand is increasing for the designers and product manufacturers to produce devices that generate lower current distortion, and for end users to select and purchase high power quality devices. These actions have already been started in many countries, as reflected by improvements in fluorescent lamp ballasts, inclusion of filters with energy saving lamps, improved PWM adjustable-speed drive controls, high power quality battery chargers, switch-mode power supplies, and uninterruptible power sources.

### 1.8.2 Harmonic Cancellation

There are some relatively simple techniques that use transformer connections to employ phase-shifting for the purpose of harmonic cancellation, including [10]:

- delta-delta and delta-wye transformers (or multiple phase-shifting transformers) for supplying harmonic producing loads in parallel (resulting in twelve-pulse rectifiers) to eliminate the 5th and 7th harmonic components,
- transformers with delta connections to trap and prevent triplen (zero-sequence) harmonics from entering power systems,
- transformers with zigzag connections for cancellation of certain harmonics and to compensate load imbalances,

- other phase-shifting techniques to cancel higher harmonic orders, if required, and
- canceling effects due to diversity [57–59] have been discovered.

### 1.8.3 Dedicated Line or Transformer

Dedicated (isolated) lines or transformers are used to attenuate both low- and high-frequency electrical noise and transients as they attempt to pass from one bus to another. Therefore, disturbances are prevented from reaching sensitive loads and any load-generated noise and transients are kept from reaching the remainder of the power system. However, some common-mode and differential noise can still reach the load. Dedicated transformers with (single or multiple) electrostatic shields are effective in eliminating common-mode noise.

Interharmonics (e.g., caused by induction motor drives) and voltage notching (e.g., due to power electronic switching) are two examples of problems that can be reduced at the terminals of a sensitive load by a dedicated transformer. They can also attenuate capacitor switching and lightning transients coming from the utility system and prevent nuisance tripping of adjustable-speed drives and other equipment. Isolated transformers do not totally eliminate voltage sags or swells. However, due to the inherent large impedance, their presence between PCC and the

source of disturbance (e.g., system fault) will lead to relatively shallow sags.

An additional advantage of dedicated transformers is that they allow the user to define a new ground reference that will limit neutral-to-ground voltages at sensitive equipment.

#### 1.8.3.1 Application Example 1.6: Interharmonic Reduction by Dedicated Transformer

Figure E1.6.1 shows a typical distribution system with linear and nonlinear loads. The nonlinear load (labeled as "distorting nonlinear load") consists of two squirrel-cage induction motors used as prime movers for chiller-compressors for a building's air-conditioning system. This load produces interharmonic currents that generate interharmonic voltage drops across the system's impedances resulting in the interharmonic content of the line-to-line voltage of the induction motors as given by Table E1.6.1. Some of the loads are very sensitive to interharmonics and these must be reduced at the terminals of sensitive loads. These loads are labeled as "sensitive loads."

Three case studies are considered:

- Case #1: Distorting nonlinear load and sensitive loads are fed from the same pole transformer (Fig. E1.6.2).

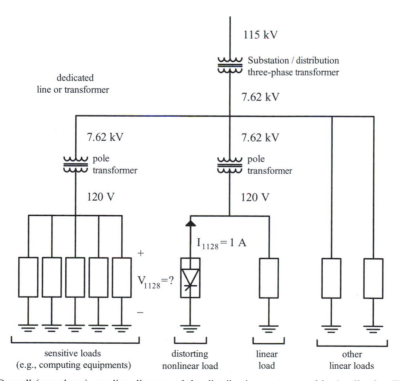

**FIGURE E1.6.1** Overall (per phase) one-line diagram of the distribution system used in Application Example 1.6.

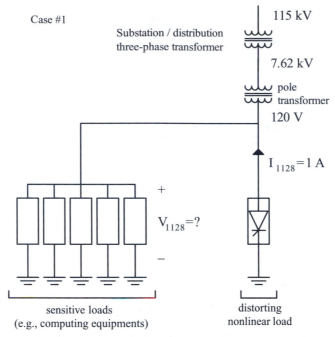

**FIGURE E1.6.2** Case #1 of Application Example 1.6: distorting nonlinear load and sensitive loads are fed from same pole transformer.

**TABLE E1.6.1** Interharmonics of Phase Current and Line-to-Line Voltage Generated by a Three-Phase Induction Motor [34]

| Interharmonic $f_h$ (Hz) | Interharmonic amplitude of phase current (%) | Interharmonic amplitude of line-to-line voltage (%) |
|---|---|---|
| 1128 | 7 | 0.40 |
| 1607 | 10 | 0.40 |
| 1730 | 10 | 0.55 |

- Case #2: A dedicated 1:1 isolation transformer is used between the distorting nonlinear load and sensitive loads (Fig. E1.6.3).
- Case #3: A dedicated 7.62 kV to 120 V pole transformer is used between the distorting nonlinear load and sensitive loads (Fig. E1.6.4).

### 1.8.4 Optimal Placement and Sizing of Capacitor Banks

It is well known that proper placement and sizing of shunt capacitor banks in distorted networks can result in reactive power compensation, improved voltage regulation, power factor correction, and power/energy loss reduction. The capacitor placement problem consists of determining the optimal numbers, types, locations, and sizes of capacitor banks such that minimum yearly cost due to peak power/energy losses and cost of capacitors is achieved, while the operational constraints are maintained within required limits.

Most of the reported techniques for capacitor placement assume sinusoidal operating conditions. These methods include nonlinear programming, and near global methods (genetic algorithms, simulated annealing, tabu search, artificial neural networks, and fuzzy theory). All these approaches ignore the presence of voltage and current harmonics [60, 61].

Optimal capacitor bank placement is a well-researched subject. However, limited attention is given to this problem in the presence of voltage and current harmonics. Some publications have taken into account the presence of distorted voltages for solving the capacitor placement problem. These investigations include exhaustive search, local variations, mixed integer and nonlinear programming, heuristic methods for simultaneous capacitor and filter placement, maximum sensitivities selection, fuzzy theory, and genetic algorithms.

According to newly developed investigations based on fuzzy and genetic algorithms [60, 61], proper placement and sizing of capacitor banks in power systems with nonlinear loads can result in lower system losses, greater yearly benefits, better voltage profiles, and prevention of harmonic parallel resonances, as well as improved power quality. Simulation results for the standard 18-bus IEEE

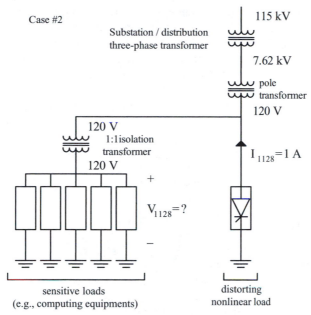

**FIGURE E1.6.3** Case #2 of Application Example 1.6: use of an isolation transformer with a turns ratio 1 : 1 between distorting (nonlinear) load and sensitive loads.

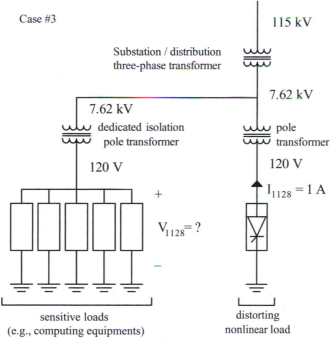

**FIGURE E1.6.4** Case #3 of Application Example 1.6: use of a dedicated (isolation transformer with turns ratio 7620 : 120) pole transformer between distorting nonlinear load and sensitive loads.

distorted distribution system show that proper placement and sizing of capacitor banks can limit voltage and current harmonics and decrease their THDs to the recommended levels of IEEE-519, without application of any dedicated high-order passive or active filters. For cases where the construction of new capacitor bank locations is not feasible, it is possible to perform the optimization process without defining any new locations. Therefore, reexamining capacitor bank sizes and locations before taking any major steps for power quality mitigation is highly recommended.

Detailed analyses for optimal sizing and placement of capacitor banks in the presence of harmonics and nonlinear loads are presented in Chapter 10.

### 1.8.5 Derating of Power System Devices

Power system components must be derated when supplying harmonic loads. Commercial buildings have drawn the most attention in recent years due to the increasing use of nonlinear loads. According to the IEEE dictionary, derating is defined as "the intentional reduction of stress/strength ratio (e.g., real or apparent power) in the application of an item (e.g., cables, transformer, electrical machines), usually for the purpose of reducing the occurrence of stress-related failure (e.g., reduction of lifetime due to increased temperature beyond rated temperature)." As discussed in Section 1.5, harmonic currents and voltages result in harmonic losses of magnetic devices, increasing their temperature rise [62]. This rise beyond the rated value results in a reduction of lifetime, as will be discussed in Chapter 6.

There are several techniques for determining the derating factors (functions) of appliances for nonsinusoidal operating conditions (as discussed in Chapter 2), including:

- from tables in standards and published research (e.g., ANSI/IEEE Std C57.110 [63] for transformer derating),
- from measured (or computed) losses,
- by determining the K-factor, and
- based on the $F_{HL}$-factor.

### 1.8.6 Harmonic Filters, APLCs, and UPQCs

One means of ensuring that harmonic currents of nonlinear components will not unduly interact with the remaining part of the power system is to place filters near or close to nonlinear loads. The main function of a filter is either to bypass harmonic currents, block them from entering the power system, or compensate them by locally supplying harmonic currents. Due to the lower impedance of the filter in comparison to the impedance of the system, harmonic currents will circulate between the load and the filter and do not affect the entire system; this is called series resonance. If other frequencies are to be controlled (e.g., that of arc furnaces), additional tuned filters are required.

Harmonic filters are broadly classified into passive, active, and hybrid structures. These filters can only compensate for harmonic currents and/or harmonic voltages at the installed bus and do not consider the power quality of other buses. New generations of active filters are active-power line conditioners that are capable of minimizing the power quality of the entire system.

Passive filters are made of passive components (inductance, capacitance, and resistance) tuned to the harmonic frequencies that are to be attenuated. The values of inductors and capacitors are selected to provide low impedance paths at the selected frequencies. Passive filters are generally designed to remove one or two harmonics (e.g., the 5th and 7th). They are relatively inexpensive compared with other means for eliminating harmonic distortion, but also suffer from some inherent limitations, including:

- Interactions with the power system;
- Forming parallel resonance circuits with system impedance (at fundamental and/or harmonic frequencies). This may result in a situation that is worse than the condition being corrected. It may also result in system or equipment failure;
- Changing characteristics (e.g., their notch frequency) due to filter parameter variations;
- Unsatisfactory performance under variations of nonlinear load parameters;
- Compensating a limited number of harmonics;
- Not considering the power quality of the entire system; and
- Creating parallel resonance. This resonance frequency must not necessarily coincide with any significant system harmonic. Passive filters are commonly tuned slightly lower than the attenuated harmonic to provide a margin of safety in case there are some changes in system parameters (due to temperature variations and/or failures). For this reason filters are added to the system starting with the lowest undesired harmonic. For example, installing a seventh-harmonic filter usually requires that a fifth-harmonic filter also be installed.

Designing passive filters is a relatively simple but tedious matter. For the proper tuning of passive filters, the following steps should be followed:

- Model the power system (including nonlinear loads) to indicate the location of harmonic sources and the orders of the injected harmonics. A harmonic power (load) flow algorithm (Chapter 7) should be used; however, for most applications with a single dominating harmonic source, a simplified equivalent model and hand calculations are adequate;

- Place the hypothetical harmonic filter(s) in the model and reexamine the system. Filter(s) should be properly tuned to dominant harmonic frequencies; and
- If unacceptable results (e.g., parallel resonance within system) are obtained, change filter location(s) and modify parameter values until results are satisfactory.

In addition to power quality improvement, harmonic filters can be configured to provide power factor correction. For such cases, the filter is designed to carry resonance harmonic currents, as well as fundamental current.

Active filters rely on active power conditioning to compensate for undesirable harmonic currents. They actually replace the portion of the sine wave that is missing in the nonlinear load current by detecting the distorted current and using power electronic switching devices to inject harmonic currents with complimentary magnitudes, frequencies, and phase shifts into the power system. Their main advantage over passive filters is their fine response to changing loads and harmonic variations. Active filters can be used in very difficult circumstances where passive filters cannot operate successfully because of parallel resonance within the system. They can also take care of more than one harmonic at a time and improve or mitigate other power quality problems such as flicker. They are particularly useful for large, distorting nonlinear loads fed from relatively weak points of the power system where the system impedance is relatively large. Active filters are relatively expensive and not feasible for small facilities.

Power quality improvement using filters, optimal placement and sizing of shunt capacitors, and unified power quality conditioners (UPQCs) are discussed in Chapters 9, 10, and 11, respectively.

### 1.8.6.1 Application Example 1.7: Hand Calculation of Harmonics Produced by Twelve-Pulse Converters

Figure E1.7.1 shows a large industrial plant such as an oil refinery or chemical plant [64] being serviced from a utility with transmission line-to-line voltage of 115 kV. The demand on the utility system is 50 MVA and 50% of its load is a twelve-pulse static power converter load.

Table E1.7.1 lists the harmonic currents ($I_h$) given in pu of the fundamental current based on the com-

**TABLE E1.7.1** Harmonic Current ($I_h$) Generated by Six-Pulse and Twelve-Pulse Converters [64] Based on $X_c^h = 0.12$ pu and $\alpha = 30°$

| Harmonic order ($h$) | $I_h$ for 6-pulse converter (pu) | $I_h$ for 12-pulse converter (pu) |
|---|---|---|
| 1 | 1.000 | 1.000 |
| 5 | 0.192 | **0.0192** |
| 7 | 0.132 | **0.0132** |
| 11 | 0.073 | 0.073 |
| 13 | 0.057 | 0.057 |
| 17 | 0.035 | **0.0035** |
| 19 | 0.027 | **0.0027** |
| 23 | 0.020 | 0.020 |
| 25 | 0.016 | 0.016 |
| 29 | 0.014 | **0.0014** |
| 31 | 0.012 | **0.0012** |
| 35 | 0.011 | 0.011 |
| 37 | 0.010 | 0.010 |
| 41 | 0.009 | **0.0009** |
| 43 | 0.008 | **0.0008** |
| 47 | 0.008 | 0.008 |
| 49 | 0.007 | 0.007 |

mutating reactance $X_c^h = 0.12$ pu and the firing angle $\alpha = 30°$ of six-pulse and twelve-pulse converters. In an ideal twelve-pulse converter, the magnitude of some current harmonics (bold in Table E1.7.1) is zero. However, for actual twelve-pulse converters, the magnitudes of these harmonics are normally taken as 10% of the six-pulse values [64].

### 1.8.6.2 Application Example 1.8: Filter Design to Meet IEEE-519 Requirements

Filter design for Application Example 1.7 will be performed to meet the IEEE-519 requirements. The circuit of Fig. E1.7.1 is now augmented with a passive filter, as shown by Fig. E1.8.1.

### 1.8.6.3 Application Example 1.9: Several Users on a Single Distribution Feeder

Figure E1.9.1 shows a utility distribution feeder that has four users along a radial feeder [64]. Each user sees a different value of short-circuit impedance or system size. Note that

$$S_{SC} = MVA_{SC} = \frac{10\text{MVA}}{Z_{sys}[\text{pu at 10MVA base}]}.$$

There is one type of transformer $(\Delta - Y)$; therefore, only six-pulse static power converters are used.

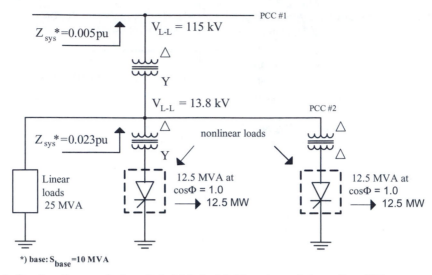

**FIGURE E1.7.1** One-line diagram of a large industrial plant fed from transmission voltage [64].

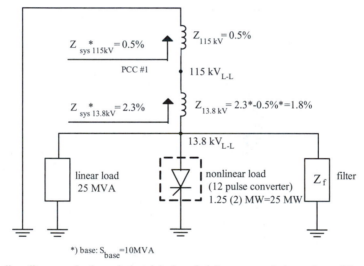

**FIGURE E1.8.1** One-line diagram of a large industrial plant fed from transmission voltage (Fig. E1.7.1) with a passive filter placed at PCC #2.

## 1.9 SUMMARY

The focus of this chapter has been on definition, measures, and classification of electric power quality as well as related issues that will be covered in the following chapters. Power quality can be defined as "the measure, analysis, and improvement of the bus voltage to maintain a sinusoidal waveform at rated voltage and frequency." Main causes of disturbances and power quality problems are unpredictable events, the electric utility, the customer, and the manufacturer.

The magnitude–duration plot can be used to classify power quality events, where the voltage magni-tude is split into three regions (e.g., interruption, undervoltage, and overvoltage) and the duration of these events is split into four regions (e.g., very short, short, long, and very long). However, IEEE stan-dards use several additional terms to classify power quality events into seven categories including: tran-sient, short-duration voltage variation, long-duration voltage variation, voltage imbalance, waveform dis-tortion, voltage fluctuation (and flicker), and power–frequency variation. Main sources for the formulations and measures of power quality are IEEE Std 100, IEC Std 61000-1-1, and CENELEC Std EN 50160. Some of the main detrimental effects of poor power

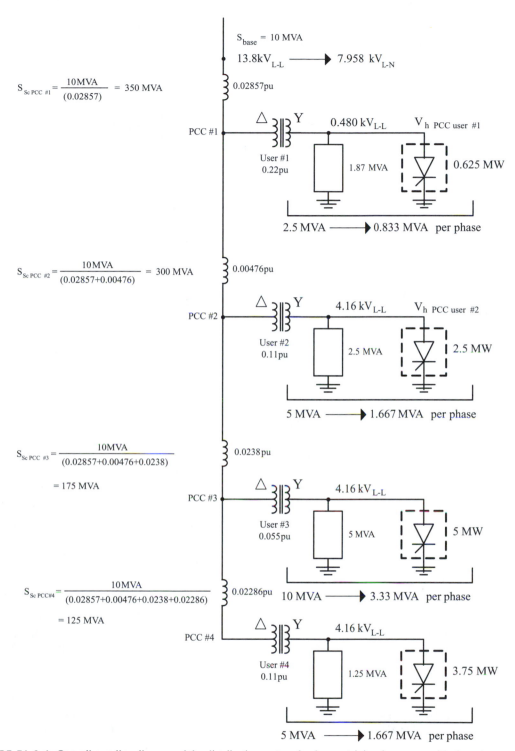

$$S_{Sc\,PCC\,\#1} = \frac{10\,MVA}{(0.02857)} = 350\,MVA$$

$$S_{Sc\,PCC\,\#2} = \frac{10\,MVA}{(0.02857 + 0.00476)} = 300\,MVA$$

$$S_{Sc\,PCC\,\#3} = \frac{10\,MVA}{(0.02857 + 0.00476 + 0.0238)} = 175\,MVA$$

$$S_{Sc\,PCC\#4} = \frac{10\,MVA}{(0.02857 + 0.00476 + 0.0238 + 0.02286)} = 125\,MVA$$

**FIGURE E1.9.1** Overall one-line diagram of the distribution system feeder containing four users with six-pulse converters [64].

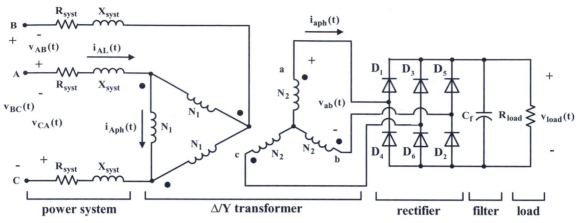

**FIGURE P1.1** Connection of a delta/wye three-phase transformer with a diode rectifier, filter, and load.

quality include increase or decrease of the fundamental voltage component, additional losses, heating, and noise, decrease of appliance and equipment lifetime, malfunction and failure of components, controllers, and loads, resonance and ferroresonance, flicker, harmonic instability, and undesired (harmonic, subharmonic, and interharmonic) torques.

Documents for control of power quality come in three levels of applicability and validity: guidelines, recommendations, and standards. IEEE-Std 519 and IEC 61000 (or EN 61000) are the most commonly used references for power quality in the United States and Europe, respectively.

Three techniques are used for harmonic analysis: time-domain simulation, frequency (harmonic)-domain modeling, and iterative procedures.

Many mitigation techniques for controlling power quality have been proposed, including high power quality equipment design, harmonic cancellation, dedicated line or transformer, optimal placement and sizing of capacitor banks, derating of devices, harmonic filters (passive, active, hybrid), and custom-build power devices. The practice is that if at PCC harmonic currents are not within the permissible limits, the consumer with the nonlinear load must take some measures to comply with standards. However, if harmonic voltages are above recommended levels—and the harmonic currents injected comply with standards – the utility will have to take appropriate actions to improve the power quality.

Nine application examples with solutions are provided for further clarifications of the presented materials. The reader is encouraged to read the overview of the text given in the preface before delving further into the book.

## 1.10 PROBLEMS

### Problem 1.1: Delta/wye Three-Phase Transformer Configuration

a) Perform a PSpice analysis for the circuit of Fig. P1.1, where a three-phase diode rectifier with filter (e.g., capacitor $C_f$) serves a load ($R_{load}$). You may assume ideal transformer conditions. For your convenience you may assume $(N_1/N_2) = 1$, $R_{syst} = 0.01\ \Omega$, $X_{syst} = 0.05\ \Omega$ @ $f = 60$ Hz, $v_{AB}(t) = \sqrt{2}600V \cos \omega t$, $v_{BC}(t) = \sqrt{2}600V \cos(\omega t - 120°)$, $v_{CA}(t) = \sqrt{2}600V \cos(\omega t - 240°)$, ideal diodes $D_1$ to $D_6$, $C_f = 500\ \mu F$, and $R_{load} = 10\ \Omega$. Plot one period of either voltage or current after steady state has been reached as requested in parts $b$ to $e$.

b) Plot and subject the line-to-line voltages $v_{AB}(t)$ and $v_{ab}(t)$ to Fourier analysis. Why are they different?

c) Plot and subject the input line current $i_{AL}(t)$ of the delta primary to a Fourier analysis. Note that the input line currents of the primary delta do not contain the 3rd, 6th, 9th, 12th, . . ., that is, harmonic zero-sequence current components.

d) Plot and subject the phase current $i_{Aph}(t)$ of the delta primary to a Fourier analysis. Why do the phase currents of the primary delta not contain the 3rd, 6th, 9th, 12th, . . ., that is, harmonic zero-sequence current components?

e) Plot and subject the output current $i_{aph}(t)$ of the wye secondary to a Fourier analysis. Why do the output currents of the secondary wye not contain the 3rd, 6th, 9th, 12th, . . ., that is, zero-sequence current components?

## Problem 1.2: Voltage Phasor Diagrams of a Three-Phase Transformer in Delta/Zigzag Connection

Figure P1.2 depicts the so-called delta/zigzag configuration of a three-phase transformer, which is used for supplying power to unbalanced loads and three-phase rectifiers. You may assume ideal transformer conditions. Draw a phasor diagram of the primary and secondary voltages when there is no load on the secondary side. For your convenience you may assume $(N_1/N_2) = 1$. For balanced phase angles 0°, 120°, and 240° of voltages and currents you may use hexagonal paper.

## Problem 1.3: Current Phasor Diagrams of a Three-Phase Transformer in Delta/Zigzag Connection With Line-To-Line Load

The delta/zigzag, three-phase configuration is used for feeding unbalanced loads and three-phase rectifiers. You may assume ideal transformer conditions. Even when only one line-to-line load (e.g., $R_{load}$) of the secondary is present as indicated in Fig. P1.3, the primary line currents $\tilde{I}_{LA}$, $\tilde{I}_{LB}$, and $\tilde{I}_{LC}$ will be balanced because the line-to-line load is distributed to all three (single-phase) transformers. This is the advantage of a delta/zigzag configuration. If there is a resistive line-to-line load on the secondary side

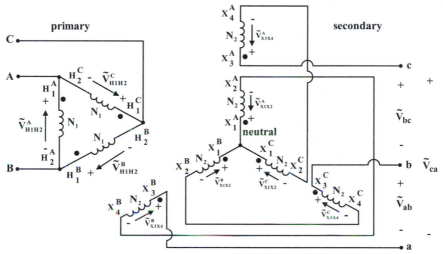

**FIGURE P1.2** Connection of a delta/zigzag, three-phase transformer with the definition of primary and secondary voltages.

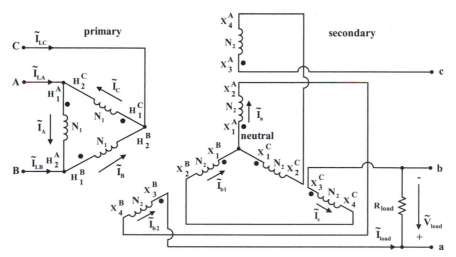

**FIGURE P1.3** Connection of a delta/zigzag, three-phase transformer with the definition of current for line-to-line load.

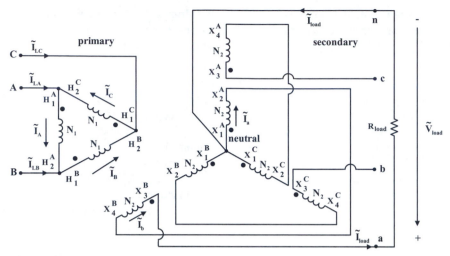

**FIGURE P1.4** Connection of a delta/zigzag, three-phase transformer with the definition of current for line-to-neutral load.

($|\tilde{I}_{load}| = 10$ A) present as illustrated in Fig. P1.3, draw a phasor diagram of the primary and secondary currents as defined in Fig. P1.3. For your convenience you may assume that the same voltage definitions apply as in Fig. P1.2 and ($N_1/N_2) = 1$. For balanced phase angles 0°, 120°, and 240° of voltages and currents you may use hexagonal paper.

### Problem 1.4: Current Phasor Diagrams of a Three-Phase Transformer in Delta/Zigzag Connection with Line-To-Neutral Load

Repeat the analysis of Problem 1.3 if there is a resistive line-to-neutral load on the secondary side ($|\tilde{I}_{load}| = 10$ A) present, as illustrated in Fig. P1.4; that is, draw a phasor diagram of the primary and secondary currents as defined in Fig. P1.4. For your convenience you may assume that the same voltage definitions apply as in Fig. P1.2 and ($N_1/N_2) = 1$. In this case the load is distributed to two (single-phase) transformers. For balanced phase angles 0°, 120°, and 240° of voltages and currents you may use hexagonal paper.

### Problem 1.5: Current Phasor Diagrams of a Three-Phase Transformer in Delta/Zigzag Connection with Three-Phase Unbalanced Load

Repeat the analysis of Problem 1.3 if there is a resistive unbalanced load on the secondary side ($|\tilde{I}_{load_a}| = 30$ A, $|\tilde{I}_{load_b}| = 20$ A, $|\tilde{I}_{load_c}| = 10$ A) present as illustrated in Fig. P1.5; that is, draw a phasor diagram of the primary and secondary currents as defined in Fig. P1.5. For your convenience you may assume that the same voltage definitions apply as in Fig. P1.2 and

($N_1/N_2) = 1$. For balanced phase angles 0°, 120°, and 240° of voltages and currents you may use hexagonal paper.

### Problem 1.6: Delta/Zigzag Three-Phase Transformer Configuration without Filter

Perform a PSpice analysis for the circuit of Fig. P1.6 where a three-phase diode rectifier without filter (e.g., capacitance $C_f = 0$) serves $R_{load}$. You may assume ideal transformer conditions. For your convenience you may assume ($N_1/N_2) = 1$, $R_{syst} = 0.01$ Ω, $X_{syst} = 0.05$ Ω @ $f = 60$ Hz, $v_{AB}(t) = \sqrt{2}600V \cos \omega t$, $v_{BC}(t) = \sqrt{2}600V \cos(\omega t - 120°)$, $v_{CA}(t) = \sqrt{2}600V \cos(\omega t - 240°)$, ideal diodes $D_1$ to $D_6$, and $R_{load} = 10$ Ω. Plot one period of either voltage or current after steady state has been reached as requested in the following parts.

a) Plot and subject the line-to-line voltages $v_{AB}(t)$ and $v_{ab}(t)$ to a Fourier analysis. Why are they different?

b) Plot and subject the input line current $i_{AL}(t)$ of the delta primary to a Fourier analysis. Note that the input line currents of the primary delta do not contain the 3rd, 6th, 9th, 12th,..., that is, harmonic zero-sequence current components.

c) Plot and subject the phase current $i_{Aph}(t)$ of the delta primary to a Fourier analysis. Why do the phase currents of the primary delta not contain the 3rd, 6th, 9th, 12th, ..., that is, harmonic zero-sequence current components?

d) Plot and subject the output current $i_{aph}(t)$ of the zigzag secondary to a Fourier analysis. Why do

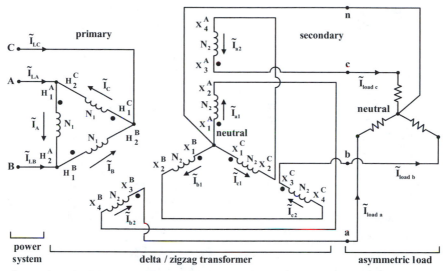

**FIGURE P1.5** Connection of a delta/zigzag, three-phase transformer with the definition of currents for three-phase unbalanced line-to-neutral loads.

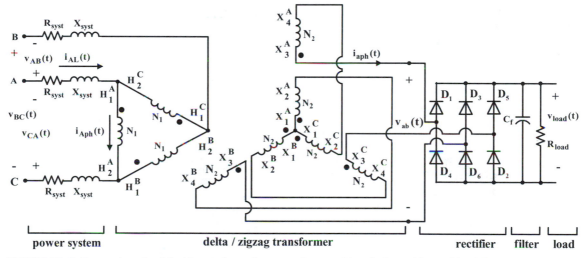

**FIGURE P1.6** Connection of a delta/zigzag, three-phase transformer with a diode rectifier and load $R_{load}$.

the output currents of the secondary zigzag not contain the 3rd, 6th, 9th, 12th, ..., that is, harmonic zero-sequence current components?

e) Plot and subject the output voltage $v_{load}(t)$ to a Fourier analysis.

## Problem 1.7: Delta/Zigzag Three-Phase Transformer Configuration with Filter

Perform a PSpice analysis for the circuit of Fig. P1.6 where a three-phase diode rectifier with filter (e.g., capacitance $C_f = 500\ \mu\text{F}$) serves the load $R_{load}$. You may assume ideal transformer conditions. For your convenience you may assume $(N_1/N_2) = 1$, $R_{syst} = 0.01\ \Omega$, $X_{syst} = 0.05\ \Omega$ @ $f = 60$ Hz, $v_{AB}(t) =$

$\sqrt{2}600V \cos\omega t$, $\qquad v_{BC}(t) = \sqrt{2}600V \cos(\omega t - 120°)$, $v_{CA}(t) = \sqrt{2}600V \cos(\omega t - 240°)$, ideal diodes $D_1$ to $D_6$, and $R_{load} = 10\ \Omega$. Plot one period of either voltage or current after steady state has been reached as requested in the following parts.

a) Plot and subject the line-to-line voltages $v_{AB}(t)$ and $v_{ab}(t)$ to a Fourier analysis. Why are they different?

b) Plot and subject the input line current $i_{AL}(t)$ of the delta primary to a Fourier analysis. Note that the input line currents of the primary delta do not contain the 3rd, 6th, 9th, 12th, ..., that is, harmonic zero-sequence current components.

c) Plot and subject the phase current $i_{Aph}(t)$ of the delta primary to a Fourier analysis. Why do the phase currents of the primary delta not contain the 3rd, 6th, 9th, 12th, ..., that is, zero-sequence current components?

d) Plot and subject the output current $i_{aph}(t)$ of the zigzag secondary to a Fourier analysis. Why do the output currents of the secondary zigzag not contain the 3rd, 6th, 9th, 12th, ..., that is, harmonic zero-sequence current components?

e) Plot and subject the output voltage $v_{load}(t)$ to a Fourier analysis.

## Problem 1.8: Transient Performance of a Brushless Dc Motor Fed by a Fuel Cell

Replace the battery (with the voltage $V_{DC} = 300$ V) of Fig. E1.3.1 by the equivalent circuit of the fuel cell as described in Fig. 2 of [67]. You may assume that in Fig. 2 of [67] the voltage $E = 300 \pm 30$ V, where the superimposed rectangular voltage $\pm 30$ V has a period of $T_{\pm 30\,V} = 1$ s. The remaining parameters of the fuel cell equivalent circuit can be extrapolated from Table III of [67]. Repeat the analysis as requested in Application Example 1.3.

## Problem 1.9: Transient Performance of an Inverter Feeding into Three-Phase Power System When Supplied by a Fuel Cell

Replace the DC source (with the voltage $V_Q = 390$ V) of Fig. E1.5.1 by the equivalent circuit of the fuel cell as described in Fig. 2 of [67]. You may assume that in Fig. 2 of [67] the voltage $E = 390 \pm 30$ V, where the superimposed rectangular voltage $\pm 30$ V has a period of $T_{\pm 30V} = 1$ s. The remaining parameters of the fuel cell equivalent circuit can be extrapolated from Table III of [67]. Repeat the analysis as requested in Application Example 1.5.

## Problem 1.10: Suppression of Subharmonic of 30 Hz with a Dedicated Transformer

The air-conditioning drive (compressor motor) generates a subharmonic current of $I_{30Hz} = 1$ A due to spatial harmonics (e.g., selection of number of slots, rotor eccentricity). A sensitive load fed from the same pole transformer is exposed to a terminal voltage with the low beat frequency of 30 Hz. A dedicated transformer can be used to suppress the 30 Hz component from the power supply of the sensitive load (see Fig. E1.6.1 and Figs. E1.6.4 to E1.6.6).

The parameters of the single-phase pole transformer at 60 Hz are $X_{sP} = X'_{pP} = 0.07$ Ω, $R_{sP} = R'_{pP} = 0$.

a) Draw an equivalent circuit of the substation transformer (per phase) and that of the pole transformer.

b) Find the required leakage primary and secondary leakage inductances $L_{pD}$ and $L_{sD}$ of the substation distribution transformer (per phase) for $R_{pD} = R_{sD} = 0$ such that the subharmonic voltage across the sensitive load $v_{30Hz} \leq 1$ mV provided $L_{sD} = L'_{pD}$, where the prime refers the inductance $L$ of the primary to the secondary side of the distribution transformer.

c) Without using a dedicated transformer, design a passive filter such that the same reduction of the subharmonic is achieved.

## Problem 1.11: Harmonic Currents of a Feeder

For Application Example 1.9 (Fig. E1.9.1), calculate the harmonic currents associated with users #3 and #4. Are they within the permissible power quality limits of IEEE-519?

## Problem 1.12: Design a Filter So That the Displacement (Fundamental Power) Factor

$\cos \Phi_{1\,with\,filter}^{total}$ will be 0.9 lagging (consumer notation) $\leq \cos \varphi_{1\,with\,filter}^{total} \leq 1.0$.

Figure P1.12.1 shows the one-line diagram of an industrial plant being serviced from a utility transmission voltage at $13.8$ kV$_{L-L}$. The total power demand on the utility system is 5 MVA: 3 MVA is a six-pulse static power converter load (three-phase rectifier with firing angle $\alpha = 30°$, note $\cos \Phi_1^{nonlinear} = 0.955 \cos \alpha$ lagging), while the remaining 2 MVA is a linear (induction motor) load at $\cos \Phi_1^{linear} = 0.8$ lagging (inductive) displacement (fundamental power) factor. The system impedance is $Z_{syst} = 10\%$ referred to a 10 MVA base.

a) Calculate the short-circuit apparent power $S_{SC}$ at PCC.

b) Find the short-circuit current $I_{scphase}$.

c) Before filter installation, calculate the displacement (fundamental power) factor $\cos \varphi_{1\,without\,filter}^{total}$, where $\varphi_{1\,without\,filter}^{total}$ is the angle between the fundamental voltage $\tilde{V}_{phase} = V_{phase} \angle 0°$ and the total fundamental phase current. Hint: For calculation of $\tilde{I}_{totalphase}$ you may:
- use a (per-phase) phasor diagram and perform calculations using the cosine law (see Fig. P1.12.2): $a^2 = b^2 + c^2 - 2 \cdot b \cdot c \cdot \cos(\alpha)$
- draw the phasor diagram to scale and find $\tilde{I}_{totalphase}$ by graphical means, or

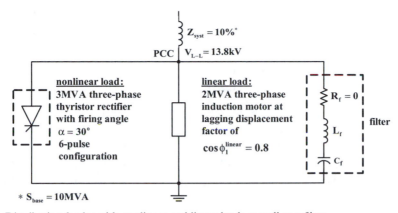

**FIGURE P1.12.1** Distribution feeder with nonlinear and linear loads as well as a filter.

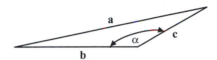

**FIGURE P1.12.2** The application of cosine law.

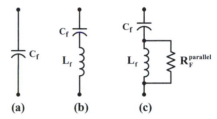

**FIGURE P1.12.3** Types of filters under consideration ($C_F = 32.36\ \mu F$, $L_F = 6.04$ mH, $R_F^{parallel} = 500\ \Omega$).

- draw the phasor diagram and use complex calculations.

d) Calculate harmonic currents and voltages without filter.

e) Design a passive LC filter at the point of common coupling (PCC) such that the fundamental (60 Hz) current through the filter is $\tilde{I}_F = j100$A. In your filter design calculation you may neglect the influence of the ohmic resistance of the filter ($R_F = 0$). Tune the filter to the 6th harmonic: this will lead to two equations and two unknowns ($L_F$ and $C_F$).

f) Calculate the displacement (fundamental power) factor $\cos\ \varphi_{1\ with\ filter}^{total}$ after the filter has been installed. Is this filter design acceptable from a displacement (fundamental power) factor point of view?

g) After the filter has been installed ($L_F = 6.04$ mH, $C_F = 32.36\ \mu F$, $R_F = 0$), compute $\rho_{system}^{5th}$ and $\rho_{system}^{7th}$. These two values provide information about resonance conditions within the feeder. What type of resonance exists?

h) Calculate the harmonic currents and voltages with filter.

i) Is there any advantage of using the $L_F C_F$ filter (see Fig. P1.12.3b) as compared to that of Fig. P1.12.3a? What is the effect on filtering if a resistance $R_F^{parallel}$ is connected in parallel with the inductor $L_F$ (see Fig. P1.12.3c)?

### Problem 1.13: Passive Filter Calculations as Applied to A Distribution Feeder with One User Including a Twelve-Pulse Static Power Converter Load

Figure P1.13 shows the one-line diagram of a large industrial plant being serviced from a utility transmission voltage at 13.8 kV$_{L-L}$. The demand on the utility system is 50 MVA and 50% of its load is a 12-pulse static power converter load. For a system impedance $Z_{syst} = 2.3\%$ referred to a 10 MVA base and a short-circuit current to load current ratio of $R_{sc} = 8.7$, design a passive RLC filter at the point of common coupling (PCC) such that the injected current harmonics and the resulting voltage harmonics at PCC are within the limits of IEEE-519 as proposed by the paper of Duffey and Stratford [64]: this paper shows that (without filter) the 11th, 13th, 23rd, 25th, 35th, 37th, 47th, and 49th current harmonics do not satisfy the limits of IEEE-519, and the 11th and 13th harmonic voltages violate the guidelines of IEEE-519 as well.

a) For your design you may assume that an inductor with $L = 1$ mH and $R = 0.10\ \Omega$ is available. Is this inductor suitable for such an RLC filter design?

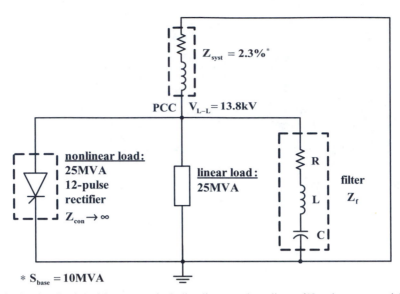

**FIGURE P1.13** Distribution feeder with one user including linear and nonlinear (12-pulse converter) load.

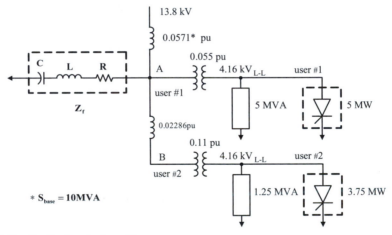

**FIGURE P1.14** Utility distribution feeder with two users.

Compute the fundamental (60 Hz) current through the RLC filter and compare this current with the total load current.

b) Calculate the harmonic currents and the harmonic voltages (5th to 19th) at PCC after the filter has been installed.

### Problem 1.14: Passive Filter Calculations as Applied to a Distribution Feeder with Two Users Each with a Six-Pulse Static Power Converter Load

Figure P1.14 shows a utility distribution feeder that has two users along the radial feeder. Each user sees a different value of short-circuit or system size.

a) For each user compute from the plant specifications apparent short-circuit power $S_{sc}$, short-circuit phase current $I_{scph}$, apparent load power $S_l$, load phase current $I_{lph}$, load phase current of static power converter $I_{lspc}$, and short-circuit ratio $R_{sc}$.

b) Determine the harmonic currents (in amperes and %) injected into the system due to the static power converter loads (up to 19th harmonic).

c) Compute the harmonic voltages (in volts and %) induced at the 13.8 kV bus due to the harmonic currents transmitted (up to 19th harmonic).

d) Design an RLC filter tuned to the frequency of the current with the largest harmonic amplitude. You may assume that an inductor with $R = 0.10 \, \Omega$ and $L = 1$ mH is available. The filter is

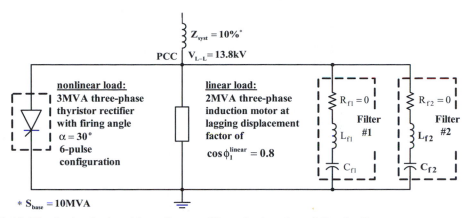

**FIGURE P1.15** Distribution feeder with nonlinear and linear loads and two LC series filters.

to be installed at the 13.8 kV bus next to user #1.

e) Recompute the current harmonics (up to 19th harmonic) transmitted into the system and their associated harmonic voltages at the 13.8 kV bus after the filter has been installed.

f) Are the harmonic currents and voltages at user #1 and user #2 within the limits of IEEE-519 as recommended by the paper of Duffey and Stratford [64]?

## Problem 1.15: Design Two Series LC Filters So That the Total Displacement Power Factor cos $\Phi_{1 \text{ with filter}}^{total}$ Will Be 0.9 lagging $\leq$ cos $\varphi_{1 \text{with filter}}^{total} \leq$ 1.0, and the Recommendations According To IEEE-519 Are Satisfied

Figure P1.15 shows the one-line diagram of an industrial plant being serviced from a utility transmission voltage at 13.8 kV$_{L-L}$. The total power demand on the utility system is 5 MVA: 3 MVA is a six-pulse static power converter load (three-phase rectifier with firing angle $\alpha = 30°$, note cos $\Phi_1^{nonlinear} = 0.955$ cos $\alpha$ lagging), while the remaining 2 MVA is a linear (induction motor) load at cos $\Phi_1^{linear} = 0.8$ lagging (inductive) displacement power factor. The system impedance is $Z_{syst} = 10\%$ referred to a $S_{base} = 10$ MVA base.

a) Before any filter installation, the displacement power factor is cos $\varphi_{1 \text{ with filter}}^{total} = 0.826$ or $\phi_{1 \text{ without filter}}^{total} = 34.2°$ lagging. The nonlinear load

current is $|\tilde{I}_{nonlinear \_load}| = 125.5$ A, the linear load current is $|\tilde{I}_{linear \_load}| = 83.66$ A, and the total load current is $|\tilde{I}_{total \_load}| = 209.1$ A. Verify these data.

b) Calculate short-circuit apparent power $S_{SC}$ at the point of common coupling (PCC).

c) Find short-circuit current $I_{scphase}$ and the short-circuit current ratio $R_{sc}$.

d) If no filter is employed the harmonic currents $I_h$ exceed IEEE-519 limits. Calculate the harmonic currents and voltages at PCC without filter.

e) To comply with IEEE-519 it is recommended to install two filters: one tuned at the 5th and the other one tuned at the 11th harmonic. Design two passive, series LC filters at PCC such that the fundamental (60 Hz) current through each of the filters is $\tilde{I}_{F1} = \tilde{I}_{F2} = \tilde{I}_F = j50$ A. In your filter design calculation you may neglect the influence of the ohmic resistances of the two filters ($R_{F1} = R_{F2} = R_F = 0$). As a function of the filter impedance of filter #1, $Z_{F1\_h}$, and the filter impedance of filter #2, $Z_{F2\_h}$, you may calculate an equivalent filter impedance $Z_{Fequivalent\_h}$ which can be used for the calculation of the parameter $\rho_{sys\_h}$. Make sure there are no parallel resonances.

f) Calculate the harmonic currents $I_{sys5}$ through $I_{sys19}$ after the two filters have been installed. Are the harmonic currents after these two filters have been installed below the recommended IEEE-519 limits?

g) Calculate the harmonic voltages $V_h$ at PCC. Do they meet IEEE-519 recommendations?

## 1.11 REFERENCES

1) *IEEE Standard 519*, IEEE recommended practices and requirements for harmonic control in electric power systems, IEEE-519, 1992.

2) IEC 61000-3-2 (2001-10) Consolidated Edition. Electromagnetic compatibility (EMC) – Part 3-2: Limits for harmonic current emissions.

3) Shepherd, W.; and Zand, P.; *Energy Flow and Power Factor in Nonsinusoidal Circuits*, Cambridge University Press, 1979.

4) Arrillaga, J.; Bradley, D.A.; and Bodeger, P.S.; *Power System Harmonics*, John Wiley & Sons, 1985.

5) Heydt, G.T.; *Electric Power Quality*, Star in a Circle Publications, 1991.

6) Dugan, R.C.; McGranaghan, M.F.; and Beaty, H.W.; *Electrical Power Systems Quality*, McGraw-Hill, 1996.

7) Arrillaga, J.; Smith, B.C.; Watson, N.R.; and Wood, A.R.; *Power Systems Harmonic Analysis*, John Wiley & Sons, 1997.

8) Schlabbach, J.; Blume, D.; and Stephanblome, T.; *Voltage Quality in Electrical Power Systems*, Originally Published in German by VDE-Verlag in 1999, English Edition by The Institution of Electrical Engineers, UK, 2001.

9) Arrillaga, J.; Watson, N.R.; and Chen, S.; *Power System Quality Assessment*, John Wiley & Sons, 2000.

10) Sankaran, C.; *Power Quality*, CRC Press, 2002.

11) Bollen, M.H.J.; *Understanding Power Quality Problems*, IEEE Press Series on Power Engineering, 2000.

12) Kennedy, B.W.; *Power Quality Primer,* McGraw-Hill, 2000.

13) http://hesperia.gsfc.nasa.gov/sftheory/.

14) Edris, A.; "FACTS technology development: an update," *IEEE Power Engineering Review*, Vol. 20, Issue 3, March 2000, pp. 4–9.

15) Nozari, F.; and Patel, H.S.; "Power electronics in electric utilities: HVDC power transmission systems," *Proceedings of the IEEE*, Vol. 76, Issue 4, April 1988, pp. 495–506.

16) Fuchs, E.F.; "Electromagnetic field (EMF) generation and corona effects in transmission lines", *RMEL Spring Conference*, Denver, May 15–18, 2005.

17) Xiao Hui; Zeng Xiangjun; Fan Shaosheng; Wu Xuebin; and Kuang Lang; "Medium-voltage power line carrier communication system," *PowerCon 2004. 2004 International Conference on Power System Technology, 2004,* Vol. 2, 21–24 Nov. 2004, pp. 1536–1539.

18) Stahlkopf, K.; "Dream jobs 2004, engineer in paradise," *IEEE Spectrum*, February 2004, p. 32.

19) Moore, T.; "Sharpening the focus on EMF research," *IEEE Power Engineering Review*, Vol. 12, Issue 8, Aug. 1992, pp. 6–9.

20) Bhargava, B.; "Arc furnace flicker measurements and control," *IEEE Transactions on Power Delivery,* Vol. 8, Issue 1, Jan. 1993, pp. 400–410.

21) Shoup, D.J.; Paserba, J.J.; and Taylor, C.W.; "A survey of current practices for transient voltage dip/sag criteria related to power system stability," *IEEE PES Power Systems Conference and Exposition 2004,* 10–13 Oct. 2004, Vol. 2, pp. 1140–1147.

22) Standard ANSI C84.1.

23) Standard IEEE-1159.

24) Standard EN-50160.

25) Standard IEEE-1250.

26) Kappenman, J.G.; and Albertson, V.D.; "Bracing for the geomagnetic storms," *IEEE Spectrum*, Vol. 27, Issue 3, March 1990, pp. 27–33.

27) Fuchs, E.F.; You, Y.; and Lin, D.; *Development and validation of GIC transformer models*, Final report prepared for Martin Marietta Energy Systems, Inc., Oak Ridge, Tennessee 37831-6501, Research Project Contract #19-SK205V, June 1996.

28) You, Y.; Fuchs, E.F.; Lin, D.; and Barnes, P.R.; "Reactive power demand of transformers with DC bias," *IEEE Industry Applications Magazine*, July/August 1996, pp. 45–52.

29) Fuchs, E.F.; and Fardoun, A.A.; *Simulation of controller-motor drive system using SPICE*, Final report prepared for Unique Mobility, Inc., Contract UNIQUE 153-7065-1, Englewood, Colorado 80110, January 1991.

30) Grady, W.; Fuchs, E.F.; Lin, D.; and Stensland, T.; *The potential effects of single-phase power electronic-based loads on power system distortion and losses,* Vol. 2, Final report #1000655 prepared for the EPRI-PEAC Corporation, September 2003.

31) Fuchs, E.F.; *Investigations on the impact of voltage and current harmonics on end-use devices and their protection,* Summary Report DE-RA-50150-23 prepared for the U.S. Department of Energy under contract No. DE-AC02-80RA 50150, January 1987.

32) Fuchs, E.F.; *Power quality in power systems and electric machines*, ECEN 5787 Course Notes, Department of Electrical and Computer Engineering, University of Colorado at Boulder, Fall 2005.

33) Fuchs, E.F.; Roesler, D.J.; and Alashhab, F.S.; "Sensitivity of home appliances to harmonics and fractional harmonics of the power system's voltage, Part I: transformers and induction machines," *International Conference on Harmonics in Power Systems*, Worcester, Massachusetts, October 22–23, 1984.

34) Fuchs, E.F.; Roesler, D.J.; and Kovacs, K.P.; "Sensitivity of home appliances to harmonics and fractional harmonics of the power system's voltage, Part II: television sets, induction watthour meters and universal machines," *International Conference on Harmonics in Power Systems*, Worcester, Massachusetts, October 22–23, 1984.

35) Fuller, J.F.; Fuchs, E.F.; and Roesler, D.J.; "Influence of harmonics on power system distribution protection," *IEEE Transactions on Power Delivery*, April 1988, Vol. TPWRD-3, No. 2, pp. 546–554.

36) Standard IEC 61000-4-7.

37) Dennhardt, A.; *Grundzüge der Tonfrequenz-Rundsteuertechnik und Ihre Anwendung*, Verlags-und Wirtschaftsgesellschaft der Elektrizitätswerke m.b.H., Frankfurt 70-Stesemannallee 23, 1971.

38) ANSI C84.1-1982.

39) Varma, R.K.; Mathur, R.M.; Rogers, G.J.; and Kundur, P.; "Modeling effects of system frequency variation in long-term stability studies," *IEEE Transactions on Power Systems*, Vol. 11, Issue 2, May 1996, pp. 827–832.

40) Standard IEEE Std 100.

41) Standard CENELEC Std EN 50160.

42) Fuchs, E.F.; and Lin, D.; *Testing of Power Planners at the University of Colorado*, Final report, Boulder, Colorado, February 11, 1996.

43) Batan, T; *Real-Time Monitoring and Calculation of the Derating of Single-Phase transformers Under Nonsinusoidal Operation*, Ph.D. Thesis, University of Colorado at Boulder, April 8, 1999.

44) Yildirim, D.; Fuchs, E.F.; and Batan, T.; "Test results of a 20 kW, direct-drive, variable-speed wind power plant," *Proceedings of the International Conference on Electrical Machines, Istanbul*, Turkey, September 2–4, 1998, pp. 2039–2044.

45) Fuchs, E.F.; Masoum, M.A.S.; and Ladjevardi, M.; "Effects on distribution feeders from electronic loads based on future peak-load growth, Part I: Measurements," *Proceedings of the IASTED International Conference on Power and Energy Systems (EuroPES 2005)*, Paper # 468-019, Benalmadena, Spain, June 15–17, 2005.

46) Agrawal, B.L.; and Farmer, R.G.; "Effective damping for SSR analysis of parallel turbine-generators," *IEEE Transactions on Power Systems*, Vol. 3, Issue 4, Nov. 1988 pp. 1441–1448.

47) Wiik, J.; Gjefde, J.O.; and Gjengedal, T.; "Impacts from large scale integration of wind energy farms into weak power systems," *PowerCon 2000, Proceedings of International Conference on Power System Technology, 2000*. Vol. 1, 4–7 Dec. 2000, pp. 49–54.

48) Clarke, E.; *Circuit Analysis of AC Power Systems*, Vol. II, John Wiley & Sons, 1950.

49) Fuchs, E.F.; and You, Y.; "Measurement of ($\lambda$–i) characteristics of asymmetric three-phase transformers and their applications," *IEEE Transactions on Power Delivery*, Vol. 17, No. 4, October 2002, pp. 983–990.

50) Budeanu, C.I.; "Puissances Reactive et Fictive," *Inst. Romain de l'Energie*, Bucharest, 1927.

51) Yildirim, D.; Commissioning *of a 30kVA Variable-Speed, Direct-Drive Wind Power Plant*, Ph.D. Thesis, University of Colorado at Boulder, February 14, 1999.

52) Fuchs, E.F.; Fei, R.; and Hong J.; *Harmonic Losses in Isotropic and Anisotropic Transformer Cores at Low Frequencies (60–3,000 Hz)*, Final Report prepared for Martin Marietta Energy Systems, Inc., Contract No. 19X-SD079C, January 1991.

53) de Abreu, J.P.G.; and Emanuel, A. E.; "The need to limit subharmonic injection", *Proceedings of the 9th International Conference on Harmonics and Quality of Power*, Vol. I, October 1–4, 2000, pp. 251–253.

54) Standard IEC 61000.

55) Vladimirescu, A.; "SPICE – the third decade," *Proceedings of the 1990 Bipolar Circuits and Technology Meeting, 1990*, 17–18 Sept. 1990 pp. 96–101.

56) Long, W.; Cotcher, D.; Ruiu, D.; Adam, P.; Lee, S.; and Adapa, R.; "EMTP – a powerful tool for analyzing power system transients," *IEEE Computer Applications in Power*, Vol. 3, Issue 3, July 1990, pp. 36–41.

57) Mansoor, A.; Grady, W.M.; Chowdhury, A.H.; and Samotyi, M.J.; "An investigation of harmonics attenuation and diversity among distributed single-phase power electronic loads," *IEEE Transactions on Power Delivery*, Vol. 10, No. 1, January 1995, pp. 467–473.

58) Mansoor, A.; Grady, W.M.; Thallam, R.S.; Doyle, M.T.; Krein, S.D.; and Samotyj, M.J.; "Effect of supply voltage harmonics on the input current of single-phase diode bridge rectifier loads," *ibid.*, Vol. 10, No. 3, July 1995, pp. 1416–1422.

59) Mansoor, A.; Grady, W.M.; Staats, P.T.; Thallam, R.S.; Doyle, M.T.; and Samotyj, M.J.; "Predicting the net harmonic currents produced by large numbers of distributed single-phase computer loads," *ibid.*, Vol. 10, No. 4, 1995, pp. 2001–2006.

60) Masoum, M.A.S.; Jafarian, A.; Ladjevardi, M.; Fuchs, E.F.; and Grady, W.M.; "Fuzzy approach for optimal placement and sizing of capacitor banks in the presence of harmonics," *IEEE Transactions on Power Delivery*, Vol. 19, No. 2, April 2004, pp. 822–829.

61) Masoum, M.A.S.; Ladjevardi, M.; Jafarian, A.; and Fuchs, E.F.; "Optimal placement, replacement and sizing of capacitor banks in distorted distribution networks by genetic algorithms," *IEEE Transactions on Power Delivery*, Vol. 19, No. 4, October 2004, pp. 1794–1801.

62) Fuchs, E.F.; "Are harmonic recommendations according to IEEE 519 and CEI/IEC too restrictive?" *Proceedings of Annual Frontiers of Power Conference,* Oklahoma State University, Stillwater, Oklahoma, Oct. 28–29, 2002, pp. IV-1 to IV-18.

63) Standard ANSI/IEEE Std C57.110.

64) Duffey, C.K.; and Stratford, R.P.; "Update of harmonic standard IEEE-519: IEEE recommended practices and requirements for harmonic control in power systems", *IEEE Transactions on Industry Applications*, Vol. 25, No. 6, Nov./Dec. 1989, pp. 1025–1034.

65) www.ballard.com.

66) "The road ahead for EV batteries," ERPI Journal, March/April 1996.

67) Wang, C.; Nehrir, M.H.; and Shaw, S.R.; "Dynamic models and model validation for PEM fuel cells using electrical circuits," *IEEE Transactions on Energy Conversion*, Vol. 20, No. 2, June 2005, pp. 442–451.

## 1.12 ADDITIONAL BIBLIOGRAPHY

68) Fuchs, E.F.; You, Y.; and Roesler, D.J.; "Modeling, simulation and their validation of three-phase transformers with three legs under DC bias," *IEEE Transactions on Power Delivery*, Vol. 14, No. 2, April 1999, pp. 443–449.

69) Fardoun, A.A.; Fuchs, E.F.; and Huang, H.; "Modeling and simulation of an electronically commutated permanent-magnet machine drive system using SPICE," *IEEE Transactions on Industry Applications*, July/August 1994, Vol. 30, No. 4, July/August 1994, pp. 927–937.

70) Fuchs, E.F.; Lin, D.; and Martynaitis, J.; "Measurement of three-phase transformer derating and reactive power demand under nonlinear loading conditions," *IEEE Transactions on Power Delivery,* Vol. 21, Issue 2, April 2006, pp. 665–672.

71) Fuchs, E.F.; Roesler, D.J.; and Kovacs, K.P.; "Aging of electrical appliances due to harmonics of the power system's voltage," *IEEE Transactions on Power Delivery*, July 1986, Vol. TPWRD-1, No. 3, pp. 301–307.

72) Fuchs, E.F.; and Hanna, W.J.; "Measured efficiency improvements of induction motors with thyristor/triac controllers," *IEEE Transactions on Energy Conversion*, Vol. 17, No. 4, December 2002, pp. 437–444.

73) Lin, D.; and Fuchs, E.F.; "Real-time monitoring of iron-core and copper losses of three-phase transformer under (non)sinusoidal operation," *IEEE Transactions on Power Delivery*, Vol. 21, No. 3, July 2006, pp. 1333–1341.

74) Yildirim, D.; and Fuchs, E.F.; "Commentary to various formulations of distortion power D," *IEEE Power Engineering Review*, Vol. 19, No. 5, May 1999, pp. 50–52.

75) Fuchs, E.F.; and Fei, R.; "A New computer-aided method for the efficiency measurement of low-loss transformers and inductors under nonsinusoidal operation," *IEEE Transactions on Power Delivery*, January 1996, Vol. PWRD-11, No. 1, pp. 292–304.

76) Fuchs, E.F.; Roesler, D.J.; and Masoum, M.A.S.; "Are harmonic recommendations according to IEEE and IEC too restrictive?" *IEEE Transactions on Power Delivery*, Vol. 19, No. 4, Oct. 2004, pp. 1775–1786.

# Harmonic Models of Transformers

Extensive application of power electronics and other nonlinear components and loads creates single-time (e.g., spikes) and periodic (e.g., harmonics) events that could lead to serious problems within power system networks and its components (e.g., transformers). Some possible impacts of poor power quality on power transformers are

- saturating transformer core by changing its operating point on the nonlinear $\lambda - i$ curve,
- increasing core (hysteresis and eddy current) losses and possible transformer failure due to unexpected high losses associated with hot spots,
- increasing (fundamental and harmonic) copper losses,
- creating half-cycle saturation in the event of even harmonics and DC current,
- malfunctioning of transformer protective relays,
- aging and reduction of lifetime,
- reduction of efficiency,
- derating of transformers,
- decrease of the power factor,
- generation of parallel (harmonic) resonances and ferroresonance conditions, and
- deterioration of transformers' insulation near the terminals due to high voltage stress caused by lightning and pulse-width-modulated (PWM) converters.

These detrimental effects call for the understanding of how harmonics affect transformers and how to protect them against poor power quality conditions. For a transformer, the design of a harmonic model is essential for loss calculations, derating, and harmonic power flow analysis.

This chapter investigates the behavior of transformers under harmonic voltage and current conditions, and introduces harmonic transformer models suitable for loss calculations, harmonic power flow analysis, and computation of derating functions. After a brief introduction of the conventional (sinusoidal) transformer model, transformer losses with emphasis on the impacts of voltage and current harmonics are presented. Several techniques for the computation of single-phase transformer derating

factors (functions) are introduced. Thereafter, a survey of transformer harmonic models is given. In subsequent sections, the issues of ferroresonance, geomagnetically induced currents (GICs), and transformer grounding are presented. A section is also provided for derating of three-phase transformers.

## 2.1 SINUSOIDAL (LINEAR) MODELING OF TRANSFORMERS

Transformer simulation under sinusoidal operating conditions is a well-researched subject and many steady-state and transient models are available. However, transformer cores are made of ferromagnetic materials with nonlinear B–H ($\lambda - i$) characteristics. They exhibit three types of nonlinearities that complicate their analysis: saturation effect, hysteresis (major and minor) loops, and eddy currents. These phenomena result in nonsinusoidal flux, voltage and current waveforms on primary and secondary sides, and additional copper (due to current harmonics) and core (due to hysteresis loops and eddy currents) losses at fundamental and harmonic frequencies. Linear techniques for transformer modeling neglect these nonlinearities (by assuming a linear $\lambda - i$ characteristic) and use constant values for the magnetizing inductance and the core-loss resistance. Some more complicated models assume nonlinear dependencies of hysteresis and eddy-current losses with fundamental voltage magnitude and frequency, and use a more accurate equivalent value for the core-loss resistance. Generally, transformer total core losses can be approximated as

$$P_{fe} = P_{hys} + P_{eddy} = K_{hys} (B_{max})^s f + K_{eddy}(B_{max})^2 f^2, \quad (2-1)$$

where $P_{hys}$, $P_{eddy}$, $B_{max}$, and $f$ are hysteresis losses, eddy-current losses, maximum value of flux density, and fundamental frequency, respectively. $K_{hys}$ is a constant for the grade of iron employed and $K_{eddy}$ is the eddy-current constant for the conductive material. S is the Steinmetz exponent ranging from 1.5 to 2.5 depending on the operating point of transformer core.

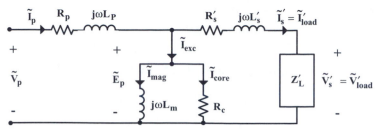

**FIGURE 2.1** Linear single-phase, steady-state transformer model for sinusoidal analysis.

Transient models are used for transformer simulation during turning-on (e.g., inrush currents), faults, and other types of disturbances. They are based on a system of time-dependent differential equations usually solved by numerical algorithms. Transient models require a considerable amount of computing time. Steady-state models mostly use phasor analysis in the frequency domain to simulate transformer behavior, and require less computing times than transient models. Transformer modeling for sinusoidal conditions is not the main objective of this chapter. However, Fig. 2.1 illustrates a relatively simple and accurate frequency-based linear model that will be extended to a harmonic model in Section 2.4. In this figure $R_c$ is the core-loss resistance, $L_m$ is the (linear) magnetizing inductance, and $R_p$, $R'_s$, $L_p$, and $L'_s$ are transformer primary and secondary resistances and inductances, respectively. Superscript $'$ is used for quantities referred from the secondary side to the primary side of the transformer.

The steady-state model of Fig. 2.1 is not suitable for harmonic studies since constant values are assumed for the magnetizing inductance and the core-loss resistance. However, this simple and practical frequency-based model generates acceptable results if a transformer were to operate in the linear region of the $\lambda - i$ characteristic, and the harmonic frequency is taken into account.

## 2.2 HARMONIC LOSSES IN TRANSFORMERS

Losses due to harmonic currents and voltages occur in windings because of the skin effect and the proximity effect. It is well known that harmonic current $i_h(t)$ and harmonic voltage $v_h(t)$ must be present in order to produce harmonic losses

$$p_h(t) = i_h(t) \cdot v_h(t). \tag{2-2}$$

If either $i_h(t)$ or $v_h(t)$ are zero then $p_h(t)$ will be zero as well. Harmonic losses occur also in iron cores due to hysteresis and eddy-current phenomena. For linear (B–H) characteristics of iron cores, the losses

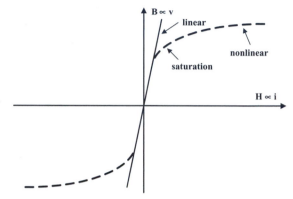

**FIGURE 2.2** Linear and nonlinear iron-core characteristics.

are dependent on fundamental and harmonic amplitudes only, whereas for nonlinear iron-core (B–H) characteristics (Fig. 2.2) the phase shift between harmonic voltage and fundamental voltage is important as well. For example, a magnetizing current with maximum peak-to-peak values results in larger maximum flux densities than a magnetizing current with minimum peak-to-peak values. Proximity losses in windings and (solid) conducting parts of a device (e.g., frame) occur due to the relative location between the various windings and conductive parts.

### 2.2.1 Skin Effect

If a conductor with cross section $a_{cond}$ conducts a DC current $I_{DC}$, the current density $j_{DC} = I_{DC}/a_{cond}$ is uniform within the conductor and a resistance $R_{DC}$ can be assigned to the conductor representing the ratio between the applied voltage $V_{DC}$ and the resulting current $I_{DC}$, that is,

$$R_{DC} = \frac{V_{DC}}{I_{DC}}. \tag{2-3}$$

For (periodic) AC currents $i_{ACh}(t)$, the current flows predominantly near the surface of the conductor and the current density $j_{ACh}$ is nonuniform within the conductor (Fig. 2.3). In general,

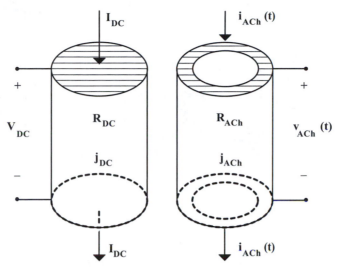

**FIGURE 2.3** DC resistance $R_{DC}$ versus AC resistance $R_{ACh}$.

$$R_{DC} < R_{ACh}.$$

The higher the order h of the harmonic current $i_{ACh}(t)$, the larger is the skin effect.

### 2.2.2 Proximity Effect

The AC current distribution within a conductor depends on the current distribution of neighboring conductors or conducting parts. The AC field $\vec{H}$ of a single conductor in free space consists of circles symmetric to the axis of the conductor.

$$\text{Ampere's law: } \int \vec{H} \cdot d\vec{l} = \int_S \vec{J} \cdot d\vec{s}. \qquad (2\text{-}4)$$

If there are two conductors or more, the circular fields will be distorted and the resulting eddy-current losses within the conductor will differ from those of single conductors. In Figs. 2.4 and 2.5, H stands for $|\vec{H}|$.

### 2.2.3 Magnetic Iron-Core (Hysteresis and Eddy-Current) Losses

All magnetic materials exhibit saturation and hysteresis properties: there are major loops and minor loops (Fig. 2.6) and such characteristics are multivalued for either a single H or a single B value.

Therefore, even if there is a sinusoidal input voltage to a magnetic circuit and no minor B–H loops are assumed, a nonsinusoidal current and increased losses are generated. For nonsinusoidal excitation and/or nonlinear loads, excessive magnetic and (fundamental and harmonic) copper losses

may result that could cause transformer failure. In most cases hysteresis losses can be neglected because they are relatively small as compared with copper losses. However, the nonlinear saturation behavior must be taken into account because all transformers and electric machines operate for economic reasons beyond the knee of saturation.

Faraday's law

$$e(t) = \frac{d\lambda(t)}{dt}, \qquad (2\text{-}5)$$

or

$$\lambda(t) = \int e(t)dt, \qquad (2\text{-}6)$$

is valid, where $e(t)$ is the induced voltage of a winding residing on an iron core, $\lambda(t) = N\Phi(t)$ are the flux linkages, $N$ is the number of turns of a winding, and $\Phi(t)$ is the flux linked with the winding.

Depending on the phase shift of a harmonic voltage with respect to the fundamental voltage the resulting wave shape of the flux linkages will be different and not proportional to the resulting voltage wave shape due to the integral relationship between flux linkages $\lambda(t)$ and induced voltage $e(t)$ [1–6].

#### 2.2.3.1 Application Example 2.1: Relation between Voltages and Flux Linkages for 0° Phase Shift between Fundamental and Harmonic Voltages

Third harmonic voltage $e_3(t)$ is in phase (0°) with the fundamental voltage $e_1(t)$ resulting in a maximum

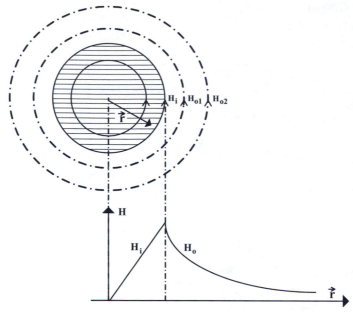

**FIGURE 2.4** Magnetic field strength inside and outside of a conductor.

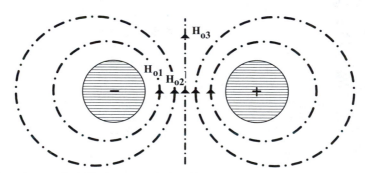

**FIGURE 2.5** Distortion of magnetic fields due to proximity effect.

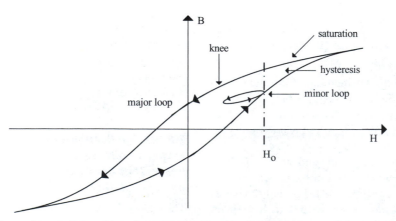

**FIGURE 2.6** Nonlinear characteristics of transformer core with major and minor hysteresis loops.

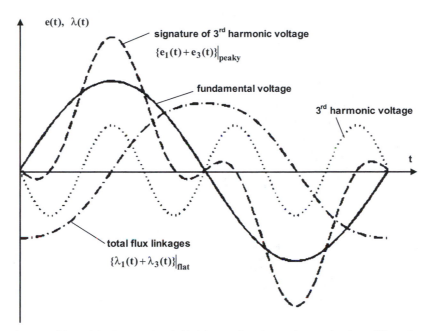

**FIGURE E2.1.1** Superposition of fundamental and third harmonic voltages that are in phase (0°, peak-to-peak voltage is maximum).

peak-to-peak value for the total nonsinusoidal voltage. According to Eq. 2-6, the peak-to-peak value of the generated $\lambda(t)$ is minimum, as illustrated by Fig. E2.1.1.

**Rule.** A "peaky" (peak-to-peak is maximum) voltage $\{e_1(t) + e_3(t)\}$ results in "flat" (peak-to-peak is minimum) flux linkages.

### 2.2.3.2 Application Example 2.2: Relation between Voltages and Flux Linkages for 180° Phase Shift between Fundamental and Harmonic Voltages

The third harmonic voltage $e_3(t)$ is out of phase (180°) with the fundamental voltage $e_1(t)$. In this case, the total voltage and the resulting flux linkages have minimum and maximum peak-to-peak values, respectively (Fig. E2.2.1).

**Rule.** A "flat" (peak-to-peak is minimum) voltage $\{e_1(t) + e_3(t)\}$ results in "peaky" (peak-to-peak is maximum) flux linkages.

**Generalization.** For higher harmonic orders, similar relations between the nonsinusoidal waveforms of induced voltage and flux linkages exist; however, there is an alternating behavior as demonstrated in Table E2.2.1.

This alternating behavior (due to the integral relationship between $e(t)$ and $\lambda(t)$) influences the

**TABLE E2.2.1** Phase Relations between Induced Voltages and Flux Linkages when a Harmonic is Superposed with the Fundamental

| Harmonic order | Nonsinusoidal voltage $e_1(t) + e_h(t)$ | Nonsinusoidal flux linkage $\lambda_1(t) + \lambda_h(t)$ |
|---|---|---|
| $h = 3$ | $\{e_1(t) + e_3(t)\}$ peaky[a] | $\{\lambda_1(t) + \lambda_3(t)\}$ flat[b] |
|  | $\{e_1(t) + e_3(t)\}$ flat | $\{\lambda_1(t) + \lambda_3(t)\}$ peaky |
| $h = 5$ | $\{e_1(t) + e_5(t)\}$ peaky | $\{\lambda_1(t) + \lambda_5(t)\}$ peaky |
|  | $\{e_1(t) + e_5(t)\}$ flat | $\{\lambda_1(t) + \lambda_5(t)\}$ flat |
| $h = 7$ | $\{e_1(t) + e_7(t)\}$ peaky | $\{\lambda_1(t) + \lambda_7(t)\}$ flat |
|  | $\{e_1(t) + e_7(t)\}$ flat | $\{\lambda_1(t) + \lambda_7(t)\}$ peaky |
| $h = 9$ | $\{e_1(t) + e_9(t)\}$ peaky | $\{\lambda_1(t) + \lambda_9(t)\}$ peaky |
|  | $\{e_1(t) + e_9(t)\}$ flat | $\{\lambda_1(t) + \lambda_9(t)\}$ flat |
| ⋮ | ⋮ | ⋮ |

[a]Maximum peak-to-peak value.
[b]Minimum peak-to-peak value.

iron-core (magnetic) losses significantly. Therefore, it is possible that a nonsinusoidal voltage results in less iron-core losses than a sinusoidal wave shape [2]. Note that the iron-core losses are a function of the maximum excursions of the flux linkages (or flux densities $B_{max}$).

### 2.2.4 Loss Measurement

For low efficiency ($\eta < 97\%$) devices the conventional indirect loss measurement approach, where the losses $P_{loss}$ are the difference between measured input power $P_{in}$ and measured output power $P_{out}$, is acceptable. However, for high efficiency ($\eta \geq 97\%$)

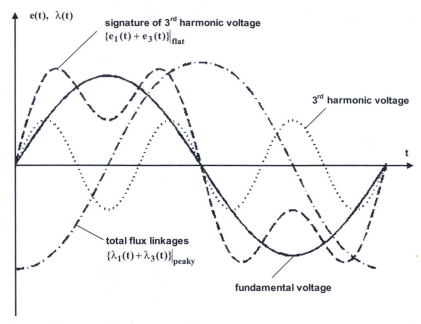

**FIGURE E2.2.1** Superposition of fundamental and third harmonic voltages that are out of phase (180°, peak-to-peak voltage is minimum).

devices the indirect loss measurement approach yields losses that have a large error.

### 2.2.4.1 Indirect Loss Measurement

Consider the two-port system shown in Fig. 2.7:

$$P_{loss} = P_{in} - P_{out}, \qquad (2\text{-}7)$$

or

$$P_{loss} = \frac{1}{T}\int_0^T v_1 i_1 dt - \frac{1}{T}\int_0^T v_2' i_2' dt. \qquad (2\text{-}8)$$

This is the conventional and relatively simple (indirect) technique used for loss measurements of most low-efficiency electric devices under sinusoidal and/or nonsinusoidal operating conditions.

### 2.2.4.2 Direct Loss Measurement

The two port system of Fig. 2.7 can be specified in terms of series and shunt impedances [1–6], as illustrated in Fig. 2.8.

The powers dissipated in $Z_{series1}$ and $Z'_{series1}$ are

$$P_{series} = \frac{1}{T}\int_0^T p_{series}(t)dt. \qquad (2\text{-}9)$$

The power dissipated in $Z_{shunt}$ is

$$P_{shunt} = \frac{1}{T}\int_0^T p_{shunt}(t)dt. \qquad (2\text{-}10)$$

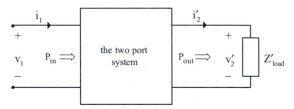

**FIGURE 2.7** Voltage and current definitions for a two-port system.

Therefore,

$$P_{loss} = P_{series} + P_{shunt} \qquad (2\text{-}11)$$

### 2.2.4.3 Application Example 2.3: Application of the Direct-Loss Measurement Technique to a Single-Phase Transformer

The direct-loss measurement technique is used to measure losses of a single-phase transformer as illustrated in Fig. E2.3.1 [7].

## 2.3 DERATING OF SINGLE-PHASE TRANSFORMERS

According to the IEEE dictionary, derating is defined as "the intentional reduction of stress/strength ratio (e.g., real or apparent power) in the application of an item (e.g., transformer), usually for the purpose of reducing the occurrence of stress-related failure (e.g., reduction of lifetime of transformer due to

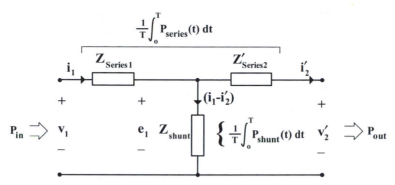

**FIGURE 2.8** Series and shunt impedances of a two-port system.

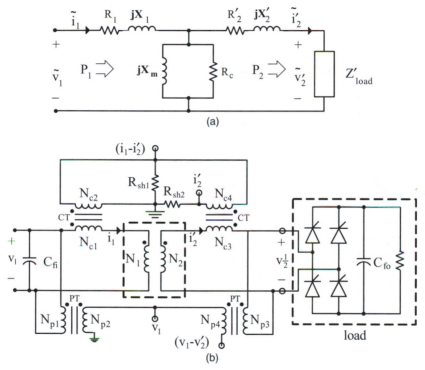

**FIGURE E2.3.1** Application of direct-loss measurement technique to a single-phase transformer; (a) equivalent circuit, (b) measuring circuit.

increased temperature beyond rated temperature)." As has been discussed in Section 2.2 harmonic currents and voltages result in harmonic losses increasing the temperature rise. This rise beyond its rated value results in a reduction of lifetime [10] as will be discussed in Chapter 6. For transformers two derating parameters can be defined:

- reduction in apparent power rating (RAPR), and
- real power capability (RPC).

Although the first one is independent of the power factor, the second one is strongly influenced by it.

### 2.3.1 Derating of Transformers Determined from Direct-Loss Measurements

The direct-loss measurement technique of Section 2.2.4.2 is applied and the reduction of apparent power (RAPR) is determined such that for any given total harmonic distortion of the current (THD$_i$), the rated losses of the transformer will not be exceeded.

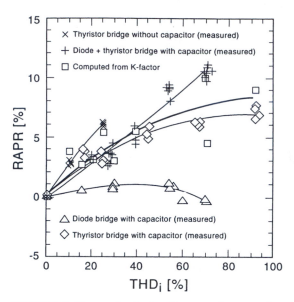

**FIGURE 2.9** Measured reduction in apparent power rating (RAPR) of 25 kVA single-phase pole transformer as a function of total harmonic current distortion (THD$_i$) where 3rd and 5th current harmonics are dominant. Calculated values from K-factor [12].

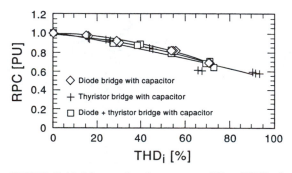

**FIGURE 2.10** Measured real power capability (RPC) of 25 kVA single-phase pole transformer as a function of total harmonic current distortion (THD$_i$) [12].

Figures 2.9 and 2.10 show the RAPR and RPC as a function of the THD$_i$.

## 2.3.2 Derating of Transformers Determined from the K-Factor

It is well known that power system transformers must be derated when supplying nonlinear loads. Transformer manufacturers have responded to this problem by developing transformers rated for 50% or 100% nonlinear load. However, the impact of nonlinear loads on transformers greatly depends on the nature and the harmonic spectrum caused by the nonlinear load, which is not considered by the manufacturers. The K-factor rating is an alternative tech-

nique for transformer derating that includes load characteristics [11, 13].

The assumptions used in calculating K-factor can be found in ANSI/IEEE C57.110, IEEE Recommended Practice for Establishing Transformer Capability when Supplying Nonsinusoidal Load Currents [13]. This recommendation calculates the harmonic load current that causes losses equivalent to the rated (sinusoidal) load current so that the transformer's "hot spot" temperature will not be exceeded. The hot spot is assumed to be where eddy-current losses are the greatest and the coolant (oil) is hottest (e.g., at the inner low-voltage winding).

The load loss (LL) generating the hot spot temperature is assumed to consist of $I^2R$ and eddy-current $P_{EC}$ losses:

$$P_{LL} = I^2R + P_{EC} \quad \text{(watts)}. \qquad (2\text{-}12)$$

The eddy current losses (Eq. 2–12) are proportional to the square of the current and frequency:

$$P_{EC} = K_{EC}I^2h^2, \qquad (2\text{-}13)$$

where $K_{EC}$ is a proportionality constant. Under rated (sinusoidal) conditions, the load loss resulting in the hot spot temperature can be expressed as

$$P_{LL-R} = I^2R^2 + P_{EC-R}, \qquad (2\text{-}14)$$

where $P_{EC-R}$ is the eddy-current loss resulting in the hot spot temperature under rated (sinusoidal) operating conditions. In per unit of the rated $I^2R$ loss, Eq. 2-14 becomes

$$P_{LL-R} = 1 + P_{EC-R} \quad \text{(pu)}. \qquad (2\text{-}15)$$

Therefore, the transformer load loss under harmonic conditions is

$$P_{LL} = \sum_{h=1}^{h\,max} I_h^2 + \left( \sum_{h=1}^{h\,max} I_h^2 h^2 \right) P_{EC-R} \quad \text{(pu)}. \qquad (2\text{-}16)$$

The first and second terms on the right side of the above equation represent the $I^2R$ loss and the eddy-current loss, respectively. Setting $P_{LL} = P_{LL-R}$ gives

$$1 + P_{EC-R} = \sum_{h=1}^{h\,max} I_h^2 + \left( \sum_{h=1}^{h\,max} I_h^2 h^2 \right) P_{EC-R} \quad \text{(pu)}. \qquad (2\text{-}17)$$

Now, if we define the K-factor as follows [13]:

$$K = \frac{\displaystyle\sum_{h=1}^{h\,max} I_h^2 h^2}{I_R^2} \quad \text{(pu)}, \qquad (2\text{-}18)$$

then

$$1 + P_{EC-R} = \sum_{h=1}^{h\,\text{max}} I_h^2 + K \left( \frac{\sum_{h=1}^{h\,\text{max}} I_h^2}{\sum_{h=1}^{h\,\text{max}} I_h^2} I_R^2 \right) P_{EC-R} \quad \text{(pu).}$$

(2-19)

Solving for $\sum_{h=1}^{h\,\text{max}} I_h^2$,

$$\sum_{h=1}^{h\,\text{max}} I_h^2 = \frac{1 + P_{EC-R}}{1 + K \dfrac{I_R^2}{\sum_{h=1}^{h\,\text{max}} I_h^2} (P_{EC-R})} \quad \text{(pu).} \quad (2\text{-}20)$$

Therefore, the maximum amount of rms harmonic load current that the transformer can deliver is

$$I_{\text{max}}^{\text{pu}} = \sqrt{\frac{1 + P_{EC-R}}{1 + K \dfrac{I_R^2}{\sum_{h=1}^{h\,\text{max}} I_h^2} [(P_{EC-R})]}} \quad \text{(pu).} \quad (2\text{-}21)$$

Using the K-factor and transformer parameters, the maximum permissible rms current of the transformer can also be defined as follows [11]:

$$I_{\text{max}}^{\text{p.u.}} = \sqrt{\frac{R_{DC} + R_{EC-R}(1-K) - \dfrac{(\Delta P_{fe} + \Delta P_{OSL})}{I_R^2}}{R_{DC}}}. \quad (2\text{-}22)$$

$R_{DC} = R_{DC\text{primary}} + R'_{DC\text{secondary}}$ is the total DC resistance of a transformer winding. $R_{EC-R}$ is the rated pu additional resistance due to eddy currents. In addition,

$$\Delta P_{fe} = \sum_{h=1}^{h\,\text{max}} P_{feh} - P_{feR},$$

and

$$\Delta P_{OSL} = \sum_{h=1}^{h\,\text{max}} P_{OSLh} - P_{OSLR}.$$

$\Delta P_{fe}$ is the difference between the total iron core losses (including harmonics) and the rated iron core losses without harmonics. $\Delta P_{OSL}$ is the difference between the total other stray losses (OSL) including harmonics and the rated other stray losses without harmonics.

The reduction of apparent power (RAPR) is

$$\text{RAPR} = 1 - (V_{2rms}^{\text{nonlinear}}/V_{2rms}^{\text{rat}}) I_{\text{max}}^{\text{p.u.}}, \quad (2\text{-}23)$$

where $V_{2rms}^{\text{nonlinear}}$ and $V_{2rms}^{\text{rat}}$ are the total rms value of the secondary voltage including harmonics and the rated rms value of the secondary winding without harmonics, respectively. All above parameters are defined in [11].

### 2.3.3 Derating of Transformers Determined from the $F_{HL}$-Factor

The K-factor has been devised by the Underwriters Laboratories (UL) [13] and has been recognized as a measure for the design of transformers feeding nonlinear loads demanding nonsinusoidal currents. UL designed transformers are K-factor rated. In February 1998 the IEEE C57.110/D7 [14] standard was created, which represents an alternative approach for assessing transformer capability feeding nonlinear loads. Both approaches are comparable [15].

The $F_{HL}$-factor is defined as

$$F_{HL} = \frac{P_{EC}}{P_{EC-O}} = \frac{\sum_{h=1}^{h\,\text{max}} I_h^2 h^2}{\sum_{h=1}^{h\,\text{max}} I_h^2}. \quad (2\text{-}24)$$

Thus the relation between K-factor and $F_{HL}$-factor is

$$K = \left( \frac{\sum_{h=1}^{h\,\text{max}} I_h^2}{I_R^2} \right) F_{HL}, \quad (2\text{-}25)$$

or the maximum permissible current is [15]

$$I_{\text{max}}^{\text{p.u.}} = \sqrt{\frac{R_{DC} + R_{EC-R} - \dfrac{\Delta P_{Fe} + \Delta P_{osl}}{I_R^2}}{R_{DC} + F_{HL} R_{EC-R}}}, \quad (2\text{-}26)$$

and the reduction in apparent power rating (RAPR) is

$$\text{RAPR} = 1 - \left( \frac{V_{2rms}^{\text{nonlinear}}}{V_{2rms}^{\text{rat}}} \right) I_{\text{max}}^{\text{p.u.}}. \quad (2\text{-}27)$$

The RAPR as a function of $F_{HL}$ is shown in Fig. 2.11.

### 2.3.3.1 Application Example 2.4: Sensitivity of K- and $F_{HL}$-Factors and Derating of 25 kVA Single-Phase Pole Transformer with Respect to the Number and Order of Harmonics

The objective of this example is to show that the K- and $F_{HL}$-factors are a function of the number and order of harmonics considered in their calculation.

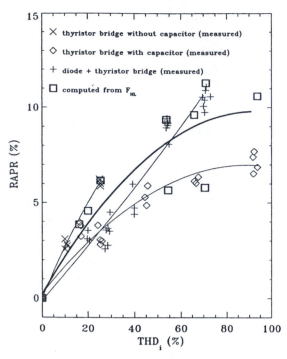

**FIGURE 2.11** Measured reduction in apparent power rating (RAPR) of 25 kVA single-phase pole transformer as a function of total harmonic current distortion (THD$_i$) where 3rd and 5th current harmonics are dominant. Calculated values from $F_{HL}$-factor [12].

**TABLE E2.5.1** Current Spectrum of the Air Handler of a Variable-Speed Drive with (Medium or) 50% of Rated Output Power

| $I_{DC}$ (A) | $I_1$ (A) | $I_2$ (A) | $I_3$ (A) | $I_4$ (A) | $I_5$ (A) | $I_6$ (A) |
|---|---|---|---|---|---|---|
| −0.82 | 10.24 | 0.66 | 8.41 | 1.44 | 7.27 | 1.29 |
| $I_7$ (A) | $I_8$ (A) | $I_9$ (A) | $I_{10}$ (A) | $I_{11}$ (A) | $I_{12}$ (A) | $I_{13}$ (A) |
| 4.90 | 0.97 | 3.02 | 0.64 | 1.28 | 0.48 | 0.57 |
| $I_{14}$ (A) | $I_{15}$ (A) | $I_{16}$ (A) | $I_{17}$ (A) | $I_{18}$ (A) | $I_{19}$ (A) | $I_{20}$ (A) |
| 0.53 | 0.73 | 0.57 | 0.62 | 0.31 | 0.32 | 0.14 |
| $I_{21}$ (A) | $I_{22}$ (A) | $I_{23}$ (A) | $I_{24}$ (A) | $I_{25}$ (A) | $I_{26}$ (A) | $I_{27}$ (A) |
| 0.18 | 0.29 | 0.23 | 0.39 | 0.34 | 0.39 | 0.28 |

sured current spectrum of the air handler of a variable-speed drive with (medium or) 50% of rated output power.

a) Plot the input current wave shape of this variable-speed drive and its Fourier approximation.

b) Compute the K- and $F_{HL}$-factors and their associated derating of the 25 kVA single-phase pole transformer feeding variable-speed drives of central residential air-conditioning systems. This assumption is a worst-case condition, and it is well-known that pole transformers also serve linear loads such as induction motors of refrigerators and induction motors for compressors of central residential air-conditioning systems.

## 2.4 NONLINEAR HARMONIC MODELS OF TRANSFORMERS

Appropriate harmonic models of all power system components including transformers are the basis of harmonic analysis and loss calculations. Harmonic models of transformers are devised in two steps: the first is the construction of transformer harmonic model, which is mainly characterized by the analysis of the core nonlinearity (due to saturation, hysteresis, and eddy-current effects), causing nonsinusoidal magnetizing and core-loss currents. The second step involves the relation between model parameters and harmonic frequencies. In the literature, many harmonic models for power transformers have been proposed and implemented. These models are based on one of the following approaches:

- time-domain simulation [17–25],
- frequency-domain simulation [26–29],
- combined frequency- and time-domain simulation [30, 31], and
- numerical (e.g., finite-difference, finite-element) simulation [1, 2, 6, 32–34].

The harmonic content of a square wave with magnitude 1.00 pu is known.

a) Develop a Fourier analysis program for any periodic function resulting in the DC offset and the amplitudes of any even and odd harmonics involved.

b) Apply this program to a square wave with magnitude 1.00 pu by taking into account the Nyquist criterion.

c) Plot the original square wave and its Fourier approximation.

d) Compute the K- and $F_{HL}$-factors by taking various numbers and orders of harmonics into account.

### 2.3.3.2 Application Example 2.5: K- and $F_{HL}$-Factors and Their Application to Derating of 25 kVA Single-Phase Pole Transformer Loaded by Variable-Speed Drives

The objective of this example is to apply the computational procedure of Application Example 2.4 to a 25 kVA single-phase pole transformer loaded only by variable-speed drives of central residential air-conditioning systems [16]. Table E2.5.1 lists the mea-

This section starts with the general harmonic model for a power transformer and the simulation techniques for modeling its nonlinear iron core. The remainder of the section briefly introduces the basic concepts and equations involving each of the above-mentioned modeling techniques.

### 2.4.1 The General Harmonic Model of Transformers

Figure 2.12 presents the physical model of a single-phase transformer. The corresponding electrical and magnetic equations are

$$v_p(t) = R_p i_p(t) + L_p \frac{di_p(t)}{dt} + e_p(t), \quad (2\text{-}28)$$

$$v_s(t) = R_s i_s(t) + L_s \frac{di_s(t)}{dt} + e_s(t), \quad (2\text{-}29)$$

$$\Phi_m = BA, \quad (2\text{-}30)$$

$$B = f_{\text{nonlinear}}(H), \quad (2\text{-}31)$$

$$N_p i_p(t) + N_s i_s(t) = H\ell, \quad (2\text{-}32)$$

where

- $R_p$ and $R_s$ are the resistances of the primary and the secondary windings, respectively,
- $L_p$ and $L_s$ are the leakage inductances of primary and secondary windings, respectively,
- $e_p(t) = N_p \dfrac{d\Phi_m}{dt}$ and $e_s(t) = N_s \dfrac{d\Phi_m}{dt}$ are the induced voltages of the primary and the secondary windings, respectively,

- $N_p$ and $N_s$ are the number of turns of the primary and the secondary windings, respectively,
- $B$, $H$ and $\Phi_m$ are the magnetic flux density, the magnetic field intensity, and the magnetic flux in the iron core of the transformer, respectively, and
- $A$ and $\ell$ are the effective cross section and length of the integration path of transformer core, respectively.

Dividing Eq. 2-32 by $N_p$ yields

$$i_p(t) + \frac{N_s}{N_p} i_s(t) = i_{exc}(t), \quad (2\text{-}33)$$

where $i_{exc}(t) = H\ell/N_p$ is the transformer excitation (no-load) current, which is the sum of the magnetizing ($i_{mag}(t)$) and core-loss ($i_{core}(t)$) currents. Combining Eqs. 2-30 and 2-31, it is clear that the no-load current is related to the physical parameters, that is, the magnetizing curve (including saturation and hysteresis) and the induced voltage.

Based on Eqs. 2-28, 2-29, and 2-33, the general harmonic model of transformer is obtained as shown in Fig. 2.13. There are four dominant characteristic parameters:

- winding resistance,
- leakage inductance,
- magnetizing current, and
- core-loss current.

Some models assume constant values for the primary and the secondary resistances. However, most references take into account the influence of skin effects and proximity effects in the harmonic model. Since primary ($\Phi_{p\ell}$) and secondary ($\Phi_{s\ell}$) leakage fluxes mainly flow across air, the primary and the secondary leakage inductances can be assumed to be constants. The main difficulty arises in the computation of the magnetizing and core-loss currents, which are the main sources of harmonics in power transformers.

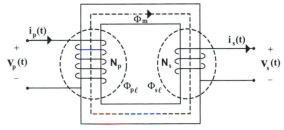

**FIGURE 2.12** Physical model of a single-phase transformer.

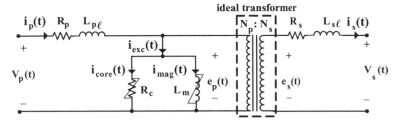

**FIGURE 2.13** General harmonic model of a transformer.

### 2.4.2 Nonlinear Harmonic Modeling of Transformer Magnetic Core

Accurate transformer models incorporate nonlinear saturation and hysteresis phenomena. Numerous linear, piecewise linear, and nonlinear models are currently available in the literature for the representation of saturation and hysteresis effects of transformer cores. Most models are based on time-domain techniques that require considerable computing time; however, there are also a few frequency-based models with acceptable degrees of accuracy. It is the purpose of this section to classify these techniques and to introduce the most widely used models.

#### 2.4.2.1 Time-Domain Transformer Core Modeling by Multisegment Hysteresis Loop

The transformer core can be accurately modeled in the time domain by simulating its $i - \lambda$ characteristic including the major hysteresis loop (with or without minor loops), which accounts for all core effects: hysteresis loss, eddy-current loss, saturation, and magnetization (Fig. 2.14). The $i - \lambda$ loop is divided into a number of segments and each segment is approximated by a parabola [20], a polynomial [31], a hyperbolic [25], or other functions. The functions expressing the segments must be defined so that $di/d\lambda$ (or $dH/dB$) is continuous in the entire defined region of the $i - \lambda$ (or $B—H$) plane.

As an example, two typical $i - \lambda$ characteristics are shown in Fig. 2.15. Five segments are used in Fig. 2.15a to model the $i - \lambda$ loop [20]. The first three are approximated by polynomials of the 13th order while the fourth and the fifth segments (in the positive and negative saturation regions) are represented by parabolas. The expression for describing the four-segment loop of Fig. 2.15b is [31]

$$i_{exc} = A + Be^{C\lambda} \quad \text{segment I}$$
$$i_{exc} = D + Ee^{F\lambda} \quad \text{segment II}$$
$$i_{exc} = -D - Ee^{-F\lambda} \quad \text{segment III.}$$
$$i_{exc} = -A - Be^{-C\lambda} \quad \text{segment IV} \quad (2\text{-}34)$$

There is a major difficulty with time-domain approaches. For a given value of maximum flux linkage, $\lambda_{max}$ the loop is easily determined experimentally; however, for variable $\lambda_{max}$ the loop not only changes its size but also its shape. Since these changes are particularly difficult to predict, the usual approach is to neglect the variation in shape and to assume linear changes in size. This amounts to scaling the characteristics in the $\lambda$ and i directions for different values of $\lambda_{max}$.

This is a fairly accurate model for the transformer core; however, it requires considerable computing time. Some more sophisticated models also include minor hysteresis loops in the time-domain analysis [27].

#### 2.4.2.2 Frequency- and Time-Domain Transformer Core Modeling by Saturation Curve and Harmonic Core-Loss Resistances

In these models, transformer saturation is simulated in the time domain while eddy-current losses and hysteresis are approximated in the frequency domain. If the voltage that produces the core flux is sinusoidal with rms magnitude of E and frequency of $f$, Eq. 2-1 may be rewritten as [31]

$$P_{fe} = P_{hys} + P_{eddy} = k_{hys}f^{1-S}E^S + k_{eddy}E^2, \quad (2\text{-}35)$$

where $k_{hys} \neq K_{hys}$ and $k_{eddy} \neq K_{eddy}$.

Although the application of superposition is incorrect for nonlinear circuits, in some cases it may be applied cautiously to obtain an approximate solution. If superposition is applied, the core loss due to individual harmonic components may be defined to model both hysteresis and eddy-current losses by dividing Eq. 2-35 by the square of the rms harmonic voltages. We define the conductance $G_{eddy}$ (accounting for eddy-current losses) and harmonic conductance $G_{hys}^{(h)}$ (accounting for hysteresis losses at harmonic frequencies) as [31]

$$G_{eddy} = k_{eddy}, \quad (2\text{-}36a)$$

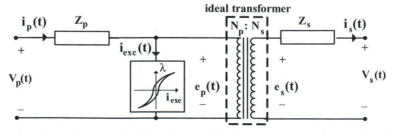

**FIGURE 2.14** Time-domain nonlinear model of transformer with the complete $i - \lambda$ characteristic [30].

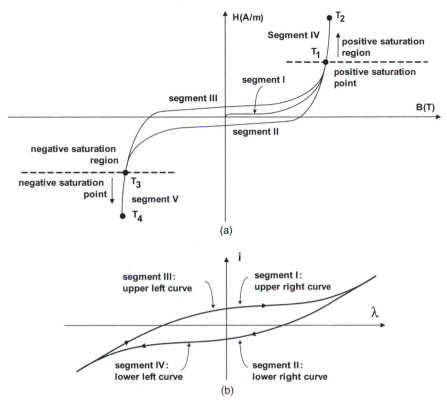

**FIGURE 2.15** Time-domain nonlinear model of transformer core with the complete hysteresis loop; (a) using five segments [20], (b) using four segments [31].

$$G_{hys}^{(h)} = k_{hys} \, (hf)^{1-S} E^{S-2}, \qquad (2\text{-}36b)$$

where $f$ and $h$ are the fundamental frequency and harmonic order, respectively. Therefore, eddy current $(P_{eddy})$ and hysteresis $(P_{hys})$ losses are modeled by a constant conductance $(G_{eddy})$ and harmonic conductances $(G_{hys}^{(h)})$, respectively, and the transformer $i - \lambda$ characteristic is approximated by a single-valued saturation curve, as shown in Fig. 2.16 [31]. There are basically two main approaches for modeling the transformer saturation curve: piecewise linear inductances and incremental inductance [19], as shown by Fig. 2.16b,c. The incremental approach uses a polynomial [31, 30], arctangent [22, 23], or other functions to model the transformer magnetizing characteristic. The incremental reluctances (or inductances) are then obtained from the slopes of the inductance-magnetizing current characteristic.

### 2.4.2.3 Time-Domain Transformer Coil Modeling by Saturation Curve and a Constant Core-Loss Resistance

Due to the fact that the effect of hysteresis in transformers is much smaller than that of the eddy current,

and to simplify the computation, most models assume a constant value for the magnetizing conductance [22, 23, 30]. As mentioned before, the saturation curve may be modeled using a polynomial [30, 29], arctangent [22, 23], or other functions. The slightly simplified transformer model is shown in Fig. 2.17.

### 2.4.2.4 Frequency-Domain Transformer Coil Modeling by Harmonic Current Sources

Some frequency-based algorithms use harmonic current sources $(I_{core}^{(h)}$ and $I_{mag}^{(h)})$ to model the nonlinear effects of eddy current, hysteresis, and saturation of the transformer core, as shown in Fig. 2.18. The magnitudes and phase angle of the current sources are upgraded at each step of the iterative procedure. There are different methods for upgrading (computing) magnitudes and phase angles of current sources at each step of the iterative procedure. Some models apply the saturation curve to compute $I_{mag}^{(h)}$ in the time domain, and use harmonic loss-density and so called phase-factor functions to compute $I_{core}^{(h)}$ in the frequency domain [1–4, 20]. These currents could also be approximately computed in the frequency domain as explained in the next section.

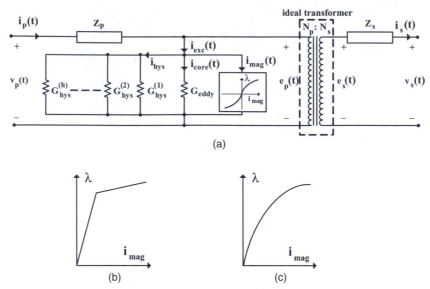

**FIGURE 2.16** Frequency- and time-domain nonlinear model of transformer core with harmonic conductances and a single-valued saturation curve; (a) transformer model [31], (b) the piecewise linear model of saturation curve, (c) the incremental model of saturation curve [19].

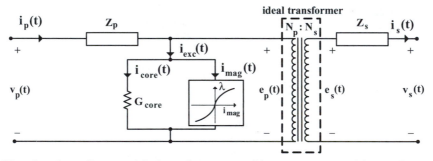

**FIGURE 2.17** Time-domain nonlinear model of transformer core with a constant magnetizing conductance and a single-valued ($\lambda - i$) characteristic [22].

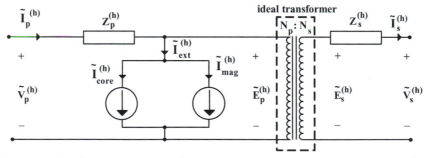

**FIGURE 2.18** Frequency-domain nonlinear model of transformer core with harmonic current sources [1–4].

### 2.4.2.5 Frequency-Domain Transformer Coil Modeling by Describing Functions

The application of describing functions for the computation of the excitation current in the frequency domain was proposed and implemented for a single-phase transformer under no-load conditions [26]. The selected describing function, which assumes

sinusoidal flux linkages and a piecewise linear hysteresis loop, is relatively fast and computes the nonsinusoidal excitation current with an acceptable degree of accuracy. The concept of describing functions as applied to the nonlinear (symmetric) network N is demonstrated in Fig. 2.19, where a sinusoidal input function $\lambda(t) = E \sin(\omega t)$ results in a nonsinusoidal output function $i_e(t)$. Neglecting the influence of harmonics, $i_e(t)$ can be approximated as

$$i_e(t) \approx A_1 \cos(\omega t) + B_1 \sin(\omega t) = C_1 \sin(\omega t + \phi)$$

$$A_1 = \frac{1}{\pi} \int_{-\pi}^{\pi} i_e(t) \cos(\omega t) d(\omega t)$$

$$B_1 = \frac{1}{\pi} \int_{-\pi}^{\pi} i_e(t) \sin(\omega t) d(\omega t). \quad (2\text{-}37)$$

In Eq. 2-37, $A_1$ and $B_1$ are the fundamental coefficients of the Fourier series representation of $i_e(t)$ and its harmonic components are neglected. Also, $C_1 \angle \phi = B_1 + jA_1$.

Based on Eq. 2-37 and Fig. 2.19, the describing function (indicating the gain of nonlinearity) is defined as

$$N(E, \omega) = \frac{B_1 + jA_1}{E} = \frac{C \angle \phi}{E}. \quad (2\text{-}38)$$

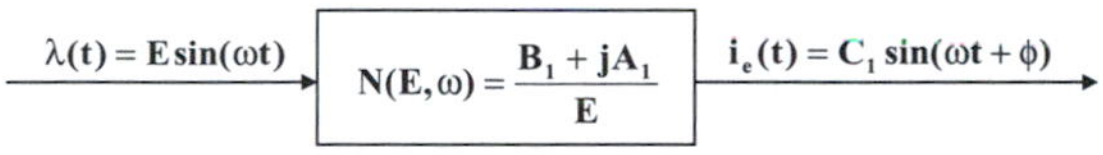

$$\lambda(t) = E\sin(\omega t) \longrightarrow \boxed{N(E,\omega) = \frac{B_1 + jA_1}{E}} \longrightarrow i_e(t) = C_1 \sin(\omega t + \phi)$$

**FIGURE 2.19** Expression of describing function for a nonlinear network N [26].

Figure 2.20 illustrates a typical $i - \lambda$ characteristic where a hysteresis loop defines the nonlinear relationship between the input sinusoidal flux linkage and the output nonsinusoidal exciting current. In order to establish a describing function for the $i - \lambda$ characteristic, the hysteresis loop is divided into three linear pieces with the slopes $1/k$, $1/k_1$, and $1/k_2$ as shown in Fig. 2.21.

Comparing the corresponding gain at different time intervals to both input flux-linkage waveform and output exciting current waveform during the same period allows us to calculate the value of $N(E, \omega)$ in three different time intervals, as illustrated by Eq. 2–38 and Fig. 2.21. Therefore, if input $\lambda(t) = E \sin(\omega t)$ is known, $i_e(t)$ can be calculated as follows:

$$i_e(t) = \begin{cases} \dfrac{1}{k}[E\sin(\omega t)] + D & 0 < \omega t < \beta \\[2mm] \dfrac{1}{k_1}[E\sin(\omega t) - M] + \dfrac{M}{k} + D & \beta < \omega t > \dfrac{\pi}{2} \\[2mm] \dfrac{1}{k_2}[E\sin(\omega t) - M] + \dfrac{M}{k} - D & \dfrac{\pi}{2} < \omega t < \pi - \beta \\[2mm] \dfrac{1}{k}[E\sin(\omega t)] - D & \pi - \beta < \omega t > \pi \end{cases}$$

$$(2\text{-}39)$$

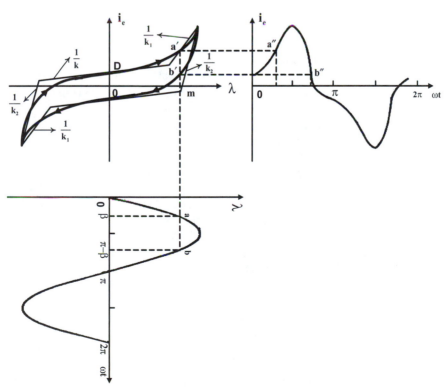

**FIGURE 2.20** Excitation current waveform for a transformer core [26].

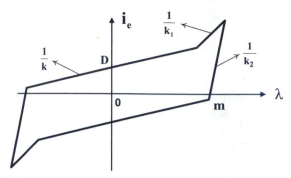

**FIGURE 2.21** Piecewise linear function approximation of hysteresis loop defining transformer describing function [26].

where $\beta = \sin^{-1}$ (M/E) is the angle corresponding to the knee point of the hysteresis loop in radians.

Applying Fourier series to Eq. 2-39, the harmonic components of the excitation current ($I_{\text{exe}}^{(h)}$) can be determined. Active (real) and reactive (imaginary) parts of the excitation current correspond to harmonics of the core-loss ($I_{\text{core}}^{(h)}$) and the magnetizing ($I_{\text{mag}}^{(h)}$) currents, respectively.

### 2.4.3 Time-Domain Simulation of Power Transformers

Time-domain techniques use analytical functions to model transformer primary and secondary circuits and core characteristics [17–25]. Saturation and hysteresis effects, as well as eddy-current losses, are included with acceptable degrees of accuracy. These techniques are mostly used for the electromagnetic transient analysis (such as inrush currents, overvoltages, geomagnetically induced currents, and out-of-phase synchronization) of multiphase and multilimb core-type power transformers under (un)balanced (non)sinusoidal excitations with (non)linear loads. The main limitation of time-domain techniques is the relatively long computing time required to solve the set of differential equations representing transformer dynamic behavior. They are not usually used for steady-state analyses. Harmonic modeling of power transformers in the time domain are performed by some popular software packages and circuit simulators such as EMTP [24] and DSPICE (Daisy's version of circuit simulator SPICE [25]). The electromagnetic mathematical model of multiphase, multilimb core-type transformer is obtained by combining its electric and the magnetic circuits [35–37]. The principle of duality is usually applied to simplify the magnetic circuit. Figure 2.22 illustrates the topology and the electric

equivalent circuit for the general case of a five-limb transformer, from which other configurations such as three-phase, three-limb, and single-phase ones can be derived. The open ends of the nonlinear multiport inductance matrix L (or its inverse, the reluctance matrix R) allow the connection for any electrical configuration of the source and the load at the terminals of the transformer.

Most time-domain techniques are based on a set of differential equations defining transformer electric and magnetic behaviors. Their computational effort involves the numerical integration of ordinary differential equations (ODEs), which is an iterative and time-consuming process. Other techniques use Newton methodology to accelerate the solution [22, 23]. Transformer currents and/or flux linkages are usually selected as variables. Difficulties arise in the computation and upgrading of magnetic variables (e.g., flux linkages), which requires the solution of the magnetic circuit or application of the nonlinear hysteresis characteristics, as discussed in Section 2.4.2.

In the next section, time-domain modeling based on state-space formulation of transformer variables is explained. Either transformer currents and/or flux linkages may be used as the state variables. Some models [22, 23] prefer flux linkages since they change more slowly than currents and more computational stability is achieved.

#### 2.4.3.1 State-Space Formulation

The state equation for an m-phase, n-winding transformer in vector form is [19]

$$\vec{V} = \vec{R}\vec{I} + \vec{L}\frac{d\vec{I}}{dt} + \frac{d\vec{\Psi}}{dt}, \qquad (2\text{-}40)$$

where $\vec{V}$, $\vec{I}$, $\vec{R}$, $\vec{L}$, and $\vec{\Psi}$ are the terminal voltage vector, the current vector, the resistance matrix, the leakage inductance matrix, and the flux linkage vector, respectively.

The flux linkage vector can be expressed in terms of the core flux vector by

$$\vec{\Psi} = \vec{W}\vec{\Phi} \quad \text{and} \quad \frac{d\vec{\Psi}}{dt} = \vec{W}\frac{d\vec{\Phi}}{dt}, \qquad (2\text{-}41)$$

where $\vec{W}$ is the transformation ratio matrix (number of turns) and $\vec{\Phi}$ is the core-flux vector.

In general, the core fluxes are nonlinear functions of the magnetomotive forces ($\vec{F}$), therefore, $d\vec{\Phi}/dt$ can be expressed as

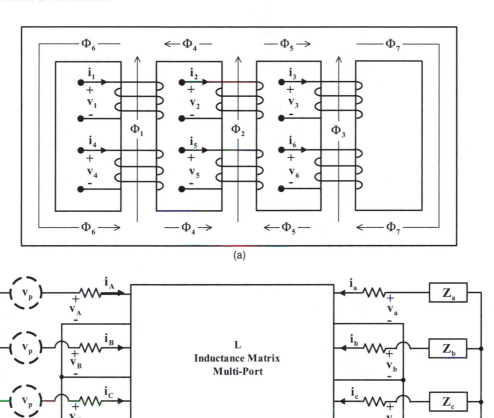

(a)

(b)

**FIGURE 2.22** Time-domain harmonic model of power transformers; (a) general topology for the three-phase, five-limb structure, (b) equivalent circuit [23].

$$\frac{d\vec{\Phi}}{dt} = \frac{d\vec{\Phi}}{d\vec{F}}\frac{d\vec{F}}{dt}, \qquad (2\text{-}42)$$

where $d\vec{\Phi}/d\vec{F}$ is a m × m Jacobian matrix. The magnetomotive force vector can be expressed in terms of the terminal current vector by

$$\vec{F} = \vec{C}_W \vec{I}, \qquad (2\text{-}43)$$

where $\vec{C}_W$ is a matrix that can be determined from matrix $\vec{W}$ and the configuration of the transformer.

Substituting Eqs. 2-42 and 2-43 into Eq. 2-40, we obtain

$$\vec{V} = \vec{R}\vec{I} + \vec{L}\frac{d\vec{I}}{dt} + \vec{W}\frac{d\vec{\Phi}}{d\vec{F}}\vec{C}_W\frac{d\vec{I}}{dt}. \qquad (2\text{-}44)$$

Defining the nonlinear incremental (core) inductance

$$\vec{L}_d = \vec{W}\frac{d\vec{\Phi}}{d\vec{F}}\vec{C}_W, \qquad (2\text{-}45)$$

the transformer state equation is finally expressed as follows:

$$\vec{V} = \vec{R}\vec{I} + (\vec{L} + \vec{L}_d)\frac{d\vec{I}}{dt}. \qquad (2\text{-}46)$$

Equations 2-45 and 2-46 are the starting point for all modeling techniques based on the decoupling of magnetic and electric circuits. The basic difficulty is the calculation of the elements of the Jacobian $d\vec{\Phi}/d\vec{F}$ at each integration step. The incorporation of nonlinear effects (such as magnetic saturation and hysteresis) and the computation of $d\vec{\Phi}/d\vec{F}$ are performed by appropriate modifications of the differential and algebraic equations (Eqs. 2-40 to 2-46). As discussed in Section 2.4.2, numerous possibilities are available for accurate representation of transformer saturation and hysteresis. However, there is a trade-off between accuracy and computational speed of the solution. Figure 2.23 shows the flowchart of the nonlinear iterative algorithm for transformer modeling based on Eqs. 2-40 to 2-46, where elements of $d\vec{\Phi}/d\vec{F}$ can be derived from the solution of transformer magnetic circuit with a piecewise linear (Fig. 2.16b) or an incremental (Fig. 2.16c) magnetizing characteristic [19].

### 2.4.3.2 Transformer Steady-State Solution from the Time-Domain Simulation

Conventional time-domain transformer models based on the brute force (BF) procedure [22, 23] are not usually used for the computation of the periodic steady-state solution because of the computational effort involved requiring the numerical integration of ODEs until the initial transient decays. This drawback is overcome with the introduction of numerical

differentiation (ND) and Newton techniques to enhance the acceleration of convergence [22, 23].

### 2.4.4 Frequency-Domain Simulation of Power Transformers

Frequency- or harmonic-domain simulation techniques use Fourier analysis to model the transformer's primary and secondary parameters and to approximate core nonlinearities [26–29]. Most frequency-based models ignore hysteresis and eddy currents or use some measures (e.g., the concept of describing functions as discussed in Section 2.4.2.5) to approximate them; therefore, they are not usually as accurate as time-domain models. Harmonic-based simulation techniques are relatively fast, which makes them fine candidates for steady-state analysis of power transformers. Most (but not all) frequency-domain analyses assume balanced three-phase operating conditions and single-limb transformers (bank of single-phase transformers).

Figure 2.24 shows a frequency-domain transformer model in the form of the harmonic Norton equivalent circuit. Standard frequency-domain analyses are used to formulate the transformer's electric circuits whereas frequency-based simulation techniques are required to model magnetic circuits. Figure 2.25 shows the flowchart of the nonlinear iterative algorithm for transformer modeling in the frequency domain. At each step of the algorithm, harmonic current sources $I_{core}^{(h)}$ and $I_{mag}^{(h)}$ are upgraded using a frequency-based model for the transformer core, as discussed in Sections 2.4.2.4 and 2.4.2.5. Some references use the harmonic lattice representation of multilimb, three-phase transformers to define a harmonic Norton equivalent circuit [27]. The main limitations of this approach, as with all frequency-based simulation techniques, are the approximations used for modeling transformer magnetic circuits.

### 2.4.5 Combined Frequency- and Time-Domain Simulation of Power Transformers

To overcome the limitations of frequency-based techniques and to preserve the advantages of the

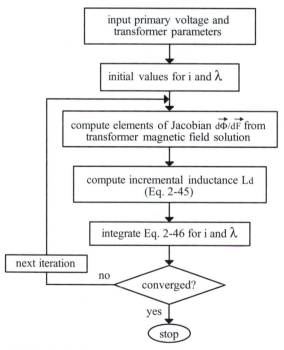

**FIGURE 2.23** Flowchart of the time-domain iterative algorithm for transformer modeling [19].

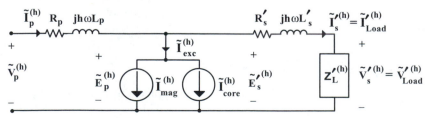

**FIGURE 2.24** Frequency-domain harmonic model of power transformer [1–4].

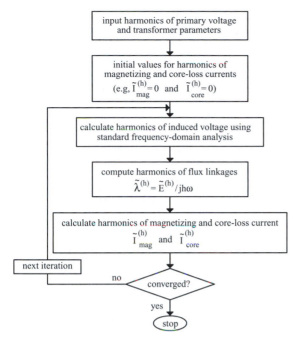

**FIGURE 2.25** Flowchart of the frequency-domain iterative algorithm for transformer modeling.

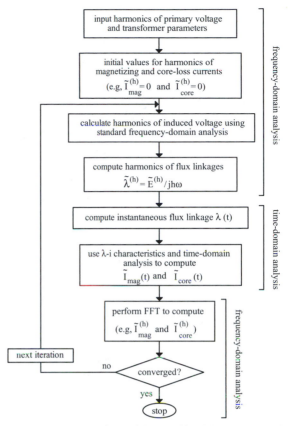

**FIGURE 2.26** Flowchart of the combined frequency- and time-domain iterative algorithm for transformer modeling [1–4].

time-domain simulation, most transformer models use a combination of frequency-domain (to model harmonic propagation on primary and secondary sides) and time-domain (to model core nonlinearities) analyses to achieve the steady-state solution with acceptable accuracy and speed [30, 31, 1–4]. Thus, accurate modeling of saturation, hysteresis, and eddy-current phenomena are possible. Combined frequency- and time-domain models are fine candidates for steady-state analysis of power transformers. Figure 2.26 shows the flowchart of the nonlinear iterative algorithm for transformer modeling in the frequency and time domains. Transformer electric circuits are models for standard frequency-domain techniques whereas the magnetic analysis is accurately performed in the time domain. At each step of the algorithm, instantaneous flux linkages $(\lambda(t))$ and transformer $i - \lambda$ characteristics are used to compute the instantaneous magnetizing $(i_{mag}(t))$ and core-loss $(i_{core}(t))$ currents, as discussed in Sections 2.4.2.1 to 2.4.2.4. Fast Fourier transformation (FFT) is performed to compute and upgrade harmonic current sources $\tilde{I}_{core}^{(h)}$ and $\tilde{I}_{mag}^{(h)}$.

## 2.4.6 Numerical (Finite-Difference, Finite-Element) Simulation of Power Transformers

These accurate and timely simulation techniques are based on two- or three-dimensional modeling of the anisotropic power transformer [1–4, 6, 32–34]. They require detailed information about transformer size, structure, and materials. The anisotropic transformer's magnetic core is divided into small (two- or three-dimensional) meshes or volume elements, respectively. Within each mesh (element), isotropic materials and linear B–H characteristics are assumed. The resulting linear equations are solved using matrix solution techniques. Mesh equations are related to each other by boundary conditions. With these nonlinear and time-consuming techniques, simultaneous consideration of most transformer nonlinearities are possible. Finite element/difference techniques are the most accurate tools for transformer modeling under sinusoidal and nonsinusoidal operating conditions. Numerical (e.g., finite-element and finite-difference) software packages are available for the simulation of electromagnetic phenomena. However, numerical techniques are not very popular since they require much detailed information about transformer size, air gaps, and materials that makes their application a tedious and time-consuming process.

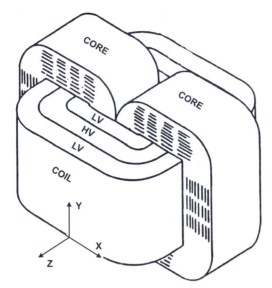

**FIGURE 2.27** Construction of 25 kVA single-phase pole transformer [36].

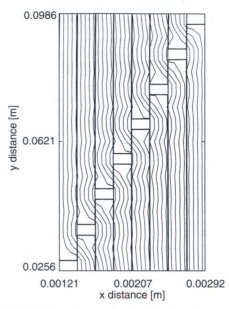

**FIGURE 2.29** Sample strip for mixed-grain region showing flux paths for an excitation of 10.12 At. One flux tube contains 0.000086 Wb/m [36].

ROLLING DIRECTION AND FLUX PATH

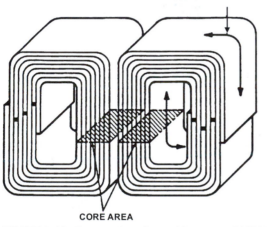

CORE AREA

**FIGURE 2.28** Cross section of wound iron-core of 25 kVA pole transformer [36].

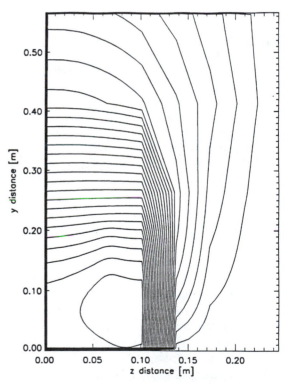

**FIGURE 2.30** Flux distribution in y–z plane for an excitation of 10.12 At. One flux tube contains 0.00000138 Wb/m [36].

Figure 2.27 illustrates the construction of a single-phase 25 kVA pole transformer as it is widely employed in residential single-phase systems. The butt-to-butt air gaps and the interlamination gaps of the wound core of Fig. 2.27 are illustrated in Fig. 2.28. These air gaps, grain orientation of the electrical steel (anisotropy), and saturation of the iron core require the calculation of representative reluctivities as indicated in Fig. 2.29. These representative reluctivity data are used for the computation of the end-region (y–z plane) and x–y planes of the single-phase transformer as depicted in Figs. 2.30 and 2.31,

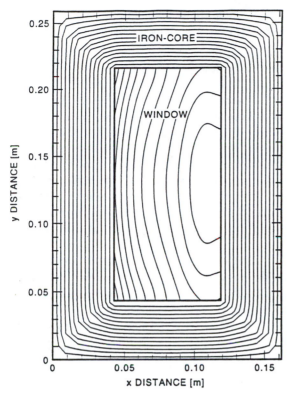

**FIGURE 2.31** Flux distribution in x – y plane for an excitation of 920 At. One flux tube inside window contains 0.000024 Wb/m and one flux tube in iron contains 0.00502 Wb/m [36].

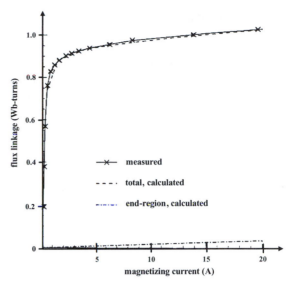

**FIGURE 2.32** Measured and calculated $\lambda - i$ characteristics. Total characteristic is that of the end region (y – z) plus that of the iron-core region (x – y) [36].

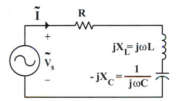

**FIGURE 2.33** Linear resonant circuit.

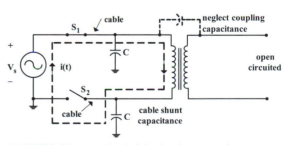

**FIGURE 2.34** An unloaded single-phase transformer connected to power system via cable.

respectively. Measured and computed flux-linkage versus current characteristics are presented in Fig. 2.32. Detailed analyses are given in references [1–4] and [36].

## 2.5 FERRORESONANCE OF POWER TRANSFORMERS

Before discussing ferroresonance phenomena, it may be good to review resonance as it occurs in a linear RLC circuit, as shown in Fig. 2.33. The current is

$$|\tilde{I}| = \frac{|\tilde{V}_s|}{\sqrt{R^2 + (X_L - X_C)^2}}. \qquad (2\text{-}47)$$

From the above equation we observe that whenever the value of the capacitive reactance $X_C$ equals the value of the inductive reactance $X_L$, the current in the circuit only is limited by the resistance $R$. Very often the value of R can be very small compared with $X_L$ or $X_C$, in which case the current can reach very high levels (resonant current).

Now let us look at a simple power circuit, consisting of a transformer connected to a power system as shown by Fig. 2.34; its single-phase equivalent circuit is illustrated in Fig. 2.35. Even though switch $S_2$ is open, a current $i(t)$ can still flow because the cable shunt capacitance C closes the circuit. The transformer is assumed to be unloaded but energized. The magnetizing current $i(t)$ has to return to its source via the cable shunt capacitance C. Therefore,

**Resonance occurs whenever $X_L$ and $X_C$ are close to each other.**

In power systems, the variations of $X_L$ are mainly due to the nonlinear magnetizing inductance of transformers, as shown in Fig. 2.36, and most changes in $X_C$ are caused by switching capacitor banks and cables. The capacitance between two charged plates is illustrated in Fig. 2.37.

## 2.5.1 System Conditions Susceptible (Contributive, Conducive) to Ferroresonance

In the previous section, a single-phase circuit was used to explain in a simplified manner the problem of ferroresonance. Here, three-phase systems will be covered because they are the basic building block of a power system and, as will be shown, they are the most prone to this problem.

Now that the principle of ferroresonance has been explained, the next important question is what conditions in an actual power system are contributive (conducive) to ferroresonance? Some conditions may not by themselves initiate the problem, but they certainly are prerequisites (necessary) so that when other conditions are present in a power system, ferroresonance can occur. The purpose of the next section is to prevent these prerequisite conditions.

## 2.5.2 Transformer Connections and Single-Phase (Pole) Switching at No Load

The transformer is the basic building block in stepping up or down voltages and currents in a power system. It is also one of the basic elements of a ferroresonant circuit. The two most widely used transformer connections are the wye and delta. Each one has its advantages and disadvantages, and depending on the particular application and engineering philosophies, one or the other or a combination of both will be used. First we will look at the delta connected (at primary) transformer, as shown in Fig. 2.38. Figure 2.38a shows a simple power system. The source of power is represented by a grounded equivalent power source because normally the system is grounded at the source and/or at some other points in the power system. The connection between the source and the transformer is made by an (underground) cable having a capacitance C.

There are four prerequisites for ferroresonance:

- The first prerequisite for ferroresonance is the employment of cables with relatively large capacitances. The transformer primary is delta connected or in ungrounded wye connection, while the secondary can be either grounded or ungrounded wye or delta connected. This system is shown in Fig. 2.38a and the transformer is assumed to be unloaded (load provides damping).
- The second prerequisite for ferroresonance is the transformer connection.
- The third prerequisite for ferroresonance is an unloaded transformer. In order to simplify this discussion, the coupling capacitances of the transformer (from phase to ground, phase to phase, and from primary to secondary) will be assumed to be low enough compared with the shunt capacitance

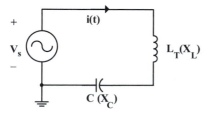

**FIGURE 2.35** Equivalent circuit of Fig. 2.34.

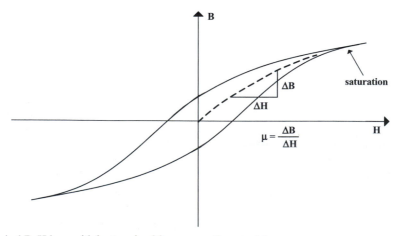

**FIGURE 2.36** Typical B–H loop with hysteresis of ferromagnetic materials.

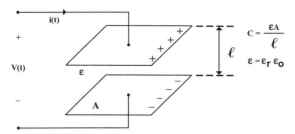

**FIGURE 2.37** Capacitance between two charged parallel plates.

of the primary line (cables in populated areas) so that they can be neglected. However, it must be noted that at voltage levels above 15 kV the coupling capacitances (increased insulation) of the transformer become more important so that their contribution to the ferroresonant circuit cannot be neglected.

- The fourth prerequisite is when either one or two phases are connected; then these circuit configurations are susceptible to ferroresonance. When the

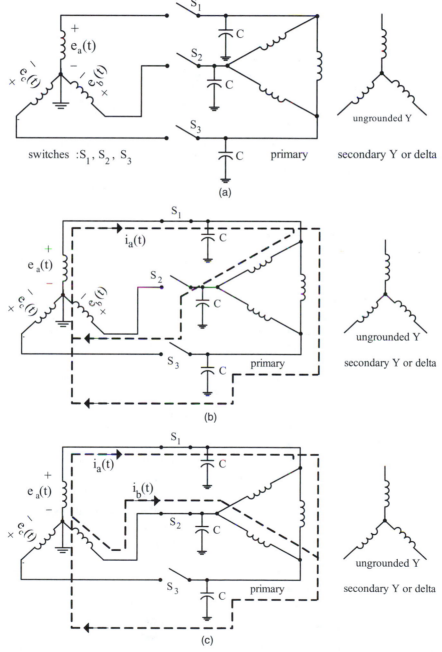

**FIGURE 2.38** Simple power system; the unloaded secondary can be either wye or delta (not shown) connected. (a) Transformer deenergized; (b) one phase energized: ferroresonance is likely to occur; (c) two phases energized: ferroresonance is likely to occur.

third phase is energized, the system will be balanced and the current will flow directly to and from the transformer, bypassing the capacitance to ground.

Adding a load to the secondary side of the transformer changes the ferroresonant circuit considerably: the effective load impedance $Z_L$ will be reflected back into the primary side by the square of the turns ratio of the primary ($N_1$) to the secondary ($N_2$). This reflected impedance $Z_L' = (N_1/N_2)^2 Z_L$ will in turn limit considerably the current in the ferroresonant circuit, thus reducing the possibility of this phenomenon significantly.

Now let us look at a transformer with its primary winding connected ungrounded wye as shown in Fig. 2.39. Note that the source is shown grounded and that the unloaded secondary side of the transformer can be connected either ungrounded wye or delta. When one (or two) phase(s) is (are) energized, a closed loop is formed, consisting of transformer windings and the capacitance of the primary conductor (cable) to ground. The resulting large nonsinusoidal magnetizing current can induce ferroresonant overvoltages in the circuit generating large losses within the transformer. As in the previous case, a secondary load can greatly reduce the probability of ferroresonance.

A single-phase transformer connected phase-to-phase to a grounded supply is also likely to become ferroresonant when one of the phases is opened or closed, as shown in Fig. 2.40. If on the other hand, the transformer is connected phase to ground to a grounded source (Fig. 2.41), the circuit is not likely to become ferroresonant when the hot line ($S_1$) is opened.

The mere presence of the ferroresonant circuit does not necessarily mean that the circuit involved will experience overvoltages due to this problem. The ratio of the capacitive reactance $X_C$ to the transformer magnetizing reactance $X_t$ of the ferroresonant circuit ($X_C/X_t$) will determine whether the circuit will experience overvoltages or not. These tests have been generalized to cover a wide range of transformer sizes (15 kVA to 15 MVA) and line-to-ground (line-to-neutral) voltages ranging from 2.5 to 19.9 kV, and it has been determined that for three-phase core type banks, they will experience overvoltages above 1.1–1.25 pu when the ratio $X_C/X_t < 40$, where $X_t = \omega L$, $X_C = 1/\omega C$ or $\left(\dfrac{1}{\omega^2 LC}\right) < 40$.

Note that a three-limb, three-phase transformer has a smaller zero-sequence reactance than that of a

bank of three single-phase transformers or that of a five-limb, three-phase transformer [38–41].

For banks of single-phase transformers, a ratio $X_C/X_t < 25$ has been found to produce overvoltages when the ferroresonant circuit is present.

### 2.5.2.1 Application Example 2.6: Susceptibility of Transformers to Ferroresonance

Assuming linear $\lambda$-$i$ characteristics with $\lambda = 0.63$ Wb-turns at 0.5 A, we have

$$L = \frac{\lambda}{i} = \frac{0.63\,\text{Wb}-\text{turns}}{0.5\,\text{A}} = 1.27\,\text{H},$$

$$C = 100\ \mu\text{F},$$

$$\omega = 377\ \text{rad/s},$$

$$\frac{1}{\omega^2 LC} = \frac{1}{(377)^2 (1.27)(100)(10^6)} = 0.055.$$

Therefore, $X_C/X_t = 0.055 < 40$, that is, ferroresonance is likely to occur.

### 2.5.3 Ways to Avoid Ferroresonance

Now that the factors conducive to ferroresonance have been identified, the next question is how to avoid the problem or at least try to mitigate its effects as much as possible. There are a few techniques to limit the probability of ferroresonance:

- *The grounded-wye/grounded-wye transformer.* When the neutral of the transformer is grounded, the capacitance to ground of the line (cable) is essentially bypassed, thus eliminating the series LC circuit that can initiate ferroresonance. It is recommended that the neutral of the transformer be solidly grounded to avoid instability in the neutral of the transformer (grounding could be done indirectly via a "Peterson coil" to limit short-circuit currents). When the neutral of a wye/wye transformer is grounded, other things can happen too: if the grounded three-phase bank is not loaded equally on its three phases, the currents flowing in the bank and consequently the fluxes will not cancel out to zero. If the bank is a three-limb core transformer, the resultant flux will be forced to flow outside the core. Under these conditions, this flux will be forced to flow in the tank of the transformer, thus causing a current to circulate in it, which can cause excessive tank overheating if the imbalance is large enough. Another approach is to use banks made up of three single-phase units or three-phase, five-limb core transformers. When single-phase units are used, the flux for each phase

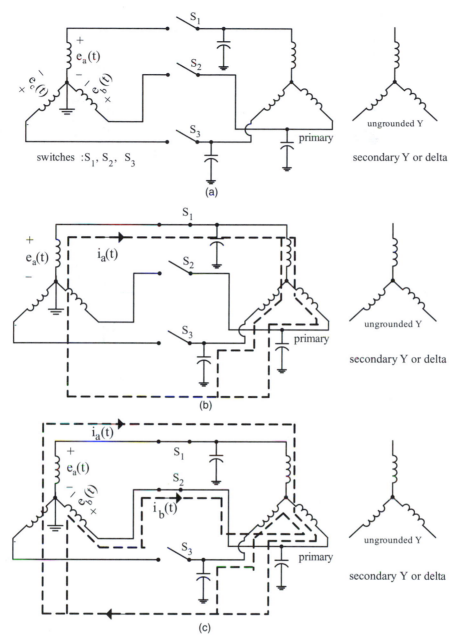

**FIGURE 2.39** The unloaded secondary can be either wye or delta (not shown) connected. (a) Transformer is de-energized; (b) one phase energized: ferroresonance is likely; (c) two phases energized: ferroresonance is likely.

will be restricted to its respective core. In the case of the five-limb core, the two extra limbs provide the path for the remaining flux due to imbalance.

- *Limiting the cable capacitance by limiting the length of the cable.* This is not a practical technique.
- *Fast three-phase switching to avoid longer one- and two-phase connections.* This is an effective but expensive approach
- Use of cables with low capacitance [42–44].

### 2.5.3.1 Application Example 2.7: Calculation of Ferroresonant Currents within Transformers

The objective of this example is to study the (chaotic) phenomenon of ferroresonance in transformers using Mathematica or MATLAB.

Consider a ferroresonant circuit consisting of a sinusoidal voltage source $v(t)$, a (cable) capacitance $C$, and a nonlinear (magnetizing) impedance

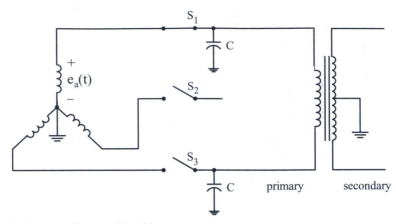

**FIGURE 2.40** Single-phase transformer with cable.

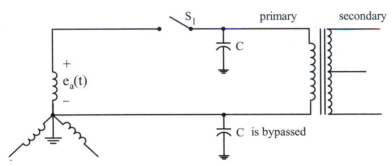

**FIGURE 2.41** Single-phase transformer with phase to ground: ferroresonance is unlikely.

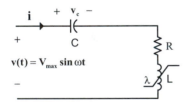

**FIGURE E2.7.1** Ferroresonant nonlinear circuit.

(consisting of a resistance $R$ and an inductance $L$) connected in series, as shown in Fig. E2.7.1. If the saturation curve (flux linkages $\lambda$ as a function of current $i$) is represented by the cubic function $i = a\lambda^3$, the following set of first-order nonlinear differential equations can be formulated:

$$i = a\lambda^3 \Rightarrow \begin{cases} \dfrac{d\lambda}{dt} = V_{max}\sin\omega t - v_c - aR\lambda^3 \\ \dfrac{dv_c}{dt} = \dfrac{a\lambda^3}{C} \end{cases},$$

or in an equivalent manner, one can formulate the second-order nonlinear differential equation

$$\frac{d^2\lambda}{dt^2} + k\lambda^2\frac{d\lambda}{dt} + \lambda^3 = b\cos\omega t,$$

where $k = 3aR$, $b = V_{max}\omega$, and $C = a$.

a) Show that the above equations are true for the circuit of Fig. E2.7.1.

b) The system of Fig. E2.7.1 displays for certain values of the parameters $b = V_{max}\omega$ and $k = 3aR$ ($\omega = 2\pi f$ where $f = 60$ Hz, $R = 0.1\ \Omega$, $C = 100\ \mu F$, $a = C$, low voltage $V_{maxlow} = \sqrt{2} \cdot 1000$ V, and high voltage $V_{maxhigh} = \sqrt{2} \cdot 20000$ V), which characterize the source voltage and the resistive losses of the circuit, respectively, chaotic behavior as described by [45]. For zero initial conditions – using Mathematica or MATLAB – compute numerically $\lambda$, $d\lambda/dt$, and $i$ as a function of time from $t_{start} = 0$ to $t_{end} = 0.5$ s.

c) Plot $\lambda$ versus $d\lambda/dt$ for the values as given in part b.

d) Plot $\lambda$ and $d\lambda/dt$ as a function of time for the values as given in part b.

e) Plot $i$ as a function of time for the values as given in part b.

f) Plot voltage dλ/dt versus *i* for the values as given in part *b*.

## 2.6 EFFECTS OF SOLAR-GEOMAGNETIC DISTURBANCES ON POWER SYSTEMS AND TRANSFORMERS

Solar flares and other solar phenomena can cause transient fluctuations in the earth's magnetic field (e.g., $B_{earth} = 0.6$ gauss $= 60 \, \mu T$). When these fluctuations are severe, they are known as geomagnetic storms and are evident visually as the aurora borealis in the Northern Hemisphere and as the aurora australis in the Southern Hemisphere (Fig. 2.42).

These geomagnetic field changes induce an earth-surface potential (ESP) that causes geomagnetically induced currents (GICs) to flow in large-scale 50 or 60 Hz electric power systems. The GICs enter and exit the power system through the grounded neutrals of wye-connected transformers that are located at opposite ends of a long transmission line. The causes and nature of geomagnetic storms and their resulting effects in electric power systems will be described next.

### 2.6.1 Application Example 2.8: Calculation of Magnetic Field Strength $\vec{H}$

The magnetic field strength $\vec{H}$ of a wire carrying 100 A at a distance of $R = 1$ m (Fig. E2.8.1) is

$$|\vec{H}| = \frac{I}{2\pi R} = \frac{100 \, \text{A}}{2\pi} \frac{}{1 \text{m}} = 15.92 \, \text{A/m}$$

or

$$|\vec{B}| = \mu_0 |\vec{H}| = 20 \mu T.$$

### 2.6.2 Solar Origins of Geomagnetic Storms

The solar wind is a plasma of protons and electrons emitted from the sun due to

- solar flares,
- coronal holes, and
- disappearing filaments.

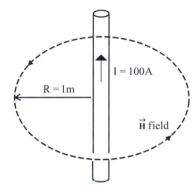

**FIGURE E2.8.1** $\vec{H}$ field of a single conductor in free space.

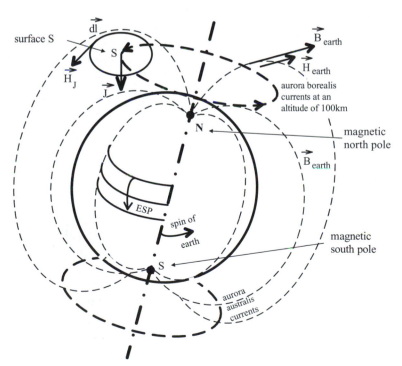

**FIGURE 2.42** Aurora borealis and aurora australis.

The solar wind particles interact with the earth's magnetic field in a complicated manner to produce auroral currents that follow circular paths (see Fig. 2.42) around the geomagnetic poles at altitudes of 100 km or more. These auroral currents produce fluctuations in the earth's magnetic field ($\Delta B = \pm 500$ nT $= \pm 0.5\ \mu$T) that are termed geomagnetic storms when they are of sufficient severity.

### 2.6.3 Sunspot Cycles and Geomagnetic-Disturbance Cycles

On average, solar activity as measured by monthly sun spot numbers follows an 11-year cycle [46–53]. The last sunspot cycle 23 had its minimum in September 1997, and its peak in 2001–2002. The largest geomagnetic storms are most likely due to flare and filament eruption events and could occur at any time during the cycle; the severe geomagnetic storm on March 13, 1989 (Hydro Quebec system collapsed with 21.5 GW) was evidently a striking example.

### 2.6.4 Earth-Surface Potential (ESP) and Geomagnetically Induced Current (GIC)

The auroral currents that result from solar-emitted particles interact with the earth's magnetic field and produce fluctuations in the earth's magnetic field $B_{earth}$. The earth is a conducting sphere, admittedly nonhomogeneous, that experiences (or at least portions of it) a time-varying magnetic field. The portions of the earth that are exposed to the time-varying magnetic field will have electric potential gradients (voltages) induced, which are termed earth-surface potentials (ESPs). Analytical methods have been developed to estimate the ESP based on geomagnetic field fluctuation data and a multilayered earth conductivity model: values in the range of 1.2–6 V/km or 2–10 V/mile can be obtained during severe geomagnetic storms in regions of low earth conductivity. Low earth conductivity (high resistivity) occurs in regions of igneous rock geology (e.g., Rocky Mountain region). Thus, power systems located in regions of igneous rocky geology are the most susceptible to geomagnetic storm effects.

Electric power systems become exposed to the ESP through the grounded neutrals of wye-connected transformers that may be at the opposite ends of long transmission lines. The ESP acts eventually as an ideal voltage source impressed across the grounded-neutral points and because the ESP has a frequency of one to a few millihertz, the resulting GICs can be determined by dividing the ESP by the

equivalent DC resistance of the paralleled transformer winding and line conductors between the two neutral grounding points. The GIC is a quasi-direct current, compared to 50 or 60 Hz, and GIC values in excess of 100 A have been measured in transformer neutrals [46–53].

Ampere's law is valid for the calculation of the aurora currents of Fig. 2.42.

$$f_C \vec{H}_J \cdot d\vec{l} = \int_S \vec{J} \cdot d\vec{S}. \qquad (2\text{-}48)$$

### 2.6.5 Power System Effects of GIC

GIC must enter and exit power systems through the grounded-neutral connections of wye-connected transformers or autotransformers. The per-phase GIC can be many times larger in magnitude than the rms AC magnetizing current of a transformer. The net result is a DC bias in the transformer core flux, resulting in a high level of half-cycle saturation [40, 54]. This half-cycle saturation of transformers operating within a power system is the source of nearly all operating and equipment problems caused by GICs during magnetic storms (reactive power demand and its associated voltage drop). The direct consequences of the half-cycle transformer saturation are

- The transformer becomes a rich source of even and odd harmonics;
- A great increase in inductive power (VAr) drawn by the transformer occurs resulting in an excessive voltage drop; and
- Large stray leakage flux effects occur with resulting excessive localized (e.g., tank) heating.

There are a number of effects due to the generation of high levels of harmonics by system power transformers, including

- overheating of capacitor banks,
- possible misoperation of relays,
- sustained overvoltages on long-line energization,
- higher secondary arc currents during single-pole switching,
- higher circuit breaker recovery voltage,
- overloading of harmonic filters of high voltage DC (HVDC) converter terminals, and
- distortion of the AC voltage wave shape that may result in loss of power transmission.

The increased inductive VArs drawn by system transformers during half-cycle saturation are sufficient to cause intolerable system voltage depression, unusual swings in MW and MVAr flow on transmis-

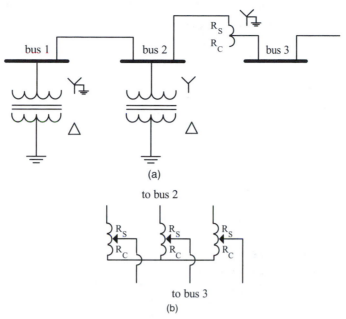

(a)

(b)

**FIGURE 2.43** (a) Sample network to illustrate system and ESP modeling, (b) autotransformer as used in (a).

sion lines, and problems with generator VAr limits on some instances.

### 2.6.6 System Model for Calculation of GIC

The ESP and GIC due to geomagnetic storms have frequencies in the millihertz range, and appear as quasi-DC in comparison to 50 and 60 Hz power system frequencies. Thus the system model is basically a DC conducting path model, in which the ESP voltage sources, in series with station ground-mat resistances, transformer resistances, and transmission line resistances, are impressed between the neutral grounds of all grounded-wye transformers.

Although there are similarities between the paths followed by GIC and zero-sequence currents in a power system, there are also important differences in the connective topography representing transformers in the two instances. The GIC transformer models are not concerned with leakage reactance values, but only with paths through the transformer that could be followed by DC current. Thus, the zero-sequence network configuration of an AC transmission line grid cannot be used directly to determine GIC. Transmission lines can be modeled for GIC determination by using their positive-sequence resistance values with a small correction factor to account for the differences between AC and DC resistances due to skin, proximity, and mag-

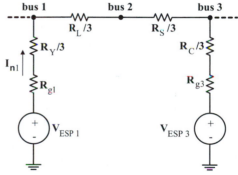

**FIGURE 2.44** DC model of sample network (Fig. 2.43) with ESP sources for buses $i = 1$, 2, and 3.

netic loss effects. A simple three-phase network is shown in Fig. 2.43, and the DC equivalent for determining GIC is shown by Fig. 2.44 [46–54]. The symbols used are defined as follows:

$R_L \equiv$ line resistance per phase,

$R_y \equiv$ resistance of grounded wye-connected winding per phase,

$R_S \equiv$ autotransformer series winding resistance per phase,

$R_C \equiv$ autotransformer common winding resistance per phase,

$R_{gi} \equiv$ grounding resistance of bus $i$,

$I_{ni} \equiv$ neutral GIC at bus $i$, and

$V_{ESPi} \equiv$ earth surface potential at bus $i$.

### 2.6.7 Mitigation Techniques for GIC

Much work has been done [46–54] after some power system failures have occurred during the past 20 years. The most important finding is that three-limb, three-phase transformers with small zero-sequence reactances – requiring a relatively large air gap between iron core and solid iron tank – are least affected by geomagnetically induced currents.

### 2.6.8 Conclusions Regarding GIC

Geomagnetic storms are naturally occurring phenomena that can adversely affect electric power systems. Only severe storms produce power system effects. Power systems in northern latitudes that are in regions of high geological resistivity (igneous rock formations) are more susceptible. Three limb, three-phase transformers with large gaps between iron core and tank permit GICs to flow but suppress their detrimental effects such as VAr demand [37–39]. The mitigation techniques of Section 2.6.7 have been implemented and for this reason now a major collapse of power systems due to geomagnetic storms can be avoided. This is an example of how research can help increase the reliability of power systems.

## 2.7 GROUNDING

Proper power system grounding is very essential for personnel safety, equipment protection, and power continuity to electrical loads. Special attention should be given to grounding location and means for minimizing circulating ground currents, as well as ground-fault sensing. Electrical power systems require two types of grounding, namely, system grounding and equipment grounding; each has its own functions [55]. The IEEE Green Book, IEEE Recommended Practice for Grounding of Industrial and Commercial Power Systems, ANSI/IEEE C114.1, Std.142, is the authoritative reference on grounding techniques [56].

### 2.7.1 System Grounding

System and circuit conductor grounding is an intentional connection between the electric system conductors and grounding electrodes that provides an effective connection to ground (earth). The basic objectives being sought are to

- limit overvoltages due to lightning,
- limit line surges, or unintentional contact with high-voltage lines,
- stabilize the voltage to ground during normal operation, and

- facilitate overcurrent device operation in case of ground faults.

#### 2.7.1.1 Factors Influencing Choice of Grounded or Ungrounded System

Whether or not to ground an electrical system is a question that must be faced sometime by most engineers charged with planning electrical distribution. A decision in favor of a grounded system leads then to the question of how to ground.

**Definitions**
- *Solidly grounded*: no intentional grounding impedance.
- *Effectively grounded*: grounded through a grounding connection of sufficiently low impedance (inherent or intentionally added or both) such that ground faults that may occur cannot build up voltages in excess of limits established for apparatus, circuits, or systems so grounded. $R_0 < X_1$, $X_0 \leq 3X_1$, where $X_1$ is the leakage reactance, and $X_0$ and $R_0$ are the grounding reactance and resistance, respectively.
- *Reactance grounded*: $X_0 \leq 10X_1$.
- *Resistance grounded*: $R_0 \geq 2X_0$.
- *High-resistance grounded*: ground fault up to 10 A is permitted.
- *Grounded for serving line-to-neutral loads*: $Z_0 \leq Z_1$.
- *Ungrounded*: no intentional system grounding connection.

**Service Continuity.** For many years a great number of industrial plant distribution systems have been operated ungrounded (Fig. 2.45) at various voltage levels. In most cases this has been done with the thought of gaining an additional degree of service continuity. The fact that any one contact occurring between one phase of the system and ground is unlikely to cause an immediate outage to any load may represent an advantage in many plants (e.g.,

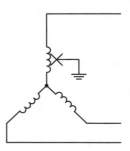

**FIGURE 2.45** Ungrounded system with one fault.

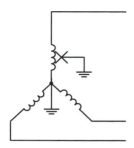

**FIGURE 2.46** Grounded system with one fault.

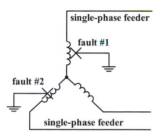

**FIGURE 2.47** Ungrounded system with two faults.

submarines), varying in its importance according to the type of plant.

**Grounded Systems.** In most cases systems are designed so that circuit protective devices will remove a faulty circuit from the system regardless of the type of fault. A phase-to-ground fault generally results in the immediate isolation of the faulted circuit with the attendant outage of the loads on that circuit (Fig. 2.46). However, experience has shown in a number of systems that greater service continuity may be obtained with grounded-neutral than with ungrounded-neutral systems.

**Multiple Faults to Ground.** Although a ground fault on one phase of an ungrounded system generally does not cause a service interruption, the occurrence of a second ground fault on a different phase before the first fault is cleared does result in an outage (Fig. 2.47). If both faults are on the same feeder, that feeder will be opened. If the second fault is on a different feeder, both feeders may be deenergized.

The longer a ground fault is allowed to remain on an ungrounded system, the greater is the likelihood of a second ground fault occurring on another phase resulting in an outage. The advantage of an ungrounded system in not immediately dropping load upon the occurrence of a ground fault may be largely destroyed by the practice of ignoring a ground fault until a second one occurs and repairs are required to restore service. With an ungrounded

system it is extremely important that an organized maintenance program be provided so that first ground fault is located and removed as soon as possible after its occurrence.

**Safety.** Many of the hazards to personnel and property existing in some industrial electrical systems are the result of poor or nonexisting grounding of electrical equipment and metallic structures. Although the subject of equipment grounding is treated in the next section, it is important to note that regardless of whether or not the system is grounded, safety considerations require thorough grounding of equipment and structures.

Proper grounding of a low-voltage (600 V or less) distribution system may result in less likelihood of accidents to personnel than leaving the system supposedly ungrounded. The knowledge that a circuit is grounded generally will result in greater care on the part of the workman.

It is erroneous to believe that on an ungrounded system a person may contact an energized phase conductor without personal hazard. As Fig. 2.48 shows, an ungrounded system with balanced phase-to-ground capacitance has rated line-to-neutral voltages existing between any phase conductor and ground. To accidentally or intentionally contact such a conductor may present a serious, perhaps lethal shock hazard in most instances.

During the time a ground fault remains on one phase of an ungrounded system, personnel contacting one of the other phases and ground are subjected to a voltage 1.73 (that is, $\sqrt{3}$) times that which would be experienced on a solidly neutral-grounded system. The voltage pattern would be the same as the case shown in Fig. 2.48c.

**Power System Overvoltages.** Some of the more common sources of overvoltages on a power system are the following:

- lightning (see Application Examples 2.9 and 2.10),
- switching surges,
- capacitor switching,
- static,
- contact with high-voltage system,
- line-to-ground faults.
- resonant conditions, and
- restriking ground faults.

**Lightning.** Most industrial systems are effectively shielded against direct lightning strikes (Fig. 2.49).

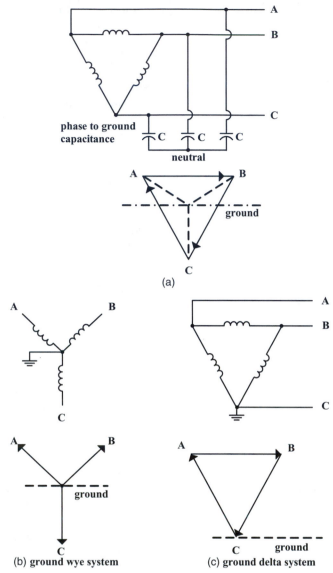

**FIGURE 2.48** (a) Ungrounded balanced system (circuit and phasor diagram) indirectly grounded by the neutral of capacitor bank (neutral is on ground potential). (b, c) Diagrams showing circuit and voltages to ground under steady-state conditions.

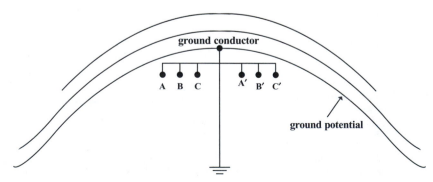

**FIGURE 2.49** Conductor configuration on transmission tower.

Many circuits are either underground in ducts or within grounded metal conduits or raceways. Even open-wire circuits are often shielded by adjacent metallic structures and buildings. Large arresters applied at the incoming service limit the surge voltages within the plant resulting from lightning strikes to the exposed service lines. Other arrester applications may be necessary within the plant to protect low impulse strength devices such as rotating machines and computers.

**Switching Surges.** Normal switching operation in the system (e.g., capacitor switching) can cause overvoltages. These are generally not more than three times the normal voltage and of short duration.

### 2.7.1.2 Application Example 2.9: Propagation of a Surge through a Distribution Feeder with an Insulator Flashover

This example illustrates the performance of a transmission system when the current of a lightning strike is large enough to cause flashover of tower insulator strings [57]. A lightning strike with the rise time $\tau_1 = 1.0\ \mu\text{sec}$, the fall time $\tau_2 = 5.0\ \mu\text{s}$, and the peak magnitude of current waveform $I_0 = 15.4\ \text{kA}$ strikes a 138 kV transmission line at node 1 (see Fig. E2.9.1). The span lengths of this transmission line are nonuniform and are given by NL (normalized length). Both ends of the transmission line are terminated by $Z_0 = 300\ \Omega$.

The transmission line has steel towers with 10 porcelain insulators. The tower-surge impedance consists of a 20 Ω resistance in series with 20 μH of inductance. The model used to simulate the first two insulators is shown in Fig. E2.9.1. Note that the transmission line can be simulated by the PSpice (lossless) transmission line model $T$, where $Z_0 = 300\ \Omega$, $f = 60\ \text{Hz}$, and $NL$ as indicated in Fig. E2.9.1. The current-controlled switch can be simulated by $W$, where $I_{ON} = 1\ \text{A}$.

a) Calculate the four span lengths in meters.
b) Calculate (using PSpice) and plot the transmission line voltage (in MV) as a function of time (in μs) at the point of strike, at the first insulator, and at the second insulator.

### 2.7.1.3 Application Example 2.10: Lightning Arrester Operation

This problem illustrates the performance of a lightning arrester [57]. The same basic parameters as for the lightning strike employed in Application Example 2.9 are used in this example. The only changes are in the value of $I_0$ (in this problem a value of 7.44 kA is used) and in the voltage of the second lightning arrester ($V_{D2} = 600\ \text{kV}$). This value is small enough that flashover of the insulator string will not occur; however, it is large enough to cause the lightning arrester to operate due to $V_{D2} = 600\ \text{kV}$. The system being simulated is shown in Fig. E2.10.1.

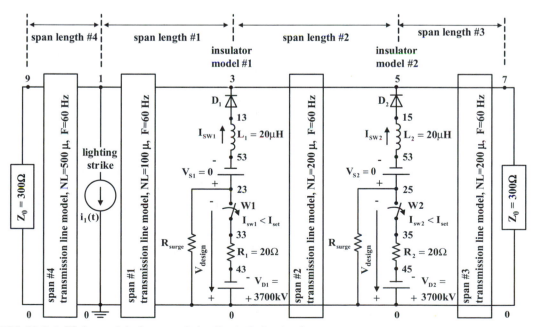

**FIGURE E2.9.1** PSpice model of a transmission line including insulators.

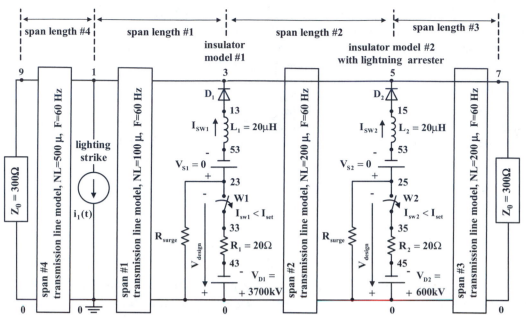

**FIGURE E2.10.1** PSpice model of a transmission line including insulator and lightning arrester.

a) Calculate and plot the transmission line voltage (in MV) as a function of time (in $\mu$s) at the point of strike, at the insulator, and at the lightning arrester.

b) Calculate and plot the currents at the point of strike, at the insulator, and at the lightning arrester (in kA) as a function of time (in $\mu$s).

## 2.7.2 Equipment Grounding

All exposed metal parts of electrical equipment (including generator frames, mounting bases, electrical instruments, and enclosures) should be bonded to a grounding electrode.

**Basic Objectives.** Equipment grounding, in contrast with system grounding, relates to the manner in which nonelectrical conductive material, which encloses energized conductors or is adjacent thereto, is to be interconnected and grounded. The basic objectives being sought are the following:

• to maintain a low potential difference between nearby metal members, and thereby protect people from dangerous electric-shock-voltage exposure in a certain area,

• to provide current-carrying capability, both in magnitude and duration, adequate to accept the ground-fault current permitted by the overcurrent protection system without creating a fire or explosive hazard to buildings or contents,

• to provide an effective electrical path over which ground-fault currents can flow without creating a fire or explosive hazard, and

• to contribute to superior performance of the electrical system (e.g., suppression of differential and common-mode electrical noise).

Investigations have pointed out that it is important to have good electric junctions between sections of conduit or metal raceways that are used as equipment grounding paths, and to assure adequate cross-sectional areas and conductivity of these grounding paths. Where systems are solidly grounded, equipment grounding conductors are bonded to the system grounded conductor and the grounding electrode conductor at the service equipment and at the source of a separately derived system in accordance with the National Electrical Code (NEC) Article 250 [55].

## 2.7.3 Static Grounding

The British standard [59] regarding grounding states that the most common source of danger from static electricity is the retention of charge on a conductor, because virtually all the stored energy can be released in a single spark to earth or to another conductor. The accepted method of avoiding the hazard is to connect all conductors to each other and to earth via electrical paths with resistances sufficiently low to permit the relaxation of the charges. This was stated in the original British standard (1980) and is carried out worldwide [60].

**Purpose of Static Grounding.** The accumulation of static electricity, on materials or equipment being handled or processed by operating personnel, introduces a potentially serious hazard in any area where combustible gases, dusts, or fibers are present. The discharge of the accumulation of static electricity from an object to ground or to another ungrounded object of lower potential is often the cause of a fire or an explosion if it takes place in the presence of readily combustible materials or combustible vapor and air mixtures.

### 2.7.4 Connection to Earth

*Resistance to Earth.* The most elaborate grounding system that can be designed may prove to be inadequate unless the connection of the system to the earth is adequate and has a low resistance. It follows, therefore, that the earth connection is one of the most important parts of the whole grounding system. It is also the most difficult part to design and to obtain. The perfect connection to earth should have zero resistance, but this is impossible to obtain. Ground resistance of less than 1 Ω can be obtained, although this low resistance may not be necessary in many cases. Since the desired resistance varies inversely with the fault current to ground, the larger the fault current, the lower must be the resistance [61, 62]. Figure 2.50a illustrates the deformation of grounding conductor due to large lightning currents.

### 2.7.5 Calculation of Magnetic Forces

Considerable work [63, 64] has been done concerning the calculation of electromagnetic forces based on closed-form analysis. These methods can be applied to simple geometric configurations only. In many engineering applications, however, three-dimensional geometric arrangements must be analyzed. Based on the theory developed by Lawrenson [65] filamentary conductor configurations can be analyzed. Figure 2.50b illustrates the force density distribution for a filamentary 90° bend at a current of $I = 100$ kA. Three-dimensional arrangements are investigated in [66]. Force calculations based on the so-called Maxwell stress are discussed in Chapter 4.

### 2.8 MEASUREMENT OF DERATING OF THREE-PHASE TRANSFORMERS

Measuring the real and apparent power derating of three-phase transformers is desirable because additional losses due to power quality problems (e.g., harmonics, DC excitation) can be readily limited

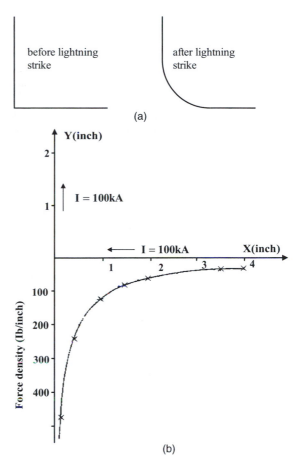

**FIGURE 2.50** (a) Deformation of grounding conductor due to lightning strike. (b) Force density distribution as it occurs along the $x$ axis for a filamentary 90° bend at a current of $I = 100$ kA. The force density distribution along the $y$ axis is not shown but is similar to that along the $x$ axis.

before any significant damage due to additional temperature rises occurs [8, 9]. Measuring transformer losses from the input power minus the output power in real time is inaccurate because the losses are the difference of two large values; this indirect approach results in maximum errors in the losses of more than 60% for high-efficiency ($\eta > 98\%$) transformers [7]. The usually employed indirect method consisting of no-load (iron-core loss) and short-circuit (copper loss) tests [67] cannot be performed online while the transformer is partially or fully loaded. IEEE Recommended Practice C57.110 computes the transformer derating based on $R_h$ for various harmonics $h$, which is derived from the DC winding resistance and the rated load loss [14]. Kelly *et al.* [68] describe an improved measuring technique of the equivalent effective resistance $R_T$ as a function of frequency $f$ of single-phase transformers, which allows the direct calculation of transformer loss at harmonic frequencies from $f = 10$ Hz up to 100 kHz. This equivalent effective resistance takes into account the total losses of the transformer: the copper losses plus the iron-core losses. Based on the fact that the iron-core losses do not depend on harmonic currents, but depend on harmonic voltages (amplitudes and phase shifts) [69], the total losses determined by [68] are not accurate. In addition, temperature-dependent operating conditions cannot be considered in [68]. Mocci [70] and Arri *et al.* [71] present an analog measurement circuit to directly measure the total losses for single- and ungrounded three-phase transformers. However, employment of many potential transformers (PTs) and current transformers (CTs) in the three-phase transformer measuring circuits [71] decreases the measurement accuracy.

This section presents a direct method for measuring the derating of three-phase transformers while transformers are operating at any load and any arbitrary conditions. The measuring circuit is based on PTs, CTs, shunts, voltage dividers or Hall sensors [72], A/D converter, and computer (microprocessor). Using a computer-aided testing (CAT) program [7, 73], losses, efficiency, harmonics, derating, and

wave shapes of all voltages and currents can be monitored within a fraction of a second. The maximum errors in the losses are acceptably small and depend mainly on the accuracy of the sensors used.

### 2.8.1 Approach

#### 2.8.1.1 Three-Phase Transformers in Δ-Δ or Y-Y Ungrounded Connection

For Δ-Δ or ungrounded Y-Y connected three-phase transformers (see Figs. 2.51 and 2.52 for using shunts, voltage dividers, and optocouplers), one obtains with the application of the two-wattmeter method at the input and output terminals by inspection:

$$P_{in} - P_{out} = \frac{1}{T}\int_0^T [(v_{AC}i_A + v_{BC}i_B) - (v'_{ac}i'_a + v'_{bc}i'_b)]dt,$$
$$= \frac{1}{T}\int_0^T [v_{AC}(i_A - i'_a) + v_{BC}(i_B - i'_b) + (v_{AC} - v'_{ac})i'_a + (v_{BC} - v'_{bc})i'_b]dt.$$

$$(2\text{-}49)$$

The term $p_{fe}(t) = v_{AC}(i_A - i'_a) + v_{BC}(i_B - i'_b)$ represents the instantaneous iron-core loss, and $p_{cu}(t) = (v_{AC} - v'_{ac})i'_a + (v_{BC} - v'_{bc})i'_b$ is the instantaneous copper loss.

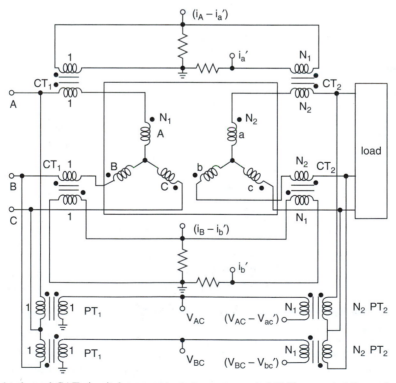

**FIGURE 2.51** Eight-channel CAT circuit for accurate Δ-Δ or ungrounded Y-Y connected three-phase transformer loss monitoring using PTs and CTs.

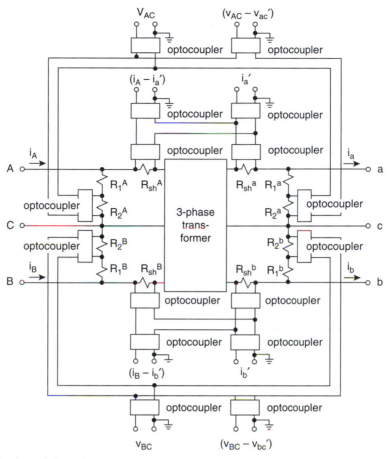

**FIGURE 2.52** Eight-channel CAT circuit for accurate $\Delta$–$\Delta$ or ungrounded $Y$–$Y$ connected three-phase transformer loss monitoring using shunts, voltage dividers, and optocouplers.

### 2.8.1.2 Three-Phase Transformers in $\Delta$–$Y$ Connection

Figure 2.53 illustrates the application of the digital data-acquisition method to a $\Delta$–$Y$ connected transformer bank.

For ungrounded $\Delta$–$Y$ connected three-phase transformers, the currents of the $Y$ side must be referred to the line currents of the $\Delta$ side, as shown in Fig. 2.53. The loss of the transformer is given by

$$P_{in} - P_{out} = \frac{1}{T}\int_0^T [(v_{AC}i_{Al} + v_{BC}i_{Bl}) - (v'_{an}i'_{al} - v'_{cn}v'_{bl})]dt,$$

(2-50)

$$= \frac{1}{T}\int_0^T [v_{AC}(i_{Al} - i'_{al}) + v_{BC}(i_{Bl} - i'_{bl}) + (v_{AC} - v'_{an})i'_{al} + (v_{BC} + v'_{cn})i'_{bl}]dt,$$

(2-51a)

where $i_{Al}$ and $i_{Bl}$ are input line currents of phases $A$ and $B$, respectively; $i_{al}' = (i_a' - i_b')$ and $i_{bl}' = (i_b' - i_c')$

are the two-wattmeter secondary currents; $v_{an}'$ and $v_{cn}'$ are the output line-to-neutral voltages of phases $a$ and $c$, respectively.

Because the neutral of the $Y$-connected secondary winding ($N$) is not accessible, the secondary phase voltages are measured and referred to the neutral $n$ of the $Y$-connected PTs (see Fig. 2.53). This does not affect the accuracy of loss measurement, which can be demonstrated below. The output power is

$$P_{out} = \frac{1}{T}\int_0^T (v'_{aN}i'_a + v'_{bN}i'_b + v'_{cN}i'_c)dt,$$

$$= \frac{1}{T}\int_0^T [(v'_{an} + v'_{nN})i'_a + (v'_{bn} + v'_{nN})i'_b + (v'_{cn} + v'_{nN})i'_c]dt,$$

$$= \frac{1}{T}\int_0^T [v'_{an}i'_a + v'_{bn}i'_b + v'_{cn}i'_c + v'_{nN}(i'_a + i'_b + i'_c)]dt,$$

(2-51b)

where $N$ denotes the neutral of the $Y$-connected secondary winding. Because $i_a' + i_b' + i_c' = 0$, the measured output power referred to the neutral $n$ of PTs

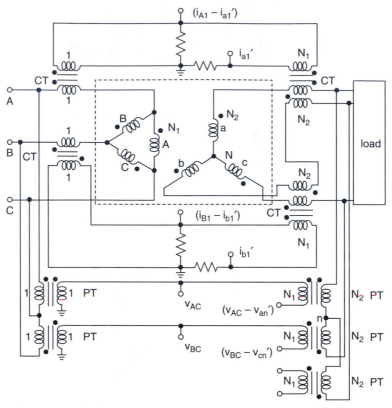

**FIGURE 2.53** Eight-channel CAT circuit for accurate $\Delta$–$Y$ three-phase transformer loss monitoring using PTs and CTs.

is the same as that referred to the neutral $N$ of the secondary winding.

### 2.8.1.3 Accuracy Requirements for Instruments

The efficiencies of high-power electrical apparatus such as single- and three-phase transformers in the kVA and MVA ranges are 97% or higher. For a 15 kVA three-phase transformer with an efficiency of 97.02% at rated operation, the total losses are $P_{\text{Loss}} = 409.3$ W, copper and iron-core losses $P_{cu} = 177.1$ W and $P_{fe} = 232.2$ W at $V_{L\text{-}L} = 234.2$ V and $I_L = 32.4$ A, respectively, as listed in Table 2.7a.

The determination of the losses from voltage and current differences as described in Fig. 2.51 – where differences are calibrated – greatly reduces the maximum error in the loss measurement. The series voltage drop and exciting current at rated operation referred to the primary of the 15 kVA, 240 V/240 V three-phase transformer are $(v_{AC} - v_{ac}') = 3.56$ V, $(i_A - i_a') = 2.93$ A, respectively (Table 2.7a). The instruments and their error limits are listed in Table 2.1. In Table 2.1, $v_{AC}$, $v_{AC} - v_{ac}'$, $i_A - i_a'$, and $i_a'$ stand

**TABLE 2.1** Instruments and Their Error Limits for Figs. 2.51 and 2.52

| Instrument (Fig. 2.51) | Instrument (Fig. 2.52) | Rating | Error limits |
|---|---|---|---|
| $\text{PT}_1$ | $\text{VD}_A$ | 240 V/240 V | $\varepsilon_{\text{PT1}} = 234$ mV |
| $\text{PT}_2$ | $\text{VD}_a$ | 240 V/240 V | $\varepsilon_{\text{PT2}} = 234$ mV |
| $\text{CT}_1$ | $\text{SH}_A$ | 30 A/5 A | $\varepsilon_{\text{CT1}} = 5.4$ mA |
| $\text{CT}_2$ | $\text{SH}_a$ | 30 A/5 A | $\varepsilon_{\text{CT2}} = 5.4$ mA |
| $v_{AC}$ | $v_{AC}$ | 300 V | $\varepsilon_{\text{V1}} = 234$ mV |
| $v_{AC} - v_{ac}'$ | $v_{AC} - v_{ac}'$ | 10 V | $\varepsilon_{\text{V2}} = 3.56$ mV |
| $i_A - i_a'$ | $i_A - i_a'$ | 5 A | $\varepsilon_{\text{A1}} = 0.49$ mA |
| $i_a'$ | $i_a'$ | 5 A | $\varepsilon_{\text{A2}} = 5.4$ mA |

for the relevant calibrated voltmeters and ammeters. Because all voltage and current signals are sampled via PTs, CTs (or optocouplers), and current shunts, the error limits for all instruments are equal to the product of the actually measured values and their relative error limits, instead of full-scale errors, as shown in Table 2.1. All error limits are referred to the meter side.

In Fig. 2.52, the voltage divider ($R_1^A$, $R_2^A$) combined with an optocoupler ($v_{AC}$) emulates the function of a PT without hysteresis. The optocoupler can alter the amplitude of a signal and provides isolation without affecting the phase shift of the signal as it is corrupted by PTs. The current shunt ($R_{sh}{}^A$) and optocouplers ($i_A$, $i_a'$, $i_A - i_a'$) emulate that of a CT without hysteresis and parasitic phase shift. The prime ($'$) indicates that $i_a'$ is about of the same magnitude as $i_A$; this is accomplished by adjustment of the amplifier gain(s) of the optocoupler(s). *VD* and *SH* in Table 2.1 stand for voltage divider and shunt, respectively.

The line-to-line voltage is measured with the maximum error of (taking into account the maximum errors of $PT_1$ and voltmeter)

$$\frac{\Delta v_{AC}}{v_{AC}} = \frac{\pm \varepsilon_{PT1} \pm \varepsilon_{V1}}{v_{AC}} = \frac{\pm 0.234 \pm 0.234}{234} = \pm 0.002. \quad (2\text{-}52)$$

The current difference is measured with the maximum error of

$$\frac{\Delta(i_A - i_a')}{(i_A - i_a')} = \frac{\pm \varepsilon_{CT1} \pm \varepsilon_{CT2} \pm \varepsilon_{A1}}{(i_A - i_a')/6}$$
$$= \frac{(\pm 5.4 \pm 5.4 \pm 0.49)\,\text{mA}}{0.49\,\text{A}} = \pm 0.023. \quad (2\text{-}53)$$

Therefore, the loss component $v_{AC}(i_A - i_a')$ in Eq. 2-49 is measured with the maximum error of

$$\frac{\Delta v_{AC}}{v_{AC}} + \frac{\Delta(i_A - i_a')}{(i_A - i_a')} = \pm 0.025. \quad (2\text{-}54)$$

The series voltage drop is measured with the maximum error of

$$\frac{\Delta(v_{AC} - v_{ac}')}{(v_{AC} - v_{ac}')} = \frac{\pm \varepsilon_{PT1} \pm \varepsilon_{PT2} \pm \varepsilon_{V2}}{(v_{AC} - v_{ac}')}$$
$$= \frac{(\pm 234 \pm 234 \pm 3.56)\,\text{mV}}{3.56\,\text{V}} = \pm 0.1325.$$
$$(2\text{-}55)$$

The output current is measured with the maximum error of

$$\frac{\Delta i_a'}{i_a'} = \frac{\pm \varepsilon_{CT2} \pm \varepsilon_{A2}}{i_a'/6} = \frac{\pm 0.0054 \pm 0.0054}{5.4} = \pm 0.002$$
$$(2\text{-}56)$$

and the loss component $(v_{AC} - v_{ac}')i_a'$ in Eq. 2.49 is measured with the maximum error of

$$\frac{\Delta i_a'}{i_a'} + \frac{\Delta(v_{AC} - v_{ac}')}{(v_{AC} - v_{ac}')} = \pm 0.1345. \quad (2\text{-}57)$$

Thus, the total loss is measured with the maximum error of

$$\frac{\Delta P_{\text{Loss}}}{P_{\text{Loss}}} = \frac{\Delta P_{cu} + \Delta P_{Fe}}{P_{\text{Loss}}}$$
$$= 2\left[\frac{\pm 0.1345 \times 177.1 \pm 0.025 \times 232.2}{409.3}\right]$$
$$= \pm 0.145 = \pm 14.5\%. \quad (2\text{-}58)$$

The above error analysis employs PTs and CTs. If these devices generate too-large errors because of hysteresis, voltage dividers (e.g., $VD_A$) and shunts (e.g., $SH_A$) combined with optocouplers can be used, as indicated in Fig. 2.52 [74]. A similar error analysis using shunts, dividers, and optocouplers leads to the same maximum error in the directly measured losses, provided the same standard maximum errors (0.1%) of Table 2.1 are assumed. The factor 2 in Eq. 2-58 is employed because loss components in Eq. 2-54 and Eq. 2-57 are only half of those in Eq. 2-49.

### 2.8.1.4 Comparison of Directly Measured Losses with Results of No-Load and Short-Circuit Tests

A computer-aided testing electrical apparatus program (CATEA) [73] is used to monitor the iron-core and copper losses of three-phase transformers. The nameplate data of the tested transformers are given in Appendixes 5.1 to 5.4. The results of the on-line measurement of the iron-core and copper losses for a $Y$–$Y$ connected 9 kVA three-phase transformer (consisting of three 3 kVA single-phase transformers, Appendix 5.1) are given in Table 2.2 for sinusoidal rated line-to-line voltages of 416 V, where the direct (online) measurement data are compared with those of the indirect (separate open-circuit and short-circuit tests) method. The iron-core loss of the indirect method is larger than that of the direct method because the induced voltage of the former is larger than that of the latter. The copper loss of the indirect method is smaller than that of the direct method because the input current of the former (which is nearly the same as the output current) is smaller than that of the latter.

**TABLE 2.2** Measured Iron-Core and Copper Losses of 9 kVA $Y$–$Y$ Connected Transformer ($V_{LL} = 416$ V)

| Losses | Direct method [7, 73] | Indirect method [67] |
|---|---|---|
| Iron-core loss | 98.4 W | 114.1 W |
| Copper loss | 221.1 W | 199.3 W |
| Total losses | 319.5 W | 313.4 W |

## 2.8.2 A 4.5 kVA Three-Phase Transformer Bank #1 Feeding Full-Wave Rectifier

A 4.5 kVA, 240 V/240 V, Δ–Δ connected three-phase transformer (Fig. 2.51) consisting of three single-phase transformers (bank #1, Appendix 5.2) is used to feed a full-wave diode rectifier (see Appendix 5.5) with an LC filter connected across the resistive load (see Fig. 6-5 of [75]). In Table 2.3, measured data are compared with those of linear load condition. The measured wave shapes of input voltage ($v_{AC}$), exciting current ($i_A - i_a'$), series voltage drop ($v_{AC} - v_{ac}'$), and output current ($i_a'$) of phase A are shown in Fig. 2.54a,b for linear and nonlinear load conditions. The total harmonic distortion $THD_i$, $K$-factor, and harmonic components of the output current are listed in Table 2.4.

Table 2.3 compares measured data and shows that the transformer is operated at nonlinear load with about the same losses occurring at linear load (261.3 W). With the apparent power derating definition

$$^{\text{derating}}S = \frac{V_{rms}\,I_{rms}}{V_{rms}^{\text{rated}}\,I_{rms}^{\text{rated}}}, \qquad (2\text{-}59)$$

the derating at the nonlinear load of Table 2.3 is 99%. The real power delivered to the nonlinear load is 91.4% of that supplied at linear load.

**TABLE 2.3** Measured Data of 4.5 kVA Transformer Bank #1

|  | Linear load (Fig. 2.54a) | Nonlinear load (Fig. 2.54b) |
|---|---|---|
| $v_{AC}$ | 237.7 V$_{rms}$ | 237.7 V$_{rms}$ |
| $i_A - i_a'$ | 1.10 A$_{rms}$ | 1.12 A$_{rms}$ |
| $v_{AC} - v_{ac}'$ | 8.25 V$_{rms}$ | 8.19 V$_{rms}$ |
| $i_a'$ | 11.09 A$_{rms}$ | 10.98 A$_{rms}$ |
| Iron-core loss | 100.7 W | 103.9 W |
| Copper loss | 161.3 W | 157.6 W |
| Total loss | 261.3 W | 261.5 W |
| Output power | 4532 W | 4142 W |
| Efficiency | 94.55% | 94.09% |

## 2.8.3 A 4.5 kVA Three-Phase Transformer Bank #2 Supplying Power to Six-Step Inverter

A 4.5 kVA transformer bank #2 (Appendix 5.3) supplies power to a half-controlled six-step inverter (Appendix 5.6), which in turn powers a three-phase induction motor. The motor is controlled by adjusting the output current and frequency of the inverter. The transformer is operated at rated loss at various motor speeds. Rated loss of bank #2 is determined by a linear (resistive) load at rated operation. The iron-core and copper losses are measured separately and are listed in Table 2.5a. Measured wave shapes of input voltage, exciting current, series voltage drop, and output current of phase A are shown in Fig. 2.55a,b for linear and nonlinear conditions. The output current includes both odd and even harmonics due to the half-controlled input rectifier of the six-step inverter. Dominant harmonics of input voltage and output current are listed for different motor speeds in Table 2.5b. The total harmonic distortion of the input voltage and output current as well as the K-factor are listed in Table 2.5c for the speed conditions of Tables 2.5a and 2.5b.

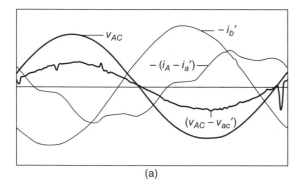

(a)

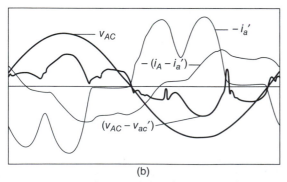

(b)

**FIGURE 2.54** (a) Measured wave shapes of 4.5 kVA three-phase transformer bank #1 feeding linear load (see rms values of Table 2.3). (b) Measured wave shapes of 4.5 kVA three-phase transformer bank #1 feeding full-wave diode rectifier load (see rms values of Tables 2.3 and 2.4).

**TABLE 2.4** Output Current Harmonic Components, THD$_i$, and K-Factor (Fig. 2.54b)

| $i_h$ (A$_{rms}$) | | | | | | |
|---|---|---|---|---|---|---|
| $h = 1$ | $h = 5$ | $h = 7$ | $h = 11$ | $h = 13$ | THD$_i$ (%) | $K$ |
| 10.4 | 3.55 | 0.81 | 0.77 | 0.24 | 36.34 | 4.58 |

**TABLE 2.5a** Measured Iron-Core and Copper Losses of 4.5 kVA Δ–Δ Connected Transformer Bank #2 Feeding a Six-Step Inverter at Various Motor Speeds (Fig. 2.55b)

| Motor speed (%)[a] | Input voltage ($V_{rms}$) | Output current ($A_{rms}$) | Core loss (W) | Copper loss (W) | Total loss (W) | Output power (W) |
|---|---|---|---|---|---|---|
| [b] | 237.8 | 11.4 | 65.7 | 154.9 | 220.6 | 4542 |
| 100 | 237.9 | 11.1 | 69.6 | 150.5 | 220.2 | 3216 |
| 90 | 238.1 | 11.5 | 62.9 | 158.8 | 221.7 | 2913 |
| 80 | 238.3 | 11.2 | 65.6 | 153.3 | 218.9 | 2398 |
| 70 | 238.3 | 11.3 | 65.7 | 154.8 | 220.4 | 2081 |
| 60 | 238.2 | 11.4 | 64.5 | 162.2 | 226.7 | 1862 |
| 50 | 238.4 | 10.8 | 69.2 | 151.3 | 220.6 | 1530 |
| 40 | 238.6 | 11.1 | 68.5 | 151.2 | 219.7 | 1308 |

[a]Refers to rated speed of 1145 rpm.
[b]Linear resistive load.

**TABLE 2.5b** Input Voltage (in rms Volts) and Output Current (in rms Amperes) Harmonics of Phase A (Fig. 2.55a,b)

| h | | 1 | 2 | 4 | 5 | 7 | 8 | 10 |
|---|---|---|---|---|---|---|---|---|
| Linear load | $v_{ACh}$ | 236.9 | 0.42 | 0.48 | 2.96 | 1.48 | 0.20 | 0.40 |
|  | $i_{ah}'$ | 11.24 | 0.04 | 0.02 | 0.12 | 0.08 | 0.00 | 0.00 |
| 100% speed | $v_{ACh}$ | 237.2 | 1.37 | 0.49 | 2.88 | 1.58 | 0.06 | 0.30 |
|  | $i_{ah}'$ | 9.05 | 5.91 | 1.08 | 0.56 | 0.30 | 0.32 | 0.20 |
| 90% speed | $v_{ACh}$ | 237.4 | 1.58 | 0.53 | 2.57 | 1.60 | 0.10 | 0.20 |
|  | $i_{ah}'$ | 9.06 | 6.83 | 1.87 | 0.54 | 0.54 | 0.16 | 0.30 |
| 80% speed | $v_{ACh}$ | 237.5 | 1.67 | 0.43 | 2.91 | 1.63 | 0.02 | 0.30 |
|  | $i_{ah}'$ | 8.78 | 6.82 | 2.47 | 0.69 | 0.67 | 0.43 | 0.20 |
| 70% speed | $v_{ACh}$ | 237.6 | 1.67 | 0.41 | 2.78 | 1.50 | 0.13 | 0.20 |
|  | $i_{ah}'$ | 8.64 | 6.91 | 2.82 | 0.99 | 0.68 | 0.60 | 0.20 |
| 60% speed | $v_{ACh}$ | 237.3 | 1.71 | 0.28 | 3.14 | 1.21 | 0.14 | 0.20 |
|  | $i_{ah}'$ | 8.59 | 6.90 | 3.20 | 1.29 | 0.68 | 0.74 | 0.20 |
| 50% speed | $v_{ACh}$ | 237.8 | 1.73 | 0.35 | 3.18 | 1.30 | 0.21 | 0.00 |
|  | $i_{ah}'$ | 8.09 | 6.57 | 3.52 | 1.67 | 0.54 | 0.79 | 0.50 |
| 40% speed | $v_{ACh}$ | 237.7 | 1.73 | 0.47 | 3.08 | 1.67 | 0.30 | 0.30 |
|  | $i_{ah}'$ | 7.78 | 6.66 | 3.70 | 2.17 | 0.48 | 0.82 | 0.80 |

**TABLE 2.5c** Measured THD-Values and K-Factor (Fig. 2.55a,b)

| Condition | $THD_v$ | $THD_i$ | K-factor |
|---|---|---|---|
| Linear load | 1.60% | 1.41% | 1.01 |
| 100% speed | 1.62% | 67.19% | 2.48 |
| 90% speed | 1.62% | 78.76% | 2.76 |
| 80% speed | 1.67% | 83.65% | 3.25 |
| 70% speed | 1.58% | 87.99% | 3.72 |
| 60% speed | 1.71% | 90.86% | 4.40 |
| 50% speed | 1.73% | 96.05% | 5.53 |
| 40% speed | 1.88% | 104.05% | 6.69 |

**TABLE 2.6** Measured Data of 15 kVA Three-Phase Transformer with Resonant Rectifier Load (Fig. 2.56a,b)

| | Linear load | Resonant rectifier |
|---|---|---|
| $v_{AC}$ | 215.7 $V_{rms}$ | 215.6 $V_{rms}$ |
| $(i_A - i_a')$ | 2.39 $A_{rms}$ | 2.59 $A_{rms}$ |
| $(v_{AC} - v_{ac}')$ | 4.42 $V_{rms}$ | 4.8 $V_{rms}$ |
| $i_a'$ | 27.00 $A_{rms}$ | 26.94 $A_{rms}$ |
| Iron-core loss | 42.0 W | 49.8 W |
| Copper loss | 191.9 W | 192.2 W |
| Total loss | 233.9 W | 242 W |
| Output power | 10015 W | 7859 W |
| Efficiency | 97.72% | 97.01% |

## 2.8.4 A 15 kVA Three-Phase Transformer Supplying Power to Resonant Rectifier

A 15 kVA, 240 V/240 V, Δ–Δ connected three-phase transformer (Appendix 5.4) is used to supply power to a resonant rectifier [12] (Appendix 5.7). The transformer is operated with the resonant rectifier load at the same total loss generated by a three-phase linear (resistive) load. Measured data are compared in Table 2.6. Measured wave shapes of input voltage, exciting current, series voltage drop, and output current of phase A are depicted in Fig. 2.56a,b. The fundamental phase shift between output transformer line-to-line voltage $v_{AC}'$ and phase current $i_a'$ of the

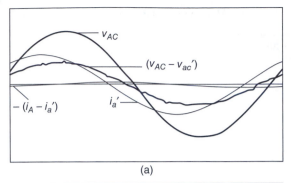

(a)

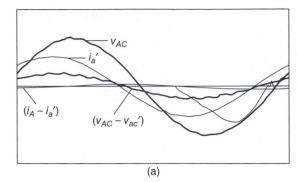

(a)

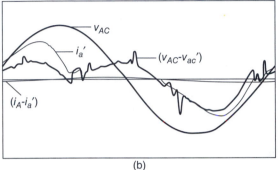

(b)

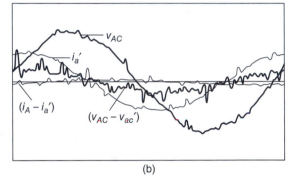

(b)

**FIGURE 2.55** (a) Measured wave shapes of 4.5 kVA three-phase transformer bank #2 feeding linear load (see rms values of Tables 2.5a and 2.5b). (b) Measured wave shapes of 4.5 kVA three-phase transformer bank #2 feeding half-controlled six-step inverter (see rms values of Tables 2.5a and 2.5b).

**FIGURE 2.56** (a) Measured wave shapes of 15 kVA three-phase transformer feeding linear load (see rms values of Table 2.6). (b) Measured wave shapes of 15 kVA three-phase transformer feeding resonant rectifier (see rms values of Table 2.6).

resonant rectifier is 67.33°, and the output displacement factor (within transformer phase) is therefore $\cos(67.33° - 30°) = 0.795$. The fundamental phase shift between output line-to-line voltage and phase current for the linear resistive load is 30.95°, and the output displacement factor (within transformer phase) is, therefore, $\cos(30.95° - 30°) \approx 1.0$. Note that the wave shapes of the output currents of Fig. 2.56a,b are about sinusoidal.

If the transformer with the resonant rectifier load is operated (see Table 2.6) at about the same loss as linear load (233.9 W), the output current of the transformer with the resonant rectifier load is 26.94 A, which corresponds to the copper loss of 192.2 W = (242 W − 49.8 W). Therefore, the transformer apparent power derating of the transformer for the nonlinear load is 99.7%. The real power supplied to the nonlinear load is 78.5% of that of the linear load.

### 2.8.5 A 15 kVA Three-Phase Transformer Bank Absorbing Power from a PWM Inverter

The same transformer of Section 2.8.4 absorbs power from a PWM inverter [12] (Appendix 5.8) and sup-

**TABLE 2.7a** Measured Data of 15 kVA Three-Phase Transformer Connected to PWM Inverter (Fig. 2.57a,b,c)

|  | Linear load of transformer | Case #1 | Case #2 |
|---|---|---|---|
| $v_{AC}$ | 234.2 V$_{rms}$ | 234.4 V$_{rms}$ | 248.2 V$_{rms}$ |
| $(i_A - i_a')$ | 2.93 A$_{rms}$ | 3.30 A$_{rms}$ | 3.82 A$_{rms}$ |
| $(v_{AC} - v_{ac}')$ | 3.56 V$_{rms}$ | 3.86 V$_{rms}$ | 4.28 V$_{rms}$ |
| $i_a'$ | 32.4 A$_{rms}$ | 33.3 A$_{rms}$ | 28.2 A$_{rms}$ |
| Iron-core loss | 232.2 W | 254.6 W | 280.5 W |
| Copper loss | 177.1 W | 172.6 W | 130.4 W |
| Total loss | 409.3 W | 427.2 W | 410.9 W |
| Output power | 13337 W | 13544 W | 12027 W |
| Efficiency | 97.02% | 96.94% | 96.07% |

plies power to a resistive load (case #1) and to a utility system (case #2). The transformer losses are also measured when supplying a linear resistive load fed from sinusoidal power supply (linear load). All measured data are compared in Table 2.7a.

Measured wave shapes of input voltage, exciting current, series voltage drop, and output current of phase A are shown in Fig. 2.57a,b,c. The total harmonic distortion $THD_i$, $K$-factor, and harmonic

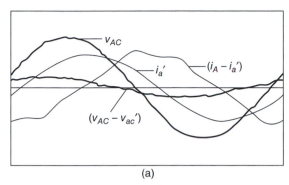

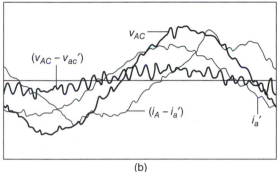

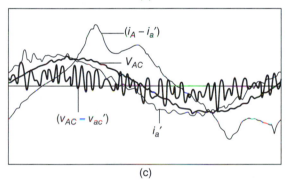

**FIGURE 2.57** (a) Measured wave shapes of 15 kVA three-phase transformer feeding linear load (see rms values of Table 2.7a). (b) Measured wave shapes of 15 kVA three-phase transformer fed by a PWM inverter (case #1) (see rms values of Tables 2.7a and 2.7b). (c) Measured wave shapes of 15 kVA three-phase transformer fed by a PWM inverter (case #2) (see rms values of Tables 2.7a and 2.7b).

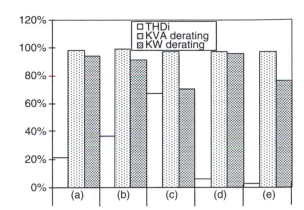

**FIGURE 2.58** Total harmonic distortion of current ($THD_i$), apparent power (kVA) derating, and real power (kW) derating for uncontrolled (a, b), half-controlled (c), and controlled (d, e) converter loads.

**TABLE 2.7b** Output Current Harmonics, $THD_i$, and K-Factor (Fig. 2.57b,c)

| | $I_h(A)$ | | | | | $THD_i$ | |
| Case | $h=1$ | $h=5$ | $h=7$ | $h=11$ | $h=13$ | (%) | K |
|---|---|---|---|---|---|---|---|
| **#1** | 32.8 | 0.53 | 0.87 | 0.12 | 0.19 | 5.84 | 1.7 |
| **#2** | 28.2 | 1.15 | 1.36 | 0.19 | 0.38 | 10.0 | 2.7 |

amplitudes of the transformer output current are listed in Table 2.7b.

Figure 2.58 summarizes the total harmonic distortion ($THD_i$), apparent power (kVA) derating, and real power (kW) derating for uncontrolled (a, b), half-controlled (c), and controlled (d, e) converter loads of transformers. In particular the graphs of Fig. 2.58 can be identified [75] as follows:

a) 25 kVA single-phase pole transformer [7, 11, 15, 75, 76] feeding uncontrolled full wave rectifier load,

b) 4.5 kVA three-phase transformer feeding uncontrolled full wave rectifier load,

c) 4.5 kVA three-phase transformer feeding half-controlled rectifier load,

d) 15 kVA three-phase transformer absorbing power from PWM inverter (14 kW), and

e) 15 kVA three-phase transformer feeding resonant rectifier load (8 kW).

### 2.8.6 Discussion of Results and Conclusions

#### 2.8.6.1 Discussion of Results

A new approach for the measurement of the derating of three-phase transformers has been described and applied under nonsinusoidal operation. It extends the measurement approach of single-phase transformers [7, 73, 12, 75, 76, 77] to three-phase transformers [75, 76, 78].

The apparent power (kVA) derating (Eq. 2-59) of three-phase transformers is not greatly affected by the $THD_i$. Even for $THD_i$ values of about 70%, derating is about 99%.

The real power (kW) derating is greatly affected (see Fig. 2.58) by the current wave shape generated by solid-state converters, in particular by the phase shift of the fundamental current component. Therefore, inverters and rectifiers should be designed such that they supply and draw power, respectively, at a displacement (power) factor of about 1.

Three-phase transformers have similar derating properties as single-phase transformers [7, 12, 73, 75, 76, 77]. The maximum error of the directly measured losses is about 15%, which compares favorably with the maximum error of more than 60% [7, 73] for loss measurement based on the difference between input and output powers as applied to high-efficiency ($\eta > 98\%$) transformers.

Transformers of the same type may have significantly different iron-core losses as measured in Table 2.3 (100.7 W) and Table 2.5a (65.7 W). Small transformers (kVA range) have relatively small wire cross sections resulting in small skin-effect losses. Large transformers (MW range) have aluminum secondary windings with relatively large wire cross sections resulting in relatively large skin-effect losses. For this reason substation transformers can be expected to have larger apparent power derating than the ones measured in this section. Unfortunately, transformers in the MW range cannot be operated in a laboratory under real-life conditions. Therefore, it is recommended that utilities sponsor the application of the method of this section and permit on-site measurements.

### 2.8.6.2 Comparison with Existing Techniques

The maximum error of the directly measured losses is about 15% (using potential and current transformers), which compares favorably with the maximum error of more than 60% [7, 73] (employing shunts and voltage dividers) for loss measurement based on the difference between input and output powers as applied to high-efficiency ($\eta > 98\%$) transformers.

The technique of [68] uses the premeasured transformer effective resistance $R_T$ as a function of frequency $f$ to calculate transformer total losses for various harmonic currents. This method can be classified as an indirect method because the transformer losses are obtained by computation, instead of direct measurement. In addition, the approach of [68] neglects the fact that the iron-core losses are a function of the harmonic phase shift [69]; in other words, the $R_T$ values are not constant for any given harmonic current amplitude but vary as a function of the harmonic voltage amplitude and phase shift as well. Finally, temperature-dependent operating conditions, for example, cannot be considered in [68]. For the above reasons, the method of [68] must be validated by some direct measurements, such as the method presented in this section.

The method of Mocci [70] and Arri *et al.* [71] has not been practically applied to three-phase transformers. The presented measurement circuit for three-phase transformers [71] uses too many instrument transformers (e.g., nine CTs and nine PTs), and therefore, the measured results will be not as accurate as those based on the measurement circuits of this section, where only four CTs and five PTs are used as shown in Fig. 2.53.

## 2.9 SUMMARY

This chapter addresses issues related to modeling and operation of transformers under (non)sinusoidal operating conditions. It briefly reviews the operation and modeling of transformers at sinusoidal conditions. It highlights the impact of harmonics and poor power quality on transformer (fundamental and harmonic) losses and discusses several techniques for derating of single- and three-phase transformers including $K$-factor, $F_{HL}$-factor, and loss measurement techniques. Several transformer harmonic models in the time domain and/or frequency domain are introduced. Important power quality problems related to transformers such as ferroresonance and geomagnetically induced currents (GICs) are explored. Different techniques for system and transformer grounding are explained. Many application examples explaining nonsinusoidal flux, voltage and current, harmonic (copper and iron-core) loss measurements, derating, ferroresonance, magnetic field strength, propagation of surge, and operation of lightning arresters are presented for further exploration and understanding of the presented materials.

## 2.10 PROBLEMS

### Problem 2.1 Pole Transformer Supplying a Residence

Calculate the currents flowing in all branches as indicated in Fig. P2.1 if the pole transformer is loaded as shown. For the pole transformer you may assume ideal transformer relations. Note that such a single-phase configuration is used in the pole transformer (mounted on a pole behind your house) supplying residences within the United States with electricity. What are the real (P), reactive (Q), and apparent (S) powers supplied to the residence?

### Problem 2.2 Analysis of an Inductor with Three Coupled Windings

a) Draw the magnetic equivalent circuit of Fig. P2.2.

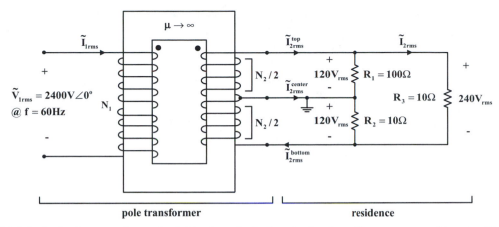

FIGURE P2.1 Single-phase transformer loaded by three single-phase loads, where $R_1$ represents a refrigerator, $R_2$ a microwave oven, and $R_3$ an electric stove.

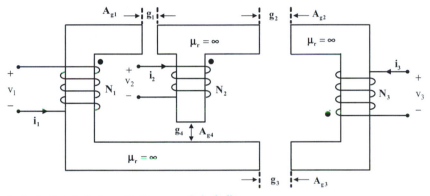

FIGURE P2.2 Analysis of an inductor with three coupled windings.

b) Determine all self- and mutual inductances. Assuming that the winding resistances are known draw the electrical equivalent circuit of Fig. P2.2.

## Problem 2.3 Derating of Transformers Based on K-Factor

A substation transformer ($S = 10$ MVA, $V_{L-L} = 5$ kV, $\%P_{EC-R} = 10$) supplies a six-pulse rectifier load with the harmonic current spectrum as given by Table X of the paper by Duffey and Stratford [82].

a) Determine the $K$-factor of the rectifier load.
b) Calculate the derating of the transformer for this nonlinear load.

## Problem 2.4 Derating of Transformers Based on $F_{HL}$-Factor

A substation transformer ($S = 10$ MVA, $V_{L-L} = 5$ kV, $\%P_{EC-R} = 10$) supplies a six-pulse rectifier load with the harmonic current spectrum as given by Table X of the paper by Duffey and Stratford [82].

a) Determine the $F_{HL}$-factor of the rectifier load.
b) Calculate the derating of the transformer for this nonlinear load.

## Problem 2.5 Derating of Transformers Based on K-Factor for a Triangular Current Wave Shape

a) Show that the Fourier series representation for the current waveform in Fig. P2.5 is

$$i(t) = \frac{8I_m}{\pi^2} \sum_{h=1,3,5,\ldots}^{\infty} \left(\frac{1}{h^2}\right) \sin\left(\frac{h\pi}{2}\right) \sin h\omega_0 t.$$

b) Approximate the wave shape of Fig. P2.5 in a discrete manner, that is, sample the wave shape using 36 points per period and use the program of Appendix 2 to find the Fourier coefficients.

c) Based on the Nyquist criterion, what is the maximum order of harmonic components $h_{max}$ you can compute in part b? Note that for $h_{max}$ the approximation error will be a minimum.

d) Calculate the $K$-factor and the derating of the transformer when supplied with a triangular current wave shape, and $\%P_{EC-R} = 10$. The rated current $I_R$ corresponds to the fundamental component of the triangular current.

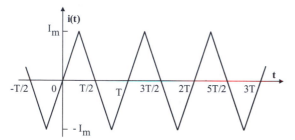

**FIGURE P2.5** Triangular current wave shape.

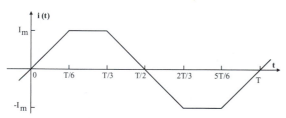

**FIGURE P2.6** Trapezoidal current wave shape.

## Problem 2.6 Derating of Transformers Based on K-Factor for a Trapezoidal Current Wave Shape

a) Show that the Fourier series representation for the current waveform in Fig. P2.6 is

$$i(t) = \frac{12I_m}{\pi^2} \sum_{h=1,3,5,\ldots}^{\infty} \left(\frac{1}{h^2}\right)\sin\left(\frac{h\pi}{3}\right)\sin h\omega_0 t.$$

b) Approximate the wave shape of Fig. P2.6 in a discrete manner, that is, sample the wave shape using 36 points per period and use the program of Appendix 2 to find the Fourier coefficients.

c) Based on the Nyquist criterion, what is the maximum order of harmonic components $h_{max}$ you can compute in part b? Note that for $h_{max}$ the approximation error will be a minimum.

d) Calculate the $K$-factor and the derating of the transformer when supplied with a trapezoidal current wave shape, and $\%P_{EC-R} = 10$. The rated current $I_R$ corresponds to the fundamental component of the trapezoidal current.

## Problem 2.7 Force Calculation for a Superconducting Coil for Energy Storage

Design a magnetic storage system which can provide for about 6 minutes a power of 100 MW, that is, an energy of 10 MWh. A coil of radius $R = 10$ m, height $h = 6$ m, and $N = 3000$ turns is shown in Fig. P2.7. The magnetic field strength $H$ inside such a coil is axially directed, essentially uniform, and equal to $H = Ni/h$. The coil is enclosed by a superconducting box of 2

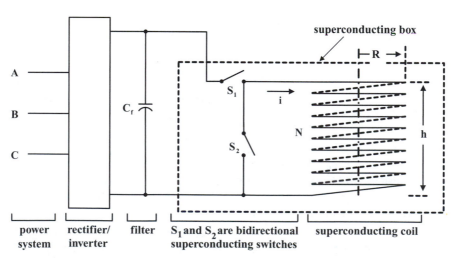

**FIGURE P2.7** Superconducting coil for energy storage.

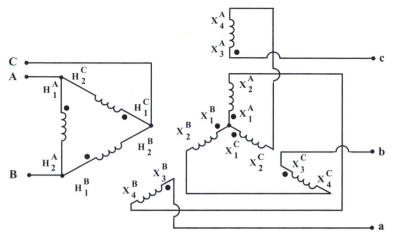

**FIGURE P2.8** Delta/zigzag three-phase transformer configuration.

times its volume. The magnetic field outside the superconducting box can be assumed to be negligible and you may assume a permeability of free space for the calculation of the stored energy.

a) Calculate the radial pressure in newtons per square meter acting on the sides of the coil for constant current $i = I_0 = 8000$ A.

b) Calculate the energy $E_{stored}$ stored in the magnetic field (either in terms of $H$ or $B$).

c) Determine the radial pressure $p_{radial}$ acting on the coil.

d) Devise an electronic circuit that enables the controlled extraction of the energy from this storage device. You may use two bidirectional superconducting switches $S_1$ and $S_2$, which are controlled by a magnetic field via the quenching effect. $S_1$ is ON and $S_2$ is OFF for charging the superconducting coil. $S_1$ is OFF and $S_2$ is ON for the storage time period. $S_1$ is ON and $S_2$ is OFF for discharging the superconducting coil.

## Problem 2.8 Measurement of Losses of Delta/Zigzag Transformers

Could the method of Section 2.8 also be applied to a Δ/zigzag three-phase transformer as shown in Fig. P2.8? This configuration is of interest for the design of a reactor [67] where the zero-sequence (e.g., triplen) harmonics are removed or mitigated, and where through phase shifting the 5th and 7th harmonics [83] are canceled.

## Problem 2.9 Error Calculations

For a 25 kVA single-phase transformer of Fig. P2.9 with an efficiency of 98.46% at rated operation, the

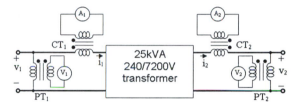

**FIGURE P2.9** Instrumentation for single-phase transformers employing conventional approach for loss measurement.

**TABLE P2.9** Instruments and Their Error Limits of Fig. P2.9

| Instrument | Full scale | Full-scale errors |
|---|---|---|
| $PT_1$ | 240 V/240 V | $\varepsilon_{PT1} = 0.24$ V |
| $PT_2$ | 7200 V/240 V | $\varepsilon_{PT2} = 0.24$ V |
| $CT_1$ | 100 A/5 A | $\varepsilon_{CT1} = 5$ mA |
| $CT_2$ | 5 A/5 A | $\varepsilon_{CT2} = 5$ mA |
| $V_1$ | 300 V | $\varepsilon_{V1} = 0.3$ V |
| $V_2$ | 300 V | $\varepsilon_{V2} = 0.3$ V |
| $A_1$ | 5 A | $\varepsilon_{A1} = 5$ mA |
| $A_2$ | 5 A | $\varepsilon_{A2} = 5$ mA |

total losses are $P_{loss} = 390$ W, where the copper and iron-core losses are $P_{cu} = 325$ W and $P_{fe} = 65$ W, respectively. Find the maximum measurement errors of the copper $P_{cu}$ and iron-core $P_{fe}$ losses as well as that of the total loss $P_{loss}$ whereby the instrument errors are given in Table P2.9.

## Problem 2.10 Delta/Wye Three-Phase Transformer Configuration with Six-Pulse Midpoint Rectifier Employing an Interphase Reactor

a) Perform a PSpice analysis for the circuit of Fig. P2.10, where a three-phase thyristor rectifier

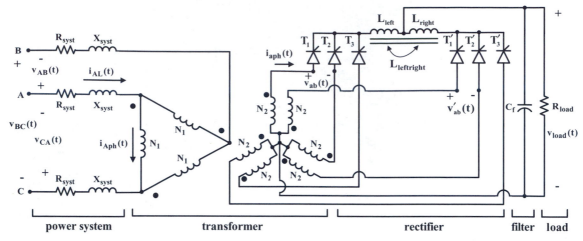

**FIGURE P2.10** Connection of a delta/wye three-phase transformer configuration with six-pulse midpoint rectifier employing an interphase reactor.

with filter (e.g., capacitor $C_f$) serves a load ($R_{load}$). You may assume ideal transformer conditions. For your convenience you may assume $(N_1/N_2) = 1$, $R_{syst} = 0.01\ \Omega$, $X_{syst} = 0.05\ \Omega$ @ $f = 60$ Hz, $v_{AB}(t) = \sqrt{2}600V\cos\omega t$, $v_{BC}(t) = \sqrt{2}$ $600V\cos(\omega t - 120°)$, and $v_{CA}(t) = \sqrt{2}600V\cos(\omega t - 240°)$. Each of the six thyristors can be modeled by a self-commutated electronic switch and a diode in series, as is illustrated in Fig. E1.1.2 of Application Example 1.1 of Chapter 1. Furthermore, $C_f = 500\ \mu F$, $L_{left} = 5$ mH, $L_{right} = 5$ mH, and $R_{load} = 10\ \Omega$. You may assume that the two inductors $L_{left}$ and $L_{right}$ are not magnetically coupled, that is, $L_{leftright} = 0$. Plot one period of either voltage or current after steady state has been reached for firing angles of $\alpha = 0°$, 60°, and 120°, as requested in parts b to e. Print the input program.

b) Plot and subject the line-to-line voltages $V_{AB}(t)$, $V_{ab}(t)$, and $V'_{ab}(t)$ to Fourier analysis. Why are they different?

c) Plot and subject the input line current $i_{AL}(t)$ of the delta primary to a Fourier analysis. Note that the input line currents of the primary delta do not contain the 3rd, 6th, 9th, 12th, ..., that is, harmonic zero-sequence current components.

d) Plot and subject the phase current $i_{Aph}(t)$ of the delta primary to a Fourier analysis. Note that the phase currents of the primary delta contain the 3rd, 6th, 9th, 12th, ..., that is, harmonic zero-sequence current components.

**TABLE P2.11** Measurement Values of the Input Quantities $\tilde{V}_t$, $\tilde{I}_t$, and $\Phi$ Obtained from Five Sets of Simultaneous Observations

| Set number k | Input quantities | | |
|---|---|---|---|
|  | $|\tilde{V}_t|$ (V) | $|\tilde{I}_t|$ (mA) | $\Phi$ (radians) |
| 1 | 5.006 | 19.664 | 1.0457 |
| 2 | 4.993 | 19.638 | 1.0439 |
| 3 | 5.004 | 19.641 | 1.0468 |
| 4 | 4.991 | 19.684 | 1.0427 |
| 5 | 4.998 | 19.679 | 1.0433 |

e) Plot and subject the output voltage $v_{load}(t)$ to a Fourier analysis.

f) How do your results change if the two inductors $L_{left}$ and $L_{right}$ are magnetically coupled, that is, $L_{leftright} = 4.5$ mH?

## Problem 2.11 Determination of Measurement Uncertainty

Instead of maximum errors, the U.S. Guide to the Expression of Uncertainty in Measurement [84] recommends the use of uncertainty. As an example the uncertainty and the associated correlation coefficients of the simultaneous measurement of resistance and reactance of an inductor having the inductance $L$ and the resistance $R$ are to be calculated. The five measurement sets of Table P2.11 are available for an inductor consisting of the impedance $Z = R + jX$, where terminal voltage $\tilde{V}_t$, current $\tilde{I}_t$, and phase-shift angle $\Phi$ are measured.

With $R = (|\tilde{V}_t||\tilde{I}_t|)\cos\Phi$, $X = (|\tilde{V}_t||\tilde{I}_t|)\sin\Phi$, $Z = (|\tilde{V}_t||\tilde{I}_t|)$, and $Z^2 = R^2 + X^2$ one obtains two independent output quantities – called measurands.

a) Calculate the arithmetic mean of $|\tilde{V}_t|$, $|\tilde{I}_t|$, and $\Phi$.

b) Determine the experimental standard deviation of mean of $s(\bar{V})$, $s(\bar{I})$, and $s(\bar{\Phi})$.

c) Find the correlation coefficients $r(|\tilde{V}_t|, |\tilde{I}_t|)$, $r(|\tilde{V}_t|, \Phi)$, and $r(|\tilde{V}_t|, \Phi)$.

d) Find the combined standard uncertainty $\mu \equiv u_c(y_\ell)$ of result of measurement where $\ell$ is the measurand index.

e) Determine the correlation coefficients $r(R/X)$, $r(R/Z)$, and $r(X/Z)$.

## 2.11 REFERENCES

1) Fuchs, E.F.; Masoum, M.A.S.; and Roesler, D.; "Large signal nonlinear model of anisotropic transformers for nonsinusoidal operation, part I: $\lambda$–$i$ characteristic," *IEEE Transactions on Power Delivery*, Vol. 6, No. 1, pp. 174–186, 1991.

2) Masoum, M.A.S.; Fuchs, E.F.; and Roesler, D.; "Large signal nonlinear model of anisotropic transformers for nonsinusoidal operation, part II: magnetizing and core loss currents," *IEEE Transactions on Power Delivery*, Vol. 6, No. 4, pp. 1509–1516, 1991.

3) Masoum, M.A.S.; Fuchs, E.F.; and Roesler, D.; "Impact of nonlinear loads on anisotropic transformers," *IEEE Transactions on Power Delivery*, Vol. 6, No. 4, pp. 1781–1788, 1991.

4) Masoum, M.A.S.; and Fuchs, E.F.; "Transformer magnetizing current in harmonic power flow," *IEEE Transactions on Power Delivery*, Vol. 9, No. 1, pp. 1020, 1994.

5) Fuchs, E.F.; and Fei, R.; "A new computer-aided method for the efficiency measurement of low-loss transformers and inductors under nonsinusoidal operation," *IEEE Transactions on Power Delivery*, Vol. PWRD-11, No. 1, pp. 292–304, 1996.

6) Stensland, T.; Fuchs, E.F.; Grady, W.M.; and Doyle, M.; "Modeling of magnetizing and core-loss currents in single-phase transformers with voltage harmonics for use in power flow," *IEEE Transactions on Power Delivery*, Vol. 12, No. 2, pp. 768–774, 1997.

7) Fuchs, E.F.; Yildirim, D.; and Batan, T.; "Innovative procedure for measurement of losses of transformers supplying nonsinusoidal loads," *IEE Proceedings – Generation, Transmission and Distribution*, Vol. 146, No. 6, November 1999, IEE Proceedings online no. 19990684.

8) Fuchs, E.F.; Lin, D.; and Martynaitis, J.; "Measurement of three-phase transformer derating and reactive power demand under nonlinear loading conditions," *IEEE Transactions on Power Delivery*, Vol. 21, Issue 2, pp. 665–672, 2006.

9) Lin, D.; and Fuchs, E.F.; "Real-time monitoring of iron-core and copper losses of three-phase transformers under (non)sinusoidal operation," *IEEE Transactions on Power Delivery*, Vol. 21, No. 3, pp. 1333–1341, July 2006.

10) Montsinger, V.M.; "Loading transformer by temperature," *Trans. AIEE*, Vol. 49, 1930, p. 776.

11) Fuchs, E.F.; Yildirim, D.; and Grady, W.M.; "Measurement of eddy-current loss coefficient $P_{EC\text{-}R}$, derating of single-phase transformers, and comparison with K-factor approach," *IEEE Transactions on Power Delivery*, Vol. 15, No. 1, January 2000, pp. 148–154, and Vol. 15, No. 4, October 2000, pp. 1331–1333 (discussions/closure) and p. 1357.

12) Yildirim, D.; "Commissioning of 30 kVA variable-speed direct-drive wind power plant," Ph.D. Thesis, University of Colorado, Boulder, May 1999.

13) Standard UL 1561, "Dry-type general purpose and power transformers," April 22, 1994.

14) Standard ANSI/IEEE C57.110, 1998, "Recommended practice for establishing transformer capability when supplying nonsinusoidal load currents," February 1998.

15) Yildirim, D.; and Fuchs, E.F.; "Measured transformer derating and comparison with harmonic loss factor $F_{HL}$ approach," *IEEE Transactions on Power Delivery*, Vol. 15, No. 1, January 2000, pp. 186–191, and Vol. 15, No. 4, October 2000, pp. 1328–1329 (discussions/closure) and p. 1357.

16) Fuchs, E.F.; Masoum, M.A.S.; and Ladjevardi, M.; "Effects on distribution feeders from electronic loads based on future peak-load growth, part I: measurements," *Proceedings of the IASTED International Conference on Power and Energy Systems (EuroPES 2005)*, Paper #468-019, Benalmadena, Spain, June 15–17, 2005.

17) Dommel, H.; "Digital computer solution of electromagnetic transients in single and multiphase network," *IEEE Transactions on Power Apparatus and Systems*, Vol. PAS-88, pp. 487–493, 1974.

18) Dommel, H.; "Transformer models in the simulation of electromagnetic transient," *5th Power Systems Computation Conference*, Paper 3.1/4, pp. 1–16, Cambridge, England, September 1–5, 1975.

19) Hatziantoniu, C.; Galanos, G.D.; and Milias-Argitis, J.; "An incremental transformer model for the study of harmonic overvoltages in weak AC/DC systems," *IEEE Transactions on Power Delivery*, Vol. 3, No. 3, pp. 1111–1121, 1988.

20) Dolinar, D.; Pihler, J.; and Grcar, B.; "Dynamic model of a three-phase power transformer," *IEEE Transactions on Power Delivery*, Vol. 8, No. 4, pp. 1811–1819, 1993.

21) Leon, F.; and Semlyen, A.; "Complete transformer model for electromagnetic transients," *IEEE Transactions on Power Delivery*, Vol. 9, No. 1, pp. 231–239, 1994.

22) Garcia, S.; Medina, A.; and Perez, C.; "A state space single-phase transformer model incorporating nonlinear phenomena of magnetic saturation and hysteresis for transient and periodic steady-state analysis," *IEEE Power Engineering Society Summer Meeting*, 2000, Vol. 4, 16–20 July 2000, pp. 2417–2421.

23) Garcia, S.; and Medina, A.; "A state space three-phase multilimb transformer model in time domain: fast periodic steady-state analysis," *IEEE Power Engineering Society Summer Meeting*, 2001, Vol. 3, 15–19 July 2001, pp. 1859–1864.

24) *Reference Manual for Electromagnetic Transient Program (EMTP)*, Bonneville Power Administration, Portland, Oregon.

25) Chan, J.H.; Vladimirescu, A.; Gao, X.C.; Liebmann, P.; and Valainis, J.; "Nonlinear transformer model for circuit simulation," *IEEE Transactions on Computer-Aided Design*, Vol. 10, No. 4, pp. 476–482, 1991.

26) Huang, S.R.; Chung, S.C.; Chen, B.N.; and Chen, Y.H.; "A harmonic model for the nonlinearities of single-phase transformer with describing functions," *IEEE Transactions on Power Delivery*, Vol. 18, No. 3, pp. 815–820, 2003.

27) Medina, A.; and Arrillaga, J.; "Generalized modeling of power transformers in the harmonic domain," *IEEE Transactions on Power Delivery*, Vol. 7, No. 3, pp. 1458–1465, 1992.

28) Semlyen, A.; Acha, E.; and Arrillaga, J.; "Newton-type algorithms for the harmonic phasor analysis of nonlinear power circuits in periodical steady state with special reference to magnetic nonlinearities," *IEEE Transactions on Power Delivery*, Vol. 3, No. 3, pp. 1090–1099, 1988.

29) Pedra, J.; Corcoles, F.; Sainz , L.; and Lopez, R.; "Harmonic nonlinear transformer modeling," *IEEE Transactions on Power Delivery*, Vol. 19, No. 2, pp. 884–890, 2004.

30) Dugui, W.; and Zheng, X.; "Harmonic model of transformer," *Proceedings of 1998 International Conference on Power System Technology, POWERCON '98*, Vol. 2, 18–21 Aug. 1998, pp. 1045–1049.

31) Greene, J.D.; and Gross, C.A.; "Nonlinear modeling of transformers," *IEEE Transactions on Industry Applications*, Vol. 24, No. 3, pp. 434–438, 1988.

32) Yamada, S.; Bessho, K.; and Lu, J.; "Harmonic balance finite element method applied to nonlinear AC magnetic analysis," *IEEE Transactions on Magnetics*, Vol. 25, No. 4, pp. 2971–2973, 1989.

33) Silvester, P.; and Chari, M.V.K.; "Finite element solution of saturable magnetic field problem," *IEEE Transactions on Power Apparatus and Systems*, Vol. PAS-89, No. 7, pp. 1642–1649, 1989.

34) Enokizono, M.; and Soda, N.; "Finite element analysis of transformer model core with measured reluctivity tensor," *IEEE Transactions on Magnetics*, Vol. 33, Issue 5, Part 2, Sept. 1997, pp. 4110–4112.

35) Masoum, M.A.S.; *Generation and Propagation of Harmonics in Power System Feeders Containing Nonlinear Transformers and Loads*, Ph.D. Thesis, University of Colorado at Boulder, April 1991.

36) Stensland, T.D.; *Effects of Voltage Harmonics on Single-Phase Transformers and Induction Machines Including Pre-Processing for Power Flow*, MS Thesis, University of Colorado at Boulder, Colorado, 1995.

37) Marti, J.R.; Soudack, A.C.; "Ferroresonance in power systems: fundamental solutions," *IEEE Proceedings C, Generation, Transmission and Distribution* [see also *IEE Proceedings-Generation, Transmission and Distribution*], Vol. 138, Issue 4, July 1991, pp. 321–329.

38) Yunge Li; Wei Shi; Rui Qin; and Jilin Yang; "A systematical method for suppressing ferroresonance at neutral-grounded substations," *IEEE Transactions on Power Delivery*, Vol. 18, Issue 3, July 2003, pp. 1009–1014.

39) Fuchs, E.F.; You, Y.; and Roesler, D. J.; "Modeling, simulation and their validation of three-phase transformers with three legs under DC bias," *IEEE Transactions on Power Delivery*, Vol. 14, No. 2, April 1999, pp. 443–449.

40) Fuchs, E.F.; and You, Y.; "Measurement of $(\lambda–i)$ characteristics of asymmetric three-phase transformers and their applications," *IEEE Transactions on Power Delivery*, Vol. 17, No. 4, October 2002, pp. 983–990.

41) Young, F.S.; Schmid, R.L; and Fergestad, P.I.; "A laboratory investigation of ferroresonance in cable-connected transformers," *IEEE Transactions on Power Apparatus and Systems*, Vol. PAS-87, Issue 5, May 1968, pp. 1240–1249.

42) Koch, H. J.; "Gas-insulated transmission line (GIL)" 0-7803-7989-6/03/$17.00 copyright 2003 IEEE, Panel Discussion: Gas-Insulated Transmission Lines, Proceedings of the IEEE PES General Meeting, Toronto, pp. 2480–2483.

43) Koch, H.; "Gas-insulated line (GIL) of the 2nd generation," *AC-DC Power Transmission*, 28–30 November 2001, Conference Publication No. 485 copyright IEEE 2001, pp. 39–43.

44) Koch, H.; and Hillers, T.; "Second-generation gas-insulated line," *Power Engineering Journal*, June 2002, pp. 111–116.

45) Kieny, C.; Le Roy, G.; and Sbai, A.; "Ferroresonance study using Galerkin method with pseudo-arclength continuation method," *IEEE Transactions on Power Delivery*, Vol. 6, Issue 4, Oct. 1991, pp. 1841–1847.

46) Lahtinen, M.; and Elovaara, J.; "GIC occurrences and GIC test for 400 kV system transformer," *IEEE Transactions on Power Delivery*, Vol. 17, Issue 2, April 2002, pp. 555–561.

47) Boteler, D.H.; Shier, R.M.; Watanabe, T.; Horita, R.E.; "Effects of geomagnetically induced currents in the BC Hydro 500 kV system," *IEEE Transactions on Power Delivery*, Vol. 4, Issue 1, Jan. 1989, pp. 818–823.

48) Kappenman, J.G.; Norr, S.R.; Sweezy, G.A.; Carlson, D.L.; Albertson, V.D.; Harder, J.E.; and Damsky, B.L.; "GIC mitigation: a neutral blocking/bypass device to prevent the flow of GIC in power systems," *IEEE Transactions on Power Delivery*, Vol. 6, Issue 3, July 1991, pp. 1271–1281.

49) Gish, W.B.; Feero, W.E.; and Rockefeller, G.D.; "Rotor heating effects from geomagnetic induced currents," *IEEE Transactions on Power Delivery*, Vol. 9, Issue 2, April 1994, pp. 712–719.

50) Bozoki, B.; Chano, S.R.; Dvorak, L.L.; Feero, W.E.; Fenner, G.; Guro, E.A.; Henville, C.F.; Ingleson, J.W.; Mazumdar, S.; McLaren, P.G.; Mustaphi, K.K.; Phillips, F.M.; Rebbapragada, R.V.; and Rockefeller, G.D.; "The effects of GIC on protective relaying," *IEEE Transactions on Power Delivery*, Vol. 11, Issue 2, April 1996, pp. 725–739.

51) Dickmander, D.L.; Lee, S.Y.; Desilets, G.L.; and Granger, M.; "AC/DC harmonic interactions in the presence of GIC for the Quebec–New England Phase II HVDC transmission," *IEEE Transactions on Power Delivery*, Vol. 9, Issue 1, Jan. 1994, pp. 68–78.

52) Walling, R.A.; and Khan, A.N.; "Characteristics of transformer exciting-current during geomagnetic disturbances," *IEEE Transactions on Power Delivery*, Vol. 6, Issue 4, Oct. 1991, pp. 1707–1714.

53) You, Y.; Fuchs, E.F.; Lin, D.; and Barnes, P.R.; "Reactive power demand of transformers with DC bias," *IEEE Industry Applications Society Magazine*, Vol. 2, No. 4, July/August 1996, pp. 45–52.

54) Fardoun, A.A.; Fuchs, E.F.; and Masoum, M.A.S.; "Experimental analysis of a DC bucking motor blocking geomagnetically induced currents," *IEEE Transactions on Power Delivery*, January 1994, Vol. TPWRD-9, pp. 88–99.

55) Castenschiold, R.; and Johnson, G.S.; "Proper grounding of on-site electrical power systems," *IEEE Industry Applications Society Magazine*, Vol. 7, Issue 2, March–April 2001, pp. 54–62.

56) American National Standard ANSI/IEEE C114.1, *IEEE recommended practice for grounding of industrial and commercial power systems, the IEEE green book*, 1973).

57) Kraft, L.A.; "Modelling lightning performance of transmission systems using PSpice," *IEEE Transactions on Power Systems*, Vol. 6, Issue 2, May 1991, pp. 543–549.

58) Johnk, C.T.A.; *Engineering Electromagnetic Fields and Waves*, John Wiley & Sons, New York, NY, 1975.

59) *British Standard Code of Practice for Control of Undesirable Static Electricity*, pt. 1, sec. 4, 1991.

60) Luttgens, G.; "Static electricity hazards in the use of flexible intermediate bulk containers (FIBCs)," *IEEE Transactions on Industry Applications*, Vol. 33, Issue 2, pp. 444–446, 1997.

61) Pillai, P.; Bailey, B.G.; Bowen, J.; Dalke, G.; Douglas, B.G.; Fischer, J.; Jones, J.R.; Love, D.J.; Mozina, C.J.; Nichols, N.; Normand, C.; Padden, L.; Pierce, A.; Powell, L.J.; Shipp, D.D.; Stringer, N.T.; and Young, R.H.; "Grounding and ground fault protection of multiple generator installations on medium-voltage industrial and commercial systems-Part 2: Grounding methods working group report," *IEEE Transactions on Industry Applications*, Vol. 40, Issue 1, Jan.–Feb. 2004, pp. 17–23.

62) Meliopoulos, A.P.S.; and Dunlap, J.; "Investigation of grounding related problems in AC/DC converter stations," *IEEE Transactions on Power Delivery*, Vol. 3, Issue 2, April 1988, pp. 558–567.

63) Frick, C.W.; "Electromagnetic forces on conductors with bends, short lengths and cross-overs," *General Electric Review*, vol. 36, no. 5.

64) Dwight, H.B.; "Calculation of magnetic forces in disconnecting switches," *A.I.E.E. Trans.*, vol. 39, 1920, p. 1337.

65) Lawrenson, P.J.; "Forces on turbogenerator end windings," *Proc. IEE*, vol. 112, no. 6, June 1965.

66) Al-Barrak, A. S.; "Electromagnetic forces and bending moments on conductors of arbitrary polygonal filament and finite shape," Independent Study, University of Colorado, Boulder, 1979.

67) *Electrical Transmission and Distribution Reference Book*, Westinghouse Electric Corporation, 1964.

68) Kelly, A.W.; Edwards, S.W.; Rhode, J.P.; and Baran, M.; "Transformer derating for harmonic currents: a wideband measurement approach for energized transformers," *Proceedings of the Industry Applications Conference*, 1995, Thirtieth IAS Annual Meeting, IAS 1995. Vol. 1, Oct. 8–12, 1995, pp. 840–847.

69) Masoum, M.A.S.; and Fuchs, E.F.; "Derating of anisotropic transformers under nonsinusoidal operating conditions," *Electrical Power and Energy Systems*, Vol. 25 (2003), pp. 1–12.

70) Mocci, F.; "Un nuovo methodo per la determinazione della potenza assorbita dai doppi bipoli," *L'Energia Elettrica*, No. 7–8, 19989, pp. 323–329.

71) Arri, E.; Locci, N.; and Mocci, F.; "Measurement of transformer power losses and efficiency in real working conditions," *IEEE Transactions on Instrumentation and Measurement*, Vol. 40, No. 2, April 1991, pp. 384–387.

72) *Current and Voltage Transducer Catalog*, Third Edition, LEM U.S.A., Inc., 6643 West Mill Road, Milwaukee, WI 53218.

73) Lin, D.; Fuchs, E.F.; and Doyle, M.; "Computer-aided testing of electrical apparatus supplying nonlinear loads," *IEEE Transactions on Power Systems*, Vol. 12, No. 1, February 1997, pp. 11–21.

74) Rieman, S.; "An optically isolated data acquisition system," Independent Study, University of Colorado, Boulder, December 1997.

75) Grady, W.M.; Fuchs, E.F.; Lin, D.; and Stensland, T.D.; "The potential effects of single-phase power electronic-based loads on power system distortion and losses, Vol. 2: Single-phase transformers and induction motors," Technical Report published by the Electric Power Research Institute (EPRI), Palo Alto, CA, 2003, 1000655, September 2003.

76) Fuchs, E.F.; "Transformers, liquid filled," *Encyclopedia of Electrical and Electronics Engineering 2000*, edited by John G. Webster, John Wiley & Sons, 605 Third Avenue, New York, NY 10158, paper No. 934C, 40 pages, http://www.interscience.wiley.com:38/eeee/32/6132/w.6132-toc.html.

77) Batan, T.; Discussion of "Measurement of eddy-current loss coefficient $P_{EC-R}$, derating of single phase transformers, and comparison with $K$-factor approach," *IEEE Transactions on Power Delivery*, Vol. 15, issue 4, Oct. 2000, pp. 1331–1333.

78) Fuchs, E.F.; You, Y.; and Lin, D.; "Development and validation of GIC transformer models," Final Report for Contract 19X-SK205V, Prepared for Martin Marietta Energy Systems, Inc., Oak Ridge, TN, June 1996.

79) "High Speed Bipolar Monolithic Sample/Hold Amplifier," Burr-Brown, SHC5320.

80) DAS-50 User's Guide, Keithley MetraByte Corporation, 440 Myles Standish Boulevard, Taunton, MA 02780.

81) Fuchs, E.F.; Fei, R.; and Hong, J.; "Harmonic losses in isotropic and anisotropic transformer cores at low frequencies (60–3,000 Hz)," Final Report for Contract No. 19X-SD079C. Prepared for Martin Marietta Energy Systems, Inc., by University of Colorado, Boulder, January 1991.

82) Duffey, C.K.; and Stratford, R.P.; "Update of harmonic standard IEEE-519: IEEE recommended practices and requirements for harmonic control in electric power systems," *IEEE Transactions on Industry Applications*, Vol. 25, Issue 6, Nov.–Dec. 1989, pp. 1025–1034.

83) "Neutral Current Eliminator (NCE) and Combined Neutral Current Eliminator (CNCE)," *MIRUS International Inc.* http://mirusinternational.com.

84) "American National Standard for Calibration—U.S. Guide to the Expression of Uncertainty in Measurement," *NCSL International*, ANCI/NCSL Z540-2-1997 (r2002).

## 2.12 ADDITIONAL BIBLIOGRAPHY

85) Hesterman, B.; "Time-domain $K$-factor computation methods," *Proc. PCIM 1994, 29th Int. Power Conference*, Sept. 1994, pp. 406–417.

86) Fuchs, E.F.; Roesler, D.J.; and Kovacs, K.P.; "Aging of electrical appliances due to harmonics of the power system's voltage," *IEEE Transactions on Power Delivery*, July 1986, Vol. TPWRD-1, No. 3, pp. 301–307.

87) Fuchs, E.F.; Roesler, D.J.; and Alashhab, F.S.; "Sensitivity of electrical appliances to harmonics and fractional harmonics of the power system's voltage, part I," *IEEE Transactions on Power Delivery*, April 1987, Vol. TPWRD-2, No. 2, pp. 437–444.

88) Fuchs, E.F.; Roesler, D.J.; and Kovacs, K.P.; "Sensitivity of electrical appliances to harmonics and fractional harmonics of the power system's voltage, part II," *IEEE Transactions on Power Delivery*, April 1987, Vol. TPWRD-2, No. 2, pp. 445–453.

89) Masoum, M.A.S.; and Fuchs, E.F.; "Transformer magnetizing current and core losses in harmonic power flow," *IEEE Transactions on Power Delivery*, January 1994, Vol. PWRD-9, pp. 10–20.

90) Chowdhury, A.H.; Grady, W.M.; and Fuchs, E.F.; "An investigation of harmonic characteristics of transformer excitation current under nonsinusoidal supply voltage," *IEEE Transactions on Power Delivery*, Vol. 14, No. 2, April 1999, pp. 450–458.

91) Fuchs, E.F.; Roesler, D.J.; and Masoum, M.A.S.; "Are harmonic recommendations according to IEEE and IEC too restrictive?" *IEEE Transactions on Power Delivery*, Vol. 19, No. 4, October 2004, pp. 11775–11786.

92) Fuchs, E.F.; and Masoum, M.A.S.; "Suppression of harmonic distortion in power systems due to geomagnetically induced currents (GICs) through enforcing gic balance in all phases of a system," Invention Disclosure, University of Colorado, Boulder, June 1, 1992.

93) Fuchs, E.F.; "A new method for measuring the efficiency of low-loss electrical apparatus," Invention Disclosure, University of Colorado, Boulder, February, 1994.

94) You, Y.; and Fuchs, E.F.; "Means for suppressing harmonic distortion due to DC bias currents/fluxes in electric circuits containing any three-phase transformers through balancing and canceling of DC MMF/flux of every phase," Invention Disclosure, University of Colorado, Boulder, June, 1994.

95) You, Y.; and Fuchs, E.F.; "High AC impedance, low DC resistance inductor," Invention Disclosure, University of Colorado, Boulder, June, 1994.

96) Fuchs, E.F.; and Lin, D.; "On-line monitoring of iron-core and copper losses of single and three-phase transformers under (non)sinusoidal operation at any load," Invention Disclosure, University of Colorado, Boulder, January 1995.

97) Fuchs, E.F.; and Yildirim, D.; "On-line technique for separate measurement of copper loss, iron-core loss and (rated) eddy-current loss PEC-R of transformers supplying nonlinear loads," Invention Disclosure, University of Colorado, Boulder, April 1998.

# 3
CHAPTER

# Modeling and Analysis of Induction Machines

Before the development of the polyphase AC concept by Nikola Tesla [1] in 1888, there was a competition between AC and DC systems. The invention of the induction machine in the late 1880s completed the AC system of electrical power generation, transmission, and utilization. The Niagara Falls hydroplant was the first large-scale application of Tesla's polyphase AC system. The theory of single- and three-phase induction machines was developed during the first half of the twentieth century by Steinmetz [2], Richter [3], Kron [4], Veinott [5], Schuisky [6], Bödefeld [7], Alger [8], Fitzgerald *et al.* [9], Lyon [10], and Say [11] – just to name a few of the hundreds of engineers and scientists who have published in this area of expertise. Newer textbooks and contributions are by Match [12], Chapman [13], and Fuchs *et al.* [14, 15]. In these works mostly balanced steady-state and transient performances of induction machines are analyzed. Most power quality problems as listed below are neglected in these early publications because power quality was not an urgent issue during the last century.

Today, most industrial motors of one horsepower or larger are three-phase induction machines. This is due to their inherent advantages as compared with synchronous motors. Although synchronous motors have certain advantages – such as constant speed, generation of reactive power (leading power factor based on consumer notation of current) with an overexcited field, and low cost for low-speed motors – they have the constraints of requiring a DC exciter, inflexible speed control, and high cost for high-speed motors. The polyphase induction motors, however, have certain inherent advantages, including these:

- Induction motors require no exciter (no electrical connection to the rotor windings except for some doubly excited machines);
- They are rugged and relatively inexpensive;
- They require very little maintenance;
- They have nearly constant speed (slip of only a few percent from no load to full load);
- They are suitable for explosive environments;
- Starting of motors is relatively easy; and

- Operation of three-phase machines within a single-phase system is possible.

The main disadvantages of induction motors are:

- complicated speed control,
- high starting current if no starters are relied on, and
- low and lagging (based on consumer notation of current) power factor at light loads.

For most applications, however, the advantages far outweigh the disadvantages. In today's power systems with a large number of nonlinear components and loads, three-phase induction machines are usually subjected to nonsinusoidal operating conditions. Conventional steady-state and transient analyses do not consider the impact of voltage and/or current harmonics on three-phase induction machines. Poor power quality has many detrimental effects on induction machines due to their abnormal operation, including:

- excessive voltage and current harmonics,
- excessive saturation of iron cores,
- static and dynamic rotor eccentricities,
- one-sided magnetic pull due to DC currents,
- shaft fluxes and associated bearing currents,
- mechanical vibrations,
- dynamic instability when connected to weak systems,
- premature aging of insulation material caused by cyclic operating modes as experienced by induction machines for example in hydro- and wind-power plants,
- increasing core (hysteresis and eddy-current) losses and possible machine failure due to unacceptably high losses causing excessive hot spots,
- increasing (fundamental and harmonic) copper losses,
- reduction of overall efficiency,
- generation of inter- and subharmonic torques,
- production of (harmonic) resonance and ferroresonance conditions,
- failure of insulation due to high voltage stress caused by rapid changes in supply current and lightning surges,

- unbalanced operation due to an imbalance of power systems voltage caused by harmonics, and
- excessive neutral current for grounded machines.

These detrimental effects call for the understanding of the impact of harmonics on induction machines and how to protect them against poor power quality conditions. A harmonic model of induction motors is essential for loss calculations, harmonic torque calculations, derating analysis, filter design, and harmonic power flow studies.

This chapter investigates the behavior of induction machines as a function of the harmonics of the power system's voltages/currents, and introduces induction machine harmonic concepts suitable for loss calculations and harmonic power flow analysis. After a brief review of the conventional (sinusoidal) model of the induction machine, time and space harmonics as well as forward- and backward-rotating magnetic fields at harmonic frequencies are analyzed. Some measurement results of voltage, current, and flux density waveforms and their harmonic components are provided. Various types of torques generated in induction machines, including fundamental and harmonic torques as well as inter- and subharmonic torques, are introduced and analyzed. The interaction of space and time harmonics is addressed. A section is dedicated to voltage-stress winding failures of AC machines fed by voltage-source and current-source PWM inverters. Some available harmonic models of induction machines are briefly introduced. The remainder of the chapter includes discussions regarding rotor eccentricity, classification of three-phase induction machines, and their operation within a single-phase system. This chapter contains a number of application examples to further clarify the relatively complicated issues.

## 3.1 COMPLETE SINUSOIDAL EQUIVALENT CIRCUIT OF A THREE-PHASE INDUCTION MACHINE

Simulation of induction machines under sinusoidal operating conditions is a well-researched subject and many transient and steady-state models are available. The stator and rotor cores of an induction machine are made of ferromagnetic materials with nonlinear B–H ($\lambda - i$) characteristics. As mentioned in Chapter 2, magnetic coils exhibit three types of nonlinearities that complicate their analysis: saturation effects, hysteresis loops, and eddy currents. These phenomena result in nonsinusoidal flux, voltage and current waveforms in the stator and rotor windings, and additional copper (due to current harmonics)

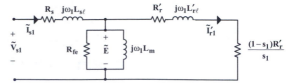

**FIGURE 3.1** Complete linear per-phase equivalent circuit of a three-phase induction machine for sinusoidal analysis.

and core (due to hysteresis loops and eddy currents) losses at fundamental and harmonic frequencies. Under nonsinusoidal operating conditions, the stator magnetic field will generate harmonic rotating fields that will produce forward- and backward-rotating magnetomotive forces (mmfs); in addition to the time harmonics (produced by the nonsinusoidal stator voltages), space harmonics must be included in the analyses. Another detrimental effect of harmonic voltages at the input terminals of induction machines is the generation of unwanted harmonic torques that are superimposed with the useful fundamental torque, producing vibrations causing a deterioration of the insulation material and rotor copper/aluminum bars. Linear approaches for induction machine modeling neglect these nonlinearities (by assuming linear $\lambda - i$ characteristics and sinusoidal current and voltage waveforms) and use constant values for the magnetizing inductance and the core-loss resistance.

Sinusoidal modeling of induction machines is not the main subject of this chapter. However, Fig. 3.1 illustrates a relatively simple and accurate frequency-based linear model that will be extended to a harmonic model in Sections 3.4 to 3.9. In this figure subscript 1 denotes the fundamental frequency ($h = 1$); $\omega_1$ and $s_1$ are the fundamental angular frequency (or velocity) and the fundamental slip of the rotor, respectively. $R_{fe}$ is the core-loss resistance, $L_m$ is the (linear) magnetizing inductance, $R_s$, $R'_r$, $L_{s\ell}$, and $L'_{r\ell}$ are the stator and the rotor (reflected to the stator) resistances and leakage inductances, respectively [3–15]. The value of $R_{fe}$ is usually very large and can be neglected. Thus, the simplified sinusoidal equivalent circuit of an induction machine results, as shown in Fig. 3.2.

Replacing the network to the left of line a–b by Thevenin's theorem (TH) one gets

$$\left| \tilde{V}_{s1TH} \right| = \frac{\left| \tilde{V}_{s1} \right| (\omega_1 L_m) \angle \theta_{TH}}{\sqrt{R_s^2 + (\omega_1 L_{s\ell} + \omega_1 L_m)^2}}, \quad \text{(3-1a)}$$

$$\theta_{TH} = \frac{\pi}{2} - \tan^{-1} \left( \frac{\omega_1 L_{s\ell} + \omega_1 L_m}{R_s} \right), \quad \text{(3-1b)}$$

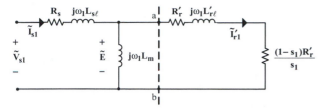

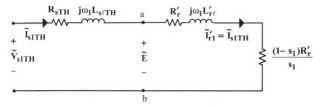

FIGURE 3.2 Simplified per-phase equivalent circuit of a three-phase induction machine for sinusoidal analysis.

FIGURE 3.3 Thevenin adjusted equivalent circuit of a three-phase induction machine for fundamental frequency ($h = 1$).

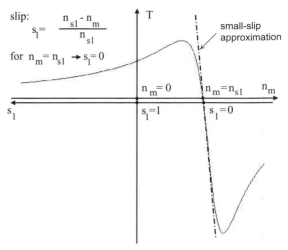

FIGURE 3.4 Fundamental electrical torque-speed characteristics for a three-phase induction machine.

$$R_{sTH} = \frac{(\omega_1 L_m)^2 R_s}{R_s^2 + (\omega_1 L_{s\ell} + \omega_1 L_m)^2}, \qquad (3\text{-}1c)$$

$$X_{s\ell TH} = \omega_1 L_{s\ell TH} =$$
$$\frac{(\omega_1 L_m) R_s^2 + (\omega_1 L_{s\ell})(\omega_1 L_m)(\omega_1 L_{s\ell} + \omega_1 L_m)}{R_s^2 + (\omega_1 L_{s\ell} + \omega_1 L_m)^2}. \qquad (3\text{-}1d)$$

Using Eq. 3-1, the Thevenin adjusted equivalent circuit is shown in Fig. 3.3.

The fundamental ($h = 1$) slip is

$$s_1 = \frac{n_{s1} - n_m}{n_{s1}} = \frac{\omega_{s1} - \omega_m}{\omega_{s1}}, \qquad (3\text{-}2)$$

where $n_{s1}$ is the synchronous speed of the stator field, $n_m$ is the (mechanical) speed of the rotor, and $\omega_{s1}$, $\omega_m$ are the corresponding angular velocities (or frequencies), respectively. Note that subscript 1 denotes fundamental frequency.

For a p-pole machine,

$$\omega_{s1} = \left(\omega_1 \Big/ \frac{p}{2}\right) = 2\pi n_{s1}^{rps} = 2\pi \frac{n_{s1}^{rpm}}{60}, \qquad (3\text{-}3a)$$

$$\omega_m = 2\pi n_m^{rps} = 2\pi \frac{n_m^{rpm}}{60}, \qquad (3\text{-}3b)$$

where $\omega_1 = 2\pi \, f_1 = 377$ rad/s is the fundamental angular velocity (frequency).

For 

| | |
|---|---|
| $p = 2$ | $n_{s1}^{rpm} = 3600$ rpm at $\omega_1 = 2\pi \, f_1 = 377$ rad/s |
| $p = 4$ | $n_{s1}^{rpm} = 1800$ rpm |
| $p = 6$ | $n_{s1}^{rpm} = 1200$ rpm |
| $p = 8$ | $n_{s1}^{rpm} = 900$ rpm. |

The torque of a three-phase ($q_1 = 3$) induction machine is (if $V_{s1TH} = |\tilde{V}_{s1TH}|$)

$$T = \frac{1}{\omega_{s1}} \frac{q_1 V_{s1TH}^2 \dfrac{R_r'}{s_1}}{\left(R_{sTH} + \dfrac{R_r'}{s_1}\right)^2 + (\omega_1 L_{s\ell TH} + \omega_1 L_{r\ell}')^2} \qquad (3\text{-}4a)$$

or

$$T = \frac{1}{\omega_{s1}} \frac{q_1 V_{s1TH}^2 R_r'}{s_1 \left[\dfrac{(R_{sTH}s_1 + R_r')^2}{s_1^2} + (\omega_1 L_{s\ell TH} + \omega_1 L_{r\ell}')^2\right]} \tag{3-4b}$$

or

$$T = \frac{1}{\omega_{s1}} \frac{q_1 V_{s1TH}^2 R_r'}{\dfrac{(R_{sTH}s_1 + R_r')^2}{s_1} + (\omega_1 L_{s\ell TH} + \omega_1 L_{r\ell}')^2 s_1} \qquad (3\text{-}4c)$$

or

$$T = \frac{1}{\omega_{s1}} \frac{q_1 V_{s1TH}^2 R_r' s_1}{(R_{sTH}s_1 + R_r')^2 + (\omega_1 L_{s\ell TH} + \omega_1 L_{r\ell}')^2 s_1^2}. \qquad (3\text{-}4d)$$

For small values of $s_1$ we have

$$T \approx \frac{1}{\omega_{s1}} \left(\frac{q_1 V_{s1TH}^2 s_1}{R_r'}\right) \qquad s_1 = \text{small}. \qquad (3\text{-}5)$$

This relation is called the small-slip approximation for the torque.

Based on Eq. 3-4, the fundamental torque-speed characteristic of a three-phase induction machine is shown in Fig. 3.4.

Mechanical output power (neglecting iron-core losses, friction, and windage losses) is

$$P_{out} = T \cdot \omega_m. \qquad (3\text{-}6)$$

The air-gap power is (Fig. 3.5)

$$P_{gap} = q_1 |\tilde{I}'_{r1}|^2 \frac{R'_r}{s_1}, \qquad (3\text{-}7)$$

where $|\tilde{I}'_{r1}| = |\tilde{I}_{s1TH}|$.

The loss distribution within the machine is given in Fig. 3.6. Note that

$$P_{out} = P_{gap} - P_{rloss} = P_{in} - P_{sloss} - P_{rloss}$$

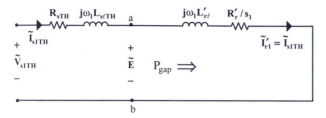

**FIGURE 3.5** Definition of air-gap power.

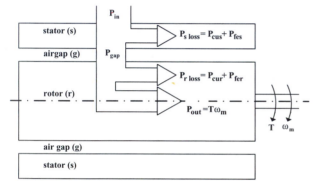

**FIGURE 3.6** Loss breakdown within induction machine neglecting frictional and windage losses.

and

$$P_{out} = T \cdot \omega_m.$$

The complete fundamental torque-speed relation detailing braking, motoring, and generation regions is shown in Fig. 3.7, where $n_{s1}$ denotes the synchronous (mechanical) speed of the fundamental rotating magnetic field.

### 3.1.1 Application Example 3.1: Steady-State Operation of Induction Motor at Undervoltage

A $P_{out} = 100$ hp, $V_{L\text{-}L\_rat} = 480$ V, $f = 60$ Hz, $p = 6$ pole, three-phase induction motor runs at full load and rated voltage with a slip of 3%. Under conditions of stress on the power system, the line-to-line voltage drops to $V_{L\text{-}L\_low} = 430$ V. If the load is of the constant torque type, compute for the lower voltage:

a) Slip $s_{low}$ (use small-slip approximation).
b) Shaft speed $n_{m\_low}$ in rpm.
c) Output power $P_{out\_low}$.
d) Rotor copper loss $P_{cur\_low} = I'^2_r R'_r$ in terms of the rated rotor copper loss at rated voltage.

### 3.1.2 Application Example 3.2: Steady-State Operation of Induction Motor at Overvoltage

Repeat Application Example 3.1 for the condition when the line-to-line voltage of the power supply system increases to $V_{L\text{-}L\_high} = 510$ V.

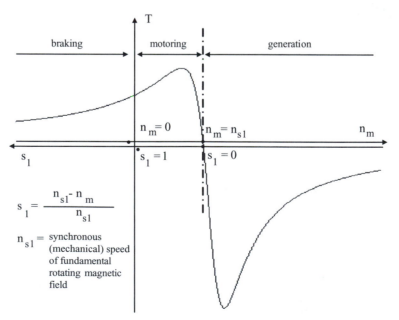

**FIGURE 3.7** Complete fundamental torque-speed relation detailing braking, motoring, and generation.

### 3.1.3 Application Example 3.3: Steady-State Operation of Induction Motor at Undervoltage and Under-Frequency

Repeat Application Example 3.1 for the condition when the line-to-line voltage of the power supply system and the frequency drop to $V_{L\text{-}L\_\text{low}} = 430$ V and $f_{\text{low}} = 59$ Hz, respectively.

## 3.2 MAGNETIC FIELDS OF THREE-PHASE MACHINES FOR THE CALCULATION OF INDUCTIVE MACHINE PARAMETERS

Numerical approaches such as the finite-difference and finite-element methods [16] enable engineers to compute no-load and full-load magnetic fields and those associated with short-circuit and starting conditions, as well as fields for the calculation of stator and rotor inductances/reactances. Figures 3.8 and 3.9

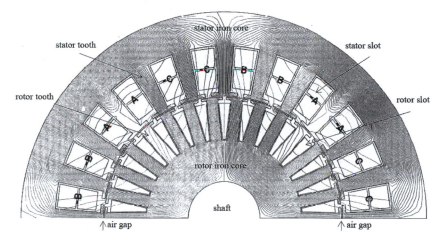

**FIGURE 3.8** No-load field of four-pole, 800 W, three-phase induction motor. One flux tube contains a flux per unit length of 0.0005 Wb/m.

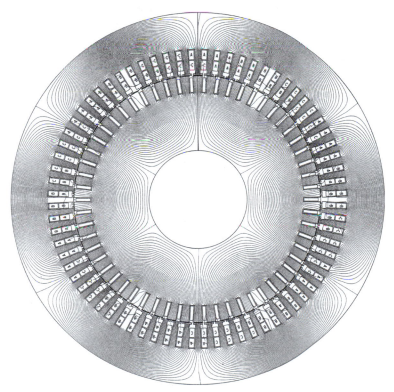

**FIGURE 3.9** No-load field of six-pole, 75 kW, three-phase induction motor. One flux tube contains a flux per unit length of 0.0006 Wb/m.

represent the no-load fields of four- and six-pole induction machines [17, 18]. Figure 3.10a–e illustrates radial forces generated as a function of the rotor position. Such forces cause audible noise and vibrations. The calculation of radial and tangential magnetic forces is discussed in Chapter 4 (Section 4.2.14), where the concept of the "Maxwell stress" is employed. Figures 3.11 to 3.13 represent unsaturated stator and rotor leakage fields and the associated field during starting of a two-pole induction motor. Figures 3.14 and 3.15 represent saturated stator and rotor leakage fields, respectively, and Fig. 3.16 depicts the associated field during starting of a two-pole induction machine. The starting current and starting

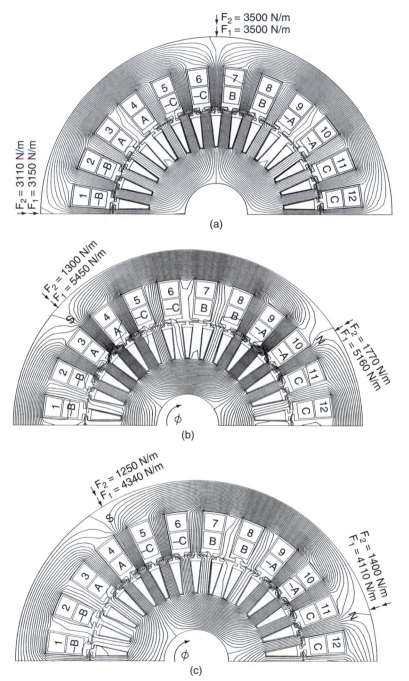

**FIGURE 3.10** Flux distribution and radial stator core forces at no load and rated voltage for (a) rotor position #1, (b) rotor position #2, (c) rotor position #3, (d) rotor position #4, and (e) rotor position #5.

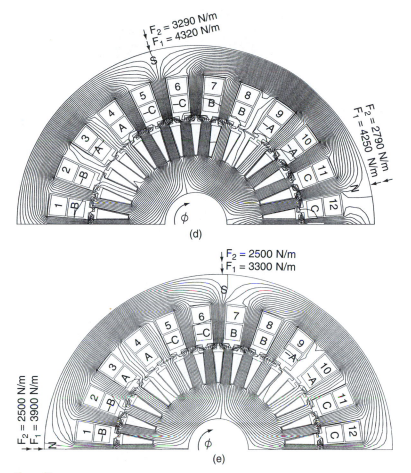

FIGURE 3.10 (continued)

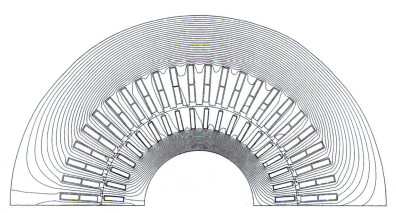

**FIGURE 3.11** Field for the determination of the unsaturated stator leakage flux of a 3.4 MW, two-pole, three-phase induction motor. One flux tube contains a flux per unit length of 0.005 Wb/m [19].

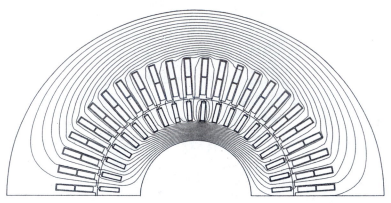

**FIGURE 3.12** Field for the determination of the unsaturated rotor leakage flux of a 3.4 MW, two-pole, three-phase induction motor. One flux tube contains a flux per unit length of 0.015 Wb/m [19].

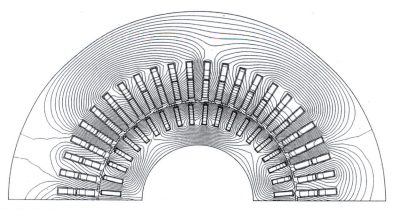

**FIGURE 3.13** Field distribution (first approximation) at starting with rated voltage of a 3.4 MW, two-pole, three-phase induction motor. One flux tube contains a flux per unit length of 0.005 Wb/m [19].

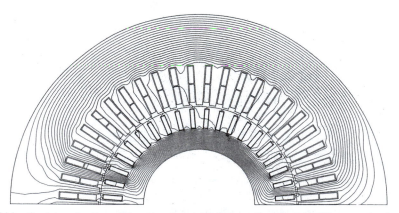

**FIGURE 3.14** Field for the determination of the saturated stator leakage flux of a 3.4 MW, two-pole, three-phase induction motor. One flux tube contains a flux per unit length of 0.15 Wb/m [19].

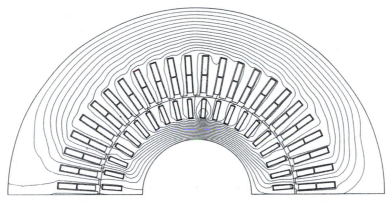

**FIGURE 3.15** Field for the determination of the saturated rotor leakage flux of a 3.4 MW, two-pole, three-phase induction motor. One flux tube contains a flux per unit length of 0.005 Wb/m [19].

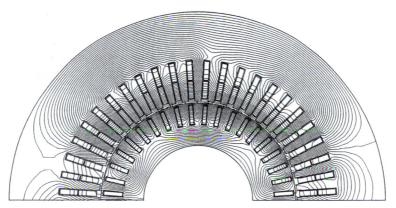

**FIGURE 3.16** Field distribution (second approximation) during starting with rated voltage of a 3.4 MW, two-pole, three-phase induction motor. One flux tube contains a flux per unit length of 0.005 Wb/m [19].

torque as a function of the terminal voltage are shown in Fig. 3.17 [19]. This plot illustrates how saturation influences the starting of an induction motor. Note that the linear (hand) calculation results in lower starting current and torque than the numerical solution.

Any rotating machine design is based on iterations. No closed form solution exists because of the nonlinearities (e.g., iron-core saturation) involved. In Fig. 3.13 the field for the first approximation, where saturation is neglected and a linear B–H characteristic is assumed, permits us to calculate stator and rotor currents for which the starting field can be computed under saturated conditions assuming a nonlinear B–H characteristic as depicted in Fig. 3.16. For the reluctivity distribution caused by the saturated short-circuit field the stator (Fig. 3.14) and rotor (Fig. 3.15) leakage reactances can be

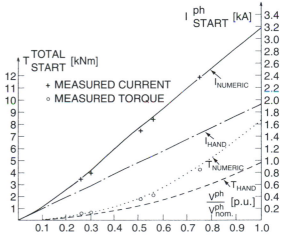

**FIGURE 3.17** Starting currents and torques as a function of terminal voltage for a 3.4 MW, three-phase, induction motor [19].

recomputed, leading to the second approximation as indicated in Fig. 3.16. In practice a few iterations are sufficient to achieve a satisfactory solution for the starting torque as a function of the applied voltage as illustrated in Fig. 3.17. It is well known that during starting saturation occurs only in the stator and rotor teeth and this is the reason why Figs. 3.13 and 3.16 are similar.

## 3.3 STEADY-STATE STABILITY OF A THREE-PHASE INDUCTION MACHINE

The steady-state stability criterion is [20]

$$\frac{\partial T_L}{\partial \omega_m} > \frac{\partial T}{\partial \omega_m}, \qquad (3\text{-}8)$$

where $T_L$ is the load torque and $T$ is the motor torque. Using this equation, the stable steady-state operating point of a machine for any given load-torque characteristic can be determined.

### 3.3.1 Application Example 3.4: Unstable and Stable Steady-State Operation of Induction Machines

Figure E3.4.1 shows the torque–angular velocity characteristic of an induction motor $(T - \omega_m)$ supplying a mechanical constant torque load $(T_L)$. Determine the stability of the operating points $Q_1$ and $Q_2$.

### 3.3.2 Application Example 3.5: Stable Steady-State Operation of Induction Machines

Figure E3.5.1 shows the torque–angular velocity relation $(T - \omega_m)$ of an induction motor supplying a mechanical load with nonlinear (parabolic) torque $(T_L)$. Determine the stability of the operating point $Q_1$.

### 3.3.3 Resolving Mismatch of Wind-Turbine and Variable-Speed Generator Torque–Speed Characteristics

It is well known that the torque–speed characteristic of wind turbines (Fig. 3.18) and commonly available variable-speed drives employing field weakening (Fig. 3.19a) do not match because the torque of a wind turbine is proportional to the square of the speed and that of a variable-speed drive is inversely proportional to the speed in the field-weakening region. One possibility to mitigate this mismatch is an electronic change in the number of poles p or a change of the number of series turns of the stator winding [21–23]. The change of the number of poles is governed by the relation

$$\frac{E \cdot p}{f} = 4.44 B_{max} \cdot N_{rat} \cdot 4 \cdot R \cdot L,$$

where $B_{max}$ is the maximum flux density at the radius $R$ of the machine, $N_{rat}$ is the rated number of series turns of a stator phase winding, and $L$ is the axial iron-core length of the machine.

The change of the number of turns N is governed by

$$\frac{E}{f \cdot N} = 4.44 B_{max} \cdot 4 \cdot R \cdot L/p.$$

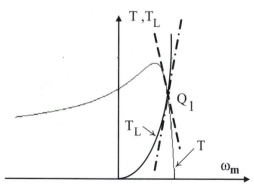

**FIGURE E3.5.1** Steady-state stability of induction machine with nonlinear load torque.

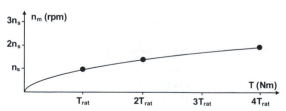

**FIGURE 3.18** Torque–speed characteristic of a wind turbine $T \propto n_m^2$.

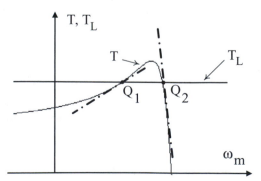

**FIGURE E3.4.1** Steady-state stability of induction machine with constant load torque.

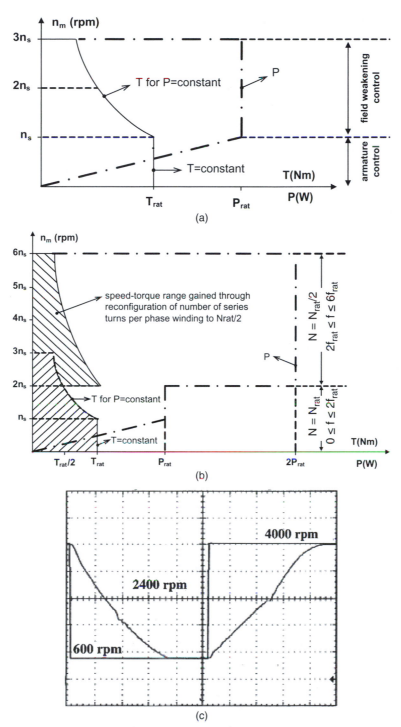

**FIGURE 3.19** (a) Variable-speed drive torque–speed characteristic without winding switching $T \propto 1/n_m$ in field-weakening region. The full line corresponds to the torque T and the dashed-dotted line to the power P. (b) Variable-speed drive torque–speed characteristic by changing the number of series turns of the stator winding from $N_{rat}$ to $N_{rat}/2$. The full line corresponds to the torque T and the dashed-dotted line to the power P. (c) Measured variable-speed drive response with winding switching from pole number $p_1$ to $p_2$ resulting in a speed range from 600 to 4000 rpm [21]. One horizontal division corresponds to 200 ms. One vertical division corresponds to 800 rpm.

The introduction of an additional degree of freedom – either by a change of p or N – permits an extension of the constant flux region. Adding this degree of freedom will permit wind turbines to operate under stalled conditions at all speeds, generating maximum power at a given speed with no danger of a runaway. This will make the blade-pitch control obsolete – however, a furling of the blades at excessive wind velocities must be provided – and wind turbines become more reliable and less expensive due to the absence of mechanical control. Figure 3.19b illustrates the speed and torque increase due to the change of the number of turns from $N_{rat}$ to $N_{rat}/2$, and Fig. 3.19c demonstrates the excellent dynamic performance of winding switching with solid-state switches from number of poles $p_1$ to $p_2$ [21]. The winding reconfiguration occurs near the horizontal axis (small slip) at 2400 rpm.

## 3.4 SPATIAL (SPACE) HARMONICS OF A THREE-PHASE INDUCTION MACHINE

Due to imperfect (e.g., nonsinusoidal) winding distributions and due to slots and teeth in stator and rotor, the magnetomotive forces (mmfs) of an induction machine are nonsinusoidal.

For sinusoidal distribution one obtains the mmfs

$$F_a = \{A_1 \cos\theta\} i_a$$
$$F_b = \{A_1 \cos(\theta - 120°)\} i_b. \qquad (3-9)$$
$$F_c = \{A_1 \cos(\theta - 240°)\} i_c$$

Note that $\cos(\theta - 240°) = \cos(\theta + 120°)$. For nonsinusoidal mmfs (consisting of fundamental, third, and fifth harmonics) one obtains

$$F_a = \{A_1 \cos\theta + A_3 \cos 3\theta + A_5 \cos 5\theta\} i_a(t)$$
$$F_b = \{A_1 \cos(\theta - 120°) + A_3 \cos 3(\theta - 120°) + A_5 \cos 5(\theta - 120°)\} i_b(t), \qquad (3-10)$$
$$F_c = \{A_1 \cos(\theta - 240°) + A_3 \cos 3(\theta - 240°) + A_5 \cos 5(\theta - 240°)\} i_c(t)$$

where

$$i_a(t) = I_m \cos \omega_1 t$$
$$i_b(t) = I_m \cos(\omega_1 t - 120°). \qquad (3-11)$$
$$i_c(t) = I_m \cos(\omega_1 t - 240°)$$

The magnetomotive force (mmf) originates in the phase belts a–a′, b–b′, and c–c′ as shown in Fig. 3.20. The total mmf is

$$F_{tot}(\theta, t) = F_a + F_b + F_c, \qquad (3-12)$$

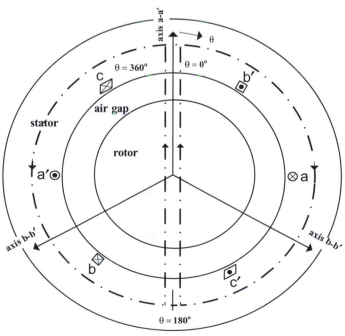

**FIGURE 3.20** Definition of phase belts and their associated axes in an induction machine.

where

$$F_a = \{A_1 \cos\theta + A_3 \cos 3\theta + A_5 \cos 5\theta\} I_m \cos\omega_1 t \quad (3\text{-}13a)$$

$$F_a = A_1 I_m \cos\theta \cos\omega_1 t + A_3 I_m \cos 3\theta \cos\omega_1 t + A_5 I_m \cos 5\theta \cos\omega_1 t$$

Note that $\cos x \cos y = \tfrac{1}{2}\{\cos(x-y) + \cos(x+y)\}$; therefore,

$$F_a = \frac{A_1 I_m}{2}\{\cos(\theta - \omega_1 t) + \cos(\theta + \omega_1 t)\}$$
$$+ \frac{A_3 I_m}{2}\{\cos(3\theta - \omega_1 t) + \cos(3\theta + \omega_1 t)\} \quad (3\text{-}13b)$$
$$+ \frac{A_5 I_m}{2}\{\cos(5\theta - \omega_1 t) + \cos(5\theta + \omega_1 t)\}.$$

Correspondingly,

$$F_b = \frac{A_1 I_m}{2}\{\cos(\theta - \omega_1 t) + \cos(\theta + \omega_1 t - 240°)\}$$
$$+ \frac{A_3 I_m}{2}\{\cos(3\theta - \omega_1 t - 240°) + \cos(3\theta + \omega_1 t$$
$$- 480°)\} + \frac{A_5 I_m}{2}\{\cos(5\theta - \omega_1 t - 480°)$$
$$+ \cos(5\theta + \omega_1 t - 720°)\}.$$
$$(3\text{-}13c)$$

$$F_c = \frac{A_1 I_m}{2}\{\cos(\theta - \omega_1 t) + \cos(\theta + \omega_1 t + 240°)\}$$
$$+ \frac{A_3 I_m}{2}\{\cos(3\theta - \omega_1 t + 240°) + \cos(3\theta + \omega_1 t$$
$$+ 480°)\} + \frac{A_5 I_m}{2}\{\cos(5\theta - \omega_1 t + 480°)$$
$$+ \cos(5\theta + \omega_1 t + 720°)\}.$$
$$(3\text{-}13d)$$

Therefore, with Eq. 3-12 the total mmf is simplified:

$$F_{\text{tot}} = \frac{3}{2} A_1 I_m \cos(\omega_1 t - \theta) + \frac{3 A_5}{2} I_m \cos(\omega_1 t + 5\theta) \quad (3\text{-}14)$$

The angular velocity of the fundamental mmf $\dfrac{d\theta_1}{dt}$ is

$$\omega_1 \cdot t - \omega_1 = 0,$$

$$\theta_1 = \omega_1 \cdot t,$$

$$\frac{d\theta_1}{dt} = \omega_1$$

and the angular velocity of the 5th space harmonic mmf $\dfrac{d\theta_5}{dt}$ is

$$\omega_1 \cdot t + 5\theta_5 = 0,$$

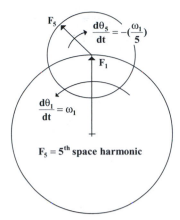

**FIGURE 3.21** Fundamental and 5th spatial harmonic of an induction machine rotating in opposite directions (see Eq. 3-14). Note: $F_1$ is rotating in forward (+) or counterclockwise direction and $F_5$ is rotating in backward (–) or clockwise direction.

**TABLE 3.1** Positive-, Negative-, and Zero-Sequence Spatial Harmonics

| Spatial-harmonic sequence | + | – | 0 |
|---|---|---|---|
| Spatial-harmonic order | 1 | 2 | 3 |
| | 4 | 5 | 6 |
| | 7 | 8 | 9 |
| | 10 | 11 | 12 |
| | 13 | 14 | 15 |
| | . | . | . |

$$5\theta_5 = -\omega_1 \cdot t,$$

$$\frac{d\theta_5}{dt} = -\frac{\omega_1}{5}$$

Graphical representation of spatial harmonics (in the presence of fundamental, third, and fifth harmonics mmfs) with phasors is shown in Fig. 3.21. Note that the 3rd harmonic mmf cancels (is equal to zero).

Similar analysis is performed to determine the rotating directions of each individual space harmonic, and the positive-, negative-, and zero-sequence harmonic orders can be defined, as listed in Table 3.1.

Note that:

***Even and triplen harmonics are normally not present in a balanced three-phase system!***

In general one can write for spatial harmonics

$$\frac{d\theta_h}{dt} = \left(\begin{matrix} + \\ - \\ 0 \end{matrix}\right) \frac{\omega_1}{h}, \quad (3\text{-}15)$$

where +, –, and 0 are used for positive-, negative-, and zero-sequence space harmonic orders,

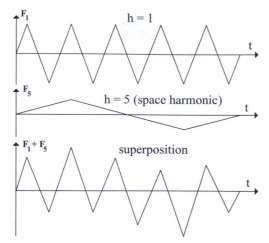

**FIGURE 3.22** Superposition of fundamental mmf with 5th space harmonic resulting in amplitude modulation of fundamental of 60 Hz with 12 Hz.

respectively. Figure 3.22 illustrates the superposition of fundamental mmf with 5th space harmonic resulting in amplitude modulation where fundamental of 60 Hz is modulated with 12 Hz.

## 3.5 TIME HARMONICS OF A THREE-PHASE INDUCTION MACHINE

A three-phase induction machine is excited by balanced three-phase $f_1 = 60$ Hz currents containing a fifth time harmonic. The equations of the currents are for $\omega_1 = 2\pi f_1$:

$$i_a(t) = \{I_{m1} \cos \omega_1 t + I_{m5} \cos 5\omega_1 t\}$$

$$i_b(t) = \{I_{m1} \cos(\omega_1 t - 120°) + I_{m5} \cos(5\omega_1 t - 120°)\}. \tag{3-16}$$

$$i_c(t) = \{I_{m1} \cos(\omega_1 t + 120°) + I_{m5} \cos(5\omega_1 t + 120°)\}$$

Note that the three-phase system rotates in a clockwise (cw) manner; that is, in a mathematically negative sense. Assume that the winding has been designed to eliminate all spatial harmonics. Thus for phase a the mmf becomes

$$F_a = A_1 \cos \theta i_a(t) \tag{3-17}$$

Correspondingly,

$$F_a = A_1 \cos(\theta)\{I_{m1} \cos(\omega_1 t) + I_{m5} \cos(5\omega_1 t)\}$$

$$F_b = A_1 \cos(\theta - 120°)\{I_{m1} \cos(\omega_1 t - 120°) + I_{m5} \cos(5\omega_1 t - 120°)\}. \tag{3-18}$$

$$F_c = A_1 \cos(\theta + 120°)\{I_{m1} \cos(\omega_1 t + 120°) + I_{m5} \cos(5\omega_1 t + 120°)\}$$

The total mmf is

$$F_{tot} = F_a + F_b + F_c, \tag{3-19}$$

expanded:

$$F_a = \left(\frac{A_1 I_{m1}}{2}\right)\cos(\theta - \omega_1 t) + \left(\frac{A_1 I_{m1}}{2}\right)\cos(\theta + \omega_1 t)$$
$$+ \left(\frac{A_1 I_{m5}}{2}\right)\cos(\theta - 5\omega_1 t) + \left(\frac{A_1 I_{m5}}{2}\right)$$
$$\cos(\theta + 5\omega_1 t),$$

$$F_b = \left(\frac{A_1 I_{m1}}{2}\right)\cos(\theta - \omega_1 t) + \left(\frac{A_1 I_{m1}}{2}\right)$$
$$\cos(\theta + \omega_1 t - 240°) + \left(\frac{A_1 I_{m5}}{2}\right)\cos(\theta - 5\omega_1 t)$$
$$+ \left(\frac{A_1 I_{m5}}{2}\right)\cos(\theta + 5\omega_1 t - 240°),$$

$$F_c = \left(\frac{A_1 I_{m1}}{2}\right)\cos(\theta - \omega_1 t) + \left(\frac{A_1 I_{m1}}{2}\right)$$
$$\cos(\theta + \omega_1 t + 240°) + \left(\frac{A_1 I_{m5}}{2}\right)\cos(\theta - 5\omega_1 t)$$
$$+ \left(\frac{A_1 I_{m5}}{2}\right)\cos(\theta + 5\omega_1 t + 240°),$$

$$\tag{3-20}$$

or

$$F_{tot} = \left(\frac{A_1 I_{m1}}{2}\right)\cos(\theta - \omega_1 t) + \left(\frac{A_1 I_{m5}}{2}\right)\cos(\theta - 5\omega_1 t),$$

$$\tag{3-21}$$

The angular velocity of the fundamental $\dfrac{d\theta_1}{dt}$ is

$$\theta_1 - \omega_1 \cdot t = 0,$$

$$\frac{d\theta_1}{dt} = \omega_1,$$

and the angular velocity of the fifth time harmonic $\dfrac{d\theta_5}{dt}$ is

$$\theta_5 - 5\omega_1 \cdot t = 0,$$

$$\frac{d\theta_5}{dt} = 5\omega_1.$$

For the current system (consisting of the fundamental and 5th harmonic components where the 5th harmonic system rotates in counterclockwise direction)

$$i_a(t) = \{I_{m1} \cos \omega_1 t + I_{m5} \cos 5\omega_1 t\}$$

$$i_b(t) = \{I_{m1} \cos(\omega_1 t - 120°) + I_{m5} \cos(5\omega_1 t + 120°)\}, \tag{3-22}$$

$$i_c(t) = \{I_{m1} \cos(\omega_1 t + 120°) + I_{m5} \cos(5\omega_1 t - 120°)\}$$

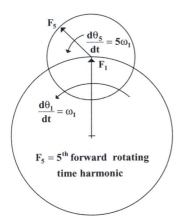

**FIGURE 3.23** Forward (+) rotating fundamental mmf of an induction machine superposed with forward (+) rotating 5th time harmonic rotating in the same direction.

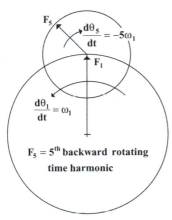

**FIGURE 3.24** Forward (+) rotating fundamental mmf of an induction machine superposed with backward (−) rotating 5th time harmonic rotating in the opposite direction.

the fifth time harmonic has the angular velocity

$$\frac{d\theta_5}{dt} = -5\omega_1.$$

Phasor representation for forward rotating fundamental and forward rotating 5th harmonic current systems is given in Fig. 3.23. Phasor representation for forward rotating fundamental and backward rotating 5th harmonic current systems is shown in Fig. 3.24.

Similar analysis can be performed to determine the rotating directions of each individual time harmonic and to define the positive-, negative-, and zero-sequence harmonic orders, as listed in Table 3.2. As with the space harmonics, even and triplen harmonics are normally not present, provided the system is balanced.

**TABLE 3.2** Positive-, Negative-, and Zero-Sequence Time Harmonics

| Time-harmonic sequence | + | − | 0 |
|---|---|---|---|
| Time-harmonic order | 1 | 2 | 3 |
| | 4 | 5 | 6 |
| | 7 | 8 | 9 |
| | 10 | 11 | 12 |
| | 13 | 14 | 15 |
| | ⋮ | ⋮ | ⋮ |

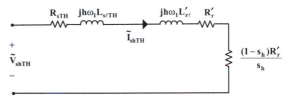

**FIGURE 3.25** Thevenin (TH) adjusted equivalent circuit of an induction machine for the hth harmonic. It is assumed that the resistive and inductive circuit parameters are independent of h.

In general one can write for time harmonics

$$\frac{d\theta_h}{dt} = \left(\begin{matrix}+\\-\\0\end{matrix}\right)h\omega_1, \tag{3-23}$$

where +, −, and 0 are used for positive-, negative-, and zero-sequence time harmonic orders, respectively.

Therefore, time harmonics voltages have an important impact on induction machines. Forward- and backward-rotating fields are produced by positive-, negative-, and zero-sequence harmonics that produce harmonic shaft torques.

### 3.6 FUNDAMENTAL AND HARMONIC TORQUES OF AN INDUCTION MACHINE

Starting with the Thevenin-adjusted circuit as illustrated in Fig. 3.25, one obtains the current of the hth harmonic

$$\tilde{I}_{shTH} = \frac{\tilde{V}_{shTH}}{\left(R_{sTH} + \dfrac{R'_r}{s_h}\right) + jh\omega_1(L_{s\ell TH} + L'_{r\ell})}. \tag{3-24}$$

Therefore, the torque for the hth harmonic is

$$T_{eh} = \frac{1}{\omega_{sh}} \frac{q_1 V_{shTH}^2 \dfrac{R'_r}{s_h}}{\left(R_{sTH} + \dfrac{R'_r}{s_h}\right)^2 + (h\omega_1)^2(L_{s\ell TH} + L'_{r\ell})^2}. \tag{3-25}$$

Similarly, the fundamental (h = 1) torque is

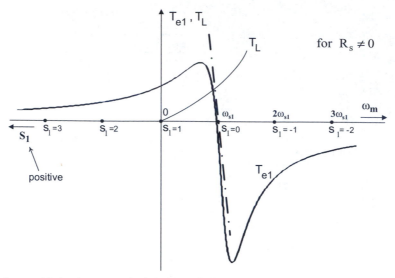

**FIGURE 3.26** Fundamental induction motor ($T_{e1}$) and load ($T_L$) torques as a function of angular velocity and slip $s_1$.

$$T_{e1} = \frac{1}{\omega_{s1}} \frac{q_1 V_{s1TH}^2 \dfrac{R_r'}{s_1}}{\left(R_{sTH} + \dfrac{R_r'}{s_1}\right)^2 + (\omega_1)^2 (L_{s\ell TH} + L_{r\ell}')^2}. \quad (3\text{-}26)$$

### 3.6.1 The Fundamental Slip of an Induction Machine

The fundamental slip is

$$s_1 = \frac{\omega_{s1} - \omega_m}{\omega_{s1}} \quad (3\text{-}27)$$

or

$$s_1 = 1 - \frac{\omega_m}{\omega_{s1}} \quad \Rightarrow \quad \frac{\omega_m}{\omega_{s1}} = 1 - s_1$$

where $\omega_{s1} = \dfrac{\omega_1}{p/2}$ is the (mechanical) synchronous fundamental angular velocity and $\omega_m$ is the mechanical angular shaft velocity.

The fundamental torque referred to fundamental slip $s_1$ is shown in Fig. 3.26 where $T_{e1}$ is the machine torque and $T_L$ is the load torque. If $R_s = 0$ then $T_{e1}$ is symmetric to the point at $(T_{e1} = 0/s_1 = 0)$ or $(T_{e1} = 0/\omega_{s1})$.

### 3.6.2 The Harmonic Slip of an Induction Machine

The harmonic slip (without addressing the direction of rotation of the harmonic field) is defined as

$$s_h = \frac{h\omega_{s1} - \omega_m}{h\omega_{s1}},$$

where $\omega_{s1} = (\omega_1)/(p/2)$ and $\omega_1$ is the electrical angular velocity, $\omega_1 = 2\pi f_1$ and $f_1 = 60$ Hz.

To include the direction of rotation of harmonic mmfs, in the following we assume that the fundamental rotates in forward direction, the 5th in backward direction, and the 7th in forward direction (see Table 3.2 for positive-, negative-, and zero-sequence components).

For motor operation $\omega_m < \omega_{s1}$; thus for the 5th harmonic component, one obtains

$$s_5 = \frac{-5\omega_{s1} - \omega_m}{-5\omega_{s1}},$$

where $(-5\omega_{s1})$ means rotation in backward direction, or

$$s_5 = \frac{5\omega_{s1} + \omega_m}{5\omega_{s1}}.$$

Note that in this equation $5\omega_{s1}$ is the base (reference) angular velocity.

Correspondingly, one obtains for the forward rotating 7th harmonic:

$$s_7 = \frac{7\omega_{s1} - \omega_m}{7\omega_{s1}},$$

where $7\omega_{s1}$ is the base.

Therefore, the harmonic slip is

$$s_h = \frac{h\omega_{s1} - (\pm)\omega_m}{h\omega_{s1}}, \quad \begin{array}{l} (+) : \text{for forward rotating field} \\ (-) : \text{for backward rotating field} \end{array}.$$

$$(3\text{-}28)$$

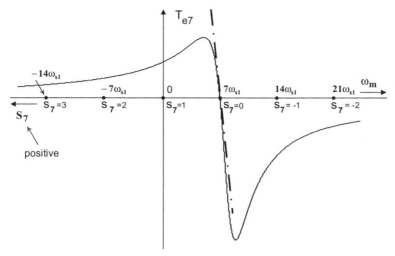

**FIGURE 3.27** Induction motor torque for 7th time harmonic as a function of angular velocity and slip $s_7$.

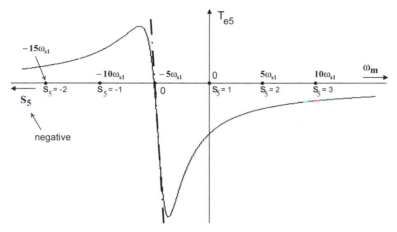

**FIGURE 3.28** Induction motor torque for 5th time harmonic as a function of angular velocity and slip $s_5$.

Harmonic torques $T_{e7}$ and $T_{e5}$ (as a function of the angular velocity $\omega_m$ and the harmonic slips $s_7$ and $s_5$) are depicted in Figs. 3.27 and 3.28, respectively.

### 3.6.3 The Reflected Harmonic Slip of an Induction Machine

If $\omega_{s1}$ is taken as the reference angular velocity, then for the backward rotating fifth harmonic one obtains

$$s_5^{(1)} = \frac{5\omega_{s1} + \omega_m}{\omega_{s1}}, \qquad (3\text{-}29)$$

where $s_5^{(1)}$ is the reflected fifth harmonic slip.

Correspondingly, if one takes $\omega_{s1}$ as reference then for the forward rotating seventh harmonic one gets

$$s_7^{(1)} = \frac{7\omega_{s1} - \omega_m}{\omega_{s1}},$$

where $s_7^{(1)}$ is the reflected seventh harmonic slip.

In general, the reflected harmonic slip is

$$s_h^{(1)} = \frac{h\omega_{s1} - (\pm)\omega_m}{\omega_{s1}}, \quad \begin{array}{l} (+): \text{for forward rotating field} \\ (-): \text{for backward rotating field} \end{array}.$$

$$(3\text{-}30)$$

To understand the combined effects of fundamental and harmonic torques, it is more convenient to plot harmonic torques ($T_{eh}$) as a function of the reflected harmonic slips ($s_h^{(1)}$):

- The harmonic torque $T_{e7}$ reflected to harmonic slip $s_7^{(1)} = \dfrac{7\omega_{s1} - \omega_m}{\omega_{s1}}$ is depicted in Fig. 3.29; and

- The harmonic torque $T_{e5}$ reflected to harmonic slip $s_5^{(1)} = \dfrac{5\omega_{s1} + \omega_m}{\omega_{s1}}$ is shown in Fig. 3.30.

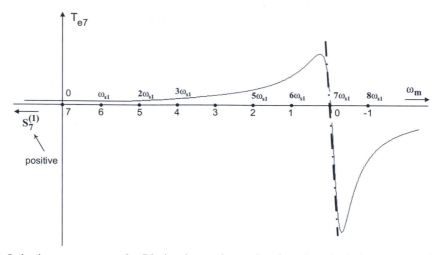

**FIGURE 3.29** Induction motor torque for 7th time harmonic as a function of mechanical angular velocity and reflected harmonic slip $s_7^{(1)}$.

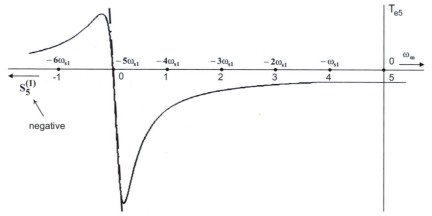

**FIGURE 3.30** Induction motor torque for 5th time harmonic as a function of mechanical angular velocity and reflected harmonic slip $s_5^{(1)}$.

### 3.6.4 Reflected Harmonic Slip of an Induction Machine in Terms of Fundamental Slip

The relation between $s_1$ (fundamental slip) and $s_h^{(1)}$ (reflected harmonic slip) is

$$s_h^{(1)} = \frac{h\omega_{s1} - \omega_m}{\omega_{s1}} = \frac{\omega_{s1} - \omega_m}{\omega_{s1}} + \frac{(h-1)\omega_{s1}}{\omega_{s1}}$$

$$= 1 - \frac{\omega_m}{\omega_{s1}} + h - 1 = h - (1 - s_1).$$

Therefore, the reflected harmonic slip, in terms of fundamental slip, is

$$s_h^{(1)} = s_1 + (h - 1). \qquad (3\text{-}31\text{a})$$

For the backward rotating 5th harmonic one gets

$$s_5^{(1)} = -s_1 + 6 \text{ thus } s_1 = -s_5^{(1)} + 6,$$

and for the forward rotating 7th harmonic

$$s_7^{(1)} = s_1 + 6 \text{ thus } s_1 = s_7^{(1)} - 6.$$

Note that

$$s_5^{(1)} = \frac{5\omega_{s1} + \omega_m}{\omega_{s1}}$$

In general, the reflected harmonic slip $s_h^{(1)}$ for the forward and the backward rotating harmonics is a linear function of the fundamental slip $s_1$:

$$s_h^{(1)} = s_1 + (h - 1) \text{ for forward – rotating field}$$
$$s_h^{(1)} = -s_1 + (h + 1) \text{ for backward – rotating field.}$$
$$(3\text{-}31\text{b})$$

Superposition of fundamental ($h = 1$), fifth harmonic ($h = 5$), and seventh harmonic ($h = 7$) torque $T_e = T_{e1} + T_{e5} + T_{e7}$ is illustrated in Fig. 3.31. At $\omega_m = \omega_{mrated}$ the total electrical torque $T_e$ is identical to the load torque $T_L$ or $T_{e1} + T_{e5} + T_{e7} = T_L$, where $T_{e1}$ is a motoring torque, $T_{e7}$ is a motoring torque, and $T_{e5}$ is a

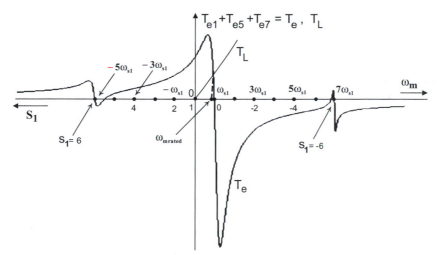

**FIGURE 3.31** Induction motor torque for fundamental, 5th, and 7th time harmonics as a function of angular velocity and slip $s_1$.

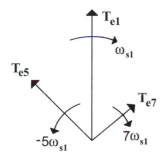

**FIGURE 3.32** Graphical phasor representation of superposition of torques as given in Fig. 3.31.

braking torque. Figure 3.32 shows the graphical phasor representation of superposition of torques as given in Fig. 3.31.

### 3.6.5 Reflected Harmonic Slip of an Induction Machine in Terms of Harmonic Slip

In this section, the relation between the reflected harmonic slip $s_h^{(1)}$ and the harmonic slip $s_h$ is determined as $s_h^{(1)} = f(s_h)$.

The harmonic slip (without addressing the direction of rotation of the harmonic field) is

$$s_h = \frac{h\omega_{s1} - \omega_m}{h\omega_{s1}} = \frac{(h-1)\omega_{s1}}{h\omega_{s1}} + \frac{\omega_{s1} - \omega_m}{h\omega_{s1}}$$

with

$$s_1 = \frac{\omega_{s1} - \omega_m}{\omega_{s1}}.$$

It follows that

$$s_h = \frac{(h-1)\omega_{s1}}{h\omega_{s1}} + \frac{s_1}{h}.$$

With $s_h^{(1)} = s_1 + (h-1)$, one obtains for forward rotating harmonics

$$s_1 = s_h^{(1)} - (h-1)$$

or

$$s_h = \frac{(h-1)\omega_{s1}}{h\omega_{s1}} + \frac{s_h^{(1)} - (h-1)}{h} = \frac{(h-1)}{h} + \frac{s_h^{(1)}}{h} - \frac{(h-1)}{h}$$

$$= \frac{s_h^{(1)}}{h}.$$

Similar analysis can be used for backward-rotating harmonics. Therefore, $s_h^{(1)}$ is a linear function of $s_h$:

$$s_h^{(1)} = h s_h \text{ for forward- and backward-rotating fields.} \tag{3-32}$$

### 3.7 MEASUREMENT RESULTS FOR THREE- AND SINGLE-PHASE INDUCTION MACHINES

Measurement techniques for the determination of the (copper and iron-core) losses of induction machines are described in [24]. The losses of harmonics of single-phase induction motors are measured in [25]. Typical input current wave shapes for single- and three-phase induction motors fed by thyristor converters are depicted in Figs. 3.33 and 3.34 [26]. In addition to semiconductor converters, harmonics can be generated by the DC magnetization of transformers [27]. A summary of the heating effects due to harmonics on induction machines is given in [28]. The generation of sub- and interharmonic torques by induction machines is presented in [29]. Power quality of machines relates to harmonics of the currents and voltages as well as to those of flux densities. The latter are more difficult to measure because sensors (e.g., search coils, Hall devices) must be implanted in the core. Nevertheless, flux density harmonics are important from an acoustic noise [30] and vibration [31–33] point of view. Excessive vibrations may lead to a reduction of lifetime of the machine. Single-phase induction machines are one of the more complicated electrical apparatus due to the existence

of forward- and backward-rotating fields. The methods as applied to these machines are valid for three-phase machines as well, where only forward fundamental rotating fields exist.

### 3.7.1 Measurement of Nonlinear Circuit Parameters of Single-Phase Induction Motors

The optimization of single- and three-phase induction machines is based on sinusoidal quantities neglecting the influence of spatial and time harmonics in flux densities, voltages, and currents [34, 35]. At nonsinusoidal terminal voltages, the harmonic losses of single-phase induction motors impact their efficiency [25]. The computer-aided testing of [24] will be relied on to measure nonlinear circuit parameters and flux densities in stator teeth and yokes as well as in the air gap of permanent-split-capacitor (PSC) motors. Detailed test results are presented in [36]. One concludes that tooth flux density wave shapes can be nearly sinusoidal although the exciting or magnetizing currents are nonsinusoidal.

A computer-aided testing (CAT) circuit and the computer-aided testing of electrical apparatus program (CATEA) are used in [24] to measure non-linear parameters. The CAT circuit is shown in Fig. 3.35, where four-channel signals $v_m$, $i_m$, $v_a$, and $i_a$ corresponding to main (m) and auxiliary (a) phase voltages and currents, respectively, are sampled by a personal computer via a 12-bit A/D converter.

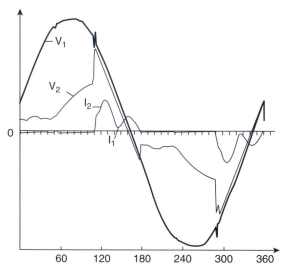

**FIGURE 3.33** Waveforms of input and output voltages and currents of a thyristor/triac controller feeding a 2 hp single-phase induction motor at no load. $V_1$, $I_1$ and $V_2$, $I_2$ are input voltage/current and output voltage/current of controller, respectively [26].

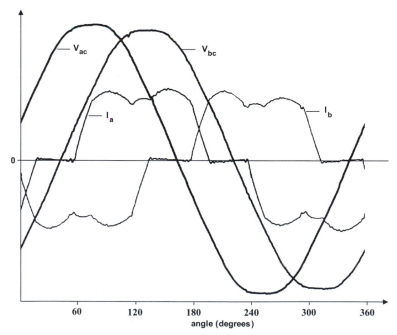

**FIGURE 3.34** Waveforms of input voltages $V_{ac}$, $V_{bc}$ and input currents $I_a$, $I_b$ of 7.5 hp three-phase induction motor at no load fed by a thyristor/triac controller [26].

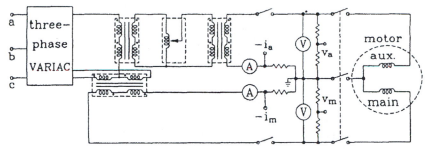

**FIGURE 3.35** CAT circuit for measuring of nonlinear circuit parameters of single-phase induction machines.

**TABLE 3.3** Three Fractional Horsepower PSC Single-Phase Induction Motors Serving as Prime Movers for Window Air Conditioners

| | |
|---|---|
| Single-phase motor R333MC | Capacitor: 15 $\mu$F, 370 V, output power: 1/2 hp |
| Single-phase motor Y673MG1 | Capacitor: 17.5 $\mu$F, 366 V, output power: 3/4 hp |
| Single-phase motor 80664346 | Capacitor: 25 $\mu$F, 370 V, output power: 1/2 hp |

Three (see Table 3.3) fractional horsepower PSC single-phase induction motors serving as prime movers for window air conditioners are tested and their parameters listed.

All motors to be tested are operated at minimum load, where input and output lines at the compressor are open to air. Measurements are conducted at an ambient temperature of 22°C. The linear stator leakage impedance of a motor is measured by removing the rotor from the stator bore and applying various terminal voltage amplitudes to the stator windings. When the main-phase parameters are measured, the switch leading to the auxiliary phase must be open, and vice versa. From the sampled data one obtains the fundamental voltages and currents of the main- and auxiliary-phase windings through Fourier analysis.

**Stator Impedance.** The stator input resistance and reactance are derived via linear regression. The stator leakage reactance is obtained by subtracting the reactance corresponding to leakage flux inside the bore of the motor from its input reactance [25].

**Rotor Impedance.** The rotor leakage impedance is frequency dependent and nonlinear because of the closed-rotor slots. The rotor impedance, referred to the main-phase winding, is measured by passing varying amplitudes of current – from about 5 to 150% of rated current – through the main-phase winding under locked-rotor conditions. The total

input resistance is derived from linear regression and the rotor resistance at rated frequency is obtained by subtracting the main-phase stator winding resistance from the total input resistance. The rotor resistance, taking the skin effect into account, can be obtained from [25]:

$$R_r = \phi(\xi)R_{r0}, \tag{3-33}$$

$$\phi(\xi) = \xi \frac{\sin h(\xi) + \sin(2\xi)}{\cos h(2\xi) - \cos(2\xi)}, \tag{3-34a}$$

where

$$\xi = k_r \sqrt{f}. \tag{3-34b}$$

In Eq. 3-34b, $f$ is the frequency of the rotor current. $k_r$ is the skin-effect coefficient derived from the height of the rotor slot $h_r$ and the resistivity of the bar $\rho_r$:

$$k_r = h_r \sqrt{\frac{\pi \mu_0}{\rho_r}}. \tag{3-34c}$$

In Eq. 3-33, $R_{r0}$ is the DC rotor resistance that can be derived from the rotor AC resistance at rated frequency and its skin-effect coefficient. The nonlinear rotor leakage reactance as a function of the amplitude of the rotor current – which can be described by a $(\lambda - i)$ characteristic – is obtained by subtracting the stator main-phase winding leakage reactance from the total input reactance. A linear rotor leakage reactance – which can be used as an approximation of the nonlinear rotor leakage reactance – is also derived from a linear regression.

**Magnetizing Impedance.** The magnetizing $(\lambda - i)$ characteristic referred to the stator main-phase winding is measured by supplying varying amplitudes of voltage to the main-phase winding at no-load operation. The forward induced voltage is determined by subtracting from the input voltage the voltage drop across the stator impedance and the backward induced voltage (the voltage drop across the rotor backward equivalent impedance); note that in [25, Fig. 7a], the main-phase forward circuit, and [25, Fig. 7c], the main-phase backward circuit, are connected in series, whereby $h = 1$. Thereafter, the magnetizing impedance is derived from the ratio of the forward induced voltage and the input current in the frequency (fundamental) domain. The magnetizing reactance as a function of the amplitude of the forward induced voltage is derived from the imaginary part of the (complex) magnetizing impedance.

The $(\lambda - i)$ characteristics are given in the frequency domain, which means that both $\lambda$ (flux linkages) and $i$ (current) have fundamental amplitudes only. The fundamental amplitudes of either flux linkages or currents can be derived from each other via such characteristics.

**Iron-Core Resistance.** The resistance corresponding to the iron-core loss is determined as follows. The iron-core and the mechanical frictional losses are computed by subtracting from the input power at no load

1. the copper loss of the main-phase winding at no load,
2. the backward rotor loss at no load, and
3. the forward air-gap power responsible for balancing the backward rotor loss.

Thereafter, the iron-core and the mechanical frictional losses, the total value of which is a linear function of the square of the forward induced voltage, are separated by linear regression. The equivalent resistance corresponding to the iron-core loss is derived from the slope of the regression line.

**Turns Ratio between the Turns of the Main- and Auxiliary-Phase Windings.** The ratio between the turns of the main- and auxiliary-phase windings $|\tilde{k}_{am}|$ is also an important parameter for single-phase induction motors. The main- and auxiliary-phase windings of some motors are not placed in spatial quadrature; therefore, an additional parameter $\theta$, the spatial angle between the axes of the two stator windings, must be determined. These two parameters ($|\tilde{k}_{am}|$, $\theta$) are found by measuring the main-phase induced voltage and current as well as the auxiliary-phase induced voltage when the motor is operated at no load. As has been mentioned above, the main-phase forward (f) and backward (b) induced voltages in the frequency domain, $\tilde{E}_f$ and $\tilde{E}_b$, respectively, can be obtained from the input voltage and current. If

$$\tilde{k}_{am} = |\tilde{k}_{am}|e^{j\theta} \qquad (3\text{-}35)$$

is used to represent this complex turns ratio, then the auxiliary-phase induced voltage is given by

$$\tilde{E}_a = \tilde{k}_{am}\tilde{E}_f + \tilde{k}_{am}^*\tilde{E}_b. \qquad (3\text{-}36)$$

Therefore, the complex ratio $|\tilde{k}_{am}|$ can be iteratively computed from

$$\tilde{k}_{am} = (\tilde{E}_a - \tilde{k}_{am}^*\tilde{E}_b)/\tilde{E}_f, \qquad (3\text{-}37)$$

where $\tilde{k}_{am}^*$ is the complex conjugate of $\tilde{k}_{am}$. Measured parameters of PSC motors are given in Section 3.7.2.

### 3.7.1.1 Measurement of Current and Voltage Harmonics

Voltage and current harmonics can be measured with state-of-the-art equipment [24–26]. Voltage and current waveshapes representing input and output voltages or currents of thyristor or triac controllers for induction motors are presented in Figs. 3.33 and 3.34. Here it is important that DC currents due to the asymmetric gating of the switches do not exceed a few percent of the rated controller current; DC currents injected into induction motors cause braking torques and generate additional losses or heating [26, 28, 41].

### 3.7.1.2 Measurement of Flux-Density Harmonics in Stator Teeth and Yokes (Back Iron)

Flux densities in the teeth of rotating machines are of the alternating type, whereas those in the yokes (back iron) are of the rotating type. The loss characteristics for both types are different and these can be requested from the electrical steel manufacturers. The motors to be tested are mounted on a testing frame, and measurements are performed under no load or minimum load with rated run capacitors. Four stator search coils with $N = 3$ turns each are employed to indirectly sense the flux densities of the teeth and yokes located in the axes of the main- and auxiliary-phase windings (see Fig. 3.36). The induced voltages in the four search coils are measured with two different methods: the computer sampling and the oscilloscope methods.

**Measurement via Computer Sampling.** The computer-aided testing circuit and program [24] are relied on to measure the induced voltages of the four search coils of Fig. 3.36. The CAT circuit is shown in

Figs. 3.36 and 3.37, where eight channel signals ($v_{in}$, $i_{in}$, $i_m$, $i_a$, $e_{mt}$, $e_{at}$, $e_{my}$, and $e_{ay}$ corresponding to input voltage and current, main- and auxiliary-phase currents, main- and auxiliary-phase tooth search-coil induced voltages, and main- and auxiliary-phase yoke (back iron)-search coil induced voltages, respectively), are sampled [36].

The flux linkages of the search coils are defined as

$$\lambda(t) = \int e(t)dt, \qquad (3\text{-}38)$$

and numerically obtained from

$$\lambda_0 = 0 \text{ and } \lambda_{i+1} = \lambda_i + \Delta t(e_{i+1} + e_i)/2,$$
$$\text{for } i = 0, 1, 2, \ldots, (n-1). \qquad (3\text{-}39)$$

The average or DC value is

$$\lambda_{ave} = \frac{1}{n}\left(\sum_{i=0}^{n-1} \lambda_i\right), \qquad (3\text{-}40)$$

or the AC values are

$$\lambda_i = \lambda_i - \lambda_{ave}, \qquad \text{for } i = 0, 1, 2, 3, \ldots, n, \quad (3\text{-}41)$$

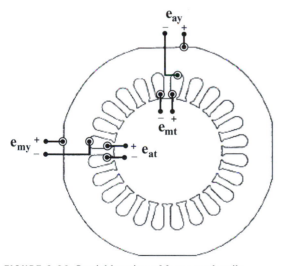

**FIGURE 3.36** Spatial location of four search coils.

where $e(t)$ of Eq. 3-38 is the induced voltage measured in a search coil and $n$ in Eqs. 3-39 to 3-41 is the number of sampled points. The maximum flux linkage $\lambda_{max}$ can be obtained from $\lambda_i$. The maximum flux density is given by

$$B_{max} = \lambda_{max}/(N \cdot s), \qquad (3\text{-}42)$$

where the number of turns of each search coil is $N = 3$ and $s$ is given for a stator tooth and yoke by the following equations:

$$s = k_{fe}w_t l, \qquad (3\text{-}43)$$

$$s = k_{fe}h_y l, \qquad (3\text{-}44)$$

where $w_t$, $h_y$, $l$, and $k_{fe}$ are the width of the (parallel) stator teeth (where two search coils reside), the height of the yoke or stator back iron (where two search coils reside), the length of the iron core, and the iron-core stacking factor, respectively. Measured data of PSC motors at rated voltage are presented in Section 3.7.2.

The maximum flux densities in stator teeth and yoke at the axes of the main- and auxiliary-phase windings are measured for various voltage amplitudes by recording the induced voltages in the four search coils. A digital oscilloscope is used to plot the induced voltages of the search coils. These waveforms are then sampled either by hand or by computer based on 83 points per period. Complex Fourier series components $\tilde{C}_h$ are computed – taking into account the Nyquist criterion – and the integration is performed in the complex domain as follows:

$$e(t) = \sum_{h=1}^{27} \tilde{C}_h e^{jh\omega t}, \qquad (3\text{-}45)$$

$$\lambda(t) = \sum_{h=1}^{27} \frac{1}{jh\omega} \tilde{C}_h e^{jh\omega t}. \qquad (3\text{-}46)$$

The maximum flux linkage $\lambda_{max}$ is found from $\lambda(t)$ and the maximum flux densities are determined from Eqs. 3-42 to 3-44.

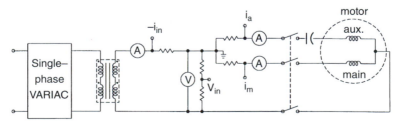

**FIGURE 3.37** CAT circuit for flux density measurement.

### 3.7.2 Application Example 3.6: Measurement of Harmonics within Yoke (Back Iron) and Tooth Flux Densities of Single-Phase Induction Machines

The three permanent-split capacitor (PSC) motors (R333MC, Y673MG1, and 80664346) of Table 3.3 are subjected to the tests described in the prior subsections of Section 3.7, and pertinent results will be presented. These include stator, rotor, and magnetizing parameters, and tables listing no-load input impedances of the main and auxiliary phases (Tables E3.6.1 and E3.6.2). The bold data of the tables correspond to rated main- and auxiliary-phase voltages at no load. The rated auxiliary-phase voltage is obtained by multiplying the rated main-phase voltage with the measured turns ratio $|\tilde{k}_{am}|$. $L_m$ and $L_a$ are inductances of the main- and auxiliary-phase windings, respectively.

### 3.8 INTER- AND SUBHARMONIC TORQUES OF THREE-PHASE INDUCTION MACHINES

In Sections 3.4 and 3.5 the magnetomotive forces (mmfs) of spatial and time harmonics including the

**TABLE E3.6.1** Main-Phase (m) Input Impedance at Single-Phase No-Load Operation as a Function of Input Voltage for R333MC

| $V_m$ (V) | $R_m$ (Ω) | $X_m$ (Ω) | $Z_m$ (Ω) | $L_m$ (mH) |
|---|---|---|---|---|
| 250.3 | 9.65 | 46.6 | 47.6 | 126 |
| **229.9** | **11.2** | **56.9** | **58.0** | **154** |
| 210.3 | 13.2 | 67.8 | 69.1 | 183 |
| 189.8 | 15.1 | 77.3 | 78.7 | 209 |
| 170.6 | 17.0 | 82.4 | 84.1 | 223 |
| 149.5 | 18.6 | 85.4 | 87.4 | 232 |
| 129.8 | 20.5 | 86.3 | 88.7 | 235 |
| 109.2 | 22.6 | 86.7 | 89.5 | 238 |
| 89.9 | 22.5 | 87.8 | 90.6 | 240 |
| 69.4 | 26.3 | 87.0 | 90.9 | 241 |

**TABLE E.3.6.2** Auxiliary-Phase (a) Input Impedance at Single-Phase No-Load Operation as a Function of Input Voltage for R333MC

| $V_a$ (V) | $R_a$ (Ω) | $X_a$ (Ω) | $Z_a$ (Ω) | $L_a$ (mH) |
|---|---|---|---|---|
| **298.88** | **30.6** | **94.7** | **99.5** | **264** |
| 279.9 | 33.6 | 106.0 | 111.2 | 295 |
| 259.3 | 36.8 | 117.9 | 123.5 | 328 |
| 239.8 | 39.1 | 126.2 | 132.1 | 351 |
| 219.4 | 41.1 | 132.5 | 138.7 | 368 |
| 199.0 | 43.8 | 135.6 | 142.5 | 378 |
| 179.3 | 45.6 | 137.4 | 144.7 | 384 |
| 160.0 | 46.7 | 138.6 | 146.3 | 388 |
| 139.1 | 49.9 | 139.0 | 147.6 | 392 |
| 120.8 | 50.3 | 140.2 | 148.9 | 395 |
| 100.1 | 54.3 | 139.3 | 149.5 | 397 |

fundamental fields were derived, respectively. It can be shown that harmonic fields (where $h$ = integer) generate very small asynchronous torques as compared with the fundamental torque. Sub- and interharmonic ($0 \le h \le 1$) torques, however, can generate parasitic torques near zero speed – that is, during the start-up of an induction motor. The formulas derived for harmonic torques are valid also for sub- and interharmonic torques. The following subsections confirm that sub- and interharmonic torques can have relatively large amplitudes if they originate from constant-voltage sources and not from constant-current sources.

### 3.8.1 Subharmonic Torques in a Voltage-Source-Fed Induction Motor

The equivalent circuit of a voltage-source-fed induction motor is shown in Fig. 3.38a for the fundamental ($h = 1$) and for the harmonic of order h. Using Thevenin's theorem (subscript TH), the magnetizing branch can be eliminated (Fig. 3.38b) and one obtains the following relation for the electrical torque of the hth harmonic where $V_{shTH} = |\tilde{V}_{shTH}|$:

$$T_{eh} = \frac{1}{\omega_{msh}} \frac{q_1 V_{shTH}^2 R_r'/s_h}{(R_{shTH} + R_r'/s_h)^2 + (h\omega_1(L_{s\ell TH} + L_{r\ell}'))^2}.$$

$$(3\text{-}47a)$$

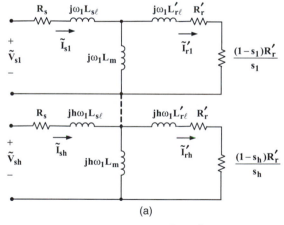

(a)

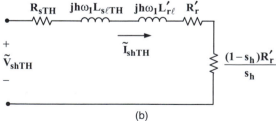

(b)

**FIGURE 3.38** (a) Per-phase equivalent circuit of three-phase induction machine for fundamental ($h = 1$) and hth harmonic fed by voltage source. (b) Thevenin (TH) adjusted per-phase equivalent circuit.

The Thevenin-adjusted machine parameters are

$$V_{shTH} = \frac{V_{sh}(h\omega_1 L_m)\angle\theta_{hTH}}{\sqrt{R_s^2 + (h\omega_1 L_{s\ell} + h\omega_1 L_m)^2}} \quad (3\text{-}47b)$$

with

$$\theta_{hTH} = \frac{\pi}{2} - \tan^{-1}\left[\frac{h\omega_1 L_{s\ell} + h\omega_1 L_m}{R_s}\right], \quad (3\text{-}47c)$$

$$R_{shTH} = \frac{h^2(\omega_1 L_m)^2 R_s}{R_s^2 + (h\omega_1 L_{s\ell} + h\omega_1 L_m)^2}, \quad (3\text{-}47d)$$

$$(h\omega_1 L_{s\ell TH}) =$$
$$\frac{h(\omega_1 L_m)R_s^2 + h^2(\omega_1 L_{s\ell})(\omega_1 L_m)[h(\omega_1 L_{s\ell}) + h(\omega_1 L_m)]}{R_s^2 + (h\omega_1 L_{s\ell} + h\omega_1 L_m)^2}.$$
$$(3\text{-}47e)$$

The mechanical synchronous angular velocity is

$$\omega_{msh} = \frac{h\omega_1}{(p/2)}, \quad (3\text{-}48)$$

where $\omega_1$ is the electrical synchronous angular velocity, $p$ is the number of poles, and $q_1$ is the number of phases (e.g., $q_1 = 3$).

In order to obtain the maximum or breakdown torque one matches the impedances of the Thevenin-adjusted equivalent circuit at $s_h = s_{h\max}$:

$$R_{rh}^2 = \left(\frac{R_r'}{s_{h\max}}\right)^2 = R_{shTH}^2 + [h\omega_1(L_{s\ell TH} + L_{r\ell}')]^2, \quad (3\text{-}49a)$$

and the breakdown slip for the hth harmonic becomes

$$s_{h\max} = \frac{R_r'}{\pm\sqrt{R_{shTh}^2 + (h\omega_1(L_{s\ell TH} + L_{r\ell}'))^2}}. \quad (3\text{-}49b)$$

The maximum motoring (*mot*) positive torque and the maximum generator (*gen*) negative torque for the time harmonic of hth order are

$$T_{eh\max}^{mot} = \frac{1}{\omega_{msh}} \frac{q_1 V_{shTH}^2 R_r'/|s_{h\max}|}{(R_{shTH} + R_r'/|s_{h\max}|)^2 + (h\omega_1(L_{s\ell TH} + L_{r\ell}'))^2}, \quad (3\text{-}50a)$$

$$T_{eh\max}^{gen} = -\frac{1}{\omega_{msh}} \frac{q_1 V_{shTH}^2 R_r'/|s_{h\max}|}{(R_{shTH} - R_r'/|s_{h\max}|)^2 + (h\omega_1(L_{s\ell TH} + L_{r\ell}'))^2}. \quad (3\text{-}50b)$$

The generator (*gen*) electrical torque reaches its absolute (*abs*) maximum, if

$$R_{rhTH} = R_r'/|s_{h\max}|, \quad (3\text{-}51)$$

that is,

$$T_{eh\max}^{gen\,abs} = -\frac{1}{\omega_{msh}} \frac{q_1 V_{shTH}^2 R_r'/|s_{h\max}|}{(h\omega_1(L_{s\ell TH} + L_{r\ell}'))^2}$$
$$= -\frac{(p/2)}{(h\omega_1)} \frac{q_1 V_{shTH}^2 R_{rhTH}}{(h\omega_1(L_{s\ell TH} + L_{r\ell}'))^2}. \quad (3\text{-}52)$$

### 3.8.2 Subharmonic Torques in a Current-Source-Fed Induction Motor

The equivalent circuits of a current-source-fed induction motor are shown in Fig. 3.39a,b for the fundamental ($h = 1$) and for the harmonic of order h.

The torque of the hth time harmonic is, from Fig. 3.39b,

$$T_{eh} = \frac{1}{\omega_{msh}} \frac{q_1 V_{sh}^2 R_r'/s_h}{(R_r'/s_h)^2 + (h\omega_1(L_m + L_{r\ell}'))^2}, \quad (3\text{-}53)$$

where $\tilde{V}_{sh} = jh\omega_1 L_m \tilde{I}_{sh}$.

The breakdown slip is

$$s_{h\max} = \pm R_r'/(h\omega_1(L_m + L_{r\ell}')). \quad (3\text{-}54)$$

Two application examples will be presented where an induction motor is first fed by a voltage

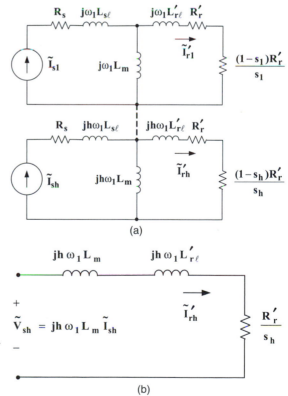

(a)

(b)

**FIGURE 3.39** (a) Per-phase equivalent circuit of three-phase induction machine for fundamental ($h = 1$) and hth harmonic fed by current source. (b) Equivalent circuit of three-phase induction machine for the hth harmonic taking into account the magnetizing branch.

source and then by a current source containing a forward rotating subharmonic component of 6 Hz ($h = 0.1$).

### 3.8.3 Application Example 3.7: Computation of Forward-Rotating Subharmonic Torque in a Voltage-Source-Fed Induction Motor

Terminal voltages of a three-phase induction motor contain the forward-rotating subharmonic of 6 Hz at an amplitude of $V_{0.1} = 5\%$. The fundamental voltage is $V_1 = 240$ V and the parameters of a three-phase induction machine are as follows:

$$240/416 \text{ V}, \Delta/Y \text{ connected}, f = 60 \text{ Hz},$$
$$n_{\text{rated}} = 1738 \text{ rpm}, p = 4 \text{ poles},$$

$$R_s = 7.0 \ \Omega, \omega_1 L_{sl} = 8 \ \Omega, \omega_1 L_m = 110 \ \Omega,$$
$$R'_r = 5 \ \Omega, \omega_1 L'_{r\ell} = 7 \ \Omega. \qquad \text{(E3.7-1)}$$

Compute and plot the sum of the fundamental and subharmonic torques ($T_{e1} + T_{e0.1}$) as a function of the fundamental slip $s_1$.

### 3.8.4 Application Example 3.8: Rationale for Limiting Harmonic Torques in an Induction Machine

Based on results of the previous application example, explain why low-frequency subharmonic voltages at the terminal of induction machines should be limited.

### 3.8.5 Application Example 3.9: Computation of Forward-Rotating Subharmonic Torque in Current-Source-Fed Induction Motor

Terminal currents of a three-phase induction motor contain a forward-rotating subharmonic of 6 Hz at an amplitude $I_{0.1} = 5\%$ of the fundamental $I_1 = 1.73$ A. The parameters of a three-phase induction machine are as follows:

$$3 \text{ A}/1.73 \text{ A}, \Delta/Y \text{ connected}, f = 60 \text{ Hz},$$
$$n_{\text{rat}} = 1773 \text{ rpm}, p = 4 \text{ poles},$$

$$R_s = 7.0 \ \Omega, \omega_1 L_{sl} = 5 \ \Omega, \omega_1 L_m = 110 \ \Omega,$$
$$R'_r = 5 \ \Omega, \omega_1 L'_{r\ell} = 4 \ \Omega. \qquad \text{(E3.9-1)}$$

Compute the fundamental torque ($T_{e1}$) as a function of the fundamental slip ($s_1$) and subharmonic torque ($T_{e0.1}$) as a function of the subharmonic slip ($s_{0.1}$).

### 3.9 INTERACTION OF SPACE AND TIME HARMONICS OF THREE-PHASE INDUCTION MACHINES

For mmfs with a fundamental and a spatial harmonic of order H one obtains

$$F_a = \{A_1 \cos \theta + A_H \cos H\theta\} i_a(t)$$
$$F_b = \{A_1 \cos(\theta - 120°) + A_H \cos H(\theta - 120°)\} i_b(t) \qquad \text{(3-55)}$$
$$F_c = \{A_1 \cos(\theta - 240°) + A_H \cos H(\theta - 240°)\} i_c(t),$$

where the currents consist of a fundamental and a time harmonic of order $h$:

$$i_a(t) = \{I_{m1} \cos \theta + I_{mh} \cos(h\omega_1 t)\}$$
$$i_b(t) = \{I_{m1} \cos(\theta - 120°) + I_{mh} \cos(h\omega_1 t - 120°)\} \qquad \text{(3-56)}$$
$$i_c(t) = \{I_{m1} \cos(\theta - 240°) + I_{mh} \cos(h\omega_1 t - 240°)\},$$

In Eq. 3-55, $A_1$, $A_H$, $\theta$, and H are fundamental mmf amplitude, harmonic mmf amplitude, mechanical (space) angular displacement, and spatial harmonic order, respectively. In Eq. 3-56, $I_{m1}$, $I_{mh}$, $\omega_1$, and $h$ are fundamental current amplitude, harmonic current amplitude, electrical angular velocity, and (time) harmonic order, respectively. The total mmf becomes $F_{\text{total}} = F_a + F_b + F_c$,

$$
\begin{aligned}
F_{\text{total}} = {} & \frac{3A_1 I_{m1}}{2}\cos(\theta - \omega_1 t) + \frac{3A_1 I_{mh}}{2}\cos(\theta - h\omega_1 t) \\
& + \frac{A_H I_{m1}}{2}\{\cos(H\theta - \omega_1 t) \\
& + \cos(H\theta + (1-H)120° \\
& - \omega_1 t) + \cos(H\theta - (1-H)120° - \omega_1 t)\} \\
& + \frac{A_H I_{m1}}{2}\{\cos(H\theta + \omega_1 t) \\
& + \cos(H\theta - (1+H)120° \\
& + \omega_1 t) + \cos(H\theta + (1+H)120° + \omega_1 t)\} \\
& + \frac{A_H I_{mh}}{2}\{\cos(H\theta - h\omega_1 t) \\
& + \cos(H\theta + (1-H)120° \\
& - h\omega_1 t) + \cos(H\theta - (1-H)120° - h\omega_1 t)\} \\
& + \frac{A_H I_{mh}}{2}\{\cos(H\theta + h\omega_1 t) \\
& + \cos(H\theta - (1+H)120° \\
& + h\omega_1 t) + \cos(H\theta + (1+H)120° + h\omega_1 t)\}
\end{aligned}
\tag{3-57}
$$

The evaluation of Eq. 3-57 depends on the amplitudes $A_1$, $A_H$, $I_{m1}$, and $I_{mh}$ as well as on the harmonic orders $h$ and $H$.

### 3.9.1 Application Example 3.10: Computation of Rotating MMF with Time and Space Harmonics

Compute the mmf and its angular velocities for an induction machine with the following time and space harmonics:

(a) $A_1 = 1$ pu, $A_H = 0.05$ pu, $I_{m1} = 1$ pu, $I_{mh} = 0.05$ pu, $H = 5$, and $h = 5$.

(b) $A_1 = 1$ pu, $A_H = 0.05$ pu, $I_{m1} = 1$ pu, $I_{mh} = 0.05$ pu, $H = 13$, and $h = 0.5$.

### 3.9.2 Application Example 3.11: Computation of Rotating MMF with Even Space Harmonics

Nonideal induction machines may generate even space harmonics. For $A_1 = 1$ pu, $A_H = 0.05$ pu, $I_{m1} = 1$ pu, $I_{mh} = 0.05$ pu, $H = 12$, and $h = 0.1$, compute the resulting rotating mmf and its angular velocities.

### 3.9.3 Application Example 3.12: Computation of Rotating MMF with Noninteger Space Harmonics

Nonideal induction machines may generate noninteger space harmonics. For $A_1 = 1$ pu, $A_H = 0.05$ pu, $I_{m1} = 1$ pu, $I_{mh} = 0.05$ pu, $H = 13.5$, and $h = 0.1$, compute the resulting rotating mmf and its angular velocities.

Application Examples 3.10 to 3.12 illustrate how sub- and noninteger harmonics are generated by induction machines. Such harmonics may cause the malfunctioning of protective (e.g., under-frequency) relays within a power system [42].

The calculation of the asynchronous torques due to harmonics and inter- and subharmonics is based on the equivalent circuit without taking into account the variation of the differential leakage as a function of the numbers of stator slots and rotor slots. It is well known that certain stator and rotor slot combinations lead to asynchronous and synchronous harmonic torques in induction machines. A more comprehensive treatment of such parasitic asynchronous and synchronous harmonic torques is presented in [30, 43–69]. Speed variation of induction motors is obtained either by voltage variation with constant frequency or by variation of the voltage and the fundamental frequency using inverters as a power source. Inverters have somewhat nonsinusoidal output voltages or currents. Thus parasitic effects like additional losses, dips in the torque–speed curve, output power derating, magnetic noise, oscillating torques, and harmonic line currents are generated. It has already been shown [43–45] that the multiple armature reaction has to be taken into account if effects caused by delta-connection of the stator windings, parallel winding branches, stator current harmonics, or damping of the air-gap field are to be considered. Moreover, the influence of the slot openings [46] and interbar currents [30, 47, 48, 68] in case of slot skewing is important. Compared with conventional methods [25, 30,

68] the consideration of multiple armature reactions requires additional work consisting of the solution of a system of equations of the order from 14 to 30 for a three-phase induction machine and the summation of the air-gap inductances. The presented theory [69] is based on Fourier analysis and does not employ any transformation. As an alternative to the analytical method described [69] two-dimensional numerical methods (e.g., finite-element method) with time stepping procedures are available [49–51]. These have the advantages of analyzing transients, variable permeability, and eddy currents in solid iron parts, but have the disadvantages of grid generation, discretization errors, and preparatory work and require large computing CPU time.

## 3.10 CONCLUSIONS CONCERNING INDUCTION MACHINE HARMONICS

One can draw the following conclusions:

- Three-phase induction machines can generate asynchronous as well as synchronous torques depending on their construction (e.g., number of stator and rotor slots) and the harmonic content of their power supplies [25–69].
- The relations between harmonic slip $s_h$ and the harmonic slip referred to the fundamental slip $s_h^{(1)}$ are the same for forward- and backward-rotating harmonic fields.
- Harmonic torques $T_{eh}$ (where $h$ = integer) are usually very small compared to the rated torque $T_{e1}$.
- Low-frequency and subharmonic torques $(0 \leq h \leq 1)$ can be quite large even for small low-frequency and subharmonic voltage amplitudes (e.g., 5% of fundamental voltage). These low-frequency or subharmonics may occur in the power system or at the output terminals of voltage-source inverters.
- Low-frequency and subharmonic torques $(0 \leq h \leq 1)$ are small even for large low-frequency harmonic current amplitudes (e.g., 5% of fundamental current) as provided by current sources.
- References [25], [28], and [39–41] address the heating of induction machines due to low-frequency and subharmonics, and reference [42] discusses the influence of inter- and subharmonics on under-frequency relays, where the zero crossing of the voltage is used to sense the frequency. Because of these negative influences the low-frequency, inter- and subharmonic voltage amplitudes generated in the power system due to inter- and subharmonic currents must be limited to below 0.1% of the rated voltage.

- Time and spatial harmonic voltages or currents may generate sufficiently large mmfs to cause significant harmonic torques near zero speed.
- The interaction of time and space harmonics generates sufficiently small mmfs so that start-up of an induction motor will not be affected.

## 3.11 VOLTAGE-STRESS WINDING FAILURES OF AC MOTORS FED BY VARIABLE-FREQUENCY, VOLTAGE- AND CURRENT-SOURCE PWM INVERTERS

Motor winding failures have long been a perplexing problem for electrical engineers. There are a variety of different causes for motor failure: heat, vibration, contamination, defects, and voltage stress. Most of these are understood and can be controlled to some degree by procedure and/or design. Voltage (stress) transients, however, have taken on a new dimension with the development of variable-frequency converters (VFCs), which appear to be more intense than voltage stress on the motor windings caused by motor switching or lightning. This section will use induction motors as the baseline for explaining this phenomenon. However, it applies to transformers and synchronous machines as well.

Control and efficiency advantages of variable-frequency drives (VFDs) based on power electronic current- or voltage-source inverters lead to their prolific employment in utility power systems. The currents and voltages of inverters are not ideally sinusoidal but have small current ripples that flow in cables, transformers, and motor or generator windings. As a result of cable capacitances and parasitic capacitances this may lead to relatively large voltage spikes (see Application Example 3.14). Most short-circuits are caused by a breakdown of the insulation and thus are initiated by excessive voltage stress over some period of time. This phenomenon must be analyzed using traveling wave theory because the largest voltage stress does not necessarily appear at the input terminal but can occur because of the backward (reflected) traveling wave somewhere between the input and output terminals. An analysis of this phenomenon involving pulse-width-modulated (PWM) inverters has not been done, although some engineers have identified this problem [70–74]. It is important to take into account the damping (losses) within the cables, windings, and cores because the voltage stress caused by the repetitive nature of PWM depends on the superposition of the forward- and backward-traveling waves, the local magnitudes of which are dependent on the damping impedances of the cable, winding, and core.

With improved semiconductor technology in VFDs and pulse-width modulators, the low-voltage induction motor drive has become a very viable alternative to the DC motor drive for many applications. The performance improvement, along with its relatively low cost, has made it a major part of process control equipment throughout the industry. The majority of these applications are usually under the 2 kV and the 1 MW range. Even with advancements in technology, the low-voltage induction motor experiences failures in the motor winding. Although there are differences between medium-voltage and high- and low-voltage induction motor applications, there are similarities between their failures.

The medium-voltage induction motor [75] experiences turn-to-turn failures when exposed to surge transients induced by switching or lightning. Most research shows that the majority of failures that occur in the last turn of the first coil prior to entering the second coil of the windings are due to (capacitor) switching transients. One might not expect to see turn-to-turn failures in small induction machines because of their relatively low voltage levels compared to their high insulation ratings, but this is not the case. Turn-to-turn failures are being experienced in all turns of the first coil of the motor.

As indicated above, most short-circuits in power systems occur because of insulation failure. The breakdown of the insulation depends on the voltage stress integrated over time. Although the time component cannot be influenced, the voltage stress across a dielectric can be reduced. Some preliminary work has been done in exploring this wave-propagation phenomenon where forward- and backward- traveling waves superimpose; here, the maximum voltage stress does not necessarily occur at the terminal of a winding, but somewhere in between the two (input, output) terminals. Because of the repetitive nature of PWM pulses, the damping due to losses in the winding and iron core cannot be neglected. The work by Gupta *et al.* [75] and many others [70–74 and 76–78] neglects the damping of the traveling waves; therefore, it cannot predict the maximum voltage stress of the winding insulation. Machine, transformer windings, and cables can be modeled by leakage inductances, resistances, and capacitances. Current-source inverters generate the largest voltage stress because of their inherent current spikes resulting in a large L(di/dt).

The current model [75] of medium-voltage induction motors is based on the transmission line theory neglecting damping effects in windings and cores: since the turns of the coil are in such close proximity to each other, they form sections of a transmission line. As a single steep front wave, or surge, travels through the coil's turns, it will encounter impedance mismatches, usually at the entrance and exit terminals of the coil, as well as at the transitions of the uniform parts (along the machine axial length) to the end winding parts. At these points of discontinuity, the wave will go through a transition, where it can be increased up to 2 per unit. With repetitive PWM pulses this is not the case: simulations have shown voltage increases to be as high as 20 times its source (rated) value due to the so-called swing effect. These increases in magnitude cause turn-to-turn failures in the windings. In dealing with medium-voltage induction motors [75], the transmission line theory neglecting any damping has been quite accurate in predicting where the failures are occurring for a single-current surge. This is not the case with low- and high-voltage induction motors fed by VFDs.

Existing theoretical results can be summarized as follows:

- The existing model [75–78] based on transmission line theory neglects the damping in the winding and iron core and, therefore, is not suitable for repetitive PWM switching simulation where the swing effect (cumulative increase of voltage magnitudes because of repetitive switching such that a reinforcement occurs due to positive feedback caused by winding parameter discontinuities) becomes important.
- A new equivalent circuit of induction motors based on a modified transmission line theory employing resistances, coupling capacitances, and inductances including the feeding cable has been developed, as is demonstrated in the following examples.
- The case of lossless transmission line segments modeling one entire phase of an induction motor with a single-current pulse for input agrees with the findings of the EPRI study [75] and other publications [76–78].
- Forward- and backward-traveling waves superimpose and generate different voltage stresses in coils. Voltage stresses will be different at different locations in the winding.
- Voltage stress appears to be dependent on PWM switching frequency, duty cycle, and motor parameters.
- Voltage stress is also dependent on the wave shape of the voltage/current surge. The wave shape is characterized by the rise, fall, and duration times as well as switching frequency.

- Current-source inverters (CSIs) produce larger voltage stress than voltage-source inverters (VSIs).
- Filters intended to reduce transient voltage damage to a motor may not be able to cover the entire frequency range of the inverter for a variable-frequency drive (VFD).
- Increases in voltage magnitude cause turn-to-turn and coil-to-frame failures in the windings.
- Increasing insulation at certain portions of a turn will help.
- Motor winding parameters are somewhat nonuniform due to the deep motor slots. Inductance and capacitance change along the coil (e.g., the end and slot regions of the coil).

### 3.11.1 Application Example 3.13: Calculation of Winding Stress Due to PMW Voltage-Source Inverters

Voltage and current ripples due to PWM inverters cause voltage stress in windings of machines and transformers. The configuration of Fig. E3.13.1 represents a simple winding with two turns consisting of eight segments. Each segment can be represented by an inductance $L_i$, a resistance $R_i$, and a capacitance to ground (frame) $C_i$, and there are four interturn capacitances $C_{ij}$ and inductances $L_{ij}$. Figure E3.13.2 represents a detailed equivalent circuit for the configuration of Fig. E3.13.1.

a) To simplify the analysis neglect the capacitances $C_{ij}$ and the inductances $L_{ij}$. This leads to the circuit of Fig. E3.13.3, where the winding is fed by a PWM voltage source (see Fig. E3.13.4). One obtains the 16 differential equations (Eqs. E3.13-1a,b to E3.13-8a,b) as listed below.

$$\frac{di_1}{dt} = -\frac{R_1}{L_1}i_1 - \frac{v_1}{L_1} + \frac{v_s(t)}{L_1}, \quad \text{(E3.13-1a)}$$

$$\frac{dv_1}{dt} = \frac{i_1}{C_1} - \frac{G_1}{C_1}v_1 - \frac{i_2}{C_1}, \quad \text{(E3.13-1b)}$$

$$\frac{di_2}{dt} = -\frac{R_2}{L_2}i_2 - \frac{v_2}{L_2} + \frac{v_1}{L_2}, \quad \text{(E3.13-2a)}$$

$$\frac{dv_2}{dt} = \frac{i_2}{C_2} - \frac{G_2}{C_2}v_2 - \frac{i_3}{C_2}, \quad \text{(E3.13-2b)}$$

$$\frac{di_3}{dt} = -\frac{R_3}{L_3}i_3 - \frac{v_3}{L_3} + \frac{v_2}{L_3}, \quad \text{(E3.13-3a)}$$

$$\frac{dv_3}{dt} = \frac{i_3}{C_3} - \frac{G_3}{C_3}v_3 - \frac{i_4}{C_3}, \quad \text{(E3.13-3b)}$$

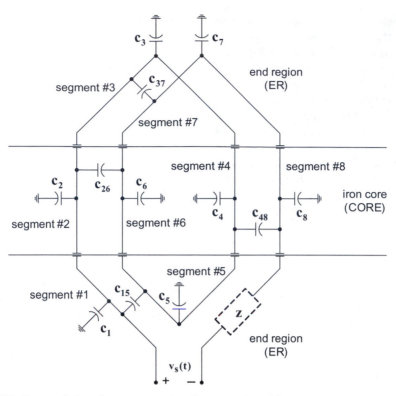

**FIGURE E3.13.1** Winding consisting of two turns, and each turn consists of four segments.

$$\frac{di_4}{dt} = -\frac{R_4}{L_4}i_4 - \frac{v_4}{L_4} + \frac{v_3}{L_4}, \quad \text{(E3.13-4a)}$$

$$\frac{dv_4}{dt} = \frac{i_4}{C_4} - \frac{G_4}{C_4}v_4 - \frac{i_5}{C_4}, \quad \text{(E3.13-4b)}$$

$$\frac{di_5}{dt} = -\frac{R_5}{L_5}i_5 - \frac{v_5}{L_5} + \frac{v_4}{L_5}, \quad \text{(E3.13-5a)}$$

$$\frac{dv_5}{dt} = \frac{i_5}{C_5} - \frac{G_5}{C_5}v_5 - \frac{i_6}{C_5}, \quad \text{(E3.13-5b)}$$

$$\frac{di_6}{dt} = -\frac{R_6}{L_6}i_6 - \frac{v_6}{L_6} + \frac{v_5}{L_6}, \quad \text{(E3.13-6a)}$$

$$\frac{dv_6}{dt} = \frac{i_6}{C_6} - \frac{G_6}{C_6}v_6 - \frac{i_7}{C_6}, \quad \text{(E3.13-6b)}$$

$$\frac{di_7}{dt} = -\frac{R_7}{L_7}i_7 - \frac{v_7}{L_7} + \frac{v_6}{L_7}, \quad \text{(E3.13-7a)}$$

$$\frac{dv_7}{dt} = \frac{i_7}{C_7} - \frac{G_7}{C_7}v_7 - \frac{i_8}{C_7}, \quad \text{(E3.13-7b)}$$

$$\frac{di_8}{dt} = -\frac{R_8}{L_8}i_8 - \frac{v_8}{L_8} + \frac{v_7}{L_8}, \quad \text{(E3.13-8a)}$$

$$\frac{dv_8}{dt} = \frac{i_8}{C_8} - v_8\left(\frac{1}{ZC_8} + \frac{G_8}{C_8}\right). \quad \text{(E3.13-8b)}$$

The parameters are

$R_1 = R_2 = R_3 = R_4 = R_5 = R_6 = R_7 = R_8 = 25\ \mu\Omega$;
$L_1 = L_3 = L_5 = L_7 = 1$ mH, $L_2 = L_4 = L_6 = L_8 = 10$ mH;
$C_1 = C_3 = C_5 = C_7 = 0.7$ pF, $C_2 = C_4 = C_6 = C_8 = 7$ pF,
$C_{15} = C_{37} = 0.35$ pF, $C_{26} = C_{48} = 3.5$ pF;
$G_1 = G_3 = G_5 = G_7 = 1/(5000\ \Omega)$,
$G_2 = G_4 = G_6 = G_8 = 1/(500\ \Omega)$; $L_{15} = L_{37} = 0.005$ mH,
$L_{26} = L_{48} = 0.01$ mH.

$Z$ is either $1\ \mu\Omega$ (short-circuit) or $1\ M\Omega$ (open-circuit). As input, assume the PWM voltage-step function $v_s(t)$ (illustrated in Fig. E3.13.4).

Using Mathematica or MATLAB compute all state variables for the time period from $t = 0$ to 2 ms. Plot the voltages $(v_s - v_1)$, $(v_1 - v_5)$, $(v_2 - v_6)$, $(v_3 - v_7)$, and $(v_4 - v_8)$.

b) For the detailed equivalent circuit of Fig. E3.13.2 establish all differential equations in a similar manner as has been done in part a. As input, assume a PWM voltage-step function $v_s(t)$ (illustrated in Fig. E3.13.4).

Using Mathematica or MATLAB compute all state variables for the time period from $t = 0$ to 2 ms. Plot the voltages $(v_s - v_1)$, $(v_1 - v_5)$, $(v_2 - v_6)$, $(v_3 - v_7)$, and $(v_4 - v_8)$ as in part a.

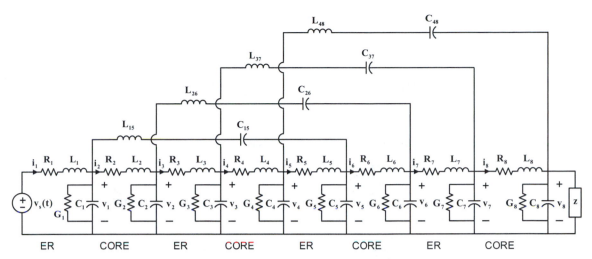

**FIGURE E3.13.2** Detailed equivalent circuit of Fig. E3.13.1.

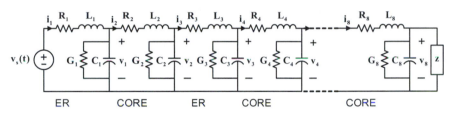

**FIGURE E3.13.3** Simplified equivalent circuit of Fig. E3.13.1.

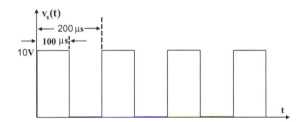

**FIGURE E3.13.4** PWM voltage source $v_s(t)$.

### 3.11.2 Application Example 3.14: Calculation of Winding Stress Due to PMW Current-Source Inverters

Voltage and current ripples due to PWM current-source inverters cause voltage stress in windings of machines and transformers. The configuration of Fig. E3.14.1 represents a simple winding with two turns. Each winding can be represented by four segments having an inductance $L_i$, a resistance $R_i$, a capacitance to ground (frame) $C_i$, and some interturn capacitances $C_{ij}$ and inductances $L_{ij}$. Figure E.3.14.2 represents a detailed equivalent circuit for the configuration of Fig. E3.14.1.

a) To simplify the analysis neglect the capacitances $C_{ij}$ and the inductances $L_{ij}$. This leads to the circuit of Fig. E3.14.3, where the winding is fed by a PWM current source. One obtains the 15 differential equations of Eqs. E3.14-1b to E3.14-8a,b as listed below. In this case the current through $L_1$ is given (e.g., $i_s(t)$) and, therefore, there are only 15 differential equations, as compared with Application Example 3.13.

$$\frac{dv_1}{dt} = \frac{i_s}{C_1} - \frac{G_1}{C_1}v_1 - \frac{i_2}{C_1}, \quad \text{(E3.14-1b)}$$

$$\frac{di_2}{dt} = -\frac{R_2}{L_2}i_2 - \frac{v_2}{L_2} + \frac{v_1}{L_2}, \quad \text{(E3.14-2a)}$$

$$\frac{dv_2}{dt} = \frac{i_2}{C_2} - \frac{G_2}{C_2}v_2 - \frac{i_3}{C_2}, \quad \text{(E3.14-2b)}$$

$$\frac{di_3}{dt} = -\frac{R_3}{L_3}i_3 - \frac{v_3}{L_3} + \frac{v_2}{L_3}, \quad \text{(E3.14-3a)}$$

$$\frac{dv_3}{dt} = \frac{i_3}{C_3} - \frac{G_3}{C_3}v_3 - \frac{i_4}{C_3}, \quad \text{(E3.14-3b)}$$

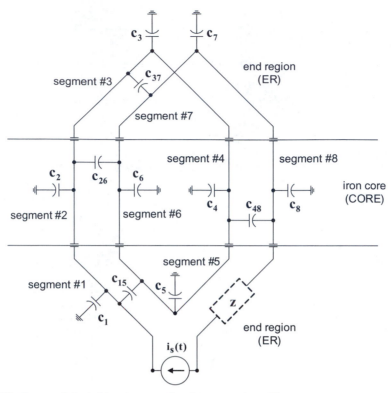

**FIGURE E3.14.1** Winding consisting of two turns, and each turn consists of four segments.

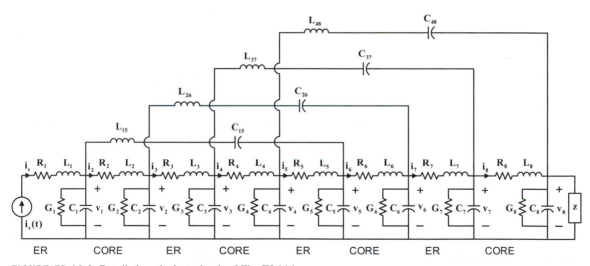

**FIGURE E3.14.2** Detailed equivalent circuit of Fig. E3.14.1.

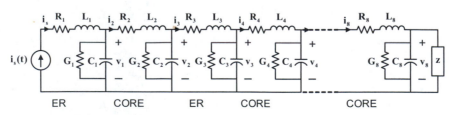

**FIGURE E3.14.3** Simplified equivalent circuit of Fig. E3.14.1.

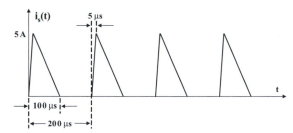

**FIGURE E3.14.4** PWM current source $i_s(t)$.

$$\frac{di_4}{dt} = -\frac{R_4}{L_4}i_4 - \frac{v_4}{L_4} + \frac{v_3}{L_4}, \qquad \text{(E3.14-4a)}$$

$$\frac{dv_4}{dt} = \frac{i_4}{C_4} - \frac{G_4}{C_4}v_4 - \frac{i_5}{C_4}, \qquad \text{(E3.14-4b)}$$

$$\frac{di_5}{dt} = -\frac{R_5}{L_5}i_5 - \frac{v_5}{L_5} + \frac{v_4}{L_5}, \qquad \text{(E3.14-5a)}$$

$$\frac{dv_5}{dt} = \frac{i_5}{C_5} - \frac{G_5}{C_5}v_5 - \frac{i_6}{C_5}, \qquad \text{(E3.13-5b)}$$

$$\frac{di_6}{dt} = -\frac{R_6}{L_6}i_6 - \frac{v_6}{L_6} + \frac{v_5}{L_6}, \qquad \text{(E3.14-6a)}$$

$$\frac{dv_6}{dt} = \frac{i_6}{C_6} - \frac{G_6}{C_6}v_6 - \frac{i_7}{C_6}, \qquad \text{(E3.14-6b)}$$

$$\frac{di_7}{dt} = -\frac{R_7}{L_7}i_7 - \frac{v_7}{L_7} + \frac{v_6}{L_7}, \qquad \text{(E3.14-7a)}$$

$$\frac{dv_7}{dt} = \frac{i_7}{C_7} - \frac{G_7}{C_7}v_7 - \frac{i_8}{C_7}, \qquad \text{(E3.14-7b)}$$

$$\frac{di_8}{dt} = -\frac{R_8}{L_8}i_8 - \frac{v_8}{L_8} + \frac{v_7}{L_8}, \qquad \text{(E3.14-8a)}$$

$$\frac{dv_8}{dt} = \frac{i_8}{C_8} - v_8\left(\frac{1}{ZC_8} + \frac{G_8}{C_8}\right). \qquad \text{(E3.14-8b)}$$

The parameters are

$R_1 = R_2 = R_3 = R_4 = R_5 = R_6 = R_7 = R_8 = 25\ \mu\Omega;$
$L_1 = L_3 = L_5 = L_7 = 1\ \text{mH},\ L_2 = L_4 = L_6 = L_8 = 10\ \text{mH};$
$C_1 = C_3 = C_5 = C_7 = 0.7\ \text{pF},\ C_2 = C_4 = C_6 = C_8 = 7\ \text{pF},$
$C_{15} = C_{37} = 0.35\ \text{pF},\ C_{26} = C_{48} = 3.5\ \text{pF};$
$G_1 = G_3 = G_5 = G_7 = 1/(5000\ \Omega),$
$G_2 = G_4 = G_6 = G_8 = 1/(500\ \Omega);$ and
$L_{15} = L_{37} = 0.005\ \text{mH},\ L_{26} = L_{48} = 0.01\ \text{mH}.$

Z is either $1\ \mu\Omega$ (short-circuit) or $1\ \text{M}\Omega$ (open-circuit). Assume a PWM current-step function $i_s(t)$ as indicated in Fig. E3.14.4.

Using Mathematica or MATLAB compute all state variables for the time period from $t = 0$ to 2 ms.

Plot the voltages $(v_1 - v_5)$, $(v_2 - v_6)$, $(v_3 - v_7)$, and $(v_4 - v_8)$.

b) For the detailed equivalent circuit of Fig. E3.14.2 establish all differential equations in a similar manner as has been done in part a. As input assume a PWM current-step function $i_s(t)$ as indicated in Fig. E3.14.4.

Using Mathematica or MATLAB compute all state variables for the time period from $t = 0$ to 2 ms. Plot the voltages $(v_1 - v_5)$, $(v_2 - v_6)$, $(v_3 - v_7)$, and $(v_4 - v_8)$ as in part a.

### 3.12 NONLINEAR HARMONIC MODELS OF THREE-PHASE INDUCTION MACHINES

The aim of this section is to introduce and briefly discuss some harmonic models for induction machines available in the literature. Detailed explanations are given in the references cited.

#### 3.12.1 Conventional Harmonic Model of an Induction Motor

Figures 3.3 and 3.25 represent the fundamental and harmonic per-phase equivalent circuits of a three-phase induction motor, respectively, which are used for the calculation of the steady-state performance. The rotor parameters may consist of the equivalent rotor impedance taking into account the cross-path resistance of the rotor winding, which is frequency dependent because of the skin effect in the rotor bars. The steady-state performance is calculated by using the method of superposition based on Figs. 3.3 and 3.25. In general, for the usual operating speed regions, the fundamental slip is very small and, therefore, the harmonic slip, $s_h$, is close to unity.

#### 3.12.2 Modified Conventional Harmonic Model of an Induction Motor

In order to improve the loss prediction of the conventional harmonic equivalent circuit of induction machines (Fig. 3.25), the core loss and stray-core losses associated with high-frequency leakage fluxes are taken into account. Stray-load losses in induction machines are considered to be important – due to leakage – when induction machines are supplied by nonsinusoidal voltage or current sources. This phenomenon can be accounted for by a modified loss model, as shown in Fig. 3.40 [79].

**Inclusion of Rotor-Core Loss.** Most publications assume negligible rotor-core losses due to the very

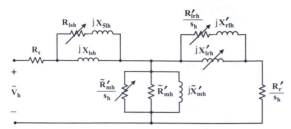

**FIGURE 3.40** Modified harmonic loss model of induction motor including separate resistances for rotor- and stator-core losses, as well as eddy currents due to leakage fluxes [79].

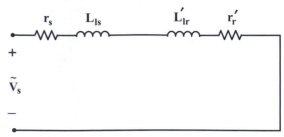

**FIGURE 3.41** Simplified conventional harmonic model of induction motor [80].

low frequency (e.g., slip frequency) of the magnetic field within the rotor core. To take into account the rotor-iron losses – associated with the magnetizing path – a slip-dependent resistance $R'_{mh}$ is connected in parallel with the stator-core loss resistance (Fig. 3.40).

**Iron Loss Associated with Leakage Flux Paths (Harmonic Stray Loss).** Resistances in parallel with the primary and the secondary leakage inductances ($R_{lsh}$, $R'_{lrh}$) in the harmonic equivalent circuit of Fig. 3.40 represent harmonic losses, referred to as stray losses. Because almost all stray losses result from hysteresis and eddy-current effects, these equivalent parallel resistors are assumed to be frequency dependent.

### 3.12.3 Simplified Conventional Harmonic Model of an Induction Motor

As the mechanical speed is very close to the synchronous speed, the rotor slip with respect to the harmonic mmfs will be close to unity (e.g., about 6/5 and 6/7 for the 5th and 7th harmonics, respectively). Therefore, the resistance $(1 - s_h)R'_r/s_h$ will be negligibly small in the harmonic model of Fig. 3.25, and the equivalent harmonic circuit may be reduced to that of Fig. 3.41. Only the winding resistances and the leakage inductances are included; the magnetizing branches are ignored because the harmonic voltages can be assumed to be relatively small, and at harmonic frequencies the core-flux density is small as well as compared with fundamental flux density. It should be noted that the winding resistances and leakage inductances are frequency dependent.

### 3.12.4 Spectral-Based Harmonic Model of an Induction Machine with Time and Space Harmonics

Figure 3.42 shows a harmonic equivalent circuit of an induction machine that models both time and

space harmonics [81]. The three windings may or may not be balanced as each phase has its own per phase resistance $R_{equj}$, reactance $X_{equj}$, and speed-dependent complex electromagnetic force emf $e_j$ (where $j = 1, 2, 3$ represents the phase number). To account for the core losses, two resistances in parallel are added to each phase: $R_{c1}$ representing the portion of losses that depends only on phase voltages and $(\omega_b/\omega_s)R_{c2}$ representing losses that depend on the frequency and voltage of the stator ($\omega_b$ is the rated stator frequency whereas $\omega_s$ is the actual stator frequency).

To account for space-harmonic effects, the spectral approach modeling technique is used. This concept combines effects of time harmonics (from the input power supply) and space harmonics (from machine windings and magnetic circuits), bearing in mind that no symmetrical components exist in both quantities.

The equivalent circuit of Fig. 3.42 has several branches (excluding the fundamental and the core losses), where each branch per phase is associated with one harmonic of the power supply (impedance) or one frequency for the space distribution (current source). To evaluate various parameters, the spectral analysis of both time-domain stator voltages ($v_1$, $v_2$, $v_3$) and currents ($i_1, i_2, i_3$) are required. The conventional blocked-rotor, no-load, and loaded shaft tests can be performed at fundamental frequency to evaluate phase resistance $R_{equj}$, reactance $X_{equj}$, complex emf $e_j$, $R_{c1}$, and $R_{c2}$.

The harmonic stator voltage ($V_{fij}$) and current ($i_{fij}$) with their respective magnitudes and phase angles need to be measured to determine harmonic impedances (e.g., $R_{fij} + jX_{fij}$ is the ratio of the two phasors).

For the current sources coming from the space harmonics, one has to differentiate between power supply harmonics and space harmonics and those which are solely space-harmonic related. This can be

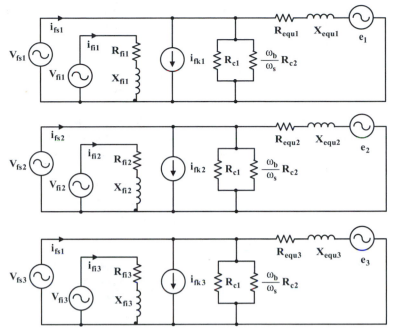

**FIGURE 3.42** Spectral-based equivalent circuit of induction motor with space-harmonic, time–harmonic, and core losses [81].

done using a blocked-rotor test or a long-time starting test (with the primary objective of obtaining data at zero speed). Voltage and current measurements in the time domain allow the evaluation of the effects only related to the power supply harmonics in the frequency domain. Determination of the space-harmonics effects is made possible by utilizing the principle of superposition in steady state at a nonzero speed.

In the model of Fig. 3.42:

- Effects of both time and space harmonics are included;
- $R_{fij} + jX_{fij} = V_{fij}/i_{fij}$ originates from phase voltage time harmonics (fi is the ith harmonic);
- $i_{fkj}$ originates from the space harmonics (subscript fk is used for space harmonics);
- The skin effect is ignored, but could be easily included; and
- Core losses related to power supply harmonics except the fundamental are included in the harmonic circuit.

To introduce space harmonics into the model, current sources with the phasor notation $i_{fkj}$ are used which are only linked to the load condition and, as a result, to the speed of the rotor. In most cases, the space-harmonics frequency $fk$ is also a multiple of the power supply frequency $f_s$. It is understood that

the space harmonics will affect electromagnetic torque calculations, which is not shown in the equivalent circuit.

Details of this equivalent circuit are presented in reference [81].

## 3.13 STATIC AND DYNAMIC ROTOR ECCENTRICITY OF THREE-PHASE INDUCTION MACHINES

The fundamentals of rotor eccentricity are described by Heller and Hamata [30]. One differentiates between static and dynamic rotor eccentricity. The case of static rotor eccentricity occurs if the rotor axis and the stator bore axes do not coincide and the rotor is permanently shifted to one side of the stator bore. Dynamic rotor eccentricity occurs when the rotor axis rotates around a small circle within the stator bore. The most common forms are combinations of static and dynamic rotor eccentricities. This is so because any rotor cannot be perfectly (in a concentric manner) mounted within the stator bore of an electric machine. The electric machines most prone to the various forms of rotor eccentricity are induction machines due to their small air gaps. The air gaps of small induction machines are less than 1 mm, and those of large induction machines are somewhat larger than 1 mm. Rotor eccentricity originates therefore in the mounting tolerances of the radial bearings. The main effects of rotor eccentricity are noninteger harmonics, one-sided magnetic pull, acoustic noise, mechanical vibrations, and additional losses within the machine. Simulation, measurement methods, and detection systems [82–87] for the various forms of rotor eccentricities have been developed.

Some phenomena related to rotor imbalance are issues of shaft flux and the associated bearing currents [88, 89]. The latter occur predominantly due to PWM inverters and because of mechanical and electrical asymmetries resulting in magnetic imbalance.

## 3.14 OPERATION OF THREE-PHASE MACHINES WITHIN A SINGLE-PHASE POWER SYSTEM

Very frequently there are only single-phase systems available; however, relatively large electrical drives are required. Examples are pump drives for irrigation applications which require motors in the 40 hp range. As it is well known, single-phase induction motors in this range are hardly an off-the-shelf item; in addition such motors are not very efficient and are relatively expensive as compared with three-phase induction motors. For this reason three-phase induction motors are used together with a network – consisting of capacitors – which converts a single-phase system to an approximate three-phase system [90, 91].

## 3.15 CLASSIFICATION OF THREE-PHASE INDUCTION MACHINES

The rotor winding design of induction machines varies depending on their application. There are two types: wound-rotor and squirrel-cage induction machines. Most of the manufactured motors are of the squirrel-cage type. According to the National Electrical Manufacturers Association (NEMA), one differentiates between the classes A, B, C, and D as is indicated in the torque–speed characteristics of Fig. 3.43. Class A has large rotor bars sections near the surface of the rotor resulting in low rotor resistance $R'_r$ and an associated low starting torque. Most induction generators are not required to produce a

significant starting torque and therefore class A is acceptable for such applications. Class B has large and deep rotor bars, that is, the skin effect cannot be neglected during start-up. Classes A and B have similar torque–speed characteristics. Class C is of the double-cage rotor design: due to the skin effect the current flows mostly in the outer cage during start-up and produces therefore a relatively large starting torque. Class D has small rotor bars near the rotor surface and produces a large starting torque at the cost of a low efficiency at rated operation.

The equivalent circuit of a single-squirrel cage induction machine is given in Fig. 3.1. That of a double squirrel-cage type is shown in Fig. 3.44 [12, 92].

In the double-cage winding $\omega_1 L'_{r1r2}$ is the mutual reactance between cage $r_1$ and cage $r_2$. $R'_{r1}$ and $R'_{r2}$ are the resistances of cages $r_1$ and $r_2$, respectively. $R'_e$ and $\omega_1 L'_e$ are the end-winding resistance and the end-

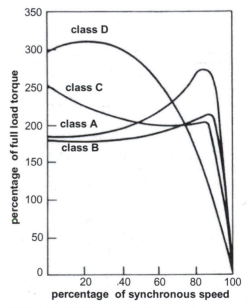

**FIGURE 3.43** NEMA classification of squirrel-cage induction machines.

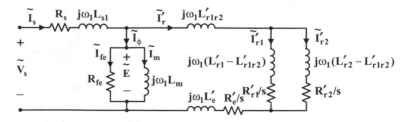

**FIGURE 3.44** Equivalent circuit of double-cage induction machine.

winding leakage reactance of the end rings, respectively, if they are common to both cages. $\omega_1 L'_{r1}$ and $\omega_1 L'_{r2}$ are the leakage reactances of cages $r_1$ and $r_2$, respectively. The induced voltage is

$$\tilde{E} = \tilde{I}'_{r1}\left(\frac{R'_{r1}}{s} + j\omega_1 L'_{r1}\right) + \tilde{I}'_{r2}j\omega_1 L'_{r1r2} + \tilde{I}'_{r}\left(\frac{R'_e}{s} + j\omega_1 L'_e\right).$$

(3-58)

With

$$\tilde{I}'_r = \tilde{I}'_{r1} + \tilde{I}'_{r2}$$

(3-59)

one obtains alternatively

$$\tilde{E} = \tilde{I}'_{r1}\left[\frac{R'_{r1}}{s} + j\omega_1(L'_{r1} - L'_{r1r2})\right] + I'_r\left(j\omega_1 L'_{r1r2} + \frac{R'_e}{s} + j\omega_1 L'_e\right),$$

(3-60)

and

$$\tilde{E} = \tilde{I}'_{r2}\left[\frac{R'_{r2}}{s} + j\omega_1(L'_{r2} - L'_{r1r2})\right] + I'_r\left(j\omega_1 L'_{r1r2} + \frac{R'_e}{s} + j\omega_1 L'_e\right).$$

(3-61)

The resultant rotor impedance is now

$$\frac{\tilde{E}}{\tilde{I}'_r} = \frac{R'_r(s)}{s} + j\omega_1 L'_r(s)$$

$$= \frac{\left[\frac{R'_{r1}}{s} + j\omega_1(L'_{r1} - L'_{r1r2})\right]\left[\frac{R'_{r2}}{s} + j\omega_1(L'_{r2} - L'_{r1r2})\right]}{\left[\frac{R'_{r1}}{s} + j\omega_1(L'_{r1} - L'_{r1r2})\right] + \left[\frac{R'_{r2}}{s} + j\omega_1(L'_{r2} - L'_{r1r2})\right]}$$

$$+ \left(j\omega_1 L'_{r1r2} + \frac{R'_e}{s} + j\omega_1 L'_e\right)$$

(3-62)

In Eq. 3-62 the resulting rotor resistance $R'_r(s)$ and leakage inductance $L'_r(s)$ are slip ($s$) dependent. For an induction motor to be very efficient at rated slip the rotor resistance must be small. However, such a machine has a low starting torque. If the rotor resistance is relatively large the starting torque will be large, but the efficiency of such a machine will be very low. A double-squirrel cage solves this problem through a slip-dependent rotor resistance: at starting the currents mainly flow through the outer squirrel-cage winding having a large resistance, and at rated operation the currents flow mainly through the inner squirrel-cage winding. Triple- or quadruple-squirrel cage windings are employed for special applications only.

## 3.16 SUMMARY

This chapter addresses first the normal operation of induction machines and the calculation of machine parameters, which are illustrated by flux plots. Power quality refers to the effects caused by induction machines due to their abnormal operation or by poorly designed machines. It discusses the influences of time and space harmonics on the rotating field, and the influence of current, voltage, and flux density harmonics on efficiency. This chapter highlights the detrimental effects of inter- and subharmonics on asynchronous and synchronous torque production, and the voltage stress induced by current source inverters and lightning surges. The applicability of harmonic models to three-phase induction machines is addressed. Static and dynamic rotor eccentricity, bearing currents, the operation of three-phase motors within a single-phase system, and the various classes and different equivalent circuits of induction machines are explained. Many application examples refer to the stable or unstable steady-state operation, the operation at under- or overvoltage and under- or over-frequency and combinations thereof. Calculation of asynchronous (integer, inter-, sub-) harmonic torques is performed in detail, whereas the generation of synchronous torque is mentioned. The calculation of the winding stress due to current and voltage-source inverters is performed with either Mathematica or MATLAB.

## 3.17 PROBLEMS

### Problem 3.1: Steady-State Stability

Figure P3.1a,b illustrates the torque–speed characteristics of a 10 kW three-phase induction motor with different rotor slot skewing with load torques $T_{L1}$, $T_{L2}$, and $T_{L3}$. Comment on the steady-state stability of the equilibrium points A, B, C, D, E, F, G, H, I, and J in Fig. P3.1a and A, B, C, D, and E in Fig. P3.1b.

### Problem 3.2: Phasor Diagram of Induction Machine at Fundamental Frequency and Steady-State Operation

Draw the phasor diagram for the equivalent circuit of Fig. P3.2. You may use the consumer reference system.

### Problem 3.3: Analysis of an Induction Generator for a Wind-Power Plant

A $P_{out} = 3.5$ MW, $V_{L-L} = 2400$ V, $f = 60$ Hz, $p = 12$ poles, $n_{m\text{-generator}} = 618$ rpm, Y-connected squirrel-cage wind-power induction generator has the following parameters per phase referred to the stator: $R_S = 0.1\ \Omega$, $X_S = 0.5\ \Omega$, $X_m \rightarrow \infty$ (that is, $X_m$ can be ignored), $R'_r = 0.05\ \Omega$, and $X'_r = 0.2\ \Omega$. You may use

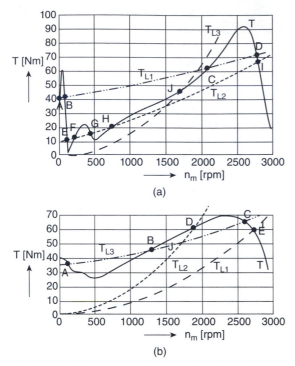

(a)

(b)

**FIGURE P3.1** Torque–speed characteristic $T$–$n_m$ of 10 kW three-phase induction motor with three load characteristics $T_{L1}$–$n_m$, $T_{L2}$–$n_m$, and $T_{L3}$–$n_m$ with (a) half a rotor-slot pitch skewing and (b) full rotor-slot pitch skewing.

the equivalent circuit of Fig. P3.2 with $X_m$ and $R_{fe}$ large so that they can be ignored. Calculate:

a) Synchronous speed $n_s$ in rpm.
b) Synchronous angular velocity $\omega_s$ in rad/s.
c) Slip $s$.
d) Rotor current $\tilde{I}'_r$.
e) (Electrical) generating torque $T_{gen}$.
f) Slip $s_m$ where the maximum torque occurs.
g) Maximum torque $T_{max}$ in generation region.

## Problem 3.4: E/f Control of a Squirrel-Cage Induction Motor [93]

The induction machine of Problem 3.3 is now used as a motor at $n_{m\text{-motor}} = 582$ rpm. You may use the equivalent circuit of Fig. P3.2 with $X_m$ and $R_{fe}$ large so that they can be ignored. Calculate for E/f control:

a) Induced voltage $\tilde{E}_{rated}$.
b) Electrical torque $T_{rated}$.
c) The frequency $f_{new}$ at $n_{m\text{-new}} = 300$ rpm.
d) The new slip $s_{new}$.

## Problem 3.5: V/f Control of a Squirrel-Cage Induction Motor [93]

The induction machine of Problem 3.3 is now used as a motor at $n_{m\text{-motor}} = 582$ rpm. You may use the

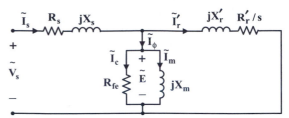

**FIGURE P3.2** Equivalent circuit of induction machine.

equivalent circuit of Fig. P3.2 with $X_m$ and $R_{fe}$ large so that they can be ignored. Calculate for V/f control:

a) Electrical torque $T_{rated}$.
b) The frequency $f_{new}$ at $n_{m\text{-new}} = 300$ rpm.
c) The new slip $s_{new}$.

## Problem 3.6: Steady-State Operation at Undervoltage

A $P_{out} = 100$ hp, $V_{L\text{-}L\_rat} = 480$ V, $f = 60$ Hz, $p = 6$ pole three-phase induction motor runs at full load and rated voltage with a slip of 3%. Under conditions of stress on the power supply system, the line-to-line voltage drops to $V_{L\text{-}L\_low} = 450$ V. If the load is of the constant torque type, compute for the lower voltage:

a) The slip $s_{low}$ (use small-slip approximation).
b) The shaft speed $n_{m\_low}$ in rpm.
c) The output power $P_{out\_low}$.
d) The rotor copper loss $P_{cur\_low} = I'^2_r R'_r$ in terms of the rated rotor copper loss at rated voltage.

## Problem 3.7: Steady-State Operation at Overvoltage

Repeat Problem 3.6 for the condition when the line-to-line voltage of the power supply system increases to $V_{L\text{-}L\_high} = 500$ V.

## Problem 3.8: Steady-State Operation at Undervoltage and Under-Frequency

A $P_{out\_rat} = 100$ hp, $V_{L\text{-}L\_rat} = 480$ V, $f = 60$ Hz, $p = 6$ pole three-phase induction motor runs at full load and rated voltage and frequency with a slip of 3%. Under conditions of stress on the power supply system the terminal voltage and the frequency drop to $V_{L\text{-}L\_low} = 450$ V and $f_{low} = 59.5$ Hz, respectively. If the load is of the constant torque type, compute for the lower voltage and frequency:

a) The slip (use small-slip approximation).
b) The shaft speed in rpm.
c) The hp output.

d) The rotor copper loss $P_{\text{cur\_low}} = I_r'^2 R_r'$ in terms of the rated rotor copper loss at rated frequency and voltage.

## Problem 3.9: Steady-State Operation at Overvoltage and Over-Frequency

A $P_{\text{out\_rat}} = 100$ hp, $V_{L\text{-}L\_\text{rat}} = 480$ V, $f = 60$ Hz, $p = 6$ pole three-phase induction motor runs at full load and rated voltage and frequency with a slip of 3%. Assume the terminal voltage and the frequency increase to $V_{L\text{-}L\_\text{low}} = 500$ V and $f_{\text{high}} = 60.5$ Hz, respectively. If the load is of the constant torque type, compute for the lower voltage and frequency:

a) The slip (use small-slip approximation).
b) The shaft speed in rpm.
c) The hp output.
d) The rotor copper loss $P_{\text{cur\_low}} = I_r'^2 R_r'$ in terms of the rated rotor copper loss at rated frequency and voltage.

## Problem 3.10: Calculation of the Maximum Critical Speeds $n_{\text{mcritical}}^{\text{motoring}}$ and $n_{\text{mcritical}}^{\text{regeneration}}$ for Motoring and Regeneration, Respectively [93]

A $V_{L\text{-}L\_\text{rat}} = 480$ V, $f_{\text{rat}} = 60$ Hz, $n_{\text{mrat}} = 1760$ rpm, 4-pole, Y-connected squirrel-cage induction machine has the following parameters per phase referred to the stator: $R_S = 0.2\ \Omega$, $X_S = 0.5\ \Omega$, $X_m = 20\ \Omega$, $R_r' = 0.1\ \Omega$, $X_r' = 0.8\ \Omega$. Use the equivalent circuit of Fig. P3.10.1.

A hybrid car employs the induction machine with the above-mentioned parameters for the electrical part of the drive.

a) Are the characteristics of Figure P.3.10.2 (for a < 1) either E/f control or V/f control?
b) Compute the rated electrical torque $T_{\text{rat}}$, the rated angular velocity $\omega_{\text{mrat}}$ and the rated output power $P_{\text{rat}}$ (see operating point $Q_1$ of Figure P3.10.2).
c) Calculate the maximum critical motoring speed $n_{\text{mcritical}}^{\text{motoring}}$, whereby rated output power $(P_{\text{rat}} = T_{\text{rat}} \cdot \omega_{\text{mrat}})$ is delivered by the machine at the shaft (see operating point $Q_2$ of Fig. P3.10.2).

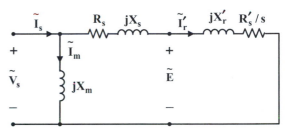

**FIGURE P3.10.1** Equivalent circuit.

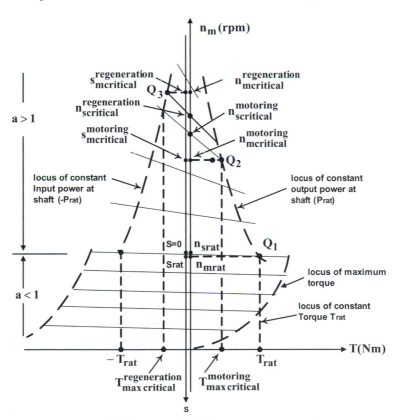

**FIGURE P3.10.2** Definition of $n_{m\_\text{rat}}$, $n_{\text{mcritical}}^{\text{motoring}}$, and $n_{\text{mcritical}}^{\text{regeneration}}$. Note that only the linear portions of the torque–speed characteristics are shown.

d) Calculate the maximum critical speed $n_{mcritical}^{regeneration}$, whereby rated power is absorbed at the shaft by the machine (see operating point $Q_3$ of Fig. P3.10.2).

e) In order to increase the critical motoring speed to about 3 times the rated speed $n_{mrat}$ an engineer proposes to decrease $X_s$ and $X_r'$ by a factor of two. Will this accomplish the desired speed increase? Note that the resistances $R_s$ and $R_r'$ are so small that no significant speed increase can be expected by changing them.

## Problem 3.11: Torque Characteristic of Induction Machine with Fundamental and Subharmonic Fields

The equivalent circuit of a voltage-source-fed induction motor is shown in Fig. 3.38. The fundamental angular frequency of the voltage source is $\omega_1$ and the time-harmonic angular frequency is $h\omega_1$ (caused by imperfections of the voltage-source inverter).

The parameters of the three-phase induction motor are as follows: $V_{L-L} = 460$ V, Y-connected, $p = 2$ poles, $n_{rated} = 3528$ rpm, $f = 60$ Hz, $R_s = 1\ \Omega$, $\omega_1 L_{sl} = 2\ \Omega$, $\omega_1 L_m = \infty$ (you may neglect the magnetizing branch), $R_r' = 1\ \Omega$, $\omega_1 L_{rl}' = 2\ \Omega$.

a) Compute a few points (e.g., $T_{e1}(s_1 = 0)$, $T_{e1}(s_1 = s_{1max+})$, $T_{e1}(s_1 = s_{1max-})$, $T_{e1}(s_1 = 1.0)$) and sketch the fundamental torque characteristic $T_{e1} = f(s_1)$ in Nm, where $s_1$ is the slip of the rotor with respect to the fundamental positively rotating (forward rotating, positive sequence) field.

b) Provided a negatively rotating (backward rotating, negative sequence) voltage time subharmonic of 30 Hz ($h = 0.5$) is present with a magnitude of $V_{s0.5}/V_{s1} = 0.10$ compute a few points (e.g., $T_{e0.5}(s_{0.5} = 0)$, $T_{e0.5}(s_{0.5} = s_{0.5max+})$, $T_{e0.5}(s_{0.5} = s_{0.5max-})$, $T_{e0.5}(s_{0.5} = 1.0)$) of this subharmonic torque characteristic $T_{e0.5} = f(s_{0.5})$ in Nm.

c) Sketch the subharmonic torque characteristic $T_{e0.5} = f(s_{0.5})$ in Nm.

d) Sketch the total torque characteristics $(T_{e1} + T_{e0.5}) = f(s_1)$ in Nm.

e) What is the starting torque (at $s_1 = 1$) without and with this ($h = 0.5$) time subharmonic?

## Problem 3.12: Voltage Stress due to Current Ripples in Stator Windings of an Induction Motor Fed by a Current-Source Inverter Through a Cable of 10 m Length

Voltage and current ripples due to PWM current-source inverters cause voltage stress in windings of machines. The configuration of Fig. E3.14.1 repre-

sents the first two turns of the winding of an induction motor. Each turn can be represented by four segments having an inductance $L_i$, a resistance $R_i$, a capacitance to ground (frame) $C_i$, and some interturn capacitances $C_{ij}$ and inductances $L_{ij}$. Figure E.3.14.2 represents a detailed equivalent circuit for the configuration of Fig. E3.14.1. To simplify the analysis neglect the capacitances $C_{ij}$ and the inductances $L_{ij}$. This leads to the circuit in Fig. E.3.14.3, where the winding is fed by a PWM current source. One obtains the 15 differential equations as listed below.

$$\frac{dv_1}{dt} = \frac{i_s}{C_1} - \frac{G_1}{C_1}v_1 - \frac{i_2}{C_1},$$

$$\frac{di_2}{dt} = -\frac{R_2}{L_2}i_2 - \frac{v_2}{L_2} + \frac{v_1}{L_2},$$

$$\frac{dv_2}{dt} = \frac{i_2}{C_2} - \frac{G_2}{C_2}v_2 - \frac{i_3}{C_2},$$

$$\frac{di_3}{dt} = -\frac{R_3}{L_3}i_3 - \frac{v_3}{L_3} + \frac{v_2}{L_3},$$

$$\frac{dv_3}{dt} = \frac{i_3}{C_3} - \frac{G_3}{C_3}v_3 - \frac{i_4}{C_3},$$

$$\frac{di_4}{dt} = -\frac{R_4}{L_4}i_4 - \frac{v_4}{L_4} + \frac{v_3}{L_4},$$

$$\frac{dv_4}{dt} = \frac{i_4}{C_4} - \frac{G_4}{C_4}v_4 - \frac{i_5}{C_4},$$

$$\frac{di_5}{dt} = -\frac{R_5}{L_5}i_5 - \frac{v_5}{L_5} + \frac{v_4}{L_5},$$

$$\frac{dv_5}{dt} = \frac{i_5}{C_5} - \frac{G_5}{C_5}v_5 - \frac{i_6}{C_5},$$

$$\frac{di_6}{dt} = -\frac{R_6}{L_6}i_6 - \frac{v_6}{L_6} + \frac{v_5}{L_6},$$

$$\frac{dv_6}{dt} = \frac{i_6}{C_6} - \frac{G_6}{C_6}v_6 - \frac{i_7}{C_6},$$

$$\frac{di_7}{dt} = -\frac{R_7}{L_7}i_7 - \frac{v_7}{L_7} + \frac{v_6}{L_7},$$

$$\frac{dv_7}{dt} = \frac{i_7}{C_7} - \frac{G_7}{C_7}v_7 - \frac{i_8}{C_7},$$

$$\frac{di_8}{dt} = -\frac{R_8}{L_8}i_8 - \frac{v_8}{L_8} + \frac{v_7}{L_8},$$

$$\frac{dv_8}{dt} = \frac{i_8}{C_8} - v_8\left(\frac{1}{ZC_8} + \frac{G_8}{C_8}\right).$$

The parameters are

$R_1 = R_2 = R_3 = R_4 = R_5 = R_6 = R_7 = R_8 = 25\ \mu\Omega$;
$L_1 = L_3 = L_5 = L_7 = 1$ mH, and
$L_2 = L_4 = L_6 = L_8 = 10$ mH; $C_1 = 0.2\ \mu$F for the cable
of 10 m length; $C_3 = C_5 = C_7 = 0.7$ pF,
$C_2 = C_4 = C_6 = C_8 = 7$ pF, $C_{15} = C_{37} = 0.35$ pF,
$C_{26} = C_{48} = 3.5$ pF; $G_1 = G_3 = G_5 = G_7 = 1/(5000\ \Omega)$,
$G_2 = G_4 = G_6 = G_8 = 1/(500\ \Omega)$; and
$L_{15} = L_{37} = 0.005$ mH, $L_{26} = L_{48} = 0.01$ mH.

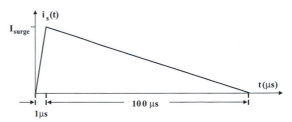

**FIGURE P3.13** Lightning surge wave shape according to IEC [94].

Z is either $1\ \mu\Omega$ (short-circuit) or $1\ M\Omega$ (open-circuit).

Assume a PWM triangular current-step function $i_s(t)$ as indicated in Fig. E3.14.4. Using Mathematica or MATLAB compute all state variables for the time period from $t = 0$ to 2 ms. Plot the voltages $(v_1 - v_5)$, $(v_2 - v_6)$, $(v_3 - v_7)$, and $(v_4 - v_8)$.

## Problem 3.13: Voltage Stress within the Stator Winding of a Large Induction Hydrogenerator Due to Lightning Surge

Lightning surges cause voltage stress in windings of machines. The configuration of Fig. E3.14.1 represents the first two turns of the winding of an induction generator operating without transformer on the 2400 $V_{L\text{-}L}$ power system. Each turn can be represented by four segments having an inductance $L_i$, a resistance $R_i$, a capacitance to ground (frame) $C_i$, and some interturn capacitances $C_{ij}$ and inductances $L_{ij}$. Figure E.3.14.2 represents a detailed equivalent circuit for the configuration of Fig. E3.14.1. To simplify the analysis neglect the capacitances $C_{ij}$ and the inductances $L_{ij}$. This leads to the circuit Fig. E3.14.3. One obtains the 15 differential equations as listed in Problem 3.12.

The parameters are

$R_1 = R_2 = R_3 = R_4 = R_5 = R_6 = R_7 = R_8 = 25\ \mu\Omega$;
$L_1 = L_3 = L_5 = L_7 = 1$ mH, and
$L_2 = L_4 = L_6 = L_8 = 10$ mH; $C_1 = C_3 = C_5 = C_7 = 7$ pF,
$C_2 = C_4 = C_6 = C_8 = 70$ pF, $C_{15} = C_{37} = 3.5$ pF,
$C_{26} = C_{48} = 35$ pF; $G_1 = G_3 = G_5 = G_7 = 1/(5000\ \Omega)$,
$G_2 = G_4 = G_6 = G_8 = 1/(500\ \Omega)$; and
$L_{15} = L_{37} = 0.005$ mH, $L_{26} = L_{48} = 0.01$ mH.

Z is either $1\ \mu\Omega$ (short-circuit) or $1\ M\Omega$ (open-circuit), $I_{surge} = 10,000$ A.

Assume a lightning surge function $i_s(t)$ as indicated in Fig. P3.13. Using Mathematica or MATLAB compute all state variables for the time period from $t = 0$ to 0.2 ms. Plot the voltages $(v_1 - v_5)$, $(v_2 - v_6)$, $(v_3 - v_7)$, and $(v_4 - v_8)$.

## Problem 3.14: Harmonics of Current and Voltage Wave Shapes

Determine the Fourier coefficients of the currents and voltages of Figs. 3.33, 3.34, and E3.6.7a,b. Show that for a given sampling rate the Nyquist criterion is satisfied which in turn means that the approximation error of the reconstructed function is minimum.

## 3.18 REFERENCES

1) Neidhoefer, G; "Early three-phase power," IEEE Power & Energy, vol. 5, no. 5, September/October 2007, pp. 88–102.

2) Steinmetz, C.P.; "Steinmetz on induction-motor discussion in New York," *AIEE Trans.*, February 1898, pp. 106–110.

3) Richter, R.; *Elektrische Maschinen*, Verlag von Julius Springer, Berlin, vol. 1, 1924.

4) Kron, G.; "The generalized theory of electrical machinery," *AIEE Transactions*, vol. 49, April 1930, pp. 666–683.

5) Veinott, C.G.; "Performance calculation of induction motors," *AIEE Transactions*, 1932, pp. 743–755.

6) Schuisky, W.; *Berechnung elektrischer Maschinen*, Springer-Verlag, Wien, 1960.

7) Bödefeld, T.; and Sequenz, H.; *Elektrische Maschinen*, Springer-Verlag, Wien, 1962.

8) Alger, P.L.; *The Nature of Polyphase Induction Machines*, John Wily & Sons, Inc., New York, 1951.

9) Fitzgerald, A.E.; Kingsley, C.; and Umans, S.D.; *Electric Machinery*, Fifth Edition, McGraw-Hill Publishing Company, New York, 1990.

10) Lyon, W.V.; *Transient Analysis of Alternating Current Machinery*, John Wiley & Sons, Inc., New York, 1954.

11) Say, M.G.; *Alternating Current Machines*, John Wiley & Sons, New York, 1983.

12) Match, L.W.; and Morgan, J.D.; *Electromagnetic and Electromechanical Machines*, Third Edition, Harper & Row Publishers, New York, 1986.

13) Chapman, S.J.; *Electric Machinery Fundamentals*, Third Edition, McGraw Hill Boston, 1999.

14) Fuchs, E.F.; Appelbaum, J.; Khan, I.A.; Höll, J.; and Frank, U.V.; *Optimization of Induction Motor Efficiency, Vol. 1: Three-Phase Induction Motors*, Electric Power Research Institute, EPRI EL-4152, 1985.

15) Fuchs, E.F.; Huang, H.; Vandenput, A.J.; Appelbaum, J.; Zak, Z.; and Erlicki, M.S.; *Optimization of Induction Motor Efficiency, Vol. 2: Single-Phase Induction Motors*, Electric Power Research Institute, EPRI EL-4152, 1987.

16) Fuchs, E.F.; and McNaughton, G.A.; "Comparison of first-order finite difference and finite element algorithms for the analysis of magnetic fields, part I: theoretical analysis," *IEEE Transactions on Power Apparatus and Systems*, May 1982, vol. PAS-101, no. 5, pp. 1170–1180, and "Comparison of first-order finite difference and finite element algorithms for the analysis of magnetic fields, part II: numerical results," *IEEE Transactions on Power Apparatus and Systems*, May 1982, vol. PAS-101, no. 5, pp. 1181–1201.

17) Fuchs, E.F.; Chang, L.H.; and Appelbaum, J.; "Magnetizing current, iron losses and forces of three-phase induction machines at sinusoidal and nonsinusoidal terminal voltages, part I: analysis," *IEEE Transactions on Power Apparatus and Systems*, November 1984, vol. PAS-103, no. 11, pp. 3303–3312.

18) Fuchs, E.F.; Roesler, D.J.; and Chang, L.H.; "Magnetizing current, iron losses and forces of three-phase induction machines at sinusoidal and nonsinusoidal terminal voltages, part II: results," *IEEE Transactions on Power Apparatus and Systems*, November 1984, vol. PAS-103, no. 11, pp. 3313–3326.

19) Fuchs, E.F.; Poloujadoff, M.; and Neal, G.W.; "Starting performance of saturable three-phase induction motors," *IEEE Transactions on Energy Conversion*, September 1988, vol. EC- 3, no. 3, pp. 624–635.

20) Leonhard, W.; *Control of Electrical Drives*, Springer-Verlag, Berlin, 1985.

21) Schraud, J.; "Control of an induction motor by electronically switching its windings with different pole numbers to extend the speed range at constant power," Independent Study, University of Colorado, Boulder, May 1999.

22) Fuchs, E.F.; Schraud, J.; and Fuchs, F.S.; "Analysis of critical-speed increase of induction machines via winding reconfiguration with solid-state switches," *IEEE Transactions on Energy Conversion*, paper no. TEC-00375-2005.

23) Schraud, J.; Fuchs, E.F.; and Fuchs, H.A.; "Experimental verification of critical-speed increase of induction machines via winding reconfiguration with solid-state switches," *IEEE Transactions on Energy Conversion*, paper no. TEC-00376-2005.

24) Lin, D.; Fuchs, E.F.; and Doyle, M.; "Computer-aided testing program of electrical apparatus supplying nonlinear loads," *IEEE Transactions on Power Systems*, vol. 12, no. 1, February 1997, pp. 11–21.

25) Lin, D.; Batan, T.; Fuchs, E.F.; and Grady, W.M.; "Harmonic losses of single-phase induction motors under nonsinusoidal voltages," *IEEE Transactions on Energy Conversion*, vol. 11, no. 2, June 1996, pp. 273–286.

26) Fuchs, E.F.; and Lin, D.; *Testing of Power Planners at the University of Colorado*, Final report, Boulder, Colorado, February 11, 1996.

27) Fuchs, E.F.; You, Y.; and Roesler, D.J.; "Modeling, simulation and their validation of three-phase transformers with three legs under DC bias," *IEEE*

*Transactions on Power Delivery*, vol. 14, no. 2, April 1999, pp. 443–449.

28) Fuchs, E.F.; Roesler, D.J.; and Masoum, M.A.S.; "Are harmonic recommendations according to IEEE and IEC too restrictive?" *IEEE Transactions on Power Delivery*, vol. 19, no. 4, October 2004, pp. 1775–1786.

29) Fuchs, E.F.; ECEN 5787, Power Quality in Power Systems and Electric Machines, Course Notes, University of Colorado, 2006.

30) Heller, B.; and Hamata, V.; *Harmonic Field Effects in Induction Machines*, Academia, Publishing House of Czechoslovak Academy of Sciences, Prague, 1977.

31) Verma, S.P.; and Balan, A.; "Measurement techniques for vibration and acoustic noise of electrical machines," *Proceedings of the Sixth International Conference on Electrical Machines and Drives (Conf. Publ. No. 376)*, 8–10 Sept. 1993, pp. 546–551.

32) Verma, S.P.; and Li, W.; "Measurement of vibrations and radiated acoustic noise of electrical machines," *Proceedings of the Sixth International Conference on Electrical Machines and Systems*, Volume 2, 9–11 Nov. 2003, pp. 861–866.

33) Verma, S.P.; and Balan, A.; "Determination of radial-forces in relation to noise and vibration problems of squirrel-cage induction motors," *IEEE Transactions on Energy Conversion*, vol. 9, issue 2, June 1994, pp. 404–412.

34) Huang, H.; Fuchs, E.F.; Zak, Z.; and White, J.C.; "Optimization of single-phase induction motor design, parts I and II," *IEEE Transactions on Energy Conversion*, vol. EC-3, no. 2, June 1988, pp. 349–356 and pp. 357–366.

35) Fei, R.; Fuchs E.F.; and Huang, H.; "Comparison of two optimization techniques as applied to three-phase induction motor design," *IEEE Transactions on Energy Conversion*, vol. EC-4, December 1989, pp. 651–660.

36) Fuchs, E.F.; Lin D.; and Yildirim, D.; "Test results of prior art permanent-split-capacitor (PSC) motors," *Final Report*, University of Colorado, Boulder, April 20, 1996.

37) Fuchs, E.F.; Huang, H.; Vandenput, A.J.; Höll, J.; Zak, Z.; Appelbaum, J.; and Erlicki, M.; "Optimization of induction motor efficiency, vol. 2: single-phase induction motors," *Publication of the Electric Power Research Institute*, Palo Alto, California, EPRI EL-4152-CCM, May 1987, 450 pages.

38) Fuchs, E.F.; Chang, L.H.; Roesler, D.J.; and Appelbaum, J.; "Magnetizing current, iron losses and forces of three-phase induction machines at sinusoidal and nonsinusoidal terminal voltages, Parts I and II," *IEEE Transactions on Power Apparatus and Systems*, vol. PAS-103, no. 11, pp. 3303–3312 and 3313–3326.

39) Fuchs, E.F.; Roesler, D.J.; and Alashhab, F.S.; "Sensitivity of electrical appliances to harmonics and fractional harmonics of the power system's voltage, part I," *IEEE Transactions on Power Delivery*, April 1987, vol. TPWRD-2, no. 2, pp. 437–444.

40) Fuchs, E.F.; Roesler, D.J.; and Kovacs, K.P.; "Sensitivity of electrical appliances to harmonics and fractional harmonics of the power system's voltage, part II," *IEEE Transactions on Power Delivery*, April 1987, vol. TPWRD-2, no. 2, pp. 445–453.

41) de Abreu, J.P.G.; and Emanuel, A.E.; "The need to limit subharmonic injection," *Proceedings of the 9th International Conference on Harmonics and Quality of Power, vol. I*, October 1–4, 2000, pp. 251–253.

42) Fuller, J.F.; Fuchs, E.F.; and Roesler, D.J.; "Influence of harmonics on power system distribution protection," *IEEE Transactions on Power Delivery*, April 1988, vol. TPWRD-3, no. 2, pp. 546–554.

43) Oberretl, K.; "The field-harmonic theory of the squirrel cage motor taking multiple armature reaction and parallel winding branches into account," *Archiv für Elektrotechnik*, 1965, 49, pp. 343–364.

44) Oberretl, K.; "Field-harmonic theory of slip-ring motor taking multiple armature reactions into account," *Proc. IEE*, 117(1970), pp. 1667–1674.

45) Oberretl, K.; "Influence of parallel winding branches, delta connection, coil pitching, slot opening, and slot skew on the starting torque of squirrel-cage motors," *Elektrotech. Z.*, 1965, 86A, pp. 619–627.

46) Oberretl, K.; "General field-harmonic theory for three-phase, single-phase and linear motors with squirrel-cage rotor, taking multiple armature reaction and slot openings into account," *Archiv für Elektrotechnik*, 76 (1993), Part I: Theory and method of calculation, pp. 111–120, and Part II: Results and comparison with measurements, pp. 201–212.

47) Wagner, W.; "Calculation of three-phase induction motors with squirrel-cage, taking multiple armature reaction, slot openings and inter-bar currents into account," Doctoral Dissertation, Universität Dortmund (1986).

48) Serrano-Iribarnegaray, L.; and Martinez-Roman, J.; "Critical review of the analytical approaches accounting for inter-bar currents and experimental study of aging in two-speed motors for elevator drives," *Proc. IEE*, vol. 152 (2005), pp. 72–80.

49) Tenhunen, A.; and Arkio, A.; "Modeling of induction machines with skewed rotor slots," *IEE Proc. Electr. Power Appl.*, vol. 148 (2001), pp. 45–50.

50) Gottkehaskamp, R.; "Nonlinear analysis of induction machines in transient and steady state using finite-element and time-stepping method accounting for solid-iron rotor with squirrel cage," Doctoral Dissertation, University of Dortmund (1992).

51) Williamson, S.; Lim, L.H.; and Smith, A.C.; "Transient analysis of cage-induction motors using finite elements," *IEEE Transactions on Magnetics*, vol. 26 (1990), pp. 941–944.

52) Oberretl, K.; "Dependence of rotor bar currents from rotor position in squirrel-cage motors," *International Conference on Electrical Machines*, Athèns, Sept. 1980, pp. 1761–1767.

53) Oberretl, K.; "Distribution of currents in squirrel-cage rotors dependent on rotor position," *Archiv für Elektrotechnik*, 70 (1987), pp. 217–225.

54) Oberretl, K.; "Iron losses, flux pulsation and magnetic slot wedges in squirrel-cage motors," *Electr. Engineering*, 82 (2000), pp. 301–311.

55) Oberretl, K.; "New facts about parasitic torques in squirrel-cage induction motors," *Bulletin Oerlikon*, no. 389/390, 1969, pp. 130–155, in English, French, German.

56) Oberretl, K.; "Parasitic synchronous torques and oscillations in squirrel-cage induction motors, influence of transients and iron saturation," *Archiv für Elektrotechnik*, 77 (1994), Part I: Steady state, pp. 179–190, and Part II: Rapid running-up, field-dependent permeability, pp. 277–288.

57) Oberretl, K.; "13 rules to minimize stray load losses in induction motors," *Bulletin Oerlikon*, no. 389/390, 1969, pp. 1–11, in English, French, German.

58) Oberretl, K.; "Tooth breakage and tooth forces in asynchronous motors," *Electr. Engineering*, 80 (1997), pp. 309–323.

59) Frohne, H.; "Über die primären Bestimmungsgrössen der Lautstärke bei Asynchronmotoren," Doctoral Dissertation, University of Hannover (1959).

60) Jordan, H.; "Geräuscharme Electromotoren," Girardet, Essen (1950).

61) Narolski, B.; "Beiträge zur Berechnung des magnetischen Geräusches bei Asynchronmotoren," *Acta Technica CSAV* (1965), pp. 156–171.

62) Üner, Z.; "Über die Ermittlung der Lautstärke des magnetischen Lärms von Drehstrommotoren," Doctoral Dissertation, University of Hannover (1964).

63) Girgis, R.S.; and Verma, S.P.; "Resonant frequencies and vibration behavior of electrical machines as affected by teeth, windings, frame and laminations," *IEEE Transactions on PAS*, 1979, pp. 1446–1455.

64) Kuhl, W.; "Messungen zu den Theorien der Eigenschwingungen von Kreisringen beliebiger Wandstärke," *Akust. Zeitschr.*, Bd. 7 (1942), pp. 125–152.

65) Beranek, L.; *Acoustics*, McGraw-Hill, New York (1954).

66) Ellison, A.J.; and Moore, C.J.; "Calculation of acoustic power radiated by short electric machines," *Acustica* (1969), vol. 21, pp. 10–15.

67) Zhu, Z.Q.; and Howe, D.; "Improved methods for prediction of electromagnetic noise radiated by electrical machines," *Proc. IEE* (1994), vol. 1451, pp. 199–120.

68) Williamson, S.; and Smith, A.C.; "Equivalent circuits for cage induction motors with inter-bar currents," *IEEE Proc. Electric Power App.*, vol. 149 (2002).

69) Oberretl, K.; "Losses, torques and magnetic noise in induction motors with static converter supply taking multiple armature reaction and slot openings into account," *Electric Power Applications, IET*, vol. 1, no. 4, pp. 517–531, July 2007.

70) Oyegoke, B.S.; "Effect of the overhang capacitance on the fast-rising surge distribution in winding of a large motor," *IEE Proc. Sci. Meas. Technol.*, 1999. 146, no. 6, November 1999, pp. 299–303.

71) Kaufhold, M.; et al.; "Electrical stress and failure mechanism of the winding insulation in PWM-inverter-fed low-voltage induction motors," *IEEE Transactions on Industrial Electronics*, vol. 47, no. 3, April 2000, pp. 396–402.

72) Wang, J.; et al.; "Insulation fault detection in a PWM controlled induction motor – experimental design and preliminary results," Paper No. 0-7803-6, 2000 IEEE.

73) Bonnet, A.H.; "Analysis of the impact of pulse-width modulated inverter voltage waveforms on AC induction motors," *IEEE Transactions on Industry Applications*, vol. 32, no. 2, March/April 1996, pp. 386–392.

74) Penrose, H.W.; "Electric motor repair for low voltage induction motors in PWM inverter duty environments," *Proceedings of Electrical Insulation and Electrical Manufacturing and Coil Winding Conference (Cat. No. 97CH36075)*, pp. 841–848.

75) Gupta, B.K.; et al.; "Turn insulation capability of large AC motors, parts 1, 2, 3," *IEEE Transactions on Energy Conversion*, vol. EC-2, no. 4, December 1987, pp. 658–665, pp. 666–673, pp. 674–679.

76) Narang, A.; et al.; "Measurement and analysis of surge distribution in motor windings," *IEEE*

*Transactions on Energy Conversion*, vol. 4, no. 1 , March 1989, pp. 126–134.

77) Guardado, J.L.; and Cornick, K.J.; "The effects of coil parameters on the distribution of steep-fronted surges in machine windings," *IEEE Transactions on Energy Conversion*, vol. 7, no. 3, September 1992, pp. 552–556.

78) Guardado, J.L.; and Cornick, K.J.; "A computer model for calculating steep-fronted surge distribution in machine windings," *IEEE Transactions on Energy Conversion*, vol. 4, no. 1, March 1989, pp. 95–101.

79) Kinnares, V.; Jaruwanchai, P.; Suksawat, D.; and Pothivejkul, S.; "Effect of motor parameter changes on harmonic power loss in PWM fed induction machines," *IEEE International Conference on Power Electronic and Drive System, PEDS '99*, July 1999, Hong Kong.

80) Smith, K.; and Ran, L.; "A time domain equivalent circuit for the inverter-fed induction motor," *9th International Conference on Electrical Machines and Drives*, No. 486, IEE, Sept. 1999, Pages: 1–5.

81) Capolino, G.A.; and Henao, H.; "A new model for three-phase induction machine diagnosis using a simplified spectral approach," Industry Applications Conference, 2001. *Thirty-Sixth IAS Annual Meeting, Conference Record of the 2001 IEEE*, vol. 3, 30 Sept.–4 Oct. 2001, pp. 1558–1563.

82) Hurst, K.D.; and Habetler, T.G.; "Sensorless speed measurement using current harmonic spectral estimation in induction machine drives," *IEEE Transactions on Power Electronics*, vol. 11, no. 1, Jan. 1996, pp. 66–73.

83) Fruchtenicht, S.; Pittius, E.; and Seinsch, H.O.; "A diagnostic system for three-phase asynchronous machines," *Fourth International Conference on Electrical Machines and Drives*, 13–15 Sept. 1989, pp. 163–171.

84) Dorrell, D.G.; Thomson, W.T.; and Roach, S.; "Combined effects of static and dynamic eccentricity on airgap flux waves and the application of current monitoring to detect dynamic eccentricity in 3-phase induction motors," *Seventh International Conference on Electrical Machines and Drives*, 11–13 Sept. 1995, pp. 151–155.

85) Wolbank, T.M.; Macheiner, P.; Machl, J.L.; and Hauser, H.; "Detection of rotor eccentricity caused air gap asymmetries in inverter fed AC machines based on current step response," *The Fifth International Conference on Power Electronics and Drive Systems 2003, PEDS 2003*, vol. 1, 17–20 Nov. 2003, pp. 468–473.

86) Toliyat, H.A.; Arefeen, M.S.; and Parlos, A.G.; "A method for dynamic simulation of air-gap eccentricity in induction machines," *IEEE Transactions on Industry Applications*, vol. 32, no. 4, July–Aug. 1996, pp. 910–918.

87) Kral, C.; Habetler, T.G.; and Harley, R.G.; "Detection of mechanical imbalances of induction machines without spectral analysis of time-domain signals," *IEEE Transactions on Industry Applications*, vol. 40, no. 4, July–Aug. 2004, pp. 1101–1106.

88) Zitzelsberger, J.; and Hofmann, W.; "Effects of modified modulation strategies on bearing currents and operational characteristics of AC induction machines," *Industrial Electronics Society, 2004. IECON 2004. 30th Annual Conference of IEEE*, vol. 3, 2–6 Nov. 2004, pp. 3025–3030.

89) Chen, S.; Lipo, T.A.; Fitzgerald, D.; "Source of induction motor bearing currents caused by PWM inverters," *IEEE Transactions on Energy Conversion*, vol. 11, no. 1, March 1996, pp. 25–32.

90) Smith, O.J.M.; "Large low-cost single-phase SEMIHEX™ motors," *IEEE Transactions on Energy Conversion*, vol. 14, no. 4, Dec. 1999, pp. 1353–1358.

91) Smith, O.J.M.; "High-efficiency single-phase motor," *IEEE Transactions on Energy Conversion*, vol. 7, no. 3, Sept. 1992, pp. 560–569.

92) Barnes, M.L.; and Gross, C.A.; "Comparison of induction machine equivalent circuit models," *Proceedings of the Twenty-Seventh Southeastern Symposium on System Theory*, 1995, 12–14 March 1995, pp. 14–17.

93) Dubey, G.K.; *Power Semiconductor Controlled Drives*, Prentice Hall, Englewood Cliffs, New Jersey, 1989.

94) High voltage test techniques, IEC Publication 60060-2.

## 3.19 ADDITIONAL BIBLIOGRAPHY

95) Fuchs, E.F.; and Hanna, W.J.; "Measured efficiency improvements of induction motors with thyristor/ triac controllers," *IEEE Transactions on Energy Conversion*, vol. 17, no. 4, December 2002, pp. 437–444.

96) Kovacs, P.K.; *Transient Phenomena in Electrical Machines*, Akademiai Kiado, Budapest, Hungary, 1984.

97) Fuchs, E.F.; "Speed control of electric machines based on the change of the number of series turns of windings and the (induced) voltage/frequency ratio," Invention Disclosure, University of Colorado, Boulder, July 1999.

98) Fuchs, E.F.; "Elimination of blade-pitch control by operating variable-speed wind turbines under stalled conditions at any rotational speed," Invention Disclosure, University of Colorado, Boulder, February 2007.

99) Fuchs, E.F.; "Integrated starter/alternator with wide torque-speed range for automotive applications on land, water, and in air," Invention Disclosure, University of Colorado, Boulder, February 2007.

100) Fuchs, E.F.; "Electric integrated motor/generator drive for hybrid/electric automotive applications on land, water, and in air," Invention Disclosure, University of Colorado, Boulder, February 2007.

# Modeling and Analysis of Synchronous Machines

Power quality problems of synchronous machines can be of the following types due to abnormal operation:

- unbalanced load,
- torques during faults such as short-circuits (e.g., balanced three-phase short-circuit, line-to-line short-circuit), out-of-phase synchronization, unbalanced line voltages, reclosing,
- winding forces during abnormal operation and faults,
- excessive saturation of iron cores,
- excessive voltage and current harmonics,
- harmonic torques,
- mechanical vibrations and hunting,
- static and dynamic rotor eccentricities,
- bearing currents and shaft fluxes,
- insulation stress due to nonlinear sources (e.g., inverters) and loads (e.g., rectifiers),
- dynamic instability when connected to weak systems, and
- premature aging of insulation material caused by cyclic operating modes as experienced by machines, for example, in pumped-storage and wind-power plants.

The theory of synchronous machines under load was developed during the first half of the twentieth century by Blondel [1], Doherty and Nickle [2, 3], Park [4, 5], Kilgore [6], Concordia [7], and Lyon [8] – just to name a few of the hundreds of engineers and scientists who have published in this area of expertise. In these works mostly balanced steady-state, transient, and subtransient performances of synchronous machines are analyzed. Most power quality problems as listed above are neglected in these early publications because power quality was not an issue during the last century. However, the asymmetric properties of synchronous machines are well known, resulting in an infinite series of even-current harmonics in the rotor and an infinite series of odd-current harmonics in the stator. An extremely asymmetric machine is the single-phase synchronous machine. Such machines are still used today – albeit with a very strong damper winding (amortisseur) in order to attenuate the higher harmonics within the machine. This attenuation of harmonics requires large amortisseurs and results in very large machines; that is, the volume per generated power is large. Examples are the 16 2/3 Hz generator of the German railroad system [9].

Application of power electronic devices, nonlinear loads, and distributed generation (DG) due to renewable energy sources in interconnected and islanding power systems causes concerns with respect to the impact of harmonics and poor power quality on the performance and stability of synchronous generators. Although standards generally accept small to medium-sized distorting loads, for large nonlinear loads a detailed harmonic flow study is desirable. For strong power systems, studies do not extend beyond the substation level, but for weak power systems the analysis may need to include generators of larger generating plants. For the analysis of such systems and for the optimal design of newly developed DG and isolated systems as well as synchronous motor drives, accurate harmonic models of synchronous machines are required. An example for the latter is the connection of wind-farm generators (without power electronic interface) to a weak power system.

The synchronous machine is an essential component of a power system; it allows conversion from mechanical to electrical energy. It is a device that works in synchronism with the rest of the electrical network. Several frames of reference have been used to model synchronous machine operation. The first and still the most widely used model is based on the concept of the $dq0$-coordinate system [4, 5]. The synchronous machine has also been represented in $\alpha\beta0$ coordinates [10] to allow a natural transition between $abc$ and $dq0$ coordinates. Detailed models of the synchronous machine have also been developed for harmonic analysis [11–13].

All these models, however, cannot accurately describe the transient and steady-state unbalanced operation of a synchronous machine unless transient and subtransient parameters are introduced. A machine model in the $abc$-coordinate system can naturally simulate these abnormal operating conditions, because it is based on a realistic representation that can take into account the explicit time-varying

**155**

nature of the stator inductances and that of the mutual stator–rotor inductances, as well as spatial harmonic effects. In one of the first models in the *a,b,c* time domain [14], the analysis illustrates the advantages of this *abc*-coordinate system (as applied to synchronous machine representation) compared with the models based on *dq*0 and *αβ*0 coordinates. For example, a model in *abc* coordinates is used for the dynamic analysis of a three-phase synchronous generator feeding a static converter for high-voltage DC (HVDC) transmission [15]: harmonic terms up to the fourth order are introduced in the stator-inductance matrix. In more recent contributions [16, 17], *a,b,c* time-domain models of a synchronous machine are proposed where saturation effects are incorporated.

This chapter reviews the electrical and mechanical equations related to synchronous machines, and their conventional model in *dq*0 coordinates – which is suitable for sinusoidal operating conditions – is presented. It investigates the behavior of synchronous machines under faults (e.g., balanced and unbalanced short-circuits, out-of-phase synchronization, reclosing) and the influence of harmonics superimposed with the fundamental quantities, and it introduces various harmonic models of synchronous machines.

## 4.1 SINUSOIDAL STATE-SPACE MODELING OF A SYNCHRONOUS MACHINE IN THE TIME DOMAIN

A synchronous machine is a complicated electromagnetic device, and it is very important for the operation of power systems. Detailed models are needed to analyze its behavior under different (e.g., steady-state, transient, subtransient, imbalance, under the influence of harmonics) operating conditions and to understand their impact on the power system. Before introducing the conventional *dq*0 model of synchronous machines for sinusoidal operating conditions, the associated electrical and mechanical equations and magnetic nonlinearities are presented.

### 4.1.1 Electrical Equations of a Synchronous Machine

Based on the stator and rotor equivalent circuits, voltage equations can be obtained in terms of flux linkages and winding resistances [17]. According to Faraday's and Kirchhoff's laws

$$v = Ri + \frac{d\Psi}{dt}. \tag{4-1}$$

Neglecting saturation the flux linkages are proportional to the currents; thus

$$\Psi = Li. \tag{4-2}$$

Substitution of Eq. 4-2 into Eq. 4-1 yields – after solving for $d\Psi/dt$ – the differential equation

$$\frac{d\Psi}{dt} = v - RL^{-1}\Psi. \tag{4-3}$$

However, it is required to take into account the interaction between self- and mutual inductances of the windings residing on the stator and rotor members. Thus, a set of differential equations can be written in matrix form for the stator and rotor circuits as

$$v_{abc} = R_S i_{abc} + p\Psi_{abc}, \tag{4-4}$$

$$v_{fdq} = R_R i_{fdq} + p\Psi_{fdq}, \tag{4-5}$$

where subscripts *abc* and *fdq* represent the stator (e.g., *abc* components) and rotor (e.g., field, *d*- and *q*-axes components), respectively, and *p* is the differential operator.

The matrix equation for the flux linkages is

$$\begin{bmatrix} \Psi_{abc} \\ \Psi_{fdq} \end{bmatrix} = [L] \begin{bmatrix} i_{abc} \\ i_{fdq} \end{bmatrix}. \tag{4-6}$$

In Eqs. 4-1 to 4-6:

$$p\Psi = p[\Psi_{abc} \quad \Psi_{fdq}]^T = p[\Psi_a \quad \Psi_b \quad \Psi_c \quad \Psi_f$$
$$\Psi_{kd1} \quad \cdots \quad \Psi_{kdn} \Psi_{kq1} \cdots \Psi_{kqn}]^T, \tag{4-7}$$

$$v = [v_{abc} \quad v_{fdq}]^T = [v_a \quad v_b \quad v_c \quad v_f \quad v_{kd1} \quad \cdots$$
$$v_{kdn} \quad v_{kq1} \quad \cdots \quad v_{kqn}]^T, \tag{4-8}$$

$$i = [i_{abc} \quad i_{fdq}]^T = [i_a \quad i_b \quad i_c \quad i_f \quad i_{kd1} \quad \cdots \quad i_{kdn} \quad i_{kq1}$$
$$\cdots \quad i_{kqn}]^T, \tag{4-9}$$

$$R = \text{diag}[R_a \quad R_b \quad R_c \quad R_f \quad R_{kd1} \quad \cdots \quad R_{kdn} \quad R_{kq1}$$
$$\cdots \quad R_{kqn}]^T, \tag{4-10}$$

where *kdn* and *kqn* are the number of damper windings of the *d*- and *q*-axes, respectively. The inductance matrix *L* has the form

$$[L] = \begin{bmatrix} L_{SS} & L_{SR} \\ L_{RS} & L_{RR} \end{bmatrix}. \tag{4-11}$$

$L_{SS}$ are stator self-inductances, $L_{SR}$ are stator–rotor mutual inductances, where $L_{SR} = [L_{RS}]^T$, and $L_{RR}$ are rotor self-inductances. $T$ indicates the transpose of a matrix. Note that $L_{SS}$, $L_{SR}$, and $L_{RS}$ consist of time-varying inductances representing the interrelationships between the stator and rotor windings as a function of the rotor position angle $\theta$.

Therefore, the inductance matrix is a function of time due to $\theta$ which itself is a function of time (e.g., $\theta = \omega_r t + \delta - \pi/2$, where $\omega_r$ and $\delta$ are rotor angular velocity and rotor angle, respectively):

$$
L = \begin{bmatrix}
L_{aa} & L_{ab} & L_{ac} & L_{af} & L_{akd} & L_{akq} & L_{akD} & L_{akQ} \\
L_{ba} & L_{bb} & L_{bc} & L_{bf} & L_{bkd} & L_{bkq} & L_{dkD} & L_{bkQ} \\
L_{ca} & L_{cb} & L_{cc} & L_{cf} & L_{ckd} & L_{ckq} & L_{ckD} & L_{ckQ} \\
L_{fa} & L_{fb} & L_{fc} & L_{ff} & L_{fkd} & 0 & L_{fkD} & 0 \\
L_{kda} & L_{kdb} & L_{kdc} & L_{kdf} & L_{kkd} & 0 & L_{kdkD} & 0 \\
L_{kqa} & L_{kqb} & L_{kqc} & 0 & 0 & L_{kkq} & 0 & L_{kqkQ} \\
L_{kDa} & L_{kDb} & L_{kDc} & L_{kDf} & L_{kDkd} & 0 & L_{kkD} & 0 \\
L_{kQa} & L_{kQb} & L_{kQc} & 0 & 0 & L_{kQkq} & 0 & L_{kkQ}
\end{bmatrix}.
$$
(4-12a)

In the above equations, only two damper windings along the $d$- and $q$-axes are assumed and [17]:

- The entries of $L_{RR}$ represent – neglecting saturation – constant inductances, independent of the rotor position $\theta$. $L_{RR}$ has the form

$$
L_{RR} = \begin{bmatrix}
L_{ff} & L_{fkd} & 0 & L_{fkD} & 0 \\
L_{kdf} & I_{kkd} & 0 & L_{kdkD} & 0 \\
0 & 0 & L_{kkq} & 0 & L_{kqkQ} \\
L_{kDf} & L_{kDkd} & 0 & L_{kkD} & 0 \\
0 & 0 & L_{kQkq} & 0 & L_{kkQ}
\end{bmatrix}.
$$
(4-12b)

- Considering even harmonics ($h$) up to the $n$th order, $L_{SS}$ has the form

$$
L_{SS} = \begin{bmatrix}
L_{aa0} + \sum_{h=2}^{n} L_{aa_h} \cos h\theta & -L_{ab_0} - \sum_{h=2}^{n} L_{ab_h} \cos h(\theta + \pi/6) & -L_{ac_0} - \sum_{h=2}^{n} L_{ac_h} \cos h(\theta + 5\pi/6) \\
-L_{ba_0} - \sum_{h=2}^{n} L_{ba_h} \cos h(\theta + \pi/6) & L_{bb_0} + \sum_{h=2}^{n} L_{bb_h} \cos h(\theta - 2\pi/3) & -L_{bc_0} - \sum_{h=2}^{n} L_{bc_h} \cos h(\theta - \pi/2) \\
-L_{ac_0} - \sum_{h=2}^{n} L_{ac_h} \cos h(\theta + 5\pi/6) & -L_{cb_0} - \sum_{h=2}^{n} L_{cb_h} \cos h(\theta - \pi/2) & L_{cc_0} + \sum_{h=2}^{n} L_{cc_h} \cos h(\theta + 2\pi/3)
\end{bmatrix}.
$$
(4-12c)

- Considering odd harmonics ($h$) up to the mth order, $L_{SR}$ has the form

$$
L_{SR} = \begin{bmatrix}
\sum_{h=1}^{m} M_{f_h} \cos h\theta & \sum_{h=1}^{m} M_{kd_h} \cos h\theta & -\sum_{h=1}^{m} M_{kq_h} \cos h\theta & \sum_{h=1}^{m} M_{kD_h} \cos h\theta & -\sum_{h=1}^{m} M_{kQ_h} \cos h\theta \\
\sum_{h=1}^{m} M_{f_h} \cos h(\theta - 2\pi/3) & \sum_{h=1}^{m} M_{kd_h} \cos h(\theta - 2\pi/3) & -\sum_{h=1}^{m} M_{kq_h} \cos h(\theta - 2\pi/3) & \sum_{h=1}^{m} M_{kD_h} \cos h(\theta - 2\pi/3) & -\sum_{h=1}^{m} M_{kQ_h} \cos h(\theta - 2\pi/3) \\
\sum_{h=1}^{m} M_{f_h} \cos h(\theta + 2\pi/3) & \sum_{h=1}^{m} M_{kd_h} \cos h(\theta + 2\pi/3) & -\sum_{h=1}^{m} M_{kq_h} \cos h(\theta + 2\pi/3) & \sum_{h=1}^{m} M_{kD_h} \cos h(\theta + 2\pi/3) & -\sum_{h=1}^{m} M_{kQ_h} \cos h(\theta + 2\pi/3)
\end{bmatrix}.
$$
(4-12d)

Combining Eqs. 4-3 to 4-12 results in a unified representation for the electrical part of the synchronous machine having the form

$$p\Psi = v - RL^{-1}\Psi.$$
(4-13)

It is important to note that, for the computation of the state variables with Eq. 4-13, the inverse of the time-varying inductance matrix must be obtained at each step of integration.

The electric torque is expressed in terms of the stator currents and flux linkages as

$$T_e = p\frac{2}{3\sqrt{3}}[\Psi_a(i_b - i_c) + \Psi_b(i_c - i_a) + \Psi_c(i_a - i_b)],$$
(4-14)

where $p$ is the number of poles.

### 4.1.2 Mechanical Equations of Synchronous Machine

The second-order mechanical equation of a synchronous machine can be decomposed into two first-order differential equations: one for the mechanical angular velocity of the rotor $\omega_r$ and the other for the mechanical rotor angle $\delta$. The constant $k_D$ is usually incorporated into the angular velocity equation to add a damping component that is proportional to the difference of angular velocities $\omega_r$ and $\omega_B$:

$$\frac{d\omega_r}{dt} = \frac{\omega_B}{2H}(T_m - T_e - k_D(\omega_r - \omega_B)). \quad (4\text{-}15)$$

$$\frac{d\delta}{dt} = \omega_r - \omega_B, \quad (4\text{-}16)$$

where

$$H = \frac{1}{2}\frac{J\omega_B^2}{VA_{\text{base}}}\text{ seconds.} \quad (4\text{-}17)$$

In the above equations $\omega_B$ is the rated angular velocity of the rotor, $k_D$ is the damping factor (pu torque/pu angular velocity), $H$ is the unit inertia constant (watt-s/VA at rated angular velocity), $J$ is the axial combined (rotor and prime mover) moment of inertia (kg m$^2$), and $VA_{\text{base}}$ is the apparent power base $S_{\text{base}}$.

### 4.1.3 Magnetic Saturation of Synchronous Machine

This section introduces a simple approach to approximately include magnetic saturation in the synchronous machine equations [17]. The nonlinear effect of saturation in a synchronous machine is modeled according to the following algorithm, which is based on modifying the stator currents in $dq0$ coordinates.

- For the input stator currents and fluxes (e.g., $i_a$, $i_b$, $i_c$, and $\Psi_a$, $\Psi_b$, $\Psi_c$) apply the $dq0$ transformation to obtain $i_d$, $i_q$, $i_o$, and $\Psi_d$, $\Psi_q$, $\Psi_o$:

$$\begin{bmatrix} i_d \\ i_q \\ i_o \end{bmatrix} = [T_{\text{PARK}}]\begin{bmatrix} i_a \\ i_b \\ i_c \end{bmatrix}$$
$$\begin{bmatrix} \Psi_d \\ \Psi_q \\ \Psi_o \end{bmatrix} = [T_{\text{PARK}}]\begin{bmatrix} \Psi_a \\ \Psi_b \\ \Psi_c \end{bmatrix}, \quad (4\text{-}18a)$$

where the Park transformation is

$$[T_{\text{PARK}}] = \sqrt{\frac{2}{3}}\begin{bmatrix} \cos\theta & \cos(\theta - 120°) & \cos(\theta + 120°) \\ -\sin\theta & -\sin(\theta - 120°) & -\sin(\theta + 120°) \\ 1/\sqrt{2} & 1/\sqrt{2} & 1/\sqrt{2} \end{bmatrix}.$$
$$(4\text{-}18b)$$

- Calculate the $dq$-components of the total mmf in the machine as

$$I_d = i_d + i_f + \sum_{j=1}^{kdn} i_{kdj}$$
$$I_q = i_q + \sum_{j=1}^{kqn} i_{kqj}, \quad (4\text{-}19)$$

where $kdn$ and $kqn$ are the number of damper windings of $d$- and $q$-axes, respectively.

- For given saturation characteristics of the $d$- and $q$-axes, saturated mmfs ($I'_d$ and $I'_q$) are calculated based on a polynomial approximation, for example,

$$I'_d = \alpha_d\Psi_d + \beta_d\Psi_d^n$$
$$I'_q = \alpha_q\Psi_q + \beta_q\Psi_q^m. \quad (4\text{-}20)$$

- The relationship between the saturated mmfs of Eq. 4-20 and linear mmfs of Eq. 4-19 are

$$K_d = I'_d / I_d$$
$$K_q = I'_q / I_q. \quad (4\text{-}21)$$

- Stator and rotor currents are now adjusted for saturation, as follows:

$$\begin{aligned} i'_a &= K_d i_a, \quad i'_b = K_d i_b, \quad i'_c = K_d i_c \\ i'_f &= K_d i_f \\ i'_{kd} &= K_d i_{kd}, \quad i'_{kD} = K_d i_{kD} \\ i'_{kq} &= K_q i_{kq}, \quad i'_{kQ} = K_q i_{kQ} \end{aligned} \quad (4\text{-}22)$$

### 4.1.4 Sinusoidal Model of a Synchronous Machine in $dq0$ Coordinates

Several reference frames can be employed to represent synchronous machine operation. The first and most widely used model is based on the concept of the ideal synchronous machine represented by fictitious $dq0$ coordinates attached to the rotating rotor reference frame [4, 5], as shown in Fig. 4.1. This model assumes balanced operating conditions at the input terminals and within the machine. Synchronous machines can be also represented in $\alpha\beta0$ coordinates [10] to allow for a natural transition between $abc$ and $dq0$ coordinates.

### 4.2 STEADY-STATE, TRANSIENT, AND SUBTRANSIENT OPERATION

Numerical methods such as the finite-difference and finite-element methods [18] enable the engineer to

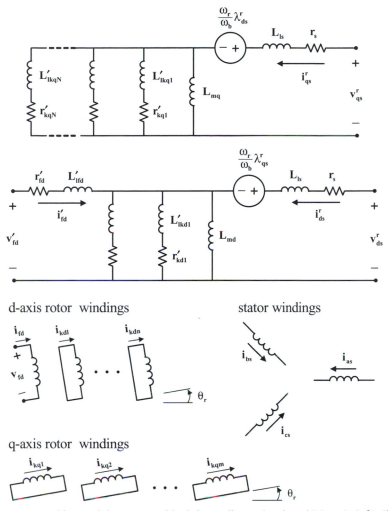

**FIGURE 4.1** Synchronous machine model represented in dq0 coordinates for sinusoidal analysis [1, 2].

compute no-load and full-load magnetic fields and those associated with short-circuit and starting conditions, as well as fields for the calculation of leakage, synchronous, transient, and subtransient inductances or reactances. Figures 4.2, 4.3, and 4.4 represent the no-load, short-circuit, and full-load fields, respectively, of a four-pole generator [19, 20]. Figure 4.3 illustrates the winding forces during steady-state short-circuit conditions. These forces were computed based on the "Maxwell stress" discussed later in this chapter. Figure 4.2b [21] shows that at rated voltage only a small amount of flux (e.g., 1.4 mT) enters the space between the iron core (stator back iron) and the stator frame, and the eddy currents induced in the solid key bars are negligible. Figures 4.5 and 4.6 illustrate synchronous linear $L_d$, $L_q$ inductances or synchronous $X_d$, $X_q$ reactances [19, 20, 22] of a two-

pole generator. Figures 4.5 and 4.6 do not indicate any coupling between the $d$- and $q$-axes. Such a coupling is due to the saturation and is called cross-coupling [23], which will be neglected in subsequent sections. Figure 4.7 pertains to the calculation of the transient inductance $L_d'$ or transient reactance $X_d'$ [19, 22]. Figure 4.8 illustrates the field of a two-pole generator at full load.

In general, for synchronous machines with a field-current excitation in the d-axis the following inequalities are valid: $X_d \geq X_q$, $X_d' \leq X_q'$, and $X_d'' \leq X_q''$. However, for inverter-fed synchronous machines with permanent-magnet excitation – also called brushless DC machines – as used in high-efficiency, variable-speed drives [24, 25], due to the lack of field and damper windings in such machines, the transient and subtransient reactances are not defined. Figures

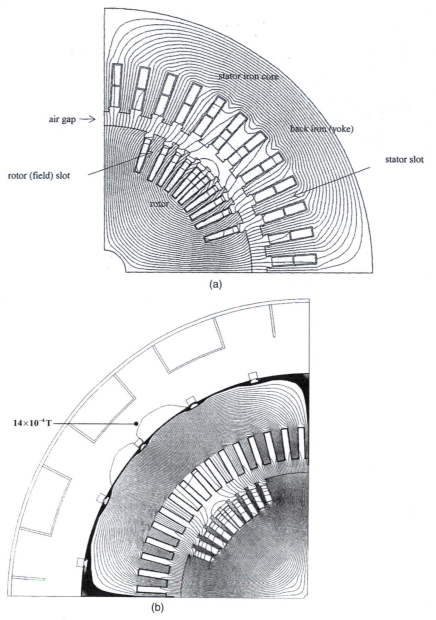

(a)

(b)

**FIGURE 4.2** (a) No-load field of four-pole synchronous generator. (b) No-load field of four-pole synchronous generator including leakage fields outside the iron core [21].

4.9 and 4.10 depict no-load and full-load fields of a permanent-magnet machine, respectively, whereas Fig. 4.11 illustrates the field for the calculation of the stator self-inductance (e.g., $L_q$). In permanent-magnet machines the leakage is defined as the leakage flux between two stator winding phases and not the leakage flux between the stator and the rotor windings (which do not exist) as it is normally defined for other types of machines. In addition $L_q$ is larger than $L_d$, and there is a significant amount of cogging [26].

### 4.2.1 Definition of Transient and Subtransient Reactances as a Function of Leakage and Mutual Reactances

The $dq0$ model requires the availability of values for the synchronous ($X_d$, $X_q$), transient ($X_d'$, $X_q'$), and subtransient ($X_d''$, $X_q''$) reactances of synchronous machines. Most turbogenerators have field windings in the $d$-axis only and none in the $q$-axis, therefore $X_q' \approx X_q$. The transient reactance $X_d'$ can be defined in terms of leakage and mutual reactances, as depicted in Fig. 4.12:

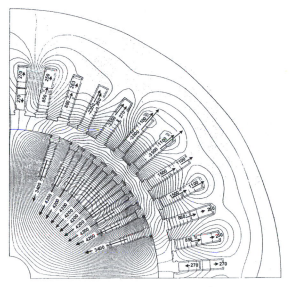

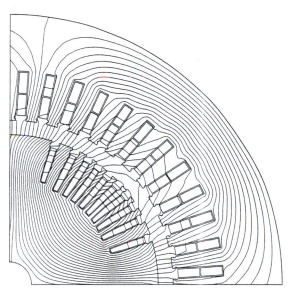

**FIGURE 4.3** Short-circuit field and force (measured in N/m, see Section 4.2.14) distribution of four-pole synchronous generator. The arrows indicate the direction and value (length of arrow) of the winding forces developed due to balanced steady-state, three-phase short-circuit.

**FIGURE 4.4** Full-load field of four-pole synchronous generator.

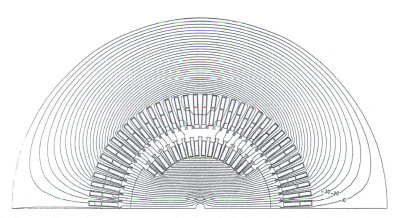

**FIGURE 4.5** Field of two-pole synchronous generator for the calculation of either $L_d$ or $X_d$.

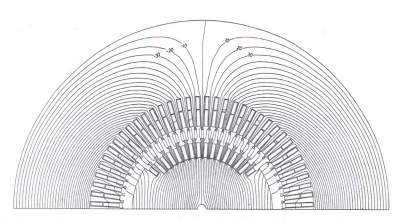

**FIGURE 4.6** Field of two-pole synchronous generator for the calculation of either $L_q$ or $X_q$.

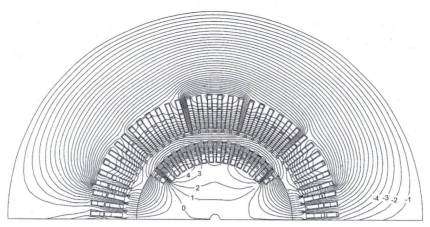

**FIGURE 4.7** Field of two-pole synchronous generator for the calculation of either $L_d'$ or $X_d'$.

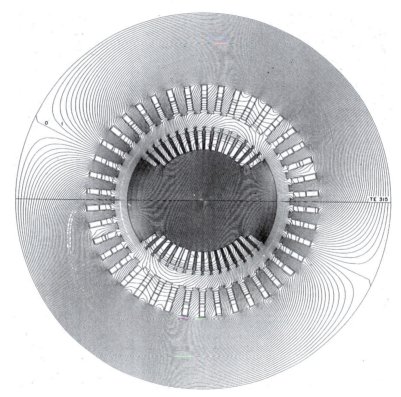

**FIGURE 4.8** Magnetic field of a two-pole, three-phase synchronous machine under full load [27].

$$X_d' = X_{al} + \frac{X_{af}X_{fl}}{X_{af} + X_{fl}}, \qquad (4\text{-}23)$$

where $X_{al}$ is the amature (stator) leakage reactance, $X_{fl}$ is the field leakage reactance, and $X_{af}$ is the mutual reactance between armature and field circuits.

Correspondingly, one obtains for the subtransient reactance $X_d''$ (Fig. 4.13):

$$X_d'' = X_{al} + \frac{X_{af}X_{fl}X_{dl}}{X_{af}X_{fl} + X_{af}X_{dl} + X_{dl}X_{fl}}, \quad (4\text{-}24)$$

where $X_{dl}$ is the d-axis damper winding leakage reactance.

Representative per-unit values for a typical turbogenerator are

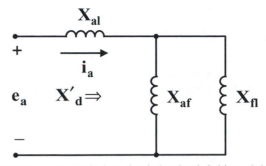

FIGURE 4.12 Equivalent circuit for the definition of the transient reactance $X'_d$.

**FIGURE 4.9** Flux distribution at no load of a permanent-magnet machine.

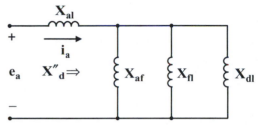

FIGURE 4.13 Equivalent circuit for the definition of the subtransient reactance $X''_d$.

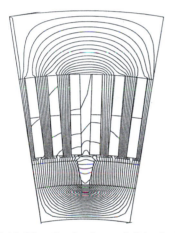

**FIGURE 4.10** Flux distribution at full load of a permanent-magnet machine.

$$X_d = 1.6, \; X_q = 1.4, \; X'_d = 0.2, \; X''_d = 0.15. \quad (4\text{-}25)$$

In general, $1 \text{ pu} \leq X_d \leq 2 \text{ pu}$, $1 \text{ pu} \leq X_q \leq 2 \text{ pu}$, $0.2 \text{ pu} \leq X'_d \leq 0.4 \text{ pu}$, $0.1 \text{ pu} \leq X''_d \leq 0.3 \text{ pu}$, and the values of $X'_q$ and $X''_q$ depend on the field and damper winding design within the $q$-axis, respectively. Note that the value of $X''_q$ depends on the rotor construction and the damper winding associated with the q-axis: if there are no additional damper bars located on the rotor pole, then $X''_q$ is relatively large and in the neighborhood of $X_q$. In the presence of damper windings in the q-axis $X''_q$ is somewhat larger than $X''_d$ because the damper winding (e.g., solid rotor or embedded damper bars on the rotor pole) of the $q$-axis is less effective than that of the d-axis. The rationale for selecting the above ranges for the various reactances lies in the stability requirements for generators when operating on the power system: machines with small reactances exhibit better stability than machines with large reactances; or in other words, machines with small reactances or inductances have smaller time constants than those with large reactances or inductances. Small reactances require a large air gap and a large field current for setting up the maximum flux density across the air gap of about $B_{\max} = 0.7 \; T$. The large air gap makes

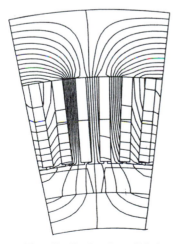

**FIGURE 4.11** Flux distribution for self-inductance calculation ($L_q$) of a permanent-magnet machine. $\phi = Li$

machines less efficient as compared to those with small air gaps. One has to weigh the benefits of a more stable but less efficient machine versus that with less stability but greater efficiency. In addition, machines with small reactances result in larger short-circuit currents and larger transient torques and winding forces than machines with larger reactances.

### 4.2.2 Phasor Diagrams for Round-Rotor Synchronous Machines

There are two ways of drawing equivalent circuits and phasor diagrams of synchronous machines. The first one is based on the so-called consumer system, where the terminal current flows into the equivalent circuit, and the second one is the generator system, where the terminal current flows out of the equivalent circuit. The first one is mostly employed by engineers concerned with drives, and the latter one is mostly used by power system engineers who are concerned with generation issues.

#### 4.2.2.1 Consumer (Motor) Reference Frame

Figure 4.14a depicts the equivalent circuit of a round-rotor synchronous machine based on the consumer (motor) reference current system. Figure 4.14b illustrates the corresponding phasor diagram for overexcited operation, and Fig. 4.14c pertains to underexcited operation. Note that the polarity of the voltages is indicated by + and − signs as well as by arrows, where the head of the arrow coincides with the + sign and the tail of the arrow with the − sign. Such an arrow notation makes it easier to draw phasor diagrams. The phasor diagrams are not to scale because the ohmic voltage drop $\tilde{V}_R$ is normally much smaller than the reactive voltage drop $\tilde{V}_X$.

#### 4.2.2.2 Generator Reference Frame

Figure 4.15a depicts the equivalent circuit of a round-rotor synchronous machine based on the generator reference current system. Figure 4.15b,c illustrate the corresponding phasor diagram for overexcited and underexcited operation, respectively.

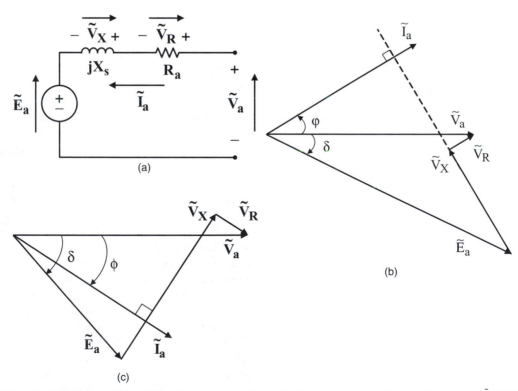

**FIGURE 4.14** (a) Equivalent circuit based on consumer (motor) reference frame for the terminal current $\tilde{I}_a$. (b) Phasor diagram based on consumer (motor) reference frame for leading (overexcited) terminal current $\tilde{I}_a$. (c) Phasor diagram based on consumer (motor) reference frame for lagging (underexcited) terminal current $\tilde{I}_a$.

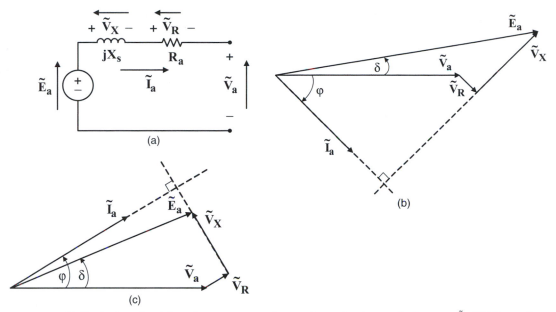

**FIGURE 4.15** (a) Equivalent circuit based on generator reference frame for the terminal current $\tilde{I}_a$. (b) Phasor diagram based on generator reference frame for lagging (overexcited) terminal current $\tilde{I}_a$. (c) Phasor diagram based on generator reference frame for leading (underexcited) terminal current $\tilde{I}_a$.

### 4.2.2.3 Similarities between Synchronous Machines and Pulse-Width-Modulated (PWM) Current-Controlled, Voltage-Source Inverters

Inverters are electronic devices transforming voltages and currents from a DC source to an equivalent AC source. Figure 4.16a illustrates the actual circuit of an inverter, where the input voltage is a DC voltage $V_d$ and the output voltage is an AC voltage $\tilde{V}_{load}$. It is assumed that the PWM switching is lossless. In Fig. 4.16b the DC voltage ($V_d/2$) is transformed to the AC side and represented by the phasor ($\tilde{V}_d/2$), which makes the equivalent circuit of Fig. 4.16b similar to that of a round-rotor synchronous machine, as depicted in Fig. 4.15a. The relation between the (fundamental) output voltage $|\tilde{V}_{load}|$ of the inverter and the input voltage $\tilde{V}_d/2$ (transformed to the AC side) is given [28] by

$$|\tilde{V}_{load}| = m \cdot \frac{(V_d/2)}{\sqrt{2}}, \qquad (4\text{-}26)$$

where $m \leq 1$ is the modulation index of the (sinusoidal) PWM. This relation appears to hold for operation around unity displacement (fundamental power) factor only [29, 30]. In these references it is shown that at lagging (overexcited, delivering reactive power to the grid) displacement power factor a much

higher input voltage $V_d$ than specified by Eq. 4-26 is required, whereas for leading (underexcited, absorbing reactive power from the grid) displacement power factor a smaller input voltage $V_d$ is sufficient. For example, although for unity displacement factor $V_{d\_unity\_pf} = 400$ V is acceptable, a much higher input voltage, e.g., $V_{d\_lagging\_pf} = 600$ V is required for displacement factors larger than $\cos \varphi = 0.8$ lagging (overexcited) [29, 30] at $\tilde{V}_{load} = 139$ V.

### 4.2.2.4 Phasor Diagram of a Salient-Pole Synchronous Machine

The $d$- and $q$-equivalent circuits and the associated phasor diagram of a (balanced) salient-pole synchronous machine are given in Fig. 4.17a,b,c. Note there is no zero-sequence equivalent circuit, which becomes important for unbalanced operation (e.g., asymmetrical short-circuits) of synchronous machines only. Figure 4.17c represents the phasor diagram without cross-coupling between the $d$- and $q$-axes due to saturation. The relation between voltages, currents, and reactances is given by the phasor diagram of Fig. 4.18 including cross-coupling of $d$- and $q$-axes parameters [20, 23]. This cross-coupling is due to saturation of the machine and will be neglected in subsequent sections.

According to [7] the following relations are valid:

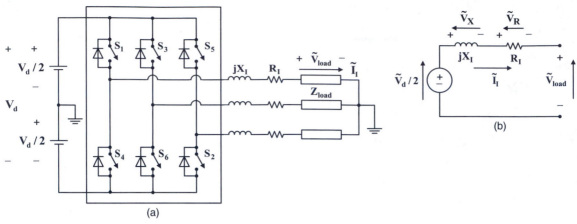

**FIGURE 4.16** (a) Equivalent circuit for inverter (feeding a load) where input voltage $V_d$ is referred to the primary side. (b) Equivalent circuit for inverter where input voltage $\tilde{V}_d/2$ is referred to the secondary side.

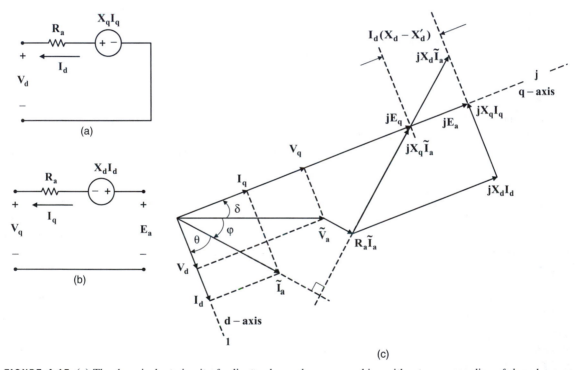

**FIGURE 4.17** (a) The d-equivalent circuit of salient-pole synchronous machine without cross-coupling of $d$- and $q$-axes. (b) The q-equivalent circuits of salient-pole synchronous machine without cross-coupling of $d$- and q-axes. (c) Phasor diagram of salient-pole synchronous machine without cross-coupling of $d$- and $q$-axes.

Terminal voltage component in d-axis as a function of the fluxes and currents in $d$- and $q$-axes:

$$V_d = -\Psi_q - R_a I_d = X_q I_q - R_a I_d. \qquad (4\text{-}27)$$

Terminal voltage component in $q$-axis:

$$V_q = \Psi_d - R_a I_q = E_a - X_d I_d - R_a I_q. \qquad (4\text{-}28)$$

Eqs. 4-27 and 4-28 solved for $I_d$ and $I_q$ yield

$$I_d = \frac{-R_a V_d + X_q (E_a - V_q)}{X_d X_q + R_a^2}, \qquad (4\text{-}29)$$

$$I_q = \frac{X_d V_d + R_a (E_a - V_q)}{X_d X_q + R_a^2}, \qquad (4\text{-}30)$$

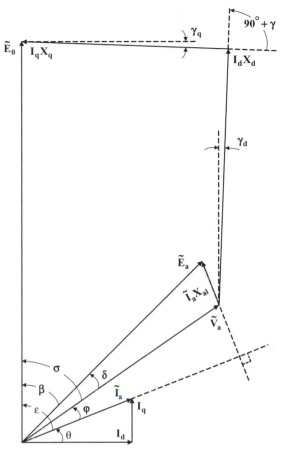

**FIGURE 4.18** Phasor diagram of salient-pole synchronous machine with cross-coupling of d- and q-axes [19, 20].

which may be written in terms of the terminal voltage $\tilde{V}_a$:

$$I_d = \frac{-R_a|\tilde{V}_a|\sin\delta + X_q(E_a - |\tilde{V}_a|\cos\delta)}{X_d X_q + R_a^2}, \quad (4\text{-}31)$$

$$I_q = \frac{X_d|\tilde{V}_a|\sin\delta + R_a(E_a - |\tilde{V}_a|\cos\delta)}{X_d X_q + R_a^2}, \quad (4\text{-}32)$$

where

$$|\tilde{V}_a| = \sqrt{V_d^2 + V_q^2}, \quad (4\text{-}33)$$

$$|\tilde{I}_a| = \sqrt{I_d^2 + I_q^2}. \quad (4\text{-}34)$$

### 4.2.3 Application Example 4.1: Steady-State Analysis of a Nonsalient-Pole (Round-Rotor) Synchronous Machine

A three-phase ($m = 3$), four-pole ($p = 4$) nonsalient-pole (round-rotor) synchronous machine has the parameters $X_S = 2$ pu and $R_a = 0.05$ pu. It is operated at (the phase voltage) $V_a = 1$ pu at an overexcited (lagging current based on the generator reference system, where the current flows out of the machine) displacement power factor of $\cos\varphi = 0.8$ lagging, and per-phase current $|\tilde{I}_a| = 1$ pu. The base (rated phase) values are $|\tilde{V}_a|_{\text{rated}} = V_{\text{base}} = 24$ kV, $|\tilde{I}_a|_{\text{rated}} = I_{\text{base}} = 1.4$ k A, and the base impedance is $Z_{\text{base}} = V_{\text{base}}/I_{\text{base}} = 17.14\ \Omega$.

1. Draw the per-phase equivalent circuit of this machine.
2. What is the total rated apparent input power $S$ (expressed in MVA)?
3. What is the total rated real input power $P$ (expressed in MW)?
4. Draw a per-phase phasor diagram with the voltage scale of $1.0$ pu $\equiv 3$ inches, and the current scale of $1.0$ pu $\equiv 2.5$ inches.
5. From this phasor diagram determine the per-phase induced voltage $|\tilde{E}_a|$ and the torque angle $\delta$.
6. Calculate the rated speed (in rpm) of this machine at $f = 60$ Hz.
7. Calculate the angular velocity $\omega_S$ (in rad/s).
8. Calculate the total (approximate) shaft power $P_{\text{shaft}} \approx 3(E_a \cdot V_a \cdot \sin\delta)/X_S$.
9. Find the shaft torque $T_{\text{shaft}}$ (in Nm).
10. Determine the total ohmic loss of the motor $P_{\text{loss}} = 3 \cdot R_a|\tilde{I}_a|^2$.
11. Repeat the above analysis for $\cos\varphi = 0.8$ underexcited (leading current based on generator notation) displacement power factor and $|\tilde{I}_a| = 0.5$ pu.

### 4.2.4 Application Example 4.2: Calculation of the Synchronous Reactance $X_S$ of a Cylindrical-Rotor (Round-Rotor, Nonsalient-Pole) Synchronous Machine

A $p = 2$ pole, $f = 60$ Hz synchronous machine (generator, alternator, motor) is rated $S_{\text{rat}} = 16$ MVA at a lagging (generating reference system) displacement power factor $\cos\varphi = 0.8$, a line-to-line terminal voltage of $V_{L\text{-}L} = 13,800$ V, and a (one-sided) air-gap length of $g = 0.0127$ m. The stator has 48 slots and has a three-phase double-layer, 60° phase-belt winding with 48 armature coils having a short pitch of 2/3. Each coil consists of one turn $N_{\text{coil}} = 1$. Sixteen stator coils are connected in series, that is, the number of series turns per phase of the stator (st) are $N_{st\,\text{phase}} = 16$ (see Fig. E4.2.1). The maximum air-gap flux density at no load (stator current $I_{st\,\text{phase}} = 0$) is $B_{st\,\text{max}} = 1$ T.

There are eight field coils per pole (see Fig. E4.2.2) and they are pitched 44–60–76–92–124–140–156–172 degrees, respectively, as indicated in Fig. E4.2.2. Each field coil has 18 turns. The developed rotor (field) magnetomotive force (*mmf*) $F_r$ is depicted in Fig. E4.2.3. Note that the field *mmf* is approximately sinusoidal.

a) Calculate the distribution (breadth) factor of the stator winding $k_{st\,b}$ ($=k_{st\,d}$).

b) Calculate the fundamental pitch factor of the stator winding $k_{st\,p1}$.

c) Calculate the pitch factor of the rotor (*r*) winding $k_{r\,p}$.

d) Determine the stator flux $\varphi_{stmax}$.

e) Compute the area ($area_p$) of one pole.

f) Compute the field (*f*) current $I_{fo}$ required at no load.

g) Determine the synchronous (*s*) reactance (per phase value) $X_S$ of this synchronous machine in ohms.

h) Find the base impedance $Z_{base}$ expressed in ohms.

i) Express the synchronous reactance $X_S$ in per unit (pu).

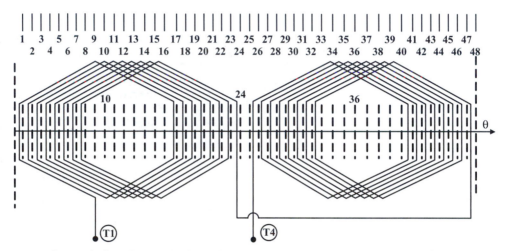

**FIGURE E4.2.1** Stator winding of two-pole, three-phase synchronous machine.

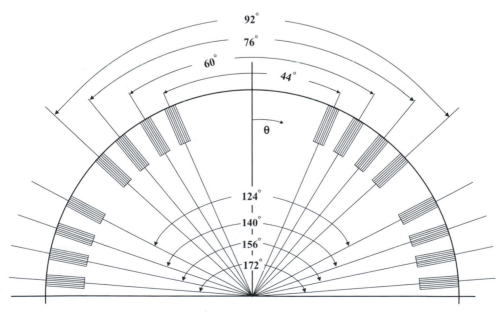

**FIGURE E4.2.2** Rotor winding of two-pole, three-phase synchronous machine.

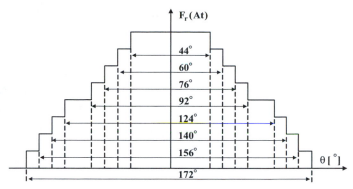

**FIGURE E4.2.3** Rotor mmf $F_r$ of two-pole, three-phase synchronous machine.

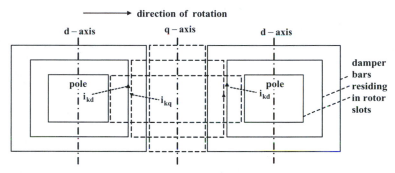

**FIGURE E4.4.1** Decomposition of amortisseur winding of a two-pole generator [40]. In this graph only six damper windings are shown: three in the d-axis and three in the $q$-axis. Figures E4.4.2 to E4.4.4 and E4.4.11 to E4.4.13 provide details about the location of the 16 damper bars within the rotor slots.

### 4.2.5 Application Example 4.3: *dq*0 Modeling of a Salient-Pole Synchronous Machine

Engineers are most concerned about the torques generated by the machine under abnormal operating conditions such as

- balanced three-phase short-circuit,
- line-to-line short-circuit,
- line-to-neutral short-circuit,
- out-of-phase synchronization,
- synchronizing and damping torques,
- reclosing, and
- stability.

Linear (not saturation-dependent) reactances are assumed and amortisseur (damper) windings are neglected except for some out-of-phase synchronization results and for reclosing events. Most machines have very weak damper windings and, therefore, this assumption is justified.

### 4.2.6 Application Example 4.4: Calculation of the Amortisseur (Damper Winding) Bar Losses of a Synchronous Machine during Balanced Three-Phase Short-Circuit, Line-to-Line Short-Circuit, Out-of-Phase Synchronization, and Unbalanced Load Based on the Natural abc Reference System [40]

For the analysis of subtransient phenomena, a simulation of the damper-winding system of alternators is of utmost importance. Concordia [7] expanded the usual d- and q- axes decomposition by introducing more than two (e.g., $d$ and $q$) damper winding systems. In this application example the amortisseur is approximated by 16 damper windings in the $d$- and $q$-axes as shown in Fig. E4.4.1, that is, 8 damper windings in the $d$-axis and 8 damper windings in the $q$-axis. Damper bars reside in the rotor slots and some bars are embedded in the solid rotor pole zone. To simplify this analysis it is assumed that the rotor is laminated; that means no eddy currents are induced, saturation is neglected, and there are no damper bars embedded in the pole zone.

### 4.2.7 Application Example 4.5: Measured Voltage Ripple of a 30 kVA Permanent-Magnet Synchronous Machine, Designed for a Direct-Drive Wind-Power Plant

A permanent-magnet generator is to be designed [29, 30] for a variable-speed drive of a wind-power plant without any mechanical gear. The rated speed range is from 60 to 120 rpm. Line-to-neutral voltages $V_{L-N}$ and line-to-line voltages $V_{L-L}$ are recorded at no load and under rated load.

The voltage wave shapes are subjected to a Fourier analysis. Figures E4.5.1 and E4.5.2 depict the line-to-neutral voltages at no load and rated load, respectively, and Tables E4.5.1 and E4.5.2 represent their Fourier coefficients. Figures E4.5.3 and E4.5.4 represent the line-to-line voltages at no load and at rated load, and Tables E4.5.3 and E4.5.4 list their spectra. To minimize the size of this low-speed generator, the stator pitch $k_p$ and distribution $k_d$ winding factors are chosen relatively high $(k_p \cdot k_d = 0.8)$ resulting in a somewhat nonsinusoidal voltage wave shape. The generator feeds a rectifier and for this

reason the nonsinusoidal voltage wave shape does not matter, except it increases the generator losses within acceptable limits.

### 4.2.8 Application Example 4.6: Calculation of Synchronous Reactances $X_d$ and $X_q$ from Measured Data Based on Phasor Diagram

Synchronous reactances are defined for fundamental frequency. Subjecting the voltage and current to a Fourier series yields the following data: line-to-neutral terminal voltage $V_{(l-n)}$, phase current $I_{(l-n)}$, instead of the induced voltage $E$ the no-load voltage

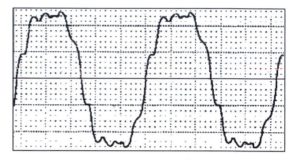

**FIGURE E4.5.2** Measured line-to-neutral $V_{L-N}$ voltage wave shape of permanent-magnet generator at load designed for wind-power application [44].

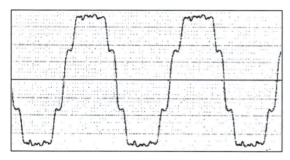

**FIGURE E4.5.1** Measured line-to-neutral $V_{L-N}$ voltage wave shape of permanent-magnet generator at no load designed for wind-power application [44].

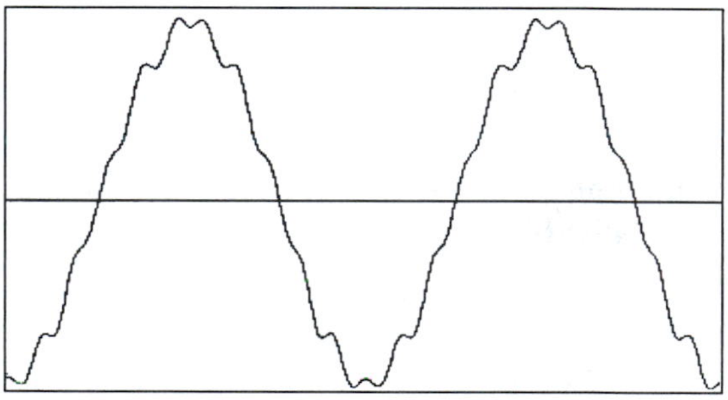

**FIGURE E4.5.3** Measured line-to-line $V_{L-L}$ voltage wave shape of permanent-magnet generator at no load designed for wind-power application [44].

**TABLE E4.5.1** Harmonic Spectrum of Fig. E4.5.1

| Harmonic order | Harmonic magnitude (%) | Harmonic order | Harmonic magnitude (%) |
|---|---|---|---|
| 1 | 100 | 7 | 3 |
| 2 | 0 | 8 | 0 |
| 3 | 11 | 9 | 8 |
| 4 | 0 | 10 | 0 |
| 5 | 1 | 11 | 7 |
| 6 | 0 | 12 | 0 |

**TABLE E4.5.2** Harmonic Spectrum of Fig. E4.5.2

| Harmonic order | Harmonic magnitude (%) | Harmonic order | Harmonic magnitude (%) |
|---|---|---|---|
| 1 | 100 | 7 | 2 |
| 2 | 0 | 8 | 0 |
| 3 | 14 | 9 | 7 |
| 4 | 0 | 10 | 0 |
| 5 | 1 | 11 | 5 |
| 6 | 0 | 12 | 0 |

$V_{(l-n)\_no\text{-}load} \approx E$ can be used for a permanent-magnet machine if saturation is not dominant. The displacement power factor angle $\varphi$ and the torque angle $\delta$ can be determined from an oscilloscope recording and from a stroboscope, respectively. The ohmic resistance $R$ can be measured as a function of the machine temperature. The relationship between these quantities is given by the phasor diagram of Fig. E4.6.1.

From the above phasor diagram one can derive the following relations for the synchronous reactances $X_q$ and $X_d$.

**TABLE E4.5.3** Harmonic Spectrum of Fig. E4.5.3

| Harmonic order | Harmonic magnitude (%) | Harmonic order | Harmonic magnitude (%) |
|---|---|---|---|
| 1 | 100 | 7 | 3 |
| 2 | 1 | 8 | 0 |
| 3 | 0 | 9 | 0 |
| 4 | 1 | 10 | 0 |
| 5 | 1 | 11 | 6 |
| 6 | 0 | 12 | 0 |

**TABLE E4.5.4** Harmonic Spectrum of Fig. E4.5.4

| Harmonic order | Harmonic magnitude (%) | Harmonic order | Harmonic magnitude (%) |
|---|---|---|---|
| 1 | 100 | 7 | 3 |
| 2 | 1 | 8 | 0 |
| 3 | 1 | 9 | 0 |
| 4 | 0 | 10 | 0 |
| 5 | 1 | 11 | 6 |
| 6 | 0 | 12 | 0 |

$$X_q = \frac{V_{(l-n)}\sin\delta + RI_{(l-n)}\sin\delta\cos l + RI_{(l-n)}\cos\delta\sin l}{I_{(l-n)}\cos l\cos\delta - I_{(l-n)}\sin l\sin\delta},$$

$$(E4.6\text{-}1)$$

$$X_d = \frac{E - V_{(l-n)}\cos\delta - RI_{(l-n)}\cos(\delta - l)}{I_{(l-n)}\cos(90° - l - \delta)}. \quad (E4.6\text{-}2)$$

An application of these relations is given in [44].

### 4.2.9 Application Example 4.7: Design of a Low-Speed 20 kW Permanent-Magnet Generator for a Wind-Power Plant

A $P = 20\,\text{kW}$, $V_{L\text{-}L} = 337\,\text{V}$ permanent-magnet generator has the cross sections of Figs. E4.7.1 and E4.7.2. The number of poles is $p = 12$ and its rated speed is $n_{rat} = n_S = 60\,\text{rpm}$, where $n_S$ is the synchronous speed. The B–H characteristic of the neodymium–iron–boron (NdFeB) permanent-magnet material is shown in Fig. E4.7.3.

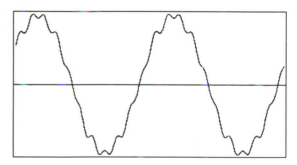

**FIGURE E4.5.4** Measured line-to-line $V_{L\text{-}L}$ voltage wave shape of permanent-magnet generator at rated load designed for wind-power application [44].

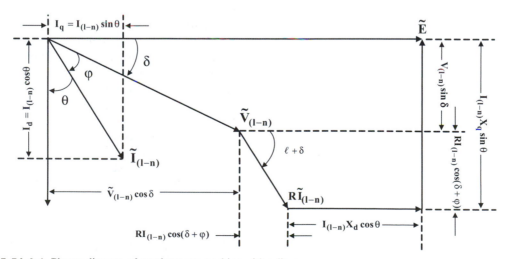

**FIGURE E4.6.1** Phasor diagram of synchronous machine with saliency.

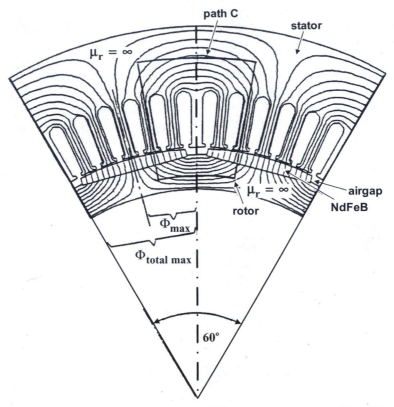

**FIGURE E4.7.1** Cross section of one period (two pole pitches) of permanent-magnet machine with no-load field.

a) Calculate the frequency f of the stator voltages and currents.

b) Apply Ampere's law to path C (shown in Figs. E4.7.1 and E4.7.2) assuming the stator and rotor iron cores are ideal ($\mu_r \to \infty$).

c) Apply the continuity of flux condition for the areas $A_m$ and $A_g$ perpendicular to path C (two times the one-sided magnet ($m$) length $l_m$, and two times the one-sided air gap ($g$) length $l_g$, respectively). Furthermore $l_m = 7$ mm, $l_g = 1$ mm, $A_m = 11,668$ mm$^2$, and $A_g = 14,469$ mm$^2$.

d) Provided the relative permeability of the stator and rotor cores is $\mu_r \to \infty$, compute the load line of this machine. Plot the load line in a figure similar to that of Fig. E4.7.3. Is the permanent-magnet material (NdFeB) operated at the point of the maximum energy product?

e) What is the recoil permeability $\mu_R$ of the NdFeB material?

f) Compute the fluxes $\Phi_{max}$ and $\Phi_{totalmax} = 2\Phi_{max}$, provided there are $N = 360$ turns per stator phase, the (pitch and distribution) winding factor is $k_w = 0.8$, and the iron-core stacking factor is $k_s = 0.94$.

g) Determine the diameter $D$ of stator wire for a copper-fill factor of $k_{cu} = 0.62$ and a stator slot cross section of $A_{st\_slot} = 368$ mm$^2$.

### 4.2.10 Application Example 4.8: Design of a 10 kW Wind-Power Plant Based on a Synchronous Machine

Design a wind-power plant (Fig. E4.8.1) feeding $P_{out\_transformer} = 10$ kW into the distribution system at a line-to-line voltage of $V_{L\text{-}L\text{ system}} = 12.47$ kV = $V_{L\text{-}L\_secondary\ of\ transformer}$. The wind-power plant consists of a $Y - \Delta$ three-phase (ideal) transformer connected between three-phase PWM inverter and power system, where the $Y$ is the primary and $\Delta$ is the secondary of the ideal transformer (Fig. E4.8.2). The input voltage of the PWM inverter (see Chapter 1, Fig. E1.5.1) is $V_{DC} = 600$ V, where the inverter delivers an output AC current $I_{inverter}$ at unity displacement power factor $\cos\varphi = 1$ to the primary ($Y$) of the $Y - \Delta$ transformer, and the inverter output line-to-line voltage is $V_{L\text{-}L\_inverter} = 240$ V = $V_{L\text{-}L\_primary\ of\ transformer}$.

The three-phase rectifier (see Chapter 1, Fig. E1.2.1) with one self-commuted PWM-operated switch (IGBT) is fed by a three-phase synchronous generator. The input line-to-line voltages of the three-phase rectifier is $V_{L\text{-}L\ rectifier} = 480$ V = $V_{L\text{-}L\_generator}$. Design the mechanical gear between wind turbine and generator (with $p = 2$ poles at $n_S = 3600$ rpm, utilization factor $C = 1.3$ kWmin/m$^3$,

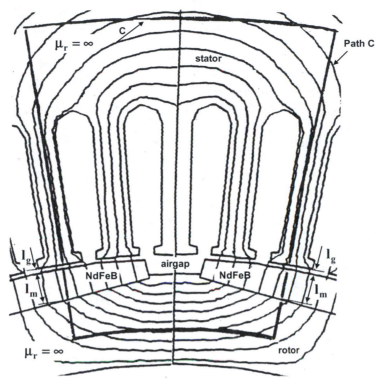

**Figure E4.7.2** Magnified cross section of area enclosed by path C of a pole pitch of permanent-magnet machine (see Fig. E4.7.1).

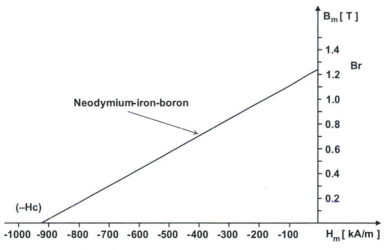

**FIGURE E4.7.3** $B_m$–$H_m$ characteristic of neodymium–iron–boron (NdFeB) permanent-magnet material.

$D_{rotor} = 0.2$ m, the leakage reactance $X_{sleakage} = 2\pi f L_{sleakage}$ is 10% of the synchronous reactance $X_S = 1.5$ pu, maximum flux density $B_{max} = 0.7$ T, iron-stacking factor $k_{fe} = 0.95$, stator winding factor $k_S = 0.8$, rotor winding factor $k_r = 0.8$, and torque angle $\delta = 30°$) so that a wind turbine can operate at $n_{turbine} = 30$ rpm. Lastly, the wind turbine – using the famous Lanchester-Betz limit [45] for the maximum energy efficiency of a wind turbine as a guideline – is to be designed for the rated wind velocity of $v = 10$ m/s at an altitude of 1600 m and a coefficient of performance (actual efficiency) $c_p = 0.3$. You may assume that all components (transformer, inverter, rectifier, generator, mechanical gear) except the wind turbine have each an efficiency of $\eta = 95\%$.

a) Based on the given transformer output power of $P_{out\_transformer} = 10$ kW compute the output powers of inverter, rectifier, generator, gear, and wind turbine.

b) For the circuit of Fig. E4.8.1 determine the transformation ratio $(N_Y/N_\Delta)$ of the $Y - \Delta$ transformer, where the $Y$ is the primary (inverter side) and $\Delta$ the secondary (power system side).

c) What is the phase shift between $\tilde{V}_{L\text{-}L \text{ system}}$ and $\tilde{V}_{L\text{-}L \text{ inverter}}$?

d) What is the phase shift between $\tilde{I}_{L \text{ system}}$ and $\tilde{I}_{L \text{ inverter}}$?

e) Use sinusoidal PWM to determine for the circuit of Fig. E1.5.1 of Chapter 1 the inverter output current $\tilde{I}_{inverter}$ which is in phase with $\tilde{V}_{L\text{-}N\_invert\text{-}er} = 138.57$ V, that is, a (unity) displacement power factor $\cos \varphi = 1$ at an inverter switching frequency of $f_{sw} = 7.2$ kHz. You may assume $R_{system} = 50$ m$\Omega$, $X_{system} = 0.1$ $\Omega$, $R_{inverter} = 10$ m$\Omega$, and $X_{inverter} = 0.377$ $\Omega$ at $f = 60$ Hz.

f) For the circuit of Fig. E1.2.1 of Chapter 1 and a duty cycle of $\delta = 50\%$, compute the input current of the rectifier $I_{rectifier}$.

g) Design the synchronous generator for the above given data provided it operates at unity displacement power factor.

h) Design the mechanical gear.

i) Determine the radius of the wind turbine blades for the given conditions.

### 4.2.11 Synchronous Machines Supplying Nonlinear Loads

Frequently synchronous machines or permanent-magnet machines are used together with solid-state converters, that is, either rectifiers as a load [46–49] or inverters feeding the machine. The question arises how much current distortion can be permitted in order to prevent noise and vibrations as well as over-heating? It is recommended to limit the total har-

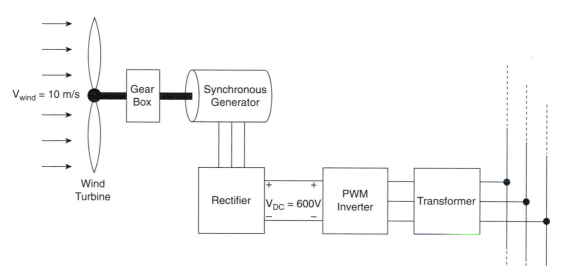

**FIGURE E4.8.1** Block diagram of wind-power plant.

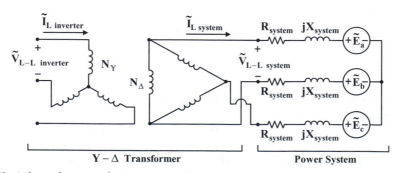

**FIGURE E4.8.2** $Y - \Delta$ three-phase transformer connected to power system.

monic current distortion of the phase current to $THD_i \leq 5$–$20\%$ as specified by recommendation IEEE-519 (see Chapter 1). The 5% limit is for very large machines and the 20% limit is for small machines. The calculation of the synchronous, transient, and subtransient reactances based on numerical field calculations is addressed in [50]. Synchronous machines up to 5 MW are used in variable-speed drives for wind-power plants. Enercon [51] uses synchronous generators with a large pole number; this enables them to avoid any mechanical gear within the wind-power train. The AC output is rectified and an inverter supplies the wind power to the grid. Hydrogenerators are another example where a large number of poles are used.

### 4.2.12 Switched-Reluctance Machine

Switched-reluctance machines [52–55] have the advantage of simplicity: such machines consist of a salient pole (solid) rotor member without any winding and the stator member carries concentrated coils that are excited by a solid-state converter. The only disadvantages of such machines are the encoder required to sense the speed of the rotor and the acoustic noise emanating form the salient-pole rotor. The windage losses of such machines can be considerable. Figure 4.19a,b,c,d illustrates how the field of a switched-reluctance machine changes as a function of the rotor position.

### 4.2.13 Some Design Guidelines for Synchronous Machines

In order for rotating machines to work properly and not to cause power quality problems, limits for maximum flux and current densities must be met so that neither excessive saturation nor heating occurs. These limits are somewhat different for induction (see Chapter 3) and synchronous machines.

#### 4.2.13.1 Maximum Flux Densities

Table 4.1 lists recommended maximum flux densities within the stator and rotor of synchronous machines.

#### 4.2.13.2 Recommended Current Densities

Table 4.2 lists recommended current densities within the stator and rotor windings of synchronous machines. These current densities depend on the cooling methods applied.

#### 4.2.13.3 Relation between Induced $E_{phase}$ and Terminal $V_{phase}$ Voltages

The relationship between induced voltage and terminal voltage is for motor operation approximately $E_{\mathrm{phase}}/V_{\mathrm{phase}} \approx 0.95$ and for generator operation $E_{\mathrm{phase}}/V_{\mathrm{phase}} \approx 1.05$.

#### 4.2.13.4 Iron-Core Stacking Factor and Copper-Fill Factor

The iron-core stacking factor depends on the lamination thicknesses. An approximate value for most commonly used designs is $k_{fe} \approx 0.95$.

The copper-fill factor depends on the winding cross section (e.g., round, square) and the wire diameter. An approximate value for most commonly used designs is $k_{cu} \approx 0.70$.

### 4.2.14 Winding Forces during Normal Operation and Faults

In power apparatus magnetic forces are often large enough to cause unwanted noise or disabling damage. For instance, in [56] the problem of stator bar vibrations causing mechanical wear in hydrogenerators is addressed. Useful methods for reducing the vibrations, which are caused by magnetic forces, are suggested. Along with the suggestions a force calculation

**TABLE 4.1** Guidelines for Maximum Flux Densities (Measured in Teslas) in Synchronous Machines

| Stator back iron | Stator teeth, maximum value | Stator teeth, middle of tooth height | Solid rotor | Solid rotor teeth, near winding | Solid rotor teeth, not near winding | Solid rotor back iron, forged iron | Solid rotor back iron, cast iron |
|---|---|---|---|---|---|---|---|
| 1–1.4 | 1.6–1.8 | 1.35–1.55 | 1.2–1.5 | <2.4 | 1.4–1.6 | 1–1.4 | 0.7 |

**TABLE 4.2** Guidelines for Current Densities (Measured in A/mm$^2$) in Windings of Synchronous Machines

| Without forced air cooling | With forced air cooling (e.g., ventilator) | Hydrogen or indirect water cooling | Direct water cooling |
|---|---|---|---|
| 1–2 | 2–4 | 4–8 | 8–16 |

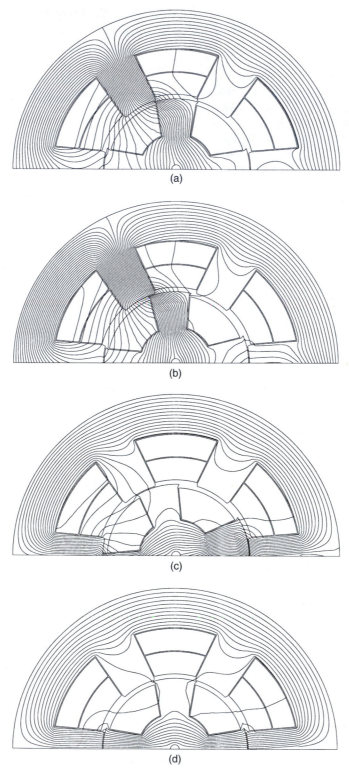

**FIGURE 4.19** Field of switched-reluctance machine for (a) rotor position #1 (0 mechanical degrees), rotor position #2 (5 mechanical degrees), (c) rotor position #3 (15 mechanical degrees), and (d) rotor position #4 (30 mechanical degrees).

formula proposed in 1931 is given. The surface integral method recommended in this section is an improvement over such formulas since it uses magnetic field calculation, which accounts for magnetic saturation of the iron cores.

### 4.2.14.1 Theoretical Basis

Several approaches may be taken to arrive at the expressions necessary for the surface integral force calculations. Using a stress of Maxwell and energy considerations, Stratton [57] formally derives this expression for the total force on a nonferromagnetic body:

$$\vec{F} = \int_{S1} [\mu \vec{H}(\vec{H} \cdot \vec{n}) - \frac{\mu}{2} H^2 \vec{n}] da. \qquad (4\text{-}35)$$

In Eq. 4-35 $S_1$ is an arbitrary surface surrounding the body. Equivalent expressions are given by Carpenter [58]. He begins by assuming that "any distribution of poles or currents, or both, which, when put in place of a piece of magnetized iron, reproduces the magnetic-field condition at all points outside the iron, must experience the same total mechanical force as a part which it replaces." One special distribution consists of poles and currents existing on a surface surrounding the body of interest. The force on the body is thus modeled by the force densities

$$F_t = \mu_o H_n H_t, \qquad (4\text{-}36a)$$

$$F_n = \frac{1}{2} \mu_o (H_n^2 - H_t^2), \qquad (4\text{-}36b)$$

in the tangential and normal directions, respectively. Surface integrals of these densities then give the total force on the object.

A third treatment of the subject is made by Reichert [59]. He speaks of a quantity $\vec{p}$, the surface stress. Integration of stress,

$$\vec{p} = \frac{1}{\mu_o} (\vec{n} \cdot \vec{B}) \vec{B} - \frac{1}{2\mu_o} B^2 \vec{n}, \qquad (4\text{-}37)$$

is said to give the total force on the body. This force calculation is equivalent to Stratton's and Carpenter's. However, Reichert goes on to adapt the stress to numerical methods.

Required characteristics of the above integration surfaces, which are clarified by Reichert, turn out to limit the possible application of the method. The necessary characteristics are

- Surface must be in air;
- An exception may be made to the above rule by creating an artificial air gap. This may be done if the artificial air gap does not cause a significant change in the magnetic field; and
- Surface may be any that fully encloses the body and follows the above rules.

Especially note that only the total force or torque acting on the body is given by the integration. No information can be obtained about the actual distribution of the force within the body. Figure 4.20 shows forces calculated with the above method.

## 4.3 HARMONIC MODELING OF A SYNCHRONOUS MACHINE

Detailed nonlinear models of a synchronous machine in the frequency domain have been developed for harmonic analysis [11–13]. These models, however, cannot accurately describe the transient and steady-state unbalanced operation of a synchronous machine under nonsinusoidal voltage and current conditions. A machine model in phase (*abc*) coordinates can naturally reproduce these abnormal operation conditions because it is based on a realistic representation that can take into account the explicit time-varying nature of the stator and the mutual stator–rotor inductances, as well as the spatial (space) harmonic effects. Harmonic modeling of synchronous machines is complicated by the following factors:

- *Frequency-Conversion Process.* Synchronous machines may experience a negative-sequence current in their armature (stator), e.g., under unbalanced three-phase conditions. This current may induce a second-order harmonic of this negative-sequence armature current in the rotor. The rotor harmonic in turn may induce a third-order harmonic of this negative-sequence current in the armature (mirror action), and so on. This frequency conversion causes the machine itself to internally generate harmonics, and therefore complicates the machine's reaction to external harmonics imposed by power sources.
- *Saturation Effects.* Saturation affects the machine's operating point. Saturation effects interact with the frequency-conversion process and cause a cross-coupling between the *d*- and *q*-axes [23].
- *Machine Load Flow (or System) Constraints.* Synchronous machine harmonic models are incorporated into harmonic load flow programs and must satisfy constraints imposed by the load flow program.

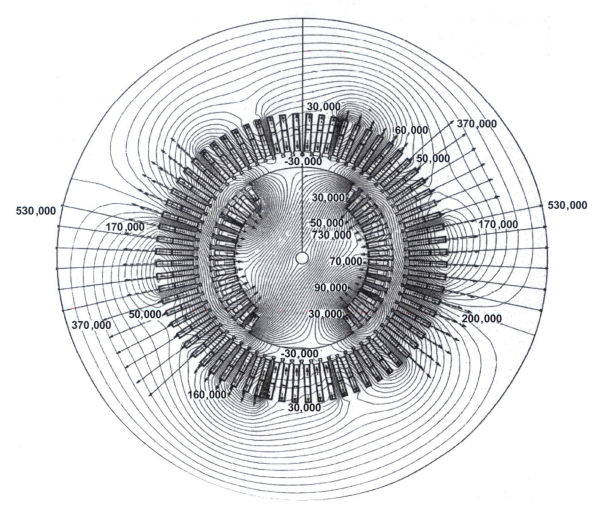

**FIGURE 4.20** Flux and force (measured in N/m) distribution in a two-pole turbogenerator during a sudden three-phase short-circuit. The arrows indicate the direction and value (length of arrow) of the winding forces developed due to balanced subtransient, three-phase short-circuit.

- *Inverter-Fed or Rectifier-Loaded Machines.* Examples are drives employed by the French railroad system, hybrid automobile drives that predominantly are based on brushless DC machines [24, 25], wind-power applications [51], and rotating rectifiers for synchronous generators [49].

In one of the first models in the phase (*abc*) domain, the analysis of [14] discusses the advantages of this synchronous machine representation in *abc* coordinates with respect to the models based on *dq*0 and *αβ*0 coordinates. A model using *abc* coordinates is employed [15] for the dynamic analysis of a three-phase synchronous generator connected to a static converter for high-voltage DC (HVDC) transmission, where harmonic terms up to the fourth

order are introduced in the stator-inductance matrix. In more recent contributions [16, 17] phase-domain models of the synchronous machine are employed and nonlinear saturation effects are incorporated.

### 4.3.1 Model of a Synchronous Machine as Applied to Harmonic Power Flow

A well-known harmonic model of a synchronous machine is based on the negative-, positive-, and zero-sequence reactances [60–62]

$$
\begin{aligned}
X_1(h) &= jhX_1 \\
X_2(h) &= jhX_2, \\
X_o(h) &= jhX_0
\end{aligned}
\tag{4-38}
$$

where $h$ is the order of harmonics and $X_1$, $X_2$, $X_0$ are positive-, negative-, and zero-sequence reactances of the machine at fundamental frequency, respectively.

Losses may be included by adding a resistor, as shown in Fig. 4.21. This model of a synchronous machine is used in harmonic (balanced) power flow analysis.

#### 4.3.1.1 Definition of Positive-, Negative-, and Zero-Sequence Impedances/Reactances

Symmetrical components [10] rely on positive-, negative- and zero-sequence impedances. Figure 4.22 gives a summary for the definition of these compo-

nents [62]. Figure 4.22a represents the three-phase circuit and Fig. 4.22b shows the per-phase equivalent circuit for the positive sequence. Figures 4.22c,d detail the circuits for the negative components, and

**FIGURE 4.21** Harmonic equivalent circuit of synchronous machine based on negative-sequence impedance.

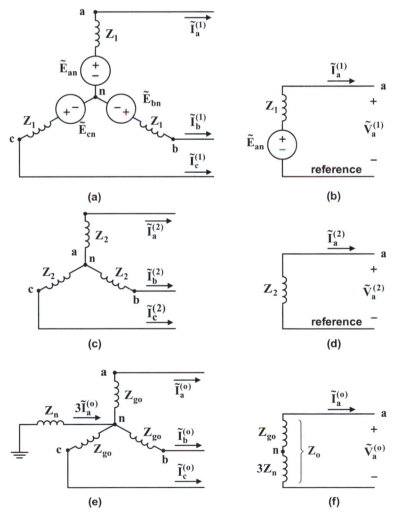

**FIGURE 4.22** Definition of symmetrical components: (a) positive-sequence voltages, currents, and impedances; (b) per-phase positive-sequence circuit; (c) negative-sequence voltages, currents, and impedances; (d) per-phase negative-sequence circuit; (e) zero-sequence voltages, currents, and impedances; (f) per-phase zero-sequence circuit.

Figs. 4.22e,f represent the circuits for the zero-sequence components.

### 4.3.1.2 Relations between Positive-, Negative-, and Zero-Sequence Reactances and the Synchronous, Transient, and Subtransient Reactances

The following relations exist between sequence, transient, and subtranient reactances:

- The positive-sequence reactance $X_1$ is identical with the synchronous reactance $X_s$, which is $X_d = X_q$ for an (ideal) round-rotor turbogenerator.
- The negative-sequence reactance $X_2$ is identical with the subtransient reactance $X_d''$.
- The zero-sequence reactance $X_0$ depends on the stator winding pitch and it varies from 0.1 to 0.7 of $X_d''$.

Therefore,

$$\begin{cases} X_1 = X_s (= X_d = X_q, \text{ for an ideal round} \\ \quad -\text{rotor turbogenerator}) \\ X_2 = X_d'' \\ X_0 = 0.1 X_d'' \text{ to } 0.7 X_d'' \end{cases} \quad (4\text{-}39)$$

### 4.3.2 Synchronous Machine Harmonic Model Based on Transient Inductances

The simple model of Fig. 4.21 is inappropriate for harmonic modeling of a synchronous machine when detailed information (e.g., harmonic torque) must be extracted. To improve this model three cases are considered, where the synchronous machine is fed by

- harmonic currents,
- harmonic voltages, and
- combination of current and voltage harmonics (as expected in actual power systems).

**Current-Source (Containing Harmonics) Fed Synchronous Machine.** The stator voltages that result when balanced harmonic currents of order $h$ are applied to a three-phase synchronous machine are determined. The synchronous machine has a field winding in the $d$-axis residing on the round rotor and no subtransient properties. In complex exponential notation the applied currents are

$$\begin{aligned} i_a &= i_h \exp j[h\omega t + \angle i_h] \\ i_b &= i_h \exp j[h\omega t + \angle i_h \mp 2\pi/3], \quad (4\text{-}40) \\ i_c &= i_h \exp j[h\omega t + \angle i_h \pm 2\pi/3] \end{aligned}$$

where the upper signs in Eq. 4-40 apply to the positive-phase sequence. Expressions for voltage, current, and flux may be obtained by taking the real part of each relevant complex equation. The $a$-phase stator voltage is according to [60]

$$\begin{aligned} v_a &= i_h jh\omega \left( \frac{L_d' + L_q'}{2} \right) \exp j[h\omega t + \angle i_h] + i_h \left( \frac{L_d' - L_q'}{2} \right) \\ & \quad j[(h \mp 2)\omega \exp j(h \mp 2)\omega t + \angle i_h \mp 2\alpha]. \end{aligned} \quad (4\text{-}41)$$

Using a similar analysis the $b$-phase voltage is

$$\begin{aligned} v_b &= i_h jh\omega \left( \frac{L_d' + L_q'}{2} \right) \exp j[h\omega t \mp 2\pi/3 + \angle i_h] \\ & \quad + i_h j(h \mp 2)\omega \left( \frac{L_d' - L_q'}{2} \right) \exp j[(h \mp 2)\omega t \mp \\ & \quad 2\alpha \pm 2\pi/3 + \angle i_h] \end{aligned} \quad (4\text{-}42)$$

Comparing Eqs. 4-41 and 4-42 it can be noted that the additional harmonic voltage term has the opposite phase sequence than that of the applied current. For example, if the applied current of order $h = 7$ has a positive-phase sequence, then the additional harmonic voltage component has the order 5 with a negative-phase sequence. This is the usual situation for a balanced system. If, however, the applied current of order $h = 7$ has negative-phase sequence, then the additional harmonic voltage component has the order 9 with a positive-phase sequence, that is, not a zero-sequence as is commonly expected. Consequently, the phase voltage set will not be balanced. The analysis is also valid for the application of a negative-sequence current of fundamental frequency; in this case the additional component has the harmonic order 3 and has positive-phase sequence [63].

**Voltage-Source (Containing Harmonics) Fed Synchronous Machine.** If a balanced set of harmonic voltages is applied to a synchronous machine, an additional harmonic current with harmonic orders $(h \mp 2)$ – based on the results of the prior section – can be expected. The current solution is

$$\begin{aligned} i_a &= i_h \exp j[h\omega t + \angle i_h] + \bar{i}_h \exp j[(h \mp 2)\omega t + \angle \bar{i}_h] \\ i_b &= i_h \exp j[h\omega t + \angle i_h \mp 2\pi/3] + \bar{i}_h \exp j[(h \mp 2)\omega t \\ & \quad + \angle \bar{i}_h \pm 2\pi/3] \\ i_c &= i_h \exp j[h\omega t + \angle i_h \pm 2\pi/3] + \bar{i}_h \exp j[(h \mp 2)\omega t \\ & \quad + \angle \bar{i}_h \pm 2\pi/3] \end{aligned} \quad (4\text{-}43)$$

The quantities with a bar ($^-$) are the current components with the frequency $(h \mp 2)\omega$. The phase sequence of these components is the opposite of that of the applied voltage with frequency $h\omega$. Using the results of the prior section, the a-phase voltage is

$$v_a = i_h jh\omega\left(\frac{L'_d + L'_q}{2}\right)\exp j[h\omega t + \angle i_h] + i_h j(h \mp 2)$$

$$\omega\left(\frac{L'_d - L'_q}{2}\right)\exp j[(h \mp 2)\omega t \mp 2\alpha + \angle i_h]$$

$$+ \bar{i}_h j(h \mp 2)\omega\left(\frac{L'_d + L'_q}{2}\right)\exp j[(h \mp 2)\omega t + \angle \bar{i}_h]$$

$$+ \bar{i}_h jh\omega\left(\frac{L'_d - L'_q}{2}\right)\exp j[h\omega t \mp 2\alpha - \angle \bar{i}_h]$$

$$\text{(4-44)}$$

For an applied voltage

$$V_a = V_h \exp jh\omega t. \qquad \text{(4-45)}$$

Comparing Eqs. 4-44 and 4-45 results in

$$v_h = i_h jh\omega\left(\frac{L'_d + L'_q}{2}\right) + \bar{i}_h jh\omega\left(\frac{L'_d - L'_q}{2}\right) \qquad \text{(4-46)}$$

$$\angle i_h = -\pi/2$$

$$0 = +i_h(h \mp 2)\omega\left(\frac{L'_d - L'_q}{2}\right) + \bar{i}_h(h \mp 2)\omega\left(\frac{L'_d + L'_q}{2}\right)$$

$$\angle \bar{i}_h = \mp 2\alpha + \angle i_h. \qquad \text{(4-47)}$$

Eqs. 4-46 and 4-47 may be rearranged to give

$$v_h = jh\omega i_h\left(\frac{2L'_d L'_q}{L'_d + L'_q}\right). \qquad \text{(4-48)}$$

Comparing Eqs. 4-42 and 4-48, one notes that the effective inductance at the applied frequency is different for the voltage- and current-source-fed machines.

If harmonic voltages are applied at both frequencies $h\omega$ and $(h \mp 2)\omega$, then Eqs. 4-46 and 4-47 can be modified by the inclusion of $\bar{V}_h$ and $\bar{V}_h$. The currents at both frequencies will be affected.

**Synchronous Machines Fed by a Combination of Harmonic Voltage and Current Sources.** When a synchronous machine is subjected to a harmonic voltage disturbance at frequency $h\omega$, harmonic-current components are drawn at $h\omega$ and at the associated frequency $(h \mp 2)\omega$. The upper sign applies to the positive- and the lower sign to the negative-phase sequence. Current flow at the associated frequency

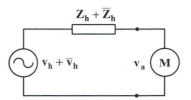

**FIGURE 4.23** System in general case: balanced representation.

occurs because the machine is a time-varying electrical system. In a similar manner, a synchronous machine fed by harmonic currents at frequency $h\omega$ develops a voltage across its terminal at both the applied and the associated frequency $(h \mp 2)\omega$. Consequently the machine cannot be modeled by impedances defined by a single harmonic frequency [60].

In the general case, a machine will be subjected to applied voltages $v_h$ and $\bar{v}_h$ via system impedances $z_h$ and $\bar{z}_h$ as shown in Fig. 4.23.

Now a linearized Thevenin model can be employed for the system (left-hand side of Fig. 4.23). The applied harmonic voltages $v_h$ and $\bar{v}_h$ are assumed to have opposite phase sequences: one is of positive and the other of negative sequence. This is the condition that would apply if the distorting load responsible for the harmonic disturbance drew balanced currents. Consequently, a single-phase model may be used. Voltage equations for the general case are obtained by applying a voltage mesh equation at both frequencies $h\omega$ and $(h \mp 2)\omega$. The voltage components applying across the machine are those given by Eq. 4-44 in response to the currents given by Eq. 4-43. Voltage equations suggest the equivalent circuit shown in Fig. 4.24. The interaction between the two sides of the circuit is represented by a mutual inductance, which is a simplification. This mutual inductance is nonreciprocal, and the frequency difference between the two sides is ignored. This model may be used to calculate harmonic current flow.

The results presented here show that a synchronous machine cannot be modeled by one impedance (Fig. 4.21) if voltage sources have voltage components with several frequencies. In the case where one harmonic voltage source is large and the other small then $v_h \approx 0$, and the apparent machine impedance to current flow is linear and time-varying:

$$Z_h = \frac{jh\omega(L'_d + L'_q)}{2} + \frac{(h \mp 2)h\omega^2\left(\frac{L'_d - L'_q}{2}\right)^2}{\xi_h}. \qquad \text{(4-49)}$$

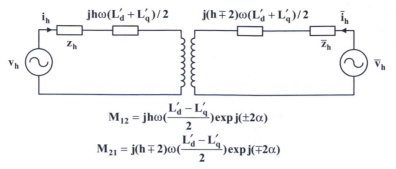

**FIGURE 4.24** Harmonic equivalent circuit of synchronous machine based on transient inductances [60].

**TABLE E4.9.1** Measured Current Spectrum of a 5 kVA, Four-Pole Synchronous Machine Excited by Positive- and Negative-Sequence 350 Hz Voltages

| Positive-phase sequence | | | Negative-phase sequence | | |
|---|---|---|---|---|---|
| Frequency (Hz) | Harmonic order | RMS current (mA) | Frequency (Hz) | Harmonic order | RMS current (mA) |
| 50 | 1 | 19.7 | 50 | 1 | 26.3 |
| 250 | 5 | 143.1 | 250 | 5 | 6.6 |
| 350 | 7 | 324.0 | 350 | 7 | 329.2 |
| 450 | 9 | 0.8 | 450 | 9 | 141.4 |
| 550 | 11 | 8.2 | 550 | 11 | 0.1 |

If several harmonic frequency sources exist, the total current flow can be obtained by the principle of superposition when applied to each harmonic h and its associated harmonics $(h \mp 2)\omega$. Because of the variation of the flux-penetration depth with frequency the governing machine impedances will be a function of frequency.

### 4.3.2.1 Application Example 4.9: Measured Current Spectrum of a Synchronous Machine

To validate the preceding theory, a series of measurements is recorded for a 5 kVA, 415 V, four-pole synchronous machine [60]. A 350 Hz ($h = 7$) three-phase voltage is applied to the machine and the current spectrum is measured by a spectrum analyzer for both negative- and positive-phase sequence sources (Table E4.9.1). These results are in agreement with the above-outlined theory; that is, for $h = 7$ (positive-sequence voltage excitation) a large current signal is generated at the 5th current harmonic, and for the positive-sequence voltage excitation for $h = 7$ a strong signal occurs at the 9th current harmonic with negative sequence.

In addition, the spectrum shows a frequency range that is associated with the nonsinusoidal inductance variation. The flux waveforms arising from induced damper currents exhibit significant harmonic content, and the frequency range of the spectrum is broader than one would expect based on the nonsinusoidal

inductance variation measured under static conditions.

### 4.3.3 Synchronous Machine Model with Harmonic Parameters

A synchronous machine model based on modified $dq0$ equations – frequently used in harmonic load flow studies – is presented next [61]. From the physical interpretation of this model one notices that this machine equivalent circuit is asymmetric. This model is then incorporated with the extended "decoupling-compensation" network used in harmonic power flow analysis, and it is now extended to problems including harmonics caused by the asymmetry of transmission lines, asymmetry of loads, and nonlinear elements like static VAr compensators (SVCs). This synchronous machine model has the form

$$\begin{aligned}
I_1(h) &= y_{11}(h-1)V_1(h) + \Delta I_1(h) \\
\Delta I_1(h) &= y_{12}(h-1)V_2(h-2) \\
I_2(h) &= y_{22}(h+1)V_2(h) + \Delta I_2(h) \\
\Delta I_2(h) &= y_{21}(h+1)V_1(h+2) \\
I_o(h) &= y_{00}(h)V_0(h)
\end{aligned} \qquad (4\text{-}50)$$

In these equations, the subscripts 1, 2, 0 identify positive-, negative-, and zero-sequence quantities, respectively; $y_{11}(h)$, $y_{12}(h)$, $y_{21}(h)$, $y_{22}(h)$, and $y_{00}(h)$ are harmonic parameters of the machine obtained from the dq0-equation model. In accordance with

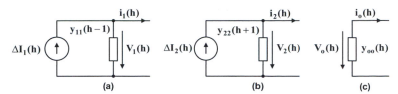

**FIGURE 4.25** Synchronous machine equivalent circuits with harmonic parameters; (a) positive sequence, (b) negative sequence, (c) zero sequence.

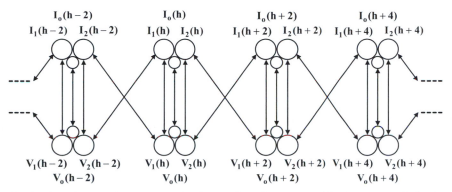

**FIGURE 4.26** Derivation of harmonic currents and voltages of synchronous machine in a symmerical reference frame.

Eq. 4-50 the equivalent circuits of the machine can be assembled based on the different sequence networks as depicted in Fig. 4.25. This is a decoupled harmonic model since $\Delta I_1(h)$ and $\Delta I_2(h)$ are decoupled-harmonic current sources. Equivalent circuits of different sequences and different harmonic orders are decoupled and can be incorporated separately with the corresponding power network models.

The physical interpretation of this harmonic-decoupled model is that

- The hth order positive-sequence current flowing in the machine armature is not entirely caused by the hth order positive-sequence voltage, but will generate in addition a $(h-2)$th order negative-sequence voltage at the machine terminal;
- The hth order negative-sequence current flowing in the machine armature is not only caused by the same order negative-sequence voltage, but will also generate a $(h+2)$th order positive-sequence voltage at the machine terminal; and
- The zero-sequence harmonic currents are relevant to the same order zero-sequence harmonic voltages only.

Such a phenomenon, stemming from the asymmetry of the machine rotor, is depicted in Fig. 4.26.

The reason why there is always a difference of the order between the positive- and negative-sequence harmonic current components and their corresponding admittances in Eq. 4-50 is because these admittances are transformed by the $dq0$ equations. The reference frame of the $dq0$ components is attached to the rotor, while the reference frame of symmetrical components remains stationary in space.

This mathematical model of a synchronous machine (Fig. 4.25) can be used in harmonic load flow studies to investigate the impact of a machine's asymmetry and nonlinearity. Moreover, decoupled equivalent circuits of different sequences and harmonics can be incorporated separately with the corresponding power network models.

### 4.3.3.1 Application Example 4.10: Harmonic Modeling of a 24-Bus Power System with Asymmetry in Transmission Lines

The 24-bus sample system shown in Fig. E4.10.1 is used in this example [61]. Parameters of the system and generators are provided in [61, 64]. Simulation results will be compared for the simple harmonic model (Fig. 4.21, Eq. 4-38) and the model with harmonic parameters (Fig. 4.24, Eq. 4-50).

Two cases are investigated, that is, harmonics caused by the asymmetry of transmission lines (or loads) and by nonlinear SVC devices (see Application Example 4.11). Harmonic admittances of synchronous machines are computed and compared in

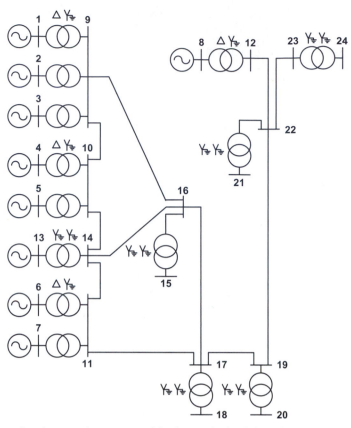

**FIGURE E4.10.1** Twenty-four-bus sample system used for harmonic simulation of synchronous machine.

[61]. They remain unchanged with the change of external network conditions. The imaginary components of admittances for the two models are very similar. This may be one of the reasons why the simple model is frequently employed.

All transmission lines (although transposed) are asymmetric, whereas all loads are symmetric. Decoupled positive- and negative-harmonic current sources of the equivalent circuit ($\Delta I_1(h)$ and $\Delta I_2(h)$ in Fig. 4.24 and Eq. 4-50) are computed and listed in [61] for generators 1, 6, and 8. The harmonic frequency voltage components at buses 1, 12, and 16 are also given in the same reference.

### 4.3.3.2 Application Example 4.11: Harmonic Modeling of a 24-Bus Power System with a Nonlinear Static VAr Compensator (SVC)

All transmission lines and loads in Fig. E4.10.1 are assumed to be symmetric. A static VAr compensator (SVC) device is connected at the additional bus 25 via a step-up transformer placed at bus 14. The

thyristor-controlled reactor (TCR) part of this SVC is a six-pulse thyristor controlled Δ-connected reactor [61] that injects odd harmonics into the system (Table E4.11.1). Harmonic voltages at buses 1, 12, and 14 are given in Table E4.11.2. For comparison, the negative-sequence fifth harmonic voltages (computed with models of Figs. 4.21 and 4.25) are listed in Table E4.11.3. It can be seen that harmonic voltages at those buses near the harmonic source (bus 25) are higher than those of others.

### 4.3.4 Synchronous Machine Harmonic Model with Imbalance and Saturation Effects

This section introduces a synchronous machine harmonic model [12] that incorporates both frequency conversions and saturation effects under various machine load-flow constraints (e.g., unbalanced operation). The model resides in the frequency domain and can easily be incorporated into harmonic load-flow programs.

To model these effects and also to maintain an equivalent-circuit representation, a three-phase harmonic Norton equivalent circuit (Fig. 4.27) is developed with the following equations [12]:

$$[I_{km}(h)] = [Y(h)]([V_h(h)] - [V_m(h)] - [E(h)]) + [I_{nl}(h)], \quad (4\text{-}51a)$$

$$[E_{(h)}] = 0, \text{ if } h \neq 1$$

$$[I_{km}] = [I_{km-a} \quad I_{km-b} \quad I_{km-c}]^T$$

$$[V_k] = [V_{k-a} \quad V_{k-b} \quad V_{k-c}]^T$$

$$[V_m] = [V_{m-a} \quad V_{m-b} \quad V_{m-c}]^T \quad (4\text{-}51b)$$

$$[E] = [E_p \quad a^2 E_p \quad a E_p]^T$$

$$[I_{nl}] = [I_{nl-a} \quad I_{nl-b} \quad I_{nl-c}]^T$$

$$a = \exp(-j2\pi/3)$$

where $h$ is the harmonic order. In the model of Fig. 4.27, nonlinear effects are represented by a harmonic current source ($[I_{nl}(h)]$) that includes the harmonic effects stemming from frequency conversion ($[I_f(h)]$) and those from saturation ($[I_s(h)]$):

$$[I_{nl}(h)] = [I_f(h)] + [I_S(h)]. \quad (4\text{-}52)$$

This harmonic model reduces to the conventional, balanced three-phase equivalent circuit of a synchronous machine if $[I_{nl}(h)]$ is zero and $[Y(h)]$ is computed according to reference [65]. Since $[I_{nl}(h)]$ is a known current source (determined from the machine load-flow conditions), the model is linear and decoupled from a harmonic point of view. Therefore, it can be solved with network equations in a way similar to the traditional machine model.

**TABLE E4.11.1** Harmonic Current Injections of SVC

| $h$ | Positive-sequence voltage magnitude (u) | Negative-sequence voltage magnitude (pu) |
|---|---|---|
| 5 | 0 | $0.9440 \times 10^{-2}$ |
| 7 | $0.7897 \times 10^{-2}$ | 0 |
| 9 | 0 | 0 |
| 11 | 0 | $0.4288 \times 10^{-2}$ |
| 13 | $0.2581 \times 10^{-2}$ | 0 |

**TABLE E4.11.2** Simulation Results for the 24-Bus System (Fig. E4.10.1): Harmonic Voltage Magnitudes at Some Buses

| $h$ | Positive-sequence voltage magnitude (pu) | | | Negative-sequence voltage magnitude (pu) | | |
|---|---|---|---|---|---|---|
| | Bus 1 | Bus 12 | Bus 14 | Bus 1 | Bus 12 | Bus 14 |
| 5 | 0 | 0 | 0 | $0.151 \times 10^{-3}$ | $0.238 \times 10^{-3}$ | $0.302 \times 10^{-3}$ |
| 7 | $0.156 \times 10^{-3}$ | $0.312 \times 10^{-3}$ | $0.317 \times 10^{-3}$ | 0 | 0 | 0 |
| 9 | 0 | 0 | 0 | 0 | 0 | 0 |
| 11 | 0 | 0 | 0 | $0.194 \times 10^{-3}$ | $0.216 \times 10^{-3}$ | $0.334 \times 10^{-3}$ |
| 13 | $0.101 \times 10^{-3}$ | $0.330 \times 10^{-4}$ | $0.136 \times 10^{-3}$ | 0 | 0 | 0 |

**TABLE E4.11.3** Simulation Results for the 24-Bus System (Fig. E4.10.1): Negative-Sequence Fifth Harmonic Voltages at Different Buses

| Bus number | Negative-sequence 5th harmonic voltage magnitude (pu) | | Bus number | Negative-sequence 5th harmonic voltage magnitude (pu) | |
|---|---|---|---|---|---|
| | Simple model (Fig. 4.21) | New model (Fig. 4.25) | | Simple model (Fig. 4.21) | New model (Fig. 4.25) |
| 1 | $0.161 \times 10^{-3}$ | $0.151 \times 10^{-3}$ | 14 | $0.302 \times 10^{-3}$ | $0.302 \times 10^{-3}$ |
| 2 | $0.161 \times 10^{-3}$ | $0.151 \times 10^{-3}$ | 15 | $0.247 \times 10^{-3}$ | $0.246 \times 10^{-3}$ |
| 3 | $0.161 \times 10^{-3}$ | $0.151 \times 10^{-3}$ | 16 | $0.292 \times 10^{-3}$ | $0.296 \times 10^{-3}$ |
| 4 | $0.184 \times 10^{-3}$ | $0.184 \times 10^{-3}$ | 17 | $0.279 \times 10^{-3}$ | $0.279 \times 10^{-3}$ |
| 5 | $0.184 \times 10^{-3}$ | $0.184 \times 10^{-3}$ | 18 | $0.265 \times 10^{-3}$ | $0.269 \times 10^{-3}$ |
| 6 | $0.185 \times 10^{-3}$ | $0.179 \times 10^{-3}$ | 19 | $0.276 \times 10^{-3}$ | $0.276 \times 10^{-3}$ |
| 7 | $0.185 \times 10^{-3}$ | $0.179 \times 10^{-3}$ | 20 | $0.224 \times 10^{-3}$ | $0.224 \times 10^{-3}$ |
| 8 | $0.141 \times 10^{-3}$ | $0.147 \times 10^{-3}$ | 21 | $0.200 \times 10^{-3}$ | $0.208 \times 10^{-3}$ |
| 9 | $0.259 \times 10^{-3}$ | $0.257 \times 10^{-3}$ | 22 | $0.237 \times 10^{-3}$ | $0.246 \times 10^{-3}$ |
| 10 | $0.269 \times 10^{-3}$ | $0.268 \times 10^{-3}$ | 23 | $0.233 \times 10^{-3}$ | $0.241 \times 10^{-3}$ |
| 11 | $0.281 \times 10^{-3}$ | $0.279 \times 10^{-3}$ | 24 | $0.194 \times 10^{-3}$ | $0.201 \times 10^{-3}$ |
| 12 | $0.229 \times 10^{-3}$ | $0.238 \times 10^{-3}$ | 25 | $0.119 \times 10^{-3}$ | $0.119 \times 10^{-3}$ |
| 13 | $0.268 \times 10^{-3}$ | $0.268 \times 10^{-3}$ | | | |

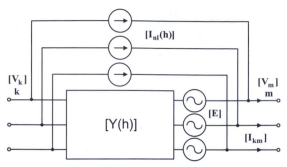

**Figure 4.27** Synchronous machine harmonic model with imbalance and saturation effects [12].

To include the usual load-flow conditions, the following constraints are specified at the fundamental frequency ($h = 1$):

- **Slack machine**: The constraints are the magnitude and the phase angle of positive-sequence voltage at the machine terminals:

$$[T]([V_k] - [V_m]) = V_{\text{specified}}$$
$$[T] = (1/3)[1 \quad a \quad a^2]. \tag{4-53}$$

- **PV machine**: The constraints are the three-phase real power output and the magnitude of the positive-sequence voltage at machine terminals, where superscript $H$ denotes conjugate transpose:

$$\text{Real}\{-[I_{km}]^H([V_k] - [V_m])\} = P_{\text{specified}}, \tag{4-54}$$

$$|[T]([V_k] - [V_m])| = V_{\text{specified}}. \tag{4-55}$$

- **PQ machine**: The constraints are the three-phase real and the three-phase reactive power outputs:

$$-[I_{km}]^H([V_k] - [V_m]) = (P + jQ)_{\text{specified}}. \tag{4-56}$$

These constraint equations and the machine structure equation (Eq. 4-51a) jointly define a three-phase machine model with nonlinear effects. Combining these equations with other network equations leads to multiphase load flow equations for the entire system [65]. The system is then solved by the Newton–Raphson method for each frequency. Detailed analysis and demonstration of Newton–Raphson based harmonic load flow is presented in Chapter 7.

In the model of Fig. 4.27, $[I_{nl}(h)]$ is not known and will be computed using the following iterative procedure:

- **Step 1. Initialization**
Set the harmonic current source $[I_{nl}(h)]$ to zero.

- **Step 2. Network Load-Flow Solution**
Replace the synchronous machines with their harmonic Norton equivalent circuits (Fig. 4.27) and solve the network equations for the fundamental and harmonic frequencies.

- **Step 3. Harmonic Current-Source Computation**
Use the newly obtained network voltages and currents (from Step 2) to update $[I_{nl}(h)]$ for the machine equivalent circuit.

- **Step 4. Convergence**
If the computed $[I_{nl}(h)]$ values are sufficiently close to the previous ones stop; otherwise, go to Step 2.

The network load-flow solution and the harmonic iteration processes can be performed by any general purpose harmonic load-flow program as explained in Chapter 7 and reference [65]. Therefore, $[I_{nl}(h)]$ needs to be computed (Step 3). This will require machine harmonic models in the dq0 and abc coordinates.

### 4.3.4.1 Synchronous Machine Harmonic Model Based on dq0 Coordinates

To incorporate the synchronous machine model (Fig. 4.27) in a general purpose harmonic analysis program, reasonable assumptions and simplifications according to the general guidelines indicated in references [66] and [67] are made. Under these assumptions, the following dq0 transformation is used to transfer the machine quantities from abc coordinates into rotating dq0 coordinates [12]:

$$[v_d \quad v_q \quad v_o]^T = [P]^{-1}[v_a \quad v_b \quad v_c]^T, \tag{4-57}$$

where

$$[P]^{-1} = \sqrt{2/3} \begin{bmatrix} \cos\theta & \cos(\theta - 2\pi/3) & \cos(\theta + 2\pi/3) \\ -\sin\theta & -\sin(\theta - 2\pi/3) & -\sin(\theta + 2\pi/3) \\ \sqrt{1/2} & \sqrt{1/2} & \sqrt{1/2} \end{bmatrix}. \tag{4-58}$$

In this equation $\theta = \omega t + \delta$ and $\delta$ is the angle between $d$-axis and the real axis of the network phasor reference frame. Defining

$$[D] = \begin{bmatrix} 1 & e^{-j2\pi/3} & e^{-j4\pi/3} \\ -j & -je^{-j2\pi/3} & -je^{-j4\pi/3} \\ 0 & 0 & 0 \end{bmatrix} e^{j\delta}/\sqrt{6}$$

$$[D_o] = \begin{bmatrix} 0 & 0 & 0 \\ 0 & 0 & 0 \\ \sqrt{1/3} & \sqrt{1/3} & \sqrt{1/3} \end{bmatrix}, \tag{4-59}$$

matrix $[P]^{-1}$ can be simplified as

$$[P]^{-1} = [D]e^{j\alpha t} + [D]^C e^{-j\alpha t} + [D_o]. \qquad (4\text{-}60)$$

Because the $dq0$ transformation is an orthogonal transformation, it follows that

$$[P] = ([P]^{-1})^T = [D]^T e^{j\alpha t} + [D]^H e^{-j\alpha t} + [D_o]^T, \qquad (4\text{-}61)$$

where superscripts $T$, $C$, and $H$ represent transpose, conjugate, and conjugate transpose, respectively.

Using these equations for the $dq0$ transformation and the generator reference convention, a synchronous machine is represented in the $dq0$ coordinates as

$$[v_{park}] = -[R][i_{park}] - \frac{d}{dt}[\Psi_{park}] + [F][\Psi_{park}], \qquad (4\text{-}62)$$
$$[\Psi_{park}] = [L][i_{park}]$$

where

$$[v_{park}] = [v_d \quad v_q \quad v_o \quad v_f \quad v_g \quad v_D \quad v_Q]^T$$

$$[i_{park}] = [i_d \quad i_q \quad i_o \quad i_f \quad i_g \quad i_D \quad i_Q]^T,$$

$$[R] = [r_a \quad r_a \quad r_a \quad r_f \quad r_g \quad r_D \quad r_Q]^T$$

$$[F] = \begin{bmatrix} 0 & -\omega & 0 & 0 & 0 & 0 & 0 \\ \omega & 0 & 0 & 0 & 0 & 0 & 0 \\ 0 & 0 & 0 & 0 & 0 & 0 & 0 \\ 0 & 0 & 0 & 0 & 0 & 0 & 0 \\ 0 & 0 & 0 & 0 & 0 & 0 & 0 \\ 0 & 0 & 0 & 0 & 0 & 0 & 0 \\ 0 & 0 & 0 & 0 & 0 & 0 & 0 \end{bmatrix},$$

$$[L] = \begin{bmatrix} L_d & 0 & 0 & M_{df} & 0 & M_{dD} & 0 \\ 0 & L_q & 0 & 0 & M_{gg} & 0 & M_{gQ} \\ 0 & 0 & Lo & 0 & 0 & 0 & 0 \\ M_{df} & 0 & 0 & L_{ff} & 0 & M_{fD} & 0 \\ 0 & M_{qg} & 0 & 0 & L_{gg} & 0 & M_{gQ} \\ M_{dD} & 0 & 0 & M_{fD} & 0 & L_{DD} & 0 \\ 0 & M_{qQ} & 0 & 0 & M_{gQ} & 0 & L_{QQ} \end{bmatrix}.$$

In these equations, subscripts $f$ and $D$ represent the field and damping windings of the $d$-axis, and subscripts $g$ and $Q$ indicate the field and damper windings of the $q$-axis.

Substituting Eq. 4-62 into Eq. 4-61 and defining the differential operator $p = \dfrac{d}{dt}$:

$$[v_{park}] = -[R][i_{park}] - p([L][i_{park}]) + [F][L][i_{park}],$$
$$= (-[R] - p[L] + [F][L])[i_{park}] = [Z(p)][i_{park}] \qquad (4\text{-}63)$$

where $[Z(p)] = -[R] - p(L) = [F][L]$ is the operation impedance matrix.

The rotor angular velocity ($\omega$) is constant during steady-state operation; therefore, Eq. 4-63 represents a linear time-invariant system and can be written in the frequency domain for each harmonic $h\omega$, as follows:

$$[V_{park}(h)] = [Z(h)][I_{park}(h)]. \qquad (4\text{-}64)$$

For the harmonic quantities $h \neq 0$ (e.g., excluding DC), the $f$, $g$, $D$, and $Q$ windings are short-circuited; therefore, Eq. 4-64 can be rewritten as

$$\begin{bmatrix} V_d \\ V_q \\ V_o \\ 0 \\ 0 \\ 0 \\ 0 \end{bmatrix} = \begin{bmatrix} & \vdots & \\ Z_{11}(h) & \vdots & Z_{12}(h) \\ & \vdots & \\ \cdots & \cdots & \cdots & \vdots & \cdots & \cdots & \cdots \\ & \vdots & \\ Z_{21}(h) & \vdots & Z_{22}(h) \\ & \vdots & \end{bmatrix} \begin{bmatrix} I_d \\ I_q \\ I_o \\ I_f \\ I_g \\ I_D \\ I_Q \end{bmatrix}. \qquad (4\text{-}65)$$

Solving for the $dq0$ currents as a function of the $dq0$ voltages, we have [12]:

$$[I_{dqo}(h)] = [Z_{11}(h) - Z_{12}(h)Z_{22}^{-1}(h)Z_{21}(h)]^{-1}[V_{dqo}(h)]$$
$$= [Y_{dqo}(h)][V_{dqo}(h)], \qquad (4\text{-}66)$$

where

$$[Y_{dqo}(h)] = [Z_{11}(h) - Z_{12}(h)Z_{22}^{-1}(h)Z_{21}(h)]^{-1}$$
$$[I_{dqo}(h)] = [I_d \quad I_q \quad I_o]^T$$
$$[V_{dqo}(h)] = [V_d \quad V_q \quad V_o]^T$$

Equation 4-66 (consisting of a simple admittance matrix relating the $dq0$ components) defines the harmonic machine model in the $dq0$ coordinates.

For the DC components (e.g., $h = 0$), the voltage of the f-winding is no longer zero. Applying similar procedures, the results will be similar to Eq. 4-66, except that there is an equivalent DC voltage source (which is a function of the DC excitation voltage $v_f$) in series with the $[Y_{dq0}]$ matrix.

### 4.3.4.2 Synchronous Machine Harmonic Model Based on abc Coordinates

To interconnect the synchronous machine model of Eq. 4-66 with the rest of the system, the $dq0$ voltages and currents must be converted to the corresponding

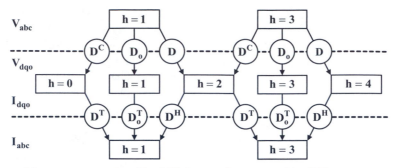

**Figure 4.28** Process of frequency conversion (Eq. 4-70) in a synchronous machine [12].

phase ($abc$) quantities using Park's transformation. Assume that the machine voltage $[v_{abc}(t)]$ vector consists of a single harmonic component of order $h$:

$$[v_{abc}(t)] = \text{Real}\{\sqrt{2}[V_{abc}(h)]e^{jh\omega t}\}. \qquad (4\text{-}67)$$

The corresponding $dq0$-voltage vector is obtained by Park's transformation (Eq. 4-58) as

$$
\begin{aligned}
[v_{dqo}(t)] &= [P]^{-1}[V_{abc}(t)] = \text{Real}\{[P]^{-1}\sqrt{2}[V_{abc}(h)]e^{jh\omega t}\} \\
&= \sqrt{2}\text{Real}\{[D][V_{abc}(h)]e^{j(h+1)\omega t} + [D]^C. \\
&\quad [V_{abc}(h)]e^{j(h-1)\omega t} + [D_o][V_{abc}(h)]e^{jh\omega t}\}
\end{aligned}
$$
$$(4\text{-}68)$$

Therefore, one harmonic $abc$ voltage introduces three harmonics in terms of $dq0$ coordinates. The machine's $dq0$ model can now be used to obtain the $dq0$ currents. From Eq. 4-66:

$$
\begin{aligned}
[i_{dqo}(t)] &= \sqrt{2}\text{Real}\{[Y_{dqo}(h+1)][D][V_{abc}(h)]e^{j(h+1)\omega t} \\
&\quad + [Y_{dqo}(h-1)][D]^C[V_{abc}(h)]e^{j(h-1)\omega t} \\
&\quad + [Y_{dqo}(h)][D_o][V_{abc}(h)]e^{jh\omega t}\}.
\end{aligned}
$$
$$(4\text{-}69)$$

Transferring Eq. 4-69 back into the $abc$ coordinates (using the inverse Park transformation), and with the condition $[D]^T[D_0] = 0$, we get

$$
\begin{aligned}
[i_{abc}(t)] &= [P][i_{dqo}(t)] \\
&= \sqrt{2}\text{Real}\{[D_o^T Y_{dqo}(h)D_o + D^T Y_{dqo}(h-1)D^C \\
&\quad + D^H Y_{dqo}(h+1)D][V_{abc}(h)]e^{jh\omega t} \\
&\quad + [D^H Y_{dqo}(h-1)D^C][V_{abc}(h)]e^{j(h-2)\omega t} \\
&\quad + [D^T Y_{dqo}(h+1)D][V_{abc}(h)]e^{j(h+2)\omega t}\}.
\end{aligned}
$$
$$(4\text{-}70)$$

Based on this equation, the harmonic voltage of order $h$, $[V_{abc}(h)]$, can generate harmonic currents of order $(h-2)$ and $(h+2)$ in $abc$ coordinates (Fig.

4.28). This is known as frequency conversion. Note that (Fig. 4.28)

- Negative-sequence voltage $[D][V_{abc}(h)]$ generates harmonic currents of order $(h+2)$ and $h$.
- Positive-sequence voltage $[D]^C[V_{abc}(h)]$ generates harmonic currents of order $(h-2)$ and $h$.
- Zero-sequence voltage $[D_o][V_{abc}(h)]$ generates only the harmonic current of the same order.
- There are three particular harmonic voltages that can generate harmonic currents of the same order.

All these observations are consistent with the revolving magnetic field theory [68].

### 4.3.4.3 Computation of Synchronous Machine Injected Harmonic Currents [$I_{nl}(h)$]

The current source $[I_{nl}(h)]$ of the synchronous machine harmonic model (Fig. 4.27) consists of two components: $[I_f(h)]$ due to the frequency-conversion phenomena and $[I_s(h)]$ due to the saturation effects. These currents will be computed separately.

**Injected Harmonic Current Due to Frequency Conversion [$I_f(h)$].** As mentioned before, Eq. 4-70 also indicates that there are three particular harmonic voltages that can generate harmonic currents of the same order. Assuming that the voltage vector $[v_{abc}(t)]$ includes all harmonic components, the harmonic terminal voltages that result in currents of the same harmonic order $h$ can be grouped and described in phasor form as [12]

$$
\begin{aligned}
[I_{abc}(h)] &= [D^T Y_{dqo}(h-1)D^C + D^H Y_{dqo}(h+1)D \\
&\quad + D^T_o Y_{dqo}(h)D_o][V_{abc}(h)] \\
&\quad + [D^H Y_{dqo}(h+1)D^C][V_{abc}(h+2)] \\
&\quad + [D^T Y_{dqo}(h-1)D][V_{abc}(h-2)].
\end{aligned}
$$
$$(4\text{-}71)$$

If we substitute

$$[Y(h)] = [D^T Y_{dqo}(h-1)D^C + D^H Y_{dqo}(h+1)D$$
$$+ D^T_o Y_{dqo}(h)D_o], \tag{4-72}$$

$$[I_f(h)] = [D^H Y_{dqo}(h+1)D^C][V_{abc}(h+2)]$$
$$+ [D^T Y_{dqo}(h-1)D][V_{abc}(h-2)], \tag{4-73}$$

then Eq. 4-71 can be rewritten as

$$[I_{abc}(h)] = [Y(h)][V_{abc}(h)] + [I_f(h)]. \tag{4-74}$$

This equation defines the machine model shown in Fig. 4.27, where

- $[I_{abc}(h)] = [I_{km}(h)]$.
- $[V_{abc}(h)] = [V_k(h)] - [V_m(h)]$.
- $[Y(h)]$ correlates harmonic voltages to harmonic currents of the same order.
- $[I_f(h)]$ describes the harmonic coupling due to frequency-conversion effects. This current vector is generated from the conversion of voltages of different harmonic orders.

Following a similar approach, the equivalent DC voltage source of the $dq0$-machine model at harmonic order $h = 0$ becomes a set of positive-sequence fundamental frequency voltage sources in $abc$ coordinates. As explained earlier, the values of these voltages are determined in conjunction with load-flow constraints.

Therefore, matrix operations are sufficient to compute the admittance matrix $[Y(h)]$ and the equivalent current source $[I_f(h)]$ needed for the harmonic machine model:

- Compute $[Y(h)]$ according to Eq. 4-72 whenever the machine admittance matrix needs to be added to the network admittance matrix in the frequency scan process; and
- Compute $[I_f(h)]$ according to Eq. 4-73 where $[V_{abc}(h)]$ contains the machine voltages obtained from the load-flow solution of the previous iteration.

### Injected Harmonic Currents Due to Saturation $[I_s(h)]$.
Saturation of the stator and rotor cores of a synchronous machine has a significant impact on the machine's operation and the corresponding operating point [68]. Other saturation factors (e.g., cross-coupling between $d$- and $q$-axes) are assumed to be negligible and are not considered in the model of Fig. 4.27.

It is shown in [12] that the influence of saturation can be represented as a current source in parallel with the $dq0$-machine model (Eq. 4-66):

$$[I_{dqo}(h)] = [Y_{dqo}(h)][V_{dqo}(h)] + [I_{dqo-s}(h)], \tag{4-75}$$

where $[I_{dq0-s}(h)]$ is a known current vector representing the effect of saturation. For a given set of machine terminal voltages, vector $[I_{dq0-s}(h)]$ is obtained through a process of subiterations [12].

With saturation effects included in $dq0$ coordinates, Park's transformation can now be applied to obtain the machine's model in $abc$ coordinates. The processing of the first part of Eq. 4-75 has been explained in the previous section. The second part (e.g., $[I_{dq0-s}(h)]$) will be discussed next.

Using Park transformation matrix (Eq. 4-61), the $dq0$-harmonic currents of order $h$ appear in $abc$ coordinates as

$$[i_{abc-s}(t)] = \text{Real}\{[P]\sqrt{2}[I_{dqo-s}(h)]e^{jh\omega t}\}$$
$$= \text{Real}\{\sqrt{2}[D^T e^{j(h+1)\omega t} + D^H e^{j(h-1)\omega t}$$
$$+ D_o e^{jh\omega t}][I_{dqo-s}(h)]\}. \tag{4-76}$$

It can be seen that the $dq0$ currents $[I_{dq0-s}(h)]$ also introduce three separate harmonic components of order $(h-1)$, $h$, and $(h+1)$ in the $abc$ coordinates. Assuming that $[I_{dqo-s}(h)]$ includes harmonics of all orders, the particular $dq0$ harmonics that produce harmonic currents of the same order in $abc$ coordinates can be grouped together in phasor form as

$$[I_s(h)] = [D]^T[I_{dqo-s}(h-1)] + [D]^H[I_{dqo-s}(h+1)]$$
$$+ [D_o][I_{dqo-s}(h)]. \tag{4-77}$$

This is the current that is needed to represent the effects of saturation in $abc$ coordinates.

The above equation defines the final machine model used for harmonic load-flow studies. The approach of computing $[I_s(h)]$ can be summarized as follows:

- Compute the saturation current $[I_{dq0-s}(h)]$ using the subiteration process described in reference [12] or any other desired approach; and
- Compute $[I_s(h)]$ according to Eq. 4-77 for all harmonics of interest.

### Total Injected Harmonic Current $[I_{nt}(h)]$.
The complete machine model with both the frequency-conversion and saturation effects included can now be described by

$$[I_{abc}(h)] = [Y(h)][V_{abc}(h)] + [I_f(h)] + [I_s(h)]. \tag{4-78}$$

**TABLE E4.12.1** Values of Negative-Sequence Impedance for Synchronous Machine [12]

| Test type | Exact[a] | Harmonic truncation[b] | | |
| --- | --- | --- | --- | --- |
| | | $h = 1$ | $h = 1, 3$ | $h = 1, 3, 5$ |
| I-definition | $0.0655 + j0.3067$ | $0.0857 + j0.2165$ | $0.0655 + j0.3066$ | $0.0655 + j0.3066$ |
| V-definition | $0.0858 + j0.2165$ | $0.0856 + j0.2166$ | $0.0859 + j0.2164$ | $0.0859 + j0.2164$ |

[a]Based on analytical formulas.

[b]$h = 1$: only the fundamental frequency component included; $h = 1, 3$: $h = 1$ modeling plus the third harmonic; $h = 1, 3, 5$: $h = 1, 3$ modeling plus the fifth harmonic.

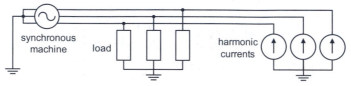

**Figure E4.13.1** Single machine test system used in application problems [12].

#### 4.3.4.4 Application Example 4.12: Effect of Frequency Conversion on Synchronous Machine Negative-Sequence Impedance

The negative-sequence impedance of a synchronous machine is important for simulation under unbalanced conditions. Its value is different depending on whether negative-sequence currents or negative-sequence voltages are used to determine the impedance [68]. This inconsistency is caused by the frequency conversion process. In this example the harmonic model of Fig. 4.27 (implemented in a multiphase harmonic load flow [65]) is used to run two test conditions that are commonly accepted to define the machine's negative-sequence impedance:

- I-definition: synchronous machine is connected to a negative-sequence current source and the negative-sequence terminal voltage is recorded; and
- V-definition: synchronous machine is connected to a negative-sequence voltage source and the negative-sequence terminal current is recorded.

The impedance resulting from these tests is the ratio of the fundamental frequency voltage to the fundamental frequency current obtained by harmonic analysis. The program was run for three different levels of harmonic truncation (Table E4.12.1).

The following observations can be derived from this table:

- The difference between the machine impedances calculated from the I- and V-definitions is caused by the 3rd harmonic. The contribution of the 5th harmonic is negligible;

- If harmonics are not included, the proposed machine model corresponds to the impedance of V-definition;
- In the general case of unbalanced machine operation, the machine's negative-sequence impedance is neither the one from the V-definition nor the one from the I-definition. The correct machine response can only be fully evaluated with harmonics included. The combination of the load-flow constraints with harmonic solutions is therefore the best approach in such situations; and
- Agreement between simulation and exact results justifies the validity of the synchronous machine harmonic model (Fig. 4.27).

#### 4.3.4.5 Application Example 4.13: Effect of Imbalance on Power Quality of Synchronous Machines

A profile of the harmonic content of the synchronous machine currents and voltages is calculated (using the harmonic model of Fig. 4.27) for two test machines. The voltage and current harmonics in a machine will be influenced by the conditions imposed by the external network. To simplify the interpretation of the results, a straightforward operating condition with grounded Y synchronous machines and unbalanced R, L, and C loads is considered (Fig. E4.13.1 without the harmonic current sources). The loads are adjusted so that the machine's current imbalance at fundamental frequency is around the normal tolerance of 10%. The results (using two synchronous machines with different positive- and negative-sequence currents) are shown in Table E4.13.1 for the voltage harmonics

**TABLE E4.13.1** Harmonic Profiles of Two Synchronous Machines (Unbalanced Operating Conditions) [12]

| Machine | Item | h = 1 | | h = 3 | | h = 5 | | h = 7 |
|---|---|---|---|---|---|---|---|---|
| | | Magnitude | Angle | Magnitude | Angle | Magnitude | Angle | Magnitude |
| No. 1 | $V_a$ | 1.103 | 1.1 | 0.0148 | 88.3 | 0.0002 | 148.4 | 0 |
| $NP^a = 9.35\%$ | $V_b$ | 1.025 | −121.5 | 0.0131 | −36.5 | 0.0002 | 23.8 | 0 |
| | $V_c$ | 1.022 | 120.2 | 0.0131 | −151.5 | 0.0002 | −90.2 | 0 |
| | $I_{f\text{-}a}$ | 0.027 | 167.3 | 0.0225 | −152.5 | 0.0002 | −99.5 | 0 |
| | $I_{f\text{-}b}$ | 0.027 | −72.7 | 0.0225 | 87.4 | 0.0002 | 140.4 | 0 |
| | $I_{f\text{-}c}$ | 0.027 | 47.3 | 0.0225 | −32.4 | 0.0002 | 20.4 | 0 |
| No. 2 | $V_a$ | 1.080 | 1.3 | 0.0345 | 92.1 | 0.0011 | 157.2 | 0 |
| $NP^a = 12.41\%$ | $V_b$ | 1.041 | −121.6 | 0.0313 | −33.0 | 0.0009 | 32.4 | 0 |
| | $V_c$ | 1.030 | 120.3 | 0.0315 | −148.2 | 0.0010 | −82.7 | 0 |
| | $I_{f\text{-}a}$ | 0.025 | 155.2 | 0.0704 | −151.3 | 0.0016 | −94.8 | 0 |
| | $I_{f\text{-}b}$ | 0.025 | −84.7 | 0.0696 | 88.3 | 0.0016 | 145.2 | 0 |
| | $I_{f\text{-}c}$ | 0.025 | 35.2 | 0.0696 | −30.9 | 0.0016 | 25.2 | 0 |

[a] NP: ratio of the negative- to positive-sequence current at fundamental frequency.

**TABLE E4.14.1** Third Harmonic Currents in Grounded Y- and Δ-Connected Synchronous Generators [12]

| | | Phase A | | Phase B | | Phase C | |
|---|---|---|---|---|---|---|---|
| | | Magnitude | Angle | Magnitude | Angle | Magnitude | Angle |
| $V_3$ | Y−g | 0.0148 | 88.3 | 0.0130 | −36.5 | 0.0131 | −151.5 |
| | Δ | 0.0091 | 81.7 | 0.0071 | −40.7 | 0.0078 | −168.4 |
| $I_3$ | Y−g | 0.0090 | 48.3 | 0.0125 | −71.7 | 0.0108 | 170.1 |
| | Δ | 0.0055 | 41.6 | 0.0068 | −76.0 | 0.0065 | 153.3 |

and for the frequency-conversion-induced equivalent current source $I_f(h)$. Significant harmonics are observed even for cases when the machines are operated within an acceptable range of imbalance. The effects of these harmonics could easily be magnified if voltage resonance takes place. Note that the harmonic magnitudes decrease quickly for higher orders. From the results of Tables E4.12.1 and E4.13.1, it may be concluded that only 3rd and 5th harmonics need to be included in synchronous machine's representation for harmonic studies.

#### 4.3.4.6 Application Example 4.14: Effect of Delta Connection on Power Quality of Synchronous Machines

It is believed that the flow of the third harmonic in lines can be prevented by connecting the three-phase equipment in delta. This assumption is tested here by assuming that the winding of the first synchronous generator (Application Example 4.13) is to be connected in delta. The results of this test (third harmonic currents for the Δ and Y-connected machines) are presented in Table E4.14.1. Even though the third harmonic components are about 40% smaller

in the Δ connection, they do not cancel completely. The reason is

*The third harmonic generated by the frequency-conversion process in a synchronous machine is of negative sequence and not of zero sequence, and therefore cannot be eliminated by a Δ connection.*

#### 4.3.4.7 Application Example 4.15: Effect of Saturation on Power Quality of Synchronous Machines

In the model of Fig. 4.27, the saturation curve is approximated in a piecewise-linear manner between the data points (MMF,$\lambda$) = (0.9,0.9),(1.5,1.2)(2.7,2.8). Table E4.15.1 lists the harmonic profiles of the machine with and without saturation. There are large differences between the cases with and without saturation (the last column of Table E4.15.1). The differences in the phase angles are even more noticeable. These results suggest that

*Saturation can have an important influence on harmonic power-flow analysis.*

**TABLE E4.15.1** Harmonic Profiles of Two Synchronous Machines with and without Saturation (under Unbalanced Operating Conditions) [12]

| | | $h = 1$ | | $h = 3$ | | $h = 5$ | | |
|---|---|---|---|---|---|---|---|---|
| | | Magnitude | Phase | Magnitude | Phase | Magnitude | Phase | Difference (%)[a] |
| NS[b] | $V_a$ | 1.103 | 1.1 | 0.0148 | 88.3 | 0.00020 | 148.4 | |
| SAT | $V_a$ | 1.103 | 1.1 | 0.0120 | 76.3 | 0.00014 | 115.8 | 19.0 |
| NS | $V_b$ | 1.025 | −121.5 | 0.0130 | −36.5 | 0.00017 | 23.8 | |
| SAT | $V_b$ | 1.025 | −121.5 | 0.0106 | −48.4 | 0.00012 | −8.8 | 18.6 |
| NS | $V_c$ | 1.022 | 120.2 | 0.0131 | −151.5 | 0.00017 | −90.2 | |
| SAT | $V_c$ | 1.022 | 120.2 | 0.0106 | −163.7 | 0.00013 | −122.9 | 18.6 |
| NS | $I_a$ | 0.0027 | 167.3 | 0.0225 | −152.5 | 0.00022 | −99.5 | |
| SAT | $I_a$ | 0.0022 | 168.5 | 0.0182 | −164.5 | 0.00014 | −131.3 | 19.0 |
| NS | $I_b$ | 0.0027 | −72.7 | 0.0224 | 87.4 | 0.00022 | 140.4 | |
| SAT | $I_b$ | 0.0022 | −71.4 | 0.0182 | 75.5 | 0.00014 | 108.6 | 18.6 |
| NS | $I_c$ | 0.0027 | 47.3 | 0.0225 | −32.4 | 0.00022 | 20.4 | |
| SAT | $I_c$ | 0.0022 | 48.5 | 0.0182 | −44.6 | 0.00014 | −11.3 | 18.7 |

[a]Relative difference of the 3rd harmonic magnitude.

[b]NS: without saturation; SAT: with saturation.

**TABLE E4.16.1** Voltages of Machine 1 (Defined in Table E4.13.1) in the Presence of Nonlinear Loads (Fig. E4.13.1) [12]

| | Phase A | | Phase B | | Phase C | | |
|---|---|---|---|---|---|---|---|
| | Magnitude | Phase | Magnitude | Phase | Magnitude | Phase | Case study |
| $h = 1$ | 1.099 | 0.2 | 1.023 | −120.8 | 1.028 | 120.5 | 1 |
| | 1.100 | 0.3 | 1.024 | −120.9 | 1.026 | 120.5 | 2 |
| | 1.000 | 0.3 | 1.024 | −120.8 | 1.026 | 120.5 | 3 |
| $h = 3$ | 0.0607 | 69.4 | 0.0304 | −171.5 | 0.0536 | −68.4 | 1 |
| | 0.0777 | 67.3 | 0.0283 | −155.3 | 0.0572 | −82.9 | 2 |
| | 0.724 | 66.4 | 0.0295 | −156.7 | 0.0526 | −79.7 | 3 |
| $h = 5$ | 0.0336 | 69.7 | 0.0450 | −59.2 | 0.0379 | −173.5 | 1 |
| | 0.0445 | 80.1 | 0.0539 | −50.0 | 0.0458 | −173.6 | 2 |
| | 0.0417 | 76.8 | 0.0517 | −52.9 | 0.0439 | −177.4 | 3 |

### 4.3.4.8 Application Example 4.16: Impact of Nonlinear Loads on Power Quality of Synchronous Machines

The effects of nonlinearities of synchronous machines are further investigated by including nonlinear loads, represented as 3rd and 5th current injections to the generator-load system (Fig. E4.13.1). Three case studies are performed:

- *Case 1:* with no machine nonlinearities,
- *Case 2:* with only frequency conversion, and
- *Case 3:* with both frequency conversion and saturation.

The results (Table E4.16.1) further confirm the need for detailed modeling of machine's nonlinearities.

### 4.3.5 Static- and Dynamic-Rotor Eccentricities Generating Current and Voltage Harmonics

Most synchronous machines have relatively large radial air-gap lengths, which range from a few milli-

meters to tens of centimeters. This is so because the synchronous reactances ($X_d$, $X_q$) must be small enough to guarantee a good dynamic (transient, subtransient) behavior. Although permanent-magnet machines have a mechanical air gap of less than one millimeter up to a few millimeters, their electrical air gap is always larger than one millimeter because the radial air-gap height (length) of the permanent magnet behaves magnetically like air with the permeability of free space $\mu_o$ because the recoil permeability $\mu_R$ of excellent permanent-magnetic material (e.g., NdFeB) is about $\mu_R \approx 1.05\mu_o$. Therefore, an imperfect mechanical mounting – resulting in a static-rotor eccentricity – or a bent shaft – resulting in dynamic-rotor eccentricity – will not significantly alter the magnetic coupling of the stator windings and the rotating magnetic flux will be about circular. Here one should mention that permanent-magnet machines cannot be easily assembled for ratings larger than a few tens of kilowatts because a mounting gear will be required, which maintains a uniform concentric air gap during the mounting of the rotor.

The modified winding-function approach (MWFA) accounting for all space harmonics has been used for the calculations of machine inductances under rotor eccentricities [69]. Machines can fail due to air-gap eccentricity caused by mechanical problems including the following:

- shaft deflection,
- inaccurate positioning of the rotor with respect to the stator,
- worn bearings,
- stator-core movement, and
- bent rotor shaft.

Electrical asymmetries contribute to harmonic generation such as

- Damper windings (amortisseur) are incomplete, that is, the amortisseur has a different construction between poles as compared with that on the pole d-axis; or
- Eddy currents in the pole faces of salient-pole machines contribute to subtransient response;
- Heating of the rotor due to nonuniform amortisseur and nonuniform eddy-current generation around the rotor circumference may cause shaft bending/deflection; and
- Incomplete shielding of the field winding of a machine fitted with nonuniform amortisseurs and subjected to harmonic disturbance.

Nonsinusoidal air-gap flux wave effects – due to any of the above-listed reasons – can be accounted for by representing the mutual inductances between the field and stator windings, and those between the stator windings, by trigonometric series as a function of the rotor angle $\theta$. It is found that the inclusion of nonsinusoidal inductance variation causes a broad spectrum such that injection of a single-harmonic current results in voltages containing all odd harmonic orders including the fundamental. The extent to which the energy is spread across the spectrum depends on the pole shape, the numbers of stator and rotor slots, and the amortisseur design. The spectral spread is a second-order effect compared to the generation of the associated voltage at frequency ($h \pm 2$). Harmonic behavior is also affected by nonsinusoidal inductance variation with rotor position and by the extent to which induced damper winding currents shield the field winding from the harmonic-gap flux waves. Rotor eccentricity in machines can be of two forms:

- static air gap eccentricity, and
- dynamic air gap eccentricity.

With the static air-gap eccentricity, the position of the minimum radial air-gap length is fixed in space and the center of rotation and that of the rotor are the same. Static eccentricity can be caused by oval stator cores or by the incorrect relative mounting of stator and rotor. In case of dynamic air-gap eccentricity, the center of rotation and that of the rotor are not the same and the minimum air gap rotates with the rotor. Therefore, dynamic eccentricity is both time and space dependent. Dynamic eccentricity can be caused by misalignment of bearings, mechanical resonance at critical speeds, a bent rotor shaft, and wear of bearings [69].

MATLAB/Simulink [70] can be used for the simulation of machine variables. All machine inductances employed for the simulation are expressed in their Fourier series based on the MWFA. Two different cases, namely, a noneccentric and an eccentric rotor case, for this analysis are investigated. In the first case, it is assumed that dynamic air-gap eccentricity is not present, and in the second one 50% dynamic air-gap eccentricity is introduced to investigate the effect of the eccentric rotor on stator-current signatures. To analyze the stator-current signatures of these two cases, FFT (fast Fourier transform) of the current signals are performed for all cases. The stator currents show the existence of the 5th (300 Hz), 7th (420 Hz), 11th (660 Hz), 13th (780 Hz), 17th (1020 Hz), and 19th (1140 Hz) harmonics even when the rotor is not eccentric. Implementing 50% eccentricity will cause the stator and rotor current induced harmonics to increase when compared to those generated without rotor eccentricity as is listed in Table 4.3.

The stator current harmonics exist because the interaction of the magnetic fields caused by both the stator and rotor windings will produce harmonic fluxes that move relative to the stator and they induce corresponding current harmonics in the stationary stator windings. These harmonic fluxes in the air gap will increase as the rotor dynamic eccentricity increases, and consequently the current harmonic content increases too. The 3rd harmonic and its multiples are assumed not to exist in the stator windings because it is a three-phase system with no neutral

**TABLE 4.3** Relative Percentage of Stator Harmonics Due to 50% Dynamic Eccentricity

| 5th | 7th | 11th | 13th | 17th | 19th |
|-------|-------|-------|-------|-------|-------|
| 22.8% | 12.4% | 20.9% | 28.4% | 47.1% | 36.9% |

connection. However, third-harmonic components are induced in the rotor-field winding. Because the stator and rotor-current signatures of synchronous machines have changed, either or both signatures can be utilized for detecting dynamic eccentricity. However, it is more practical to implement the stator-current signature analysis for the condition monitoring than the rotor-current signature analysis because some synchronous machines have brushless excitation [49], which will make the rotor current inaccessible. In summary, one can conclude that synchronous machines are not very prone to the effects caused by rotor eccentricity due to their large air gaps.

### 4.3.6 Shaft Flux and Bearing Currents

The generation of shaft flux [71] due to mechanical and magnetic asymmetries as well as solid-state switching is an important issue for large synchronous generators. Bearing currents are predominantly experienced in PWM AC drives [72–76] due to the high-frequency switching ripples resulting in bearing electroerosion. Techniques for measurement of parameters related to inverter-induced bearing currents are presented in [77].

### 4.3.7 Conclusions

The most important conclusions of this section are

- Synchronous machines generate harmonics due to
  1. frequency conversion,
  2. saturation, and
  3. unbalanced operation.
- Current harmonics due to nonlinear loads should be limited based on IEEE-519 [78].
- Unbalanced load-flow analysis without taking into account harmonics is not correct due to the ambiguous value of the negative-sequence impedance of synchronous machines. Correct results can only be obtained by including harmonics in a three-phase, load-flow approach.
- Only 3rd and 5th harmonics need to be included for the machine's representation in harmonic analysis. The 3rd harmonic is of the negative-sequence and not of the zero-sequence type. Hence, it cannot be eliminated by Δ transformer connections. The same is true for transformer applications with DC current bias [79].
- Machine saturation can have noticeable effects on the harmonic distribution. These effects are more significant with respect to the harmonic phase angles.

- Rotor eccentricity has a minor influence on harmonic generation due to the large air gap of synchronous machines.
- Shaft fluxes and bearing currents can have detrimental impacts.

## 4.4 SUMMARY

In the past and at present synchronous generators represent an integral part of a power system. Although the tendency exists to move from a central power station approach to a distributed generation (DG) system with renewable energy sources in the megawatt range, there will be always the requirement of frequency control, which is best performed by a central power station in the gigawatt range that can absorb load-flow fluctuations. Renewable plants will be mostly operated at their maximum power output and therefore cannot provide additional (real or reactive) power if the load increases beyond their assigned output powers. In addition, such plants have an intermittent output, which cannot be controlled by dispatch and control centers. This is to say that also in future power systems the power quality will depend on the reliable operation of synchronous generators.

This chapter starts out with the introduction of the synchronous machine model based on $dq0$ coordinates. To visualize steady-state saturated magnetic fields, numerical solutions are presented for no-load, full-load, and short-circuit conditions. Fields for permanent-magnet machines and switched-reluctance machines are presented as well. Thereafter, the two possible reference systems (e.g., consumer and generator) are addressed. The application examples relate to the calculation of synchronous reactances, the investigation of various fault conditions including reclosing, and the calculation of the amortisseur current for subtransient faults such as line-to-line, line-to-neutral, and balanced three-phase short circuits. The design of synchronous machines and permanent-magnet machines for wind-power and hybrid drive applications, respectively, is discussed. Design guidelines for synchronous machines are presented, and the calculation of magnetic forces – based on the Maxwell stress – is included. The performance of synchronous machines under the influence of harmonics is explained based on models used in application examples. The employment of such models for harmonic power flow analyses is mandatory. Finally, approaches for analyzing static and dynamic eccentricities, shaft fluxes, and bearing currents are outlined.

## 4.5 PROBLEMS

### Problem 4.1: Steady-State Analysis of a Nonsalient-Pole (Round-Rotor) Synchronous Machine Using Phasor Diagrams

A nonsalient-pole synchronous machine has the following data: $X_S = (X_d = X_q) = 1.8$ pu, $R_a = 0.01$ pu, $I_{phase} = 1.0$ pu and $V_{L-N} = 1.0$ pu.

a) Sketch the equivalent circuit using the consumer reference system.
b) Draw the phasor diagram (to scale) based on the consumer reference system for a lagging (underexcited) displacement power factor of $\cos \varphi = 0.90$.
c) Sketch the equivalent circuit based on the generator reference system.
d) Draw the phasor diagram (to scale) employing the generator reference system for a lagging (overexcited) displacement power factor of $\cos \varphi = 0.90$.

### Problem 4.2: Steady-State Analysis of a Salient-Pole Synchronous Machine Using Phasor Diagram [7]

A salient-pole synchronous machine has the following data: $X_d = 1.8$ pu, $X_q = 1.5$ pu, $R_a = 0.01$ pu, $I_{phase} = 1.0$ pu, $V_{L-N} = 1.0$ pu.

a) Sketch the equivalent circuit based on the consumer reference system.
b) Draw the phasor diagram (to scale) using the consumer reference system for a lagging (underexcited) displacement power factor of $\cos \varphi = 0.70$.
c) Sketch the equivalent circuit based on the generator reference system.
d) Draw the phasor diagram (to scale) employing the generator reference system for a lagging (overexcited) displacement power factor of $\cos \varphi = 0.70$.

### Problem 4.3: Phasor Diagram of a Synchronous Generator With Two Displaced Stator Windings [80]

Synchronous machines with two by 30° displaced stator windings (winding #1 and winding #2) are used to improve the voltage/current wave shapes so that the output voltages of a generator become more sinusoidal. This is similar to the employment of 12-pulse rectifiers versus 6-pulse rectifiers. To accom-

plish this, stator winding #1 feeds a $Y$–$Y$ connected transformer and stator winding #2 supplies a $\Delta$–$Y$ connected transformer as illustrated in Fig. P4.3.1. For a $S = 1200$ MVA, $V_{L-L} = 24$ kV, $f = 60$ Hz, $p = 2$ turbogenerator the machine parameters are in per unit (pu): $X_d = 1.05$, $X_{12d} = 0.80$, $X_q = 0.95$, $X_{12q} = 0.73$, $X_{ffd} = 2.10$, $k_o = 0.5$, $k_1 = 0.638$, $r = R_a = 0.00146$, $r_f = R_f = 0.0007$.

Draw the phasor diagram (to scale) for rated operation and a displacement power factor of $\cos \varphi = 0.8$ lagging (overexcited, generator reference system).

### Problem 4.4: Balanced Three-Phase Short-Circuit Currents of a Synchronous Machine Neglecting Influence of Amortisseur

A three-phase synchronous generator is initially unloaded and has a rated excitation so that $E_f = 1.0$ pu. At time $t = 0$, a three-phase short-circuit is applied. The machine parameters are

direct ($d$) axis: $X_d = 1.2$ pu, $X_f = 1.1$ pu, $X_m = 1.0$ pu, $R_a = 0.005$ pu, $R_f = 0.0011$ pu;

quadrature ($q$) axis: $X_q = 0.8$ pu, $X_m = 0.6$ pu, $R_a = 0.005$ pu.

a) Plot the first-half cycle of the torque $T_{3phase\_short-circuit}$ due to the three-phase short-circuit neglecting damping and assuming that there is no amortisseur.
b) Plot the torque $T_{3phase\_short-circuit}$ as a function of the time angle $\varphi = \omega t$.
c) At which angle $\varphi_{max}$ (measured in degrees) occurs the maximum torque $T_{max}$ (measured in per unit)?

### Problem 4.5: Torque $T_{L-L}$ and Induced Voltage $e_a$ During the First-Half Cycle of a Line (Line b)-to-Line (Line c) Short-Circuit of a Synchronous Machine Neglecting Influence of Amortisseur

a) For the machine parameters of Problem 4.4 calculate and plot the line-to-line torque $T_{L-L}$ for the first half-cycle when phases b and c are short-circuited, and determine at which angle $\varphi_{max}$ (measured in degrees) the maximum torque (measured in per unit) occurs.
b) Plot the induced open-circuit voltage of phase a, that is, $e_a$ (measured in per unit) as a function of

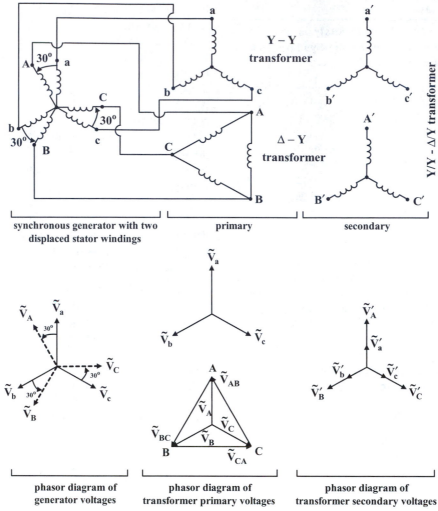

**Figure P4.3** Generator-transformer interconnection with two displaced stator windings and a Y/Y–Δ/Y transformer.

the phase angle $\varphi$ (measured in degrees) for the first half-cycle.

## Problem 4.6: Line-to-Neutral Short-Circuit Current of a Synchronous Machine Neglecting Influence of Amortisseur

a) For the machine parameters of Problem 4.4 calculate the total (fundamental and harmonics) line-to-neutral short-circuit current $i_a$ for the first half-cycle when phase a and the neutral are short-circuited, and determine at which angle $\varphi_{max}$ (measured in degrees) the maximum current $I_{L\text{-}N\_max}$ (measured in per unit) occurs.

b) Plot the fundamental short-circuit current of phase a $i_{a1}$ (measured in per unit) for the first half-cycle as a function of the phase angle $\varphi$ (measured in degrees).

## Problem 4.7: Out-of-Phase Synchronization Torque of a Synchronous Machine Neglecting Influence of Amortisseur

a) For the machine parameters of Problem 4.4 calculate and plot for the first half-cycle the out-of-phase synchronizing torque.

b) Determine at which angle $\varphi_{max}$ (measured in degrees) the maximum torque (measured in per unit) occurs.

## Problem 4.8: Synchronizing and Damping Torques of a Synchronous Machine Neglecting Influence of Amortisseur

For the machine parameters of Problem 4.4 calculate and plot for the first half-cycle the steady-state torque $T_0$, the synchronizing torque $T_S$, and the damping torque $T_D$. The per-unit transient time

**TABLE P4.9** Starting and Accelerating Torque of a Synchronous Machine Developed by Induction-Motor Action as a Function of Slips

| s (pu) | 1.0 | 0.9 | 0.8 | 0.7 | 0.6 | 0.5 | 0.4 | 0.3 | 0.2 | 0.1 | 0 |
|---|---|---|---|---|---|---|---|---|---|---|---|
| $T_{\text{starting}}$ (pu) | 0.0172 | | | | 0.0685 | | | | | | |

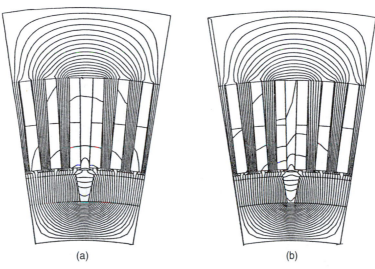

(a)                 (b)

**Figure P4.10** No-load (a) and full-load (b) flux patterns of permanent-magnet DC motor where the flux within one tube is $\Delta A = 0.0002$ Wb/m.

constant of the $d$-axis is $T'_d = 242$ pu and the per unit voltages $V = V_0 = V_{l\text{-}n} = V_{l\text{-}n0} = 1.0$ pu may be used.

## Problem 4.9: Starting and Accelerating Torque of a Synchronous Machine Developed by Induction Motor Action Neglecting Influence of Amortiseur

The starting of synchronous generators without amortisseur (damper) winding by induction-motor action is not advisable, as will be demonstrated in this problem. Calculate for a synchronous machine with the parameters $X_d = 1.2$ pu, $X_q = 0.8$ pu, $X_f = 1.1$ pu, $X_m = 1.0$ pu, $R_a = 0.005$ pu, and $R_f = 0.0011$ pu the starting and accelerating torque $T_{\text{starting}}$ as a function of the slip $s$. To do this, start out from the $dq0$-coordinate system and transform it to the fbo [81] coordinate system. Table P4.9.1 presents some of the results. It is the task of this problem to fill in the missing values of this table.

As can be seen from the few entries of this table, a synchronous machine cannot develop any significant induction motor torque without amortisseur winding. Most large synchronous generators have a very rudimentary amortisseur and must not be started by induction-motor action. Small synchronous machines with ratings of less than 1 MW can be

equipped with a full amortisseur winding and can be started by induction-motor action.

## Problem 4.10: Design of a Brushless DC (Permanent-Magnet) Motor Drive [24]

For a hybrid automobile drive a permanent-magnet motor – as a part of a brushless DC machine – is to be analyzed with the following data:

Motor rating
output power $P = 50$ kW
efficiency $\eta = 0.95$

indirect water cooling

$V_{L\text{-}L} = 240$ V, $f = 400$ Hz

$\cos \varphi = 0.9$ lagging
(consumer system)

Stator
axial length = 300 mm
outer diameter
  $D_{\text{sout}} = 244$ mm
inner diameter
  $D_{\text{sin}} = 170.4$ mm
number of slots in stator
120
number of poles 20

Rotor
outer diameter $D_{\text{rout}} = 169.4$ mm
inner diameter $D_{\text{rin}} = 140.2$ mm

An encoder is used to sense the rotor speed. For your design it will be helpful to rely on the flux patterns and machine cross section of Fig. P4.10a,b. It is recommended to use as permanent-magnet material NdFeB.

a) Design the stator winding.
b) What is the rated synchronous speed $n_{s\_rat}$?
c) Compute the (maximum) height (radial one-sided length) of the permanent magnets $h_{mag\_max}$.
d) Calculate the maximum flux density of this machine.
e) Determine the leakage inductance $L_\ell$ (from stator phase a to stator phase b) and the winding resistance $R_a$ of one stator phase.
f) Determine the current density $J_{stator}$, provided the stator winding copper fill factor is $k_{cu} = 0.7$, and the average stator-slot width is about the same as the average stator-tooth width.

## Problem 4.11: Fourier Analyses of Terminal Currents of a Brushless DC Motor (Permanent-Magnet Motor with Inverter)

Find the Fourier coefficients of the current wave shapes of Fig. P4.11 that represent the input MOSFET current (upper wave shape) and the current of a brushless DC motor (lower wave shape). Select a certain sampling interval, e.g., $\Delta t = 0.025$ ms, perform a Fourier analysis, and show that for a minimum approximation error the Nyquist criterion is satisfied.

## Problem 4.12: Steady-State Frequency Variation within an Interconnected Power System as a Result of Two Load Changes [82]

A block diagram of two interconnected areas of a power system (e.g., area #1 and area #2) is shown in Fig. P4.12. The two areas are connected by a single transmission line. The power flow over the transmission line will appear as a positive load to one area and an equal but negative load to the other, or vice versa, depending on the direction of power flow.

a) For steady-state operation show that with $\Delta\omega_1 = \Delta\omega_2 = \Delta\omega$ the change in the angular velocity (which is proportional to the frequency $f$) is

$$\Delta\omega = \text{function of}$$
$$(\Delta P_{L1}, \Delta P_{L2}, D_1, D_2, R_1, R_2) \tag{P12-1}$$

and

$$\Delta P_{tie} = \text{function of the same parameters as in Eq. P12-1.} \tag{P12-2}$$

b) Determine values for $\Delta\omega$ (Eq. P12-1), $\Delta P_{tie}$ (Eq. P12-2), and the new frequency $f_{new}$, where the nominal (rated) frequency is $f_{rated} = 60$ Hz, for these parameters: $R_1 = 0.05$ pu, $R_2 = 0.1$ pu, $D_1 = 0.8$ pu, $D_2 = 1.0$ pu, $\Delta P_{L1} = 0.2$ pu, $\Delta P_{L2} = -0.3$ pu.
c) For a base apparent power $S_{base} = 1000$ MVA compute the power flow across the transmission line.
d) How much is the load increase in area #1 ($\Delta P_{mech1}$) and area #2 ($\Delta P_{mech2}$) due to the two load steps?
e) How would you change $R_1$ and $R_2$ in case $R_1$ is a wind or solar power plant operating at its maximum power point (and cannot accept any significant load increase due to the two load steps) and $R_2$ is a coal fired plant that can be overloaded?

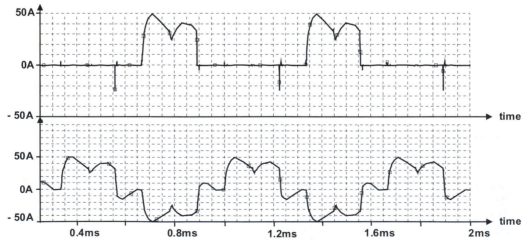

**Figure P4.11** Current wave shape of the MOSFET current (upper wave shape), and the input current of a brushless dc-motor (lower wave shape).

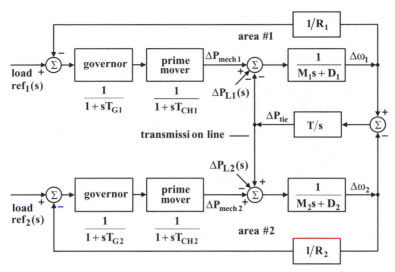

**Figure P4.12** Interconnected power system consisting of area #1 and area #2.

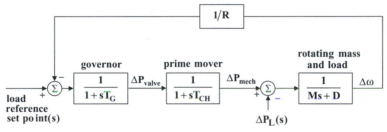

**Figure P4.13** Block diagram of governor, prime mover, and rotating mass and load.

## Problem 4.13: Frequency Control of an Isolated Power Plant (Islanding Operation)

Figure P4.13 illustrates the block diagram of governor, prime mover (steam turbine), and rotating mass and load of a turbogenerator set. For the frequency change ($\Delta\omega$) per generator output power change ($\Delta P$),

$$R = \frac{\Delta\omega}{\Delta P}\,\text{pu} = 0.01\,\text{pu},$$

the load change ($\Delta PL$) per frequency change

$$D = \frac{\Delta P_L}{\Delta\omega} = 0.8\,\text{pu},$$

step load change $\Delta P_L(s) = \dfrac{\Delta P_L}{s}\,\text{pu} = \dfrac{0.2}{s}\,\text{pu}$,

angular momentum of steam turbine and generator set $M = 4.5$,
base apparent power $S_{base} = 500\,\text{MVA}$,
governor time constant $T_G = 0.01\,\text{s}$,
valve charging time constant $T_{CH} = 1.0\,\text{s}$,
load reference set point = 1.0.

a) Derive for Fig. P4.13 $\Delta\omega_{\text{steady state}}$ by applying the final value theorem. You may assume the load reference set point(s) = 0 and $\Delta P_L(s) = \dfrac{\Delta P_L}{s}\,\text{pu} = \dfrac{0.2}{s}\,\text{pu}$, where $s$ is the Laplace operator.

b) List the ordinary differential equations and the algebraic equations of the block diagram of Fig. P4.13.

c) Use either Mathematica or MATLAB to establish steady-state conditions by imposing a step function for load reference set point $(s) = \dfrac{1}{s}\,\text{pu}$ and run the program with a zero step-load change $\Delta P_L = 0$ for 5 s. After 5 s impose a step load change of $\Delta P_L(s) = \dfrac{\Delta P_L}{s}\,\text{pu} = \dfrac{0.2}{s}\,\text{pu}$ to find the transient response of $\Delta\omega(t)$.

d) Use Mathematica or MATLAB to establish steady-state conditions by imposing a step function for load reference set point $\Delta P_L(s) = \dfrac{\Delta P_L}{s}\,\text{pu}$ $= \dfrac{-0.2}{s}\,\text{pu}$ and run the program with a zero step-load change $\Delta P_L = 0$ for 5 s. After 5 s impose a step load change of $\Delta P_L (s) = \dfrac{\Delta P_L}{s}$ pu $= \dfrac{-0.2}{s}$ pu to find the transient response of $\Delta\omega(t)$.

### Problem 4.14 Frequency Control of an Interconnected Power System Broken into Two Areas, Each Having One Generator

Figure P4.12 shows the block diagram of two generators interconnected by a tie line (transmission line). Data of generation set (steam turbine and generator) #1:

for frequency change ($\Delta\omega_1$) per generator output

power change ($\Delta P_1$): $R_1 = \dfrac{\Delta\omega_1}{\Delta P_1}$ pu $= 0.01$ pu,

the load change ($\Delta P_{L1}$) per frequency change

($\Delta\omega_1$): $D_1 = \dfrac{\Delta P_{L1}}{\Delta\omega_1} = 0.08$ pu,

step-load change: $\Delta P_{L1}(s) = \dfrac{\Delta P_{L1}}{s}$ pu $= \dfrac{0.2}{s}$ pu

angular momentum of steam turbine and generator set: $M_1 = 4.5$,
base apparent power: $S_{base} = 500$ MVA,
governor time constant: $T_{G1} = 0.01$ s,
valve charging time constant: $T_{CH1} = 1.0$ s,
(load reference set point)$_1 = 0.8$ pu.

Data of generation set (steam turbine and generator) #2:

for frequency change ($\Delta\omega_2$) per change in generator

output power ($\Delta P_2$): $R_2 = \dfrac{\Delta\omega_2}{\Delta P_2}$ pu $= 0.02$ pu,

the load change ($\Delta P_{L2}$) per frequency change ($\Delta\omega_2$):

$D_2 = \dfrac{\Delta P_{L2}}{\Delta\omega_2} = 1.0$ pu,

step load change: $\Delta P_{L2}(s) = \dfrac{\Delta P_{L2}}{s}$ pu $= \dfrac{-0.2}{s}$ pu,

angular momentum of steam turbine and generator set: $M_2 = 6$,
base apparent power: $S_{base} = 500$ MVA,
governor time constant: $T_{G2} = 0.02$ s,
valve charging time constant: $T_{CH1} = 1.5$ s,
(load reference set point)$_2 = 0.8$ pu.

Data for tie line:

$$T = \dfrac{377}{X_{tie}} \text{ with } X_{tie} = 0.2 \text{ pu},$$

a) List the ordinary differential equations and the algebraic equations of the block diagram of Fig. P4.12.

b) Use either Mathematica or MATLAB to establish steady-state conditions by imposing a step function for load reference set point(s)$_1 = $ load

ref$_1 (s) = \dfrac{0.8}{s}$ pu, load reference set point (s)$_2 = $ load

ref$_2 (s) = \dfrac{0.8}{s}$ pu and run the program with a zero

step-load changes $\Delta P_{L1} = 0$, $\Delta P_{L2} = 0$ for 10 s. After additional 5 s impose step-load change $\Delta P_{L1}(s) = \dfrac{\Delta P_{L1}}{s}$ pu $= \dfrac{0.2}{s}$ pu, and after additional

2 s impose $\Delta P_{L2}(s) = \dfrac{\Delta P_{L2}}{s}$ pu $= \dfrac{-0.2}{s}$ pu to find the

transient responses $\Delta\omega_1(t)$ and $\Delta\omega_2(t)$.

### Problem 4.15 Determination of the sequence-component equivalent circuits and matrices $Z^{(1)}_{bus}$, $Z^{(2)}_{bus}$, $Z^{(0)}_{bus}$, line-to-ground (neutral), and line-to-line fault currents at bus 2 of Fig. P4.15 [83]

Two synchronous generators are connected through three-phase transformers to a transmission line, as shown in Fig. P4.15. The ratings and reactances of the generators and transformers are

1. Generators 1 and 2: 600 MVA, 34.5 kV;

$$X''_d = X_{1\_gen} = X_{2\_gen} = 25\% = 0.25 \text{ pu}$$

$$X_{o\_gen} = 10\% = 0.1 \text{ pu, and}$$
$$X_{n\_grounding\_coil} = 8\% = 0.08 \text{ pu}.$$

2. Transformer $T_1$ 600 MVA, (34.5 kV connected in $\Delta$)/(345 kV connected in $Y$); $X_{transformer\ leakage} = 10\% = 0.10$ pu. Note that the $Y$ of transformer $T_1$ is grounded.
3. Transformer $T_2$ 600 MVA, (34.5 kV connected in $\Delta$)/(345 kV connected in $Y$); $X_{transformer\ leakage} = 20\% = 0.20$ pu. Note that the $Y$ of transformer $T_2$ is not grounded.
4. On a chosen base of $S_{base} = 600$ MVA, $V_{L-L}$ $_{base} = 345$ kV in the transmission-line circuit, the line reactances are $X_{1line} = X_{2line} = 10\% = 0.10$ pu and $X_{0line} = 30\% = 0.30$ pu.
   a) Draw each of the three (positive-, negative-, zero-) sequence networks.
   b) Determine the zero-, positive-, and negative-sequence bus impedance (reactance) matrices by means of the $Z_{bus}$ building algorithm, as described in Chapter 8 of [83].
   c) Apply the matrix reduction technique by Kron, as outlined in Chapter 7 of [83].
   d) Determine numerical value (subtransient) for the line-to-ground fault current when the fault occurs at bus 2 of Fig. P4.15.

**Figure P4.15** Single-line diagram of a two-machine system connected via transformers and transmission line.

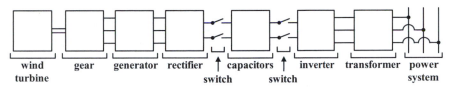

**Figure P4.16.1** Block diagram for charging and discharging supercapacitor bank.

e) Determine numerical value (subtransient) for the line-to-line fault current when the fault occurs at bus 2 of Fig. P4.15.

## Problem 4.16 Analysis of a Synchronous Machine Feeding a Rectifier

One of the major problems in using wind and solar energy is energy storage. Wind or solar energy may not be available when it is needed. A report from a wind farm in New Mexico states that the wind farm could lose as much as 60 MW within a minute. For example, there are several scenarios of how the power change of 60 MW per minute can be mitigated through complementary albeit more expensive power sources: one of them is the combination of a compressed-air power plant with a supercapacitor plant for bridging the time from when the wind power plant output decreases (60 MW per minute) to when the compressed-air energy storage (CAES) plant can take over. A CAES plant requires a start-up time of about 6 minutes. To bridge this 6 minute start-up time required for a 100 MW compressed-air power plant, supercapacitors or ultracapacitors are proposed to provide up to 100 MW during a 6 minute interval amounting to a required energy storage of 10 MWh. Inverters fed from supercapacitors can deliver power within milliseconds to the power system replacing the lost power of 60 MW per minute almost instantaneously. This combination of compressed-air power plant and supercapacitors or ultra-capacitors as bridging energy sources can be employed for peak-power operation as well as for improving power quality by preventing short brown-outs or blackouts.

Figure P4.16.1 depicts the block diagram of such a supercapacitor plant consisting of wind turbine, mechanical gear, synchronous generator, three-phase rectifier, supercapacitor bank, three-phase inverter, transformer, and power system.

a) Design a controlled three-phase rectifier in the tens of kilowatts range. Figure P4.16.2 shows the three-phase rectifier of Fig. P4.16.1 with six diodes and one self-commutated switch, an insulated-gate bipolar transistor (IGBT). The nominal phase voltages of the generator can be assumed to be $v_{an} = 1200$ V sin $\omega t$, $v_{bn} = 1200$ V sin$(\omega t - 120°)$, and $v_{cn} = 1200$ V sin$(\omega t - 240°)$. The IGBT is gated with a switching frequency of 3 kHz and you may assume a duty cycle of $\delta = 50\%$. $C_{f1} = 200$ $\mu F$, $L_{f1} = 90$ mH, $C_{f2} = 50$ $\mu F$, ideal diodes $D_1$ to $D_6$ and $D_f$, $v_{gs} = 40$ V magnitude, $R_{sd} = R_{ss} = 10$ $\Omega$, $C_{sd} = C_{ss} = 0.1$ $\mu F$, $C_f = 1000$ $\mu F$, $L_f = 1$ mH, and $R_{load} = 10$ $\Omega$.

1) Perform a PSpice analysis and plot output voltage $v_{load}(t)$, generator phase current $i_{aphase}$, and the generator line-to-line voltage $v_{ab}$.
2) Subject the generator phase current to a Fourier analysis.
3) Determine the displacement power factor of the generator.
4) Determine the power factor of the generator.

b) Design PWM (pulse-width-modulated) three-phase current controlled voltage-source inverter feeding power into the utility system via $Y_{grounded}/\Delta$ transformer in the tens of kilowatts range. The inverter circuit of Fig. P4.16.3 is to be analyzed with PSpice where the DC voltage is $V_{dc} = 600$ V, AC output voltage of the inverter is $V_{acL-L} = 360$ V, and the switching frequency of the IGBTs is $f_{switch} = 7.2$ kHz. $C_{filter} = 1000$ $\mu F$, $L_w = 1$ mH, $R_w = 10$ m$\Omega$, $L_f = 45$ $\mu H$, $C_f = 10.3$ $\mu F$, $R_f = 10$ m$\Omega$, $L_{syst} = 265$ $\mu H$, $R_{syst} = 50$ m$\Omega$. The power system phase voltages are $v_{agrid}(t) = 7200$ V sin $\omega t$, $v_{bgrid}(t) = 7200$ V sin$(\omega t - 120°)$, and $v_{cgrid}(t) = 7200$ V sin$(\omega t - 240°)$. Note the voltages $v_{asyst}(t)$,

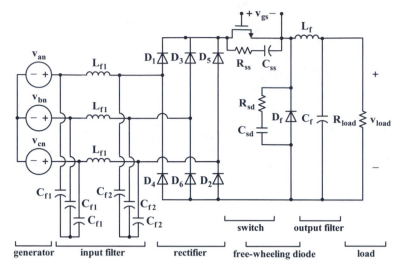

**Figure P4.16.2** Controlled three-phase rectifier.

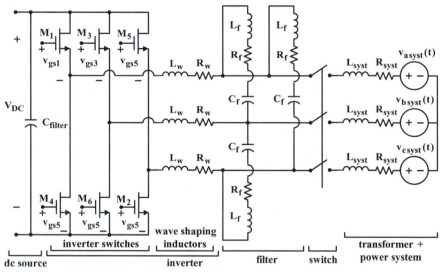

**Figure P4.16.3** PWM (pulse-width-modulated) three-phase current controlled voltage-source inverter.

$v_{bsyst}(t)$, and $v_{csyst}(t)$ are the power system voltages $v_{agrid}(t)$, $v_{bgrid}(t)$, and $v_{cgrid}(t)$ referred to the Y- or inverter side of the $Y/\Delta$ transformer.

1) Determine the transformation ratio of the $Y/\Delta$ transformer between inverter and power system; you may assume ideal transformer conditions.
2) Using PSpice calculate and plot input and output voltages and currents of the inverter, provided the output voltage is a sinusoid as given by the power system.
3) Subject the output current of the inverter to a Fourier analysis.

4) Determine the overall costs of this power plant if a specific cost of $4000/kW installed capacity is assumed. Note that a coal-fired plant has a specific cost of $2000/kW installed capacity.

## Problem 4.17 Voltage Stress in a Synchronous Motor Winding Fed Through a Cable Of 10 m Length by a PWM Current-Source Inverter

Voltage and current ripples due to PWM current-source inverters cause voltage stress in windings of machines. The configuration of Fig. E3.14.1 represents the first two turns of the winding of an induction motor, which is assumed to be identical to the winding of a synchronous motor. Each turn can be represented by four segments having an inductance $L_i$, a resistance $R_i$, a capacitance to ground (frame) $C_i$, and some interturn capacitances $C_{ij}$ and inductances $L_{ij}$. Figure E.3.14.2 represents a detailed equivalent circuit for the configuration of Fig. E.3.14.1.

To simplify the analysis neglect the capacitances $C_{ij}$ and the inductances $L_{ij}$. This leads to the circuit Fig. E.3.14.3, where the winding is fed by a PWM current source. One obtains the 15 differential equations as listed below.

$$\frac{dv_1}{dt} = \frac{i_s}{C_1} - \frac{G_1}{C_1}v_1 - \frac{i_2}{C_1},$$

$$\frac{di_2}{dt} = -\frac{R_2}{L_2}i_2 - \frac{v_2}{L_2} + \frac{v_1}{L_2},$$

$$\frac{dv_2}{dt} = \frac{i_2}{C_2} - \frac{G_2}{C_2}v_2 - \frac{i_3}{C_2},$$

$$\frac{di_3}{dt} = -\frac{R_3}{L_3}i_3 - \frac{v_3}{L_3} + \frac{v_2}{L_3},$$

$$\frac{dv_3}{dt} = \frac{i_3}{C_3} - \frac{G_3}{C_3}v_3 - \frac{i_4}{C_3},$$

$$\frac{di_4}{dt} = -\frac{R_4}{L_4}i_4 - \frac{v_4}{L_4} + \frac{v_3}{L_4},$$

$$\frac{dv_4}{dt} = \frac{i_4}{C_4} - \frac{G_4}{C_4}v_4 - \frac{i_5}{C_4},$$

$$\frac{di_5}{dt} = -\frac{R_5}{L_5}i_5 - \frac{v_5}{L_5} + \frac{v_4}{L_5},$$

$$\frac{dv_5}{dt} = \frac{i_5}{C_5} - \frac{G_5}{C_5}v_5 - \frac{i_6}{C_5},$$

$$\frac{di_6}{dt} = -\frac{R_6}{L_6}i_6 - \frac{v_6}{L_6} + \frac{v_5}{L_6},$$

$$\frac{dv_6}{dt} = \frac{i_6}{C_6} - \frac{G_6}{C_6}v_6 - \frac{i_7}{C_6},$$

$$\frac{di_7}{dt} = -\frac{R_7}{L_7}i_7 - \frac{v_7}{L_7} + \frac{v_6}{L_7},$$

$$\frac{dv_7}{dt} = \frac{i_7}{C_7} - \frac{G_7}{C_7}v_7 - \frac{i_8}{C_7},$$

$$\frac{di_8}{dt} = -\frac{R_8}{L_8}i_8 - \frac{v_8}{L_8} + \frac{v_7}{L_8},$$

$$\frac{dv_8}{dt} = \frac{i_8}{C_8} - v_8\left(\frac{1}{ZC_8} + \frac{G_8}{C_8}\right).$$

The parameters are

$R_1 = R_2 = R_3 = R_4 = R_5 = R_6 = R_7 = R_8 = 25~\mu\Omega$; $L_1 = L_3 = L_5 = L_7 = 1$ mH, and $L_2 = L_4 = L_6 = L_8 = 10$ mH; $C_1 = 0.2~\mu F$ for the cable of 10 m length; $C_3 = C_5 = C_7 = 0.7$ pF, $C_2 = C_4 = C_6 = C_8 = 7$ pF, $C_{15} = C_{37} = 0.35$ pF, $C_{26} = C_{48} = 3.5$ pF; $G_1 = G_3 = G_5 = G_7 = 1/(5000~\Omega)$, $G_2 = G_4 = G_6 = G_8 = 1/(500~\Omega)$; and $L_{15} = L_{37} = 0.005$ mH, $L_{26} = L_{48} = 0.01$ mH. $Z$ is either 1 $\mu\Omega$ (short-circuit) or 1 M$\Omega$ (open circuit).

Assume a PWM triangular current-step function $i_s(t)$, as indicated in Fig. E.3.14.4. Using Mathematica or MATLAB compute all state variables for the time period from $t = 0$ to 2 ms. Plot the voltages $(v_1 - v_5)$, $(v_2 - v_6)$, $(v_3 - v_7)$, and $(v_4 - v_8)$.

## 4.6 REFERENCES

1) Blondel, A.; "Application de la methode des deux reactions a l'etude de phenomenes oscillatoires des alternateurs couples," *Rev. Gen. Electr.*, vol. 13, pp. 235, 275, 331, 387, 515, 1923.

2) Doherty, R.E.; and Nickle, C.A.; "Synchronous machines I and II, an extension of Blondel's two-reaction theory," *Trans. AIEE*, vol. 45, pp. 912–927, 1926.

3) Doherty, R.E.; and Nickle, C.A.; "Synchronous machines III," *Trans. AIEE*, vol. 46, p. 1, 1927.

4) Park, R.H.; "Two-reaction theory of synchronous machines, generalized method of analysis, part I," *Trans. AIEE*, vol. 48, pp. 716–730, 1929.

5) Park, R.H.; "Two-reaction theory of synchronous machines II," *Trans. AIEE*, vol. 52, pp. 352–355, 1933.

6) Kilgore, L.A.; "Effects of saturation on machine reactances," *Trans. AIEE*, vol. 54, pp. 545–550, 1930.

7) Concordia, C.; *Synchronous Machines Theory and Performance*, John Wiley & Sons, Inc., New York, 1951.

8) Lyon, W.V.; *Transient Analysis of Alternating Current Machinery*, John Wiley & Sons, Inc., New York, 1954.

9) http://www.fes-muenchen.de/rund/herbst.99/walchsee.htm

10) Clarke, E.; "Circuit analysis of AC power systems," vol. II, John Wiley & Sons, 1950.

11) Roark, J.D; and Fross, C.A.; "Unbalanced synchronous machine analysis using frequency domain methods," *Proceeding IEEE Power Engineering Society Summer Meeting*, Los Angeles, CA, 1978, paper 78SM S24-1.

12) Xu, W.W.; Dommel, H.W.; and Marti, J.R.; "A synchronous machine model for three-phase harmonic analysis and EMTP initialization," *IEEE Transactions on Power Systems*, 1991, 6, (4), pp. 1530–1538.

13) Medina, A.; Arrillaga, J.; and Eggleston, J.F.; "A synchronous machine model in the harmonic domain," *Proceeding International Conference on Electrical Machines*, Manchester, UK, September 1992, pp. 647–651.

14) Subramaniam, P.; and Malik, P.P.; "Digital simulation of a synchronous generator in direct-phase coordinates," *Proc. Inst. Electr. Eng.*, 1971, 118(1), pp. 153–106.

15) Arrillaga, J.; Campos-Barros, J.G.; and Al-Kashali, H.J.; "Dynamic modeling of single generators connected to HVDC converters," *IEEE Transactions on Power Apparatus and Systems,* 1978, 97(4), pp. 1018–1029.

16) Marti, J.R.; and Louie, K.W.; "A phase-domain synchronous generator model including saturation effects," *IEEE Transactions on Power Apparatus and Systems*, 1997, 12(1), pp. 222–229.

17) Rodriguez, O.; and Medina, A.; "Fast periodic steady-state solution of a synchronous machine model incorporating the effects of magnetic saturation and hysteresis," *Presented at IEEE PES Winter Meeting*, vol. 3, pp. 1431–1436, 28 January–1 February, Columbus, OH, 2001.

18) Fuchs, E.F.; and McNaughton, G.A.; "Comparison of first-order finite difference and finite element algorithms for the analysis of magnetic fields, part I: theoretical analysis," *IEEE Transactions on Power Apparatus and Systems*, May 1982, vol. PAS-101, no. 5, pp. 1170–1180, and "Comparison of first-order finite difference and finite element algorithms for the analysis of magnetic fields, part II: numerical results," *IEEE Transactions on Power Apparatus and Systems*, May 1982, vol. PAS-101, no. 5, pp. 1181–1201.

19) Erdelyi, E.A.; and Fuchs, E.F.; "Non-linear electrical generator study final report on phase "B", part II non-linear synchronous alternators," Contract No. N00014-67-A-405-0003, Project No. 097-374/5-19-70, Code 473, University of Colorado, July 1972.

20) Fuchs, E.F.; and Erdelyi, E.A.; "Nonlinear theory of turbogenerators, part I: magnetic fields at no-load and balanced loads," *IEEE Transactions on Power Apparatus and Systems*, March/April 1973, vol. PAS-92, no. 2, pp. 583–591, and "Nonlinear theory of turbogenerators, part II: load-dependent synchronous reactances," *IEEE Transactions on Power Apparatus and Systems*, March/April 1973, vol. PAS-92, no. 2, pp. 592–599.

21) Fuchs, E.F.; and Senske, K.; "Comparison of iterative solutions of the finite difference method with measurements as applied to Poisson's and the diffusion equations," *IEEE Transactions on Power Apparatus and Systems*, August 1981, vol. PAS-100, no. 8, pp. 3983–3992.

22) Fuchs, E.F.; "Lastabhängige transiente Reaktanzen von gesättigten Turbogeneratoren," *Archiv für Elektrotechnik*, vol. 55, 1973, pp. 263–273.

23) El-Serafi, A.M.; Abdallah, A.S.; El-Sherbiny, M.K.; and Badawy, E.H.; "Experimental study of the saturation and the cross-magnetizing phenomenon in saturated synchronous machines," *IEEE Transactions on Energy Conversion*, vol. 3, issue 4, Dec. 1988, pp. 815–823.

24) Fuchs, E.F.; and Fardoun, A.A.; "Simulation of Controller-Motor drive System Using PSpice," Final Report prepared for Unique Mobility, Inc., 3700 S.

Jason Street, Englewood, Colorado, 80110, University of Colorado, January 1991.

25) Fardoun, A.A.; Fuchs, E.F.; and Huang, H.; "Modeling and simulation of an electronically commutated permanent-magnet machine drive system using SPICE," *IEEE Transactions on Industry Applications*, vol. 30, no. 4, July/August 1994, pp. 927–937.

26) Muljadi, E.; and Green J.; "Cogging torque reduction in a permanent magnet wind turbine generator, *Proceedings of the 21st American Society of Mechanical Engineers Wind Energy Symposium*, Reno, Nevada, January 14–17, 2002.

27) Fuchs, E.F.; and Pohl, G.; "Computer generated polycentric grid design and a novel dynamic acceleration of convergence for the iterative solution of magnetic fields based on the finite difference method," *IEEE Transactions on Power Apparatus and Systems*, August 1981, vol. PAS-100, no. 8, pp. 3911–3920.

28) Dubey, G.K.; *Power Semiconductor Controlled Drives*, Prentice Hall, Englewood Cliffs, New Jersey, 1989.

29) Yildirim, D.; "Commissioning of a 30 kVA variable-speed, direct-drive wind power plant," Ph.D. Thesis, University of Colorado, Boulder, Colorado, 1999.

30) Yildirim, D.; Fuchs, E.F.; and Batan, T.; "Test results of a 20 kW, direct-drive, variable-speed wind power plant," *Proceedings of the International Conference on Electrical Machines,* Istanbul, Turkey, September 2–4, 1998, pp. 2039–2044.

31) Palmer, H.; EE 673, "Advanced synchronous machinery," Class Notes, University of Colorado, 1967.

32) Fuchs, E.F.; "Parallelschalteinrichtungen," Diplomar-beit, Technical University of Stuttgart, Germany, 1967.

33) Fuchs, E.F.; "Automatic paralleling of power systems," *Proceedings of the 1969 Midwest power Symposium*, University of Minnesota, October 1969.

34) Fuchs, E.F., "Automatic paralleling of power systems," *Proceedings of the Third Pan American Congress of Mechanical, Electrical, and Allied Engineering Branches*, San Juan, Puerto Rico, September 6–13, 1969, Paper 30.

35) Joyce, J.S.; Kulig, T.; and Lambrecht, D.; "Torsional fatigue of turbine-generator shafts caused by different electrical system faults and switching operations," *IEEE Trans. on Power Apparatus and Systems*, vol. PAS-97, 1978, pp. 1965–1977.

36) Joyce, J.S.; Kulig, T.; and Lambrecht, D.; "The impact of high-speed reclosure of single and multiphase system faults on turbine-generator shaft

torsional fatigue," *IEEE Transactions on Power Apparatus and Systems*, vol. PAS-99, 1980, pp. 1764–1779.

37) El-Serafi, A.M.; and Faried, S.O.; "Effect of sequential reclosure of multi-phase system faults on turbine-generator shaft torsional torques," *IEEE Transactions on Power Systems,* vol. 6, issue 4, Nov. 1991, pp. 1380–1388.

38) Hammons, T.J.; and Lim, C.K.; "Probability assessment of turbine-generator shaft torque following severe disturbances on the system supply," *IEEE Transactions on Energy Conversion*, vol.14, issue 4, Dec. 1999, pp. 1115–1123.

39) Ledesma, P.; and Usaola, J.; "Minimum voltage protections in variable speed wind farms," *2001 IEEE Porto Power Tech Proceedings*, vol. 4, 10–13 Sept. 2001, 6 pp.

40) Fuchs, E.F.; "Industrial experience with the relax-ation solutions of the quasi-Poissonian differential equation," *Proceedings of International Conference on Electrical Machines*, Vienna, Austria, September 1315, 1976, pp. D20-1 to 10.

41) Kulig, T.S.; Buckley, G.W.; Lambrecht, D.; and Liese, M.; "A new approach to determine transient generator winding and damper currents in case of internal and external faults and abnormal operation, part 1: fundamentals," *Paper presented at IEEE Winter Meeting*, New Orleans, 1986.

42) Kulig, T.S.; Buckley, G.W.; Lambrecht, D.; and Liese, M.; "A new approach to determine transient generator winding and damper currents in case of internal and external faults and abnormal operation, part 2: analysis," *Paper presented at IEEE Winter Meeting*, New Orleans, 1986.

43) Kulig, T.S.; Buckley, G.W.; Lambrecht, D.; and Liese, M.; "A new approach to determine transient generator winding and damper currents in case of internal and external faults and abnormal operation, part 3: results," *IEEE Transactions on Energy Conversion*, vol. 5, issue 1, March 1990, pp. 70–78.

44) Fingersh, L.J.; "The testing of a low-speed, perma-nent-magnet generator for wind power applications," M.S. Thesis, University of Colorado, Boulder, 1995.

45) Bergey, K.H.; "The Lanchester-Betz limit," *J. Energy*, vol. 3, no. 6, Nov.–Dec. 1979, pp. 382–384.

46) Franklin, P.W.; "Theory of the three-phase, salient-pole type generator with bridge rectified output, parts I and II," *IEEE Transactions on Power Apparatus and Systems*, vol. PAS-91, Sept./Oct. 1972, pp. 1960–1975.

47) Jatskevich, J.; Pekarek, S.D.; and Davoudi, A.; "Parametric average-value model of synchronous machine-rectifier systems," *IEEE Transactions on*

*Energy Conversion*, vol. 21, issue 1, March 2006, pp. 9–18.

48) Jadric, I.; Borojevic, D.; and Jadric, M.;"Modeling and control of a synchronous generator with an active DC load," *IEEE Transactions on Power Electronics* vol. 15, issue 2, March 2000, pp. 303–311.

49) Abolins, A.; Achenbach, H.; and Lambrecht, D.; "Design and performance of large four-pole turbogenerators with semiconductor excitation for nuclear power stations," *Cigre-Report*, 1972 Session, August 28–September, Paper No 11-04.

50) Fuchs, E.F.; "Numerical determination of synchronous, transient, and subtransient reactances of a synchronous machine," Ph.D. Thesis, University of Colorado, Boulder, Colorado, 1970.

51) http://www.enercon.de/en/_home.htm

52) Batan, T.; "Real-time monitoring and calculation of the derating of single-phase transformers under non-sinusoidal operation," Ph.D. Thesis, University of Colorado, Boulder, 1999.

53) Gallegos-Lopez, G.; Kjaer, P.C.; and Miller, T.J.E.; "A new sensorless method for switched reluctance motor drives," *IEEE Transactions on Industry Applications*, vol. 34, issue 4, July–Aug. 1998, pp. 832–840.

54) Staton, D.A.; Soong, W.L.; and Miller, T.J.E.; "Unified theory of torque production in switched reluctance and synchronous reluctance motors," *IEEE Transactions on Industry Applications*, vol. 31, issue 2, March–April 1995, pp. 329–337.

55) Miller, T.J.E.; "Faults and unbalance forces in the switched reluctance machine," *IEEE Transactions on Industry Applications*, vol. 31, issue 2, March–April 1995, pp. 319–328.

56) Forster, J.A.; and Klataske, L.F.; "IEEE Working group report of problems with hydrogenerator thermoset stator windings," *IEEE Transactions on Power Apparatus and Systems*, vol. PAS-100, no. 7, July 1981, pp. 3292–3300.

57) Stratton, J.A.; *Electromagnetic Theory*, McGraw-Hill Book Company, Inc., New York and London, pp. 153–155, 1941.

58) Carpenter, C.J.; "Surface-integral methods of calculating forces on magnetized iron parts," *Proc. IEE*, 106C, 1960, pp. 19–28.

59) Reichert, K.; Freundl, H.; and Vogt, W.; "The calculation of forces and torques within numerical magnetic field calculation methods," *Proceedings of Compumag Conference*, London, 1976.

60) Hart, P.M.; and Bonwick, W.J.; "Harmonic modeling of synchronous machine," *IEE Proceedings on Electric Power Applications*, vol. 135, Proceedings B, no. 2, pp. 5258, March 1988.

61) Chen, H.; Long, Y.; and Zhang, X.P.; "More sophisticated synchronous machine model and the relevant harmonic power flow study," *IEE Proceedings, Transactions on ac Distribution*, vol. 146, no. 3, May 1999.

62) Grainger, J.J.; and Stevenson, W.D.; *Power System Analysis*, McGraw-Hill, Inc., New York, 1994.

63) Kimbark, E.W.; *Direct Current Transmission*, vol. 1, Wiley–Interscience, New York, 1971.

64) Zhang, X.-P.; and Chen, H.; "Asymmetrical three-phase load-flow study based on symmetrical component theory," *IEEE Proc. C*, 1994, 141, (3), pp. 248–252.

65) Xu, W.; Marti J.R.; and Dommel, H.W.; "A multiphase harmonic load flow solution technique," *IEEE Transactions on Power Systems*, vol. 6, issue 1, Feb. 1991, pp. 174–182.

66) CIGRE–Working Group 36-05, "Harmonics, characteristic parameters, methods of study, estimates of existing values in the network," *Electra*, no. 77, pp. 35–54, July 1981.

67) Dommel, H.W.; *Electromagnetic transients program reference manual* (*EMTP theory book*), Prepared for Bonneville Power Administration by the Dept. of Electrical Engineering, University of British Columbia, Canada, Aug. 1986.

68) Smith, R.T.; *Analysis of Electrical Machines*, New York, Pergamon Press, 1982.

69) Al-Nuaim, N.A.; and Toliyat, H.A.; "A novel method for modeling dynamic air-gap eccentricity in synchronous machines based on modified winding function theory," *IEEE Transactions on Energy Conversion*, vol. 13, pp. 156–162, June 1998.

70) MATLAB, Version 4.2c.1, and Simulink, Version 1.3c, Mathematical packages by the MathWorks, Inc., Natick, MA.

71) Ammann, C.; Reichert, K.; Joho, R.; and Posedel, Z.; "Shaft voltages in generators with static excitation systems – problems and solution," *IEEE Transactions on Energy Conversion*, vol. 3, issue 2, June 1988, pp. 409–419.

72) Busse, D.; Erdman, J.; Kerkman, R.J.; Schlegel, D.; and Skibinski, G.; "System electrical parameters and their effects on bearing currents," *IEEE Transactions on Industry Applications*, vol. 33, issue 2, March–April 1997, pp. 577–584.

73) Muetze, A.; and Binder, A.; "Calculation of circulating bearing currents in machines of inverter-based drive systems," *Conference Record of the 2004 IEEE Industry Applications Conference, 2004, 39th IAS Annual Meeting*, vol. 2, 3–7 Oct. 2004, pp. 720–726.

74) Maki-Ontto, P.; and Luomi, J.; "Bearing current prevention of converter-fed ac machines with a

conductive shielding in stator slots," *Proceedings of the IEEE International Electric Machines and Drives Conference*, 2003, vol. 1, 1–4 June 2003, pp. 274–278.

75) Busse, D.; Erdman, J.; Kerkman, R.J.; Schlegel, D.; and Skibinski, G.; "Bearing currents and their relationship to PWM drives," *IEEE Transactions on Power Electronics,* vol. 12, issue 2, March 1997, pp. 243–252.

76) Hyypio, D.; "Mitigation of bearing electro-erosion of inverter-fed motors through passive common-mode voltage suppression," *IEEE Transactions on Industry Applications*, vol. 41, issue 2, March–April 2005, pp. 576–583.

77) Muetze, A.; and Binder, A.; "Techniques for measurement of parameters related to inverter-induced bearing currents," *Industry Applications Conference, 2005, 40th IAS Annual Meeting, Conference Record of the 2005*, vol. 2, 2–6 Oct. 2005, pp. 1390–1397.

78) *IEEE Standard 519*, "IEEE recommended practices and requirements for harmonic control in electric power systems," IEEE-519, 1992.

79) Fuchs, E.F.; You, Y.; and Roesler, D.J.; "Modeling, simulation and their validation of three-phase transformers with three legs under DC bias," *IEEE Transactions on Power Delivery*, vol. 14, no. 2, April 1999, pp. 443–449.

80) Fuchs, E.F.; and Rosenberg, L.T.; "Analysis of an alternator with two displaced stator windings," *IEEE Transactions on Power Apparatus and Systems*, November/December 1974, vol. PAS-93, no. 6, pp. 1776–1786.

81) White D.C.; and Woodson, H.H.; *Electrical Energy Conversion*, John Wiley & Sons, Inc., New York, 1959.

82) Wood, A. J.; and Wollenberg, B.F.; *Power Generation Operation and Control*, John Wiley & Sons, Inc., 1984.

83) Grainger, J.J.; and Stevenson, Jr., W.D.; *Power System Analysis*, McGraw-Hill, Inc. 1994.

## 4.7 ADDITIONAL BIBLIOGRAPHY

84) Sarma, M.S.; *Electric Machines – Steady-State and Dynamic Performance*, 2nd edition, West Publishing Company, Minneapolis/St. Paul, 1994.

85) Fuchs, E.F.; and Fardoun, A.A.; "Architecture of a variable-speed, direct-transmission wind power plant of wind-farm size," Invention Disclosure, University of Colorado, Boulder, February 1994.

86) Fuchs, E.F.; and Gregory, B.; "Control of a wind power plant with variable displacement (or power factor) angle and waveshape of AC current fed into the power system," Invention Disclosure, University of Colorado, Boulder, February 1994.

87) Fuchs, E.F.; and Yildirim, D.; "Transverse electric flux machines with phase numbers higher than two," Invention Disclosure, University of Colorado, Boulder, January 1995.

88) Fuchs, E.F.; and Yildirim, D.; "Permanent-magnet reluctance machine for high torques at low rated speeds and low weight based on alternating magnetic field theory," Invention Disclosure, University of Colorado, Boulder, April 1997.

89) Fuchs, E.F.; and Yildirim, D.; "Pulse-width-modulated (PWM) current-controlled inverter with reduced switching and filtering losses," Invention Disclosure, University of Colorado, Boulder, July 1999.

# 5

# Interaction of Harmonics with Capacitors

Capacitors are extensively used in power systems for voltage control, power-factor correction, filtering, and reactive power compensation. With the proliferation of nonlinear loads and the propagation of harmonics, the possibility of parallel/series resonances between system and capacitors at harmonic frequencies has become a concern for many power system engineers.

Since the 1990s, there has been an increase of nonlinear loads, devices, and control equipment in electric power systems, including electronic loads fed by residential and commercial feeders, adjustable-speed drives and arc furnaces in industrial networks, as well as the newly developing distributed generation (DG) sources in transmission and distribution systems. This has led to a growing presence of harmonic disturbances and has deteriorated the quality of electric power. Moreover, some nonlinear loads and power electronic control equipment tend to operate at relatively low power factors, causing poor voltage regulation, increasing line losses, and forcing power plants to supply more apparent power. The conventional and practical procedure for overcoming these problems, as well as compensating reactive power, are to install (fixed and/or switching) shunt capacitor banks on either the customer or the utility side of a power system. Capacitor banks are also used in power systems as reactive power compensators and tuned filters. Recent applications are power system stabilizers (PSS), flexible AC transmission systems (FACTS), and custom power devices, as well as high-voltage DC (HVDC) systems.

Misapplications of power capacitors in today's modern and complicated industrial distribution systems could have negative impacts on both the customers (sensitive linear and nonlinear loads) and the utility (equipment), including these:

- amplification and propagation of harmonics resulting in equipment overheating, as well as failures of capacitor banks themselves,
- harmonic parallel resonances (between installed capacitors and system inductance) close to the frequencies of nearby harmonic sources,
- unbalanced system conditions which may cause maloperation of (ground) relays, and
- additional power quality problems such as capacitor in-rush currents and transient overvoltages due to capacitor-switching actions.

Capacitor or frequency scanning is usually the first step in harmonic analysis for studying the impact of capacitors on system response at fundamental and harmonic frequencies. Problems with harmonics often show up at capacitor banks first, resulting in fuse blowing and/or capacitor failure. The main reason is that capacitors form either series or parallel resonant circuits, which magnify and distort their currents and voltages. There are a number of solutions to capacitor related problems:

- altering the system frequency response by changing capacitor sizes and/or locations,
- altering source characteristics, and
- designing harmonic filters.

The last technique is probably the most effective one, because properly tuned filters can maintain the primary objective of capacitor application (e.g., power-factor correction, voltage control, and reactive power compensation) at the fundamental frequency, as well as low impedance paths at dominant harmonics.

This chapter starts with the main applications of capacitors in power systems and continues by introducing some power quality issues associated with capacitors. A section is provided for capacitor and frequency scanning techniques. Feasible operating regions for safe operation of capacitors in the presence of voltage and current harmonics are presented. The last section of this chapter discusses equivalent circuits of capacitors.

## 5.1 APPLICATION OF CAPACITORS TO POWER-FACTOR CORRECTION

The application of capacitor banks in transmission and distribution systems has long been accepted as a necessary step in the design of utility power systems. Design considerations often include traditional factors such as voltage and reactive power (VAr) control, power-factor correction, and released capacity. More recent applications concern passive and active filtering, as well as parallel and series (active and reactive) power compensation. Capacitors are also incorporated in many PSS, FACTS, and custom power devices, as well as HVDC systems.

An important application of capacitors in power systems is for power-factor correction. Poor power factor has many disadvantages:

- degraded efficiency of distribution power systems,
- decreased capacity of transmission, substation, and distribution systems,
- poor voltage regulation, and
- increased system losses.

Many utility companies reward customers who improve their power factor (PF) and penalize those who do not meet the prescribed PF requirements. There are a number of approaches for power factor improvement:

- Synchronous condensers (e.g., overexcited synchronous machines with leading power factor) are used in transmission systems to provide stepless (continuous) reactive power compensation for optimum transmission line performance under changing load conditions. This technique is mostly recommended when a large load (e.g., a drive) is added to the system;
- Shunt-connected capacitor banks are used for stepwise (discrete) full or partial compensation of system inductive power demands [1–2]; and
- Static VAr compensators and power converter-based reactive power compensation systems are employed for active power-factor correction (PFC). These configurations use power electronic switching devices to implement variable capacitive or inductive impedances for optimal power-factor correction. Many PFC configurations have been proposed [3–5] for residential and commercial loads, including passive, single-stage, and double-stage topologies for low (below 300 W), medium (300–600 W) and high (above 600 W) power ratings, respectively.

The capacitor-bank approach is most widely used owing to its low installation cost and large capacity [2]. Active power-factor circuits are relied on for relatively small power rating applications and are not a subject of this chapter.

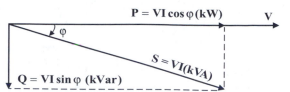

**FIGURE 5.1** Definition of lagging displacement power factor (consumer notation).

### 5.1.1 Definition of Displacement Power Factor

Displacement power factor (DPF) is the ratio of the active or real fundamental power P (measured in W or kW) to the fundamental apparent power S (measured in VA or kVA). The reactive power Q (measured in VAr or kVAr) supplied to inductive devices is the vector difference between the real and apparent powers. The DPF is the cosine of the angle between these two quantities. It reflects how efficiently a facility uses electricity by comparing the amount of useful work that is extracted from the electrical apparent power supplied. Of course, there are many other measures of energy efficiency. The relationship between S, P, and Q is defined by the power triangle of Fig. 5.1.

$$S^2 = \sqrt{P^2 + Q^2}. \tag{5-1}$$

The displacement power factor, determined from $\varphi = \tan^{-1}(Q/P)$, measures the displacement angle between the fundamental components of the phase voltage and phase current:

$$\text{displacement power factor} = DPF = \cos\varphi = P/S. \tag{5-2}$$

The DPF varies between zero and one. A value of zero means that all power is supplied as reactive power and no useful work is accomplished. Unity DPF means that all of the power consumed by a facility goes to produce useful work, such as resistive heating and incandescent lighting.

Electric power systems usually experience lagging displacement power factors – based on the consumer notation of current – because a significant part consists of reactive devices that employ coils and requires reactive power to energize their magnetic circuits. Generally, motors that are operated at or near full nameplate ratings will have high DPFs (e.g., 0.90 to 0.95) whereas the same motor under no-load and/or light load conditions may exhibit a DPF in the range of 0.30. For example, motors in large hydraulic machines such as plastic molding machines mostly

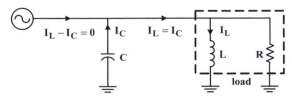

**FIGURE 5.2** Current paths demonstrating the essential features of power-factor correction.

operate under lightly loaded conditions, contributing to an overall power factor of about 0.60.

The reactive power absorbed by electrical equipment such as transformers, electric motors, welding units, and static converters increases the currents of generators, transmission lines, utility transformers, switchgear, and cables. Electric utility companies must supply the entire apparent power demand. Because a customer only extracts useful work from the real power component, a high displacement power factor is important. Therefore, in most power systems lagging (underexcited) DPFs (larger than 0.8) are acceptable, and leading (overexcited) power factors should be avoided because they may cause resonance conditions within the power system.

Inductive loads with low DPFs require the generators and transmission or distribution systems to supply reactive current with the associated power losses and increased voltage drops. If a shunt capacitor bank is connected across the load, its (capacitive) reactive current component $I_C$ can cancel the load (inductive) reactive current component $I_L$. If the bank is sufficiently large (e.g., $I_C = I_L$), it will supply all reactive power, and there will be no reactive current flow entering the power system as is indicated in Fig. 5.2.

The DPF can be measured either by a direct-reading $\cos\varphi$ meter indicating an instantaneous value or by a recording VAr meter, which allows recording of current, voltage, and power factor over time. Readings taken over an extended period of time (e.g., seconds, minutes) provide a useful means of estimating an average value of power factor for an installation.

### 5.1.2 Total Power Factor in the Presence of Harmonics

Equations 5-1 and 5-2 assume that system loads have linear voltage–current characteristics and harmonic distortions are not significant. Harmonic voltage and current distortions will change the expression for the total apparent power and the total power factor (TPF). Consider a voltage v(t) and a current i(t)

expressed in terms of their rms harmonic components:

$$v(t) = V_0 + \sum_{h=1}^{\infty} \sqrt{2} V_h \sin(h\omega_1 t + \theta_h^v), \quad (5\text{-}3a)$$

$$i(t) = I_0 + \sum_{h=1}^{\infty} \sqrt{2} I_h \sin(h\omega_1 t + \theta_h^i). \quad (5\text{-}3b)$$

The active (real, average) and reactive power are given by

$$P_{\text{total}} = V_o I_o + \sum_{h=1}^{\infty} V_h I_h \cos(\theta_h^v - \theta_h^i), \quad (5\text{-}4a)$$

$$Q_{\text{total}} = \sum_{h=1}^{\infty} V_h I_h \sin(\theta_h^v - \theta_h^i), \quad (5\text{-}4b)$$

and the apparent (voltampere) power is

$$S_{\text{total}} = \sqrt{(\sum_{h=0}^{\infty} I_h^2)(\sum_{h=0}^{\infty} V_h^2)}. \quad (5\text{-}4c)$$

Therefore, for nonsinusoidal cases, Eq. 5-1 does not hold and must be replaced by

$$D = \sqrt{S_{\text{total}}^2 - P_{\text{total}}^2 - Q_{\text{total}}^2}, \quad (5\text{-}5)$$

where D is called the distortion power.

According to Eq. 5-5, the total power factor is lower than the DPF (Eq. 5-2) in the presence of harmonic distortion of nonlinear devices such as solid-state or switched power supplies, variable-speed AC drives, and DC drives. Harmonic distortion essentially converts a portion of the useful energy into high-frequency energy that is no longer useful to most devices and is ultimately lost as heat. In this way, harmonic distortion further reduces the power factor. In the presence of harmonics, the total power factor is defined as

$$\text{total power factor} = TPF = \cos\theta = \frac{P_{\text{total}}}{S_{\text{total}}}, \quad (5\text{-}6)$$

where $P_{\text{total}}$ and $S_{\text{total}}$ are defined in Eq. 5-4. Since capacitors only provide reactive power at the fundamental frequency, they cannot correct the power factor in the presence of harmonics. In fact, improper capacitor sizing and placements can decrease the power factor by generating harmonic resonances, which increase the harmonic content of system voltage and current. $DPF = \cos \varphi$ (with $\varphi = \tan^{-1}(Q/P)$) is always greater than $TPF = \cos \theta$ (with $\theta = \cos^{-1}(P_{total}/S_{total})$) when harmonics are present. Displacement power factor is still very important to most industrial customers because utility billing with respect to power factor is almost universally based on the displacement power factor.

### 5.1.2.1 Application Example 5.1: Computation of Displacement Power Factor (DPF) and Total Power Factor (TPF)

Strong power systems have very small system impedances (e.g., $R_{syst} = 0.001\ \Omega$, and $X_{syst} = 0.005\ \Omega$) whereas weak power systems have fairly large system impedances (e.g., $R_{syst} = 0.1\ \Omega$, and $X_{syst} = 0.5\ \Omega$). Power quality problems are mostly associated with weak systems. Unfortunately, distributed generation (DG) inherently leads to weak systems because the source impedances of small generators are large, that is, they cannot supply a large current (ideally infinitely large) during transient operation. This application example studies the power-factor correction for a relatively weak power system where the difference between the displacement power factor and the total power factor is large due to the relatively large amplitudes of voltage and current harmonics.

a) Perform a PSpice analysis for the circuit of Fig. E5.1.1, where a three-phase thyristor rectifier – fed by a Y/Y transformer with turns ratio $(N_p/N_s = 1)$ – via a filter (e.g., capacitor $C_f$) serves a load ($R_{load}$). You may assume $R_{syst} = 0.05\ \Omega$, $X_{syst} = 0.1\ \Omega$, $f = 60$ Hz, $v_{AN}(t) = \sqrt{2} \cdot 346\ \text{V} \cos \omega t$, $v_{BN}(t) = \sqrt{2} \cdot 346\ \text{V} \cos(\omega t - 120°)$, and $v_{CN}(t) = \sqrt{2} \cdot 346\ \text{V} \cos(\omega t - 240°)$. Thyristors $T_1$ to $T_6$ can be modeled by MOSFETs in series with diodes, $C_f = 500\ \mu\text{F}$, $R_{load} = 10\ \Omega$, and a firing angle of $\alpha = 0°$. Plot one period of voltage and current

after steady state has been reached as requested in parts b and c.

b) Plot and subject the line-to-neutral voltage $v_{an}(t)$ to a Fourier analysis.

c) Plot and subject the phase current $i_a(t)$ to a Fourier analysis.

d) Repeat parts a to c for $\alpha = 10°$, $20°$, $30°$, $40°$, $50°$, $60°$, $70°$, and $80°$.

e) Calculate the displacement power factor (DPF) and the total power factor (TPF) based on the phase shifts of the fundamental and harmonics between $v_{an}(t)$ and $i_a(t)$ for all firing angles and plot the DPF and TPF as a function of $\alpha$.

f) Determine the capacitance (per phase) $C_{bank}$ of the power-factor correction capacitor bank such that the displacement power factor as seen by the power system is for $\alpha = 60°$ about equal to $DPF = 0.90$ lagging.

g) Explain why the total power factor cannot be significantly increased by the capacitor bank. Would the replacement of the capacitor bank by a three-phase tuned ($R_{fbank}$, $L_{fbank}$, and $C_{fbank}$) filter be a solution to this problem?

### 5.1.3 Benefits of Power-Factor Correction

Improving a facility's power factor (either DPF or TPF) not only reduces utility power factor surcharges, but it can also provide lower power consumption and offer other advantages such as reduced demand charges, increased load-carrying capabilities in existing lines and circuits, improved voltage profiles, and reduced power system losses. The overall results are lower costs to consumers and the utility alike, as summarized below:

- **Increased Load-Carrying Capabilities in Existing Circuits.** Installing a capacitor bank at the end of a feeder (near inductive loads) improves the power factor and reduces the current carried by

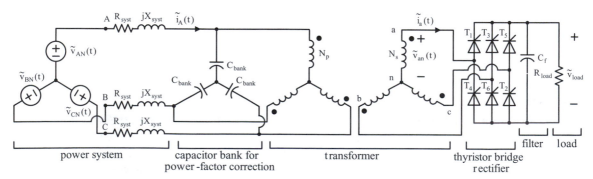

**FIGURE E5.1.1** Connection of a wye/wye three-phase transformer with a thyristor rectifier, filter, load, and a bank of power-factor correction capacitors.

the feeder. This may allow the circuit to carry additional loads and save costs for upgrading the network when extra capacity is required. In addition, the lower current flow reduces resistive losses in the circuit.

- *Improved Voltage Profile.* A lower power factor requires a higher current flow for a given load. As the line current increases, the voltage drop in the conductor increases, which may result in a lower voltage at the load. With an improved power factor, the voltage drop in the conductor is reduced.

- *Reduced Power-System Losses.* In industrial distribution systems, active losses vary from 2.5 to 7.5% of the total load measured in kWh, depending on hours of full-load and no-load plant operation, wire size, and length of the feeders. Although the financial return from conductor loss reduction alone is seldom sufficient to justify the installation of capacitors, it is an attractive additional benefit, especially in older plants with long feeders. Conductor losses are proportional to the current squared, whereas the current is reduced in direct proportion to the power-factor improvement; therefore, losses are inversely proportional to the square of the power factor.

- *Release of Power-System Capacity.* When capacitor banks are in operation, they furnish magnetizing current for electromagnetic devices (e.g., motors, transformers), thus reducing the current demand from the power plant. Less current means less apparent power, or less loading of transformers and feeders. This means capacitors can be used to reduce system overloading and permit additional load to be added to existing facilities. Release of system capacity by power-factor improvement and especially with capacitors has become an important practice for distribution engineers.

- *Reducing Electricity Bills.* To encourage more efficient use of electricity and to allow the utility to recoup the higher costs of providing service to customers with low power factor demanding a high amount of current, many utilities require that the consumers improve the power factors of their installations. This may be in the form of power-factor penalty (e.g., adjusted demand charges or an overall adjustment to the bill) or a rate structure (e.g., the demand charges or power rates are based on current or apparent power). Power-factor penalties are usually imposed on larger commercial and industrial customers when their power factors fall below a certain value (e.g., 0.90 or 0.95). Customers with current or apparent power-based

demand charges in effect pay more whenever their power factor is less than about unity. More and more utilities are introducing power-factor penalties (or apparent power-based demand charges) into their rate structures, in part to comply with provisions of the Clean Air Act and deregulation, and to some extent in response to growing competition in a newly deregulated power market. Without these surcharges there would be no motivation for consumers to install power-factor correction facilities. Against the financial advantages of reduced billing, the consumer must balance the cost of purchasing, installing, and maintaining the power-factor-improvement capacitors. The overall result of enforcing incentives is lower costs to consumers and the utility companies. A high power factor avoids penalties imposed by utilities and makes better use of the available capacity of a power system.

Therefore, power-factor correction capacitors are usually installed in power systems to improve the efficiency at which electrical energy is delivered and to avoid penalties imposed by utilities, making better use of the available capacity of a power system.

## 5.2 APPLICATION OF CAPACITORS TO REACTIVE POWER COMPENSATION

Reactive power compensation plays an important role in the planning and improvement of power systems. Its aim is predominantly to provide an appropriate placement of the compensation devices, which ensures a satisfactory voltage profile and a reduction in power and energy losses within the system. It also maximizes the real power transmission capability of transmission lines, while minimizing the cost of compensation.

The increase of real power transmission in a particular system is restricted by a certain critical voltage level. This critical voltage depends on the reactive power support available in the system to meet the additional load at a given operating condition. Use of series and shunt compensation is one of the corrective measures suggested by various researchers. This is performed by implementing power capacitor banks or FACTS devices. Shunt-connected capacitor banks and shunt FACTS devices are implemented to produce an acceptable voltage profile, minimize the loss of the investment, and enhance the power transmission capability. Series-connected capacitors and series FACTS devices are incorporated in transmission lines to reduce losses and improve transient behavior as well as the stability of power systems.

Simultaneous compensation of losses and reactive powers is performed by series and shunt-connected FACTS devices to improve the static and dynamic performances of power systems.

Shunt capacitors applied at the receiving end of a power system supplying a load of lagging power factor have several benefits, which are the reason for their extensive applications:

- reduce the lagging component of circuit current,
- increase the voltage level at the load,
- improve voltage regulation if the capacitor units are properly switched,
- reduce $I^2R$ real power loss (measured in W) in the system due to the reduction of current,
- reduce $I^2X$ reactive power loss (measured in VAr) in the system due to the reduction of current,
- increase power factor of the source generators,
- decrease reactive power loading on source generators and circuits to relieve an overloaded condition or release capacity for additional load growth,
- by reducing reactive power load on the source generators, additional real power may be placed on the generators if turbine capacity is available, and
- to reduce demand apparent power where power is purchased. Correction to 1 pu power factor may be economical in some cases.

## 5.3 APPLICATION OF CAPACITORS TO HARMONIC FILTERING

Capacitors are widely used in passive and active harmonic filters. In addition, many static and converter-based power system components such as static VAr compensators, FACTS, and power quality and custom devices use capacitors as storage and compensation components.

Filters are the most frequently used devices for harmonic compensation. Passive and active filters are the main types of filters. Passive filters are composed of passive elements such as resistors (R), inductors (L), and capacitors (C). Passive harmonic filters are the most popular and effective mitigation method for harmonic problems. The passive filter is generally designed to provide a low-impedance path to prevent the harmonic current to enter the power system. They are usually tuned to a specific harmonic such as the 5th, 7th, 11th, etc. In addition, they provide displacement power-factor correction.

Filters are generally divided into passive, active, and hybrid filters. The hybrid structure uses a combination of active filters with passive filters. The passive part is used to reduce the overall filter

rating and improve its performance. Classifications of passive and active filters are provided in Chapter 9.

### 5.3.1 Application Example 5.2: Design of a Tuned Harmonic Filter

A harmonic filter (consisting of a capacitor in series with a tuning reactor) is to be designed in parallel to a power-factor correction (PFC) capacitor to improve the power quality (as recommended by IEEE-519 [8]) and to improve the displacement power factor from 0.42 lag to 0.97 lag (consumer notation). Data obtained from a harmonic analyzer at 30% loading are 6.35 kV/phase, 618 kW/phase, DPF = 0.42 lag, $THD_i = 30.47\%$, and fifth harmonic (250 Hz) current $I_5 = 39.25$ A.

## 5.4 POWER QUALITY PROBLEMS ASSOCIATED WITH CAPACITORS

There are resonance effects and harmonic generation associated with capacitor installation and switching. Series and parallel resonance excitations lead to increase of voltages and currents, effects which can result in unacceptable stresses with respect to equipment installation or thermal degradation. Using joint capacitor banks for power-factor correction and reactive power compensation is a very well-established approach. However, there are power quality problems associated with using capacitors for such a purpose, especially in the presence of harmonics.

### 5.4.1 Transients Associated with Capacitor Switching

Transients associated with capacitor switching are generally not a problem for utility equipment. However, they could cause a number of problems for the customers, including

- voltage increases (overvoltages) due to capacitor switching,
- transients can be magnified in a customer facility (e.g., if the customer has low-voltage power-factor correction capacitors) resulting in equipment damage or failure (due to excessive overvoltages),
- nuisance tripping or shutdown (due to DC bus overvoltage) of adjustable-speed drives or other process equipment, even if the customer circuit does not employ any capacitors,
- computer network problems, and
- telephone and communication interference.

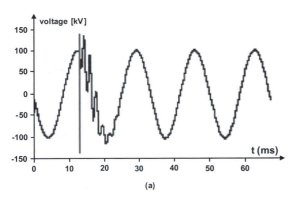

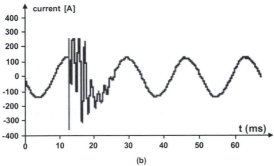

**FIGURE 5.3** Typical distribution capacitor closing transient voltage (a) and current (b) waveforms [6].

Because capacitor voltage cannot change instantaneously, energizing a capacitor bank results in an immediate drop in system voltage toward zero, followed by a fast voltage recovery (overshoot), and finally an oscillating transient voltage superimposed with the fundamental waveform (Fig. 5.3). The peak-voltage magnitude (up to 2 pu with transient frequencies of 300–1000 Hz) depends on the instantaneous system voltage at the moment of capacitor connection.

Voltage increase occurs when the transient oscillation – caused by the energization of a capacitor bank – excites a series resonance formed by the leakage inductances of a low-voltage system. The result is an overvoltage at the low-voltage bus. The worst voltage transients occur when the following conditions are met:

- The size of the switched capacitor bank is significantly larger (>10) than the low-voltage power-factor correction bank;
- The source frequency is close to the series-resonant frequency formed by the step-down transformer and the power-factor correction-capacitor bank; and

- There is relatively small damping provided by the low-voltage load (e.g., for typical industrial plant configuration or primarily motor load with high efficiency).

Solutions to the voltage increase usually involve:

- Detune the circuit by changing capacitor bank sizes;
- Switch large banks in more than one section of the system;
- Use an overvoltage control method. For example, use switching devices that do not prestrike or restrike, employ synchronous switching of capacitors, and install high-energy metal-oxide surge arrestors to limit overvoltage and protect critical equipment;
- Apply surge arresters at locations where overvoltages occur;
- Convert low-voltage power-factor correction banks into harmonic filters (e.g., detune the circuit);
- Use properly designed and tuned filters instead of capacitors;
- Install isolation or dedicated transformers or series reactors for areas of sensitive loads; and
- Rely on transient-voltage-surge suppressors.

### 5.4.2 Harmonic Resonances

Improper placement and sizing of capacitors could cause parallel and/or series resonances and tune the system to a frequency that is excited by a harmonic source [7]. In industrial power systems, capacitor banks are normally specified for PFC, filtering, or reactive-power compensation without regard to resonances and other harmonic concerns. High overvoltages could result if the system is tuned to one harmonic only that is being supplied by a nonlinear load or saturated electromagnetic device such as a transformer (e.g., second, third, fourth, and fifth harmonics result from transformer inrush currents). Moreover, the capacitive reactance is inversely proportional to frequency; therefore, harmonic currents may overload capacitors beyond their limits. Thus, capacitor banks themselves may be affected by resonance, and may fail prematurely. This may even lead to plant or feeder shutdowns.

Resonance is a condition where the capacitive reactance of a system offsets its inductive reactance, leaving the small resistive elements in the network as the only means of limiting resonant currents. Depending on how the reactive elements are arranged throughout the system, the resonance can be of a series or a parallel type. The frequency at which this offsetting effect takes place is called the resonant frequency of the system.

**Parallel Resonance.** In a parallel-resonant circuit the inductive and the capacitive reactance impedance components are in parallel to a source of harmonic current and the resistive components of the impedances are small compared to the reactive components. The presence of a capacitor (e.g., for PFC) and harmonics may create such conditions and subject the system to failure. From the perspective of harmonic sources, shunt capacitor banks appear to be in parallel with the system short-circuit reactance. The resonant frequency of this parallel combination is

$$f_r = \frac{1}{2\pi \cdot \sqrt{LC}} = f_1 \cdot \sqrt{\frac{1000 \cdot S_{sc}}{Q_{cap}}}, \qquad (5\text{-}7)$$

where $f_r$ and $f_1$ are the resonant and the fundamental frequencies, respectively. $S_{sc}$ and $Q_{cap}$ are the system short-circuit apparent power – measured in MVA – at the bus and the reactive power rating – measured in kVAr – of the capacitor, respectively. Installation of capacitors in power systems modifies the resonance frequency. If this frequency happens to coincide with one generated by the harmonic source, then excessive voltages and currents will appear, causing damage to capacitors and other electrical equipment.

**Series Resonance.** In a series-resonance circuit the inductive impedance of the system and the capacitive reactance of a capacitor bank are in series to a source of harmonic current. Series resonance usually occurs when capacitors are located toward the end of a feeder branch. From the harmonic source perspective, the line impedance appears in series with the capacitor. At, or close to, the resonant frequency of this series combination, its impedance will be very low. If any harmonic source generates currents near this resonant frequency, they will flow through the low-impedance path, causing interference in communication circuits along the resonant path, as well as excessive voltage distortion at the capacitor.

**Capacitor Bank Behaves as a Harmonic Source.** There are many capacitor banks installed in industrial and overhead distribution systems. Each capacitor bank is a source of harmonic currents of order $h$, which is determined by the system short-circuit impedance (at the capacitor location) and the capacitor size. This order of harmonic current is given by

$$h = \sqrt{\frac{X_c}{X_{sc}}} \qquad (5\text{-}8)$$

Power system operation involves many switching functions that may inject currents with different harmonic resonance frequencies. Mitigation of such harmonic sources is vital. There is no safe rule to avoid such resonant currents, but resonances above 1000 Hz will probably not cause problems except interference with telephone circuits. This means the capacitive reactive power $Q_{capacitor}$ should not exceed roughly 0.3% of the system short-circuit apparent power $S_{sc}$ at the point of connection unless control measures are taken to deal with these current harmonics. In some cases, converting the capacitor bank into a harmonic filter might solve the entire problem without sacrificing the desired power-factor improvement.

### 5.4.3 Application Example 5.3: Harmonic Resonance in a Distorted Industrial Power System with Nonlinear Loads

Figure E5.3.1a shows a simplified industrial power system ($R_{sys} = 0.06\ \Omega$, $L_{sys} = 0.11$ mH) with a power-factor correction capacitor ($C_{pf} = 1300\ \mu F$). Industrial loads may have nonlinear v–i characteristics that can be approximately modeled as (constant) decoupled harmonic current sources. The utility voltages of industrial distribution systems are often distorted due to the neighboring loads and can be approximately modeled as decoupled harmonic voltage sources (as shown in the equivalent circuit of Fig. E5.3.1b):

$$v_{sys}(t) = \sum_{h=1}^{\infty} v_{sys,h}(t) = \sum_{h=1}^{\infty} \sqrt{2} V_{sys,h} \sin(h\omega_1 t + \theta_{sys,h}^{v}),$$

$$(E5.3\text{-}1a)$$

$$i_{NL}(t) = \sum_{h=1}^{\infty} i_{NL,h}(t) = \sum_{h=1}^{\infty} \sqrt{2} I_{NL,h} \sin(h\omega_1 t + \theta_{NL,h}^{i}).$$

$$(E5.3\text{-}1b)$$

Fourier analyses of measured voltage and current waveforms of the utility system and the nonlinear loads can be used to estimate the values of $V_{sys,h}$, $\theta_{sys,h}^{v}$, $I_{NL,h}$, and $\theta_{NL,h}^{i}$. This is a practical approach because most industrial loads consist of a number of in-house nonlinear loads, and their harmonic models are not usually available.

For this example, assume

$$v_{sys}(t) = \sqrt{2}(110)\sin(\omega_1 t) + \sqrt{2}(5)\sin(5\omega_1 t) \\ + \sqrt{2}(3)\sin(7\omega_1 t)[V],$$

$$(E5.3\text{-}2a)$$

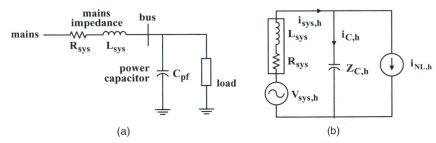

**FIGURE E5.3.1** A simplified industrial power system with capacitor and nonlinear load; (a) configuration, (b) the harmonic equivalent circuit [2] with $Z_{sys,h} = R_{sys} + jhL_{sys}$.

$$i_{NL}(t) = \sqrt{2}(1.0)\sin(\omega_1 t) + \sqrt{2}(0.3)\sin(5\omega_1 t)$$
$$+ \sqrt{2}(0.2)\sin(7\omega_1 t)[A].$$

$$(E5.3\text{-}2b)$$

Compute frequencies of the series and parallel resonances and the harmonic currents injected into the capacitor, and plot its frequency response.

### 5.4.4 Application Example 5.4: Parallel Resonance Caused by Capacitors

In the system of Fig. E5.4.1a, the source has the ratio $X/R = 10$. Assume $X/R = 4000$ (losses 0.25 W/kVAr) and $X/R = 5000$ (losses 0.2 W/kVAr) for low-voltage and medium-voltage high-efficiency capacitors, respectively. The harmonic current source is a six-pulse converter injecting harmonic currents of the order [8]

$$h = n(k + 1), \qquad (E5.4\text{-}1)$$

where $n$ is an integer (typically from 1 to 4) and k is the number of pulses (e.g., equal to 6 for a six-pulse converter). Find the resonance frequency of this circuit. Plot the frequency response and current amplification at bus 1.

### 5.4.5 Application Example 5.5: Series Resonance Caused by Capacitors

An example of a series resonance system is demonstrated in Fig. E5.5.1a. The equivalent circuit (neglecting resistances) is shown in Fig. E5.5.1b. Find the resonance frequency of this circuit. Plot the frequency response of bus 1 equivalent impedance, and the current amplification across the tuning reactor.

### 5.4.6 Application Example 5.6: Protecting Capacitors by Virtual Harmonic Resistors

Repeat Application Example 5.3 assuming a power converter is used to include a virtual harmonic resistor $R_{vir} = 0.5\ \Omega$ in series with the capacitor.

**Suggested Solutions to Resonance Problems.** As demonstrated in the above examples, harmonic currents and voltages that resonate with power system impedance are usually amplified and result in grave power quality problems such as destruction of capacitors, saturation of electromagnetic devices, and high losses and reduced lifetimes of appliances. It is difficult to come up with a single remedy for all resonance problems, as they highly depend on system configuration and load conditions. Some recommendations are

- Move resonance frequencies away from system harmonics;
- Perform system analysis and harmonic simulations before installing new capacitor banks. Optimal placement and sizing of capacitor banks in the presence of harmonic sources and nonlinear loads are highly recommended for all newly installed capacitor banks;
- Protect capacitors from harmonic destruction using damping circuits (e.g., passive or active resistors in series with the resonance circuit); and
- Use a power converter to include a virtual harmonic resistor in series with the power capacitor [2]. The active resistor only operates at harmonic frequencies to protect the capacitor. To increase system efficiency, the harmonic real power absorption of the virtual resistor can be stored on the DC capacitor of the converter, converted into fundamental real power, and regenerated back to the utility system.

## 5.5 FREQUENCY AND CAPACITANCE SCANNING

Resonances occur when the magnitude of the system impedance is extreme. Parallel resonance occurs when inductive and capacitive elements are in shunt and the impedance magnitude is a maximum. For a series-resonant condition, inductive and capacitive elements are in series and the impedance magnitude is a minimum. Therefore, the search for a resonant condition at a bus amounts to searching for extremes of the magnitude of its driving-point impedance $Z(\omega)$. There are two basic techniques to determine these extremes and the corresponding (resonant) frequencies for power systems: frequency scan and capacitive scan.

**Frequency Scan.** The procedures for analyzing power quality problems can be classified into those in the frequency domain and those in the time domain. Frequency-domain techniques are widely used for harmonic modeling and for formulations [9]. They are basically a reformulation of the fundamental load flow to include nonlinear loads and system response at harmonic frequencies.

Frequency scanning is the simplest and a commonly used approach for performing harmonic analysis, determining system resonance frequencies, and designing tuned filters. It is the computation (and plotting) of the driving-point impedance magnitude at a bus ($|Z(\omega)|$) for a range of frequencies and simply scanning the values for either a minimum or a maximum. One could also examine the phase of the driving-point impedance ($\angle Z(\omega)$) and search for its zero crossings (e.g., where the impedance is purely resistive). The frequency characteristic of impedance is called a frequency scan and the resonance frequencies are those that cause minimum and/or maximum values of the magnitude of impedance.

**Capacitance Scan.** In most power systems, the frequency is nominally fixed, but the capacitance (and/or the inductance) might change. Resonance conditions occur for values of C where the magnitude of impedance is a minimum or a maximum. Therefore, instead of sweeping the frequency, the impedance is considered as a function of C, and the capacitance is swept over a wide range of values to determine the resonance points. This type of analysis is called a capacitance scan.

It is important to note that frequency and capacitance scanning of the driving-point impedance at a bus may not reveal all resonance problems of the system. The plot of $Z(\omega)$ versus $\omega$ will only illustrate resonant conditions related to the bus under consideration. It will not detect resonances related to capacitor banks placed at locations relatively remote from the bus under study since it is not usually possible to check the driving-point impedances of all buses.

It is recommended to perform scanning and to check driving-point impedances of buses near harmonic sources (e.g., six- and twelve-pulse converters and adjustable drives) and to examine the transfer

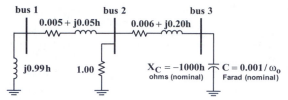

**FIGURE E5.7.1** The three-bus example system [9].

impedances between the harmonic sources and a few nearby buses.

### 5.5.1 Application Example 5.7: Frequency and Capacitance Scanning [9]

Perform frequency and capacitance scans at bus 3 of the three-bus system shown in Fig. E5.7.1. The equivalent circuit of the external power system is represented as 0.99 h at bus 1 ($h$ is the order of harmonics). The nominal design value of the shunt capacitor at bus 3 is $0.001/\omega_o$ farads and its reactance at harmonic $h$ is $(-1000/h)$.

### 5.6 HARMONIC CONSTRAINTS FOR CAPACITORS

Nonlinear loads and power electronic-based switching equipment are being extensively used in modern power systems. These have nonlinear input characteristics and emit high harmonic currents, which may result in waveform distortion, resonance problems, and amplification of voltage and current harmonics of power capacitors. Therefore, the installation of any capacitor at the design and/or operating stages should also include harmonic and power quality analyses. Some techniques have been proposed for solving harmonic problems of capacitors. Most approaches disconnect capacitor banks when their current harmonics are high. Other techniques include harmonic filters, blocking devices, and active protection circuits. These approaches use the total harmonic distortion of capacitor voltage ($THD_v$) and current ($THD_i$) as a measure of distortion level and require harmonic voltage, current, and reactive power constraints for the safe operation of capacitor banks.

The reactive power of a capacitor is given by its capacitance $C$, the rms terminal voltage $V_{rms}$, and the line frequency $f$. This reactive power is valid for sinusoidal operating conditions with rated rms terminal voltage $V_{rms\_rat}$ and rated line frequency $f_{rat}$. If the terminal voltage and the line frequency deviate from their rated values, that is, if the capacitor contains voltage and current harmonics ($V_h$ and $I_h$ for $h = 2, 3, 4, 5, \ldots$), then appropriate constraints for voltage $V_{rms}$, current $I_{rms}$, and reactive power Q must be satisfied, as indicated by IEEE [10], IEC, and European [e.g., 11, 12] standards. These constraints are examined in the following sections.

### 5.6.1 Harmonic Voltage Constraint for Capacitors

Assume harmonic voltages (Eq. 5-3a) across the terminals of a capacitor, where $V_{rms}$ is the terminal voltage. In the following relations $h$ is the harmonic order of the voltage, current, and reactive power: $h = 3^*, 5, 7, 9^*, 11, 13\ 15^*, 17, \ldots$ We have

$$V_{rms}^2 = V_1^2 + \sum_{h=2}^{\infty} V_h^2, \qquad (5\text{-}9a)$$

or

$$V_{rms} = \sqrt{V_1^2 + \sum_{h=2}^{\infty} V_h^2}. \qquad (5\text{-}9b)$$

Normalized to $V_{rms\_rat}$,

$$\frac{V_{rms}}{V_{rms\_rat}} = \sqrt{\left(V_1^2 + \sum_{h=2}^{\infty} V_h^2\right)/V_{rms\_rat}^2} \le 1.10 \text{ or } 1.15^a. \tag{5-9c}$$

### 5.6.2 Harmonic Current Constraint for Capacitors

Assume harmonic currents (Eq. 5-3b) through the terminals of a capacitor, where $I_{rms}$ is the total current:

$$I_{rms}^2 = I_1^2 + I_2^2 + I_3^2 + \ldots \qquad (5\text{-}10a)$$

For sinusoidal current components $I_1, I_2, I_3, \ldots$, one obtains with $\omega_1 = 2\pi f_1$

$$I_{rms}^2 = (\omega_1 C V_1)^2 + (2\omega_1 C V_2)^2 + (3\omega_1 C V_3)^2 + \ldots, \tag{5-10b}$$

or

$$I_{rms}^2 = \omega_1^2 C^2 (V_1^2 + 4V_2^2 + 9V_3^2 + \ldots),$$

$$V_1^2 = V_{rms}^2 - V_2^2 - V_3^2 - \ldots$$

$$I_{rms}^2 = \omega_1^2 C^2 (V_{rms}^2 + 3V_2^2 + 8V_3^2 + \ldots),$$

or

$$I_{rms}^2 = \omega_1^2 C^2 \left( V_{ms}^2 + \sum_{h=2}^{\infty} (h^2 - 1)V_h^2 \right), \qquad (5\text{-}10c)$$

$$I_{rms}^2 = \omega_1^2 C^2 V_{rms}^2 \left( 1 + \sum_{h=2}^{\infty} (h^2 - 1)\left(\frac{V_h}{V_{rms}}\right)^2 \right). \qquad (5\text{-}10d)$$

With $I_{rms\_rat} = \omega_{rat} \cdot C \cdot V_{rms\_rat}$ one can write $(\omega_{rat} = 2\pi f_{rat})$

$$\frac{I_{rms}}{I_{rms\_rat}} = \frac{f_1}{f_{rat}} \cdot \frac{V_{rms}}{V_{rms\_rat}} \sqrt{1 + \sum_{h=2}^{\infty} (h^2 - 1)\left(\frac{V_h}{V_{rms}}\right)^2} \le 1.5 \text{ or } 1.8^b. \tag{5-10e}$$

### 5.6.3 Harmonic Reactive-Power Constraint for Capacitors

In the presence of both voltage and current harmonics, the reactive power will also contain harmonics (Eq. 5-4b):

$$\begin{aligned}
Q &= Q_1 + Q_2 + Q_3 + \ldots \\
&= I_1 V_1 + I_2 V_2 + I_3 V_3 + \ldots \\
&= \omega_1 C V_1^2 + 2\omega_1 C V_2^2 + 3\omega_1 C V_3^2 + \ldots \\
&= \omega_1 C [(V_{rms}^2 - V_2^2 - V_3^2 - \ldots) + 2V_2^2 + 3V_3^2 + \ldots, \\
&= \omega_1 C [V_{rms}^2 + V_2^2 + 2V_3^2 + \ldots] \\
&= \omega_1 C [V^2 + \sum_{h=2}^{\infty} (h-1)V_h^2]
\end{aligned} \tag{5-11a}$$

and with $Q_{rat} = \omega_{rat} C V_{rms\_rat}^2$ one obtains

$$\frac{Q}{Q_{rat}} = \frac{f_1}{f_{rat}} \cdot \frac{V_{rms}^2}{V_{rms\_rat}^2} \left[ 1 + \sum_{h=2}^{\infty} (h-1)\frac{V_h^2}{V_{rms}^2} \right] \le 1.35 \text{ or } 1.45^a. \tag{5-11b}$$

---

*The triplen harmonics (which in most cases are of the zero-sequence type) are neglected for three-phase capacitor banks – these harmonics might not be zero but are usually very small. Note that not all triplen harmonics are necessarily of the zero-sequence type. All even harmonics are usually small in three- or single-phase power systems, and therefore they are neglected.

---

a The value of 1.15 for voltages and the value of 1.45 for reactive power are acceptable if the harmonic load is limited to 6 h during a 24 h period at a maximum ambient temperature of $T_{amb} = 35°C$ [13].
b The value 1.8 for harmonic currents is permissible if it is explicitly stated on the nameplate of the capacitor [13].

### 5.6.4 Permissible Operating Region for Capacitors in the Presence of Harmonics

The safe operating region for capacitors in a harmonic environment can be obtained by plotting of the corresponding constraints for voltage, current, and reactive power.

**Safe Operating Region for the Capacitor Voltage**

$$\frac{V_{rms}}{V_{rms\_rat}} = \frac{\sqrt{V_1^2 + \sum_{h=2}^{\infty} V_h^2}}{V_{rms\_rat}} \leq 1.10 \text{ or } 1.15^a, \quad (5\text{-}12a)$$

or

$$\frac{V_1}{V_{rms\_rat}} \sqrt{1 + \left(\frac{\sum_{h=2}^{\infty} V_h^2}{V_{rms\_rat}^2}\right)} = 1.10 \text{ or } 1.15^a. \quad (5\text{-}12b)$$

with $V_1 \approx V_{rms} = V_{rms\_rat}$ and relying on the lower limit in Eq. 5-12b (e.g., 1.10) one obtains

$$\sum_{h=2}^{\infty} \left(\frac{V_h^2}{V_{rms}^2}\right) = (1.1)^2 - 1 = 0.21, \quad (5\text{-}12c)$$

or the voltage constraint becomes

$$\sum_{h=2}^{\infty} \frac{V_h}{V_{rms}} = 0.458, \quad (5\text{-}12d)$$

which is independent of $h$.

**Safe Operating Region for the Capacitor Current**

$$\frac{I_{rms}}{I_{rms\_rat}} = \frac{f_1}{f_{rat}} \cdot \frac{V_{rms}}{V_{rms\_rat}} \sqrt{1 + \sum_{h=2}^{\infty} (h^2 - 1) \left(\frac{V_h}{V_{rms}}\right)^2} \quad (5\text{-}13a)$$
$$= 1.5 \text{ or } 1.8^b,$$

for $f_1 = f_{rat}$ and $V_{rms} = V_{rms\_rat}$. Relying on the lower limit in Eq. 5-13a (e.g., 1.5) one obtains

$$(1.5)^2 - 1 = \sum_{h=2}^{\infty} (h^2 - 1) \left(\frac{V_h}{V_{rms}}\right)^2 \quad (5\text{-}13b)$$

or

$$\sum_{h=2}^{\infty} \left(\frac{V_h}{V_{rms}}\right) = \sqrt{\frac{1.25}{h^2 - 1}}, \quad (5\text{-}13c)$$

which is a hyperbola.

**Safe Operating Region for the Capacitor Reactive Power**

$$\frac{Q}{Q_{rat}} = \frac{f_1}{f_{rat}} \cdot \frac{V_{rms}^2}{V_{rms\_rat}^2} \left[1 + \sum_{h=2}^{\infty} (h-1) \frac{V_h^2}{V_{rms}^2}\right] = 1.35 \text{ or } 1.45^a, \quad (5\text{-}14a)$$

for $f_1 = f_{rat}$ and $V_{rms} = V_{rms\_rat}$. Relying on the lower limit in Eq. 5-14a (e.g., 1.35) one obtains

$$\sum_{h=2}^{\infty} \left(\frac{V_h}{V_{rms}}\right) = \sqrt{\frac{0.35}{(h-1)}}, \quad (5\text{-}14b)$$

which is also a hyperbola.

If the above listed limits are exceeded for a capacitor with the rated voltage $V_{rms\_rat1}$ then one has to select a capacitor with a higher rated voltage $V_{rms\_rat2}$ and a higher rated reactive power satisfying the relation

$$Q_{rat2} = Q_{rat1} \left(\frac{V_{rms\_rat2}}{V_{rms\_rat1}}\right)^2. \quad (5\text{-}15)$$

Note all triplen and all even harmonics are small in three-phase power systems. However, triplen harmonics can be dominant in single-phase systems.

**Feasible Operating Region for the Capacitor Reactive Power.** Based on Eqs. 5-12d, 5-13c, and 5-14b, the safe operating region for capacitors in the presence of voltage and current harmonics is shown in Fig. 5.4.

### 5.6.5 Application Example 5.8: Harmonic Limits for Capacitors when Used in a Three-Phase System

The reactance of a capacitor decreases with frequency, and therefore the capacitor acts as a sink for

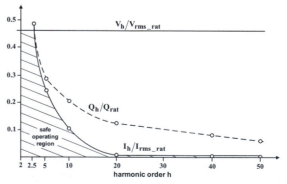

**FIGURE 5.4** Feasible operating region of capacitors with voltage, current, and reactive power harmonics (only one harmonic is present at any time).

higher harmonic currents. The effect is to increase the heating and dielectric stresses. ANSI/IEEE [10], IEC, and European [e.g., 11–13] standards list limits for voltage, currents, and reactive power of capacitor banks. These limits can be used to determine the maximum allowable harmonic levels. The result of the increased heating and voltage stress brought about by harmonics is a shortened capacitor life due to premature aging.

The following constraints must be satisfied:

$$V_{rms}/V_{rated} = \sqrt{\left(V_1^2 + \sum_{h=2}^{h\,max} V_h^2\right)/V_{rated}^2} \leq 1.10, \quad \text{(E5.8-1)}$$

$$I_{rms}/I_{rated} = (f_1/f_{rated})(V_{rms}/V_{rated})$$
$$= \sqrt{1 + \sum_{h=2}^{h\,max} (h^2 - 1)\left(\frac{V_h}{V_{rms}}\right)^2} \leq 1.3, \quad \text{(E5.8-2)}$$

$$Q/Q_{rated} = (f_1/f_{rated})(V_{rms}/V_{rated})^2$$
$$= 1 + \sum_{h=2}^{h\,max} (h-1)(V_h/V_{rms})^2 \leq 1.35, \quad \text{(E5.8-3)}$$

where $V_{rms\_rat}$ is the rated terminal voltage, $V_{rms}$ is the applied effective (rms) terminal voltage, $f_1$ is the line frequency, and $V_1$ is the fundamental (rms) voltage.

Plot for $V_1 \approx V_{rms} = V_{rms\_rat}$ and $f_1 = f_{rat}$ the loci for $V_h/V_{rms}$, where $(V_{rms}/V_{rms\_rat}) = 1.10$, $(I_{rms}/I_{rms\_rat}) = 1.3$, and $(Q/Q_{rat}) = 1.35$, as a function of the harmonic order $h$ for $5 \leq h \leq 49$ (only one harmonic is present at any time within a three-phase system).

## 5.7 EQUIVALENT CIRCUITS OF CAPACITORS

The impedance of a capacitor with conduction and dielectric losses can be represented by the two equivalent circuits of Fig. 5.5a (representing a parallel connection of capacitance $C$ and conductance $G_p$) and Fig. 5.5b (representing a series connection of capacitance $C$ and resistance $R_s$). The phasor diagrams for the admittance $G_p + j\omega C$ and the impedance $R_s - j/\omega C$ are depicted in Fig. 5.6a,b.

The loss factor $\tan \delta$ or the dissipation factor $DF = \tan \delta$ – which is frequency dependent – is defined for the parallel connection

$$\tan \delta = \frac{G_p}{\omega C}, \quad \text{(5-16)}$$

and for the series connection

$$\tan \delta = R_s \omega C. \quad \text{(5-17)}$$

Most AC capacitors have two types of losses:

- conduction loss caused by the flow of the actual charge through the dielectric, and
- dielectric loss due to the movement or rotation of the atoms or molecules of the dielectric in an alternating electric field.

The two equivalent circuits can be combined and expanded to the single equivalent circuit of Fig. 5.7, where $L_s = ESL$ = equivalent series inductance,

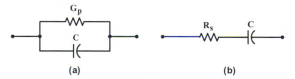

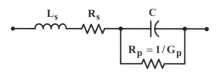

**FIGURE 5.5** Equivalent circuits of a capacitor with losses; (a) parallel connection, (b) series connection.

**FIGURE 5.7** Detailed equivalent circuit for a capacitor.

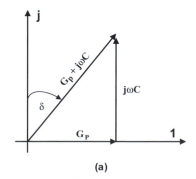

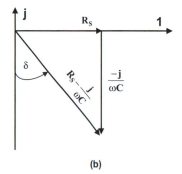

**FIGURE 5.6** Phasor diagrams for admittance $G_p + j\omega C$ and impedance $R_s - (j/\omega C)$; (a) parallel connection, (b) series connection.

$R_s = ESR$ = equivalent series resistance, and $R_p = EPR$ = equivalent parallel resistance. For most power system applications such a detailed equivalent circuit is not required. This circuit indicates that every capacitor has a self-resonant frequency, above which it becomes an inductor. $R_s$ is readily measured by applying this frequency to a capacitor, measuring voltage and current, and calculating their ratio. The resistance $R_p$ will always be much larger than the capacitive reactance at the resonant frequency and therefore this parallel resistance can be neglected at this resonant frequency.

The presence of voltage distortion increases the dielectric loss of capacitors; the total loss is for the frequency-dependent loss factor $\tan \delta_h = R_{sh}\omega_h C$ at $h_{max}$ harmonic voltages:

$$P_{C\_total} = \sum_{h=1}^{h=h_{max}} C \cdot (\tan \delta_h)\omega_h V_h^2, \quad (5\text{-}18)$$

where $\omega_h = 2 \cdot \pi \cdot f_h$ and $V_h$ is the rms voltage of the hth harmonic. For capacitors used for low-frequency applications the relation becomes

$$P_{C\_total} \approx \sum_{h=1}^{h=h_{max}} C^2 \cdot R_{sh} \cdot \omega_h^2 \cdot V_h^2. \quad (5\text{-}19)$$

For 60 Hz applications inexpensive AC capacitors can be used, whereas for power electronics applications and filters low-loss (e.g., ceramic or polypropylene) capacitors must be relied on.

### 5.7.1 Application Example 5.9: Harmonic Losses of Capacitors

For a capacitor with $C = 100 \ \mu F$, $V_{rat} = 460$ V, $R_{s1} = 0.01 \ \Omega$ (where $R_{s1}$ is the series resistance of the capacitor at fundamental ($h = 1$) frequency of $f_{rat} = 60$ Hz), compute the total harmonic losses for the harmonic spectra of Table E5.9.1 (up to and including the 19th harmonic) for the following conditions:

a) $R_{sh}$ is constant, that is, $R_{sh} = R_{s1} = 0.01 \ \Omega$.
b) $R_{sh}$ is proportional to frequency, that is, $R_{sh} = R_{s1}(f/f_{rat}) = 0.01$ h $\Omega$.
c) $R_{sh}$ is proportional to the square of frequency, $R_{sh} = R_{s1}(f/f_{rat})^2 = 0.01$ h$^2$ $\Omega$.

### 5.8 SUMMARY

Capacitors are important components within a power system: they are indispensable for voltage control, power-factor correction, and the design of filters.

**TABLE E5.9.1** Possible Voltage Spectra and Associated Capacitor Losses

| h | $\left(\dfrac{V_h}{V_{60Hz}}\right)_{1\Phi}$ (%) | $P_h$ for part a (mW) | $P_h$ for part b (mW) | $P_h$ for part c (mW) | $\left(\dfrac{V_h}{V_{60Hz}}\right)_{3\Phi}$ (%) | $P_h$ for part a (mW) | $P_h$ for part b (mW) | $P_h$ for part c (mW) |
|---|---|---|---|---|---|---|---|---|
| 1 | 100 | 3005 | 3005 | 3005 | 100 | 3005 | 3005 | 3005 |
| 2 | 0.5 | 0.3 | 0.6 | 1.2 | 0.5 | 0.3 | 0.6 | 1.2 |
| 3 | 4.0 | 40 | 120 | 360 | 2.0[a] | 10.8 | 32.4 | 97.2 |
| 4 | 0.3 | 0.4 | 1.6 | 6.4 | 0.5 | 1.2 | 4.8 | 19.2 |
| 5 | 3.0 | 67.7 | 339 | 1695 | 5.0 | 188 | 940 | 4700 |
| 6 | 0.2 | 0.43 | 2.6 | 15.6 | 0.2 | 0.43 | 2.6 | 15.6 |
| 7 | 2.0 | 58.9 | 412 | 2880 | 3.5 | 181 | 1267 | 8869 |
| 8 | 0.2 | 0.77 | 6.16 | 49.3 | 0.2 | 0.77 | 6.16 | 49.3 |
| 9 | 1.0 | 24.4 | 220 | 1980 | 0.3 | 2.19 | 19.71 | 177.4 |
| 10 | 0.1 | 0.3 | 3 | 30 | 0.1 | 0.3 | 3 | 30 |
| 11 | 1.5 | 81.9 | 900 | 9900 | 1.5 | 81.9 | 900 | 9900 |
| 12 | 0.1 | 0.433 | 5.2 | 62.4 | 0.1 | 0.433 | 5.2 | 62.4 |
| 13 | 1.5 | 114 | 1480 | 19240 | 1.0 | 114 | 1482 | 19270 |
| 14 | 0.1 | 0.59 | 8.26 | 115.6 | 0.05 | 0.15 | 2.58 | 28.81 |
| 15 | 0.5 | 16.9 | 254 | 3810 | 0.1 | 0.68 | 10.2 | 153.0 |
| 16 | 0.05 | 0.192 | 3.07 | 49.1 | 0.05 | 0.192 | 3.07 | 49.1 |
| 17 | 1.0 | 87 | 1480 | 25140 | 0.5 | 22 | 374 | 6358 |
| 18 | 0.05 | 0.244 | 4.4 | 79.2 | 0.01 | 0.0097 | 0.175 | 3.143 |
| 19 | 1.0 | 109 | 2070 | 39200 | 0.5 | 27.1 | 513 | 9750 |
| | | | All higher voltage harmonics < 0.5% | | | | | |

[a]Under certain conditions (e.g., DC bias of transformers as discussed in Chapter 2, and the harmonic generation of synchronous generators as outlined in Chapter 4) triplen harmonics are not of the zero-sequence type and can therefore exist in a three-phase system.

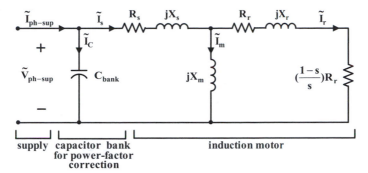

**FIGURE P5.1** Per-phase equivalent circuit of induction motor with displacement power-factor correction capacitor bank.

Their deployment may cause problems associated with capacitor switching and series resonance. Capacitor failures are induced by too large voltage, current, and reactive power harmonics. In most cases triplen (multiples of 3) and even harmonics do not exist in a three-phase system. However, there are conditions where triplen harmonics are not of the zero-sequence type and they can occur within three-phase systems. Triplen harmonics are mostly dominant in single-phase systems, whereas even harmonics are mostly negligibly small within single- and three-phase systems. In this chapter the difference between displacement power factor (DPF) – pertaining to fundamental quantities only – and the concept of the total power factor (TPF) – pertaining to fundamental and harmonic quantities – is explained. For the special case when there are no harmonics present these two factors will be identical; otherwise, the TPF is always smaller than the DPF. The TPF concept is related to the distortion power D and the DPF and TPF must be in power systems of the lagging type if consumer notation for the current is assumed. Leading power factors tend to cause oscillations and instabilities within power systems.

The losses of capacitors can be characterized by the loss factor or dissipation factor (DF) tan $\delta$, which is a function of the harmonic frequency. AC capacitor losses are computed for possible single- and three-phase voltage spectra.

## 5.9 PROBLEMS

### Problem 5.1: Calculation of Displacement Power Factor (DPF) and Its Correction

A $V_{L-L} = 460$ V, $f = 60$ Hz, $p = 6$ pole, $n_m = 1180$ rpm, Y-connected squirrel-cage induction motor has the following parameters per phase referred to the stator:

$R_S = 0.19\ \Omega$, $X_S = 0.75\ \Omega$, $X_m = 20\ \Omega$, $R'_r = 0.070\ \Omega$, $X'_r = 0.380\ \Omega$, and $R_{fe} \rightarrow \infty$.

a) Calculate the full-load current $\tilde{I}_s$ and the displacement power factor cos $\varphi$ of the motor (without capacitor bank).

b) In Fig. P5.1 the reactive current component of the motor is corrected by a capacitor bank $C_{bank}$. Find $C_{bank}$ such that the displacement power factor as seen at the input of the capacitor is cos $\varphi = 0.95$ lagging based on the consumer notation of current flow.

### Problem 5.2: Calculation of the Total Power Factor (TPF) and the DPF Correction

The following problem studies the power-factor correction for a relatively weak power system where the difference between the displacement power factor (DPF) and the total power factor (TPF) is large due to the relatively large amplitudes of voltage and current harmonics.

a) Perform a PSpice analysis for the circuit of Fig. P5.2, where a three-phase diode rectifier – fed by a Y/Y transformer with turns ratio $N_p/N_s = 1$ – is combined with a self-commutated switch (e.g., MOSFET or IGBT) and a filter (e.g., capacitor $C_f$) serving a load ($R_{load}$). You may assume $R_{syst} = 0.1\ \Omega$, $X_{syst} = 0.5\ \Omega$, $f = 60$ Hz, $v_{AN}(t) = \sqrt{2} \cdot 346$ V cos $\omega t$, $v_{BN}(t) = \sqrt{2} \cdot 346$ V cos($\omega t - 120°$), $v_{CN}(t) = \sqrt{2} \cdot 346$ V cos($\omega t - 240°$), ideal diodes $D_1$ to $D_6$, $C_f = 500\ \mu$F, $R_{load} = 10\ \Omega$, and a switching frequency of the self-commutated switch of $f_{sw} = 600$ Hz at a duty cycle of $\delta = 50\%$. Plot one period of the voltage and current after steady state has been reached as requested in parts b and c.

b) Plot and subject the line-to-neutral voltage $v_{an}(t)$ to a Fourier analysis.

c) Plot and subject the phase current $i_a(t)$ to a Fourier analysis.

d) Repeat parts a to c for $\delta = 10\%, 20\%, 30\%, 40\%, 60\%, 70\%, 80\%$, and $90\%$.

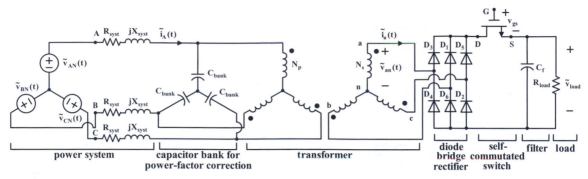

**FIGURE P5.2** Connection of a wye/wye three-phase transformer feeding a diode rectifier combined with self-commutated switch, filter, load, and a bank of power-factor correction capacitors.

e) Calculate the displacement power factor (DPF) and the total power factor (TPF) based on the phase shifts of the fundamental and harmonics between $v_{an}(t)$ and $i_a(t)$ for all duty cycles and plot the DPF and TPF as a function of $\delta$.

f) Determine the capacitance (per phase) $C_{bank}$ of the power-factor correction capacitor bank such that the displacement power factor as seen by the power systems is for $\delta = 50\%$ about equal to DPF = 0.95 lagging.

### Problem 5.3: Relation between Total and Displacement Power Factors

For the circuit of Fig. E5.1.1 (without taking into account power-factor correction capacitors), calculate with PSpice the Fourier components (magnitude and phase) of the phase voltage $v_{an}(t)$ and the phase current $i_a(t)$ up to the 10th harmonic. Compute and plot the ratio TPF/DPF as a function of the firing angle $\alpha$.

### Problem 5.4: Design of a Tuned Passive Filter

A harmonic filter (consisting of a capacitor and a tuning reactor) is to be designed in parallel to a power-factor correction (PFC) capacitor to improve the power quality (as recommended by IEEE-519 [8]) and to improve the displacement power factor *DPF* from 0.5 lag to 0.95 lag (consumer notation). Data obtained from a harmonic analyzer at 100% loading are 6.35 kV/phase, 600 kW/phase, THD$_i$ = 50%, and fifth harmonic (300 Hz) current $I_5 = 50$ A. It is desirable to reduce the THD$_i$ to 10%.

### Problem 5.5: Harmonic Resonance

Figure E5.3.1a shows a simplified industrial power system ($R_{sys} = 0.01\ \Omega$, $L_{sys} = 0.05$ mH) with a power-factor correction capacitor ($C_{pf} = 2000\ \mu F$). Indus-

trial loads may have nonlinear v–i characteristics that can be approximately modeled as (constant) decoupled harmonic current sources. The utility voltages in industrial distribution systems are often distorted due to the neighboring loads and can be approximately modeled as decoupled harmonic voltage sources:

$$v_{sys}(t) = \sqrt{2} \cdot 460 \sin(\omega_1 t) + \sqrt{2} \cdot 25 \sin(5\omega_1 t)$$
$$+ \sqrt{2} \cdot 15 \sin(7\omega_1 t)$$
$$i_{NL}(t) = \sqrt{2} \cdot 10 \sin(\omega_1 t) + \sqrt{2} \cdot 2 \sin(5\omega_1 t) \quad \text{(P5.5-1)}$$
$$+ \sqrt{2} \cdot 1 \sin(7\omega_1 t).$$

Compute frequencies of the series and parallel resonances and the harmonic currents injected into the capacitor, and plot its frequency response.

### Problem 5.6: Parallel Harmonic Resonance

In the system of Fig. E5.4.1a, the source has the ratio $X/R = 20$. Assume $X/R = 2000$ (losses 0.5 W/kVAr) and $X/R = 10,000$ (losses 0.1 W/kVAr) for low-voltage and medium-voltage high-efficiency capacitors, respectively. The harmonic current source is a six-pulse converter injecting harmonic currents of the order

$$h = n(k+1), \quad \text{(P5.6-1)}$$

where $n$ is an integer (typically from 1 to 4) and k is the number of pulses (e.g., equal to 6 for a six-pulse converter). Find the resonance frequency of this circuit. Plot the frequency response and current amplification at bus 1.

### Problem 5.7: Series Harmonic Resonance

An example of a series resonance system is demonstrated in Fig. E5.5.1a, where $S = 200$ MVA, $X/R = 20$,

$V_{\text{bus-rms}} = 7.2 \text{ kV}$,  $X_L = 1 \, \Omega$,  $Q_C = 900 \text{ kVAr}$, $V_{C\text{rms}} = 7.2 \text{ kV}$, and $I_{L\text{rms}} = 100 \text{ A}$. The equivalent circuit (neglecting resistances) is shown in Fig. E5.5.1b. Find the resonance frequency of this circuit. Plot the frequency response of the bus 1 equivalent impedance, and the current amplification across the tuning reactor.

## Problem 5.8: Protection of Capacitors by Virtual Harmonic Resistors

Repeat Application Example 5.3 assuming a power converter is used to include a virtual harmonic resistor $R_{\text{vir}} = 1.0 \, \Omega$ in series with the capacitor.

## Problem 5.9: Harmonic Current, Voltage, and Reactive Power Limits for Capacitors When Used in a Single-Phase System

The reactance of a capacitor decreases with frequency and therefore the capacitor acts as a sink for higher harmonic currents. The effect is to increase the heating and dielectric stress. ANSI/IEEE [10], IEC, and European [e.g., 11, 12] standards provide limits for voltage, currents, and reactive power of capacitor banks. This can be used to determine the maximum allowable harmonic levels. The result of the increased heating and voltage stress brought about by harmonics is a shortened capacitor life due to premature aging.

According to the nameplate of the capacitors the following constraints must be satisfied:

$$V_{\text{rms}}/V_{\text{rms-rat}} = \sqrt{\left(V_1^2 + \sum_{h=2}^{h\max} V_h^2\right)/V_{\text{rms\_rat}}^2} \le 1.15,$$

$$(P5.9\text{-}1)$$

$$I_{\text{rms}}/I_{\text{rat}} = (f_1/f_{\text{rat}})(V_{\text{rms}}/V_{\text{rat}})\sqrt{1 + \sum_{h=2}^{h\max}(h^2-1)\left(\frac{V_h}{V_{\text{rms}}}\right)^2}$$

$$\le 1.3,$$

$$(P5.9\text{-}2)$$

$$Q/Q_{\text{rat}} = (f_1/f_{\text{rat}})(V_{\text{rms}}/V_{\text{rat}})^2\left[1 + \sum_{h=2}^{h\max}(h-1)\left(\frac{V_h}{V_{\text{rms}}}\right)^2\right]$$

$$\le 1.45,$$

$$(P5.9\text{-}3)$$

where $V_{\text{rrms\_rat}}$ is the rated terminal voltage, $V_{\text{rms}}$ is the applied effective (rms) terminal voltage, $f_1$ is the line frequency, and $V_1$ is the fundamental (rms) voltage.

Plot for $V_1 \approx V_{\text{rms}} = V_{\text{rms\_rat}}$ and $f_1 = f_{\text{rat}}$ the loci for $V_h/V_{\text{rms}}$, where $(V_{\text{rms}}/V_{\text{rms\_rat}}) = 1.15$, $(I_{\text{rms}}/I_{\text{rms\_rat}}) = 1.3$,

and $(Q/Q_{\text{rat}}) = 1.45$ as a function of the harmonic order h for $3 \le h \le 49$ (only one harmonic is present at any time).

## Problem 5.10: Harmonic Losses of Capacitors

For a capacitor with $C = 100 \, \mu F$, $V_{\text{rat}} = 1000 \text{ V}$, $R_{s1} = 0.005 \, \Omega$ (where $R_{s1}$ is the series resistance of the capacitor at fundamental ($h = 1$) frequency of $f_{\text{rat}} = 60 \text{ Hz}$), compute the total harmonic losses for the harmonic spectra of Table P5.10 (up to and including the 19th harmonic) for the following conditions:

a) $R_{sh}$ is constant, that is, $R_{sh} = R_{s1} = 0.005 \, \Omega$.
b) $R_{sh}$ is proportional to frequency, that is,

$$R_{sh} = R_{s1}\left(\frac{f}{f_{\text{rat}}}\right) = 0.005 \cdot h \, \Omega.$$

c) $R_{sh}$ is proportional to the square of frequency,

$$R_{sh} = R_{s1}\left(\frac{f}{f_{\text{rat}}}\right)^2 = 0.005 \cdot h^2 \, \Omega.$$

d) Plot the calculated losses of parts a to c as a function of the harmonic order $h$.

**TABLE P5.10** Possible Voltage Spectra with High Harmonic Penetration

| $h$ | $\left(\dfrac{V_h}{V_{60\text{Hz}}}\right)_{1\Phi}$ (%) | $\left(\dfrac{V_h}{V_{60\text{Hz}}}\right)_{3\Phi}$ (%) |
|---|---|---|
| 1 | 100 | 100 |
| 2 | 2.5 | 0.5 |
| 3 | 5.71 | 1.0[a] |
| 4 | 1.6 | 0.5 |
| 5 | 1.25 | 7.0 |
| 6 | 0.88 | 0.2 |
| 7 | 1.25 | 5.0 |
| 8 | 0.62 | 0.2 |
| 9 | 0.96 | 0.3 |
| 10 | 0.66 | 0.1 |
| 11 | 0.30 | 2.5 |
| 12 | 0.18 | 0.1 |
| 13 | 0.57 | 2.0 |
| 14 | 0.10 | 0.05 |
| 15 | 0.10 | 0.1 |
| 16 | 0.13 | 0.05 |
| 17 | 0.23 | 1.5 |
| 18 | 0.22 | 0.01 |
| 19 | 1.03 | 1.0 |
| All higher harmonics < 0.2% | | |

[a]Under certain conditions (e.g., DC bias of transformers as discussed in Chapter 2, and the harmonic generation of synchronous generators as outlined in Chapter 4) triplen harmonics are not of the zero-sequence type, and they can therefore exist in a three-phase system.

## 5.10 REFERENCES

1) Wang, C.; Cheng, T.C.; Zheng, G.; Mu, Y.D.L.; Palk, B.; and Moon, M.; "Failure analysis of composite dielectric of power capacitors in distribution systems," *IEEE Transactions on Dielect. Elect. Insul.*, vol. 5, pp. 583–588, Aug. 1998.

2) Wu, J.C.; Jou, H.L.; Wu, K.D.; and Shen, N.C.; "Power converter-based method for protecting three-phase power capacitor from harmonic destruction," *IEEE Transactions on Power Delivery*, vol. 19, issue 3, pp. 1434–1441, July 2004.

3) Fernandez, A.; Sebastian, J.; Hernando, M.M.; Villegas, P.; and Garcia, J.; "Helpful hints to select a power-factor correction solution for low- and medium-power single-phase power supplies," *IEEE Transactions on Industrial Electronics*, vol. 52, issue 1, pp. 46–55, Feb. 2005.

4) Basu, S.; and Bollen, M.H.J; "A novel common power-factor correction scheme for homes and offices," *IEEE Transactions on Power Delivery*, vol. 20, issue 3, pp. 2257–2263, July 2005.

5) Wolfle, W.H.; and Hurley, W.G.; "Quasi-active power-factor correction with a variable inductive filter: Theory, design and practice," *IEEE Transactions on Power Electronics*, vol. 18, issue 1, Part 1, pp. 248–255, Jan. 2003.

6) Grebe, T.E.; "Application of distribution system capacitor banks and their impact on power quality," *IEEE Transactions on Industry Applications*, vol. 32, issue 3, pp. 714–719, May–June 1996.

7) Simpson, R.H.; "Misapplication of power capacitors in distribution systems with nonlinear loads – three case histories," *IEEE Transactions on Industry Applications*, vol. 41, no. 1, pp. 134–143, Jan.–Feb. 2005.

8) "IEEE recommended practices and requirements for harmonic control in electric power systems," *IEEE Standard 519-1992*, New York, NY, 1993.

9) Heydt, G.T.; *Electric Power Quality*, Star in a Circle Publications, USA, 1991.

10) ANSI/IEEE Standard 18-1980.

11) VDE 0560 T1 and T4.

12) DIN 48500.

13) Hoppner, A.; *Schaltanlagen, Taschenbuch für Planung, Konstruktion und Montage von Schaltanlagen*, Brown, Boveri & Cie, Aktiengesellschaft, Mannheim, Publisher: W. Girardet, Essen, Germany, 1966.

14) Dubey, G.K.; *Power Semiconductor Controlled Drives*, Prentice Hall, Englewood Cliffs, New Jersey, 1989.

## CHAPTER

# Lifetime Reduction of Transformers and Induction Machines

The total installed power capacity within the Eastern, Western, and Texan power pools of the United States is 900 GW with about 70 GW of spinning reserve. Approximately 60% of the 900 GW is consumed by induction motors and 100% passes through transformers. Similar percentages exist in most countries around the world. For this reason transformers and induction motors are important components of the electric power system.

The lifetime of any device is limited by the aging of the insulation material due to temperature: the higher the activation energy of any material, the faster the aging proceeds. Iron and copper/aluminum have low activation energies and for this reason their aging is negligible. Insulation material – either of the organic or inorganic type – is most susceptible to aging caused by temperature. If a device is properly designed then the rated temperature results in the rated lifetime. Temperature rises above rated temperature result in a decrease of lifetime below its rated value. There are a few mechanisms by which the rated lifetime can be reduced:

1. Temperature rises above the rated temperature can come about due to overload and voltage or current harmonics.
2. Lifetime can be also decreased by intermittent operation. It is well known that generators of pumped storage plants must be rewound every 15 years as compared to 40 years for generators which operate at a constant temperature.
3. Vibration within a machine due to load variations (e.g., piston compressor) can destroy the mechanical properties of conductor insulation.
4. Faults – such as short-circuits – can impact the lifetime due to excessive mechanical forces acting on the winding and their insulation.

In this chapter we are concerned with aging due to elevated temperatures caused by harmonics. The presence of current and voltage harmonics in today's power systems causes additional losses in electromagnetic components and appliances, creating substantial elevated temperature rises and decreasing lifetime of machines and transformers. Therefore, estimating additional losses, temperature rises, and aging of power system components and loads has become an important issue for utilities and end users alike.

Three different phases are involved in the estimation (or determination) of the lifetime of magnetic devices:

1. modeling and computation of the additional losses due to voltage or current harmonics,
2. determination of the ensuing temperature rises, and
3. estimation of the percentage decrease of lifetime as compared with rated lifetime [1].

The literature is rich in documents and papers that report the effect of poor power quality on losses and temperature rises of power system components and loads; however, a matter that still remains to be examined in more detail is the issue of device aging under nonsinusoidal operating conditions. Earlier papers were mostly concerned about magnetic device losses under sinusoidal operating conditions [2–13]. However, more recent research has expanded the scope to magnetic device derating under nonsinusoidal operation [14–17]. Only a few papers have addressed the subjects of device aging and economics issues caused by poor power quality [18–23].

In the first sections of this chapter the decrease of lifetime of power system components and loads such as universal motors, single- and three-phase transformers, and induction motors are estimated based on their terminal voltage harmonics. After a brief review of temperature relations, the concept of weighted harmonic factors is introduced. This is a quantity that relates the device terminal voltage harmonic amplitudes to its temperature rise. Additional temperature rises and losses due to harmonics are discussed, and thereafter the weighted harmonic factors are employed to determine the decrease of lifetime of electrical appliances. Toward the end of this chapter the time-varying nature of harmonics is addressed with respect to their measurement, summation, and propagation. The probabilistic evaluation of the economical damage – that is, cost – due to harmonic losses in industrial energy systems is explored next, and the increase of temperature in a device as a function of time as well as intermittent operation of devices concludes this chapter.

## 6.1 RATIONALE FOR RELYING ON THE WORST-CASE CONDITIONS

Field measurements show that the harmonic voltages and currents of a distribution feeder are time-varying. This is due to

1. the changes in the loads, and
2. the varying system's configuration.

This means that steady-state harmonic spectra do not exist. However, loads such as variable-speed drives for air-conditioning, rolling mills, and arc furnaces operate during their daily operating cycle for at least a few hours at rated operation. It is for this operating condition that system components (e.g., transformers, capacitors) and loads must be designed. Most electromagnetic devices such as transformers and rotating machines have thermal time constants of 1 to 3 hours – depending on their sizes and cooling mechanisms. When these components continually operate at rated power, they are exposed to steady-state current and voltage spectra resulting at an ambient temperature of $T_{amb} = 40°C$ in the maximum temperature rise $T_{risemax}$, and the resulting temperature $T_{max} = T_{amb} + T_{risemax}$ must be less or equal to the rated temperature $T_{rat}$. The worst case is therefore when the power system components operate at rated load with their associated current or voltage spectra where individual harmonic amplitudes are maximum. It is not sufficient to limit the total harmonic distortion of either voltage ($THD_v$) or current ($THD_i$) because these parameters can be the same for different spectra – resulting in different additional temperature rises, insulation stresses, and mechanical vibrations. In Sections 6.2 to 6.10 worst-case conditions will be assumed.

## 6.2 ELEVATED TEMPERATURE RISE DUE TO VOLTAGE HARMONICS

Most electric appliances use electric motors and/or transformers. In all these cases the power system's sinusoidal voltage causes ohmic and iron-core losses resulting in a temperature rise, which approaches the rated temperature rise ($T_{riserat}$) at continuous operation:

$$T_{rat} = T_{amb} + T_{riserat}, \qquad (6-1)$$

where $T_{amb} = 40°C$ is the ambient temperature. The lifetime of a magnetic device is greatly dependent on this temperature rise and the lifetime will be reduced if this rated temperature rise is exceeded over any length of time [1].

The presence of voltage harmonics in the power system's voltage causes harmonic currents in induction motors and transformers of electrical appliances, resulting in an elevated temperature rise ($T_{riserat} + \Delta T_h$) such that the device temperature is

$$T = T_{amb} + T_{riserat} + \Delta T_h, \qquad (6-2)$$

where $\Delta T_h$ is the additional temperature rise due to voltage (integer, sub-, inter-) harmonics. Through a series of studies this additional temperature rise of various electric machines and transformers was calculated and measured as a function of the harmonic amplitude $V_h$ and the phase shifts of the voltage harmonics $\varphi_h$ with respect to the fundamental [18, 19].

The insulating material of electrical apparatus, as used in electrical appliances, is of either organic or inorganic origin. The deterioration of the insulation caused by the elevated temperature rise is manifested by a reduction of the mechanical strength and/or a change of the dielectric behavior of the insulation material. This thermal degradation is best represented by the reaction rate equation of Arrhenius. Plots which will be drawn based on this equation are called Arrhenius plots [24, 25]. The slopes of these plots are proportional to the activation energy $E$ of the insulation material under consideration. Knowing the rated lifetime of insulation materials at rated temperature ($T_{rat} = T_{amb} + T_{riserat}$), and the elevated temperature rise ($\Delta T_h$) due to given amplitudes $V_h$ and phase shifts $\varphi_h$ of voltage harmonics, one will be able to estimate from the Arrhenius plot the reduction of the lifetime of an electrical appliance due to $\Delta T_h$.

*Definition: The activation energy E is the energy transmitted in the form of heat to the chemical reaction of decomposition.*

## 6.3 WEIGHTED-HARMONIC FACTORS

The harmonic voltage content (order $h$, amplitude $V_h$, and phase shift $\varphi_h$) of power system voltages varies with the type and size of harmonic generators and loads as well as with the topology of the system, and it will hardly be the same for any two networks. Since voltage harmonics result in additional losses and temperature rises ($\Delta T_h$) in electrical appliances, it would be desirable to derive one single criterion which limits – for a maximum allowable additional temperature rise – the individual amplitudes and phase shifts of the occurring harmonic voltages including their relative weight with respect to each other in contributing to the elevated temperature rise $\Delta T_h$.

Also, this criterion should be simple enough so that the additional losses, temperature rises $\Delta T_h$, and the reduction of lifetime of electrical appliances can easily be predicted as a function of the harmonic content of their terminal voltages. As most electrical appliances use transformers and motors as input devices, in this section weighted harmonic factors will be derived for single-phase and three-phase transformers and induction machines.

The ambient temperature is at the most $T_{amb} = 40°C$. The rated temperature rise depends on the class of insulation material used. $T_{rise\,rat} = 80°C$ is a commonly permitted value so that

$$T_{rat} = T_{amb} + T_{rise\,rat} = 40°C + 80°C = 120°C. \quad (6-3)$$

In transformers, iron-core sheets and copper (aluminum) conductors are insulated by paper, plastic material, or varnish.

The Swedish scientist Svante Arrhenius originated the so-called Arrhenius rule: a differential equation describing the speed of degradation or deterioration of any (organic, inorganic) material. This deterioration is not oxidation due to elevated temperatures.

The idea of a weighted harmonic factor for different magnetic devices [20] is based on the fact that the additional temperature rise $\Delta T_h$ is different for transformers, universal machines, and induction machines, although the harmonic content ($V_h$, $\varphi_h$) of the terminal voltages is the same.

### 6.3.1 Weighted-Harmonic Factor for Single-Phase Transformers

Figure 6.1 illustrates the equivalent circuit of a single-phase transformer for the fundamental ($h = 1$) and harmonic voltages of hth order.

The ohmic losses of the hth harmonic expressed in terms of the fundamental ohmic losses are

$$\frac{W_{ohmic,h}}{W_{ohmic,1}} = \left(\frac{V_{ph}}{V_{p1}}\right)^2 \left(\frac{R_{ph} + R'_{sh}}{R_{p1} + R'_{sl}}\right)(\text{RATIO}), \quad (6-4)$$

where

$$\text{RATIO} =$$

$$\frac{\left(R_{p1} + R'_{s1} + R'_{load,1}\right)^2 + \omega^2 \left(L_{p1} + L'_{s1} + L'_{load,1}\right)^2}{\left(R_{ph} + R'_{sh} + R'_{load,h}\right)^2 + (h\omega)^2 \left(L_{ph} + L'_{sh} + L'_{load,h}\right)^2}.$$

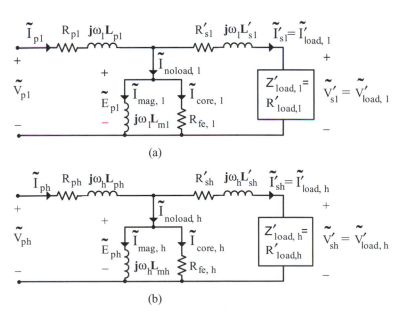

(a)

(b)

**FIGURE 6.1** Fundamental (a) and harmonic (b) equivalent circuits of single-phase transformers.

In most residential power transformers the influence of the skin effect can be neglected. However, the skin effect of an ohmic load cannot be ignored, because $R'_{\text{load},h}, R'_{\text{load},l} \gg R_{ph}, R_{p1}, R'_{sh}, R'_{s1}$. Therefore, one obtains

$$\frac{W_{\text{ohmic},h}}{W_{\text{ohmic},1}} \approx \left(\frac{V_{ph}}{V_{p1}}\right)^2 \frac{\left(R'_{\text{load},1}\right)^2 + \omega^2 \left(L_{p1} + L'_{s1} + L'_{\text{load},1}\right)^2}{\left(R'_{\text{load},h}\right)^2 + (h\omega)^2 \left(L_{ph} + L'_{sh} + L'_{\text{load},h}\right)^2}. \tag{6-5}$$

Due to $L_{ph} \approx L_{p1}$, $L'_{sh} \approx L'_{s1}$, $L'_{\text{load},h} \approx L'_{\text{load},1}$, and the fact that the load resistance $R'_{\text{load},h}$ increases less than linearly with frequency [18], the following simplified expression is considered:

$$\frac{W_{\text{ohmic},h}}{W_{\text{ohmic},1}} \approx \frac{1}{h^n}\left(\frac{V_{ph}}{V_{p1}}\right)^2, \tag{6-6}$$

where $0 < n < 2$ depends on the resistance and inductance values of the equivalent circuit of a load.

The iron-core losses of a single-phase transformer due to a voltage harmonic of order $h$ are expressed in terms of the corresponding losses caused by the fundamental as

$$\frac{W_{\text{iron},h}}{W_{\text{iron},1}} = \left(\frac{B_{\max,h}}{B_{\max,1}}\right)^2 \frac{\left\{\sigma\left(\frac{fh}{100}\right)^2 k_m^2 + \varepsilon\left(\frac{fh}{100}\right)\right\}}{\left\{\sigma\left(\frac{f}{100}\right)^2 + \varepsilon\left(\frac{f}{100}\right)\right\}}, \tag{6-7}$$

where $B_{\max,1}$ and $B_{\max,h}$ define fundamental and harmonic maximum flux densities, respectively, and $\sigma$ and $\varepsilon$ are [26] coefficients related to eddy-current and hysteresis losses, respectively. The term $k_m$ takes the reaction of the eddy currents within the laminations and between the laminations on the original field into account [18]. With $h$ $B_{\max,h} \propto E_{ph}$ and $B_{\max,1} \propto E_{p1}$, the relation becomes

$$\frac{W_{\text{iron},h}}{W_{\text{iron},1}} = \left(\frac{E_{ph}}{h E_{p1}}\right)^2 \left(\frac{R_{fe,1}}{R_{fe,h}}\right), \tag{6-8}$$

where $R_{fe,1}$ and $R_{fe,h}$ are the iron-core loss resistances for the fundamental ($h = 1$) and harmonic of order $h$, respectively.

The total losses for a harmonic of order $h$ are then determined from the sum of Eqs. 6-6 and 6-8:

$$\frac{W_{\text{total},h}}{W_{\text{total},1}} \approx \frac{1}{h^n}\left(\frac{V_{ph}}{V_{p1}}\right)^2 + \frac{1}{h^2}\left(\frac{E_{ph}}{E_{p1}}\right)^2 \left(\frac{R_{fe,1}}{R_{fe,h}}\right). \tag{6-9}$$

According to [27], the core-loss resistance ratio is $R_{fe,h}/R_{fe,1} = h^{0.6}$ and because of the leakage reactance being proportional to $h$, one obtains the inequality $(E_{ph}/E_{p1}) < (V_{ph}/V_{p1})$ or

$$\left(\frac{E_{ph}}{E_{p1}}\right)^2 = \left(\frac{V_{ph}}{V_{p1}}\right)^m, \tag{6-10}$$

where the exponent $m$ is less than 2. Therefore, for single-phase transformers Eq. 6-9 becomes

$$\frac{W_{\text{total},h}}{W_{\text{total},1}} \approx \frac{1}{h^n}\left(\frac{V_{ph}}{V_{p1}}\right)^2 + \frac{1}{h^{2.6}}\left(\frac{V_{ph}}{V_{p1}}\right)^m. \tag{6-11}$$

Depending on the relative sizes of the ohmic and iron-core losses at full load, Eq. 6-11 can be rewritten for all occurring harmonics and is called the weighted harmonic factor [20]:

$$\frac{W_{\text{total},h\Sigma}}{W_{\text{total},1}} = K_1 \sum_{h=2}^{\infty} \frac{1}{(h)^k}\left(\frac{V_{ph}}{V_{p1}}\right)^l, \tag{6-12}$$

where $W_{\text{total},h\Sigma}$ and $W_{\text{total},1}$ are the total harmonic losses and the total fundamental losses, respectively. Also $0 \le k \le 2.0$ and $0 \le l \le 2.0$. Average values for $k$ and $l$ are obtained from measurements as discussed in a later section.

### 6.3.2 Measured Temperature Increases of Transformers

Measured temperature rises in transformers and induction machines as they occur in a residential/commercial power system – together with calculated loss increases due to voltage harmonics – represent a base for the estimation of the lifetime reduction due to voltage harmonics. It is thereby assumed that for small additional temperature rises – as compared with the rated temperature rise – the additional losses are proportional to the additional temperature rise.

#### 6.3.2.1 Single-Phase Transformers

A single-phase 150 VA transformer was tested [18, 19] and Fig. 6.2a,b shows the measured temperature rises in percent of the rated temperature rises at full load for moderate (e.g., $T_{\text{amb}} = 23°C$) and high (e.g., $T_{\text{amb}} = 40°C$) ambient temperatures.

Note that if the third harmonic voltage $v_3(t)$ superposed with the fundamental voltage $v_1(t)$ produces a peak-to-peak voltage $\{v_1(t) + v_3(t)\}$ that is maximum, then the peak-to-peak flux density in the core

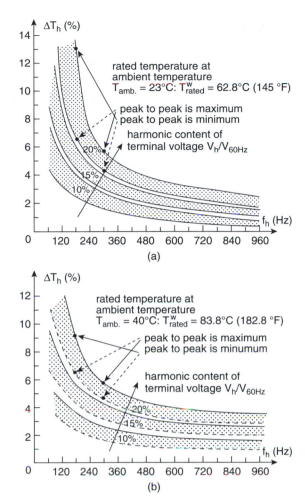

**FIGURE 6.2** Measured additional temperature rise of the 150 VA single-phase transformer winding in % of the rated temperature rise at full load at an ambient temperature of (a) 23°C and (b) 40°C as a function of the harmonic voltage amplitude, phase shift, and frequency [29, 30].

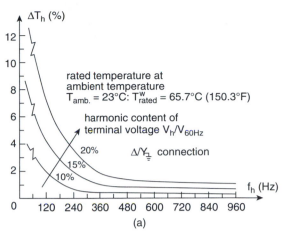

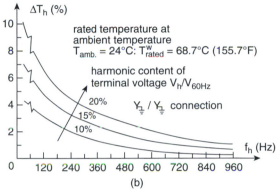

**FIGURE 6.3** (a) Measured additional temperature rise of the transformer winding in % of the rated temperature rise at full load at an ambient temperature of 23°C as a function of the harmonic voltage amplitude and frequency for either forward- or backward-rotating harmonic voltage systems, if the transformer bank is in $\Delta$/Y-grounded connection [29, 30]. (b) Measured additional temperature rise of the transformer winding in % of the rated temperature rise at full load at an ambient temperature of 24°C as a function of the harmonic voltage amplitude and frequency for either forward- or backward-rotating harmonic voltage systems, if the transformer bank is in Y-grounded/Y-grounded connection [29, 30].

$\{B_1(t) + B_3(t)\}$ is minimum. That is, the total harmonic losses are smaller than those when the peak-to-peak voltage is minimum, resulting in a maximum peak-to-peak value of the flux density. This "alternating" behavior of the harmonic losses and associated temperatures is discussed in detail in [27, 28].

### 6.3.2.2 Three-Phase Transformers

Three 150 VA single-phase transformers were assembled to $\Delta$/Y-grounded and Y-grounded/Y-grounded configurations and they were subjected to the same harmonic voltage conditions as the single-phase transformer of the preceding section. The temperatures of the transformer windings were measured at the same location as for the single-phase transformer. The tests were limited to balanced opera-

tion. However, no phase-lock loop was available and no stable relation between fundamental and harmonic voltage systems could be maintained. Therefore, the temperature data of Fig. 6.3a,b represent average values and the alternating behavior could not be observed, although it also exists in three-phase transformers.

### 6.3.3 Weighted-Harmonic Factor for Three-Phase Induction Machines

The per-phase equivalent circuits of a three-phase induction motor are shown for the fundamental ($h = 1$) and a (integer, sub-, inter-) harmonic of hth order in Fig. 6.4. In Fig. 6.4 the fundamental slip is

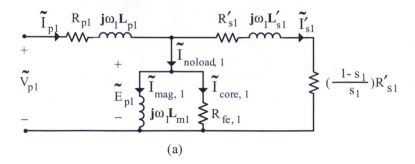

(a)

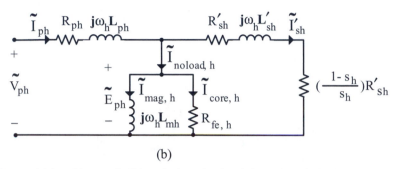

(b)

**FIGURE 6.4** Fundamental (a) and harmonic (b) equivalent circuits of three-phase induction machines.

$$s_1 = \frac{n_{s1} - n_m}{n_{s1}}, \qquad (6\text{-}13a)$$

and the harmonic slip is

$$s_h = \frac{n_{sh} - n_m}{n_{sh}}. \qquad (6\text{-}13b)$$

The ohmic losses caused by either the fundamental or a harmonic of order $h$ are proportional to the square of the respective currents. Therefore, one can write for the ohmic losses caused by voltage harmonics:

$$\frac{W_{ohmic,h}}{W_{ohmic,1}} = \left(\frac{R_{ph} + R'_{sh}}{R_{p1} + R'_{s1}}\right)\left(\frac{V_{ph}}{V_{p1}}\right)^2 \left(\frac{Z_{p1}}{Z_{ph}}\right)^2, \qquad (6\text{-}14)$$

where

$$Z_{p1} = \sqrt{\left(R_{p1} + R'_{s1}/s_1\right)^2 + \left(\omega_1 L_p + \omega_1 L'_s\right)^2}$$

and

$$Z_{ph} = \sqrt{\left(R_{ph} + R'_{sh}/s_1\right)^2 + \left(h\omega_1 L_p + h\omega_1 L'_s\right)^2}.$$

For all time harmonics ($h = 2, 3, 4, \ldots$) an induction machine operating at rated speed can be considered being at standstill because $s_1 \ll s_h$. With $R_{ph}$, $R'_{sh}/s_h \ll h\omega_1 L_p$, $h\omega_1 L'_s$, the per-unit starting impedance

$$Z_{start\,pu} = \frac{\sqrt{\left(\omega_1 L_p + \omega_1 L'_s\right)^2}}{\sqrt{\left(R_{p1} + R'_{s1}/s_1\right)^2 + \left(\omega_1 L_p + \omega_1 L'_s\right)^2}}, \qquad (6\text{-}15)$$

and the slip $s_h = \{h \mp (1 - s_1)\}/h \approx (h \mp 1)/h$, Eq. 6-14 can be rewritten as

$$\frac{W_{ohmic,h}}{W_{ohmic,1}} \approx \left(\frac{R_{ph} + R'_{sh}}{R_{p1} + R'_{s1}}\right)\left(\frac{V_{ph}}{V_{p1}}\right)^2 \frac{1}{h^2 Z_{start\,pu}^2}. \qquad (6\text{-}16)$$

The additional iron-core losses due to a voltage harmonic of order $h$ referred to the iron-core losses at the fundamental frequency are (see Fig. 6.4)

$$\frac{W_{ohmic,h}}{W_{ohmic,1}} \approx \left(\frac{V_{ph}}{V_{p1}}\right)^2 \left(\frac{Z_{fe,1}}{Z_{fe,h}}\right)^2 \frac{R_{fe,h}}{R_{fe,1}} = \left(\frac{E_{ph}}{E_{p1}}\right)^2 \frac{R_{fe,1}}{R_{fe,h}}, \qquad (6\text{-}17)$$

where

$$Z_{fe,1} = \frac{\left(j\omega_1 L_m + R_{fe,1}\right)\left(R_{p1} + j\omega_1 L_p\right)}{j\omega_1 L_m} + R_{fe,1},$$

$$Z_{fe,h} = \frac{\left(jh\omega_1 L_m + R_{fe,h}\right)\left(R_{ph} + jh\omega_1 L_p\right)}{jh\omega_1 L_m} + R_{fe,h}.$$

From [31] follows $\dfrac{R_{fe,h}}{R_{fe,1}} \approx h^{0.6}$, and therefore the iron-core losses are

$$\frac{W_{\text{ohmic},h}}{W_{\text{ohmic},1}} \approx \left(\frac{E_{ph}}{E_{p1}}\right)^2 \left(\frac{1}{h^{0.6}}\right). \qquad (6\text{-}18)$$

Summing Eqs. 6-16 and 6-18, the total iron-core losses for a harmonic of order $h$ of a three-phase induction machine become

$$\frac{W_{\text{ohmic},h}}{W_{\text{ohmic},1}} \approx \left(\frac{R_{ph}+R'_{sh}}{R_{p1}+R'_{s1}}\right)\left(\frac{V_{ph}}{V_{p1}}\right)^2 \frac{1}{h^2 Z^2_{\text{start }pu}}$$
$$+ \left(\frac{E_{ph}}{E_{p1}}\right)^2 \left(\frac{1}{h^{0.6}}\right), \qquad (6\text{-}19)$$

or for all occurring harmonics, one can show that the total harmonic losses $W_{\text{total}h\text{E}}$ relate to the total fundamental losses, as defined by the weighted harmonic factor [32]:

$$\frac{W_{\text{total}h\Sigma}}{W_{\text{total}1}} = K_2 \sum_{h=2}^{\infty} \frac{1}{(h)^k} \left(\frac{V_{ph}}{V_{p1}}\right)^l, \qquad (6\text{-}20)$$

where $0 \le k \le 2$ and $0 \le l \le 2$.

Note that the relations for single-phase transformers (Eq. 6-12) and that for a three-phase induction motor (Eq. 6-20) are identical in structure, though the exponents $k$ and $l$ may assume different values for single-phase transformers and three-phase induction motors. These exponents will be identified in the next section and it will be shown that the same structure of the weighted harmonic factor will also be valid for three-phase transformers and single-phase induction machines as well as universal machines.

Depending on the values of $k$ and $l$ one obtains different loss or temperature dependencies as a function of the frequency. Unfortunately, these dependencies can vary within wide ranges from constant to hyperbolic functions. This is so because the iron-core losses depend on the electric steel (e.g., conductivity, lamination thickness, permeability) used and the winding configurations in primary or stator and secondary or rotor. These can exhibit different conductivities (e.g., Cu, Al) or skin and proximity effects. In order to study these dependencies various $k$ and $l$ combinations will be assumed, as is depicted in Figs. 6.5 to 6.7:

- $k = 0$ and $l = 1$ results in a linear frequency-independent characteristic (Fig. 6.5).
- $k = 1$ and $l = 1$ results in a inverse frequency-dependent characteristic (Fig. 6.6).
- $k = 2$ and $l = 2$ results in a quadratic inverse frequency-dependent characteristic (Fig. 6.7).

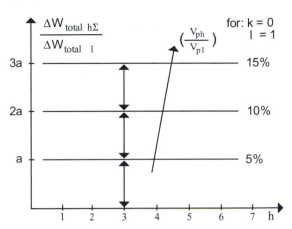

**FIGURE 6.5** Linear frequency-independent increase of harmonic losses with harmonic voltage $V_{ph}$.

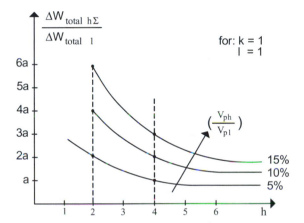

**FIGURE 6.6** Inverse frequency-dependent increase of harmonic losses with harmonic voltage $V_{ph}$.

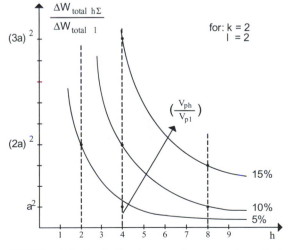

**FIGURE 6.7** Quadratic inverse frequency-dependent increase of harmonic losses with harmonic voltage $V_{ph}$.

In real transformers and induction machines combinations of these characteristics will be possible.

### 6.3.4 Calculated Harmonic Losses and Measured Temperature Increases of Induction Machines

#### 6.3.4.1 Single-Phase Induction Motors

Calculated additional losses due to voltage harmonics for the machine of Eq. 6-21 are shown for the stator in Fig. 6.8a and for the rotor in Fig. 6.8b. Measured additional temperature rises of the stator end winding and that of the squirrel-cage rotor winding at full load as a function of the harmonic voltage amplitude, phase shift, and frequency for the same single-phase induction motor are depicted in Fig. 6.9a and 6.9b, respectively. Note that the maximum

additional temperature rises are obtained when the peak-to-peak value of the terminal voltage is a maximum as a function of the superposed harmonic voltage with the fundamental voltage.

$$P = 2 \text{ hp}, \ V_t = 115/208 \text{ V}, \ I_t = 24/12 \text{ A},$$
$$f = 60 \text{ Hz}, \ \eta = 0.73, \ n_{rat} = 1725 \text{ rpm}. \quad (6\text{-}21)$$

Figure 6.8a,b shows that the additional harmonic losses due to subharmonic voltages and interharmonics below the fundamental become very large, even at low percentages and, therefore, the subharmonic voltages must be limited to below 0.5% of the fundamental. Note that at the slip frequency which corresponds to a slip of $s = 0$ the total stator losses reach a minimum (see Fig. 6.8a) because the total rotor losses are about zero (see Fig. 6.8b).

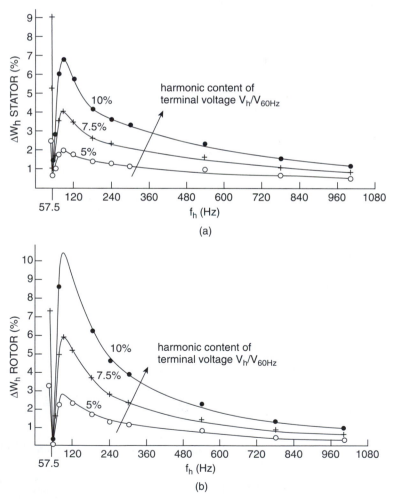

(a)

(b)

**FIGURE 6.8** Calculated total harmonic losses of (a) the stator referred to the rated losses of the stator and (b) the rotor referred to the rated losses of the rotor as a function of the harmonic frequency $h$ for the single-phase machine of Eq. 6-21 [33].

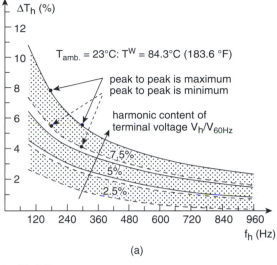

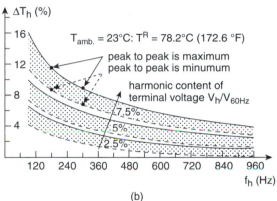

**FIGURE 6.9** (a) Measured additional temperature rise of the stator-end winding at full load as a function of the harmonic voltage amplitude, phase shift, and frequency for the single-phase machine of Eq. 6-21 [33] (referred to rated temperature rise of stator). (b) Measured additional temperature rise of the squirrel-cage rotor winding at full load as a function of the harmonic voltage amplitude, phase shift, and frequency for the single-phase machine of Eq. 6-21 [33] (referred to rated temperature rise of rotor).

### 6.3.4.2 Three-Phase Induction Motors

Figure 6.4a,b illustrates the equivalent circuit of three-phase induction machines for the fundamental and time harmonics of order $h$ (integer, sub-, and interharmonics). The total harmonic losses were calculated [18, 19] for a three-phase induction machine with the following nameplate data and equivalent circuit parameters:

$$P = 800 \text{ W}, \ p = 4 \text{ poles}, \ n_{rat} = 1738 \text{ rpm},$$
$$f = 60 \text{ Hz}, \ V_{L\text{-}L} = 220/380 \text{ V}, \ \Delta/Y \text{ connected},$$
$$I_{ph} = 2.35 \text{ A}, \ R_{s1} = 7.0 \ \Omega, \ X_{s1} = 8.0 \ \Omega,$$
$$X_{m1} = 110.0 \ \Omega, \ R'_{r1} = 4.65 \ \Omega, \ X'_{r1} = 7.3 \ \Omega. \quad (6\text{-}22)$$

Figure 6.10a shows the total harmonic stator losses referred to the total rated stator losses and Fig. 6.10b depicts the total harmonic rotor losses referred to the total rated rotor losses. In Fig. 6.10a there is a local maximum around 360 Hz; this maximum stems from the iron-core losses and its location depends on the lamination thickness of the iron-core sheets. The losses rapidly increase for subharmonics and interharmonics with decreasing frequency. The total harmonic rotor losses of Fig. 6.10b are larger for backward-rotating harmonic voltage systems (full lines) and smaller for forward-rotating harmonic systems (dashed lines). As for the stator, the rotor losses due to subharmonics and interharmonics below 60 Hz increase greatly with decreasing frequency. Note that at the slip frequency which corresponds to a slip of $s = 0$ the total stator losses reach a minimum (see Fig. 6.10a) because the total rotor losses are about zero (see Fig. 6.10b).

For the three-phase induction motor with the following nameplate data:

$$P = 2 \text{ hp}, \ q_1 = 3; \ N_o = \text{LPF}, \ f = 60 \text{ Hz},$$
$$n_{rat} = 1725 \text{ rpm}, \ V_{L\text{-}L} = 200 \text{ A}, \ I_{ph} = 7.1 \text{ A},$$
$$\text{time: continuous.} \quad (6\text{-}23)$$

the additional temperature rise of the stator end winding for forward- and backward-rotating harmonic voltage systems superimposed with a forward voltage system of fundamental frequency of 60 Hz is shown in Fig. 6.11a. Note that for subharmonic and interharmonics below 60 Hz the stator temperature increases rapidly. The corresponding temperature rises of the squirrel-cage winding of the rotor are depicted in Fig. 6.11b and 6.11c for forward- and backward-rotating harmonic voltage systems, respectively. The temperature rise due to the backward-rotating harmonic voltage system is slightly larger than that of the forward-rotating system, if the rotor temperature rise is considered. Again, the temperature rises due to sub- and interharmonics below 60 Hz are rapidly increasing with decreasing frequency.

Figures 6.10 and 6.11 illustrate that the additional temperature rises due to subharmonic and interharmonics below 60 Hz voltages become very large, even at low percentages; therefore, the subharmonic and interharmonics below 60 Hz voltage components must be limited to below 0.5% of the fundamental voltage. The sensitivity of large three-phase induction machines with respect to additional temperature rises has also been calculated in [34], and it is recommended to limit the subharmonic voltages to 0.1%.

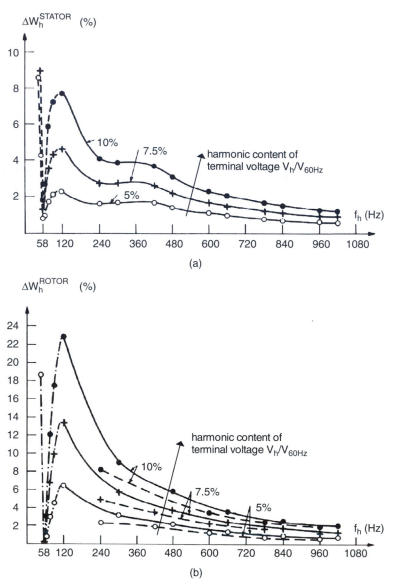

**FIGURE 6.10** (a) Calculated total harmonic stator losses referred to the total rated stator losses for the machine of Eq. 6-22 [30]. (b) Calculated total harmonic rotor losses referred to the total rated rotor losses for the machine of Eq. 6-22 [30].

## 6.4 EXPONENTS OF WEIGHTED-HARMONIC FACTORS

In references [18, 19], the functional dependencies of the additional losses and the additional temperature rises due to harmonics of the terminal voltage are calculated for linear circuits and measured for actual machines, respectively. Thereby it is assumed that the additional temperature rise is proportional to the additional losses. Inspecting these loss and temperature functions, one notes that they can be approximated by hyperbolas for $h \geq 1.0$, as shown in Fig. 6.12.

For $h < 1$ corresponding to the slip frequency where the stator field is in synchronism with the mechanical rotor no voltage $\tilde{E}$ will be induced in the rotor and the rotor currents are zero; that is, the rotor loss is zero and the total motor loss is at a minimum, as is indicated in Fig. 6.12. At very low inter- or subharmonic frequencies, say, 3 Hz, the magnetizing reactance is very small and the 3 Hz current becomes very large, resulting in large total losses and harmonic torques as discussed in Chapter 3.

**FIGURE 6.11** (a) Measured additional temperature rise of the stator end winding as a function of forward- and backward-rotating harmonic voltage systems superposed with a forward-rotating fundamental voltage system for the induction machine of Eq. 6-23 [30] (referred to rated temperature rise of stator). (b) Measured additional temperature rise of the rotor squirrel-cage winding as a function of forward-rotating harmonic voltage systems superposed with a forward-rotating fundamental voltage system for the induction machine of Eq. 6-23 [30] (referred to rated temperature rise of rotor). (c) Measured additional temperature rise of the rotor squirrel-cage winding as a function of backward-rotating harmonic voltage systems superposed with a forward-rotating fundamental voltage system for the induction machine of Eq. 6-23 [30] (referred to rated temperature rise of rotor).

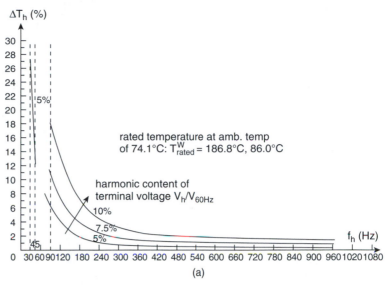

(a)

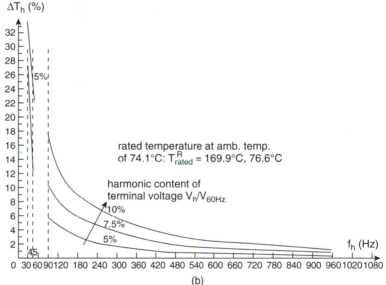

(b)

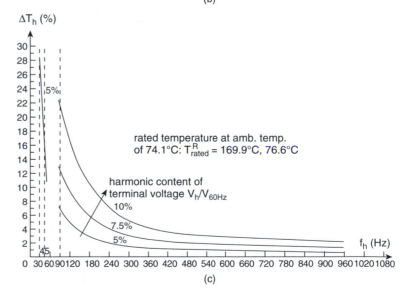

(c)

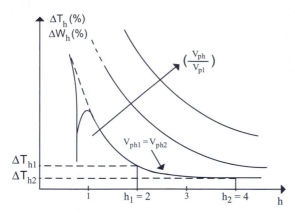

**FIGURE 6.12** Hyperbolic inverse frequency-dependent increase of harmonic losses with harmonic voltage $V_{ph}$, except at the frequency corresponding to mechanical speed (slip frequency).

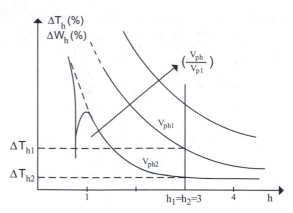

**FIGURE 6.13** Hyperbolic inverse frequency-dependent increase of harmonic losses with harmonic voltage $V_{ph}$, except at the frequency corresponding to mechanical speed (slip frequency).

**Determination of Factor k.** Knowing

$$\Delta T_h = K_1 \sum_{h=2}^{\infty} \frac{1}{(h)^k} \left( \frac{V_{ph}}{V_{p1}} \right)^l, \qquad (6\text{-}24)$$

one can determine the exponent $k$ from two given (measured) points $(h_1, \Delta T_{h1})$ and $(h_2, \Delta T_{h2})$ of Fig. 6.12. Therefore, one can write

$$k = \frac{\log \left\{ \frac{\Delta T_{h2}}{\Delta T_{h1}} \left( \frac{(V_{ph1}/V_{p1})^l}{(V_{ph2}/V_{p1})^l} \right) \right\}}{\log \left( \frac{h_1}{h_2} \right)}. \qquad (6\text{-}25)$$

From evaluation of measured points one finds the following [33]:

| | | |
|---|---|---|
| • Single- and three-phase transformers | $0.6 \leq k \leq 1.2$ | ▶ $k_{avg} = 0.90$ |
| • Single-phase induction machines | $0.5 \leq k \leq 1.2$ | ▶ $k_{avg} = 0.85$ |
| • Three-phase induction machines | $0.7 \leq k \leq 1.2$ | ▶ $k_{avg} = 0.95$ |
| • Universal machines | $0.8 \leq k \leq 1.2$ | ▶ $k_{avg} = 1.00$ |

The average values of $k = k_{avg}$ for each category are also shown.

**Determination of Factor l.** For the determination of the exponent $l$ of the weighted-harmonic factors, the functional dependencies of the additional losses and measured temperature rises on the harmonic fre-

quencies and amplitude of references [18, 19] can be used. These calculated and measured functions can be approximated by a family of hyperbolas, as shown in Fig. 6.13.

One can determine the values of the exponent $l$ from any given two points $(V_{ph1}, \Delta T_{h1})$ and $(V_{ph2}, \Delta T_{h2})$ of Fig. 6.13 for any given two harmonic frequencies of order $h_1$ and $h_2$:

$$l = \frac{\log \left\{ \frac{\Delta T_{h2}}{\Delta T_{h1}} \left( \frac{h_1}{h_2} \right)^k \right\}}{\log \left( \frac{V_{ph1}}{V_{ph2}} \right)}. \qquad (6\text{-}26)$$

From evaluation of measured points one finds the following [33]:

| | | |
|---|---|---|
| • Single- and three-phase transformers | $1.50 \leq l \leq 2.0$ | ▶ $l_{avg} = 1.75$ |
| • Single-phase induction machines | $1.0 \leq l \leq 1.80$ | ▶ $l_{avg} = 1.40$ |
| • Three-phase induction machines | $1.2 \leq l \leq 2.0$ | ▶ $l_{avg} = 1.60$ |
| • Universal machines | $1.5 \leq l \leq 2.5$ | ▶ $l_{avg} = 2.00$ |

The average values $l = l_{avg}$ for each category are also shown.

## 6.5 ADDITIONAL LOSSES OR TEMPERATURE RISES VERSUS WEIGHTED-HARMONIC FACTORS

The additional losses and temperature rises due to harmonics of the terminal voltage of transformers, induction machines, and universal machines are calculated and measured in references [18, 19]. Provided these additional losses due to such time harmonics are small as compared with the rated losses ($W_{total h\Sigma} \ll W_{total1}$), a proportionality between the additional losses and the additional temperature rises can be assumed because the cooling conditions are not significantly altered. Calculations and measurements show that the previously mentioned electromagnetic devices are sensitive to voltage harmonics in the frequency range $0 \leq f_h \leq 1500$ Hz. Harmonics of low order generate the largest additional loss. The correspondence of a given value of the weighted-harmonic factor (to an additional loss and temperature rise) leads to the additional temperature rise (or loss) versus the harmonic-factor function [20] as shown in Fig. 6.14.

With assumed values for the average exponents $k_{avg}$ and $l_{avg}$ as they apply to transformers, induction and universal machines, one can compute for given percentages of the harmonic voltages the weighted-harmonic factor and associate with it the additional losses or temperature rises.

### 6.5.1 Application Example 6.1: Temperature Rise of a Single-Phase Transformer Due to Single Harmonic Voltage

Determine the temperature rise $\Delta T_h$ of a single-phase transformer provided $V_3 = 0.10$ pu = 10%,

$T_{amb} = 40°C$, and $T_{rated} = 100°C$. Assume $k_{avg} = 0.90$ and $l_{avg} = 1.75$.

### 6.5.2 Application Example 6.2: Temperature Rise of a Single-Phase Induction Motor Due to Single Harmonic Voltage

Determine the temperature rise $\Delta T_h$ of a single-phase induction motor provided $V_3 = 0.10$ pu $\equiv 10\%$, $T_{amb} = 40°C$, and $T_{rated} = 100°C$. Assume $k_{avg} = 0.85$ and $l_{avg} = 1.40$.

Transformers, induction machines, and universal machines have different loss or temperature sensitivities with respect to voltage harmonics and, therefore, the additional temperature rises are different for all five types of devices. The additional temperature rise (loss) versus weighted-harmonic factor function indicates that the most sensitive components are single-phase induction machines, whereas the least sensitive devices are transformers with resistive load and universal machines. This is so because for any harmonic terminal voltage, a single-phase machine (and to some degree an unbalanced three-phase induction machine) develops forward- and backward-rotating harmonic fields and responds like being under short-circuit conditions due to the large slip $s_h$ of the harmonic field with respect to the rotating rotor. In transformers the resistive and inductive loads can never have zero impedance due to the nature of the frequency dependency of such loads.

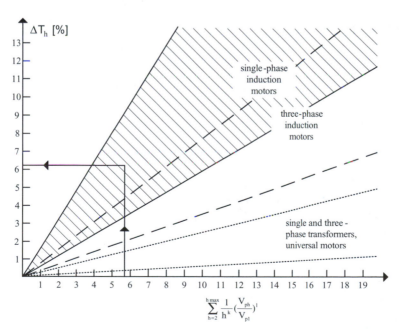

**FIGURE 6.14** Additional temperature rise (or loss) versus weighted-harmonic factor function for universal motors, single- and three-phase transformers, and induction motors [33].

## 6.6 ARRHENIUS PLOTS

A series of studies have investigated the behavior of various electric machines and transformers as they occur in a residential distribution system and are exposed to harmonics of the terminal voltage [18, 19]. The influence of such harmonics expresses itself, among others (e.g., mechanical vibration), in an elevated (additional) temperature rise of the machine windings and iron cores. The question arises how the lifetime of the machines will be affected by such additional temperature rises.

**Thermal Aging.** The insulating material of an electric apparatus as used in electrical appliances is of organic or inorganic origin. Due to the heating of these materials, caused by the loss within the machine, a deterioration of the insulating materials will occur. This deterioration is manifested either by

- lowering of the mechanical strength, and/or
- changing of the dielectric behavior of the insulating material.

It may be mentioned that not only the heat itself, but also small motions due to expansion and contraction of the wire and iron laminations, are causing deteriorations by mechanical friction. In addition, time harmonics induce small core vibrations that may aggravate the mechanical stresses. The mechanical failure of the insulation is a result of the decrease of the tensile strength or flexibility of the insulation material. The thermal lifetime of electric machines and transformers is highly dependent on the mode of utilization; there is no doubt that machines with frequently variable load are more prone to aging from a mechanical failure point of view.

All further investigations consider machines or transformers operating at constant rated load, where the chemical changes of the insulating materials are responsible only for thermal aging.

## 6.7 REACTION RATE EQUATION

For many years it has been recognized that thermal degradation of organic or inorganic materials can be best represented by the reaction rate equation [24, 25]

$$\frac{dR}{dt} = Ae^{(-E/KT)}. \qquad (6\text{-}27)$$

In this equation, $dR/dt$ is the reduction in property $R$ with respect to time, $A$ is a constant of integration,

$K$ is the gas constant or, depending on the units, the Boltzmann constant, $T$ is the absolute temperature in kelvins, and $E$ is the activation energy of the aging reaction (large $E$ leads to fast aging, small $E$ leads to slow aging).

Equation 6-27 expresses the rule of chemical reactions which was derived by Svante Arrhenius in 1880. The form of the original Arrhenius formula can be obtained from the differential equation by integration as follows:

$$dR = Ae^{-(E/KT)}dt,$$
$$\int dR = A\int e^{-(E/KT)}dt,$$
$$R = Ae^{-(E/KT)}dt,$$
$$\frac{R}{t} = Ae^{-(E/KT)},$$
$$\ln(R/t) = \ln(Ae^{-(E/KT)}),$$
$$\ln R - \ln t = \ln A - E/KT,$$

or

$$\ln t = \left(\frac{E}{K}\right)\frac{1}{T} + B. \qquad (6\text{-}28)$$

The plot $\ln t$ versus $1/T$ is a straight line with the slope $E/K$, if for the lifetime $t$ a logarithmic scale is employed, as shown in Fig. 6.15.

Equation 6-28 expresses that the logarithm of the degradation (decreased life) time $t$ is proportional to the activation energy $E$: this energy is transmitted in the form of heat to the chemical reaction or decomposition. Different kinds of materials need a different amount of heat energy to arrive at the same degree of deterioration. To obtain these data in the case of insulating materials, extensive experimental research has been done [24]. In this reference, one can find not only a list of the activation energies, but the frequency distribution of the activation energies

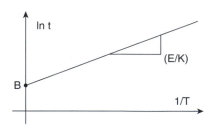

**FIGURE 6.15** Lifetime $t$ versus the inverse of the absolute temperature $T$.

of various materials. Properties monitored for these materials include flexural strength, impact strength, and dielectric strength. In Fig. 4 of [24], one notes the number of materials for each increment of 0.1 eV activation energy (see Fig. 6.16). The peak of this distribution curve occurs at about $E = 1.15$ eV.

The practical use of the Arrhenius plot and the consequences of the data of Fig. 4 of [24] will be discussed in the next section.

## 6.8 DECREASE OF LIFETIME DUE TO AN ADDITIONAL TEMPERATURE RISE

The slope of the Arrhenius plot based on

$$\ln t = \left(\frac{E}{K}\right)\frac{1}{T} + B \qquad (6\text{-}29)$$

is the most important quantity (proportional to the activation energy $E$) for our investigation; this is the only quantity that determines the aging of the insulation material. Therefore, we have to use the slope of the Arrhenius plot as a measure of aging.

If one knows from experiments two (different) points of the plot belonging to temperatures $T_1$ and $T_2$ and lifetimes $t_1$ and $t_2$, respectively, that is, $(T_1, t_1)$ and $(T_2, t_2)$, then one can obtain for a given insulation material

$$\ln t_1 - \ln t_2 = \left(\frac{E}{K}\right)\left(\frac{1}{T_1} - \frac{1}{T_2}\right). \qquad (6\text{-}30)$$

It must be noted that if the activation energy $E$ is measured in electron volts one has to use for $K$ the Boltzmann constant: $K = 1.38 \cdot 10^{-23}$ J/Kelvin; in case the activation energy is given in kilocalories per mole the constant $K$ will be the gas constant $K = R = 19.84 \cdot 10^{-4}$ kcal/mol. Note that $1$ eV $= 1.602 \cdot 10^{-19}$ J or $1$ J $= 0.624 \cdot 10^{19}$ eV.

Suppose one knows the rated lifetime $t_2$ of an apparatus and its rated (constant) temperature $T_2$ at which it is operating. The question arises to what extent the lifetime decreases provided the elevated temperature becomes $T_1 = T_2 + \Delta T$ where $\Delta T = \Delta T_h$. To answer this question one has to substitute in Eq. 6-30 $T_1$ by $T_2 + \Delta T$:

$$\ln t_1 - \ln t_2 = \left(\frac{E}{K}\right)\left(\frac{1}{T_2 + \Delta T} - \frac{1}{T_2}\right). \qquad (6\text{-}31)$$

After some manipulations, the new decreased lifetime is

$$t_1 = t_2 e^{-\left(\frac{E}{K}\right)\frac{\Delta T}{T_2(T_2 + \Delta T)}}, \qquad (6\text{-}32)$$

where $t_2$ is the rated lifetime, $T_2$ is the rated temperature in kelvins, and $\Delta T$ is the (additional) temperature rise in degrees Celsius. Several examples will illustrate the use of the above relation as applied to the calculation of the decrease of the lifetime due to the additional temperature rise caused by the additional harmonic losses. In all following examples it is assumed that the rated lifetime of the apparatus is $t_2 = 40$ years, and the steady-state rated temperature of the hottest spot is $T_2 = 100°C \equiv 273 + 100 = 373$ °Kelvin.

### 6.8.1 Application Example 6.3: Aging of a Single-Phase Induction Motor with E = 0.74 eV Due to a Single Harmonic Voltage

Determine for an activation energy of $E = 0.74$ eV the slope $E/K$ and for the additional temperature rise $\Delta T_{h=3} = 6.6°C$ (see Application Example 6.2 with $T_{amb} = 40°$, $T_2 = T_{rat} = 100°C$) the reduced lifetime of a single-phase induction motor.

### 6.8.2 Application Example 6.4: Aging of a Single-Phase Induction Motor with E = 0.51 eV Due to a Single Harmonic Voltage

Determine for an activation energy of $E = 0.51$ eV the slope $E/K$ and for the additional temperature rise $\Delta T_{h=3} = 6.6°C$ (see Application Example 6.2 with $T_{amb} = 40°$, $T_2 = T_{rat} = 100°C$) the reduced lifetime of a single-phase induction motor.

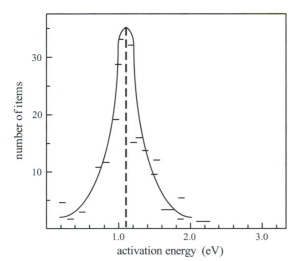

**FIGURE 6.16** Frequency distribution of activation energies of various organic and inorganic materials [24].

## 6.9 REDUCTION OF LIFETIME OF COMPONENTS WITH ACTIVATION ENERGY E = 1.1 EV DUE TO HARMONICS OF THE TERMINAL VOLTAGE WITHIN RESIDENTIAL OR COMMERCIAL UTILITY SYSTEMS

Figure 6.14 shows the additional temperature rises (or losses) in percent of the rated temperature rises (or losses) as a function of the weighted-harmonic factor. These functions are obtained from calculations and measurements [18, 19]. Note that the ambient temperature has been eliminated. For the evaluation of the reduction of the lifetime one must base all calculations on an activation energy of about $E = 1.1$ eV because of Fig. 6.16, which confirms that the majority of (insulation) materials have such a distinct activation energy.

According to Fig. 6.14, a weighted harmonic-voltage factor of $\sum_{h=2}^{h\max} \frac{1}{h^k}\left(\frac{V_{ph}}{V_{p1}}\right)^l = 5.8$ corresponds to additional temperature rises (referred to the rated temperature rises) on the average of 6.2% for single-phase induction machines, 3.2% for three-phase induction machines, and 0.85% for transformers and universal machines. These percentage values confirm that single-phase machines are very sensitive to harmonic voltages due to their forward- and backward-rotating fields and their short-circuited rotor, three-phase induction machines are sensitive to voltage harmonics due to their short-circuited rotor, transformers are not very sensitive to voltage harmonics because of their resistive load, and universal machines are not sensitive as well because the commutator transforms a voltage source to a current source where the current magnitudes are limited. A weighted-harmonic voltage factor of 5.8 results for the rated temperature $T_2 = 85°C$ at $T_{amb60Hz} \approx T_{ambh} = 23°C$ in temperature increases of 3.84°C for single-phase induction machines, 1.98°C for three-phase induction machines, and 0.53°C for transformers and universal machines. With these additional temperature increases one obtains at an activation energy of $E = 1.1$ eV – with $E/K = 12,769$ Kelvin – the decreased lifetime (using Eq. 6-32) of 31.5% for single-phase induction machines, 18% for three-phase induction machines, and 5% for transformers and universal machines.

**Conclusion.** It is believed that the weighted-harmonic voltage factor for single-phase and three-phase induction motors in the neighborhood of 5.8 represents a compromise which on the one hand promotes the installation of solid-state circuits by choosing generous permissible harmonic voltage levels, and on the other hand avoids severe detrimental reactions on the majority of the residential or commercial power system components including loads. Harmonic spectra of the residential or commercial power system voltage that satisfy for single-phase and three-phase induction motors

$$\sum_{h=2}^{h\max} \frac{1}{h^k}\left(\frac{V_{ph}}{V_{p1}}\right)^l \approx 5.8. \tag{6-33}$$

result in acceptable temperature rises as far as induction motors, transformers, and universal machines are concerned. Thus it is recommended not to rigidly fix the harmonic voltages but rather the additional temperature rises they generate.

## 6.10 POSSIBLE LIMITS FOR HARMONIC VOLTAGES

In order to illustrate the use of Eqs. 6-33 and 6-32, the harmonic voltage spectra for a single-phase and for a three-phase feeder (see Table 6.1) are proposed [33].

With average values of $k_{avg} = 0.85$ and $l_{avg} = 1.4$ for single-phase induction motors, $k_{avg} = 0.95$ and $l_{avg} = 1.6$ for three-phase induction motors, $k_{avg} = 0.90$ and $l_{avg} = 1.75$ for single-phase and three-phase transformers, and $k_{avg} = 1.0$ and $l_{avg} = 2.0$ for universal machines, one obtains for the single-phase spectrum of Table 6.1 the weighted harmonic-voltage factor

for single-phase induction motors $\sum_{h=2}^{h\max} \frac{1}{h^k}\left(\frac{V_{ph}}{V_{p1}}\right)^l \approx 5.7$,

for the three-phase spectrum of Table 6.1 for three-phase induction motors $\sum_{h=2}^{h\max} \frac{1}{h^k}\left(\frac{V_{ph}}{V_{p1}}\right)^l \approx 5.7$,

for single-phase transformers $\sum_{h=2}^{h\max} \frac{1}{h^k}\left(\frac{V_{ph}}{V_{p1}}\right)^l \approx 7.37$,

for three-phase transformers $\sum_{h=2}^{h\max} \frac{1}{h^k}\left(\frac{V_{ph}}{V_{p1}}\right)^l \approx 7.4$,

and for universal motors $\sum_{h=2}^{h\max} \frac{1}{h^k}\left(\frac{V_{ph}}{V_{p1}}\right)^l \approx 8.5$. Any other set of voltage harmonics is feasible if Eq. 6-33 is about satisfied. Based on Fig. 6.14 (with $T_2 = 85°C$, $T_{amb} = 23°C$, $E = 1.1$ eV, and rated lifetime of $t_2 = 40$ years), those harmonic factors result in the additional temperature rises and lifetime reductions of Table 6.2.

## 6.10.1 Application Example 6.5: Estimation of Lifetime Reduction for Given Single-Phase and Three-Phase Voltage Spectra with High Harmonic Penetration with Activation Energy E = 1.1 eV

Estimate the lifetime reductions of induction machines, transformers, and universal machines for the single- and three-phase voltage spectra of Table E6.5.1 and their associated lifetime reduction for an activation energy of $E = 1.1$ eV. The ambient temperature is $T_{amb} = 23°C$, the rated temperature is $T_2 = 85°C$, and the rated lifetime of $t_2 = 40$ years can be assumed.

## 6.10.2 Application Example 6.6: Estimation of Lifetime Reduction for Given Single-Phase and Three-Phase Voltage Spectra with Moderate Harmonic Penetration with Activation Energy E = 1.1 eV

Estimate the lifetime reductions for induction machines, transformers, and universal machines for the single- and three-phase voltage spectra of Table E6.6.1 and their associated lifetime reduction for an activation energy of $E = 1.1$ eV. The ambient temperature is $T_{amb} = 23°C$, the rated temperature is $T_2 = 85°C$, and the rated lifetime of $t_2 = 40$ years can be assumed.

**TABLE 6.1** Possible Voltage Spectra of Single-Phase and Three-Phase Feeders

| $h$ | $\left(\dfrac{V_h}{V_{60Hz}}\right)_{1\Phi}$ (%) | $\left(\dfrac{V_h}{V_{60Hz}}\right)_{3\Phi}$ (%) |
|---|---|---|
| 1 | 100 | 100 |
| 2 | 0.5 | 0.5 |
| 3 | 4.0 | 2.0[a] |
| 4 | 0.3 | 0.5 |
| 5 | 3.0 | 5.0 |
| 6 | 0.2 | 0.2 |
| 7 | 2.0 | 3.5 |
| 8 | 0.2 | 0.2 |
| 9 | 1.0 | 0.3 |
| 10 | 0.1 | 0.1 |
| 11 | 1.5 | 1.5 |
| 12 | 0.1 | 0.1 |
| 13 | 1.5 | 1.0 |
| 14 | 0.1 | 0.05 |
| 15 | 0.5 | 0.1 |
| 16 | 0.05 | 0.05 |
| 17 | 1.0 | 0.5 |
| 18 | 0.05 | 0.01 |
| 19 | 1.0 | 0.5 |
| | All higher harmonics < 0.5% | |

[a] Under certain conditions (e.g., DC bias of transformers as discussed in Chapter 2, and the harmonic generation of synchronous generators as outlined in Chapter 4) triplen harmonics are not of the zero-sequence type and can therefore exist in a three-phase system.

**TABLE E6.5.1** Possible Voltage Spectra with High-Harmonic Penetration

| $h$ | $\left(\dfrac{V_h}{V_{60Hz}}\right)_{1\Phi}$ (%) | $\left(\dfrac{V_h}{V_{60Hz}}\right)_{3\Phi}$ (%) |
|---|---|---|
| 1 | 100 | 100 |
| 2 | 2.5 | 0.5 |
| 3 | 5.71 | 1.0 |
| 4 | 1.6 | 0.5 |
| 5 | 1.25 | 7.0 |
| 6 | 0.88 | 0.2 |
| 7 | 1.25 | 5.0 |
| 8 | 0.62 | 0.2 |
| 9 | 0.96 | 0.3 |
| 10 | 0.66 | 0.1 |
| 11 | 0.30 | 2.5 |
| 12 | 0.18 | 0.1 |
| 13 | 0.57 | 2.0 |
| 14 | 0.10 | 0.05 |
| 15 | 0.10 | 0.1 |
| 16 | 0.13 | 0.05 |
| 17 | 0.23 | 1.5 |
| 18 | 0.22 | 0.01 |
| 19 | 1.03 | 1.0 |
| | All higher harmonics < 0.2% | |

**TABLE 6.2** Additional Temperature Rise and Associated Lifetime Reduction of Induction Motors, Transformers, and Universal Motors Due to the Harmonic Spectra of Table 6.1

| | Single-phase induction motors | Three-phase induction motors | Single-phase transformers | Three-phase transformers | Universal motors |
|---|---|---|---|---|---|
| $\Delta T_h$ (%) | 6.2 | 3.2 | 1.2 | 1.2 | 1.3 |
| $\Delta T_h$ (°C) | 3.84 | 1.98 | 0.74 | 0.74 | 0.81 |
| Lifetime reduction (%) | 31.5 | 17.8 | 7.1 | 7.1 | 7.7 |

**TABLE E6.6.1** Possible Voltage Spectra with Moderate-Harmonic Penetration

| $h$ | $\left(\dfrac{V_h}{V_{60\,Hz}}\right)_{1\Phi}$ (%) | $\left(\dfrac{V_h}{V_{60\,Hz}}\right)_{3\Phi}$ (%) |
|---|---|---|
| 1 | 100 | 100 |
| 2 | 0.5 | 0.5 |
| 3 | 3.0 | 0.5 |
| 4 | 0.3 | 0.5 |
| 5 | 2.0 | 3.0 |
| 6 | 0.2 | 0.2 |
| 7 | 1.0 | 2.5 |
| 8 | 0.2 | 0.2 |
| 9 | 0.75 | 0.3 |
| 10 | 0.1 | 0.1 |
| 11 | 1.0 | 1.0 |
| 12 | 0.1 | 0.1 |
| 13 | 0.9 | 0.85 |
| 14 | 0.1 | 0.05 |
| 15 | 0.3 | 0.1 |
| 16 | 0.05 | 0.05 |
| 17 | 0.5 | 0.3 |
| 18 | 0.05 | 0.01 |
| 19 | 0.4 | 0.2 |
| All higher harmonics < 0.2% | | |

## 6.11 PROBABILISTIC AND TIME-VARYING NATURE OF HARMONICS

As mentioned in Section 6.1, the harmonic current or voltage spectra within a distribution feeder are continually changing as a function of changes in the load and the system's condition. The rationale given for using the worst-case conditions, where loads are operating for at least a few hours at rated operation generating the "rated current/voltage spectra," is the basis for the IEEE [35] and IEC [36] harmonic guidelines and standards, respectively. Nevertheless it will be worthwhile to study the time-varying nature [37, 38] of harmonic spectra. In these publications continually changing current (THD$_i$) and voltage (THD$_v$) total harmonic distortions are plotted as a function of time. They indeed change widely from 2.5% to about 9% for THD$_i$, and from 1% to about 5% for the THD$_v$ for one site, and for another site the changes are similar in magnitude. As mentioned in prior sections, the THDs are neither a good measure for temperature rises nor for vibrations; instead harmonic components should be used in any further studies. The measured THD data are characterized by statistical measures (e.g., minimum, maximum, average or mean values, standard deviation, probability distribution), histograms or probability density functions (pdf), probability distribution functions ($PX(x)$), statistical description of sub-time

intervals, and combined deterministic or statistical description [41–45]. These publications address the harmonic summation and propagation by explaining the sum of random harmonic phasors via the representation of harmonic phasors, marginal pdf of $x$–$y$ components, pdf of sum of projections of independent phasors, and pdf of magnitude of sum of random phasors. These papers apply the methods described to a 14-bus transmission system and provide general guidelines with respect to their applicability. The probabilistic evaluation of the economic cost due to harmonic losses in industrial energy systems is presented in the next section.

## 6.12 THE COST OF HARMONICS

The cost of harmonics can originate either in

- the complicated solid-state components necessary to maintain the current or voltage harmonics at a low level, for example, switched-mode power supplies operating at unity-power factor [39], or in
- the use of simple peak-rectifiers resulting in high harmonic amplitudes [40].

Both costs are not negligibly small and must be considered in the future. The first approach appears to be favored by the IEC [36] for low-, medium-, and high-voltage power systems, whereas the latter appears to be favored by IEEE [35] for single-phase systems only.

In references [41–45] the cost of the latter approach is discussed by defining the economic damage (cost) due to harmonic losses $D = D_w + D_a$, where $D_w$ is the operating cost and $D_a$ is the aging cost. This economic damage is a probabilistic quantity because the harmonic voltage or current spectra change in a random manner. The expected value of these costs is defined as $E(D) = E(D_w) + E(D_a)$, where $E(D_w)$ is the present-worth expected value of the operating costs due to harmonic losses and $E(D_a)$ is the present-worth expected value of the aging costs due to harmonic losses. Numerical results for a 20 kV, 150 MVA industrial feeder illustrate that the economic damage for induction motors is significant and within a lifetime of 35 years amounts to about the purchase cost ($7,625) of an induction motor.

## 6.13 TEMPERATURE AS A FUNCTION OF TIME

Rotating machines, transformers, and inductors must be designed so that their rated temperatures will not be exceeded. For rotating machines the maximum torques are not a design criterion as long as the machines continue to operate. The maximum permissible temperature (hot spot) must not be exceeded, otherwise the lifetime will be reduced. Note that any machine works most efficiently and cost-effectively if the permissible (maximum) rated temperature will be reached but not exceeded at any time. At steady state machines must be able to operate at rated torque. For a short time machines can be operated above rated torque. This leads to the concept of intermittent operation discussed in a later section.

A motor can be replaced by a radiating sphere as illustrated in Fig. 6.17. During a small time increment $dt$ one can assume that the temperature of the sphere increases by the incremental temperature $d\theta$. For this reason the change in stored heat (increase) $dQ_C$ during the time $dt$ is

$$dQ_C = C \cdot d\theta, \qquad (6\text{-}34)$$

where $\theta$ is the temperature rise and $C$ is the heat-absorption capacity. Note that Eq. 6-34 is independent of time.

Due to thermal radiation a part of the stored heat will be emitted to the surrounding environment. The change of the stored heat (reduction) $dQ_A$ is time dependent:

$$dQ_A = A \cdot \theta \cdot dt, \qquad (6\text{-}35)$$

where $A$ is the heat-radiation capacity.

The sum of Eqs. 6-34 and 6-35 yields the ordinary differential equation

$$dQ = dQ_C + dQ_A = C \cdot d\theta + A \cdot \theta \cdot dt = P_{\text{loss}} \cdot dt, \qquad (6\text{-}36)$$

where $P_{\text{loss}} \cdot dt$ is the (loss) energy absorbed by the machine during the incremental time $dt$. Dividing Eq. 6-36 by $dt$ and $A$, one obtains the first-order ordinary differential equation

$$\frac{P_{\text{loss}}}{A} = \frac{C}{A} \cdot \frac{d\theta}{dt} + \theta = \tau_\theta \cdot \frac{d\theta}{dt} + \theta, \qquad (6\text{-}37)$$

where $\tau_\theta = (C/A)$ is the thermal time constant. The solution of this first-order differential equation is

$$\theta = \frac{P_{\text{loss}}}{A} + k \cdot e^{(-t/\tau_\theta)}. \qquad (6\text{-}38)$$

For $t = 0$ and $\theta = \theta_o$ (initial temperature rise) and for $t = \infty$ with $\theta = \theta_f = P_{\text{loss}}/A$ (final temperature rise) one obtains for the temperature as a function of time

$$\theta = \theta_f - (\theta_f - \theta_o) \cdot e^{(-t/\tau_\theta)}. \qquad (6\text{-}39)$$

Figure 6.18 illustrates the solution of Eq. 6-39 including its time constant. The temperature transient during a load cycle is schematically depicted in Fig. 6.19 with $\theta_f = P_{\text{loss}}/A$.

### 6.13.1 Application Example 6.7: Temperature Increase of Rotating Machine with a Step Load

An enclosed fan-cooled induction motor has a thermal time constant of $\tau_\theta = 3\,h$ and a steady-state rated temperature of $\theta_f = 120°C$ at an ambient temperature of $\theta_{\text{amb}} = 40°C$. The motor has at time $t = 0_-$ a temperature of $\theta = \theta_{\text{amb}}$, and the motor is fully loaded at $t = 0_+$. Calculate the time $t_{95\%}$ when the fully loaded motor reaches 95% of its final temperature $\theta_f = 120°C$.

### 6.14 VARIOUS OPERATING MODES OF ROTATING MACHINES

Depending on their applications, rotating machines can be subjected to various operating modes such as steady-state, short-term, steady-state with short-term, intermittent, and steady-state with intermittent operating modes.

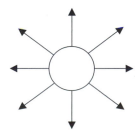

**FIGURE 6.17** Rotating machine represented by radiating sphere.

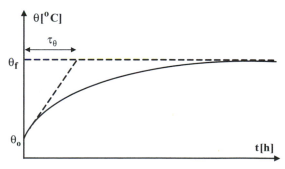

**FIGURE 6.18** Solution of Eq. 6-39 including its time constant.

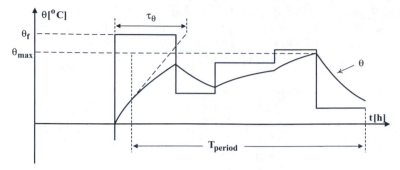

**FIGURE 6.19** Temperature transient during a load cycle.

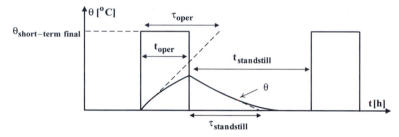

**FIGURE 6.20** Transient temperature for short-term operation.

## 6.14.1 Steady-State Operation

Steady-state temperature is reached when the operating time of the machine $t_{oper}$ is large as compared with the time constant $\tau_\theta$ of the machine. That is,

$$t_{oper} \geq (3\text{--}4) \cdot \tau_\theta. \qquad (6\text{-}40)$$

According to experience the thermal time constant for openly ventilated machines is

$$\tau_\theta \approx 1 \text{ h} \qquad (6\text{-}41)$$

and for enclosed but ventilated machines

$$\tau_\theta \approx (3\text{--}4) \text{ h}. \qquad (6\text{-}42)$$

During steady-state operation the rated output power must be delivered by the machine.

## 6.14.2 Short-Term Operation

For short-term operation one assumes that the machine cools down to the ambient temperature and its temperature rise (initial temperature rise) is $\theta_o = 0$. Fig. 6.20 illustrates the transient temperature of this operating mode.

The absence of any ventilation during standstill (time $t_{standstill}$) requires that the time constant during standstill be longer than that during operation (time $t_{oper}$), that is, $\tau_{standstill} > \tau_{oper}$. For short-term operation the times $t_{oper}$ and $t_{standstill}$ must relate to the time constants as follows:

$$t_{oper} < (3\text{--}4)\ \tau_{oper}, \qquad (6\text{-}43)$$

$$t_{standstill} < (3\text{--}4)\ \tau_{standstill}. \qquad (6\text{-}44)$$

The transient temperature is for short-term operation (see Fig. 6.20):

$$\theta_{short\text{-}term} = \theta_{rated} = \theta_{short\text{-}term\_final}\left(1 - e^{\left(-\frac{t_{oper}}{\tau_{oper}}\right)}\right). \qquad (6\text{-}45)$$

The final short-term temperature is obtained from Eq. 6-45:

$$\theta_{short\text{-}term\_final} = \frac{\theta_{rated}}{\left(1 - e^{\left(-\frac{t_{oper}}{\tau_{oper}}\right)}\right)}. \qquad (6\text{-}46)$$

with

$$\theta_{short\text{-}term\_final} \propto P_{loss\_short\text{-}term} \qquad (6\text{-}47)$$

and

$$\theta_{rated} \propto P_{loss\_rated}. \qquad (6\text{-}48)$$

It follows for the losses during short-term operation:

$$P_{\text{loss\_short-term}} = \frac{P_{\text{loss\_rated}}}{\left(1 - e^{\left(-\frac{t_{\text{oper}}}{\tau_{\text{oper}}}\right)}\right)}. \qquad (6\text{-}49)$$

For machines whose losses consist of iron-core and copper losses (e.g., induction machines) the total losses at rated operation are

$$P_{\text{loss\_rated}} = p \cdot P_{\text{rated}} + q \cdot P_{\text{rated}}. \qquad (6\text{-}50)$$

The first term of Eq. 6-50 pertains to the iron-core and the second term to the copper losses. The losses during short-term operation are

$$P_{\text{loss\_short-term}} = p \cdot P_{\text{rated}} + q \cdot P_{\text{rated}} \left(\frac{P_{\text{loss\_short-term}}}{P_{\text{rated}}}\right)^2. \qquad (6\text{-}51)$$

Introducing Eqs. 6-50 and 6-51 into Eq. 6-49 yields the ratio

$$\frac{P_{\text{loss\_short-term}}}{P_{\text{rated}}} = \sqrt{\left(1 + \frac{p}{q}\right)\frac{1}{1 - e^{\left(-\frac{t_{\text{oper}}}{\tau_{\text{oper}}}\right)}} - \frac{p}{q}}, \qquad (6\text{-}52)$$

where the ratio $p/q$ is available from the manufacturer of the machine.

### 6.14.3 Steady State with Short-Term Operation

In this case the short-term load is superposed with the steady-state load (see Fig. 6.21). The operating time $t_{\text{oper}}$ and the pause time $t_{\text{pause}}$ relate to the time constant as follows:

$$t_{\text{oper}} < (3\text{--}4) \cdot \tau_\theta, \qquad (6\text{-}53)$$

$$t_{\text{pause}} > (3\text{--}4) \cdot \tau_\theta, \qquad (6\text{-}54)$$

and the ratio between the required power during steady-state with short-term load to the rated power is

$$\frac{P_{\text{steady-state+short-time}}}{P_{\text{rated}}} = \sqrt{\frac{1}{(1 - e^{\left(\frac{t_{\text{oper}}}{\tau_\theta}\right)})}}. \qquad (6\text{-}55)$$

In Eq. 6-55 the iron-core losses are neglected $(p = 0)$.

### 6.14.4 Intermittent Operation

The mode of intermittent operation occurs most frequently. One can differentiate between two cases:

- irregular load steps (see Fig. 6.22), and
- regular load steps.

The case with irregular load steps is discussed in Application Example 6.8.

The case with regular load steps can be approximated as follows based on Fig. 6.23:

$$\frac{P_{\text{int ermittent}}}{P_{\text{rated}}} = \sqrt{\left(1 + \frac{p}{q}\right)\left(1 + \frac{t_{\text{pause}}}{t_{\text{oper}}} \cdot \frac{\tau_\theta}{\tau_{\text{standstill}}}\right) - \frac{p}{q}}. \qquad (6\text{-}56)$$

### 6.14.5 Steady State with Intermittent Operation

Steady state with superimposed periodic intermittent operation (see Fig. 6.24) does not occur frequently. The ratio between the power required for this case and the rated power is

$$\frac{P_{\text{steady-state+periodic int ermittent}}}{P_{\text{rated}}} = \sqrt{\left(1 + \frac{t_{\text{pause}}}{t_{\text{oper}}}\right)} \qquad (6\text{-}57)$$

for $t_{\text{oper}} < (3\text{--}4)\tau_\theta$ and $t_{\text{pause}} < (3\text{--}4)\tau_\theta$. In summary one can state that for the motor sizing the maximum temperature (and not the torque) is important.

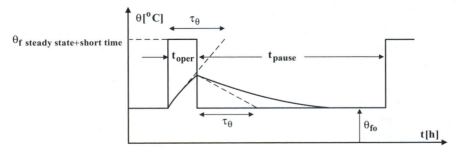

**FIGURE 6.21** Transient temperature for steady-state load superposed with short-term load.

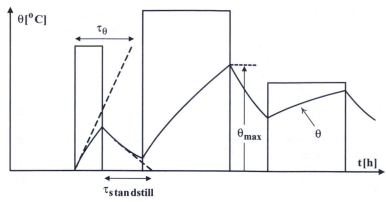

**FIGURE 6.22** Intermittent operation with irregular load steps.

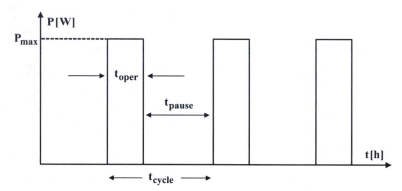

**FIGURE 6.23** Intermittent operation with regular load steps.

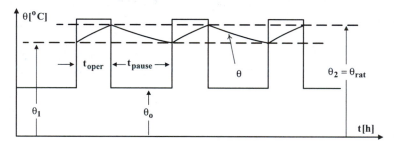

**FIGURE 6.24** Steady state with superimposed periodic intermittent operation with regular load steps.

Intermittent operation occurs in drives such as used in hybrid cars, wind-power plants, air-conditioning and refrigeration systems, and others. For this reason a detailed analysis of steady state with superposed intermittent operation will be presented in Application Example 6.8.

### 6.14.6 Application Example 6.8: Steady State with Superimposed Periodic Intermittent Operation with Irregular Load Steps

A three-phase, squirrel-cage induction motor is fed by a voltage-source inverter. The induction motor is operated at variable frequency and constant (rated) flux (that is, $|\tilde{E}|/f =$ constant) and has the following nameplate data: $V_{L-L} = 460$ V, $f = 60$ Hz, $p = 4$ poles, $n_{m\_rat} = 1720$ rpm, $P_{out\_rat} = 29.594$ kW or 39.67 hp. The stator winding is Y-connected, and the parameters per phase are $R_s = 0.5\ \Omega, R'_r = 0.2\ \Omega, X_s = X'_r = 1\ \Omega$, $X_m = 30\ \Omega$ at $f = 60$ Hz, and $R_{fe} \rightarrow \infty$. The axial moment of inertia of the motor is $J_m = 0.234$ kgm$^2$, the viscous damping coefficient $B$ can be assumed to be zero, and the inertia of the load referred to the motor shaft is $J_{load} = 5$ kgm$^2$. The steady-state/intermittent load cycle is shown in Fig. E6.8.1.

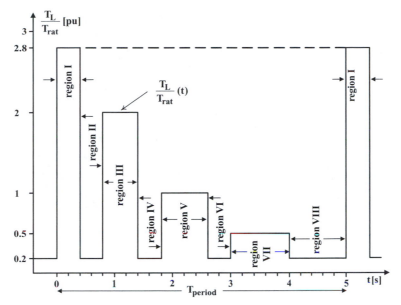

**FIGURE E6.8.1** Steady-state/intermittent load cycle.

The induction motor is operated at 60 Hz (natural torque-speed characteristic).

a) Determine the rated synchronous speed, rated synchronous angular velocity, rated angular velocity, rated slip, rated stator current $|\tilde{I}_s|$, rated rotor current $|\tilde{I}'_r|$, and rated induced voltage $|\tilde{E}|$.

b) Calculate the maximum torque $T_{max}$ at $s_m$ and the fictitious maximum torque $T_{max\_fict}$ at $s = 1$.

c) Derive from the equation of motion the slip $s(t)$ as a function of the initial slip $s(0)$, the load torque $T_L(t)$, the maximum fictitious torque $T_{max\_fict}$, the time constant $\tau_m$, and the time $t$.

d) Apply the solution for $s(t)$ to the different regions of Fig. E6.8.1 and plot $s(t)$ from $t = 0$ to the maximum time $t = 5$ s of the load-cycle time period of $T_{period} = 5$ s.

e) Calculate the torque $T(t)$ and plot it from $t = 0$ to $t = 5$ s.

f) Based on the slip function compute pointwise the stator current $|\tilde{I}_s(t)|$ and plot $|\tilde{I}_s(t)|$ from $t = 0$ to $t = 5$ s.

g) Plot $|\tilde{I}_s(t)|^2$ as a function of time from $t = 0$ to $t = 5$ s.

h) Determine the rms value of the motor (stator) current $i_s(t)$ for the entire load cycle as specified in Fig. E6.8.1; that is, determine

$$I_{s\_rms} = \sqrt{\frac{1}{T} \int_t^{t+T} |\tilde{I}_s(t)|^2 dt}.$$

i) Is this induction motor over- or underdesigned? If it is underdesigned, what is the reduction of lifetime?

### 6.14.7 Reduction of Vibrations and Torque Pulsations in Electric Machines

Vibrations and torque pulsations generate electromagnetic forces that mainly act on the end turns of stator windings. As it is well known the turns residing in the stator slots are not exposed to high magnetic fields and for this reason the magnetic forces acting on the turns within the slots are relatively small. Besides increased temperatures, vibrations and torque pulsations are a major reason for the deterioration of the insulation materials. Vibrations and pulsating torques can have their origin in motor asymmetries (e.g., rotor eccentricities, interturn faults, cogging due to permanent magnets, and a poor selection of stator and rotor slotting [47]) and loads (e.g., piston compressors). For this reason it will be important to demonstrate how torque pulsations caused by loads can be reduced to an acceptable level. This will be the subject of Application Example 6.9.

### 6.14.8 Application Example 6.9: Reduction of Harmonic Torques of a Piston-Compressor Drive with Synchronous Motor as Prime Mover

The drive motor of a piston compressor is a nonsalient-pole synchronous motor with an amortisseur. To compensate for the piston stroke an eccentric shaft is employed. The load (L) torque at the eccentric shaft consists of an average torque and two harmonic torques:

$$T_L = T_{Lavg} + T_{L1} + T_{L4}, \qquad \text{(E6.9-1)}$$

where

$$T_{Lavg} = 265.53 \text{ kNm}, \qquad \text{(E6.9-2)}$$

$$T_{L1} = 31.04 \text{ kNm} \sin(\omega_{ms}t - \pi), \qquad \text{(E6.9-3)}$$

$$T_{L4} = 50.65 \text{ kNm} \sin(4\omega_{ms} - \pi/6). \qquad \text{(E6.9-4)}$$

The drive motor and the flywheel are coupled with the eccentric shaft, and the nonsalient-pole synchronous motor has the following data: $P_{rat} = 7.5$ MW, $p = 28$ poles, $f = 60$ Hz, $V_{L\text{-}Lrat} = 6$ kV, $\cos \Phi_{rat} = 0.9$ overexcited, $\eta_{rat} = 96.9\%$, $T_{max}/T_{rat} = 2.3$.

The slip of the nonsalient pole synchronous motor when operating as an induction motor is $s_{rat} = 6.8\%$ at $T_{rat}$.

a) Calculate the synchronous speed $n_{ms}$, synchronous angular velocity $\omega_{ms}$, rated stator current $\tilde{I}_{a\_rat}$, rated torque $T_{rat}$, the torque angle $\delta$, stator current $\tilde{I}_a$ at $T_{Lavg}$ and the synchronous reactance $X_S$ (in ohms and in per unit).

b) Neglecting damping provided by induction-motor action, derive the nonlinear differential equation for the motor torque, and by linearizing this equation find an expression for the eigen angular frequency $\omega_{eigen}$ of the undamped motor.

c) Repeat part b, but do not neglect damping provided by induction motor action.

d) Solve the linearized differential equations of part b (without damping) and part c (with damping) and determine – by introducing phasors – the axial moment of inertia $J$ required so that the first harmonic motor torque will be reduced to ±4% of the average torque (Eq. E6.9-2).

e) Compare the eigen frequency $f_{eigen}$ with the pulsation frequencies occurring at the eccentric shaft.

f) State the total motor torque $T(t)$ as a function of time (neglecting damping) for one revolution of the eccentric shaft and compare it with the load torque $T_L(t)$. Why are they different?

### 6.14.9 Calculation of Steady-State Temperature Rise ΔT of Electric Apparatus Based on Thermal Networks

The losses of an electric apparatus (e.g., rotating machine, transformer, and inductor) heat up various parts of the apparatus, and the losses are dissipated via heat conduction, convection, and radiation.

**Heat Flow Related to Conduction.** Assume that the losses at end #2 with the temperature $\theta_2$ of a homogeneous body (see Fig. 6.25) are

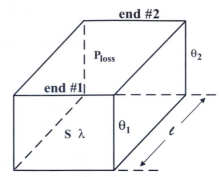

**FIGURE 6.25** Heat flow within a homogeneous body (principle of heat pipe).

$$P_{loss} = \lambda \cdot S \frac{\theta_2 - \theta_1}{\ell} = \lambda \cdot S \frac{\Delta T}{\ell}, \qquad \text{(6-58)}$$

where $\lambda$ is the thermal conductivity of a homogeneous body measured in $\left[\dfrac{W}{m^\circ C}\right]$, $S$ is the surface of the homogeneous body perpendicular to the heat flow measured in m$^2$, and $\ell$ is the length of the homogeneous body along the direction of the heat flow measured in $m$.

The temperature at end #1 $\theta_1$ is given by Eq. 6-58:

$$\Delta T = (\theta_2 - \theta_1) = \frac{\ell}{\lambda \cdot S} P_{loss} = R_{conduct} \cdot P_{loss}, \qquad \text{(6-59)}$$

where the term $\left[\dfrac{\ell}{\lambda \cdot S}\right]$ is defined as the heat conduction resistance

$$R_{conduct} = \frac{\ell}{\lambda \cdot S} \qquad \text{(6-60)}$$

**Heat Flow Related to Radiation and Convection.** The heat flow due to radiation and convection from the heat source to the cooling medium (e.g., ambient air) is proportional to the surface $A$ of the body (e.g., winding) on which the heat source resides and its temperature difference $\Delta T = (\theta_2 - \theta_1)$ with respect to the cooling medium:

$$P_{loss} = \alpha \cdot A(\theta_2 - \theta_1), \qquad \text{(6-61a)}$$

where $\alpha$ is the thermal conductivity measured in $\left[\dfrac{W}{m^2 {}^\circ C}\right]$ between the heat source and the cooling medium.

Correspondingly, one can define

$$\Delta T = (\theta_2 - \theta_1) = \frac{1}{\alpha \cdot A} P_{\text{loss}} = R_{\text{radiation/convection}} \cdot P_{\text{loss}}$$

(6-61b)

where the term $\dfrac{1}{\alpha \cdot A}$ is defined as the heat radiation/convection resistance

$$R_{\text{radiation/convection}} = \frac{1}{\alpha \cdot A}.$$

(6-62)

The application of the conduction relationship (Eq. 6-59) generates Fig. 6.26, where a linear function leads to the heat flow from a high-temperature body #2 (e.g., winding) via insulation – with the heat conduction resistance $R_{\text{conduct}} = \dfrac{\ell}{\lambda \cdot S}$ – to a low-temperature body #1 (e.g., iron core).

The application of the radiation/convection relationship (Eq. 6-61) generates Fig. 6.27, where a step function leads to the heat flow from a high-temperature body #2 (e.g., frame) – with the heat radiation/convection resistance $R_{\text{radiation/convection}} = \dfrac{1}{\alpha \cdot A}$ – to a low-temperature body #1 (e.g., ambient air). An example for this case is the interface between the motor frame and the ambient air.

Figure 6.28 depicts the heat flow as it occurs from a winding via the insulation, the iron core, the small air gap between iron core and frame, and from the frame of an electrical apparatus to the ambient air serving as cooling medium. Note that the heat

drop in the small air gap is a mixture between conduction and convection/radiation. The reason is that some of the iron core laminations have contact (heat transfer via conduction) with the frame and some have no contact (heat transfer via convection/radiation).

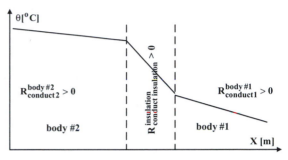

**FIGURE 6.26** Linear heat flow through conduction from a high-temperature body #2 to a low-temperature body #1 with finite thermal resistances.

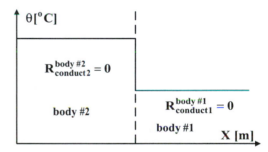

**FIGURE 6.27** Step-function heat flow through convection/radiation from a high-temperature body to a low-temperature cooling medium.

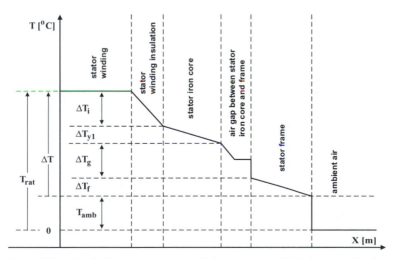

**FIGURE 6.28** Heat flow within electrical apparatus from a high-temperature body (e.g., winding) to a low-temperature cooling medium (e.g., ambient air).

### 6.14.10 Application Example 6.10: Temperature-Rise Equations for a Totally Enclosed Fan-Cooled 100 hp Motor

For a totally enclosed fan-cooled motor [48] the surface of the external frame dissipates the heat. The ribs are located along the axis of the machine where the hot air is removed by a fan. Test results show [49, 50] that the stator end windings have about the same temperature as the slot-embedded part of the stator winding. This result permits us to assume that the end winding does not dissipate heat (e.g., radiates and removes heat by convection) to the surrounding air inside the end bells of the motor, and the heat developed by the stator winding is completely transferred to the stator core. The stator iron loss is then added to the stator copper loss and the resultant heat is transferred to the stator frame via the small air gap between core and frame from which it is removed by the fan through radiation and convection. A second assumption refers to the heat of the rotor winding. It is assumed that this heat is transferred to the stator. This means that the temperature of the rotor is higher than that of the stator. The temperature of the hottest spot in the stator determines the class of winding insulation. It is assumed that this point is residing in the stator winding inside a stator slot. The temperature of the hot spot consists of the following temperature drops:

- temperature drop across stator winding insulation, $\Delta T_i$
- temperature drop across stator back iron (yoke), $\Delta T_{y1}$
- temperature drop across the small air gap between the stator back iron and the frame of the motor, $\Delta T_g$
- difference between the temperature of the frame and the ambient temperature of the air, $\Delta T_f$.

The temperature rise of the stator winding is therefore

$$\Delta T = \Delta T_i + \Delta T_{y1} + \Delta T_g + \Delta T_f. \qquad \text{(E6.10-1)}$$

The thermal network of the motor is shown in Fig. E6.10.1, where

$R_i$ is the thermal conductivity resistance of the stator slot insulation

$R_{y1}$ is the thermal conductivity resistance of the stator back iron or yoke

$R_g$ is the thermal conductivity resistance of the small air gap between the stator core and the frame of the motor

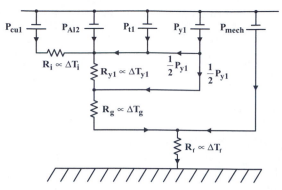

**FIGURE E6.10.1** Thermal network of a totally enclosed fan-cooled induction motor.

$R_f$ is the thermal radiation/convection resistance of the outside of the stator frame and the moving air

$P_{cu1}$ is the stator ohmic (copper) loss

$P_{Al2}$ is the rotor ohmic (aluminum) loss

$P_{t1}$ is the stator teeth (iron-core) loss

$P_{y1}$ is the stator yoke (iron-core) loss

$P_{mech}$ is the friction (bearing, windage) loss

### 6.14.11 Application Example 6.11: Temperature-Rise Equations for a Drip-Proof 5 hp Motor

The temperature rise equation is written for a drip-proof motor [48]. This equation is based on the following assumptions:

1. The ohmic losses in the embedded part of the stator winding produce a temperature drop across the slot insulation.
2. The iron-core losses of the stator teeth, the ohmic losses in the embedded part of the stator winding, and the ohmic losses of the rotor bars and half of the iron-core losses of the stator back iron generate temperature drops across the stator back iron and the motor frame. The thermal network of this drip-proof motor based on the above assumptions is depicted in Fig. E6.11.1. In this figure $P_{em}$ stands for the ohmic losses in the embedded part of the stator winding.

## 6.15 SUMMARY

1) All IEC and IEEE harmonic standards and guidelines rely on the worst-case condition where the harmonic current or voltage spectra are independent of time.

2) The weighted-harmonic voltage factor is based on the same assumption and estimates the addi-

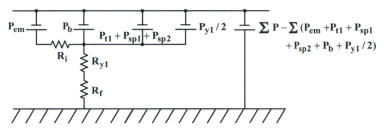

**FIGURE E6.11.1** Thermal network of a drip-proof induction motor.

tional temperature rise for rated operation of transformers, induction motors, and universal motors for the voltage spectrum with maximum individual harmonic voltage amplitudes.

3) The weighted-harmonic voltage factor

$$\sum_{h=2}^{\infty} \frac{1}{(h)^k} \left( \frac{V_h}{V_1} \right)^l \qquad (6\text{-}63)$$

described in this chapter is related to the square of the total harmonic distortion of the voltage

$$THD_v = \sqrt{\sum_{h=2}^{\infty} \left( \frac{V_h}{V_1} \right)^2} \qquad (6\text{-}64)$$

by introducing the weighting function $\left( \frac{1}{h} \right)^k$ and replacing the exponent 2 of the ratio by $\sum_{h=2}^{\infty} \frac{1}{(h)^k} \left( \frac{V_h}{V_1} \right)^2$ by the exponent $l$.

4) $\sum_{h=2}^{\infty} \frac{1}{(h)^k} \left( \frac{V_h}{V_1} \right)^l$ can be used as a measure for the losses and temperature rises of electromagnetic apparatus due to harmonic voltages.

5) Acceptable limits for the reduction of lifetime due to thermal aging requires limits for the value of the weighted-harmonic factor $\sum_{h=2}^{\infty} \frac{1}{(h)^k} \left( \frac{V_h}{V_1} \right)^l$ and indirectly limits the amplitude of the occurring voltage harmonics.

6) A weighted harmonic factor of

$$\sum_{h=2}^{\infty} \frac{1}{(h)^k} \left( \frac{V_h}{V_1} \right)^l = 5.8 \qquad (6\text{-}65)$$

is recommended.

7) It is evident that for a given weighted-harmonic voltage factor, single- and three-phase induction machines are more susceptible to aging than transformers and universal machines.

8) Strictly speaking, all current or voltage harmonic spectra are continually changing as a function of time due to changing system configuration and loads, and the assumption of time-independent spectra is a simplification leading to a first-cut approach only.

9) The cost of harmonics incurs by either purchasing more expensive solid-state equipment (e.g., PWM inverters, switched-mode power supplies), which generates a reduced amount of harmonics, or by using less-sophisticated but harmonic generating equipment (e.g., six-step inverters, diode rectifiers), which generate a greater amount of harmonics than the more expensive switched-mode equipment. In the first case the losses occur in the switched-mode equipment (e.g., efficiencies ranging from 80 to 98%), whereas in the second case of less-sophisticated solid-state equipment (e.g., efficiencies ranging from 90 to 98%) the harmonic losses incur in the loads (e.g., induction motors) and their reduction of lifetime must be accounted for due to temperature increase.

10) Design considerations for intermittent operation are presented so that no lifetime reduction occurs.

11) Most drives are subjected to pulsating loads that cause vibrations within the drive motor, in particular the motor windings. To limit vibrations to a certain value the required axial moment of inertia J of a flywheel is calculated.

12) The temperature rises for totally enclosed fan-cooled and for drip-proof induction motors are derived.

## 6.16 PROBLEMS

### Problem 6.1: Aging of Transformers and Induction Machines Due to Single Harmonic Voltage

The terminal voltage of a single-phase transformer (rated resistive load) and a single-phase induction machine (operation at rated load) contains a third-harmonic voltage component of $V_{p3}/V_{p1} = 2.5\%$, $5\%$, and $7.5\%$.

a) Determine for the worst-case condition the temperature rises $\Delta T$ for both devices, which occur due to the previously mentioned harmonic voltage components. You may assume for both devices a rated temperature of $T_{rat} = T_2 = 125°C$ at an ambient temperature of $T_{amb} = 40°C$.

b) Calculate for the temperature rises $\Delta T$ of part a the decrease of lifetime for both devices under the given three (not simultaneously occurring) harmonic voltage components for an activation energy of $E = 1.15$ eV.

## Problem 6.2: Operation of Induction Motor at Reduced Line-to-Line Voltage

A 100 hp, 480 $V_{L-L}$, 60 Hz, 6-pole three-phase induction motor runs at full load and rated voltage with a slip of 3%. Under conditions of stress on the power supply system, the line-to-line voltage drops to 420 $V_{L-L}$. If the load is of the constant torque type, then compute for the lower voltage

a) The slip (use small slip approximation).
b) The shaft speed in rpm.
c) The hp output.
d) The rotor copper loss ($I_2'^2 r_2$) in terms of the rated rotor copper loss at rated voltage.
e) Estimate the reduction of lifetime due to reduced voltage operation provided the rated temperature is $T_{rat} = 125°C$, the ambient temperature is $T_{amb} = 40°C$, rated lifetime is $t_2 = 40$ years, and $E = 1.15$ eV.

## Problem 6.3: Operation of Induction Motor at Increased Line-to-Line Voltage

Repeat Problem 6.2 for the condition when the line-to-line voltage of the power supply system increases to 520 $V_{L-L}$, and estimate the reduction of lifetime.

## Problem 6.4: Reduced Voltage and Frequency Operation of Induction Machines

A 100 hp, 480 $V_{L-L}$, 60 Hz, 6-pole three-phase induction motor runs at full load and rated voltage and frequency with a slip of 3%. Under conditions of stress on the power supply system the terminal voltage and the frequency drops to 420 $V_{L-L}$ and 54 Hz, respectively. If the load is of the constant torque type, then compute for the lower voltage and frequency:

a) The slip (use small slip approximation).
b) The shaft speed in rpm.
c) The hp output.
d) Rotor copper loss ($I_2'^2 r_2$) in terms of the rated rotor copper loss at rated frequency and voltage.
e) Estimate the reduction of lifetime for $T_{rat} = 125°C$, $T_{amb} = 40°C$, $t_2 = 40$ years, and $E = 1.15$ eV.

## Problem 6.5: Lifetime Reduction of Induction Motors and Transformers for Given Voltage Spectra

Estimate the lifetime reductions of induction machines, transformers, and universal machines for the single- and three-phase voltage spectra of Table P6.5, and their associated lifetime reductions for an activation energy of $E = 1.15$ eV. The ambient temperature is $T_{amb} = 40°C$, the rated temperature is $T_2 = 120°C$, and the rated lifetime of $t_2 = 40$ years can be assumed.

## Problem 6.6: Calculate the Stator rms Current of an Induction Motor at Steady-State Operation with Superimposed Intermittent, Irregular Load Steps

Repeat the analysis of Application Example 6.8 provided the inverter-motor drive is operated at $f = 30$ Hz with constant flux, $(E/f) = $ constant, control.

**TABLE P6.5** Possible voltage spectra with high-harmonic penetration

| $h$ | $\left(\dfrac{V_h}{V_{60\,Hz}}\right)_{1\Phi}$ (%) | $\left(\dfrac{V_h}{V_{60\,Hz}}\right)_{3\Phi}$ (%) |
|---|---|---|
| 1 | 100 | 100 |
| 2 | 0.5 | 0.5 |
| 3 | 7.0 | 0.5 |
| 4 | 1.6 | 0.5 |
| 5 | 5.0 | 7.5 |
| 6 | 0.88 | 0.2 |
| 7 | 3.0 | 5.0 |
| 8 | 0.62 | 0.2 |
| 9 | 2.0 | 0.3 |
| 10 | 0.66 | 0.1 |
| 11 | 1.0 | 4.0 |
| 12 | 0.18 | 0.1 |
| 13 | 0.8 | 3.0 |
| 14 | 0.10 | 0.05 |
| 15 | 0.10 | 0.1 |
| 16 | 0.13 | 0.05 |
| 17 | 0.23 | 1.5 |
| 18 | 0.22 | 0.01 |
| 19 | 1.03 | 1.0 |
| All higher harmonics < 0.2% | | |

## Problem 6.7: Calculate the Reduction of Harmonic Synchronous Motor Torques for a Given Time-Dependent Load

The drive motor of a pulsating load is a nonsalient-pole synchronous motor with an amortisseur. To compensate for the changing load torque as a function of time a flywheel is coupled with the shaft. The load ($L$) torque at the shaft consists of an average torque and two harmonic torques:

$$T_L = T_{\text{Lavg}} + T_{L1} + T_{L3}, \quad \text{(P6.7-1)}$$

where

$$T_{\text{Lavg}} = 260 \text{ kNm}, \quad \text{(P6.7-2)}$$

$$T_{L1} = 30 \text{ kNm} \sin(\omega_{ms} t - \pi), \quad \text{(P6.7-3)}$$

$$T_{L3} = 50 \text{ kNm} \sin(3\omega_{ms} - \pi/6). \quad \text{(P6.7-4)}$$

The nonsalient-pole synchronous motor has the following data: $P_{\text{rat}} = 5$ MW, $p = 40$ poles, $f = 60$ Hz, $V_{L\text{-}L\,\text{rat}} = 6$ kV, $\cos \Phi_{\text{rat}} = 0.8$ overexcited, $\eta_{\text{rat}} = 95\%$, $T_{\text{max}}/T_{\text{rat}} = 2.5$. The slip of the nonsalient pole synchronous motor when operating as an induction motor is $s_{\text{rat}} = 6\%$ at $T_{\text{rat}}$.

a) Calculate the synchronous speed $n_{ms}$, synchronous angular velocity $\omega_{ms}$, rated stator current $\tilde{I}_{a\_\text{rat}}$, rated torque $T_{\text{rat}}$, the torque angle $\delta$, stator current $\tilde{I}_a$ at $T_{\text{Lavg}}$ and the synchronous reactance $X_S$ (in ohms and in per unit).
b) Neglecting damping provided by induction-motor action, derive the nonlinear differential equation for the motor torque, and by linearizing this equation find an expression for the eigen angular frequency $\omega_{\text{eigen}}$ of the undamped motor.
c) Repeat part b, but do not neglect damping provided by induction motor action.
d) Solve the linearized differential equations of part b (without damping) and part c (with damping) and determine – by introducing phasors – the axial moment of inertia J required so that the first harmonic order motor torque will be reduced to ±2% of the average torque (Eq. P6.7-2).
e) Compare the eigen frequency $f_{\text{eigen}}$ with the pulsation frequencies occurring at the eccentric shaft.
f) State the total motor torque $T(t)$ as a function of time (neglecting damping) for one revolution of the eccentric shaft and compare it with the load torque $T_L(t)$. Why are they different?

## Problem 6.8: Temperature Transient of an Induction Machine

An enclosed fan-cooled induction motor has a thermal time constant of $\tau_\theta = 4$ h and a steady-state rated temperature of $\theta_f = 140°C$ at an ambient temperature of $\theta_{\text{amb}} = 40°C$. The motor has at time $t = 0-$ a temperature of $\theta = \theta_{\text{amb}}$, and the motor is fully loaded at $t = 0+$. Calculate the time $t_{95\%}$ when the fully loaded motor reaches 95% of its final temperature $\theta_f = 140°C$.

## Problem 6.9: Calculation of the Temperature Rise of a Totally Enclosed Fan-Cooled 100 hp Induction Motor

a) Based on Application Example 6.10 calculate the temperature rise $\Delta T$ of a totally enclosed fan-cooled induction motor provided the following data are given: Input power $P_{\text{in}} = 79,610$ W, output power $P_{\text{out}} = 74,600$ W, line frequency $f = 60$ Hz, thickness of stator slot insulation $\beta_i = 9 \cdot 10^{-4}$ m, thermal conductivity of stator slot insulation $\lambda_i = 0.109$ W/m°C, total area of the stator slots $S_i = 2.006$ m², thermal conductivity of the stator laminations (iron core) $\lambda_c = 47.24$ W/m°C, radial height of stator back iron $h_{y1} = 0.049$ m, cross-sectional area of the stator core at the middle of the stator back iron (stator yoke) $S_{y1} = 0.482$ m², length of the small air gap between the stator core and the frame of the motor $\beta_\delta = 3 \cdot 10^{-5}$ m, thermal conductivity of the small air gap between stator core and frame $\lambda_\delta = 0.028$ W/m°C, outside surface area of the stator frame $S_\delta = 0.563$ m², surface thermal coefficient for dissipation of heat to stationary air $\alpha_o = 14.30$ W/m²°C, surface thermal coefficient for dissipation of heat to moving air at $v = 16$ m/s $\alpha_v = 88.66$ W/m²°C, surface area in contact with stationary air $S_o = 0.841$ m², and surface area in contact with moving air $S_v = 3.36$ m². Table P6.9 details the percentage values of the various losses of the 100 hp induction motor. Note that the iron core losses of the rotor are very small due to the low frequency in the rotor member (slip frequency).
b) Determine the rated temperature of the motor $T_{\text{rat}}$ provided the ambient temperature is $T_{\text{amb}} = 40°C$.

**TABLE P6.9** Percentage Losses of a 100 hp Three-Phase Induction Motor

| $P_{\text{cu1}}$ | $P_{\text{y1}}$ | $P_{\text{t1}}$ | $P_{\text{sp1}}$ | $P_{\text{Al2}}$ | $P_{\text{sp2}}$ | $P_{\text{mech}}$ |
|---|---|---|---|---|---|---|
| 30% | 8% | 10% | 2% | 45% | 2% | 3% |

## Problem 6.10: Calculation of the Temperature Rise of a Drip-Proof 5 hp Induction Motor

a) Based on Application Example 6.11 calculate the temperature rise $\Delta T$ of a drip-proof induction motor provided the following data are given: Input power $P_{in} = 4355$ W, output power $P_{out} = 3730$ W, line frequency $f = 60$ Hz, thickness of stator slot insulation $\beta_i = 6 \cdot 10^{-4}$ m, thermal conductivity of stator slot insulation $\lambda_i = 0.148$ W/m°C, total area of the stator slots $S_i = 0.0701$ m², thermal conductivity of the stator laminations (iron core) $\lambda_c = 47.24$ W/m°C, radial height of stator back iron $h_{y1} = 0.0399$ m, cross-sectional area of the stator core at the middle of the stator back iron (stator yoke) $S_{y1} = 0.0325$ m², length of the small air gap between the stator core and the frame of the motor $\beta_\delta = 5 \cdot 10^{-6}$ m, thermal conductivity of the small air gap between stator core and frame $\lambda_\delta = 0.236$ W/m°C, outside surface area of the stator frame $S_\delta = 0.0403$ m², surface thermal coefficient for dissipation of heat to stationary air $\alpha_o = 18.70$ W/m²°C, surface thermal coefficient for dissipation of heat to moving air at $v = 4.6$ m/s $\alpha_v = 58.81$ W/m²°C, surface area in contact with stationary air $S_o = 0.181$ m², and surface area in contact with moving air $S_v = 0.696$ m². Table P6.10 details the percentage values of the various losses of the 5 hp induction motor. Note that the iron-core losses of the rotor are very small due to the low frequency in the rotor member (slip frequency).

b) Determine the rated temperature of the motor $T_{rat}$ provided the ambient temperature is $T_{amb} = 40$°C.

## Problem 6.11: Solution of the Partial Differential Equation Governing Heat Conduction

The solution of the heat equation in one spatial dimension $(x)$

$$\frac{\partial T(x,t)}{\partial t} = c^2 \frac{\partial^2 T(x,t)}{\partial x^2} \qquad \text{(P6.11-1)}$$

is based on the separation of variables as was proposed by Fourier, where $T$ is the temperature and $x$ is the spatial coordinate as illustrated, for example, in Fig. P6.11, and $t$ is the time. The boundary conditions are

$$T(0, t) = 0 \text{ and } T(\ell, t) = 0 \qquad \text{(P6.11-2)}$$

for all times $t$, and the initial condition at $t = 0$ is

$$T(x, 0) = f(x), \qquad \text{(P6.11-3)}$$

where $f(x)$ is a given function of $x$.

The temperature $T(x, t)$, is assumed separable in $x$ and $t$.

$$T(x, t) = X(x) \cdot T(t) \qquad \text{(P6.11-4)}$$

so that

$$X'' \cdot T = \frac{1}{c^2} X \cdot T', \qquad \text{(P6.11-5)}$$

or after dividing by $X(x)T(t)$

$$\frac{X''}{X} = \frac{1}{c^2} \frac{T'}{T} = k, \qquad \text{(P6.11-6)}$$

where $k$ is the separation constant. We can expect $k$ to be negative as can be seen from the time equation:

$$T' = c^2 k T \qquad \text{(P6.11-7)}$$

or after integration

$$T(t) = T_0 e^{c^2 k T}. \qquad \text{(P6.11-8)}$$

Thus $T(t)$ will increase in time if $k$ is positive and decrease in time if $k$ is negative. The increase of $T$ in time is not possible without any additional heating mechanism and therefore we assume that $k$ is negative. To force this we set $k = -p^2$. The spatial differential equation is now

$$X'' + p^2 X = 0, \qquad \text{(P6.11-9)}$$

which is similar to the harmonic motion equation with the trigonometric solution

$$X(x) = A \cos px + B \sin px. \qquad \text{(P6.11-10)}$$

Now, applying the boundary condition, we find that for

**TABLE P6.10** Percentage Losses of a 5 hp Three-Phase Induction Motor

| $P_{cu1}$ | $P_{y1}$ | $P_{t1}$ | $P_{sp1}$ | $P_{Al2}$ | $P_{sp2}$ | $P_{mech}$ |
|-----------|----------|----------|-----------|-----------|-----------|------------|
| 30% | 8% | 10% | 2% | 42% | 2% | 6% |

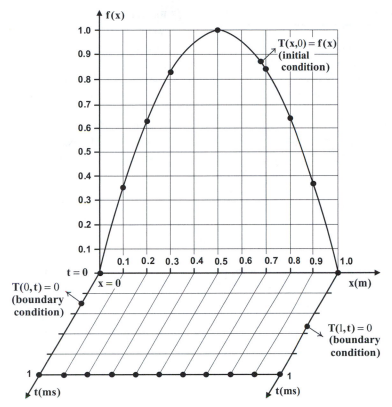

**FIGURE P6.11** Heat conduction in a homogeneous one-dimensional body of length $\ell = 1$ m.

$$X(0) = 0, \quad X(\ell) = 0. \quad \text{(P6.11-11)}$$

$A = 0$ and $\sin(p\ell) = 0$, resulting in $p\ell = n\pi$. We take the most general solution by adding together all possible solutions, satisfying the boundary conditions, to obtain

$$T(x,t) = \sum_{n=0}^{\infty} B_n \sin\frac{n\pi}{\ell} e^{\frac{n^2\pi^2 c^2}{\ell^2}t}. \quad \text{(P6.11-12)}$$

The final step is to apply the initial conditions

$$T(x,0) = \sum_{n=0}^{\infty} B_n \sin\frac{n\pi}{\ell} = f(x). \quad \text{(P6.11-13)}$$

Finding the Fourier coefficient of the above expression leads to

$$B_n = \frac{2}{\ell}\int_0^{\ell} f(x)\sin\frac{n\pi}{\ell}x\,\mathrm{d}x \quad \text{(P6.11-14)}$$

for all positive integers $n$.

Note that because every term in the solution for $T(x, t)$ has a negative exponential in it, the temperature must decrease in time, and the final solution will tend to $T = 0$.

a) For $c^2 = 1$, $\ell = 1$, that is, $\dfrac{\partial T(x,t)}{\partial t} = \dfrac{\partial^2 T(x,t)}{\partial x^2}$, the boundary conditions $T(0, t) = T(1, t) = 0$, and the initial condition $T(x, 0) = f(x) = -4x^2 + 4x$ find $T(x, t)$ for $0 < x < 1$. The initial conditions and boundary conditions are shown in Fig. P6.11 as a function of the spatial coordinate $x$ and the time $t$. Calculate and plot $T(x, t)$ as a function of $x$ for $t = 1$ ms; thereby it will be sufficient to take into account three terms of the Fourier series.

b) For $c^2 = 1$, $\ell = 1$, that is, $\dfrac{\partial T(x,t)}{\partial t} = \dfrac{\partial^2 T(x,t)}{\partial x^2}$, the boundary conditions $T(0, t) = T(1, t) = 0$, and the initial conditions $T(x, 0) = f(x) = x$ (for $0 < x < 1/2$) and $T(x, 0) = f(x) = 1 - x$ (for $1/2 < x < 1$) find $T(x, t)$. Calculate and plot $T(x, t)$ as a function of $x$ for $t = 1$ ms; thereby it will be sufficient to take into account three terms of the Fourier series.

## 6.17 REFERENCES

1) Montsinger, V.M.; "Loading transformer by temperature," *Trans. AIEE*, vol. 49, 1930, p. 776.

2) Pierce, L.W.; "Thermal considerations in specifying dry-type transformers," *IEEE Transactions on Industry Applications,* vol. 30, issue 4, July–Aug. 1994, pp. 1090–1098.

3) Yan, T.; and Zhongyan, H.; "The calculating hot-point temperature-rise of transformer windings," *ICEMS 2001, Proceedings of the Fifth International Conference on Electrical Machines and Systems 2001,* vol. 1, 18–20 Aug. 2001, pp. 220–223.

4) Pierce, L.W.; "Hottest spot temperatures in ventilated dry type transformers," *IEEE Transactions on Power Delivery*, vol. 9, issue 1, Jan. 1994, pp. 257–264.

5) Radakovic, Z.; and Feser, K.; "A new method for the calculation of the hot-spot temperature in power transformers with ONAN cooling," *IEEE Transactions on Power Delivery*, vol. 18, issue 4, Oct. 2003, pp. 1284–1292.

6) Okoro, O.I.; Weidemann, B.; and Ojo, O.; "An efficient thermal model for induction machines," 2004, 39th IAS Annual Meeting, *Conference Record of the 2004 IEEE Industry Applications Conference,* vol. 4, 3–7 Oct. 2004, pp. 2477–2484.

7) Pierce, L.W.; "Predicting liquid filled transformer loading capability," *IEEE Transactions on Industry Applications,* vol. 30, issue 1, Jan.–Feb. 1994, pp. 170–178.

8) Weekes, T.; Molinski, T.; and Swift, G.; "Transient transformer overload ratings and protection," *IEEE Electrical Insulation Magazine,* vol. 20, issue 2, Mar.–Apr. 2004, pp. 32–35.

9) Moreno, J.F.; Hidalgo, F.P.; and Martinez, M.D.; "Realisation of tests to determine the parameters of the thermal model of an induction machine," *IEEE Proceedings – Electric Power Applications*, vol. 148, issue 5, Sept. 2001, pp. 393–397.

10) Sang-Bin, L.; and Habetler, T.G.; "An online stator winding resistance estimation technique for temperature monitoring of line-connected induction machines," *IEEE Transactions on Industry Applications,* vol. 39, issue 3, May–June 2003, pp. 685–694.

11) Siyambalapitiya, D.J.T.; McLaren, P.G.; and Tavner, P.J.; "Transient thermal characteristics of induction machine rotor cage," *IEEE Transactions on Energy Conversion*, vol. 3, issue 4, Dec. 1988, pp. 849–854.

12) Duran, M.J.; and Fernandez, J.; "Lumped-parameter thermal model for induction machines," *IEEE Transactions on Energy Conversion*, vol. 19, issue 4, Dec. 2004, pp. 791–792.

13) Rajagopal, M.S.; Seetharamu, K.N.; and Ashwathna-rayana, P.A.; "Transient thermal analysis of induction motors," *IEEE Transactions on Energy Conversion*, vol. 13, issue 1, March 1998, pp. 62–69.

14) Pierce, L.W.; "Transformer design and application considerations for nonsinusoidal load currents," *IEEE Transactions on Industry Applications*, vol. 32, issue 3, May–June 1996, pp. 633–645.

15) Galli, A.W.; and Cox, M.D.; "Temperature rise of small oil-filled distribution transformers supplying nonsinusoidal load current," *IEEE Transactions on Power Delivery*, vol. 11, issue 1, Jan. 1996, pp. 283–291.

16) Picher, P.; Bolduc, L.; Dutil, A.; and Pham, V.Q.; "Study of the acceptable dc current limit in core-form power transformers," *IEEE Transactions on Power Delivery*, vol. 12, issue 1, Jan. 1997, pp. 257–265.

17) Sousa, G.C.D.; Bose, B.K.; Cleland, J.; Spiegel, R.J.; and Chappell, P.J.; "Loss modeling of converter induction machine system for variable speed drive," *Proceedings of the 1992 International Conference on Industrial Electronics, Control, Instrumentation, and Automation*, 1992, vol. 1, 9–13 Nov. 1992, pp. 114–120.

18) Fuchs, E.F.; Roesler, D.J.; and Alashhab, F.S.; "Sensitivity of electrical appliances to harmonics and fractional harmonics of the power system's voltage, part I," *IEEE Transactions on Power Delivery,* April 1987, vol. TPWRD-2, no. 2, pp. 437–444.

19) Fuchs, E.F.; Roesler, D.J.; and Kovacs, K.P.; "Sensitivity of electrical appliances to harmonics and fractional harmonics of the power system's voltage, part II," *IEEE Transactions on Power Delivery*, April 1987, vol. TPWRD-2, no. 2, pp. 445–453.

20) Fuchs, E.F.; Roesler, D.J.; and Kovacs, K.P.; "Aging of electrical appliances due to harmonics of the power system's voltage," *IEEE Transactions on Power Delivery*, July 1986, vol. TPWRD-1, no. 3, pp. 301–307.

21) Fuchs, E.F.; Roesler, D.J.; and Masoum, M.A.S.; "Are harmonic recommendations according to IEEE and IEC too restrictive?" *IEEE Transactions on Power Delivery*, vol. 19, no. 4, October 2004, pp. 11775–1786.

22) Fuchs, E.F.; Lin, D.; and Martynaitis, J.; "Measurement of three-phase transformer derating and reactive power demand under nonlinear loading conditions," *IEEE Transactions on Power Delivery,* vol. 21, issue 2, April 2006, pp. 665–672.

23) Lin, D.; and Fuchs, E. F.; "Real-time monitoring of iron-core and copper losses of three-phase transformers under (non)sinusoidal operation," *IEEE Transactions on Power Delivery*, vol. 21, issue 3, July 2006, pp. 1333–1341.

24) Dixon, R.R.; "Thermal aging predictions from an Arrhenius plot with only one data point," *IEEE Transactions on Electrical Insulation*, vol. E1-15, no. 4, 1980, p. 331.

25) http://nobelprize.org/chemistry/laureates/1903/arrhenius-bio.html.

26) Richter, R.; *Elektrische Maschinen*, Springer Verlag, Berlin, 1924, p. 157.

27) Fuchs, E.F.; Masoum, M.A.S.; and Roesler, D.J.; "Large signal nonlinear model of anisotropic transformers for nonsinusoidal operation, parts I and II," *IEEE Transactions on Power Delivery*, vol. PWRD-6, January 1991, pp. 174–186, and vol. PWRD-6, October 1991, pp. 1509–1516.

28) Fuchs, E.F.; and Fei, R.; "A new computer-aided method for the efficiency measurement of low-loss transformers and inductors under nonsinusoidal operation," *IEEE Transactions on Power Delivery*, vol. PWRD-11, January 1996, pp. 292–304.

29) Fuchs, E.F.; Alashhab, F.; Hock, T.F.; and Sen, P.K; "Impact of harmonics on home appliances," Topical Report (First Draft), prepared for the U.S. Department of Energy under contract No. DOE-RA-50150-9, June 1981, University of Colorado at Boulder.

30) Fuchs, E.F.; Chang, L.H.; Appelbaum, J.; and Moghadamnia, S.; "Sensitivity of transformer and induction motor operation to power system's harmonics," Topical Report (First Draft), prepared for the U.S. Department of Energy under contract No. DOE-RA-50150-18, April 1983, University of Colorado at Boulder.

31) Fuchs, E.F.; Chang, L.H.; and Appelbaum, J.; "Magnetizing current, iron losses and forces of three-phase induction machines at sinusoidal and nonsinu-soidal terminal voltages, part I: Analysis," *IEEE Transactions on Power Apparatus and Systems*, November 1984, vol. PAS-103, no. 11, pp. 3303–3312.

32) Cummings, P.G.; Private communication, January 4, 1984.

33) Fuchs, E.F.; "Investigations on the impact of voltage and current harmonics on end-use devices and their protection," Summary Report, prepared for the U.S. Department of Energy under contract No. DE-AC02-80RA 50150, January 1987, University of Colorado at Boulder.

34) de Abreu, J.P.G.; and Emanuel, A.E.; "The need to limit subharmonic injection," *Proceedings of the 9th International Conference on Harmonics and Quality of Power*, vol. I," October 1–4, 2000, pp. 251–253.

35) *IEEE Standard 519*, IEEE recommended practices and requirements for harmonic control in electric power systems, IEEE-519, 1992.

36) IEC 61000-3-2 (2001–10) Consolidated edition, electromagnetic compatibility (EMC) – part 3-2: limits for harmonic current emissions.

37) Baghzouz, Y.; Burch, R.F.; Capasso, A.; Cavallini, A.; Emanuel, A.E.; Halpin, M.; Imece, A.; Ludbrook, A.; Montanari, G.; Olejniczak, K.J.; Ribeiro, P.; Rios-Marcuello, S.; Tang, L.; Thaliam, R.; and Verde, P.; "Time-varying harmonics; I. Characterizing measured data," *IEEE Transactions on Power Delivery,* vol. 13, issue 3, July 1998, pp. 938–944.

38) Baghzouz, Y.; Burch, R.F.; Capasso, A.; Cavallini, A.; Emanuel, A.E.; Halpin, M.; Langella, R.; Montanari, G.; Olejniczak, K.J.; Ribeiro, P.; Rios-Marcuello, S.; Ruggiero, F.; Thallam, R.; Testa, A.; and Verde, P.; "Time-varying harmonics; II. Harmonic summation and propagation," *IEEE Transactions on Power Delivery*, vol. 17, issue 1, Jan. 2002, Page(s):279–285.

39) Erickson, R. W.; and Maksimovic, D.; *Fundamentals of Power Electronics,* Second Edition, Kluwer Academic Publishers, ISBN 0-7923-7270-0, 2000.

40) Fuchs, E.F.; Masoum, M.A.S.; and Ladjevardi, M.; "Effects on distribution feeders from electronic loads based on future peak-load growth, part I: measure-ments, " *Proceedings of the IASTED International Conference on Power and Energy Systems (EuroPES 2005),* Paper # 468-019, Benalmadena, Spain, June 15–17, 2005.

41) Carpinelli, G.; Caramia, P.; Di Vito, E.; Losi, A.; and Verde, P.; "Probabilistic evaluation of the economical damage due to harmonic losses in industrial energy system," *IEEE Transactions on Power Delivery*, vol. 11, issue 2, April 1996, pp. 1021–1031, and Correction to "Probabilistic evaluation of the economical damage due to harmonic losses in industrial energy systems," *IEEE Transactions on Power Delivery*, vol. 11, issue 3, July 1996, p. 1692.

42) Verde, P.; "Cost of harmonic effects as meaning of standard limits," *Proceedings of the Ninth Interna-tional Conference on Harmonics and Quality of Power,* vol. 1, 1–4 Oct. 2000, pp. 257–259.

43) Caramia, P.; Carpinelli, G.; Verde, P.; Mazzanti, G.; Cavallini, A.; and Montanari, G.C.; "An approach to life estimation of electrical plant components in the presence of harmonic distortion," *Proceedings of the Ninth International Conference on Harmonics and Quality of Power, 2000.* vol. 3, 1–4 Oct. 2000, pp. 887–892.

44) Caramia, P.; and Verde, P.; "Cost-related harmonic limits," *IEEE Power Engineering Society Winter Meeting, 2000.* vol. 4, 23–27 Jan. 2000, pp. 2846–2851.

45) Caramia, P.; Carpinelli, G.; Losi, A.; Russo, A.; and Verde, P.; "A simplified method for the probabilistic evaluation of the economical damage due to

harmonic losses," *Proceedings of the 8th International Conference on Harmonics And Quality of Power*, 1998. vol. 2, 14-16 Oct. 1998, Page(s): 767–776.

46) Dubey, G.K.; *Power Semiconductor Controlled Drives*, Prentice Hall, Englewood Cliffs, New Jersey, 1989.

47) Heller, B.; and Hamata,V.; *Harmonic Field Effects in Induction Machines*, Academia, Publishing House of the Czechoslovak Academy of Sciences, Prague, 1977.

48) Fuchs, E.F.; Appelbaum, J.; Khan, I.A.; Höll, J.; and Frank, U.V.; *Optimization of Induction Motor Efficiency, Volume 1: Three-Phase Induction Motors*, Final Report prepared by the University of Colorado for the Electric Power Research Institute (EPRI), Palo Alto, EL-4152-CCM, Volume 1, Research Project 1944-1, 1985.

49) Fuchs, E.F.; Huang, H.; Vandenput, A.J.; Höll, J.; Appelbaum, J.; Zak, Z.; and Ericki, M.S.; *Optimization of Induction Motor Efficiency, Volume 2: Single-Phase Induction Motors*, Final Report prepared by the University of Colorado for the Electric Power Research Institute (EPRI), Palo Alto, EL-4152-CCM, Volume 2, Research Project 1944-1, 1987.

50) Fuchs, E.F.; Höll, J.; Appelbaum, J.; Vandenput, A.J.; and Klode, H.; *Optimization of Induction Motor Efficiency, Volume 3: Experimental Comparison of Three-Phase Standard Motors with Wanlass Motors*, Final Report prepared by the University of Colorado for the Electric Power Research Institute (EPRI), Palo Alto, EL-4152-CCM, Volume 3, Research Project 1944-1, 1985.

51) http://www-solar.mcs.st-and.ac.uk/~alan/MT2003/PDE/node21.html.

# 7

# Power System Modeling under Nonsinusoidal Operating Conditions

The calculation of power system harmonic voltages and currents and analyzing power quality problems can be classified into frequency- and time-domain methods. Frequency-domain techniques are widely used for harmonic problem formulation. They are a reformulation of the fundamental load flow problem in the presence of nonlinear loads and/or distorted utility voltages. This complicated, nonlinear, and harmonically coupled problem has been solved in several ways in order to reach a compromise between simplicity and reliability of the formulation. Time-domain techniques can accurately model system nonlinearities and are mostly used for power quality problems such as transient and stability issues under nonsinusoidal operating conditions. They require more computation time than frequency-domain approaches but can accurately model nonlinearities (e.g., saturation and hysteresis), transients, and stability associated with transformers, nonlinear loads, and synchronous generators. Some techniques are based on a combination of frequency- and time-domain methods. In the frequency domain, the power system is linearly modeled while nonlinear simulation of components and loads can be accurately performed in the time domain. Hybrid time- and frequency-domain techniques result in moderate computational speed and accuracy.

This chapter starts with a review of power system matrices and the fundamental Newton–Raphson load flow algorithm. In Section 7.4, the Newton–Raphson harmonic power flow formulation is presented and explained in detail. In the last part of this chapter, a survey and classification of harmonic power flow approaches are addressed. All phasors are in per unit (pu) and all phase angles are assumed to be in radians.

## 7.1 OVERVIEW OF A MODERN POWER SYSTEM

The power system of today is a complicated interconnected network consisting of the following major sections as is illustrated in Fig. 7.1:

- generation,
- transmission and subtransmission,
- distribution, and
- linear and nonlinear loads.

**Generation.** The three-phase AC synchronous generator or alternator is one of the fundamental components of a power system consisting of two synchronously rotating fields: one field is produced by the rotor driven at synchronous speed and excited by DC current. The DC current in the rotor winding is provided by an excitation system with or without slip rings. The second field is set up through the stator windings by the three-phase armature currents.

An important element of power systems are AC generators with rotating rectifiers, known as brushless excitation systems [1]. The excitation system controls the generator voltage and indirectly controls the reactive power flow within the power system together with synchronous condensers and capacitor banks. The slip ring free rotor enables the generation of high power – from 50 MVA up to 1600 MVA – at high voltages, typically in the 30–40 kV range. Hydrogen and water cooling permit the high power density of synchronous generators, which are mostly of the 2 and 4-pole types. Hydrogenerators are of the synchronous and of the induction machine types and can have up to 40 poles [2]. The source of mechanical power for AC generators may be water turbines, steam turbines (where the steam is either generated by coal, natural gas, oil, or nuclear fuel), and gas turbines for peak-power generation. Coal or natural gas fired plants have an efficiency of about 30% [3], whereas combined-cycle plants (gas turbine in series with a steam turbine) can reach efficiencies of about 60% [4]. Today's deregulated power system also relies on distributed renewable energy resources such as solar, fuel cells, and turbines driven by wind, tides, water, and biomass generated gas. In a typical power station several generators are connected to a common point called a bus and operated in parallel to provide the total required power.

**Transformers.** Another key component of a power system is the transformer. Its role is to transform power with very high efficiency (e.g., 98%) from one

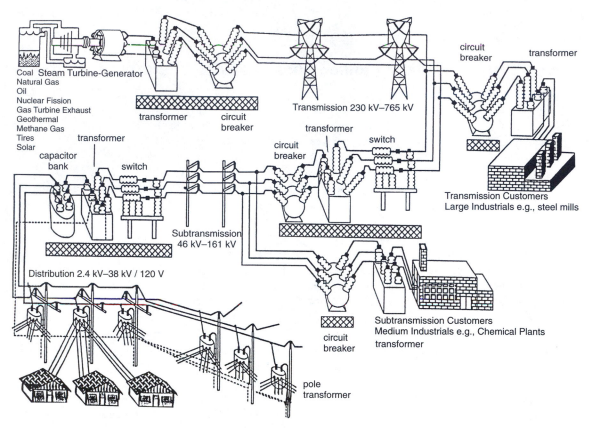

**FIGURE 7.1** Basic elements of a power system.

level of voltage to another. Using a step-up transformer of turns ratio a will reduce the secondary current by a ratio of $1/a$. This will reduce the real power loss $I^2R$ and the reactive power $I^2X$ of the transmission line, which makes the transmission of power over long distances feasible.

Due to insulation requirements and other practical design problems, terminal voltages of generators are limited to relatively low values, typically in the 30–40 kV range. Consequently, step-up transformers are used for transmission of power. At the receiving end of a transmission line, step-down transformers are employed to reduce the voltage to suitable levels for distribution or direct utilization. The power may go through several transformers between generator and the end user.

**Transmission and Subtransmission.** Overhead transmission lines permit the transfer of energy from generating units to the distribution system that serves the end users. Transmission lines also interconnect neighboring utilities to permit economic dispatch of power within regions during normal operating conditions, as well as transfer of power between regions

during emergencies. Transmission-line voltage levels above 60 kV are standardized at 69 kV, 115 kV, 138 kV, 161 kV, 230 kV, 345 kV, 500 kV, and 745 kV. Transmission voltages above 230 kV are referred to as extrahigh voltage (EHV). High-voltage transmission lines are terminated at substations.

There are three types of substations: high-voltage substations, receiving substations, and primary substations. At the primary substations, the voltage is stepped down through a transformer to a value suitable for end users. For example, large industrial customers may be served directly from the transmission system.

The segment of the transmission system that connects high-voltage substations through step-down transformers to distribution substations is called the subtransmission network. There is no clear delineation between transmission and subtransmission voltage levels. Typically, the subtransmission voltage levels range from 69 kV to 138 kV. Some large industrial customers may be served from the subtransmission system. Capacitor and reactor banks are usually installed in subtransmission substations for maintaining or controlling the transmission line voltage.

It is well known that for distribution voltages higher than 15 kV ferroresonance may occur if cables instead of overhead lines are used. To avoid this problem, gas-insulated underground cables are relied on to deliver in cities the power at high voltages.

**Distribution.** The distribution system (in the range of 4 kV to 34.5 kV) connects distribution substations to end users. Some small industrial consumers may be served directly by primary feeders. The secondary distribution network reduces the voltage for utilization by commercial and residential consumers. Either overhead transmission lines or cables are employed to deliver power to consumers. Typical secondary distribution systems operate at levels of 240/120 V (single-phase, three-wire), 208Y/120 V (three phase, four wire), and 480Y/277 V (three-phase, four-wire). The power for residential loads in the United States are derived from transformers that reduce the primary feeder voltage to 240/120 V using a single-phase, three-wire line. Distribution systems are made up of both overhead and underground lines. The growth of underground distribution has been extremely rapid, especially in new residential areas.

**Linear and Nonlinear Loads.** Power systems serve linear and nonlinear loads. Loads are divided into three key categories within a power system: industrial, commercial, and residential. Industrial loads are functions of both voltage and frequency. Induction motors form a high proportion (e.g., 60%) of these loads. Commercial and residential loads consist largely of lighting, heating, and cooling (e.g., induction motors). These loads are largely independent of frequency and consume reactive power with a total power factor of larger than TPF = 80%.

The nonlinear $v$–$i$ characteristics of nonlinear loads generate both current and voltage harmonics and call for nonsinusoidal analysis and modeling of

power systems. A typical power system consists of a few large industrial nonlinear loads (e.g., large rectifier bridges, large variable-speed drives); however, the number of small nonlinear loads (such as power electronic rectifiers, high-efficiency lighting, and small variable-speed drives) is rapidly increasing. This requires attention to such nonlinear loads during the design, planning, and operation of distribution systems.

## 7.2 POWER SYSTEM MATRICES

Frequency- and time-domain techniques employ vectors and matrices for mathematical modeling of power systems and for defining input and output variables. This section deals with the bus admittance and the Jacobian matrices, which are essential components of all power flow formulations.

### 7.2.1 Bus Admittance Matrix

Table 7.1 presents a review of circuit-component definitions and laws. Ohm's and Kirchhoff's laws can be used for matrix formulations that are useful for load flow analyses. This approach yields two matrices:

- Kirchhoff's current law (KCL) generates bus matrices, which result from the fact that the sum of currents entering a bus is zero.
- Kirchhoff's voltage law (KVL) results in loop matrices, which describe voltages in a circuit loop.

The bus matrix written in terms of admittance originates directly from the KCL applied to each circuit bus other than the reference (ground) bus.

### 7.2.1.1 Application Example 7.1: A Simple Power System Configuration

A simple power system configuration is shown in Fig. E7.1.1 where bus 1 is called the swing or

**TABLE 7.1** Review of Circuit Definitions and Laws

| Circuit-component definitions | | Circuit laws | |
|---|---|---|---|
| Resistance | $R$ | Ohm's law | $V = IZ$ |
| conductance | $G = 1/R$ | | $V = I/Y$ |
| Inductance | $L$ | Kirchhoff's voltage-loop law | $\Sigma V = 0$, leads to loop matrix |
| reactance (inductive) | $j\omega L$ | (KVL) | |
| Capacitance | $C$ | Kirchhoff's current-node law | $\Sigma I = 0$, leads to bus matrix |
| reactance (capacitive) | $1/j\omega C = -j/\omega C$ | (KCL) | |
| Impedance | $Z = R + j\omega L$ (e.g., inductor) | | |
| | $Z = R - j/\omega C$ (e.g., capacitor) | | |
| Admittance | $Y = 1/Z$ | | |
| Susceptance | $B = 1/X$ | | |

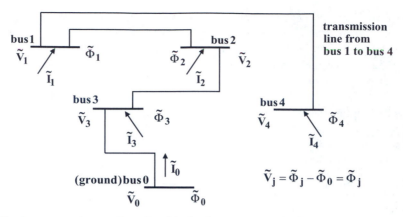

**FIGURE E7.1.1** Simple power system configuration (single-phase representation).

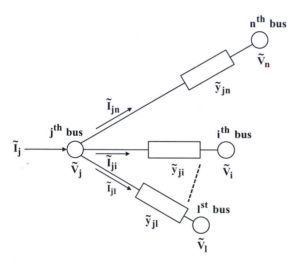

**FIGURE 7.2** Currents entering and leaving a circuit bus.

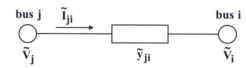

**FIGURE 7.3** Application of Ohm's law.

slack bus and bus 0 is called the reference or ground bus.

Let $j$ be an arbitrary circuit bus other than the reference bus (which is denoted as bus 0), as shown in Fig. 7.2. Then

$$\sum_{i=1}^{n} \tilde{I}_{ji} = \tilde{I}_{j}, \qquad (7\text{-}1)$$

where $\tilde{I}_{ji} \equiv$ line current, which leaves bus $j$ and enters bus $i$ via a line and $\tilde{I}_{j} \equiv$ current injected at bus $j$ from an external source or bus injection current. The tilde (~) above $\tilde{I}_{ji}$ and $\tilde{I}_{j}$ denotes a complex line current and a complex bus injection current, respectively.

Application of Ohm's law (Fig. 7.3) leads to

$$\tilde{I}_{ji} = \tilde{y}_{ji}(\tilde{V}_{j} - \tilde{V}_{i}), \qquad (7\text{-}2)$$

where $\tilde{y}_{ji}$ is the complex line admittance of a line from bus $j$ to bus $i$. The tilde notation in $\tilde{y}_{ji}$ indicates that it is a line admittance.

Certain elements of $\sum_{i=1}^{n} \tilde{I}_{ji} = \tilde{I}_{j}$ are zero:

- The current $\tilde{I}_{jj}$ is zero since no line joins bus $j$ with bus $j$ and this is reflected in $\tilde{I}_{jj} = \tilde{y}_{jj}(\tilde{V}_{j} - \tilde{V}_{j})$ because it gives zero for $\tilde{I}_{jj}$ due to $(\tilde{V}_{j} - \tilde{V}_{j}) = 0$; and
- $\tilde{I}_{ji}$ will be zero if no line exists between $j$ and $i$, the "line impedance" is infinite, and the "line admittance" $\tilde{y}_{ji}$ is zero. One can write now (introducing Eq. 7-2 into Eq. 7-1)

$$\sum_{i=1}^{n} \tilde{y}_{ji}(\tilde{V}_{j} - \tilde{V}_{i}) = \tilde{I}_{j}. \qquad (7\text{-}3)$$

Note that $\tilde{V}_{j}$ and $\tilde{V}_{i}$ are bus voltages with respect to bus 0, the reference (ground) bus. Equation 7-3 is written for $i = 1, 2, \ldots, n$ and the coefficient of voltage $\tilde{V}_{j}$ in the Kirchhoff equation at bus $j$ is $\sum_{i=1}^{n} \tilde{y}_{ji}$ whereas the coefficient of $\tilde{V}_{i}$ $(i \neq j)$ in the $j$th equation is $(-\tilde{y}_{ji})$.

All equations now read $(n \times n$ equation system) For $j = 1$:

$$\tilde{I}_{1} = \left( \sum_{\substack{i=1 \\ i \neq 1}}^{n} \tilde{y}_{1i} \right) \tilde{V}_{1} + (-\tilde{y}_{12})\tilde{V}_{2} + \ldots + (-\tilde{y}_{1n})\tilde{V}_{n}, \qquad (7\text{-}4a)$$

For $j = 2$:

$$\tilde{I}_2 = (-\tilde{y}_{21})\tilde{V}_1 + \left(\sum_{\substack{i=1 \\ i\neq 2}}^{n} \tilde{y}_{2i}\right)\tilde{V}_2 + \ldots + (-\tilde{y}_{2n})\tilde{V}_n, \quad (7\text{-}4b)$$

For $j = n$:

$$\tilde{I}_n = (-\tilde{y}_{n1})\tilde{V}_1 + (-\tilde{y}_{n2})\tilde{V}_2 + \ldots + \left(\sum_{\substack{i=1 \\ i\neq n}}^{n} \tilde{y}_{ni}\right)\tilde{V}_n. \quad (7\text{-}4c)$$

Observe the pattern of the coefficients of the bus voltages on the right-hand side of the equation system. This pattern will be generalized as the rule for the formation of a matrix of coefficients of $\tilde{V}_j$. These $n$ equations can be written in matrix form

$$\bar{I}_{\text{bus}} = \bar{Y}_{\text{bus}}\bar{V}_{\text{bus}}, \quad (7\text{-}5)$$

where $\bar{I}_{\text{bus}}$ is the bus injection vector representing the $n$ phasor injection currents, $\bar{Y}_{\text{bus}}$ is the bus admittance matrix (it is an $n \times n$ matrix of coefficients formed as indicated above), and $\bar{V}_{\text{bus}}$ is the bus voltage vector of $n$ phasor voltages referenced to bus 0:

$$\tilde{V}_n = \tilde{\Phi}_n - \tilde{\Phi}_0 = \tilde{\Phi}_n, \quad (7\text{-}6)$$

where $\tilde{\Phi}_n$ and $\tilde{\Phi}_0$ are the potentials at bus $n$ and bus 0, respectively, and $\tilde{\Phi}_0 = 0$.

If $\bar{Y}_{\text{bus}}$ is a nonsingular matrix then its inverse $(\bar{Y}_{\text{bus}})^{-1}$ can be found and there exists a solution to the matrix equation. If $\bar{Y}_{\text{bus}}$ is singular then $(\bar{Y}_{\text{bus}})^{-1}$ cannot be found and there exists no solution to the matrix equation.

$$\bar{I}_{\text{bus}} = \bar{Y}_{\text{bus}} \; \bar{V}_{\text{bus}}. \quad (7\text{-}7)$$

$$\underset{\text{row}\quad\text{column}}{\underset{n\times 1}{\uparrow}\quad \underset{n\times n}{\uparrow}\ \underset{n\times 1}{\uparrow}}$$

$$\underbrace{\qquad\qquad}_{n\times 1}$$

Note that the injection currents find a return path through bus 0 (ground). Also note that the Kirchhoff current law is not applied to bus 0 in these equations; however, the total current entering bus 0 is the negative sum of the currents $\tilde{I}_1, \tilde{I}_2, \ldots, \tilde{I}_n$. Therefore, the Kirchhoff current law at bus 0 is

$$-\sum_{i=1}^{n}\tilde{I}_j = -\left[\left(-\tilde{y}_{21} - \tilde{y}_{31} - \ldots - \tilde{y}_{n1} + \sum_{\substack{i=1 \\ i\neq 1}}^{n}\tilde{y}_{1i}\right)\tilde{V}_1 + \right.$$
$$\left.\left(-\tilde{y}_{12} - \tilde{y}_{32} - \ldots - \tilde{y}_{n2} + \sum_{\substack{i=1 \\ i\neq 2}}^{n}\tilde{y}_{2i}\right)\tilde{V}_2 + \ldots\right],$$
$$(7\text{-}8)$$

which can be readily simplified as

$$\tilde{I}_0 = -\tilde{y}_{01}\tilde{V}_1 - \tilde{y}_{02}\tilde{V}_2 - \ldots - \tilde{y}_{0n}\tilde{V}_n, \quad (7\text{-}9)$$

where $\tilde{y}_{0k}$ is the primitive admittance joining buses 0 and $k$.

Because the relation in Eq. 7-9 is a linear combination of the previous $n$ equations, it cannot be used to obtain additional information beyond that given in the $n \times n$ equation system. Inspection of the coefficients of the $\tilde{V}$ terms in the $n \times n$ equation system reveals a rule for the formation of $\bar{Y}_{\text{bus}}$. The following rules are used to "build" the bus admittance matrix referenced to bus 0.

### The Admittance Matrix Building Algorithm

- $y_{jj}$: the diagonal entries of $\bar{Y}_{\text{bus}}$ are formed by summing the admittances of lines terminating at bus $j$. Note there is no tilde on $y_{jj}$.
- $y_{ij}$: the off-diagonal entries are the negatives of the admittances of lines between buses $i$ and $j$. If there is no line between $i$ and $j$, this term is zero. Note there is no tilde on $y_{ij}$.

Usually, bus 0 is the ground bus, but other buses may be chosen as the reference (for specialized applications). Since bus 0 is commonly chosen as the reference, the notation

$$\bar{Y}_{\text{bus}} = \bar{Y}_{\text{bus}}^{(0)} \quad (7\text{-}10)$$

is employed to denote the bus admittance matrix referenced to ground. If some other bus is used as the reference, for example bus k, the following notation is used:

$$\bar{Y}_{\text{bus}}^{(k)}. \quad (7\text{-}11)$$

### 7.2.1.2 Application Example 7.2: Construction of Bus Admittance Matrix (Fig. E7.2.1)

$$-\sum_{j=1}^{4}\tilde{I}_j = -(\tilde{I}_1 + \tilde{I}_2 + \tilde{I}_3 + \tilde{I}_4) = \tilde{I}_0. \quad (E7.1\text{-}1)$$

Inspection of the $\bar{Y}_{\text{bus}}$ building algorithm reveals several properties of the bus matrix:

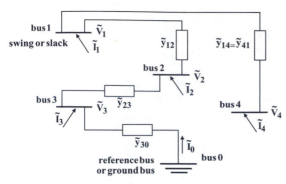

**FIGURE E7.2.1** Example for construction of bus matrix.

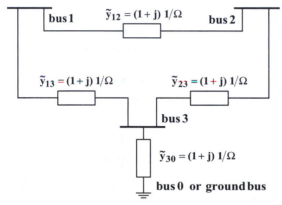

**FIGURE E7.3.1** Building of nonsingular bus admittance matrix for simple power system configuration.

- $\bar{Y}_{\text{bus}}$ is complex and symmetric;
- $\bar{Y}_{\text{bus}}$ is sparse since each bus is connected to only a few nearby buses, and this causes many $y_{ji} = 0$ entries; and
- Provided that there is a net nonzero admittance tie to the reference bus (e.g., ground), $\bar{Y}_{\text{bus}}$ is nonsingular and may be inverted to find $\bar{V}_{\text{bus}}$ from

$$\bar{I}_{\text{bus}} = \bar{Y}_{\text{bus}}\bar{V}_{\text{bus}}, \qquad (7\text{-}12)$$

or

$$\bar{V}_{\text{bus}} = (\bar{Y}_{\text{bus}})^{-1}\bar{I}_{\text{bus}}. \qquad (7\text{-}13)$$

If there are no ties to the reference bus, $\bar{Y}_{\text{bus}}$ is singular and cannot be inverted because no solution to the n × n equation system exists.

### 7.2.1.3 Application Example 7.3: Building of Nonsingular Bus Admittance Matrix (Fig. E7.3.1)

Note: Off-diagonal entries = sum of negatives of the line admittances

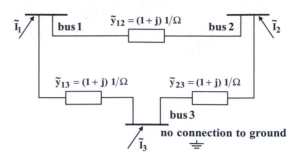

**FIGURE E7.4.1** Building of singular bus admittance matrix for simple power system configuration.

$$y_{21} = y_{12} = -(1+j)\ 1/\Omega,$$
$$y_{31} = y_{13} = -(1+j)\ 1/\Omega,$$
$$y_{32} = y_{23} = -(1+j)\ 1/\Omega,$$
Main-diagonal entry for bus 1: $y_{11} = (2+2j)\ 1/\Omega,$
Main diagonal entry for bus 2: $y_{22} = (2+2j)\ 1/\Omega,$
Main-diagonal entry for bus 3: $y_{33} = (3+3j)\ 1/\Omega,$

Therefore,

$$\bar{Y}_{\text{bus}} = \begin{bmatrix} y_{11} & y_{12} & y_{13} \\ y_{21} & y_{22} & y_{23} \\ y_{31} & y_{32} & y_{33} \end{bmatrix} = \begin{bmatrix} (2+2j) & -(1+j) & -(1+j) \\ -(1+j) & (2+2j) & -(1+j) \\ -(1+j) & -(1+j) & (3+3j) \end{bmatrix}.$$
$$(E7.3\text{-}1)$$

$\bar{Y}_{\text{bus}}$ is complex and symmetric, but not sparse.

### 7.2.1.4 Application Example 7.4: Building of Singular Bus Admittance Matrix (Fig. E7.4.1)

$$\bar{I}_{\text{bus}} = \bar{Y}_{\text{bus}}\bar{V}_{\text{bus}} = \begin{bmatrix} (2+2j) & -(1+j) & -(1+j) \\ -(1+j) & (2+2j) & -(1+j) \\ -(1+j) & -(1+j) & (2+2j) \end{bmatrix}\begin{bmatrix} \tilde{V}_1 \\ \tilde{V}_2 \\ \tilde{V}_3 \end{bmatrix},$$
$$(E7.4\text{-}1)$$

$$\tilde{I}_1 = (2+2j)\tilde{V}_1 - (1+j)\tilde{V}_2 - (1+j)\tilde{V}_3 \qquad (E7.4\text{-}2)$$

$$\tilde{I}_2 = -(1+j)\tilde{V}_1 + (2+2j)\tilde{V}_2 - (1+j)\tilde{V}_3 \qquad (E7.4\text{-}3)$$

$$\tilde{I}_3 = -(1+j)\tilde{V}_1 - (1+j)\tilde{V}_2 + (2+2j)\tilde{V}_3 \qquad (E7.4\text{-}4)$$

Add Eqs. E7.4-2 and E7.4-3:

$$(\tilde{I}_1 + \tilde{I}_2) = (1+j)\tilde{V}_1 + (1+j)\ \tilde{V}_2 - (2+2j)\tilde{V}_3$$

or

$$\underbrace{-(\tilde{I}_1 + \tilde{I}_2)}_{I_3} = -(1+j)\tilde{V}_1 - (1+j)\tilde{V}_2 + (2+2j)\tilde{V}_3$$
$$(E7.4\text{-}5)$$

Note that Eq. E7.4-2 + Eq. E7.4-3 = Eq. E7.4-5, which is the same as Eq. E7.4-4. That is, these equations are dependent. One concludes that the matrix is singular and no solution exists for Eqs. E7.4-2 to E7.4-4.

## 7.2.2 Triangular Factorization

In the United States there exist three power grids: the Eastern, Western, and Texan pools. Because of stability problems these pools are linked with small (100 MW range) DC buses. Nevertheless if even one of the pools is analyzed, the number of buses will be very large: a few thousand buses or more, depending on the detail of modeling. Fortunately, not all buses are mutually linked through a transmission line and for this reason the resulting bus admittance matrix is very sparse, that is, only about 1% of the entire matrix entries are nonzero. However, because of the large dimensions (few thousands) a direct solution (e.g., Gaussian elimination) of the equation system is not feasible and one uses the Newton–Raphson method together with the Jacobian matrix to arrive at the solution via iterations.

To reduce computer memory requirements, sparsity programming is used: it is a digital programming technique whereby sparse matrices are stored in compact forms. The usual basis of sparsity programming techniques is the storage of only nonzero entries. Two broad classes of sparsity programming techniques are

- the entry-row-column storage method, and
- the chained-data-structure method.

Applications of sparsity programming methods are primarily in $\bar{Y}_{bus}$ methods. In this chapter the Jacobian will be used; this matrix is as sparse as the $\bar{Y}_{bus}$ matrix and lends itself to sparsity programming.

In power engineering, as well as in many other branches of engineering, it is necessary to solve a simultaneous set of algebraic linear equations. The general form is the solution of

$$\bar{A} \cdot \bar{x} = \bar{b} \qquad (7\text{-}14)$$

Given are the $n$ by $n$ matrix $\bar{A}$ and the $n$-dimensional vector $\bar{b}$. Such is the case in the solution of

$$\bar{I}_{bus} = \bar{Y}_{bus} \bar{V}_{bus} \qquad (7\text{-}15)$$

or

$$\bar{V}_{bus} = [\bar{Y}_{bus}]^{-1} \bar{I}_{bus} \qquad (7\text{-}16)$$

for $\bar{V}_{bus}$ where the injection currents and the admittance matrix are given. Using the notation of $\bar{A} \cdot \bar{x} = \bar{b}$, let $\bar{A}$ be factored into two matrices called triangular factorization:

$$\bar{A} = \bar{L} \cdot \bar{U} \qquad (7\text{-}17)$$

where $\bar{L}$ is lower left triangular (e.g., $l_{rc}$ is zero for $c > r$) and $\bar{U}$ is upper right triangular (e.g., $u_{rc}$ is zero for $r > c$). Then

$$\bar{A} \cdot \bar{x} = [\bar{L} \cdot \bar{U}]\bar{x} = \bar{b} \qquad (7\text{-}18)$$

Let $(\bar{U} \cdot \bar{x})$ be a new variable $\bar{w}$, which is an $n$-dimensional vector; then

$$\bar{w} = \bar{U} \cdot \bar{x} \qquad (7\text{-}19)$$

$$\bar{L} \cdot \bar{w} = \bar{b} \qquad (7\text{-}20)$$

where

$$\bar{w} = \begin{bmatrix} w_1 \\ w_2 \\ \vdots \\ w_n \end{bmatrix}, \quad \bar{b} = \begin{bmatrix} b_1 \\ b_2 \\ \vdots \\ b_n \end{bmatrix}, \qquad (7\text{-}21\text{a,b})$$

$$\bar{L} = \begin{bmatrix} l_{11} & 0 & 0 & \cdot \\ l_{21} & l_{22} & \cdot & \cdot \\ l_{31} & l_{32} & l_{33} & \cdot \\ \cdot & \cdot & \cdot & \cdot \end{bmatrix}, \quad \bar{U} = \begin{bmatrix} u_{11} & u_{12} & u_{13} & \cdot \\ 0 & u_{22} & u_{13} & \cdot \\ 0 & 0 & u_{33} & \cdot \\ \cdot & \cdot & \cdot & \cdot \end{bmatrix}.$$

$$(7\text{-}22\text{a,b})$$

Note that $\bar{x}$ and $\bar{b}$ correspond to the $\bar{V}_{bus}$ vector and the current vector $\bar{I}_{bus}$, respectively.

### 7.2.2.1 Application Example 7.5: Matrix Multiplication

Given

$$\bar{U} = \begin{bmatrix} u_{11} & u_{12} \\ 0 & u_{22} \end{bmatrix}, \quad \bar{L} = \begin{bmatrix} l_{11} & 0 \\ l_{21} & l_{22} \end{bmatrix}. \qquad (E7.5\text{-}1\text{a,b})$$

Find entries of the matrix $\bar{L} \cdot \bar{U}$.

### 7.2.2.2 Application Example 7.6: Triangular Factorization

Given

$$\bar{U} = \begin{bmatrix} u_{11} & u_{12} \\ 0 & u_{22} \end{bmatrix}, \quad \bar{x} = \begin{bmatrix} x_1 \\ x_2 \end{bmatrix}, \qquad (E7.6\text{-}1\text{a,b})$$

show that $\bar{w} = \bar{U} \cdot \bar{x}$ is a column ($n = 2$) vector.

$$\bar{w} = \bar{U} \cdot \bar{x} = \begin{bmatrix} u_{11} & u_{12} \\ 0 & u_{22} \end{bmatrix} \begin{bmatrix} x_1 \\ x_2 \end{bmatrix} \qquad \text{(E7.6-2)}$$

$$2 \times \boxed{2 \quad 2} \times 1$$

row  column  row  column

$$\bar{w} = \begin{bmatrix} u_{11}x_1 + u_{12}x_2 \\ u_{22}x_2 \end{bmatrix}. \qquad \text{(E7.6-3)}$$

Therefore, $\bar{w}$ is a $2 \times 1$ matrix or a column vector.

**Forward Substitution.** $\bar{L} \cdot \bar{w} = \bar{b}$ is readily solved for $\bar{w}$ since the form of the equation is rather special:

$$\bar{L} = \begin{bmatrix} l_{11} & 0 & 0 & 0 & \cdot \\ l_{21} & l_{22} & 0 & 0 & \cdot \\ l_{31} & l_{32} & l_{33} & 0 & \cdot \\ \cdot & \cdot & \cdot & \cdot & \cdot \end{bmatrix} \begin{bmatrix} w_1 \\ w_2 \\ \vdots \\ w_n \end{bmatrix} = \begin{bmatrix} b_1 \\ b_2 \\ \vdots \\ b_n \end{bmatrix}. \qquad \text{(7-23)}$$

The top row of the matrix equation is solved readily for $w_1$ since the only nonzero entry on the left-hand side is $l_{11}w_1$ and therefore

$$l_{11}w_1 = b_1, \qquad \text{(7-24a)}$$

or

$$w_1 = \frac{b_1}{l_{11}}. \qquad \text{(7-24b)}$$

From row 2 one finds

$$l_{21}w_1 + l_{22}w_2 = b_2, \qquad \text{(7-24c)}$$

or

$$w_2 = \frac{b_2 - l_{21}w_1}{l_{22}}. \qquad \text{(7-24d)}$$

This process of forward substitution continues until all components of $\bar{w}$ are computed.

Having found vector $\bar{w}$, the relation $\bar{w} = \bar{U} \cdot \bar{x}$ is used to find $\bar{x}$:

$$\begin{bmatrix} w_1 \\ w_2 \\ \vdots \\ w_n \end{bmatrix} = \begin{bmatrix} u_{11} & u_{12} & u_{13} & \cdot \\ 0 & u_{22} & u_{23} & \cdot \\ 0 & 0 & u_{33} & \cdot \\ \cdot & \cdot & \cdot & \cdot \end{bmatrix} \begin{bmatrix} x_1 \\ x_2 \\ \vdots \\ x_n \end{bmatrix}. \qquad \text{(7-25)}$$

**Backward Substitution.** The bottom row is readily solved ($n = 3$) for $x_3$:

$$w_3 = u_{33}x_3, \qquad \text{(7-26a)}$$

or

$$x_3 = \frac{w_3}{u_{33}}. \qquad \text{(7-26b)}$$

The second row from the bottom yields

$$w_2 = u_{22}x_2 + u_{23}x_3 \qquad \text{(7-26c)}$$

or

$$x_2 = \frac{w_2 - u_{23}x_3}{u_{22}}. \qquad \text{(7-26d)}$$

**Rule**: The value of $x_n$ is backward substituted into the $(n-1)$st row of the above matrix equation to find $x_{n-1}$ from

$$w_{n-1} = u_{n-1\,n-1}x_{n-1} + u_{n-1\,n}x_n \qquad \text{(7-27)}$$

This process of back substitution is employed to find all other components of $\bar{x}$.

**Note**: By the process of

(*i*)   triangular factorization of $\bar{A}$,
(*ii*)  forward, and
(*iii*) backward substitution,

the solution of $\bar{A} \cdot \bar{x} = \bar{b}$ is found without inverting $\bar{A}$! At first glance, it may appear remarkable that the inversion of $\bar{A}$ may be avoided, but there is some information that lies uncalculated in the use of triangular factorization compared to a solution using inversion. If $\bar{A} \cdot \bar{x} = \bar{b}$ is to be solved for many different values of $\bar{b}$, inversion is readily used since a given vector $\bar{b}$ premultiplied by $\bar{A}^{-1}$ gives the required result $\bar{x} = \bar{A}^{-1} \cdot \bar{b}$. In such an application $\bar{A}^{-1}$ is calculated only once.

If triangular factorization is used, the forward/backward substitution process would have to be repeated for each different $\bar{b}$ vector. In power engineering applications, however, it is usually necessary to solve $\bar{A} \cdot \bar{x} = \bar{b}$ only once: the inverse of $\bar{A}$ is not really desired – only $\bar{x}$ is sought. The triangular factorization method is favored for most applications.

The success of the triangular factorization method depends on

- the speed of the factorization of $\bar{A}$ into $\bar{L}$ and $\bar{U}$, and
- the sparsity of $\bar{L}$ and $\bar{U}$.

A sparse $\bar{A}$ is of considerable interest because if the triangular matrices were sparse, the memory advantages of sparsity programming may be added to the speed advantages of triangular factorization.

**Formation of the Table of Factors.** Let $\bar{L}$ have unit (main) diagonal entries

$$\bar{L} = \begin{bmatrix} 1 & 0 & 0 & \cdot \\ l_{21} & 1 & 0 & \cdot \\ l_{31} & l_{32} & 1 & \cdot \\ \cdot & \cdot & \cdot & \cdot \end{bmatrix}, \quad \bar{U} = \begin{bmatrix} u_{11} & u_{12} & u_{13} & \cdot \\ 0 & u_{22} & u_{23} & \cdot \\ 0 & 0 & u_{33} & \cdot \\ \cdot & \cdot & \cdot & \cdot \end{bmatrix}$$

(7-28a,b)

and let $\bar{U}$ be stored superimposed onto $\bar{L}$ (omitting the writing of the unit entries of $\bar{L}$):

$$\begin{bmatrix} u_{11} & u_{12} & u_{13} & \cdot \\ l_{21} & u_{22} & u_{23} & \cdot \\ l_{31} & l_{32} & u_{33} & \cdot \\ \cdot & \cdot & \cdot & \cdot \end{bmatrix}.$$

(7-29)

This superimposed array is called the table of factors. The upper right triangle of the table is $\bar{U}$; the lower left triangle is $\bar{L}$ with the unit diagonal entries deleted.

### 7.2.3 Jacobian Matrix

The proposal to use for a partial derivative "$\partial$" instead of "d" stems from Carl Gustav Jacob Jacobi, 1804–1851, a German mathematician. At about the same time as Jacobi, the Russian mathematician Michail Vasilevich Ostrogradski (1801–1862) also introduced the Jacobian of $n$ functions $y_1, y_2, \ldots, y_n$ and $n$ unknowns $x_1, x_2, \ldots, x_n$ as follows:

$$\bar{J} = \begin{bmatrix} \dfrac{\partial y_1}{\partial x_1} & \dfrac{\partial y_1}{\partial x_2} & \cdots & \dfrac{\partial y_1}{\partial x_n} \\ \dfrac{\partial y_2}{\partial x_1} & \dfrac{\partial y_2}{\partial x_2} & \cdots & \dfrac{\partial y_2}{\partial x_n} \\ \vdots & \vdots & \vdots & \vdots \\ \dfrac{\partial y_n}{\partial x_1} & \dfrac{\partial y_n}{\partial x_2} & \cdots & \dfrac{\partial y_n}{\partial x_n} \end{bmatrix}.$$

(7-30)

It is possible to factorize arbitrary nonsingular matrices $\bar{J}$ into $\bar{L}$ and $\bar{U}$ without the need for new computer storage areas: in other words, $\bar{J}$ is destroyed and replaced by the table of factors. The table of factors is approximately as sparse as $\bar{J}$ and, therefore, the table of factors does not require additional storage.

Techniques that result in the destruction of the original data and the replacement of those data by

the solution are known as *in situ* methods. Matrix $\bar{J}$ is factored into $\bar{L}$ and $\bar{U}$ *in situ* and $\bar{J}$ is destroyed and replaced by the table of factors.

The formulas required for the calculation of the table of factors are readily found, noting that

$$\bar{L} \cdot \bar{U} = \bar{J} \; (=\bar{A}) \tag{7-31}$$

Write $\bar{L}$ and $\bar{U}$, and find the entries of $\bar{L}$ and $\bar{U}$, that is, the table of factors:

$$\begin{bmatrix} 1 & 0 & 0 & \cdot \\ l_{21} & 1 & 0 & \cdot \\ l_{31} & l_{32} & 1 & \cdot \\ \cdot & \cdot & \cdot & \cdot \end{bmatrix} \begin{bmatrix} u_{11} & u_{12} & u_{13} & \cdot \\ 0 & u_{22} & u_{23} & \cdot \\ 0 & 0 & u_{33} & \cdot \\ \cdot & \cdot & \cdot & \cdot \end{bmatrix} =$$

$$\bar{J} = \begin{bmatrix} J_{11} & J_{12} & J_{13} & \cdot \\ J_{21} & J_{22} & J_{23} & \cdot \\ J_{31} & J_{32} & J_{33} & \cdot \\ \cdot & \cdot & \cdot & \cdot \end{bmatrix}.$$

(7-32)

#### 7.2.3.1 Application Example 7.7: Jacobian Matrices

Given

$$\begin{bmatrix} 1 & 0 \\ l_{21} & 1 \end{bmatrix} \begin{bmatrix} u_{11} & u_{12} \\ 0 & u_{22} \end{bmatrix} = \begin{bmatrix} J_{11} & J_{12} \\ J_{21} & J_{22} \end{bmatrix}. \tag{E7.7-1}$$

Find $u_{11}, u_{12}, u_{22},$ and $l_{21}$

$$J_{11} = u_{11} \quad \text{or} \quad u_{11} = J_{11}$$
$$J_{12} = u_{12} \quad \text{or} \quad u_{12} = J_{12}$$
$$J_{21} = l_{21}u_{11} \quad \text{or} \quad l_{21} = J_{21}/u_{11}$$
$$J_{22} = l_{21}u_{12} + u_{22} \quad \text{or} \quad u_{22} = J_{22} - l_{21}u_{12}$$

In general,

$$u_{11} = J_{11}$$
$$u_{12} = J_{12}$$
$$\cdot \quad \cdot$$
$$\cdot \quad \cdot$$
$$u_{1c} = J_{1c}$$
$$\cdot \quad \cdot$$
$$\cdot \quad \cdot$$
$$u_{1n} = J_{1n}$$

**Forward and Backward Substitution.** The solution of a simultaneous set of algebraic linear equations by triangular factorization involves two main steps:

- formation of the table of factors, and
- forward/backward substitutions.

Having found the table of factors as indicated previously, the following relations are used to complete the solution:

$$\begin{aligned} \bar{L} \cdot \bar{w} &= \bar{b} \\ \bar{U} \cdot \bar{x} &= \bar{w} \end{aligned}. \qquad (7\text{-}33)$$

Vector $\bar{b}$ is destroyed in this process and converted *in situ* to vector $\bar{w}$. Since $l_{11}$ is unity, $w_1$ is simply $b_1$ as can be seen from the matrix solution

$$\begin{bmatrix} l_{11} & 0 & 0 & 0 & \cdot \\ l_{21} & l_{22} & 0 & 0 & \cdot \\ l_{31} & l_{32} & l_{33} & 0 & \cdot \\ \cdot & \cdot & \cdot & \cdot & \cdot \end{bmatrix} \begin{bmatrix} w_1 \\ w_2 \\ \vdots \\ w_n \end{bmatrix} = \begin{bmatrix} b_1 \\ b_2 \\ \vdots \\ b_n \end{bmatrix}. \qquad (7\text{-}34)$$

Subsequent rows of $\bar{w}$ are found using

$$w_r = b_r - \sum_{q=1}^{r-1} l_{rq} w_q$$

   ↑     ↖

created   destroyed

for $r = 1, 2, \ldots, n$

As this equation is used, $w_1$ replaces $b_1$, $w_2$ replaces $b_2$, and so on.

Having found $w_1$ a similar process is used to find $\bar{x}$:

$$\begin{bmatrix} u_{11} & u_{12} & u_{13} & \cdot \\ 0 & u_{22} & u_{23} & \cdot \\ 0 & 0 & u_{33} & \cdot \\ \cdot & \cdot & \cdot & \cdot \end{bmatrix} \begin{bmatrix} x_1 \\ x_2 \\ \vdots \\ x_n \end{bmatrix} = \begin{bmatrix} w_1 \\ w_2 \\ \vdots \\ w_n \end{bmatrix}. \qquad (7\text{-}35)$$

By inspection,

$$u_{nn} x_n = w_n \qquad (7\text{-}36)$$

or

$$x_n = \frac{w_n}{u_{nn}}. \qquad (7\text{-}37)$$

Again, *in situ* technique is used and $x_n$ replaces $w_n$.

In row $n - 1$:

$$u_{n-1\,n-1} x_{n-1} + u_{n-1\,n} x_n = w_{n-1} \qquad (7\text{-}38)$$

or

$$x_{n-1} = \frac{w_{n-1} - u_{n-1\,n} x_n}{u_{n-1\,n-1}}. \qquad (7\text{-}39)$$

Element $x_{n-1}$ replaces $w_{n-1}$. The process continues for rows $n - 2, n - 3, \ldots, 1$. In row $r$

$$x_r = \frac{w_r - \sum_{q=r+1}^{n} u_{rq} x_q}{u_{rr}}. \qquad (7\text{-}40)$$

This completes the solution.

## 7.3 FUNDAMENTAL POWER FLOW

The electric power flow problem is the most studied and documented problem in power engineering. It is the calculation of line loading given the generation and demand levels of $P$, $Q$, and $S$ measured in kW, kVAr, kVA, respectively. The transmission network is (nearly) linear, and one might expect the power flow to be a linear problem. However, because power $p = (v \cdot i)$ is a product of voltage $v$ and current $i$, the problem formulation is nonlinear even for a linear transmission network. Additional nonlinearities arise from the specification and use of complex voltages and currents. Also, there are transmission component nonlinearities that may be considered, such as tap changing transformers, in which the tap is adjusted to hold a given bus voltage at a fixed magnitude.

The power flow formulation does not consider time variations of loads, generation, or network configuration. The sinusoidal steady state is assumed and, as a result, equations are algebraic in form rather than differential.

Prior to 1940 there were a limited number of interconnected power systems and the servicing of the load was elementary since the systems were primarily radial circuits (Fig. 7.4). In more recent years, the

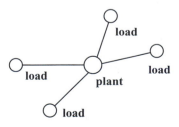

**FIGURE 7.4** Radial power systems.

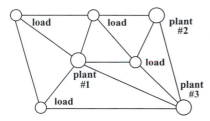

**FIGURE 7.5** Interconnected power systems.

numerous advantages of interconnection were recognized, and modern power systems are characterized by a high degree of interconnection (Fig. 7.5), for example, within the three U.S. power grids (Western, Eastern, and Texan pools). Many loop circuits with many load/generation buses and high levels of power exchange between neighboring companies exist. The latter point relates closely to interconnection advantages.

With no addition of generation capacity, it is possible to increase the generation available through interconnection. For example, in the case of an outage of a generating unit, power may be purchased from a neighboring company. The cost savings realized from lower installed capacity usually far outweigh the cost of the transmission circuits required to access neighboring companies. The analysis of large interconnected power systems generally involves the simultaneous solution of many (thousands) nonlinear algebraic equations using the Newton–Raphson approach. Like most engineering problems, the power flow problem has a set of given data and a set of quantities that must be calculated.

In this section, each complex unknown will be counted as two real unknowns, and each complex equation will be considered as two real equations. The objective of fundamental power flow is to calculate line loading for given generation and load levels ($P$, $Q$, and $S$).

### 7.3.1 Fundamental Bus Admittance Matrix

As discussed and formulated in the previous section, for an n bus system (not including the ground bus, which is the reference bus), the fundamental bus admittance matrix is an $n \times n$ matrix:

$$\bar{Y}_{\text{bus}} = \begin{bmatrix} y_{11} & y_{12} & \cdots & y_{1n} \\ y_{21} & y_{22} & \cdots & y_{2n} \\ \vdots & \vdots & \vdots & \vdots \\ y_{n1} & y_{n2} & \cdots & y_{nn} \end{bmatrix}, \quad (7\text{-}41)$$

where $f_1 = 60$ Hz and $\omega_1 = 2\pi f_1 = 377$ rad/s.

The entries of $\bar{Y}_{\text{bus}}$ are found as follows:

Main-diagonal entries:

$y_{kk} =$ sum of line admittances connected to bus $k$ ($k = 1, 2, \ldots, n$).

Off-diagonal entries:

$y_{kj} = -$(line admittance connected between buses $k$ and $j$).

Since $y_{kj} = y_{jk}$, the matrix $\bar{Y}_{\text{bus}}$ is symmetric.

### 7.3.2 Newton–Raphson Power Flow Formulation

The power (load) flow problem consists of calculations of power flows (real, reactive) and voltages (magnitude, phase) of a network for specified (demand and generation powers) terminal or bus conditions. All voltages and currents are assumed to have fundamental ($h = 1$, $f_1 = 60$ Hz) frequency components. A single-phase representation is considered to be adequate because power systems are approximately balanced within all three phases. Associated with each bus (e.g., bus $j$) are four quantities:

- real power P,
- reactive power Q,
- root-mean-squared (rms) voltage value $|\tilde{V}|$, and
- phase angle of voltage $\delta$.

Three types of buses are represented and two of the four quantities are specified at each bus:

- **Swing bus.** It is necessary to select one reference bus (called the swing or slack bus) providing the transmission losses because these are unknown until the final solution is available. The voltage amplitude and its phase angle are given at the swing bus (usually 1.0 per unit (pu) and zero radians, respectively), and all other bus voltage angles are measured with respect to the phase angle of this reference bus. The remaining buses are either PV or PQ buses.
- **Voltage controlled (PV) buses.** The real power P and the voltage magnitude (amplitude) $|\tilde{V}|$ are specified for PV buses. These are usually the generation buses at which the injected real power is specified and held fixed by turbine settings (e.g., steam entry valve). A voltage regulator holds $|\tilde{V}|$ fixed at PV buses by automatically varying the generator field excitation.
- **Load (PQ) buses.** These are mostly system loads at which real and reactive powers (P and Q, respectively) are specified.

The following notations will be used:

- For a given bus (e.g., the $j$th bus):
  - $P_j$ = real power
  - $Q_j$ = reactive power
  - $|\tilde{V}_j|$ = rms value of voltage, and
  - $\delta_j$ = angle of voltage.
- For the swing or slack bus ($j = 1$):
  - $|\tilde{V}_1| = 1.0$ pu
  - $\delta_1 = 0$ radians, and
  - It will cover all transmission line losses of the system ("slack").
- For a *PV* bus (e.g., the $j$th bus):
  - It is a generation bus
  - $P_j$ is given by turbine setting, and
  - $|V_j|$ is given by generator excitation.
- For a *PQ* bus (e.g., the $j$th bus):
  - It is a load bus,
  - $P_j$ and $Q_j$ are given by the load demand,
  - because power values are given ($p = v \cdot i$) one obtains an algebraic, nonlinear system of equations.

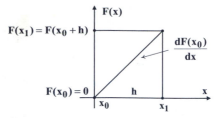

**FIGURE 7.6** Linear extrapolation from coordinate point $x_0, F(x_0)$ to $x_1, F(x_1)$.

**Fundamental Bus Voltage Vector.** For power system modeling at rated frequency, the bus voltage vector is defined as (note: the bar above the $x$ means vector, and $t$ means transpose):

$$\bar{x} = \left[\delta_2, \ |\tilde{V}_2|, \ \delta_3, \ |\tilde{V}_3|, \ \ldots \ \delta_n, \ |\tilde{V}_n|\right]^t \quad (7\text{-}42)$$

where $\delta_j$ is the phase shift of voltage at bus $j$ with respect to the swing bus voltage phase angle and $|\tilde{V}_j|$ is the voltage rms magnitude at bus $j$. Note that if bus 1 is the swing bus, $|\tilde{V}_1|$ and $\delta_1$ are known (usually 1.0 per unit and zero radians).

Although power system components are assumed to be linear in the conventional power flow formulation, the problem is a nonlinear one because power $p$ is the product of voltage $v$ and current $i$. The load flow problem requires the solution of algebraic but nonlinear equations, which are solved using the Newton–Raphson method as will be discussed next.

i) Linear approximation, one-dimensional analysis (Fig. 7.6):

$$F(x_1) = (x_1 - x_0)\frac{dF(x_0)}{dx}, \quad (7\text{-}43)$$

where $h = (x_1 - x_0)$, or

$$F(x_1) = h\frac{dF(x_0)}{dx}, \quad (7\text{-}44)$$

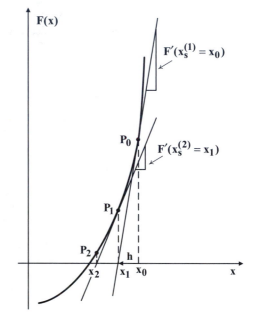

**FIGURE 7.7** Taylor series approximation for one variable, where $h = (x_1 - x_0)$.

$$\left[\frac{dF(x_0)}{dx}\right]^{-1} F(x_1) = (x_1 - x_0), \quad (7\text{-}45)$$

$$x_1 = x_0 + \left[\frac{dF(x_0)}{dx}\right]^{-1} F(x_1), \quad (7\text{-}46a)$$

or

$$x_1 = x_0 + \frac{F(x_1)}{\left[\frac{dF(x_0)}{dx}\right]}. \quad (7\text{-}46b)$$

updated guess    first guess
after first iteration

ii) Taylor series for one variable (Fig. 7.7):

$$F(x) = F(x_0) + \frac{(x - x_0)}{1!}F'(x_0) + \frac{(x - x_0)^2}{2!}F''(x_0) + \ldots$$
$$+ \frac{(x - x_0)^n}{n!}F^{(n)}(x_0) + \ldots, \quad (7\text{-}47)$$

or

$$F(x_0 + h) = F(x_0)$$
$$+ \frac{h}{1!}F'(x_0) + \frac{h^2}{2!}F''(x_0) + \ldots + \frac{h^n}{n!}F^{(n)}(x_0) + \ldots, \tag{7-48}$$

$$F(x_0 + h) \approx F(x_0) + \frac{h}{1!}F'(x_0), \tag{7-49a}$$

$$F(x_1) \approx F(x_0) + (x_1 - x_0)F'(x_0) \tag{7-49b}$$

$$\text{for } F(x_1) = 0 \quad x_1 = x_0 - \frac{F(x_0)}{F'(x_0)}. \tag{7-49c}$$

iii) Taylor series for two variables (replace "$d$" by "$\partial$", Fig. 7.6):

$$F(x_0 + h, y_0 + k) = F(x_0, y_0) + \left\{ \frac{\partial F(x_0, y_0)}{\partial x} h \right.$$
$$+ \frac{\partial F(x_0, y_0)}{\partial y} k \right\} + \frac{1}{2} \left\{ \frac{\partial^2 F(x_0, y_0)}{\partial x^2} h^2 + \right.$$
$$2 \frac{\partial^2 F(x_0, y_0)}{\partial x \partial y} hk + \frac{\partial^2 F(x_0, y_0)}{\partial y^2} k^2 \right\}$$
$$+ \frac{1}{6} \{\ldots\} + \ldots + \frac{1}{h!}\{\ldots\} + R_n, \tag{7-50}$$

where $R_n$ is the remainder.

Equation 7-50 can be written in matrix form neglecting higher order terms:

$$F(x_0 + h, y_0 + k) = F(x_0, y_0)$$
$$+ \left[ \frac{\partial F(x_0, y_0)}{\partial x} \frac{\partial F(x_0, y_0)}{\partial y} \right] \left[ \begin{matrix} h \\ k \end{matrix} \right], \tag{7-51}$$

where

$$h = (x_1 - x_0) \text{ and } k = (y_1 - y_0)$$

or, for $F(x_1, y_1) = 0$,

$$\left[ \begin{matrix} x_1 \\ y_1 \end{matrix} \right] = \left[ \begin{matrix} x_0 \\ y_0 \end{matrix} \right] - \frac{F(x_0, y_0)}{\left[ \dfrac{\partial F(x_0, y_0)}{\partial x} \dfrac{\partial F(x_0, y_0)}{\partial y} \right]}, \tag{7-52}$$

or

$$\left[ \begin{matrix} x_1 \\ y_1 \end{matrix} \right] = \left[ \begin{matrix} x_0 \\ y_0 \end{matrix} \right] - \left[ \frac{\partial F(x_0, y_0)}{\partial x} \frac{\partial F(x_0, y_0)}{\partial y} \right]^{-1} F(x_0, y_0). \tag{7-53}$$

Therefore,

$$\left[ \begin{matrix} x_1 \\ y_1 \end{matrix} \right] = \left[ \begin{matrix} x_0 \\ y_0 \end{matrix} \right] - (\bar{J})^{-1} F(x_0, y_0), \tag{7-54}$$

where the two-dimensional Jacobian $\bar{J}$ is

$$\bar{J} = \left[ \frac{\partial F(x_0, y_0)}{\partial x} \frac{\partial F(x_0, y_0)}{\partial y} \right]. \tag{7-55}$$

**Residuals of an Equation System.** Given the equation system:

$$x_1 - 2x_2 - 4x_3 = 0$$
$$3x_1 + x_2 + 7x_3 = 0.$$
$$7x_1 - x_2 + x_3 = 0$$

The approximate (guessed) solutions are $x_{10}, x_{20}, x_{30}$. Introducing approximate solutions in the equation system:

$$x_{10} - 2x_{20} - 4x_{30} = \text{residual 1}$$
$$3x_{10} + x_{20} + 7x_{30} = \text{residual 2}$$
$$7x_{10} - x_{20} + x_{30} = \text{residual 3}$$

If $\sum_{i=1}^{3} |\text{residual } i| = 0$, then solution of the equation system has been obtained.

**Fundamental Mismatch Power.** The power flow algorithm is based on the Newton–Raphson approach forcing the mismatch powers (at system buses) to zero. The mismatch powers (real and reactive powers) correspond to the residuals of the equation system as discussed above. Specifically the term "mismatch power" at bus $i$ refers to the summation of the complex powers leaving via lines connected to bus $i$ and the specified complex load (demand) power at this bus. Note that the bus itself cannot produce or dissipate power and therefore the mismatch power for a given bus must be zero. Therefore, the mismatch power vector is

$$\Delta \bar{W} = [P_2 + F_{r,2}, Q_2 + F_{i,2}, \ldots P_n + F_{r,n}, Q_n + F_{i,n}]^t, \tag{7-56}$$

where $P_j$ and $Q_j$ are the real and reactive load powers at bus $j$, respectively, and $F_{r,j}$ and $F_{i,j}$ are the real and reactive line powers computed later in this section. Figure 7.8 illustrates the definition of power at a given bus. Note the following:

- $P_2 > 0$ demand (load) of real power,
- $P_2 < 0$ generation of real power,

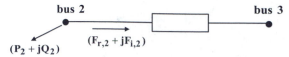

**FIGURE 7.8** Definition of powers at a bus (bus 2).

- $Q_2 > 0$ demand (inductive load) of reactive power, and
- $Q_2 < 0$ generation (capacitive load) of reactive power.

For an $n$-bus system consisting of the swing bus and PQ buses (inclusion of PV buses is considered later), the mismatch power vector has $2(n-1)$ rows corresponding to buses 2 through $n$ (bus 1 is assumed to be the swing bus). Clearly $\Delta \overline{W}$ is a function of $\overline{x}$, the bus voltage vector (Eq. 7-42).

**The Fundamental Newton–Raphson Power Flow Algorithm.** The power flow problem simplifies to finding $\overline{x}$ such that

$$\Delta \overline{W}(\overline{x}) = 0. \qquad (7\text{-}57)$$

Let $\overline{x}_s$ be the bus solution vector and expand Eq. 7-57 in a Taylor series about $\overline{x} = \overline{x}_s$ where $\overline{x}_s$ is the (guessed) solution (see Eqs. 7-46 and 7-52):

$$\Delta \overline{W}(\overline{x}) = \Delta \overline{W}(\overline{x}_s) + \overline{J}(\overline{x} - \overline{x}_s), \qquad (7\text{-}58)$$

or

$$\overline{J}^{-1}\Delta \overline{W}(\overline{x}) = \overline{J}^{-1}\Delta \overline{W}(\overline{x}_s) + (\overline{x} - \overline{x}_s), \qquad (7\text{-}59)$$

where $\overline{J}$ is the Jacobian.

With $\Delta \overline{W}(\overline{x}) = 0$ because $\overline{x}$ is the (converged) solution vector,

$$\overline{x} = \overline{x}_s - \overline{J}^{-1}\Delta \overline{W}(\overline{x}_s). \qquad (7\text{-}60)$$

Due to the linearization (neglect of higher order terms), the solution $\overline{x}_s$ of Eq. 7-60 is an approximation which only can be improved by iterations. Therefore,

$$\overline{x}^{\xi+1} = \overline{x}^{\xi} - \overline{J}^{-1}\Delta \overline{W}(\overline{x}^{\xi}) \qquad (7\text{-}61)$$

or with

$$\Delta \overline{x}^{\xi} = \overline{J}^{-1}\Delta \overline{W}(\overline{x}^{\xi}) \qquad (7\text{-}62)$$

$$\overline{x}^{\xi+1} = \overline{x}^{\xi} - \Delta \overline{x}^{\xi} \qquad (7\text{-}63)$$

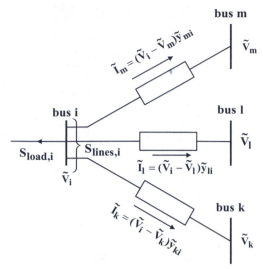

**FIGURE 7.9** Definition of load and line apparent powers as well as line currents.

where superscript $\xi$ indicates the iteration number and $\overline{J}$ is a $2(n-1) \times 2(n-1)$ matrix consisting of partial derivatives of mismatch powers with respect to the bus voltages. Note that all quantities are complex (thus the 2 in front of $(n-1)$) and the swing bus voltage (magnitude and phase angle) is known (therefore $n-1$). If $\Delta \overline{W}(\overline{x}_s) = 0$ then

$$\overline{x} = \overline{x}_s - \overline{J}^{-1}\Delta \overline{W}(\overline{x}_s) = \overline{x}_s. \qquad (7\text{-}64)$$

The evaluation of Eq. 7-61 (the solution) is done by computer using triangular factorization and forward/backward substitution. This works for a few thousands of buses quite well. In case of hand calculations, where there are just a few buses, we use matrix inversion.

### 7.3.3 Fundamental Jacobian Entry Formulas

Figure 7.9 shows a typical $i$ bus of an $n$-bus system. This $i$ bus is connected to three other buses $m$, $k$, and $l$. The load connected to bus $i$ demands a complex power $S_{\text{load},i}$.

The total complex (apparent) power leaving bus $i$ via the lines to buses $k$, $l$, $m$ is

$$S_{\text{lines},i} = \tilde{V}_i[\tilde{I}_k + \tilde{I}_l + \tilde{I}_m]^*, \qquad (7\text{-}65)$$

where* denotes the complex conjugate and $\tilde{y}_{ij}$ are the line admittances:

$$S_{\text{lines},i} = \tilde{V}_i[(\tilde{V}_i - \tilde{V}_k)\tilde{y}_{ki} + (\tilde{V}_i - \tilde{V}_l)\tilde{y}_{li} + (\tilde{V}_i - \tilde{V}_m)\tilde{y}_{mi}]^*, \qquad (7\text{-}66)$$

or

$$S_{\text{lines},i} = \tilde{V}_i[\tilde{V}_i(\tilde{y}_{ki} + \tilde{y}_{li} + \tilde{V}_{mi}) - \tilde{y}_{ki}\tilde{V}_k - \tilde{y}_{li}\tilde{V}_l - \tilde{y}_{mi}\tilde{V}_m]^*. \tag{7-67}$$

With the entries of the admittance matrix

$$y_{ii} = (\bar{Y}_{\text{bus}})_{ii} = \tilde{y}_{ki} + \tilde{y}_{li} + \tilde{y}_{mi},$$
$$y_{ki} = (\bar{Y}_{\text{bus}})_{ki} = -\tilde{y}_{ki},$$
$$y_{li} = (\bar{Y}_{\text{bus}})_{li} = -\tilde{y}_{li},$$
$$y_{mi} = (\bar{Y}_{\text{bus}})_{mi} = -\tilde{y}_{mi},$$

it follows

$$S_{\text{lines},i} = [(\tilde{Y}_{\text{bus}})_{ii}\tilde{V}_i + (\bar{Y}_{\text{bus}})_{ki}\tilde{V}_k +$$
$$(\bar{Y}_{\text{bus}})_{li}\tilde{V}_l + (\bar{Y}_{\text{bus}})_{mi}\tilde{V}_m]^*\tilde{V}_i. \tag{7-68}$$

Using the notation

$$\bar{V}_{\text{bus}} = \begin{bmatrix} \tilde{V}_i \\ \tilde{V}_k \\ \tilde{V}_l \\ \tilde{V}_m \end{bmatrix}, \tag{7-69}$$

one can write in a condensed manner

$$S_{\text{lines},i} = [(\text{row } i \text{ of } \bar{Y}_{\text{bus}})\bar{V}_{\text{bus}}]^*\tilde{V}_i. \tag{7-70}$$

Therefore, for the PQ buses we have the mismatch powers

$$\Delta W_i = S_{\text{load},i} + S_{\text{lines},i} \tag{7-71}$$

$$= S_{\text{load},i} + [(\text{row } i \text{ of } \bar{Y}_{\text{bus}})\bar{V}_{\text{bus}}]^*\tilde{V}_i, \tag{7-72}$$

where $S_{\text{load},i}$ is the load (demand) complex (apparent) power at bus $i$ ($S_{\text{load},i} = P_i + jQ_i$). Breaking the mismatch power $\Delta W_i$ into a real part $\Delta P_i$ and an imaginary part $\Delta Q_i$, we have the real and reactive mismatch powers, respectively:

$$\Delta P_i = P_i + \text{Real}\{[(\text{row } i \text{ of } \bar{Y}_{\text{bus}})\bar{V}_{\text{bus}}]^*\tilde{V}_i\} \tag{7-73}$$

and

$$\Delta Q_i = Q_i + \text{Imag}\{[(\text{row } i \text{ of } \bar{Y}_{\text{bus}})\bar{V}_{\text{bus}}]^*\tilde{V}_i\} \tag{7-74}$$

with $P_i$ and $Q_i$ being the real and reactive load (demand) powers at bus $i$.

**Inspection of Matrix Expression** (Eq. 7-70). Let us inspect the expression $\tilde{V}_i$ [(row $i$ of $\bar{Y}_{\text{bus}})\bar{V}_{\text{bus}}]^*$ (with

entries of admittance matrix in terms of line admittances, note minus signs):

$$\tilde{V}_i[(\text{row } i \text{ of } \bar{Y}_{\text{bus}})\bar{V}_{\text{bus}}]^* = \tilde{V}_i[\tilde{V}_i(\tilde{y}_{ki} + \tilde{y}_{li} + \tilde{V}_{mi}) -$$
$$\tilde{y}_{ki}\tilde{V}_k - \tilde{y}_{li}\tilde{V}_l - \tilde{y}_{mi}\tilde{V}_m]^* \tag{7-75}$$

with

$$\tilde{V}_i = V_i e^{j\delta_i},$$
$$\tilde{V}_k = V_k e^{j\delta_k},$$
$$\tilde{V}_l = V_l e^{j\delta_l},$$
$$\tilde{y}_{ki} = |\tilde{y}_{ki}|e^{j\theta_{ki}}. \tag{7-76}$$

Therefore, for the complex (apparent) power between buses k and $i$ one gets

$$\tilde{V}_i[(-\tilde{y}_{ki})\tilde{V}_k]^* = |\tilde{V}_i|e^{j\delta_i}(-|\tilde{y}_{ki}|)e^{-j\theta_{ki}}|\tilde{V}_k|e^{-j\delta_k} =$$
$$|\tilde{V}_i||\tilde{V}_k|(-|\tilde{y}_{ki}|)e^{j(-\theta_{ki}-\delta_k+\delta_i)} = F_{r,ki} + F_{i,ki}, \tag{7-77}$$

$$F_{i,ki} = V_i V_k(-|\tilde{y}_{ki}|)\cos(-\theta_{ki} - \delta_k + \delta_i), \tag{7-78}$$

and

$$F_{r,ki} = V_i V_k(-|\tilde{y}_{ki}|)\sin(-\theta_{ki} - \delta_k + \delta_i) \tag{7-79}$$

or for all buses connected to bus $i$,

$$\Delta P_i = P_i + F_{r,i} = P_i + \sum_{j=1}^{n} y_{ij} V_j V_i \cos(-\theta_{ij} - \delta_j + \delta_i) \tag{7-80}$$

and

$$\Delta Q_i = Q_i + F_{i,i} = Q_i + \sum_{j=1}^{n} y_{ij} V_j V_i \sin(-\theta_{ij} - \delta_j + \delta_i), \tag{7-81}$$

where $y_{ij}$ and $\theta_{ij}$ are the magnitude and the phase angle of the $(i,j)$th entry of the admittance matrix, respectively, and $V_j = |\tilde{V}_j|$ and $\delta_j$ are the voltage rms magnitude and phase angle at bus $j$.

The Jacobian matrix is defined as

$$\bar{J} = \begin{bmatrix} \dfrac{\partial \Delta P_2}{\partial \delta_2} & \dfrac{\partial \Delta P_2}{\partial V_2} & \cdots & \dfrac{\partial \Delta P_2}{\partial \delta_n} & \dfrac{\partial \Delta P_2}{\partial V_n} \\[2mm] \dfrac{\partial \Delta Q_2}{\partial \delta_2} & \dfrac{\partial \Delta Q_2}{\partial V_2} & \cdots & \dfrac{\partial \Delta Q_2}{\partial \delta_n} & \dfrac{\partial \Delta Q_2}{\partial V_n} \\[2mm] \vdots & \vdots & \vdots & \vdots & \vdots \\[2mm] \dfrac{\partial \Delta P_n}{\partial \delta_2} & \dfrac{\partial \Delta P_n}{\partial V_2} & \cdots & \dfrac{\partial \Delta P_n}{\partial \delta_n} & \dfrac{\partial \Delta P_n}{\partial V_n} \\[2mm] \dfrac{\partial \Delta Q_n}{\partial \delta_2} & \dfrac{\partial \Delta Q_n}{\partial \delta_n} & \cdots & \dfrac{\partial \Delta Q_n}{\partial \delta_n} & \dfrac{\partial \Delta Q_n}{\partial V_n} \end{bmatrix}, \tag{7-82a}$$

where

- The main-diagonal entries of the Jacobian are

$$\frac{\partial \Delta P_i}{\partial \delta_i} = -\sum_{\substack{j=1 \\ j \neq i}}^{n} y_{ij} V_i V_j \sin(\delta_i - \delta_j - \theta_{ij}). \qquad (7\text{-}82b)$$

- Off-diagonal entry:

$$\frac{\partial \Delta P_i}{\partial \delta_k} = y_{ik} V_i V_k \sin(\delta_i - \delta_k - \theta_{ik}) \quad k \neq i. \qquad (7\text{-}82c)$$

(Note that the negative sign cancels with negative sign due to derivative of $-\delta_k$).

- Main-diagonal entry:

$$\frac{\partial \Delta P_i}{\partial V_i} = \sum_{\substack{j=1 \\ j \neq i}}^{n} y_{ij} V_j \cos(\delta_i - \delta_j - \theta_{ij}) + 2V_i y_{ii} \cos(-\theta_{ii}).$$

$$(7\text{-}82d)$$

- Off-diagonal entry:

$$\frac{\partial \Delta P_i}{\partial V_k} = y_{ik} V_i \cos(\delta_i - \delta_k - \theta_{ik}) \quad k \neq i. \qquad (7\text{-}82e)$$

- Main-diagonal entry:

$$\frac{\partial \Delta Q_i}{\partial \delta_i} = \sum_{\substack{j=1 \\ j \neq i}}^{n} y_{ij} V_i V_j \cos(\delta_i - \delta_j - \theta_{ij}). \qquad (7\text{-}82f)$$

- Off-diagonal entry:

$$\frac{\partial \Delta Q_i}{\partial \delta_k} = -y_{ik} V_i V_k \cos(\delta_i - \delta_k - \theta_{ik}) \quad k \neq i. \qquad (7\text{-}82g)$$

- Main-diagonal entry:

$$\frac{\partial \Delta Q_i}{\partial V_i} = \sum_{\substack{j=1 \\ j \neq i}}^{n} y_{ij} V_j \sin(\delta_i - \delta j - \theta_{ij}) + 2V_i y_{ii} \sin(-\theta_{ii}).$$

$$(7\text{-}82h)$$

- Off-diagonal entry:

$$\frac{\partial \Delta Q_i}{\partial V_k} = y_{ik} V_i \sin(\delta_i - \delta_k - \theta_{ik}) \quad k \neq i. \qquad (7\text{-}82i)$$

### 7.3.4 Newton–Raphson Power Flow Algorithm

Based on the equations given, the conventional fundamental power flow algorithm is the computation

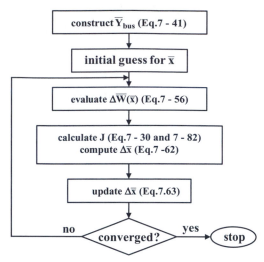

**FIGURE 7.10** Fundamental Newton–Raphson power flow algorithm.

of the bus voltage vector for a given system configuration and linear loads. The Newton–Raphson approach (Eq. 7-64) will be used to force the mismatch power to zero (Eq. 7-57). The iterative solution procedure for the fundamental Newton–Raphson power flow algorithm is as follows (Fig. 7.10):

- **Step 1:** Construct the admittance matrix $\overline{Y}_{bus}$ (Eq. 7-41).
- **Step 2:** Make an initial guess for $\overline{x}$ (e.g., 1.0 pu volts and zero radians for all bus voltages, Eq. 7-42).
- **Step 3:** Evaluate mismatch power ($\Delta \overline{W}(\overline{x})$, Eq. 7-56). If it is small enough then stop.
- **Step 4:** Establish Jacobian $\overline{J}$ (Eqs. 7-30 and 7-82) and calculate $\Delta \overline{x}^{\xi}$ (Eq. 7-62) using either matrix inversion (for problems with a few buses) or upper (lower) triangular factorization combined with backward/forward substitution.
- **Step 5:** Update the bus vector voltage (Eq. 7-63) and go to Step 3.

It is recommended to use upper (lower) triangular factorization combined with backward/forward substitution rather than matrix inversion in computing the solution because an iterative method is used where the coefficient matrix is updated from iteration to iteration.

### 7.3.5 Application Example 7.8: Computation of Fundamental Admittance Matrix

A simple (few buses) power system (Fig. E7.8.1) is used to illustrate in detail the Newton–Raphson load

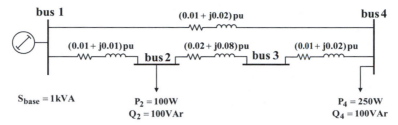

**FIGURE E7.8.1** Four-bus power systems with per unit (pu) line impedances.

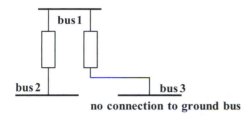

no connection to ground bus

**FIGURE 7.11** No connection to ground results in singular admittance matrix.

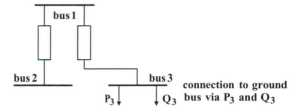

connection to ground bus via $P_3$ and $Q_3$

**FIGURE 7.12** PQ bus (bus 3) results in nonsingular Jacobian and load flow problem has a well-defined solution because the inverse of $\overline{J}$ exists, although $\overline{Y}_{bus}^{-1}$ is singular.

flow analysis. All impedances are in pu and the base apparent power is $S_{base} = 1$ kVA. Compute the fundamental admittance matrix for this system.

**Comment about Singularity**

- Considering

$$[\overline{I}_{bus}] = [\overline{Y}_{bus}]^{-1}[\overline{V}_{bus}],$$

without any connection (admittance) from any bus to ground (Fig. 7.11), $\overline{Y}_{bus}^{-1}$ is singular. However, if $\overline{Y}_{bus}$ contains load information (admittance to ground) then it will not be singular.

- Considering

$$[\Delta \overline{x}] = [\overline{J}]^{-1}[\Delta \overline{W}],$$

if $[\overline{J}]$ contains load information ($P_3$, $Q_3$, Fig. 7.12), then $[\overline{J}]$ is not singular.

### 7.3.6 Application Example 7.9: Evaluation of Fundamental Mismatch Vector

For the system of Fig. E7.8.1, assume bus 1 is the swing bus ($\delta_1 = 0$ and $|\tilde{V}_1| = 1.0$ pu). Then we can make an initial guess for bus voltage vector $\overline{x}^0 = (\delta_2, |\tilde{V}_2|, \delta_3, |\tilde{V}_3|, \delta_4, |\tilde{V}_4|)^t = (0, 1, 0, 1, 0, 1)^t$. Note, the superscript 0 means initial guess. Compute the mismatch power vector for this initial condition.

### 7.3.7 Application Example 7.10: Evaluation of Fundamental Jacobian Matrix

For the system of Fig. E7.8.1, compute the Jacobian matrix.

### 7.3.8 Application Example 7.11: Calculation of the Inverse of Jacobian Matrix

Compute the inverse of the Jacobian matrix (evaluated in Application Example 7.10) using the adjoint and determinant.

### 7.3.9 Application Example 7.12: Inversion of a 3 × 3 Matrix

In order to demonstrate the entire procedure for the computation of the inverse of a matrix the following $3 \times 3$ matrix will be used

$$\overline{A} = \begin{bmatrix} 1 & 3 & -1 \\ 2 & -4 & 2 \\ 3 & 1 & 1 \end{bmatrix}.$$

### 7.3.10 Application Example 7.13: Computation of the Correction Voltage Vector

For the system of Fig. E7.8.1, compute the correction voltage vector $\Delta \overline{x}^0$, updating the mismatch power vector, and check the convergence of the load flow algorithm (using a convergence tolerance of 0.0001).

### 7.4 NEWTON-BASED HARMONIC POWER FLOW

Classifications of solution approaches for the harmonic power (load) flow problem are summarized at the end of this chapter. In this section, the Newton-based harmonic power flow [5, 6] is selected, analyzed, and thoroughly illustrated. There are two main reasons for this selection:

- The Newton-based harmonic power flow is relatively easy to understand and to implement since it is an extension of the conventional fundamental Newton–Raphson formulation; it is based on similar terminologies and equations. It is capable of including any type of nonlinear load assuming its $v$–$i$ characteristic is available in the frequency (or time) domain.
- It was the first proposed approach for power system modeling with nonlinear loads and harmonic couplings [5, 6].

Among the limitations of the Newton-based harmonic power flow in its present form is the assumption of balanced network and load conditions.

### 7.4.1 Harmonic Bus Admittance Matrix and Power Definitions

For nonsinusoidal operating conditions power system matrices will be defined at harmonic frequencies. These matrices include the harmonic bus admittance matrix $\bar{Y}_{bus}^{(h)}$, the harmonic bus impedance matrix $\bar{Z}_{bus}^{(h)} = 1/\bar{Y}_{bus}^{(h)}$, and the harmonic Jacobian $\bar{J}^{(h)}$ matrix.

The harmonic bus admittance matrix $\bar{Y}_{bus}^{(h)}$ is identical to the fundamental bus admittance matrix $\bar{Y}_{bus}^{(1)} = \bar{Y}_{bus}$ with the difference that line admittances are evaluated at the $h$th harmonic frequency $f_h = h \cdot f_1$:

$$\bar{Y}_{bus}^{(h)} = \begin{bmatrix} y_{11}^{(h)} & y_{12}^{(h)} & \cdots & y_{1n}^{(h)} \\ y_{21}^{(h)} & y_{22}^{(h)} & \cdots & y_{2n}^{(h)} \\ \cdot & \cdot & \cdot & \cdot \\ y_{n1}^{(h)} & y_{n2}^{(h)} & \cdots & y_{nn}^{(h)} \end{bmatrix}, \quad (7\text{-}83)$$

where $\omega_h = 2\pi h f_1 = h \cdot 377$ rad/s.

**Power Definitions at Harmonic Frequencies.** Consider a voltage $v(t)$ and a current $i(t)$ expressed in terms of their rms harmonic components

$$v(t) = V_0 + \sum_{h=1}^{\infty} \sqrt{2} V_h \sin(h\omega_1 t + \theta_h^v), \quad (7\text{-}84)$$

$$i(t) = I_0 + \sum_{h=1}^{\infty} \sqrt{2} I_h \sin(h\omega_1 t + \theta_h^i). \quad (7\text{-}85)$$

Then

- the active (real, average) power is given by

$$P = V_0 I_0 + \sum_{h=1}^{\infty} V_h I_h \cos(\theta_h^v - \theta_h^i), \quad (7\text{-}86)$$

- the reactive power is given by

$$Q = \sum_{h=1}^{\infty} V_h I_h \sin(\theta_h^v - \theta_h^i), \quad (7\text{-}87)$$

- and the apparent (voltampere) power is

$$S = \sqrt{\left(\sum_{h=0}^{\infty} I_h^2\right)\left(\sum_{h=0}^{\infty} V_h^2\right)}. \quad (7\text{-}88)$$

In case of sinusoidal $v(t)$ and $i(t)$ ($h = 1$ only), we have

$$S^2 = P^2 + Q^2. \quad (7\text{-}89)$$

For nonsinusoidal cases that relation does not hold and must be replaced by

$$D = \sqrt{S^2 - P^2 - Q^2}, \quad (7\text{-}90)$$

where D is called the distortion power. Other definitions have also been proposed for harmonic powers, which are not discussed here.

**Classification of Time Harmonics.** Time harmonics can be classified into three types (Table 7.2):

- positively rotating harmonics (+),
- negatively rotating harmonics (−), and
- zero-sequence harmonics (0).

Triplen (multiples of three) harmonics ($3h$, for $h = 1, 2, 3, \ldots$) in a balanced three-phase system are predominantly of the zero-sequence type. With a few exceptions (discussed in preceding chapters), zero-sequence currents cannot flow from the network into

**TABLE 7.2** Classification of Time Harmonics

| Positive (+) sequence | Negative (−) sequence | Zero (0) sequence |
| --- | --- | --- |
| 1 | 2 | 3 |
| 4 | 5 | 6 |
| 7 | 8 | 9 |
| 10 | 11 | 12 |
| 13 | 14 | 15 |
| 16 | 17 | 18 |
| 19 | 20 | 21 |
| ⋮ | ⋮ | ⋮ |

a delta or an ungrounded wye-connected device (e.g., transformer), and therefore triplen harmonics are excluded from this analysis.

### 7.4.2 Modeling of Nonlinear and Linear Loads at Harmonic Frequencies

Before proceeding with the reformulation of the Newton–Raphson load flow to include harmonic frequencies, it is necessary to properly model linear and nonlinear loads at fundamental and harmonic frequencies.

**Harmonic Modeling of Nonlinear Loads.** The v–i relationship of nonlinear loads (such as line-commutated converters) will be modeled as coupled harmonic current sources. The injected harmonic currents of a nonlinear load ($I_{injected}^{(h)}$) at bus m will be a function of its fundamental and harmonic voltages:

$$\begin{cases} \text{Real}(I_{injected}^{(h)}) = g_{r,m}^{(h)}(\tilde{V}_m^{(1)}, \tilde{V}_m^{(5)}, \tilde{V}_m^{(7)}, \ldots, \tilde{V}_m^{(L)}, \alpha_m, \beta_m) \\ \text{Imag}(I_{injected}^{(h)}) = g_{i,m}^{(h)}(\tilde{V}_m^{(1)}, \tilde{V}_m^{(5)}, \tilde{V}_m^{(7)}, \ldots, \tilde{V}_m^{(L)}, \alpha_m, \beta_m) \end{cases},$$
$$(7\text{-}91)$$

where $\alpha_m$ and $\beta_m$ are the nonlinear load control parameters and $L$ is the maximum harmonic order considered $L = h_{max}$. The currents of Eq. 7-91 are referred to the nonlinear bus m to which the nonlinear load is connected. The nonlinearity of $g_m^{(h)}$ is due to strong couplings between harmonic currents and voltages dictated by the nonlinear load. In some cases (e.g., decoupled harmonic power flow formulation), it is convenient to ignore harmonic couplings $g_m^{(h)}(\tilde{V}_m^{(h)}, \alpha_m, \beta_m)$ and simplify the problem.

As an example, for a full-wave bridge rectifier connected to bus m (shown in Fig. 7.13), $\alpha_m$ is the firing angle of semiconductor-controlled rectifiers (SCRs), and $\beta_m$ is either the commutating impedance (inductance) or the DC voltage ($E$). Note that the equivalent circuit models a three-phase full-wave rectifier/inverter with a general load, where $L_{com}$ is

the commutating inductance, that is, the leakage inductance of the transformer, and

- for a passive load, only resistance $R$ and filter inductance $F$ are considered,
- for a DC motor drive, only $R$ and $E$ are considered ($E > 0$) when motoring, and
- for inverter operation with a DC generator, only $R$ and $E$ are considered ($E < 0$).

**Modeling Linear Loads at Harmonic Frequencies.** For the formulation of harmonic power flow, equivalent circuits of linear loads are required. At the fundamental frequency, linear loads are modeled as conventional PQ and PV buses. However, shunt admittances are used to model them at harmonic frequencies [7, 8]. The admittance of a linear load connected to bus $k$ at the hth harmonic is

$$y_k^{(h)} = \frac{(P_k^{(1)} - jQ_k^{(1)})/h}{|V_k^{(1)}|^2}. \qquad (7\text{-}92)$$

where $j$ represents the operator $\sqrt{-1}$.

Capacitor banks are modeled as a fixed shunt reactance. Transformers are approximated by linear leakage inductances, and their nonlinearities and losses due to eddy currents, hysteresis, and saturation are neglected.

### 7.4.3 The Harmonic Power Flow Algorithm (Assembly of Equations)

Power system components, loads, and generators that produce time harmonics in an otherwise harmonic-free network are termed nonlinear. Other buses such as the standard PQ and PV buses with no converter or other nonlinear devices connected are termed linear. Consider

- Bus 1 to be the conventional swing (slack) bus with specified values for voltage magnitude $|\tilde{V}_1|$ and phase angle $\delta_1$.
- Buses 2 through ($m - 1$) to be the conventional linear PQ (or PV) buses. As with the conventional power flow problem, active $P_i^{(1)}$ and reactive $Q_i^{(1)}$ powers are assumed to be known at these linear buses.
- Buses $m$ through $n$ to be nonlinear buses. At these buses, the fundamental real power $P_i^{(1)}$ and the total apparent power $S_{i,total}$ (or the total reactive power $Q_{i,total}$) are specified.

Network voltages and currents are represented by Fourier series components, and the bus voltage vector

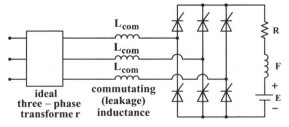

**FIGURE 7.13** Line-commutated converter with general load.

$\bar{x}$ of the fundamental power flow (Eq. 7-42) is redefined as

$$
\begin{aligned}
\bar{U} = [&\delta_2^{(1)}, |\tilde{V}_2^{(1)}|, \ldots, \delta_n^{(1)}, |\tilde{V}_n^{(1)}|, \delta_1^{(5)}, |\tilde{V}_1^{(5)}|, \ldots, \\
&\delta_n^{(5)}, |\tilde{V}_n^{(5)}|, \ldots, \delta_1^{(L)}, |\tilde{V}_1^{(L)}|, \ldots, \delta_n^{(L)}, |\tilde{V}_n^{(L)}|, \\
&\alpha_m, \beta_m, \ldots, \alpha_n, \beta_n]^t = [\bar{V}^{(1)}, \bar{V}^{(5)}, \ldots, \bar{V}^{(L)}, \bar{\Phi}]^t
\end{aligned}
$$
(7-93)

where (superscripts and subscripts denote harmonic orders and bus numbers, respectively)

Fundamental voltages are
$$\bar{V}^{(1)} = \delta_2^{(1)}, |\tilde{V}_2^{(1)}|, \ldots, \delta_n^{(1)}, |\tilde{V}_n^{(1)}|, \quad \text{(7-94a)}$$

Fifth harmonic voltages are
$$\bar{V}^{(5)} = \delta_1^{(5)}, |\tilde{V}_1^{(5)}|, \ldots, \delta_n^{(5)}, |\tilde{V}_n^{(5)}|, \quad \text{(7-94b)}$$

Lth harmonic voltages are
$$\bar{V}^{(L)} = \delta_1^{(L)}, |\tilde{V}_1^{(L)}|, \ldots, \delta_n^{(L)}, |\tilde{V}_n^{(L)}|, \quad \text{(7-94c)}$$

Nonlinear device variables vector is
$$\bar{\Phi} = \alpha_m, \beta_m, \ldots, \alpha_n, \beta_n. \quad \text{(7-95)}$$

Harmonic bus voltage magnitudes, phase angles, and the nonlinear device parameters ($\alpha_m$, $\beta_m$) are unknowns; therefore the conventional formulation (forcing the fundamental mismatch powers to zero) is not sufficient to solve the harmonic power flow problem. Three additional relations are used:

- Kirchhoff's current law for the fundamental frequency (fundamental current balance) at nonlinear buses.
- Kirchhoff's current law for each harmonic (harmonic current balance) at all buses.
- Conservation of apparent voltamperes (apparent power or voltampere balance) at nonlinear buses.

Note that Kirchhoff's current law for the fundamental frequency is not applied to linear buses since active and reactive power (mismatch power) balance has already been applied at these buses.

**Current Balance for Fundamental Frequency.** The current balance is written for fundamental frequency ($h = 1$) at the nonlinear buses as indicated in Fig. 7.14, where the nonlinear currents $g_{r,m}^{(1)}$, $g_{i,m}^{(1)}$ are referred to bus m. Therefore one obtains for the nonlinear buses (from bus m to bus n) for the fundamental current balance:

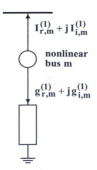

**FIGURE 7.14** Application of current balance ($I_{r,m}^{(1)} + jI_{i,m}^{(1)} + g_{r,m}^{(1)} + jg_{i,m}^{(1)} = 0$) at a nonlinear bus.

$$
\begin{bmatrix}
I_{r,m}^{(1)} \\
I_{i,m}^{(1)} \\
\cdot \\
\cdot \\
\cdot \\
I_{r,n}^{(1)} \\
I_{i,n}^{(1)}
\end{bmatrix}
= -
\begin{bmatrix}
g_{r,m}^{(1)}(\tilde{V}_m^{(1)}, \tilde{V}_m^{(5)}, \ldots, \tilde{V}_m^{(L)}, \alpha_m, \beta_m) \\
g_{i,m}^{(1)}(\tilde{V}_m^{(1)}, \tilde{V}_m^{(5)}, \ldots, \tilde{V}_m^{(L)}, \alpha_m, \beta_m) \\
\cdot \\
\cdot \\
\cdot \\
g_{r,n}^{(1)}(\tilde{V}_n^{(1)}, \tilde{V}_n^{(5)}, \ldots, \tilde{V}_n^{(L)}, \alpha_n, \beta_n) \\
g_{i,n}^{(1)}(\tilde{V}_n^{(1)}, \tilde{V}_n^{(5)}, \ldots, \tilde{V}_n^{(L)}, \alpha_n, \beta_n)
\end{bmatrix},
\quad (7\text{-}96)
$$

where $I_{r,m}^{(1)}$ and $I_{i,m}^{(1)}$ are the fundamental real and reactive line currents at the mth nonlinear bus, respectively, and $g_{r,m}^{(1)}$ and $g_{i,m}^{(1)}$ are the fundamental real and reactive injected load currents at the same bus. Note that the line and injected load currents are positive if they leave the bus.

**Current Balance at Harmonic Frequencies.** For all (linear and nonlinear) buses one can write

$$
\begin{bmatrix}
I_{r,1}^{(h)} \\
I_{i,1}^{(h)} \\
\cdot \\
\cdot \\
\cdot \\
I_{r,m-1}^{(h)} \\
I_{i,m-1}^{(h)} \\
\hline
I_{r,m}^{(h)} \\
I_{i,m}^{(h)} \\
\cdot \\
\cdot \\
\cdot \\
I_{r,n}^{(h)} \\
I_{i,n}^{(h)}
\end{bmatrix}
= -
\begin{bmatrix}
\begin{matrix} 0 \\ \cdot \\ \\ \text{(swing bus and linear buses)} \\ \\ 0 \end{matrix} \\
\hline
g_{r,m}^{(h)}(\tilde{V}_m^{(1)}, \tilde{V}_m^{(5)}, \ldots, \tilde{V}_m^{(L)}, \alpha_m, \beta_m) \\
g_{i,m}^{(h)}(\tilde{V}_m^{(1)}, \tilde{V}_m^{(5)}, \ldots, \tilde{V}_m^{(L)}, \alpha_m, \beta_m) \\
\cdot \\
\text{(nonlinear buses)} \\
\cdot \\
g_{r,n}^{(h)}(\tilde{V}_n^{(1)}, \tilde{V}_n^{(5)}, \ldots, \tilde{V}_n^{(L)}, \alpha_n, \beta_n) \\
g_{i,n}^{(h)}(\tilde{V}_n^{(1)}, \tilde{V}_n^{(5)}, \ldots, \tilde{V}_n^{(L)}, \alpha_n, \beta_n)
\end{bmatrix}
$$
(7-97)

The line currents $I_{r,m}^{(h)}$ and $I_{i,m}^{(h)}$ will be computed in the next section.

**Apparent Power or Voltampere Balance.** The third type of relation is the apparent power (voltampere) balance at each nonlinear bus $j$:

$$S_j^2 = (P_j^t)^2 + (Q_j^t)^2 + D_j^2, \qquad (7\text{-}98)$$

where $j = m, \ldots, n$ and

$$(P_j^t)^2 = \left( \sum_h P_j^{(h)} \right)^2 = (V_j^{(1)} I_j^{(1)} \cos \theta_j^{(1)} \qquad (7\text{-}99a)$$
$$+ V_j^{(5)} I_j^{(5)} \cos \theta_j^{(5)} + V_j^{(7)} I_j^{(7)} \cos \theta_j^{(7)} + \ldots)^2$$

$$(Q_j^t)^2 = \left( \sum_h Q_j^{(h)} \right)^2 = (V_j^{(1)} I_j^{(1)} \sin \theta_j^{(1)} \qquad (7\text{-}99b)$$
$$+ V_j^{(5)} I_j^{(5)} \sin \theta_j^{(5)} + V_j^{(7)} I_j^{(7)} \sin \theta_j^{(7)} + \ldots)^2$$

are the total (fundamental plus harmonics) real and reactive load powers at nonlinear buses. $D_m$ to $D_n$ can be computed from $g_{r,m}, g_{i,m}, \ldots, g_{r,n}, g_{i,n}$; therefore, distortion powers are not independent variables. Equation 7-98 is used to update (between iterations) the nonlinear-device injected powers P and Q at the nonlinear buses.

The number of unknowns for the n buses are as follows ($K$ = number of harmonics considered not including fundamental, e.g., for h = 1, 5, 7, 11 we get $K = 3$):

| | |
|---|---|
| • Fundamental bus voltage magnitudes and phases for all buses but the swing bus | $2(n-1)$ |
| • Fundamental real and reactive powers at the swing bus (losses $P_1$, $Q_1$) | 2 |
| • Harmonic (not including fundamental) voltage magnitudes and phases at all (total) $n$ buses | $2nK$ |
| • Total reactive powers $Q_j^t$ at each nonlinear bus, where $n$ is the number of all buses and $(m-1)$ the number of linear buses | $n-(m-1)$ $= (n-m+1)$ |
| • Two variables ($\alpha_j$, $\beta_j$) associated with each nonlinear bus | $2(n-m+1)$ |

| | |
|---|---|
| • Total number of unknowns | $2n(K+1) +$ $3(n-m+1)$ |

The number of available equations are as follows:

| | |
|---|---|
| • Fundamental real and reactive power balance at all $(m-1)$ linear buses but the swing bus | $2(m-2)$ |
| • Fundamental voltage magnitude and phase at the swing bus ($|\tilde{V}_1| = 1\ pu$, $\delta_1 = 0\ radians$) | 2 |
| • Fundamental real and imaginary current balance at $n-(m-1)$ nonlinear buses (at $(m-1)$ linear buses the $P_i^{(1)}$ and $Q_i^{(1)}$ are specified) | $2(n-(m-1)) = 2(n-m+1)$ |
| • Harmonic real and reactive current (except the fundamental) balance at all $n$ buses | $2nK$ |
| • Apparent voltampere balance at $n-(m-1)$ nonlinear buses to compute reactive power (if reactive power is specified at nonlinear buses, however, this apparent voltampere balance is not required at nonlinear buses) | $(n-m+1)$ |

| | |
|---|---|
| • Total number of equations | $2n(K+1) +$ $(n-m+1)$ |

There remain $2(n-m+1)$ equations to be defined. These can be the total real and reactive mismatch powers at the $n-(m-1)$ nonlinear buses:

$$\Delta P_j^{\text{nonlinear}} = P_j^t + \sum_h F_{r,j}^{(h)}, \qquad (7\text{-}100a)$$

$$\Delta Q_j^{\text{nonlinear}} = Q_j^t + \sum_h F_{i,j}^{(h)}, \qquad (7\text{-}100b)$$

where $j = m, \ldots, n$ and $P_j^t$ and $Q_j^t$ are the total (fundamental and harmonics) nonlinear load real and reactive powers at bus j, respectively. $\sum_h F_{r,j}^{(h)}$ and $\sum_h F_{i,j}^{(h)}$ (for $h = 1, 5, 7, \ldots, L$) are the total real and reactive line powers, respectively. Note that the formulas for $F_{r,j}^{(h)}$ and $F_{i,j}^{(h)}$ are identical to those for $F_{r,j}$ and $F_{i,j}$ of the fundamental power flow problem with the difference that $\bar{Y}_{\text{bus}}^{(h)}$ and the hth harmonic voltages must be used. The application of the newly defined mismatch powers at the nonlinear buses results in $2(n-m+1)$ additional equations, and the number of equations is identical to the number of unknowns.

### 7.4.4 Formulation of Newton–Raphson Approach for Harmonic Power Flow

The mismatch vector (consisting of mismatch power and mismatch currents) for harmonic power flow is defined as

$$\Delta \bar{M} = [\Delta \bar{W}, \Delta \bar{I}^{(5)}, \ldots, \Delta \bar{I}^{(L)}, \Delta \bar{I}^{(1)}]^t, \qquad (7\text{-}101)$$

where $\Delta \bar{W}$ is the mismatch power vector (Eq. 7-56) and $(\Delta \bar{I}^{(5)}, \ldots, \Delta \bar{I}^{(L)}, \Delta \bar{I}^{(1)})$ is the mismatch current vector for the harmonics including the fundamental.

In Eq. 7-101,

$$\Delta \bar{W} = [P_2^{(1)} + F_{r,2}^{(1)}, Q_2^{(1)} + F_{i,2}^{(1)}, \ldots,$$
$$P_{m-1}^{(1)} + F_{r,m-1}^{(1)}, Q_{m-1}^{(1)} + F_{i,m-1}^{(1)},$$
$$\Delta P_m^{\text{nonlinear}}, \Delta Q_m^{\text{nonlinear}}, \ldots, \Delta P_n^{\text{nonlinear}}, \Delta Q_n^{\text{nonlinear}}]^t$$
$$(7\text{-}102)$$

where $P_2^{(1)} + F_{r,2}^{(1)}$, $Q_2^{(1)} + F_{i,2}^{(1)}, \ldots$, $P_{m-1}^{(1)} + F_{r,m-1}^{(1)}$, $Q_{m-1}^{(1)} + F_{i,m-1}^{(1)}$ applies to linear buses, and $\Delta P_m^{\text{nonlinear}}$, $\Delta Q_m^{\text{nonlinear}}, \ldots, \Delta P_n^{\text{nonlinear}}, \Delta Q_n^{\text{nonlinear}}$ applies to nonlinear buses. $P_j^{(1)}$ and $Q_j^{(1)}$ are the real and reactive fundamental load powers for the linear bus $j$ (as defined by the fundamental load flow analysis), respectively; $F_{r,j}^{(1)}$ and $F_{i,j}^{(1)}$ are the (line) fundamental real and reactive powers at the linear bus $j$ (see fundamental load flow analysis), respectively.

The fundamental current mismatch is defined for nonlinear buses where all currents (e.g., line currents and nonlinear load currents) are referred to the swing bus:

$$\Delta \bar{I}^{(1)} = [I_{r,m}^{(1)} + G_{r,m}^{(1)}, I_{i,m}^{(1)} + G_{i,m}^{(1)}, \ldots,$$
$$I_{r,n}^{(1)} + G_{r,n}^{(1)}, I_{i,n}^{(1)} + G_{i,n}^{(1)}]^t, \qquad (7\text{-}103)$$

and the harmonic current mismatch is defined for linear and nonlinear buses including swing bus:

$$\Delta \bar{I}^{(h)} = [I_{r,1}^{(h)}, I_{i,1}^{(h)}, \ldots, I_{r,m-1}^{(h)}, I_{i,m-1}^{(h)}, I_{r,m}^{(h)} + G_{r,m}^{(h)}, I_{i,m}^{(h)} +$$
$$G_{i,m}^{(h)}, \ldots, I_{r,n}^{(h)} + G_{r,n}^{(h)}, I_{i,n}^{(h)} + G_{i,n}^{(h)}]^t, \qquad (7\text{-}104)$$

where the nonlinear load current components $G_{r,m}^{(h)}$ and $G_{i,m}^{(h)}$ are given (referred to swing bus) and the line current components $I_{r,m}^{(h)}$ and $I_{i,m}^{(h)}$ will be generated in the next section. The Newton–Raphson method is implemented by forcing the appropriate mismatches, $\Delta \bar{M}$, to zero using the Jacobian matrix $\bar{J}$ and obtaining appropriate correction terms $\Delta \bar{U}^{(\xi)} = \bar{U}^{(\xi)} - \bar{U}^{(\xi+1)}$, where $\xi$ represents the iteration number:

$$\Delta \bar{M} = \bar{J} \Delta \bar{U}^{(\xi)}. \qquad (7\text{-}105)$$

The Jacobian $\bar{J}$ is a $2(nK + n - 1) + 2(n - m + 1)$ by $2(nK + n - 1) + 2(n - m + 1)$ matrix, where $n$ is the total number of buses, $(m - 1)$ is the number of linear buses including the swing bus, and $K$ is the number of harmonics considered, excluding the fundamental $(h = 1)$:

$$\Delta \bar{M} = \begin{bmatrix} \Delta \bar{W} \\ \Delta \bar{I}^{(5)} \\ \Delta \bar{I}^{(7)} \\ \cdot \\ \cdot \\ \cdot \\ \Delta \bar{I}^{(L)} \\ \Delta \bar{I}^{(1)} \end{bmatrix} = \begin{bmatrix} \bar{J}^{(1)} & \bar{J}^{(5)} & \cdot & \cdot & \bar{J}^{(L)} & 0 \\ \overline{YG}^{(5,1)} & \overline{YG}^{(5,5)} & \cdot & \cdot & \overline{YG}^{(5,L)} & \bar{H}^{(5)} \\ \overline{YG}^{(7,1)} & \overline{YG}^{(7,5)} & \cdot & \cdot & \overline{YG}^{(7,L)} & \bar{H}^{(7)} \\ \vdots & \vdots & & & \vdots & \vdots \\ \overline{YG}^{(L,1)} & \overline{YG}^{(L,5)} & \cdot & \cdot & \overline{YG}^{(L,L)} & \bar{H}^{(L)} \\ \overline{YG}^{(1,1)} & \overline{YG}^{(1,5)} & \cdot & \cdot & \overline{YG}^{(1,L)} & \bar{H}^{(1)} \end{bmatrix} \begin{bmatrix} \Delta \bar{V}^{(1)} \\ \Delta \bar{V}^{(5)} \\ \Delta \bar{V}^{(7)} \\ \cdot \\ \cdot \\ \cdot \\ \Delta \bar{V}^{(L)} \\ \Delta \bar{\Phi} \end{bmatrix}. \qquad (7\text{-}106)$$

Solving the equation for the correction vector yields

$$\Delta \bar{U} = \begin{bmatrix} \Delta \bar{V}^{(1)} \\ \Delta \bar{V}^{(5)} \\ \Delta \bar{V}^{(7)} \\ \cdot \\ \cdot \\ \cdot \\ \Delta \bar{V}^{(L)} \\ \Delta \bar{\Phi} \end{bmatrix} = \begin{bmatrix} \bar{J}^{(1)} & \bar{J}^{(5)} & \cdot & \cdot & \bar{J}^{(L)} & 0 \\ \overline{YG}^{(5,1)} & \overline{YG}^{(5,5)} & \cdot & \cdot & \overline{YG}^{(5,L)} & \bar{H}^{(5)} \\ \overline{YG}^{(7,1)} & \overline{YG}^{(7,5)} & \cdot & \cdot & \overline{YG}^{(7,L)} & \bar{H}^{(7)} \\ \cdot\cdot & \cdot\cdot & \cdot\cdot & \cdot & \cdot\cdot & \vdots \\ \overline{YG}^{(L,1)} & \overline{YG}^{(L,5)} & \cdot & \cdot & \overline{YG}^{(L,L)} & \bar{H}^{(L)} \\ \overline{YG}^{(1,1)} & \overline{YG}^{(1,5)} & \cdot & \cdot & \overline{YG}^{(1,L)} & \bar{H}^{(1)} \end{bmatrix}^{-1} \begin{bmatrix} \Delta \bar{W} \\ \Delta \bar{I}^{(5)} \\ \Delta \bar{I}^{(7)} \\ \cdot \\ \cdot \\ \cdot \\ \Delta \bar{I}^{(L)} \\ \Delta \bar{I}^{(1)} \end{bmatrix}, \qquad (7\text{-}107)$$

where all elements in this matrix equation are subvectors and submatrices partitioned from $\Delta \bar{M}$, $\bar{J}$, and $\Delta \bar{U}^{(\xi)}$.

The subvectors are

- $(\Delta \overline{V}^{(1)}, \Delta \overline{V}^{(5)}, \ldots, \Delta \overline{\Phi})\, t$ = correction vector,
- $\Delta \overline{W}$ = mismatch power vector at all buses except the swing bus,
- $\Delta \overline{I}^{(1)}$ = fundamental mismatch current vector at nonlinear buses, and
- $\Delta \overline{I}^{(h)}$ = hth harmonic mismatch current vector at all buses.

The submatrices are

- $\overline{J}^{(1)}$ = Jacobian $2(n-1)$ square matrix corresponding to Jacobian of fundamental power flow written for all buses except the swing bus, and
- $\overline{J}^{(h)}$ = Jacobian of harmonic $h \neq 1$ with dimensions $2(n-1) \times 2n$, whereby there are zero entries for $(m-1)$ linear buses:

$$
\overline{J}^{(h)} = \left[ \begin{array}{c} \overline{0}_{2(m-2)\times 2n} \\[2pt] \text{(zero entries for linear buses)} \\[2pt] \text{------------------} \\[2pt] \text{partial derivatives of the h}^{\text{th}}\text{ harmonic} \\ \text{mismatch (total) } P \text{ and } Q \text{ at nonlinear} \\ \text{buses with respect to } V^{(h)} \text{ and } \delta^{(h)}. \\ \text{These are similarly formed as those for} \\ \text{the fundamental power flow.} \end{array} \right],
$$

$$(7\text{-}108)$$

where $\overline{0}_{2(m-2)\times 2n}$ denotes a $2(m-2)$ by $2n$ array of zeros and $\overline{J}^{(h)}$ is a $2(n-1)$ by $2n$ matrix.

The partial derivatives are found using the harmonic admittance matrix $\overline{Y}_{\text{bus}}^{(h)}$ and the harmonic voltages:

$$
\overline{YG}^{(h,j)} = \begin{cases} \overline{Y}^{(h,h)} + \overline{G}^{(h,h)} & \text{for } h = j \\ \overline{0} + \overline{G}^{(h,j)} & \text{for } h \neq j \end{cases} \tag{7-109}
$$

or

$$
\overline{YG}^{(h,j)} = \begin{bmatrix} \overline{Y}^{(h,h)} & \overline{G}^{(h,h)} \\ \overline{Y}^{(h,j)} \equiv 0 & \overline{G}^{(h,j)} \end{bmatrix}. \tag{7-110}
$$

- $\overline{Y}^{(h,h)}$ is an array of partial derivatives of the hth harmonic (linear) line currents with respect to the hth harmonic bus voltages,
- $\overline{G}^{(h,h)}$ is an array of partial derivatives of the hth (nonlinear) harmonic load currents with respect to the hth harmonic bus voltages, and
- $\overline{G}^{(h,j)}$ is an array of partial derivatives of the hth harmonic nonlinear load currents with respect to the jth harmonic bus voltages:

$$
\overline{G}^{(h,j)} = \left[ \begin{array}{c|ccccc} \overline{0}_{2(m-1)\times 2(m-1)} & & & \overline{0}_{2(m-1)\times 2(n-m+1)} & & \\ \hline & \dfrac{\partial G_{r,m}^{(h)}}{\partial \delta_m^{(j)}} & \dfrac{\partial G_{r,m}^{(h)}}{\partial V_m^{(j)}} & \cdot\ \cdot & 0 & 0 \\ & \dfrac{\partial G_{i,m}^{(h)}}{\partial \delta_m^{(j)}} & \dfrac{\partial G_{i,m}^{(h)}}{\partial V_m^{(j)}} & \cdot\ \cdot & 0 & 0 \\ \overline{0}_{2(n-m+1)\times 2(m-1)} & \cdot & \cdot & \cdot & \cdot & \cdot \\ & 0 & 0 & \cdot\ \cdot & \dfrac{\partial G_{r,n}^{(h)}}{\partial \delta_n^{(j)}} & \dfrac{\partial G_{r,n}^{(h)}}{\partial V_n^{(j)}} \\ & 0 & 0 & \cdot\ \cdot & \dfrac{\partial G_{i,n}^{(h)}}{\partial \delta_n^{(j)}} & \dfrac{\partial G_{i,n}^{(h)}}{\partial V_n^{(j)}} \end{array} \right], \tag{7-111}
$$

where the partial derivatives of $G_{r,m}^{(h)}$ and $G_{i,m}^{(h)}$ are referred to the swing bus. Note that $\overline{YG}^{(h,j)}$ is a $2n$ square matrix; however, for $h = 1$ only the last $2(n-m+1)$ rows exist because the fundamental current balance is not applied to linear buses. $\overline{YG}^{(1,j)}$ is a $2(n-m+1) \times 2n$ matrix and for $j = 1$ the first two columns do not exist because $|\tilde{V}_1^{(1)}|$ and $\delta_1^{(1)}$ are known, that is, $\overline{YG}^{(h,1)}$ is a $2n \times 2(n-1)$ matrix.

$\overline{H}^{(h)}$ is an array of partial derivatives of real and imaginary nonlinear load currents with respect to the firing angle $\alpha$ and the commutation impedance $\beta$:

$$\bar{H}^{(h)} = \begin{bmatrix} \overline{0}_{2(m-1)\times 2(n-m+1)} \\ \hline \dfrac{\partial G_{r,m}^{(h)}}{\partial \alpha_m} & \dfrac{\partial G_{r,m}^{(h)}}{\partial \beta_m} & . & . & 0 & 0 \\ \dfrac{\partial G_{r,m}^{(h)}}{\partial \alpha_m} & \dfrac{\partial G_{i,m}^{(h)}}{\partial \beta_m} & . & . & 0 & 0 \\ . & . & . & . & . & . \\ 0 & 0 & . & . & \dfrac{\partial G_{r,n}^{(h)}}{\partial \alpha_n} & \dfrac{\partial G_{r,n}^{(h)}}{\partial \beta_n} \\ 0 & 0 & . & . & \dfrac{\partial G_{i,n}^{(h)}}{\partial \alpha_n} & \dfrac{\partial G_{i,n}^{(h)}}{\partial \beta_n} \end{bmatrix},$$

$$(7\text{-}112)$$

where $h = 1, 5, 7, \ldots, L$. For $h = 1$ the entries are zero.

### 7.4.5 Harmonic Jacobian Entry Formulas Related to Line Currents

Figure 7.15 shows a typical nonlinear bus $i$ of an $n$ bus system. This $i$th bus is connected to three other buses, $m$, $k$, and $l$. The nonlinear device connected to bus $i$ demands a complex load current $\tilde{g}_{\text{load},i}^{(h)}$ (referred to the nonlinear bus $i$) at the hth harmonic frequency.

The total hth harmonic complex current leaving bus $i$ via the lines is

$$\tilde{I}_{\text{lines},i}^{(h)} = \tilde{I}_k^{(h)} + \tilde{I}_l^{(h)} + \tilde{I}_m^{(h)}, \qquad (7\text{-}113a)$$

$$\tilde{I}_{\text{lines},i}^{(h)} = (\tilde{V}_i^{(h)} - \tilde{V}_k^{(h)})\tilde{y}_{il}^{(h)} + (\tilde{V}_i^{(h)} - \tilde{V}_1^{(h)})\tilde{y}_{il}^{(h)} + (\tilde{V}_i^{(h)} - \tilde{V}_m^{(h)})\tilde{y}_{im}^{(h)}, \qquad (7\text{-}113b)$$

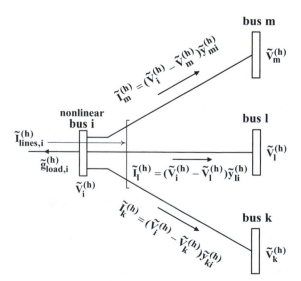

**FIGURE 7.15** Nonlinear bus $i$.

where $\tilde{y}_{ij}^{(h)}$ is the admittance of the line connecting buses $i$ and $j$ evaluated at the hth harmonic frequency. Rearranging terms gives

$$\tilde{I}_{\text{lines},i}^{(h)} = (\tilde{y}_{ik}^{(h)} + \tilde{y}_{il}^{(h)} + \tilde{y}_{im}^{(h)})\tilde{V}_i^{(h)} - \tilde{y}_{ik}^{(h)}\tilde{V}_k^{(h)} - \tilde{y}_{il}^{(h)}\tilde{V}_1^{(h)} - \tilde{y}_{im}^{(h)}\tilde{V}_m^{(h)}, \qquad (7\text{-}114a)$$

or

$$\tilde{I}_{\text{lines},i}^{(h)} = (\tilde{Y}_{\text{bus}}^{(h)})_{ii}\tilde{V}_i^{(h)} + (\tilde{Y}_{\text{bus}}^{(h)})_{ik}\tilde{V}_k^{(h)} + (\tilde{Y}_{\text{bus}}^{(h)})_{il}\tilde{V}_1^{(h)} + (\tilde{Y}_{\text{bus}}^{(h)})_{im}\tilde{V}_m^{(h)}, \qquad (7\text{-}114b)$$

where $(\tilde{Y}_{\text{bus}}^{(h)})_{ij}$ is the $(i,j)$th entry of the hth harmonic admittance matrix. Because $k$, $l$, and $m$ are the only buses connected to bus $i$ (see Fig. 7.15),

$$\tilde{I}_{\text{lines},i}^{(h)} = (\text{row } i \text{ of } \bar{Y}_{\text{bus}}^{(h)})_{ii}\bar{V}_{\text{bus}}^{(h)}, \qquad (7\text{-}115)$$

where $\bar{V}_{\text{bus}}^{(h)}$ is a complex vector of the hth harmonic bus voltages. Therefore, the mismatch current for the hth harmonic is based on the Kirchhoff current law at bus $i$:

$$\Delta \tilde{I}_i^{(h)} = \tilde{I}_{\text{lines},i}^{(h)} + \tilde{G}_{\text{load},i}^{(h)} = (\text{row } i \text{ of } \bar{Y}_{\text{bus}}^{(h)})\bar{V}_{\text{bus}}^{(h)} + \tilde{G}_{\text{load},i}^{(h)}, \qquad (7\text{-}116)$$

where $\tilde{G}_{\text{load},i}^{(h)}$ is the hth harmonic nonlinear complex load current (for bus $i$), referred to the swing bus:

$$\tilde{G}_{\text{load},i}^{(h)} = \tilde{G}_{r,i}^{(h)} + j\tilde{G}_{i,i}^{(h)} \qquad (7\text{-}117)$$

In this equation, the second index $i$ pertains to bus $i$ and the first index $i$ indicates imaginary. Breaking the mismatch current $\Delta \tilde{I}_i^{(h)}$ into a real part $\Delta I_{r,i}^{(h)}$ and an imaginary part $\Delta I_{i,i}^{(h)}$, we get

$$\Delta I_{r,i}^{(h)} = \text{Real}\{(\text{row } i \text{ of } \bar{Y}_{\text{bus}}^{(h)})\bar{V}_{\text{bus}}^{(h)}\} + G_{r,i}^{(h)}, \qquad (7\text{-}118a)$$

$$\Delta I_{i,i}^{(h)} = \text{Imag}\{(\text{row } i \text{ of } \bar{Y}_{\text{bus}}^{(h)})\bar{V}_{\text{bus}}^{(h)}\} + G_{i,i}^{(h)}, \qquad (7\text{-}118b)$$

Therefore,

$$\Delta I_{r,i}^{(h)} = I_{r,i}^{(h)} + G_{r,i}^{(h)} = \sum_{j=1}^{n} y_{ij}^{(h)}\left|\tilde{V}_j^{(h)}\right|\cos(\theta_{ij}^{(h)} + \delta_j^{(h)}) + G_{r,i}^{(h)}, \qquad (7\text{-}119a)$$

$$\Delta I_{i,i}^{(h)} = I_{i,i}^{(h)} + G_{i,i}^{(h)} = \sum_{j=1}^{n} y_{ij}^{(h)}\left|\tilde{V}_j^{(h)}\right|\sin(\theta_{ij}^{(h)} + \delta_j^{(h)}) + G_{i,i}^{(h)}, \qquad (7\text{-}119b)$$

where $y_{ij}^{(h)}$ and $\theta_{ij}^{(h)}$ are the magnitude and phase angle of the $(i,j)$th entry of the hth harmonic admittance matrix, respectively, and $|\tilde{V}_j^{(h)}|$ and $\delta_j^{(h)}$ are the hth harmonic voltage magnitude and phase angle of bus $j$ with respect to swing bus.

The entries of the Jacobian matrix corresponding to line currents are submatrices $\overline{Y}^{(h,h)}$ ($h = 1, 5, \ldots$, $L$). These matrices are defined as

$$\overline{Y}^{(h,h)} = \frac{\partial(\overline{I}^{(h)})}{\partial(\overline{V}^{(h)})}. \qquad (7\text{-}120)$$

Note that $\overline{Y}^{(h,j)} = 0$ for $h \neq j$. The zero terms are due to the fact that the v–i characteristics of the nonlinear devices are modeled in the Jacobian matrix using submatrices $\overline{G}^{(h,j)}$.

$$\overline{Y}^{(h,h)} = \begin{bmatrix} \dfrac{\partial I_{r,1}^{(h)}}{\partial \delta_1^{(h)}} & \dfrac{\partial I_{r,1}^{(h)}}{\partial V_1^{(h)}} & \cdots & \dfrac{\partial I_{r,1}^{(h)}}{\partial \delta_n^{(h)}} & \dfrac{\partial I_{r,1}^{(h)}}{\partial V_n^{(h)}} \\ \dfrac{\partial I_{i,1}^{(h)}}{\partial \delta_1^{(h)}} & \dfrac{\partial I_{i,1}^{(h)}}{\partial V_1^{(h)}} & \cdots & \dfrac{\partial I_{i,1}^{(h)}}{\partial \delta_n^{(h)}} & \dfrac{\partial I_{i,1}^{(h)}}{\partial V_n^{(h)}} \\ \cdot & \cdot & \cdots & \cdot & \cdot \\ \dfrac{\partial I_{r,n}^{(h)}}{\partial \delta_1^{(h)}} & \dfrac{\partial I_{r,n}^{(h)}}{\partial V_1^{(h)}} & \cdots & \dfrac{\partial I_{r,n}^{(h)}}{\partial \delta_n^{(h)}} & \dfrac{\partial I_{r,n}^{(h)}}{\partial V_n^{(h)}} \\ \dfrac{\partial I_{i,n}^{(h)}}{\partial \delta_1^{(h)}} & \dfrac{\partial I_{i,n}^{(h)}}{\partial V_1^{(h)}} & \cdots & \dfrac{\partial I_{i,n}^{(h)}}{\partial \delta_n^{(h)}} & \dfrac{\partial I_{i,n}^{(h)}}{\partial V_n^{(h)}} \end{bmatrix}, \qquad (7\text{-}121)$$

For $h = 1$ only the last $2(n - m + 1)$ rows exist because fundamental current balance is not performed for linear load buses. Also for $h = 1$ the first two columns of the matrix do not exist (are zero) because $|\tilde{V}_1^{(1)}|$ and $\delta_1^{(1)}$ are known and constant.

The entries of $\overline{Y}^{(h,h)}$ are computed from Eq. 7-119 by setting $G_{r,i}^{(h)}$ and $G_{i,i}^{(h)}$ to zero:

$$\frac{\partial I_{r,i}^{(h)}}{\partial \delta_j^{(h)}} = -y_{ij}^{(h)} |\tilde{V}_j^{(h)}| \sin(\theta_{ij}^{(h)} + \delta_j^{(h)}), \quad \text{for } h = 1, 5, \ldots, L,$$
$$(7\text{-}122a)$$

$$\frac{\partial I_{r,i}^{(h)}}{\partial |\tilde{V}_j^{(h)}|} = y_{ij}^{(h)} \cos(\theta_{ij}^{(h)} + \delta_j^{(h)}), \quad \text{for } h = 1, 5, \ldots, L,$$
$$(7\text{-}122b)$$

$$\frac{\partial I_{i,i}^{(h)}}{\partial \delta_j^{(h)}} = y_{ij}^{(h)} |\tilde{V}_j^{(h)}| \cos(\theta_{ij}^{(h)} + \delta_j^{(h)}), \quad \text{for } h = 1, 5, \ldots, L,$$
$$(7\text{-}122c)$$

$$\frac{\partial I_{i,i}^{(h)}}{\partial |\tilde{V}_j^{(h)}|} = y_{ij}^{(h)} \sin(\theta_{ij}^{(h)} + \delta_j^{(h)}), \quad \text{for } h = 1, 5, \ldots, L,$$
$$(7\text{-}122d)$$

where $h$ and $j$ denote harmonic order and bus number, respectively.

## 7.4.6 Newton-Based Harmonic Power Flow Algorithm

Based on the equations given, the harmonic power flow algorithm is the computation of bus voltage vector $\overline{U}$ for a given system configuration with linear and nonlinear loads. The Newton–Raphson approach (Eq. 7-107) will be used to force the mismatch vector (Eq. 7-101) to zero using the harmonic Jacobian matrix and obtaining appropriate correction terms $\Delta \overline{U}$. Therefore ($\xi$ represents the iteration number),

$$\Delta \overline{U}^\xi = J^{-1} \Delta \overline{M}(\overline{U}^\xi), \qquad (7\text{-}123)$$

$$\overline{U}^{(\xi+1)} = \overline{U}^\xi - \Delta \overline{U}^\xi. \qquad (7\text{-}124)$$

The solution procedure for the harmonic power flow algorithm is as follows (Fig. 7.16):

**Step 1:** Perform the fundamental load flow analysis (treating all nonlinear devices as linear loads) and compute an initial (approximate) value for the fundamental bus voltage magnitudes and phase angles. Make an initial guess for the harmonic bus voltage magnitudes and phase angles (e.g., 0.1 pu and 0 radians).

**Step 2:** Compute nonlinear device currents $G_{r,m}^{(h)}$ and $G_{i,m}^{(h)}$ (referred to the swing bus) for nonlinear loads.

**Step 3:** Evaluate $\Delta \overline{M}(\overline{U})$ using Eqs. 7-101, 7-102. If it is small enough then stop.

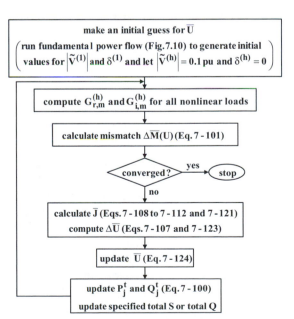

**FIGURE 7.16** Harmonic Newton-based power flow algorithm.

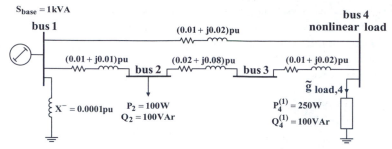

FIGURE E7.14.1 Four-bus power system with one nonlinear load bus, where $n = m = 4$, that is, there is one nonlinear bus (bus 4), two linear buses (buses 2 and 3), and the swing bus (bus 1).

**Step 4:** Evaluate $\bar{J}$ (Eqs. 7-108 to 7-112 and 7-121) and calculate $\Delta \bar{U}$ (Eqs. 7-107 and 7-123) using matrix inversion or forward/backward substitution.

**Step 5:** Update $\bar{U}$ (Eq. 7-124).

**Step 6:** Update the total (fundamental plus harmonic) powers at nonlinear buses ($P_j^t$ and $Q_j^t$ (Eq. 7-100)), and the specified total apparent power or total reactive power (Eq. 7-98), whichever is known.

**Step 7:** Go to Step 2.

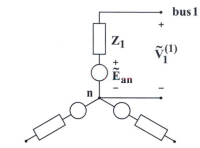

FIGURE E7.14.2 Representation of swing bus (bus 1) for fundamental ($h = 1$) frequency.

### 7.4.7 Application Example 7.14: Computation of Harmonic Admittance Matrix

The harmonic power flow algorithm will be applied to the nonlinear power system as shown in Fig. E7.14.1 where all impedances are given in pu at 60 Hz and the base is $S_{base} = 1$ kVA. In order to simplify the problem, the harmonic $v$–$i$ characteristic of the nonlinear load at bus 4 (referred to the voltage at bus 4) is given as

$$g_{r,4}^{(5)} = 0.3(V_4^{(1)})^3 \cos(3\delta_4^{(1)}) + 0.3(V_4^{(5)})^2 \cos(3\delta_4^{(5)}),$$
$$g_{i,4}^{(5)} = 0.3(V_4^{(1)})^3 \sin(3\delta_4^{(1)}) + 0.3(V_4^{(5)})^2 \sin(3\delta_4^{(5)}).$$

It will be assumed that the voltage at the swing bus (bus 1) is $|\tilde{V}_1^{(1)}| = 1.00$ pu and $\delta_1^{(1)} = 0$ rad.

The swing bus can be represented by Fig. E7.14.2 at the fundamental frequency. For the 5th harmonic, $\tilde{E}_{an}$ of Fig. E7.14.3 represents a short-circuit, and the synchronous machine impedance $Z_1$ is replaced by the (subtransient) impedance $Z_2 \approx jX^- = j0.0001$ pu at 60 Hz (see Fig. E7.14.1). In this case the subtransient reactance has been chosen to be very small in order to approximate bus 1 as an ideal bus.

Compute the fundamental and 5th harmonic admittance matrices.

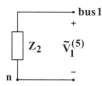

FIGURE E7.14.3 Equivalent circuit of swing bus for the 5th harmonic ($h = 5$).

### 7.4.8 Application Example 7.15: Computation of Nonlinear Load Harmonic Currents

For the system of Fig. E7.14.1, assume an initial value for the bus vector $\bar{U}^0$ and compute the fundamental and the 5th harmonic currents injected into the system by the nonlinear load. The computation of the nonlinear load harmonic currents is the initial step of the harmonic load flow algorithm.

### 7.4.9 Application Example 7.16: Evaluation of Harmonic Mismatch Vector

Using the results of Application Example 7.15, evaluate the mismatch vector for the system of Fig. E7.14.1.

### 7.4.10 Application Example 7.17: Evaluation of Fundamental and Harmonic Jacobian Submatrices

Using the results of Application Examples 7.15 and 7.16, evaluate the Jacobian matrix for the system of Fig. E7.14.1.

### 7.4.11 Application Example 7.18: Computation of the Correction Bus Vector and Convergence of Harmonic Power Flow

Using the results of Application Examples 7.15 to 7.17, evaluate the correction bus vector $\Delta \bar{U}^0$ for the system of Fig. E7.14.1. Update the mismatch vector and comment on the convergence of the harmonic load flow.

## 7.5 CLASSIFICATION OF HARMONIC POWER FLOW TECHNIQUES

Different approaches have been proposed and implemented to solve the harmonic power flow (HPF) problem [5-29]. In this section, the following criteria will be employed to classify some of these algorithms:

- Modeling technique used to simulate power system and nonlinear loads (time domain, harmonic domain).
- System condition (single-phase, three-phase, balanced, unbalanced).
- Solution approach (coupled, decoupled).

Modeling techniques include time-domain, harmonic-domain and hybrid time/harmonic-domain approaches. Time-domain approaches are based on transient analysis and have great flexibility and high accuracy; however, they usually require long computing times (especially for large power systems containing many nonlinear loads with strong harmonic couplings). Harmonic-domain approaches calculate the frequency response of power systems and require shorter computing times; however, it is usually difficult to obtain accurate frequency models for nonlinear loads [13]. Hybrid modeling techniques use a combination of frequency-domain (to limit the computing time) and time-domain (to increase the accuracy) approaches to simulate the power system and nonlinear loads, respectively [9, 14]. Therefore, comprehensive hybrid models are employed to achieve the benefits of the accuracy of the time-domain analysis and the simplicity of frequency-domain approach.

It is sufficient to use single-phase harmonic models for single-phase and balanced three-phase systems.

Unbalanced harmonic conditions are usually imposed by the system configuration, disturbances, and (non)linear loads. Unbalanced harmonic power flow solutions are relatively complicated and require long computing times and considerable memory storage.

Solution approaches may also be classified into coupled and decoupled methods [12]. Most nonlinear loads and power system components impose couplings between (time and/or space) harmonics and call for accurate coupled solution approaches. If harmonic couplings are not too strong, a decoupled approach may be implemented where separate models are used at different harmonic frequencies and superposition is employed to calculate the final solution.

Based on the above-mentioned criteria, HPF algorithms can be classified as follows:

- Newton-based coupled harmonic power flow as outlined in Section 7.4 [5, 6].
- Decoupled harmonic power flow [16, 17, 19].
- Fast harmonic power flow [17].
- Modified fast decoupled power flow [20].
- Fuzzy harmonic power flow [13].
- Probabilistic harmonic power flow [12, 22].
- Modular harmonic power flow [9, 11, 14, 23].

Table 7.3 presents a comparison of harmonic power flow algorithms. The remainder of this chapter briefly outlines the solution procedure of these algorithms.

### 7.5.1 Decoupled Harmonic Power Flow

The Newton-based harmonic power flow of Section 7.4 is very accurate because harmonic couplings are included at all frequencies. However, it might encounter convergence problems for large power systems with many nonlinear loads. For such applications, a decoupled approach is more appropriate.

At the fundamental frequency, the conventional power flow algorithm (Section 7.3) is used to model the system. At higher frequencies, harmonic couplings are neglected (to simplify the procedure, to reduce computing time, and to limit memory storage requirements) and the distorted power system is modeled by the harmonic admittance matrix. The general model of a linear load consisting of a resistance in parallel with a reactance is utilized [7, 8], and nonlinear loads are modeled with decoupled harmonic current sources. The magnitudes and phase angles of harmonic current sources depend on the nature of the nonlinear loads and can be computed

**TABLE 7.3** Classification and Comparison of Harmonic Power Flow Algorithms

| | Modeling technique | | | | | | Solution approach | | System condition | |
|---|---|---|---|---|---|---|---|---|---|---|
| | For the system | | | For nonlinear loads | | | | | | |
| Method | Time | Frequency | Hybrid | Time | Frequency | Hybrid | Coupled | Decoupled | Balanced | Unbalanced |
| Newton HPF[a] | | × | | | × | + | × | | × | |
| Decoupled HPF | | × | | | × | | | × | × | b |
| Fast HPF | | × | | | × | | | × | × | × |
| Modified fast HPF | | × | | | × | | × | × | × | |
| Fuzzy HPF | | × | | × | × | × | × | | × | |
| Probabilistic HPF | | × | | × | × | × | × | | × | |
| Modular HPF | × | × | × | × | × | × | × | | × | × |

[a]Section 7.4.

[b]The HPF algorithm may be modified to include this feature.

using analytical models or may be estimated based on measured nonsinusoidal current waveforms. This simple approach of nonlinear load modeling is very practical for most industrial loads.

The hth harmonic load admittance at bus $i$ ($y_i^{(h)}$), shunt capacitor admittance at bus $i$ ($y_{ci}^{(h)}$), and feeder admittance between buses $i$ and $i+1$ ($y_{(i,i+1)}^{(h)}$) are [7, 8, 16, 18, 19]

$$y_i^{(h)} = \frac{P_i}{\left|V_i^{(1)}\right|^2} - j\frac{Q_i}{h\left|V_i^{(1)}\right|^2}, \qquad (7\text{-}125)$$

$$y_{ci}^{(h)} = hy_{ci}^{(1)}, \qquad (7\text{-}126)$$

$$y_{i,i+1}^{(h)} = \frac{1}{R_{i,i+1} + jhX_{i,i+1}}, \qquad (7\text{-}127)$$

where $P_i$ and $Q_i$ are the fundamental active and reactive load powers at bus $i$, $y_{ci}^{(1)}$ is the fundamental frequency admittance of the capacitor bank connected to bus $i$, and the superscript $h$ indicates harmonic orders.

Nonlinear loads are modeled as decoupled harmonic current sources that inject harmonic currents into the system. The fundamental and the hth harmonic currents of the nonlinear load installed at bus $m$ with fundamental real power $P_m$ and fundamental reactive power $Q_m$ are [8, 16, 19]

$$I_m^{(1)} = [(P_m + jQ_m)/V_m^{(1)}]^*, \qquad (7\text{-}128)$$

$$I_m^{(h)} = C(h)I_m^{(1)}, \qquad (7\text{-}129)$$

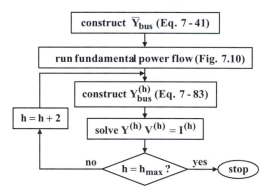

**FIGURE 7.17** Flowchart of the decoupled harmonic power flow (DHPF).

where $C(h)$ is the ratio of the hth harmonic current to its fundamental value. A conventional (fundamental) Newton–Raphson algorithm (Section 7.3) is used to solve the system at the fundamental frequency and harmonic voltages are calculated by solving the following decoupled load flow equation (derived from the node equations) [16, 17, 19]

$$Y^{(h)}V^{(h)} = I^{(h)}. \qquad (7\text{-}130)$$

The flowchart of the decoupled harmonic power flow (DHPF) based on Eqs. 7-125 to 7-130 is shown in Fig. 7.17. This simple and fast approach for the calculation of harmonic voltages and currents provides acceptable solutions as compared with Newton-based HPF (Section 7.4), which considers harmonic couplings [5, 6].

## 7.5.2 Fast Harmonic Power Flow

This method employs the equivalent current injection transformation and the forward and backward sweep techniques to solve the harmonic load flow problem [17]. For bus $i$, the complex power $S_i$ is

$$S_i = (P_i + jQ_i), \quad i = 1, \ldots, n, \qquad (7\text{-}131)$$

where $n$ is the total number of buses. The equivalent current injection ($I_i^k$) at the kth iteration is

$$I_i^k = I_i^r(V_i^k) + jI_i^i(V_i^k) = \left(\frac{P_i + jQ_i}{V_i^k}\right)*, \qquad (7\text{-}132)$$

where $V_i^k$ is the voltage of bus $i$ at the kth iteration.

For fundamental load flow, the equivalent current injection needs to be iteratively transformed (e.g., Eq. 7-132 is used at the end of each iteration to compute the equivalent current injection ($I_i^k$) from the updated voltage ($V_i^k$), active ($P_i$), and reactive ($Q_i$) powers). However, this transformation is not required for harmonic currents since they have been obtained in a suitable form based on harmonic analysis (which represents nonlinear loads with harmonic current sources).

Figure 7.18 shows part of a lateral distribution system and the associated harmonic currents. The relationships between branch currents and harmonic currents are

$$B_{jk}^{(h)} = -I_k^{(h)}, \qquad (7\text{-}133)$$

$$B_{jl}^{(h)} = -I_l^{(h)}, \qquad (7\text{-}134)$$

$$B_{ij}^{(h)} = B_{ik}^{(h)} + B_{jl}^{(h)} - I_k^{(h)}, \qquad (7\text{-}135)$$

where $B$ and $I$ are branch and injection (harmonic) currents, respectively.

It can be seen that the relationship between branch currents and harmonic (injection) currents of a radial feeder can be calculated by summing the injection currents from the receiving bus toward the sending bus of the feeder. The general form of the harmonic currents can be expressed as

$$B_{ij}^{(h)} = -I_k^{(h)} + \sum_j B_{jl}^{(h)}. \qquad (7\text{-}136)$$

On the other hand, the relationship between branch currents and bus voltages can be shown to be

$$\begin{aligned} V_j^{(h)} &= V_i^{(h)} - B_{ij}^{(h)} Z_{ij}^{(h)} \\ V_k^{(h)} &= V_j^{(h)} - B_{jk}^{(h)} Z_{jk}^{(h)} \\ V_l^{(h)} &= V_j^{(h)} - B_{jl}^{(h)} Z_{jl}^{(h)}, \end{aligned} \qquad (7\text{-}137)$$

where $Z^{(h)}$ is the equivalent impedance of the line section at the hth harmonic frequency.

It is evident for a radial distribution system that if branch currents have been calculated, then the bus voltages can be easily determined from

$$V_j^{(h)} = V_i^{(h)} - Z_{ij} B_{ij}^{(h)}. \qquad (7\text{-}138)$$

The bus voltages can be obtained by calculating the voltage from the sending bus forward to the receiving bus of the feeder. Power system components are adjusted for each harmonic order.

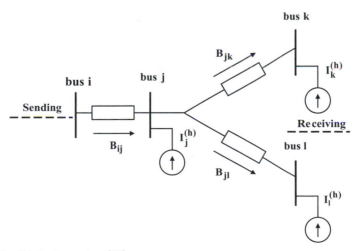

**FIGURE 7.18** Parts of a distribution system [17].

### 7.5.3 Modified Fast Decoupled Harmonic Power Flow

This method extends and generalizes the conventional fast decoupled (fundamental) power flow algorithm to accommodate harmonic sources [20]. The reduced and decoupled features of the fast decoupled algorithm is explored for the harmonic power flow application. A mathematical nonlinear model is integrated into an existing fast decoupled power flow program. This method allows separate modeling of nonlinear loads and subsequently integrating them in the main algorithm without substantially increasing computation time.

The hth harmonic bus injection current vector is defined as

$$I^{(h)} = Y^{(h)} V^{(h)}, \qquad (7\text{-}139)$$

where $Y^{(h)}$ is the hth harmonic admittance matrix. The injected current at any bus $p$ is given by

$$I_p^{(h)} = \sum_{i=1}^{n} Y_{p,i}^{(h)} V_i^{(h)}. \qquad (7\text{-}140)$$

This equation provides the change of voltage magnitude and phase angle for each harmonic as a result of the injected currents.

The problem is then solved by the Newton–Raphson method. The current injected by a nonlinear load is assumed to be constant throughout the iteration and is used to modify the solution of voltage amplitudes and angles without reinverting the Jacobian matrix. At the end of each iteration, the device parameters are adjusted to reflect the new voltage magnitudes and phase angles. The harmonic currents and injected power are then reevaluated and the process is repeated until convergence is achieved.

### 7.5.4 Fuzzy Harmonic Power Flow

All elements of Eq. 7-139 are complex valued. Therefore, it can be reformulated as [13]

$$\left[\psi^{(h)}\right]\begin{bmatrix} V_r^{(h)} \\ V_i^{(h)} \end{bmatrix} = \begin{bmatrix} I_r^{(h)} \\ I_i^{(h)} \end{bmatrix}, \qquad (7\text{-}141)$$

where $V_r^{(h)}$, $V_i^{(h)}$, $I_r^{(h)}$, and $I_i^{(h)}$ are the real and imaginary parts of the hth harmonic voltage and current, respectively. $\psi^{(h)}$ is a $2n \times 2n$ matrix restructured from $Y^{(h)}$ and $n$ is the bus number.

Fuzzy set theory is employed to represent the injection harmonic currents. For an n bus distribution system the solution is obtained by solving a set of

$4n \times 4n$ linear equations. The fuzzy solution for both harmonic magnitudes and phase angles can be individually expressed by two explicit linear equations. The solutions for any harmonic level can be easily obtained from these equations.

Harmonic sources are represented as (voltage-independent) current sources. Because harmonic penetration is a steady-state phenomenon, monitoring harmonics requires a long period of time for accurate measurements. It can be shown that the value of the harmonic currents at a bus is most likely near the center of a monitored range. Therefore, it is suitable to model the harmonic current source with a triangle fuzzy number, denoted by $\tilde{I}_h$. With a proper fuzzy model, the new fuzzy harmonic power flow equations can be expressed in the following form [13]:

$$\left[\psi^{(h)}\right]\begin{bmatrix} \tilde{V}_r^{(h)} \\ \tilde{V}_i^{(h)} \end{bmatrix} = \begin{bmatrix} \tilde{I}_r^{(h)} \\ \tilde{I}_i^{(h)} \end{bmatrix}, \qquad (7\text{-}142)$$

Note that $\psi^{(h)}$ is a constant whereas the harmonic voltages $\tilde{V}_r^{(h)}$ and $\tilde{V}_i^{(h)}$ are fuzzy numbers.

An arbitrary fuzzy number can be represented by an ordered pair of functions $(u_{min}(r), u_{max}(r)), 0 \leq r \leq 1$, which meet the following conditions:

- $u_{min}(r)$ is a bounded left continuous nondecreasing function over [0,1].
- $u_{max}(r)$ is a bounded right continuous nonincreasing function over [0,1].
- $u_{min}(r) \leq u_{max}(r), 0 \leq r \leq 1$.

The fuzzy harmonic power flow can then be expressed as follows [13]:

$$\begin{bmatrix} \psi_{11} & \psi_{12} \\ \psi_{21} & \psi_{22} \end{bmatrix} \begin{bmatrix} V_{r\,min}^h(r) \\ V_{i\,min}^h(r) \\ -V_{r\,max}^h(r) \\ -V_{i\,max}^h(r) \end{bmatrix} = \begin{bmatrix} I_{r\,min}^h(r) \\ I_{i\,min}^h(r) \\ -I_{r\,max}^h(r) \\ -I_{i\,max}^h(r) \end{bmatrix}, \qquad (7\text{-}143)$$

where $\psi_{11}$ contains the positive entries of $[\psi]$, $\psi_{12}$ represents the absolute values of the negative entries of $[\psi]$, and $[\psi] = \psi_{11} - \psi_{12}$; moreover $\psi_{12} = \psi_{21}$ and $\psi_{11} = \psi_{22}$. Equation 7-143 is linear and can then be uniquely solved for $[V_{r\,min}^{(h)}(r), V_{i\,min}^{(h)}(r), -V_{r\,max}^{(h)}(r), -V_{i\,max}^{(h)}(r)]^t$ if and only if the matrix $\begin{bmatrix} \psi_{11} & \psi_{12} \\ \psi_{21} & \psi_{22} \end{bmatrix}$ is nonsingular.

### 7.5.5 Probabilistic Harmonic Power Flow

Unavoidable uncertainties in a power system affect the input data of the modeling for the evaluation of voltage and current harmonics caused by nonlinear loads. These uncertainties are mainly due to time variations of linear load demands, network configurations, and operating modes of nonlinear loads. The variations have a random character and, therefore, are expressed using the probabilistic method. The models available for evaluating voltage distortion are

- Direct method, in which the probability functions of harmonic voltages are calculated for assigned probability functions of harmonic currents injected by nonlinear loads.
- Integrated method, in which the probability functions of the voltage and current harmonics are calculated together by properly taking into account the interactions between voltage distortions and harmonic currents.

The probabilistic method is employed for harmonic power flow analysis to describe the interactions between the supply voltage distortions and the injected harmonic currents. It transfers the deterministic approach, which is based on the solution of a nonlinear equation system, to the field of probability.

To reduce the computational effort and to overcome the convergence problem, this method linearizes the nonlinear equation describing the steady-state behavior around an expected value region. Then, the first-order approximations are used to directly establish the means and covariance matrix of the output probability density function, starting from knowledge of the means and covariance matrix of the input probability density functions.

The following vector is defined as an input parameter (the values of the vector are kept fixed and assigned as input data [21, 22]):

$$T_a = [T_1\ T_2\ T_3]^t, \qquad (7\text{-}144)$$

where $T_1$ includes the active generator power at fundamental frequency (with the exception of the slack), $T_2$ includes the active and reactive linear load powers at fundamental frequency and the total active and apparent nonlinear load powers, and $T_3$ represents the generator voltage magnitudes. In all cases, some input parameters may be affected by uncertainties; hence, a probabilistic approach is adopted.

Every probabilistic formulation requires, first, the statistical characterization of the input data and, second, the evaluation of the statistical features of the output variables of interest. The statistical characterization of the input data consists of checking

(i) which components of vector $T_a$ have to be considered as random and which ones can be kept fixed, and

(ii) what are the statistical features to characterize the randomness of the variables identified in step (i).

The most realistic approach to solve the problem is to refer to past experience, looking at both available historical data collections and technical knowledge of the electrical power system under study. According to (i), the generator voltage magnitudes can be considered deterministic if the set points of the regulator are not deliberately changed. With regard to (ii), the random nature, which appears in $T_1$ and $T_2$, is applied. A probabilistic method can be used to evaluate the statistical features of the output variable and may be based on the linearization around an expected value region of the nonlinear equation. The change of active and reactive linear loads is assumed to be Gaussian (normally distributed).

A probabilistic steady-state analysis of a harmonic power flow can be performed by either iterative probabilistic methods (e.g., Monte Carlo) or by analytical approaches. Iterative probabilistic procedures require knowledge of probability density functions of the input variables; for every random input data, a value is generated according to its proper probability density function. According to the generated input values, the operating steady-state conditions of the harmonic power flow are evaluated by solving the nonlinear equations by means of an iterative numerical method. Once convergence has been achieved, the state of the power system is completely known and the values of all the variables of interest are stored. The procedure is repeated to achieve a good estimate of probability of the output variables according to the stated accuracy. Unfortunately, iterative probabilistic procedures require a high computational effort.

An analytical approach is employed to linearize the nonlinear equations describing steady-state behavior of the power system around an expected value region. As a result, it offers a faster solution than the iterative probabilistic method. To evaluate the statistical procedure, the harmonic power flow equation is expressed as [21, 22]

$$f(N) = T_b \qquad (7\text{-}145)$$

where $N$ and $T_b$ are the output and input random vector variables, respectively. Let the vector $\mu(T_b)$ be the expected values of $T_b$. Using $\mu(T_b)$ as input data for the deterministic power flow calculation results in vector $N_o$ as solutions:

$$f(N_o) = \mu(T_b). \qquad (7\text{-}146)$$

Linearizing the equation around the point $N_o$ results in

$$N \cong N_o + J^{-1}\Delta T_b = N_o' + AT_b, \qquad (7\text{-}147)$$

where

$$\Delta T_b : Tb - \mu(T_b), \qquad (7\text{-}148)$$

$$N_o' : N_o - J^{-1}\mu(T_b). \qquad (7\text{-}149)$$

$J$ is the Jacobian matrix. Once the statistical features of the random output vector $N$ are known, the statistics of other variables on which it depends can also be evaluated.

The use of a proper linearized approach allows a drastic reduction of computational effort because the computation required practically consists of the solution of only one deterministic harmonic power flow. Since the harmonic power flow equations are linearized around an expected value of the input vector, any movement away from this region produces an error. The errors can grow with the variance of the input random vector components, with the entity linked to the nonlinear behavior of the equation system.

### 7.5.6 Modular Harmonic Power Flow

The coupled harmonic power flow calculation (Section 7.4) is established by directly involving nonlinear loads as integral parts of the solution. This makes the calculation complicated. The procedure can be simplified by detaching nonlinear device models from the main program [9, 11, 14, 23]. As most nonlinear devices can be represented by harmonic sources, detaching their models is relatively simple, and due to the iterative nature of the solution, this representation does not limit the accuracy of the results.

In modular approaches, nonlinear loads are included within the harmonic power flow as black boxes. In a black box each node has two quantities: the nodal voltage and the injected current. One quantity is specified and used to calculate the other (unknown) quantity. The procedure involves the solution of the harmonic bus voltages and using them

to compute harmonic current penetration, which is then used to obtain (updated) harmonic bus voltages.

The harmonic modeling of a nonlinear load can be obtained by either a direct or an iterative approach [11, 23]. The Newton method may be employed to achieve the solution. For the nonlinear buses, the mismatches are the difference between the calculated nodal values and the transferred values from the nonlinear device models. This formulation simplifies the calculation of the mismatch vector and Jacobian matrix.

The exact modeling of complex components (e.g., nonlinear loads) is carried out individually in the time domain for that particular component using the best-suited method. The calculations are performed in a modular fashion and separately for each component. The components are then assembled via a systemwide harmonic matrix that enables the bus voltages to be corrected so that the current mismatches $\Delta I$ at all nodes are iteratively forced to zero.

The modular harmonic power flow (MHPF) is performed iteratively. Each step consists of two stages:

(i) *Updating the periodic steady state of the individual components (black boxes) using voltage corrections from stage (ii).* The inputs to a component are the voltages at its terminals and the outputs are the terminal currents. This stage provides accurate outputs for the given inputs. The initial input voltages (at the first iteration) are usually obtained from the fundamental power flow algorithm.

(ii) *Calculation of voltage corrections for stage (i).* The currents obtained from stage (i) are combined for buses with adjacent components into the current mismatch $\Delta I$, which are expressed in the harmonic domain. These become injections into a systemwide incremental harmonic admittance matrix $Y_{\text{bus}}$, calculated in advance from such matrices for all individual components. The following equation is then solved for $\Delta V$ to be used in stage (i) to update all bus voltages:

$$\Delta I = Y_{\text{bus}}\Delta V. \qquad (7\text{-}150)$$

The approach is modular in stage (i), but at stage (ii) the voltage correction $\Delta V$ is calculated globally for the entire system. The main procedures of the MHPF algorithm are

- A fundamental power flow algorithm is used to obtain initial voltages at all nodes.

- The incremental harmonic admittance matrix $Y_{bus}$ is identified. This is first done separately for each component (black box) and then the systemwide $Y_{bus}$ matrix is assembled.
- Currents and their mismatches $\Delta I$ are obtained for each node.
- $\Delta I$ and $Y$ are used to obtain $\Delta V$ (Eq. 7-150) and the nodal voltages are finally updated for the next iteration.

The modular approach facilitates the development of new component models and the formation of a wide range of system configurations. It is also appropriate for an unbalanced system with asymmetric loads. In practice some power quality problems relate to uncharacteristic harmonic frequencies resulting from nonlinear loads and/or system asymmetries and thus requiring three-phase harmonic simulation. The solutions of MHPF algorithms are reported to be accurate as compared with those generated by the conventional (nonmodular) HPF. The computing times depend of the required accuracy of the solution at harmonic frequencies and the maximum order of harmonics.

### 7.5.7 Application Example 7.19: Accuracy of Decoupled Harmonic Power Flow

Apply the coupled (Section 7.4) and decoupled (Section 7.5.1) harmonic power flow algorithm to the distorted 18-bus system of Fig. E7.18.1 and compare simulation results including the fundamental and rms (including harmonics) voltages, distortion factor, and THD$_v$. The nonlinear load (installed at bus 5) is a 3 MW, 5 MVAr six-pulse converter. Line and load parameters are provided in [29]. For the DHPF calculations, the converter may be modeled with harmonic current sources (Table E7.18.1).

### 7.6 SUMMARY

After an overview of the interconnected power system and its components, power system matrices are introduced, in particular the bus admittance matrix. Bus-building algorithms are explained, and the concepts of singularity and nonsingularity of matrices and their physical interpretation with respect to power systems are discussed. After a brief review of matrix solution techniques under the aspects of matrix sparsity, the fundamental power

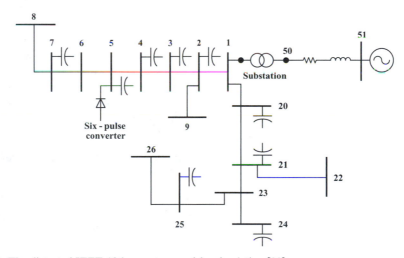

**FIGURE E7.18.1** The distorted IEEE 18-bus system used for simulation [29].

**TABLE E7.18.1** Magnitude of Harmonic Currents (as a Percentage of the Fundamental) Used to Model Six-Pulse Converters[a]

| Harmonic order | Magnitude (%) | Harmonic order | Magnitude (%) | Harmonic order | Magnitude (%) |
|---|---|---|---|---|---|
| 1 | 100 | 19 | 2.4 | 37 | 0.5 |
| 5 | 19.1 | 23 | 1.2 | 41 | 0.5 |
| 7 | 13.1 | 25 | 0.8 | 43 | 0.5 |
| 11 | 7.2 | 29 | 0.2 | 47 | 0.4 |
| 13 | 5.6 | 31 | 0.2 | 49 | 0.4 |
| 17 | 3.3 | 35 | 0.4 | | |

[a]All harmonic phase angles are assumed to be zero.

flow technique is used to explain the Newton–Raphson solution method. The harmonic power flow problem is firmly grounded on the fundamental power flow: additional circuit conditions are introduced for the harmonic power flow problem.

Numerous application examples detail for the fundamental power flow analysis the bus-building algorithm, matrix manipulations – such as multiplication and triangular factorization – formation of the Jacobian, definition of mismatch vector, and the recursive application of the Newton–Raphson algorithm. Similarly, the application examples cover the step-by-step approach for one iteration of the harmonic Newton-type solution method. The chapter concludes with a classification of existing power flow techniques.

## 7.7 PROBLEMS

### Problem 7.1: Determination of Admittance Matrices $\bar{Y}_{bus}$

a) Determine the bus admittance matrices $\bar{Y}_{bus}$ referenced to ground for Fig. P7.1a,b,c,d.
b) The system of Fig. P7.1d has no admittance connected from any one of the buses to ground. Show that its admittance matrix is singular.

### Problem 7.2: Establishing the Table of Factors

For the real matrix $\bar{A}$

$$\bar{A} = \begin{bmatrix} 1 & 5 & 1 \\ 5 & 2 & 0 \\ 1 & 0 & 1 \end{bmatrix}, \qquad \text{(P2-1)}$$

a) Find the table of factors.
b) Solve

$$\begin{bmatrix} 1 \\ 1 \\ 0 \end{bmatrix} = \bar{A} \cdot \bar{x}, \qquad \text{(P2-2)}$$

for the vector $\bar{x}$.

### Problem 7.3: Voltage Control Via Inductive Load (e.g., Reactors)

Apply the Newton–Raphson load flow analysis technique to the three-bus power system of Fig. P7.3, provided switch $S_1$ is closed and switch $S_2$ is open.

a) Find the fundamental bus admittance matrix in polar form.
b) Assume that bus 1 is the swing (slack) bus and buses 2 and 3 are PQ buses. Make an initial guess

for the bus voltage vector $\bar{x}^{(0)}$ and evaluate the initial mismatch power $\Delta \bar{W}^{(0)}$.
c) Find the Jacobian $\bar{J}^{(0)}$ for this power system configuration and compute the bus voltage correction vector $\Delta \bar{x}^{(0)}$.
d) Update the bus voltage vector $\bar{x}^{(1)} = \bar{x}^{(0)} - \Delta \bar{x}^{(0)}$ and recompute the updated mismatch power vector $\Delta \bar{W}^{(1)}$.

### Problem 7.4: Voltage Control Via Capacitive Load (e.g., Capacitors)

Repeat Problem 7.3 for the condition when switch $S_1$ is open and switch $S_2$ is closed.

### Problem 7.5: Fundamental Power Flow Exploiting the Symmetry of a Power System

Apply to the two-bus system of Fig. P7.5 the Newton–Raphson load flow analysis technique.

a) Simplify the two-bus system.
b) Find the fundamental bus admittance matrix in polar form.
c) Assume that bus 1 is the swing (slack) bus, and bus 2 is a PQ bus. Make an initial guess for the voltage vector $\bar{x}^{(0)}$ and evaluate the initial mismatch power $\Delta \bar{W}^{(0)}$.
d) Find the Jacobian matrix $\bar{J}^{(0)}$ for this power system configuration and compute the bus voltage correction vector $\Delta \bar{x}^{(0)}$.
e) Update the bus voltage vector $\bar{x}^{(1)} = \bar{x}^{0)} - \Delta \bar{x}^{(0)}$ and recompute the updated mismatch power vector $\Delta \bar{W}^{(1)}$.

### Problem 7.6: Fundamental Power Flow in a Weak Power System

The power system of Fig. P7.6 represents a weak feeder as it occurs in distributed generation (DG) with wind-power plants and photovoltaic plants as power sources. All impedances are in pu and the base apparent power is $S_{base} = 10$ MVA.

a) Compute the fundamental admittance matrix for this system.
b) Assuming bus 1 is the swing bus ($\delta_1 = 0$ and $|\tilde{V}_1| = 1.0$ pu) then we can make an initial guess for bus voltage vector $\bar{x}^0 = (\delta_2, |\tilde{V}_2|, \delta_3, |\tilde{V}_3|, \delta_4, |\tilde{V}_4|)^t = (0, 1, 0, 1, 0, 1)^t$. Note, the superscript 0 means initial guess. Compute the mismatch power vector for this initial condition.
c) Compute the Jacobian matrix.
d) Compute the inverse of the Jacobian matrix.

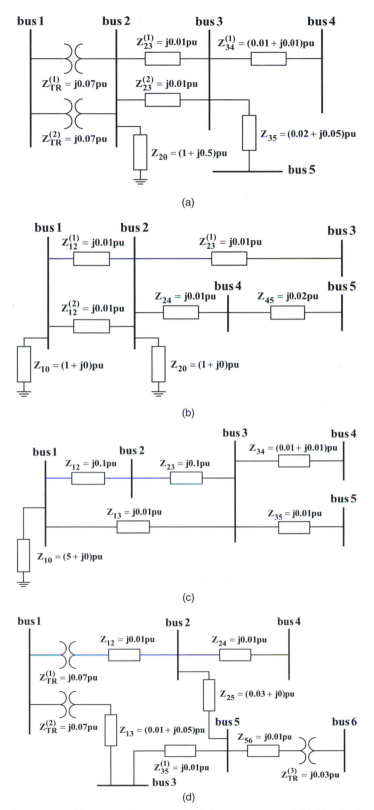

**FIGURE P7.1** (a) Five-bus system with two transformers and admittances to ground. (b) Five-bus system with admittances to ground. (c) Five-bus system with one admittance to ground. (d) Six-bus system with three transformers and no admittance to ground.

e) Compute the correction voltage vector $\Delta \overline{x}^0$, update the mismatch power vector, and check the convergence of the load flow algorithm (using a convergence tolerance of 0.0001).

## Problem 7.7: Harmonic Power Flow in a Weak Power System with One Nonlinear Bus and One Set of Harmonic Nonlinear Device Currents

The power system of Fig. P7.7 represents a weak feeder as it occurs in distributed generation (DG) with wind-power plants and photovoltaic plants as power sources. All impedances are in pu and the base apparent power is $S_{base} = 10$ MVA. The harmonic v–i characteristic of the nonlinear load at bus 4 (referred to the voltage at bus 4) is given as

$$g_{r,4}^{(5)} = 0.3(V_4^{(1)})^3\cos(3\delta_4^{(1)}) + 0.3(V_4^{(5)})^2\cos(3\delta_4^{(5)})$$
$$g_{i,4}^{(5)} = 0.3(V_4^{(1)})^3\sin(3\delta_4^{(1)}) + 0.3(V_4^{(5)})^2\sin(3\delta_4^{(5)})$$
$$(P7-1)$$

It will be assumed that the voltage at the swing bus (bus 1) is $|\tilde{V}_1^{(1)}| = 1.00$ p.u. and $\delta_1^{(1)} = 0$ rad. The fundamental power flow can be based on Fig. P7.6. For the 5th harmonic the synchronous machine impedance $Z_1$ is replaced by the (subtransient) impedance $Z_2 \approx jX^- = j0.25$ p.u. at 60 Hz. In this case the subtransient reactance has been chosen to be that of a synchronous generator in order to approximate bus 1 as a weak bus.

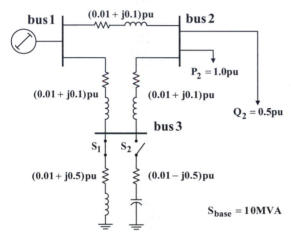

**FIGURE P7.3** Three-bus system.

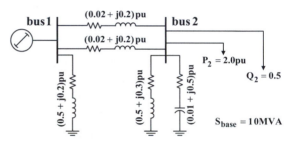

**FIGURE P7.5** Two-bus power system.

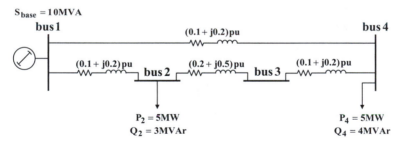

**FIGURE P7.6** Four-bus power system.

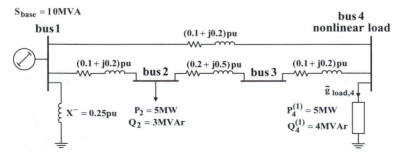

**FIGURE P7.7** Four-bus power system with one nonlinear bus and one set of harmonics.

a) Compute the fundamental and 5th harmonic admittance matrices.

b) Assume an initial value for the bus vector $\bar{U}^0$ and compute the fundamental and the 5th harmonic currents injected into the system by the nonlinear load.

c) Evaluate the mismatch vector.

d) Evaluate the Jacobian matrix.

e) Evaluate the correction vector $\Delta U^0$.

f) Update the mismatch vector and comment on the convergence of the harmonic load flow.

## Problem 7.8: Harmonic Power Flow with one Nonlinear Bus and Two Sets of Harmonic Nonlinear Device Currents

To the three-bus system of Fig. P7.8 apply the Newton–Raphson harmonic load flow analysis technique. Assume that bus 1 is the swing or slack bus, bus 2 is a linear PQ bus, and bus 3 is a nonlinear bus, where the fundamental real and reactive powers are specified. In addition, at bus 3 the real and imaginary harmonic nonlinear device currents are given as

$$g_{r,3}^{(5)} = |\tilde{V}_3^{(1)}|\cos(\delta_3^{(1)}) + |\tilde{V}_3^{(5)}|^2\cos(2\delta_3^{(5)}) + |\tilde{V}_3^{(7)}|\cos(\delta_3^{(7)}),\tag{P8-1a}$$

$$g_{i,3}^{(5)} = |\tilde{V}_3^{(1)}|\sin(\delta_3^{(1)}) + |\tilde{V}_3^{(5)}|^2\sin(2\delta_3^{(5)}) + |\tilde{V}_3^{(7)}|\sin(\delta_3^{(7)}),\tag{P8-1b}$$

$$g_{r,3}^{(7)} = |\tilde{V}_3^{(1)}|\cos(\delta_3^{(1)}) + |\tilde{V}_3^{(5)}|\cos(\delta_3^{(5)}) + |\tilde{V}_3^{(7)}|^2\cos(2\delta_3^{(7)}),\tag{P8-2a}$$

$$g_{i,3}^{(7)} = |\tilde{V}_3^{(1)}|\sin(\delta_3^{(1)}) + |\tilde{V}_3^{(5)}|\sin(\delta_3^{(5)}) + |\tilde{V}_3^{(7)}|^2\sin(2\delta_3^{(7)}).\tag{P8-2b}$$

The 5th and 7th nonlinear currents are referred to the harmonic voltages $\tilde{V}_3^{(5)}$ and $\tilde{V}_3^{(7)}$, respectively.

a) Find the fundamental and harmonic (5th and 7th) bus admittance matrices in polar form.

b) Make an initial guess for the bus vector $\bar{U}^0 = [\delta_2^{(1)}, v_2^{(1)}, \delta_3^{(1)}, v_3^{(1)}, \delta_1^{(5)}, v_1^{(5)}, \delta_2^{(5)}, v_2^{(5)}, \delta_3^{(5)}, v_3^{(5)}, \delta_1^{(7)}, v_1^{(7)}, \delta_2^{(7)}, v_2^{(7)}, \delta_3^{(7)}, v_3^{(7)}]^t$ and compute the nonlinear load device currents $G_{r,3}^{(1)}, G_{i,3}^{(1)}, G_{r,3}^{(5)}, G_{i,3}^{(5)}, G_{r,3}^{(7)}$ and $G_{i,3}^{(7)}$, which are referred to the swing bus.

c) Evaluate the initial mismatch vector $\Delta\bar{M}^0 = [\Delta\bar{W}, \Delta\bar{I}^{(5)}, \Delta\bar{I}^{(7)}, \Delta\bar{I}^{(1)}]^t$.

d) Find the Jacobian $\bar{J}^0$ for this power system configuration and compute the bus correction vector $\Delta\bar{U}^0$. The Jacobian $\bar{J}^0$ is of the form

$$\bar{J}^0 = \begin{bmatrix} \overline{J}^{(1)} & \overline{J}^{(5)} & \overline{J}^{(7)} \\ \overline{YG}^{(5,1)} & \overline{YG}^{(5,5)} & \overline{YG}^{(5,7)} \\ \overline{YG}^{(7,1)} & \overline{YG}^{(7,5)} & \overline{YG}^{(7,7)} \\ \overline{YG}^{(1,1)} & \overline{YG}^{(1,5)} & \overline{YG}^{(1,7)} \end{bmatrix}.\tag{P8-3}$$

In particular, this matrix (Eq. P8-3 or Eq. P8-4) has the following 16 unknowns: $\delta_2^{(1)}$, $V_2^{(1)}$, $\delta_3^{(1)}$, $V_3^{(1)}$, $\delta_1^{(5)}$, $V_1^{(5)}$, $\delta_2^{(5)}$, $V_2^{(5)}$, $\delta_3^{(5)}$, $V_3^{(5)}$, $\delta_1^{(7)}$, $V_1^{(7)}$, $\delta_2^{(7)}$, $V_2^{(7)}$, $\delta_3^{(7)}$, $V_3^{(7)}$, (see identification of columns of Eq. P8-4) and the following 16 equations: $\Delta P_2^{(1)}$, $\Delta Q_2^{(1)}$, $\Delta I_{r,1}^{(5)}$, $\Delta I_{i,1}^{(5)}$, $\Delta I_{r,2}^{(5)}$, $\Delta I_{i,2}^{(5)}$, $\Delta I_{r,3}^{(5)}$, $\Delta I_{i,3}^{(5)}$, $\Delta I_{r,1}^{(7)}$, $\Delta I_{i,1}^{(7)}$, $\Delta I_{r,2}^{(7)}$, $\Delta I_{i,2}^{(7)}$, $\Delta I_{r,3}^{(7)}$, $\Delta I_{i,3}^{(7)}$, $\Delta I_{r,3}^{(1)}$, $\Delta I_{i,3}^{(1)}$ (see identification of rows of Eq. P8-4). Note there are no $\bar{H}$ matrices (e.g., $\bar{H}^{(1)}$, $\bar{H}^{(5)}$ and $\bar{H}^{(7)}$) related to $\alpha_3$ and $\beta_3$. This reduces

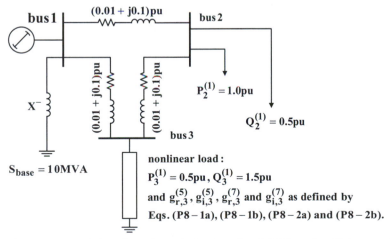

**FIGURE P7.8** Three-bus system consisting of two linear (swing bus = bus 1, PQ bus = bus 2) buses and one nonlinear bus (bus 3). The negative sequence impedance ($\approx$ subtransient reactance at 60 Hz) of the swing bus is $Z_2 = jX^- = j0.10$ pu and there are two sets of harmonics.

the equation system by two equations. In this case we can define the matrix $\bar{J}^0$ (Eq. P8-4).

e) Update the bus vector $\bar{U}^1 = (\bar{U}^0 - \Delta\bar{U}^0)$ and recompute the updated mismatch vector $\Delta\bar{M}^1$.

Hint: The top row – not part of the Jacobian – of Eq. P8-4 lists the variables:

$$[\delta_2^{(1)} \; v_2^{(1)} \; \delta_3^{(1)} \; v_3^{(1)} \; \delta_1^{(5)} \; v_1^{(5)} \; \delta_2^{(5)} \; v_2^{(5)} \; \delta_3^{(5)} \; v_3^{(5)} \; \delta_1^{(7)} \; v_1^{(7)} \; \delta_2^{(7)} \; v_2^{(7)} \; \delta_3^{(7)} \; v_3^{(7)}],$$

and the right-hand side column – not part of the Jacobian – lists the mismatch quantities which must be satisfied:

$$[\Delta P_2^{(1)} \; \Delta Q_2^{(1)} \; \Delta I_{r,1}^{(5)} \; \Delta I_{i,1}^{(5)} \; \Delta I_{r,2}^{(5)} \; \Delta I_{i,2}^{(5)} \; \Delta I_{r,3}^{(5)} \; \Delta I_{i,3}^{(5)}$$
$$\Delta I_{r,1}^{(7)} \; \Delta I_{i,1}^{(7)} \; \Delta I_{r,2}^{(7)} \; \Delta I_{i,2}^{(7)} \; \Delta I_{r,3}^{(7)} \; \Delta I_{i,3}^{(7)} \; \Delta I_{r,3}^{(1)} \; \Delta I_{i,3}^{(1)}]^t.$$

These two identifications help in defining the partial derivatives of the Jacobian. For example,

$$J_{3,5}^0 = \frac{\partial(\Delta I_{r,1}^{(5)})}{\partial(\delta_1^{(5)})}, \quad J_{3,10}^0 = \frac{\partial(\Delta I_{r,1}^{(5)})}{\partial(V_3^{(5)})}, \quad J_{8,5}^0 = \frac{\partial(\Delta I_{i,3}^{(5)})}{\partial(\delta_1^{(5)})},$$

$$J_{8,10}^0 = \frac{\partial(\Delta I_{i,3}^{(5)})}{\partial(V_3^{(5)})}, \quad J_{9,11}^0 = \frac{\partial(\Delta I_{r,1}^{(7)})}{\partial(\delta_1^{(7)})}, \quad \text{and}$$

$$J_{12,16}^0 = \frac{\partial(\Delta I_{i,2}^{(7)})}{\partial(V_3^{(7)})}.$$

| | $\delta_2^{(1)}$ | $V_2^{(1)}$ | $\delta_3^{(1)}$ | $V_3^{(1)}$ | $\delta_1^{(5)}$ | $V_1^{(5)}$ | $\delta_2^{(5)}$ | $V_2^{(5)}$ | $\delta_3^{(5)}$ | $V_3^{(5)}$ | $\delta_1^{(7)}$ | $V_1^{(7)}$ | $\delta_2^{(7)}$ | $V_2^{(7)}$ | $\delta_3^{(7)}$ | $V_3^{(7)}$ | |
|---|---|---|---|---|---|---|---|---|---|---|---|---|---|---|---|---|---|
| | ↓ | ↓ | ↓ | ↓ | ↓ | ↓ | ↓ | ↓ | ↓ | ↓ | ↓ | ↓ | ↓ | ↓ | ↓ | ↓ | |
| | 19.8 | 1.98 | −9.9 | −0.99 | 0 | 0 | 0 | 0 | 0 | 0 | 0 | 0 | 0 | 0 | 0 | 0 | $\leftarrow \Delta P_2^{(1)}$ |
| | −1.98 | 19.8 | 0.99 | −9.9 | 0 | 0 | 0 | 0 | 0 | 0 | 0 | 0 | 0 | 0 | 0 | 0 | $\leftarrow \Delta Q_2^{(1)}$ |
| | 0 | 0 | 0 | 0 | $J_{3,5}^0$ | $J_{3,6}^0$ | $J_{3,7}^0$ | $J_{3,8}^0$ | $J_{3,9}^0$ | $J_{3,10}^0$ | 0 | 0 | 0 | 0 | 0 | 0 | $\leftarrow \Delta I_{r,1}^{(5)}$ |
| | 0 | 0 | 0 | 0 | $J_{4,5}^0$ | $J_{4,6}^0$ | $J_{4,7}^0$ | $J_{4,8}^0$ | $J_{4,9}^0$ | $J_{4,10}^0$ | 0 | 0 | 0 | 0 | 0 | 0 | $\leftarrow \Delta I_{i,1}^{(5)}$ |
| | 0 | 0 | 0 | 0 | $J_{5,5}^0$ | $J_{5,6}^0$ | $J_{5,7}^0$ | $J_{5,8}^0$ | $J_{5,9}^0$ | $J_{5,10}^0$ | 0 | 0 | 0 | 0 | 0 | 0 | $\leftarrow \Delta I_{r,2}^{(5)}$ |
| | 0 | 0 | 0 | 0 | $J_{6,5}^0$ | $J_{6,6}^0$ | $J_{6,7}^0$ | $J_{6,8}^0$ | $J_{6,9}^0$ | $J_{6,10}^0$ | 0 | 0 | 0 | 0 | 0 | 0 | $\leftarrow \Delta I_{i,2}^{(5)}$ |
| | 0 | 0 | 0 | 1 | $J_{7,5}^0$ | $J_{7,6}^0$ | $J_{7,7}^0$ | $J_{7,8}^0$ | $J_{7,9}^0$ | $J_{7,10}^0$ | 0 | 0 | 0 | 0 | 0 | 1 | $\leftarrow \Delta I_{r,3}^{(5)}$ |
| $\bar{J}^0 =$ | 0 | 0 | 1 | 0 | $J_{8,5}^0$ | $J_{8,6}^0$ | $J_{8,7}^0$ | $J_{8,8}^0$ | $J_{8,9}^0$ | $J_{8,10}^0$ | 0 | 0 | 0 | 0 | 0.1 | 0 | $\leftarrow \Delta I_{i,3}^{(5)}$ |
| | 0 | 0 | 0 | 0 | 0 | 0 | 0 | 0 | 0 | 0 | $J_{9,11}^0$ | $J_{9,12}^0$ | $J_{9,13}^0$ | $J_{9,14}^0$ | $J_{9,15}^0$ | $J_{9,16}^0$ | $\leftarrow \Delta I_{r,1}^{(7)}$ |
| | 0 | 0 | 0 | 0 | 0 | 0 | 0 | 0 | 0 | 0 | $J_{10,11}^0$ | $J_{10,12}^0$ | $J_{10,13}^0$ | $J_{10,14}^0$ | $J_{10,15}^0$ | $J_{10,16}^0$ | $\leftarrow \Delta I_{i,1}^{(7)}$ |
| | 0 | 0 | 0 | 0 | 0 | 0 | 0 | 0 | 0 | 0 | $J_{11,11}^0$ | $J_{11,12}^0$ | $J_{11,13}^0$ | $J_{11,14}^0$ | $J_{11,15}^0$ | $J_{11,16}^0$ | $\leftarrow \Delta I_{r,2}^{(7)}$ |
| | 0 | 0 | 0 | 0 | 0 | 0 | 0 | 0 | 0 | 0 | $J_{12,11}^0$ | $J_{12,12}^0$ | $J_{12,13}^0$ | $J_{12,14}^0$ | $J_{12,15}^0$ | $J_{12,16}^0$ | $\leftarrow \Delta I_{i,2}^{(7)}$ |
| | 0 | 0 | 0 | 1 | 0 | 0 | 0 | 0 | 0 | 1 | $J_{13,11}^0$ | $J_{13,12}^0$ | $J_{13,13}^0$ | $J_{13,14}^0$ | $J_{13,15}^0$ | $J_{13,16}^0$ | $\leftarrow \Delta I_{r,3}^{(7)}$ |
| | 0 | 0 | 1 | 0 | 0 | 0 | 0 | 0 | 0.1 | 0 | $J_{14,11}^0$ | $J_{14,12}^0$ | $J_{14,13}^0$ | $J_{14,14}^0$ | $J_{14,15}^0$ | $J_{14,16}^0$ | $\leftarrow \Delta I_{i,3}^{(5)}$ |
| | −9.90 | −0.99 | 21.3 | 1.48 | 0 | 0 | 0 | 0 | 0 | 0 | 0 | 0 | 0 | 0 | 0 | 0 | $\leftarrow \Delta I_{r,3}^{(7)}$ |
| | −0.99 | 9.9 | −2.51 | −18.3 | 0 | 0 | 0 | 0 | 0 | 0 | 0 | 0 | 0 | 0 | 0 | 0 | $\leftarrow \Delta I_{i,3}^{(1)}$ |

$$(P8-4)$$

## 7.8 REFERENCES

1) Abolins, A.; Achenbach, H.; and Lambrecht, D.; "Design and performance of large four-pole turbogenerators with semiconductor excitation for nuclear power stations," *Cigre-Report*, 1972 Session, August 28–September, Paper No 11-04.

2) Fuchs, E.F.; "Numerical determination of synchronous, transient, and subtransient reactances of a synchronous machine," Ph.D. Thesis, University of Colorado, Boulder, 1970.

3) http://www.powerfrontiers.com/ngccplants.htm

4) http://en.wikipedia.org/wiki/Combined_cycle

5) Xia, D.; and Heydt, G.T.; "Harmonic power flow studies, part I – formulation and solution," *IEEE Transactions on Power Apparatus and System*, vol. 101, no. 6, pp. 1257–1265, 1982.

6) Xia, D.; and Heydt, G.T.; "Harmonic power flow studies, part II – implementation and practical application," *IEEE Transactions on Power Apparatus and System*, vol. 101, no. 6, pp. 1266–1270, 1982.

7) Baghzouz, Y.; and Ertem, S.; "Shunt capacitor sizing for radial distribution feeders with distorted substation voltage," *IEEE Transactions on Power Delivery*, vol. 5, no. 2, pp. 650–657, 1990.

8) Baghzouz, Y.; "Effects of nonlinear loads on optimal capacitor placement in radial feeders," *IEEE Transactions on Power Delivery*, vol. 6, no. 1, pp. 245–251, 1991.

9) Semlyen, A.; and Shlash, M.; "Principles of modular harmonic power flow methodology," *IEE Proceedings – Generation, Transmission and Distribution*, vol. 147, no. 1, pp. 1–6, 2000.

10) Arrillaga, J.; Watson, N.R.; and Bathurst, G.N.; "A multi-frequency power flow of general applicability," *IEEE Transactions on Power Delivery*, vol. 19, no. 1, pp. 342–349, 2004.

11) Bathurst, C.N.; Smith, B.C.; Watson, N.R.; and Arrillaga, J.; "A modular approach to the solution of the three-phase harmonic power-flow," *IEEE Transactions on Power Delivery*, vol. 15, no. 3, pp. 984–989, 2000.

12) Moreno Lopez de Saa, M.A.; and Usaola Garcia, J.; "Three-phase harmonic load flow in frequency and time domains," *IEE Proceedings – Electric Power Applications*, vol. 150, no. 3, pp. 295–300, 2003.

13) Hong, Y.-Y.; Lin, J.-S.; and Liu, C.-H.; "Fuzzy harmonic power flow analyses," presented at International Conference on Power System Technology, PowerCon 2000, vol. 1, 4–7 Dec. 2000, pp. 121–125.

14) Shlash, M.; and Semlyen, A.; "Efficiency issues of modular harmonic power flow," *IEE Proceedings –*

*Generation, Transmission and Distribution*, vol. 148, no. 2, pp. 123–127, 2001.

15) Williams, S.M.; Brownfield, G.T.; and Duffus, J.W.; "Harmonic propagation on an electric distribution system: field measurements compared with computer simulation," *IEEE Transactions on Power Delivery*, vol. 8, no. 2, pp. 547–552, 1993.

16) Chin, H.-C.; "Optimal shunt capacitor allocation by fuzzy dynamic programming," *Electric Power Systems Research*, vol. 35, no. 2, pp. 133–139, 1995.

17) Teng, J.-H.; and Chang, C.-Y.; "A fast harmonic load flow method for industrial distribution systems," *International Conference on Power System Technology*, vol. 3, pp. 1149–1154, December 2000.

18) Masoum, M.A.S.; Jafarian, A.; Ladjevardi, M.; Fuchs, E.F.; and Grady, W.M.; "Fuzzy approach for optimal placement and sizing of capacitor banks in the presence of harmonics," *IEEE Transactions on Power Delivery*, vol. 19, no. 2, pp. 822–829, 2004.

19) Chung, T.S.; and Leung, H.C.; "A genetic algorithm approach in optimal capacitor selection with harmonic distortion considerations," *International Journal of Electrical Power & Energy Systems*, vol. 21, no. 8, pp. 561–569, 1999.

20) Elamin, I.M.; "Fast decoupled harmonic load flow method," *Conference Record of the 1990 IEEE Industry Applications Society Annual Meeting, 1990*, 7–12 Oct. 1990, pp. 1749–1756.

21) Carpinelli, G.; Esposito, T.; Varilone, P.; and Verde, P.; "First-order probabilistic harmonic power flow," *IEE Proceedings – Generation, Transmission and Distribution*, vol. 148, no. 6, pp. 541–548, 2001.

22) Esposito, T.; Carpinelli, G.; Varilone, P.; and Verde, P.; "Probabilistic harmonic power flow for percentile evaluation," *Canadian Conference on Electrical and Computer Engineering, 2001*, vol. 2, 13–16 May 2001, pp. 831–838.

23) Bathurst, G.N.; Smith, B.C.; Watson, N.R.; and Arrillaga, J.; "A modular approach to the solution of the three-phase harmonic power-flow," *8th International Conference on Harmonics and Quality of Power, 1998*, vol. 2, 14–16 Oct. 1998, pp. 653–659.

24) Valcarel, M.; and Mayordomo, J.G.; "Harmonic power flow for unbalanced systems," *IEEE Transactions on Power Delivery*, vol. 8, no. 4, pp. 2052–2059, 1993.

25) Marinos, Z.A.; Pereira, J.L.R.; and Carneiro Jr., S.; "Fast harmonic power flow calculation using parallel processing," *IEE Proceedings – Generation, Transmission and Distribution*, vol. 141, no. 1, pp. 27–32, 1994.

26) Rios, M.S.; and Castaneda, P.R.; "Newton-Raphson probabilistic harmonic power flow through Monte Carlo simulation," *Symposium on Circuits and*

*Systems, Midwest, 1995*, vol. 2, 13–16 Aug. 1995, pp. 1297–1300.

27) Baghzouz, Y.; Burch, R.F.; Capasso, A.; Cavallini, A.; Emanuel, A.E.; Halpin, M.; Imece, A.; Ludbrook, A.; Montanari, G.; Olejniczak, K.J.; Ribeiro, P.; Rios-Marcuello, S.; Tang, L.; Thaliam, R.; and Verde, P.; "Time-varying harmonics. I. Characterizing measured data," *IEEE Transactions on Power Delivery*, vol. 13, no. 3, pp. 938–944, 1998.

28) Baghzouz, Y.; Burch, R.F.; Capasso, A.; Cavallini, A.; Emanuel, A.E.; Halpin, M.; Imece, A.; Ludbrook, A.; Montanari, G.; Olejniczak, K.J.; Ribeiro, P.; Rios-Marcuello, S.; Tang, L.; Thaliam, R.; and Verde, P.; "Time-varying harmonics. II. Harmonic summation and propagation," *IEEE Transactions on Power Delivery*, vol. 17, no. 1, pp. 279–285, 2002.

29) Grady, W.M.; Samotyj, M.J.; and Noyola, A.H.; "The application of network objective functions for actively minimizing the impact of voltage harmonics in power systems," *IEEE Transactions on Power Delivery*, vol. 7, no. 3, pp. 1379–1386, 1992.

# Impact of Poor Power Quality on Reliability, Relaying, and Security

New technical and legislative developments shape the electricity market, whereby the reliability of power systems and the profitability of the utility companies must be maintained. To achieve these objectives the evaluation of the power system and, in particular, the distribution system reliability is an important task. It enables utilities to predict the reliability level of an existing system after improvements have been achieved through restructuring and improved design and to select the most appropriate configuration among various available options.

This chapter addresses the impact of poor power quality on reliability, relaying, and security. Section 8.1 introduces reliability indices. Degradation of reliability and security due to poor power quality including the effects of harmonics and interharmonics on the operation of overcurrent and under-frequency relays, electromagnetic field (EMF) generation, corona in transmission lines, distributed generation, cogeneration, and frequency/voltage control are discussed in Section 8.2. Tools for detecting poor power quality and for improving reliability or security are treated in Sections 8.3 and 8.4, respectively. The remainder of this chapter is dedicated to techniques for improving and controlling the reliability of power systems including load shedding and load management (Section 8.5), energy-storage methods (Section 8.6), and matching the operation of intermittently operating renewable power plants with energy storage facilities (Section 8.7).

## 8.1 RELIABILITY INDICES

It is worth noting that the reliability indices as defined in the literature [1–9] are not absolute measures, but they provide relative information based on the comparison of alternative solutions. The set of reliability indices can be divided into four categories:

**Load-point indices:** average failure (fault) rate ($\lambda$), average outage time ($r$), annual outage time ($U$), repair time ($t_r$), and switching time ($t_s$).
**Customer-oriented system indices:** system average interruption frequency index (SAIFI), system average interruption duration index (SAIDI), customer average interruption duration index (CAIDI), and system average rms (variation) frequency index (SARFI, usually written as $SARFI_{\%V}$).
**Load-oriented system indices:** average service availability index (ASAI), average service unavailability index (ASUI), energy not supplied (ENS), and average energy not supplied (AENS).
**Economical evaluation of reliability:** composite customer damage function (CCDF), customer outage cost (COC), and system customer outage cost (SCOC).

In typical distribution systems, radial circuits are formed by a main feeder and some laterals as depicted in Fig. 8.1. The main feeder is supplied by a primary substation (PS or HV/MV) and connects various switching substations (SWSU)$_i$ that do not include power transformers to which the laterals are connected to supply the secondary substations (SS)$_i$. This radial configuration is simple and economical, but provides the lowest reliability level because secondary substations have only one source of supply. In Fig. 8.1 (PDu)$_i$ and (PDd)$_i$ are upstream and downstream protection devices, respectively, and (LP)$_i$ are load points.

The open-loop configuration of Fig. 8.2 is formed by two main feeders supplied by the same primary substation (PS or HV/MV). Each feeder has eight switching substations that each supply one secondary substation (SS)$_i$. No laterals are present in this configuration, but each load point (LP)$_i$ is supplied through a connection line of short length. The two feeder ends are typically disconnected by NO (normally open) switches, thus forming an open loop. Under normal conditions the NO switches separate the two main feeders of the system, but in the event of a fault they can be closed in order to supply power to loads that have been disconnected from the primary substation. In Fig. 8.2 (NOSW)$_1$ and (NOSW)$_2$ are normally open switches for the two main feeders 1 and 2, respectively.

Ring configurations are similar to open-loop schemes but are not operated in a radial manner. The main feeder is supplied at both ends by one primary substation (PS or HV/MV), but it forms a ring without normally open switches. Appropriate protection schemes allow higher reliability than for radial or open-loop configurations because one fault of the main feeder does not cause power failure at any LP. The ring configuration is shown in Fig. 8.3 whereby the number of load points (LPs), their topology, and their relative positions are the same as for the open-loop scheme.

Flower configurations are formed by ring schemes—resembling flower petals—and are supplied by the same primary substation (PS or HV/MV). Normally, the number of rings ranges from two to four for each PS. The rings belonging to different flower petals can be connected to one another through NO switches, increasing the system reliability. Figure 8.4 illustrates two flower configurations

supplying eight LPs each. Each PS is supposed to be placed at a load center. The number of LPs, their topology, and their relative positions are the same as in the cases previously considered.

The reliability evaluation of power systems is concerned besides traditional approaches [4] with the impact of automation on distribution reliability [2, 5, 7], the reliability worth for distribution planning [3], and the role of automation on the restoration of distribution systems [6]. Reliability test systems are addressed in [8] and a reliability assessment of traditional versus automated approaches is presented in [9].

### 8.1.1 Application Example 8.1: Calculation of Reliability Indices

In this example, July has been chosen because in Colorado there are short-circuit faults, demand overload, line switching, capacitor switching, and lightning strikes as indicated in Table E8.1.1. The total number of customers is $N_T = 40000$.

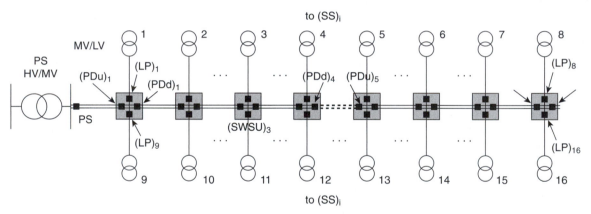

**FIGURE 8.1** Radial configuration with possible locations for protection devices (PD) at each switching substation (SWSU) with load points (LP).

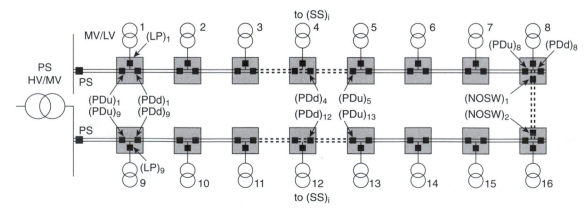

**FIGURE 8.2** Open-loop configuration with possible locations for protection devices at each switching substation.

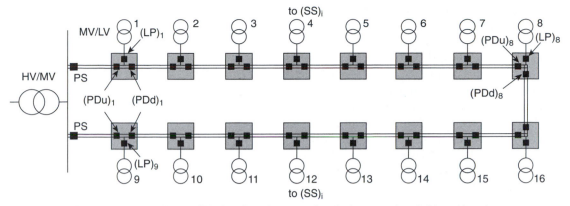

**FIGURE 8.3** Ring configuration with possible locations for protection devices at each switching substation.

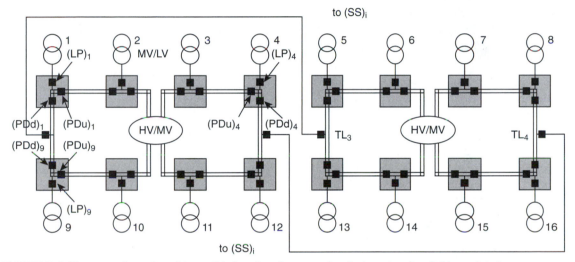

**FIGURE 8.4** Flower configuration with possible locations for protection devices at each switching substation.

**TABLE E8.1.1** Interruptions and Transients during July 2006

| Number of interruptions or transients | Interruption or transient duration | Voltage (% of rated value) | Total number of customers affected | Cause of interruption or transient |
|---|---|---|---|---|
| 4 | 4.5 h | 0 | 1,000 | Overload |
| 6 | 3 h | 0 | 1,500 | Repair |
| 25 | 1 min | 90 | 2,000 | Switching transients |
| 30 | 1 s | 130 | 3,000 | Capacitor switching |
| 50 | 0.1 s | 40 | 15,000 | Short-circuits |
| 55 | 0.1 s | 150 | 4,000 | Lightning |
| 70 | 0.01 s | 50 | 3,000 | Short-circuits |

## 8.2 DEGRADATION OF RELIABILITY AND SECURITY DUE TO POOR POWER QUALITY

### 8.2.1 Single-Time and Nonperiodic Events

The reliability and security of power systems is impaired by natural and man-made phenomena. Both types change in their severity as the power system protection develops over time. There are single-time, periodic, nonperiodic, and intermittent events. Although before the advent of interconnected systems lightning strikes [10], ice storms [11–18], ferroresonance [19], faults [20], and component failures were mainly a concern, the fact that interconnected systems are predominantly used gave rise to the influence of sunspot cycles [21] on long transmission lines built on igneous rock formations, subsynchronous resonance [22] through series compensation of long lines, nuclear explosions [23], and terrorist attacks [24–27] as well as the increase of the ambient temperature above $T_{amb} = 40°C$ due to global warming [28, 29]. It is interesting to note that utilities install ice monitors [15] on transmission lines and make an attempt to melt ice through dielectric losses [16–18]. Ferroresonance can be avoided by employing gas ($SF_6$)-filled cables [30] for distribution systems above 20 kV. The effect of sunspot cycles and those of nuclear explosions can be mitigated through the use of three-limb transformers with a large air gap between the transformer iron core and the tank [31]. The increase of the ambient temperature will necessitate in a revision of existing design standards for electric machines and transformers, and an ambient temperature of 50°C is recommended for new designs of electrical components.

## 8.2.2 Harmonics and Interharmonics Affecting Overcurrent and Under-Frequency Relay Operation [32]

The effects of nonsinusoidal voltages and currents on the performance of static under-frequency and overcurrent relays are experimentally studied in [33]. Tests are conducted for harmonics as well as sub- and interharmonics. Most single- and three-phase induction motors generate interharmonics within the rotating flux wave and the terminal currents due to the choice of the stator and rotor slots and because of air-gap eccentricity (spatial harmonics).

**Under-Frequency Relays.** The increases in operating time due to voltage harmonics from 0 to 15% of the fundamental are shown in Fig. 8.5 for a frequency set point of 58.99 Hz at frequency rate of changes of 1 Hz/s and 2 Hz/s, at a delay time of $T_D = 0$ s. Tests indicate [33] that under-frequency relays are very sensitive to interharmonics of the voltage because of the occurrence of additional zero crossings within one voltage period and, therefore, interharmonics should be limited from this point of view to less than 0.5%.

**Overcurrent Relays.** A three-phase solid-state time and instantaneous over-current relay is self-contained and can have short time, long time, instantaneous, and ground protection on the options selected. The percentage deviations from the nominal (sinusoidal) rms current values operating the overcurrent relay at lower (e.g., –0.95%) or higher

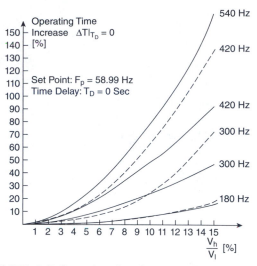

**FIGURE 8.5** Operating time increase $\Delta T |_{TD=0}$ due to voltage harmonics for the frequency set point $F_p = 58.99$ Hz. (—) Frequency rate of change of $\Delta f = 1$ Hz/s and an operating time of $T_{T1} = 60.5$ cycles or 1.008 s. (– – –) Frequency rate of change of $\Delta f = 2$ Hz/s and an operating time of $T_{T2} = 25$ cycles or 0.417 s.

(e.g., +1.43%) (nonsinusoidal) rms current values are listed for various harmonic amplitudes and phase shifts (e.g., min, max) in [33]. One notes that current harmonics cause the overcurrent device in the worst case to pick up at (100% ± 10%) of the nominal rms value based on a sinusoidal wave shape. Similar measurements reveal that the same overcurrent relay has changes at the time delay—for a given pickup rms current setting—up to 45% shorter and up to 26% longer than the nominal time delay with no harmonics. The long-time delay pickup current value is reduced for most harmonics by 20% and increased for certain harmonics by 7%. An electromechanical relay, subjected to similar tests as the solid-state relay, performed in a similar manner as the latter. With regard to the instantaneous pickup, rms current harmonics cause the overcurrent relay to pick up in the worst case at (100% + 10%) of the nominal rms value based on a sinusoidal wave shape. With regard to the time delay for given pickup rms current values, harmonics cause the hinged armature over-current relay to pick up in the worst case ($I_h/I_1 = 60\%$) at (100% + 43%) of the nominal rms value based on sinusoidal wave shape.

### 8.2.3 Power-Line Communication

Power-line carrier applications are described in [34–39]. They consist of the control of nonessential loads (load shedding) to mitigate undervoltage conditions and to prevent voltage collapse. High-frequency harmonics, interharmonics caused by solid-state converters, and voltage fluctuations caused by electric arc furnaces can cause interferences with power-line communication and control. Newer approaches for automatic meter reading [40–42] and load control are based on wireless communication. Reference [43] addresses the issue of optimal management of electrical loads.

### 8.2.4 Electromagnetic Field (EMF) Generation and Corona Effects in Transmission Lines

The minimization and mitigation of electromagnetic fields (EMFs) and corona effects, e.g., radio and television interference, coronal losses, audible noise, ozone production at or above 230/245 kV high-voltage transmission lines, is important. After addressing the generation of electric and magnetic fields, mechanisms leading to corona, the factors influencing the generation of corona, and its negative effects are explained. Solutions for the minimization of corona in newly designed transmission lines and reduction or mitigation of corona of existing lines are well known but not, however, always inexpensive. California's Public Utilities Commission initiated hearings in order to explore the possibilities of "no-cost/low-cost" EMF mitigation. A review of these ongoing hearings and their outcomes will be given. The difficulties associated with obtaining the right of way for new transmission lines lead to the upgrading of existing transmission lines with respect to higher voltages in order to increase transmission capacity. In the past, high-voltage lines with voltages of up to $V_{L-L} = 200$ kV were considered acceptable in terms of the generation of corona effects [35, 44–46]. However, the concern that electromagnetic fields may cause in addition to corona also detrimental health effects lead to renewed interest in reexamining the validity of the stated voltage limit with respect to their impact on the health of the general population. This section presents a brief summary of many publications.

#### 8.2.4.1 Generation of EMFs

**Electric Field Strength $\vec{E}$.** The electric field strength $\vec{E}$ of a single conductor having a per-unit length charge of $Q$ is

$$\vec{E} = \frac{1}{4\pi\varepsilon_o} \cdot \frac{Q}{r^2}\vec{r}, \qquad (8\text{-}1)$$

where $\varepsilon_o$ and $\vec{r}$ are the permittivity of free space and the unit vector in radial direction of a cylindrical coordinate system, respectively. Note that $\vec{E}$ reduces inversely in a quadratic manner with the radial distance $r$. For a three-phase system the three electric field components $E_a$, $E_b$, and $E_c$ must be geometrically superimposed at the point of interest. $Q$ is proportional to the transmission line voltage $V$, that is, the absolute value of $\vec{E}$ is proportional to $V$.

#### 8.2.4.2 Application Example 8.2: Lateral Profile of Electric Field at Ground Level below a Three-Phase Transmission Line

If the calculation of the electric field at ground is repeated at different points in a section perpendicular to the transmission line, the lateral profile of the transmission-line electric field is obtained. Examples of calculated lateral profiles are presented in Fig. E8.2.1. Particularly important for the line design are the maximum (worst case) field and the field at the edge of the transmission corridor. The maximum field occurs within the transmission corridor, though for flat configurations it might occur slightly outside the outer phases.

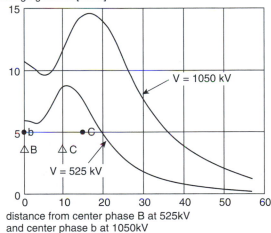

**FIGURE E8.2.1** Lateral profiles of electric field $|\vec{E}|$ at ground level. Line at $V_{L-L} = 525$ kV with $3 \times 3.3$ cm (bundle) conductors on $D = 30$ cm diameter, 45 cm spacing, spaced 10 m, and 10.6 m above ground, flat configuration (phases B and C). Line at $V_{L-L} = 1050$ kV with $8 \times 3.3$ cm (bundle) conductors on $D = 101$ cm diameter, spaced 18.3 m, and 18.3 m above ground, flat configuration (phases b and c).

Figure E8.2.2 shows universal curves using nondimensional quantities that may be used to calculate the maximum electric field at ground for a flat configuration. Results for other configurations are given in [47].

For a rated line-to-line voltage of $V_{L-L} = 525$ kV, a $3 \times 3.3$ cm conductor bundle with 45 cm spacing, phase-to-phase distance $S = 10$ m, the equivalent bundle diameter of $D = 0.3$ m, and height of the center of the bundle to ground $H = 10.6$ m the ratios $H/D = 10.6/0.3 = 35.3$, $S/H = 10/10.6 = 0.94$ result with Fig. E8.2.2 in $HE/V = 0.179$. The maximum (worst case) electric field at ground is then $|\vec{E}| = E = 0.179 \cdot 525\text{k}/10.6 = 8.8$ kV/m $= 8.8$ V/mm; this value is confirmed by Fig. E8.2.1. The presence of overhead ground wires has a negligible effect on the field at ground.

**Magnetic Field Strength $\vec{H}$.** The magnetic field strength $\vec{H}$ of a single conductor carrying a current $I$ is

$$\vec{H} = \frac{1}{2\pi} \cdot \frac{I}{r} \vec{\phi}, \qquad (8-2)$$

where $\vec{\phi}$ is the azimuthal unit vector of a cylindrical coordinate system. Note that $\vec{H}$ decreases inversely with the radial distance $r$. For a three-phase system the three magnetic field components $H_a$, $H_b$, and $H_c$ must be geometrically superimposed at the point of interest. The absolute value of the magnetic field $\vec{H}$ is proportional to $I$ [48].

### 8.2.4.3 Application Example 8.3: Lateral Profile of Magnetic Field at Ground Level under a Three-Phase Transmission Line

In most practical cases, the magnetic field in proximity to balanced three-phase lines may be calculated, considering the currents in the conductors and in the ground wires and neglecting the earth currents. As a result one obtains a magnetic field ellipse with minor and major axes. The maximum (worst case) ground level magnetic field occurs at the center line of the transmission line and is $|\vec{H}_{worst}| = H_{worst} = 14$ A/m (see Fig. E8.3.1), which corresponds to a flux density $|\vec{B}_{worst}| = B_{worst} = 0.176$ G $= 176$ mG. For comparison, the earth magnetic field is about $B_{earth} = 0.5$ G $= 500$ mG. It should be noted, however, that the earth field is not time-varying, whereas the transmission line field varies with 60 Hz frequency.

### 8.2.4.4 Mechanism of Corona

Electrical discharges are usually triggered by an electric field $|\vec{E}|$ accelerating free electrons through a gas. When these electrons acquire sufficient energy from an electric field, they may produce fresh ions by knocking electrons from atoms and molecules by collision. This process is called ionization by electron impact. The electrons multiply, as illustrated in Fig. 8.6, where secondary effects from the electrodes make the discharge self-sustaining. The initial electrons that start the ionizing process are often created by photoionization: a photon from some distant source imparts enough energy to an atom so that the atom breaks into an electron and a positively charged ion. During acceleration in the electric field, the electron collides with the atoms of nitrogen, oxygen, and other gases present. Most of these collisions are elastic collisions similar to the collision between two

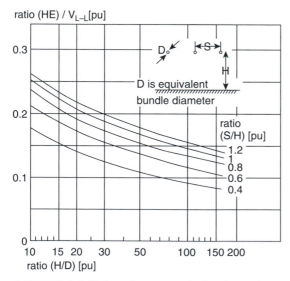

ratio (HE) / $V_{L-L}$ [pu]

D is equivalent bundle diameter

ratio (S/H) [pu]
1.2
1
0.8
0.6
0.4

ratio (H/D) [pu]

**FIGURE E8.2.2** Universal curves to calculate the maximum (worst case) electric field $|\vec{E}| = E$ at ground level for lines of flat configuration. H is the height to the center of the bundle.

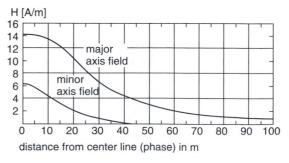

H [A/m]

major axis field

minor axis field

distance from center line (phase) in m

**FIGURE E8.3.1** Magnetic field strength $|\vec{H}| = H$ at ground level given as magnitudes of the major (real part of $|\vec{H}|$ phasor) and minor (imaginary part of $|\vec{H}|$ phasor) axes of the field ellipse. Conductor currents $i_a$, $i_b$, and $i_c$ have a magnitude of 2000 A each.

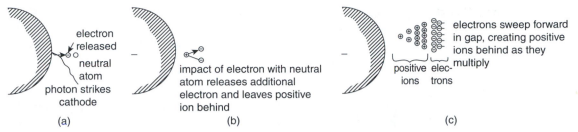

**FIGURES 8.6** Start of an electron avalanche from a negative electrode: (a) initiation, (b) formation of an electron pair, (c) multiplication of electron pairs or forming of avalanche.

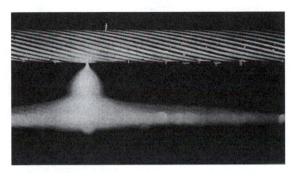

**FIGURE 8.7** Energized wet conductor with surface dirt or an insect generates plasma during discharge [47].

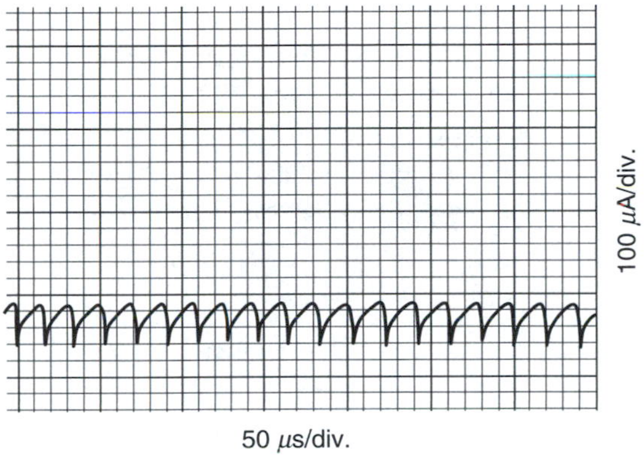

**50 μs/div.**

**FIGURE 8.8** Corona discharge at cathode resulting in 50 kHz electric noise signal [47].

billiard balls. The electron loses only a small part of its kinetic energy with each collision. Occasionally, an electron may strike an atom sufficiently hard so that excitation occurs, and the atom shifts to a higher energy state. The orbital states of one or more electrons changes, and the impacting electron loses part of its kinetic energy. Later, these excited atoms may revert to normal state, resulting in a radiation of the excess energy in the form of light (visible corona) and electromagnetic waves. An electron may also collide (recombine) with a positive ion, converting the ion to a neutral atom. Figure 8.7 shows the development of corona along an energized wet conductor, and Fig. 8.8 illustrates the frequency of corona discharge.

### 8.2.4.5 Factors Reducing the Effects of EMFs

**Active Shielding or Cancellation/Compensation of EMFs.** Active shielding is possible through compensation of electric and magnetic fields by suspending several transmission-line systems on the same towers (e.g., reverse-phased double circuit). This method is useful for newly designed transmission lines: it mitigates the generation of EMFs and reduces the amount of land used for transmission-line corridors.

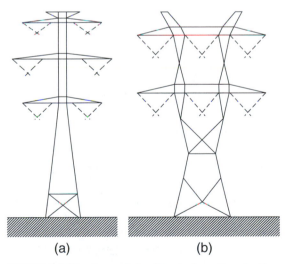

**FIGURE 8.9** Two transmission line configurations for line-to-line voltages of 345 kV, where the magnetic (as well as electric) fields are partially compensated (e.g., reverse-phased double circuit).

Figure 8.9 presents two configurations that are frequently used in practice. Within buildings and residences E. A. Leeper [49] explains how the EMFs can be compensated by active circuits.

**Passive Shielding/Mitigation using Bypassing Networks.** Passive shielding can be accomplished by using horizontal and vertical conducting grid (mesh) structures. Natural shields are provided by trees and houses. Conductive suits for linemen can protect individuals working in or near transmission corridors.

### 8.2.4.6 Factors Influencing Generation of Corona

At a given voltage, corona is determined by conductor diameter, line configuration, type of conductor (e.g., stranded), condition of its surface, and weather. Rain is by far the most important aspect of weather in increasing corona. The effect of atmospheric pressure and temperature is generally considered to modify the critical disruptive voltage (where the corona effects set in) of a conductor directly, or as the 2/3 exponent of the air density factor, $\delta$, which is given by [35]

$$\delta = \frac{17.9 \cdot (\text{barometric pressure in inches Hg})}{459 + {}^\circ\text{Fahrenheit}}.$$

$$(8\text{-}3)$$

The equation for the critical disruptive voltage is

$$V_{L-N\_o} = g_o \cdot \delta^{2/3} \cdot r \cdot m \cdot \ln\left(\frac{D}{r}\right), \qquad (8\text{-}4)$$

where

$V_{L-N\_o}$ = critical disruptive voltage in kV from line to neutral,
    $g_o$ = critical gradient in kV per centimeter,
    $r$ = radius of conductor in centimeters,
    $D$ = distance in centimeters between two conductors, and
    $m$ = surface factor.

Corona in fair weather is negligible or moderate up to a voltage near the disruptive voltage for a particular conductor. Above this voltage corona effects increase very rapidly. The calculated disruptive voltage is an indicator of corona performance. A high value of critical disruptive voltage is not the only criterion of satisfactory corona performance. Consideration should also be given to the sensitivity of the conductor to foul weather.

### 8.2.4.7 Application Example 8.4: Onset of Corona in a Transmission Line

A $V_{\text{rated}} = 345\,\text{k}V_{L-L}$ transmission line with $g_o = 21.1\,\text{kV/cm}$ (e.g., see Fig. E8.4.1b), $r = 1.6\,\text{cm}$, $D = 760\,\text{cm}$, $m = 0.84$ where at an altitude of 6000

feet the barometric pressure is 23.98 inches Hg (mercury) results at a temperature of −30°F in an air density factor of $\delta = 1.00$ and a critical disruptive voltage of $V_{L-N\_o} = 174\,\text{kV}$. That is, $(V_{L-N\_o} \cdot \sqrt{3}) = 300\,\text{k}V_{L-L} < V_{\text{rated}}$ and corona will exist on this 345 kV line—even under fair weather conditions. Under foul (e.g., dirt, rain, frost) weather conditions the critical value of $300\,\text{k}V_{L-L}$ will be further reduced. Figure E8.4.1a gives the transmission line dimensions in meters for the $V_{L-L} = 362\,\text{kV}$ class. Figure E8.4.1b presents the conductor surface gradient $g_o$ for a 362 kV class of transmission lines. $N$ is the number of phase (sub) conductors. Similar plots are given in [47] for other classes of transmission lines, e.g., aluminum-cable steel reinforced (ACSR).

The power loss associated with corona can be represented by shunt conductance. Figure E8.4.2 presents the coronal loss during fair weather conditions for different types of transmission lines. Corona loss ranges from about 0 up to 35 kW per three-phase mile. For a 500 mile transmission line the latter loss per mile value amounts to 17.5 MW. At line-to-line voltages above 230 kV, it is preferable to use more than one conductor per phase, which is known as bundling of conductors. The bundle can consist of two, three, four, or more subconductors. Figure E8.4.3 illustrates a flat configuration consisting of three bundles with six subconductors each. Bundling increases the effective radius of the line's conductor per phase and reduces the electric field strength $\bar{E}$ near the conductors, which reduces corona power loss, audible noise, radio and television interference, and the generation of ozone ($O_3$) and nitrogen oxides ($NO_x$). Another important advantage of bundling is reduced transmission line reactance, therefore increasing transmission capacity. Table E8.4.1 illustrates bundle dimensions as a function of the number of subconductors.

### 8.2.4.8 Negative Effects of EMFs [50, 51] and Corona

**Biological Effects of Electric Fields $\bar{E}$ on People and Animals.** To date, no specific biological effect of AC electric fields of the type and value applicable to transmission lines has been conclusively found and accepted by the scientific community. It is noteworthy to mention the rules established in the USSR (Union of Soviet Socialist Republics) for extrahigh voltage (EHV) substations and EHV transmission lines. For substations, limits for the duration of work

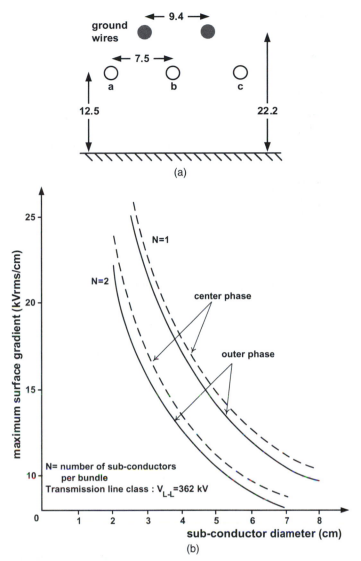

**FIGURE E8.4.1** (a) Transmission line dimensions in meters for the $V_{L-L} = 362$ kV class. (b) Conductor surface gradient $g_o$ as a function of subconductor diameter, horizontal, $V_{L-L} = 362$ kV class, where N is the number of (sub) conductors within a bundle per phase.

in a day have been established as indicated in Table 8.1.

For transmission lines, considering the infrequent and nonsystematic exposure, higher values of the electric field strength $|\vec{E}|$ are accepted:

- 20 kV/m for difficult terrain,
- 15–20 kV/m for nonpopulated region, and
- 10–12 kV/m for road crossing.

**Suggested Biological Effects of Magnetic Fields $\vec{H}$ and $\vec{B}$ on People and Animals.** The value of the magnetic field strength $\vec{H}$ at ground level close to transmission lines is of the order of 5–14 A/m corresponding to magnetic field densities of $|\vec{B}|$ from 0.08 to 0.3 G. Note that 1 gauss is equivalent to $10^{-4}$ tesla or $10^{-4}$ Wb/m$^2$ or $10^{-4}$ Vs/m$^2$ in the meter-kilogram-second-ampere (MKSA) system of units.

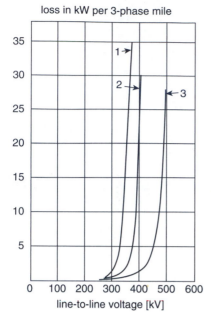

loss in kW per 3-phase mile

**FIGURE E8.4.2** Fair weather corona loss curves for smooth conductors at an air density factor $\delta = 1.0$. Curve 1: 1.4 inch hollow copper conductor (HH), $V_{L-L} = 287.5$ kV line; Curve 2: 1.1 inch HH conductor has smooth surface, $V_{L-L} = 230$ kV line; Curve 3: 1.65 inch smooth conductor has poor surface, $V_{L-L} = 230$ kV line.

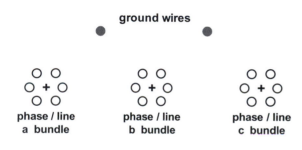

**FIGURE E8.4.3** Single-circuit horizontal (flat) line configuration.

**TABLE E8.4.1** Bundle Diameters as a Function of Number of Subconductors

| Conductors per phase | Bundle diameter (cm) |
|---|---|
| 2 | 45.7 |
| 3 | 52.8 |
| 4 | 64.6 |
| 6 | 91.4 |
| 8 | 101.6 |
| 12 | 127.0 |
| 16 | 152.4 |

**TABLE 8.1** USSR Rules for Duration of Work in Substations

| Electric field strength $|\vec{E}|$ (kV/m) | Permissible duration (minutes per day) |
|---|---|
| <5 | No restriction |
| 5–10 | 180 |
| 10–15 | 90 |
| 15–20 | 10 |
| 20–25 | 5 |

Magnetic AC fields have been reported to affect blood composition, growth, behavior, immune systems, and neural functions. However, at present, there is a lack of conclusive evidence.

**Effects of Corona.** Corona produces in the range of $V_{L-L} = 230$ kV and higher

- power loss,
- audible (acoustic) hissing sound in the vicinity of the transmission corridor,
- ozone and $NO_x$, and
- radio and television interference.

The audible noise is an environmental concern. Radio interference (RI) occurs in the AM (amplitude modulation) band. Rain and snow may produce moderate TV interference (TVI) in a low signal area. Note that voice signals are frequency modulated (FM), and video signals are amplitude modulated (AM).

### 8.2.4.9 Solutions for the Minimization of EMFs, Corona, and Other Environmental Concerns in Newly Designed Transmission Lines

Designing a transmission line with minimum environmental impact requires a study of five key factors:

1. impact of electric $\vec{E}$ and magnetic fields $\vec{H}$,
2. visual impact of the design,
3. generation of corona and its influence on radio and television interference,
4. air pollution via the generation of ozone and $NO_x$, and
5. impact of physical location.

A properly designed transmission line will have almost no perceptible negative environmental effects during fair weather, thus reducing disturbances, if they occur at all, to a small fraction of time.

**EMFs.** Although electric fields can be more readily reduced by shielding through structures with high conductivity (e.g., fences, trees), the reduction of magnetic fields is more difficult to achieve through shields with high permeability. Compensation or canceling of magnetic fields through the suspension of several transmission-line systems (e.g., reverse-phased double circuits) on the same tower as well as mounting the lines on higher towers near populated areas are frequently employed solutions. The employment of higher transmission-line voltages increases electric fields; however, they reduce the magnetic fields, which are more difficult to deal with.

**Visual Impact.** Reduction of the visual impact of a transmission line requires proper blending of the line with the countryside. Also the design and the location of supporting structures should not be objectionable to observers. Improvement in visual impact must be evaluated in conjunction with the economic penalty that the design incurs.

**Audible Noise, RI, and TVI Caused by Corona.** In the United States, there exist no regulations on a local or federal level that expressly limit the level of radio noise that a transmission line may produce. The Federal Communications Commission (FCC) places power transmission lines in the category of "incidental radiation device." As such, the FCC requires that the "device shall be operated so that the radio frequency energy that is emitted does not cause any emissions, radiation or induction which endangers the function of a radio navigation service or other safety services, or seriously degrades, obstructs or repeatedly interrupts a radio communication service" operating in accordance with FCC regulations. Figure 8.10 shows typical frequency spectra of corona-induced radio noise, and Figure 8.11 illustrates a typical lateral profile of radio noise. Figure 8.12 illustrates radio noise tolerability criteria caused by corona, one for fair weather and one for foul weather; in addition there exists a 120 Hz hum due to the changing energy content of the lines. Video signals are more affected than sound signals because they are amplitude modulated whereas the latter are frequency modulated. A reduction of the radio noise entails a reduction of corona effects (e.g., bundle conductors). Note that the number of decibels (dB) is defined by $dB = 20\log_{10}(v_{out}/v_{in})$.

**Ozone and NO$_x$ Generation.** Figure 8.13 shows an example of ozone recordings for a 1300 kV line with

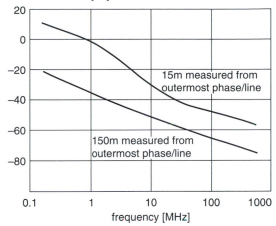

relative radio noise [dB]

**FIGURE 8.10** Typical frequency spectra of corona-produced radio noise.

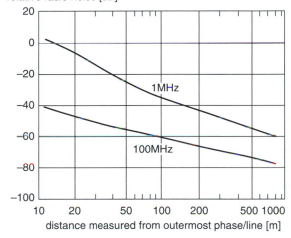

relative radio noise [dB]

**FIGURE 8.11** Typical lateral profiles of radio noise.

large corona loss of 650 kW/km. Although ozone is the primary gaseous corona product, nitrogen oxides (NO$_x$) may also be generated. The percentage of nitrogen oxide production is about one-sixth of that of ozone. The generation of ozone due to disassociation of oxygen in the corona region around the conductor is not of sufficient magnitude to be harmful. At ground level it has been found to be less than the daily natural variation. Conductor bundling is one method of reducing corona and its associated ozone generation.

**Impact of Physical Location of Transmission Lines.** The designer must also consider the environmental effects that the transmission line will have by

cumulative probability [%]

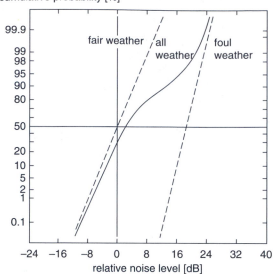

FIGURE 8.12 Typical all-weather cumulative frequency distribution of radio noise, and approximation by Gaussian distribution.

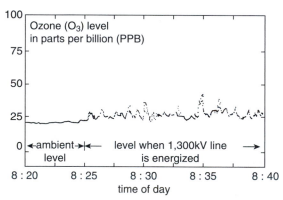

FIGURE 8.13 Detection of ozone from a 1300 kV test line. Coronal loss is 650 kW/km and wind velocity is 1.6 m/s with sensors at line height. Ozone ($O_3$) is measured in parts per billion (ppb).

its presence on certain land. These include the effect on wildlife, use of park areas, access to property, future planning, land values, and existing electromagnetic signal sources (e.g., communication transmitters). This is accomplished by rating each land parcel and selecting a route with a minimum environmental impact.

### 8.2.4.10 Economic Considerations

Corona loss of a well-designed transmission line should be below 1 kW per three-phase mile. This loss can be controlled by selecting proper configurations of bundle conductors (see Fig. E8.4.3 and Table

E8.4.1). A bundle conductor is a conductor made up of two or more subconductors, and is used as one-phase conductor. Bundle conductors are not economical at or below 200 kV, but for rated voltages of 400 kV or more they are the best solution for overhead transmission lines. The increase in transmitting capacity justifies economically the use of two-conductor bundles on 220 kV lines.

The advantages of bundle conductors are higher disruptive voltage with conductors of reasonable dimensions, reduced surge impedance and consequent higher power capabilities, and less rapid increase of corona loss and RI with increased voltage. These advantages must be weighed against increased circuit cost and increased charging apparent power (kVA). Theoretically there is an optimum subconductor separation for bundle conductors that will give the minimum crest gradient on the surface of a subconductor and hence highest disruptive voltage.

### 8.2.4.11 No-Cost/Low-Cost EMF Mitigation Hearings of PUC of California [52]

Electric fields $\vec{E}$ are created whenever power lines are energized with a voltage, whereas magnetic fields $\vec{H}$ measured in A/m (or $\vec{B}$ measured in milligauss) are created when current flows through the lines. Both electric and magnetic fields attenuate rapidly with distance from the source. Electric fields are effectively shielded by materials such as trees or buildings, whereas magnetic fields are not easily shielded by objects or materials. Therefore, concerns regarding potential power line EMF health effects arise primarily due to exposure to magnetic fields $\vec{H}$.

On January 15, 1991, the California Public Utilities Commission (CPUC) began an investigation to consider the Commission's potential role in mitigating health effects, if any, of EMFs created by

1. electric utility power lines, and
2. cellular radiotelephone facilities.

Due to the lack of scientific or medical conclusions about potential health effects from utility electric facilities and power lines, the CPUC adopted seven interim measures that help to address public concern on this subject.

- No-cost and low-cost steps to reduce EMF levels: When regulated utilities design new projects or upgrade existing facilities, approximately 4% of the project's budget may be used for reducing EMFs. The CPUC did not set specific reduction levels for EMFs. It was considered to be inappropriate to set a specific numerical standard until a scientific basis for doing so exists. No-cost measures are those steps taken in the design stage, including changes in standard practices, that will not increase the project cost but will reduce the magnetic flux density. Maximum (worst case) magnetic flux densities that could be permitted are ranging from 1 to 10 mG. Low-cost measures are those steps that will cost about 4% or less of the total project cost and will reduce the magnetic field flux density in an area (e.g., by a school, near residence) by approximately 15% or more at the edge of the right-of-way.

- New designs to reduce EMF levels: The CPUC initiated workshops to establish guidelines for new and rebuilt facilities. These guidelines incorporate alternative sites, increase the size of rights-of-way, place facilities underground, and use other suggested methods for reducing EMF levels at transmission, distribution, and substation facilities.
- Uniform residential and workplace EMF measurement methods are to be established.
- Promotion of education and research: The Department of Health Services (DHS) should manage such programs supported by public and utility funds.

CPUC-regulated utilities and municipal utilities may use ratepayer funds to pay their share of development costs for the following programs:

- EMF education,
- EMF research, and
- other research.

It has been many years since the CPUC devised the above-mentioned seven interim measures. During this time a significant body of scientific research has developed. In 2002 the DHS completed its comprehensive evaluation of existing scientific studies. Although there is still no consensus in the scientific community regarding health risks of EMF exposure, DHS reported troubling indications that EMF exposure may increase risk of certain diseases (e.g., childhood and to some extent adult leukemia, adult brain cancer, Lou Gehrig's disease) and other health problems.

### 8.2.4.12 Summary and Conclusions

The results of the many papers and books published in the area of generation of EMFs and corona by transmission lines can be summarized as follows:

- Transmission lines with line-to-line voltages of up to 200 kV do not generate any significant amount of corona; however, such lines can generate significant amounts of EMFs. For transmission lines with line-to-line voltages above 200 kV the following conclusions can be drawn: Maximum (worst case) AC electric field strength $\vec{E}$ at ground level below transmission line is up to $|\vec{E}_{worst}| = 15$ V/mm at a line-to-line voltage of 1050 kV. The maximum occurs at the edge of transmission corridor. An electric field strength of $|\vec{E}_{acceptable}| = 5$ V/mm is acceptable. Maximum (worst case) AC magnetic field strength $\vec{H}$ at ground level below transmission line is $|\vec{H}_{worst}| = 14$ A/m corresponding to a maximum flux density $|\vec{B}_{worst}| = 0.176$ G $= 176$ mG at a phase current of 2000 A. Note the DC earth field is about $B_{earth} = 0.5$ G. A flux density of $|\vec{B}_{acceptable}| = 3$ mG is acceptable.
- Corona depends on construction of transmission towers including type of conductors, e.g., hollow copper cable (e.g., HH) or aluminum cable steel reinforced (ACSR), surface of cable, type of bundle, fair or foul weather conditions, temperature, barometric pressure, and air density.
- Corona loss of 1 kW/three-phase mile is acceptable.
- Radio and television interference occurs within the 0.1 MHz to 1 GHz range.
- Audible noise ranges from 50 Hz to 120 Hz (power-frequency hum) to 10 kHz (hissing noise).
- Additional ozone ($O_3$) production is in the range of 15 ppb (parts per billion). The naturally occurring level is 25 ppb. That is, $O_3$ production by power lines is insignificant.
- Nitrogen oxide ($NO_x$) production is about one-sixth of the ozone generation. That is, $NO_x$ production by power lines is insignificant.
- Active compensation (e.g., through compensation of EMFs by arranging several transmission systems on one tower, by reducing, compensating, or canceling EMFs in houses through compensation circuits) is possible.
- Passive (e.g., horizontal or vertical mesh structures, conductive suits) and natural shielding (e.g., trees, houses) is possible.
- Overall control of EMFs and corona through transmission line (e.g., bundle conductors, reverse-phased double circuit) and tower construction (e.g., height) is feasible; it is, however, more expensive.
- For newly designed transmission lines all environmental concerns (e.g., limits for EMFs, RI, and

TVI) and technical constraints (corona loss of 1 kW/three-phase mile) can be satisfied.

- The mitigation of detrimental environmental, technical, or economic effects of existing transmission lines will be difficult to achieve at no or low cost as proposed by the Public Utilities Commission of California.

- The issue has surfaced that corona on transmission lines may create X-rays. A practical lower bound of X-rays is at least 2 keV. However, the electrical discharges of coronal activity on transmission lines can produce electron energies on the order of about 1–10 eV. This is sufficient for production of visible light but is orders of magnitude below electron energies necessary for the production of X-rays by transmission line corona [53]. That is, transmission lines do not generate X-rays.

### 8.2.5 Distributed-, Cogeneration, and Frequency/Voltage Control

Renewable energy sources such as solar and wind-power [54, 55] plants have an intermittent power output. Nevertheless it is desirable to operate them at maximum power output, for example, employment of peak-power tracker for photovoltaic plant [56]. With such peak-power and intermittent-power constraints it is impossible to operate a power system with distributed generation (DG) sources only because the drooping characteristics [57] which are responsible for load sharing require that at least one plant can deliver the additional power as requested by the utility customers, thus serving as base-load plant with spinning reserve capability. This is the reason why in Denmark [58] the control of the power system becomes more difficult as the renewable contribution approaches 30% of the total generation. This is so because the entire system exhibits properties of so-called weak systems with low short-circuit capabilities. As an alternative to a large power plant serving as a frequency leader—which can deliver sufficient additional power so that the frequency control will not be impaired—it is possible to strategically place storage peak-power plants based on pumped-hydro [59, 60] or compressed-air facilities [61, 62] that can provide peak power for several hours per day, for example, 22 h of continuous generation at 1600 MW.

Related to the availability of spinning reserve—which should be in the neighborhood of about 5–10% of the total installed power capacity—is the fast valving [63, 64] so that the time constants of the speed control circuit (e.g., governor) can be maintained within certain limits. In addition near-term [65, 66] and long-term [67, 68] load forecasts are important for the reliable operation of the power system with intermittently operating renewable energy sources. It is the task of control and dispatch centers to set the load reference points for each and every generation plant within an interconnected power system. From the following application examples one learns that renewable energy sources alone—operated at their maximum-power points [56]—cannot maintain frequency and voltage control. This is so because there must be a spinning reserve of power generation and sufficient facilities to control reactive power.

#### 8.2.5.1 Application Example 8.5: Frequency Control of an Interconnected Power System Broken into Two Areas: The First One with a 300 MW Coal-Fired Plant and the Other One with a 5 MW Wind-Power Plant

Figure P4.12 of Chapter 4 shows the block diagram of two generating plants interconnected by a tie line (transmission line). Data for generation set #1 (steam turbine and generator) are as follows:

Frequency change ($\Delta\omega_1$) per change in generator output power ($\Delta P_1$) having the droop characteristic

of $R_1 = \dfrac{\Delta\omega_1}{\Delta P_1} \text{pu} = 0.01 \text{pu}$,

The load change ($\Delta P_{L1}$) per frequency change ($\Delta\omega_1$)

resulting in $D_1 = \dfrac{\Delta P_{L1}}{\Delta\omega_1} = 0.8 \text{pu}$,

Step load change $\Delta P_{L1}(s) = \dfrac{\Delta P_{L1}}{s} \text{pu} = \dfrac{0.2}{s} \text{pu}$,

Angular momentum of steam turbine and generator set $M_1 = 4.5$,
Base apparent power $S_{base} = 500$ MVA,
Governor time constant $T_{G1} = 0.01$ s,
Valve changing (charging) time constant $T_{CH1} = 0.1$ s,
(Load reference set point)$_1 = 0.8$ pu.

Data for generation set #2 (wind turbine and generator) are as follows:

Frequency change ($\Delta\omega_2$) per change in generator output power ($\Delta P_2$) having the droop characteristic

$R_2 = \dfrac{\Delta\omega_2}{\Delta P_2} \text{pu} = 0.02 \text{pu}$,

The load change ($\Delta P_{L2}$) per frequency change ($\Delta \omega_2$)

resulting in $D_2 = \dfrac{\Delta P_{L2}}{\Delta \omega_2} = 1.0\,\text{pu}$,

Step load change $\Delta P_{L2}(s) = \dfrac{\Delta P_{L2}}{s}\,\text{pu} = \dfrac{-0.2}{s}\,\text{pu}$,

Angular momentum of wind turbine and generator set $M_2 = 6$,

Base apparent power $S_{\text{base}} = 500\,\text{MVA}$,

Governor time constant $T_{G2} = 0.02\,\text{s}$,

Valve (blade control) changing (charging) time constant $T_{CH1} = 0.75\,\text{s}$,

(Load reference set point)$_2$ = 0.8 pu.

Data for tie line: $T = \dfrac{377}{X_{tie}}$ with $X_{tie} = 0.2\,\text{pu}$.

a) List the ordinary differential equations and the algebraic equations of the block diagram of Fig. P4.12 of Chapter 4.

b) Use either Mathematica or MATLAB to establish transient and steady-state conditions by imposing a step function for load reference set point $(s)_1 = \dfrac{0.8}{s}\,\text{pu}$, load reference set point $(s)_2 = \dfrac{0.8}{s}\,\text{pu}$, and run the program with zero step load changes $\Delta P_{L1} = 0$, $\Delta P_{L2} = 0$ for 5 s. After 5 s impose the step load change $\Delta P_{L1}(s) = \dfrac{\Delta P_{L1}}{s}\,\text{pu} = \dfrac{0.2}{s}\,\text{pu}$, and after 7 s impose the step change $\Delta P_{L2}(s) = \dfrac{\Delta P_{L2}}{s}\,\text{pu} = \dfrac{-0.2}{s}\,\text{pu}$ for the droop characteristics $R_1 = 0.01\,\text{pu}$ and $R_2 = 0.02\,\text{pu}$ to find the transient and steady-state responses $\Delta \omega_1(t)$ and $\Delta \omega_2(t)$.

c) Use either Mathematica or MATLAB to establish transient and steady-state conditions by imposing a step function for load reference set point $(s)_1 = \dfrac{0.8}{s}\,\text{pu}$, load reference set point $(s)_2 = \dfrac{0.8}{s}\,\text{pu}$, and run the program with zero step load changes $\Delta P_{L1} = 0$, $\Delta P_{L2} = 0$ for 5 s. After 5 s impose step load change $\Delta P_{L1}(s) = \dfrac{\Delta P_{L1}}{s}\,\text{pu} = \dfrac{0.1}{s}\,\text{pu}$, and after 7 s impose the step load change $\Delta P_{L2}(s) = \dfrac{\Delta P_{L2}}{s}\,\text{pu} = \dfrac{-0.1}{s}\,\text{pu}$ for the droop characteristics $R_1 = 0.01\,\text{pu}$ (coal-fired plant) and $R_2 = 0.5\,\text{pu}$ (wind-power plant) to find the transient and steady-state responses $\Delta \omega_1(t)$ and $\Delta \omega_2(t)$.

### 8.2.5.2 Application Example 8.6: Frequency Control of an Interconnected Power System Broken into Two Areas: The First One with a 5 MW Wind-Power Plant and the Other One with a 5 MW Photovoltaic Plant

Figure P4.12 of Chapter 4 shows the block diagram of two generators interconnected by a tie line (transmission line).

Data for generation set #1 (wind turbine and generator) as follows:

Frequency change ($\Delta \omega_1$) per change in generator output power ($\Delta P_1$) having the droop characteristic

$R_1 = \dfrac{\Delta \omega_1}{\Delta P_1}\,\text{pu} = 0.5\,\text{pu}$,

The load change ($\Delta P_{L1}$) per frequency change ($\Delta \omega_1$)

yields $D_1 = \dfrac{\Delta P_{L1}}{\Delta \omega_1} = 0.8\,\text{pu}$,

Step load change $\Delta P_{L1}(s) = \dfrac{\Delta P_{L1}}{s}\,\text{pu} = \dfrac{0.1}{s}\,\text{pu}$,

Angular momentum of wind turbine and generator set $M_1 = 4.5$,

Base apparent power $S_{\text{base}} = 50\,\text{MVA}$,

Governor time constant $T_{G1} = 0.01\,\text{s}$,

Valve (blade control) changing (charging) time constant $T_{CH1} = 0.1\,\text{s}$,

(Load reference set point)$_1$ = 0.8 pu.

Data for generation set #2 (photovoltaic array and inverter):

Frequency change ($\Delta \omega_2$) per change in inverter output power ($\Delta P_2$) having the droop characteristic

$R_2 = \dfrac{\Delta \omega_2}{\Delta P_2}\,\text{pu} = 0.5\,\text{pu}$,

The load change ($\Delta P_{L2}$) per frequency change ($\Delta \omega_2$)

yields $D_2 = \dfrac{\Delta P_{L2}}{\Delta \omega_2} = 1.0\,\text{pu}$,

Step load change $\Delta P_{L2}(s) = \dfrac{\Delta P_{L2}}{s}\,\text{pu} = \dfrac{-0.1}{s}\,\text{pu}$,

Equivalent angular momentum $M_2 = 6$,

Base apparent power $S_{\text{base}} = 50\,\text{MVA}$,

Governor time constant $T_{G2} = 0.02\,\text{s}$,

Equivalent valve changing (charging) time constant $T_{CH2} = 0.1\,\text{s}$,

(Load reference set point)$_2$ = 0.8 pu.

Data for tie line: $T = \dfrac{377}{X_{tie}}$ with $X_{tie} = 0.2\,\text{pu}$.

Use either Mathematica or MATLAB to establish transient and steady-state conditions by imposing a step function for load reference set point$(s)_1 = \dfrac{0.8}{s}$ pu , load reference set point$(s)_2 = \dfrac{0.8}{s}$ pu and run the program with a zero step load changes $\Delta P_{L1} = 0$, $\Delta P_{L2} = 0$ for 5 s. After 5 s impose step load change $\Delta P_{L1}(s) = \dfrac{\Delta P_{L1}}{s}$ pu $= \dfrac{0.1}{s}$ pu, and after 7 s impose the step load change $\Delta P_{L2}(s) = \dfrac{\Delta P_{L2}}{s}$ pu $= \dfrac{-0.1}{s}$ pu to find the transient and steady-state responses $\Delta \omega_1(t)$ and $\Delta \omega_2(t)$.

From Application Examples 8.5 and 8.6 one draws the following conclusions:

1. Renewable energy sources stemming from solar and wind are likely to be operated at their maximum output power. This requires that the slope $R$ of the drooping characteristic of a renewable power source is large as is depicted in Figs. 8.14 and 8.15. As a consequence of the large $R$, the change in output of the renewable source results in a large frequency change $\Delta \omega$.

2. Solar, wind, and tidal power sources are intermittent in nature and their output power cannot be controlled by the dispatch center of a utility system.

3. Due to issues 1 and 2 distributed generation (DG) based on solar and wind power is not possible unless either a base load plant (e.g., gas- or coal-fired plant) or a storage plant (e.g., pumped hydro, compressed air) is ready to deliver the additional power required. That is, the base load plant or the storage plant acts as a spinning reserve plant.

4. It has been discussed in Chapter 4 (Section 4.2.2.3) that an inverter requires a larger DC input voltage for capacitive/resistive load than for inductive/resistive operation. To utilize inverter components to their fullest unity power (about resistive operation) factor operation is preferred. This means a renewable power plant cannot be very effectively employed for reactive power or voltage control.

5. The governor time constant (e.g., $T_{G1}$, $T_{G2}$) and the valve changing (charging) time constant (e.g., $T_{CH1}$, $T_{CH2}$) must be as small as possible to guarantee stability. The latter time constant is fairly large for steam, hydro, and wind-power plants. This issue has been addressed in [63, 64].

6. In summary it is not possible to operate a power system solely based on renewable power plants because they do not easily lend themselves to frequency and voltage control.

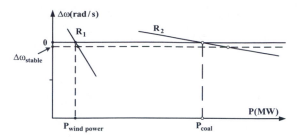

**FIGURE 8.14** Load sharing of a renewable plant (operated at maximum power with droop characteristic $R_1 = 0.5$ pu) with a coal-fired plant (operated at less than maximum power with droop characteristic $R_2 = 0.01$ pu) for a load increase of 10%, resulting in an acceptable change in angular frequency.

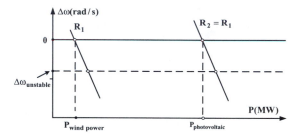

**FIGURE 8.15** Load sharing of a renewable solar plant (either photovoltaic or thermal) with a wind-power plant for a load increase of 10%. Both types of plants are operated at maximum power to maximize the renewable energy contribution within a power system both with droop characteristics of $R_1 = R_2 = 0.5$ pu, resulting in an inacceptably large change in angular frequency.

## 8.3 TOOLS FOR DETECTING POOR POWER QUALITY

### 8.3.1 Sensors

To be able to detect power quality problems one must rely on voltage and current sensors such as low-inductive voltage dividers for voltage pickup, and low-inductive shunts and Hall sensors for current pickup. Components representing the interface between sensors and measuring instruments are potential and current transformers as well as optocouplers. Voltmeters, amperemeters, oscillographs, spectrum analyzers, and network analyzers are indispensable. Special power quality tools and selection guides have been developed for detecting power quality for single- and three-phase systems, such as power quality clamp meters, power quality analyzers, voltage event recorder systems, three-phase power loggers, series three-phase power quality loggers, three-phase power quality recorders, AC

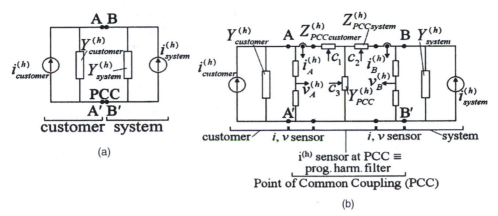

**FIGURE E8.7.1** (a) PCC without harmonic current sensor, (b) PCC with harmonic current sensor and programmable filter.

leakage current clamp meters, series three-phase power quality analyzers, high precision power analyzers, reliable power meters, multipoint power recorders, and other power quality-specific equipment. Some of them [69, 70] are useful to analyze steady-state phenomena, and some of them are transient overvoltage/under-voltage meters recording outage time, deviation from rated voltage values (e.g., sags, swells), or the number of outages within a given time period, and an instrument that distinguishes the direction of the harmonic power flow as is described in the following application example.

### 8.3.2 Application Example 8.7: Detection of Harmonic Power Flow Direction at Point of Common Coupling (PCC)

The discrimination between the contributions of the customer and the power system to generated harmonic currents at PCC is based on Fig. E8.7.1 and on references [70–84]. Figure E8.7.1a represents the equivalent circuit of one customer connected to the power system at PCC indicated by the terminals $A - A'$ and $B - B'$. To measure individual harmonic currents $i_A^{(h)}$, $i_B^{(h)}$ with respect to direction and amplitude a harmonic current sensing network (see Fig. E8.7.1b), which represents a programmable harmonic filter, is configured and connected by computer commands $c_1$, $c_2$, and $c_3$ for a fraction of a second during which measurements take place. For measuring the current $i_{customer}^{(h)}$ the impedance and admittance $z_{PCCcustomer}^{(h)}$ and $Y_{PCC}^{(h)}$, respectively, will provide a short-circuit for $i_{customer}^{(h)}(t)$ during a fraction of a second, whereas the impedance $z_{PCCsystem}^{(h)}$ is blocking $i_{system}^{(h)}(t)$. For measuring the current $i_{system}^{(h)}(t)$ the impedance and admittance $z_{PCCsystem}^{(h)}$ and $Y_{PCC}^{(h)}$, respectively, will provide a short-circuit for $i_{system}^{(h)}(t)$ during a fraction of a second, whereas the impedance $z_{PCCcustomer}^{(h)}$ is blocking $i_{customer}^{(h)}(t)$. The harmonic

impedance and admittance consists of inductors and capacitors whose very small losses—to achieve a high $Q$ of the programmable filters—will be determined as described in [83]. Current and voltage sensors (e.g., Hall sensors [81] or low-inductive current shunts and voltage dividers [82] with optocouplers) will measure the direction of the harmonic power flow as detailed in [70–74].

The main advantages of this approach are

1. Nonintrusive measurements lasting a fraction of a second. Field tests can be executed in three steps, each step lasting less than 0.1 s. In the first step the harmonic current $i_{customer}^{(h)}(t)$ will be measured by configuring the programmable filter so that $z_{PCCcustomer}^{(h)}$ and $Y_{PCC}^{(h)}$ represent a short-circuit for a harmonic customer current of hth order, while $z_{PCCsystem}^{(h)}$ represents an open-circuit for the system harmonic current of hth order. In the second step $z_{PCCsystem}^{(h)}$ and $Y_{PCC}^{(h)}$ represent a short-circuit for the system harmonic current of hth order, while $z_{PCCcustomer}^{(h)}$ represents an open circuit for the system current of hth order. In the third step the total harmonic current will be measured when $z_{PCCcustomer}^{(h)}$ and $z_{PCCsystem}^{(h)}$ are bypassed and $Y_{PCC}^{(h)} = 0$ is an open circuit.

2. Real-time measurement while system operates under load.

3. No additional dissipation occurs.

4. Very small ohmic components of inductors and capacitors can be accurately determined [83], resulting in high $Q$ of programmable harmonic filters.

5. In most harmonic standards [76, 77] the harmonic current injection from the customer installation is limited. To enforce such limits by utilities these customer-injected harmonic currents at PCC $i_{customer}^{(h)}(t)$ must be reliably and

accurately measured. This approach provides a tool for the utilities to measure customer harmonic currents. This will increase the reliability and efficiency of power system operation.

### 8.3.3 Maximum Error Analysis

The accurate online or real-time measurement of losses of circuit components such as transformers improves the reliability and security of a power system. Normally circuit components are equipped with temperature sensors (e.g., thermocouples, thermistors). This approach has two disadvantages.

1. Circuit components must be equipped with sensors during their construction, and off-the-shelf components cannot be used.
2. Temperature sensors provide information about the temperature at their location and not necessarily the temperature at the hot spot of an energy conversion device, and any temperature alarm may come too late to prevent an outage due to a fault.

The advantages of the real-time monitoring of losses of components (e.g., transformers) are that loss increases can be sensed or recorded at the time when a fault develops. In addition, the measurement components or instruments can be exchanged, thus minimizing the outage time of the transformer supplying the power system. Such a monitoring of the losses must rely on small maximum errors of the monitoring equipment.

Maximum error analysis has been extensively used in the past [70–73, 85–87]. The online measurement of transformer losses under nonsinusoidal operation is discussed in detail in [87]. The real-time monitoring of iron-core and copper losses of single- and three-phase transformers is important, particularly for transformers feeding nonlinear loads. This section devises a new digital data-acquisition method for the separate online measurement of iron-core and copper losses of three-phase transformers under any full or partial load conditions. The accuracy requirements of the instruments employed (voltage and current sensors, volt and current meters) are addressed. The maximum errors are acceptably low if voltage and current sensors with 0.1 to 0.5% errors are used.

#### 8.3.3.1 Review of Existing Methods

The separate online monitoring of the iron-core and copper losses of single- and three-phase transformers is desirable because additional losses due to

power quality problems (e.g., harmonics, DC excitation) can be readily detected before any significant damage due to additional temperature rises occurs. Those losses can accurately and economically be measured for single-phase transformers via computer-aided testing (CAT) [71, 88].

The known approaches for on-line loss measurement of high-efficiency transformers (e.g., efficiency greater than 97%) are inaccurate because they measure input and output powers and derive the loss from the difference of these two large values. The usually employed indirect method consisting of no-load (iron-core loss) and short-circuit (copper loss) tests [35] cannot be performed online while the transformer is partially or fully loaded. The same is valid for the loading back method [89].

Arri *et al.* presented an analog measurement circuit [90]. It is well known that there may occur zero-sequence currents on the Y-grounded side of a three-phase transformer, and these currents will produce corresponding losses in the transformer. The measuring circuit presented by Arri *et al.* cannot measure these zero-sequence components, and therefore the method is not suitable for transformers with grounding on any side (either primary or secondary). Moreover, Arri *et al.* rely on many instrument transformers (nine CTs and nine PTs) to convert the $\Delta$ connection to Y connection for a $Y/\Delta$ connected transformer: as a result the measuring accuracy is decreased. In [91] an analog measuring circuit with wattmeters is presented within a single-phase circuit.

This section presents a more accurate real-time method for digitally and separately measuring the steady-state iron-core and copper losses of three-phase transformers with various grounded connections while the transformers are operating at any load condition. This digital measuring circuit is based on voltage and current sensors (voltage dividers, shunts, PTs, CTs, or Hall devices), A/D converter, and personal computer. Using a computer-aided testing program (CATEA) [70] losses, efficiency, harmonics, derating, and wave shapes of all voltages and currents can be monitored within a fraction of a second.

The maximum measurement errors of the losses are acceptably small and vary in the range from 0.5 to 15%, mainly contingent on the accuracy of the voltage and current sensors used. Depending on the application, voltage dividers with optocouplers [73], current shunts with optocouplers, potential and current transformers (error < 0.1%), as well as Hall sensors [92] (error < 0.5%) can be employed.

When both voltage and current signals are obtained from Hall devices [92], the copper losses will include DC losses, if they exist. Provided voltage and current signals are generated by PT and CT sensors, the DC losses are not included in the copper losses, and must be measured by additional DC voltmeters and ammeters or sensors. In this case, the total loss consists of the iron-core, AC, and DC copper losses.

### 8.3.3.2 Approach

It is well known that for any $Y$ (ungrounded) or $\Delta$ connected three-phase windings, the two-wattmeter method is often used to measure the power based on the voltages and currents of any two phases. For a $Y$ grounded three-phase winding, an additional wattmeter is necessary for the voltage and current of the third phase, or those of the zero sequence. Therefore, for a transformer without grounding on any side (primary or secondary), power loss is measured with eight signals (voltages and currents of two phases for each side). If the primary or/and the secondary are grounded, the power loss can be measured with ten or twelve signals. This section deals with ten- and twelve-channel measuring circuits. The eight-channel measuring circuits for $Y/Y$ ungrounded

or $\Delta/\Delta$ and $\Delta$–$Y$ ungrounded three-phase transformers are discussed in [73].

A ten-channel CAT circuit applied to $\Delta$–$Y_0$ ($Y_0$ means $Y$ grounded) connected three-phase transformers is shown in Fig. 8.16. The flowchart for the ten-channel signals is shown in Fig. 8.17. The output power (at the $Y_0$ side) is

$$
\begin{aligned}
P_{\text{out}} &= \frac{1}{T}\int_0^T (v'_{an}i'_a + v'_{bn}i'_b + v'_{cn}i'_c)dt \\
&= \frac{1}{T}\int_0^T [v'_{an}i'_a + (v'_0 - v'_{an} - v'_{cn})i'_b + v'_{cn}i'_c]dt \\
&= \frac{1}{T}\int_0^T [v'_{an}i'_{al} - v'_{cn}i'_{bl} + v'_0 i'_b]dt
\end{aligned}
\tag{8-5}
$$

where $i'_{al} = (i'_a - i'_b)$, $i'_{bl} = (i'_b - i'_c)$, and $v'_0 = (v'_{an} + v'_{bn} + v'_{cn})$. The transformer loss is

$$
\begin{aligned}
P_{\text{in}} - P_{\text{out}} &= \frac{1}{T}\int_0^T [(v_{AC}i_{Al} + v_{BC}i_{Bl}) \\
&\quad - (v'_{an}i'_{al} - v'_{cn}i'_{bl} + v'_0 i'_b)]dt \\
&= \frac{1}{T}\int_0^T [v_{Ac}(i_{Al} - i'_{al}) + v_{BC}(i_{Bl} - i'_{bl}) + \\
&\quad (v_{AC} - v'_{an})i'_{al} + (v_{BC} + v'_{cn})i'_{bl} - v'_0 i'_b]dt.
\end{aligned}
\tag{8-6}
$$

The loss component $(-v'_0 i'_b)$ in Eq. 8-6 is included in the copper loss, which can be proved below:

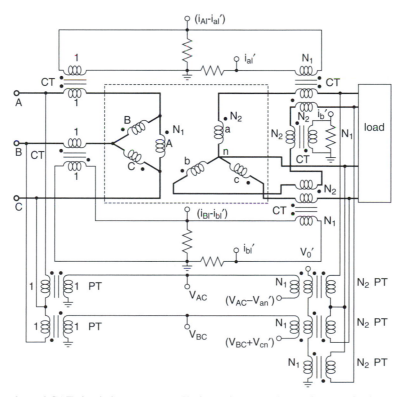

**FIGURE 8.16** Ten-channel CAT circuit for accurate $\Delta$–$Y_0$ three-phase transformer loss monitoring.

$$
\begin{aligned}
p_{cu} &= (v_{AC} - v'_{an})i'_a + (v_{CB} - v'_{cn})i'_c + (v_{BA} - v'_{bn})i'_b \\
&= (v_{AC} - v'_{an})i'_a + (-v_{BC} - v'_{cn})i'_c + \\
&\quad (v_{BC} - v_{AC} - v'_0 + v'_{an} + v'_{cn})i'_b \\
&= (v_{AC} - v'_{an})(i'_a - i'_b) + (v_{BC} + v'_{cn})(i'_b - i'_c) - v'_0 i'_b \\
&= (v_{AC} - v'_{an})i'_{al} + (v_{BC} + v'_{cn})i'_{bl} - v'_0 i'_b. \quad (8\text{-}7)
\end{aligned}
$$

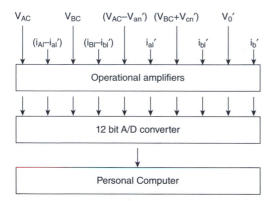

**FIGURE 8.17** Flowchart for ten-channel signals.

Note that the analog measuring circuit of [90] does not take into account zero-sequence components and, therefore, does not apply to general asymmetric operating conditions.

For $Y$–$Y_0$ connected three-phase transformers (see Fig. 8.18), the output power can be written as

$$
\begin{aligned}
P_{out} &= \frac{1}{T}\int_0^T (v'_{an}i'_a + v'_{bn}i'_b + v'_{cn}i'_c)dt \\
&= \frac{1}{T}\int_0^T [v'_{an}i'_a + v'_{bn}i'_b + v'_{cn}(i'_0 - i'_a - i'_b)]dt \\
&= \frac{1}{T}\int_0^T [(v'_{ac}i'_a + v'_{bc}i'_b + v'_{cn}i'_0)]dt \quad (8\text{-}8)
\end{aligned}
$$

where $i'_0 = (i'_a + i'_b + i'_c)$. The transformer loss can be formulated as

$$
\begin{aligned}
P_{in} - P_{out} &= \frac{1}{T}\int_0^T [(v_{AC}i_A + v_{BC}i_B) - (v'_{ac}i'_a + v'_{bc}i'_b + v'_{cn}i'_0)]dt \\
&= \frac{1}{T}\int_0^T [v_{AC}(i_A - i'_a) + v_{BC}(i_B - i'_b) - v'_{cn}i'_0 + \\
&\quad (v_{AC} - v'_{ac})i'_a + (v_{BC} - v'_{bc})i'_b]dt
\end{aligned}
$$

$$(8\text{-}9)$$

where $(-v'_{cn}i'_0)$ is included in the iron-core loss. This can be shown as follows:

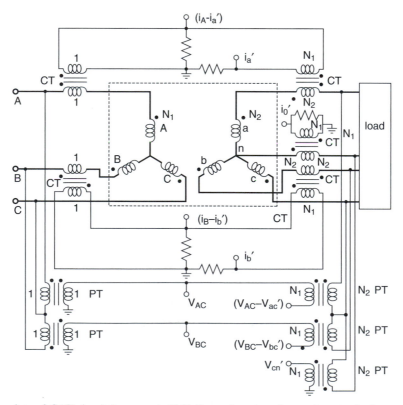

**FIGURE 8.18** Ten-channel CAT circuit for accurate $Y$–$Y_0$ three-phase transformer loss monitoring.

$$p_{fe} = v_{An}(i_A - i'_a) + v_{Bn}(i_B - i'_b) + v_{Cn}(i_C - i'_c)$$
$$= v_{An}(i_A - i'_a) + v_{Bn}(i_B - i'_b) -$$
$$v_{Cn}(i_A + i_B + i'_0 - i'_a - i'_b)$$
$$= (v_{An} - v_{Cn})(i_A - i'_a) + (v_{Bn} - v_{Cn})(i_B - i'_b) - v'_{cn}i'_0$$
$$= v_{AC}(i_A - i'_a) + v_{BC}(i_B - i'_b) - v'_{cn}i'_0. \qquad (8\text{-}10)$$

For $Y_0$–$Y_0$ connected transformers, a twelve-channel circuit is used as shown in Fig. 8.19. The loss of the transformer can be derived directly from

$$P_{in} - P_{out} = \frac{1}{T}\int_0^T [(v_{An}i_A + v_{Bn}i_B + v_{Cn}i_C) -$$
$$(v'_{an}i'_a + v'_{bn}i'_b + v'_{cn}i'_c)]dt$$
$$= \frac{1}{T}\int_0^T [v_{An}(i_A - i'_a) + v_{Bn}(i_B - i'_b) + v_{Cn}(i_C - i'_c) +$$
$$(v_{An} - v'_{an})i'_a + (v_{Bn} - v'_{bn})i'_b + (v_{Cn} - v'_{cn})i'_c]dt.$$
$$(8\text{-}11)$$

The instantaneous expression $p_{fe}(t)$ in Eq. 8-10 includes two power components: the instantaneous iron-core loss and the instantaneous exciting reactive

power. Similarly, $p_{cu}(t)$ in Eq. 8-7 includes the instantaneous copper loss and the instantaneous reactive power due to the leakage flux. At steady state, the average value of reactive power during one electric cycle is zero; therefore, the average values of $p_{fe}(t)$ and $p_{cu}(t)$ during one electric cycle reflect the steady-state iron-core and copper losses.

The instantaneous expressions of $p_{fe}(t)$ and $p_{cu}(t)$ of various three-phase transformer connections are listed in Table 8.2, where all current and voltage differences can be calibrated directly under either sinusoidal or nonsinusoidal conditions.

If the neutral line is disconnected, and $v'_0$ and $i'_b$ are not measured, Fig. 8.16 becomes the eight-channel loss-measuring circuit of $\Delta$–$Y$ connected three-phase transformers (Fig. 8.20), and the transformer loss is obtained from Eq. 8-6 with $v'_0 = 0$ [73]. Similarly, if the neutral line is disconnected, and $i'_0$ and $v_{cn}'$ are not measured, Fig. 8.18 becomes the eight-channel loss-measuring circuit of $Y$–$Y$ connected three-phase transformers, and the transformer loss is obtained

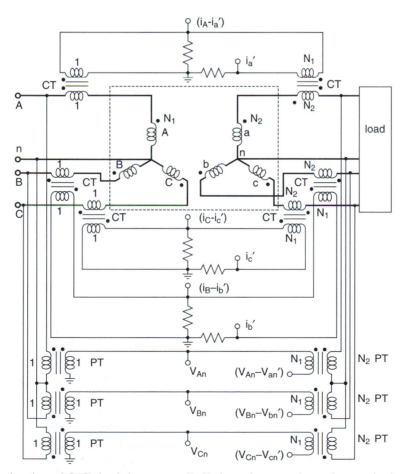

**FIGURE 8.19** Twelve-channel CAT circuit for accurate $Y_0$–$Y_0$ three-phase transformer loss monitoring.

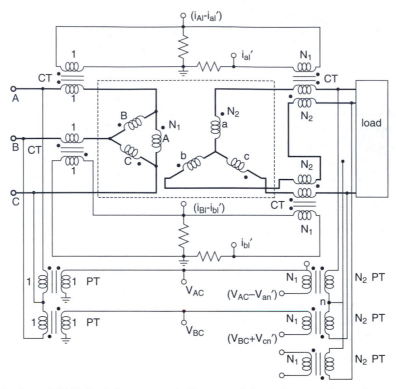

**FIGURE 8.20** Eight-channel CAT circuit for accurate $\Delta$–$Y$ connected three-phase transformer loss monitoring.

**TABLE 8.2** Iron-Core and Copper Losses of Three-Phase Transformers

| | $p_{fe}(t)$ | $p_{cu}(t)$ |
|---|---|---|
| $Y$–$Y$ (or $\Delta$–$\Delta$) | $v_{AC}(i_A - i'_a) + v_{BC}(i_B - i'_b)$ | $(v_{AC} - v'_{ac})i'_a + (v_{BC} - v'_{bc})i'_b$ |
| $\Delta$–$Y$ | $v_{AC}(i_{Al} - i'_{al}) + v_{BC}(i_{Bl} - i'_{bl})$ | $(v_{AC} - v'_{an})i'_{al} + (v_{BC} + v'_{cn})i'_{bl}$ |
| $\Delta$–$Y_0$ | $v_{AC}(i_{Al} - i'_{al}) + v_{BC}(i_{Bl} - i'_{bl})$ | $(v_{AC} - v'_{an})i'_{al} + (v_{BC} + v'_{cn})i'_{bl} - v'_0 i'_b$ |
| $Y$–$Y_0$ | $v_{AC}(i_A - i'_a) + v_{BC}(i_B - i'_b) - v'_{cn}i'_0$ | $(v_{AC} - v'_{ac})i'_a + (v_{BC} - v'_{bc})i'_b$ |
| $Y_0$–$Y_0$ | $v_{An}(i_A - i'_a) + v_{Bn}(i_B - i'_b) + v_{Cn}(i_C - i'_c)$ | $(v_{An} - v'_{an})i'_a + (v_{Bn} - v'_{bn})i'_b + (v_{Cn} - v'_{cn})i'_c$ |

**TABLE 8.3** Valid Methods for Various Connections

| Primary \ Secondary | $\Delta$ | $Y$ | $Y_0$ |
|---|---|---|---|
| $\Delta$ | Eight-channel method | Eight-channel method | Ten-channel method |
| $Y$ | Eight-channel method | Eight-channel method | Ten-channel method |
| $Y_0$ | Ten-channel method | Ten-channel method | Twelve-channel method |

from Eq. 8-9 with $i''_0 = 0$ [73]. The circuits valid for $\Delta$–$Y$, $\Delta$–$Y_0$, and $Y$–$Y_0$ connected transformers also apply for $Y$–$\Delta$, $Y_0$–$\Delta$, and $Y_0$–$Y$ connected three-phase transformers, respectively. In these cases, the capital letters A, B, and C represent the secondary, and the lower case letters a, b, and c denote the primary. The corresponding iron-core and copper losses, listed in Table 8.2, are then negative. Table 8.3 shows valid methods for three-phase transformers of various connections.

Note that in Figs. 8.16, 8.18, 8.19, and 8.20, CT and PT ratios on the primary side are $1 : 1$, whereas on the secondary side they are $N_2 : N_1$. In this way, secondary signals are referred to the primary. In

Fig. 8.20, only four CTs and five PTs are required for the loss measurement, comparing to the measurement circuit of [90] with nine CTs and nine PTs. Using fewer sensors means having higher measurement accuracy.

### 8.3.3.3 Accuracy Requirements for Instruments

The efficiencies of high-power electrical apparatus such as single- and three-phase transformers in the kVA and MVA ranges are about 98% or higher. For a 25 kVA single-phase transformer [88] with an efficiency of 98.46% at rated operation, the total losses are $P_{Loss} = 390$ W, where the copper and iron-core losses are $P_{cu} = 325$ W and $P_{fe} = 65$ W, respectively [70]. The measurement errors of losses are derived for three different configurations: the conventional, new, and back-to-back approaches.

### 8.3.3.4 Application Example 8.8: Conventional Approach $P_{Loss} = P_{in} - P_{out}$

Even if current and voltage sensors (e.g., CTs, PTs) as well as current and voltmeters with maximum error limits of 0.1% are employed, the conventional approach—where the losses are the difference between the input and output powers—always leads to relatively large maximum errors (>50%) in the loss measurements of highly efficient transformers.

The maximum error in the measured losses will be demonstrated for the 25 kVA, 240 V/7200 V single-phase transformer of Fig. E8.8.1, where the low-voltage winding is connected as primary, and rated currents are therefore $I_{2rat} = 3.472$ A, $I_{1rat} = 104.167$ A [72]. The instruments and their maximum error limits are listed in Table E8.8.1, where errors of all current and potential transformers (e.g., CTs, PTs) are referred to the meter sides. In Table E8.8.1, error limits for all ammeters and voltmeters are given

based on the full-scale errors, and those for all current and voltage transformers are derived based on the measured values of voltages and currents (see Appendix A6.1).

The maximum error of losses at unity power factor is

$$P_{Loss} = (240\,\text{V} \pm \varepsilon_{PT1} \pm \varepsilon_{V1})(5.20835\,\text{A} \pm \varepsilon_{CT1} \pm \varepsilon_{A1})20 - 30(240\,\text{V} \pm \varepsilon_{PT2} \pm \varepsilon_{V2})$$
$$(3.472\,\text{A} \pm \varepsilon_{CT2} + \varepsilon_{A2})$$
$$= 240.54\,\text{V}\,104.28\,\text{A} - 7183.8\,\text{V}\,3.4635\,\text{A}$$
$$= 202.5\,\text{W};$$

therefore,

$$\frac{\Delta P_{Loss}}{P_{Loss}} = \frac{\pm 202.5}{390} \times 100\% = \pm 51.9\%.$$

### 8.3.4 Application Example 8.9: New Approach $p_{cu} = i'_2(v_1 - v'_2)$ and $p_{fe} = v_1(i_1 - i'_2)$

The determination of the losses from voltage and current differences greatly reduces this maximum error. The copper and iron-core losses are measured based on $p_{cu} = i'_2(v_1 - v'_2)$ and $p_{fe} = v_1(i_1 - i'_2)$, respectively, as shown in Fig. E8.9.1.

The series voltage drop and exciting current at rated operation referred to the low-voltage side of the 25 kVA, 240 V/7200 V single-phase transformer are $(v_1 - v'_2) = 4.86$ V and $(i_1 - i'_2) = 0.71$ A, respectively [70, 72]. Note that in Table 2 of [70] one-half the losses of two identical 25 kVA single-phase transformers—connected back-to-back—are given while the voltage drop $(v_1 - v'_2) = 9.72$ V and the current difference $(i_1 - i'_2) = 1.41$ A pertain to the series connection of both single-phase transformers at linear load; the same applies to the nonlinear loads. The instruments and their full-scale errors are listed in Table E8.9.1.

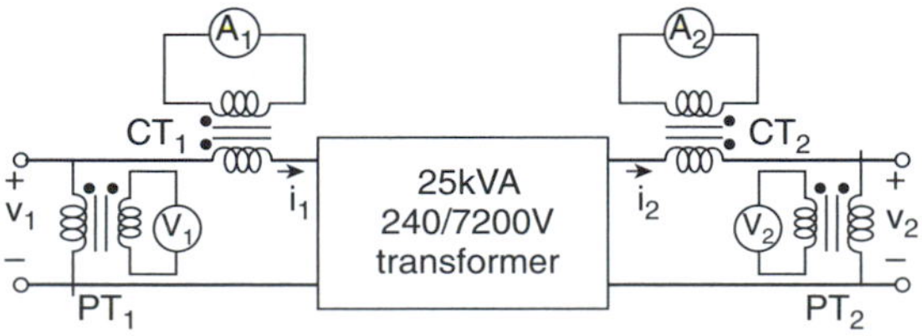

**FIGURE E8.8.1** Instrumentation for single-phase transformers employing conventional approach for loss measurement.

**TABLE E8.8.1** Instruments and Their Error Limits of Fig. E8.8.1

| Instrument | Rating | Error limits |
|---|---|---|
| PT$_1$ | 240 V/240 V | $\varepsilon_{PT1} = \pm 0.24$ V |
| PT$_2$ | 7200 V/240 V | $\varepsilon_{PT2} = \pm 0.24$ V |
| CT$_1$ | 100 A/5 A | $\varepsilon_{CT1} = \pm 5.2$ mA |
| CT$_2$ | 5 A/5 A | $\varepsilon_{CT2} = \pm 3.47$ mA |
| V$_1$ | 300 V | $\varepsilon_{V1} = \pm 0.3$ V |
| V$_2$ | 300 V | $\varepsilon_{V2} = \pm 0.3$ V |
| A$_1$ | 5 A | $\varepsilon_{A1} = \pm 5$ mA |
| A$_2$ | 5 A | $\varepsilon_{A2} = \pm 5$ mA |

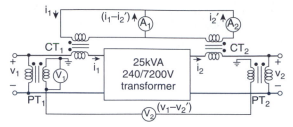

**FIGURE E8.9.1** Loss measurements of single-phase transformers from voltage and current differences.

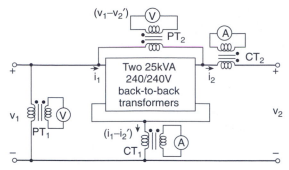

**FIGURE E8.10.1** Instrumentation for single-phase transformer employing back-to-back method of loss measurement.

**TABLE E8.9.1** Instruments and Their Error Limits of Fig. E8.9.1

| Instrument | Rating | Error limits |
|---|---|---|
| $PT_1$ | 240 V/240 V | $\varepsilon_{PT1} = \pm 0.24$ V |
| $PT_2$ | 7200 V/240 V | $\varepsilon_{PT2} = \pm 0.24$ V |
| $CT_1$ | 100 A/5 A | $\varepsilon_{CT1} = \pm 5.2$ mA |
| $CT_2$ | 5 A/5 A | $\varepsilon_{CT2} = \pm 3.47$ mA |
| $V_1$ | 300 V | $\varepsilon_{V1} = \pm 0.3$ V |
| $V_2$ | 10 V | $\varepsilon_{V2} = \pm 0.01$ V |
| $A_1$ | 100 mA | $\varepsilon_{A1} = \pm 0.1$ mA |
| $A_2$ | 5 A | $\varepsilon_{A2} = \pm 5$ mA |

**TABLE E8.10.1** Instruments and Their Error Limits of Fig. E8.10.1

| Instrument | Rating | Error limits |
|---|---|---|
| $PT_1$ | 240 V/240 V | $\varepsilon_{PT1} = \pm 0.24$ V |
| $PT_2$ | 12 V/12 V | $\varepsilon_{PT2} = \pm 9.7$ mV |
| $CT_1$ | 2 A/2 A | $\varepsilon_{CT1} = \pm 1.4$ mA |
| $CT_2$ | 100 A/5 A | $\varepsilon_{CT2} = \pm 5.2$ mA |
| $V_1$ | 300 V | $\varepsilon_{V1} = \pm 0.3$ V |
| $V_2$ | 10 V | $\varepsilon_{V2} = \pm 0.01$ V |
| $A_1$ | 2 A | $\varepsilon_{A1} = \pm 2$ mA |
| $A_2$ | 5 A | $\varepsilon_{A2} = \pm 5$ mA |

### 8.3.5 Application Example 8.10: Back-to-Back Approach of Two Transformers Simulated with CTs and PTs

If two transformers of the same type are available, the copper and iron-core losses can be measured via the back-to-back method. The testing circuit for this method is shown in Fig. E8.10.1. The series voltage drop and exciting current at rated operation referred to the low-voltage side of the two 25 kVA, 240 V/7200 V single-phase transformers are therefore 9.72 V and 1.41 A, respectively (see Table 2 of [70], where the data were measured via the back-to-back approach, and therefore the exciting currents and voltage drops are those of two transformers in series). The instruments and their full-scale errors are listed in Table E8.10.1.

### 8.3.6 Application Example 8.11: Three-Phase Transformer with DC Bias Current

As an application example, the losses of a 4.5 kVA three-phase transformer bank supplying power to resistive load under various DC bias conditions are measured. The measuring circuit for the $\Delta$–$Y_0$ connected three-phase transformer bank (see Appendix 6.2) is shown in Fig. 8.16. The transformer is operated at 75% of rated load with various balanced DC

bias currents injected on the $Y_0$ side, and the DC copper, AC copper, and iron-core losses are measured separately. Such DC currents can be generated by either unbalanced rectifiers, geomagnetic storms [31, 80, 93, 94], or nuclear explosions. Measured iron-core and copper losses vary with the DC bias current and are listed in Table E8.11.1.

Table E8.11.1 demonstrates that the DC copper loss due to the DC bias current is negligibly small. However, the AC copper losses and the iron-core losses increase at rated input voltage (240 V) about 10% and 40%, respectively, due to a DC bias current of 100% of the magnetizing current magnitude ($I_m = 1.25$ A) within the $Y_0$ connected windings; thus the efficiency of the transformer decreases from

$$\frac{3340\,\text{W}}{3340\,\text{W} + 150.3\,\text{W}} = 95.71\%$$

to

$$\frac{3293\,\text{W}}{3293\,\text{W} + 188.7\,\text{W}} = 94.6\%.$$

**TABLE E8.11.1a** Measured Iron-Core and Copper Losses of 4.5 kVA $\Delta$–$Y_0$ Connected Transformer with DC Bias for a Rated Line-to-Line Input Voltage of 242 V

| DC bias current (A) | Iron-core losses (W) | AC copper losses (W) | DC copper losses (W) | Total losses (W) | Output power (W) |
|---|---|---|---|---|---|
| 0.00 | 72.2 | 78.1 | 0.0 | 150.3 | 3340 |
| 0.56 | 79.9 | 78.9 | 0.4 | 159.2 | 3327 |
| 0.74 | 83.1 | 79.1 | 0.7 | 162.9 | 3316 |
| 1.01 | 88.7 | 81.8 | 1.3 | 171.8 | 3306 |
| 1.25 | 101.7 | 85.2 | 1.9 | 188.7 | 3293 |

**TABLE E8.11.1b** Measured Iron-Core and Copper Losses of 4.5 kVA $\Delta$–$Y_0$ Connected Transformer with DC Bias for a Below-Rated Line-to-Line Input Voltage of 230 V

| DC bias current (A) | Iron-core losses (W) | AC copper losses (W) | DC copper losses (W) | Total losses (W) | Output power (W) |
|---|---|---|---|---|---|
| 0.00 | 62.9 | 67.9 | 0.0 | 130.8 | 2958 |
| 0.55 | 68.6 | 69.3 | 0.4 | 138.3 | 2950 |
| 0.76 | 72.0 | 72.1 | 0.7 | 144.8 | 2944 |
| 1.03 | 90.7 | 72.3 | 1.3 | 164.3 | 2940 |
| 1.24 | 95.1 | 76.2 | 1.9 | 173.2 | 2925 |

**TABLE E8.11.1c** Measured Iron-Core and Copper Losses of 4.5 kVA $\Delta$–$Y_0$ Connected Transformer with DC Bias for an Above-Rated Line-to-Line Input Voltage of 254 V

| DC bias current (A) | Iron-core losses (W) | AC copper losses (W) | DC copper losses (W) | Total losses (W) | Output power (W) |
|---|---|---|---|---|---|
| 0.00 | 82.5 | 87.0 | 0.0 | 169.5 | 3663 |
| 0.56 | 88.4 | 86.7 | 0.4 | 175.5 | 3657 |
| 0.74 | 93.9 | 87.4 | 0.7 | 182.0 | 3658 |
| 1.02 | 103.0 | 90.0 | 1.7 | 194.3 | 3654 |
| 1.25 | 109.8 | 93.5 | 1.9 | 205.3 | 3645 |

## 8.3.7 Discussion of Results and Conclusions

New approaches for the real-time monitoring of three-phase transformer power loss have been described and applied under nonsinusoidal operation. Accurate (maximum error of 0.5 to 15%), inexpensive ($3,000 for PC and A/D converter) computer-aided monitoring circuits have been devised and their applications demonstrated.

The digital version of the proposed measuring circuits has the following advantages as compared with the analog circuit of Arri *et al.* [90]:

a) Measuring circuits are suitable for three-phase transformers with various commonly used connections (any combination of $Y$ ungrounded, $Y$ grounded, and $\Delta$ connections). Because the iron-core and copper losses of grounded three-phase transformers can be correctly measured, the method is also applicable for loss monitoring of power transformers under DC bias that can be generated by either unbalanced rectifiers, geomagnetic storms [31, 80, 93, 94], or nuclear explosions.

b) Fewer current and voltage transformers are used (only four CTs and five PTs are required, as shown in Fig. 8.20, comparing to nine CTs and nine PTs used in [90]), and therefore, the measurement accuracy is higher.

c) The software [70] of this monitoring approach does not rely on the discrete Fourier transform (DFT) but on a more accurate transform, which is based on functional derivatives and can also be computed via the fast Fourier transform (FFT). With such online monitoring, the iron-core and copper losses can be sampled within a fraction of a second. Moreover, current and voltage wave shapes, harmonic components, total harmonic distortions, input, output, apparent powers, and the efficiency can be dynamically displayed and updated, and it is quite straightforward to measure the transformer derating under various nonsinusoidal operations [73]. The software is compatible with some commercially available data-acquisition tools [95, 96].

Online measurement of power transformer losses has the following practical applications: apparent power derating, loading guide, prefault recognition, predictive maintenance, and transformer aging and remaining life quantification. The method proposed in this section can be used as a powerful monitoring tool in these applications. The method is suitable for any kind of transformer regardless of size, rating, type, different magnetic circuit (core or shell, three or five limbs), cooling system, power system topology (nonsinusoidal and unbalanced), etc.

### 8.3.8 Uncertainty Analysis [102]

The standard ANSI/NCSL Z540-2-1997 (R2002) recommends that the maximum error analysis as used in Section 8.3.3 be replaced by the expression of uncertainty in measurement. The rationale is that the growing international cooperation and the National Conference of Standards Laboratories International (NCSLI) have recommended that standard. The standard ANSI/NCSL Z540-2-1997 (R2002) "is an adoption of the International Organization for Standardization (ISO) Guide to the Expression of Uncertainty in Measurement to promote consistent international methods in the expression of measurement uncertainty within U.S. standardization, calibration, laboratory accreditation, and metrology services." This standard serves the following purposes:

1. It makes the comparison of measurements more reliable than the maximum error method.
2. Even when all of the known or suspected components of error have been evaluated and appropriate corrections have been applied there still remains an uncertainty about the correctness of the stated result.
3. The evaluation and expression of uncertainty should be universal and the method should be applicable to all kinds of measurements.
4. The expression of uncertainty must be internally consistent and transferable.
5. The approach of this standard has been approved by the Bureau International des Poids et Mesures (BIPM).

It is for these reasons that the measurement errors in [74] have been expressed by the uncertainty as recommended by the standard.

### 8.3.9 SCADA and National Instrument LabVIEW Software [103]

Supervisory control and data acquisition (SCADA) plays an important role in distribution system

components such as substations. There are many parts of a working SCADA system, which usually includes signal hardware (input and output), controllers, networks, human–machine interface (HMI), communications equipment, and software. The term SCADA refers to the entire central system, which monitors data from various sensors that are either in close proximity or off site. Remote terminal units (RTUs) consist of programmable logic controllers (PLCs). RTUs are designed for specific requirements but they permit human intervention; for example, in a factory, the RTU might control the setting of a conveyer belt, but the speed can be changed or overridden at any time by human intervention. Any changes or errors are usually automatically logged and/or displayed. Most often, a SCADA system will monitor and make slight changes to function optimally and they are closed-loop systems and monitor an entire system in real time. This requires

- data acquisitions including meter reading,
- checking the status of sensors, the data of which are transmitted at regular intervals depending on the system. Besides the data being used by the RTU, it is also displayed to a human who is able to interface with the system to override settings or make changes when necessary.

### 8.4 TOOLS FOR IMPROVING RELIABILITY AND SECURITY

SCADA equipment such as RTUs and PLCs are applications of telecommunications principles. Interdisciplinary telecommunications programs (ITPs) [104] combine state-of-the-art technology skills with the business, economic, and regulatory insights necessary to thrive in a world of increasingly ubiquitous communications networks. ITPs are served by faculties in economics, electrical engineering, computer science, law, business, telecommunications industry, and government. They address research in wireless, cyber-security, network protocols, telecom economics and strategy, current national policy debates, multimedia, and so forth. Key topics by ITPs are

- Security and reliability of utility systems addressing a range of security-related issues on physical and information security as well as vulnerability assessment and management. The major areas discussed include physical security and access control, asset protection, information systems analysis and protection, vulnerability, assessment, risk management, redundancy, review of existing security and reliability in power systems, new security guidelines, threat response plan, and contingency operation.

- Risk analysis and public safety review various kinds of risks ranging from floods, power outages, and shock to biological hazards. Various approaches to setting safety standards and minimizing risks are explored.

- Current, future, and basic technical concepts of telecommunications systems are studied. They include an in-depth look at basic telecom terminology and concepts, introductions to voice and data networks, signaling, modulation/multiplexing, frequency-band and propagation characteristics, spectral analysis of signals, modulation (amplitude modulation AM, frequency modulation FM, phase modulation PM, and pulse-code modulation PCM), digital coding, modulation multiplexing, detection, transmission systems, and switching systems with an introduction to different network configurations and traffic analysis.

- Data communications define data and computer communications terminology, standards, network models, routing and switching technologies, and communication and network protocols that apply to wide-area networks WANs, metropolitan-area networks MANs, community wireless networks CWNs, and local-area networks LANs. Studies therefore focus on asynchronous and synchronous wide-area networks such as frame relay and synchronous optical networks SONETs, the Internet, routers, selected Internet applications, and Ethernets.

- Applied network security examines the critical aspects of network security. A technical discussion of threats, vulnerabilities, detection, and prevention is presented. Issues addressed are cryptography, firewalls, network protocols, intrusion detection, security architecture, security policy, forensic investigation, privacy, and the law.

- The emerging technology of internet access via the distribution system is addressed [105].

The 2003 U.S.–Canada blackout is analyzed in [106] and recommendations are given for prevention of similar failures in the future. Reliability questions are addressed in [1, 107]. With the advent of distributed generation [108, 109], the control of power systems will become more complicated due to the limited short-circuit capability of renewable sources and their nonsinusoidal voltage and current wave shapes [110].

## 8.4.1 Fast Interrupting Switches and Fault-Current Limiters

Fast interrupting switches and fault-current limiters serve to maintain the stability of an interconnected system by isolating the faulty portion or by mitigating the effect of a fault on the healthy part of the power system. Fast interrupting switches are used to prevent ferroresonance due to the not quite simultaneous closing of the three phases, and to ease reclosing applications. In the first application the almost simultaneous switching of the three phases prevents the current flows through the capacitance of a cable, which is required for inducing ferroresonant currents (see Chapter 2), whereas in the latter the reclosing should occur within a few cycles as indicated in Chapter 4.

**Fast Interrupting Switches (FIS).** Fast interrupting switches [111] rely on sulfur hexafluoride ($SF_6$) as a quenching medium resulting in an increased interrupting capability of the circuit breaker. $SF_6$ is an inorganic compound; it is a colorless, odorless, nontoxic, and nonflammable gas. $SF_6$ has an extremely great influence on global warming and therefore its use should be minimized or entirely avoided. From an environmental point of view a circuit breaker is expected to use smaller amounts of $SF_6$ or to adopt as a quenching medium a gas that causes no greenhouse effect. Research is performed to substitute $SF_6$ by a greenhouse benign alternative quenching gas. The series connection of an FIS with a fault-current limiter either reduces the interrupting conditions or makes the substitution of $SF_6$ with a less effective quenching gas possible. For this reason it is important to investigate the dependence of the interrupting condition on the circuit parameters as will be done in the next section.

**Fault-Current Limiters (FCLs).** Fault-current limiters are installed in transmission and distribution systems to reduce the magnitude of the fault current and thereby mitigate the effect of the fault on the remaining healthy network. The fault-current limiter is a device having variable impedance, connected in series with a circuit breaker (CB) to limit the current under fault conditions. Several concepts for designing FCLs have been proposed: some are based on superconductors [112–114], power electronics components [115–118], polymer resistors [119], and control techniques based on conventional components [120]. FLCs not only limit the fault currents but can have the following additional functions:

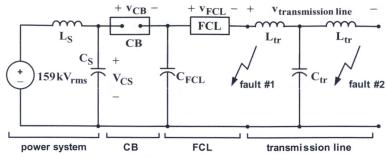

**FIGURE E8.12.1** Single-phase equivalent circuit of a circuit breaker (CB) connected in series with a fault-current limiter (FCL) with two three-phase to ground faults at different locations.

- reduction of voltage sags during short-circuits [121],
- improvement of power system stability [122, 123],
- reduction of the maximum occurring mechanical and electrical torques of a generator [124], and
- easement of the interrupting burden on circuit breakers by limiting the fault current to a desirable level.

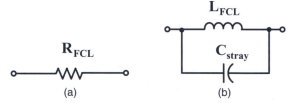

**FIGURE E8.12.2** Model for resistive (a) and inductive (b) fault current limiter.

Note that in the latter case the burden on the circuit breaker depends not only on the interrupting current but also on the transient recovery voltage (TRV) appearing across the contacts of a circuit breaker. It is conceivable that a limiting impedance of the FCL and a stray capacitance can result in a change of the TRV and may bring about a more severe interrupting duty than in the absence of an FLC. In the following section the influence of a resistive and an inductive FLC on the interrupting duty of a circuit breaker will be investigated for various fault locations. First, a fault occurring near an FCL must be addressed because it produces the maximum fault current that a circuit breaker must interrupt. In this case the TRV across the circuit breaker will be not as severe as compared with the case where the fault occurs a few kilometers from the terminals of the FCL. However, a fault a few kilometers from FCL must also be taken into account because the rate of rise of the recovery voltage (rrrv) is higher in this case than for a fault occurring near the FCL.

## 8.4.2 Application Example 8.12: Insertion of a Fault Current Limiter (FCL) in the Power System

The system voltage and frequency of the transmission system are assumed to be $V_{L-L} = 275$ kV (or $V_{L-N} = 159$ kV) and 60 Hz, respectively. A symmetrical three-phase to ground short-circuit is assumed to occur at a distance $\ell$ of 0 to 8 km from the load-side terminals of the FCL. Figure E8.12.1 shows the corresponding single-phase equivalent circuit; on the supply-side the inductance $L_s = 6.69$ mH and the capacitor $C_s = 750$ nF are given. At a fault near the circuit breaker (CB)—in the absence of an FCL—the CB is assumed to interrupt 63 kA$_{rms}$ at a rate of rise of the recovery voltage (rrrv) of 2.3 kV/μs. In the absence of an FCL the current at a distance $\ell$ of 1 km, 4 km, and 8 km is 90%, 68%, and 51% of 63 kA$_{rms}$. $C_{FCL} = 1$ nF represents the stray capacitance to ground of the FCL, and the transmission line inductance $L_{tr}$ and capacitance $C_{tr}$ per unit length are 0.8 mH/km and 15 nF/km, respectively. The FCL is assumed to produce a limiting impedance $Z_{FCL}$ of either a resistive type or an inductive type as is illustrated in Fig. E8.12.2. The resistance $R_{FCL}$ is in the range of 1 to 9 Ω and the inductance $L_{FCL}$ in the range of 2.6 to 23.9 mH; that is, for 60 Hz the resistive ($R_{FCL}$) and reactive ($2\pi f \cdot L_{FCL}$) components are about the same. In the latter case the capacitance $C_{stray}$ (10 to 100 nF) represents a stray capacitance and is connected in parallel with $L_{FCL}$. The current interruption process of the circuit breaker can be assumed to be that of an ideal switch.

### 8.4.3 Intentional Islanding, Interconnected, Redundant, and Self-Healing Power Systems

The present interconnected power system evolved after 1940 and represents an efficient however complicated energy generation, distribution, and utilization network. It is reliable but also vulnerable to sabotage, economic issues, distributed generation, and regulations as well as a lack of appropriately trained or educated engineers. Power systems can be configured in multiple ways: at one extreme we have one large power source that delivers all the required power to multiple loads over transmission lines of varying lengths. At the other extreme there are individual sources that supply the needs for each and every load. In between we have multiple sources (e.g., distributed generation) and loads of varying size that are connected by many transmission lines to supply the loads. The fundamental issue is how to balance efficiency, reliability, and security. For relatively small systems a single source serving multiple loads that are distributed over relatively small distances has the advantages of being able to share capacity as the demand varies from load to load as a function of time and to take advantage of the efficiencies of size. A single source for either one single load or several loads is the easiest to control (e.g., isochronous control) and, in many respects, the most secure, as possible failure mechnisms are associated with one source only. By connecting given loads to multiple sources (e.g., with drooping frequency-load characteristics, see Problems 4.12 to 4.14 of Chapter 4) with multiple transmission lines we increase the reliability, as the failure of any one source or transmission line can be compensated for by picking up the load from other sources or over other transmission lines. However, as the number of lines and sources becomes large the control problem becomes too large to be managed centrally. This problem of distributed control for complicated power networks is closely related to the same problem in communications, biological, economic, political, and many other systems.

In the case of distributed generation with renewable energy sources (e.g., wind, solar) additional constraints enter the control approach where the renewable sources are operated at near 100% capacity by imposing droop characteristics with steep slopes. The intermittent operation of renewable sources requires that peak-power plants are able to take over when the renewable sources are unable to generate power. In order to minimize peak-power generation (e.g., spinning reserve) capacity it will be advisable to also rely on energy storage plants (e.g., hydro, compressed air, supercapacitors).

In the following we explore the requirements necessary so that power systems can organize themselves. The information that is required to operate a conventional stand-alone power plant with isochronous control is relatively well known. However, it is much less clear what information one needs to operate a network that is faced with new and changing requirements and constraints that are brought about by new security problems, environmental regulations, distributed generation, competition, etc. Questions to be addressed by the control and dispatch center of a power system include the following: capacity at which each transmission line and power plant is operating, fuel reserves, frequency and phase of the generators, weather forecast, scheduled and nonscheduled power exchanges, times to shed noncritical loads or pick up new loads, repair times for various kinds of damage and maintenance, hierarchy of controls, and design of the system so that if a terrorist has access to a port on the network the damage he can do is limited. How does the probability that this can happen increase with the number of ports/buses on the network? It is advisable to look at the design of other complex systems that organize themselves as possible models for our power system. Systems that can be considered as role models include the Internet, multistream multiprocessor computer systems for parallel computing, and biological systems.

The systems with the most closely related set of constraints to the problem of the allocation of generation capacity and the distribution of power for varying loads with the loss of a generator or a transmission line are communications networks. These networks have a large number of sources and sinks for information that may be connected by many transmission or communication lines. Strategies for the design of these networks that take into account the cutting of communication lines or the loss of nodes such as telephone switches have been developed [125]. Typical designs for reliability include dual rings and meshes. Considerable savings can be obtained as well as an increase in reliability by the use of meshes. In order for these networks to be effective under varying levels and patterns of use, each node needs information about the traffic flow and the state of adjacent nodes on the network. If the proper information on the bandwidth and traffic are updated on a continuing basis, then the nodes can be prepared to allocate the traffic around a given fault when it occurs. Thus a given node can know at all times how it will reroute the traffic if a communication line is broken or an adjacent node is lost. This can be done without a central command system and in a sense provides a self-organizing system to correct for communication line outages or the loss of a node. Strategies used in communication theory can be effectively employed or modified to improve the reliability of power systems and their efficient operation.

Biological systems can serve as a role model for the design of power systems. Biological systems are highly adaptive and provide both repair and controls that allow them to deal with a large number of unexpected events. These systems use a hierarchy of control systems to coordinate a large number of processes. The brain typically controls the action of the body as a whole, but individual components such as the heart have local control systems that respond to demand for oxygen as communicated to it by signals from other sensory systems such as the baroreceptors. Thus there are functions that require a global response and there are others that are better dealt with locally. Additionally, different biological control systems respond with different time constants [126]. Similarly, power systems have parameters such as fuel reserves with relatively long time constants and other events such as transmission line short-circuits that need to be responded to in fractions of seconds or minutes.

Neural networks are useful ways to approach many complex pattern recognition and control problems that are at least philosophically modeled on biological systems, which use both feedback and feed-forward networks to generate adaptive responses to problems. One of the most common neural networks adjusts the connection weights, using a back propagation algorithm to adapt the network to recognize the desired pattern. This property can be used to develop a control system that has the best response to a wide variety of network problems. Feed-forward networks can be used to anticipate the power system behavior and to prevent problems such as a cascade of network overloads. Power system control and security problems can be formulated in such a way that neural networks can be used as portions of an effective control system. It is to be noted that we can now place computers in the feedback loop of a control system to perform much more complex functions than the classical control systems with a few passive elements.

The three power pools (Western, Eastern, and Texan) within the contiguous United States and their DC links operate quite effectively and reliably. This is so because there is distributed control within the transmission system and the distribution feeders: the integrity of the power pools – including the DC link between the three pools – is maintained through relaying and supervisory control. The everyday operation of power systems takes into account single power plant and single transmission line outages; contingency plans are readily available for the operators. The question arises: what must be done if major emergencies happen? We can identify four types of emergencies:

1. Failure of a major power system component (e.g., power plant, transmission line).
2. Simultaneous failure of more than one major component at different locations such as caused by terrorist attacks.
3. Outage of a power system within an entire region caused by a nuclear explosion or geomagnetically induced currents (GICs) [127].
4. Disruption of the sensing or communication network.

Although the first type of failure can be handled by the present expertise or equipment of power system operators by established contingency plans, the last three failure modes are not planned for because they do not commonly occur. Strategies and countermeasures must be devised so that the power system will be able to survive such major emergencies.

### 8.4.4 Definition of Problem

The problem at hand is multifaceted: it involves behavior, chemical, biological, economic, and engineering sciences. Although interdisciplinary in its nature, only topics involving electrical, telecommunications, and civil or architectural engineering will be addressed.

These sciences will be employed to discuss the following:

1. What can be done to prepare the power system prior to an emergency?
2. What should be done during the failure?
3. What will be done after the failure has occurred?
4. What strategies for efficiency improvements and maintaining regulatory constraints must be addressed?
5. What educational aspects must be highlighted at the undergraduate and graduate levels?

These queries will now be addressed in detail.

### 8.4.5 Solution Approach

There are several approaches for the control of the power networks. The first one draws on the techniques that have been used to control communications networks, that is, to provide each power source with the intelligence to keep track of its local environment and maintain the plans on how it would reroute power if a transmission line is cut or if an adjacent node is either dropped out of the network or has a sudden change in load. This leads to an examination of the power distribution networks to see how the network can be configured to increase its reliability, and this may well mean adding some new transmission lines that enable loads to be supplied from multiple sources. In addition it will require multiple communication links between critical loads and generating power plants. A result of this approach is to allow local nodes on the network to have a high degree of independence that lets them control themselves. This gives both the loads and the generators great flexibility. Writing the rules so that such a system is stable, economical, reliable, and secure is major.

A second approach uses biological systems as a model to develop a hierarchy of controls and information processing that will improve the overall efficiency and reliability of the system. The objective of this effort is the separation of the control problems so that the control of each function is handled by a network that functions as efficiently as possible. This means handling local problems locally and scaling up to larger regions only when it leads to improved system performance in terms of efficiency, reliability, and security. It also means finding a way to make the control system respond on the time scale needed to handle the problem; that is, there are multiple control systems that interface and transmit information between them. Biological systems are often extremely good at optimizing the distribution of controls and efficiency; they are a good starting point to look for ideas for solving related problems. For example, sharks and rays can detect electric fields as small as $10^{-7}$ V/m using a combination of antennas, gain, and parallel processing to detect fish and navigate with signals that at first glance are below the noise level. The way humans use past experience to anticipate future problems may well provide us with an approach to using neural networks as a part of the control system and preventing terrorist attacks.

**What Can Be Done to Prepare the Power System Prior to an Emergency?** This is the most important part for coping with emergencies where several power plants or transmission lines fail or where the power system of an entire region is affected. It is recommended to divide each of the three power pools into independently operating power regions (IOPRs), with distributed intelligence so that in the case of a major emergency – where short-circuits occur – fault-current limiters (FCLs) separate [121–124] within a fraction of a 60 Hz cycle the healthy grid from the faulty section. Each of the designated IOPRs must be able to operate independently and maintain frequency and voltage at about their nominal values either with isochronous or drooping frequency-load control. To achieve this, optimal reconfiguration of the power system must occur within milliseconds permitting intentional islanding operation. If necessary, optimal load shedding must take place but maintaining power to essential loads. To ease such an optimal reconfiguration, provisions must be made to relax usually strict requirements such as voltage and power quality constraints. The ability to communicate and to limit short-circuit currents within milliseconds will be an essential feature of this strategy. It is well known that nuclear explosions generate plasma that is similar to that of solar flares and therefore induce low frequency or quasi-DC currents within the high-voltage transmission system. To minimize the impact of nuclear explosions on the transmission system, three-limb, three-phase transformers with small zero-sequence impedances $Z_0$ will have to be used [31, 80, 94, 127] and DC current compensation networks [94] will have to be installed at key transformers connected to long transmission lines to enforce balanced DC and AC currents.

**What Should Be Done during the Failure?** In an emergency the fault current limiters sense overcurrents and limit these. Thereafter, reclosing switches connected in series with the FCLs will attempt to restore the connection within a few 60 Hz cycles. If the fault duration is longer than a few 60 Hz cycles then circuit breakers connected in series with the FCLs (see Fig. 8.21) will permanently open the faulty line until the source of the fault has been diagnosed. Reclosing switches are employed first because a great number of faults are temporary in their nature. This even applies to a nuclear explosion that lasts only a few seconds but generates plasma that exists for about 60 s. This plasma consists of charged particles that radiate electromagnetic waves and therefore induce low-frequency or quasi-DC currents in the long high-voltage transmission lines. If these quasi-DC currents are balanced within the three phases, the reactive power demand and its

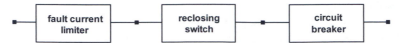

**FIGURE 8.21** Series connection of fault current limiter (FCL) and circuit breaker (CB) where reclosing switch (RCS) is in series to the FCL.

associated voltage drops are manageable if the zero-sequence impedance $Z_0$ of three-limb, three-phase transformers is small. For unbalanced DC currents the reactive power demand and its associated voltage drops are large. For this reason a temporary (seconds) interruption of transmission lines might be beneficial until the plasma has dispersed. This is analogous to a patient who will be rendered unconscious while the operation progresses. Instead of enforcing an interruption another possibility is to enforce a balance of the quasi-DC currents with DC current compensation networks [94].

**What Will Be Done after the Failure Has Occurred?** It is essential that the IOPRs can operate for a longer period of time (e.g., days) so that essential loads can be supplied. If there is a problem with maintaining frequency and voltage control then optimal load shedding will take place. The protocol for optimal load shedding must be defined before an emergency occurs. Relying on a central clock (e.g., National Institute of Standards and Technology, NIST), the paralleling of individual IOPRs can proceed to restore the integrity of each of the three power pools. An alternate strategy is to have a number of atomic clocks located throughout the networks that can provide both the frequency and phase information.

**What Strategies for Efficiency Improvements and Maintaining Regulatory Constraints Must Be Addressed?** Efficiency improvement and maintaining power quality and other regulatory constraints will be unimportant while the failure occurs. For all other times an algorithm developed operates the power system efficiently for given constraints. The development of such software combines harmonic load flow [78], power quality [32, 110, 128–130] constraints, capacitor switching, and optimization where efficiency and cost are components of the objective function [131–134].

**Reliability.** Most short-circuits in power systems occur because of insulation failure. The breakdown of the insulation depends on the voltage stress integrated over time. Although the time component cannot be influenced, the voltage stress across a die-

lectric can be reduced. With improved semiconductor technology and variable-frequency drives (VFDs) based on pulse-width modulation (PWM), the induction motor has become a viable alternative – as compared with DC motor drives – for many applications. The majority of these applications are usually under the 2 kV and the 1 MW range. Even with advancements in technology, the induction, synchronous, or permanent-magnet motors or generators experience failures of their windings due to large voltage stress when fed by PWM frequency current- or voltage-source converters. Repair work and fault analysis and research show that the majority of insulation failures occur near the last turn of the first coil prior to entering the second coil of the windings. Wave-propagation theory shows that forward- and backward-traveling waves superimpose so that the maximum voltage stress does not necessarily occur at the terminals of a winding but somewhere in between the two terminals. Because of the repetitive nature of PWM pulses the damping due to losses in the winding and iron core cannot be neglected. Work by Gupta [135–137] and many others [138–154] neglect the damping of the traveling waves; therefore it cannot predict the maximum voltage stress of the winding insulation. Figure E3.14.1 of Chapter 3 illustrates a part of a machine winding that can be modeled by inductances and capacitances – a model suitable for the traveling wave theory. This model neglecting losses has been used in [135]. By computing and measuring the maximum voltage stress as a function of PWM switching frequency and duty cycle, one finds that current-source inverters and lightning strikes generate the largest voltage stress because of their inherent current spikes resulting in a large L(di/dt).

**Demand-Side Management Programs.** In the upcoming years, utilities continue to face fundamental changes:

- Distributed generation
- Greater access by utilities and others to the transmission systems of other utilities
- Competition for retail customers from other power sources, self-generation or cogeneration, and from other utility suppliers

- Reduction in the cost of renewable resources
- Growing use of demand-side management (DSM) programs as capacity and energy resources
- Increased concern with the environmental consequences of electricity production
- Growing public opposition to construction of new power plants
- Uncertainty about future load growth, fossil-fuel prices and availability, and possible additional regulations

The traditional approach to utility planning with its narrow focus on utility-built power plants is no longer adequate. A new paradigm for utility resource planning has been promoted [155]. The new approach should account for the availability of DSM and renewable-energy technologies and the public concern with environmental qualities.

Demand-side management programs present modest but growing sources of energy and capacity resources. In 1993, DSM programs cut annual electricity sales by 1.5% and potential peak demand by 6.8% [156, 157]. DSM programs include a variety of technologies related to energy efficiency, load management, and fuel switching.

A survey of 2321 DSM programs conducted by 666 electric utilities in the United States indicated a wide range of technology alternatives, market implementation techniques, and incentive structures [158]. Of the total programs surveyed, 64% include residential customers, 50% commercial customers, 30% industrial customers, and 13% agricultural customers. Moreover, 56% of the programs emphasize peak-clipping goals and 54% feature energy-efficiency goals. Load-shifting, valley-filling, and load-growth objectives are also associated with reported DSM programs.

Load control programs are the most prevalent forms of demand-side management practiced by utilities interested in peak clipping. Three types of techniques are commonly used by utilities that would want to exercise control over customer loads:

- Direct control techniques by utilizing a communications system to remotely affect load operation. The communication technologies used include VHF radio, power liner carrier, ripple, FM-SCA radio, cable TV, and telephone.
- Local control techniques through the promotion and use of demand control equipment that operates according to local conditions. The technologies used in local control include temperature-activated cycling (mostly for the control of air conditioners), timer tripping (on pumps), and

various mechanisms that employ timers, interlocks, and/or multiple function demand limiters.
- Distributed control techniques that merge both direct and local control concepts by using a communications system to interface with and/or activate a local control device.

In the EPRI survey [158], a total of 467 load control programs have been reported by 445 different U.S. utilities. The vast majority (440) of reported programs utilize direct control technologies. Only 19 programs employ local control technologies and 8 programs use distributed controls involving interfacing between direct and local control technologies.

A crucial part of any DSM program is electrical load monitoring. Most utilities rely on revenue meters at the point of electrical entrance service to individual buildings or group of buildings. Some DSM programs may require the installation of additional meters for individual electricity-consuming devices such as lighting systems, ventilation fans, or chillers. With these additional meters, controls of targeted equipment can be performed. To make implementation of DSM programs cost-effective, there is an increased interest in monitoring techniques that allow detection and separation of individual loads from measurements performed at a single point (such as the service entrance of a building) serving several electricity-consuming devices. Such techniques have been proposed and have been shown to work relatively well for residential buildings but have not been effective for commercial buildings with irregular and variable loads [159].

Electricity production, transmission, and distribution have substantial environmental costs. Utilities use different methods to assess these impacts [160]. The simplest approach ignores externalities and assumes that compliance with all government environmental regulations yields zero externalities. A more complicated approach consists of ranking and weighting the individual air, water, and terrestrial impacts of different resource options. Some utilities utilize approaches that require quantification (such as tons of either sulfur dioxide or carbon dioxide emitted per million BTU of coal) and monetization (dollars of environmental damage per ton of either carbon dioxide or sulfur dioxide) of the emissions associated with resource options. However, since the estimates of monetized environmental costs vary so much, many utilities are reluctant to use these approaches to assess the environmental externalities of resource options.

Distribution circuits with high reliability indices (e.g., flower configuration) and the use of self-healing system configurations are more expensive than the radial, open-loop, and ring configurations. At present the distribution system contains not much intelligence if one ignores SCADA. In the future the sub-transmission and distribution system will have to be equipped with intelligent components as well so that the distributed generation can play a role even if central power stations go off-line because of faults.

### 8.4.6 Voltage Regulation, Ride-Through Capabilities of Load Components; CBEMA, ITIC Tolerance Curves, and SEMI F47 Standard

**Voltage Regulation $V_R$.** Faults (e.g., short-circuits), as discussed in Chapters 1 and 4, and environmental (e.g., lightning) events addressed in Chapter 1 cause the voltage to deviate from its rated wave shape (e.g., magnitude and waveform). The steady-state voltage deviation is measured by the voltage regulation [161]

$$V_R = \left(1 - \frac{V_{\text{rated}} - |V|}{V_{\text{rated}}}\right) \cdot 100\%,$$

where $V_R$ is the percent voltage regulation, which is 100% for $|V| = V_{\text{rated}}$. The voltages $|V|$ and $V_{\text{rated}}$ are the measured rms voltage and the rated (nominal) voltage, respectively. The voltage regulation is a measure for the strength of a bus, that is, the ability of the bus to supply current without changing its voltage amplitude. The short-circuit ratio ($SCR = I_{sc}/I_{\text{rated}}$) discussed in Chapter 1 is indicative of the strength of a power system. Renewable sources have a limited short-circuit current capability, and for this reason they lead to weak systems. Typical values of $SCR$ are in the range from 20 to 200 for residential and 1000 or more for industrial circuits. Whereas the voltage regulation characterizes steady-state voltage amplitudes, the voltage-tolerance (also called power-acceptability and equipment-immunity) curves measure the change in bus voltage during a few 60 Hz cycles.

**Voltage-Tolerance CBEMA and ITIC Curves.** The voltage-tolerance curves [162–165] are loci drawn in the bus voltage versus duration time semilogarithmic plane. The loci indicate the tolerance of a load to withstand either low or high voltages of short duration (few 60 Hz cycles). Figure 8.22 illustrates the voltage-tolerance curve established by the Compu-

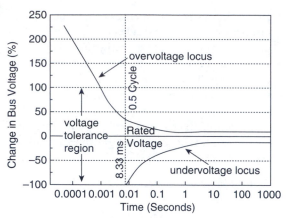

**FIGURE 8.22** CBEMA voltage-tolerance curve.

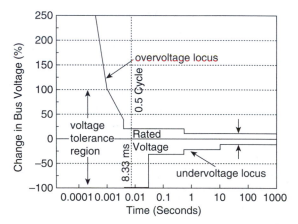

**FIGURE 8.23** ITIC voltage-tolerance curve.

ter Business Equipment Manufacturers Association (CBEMA), usually called the CBEMA curve. There are two loci: the overvoltage locus above the rated voltage value and the undervoltage locus below the rated voltage value. The linear vertical axis (ordinate) of Fig. 8.22 represents the percent change in bus voltage from its rated value. The semilogarithmic horizontal (abscissa) axis indicates the duration of the voltage disturbance either expressed in 60 Hz cycles or seconds. The steady-state value of the undervoltage locus approaches −13% below the rated voltage. Overvoltages of very short duration are tolerable if the voltage values are below the upper locus of the voltage-tolerance curve. The term "voltage tolerance region" in Fig. 8.22 identifies permissible magnitudes for transient voltage events. Voltage events are due to lightning strikes, capacitor switching, and line switching due to faults.

In 1996 the CBEMA curve was replaced by the ITIC curve of Fig. 8.23 as recommended by the Information Technology Industry Council (ITIC).

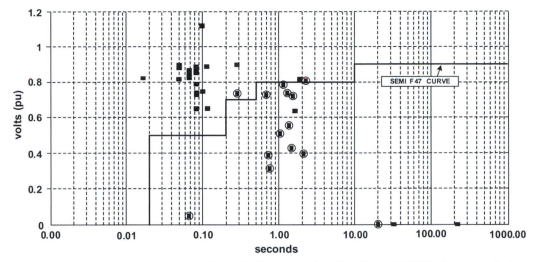

**FIGURE 8.24** Equipment ride-through standard SEMI F47 curve obtained from $V_{L-L} = 13.8$ kV voltage sags; during a time frame of 60 s 38 sags are obtained. The circled data points lead to plant outages. $V_{base} = 13.8$ kV.

**SEMI F47 Standard.** The manufacturing process for semiconductor equipment is very sensitive to voltage sags. For this reason the semiconductor industry joined forces and established criteria for permissible sags that will not affect the manufacturing process. Figure 8.24 shows measured voltage sags that have been characterized by their magnitudes and durations and are plotted along with the equipment ride-through standard for semiconductor manufacturing equipment [166–169]. The circled data in Figure 8.24 represent events that resulted in the process interruption of semiconductor manufacturing. In comparison to the CBEMA and ITIC curves, the SEMI F47 curve addresses low-voltage conditions only.

### 8.4.7 Application Example 8.13: Ride-Through Capability of Computers and Semiconductor Manufacturing Equipment

Determine for the data given in Table E8.1.1 the ride-through capability of

- computers, and
- semiconductor manufacturing equipment.

### 8.4.8 Backup, Emergency, or Standby Power Systems (Diesel-Generator Set, Batteries, Flywheels, Fuel Cells, and Supercapacitors)

Standard IEEE-446 [170] gives definitions related to emergency and standby power systems, general need guidelines, systems and hardware, maintenance, protection, grounding, and industry applications. An important variable – besides the voltage amplitude

characterized by the voltage regulation $V_R$ – for the safe operation of a power system is the frequency regulation %$R$. The frequency regulation [170] is defined by

$$\%R = \frac{(F_{n1} - F_{f1})}{F_{n1}} \cdot 100\%.$$

It represents the percentage change in frequency from steady-state no-load operation of an emergency or standby power system to steady-state full-load operation. The frequency regulation is a function of the prime mover (e.g., diesel engine) and the governing system.

Typical ratings of engine-driven generators are from 5 kW to 1 MW. The engine is mostly of the internal-combustion type – fueled with either diesel, gasoline, or natural gas. There are single- and multiple-engine generator set systems. Some of the emergency and standby power systems use steam or gas turbines. The latter ones are available beyond 3 MW. Simple inertia "ride-through" systems are available powered by induction motors or by batteries with DC machines as an interface between the DC and AC system. Parallel-supplied, parallel-redundant uninterruptible power supplies powered by diesel-driven generators are available up to a few megawatts, mostly used for critical computer loads. Battery-supplied systems with inverters are available as well. Combinations of rectifier/inverter (static), battery, and rotating uninterruptible power supplies provide emergency power within seconds.

### 8.4.9 Automatic Disconnect of Distributed Generators in Case of Failure of Central Power Station(s)

The paradigm for controlling existing power systems is based on central power stations with an appropriate spinning reserve of about 5 to 10%. At a total installed power capacity of about 800 GW within the United States, the spinning reserve is about 40 to 80 GW. The load sharing of the power plants is governed by drooping characteristics as discussed in Application Examples 8.5 and 8.6 and Problems 8.5 to 8.7. The use of renewable energy sources such as solar and wind is based on the principle of maximum power extraction. Due to the intermittent nature of such renewable sources the central power stations cannot be dispensed with or they have to be replaced by storage devices in the hundreds of gigawatts range on a nationwide basis. In the present scenario, where at least 70% of the power is generated by coal, gas, and nuclear power plants, the frequency control and to some degree the voltage (reactive power) control is governed by central power stations. Distributed generation (DG) sources can augment or displace the power provided by central power stations, but they cannot increase their output powers on demand because they are operated at the maximum output power point. If now an important central power station (having a drooping characteristic with a small slope) – which controls the frequency – shuts down due to failure then renewable sources must be disconnected from the system because they can neither control the frequency (due to their drooping characteristics with large slopes) nor can they significantly control the voltage. This is one of the drawbacks of distributed generation unless storage (e.g., pumped-storage hydro or compressed-air) plants are available on a large scale. Application Examples 8.5 and 8.6 and Problems 8.5 to 8.7 address such control issues. The Danish experience with a relatively high penetration of DG indicates that the percentage of DG can assume values in the range of 30% before weak power system effects become dominant.

### 8.5 LOAD SHEDDING AND LOAD MANAGEMENT

Load shedding is an important method to maintain the partial operation of the power system. Some utilities make agreements with customers who do not necessarily require power at all times. In case of a shortage of electrical power due to failures or overloads due to air conditioning on a very hot day, customers will be removed from the load on a temporary basis. Utilities try to entice customers to this program by offering a lower electricity rate or other reimbursement mechanisms.

### 8.6 ENERGY-STORAGE METHODS

Energy storage methods have played an important role in the past and they will even play a more important role in future power system architectures when distributed generation contributes about 30% of the entire electricity production. As long as the DG contribution is well below 30% there will be not much need for an increase in large (e.g., gigawatt range) scale storage facilities. One example for a large-scale storage plant is the Raccoon Mountain hydro pumped-storage plant commissioned around 1978 [60]. Other than hydro-storage plants there are compressed-air storage [61, 62], flywheel [171], magnetic storage [172], and supercapacitor, battery, or fuel cell [173, 174] storage.

### 8.7 MATCHING THE OPERATION OF INTERMITTENT RENEWABLE POWER PLANTS WITH ENERGY STORAGE

The operation of renewable intermittent energy sources must be matched with energy storage facilities. This is so because the wind may blow when there is no power demand and the wind may not blow when there is demand. A similar scenario applies to photovoltaic plants. It is therefore recommended to design a wind farm of, say, 100 MW power capacity together with a hydro pumped-storage plant so that the latter can absorb 100 MW during a 3–day period; that is, the pumped-storage plant must be able to store water so that the electricity storage amounts to 7.2 GWh. Similar storage capacity can be provided by compressed-air facilities where the compressed air is stored in decommissioned mines [61].

### 8.7.1 Application Example 8.14: Design of a Hydro Pumped-Storage Facility Supplied by Energy from a Wind Farm

A pumped-storage hydro-power plant (Fig. E8.14.1) is to be designed with a rated power of 250 MW and a rated energy capacity of 1500 MWh per day. It consists of an upper and a lower reservoir with a water capacity $C_{H2O}$ each. In addition there must be an emergency reserve for 625 MWh. Water evaporation must be taken into account and is 10% per year for each reservoir, and the precipitation per year is 20 inches. The maximum and minimum elevation (water level) of the upper reservoir is 1000 ft and

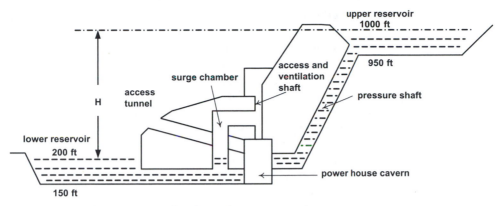

**FIGURE E8.14.1** Pumped-storage power plant for peak-power generation.

950 ft, respectively. The maximum and minimum elevations (water level) for the lower reservoir are 200 ft and 150 ft, respectively. The water turbine is of the Francis type and is coupled with a salient-pole synchronous machine with 24 poles, which can be used as a generator for generating electricity by releasing the water from the upper reservoir to the lower one, and as a motor for pumping the water from the lower reservoir to the upper one.

a) If the power efficiencies of the water turbine and the synchronous generator are $\eta_{turbine} = 0.8$ and $\eta_{synchronous\ machine} = 0.95$, respectively, compute the power required at the turbine input $P_{turbine}^{required}$.

b) Provided the head of the water is $H = 800$ ft, the frictional losses between water and pipe amount to 15%, and the water flow measured in cubic feet per second is $Q = 6000$ cfs, compute the mechanical power available in kW at the turbine input $P_{kW}$.

c) How does $P_{turbine}^{required}$ compare with $P_{kW}$?

d) Compute the specific speed $N_q$. Is the selection of the Francis turbine justified?

e) What other types of water turbines exist?

f) What is the amount of water the upper or lower reservoirs must hold to generate $E = (1500\ MWh + 625\ MWh) = 2125\ MWh$ per day during an 11.3 hour period?

g) Is the given precipitation per year sufficient to replace the evaporated water? If not, what is the required "rain-catch" area to replace the yearly water loss through evaporation?

h) The pumped-storage plant delivers the energy $E = (1500\ MWh + 625\ MWh) = 2125\ MWh$ per day for which customers pay $0.20/kWh due to peak-power generation. What is the payback period of this pumped-storage plant if the construction price is $2,000 per installed power of 1 kW, the cost for pumping is $0.03/kWh, and the interest rate is 3%?

i) Design a wind farm that can supply the hydro pumped-storage plant with sufficient energy during off-peak power times.

### 8.7.2 Application Example 8.15: Peak-Power Tracker [56] for Photovoltaic Power Plants

Next to wind power, photovoltaic systems have become an accepted renewable source of energy. Sunshine (insolation, irradiation, solar radiation) is in some regions abundant and distributed throughout the earth. The main disadvantages of photovoltaic plants are high installation cost and low energy conversion efficiency, which are caused by the nonlinear and temperature-dependent v–i and p–i characteristics, and the shadowing effect due to clouds, dust, snow, and leaves. To alleviate some of these effects many maximum power point tracking techniques have been proposed, analyzed, and implemented. They can be categorized [56] as

1. Look-up table methods: Maximum-power operating points of solar plants at different insolation and temperature conditions are measured and stored in "look-up tables." Any peak-power tracking (PPT) process will be based on recorded information. The problem with this approach is the limited amount of stored information. The nonlinear nature of solar cells and their dependency on insolation and temperature levels as well as degradation (aging) effects make it impossible to record and store all possible system conditions.

2. Perturbation and observation (P&O) methods. Measured cell characteristics (current, voltage, power) are used along with an on-line search algorithm to compute the corresponding maximum-power point, regardless of insolation, temperature, or degradation levels. The problem with this approach is the measurement errors (especially for current), which strongly affect tracker accuracy.

3. Numerical methods. The nonlinear v–i characteristics of a solar panel is modeled using mathematical or numerical equations. The model must be valid under different insolation, temperature, and degradation conditions. Based on the modeled v–i characteristics, the corresponding maximum-power points are computed as a function of cell open-circuit voltages or cell short-circuit currents under different insolation and temperature conditions.

In [56] two simple and powerful maximum-power point tracking techniques (based on numerical methods) – known as voltage-based PPT and current-based PPT – are modeled, constructed, and compared. Relying on theoretical and experimental results, the advantages and shortcomings of each technique are given and their optimal applications are classified. Unfortunately, none of the existing maximum-power tracking methods can accommodate the partial covering of the solar cells [175] through shadowing effects, and bypass diodes are one means of mitigating these detrimental effects.

## 8.8 SUMMARY

The well-established reliability indices are reviewed, and their application to frequently occurring feeder configurations within a distribution system are addressed. An estimation of electric and magnetic fields generated by transmission lines and their associated corona effects are important from an environmental and from a power quality point of view, respectively. Distributed generation (DG) – where renewable sources are a significant (e.g., 20–30%) part of the generation mix – may lead to frequency and voltage control problems. This is so because renewable sources will be operated near their maximum power range and cannot deliver large transient currents during non-steady-state conditions; that is, renewable sources such as solar and wind-power plants contribute to the properties of a weak generation system. This in turn makes a power system that consists of only intermittently operating

renewable plants inoperable because they are operated at their maximum output powers and cannot deliver additional power when demanded by the loads. In other words, the drooping characteristics of intermittently operating renewable sources have a large slope $R$, whereas base load plants such as coal-fired, natural-gas, or storage plants can have droop characteristics with small slopes. The droop characteristics with small slopes guarantee that additional power demand can be covered, through spinning reserve, as requested by the loads. This aspect is important when in the case of severe faults or terrorist attacks the interconnected power system must be separated into several independently operating power system regions where each and every part has its own frequency and voltage control resulting in intentional islanding operation. To maintain a stable operation of each independently operating power region, each one must have sufficient spinning reserve. The maximum error and uncertainty principles are introduced to estimate to total measurement error as applied to power system components. Reliable measurement errors are important for the decision making of dispatch and control centers. Fast switches and current limiters play an important role for the stable operation of a utility system because the faulty part of an interconnected system can be isolated before a domino effect sets in, which may bring down the entire system.

In systems with distributed generation – based on renewable intermittently operating energy sources – larger than 20–30%, storage plants become important to augment the spinning reserve. Storage plants can be put on line within a few (e.g., 6) minutes. This is not fast enough, however, to replace the output of wind-power plants, which may reduce as much as 60 MW per minute. Thus the concept of spinning reserve is important, where the delivery of electric energy can be increased within a few cycles. Peak-power tracking equipment such as batteries for small island photovoltaic plants and peak-power trackers for grid-connected photovoltaic plants are indispensable for maximizing the output power of intermittently operating renewable energy sources. In case of a blackout or brownout of a local power system, emergency power supplies must be employed; these can be based on the conventional diesel-generator or the fuel-cell type. The CBEMA or the newer ITIC curves will make sure that computer loads will not be affected during short outage periods because the power supplies of these components can provide sufficient energy and override a few cycles of voltage outage, reduction, increase, sags, or swells.

Lastly, various demand-side management (DSM) programs can play an important role in maintaining the reliability and security of a distribution system. These appear to be very effective with rate increases or decreases in managing peak-power conditions without requiring additional generation.

## 8.9 PROBLEMS

### Problem 8.1: Reliability, CBEMA, ITIC, and SEMI F47 Calculations

a) For interruptions listed in Table P8.1.1 calculate reliability indices (e.g., SAIDI, SAIFI, SARFI$_{\%V}$). The total number of customers is 490,000.

b) Determine whether for the interruptions of Table P8.1.1 computer (CBEMA and ITIC voltage-tolerance curves) and semiconductor manufacturing (SEMI F47 curve) equipment have a ride-through capability.

### Problem 8.2: $|\vec{E}|$ = E-field Calculation

For a $V_{L-L} = 362$ kV transmission line of flat configuration (Fig. E8.2.2) compute the $|\vec{E}|$ = E-field measured in V/mm for the transmission-line data: equivalent bundle diameter $D = 0.2$ m, phase-to-phase distance $S = 12$ m, and height of the center of the bundle to ground $H = 15$ m.

### Problem 8.3: Voltage Pick-up Through the $|\vec{H}|$ = H-field from a Transmission Line Via a Coil in the Ground

A utility requires a 60 V DC voltage source to transmit data to the control and dispatch center from a distant $V_{L-L} = 262$ kV transmission line operating at a phase rms current of $I_{phase} = 500$ A resulting at the ground level in the magnetic field strength of $H_{ground} = 1.4$ A/m or a maximum flux density of $B_{ground\_max} = 17.6$ mG. To avoid any step-down transformer connection attached to the transmission line it was decided to put a single-phase coil into the ground and rectify the induced voltage supplying DC

voltage to the transmitter's battery. The transmitter requires a battery current of $I_{DC} = 1$ A.

a) Sketch the transmission line and the location of the single-phase coil buried in the ground.

b) Assume that the rectangular single-phase coil has a length of $l = 10$ m, a width of $w = 3$ m, and a very small wire cross section. Determine the number of turns N required to obtain a rectified DC voltage of 60 V. You may assume that the magnetic field is tangential to the ground level.

### Problem 8.4: Maximum Surface Gradients $E_{L-L\_outer}$, $E_{L-L\_center}$ and Critical Disruptive Voltage $V_{L-L\_o}$, Determine the Susceptibility of a Transmission Line to Corona [35]

a) For a transmission line with the line-to-line voltage $V_{L-L} = 362$ kV and the basic geometry of the three-phase transmission line (flat configuration) of Fig. E8.4.1a where the phase spacing is $S = 7.5$ m, the subconductor diameter $d = 3$ cm, the conductor height $H = 12.5$ m, the number of conductors per bundle $N = 2$, and the bundle diameter $D_{bundle} = 45.7$ cm, determine the maximum surface gradients $E_{L-L\_outer}$ and $E_{L-L\_center}$ measured in kV$_{rms}$/cm [47].

b) Determine the crititical disruptive voltage $V_{L-L\_o}$ in kV for the maximum gradients found in part a, where the radius of a subconductor is $r = 1.5$ cm, the equivalent phase spacing (distance) for the three-phase system is $D_{equi} = 500$ cm, the surface factor $m = 0.84$, and the air density factor $\delta = 0.736$ [35].

### Problem 8.5: Frequency Variation Within an Interconnected Power System as a Result of Two Load Changes [57]

A block diagram of two interconnected areas of a power system (e.g., area #1 and area #2) is shown in Fig. P4.12 of Chapter 4. The two areas are connected

**TABLE P8.1.1** Electricity interruptions/transients

| Number of interruptions/transients | Number of interruptions/transients duration | Voltage in % of rated value | Total number of customers affected | Cause of interruption/transient |
|---|---|---|---|---|
| 10 | 4.5 h | 0 | 1000 | overload |
| 15 | 3 h | 0 | 1500 | repair |
| 60 | 1 min | 90 | 500 | switching transients |
| 90 | 1 s | 130 | 500 | capacitor switching |
| 110 | 0.1 s | 40 | 15000 | short-circuits |
| 95 | 0.1 s | 150 | 4000 | lightning |
| 130 | 0.01 s | 50 | 3000 | short-circuits |

by a single transmission line. The power flow over the transmission line will appear as a positive load to one area and an equal but negative load to the other, or vice versa, depending on the direction of power flow.

a) For steady-state operation show that with $\Delta\omega_1 = \Delta\omega_2 = \Delta\omega$ the change in the angular velocity (which is proportional to the frequency f) is

$$\Delta\omega = \text{function of } (\Delta P_{L1}, \Delta P_{L2}, D_1, D_2, R_1, R_2)$$
$$\text{(P5-1)}$$

and

$$\Delta P_{tie} = \text{function of the same parameters as in}$$
$$\text{Eq. P5-1.} \qquad \text{(P5-2)}$$

b) Determine values for $\Delta\omega$ (Eq. P5-1), $\Delta P_{tie}$ (Eq. P5-2), and the new frequency $f_{new}$, where the nominal (rated) frequency is $f_{rated} = 60$ Hz, for the parameters $R_1 = 0.05$ pu, $R_2 = 0.1$ pu, $D_1 = 0.8$ pu, $D_2 = 1.0$ pu, $\Delta P_{L1} = 0.2$ pu, and $\Delta P_{L2} = -0.3$ pu.

c) For a base apparent power $S_{base} = 1000$ MVA compute the power flow across the transmission line.

d) How much is the load increase in area #1 ($\Delta P_{mech1}$) and area #2 ($\Delta P_{mech2}$) due to the two load steps?

e) How would you change $R_1$ and $R_2$ in case $R_1$ is a wind or solar power plant operating at its maximum power point (and cannot accept any significant load increase due to the two load steps) and $R_2$ is a coal-fired plant that can supply additional load?

## Problem 8.6: Load Sharing Control of Renewable and Coal-fired Plants

This problem is concerned with the frequency and load-sharing control of an interconnected power system broken into two areas: the first one with a 50 MW wind-power plant and the other one with a 600 MW coal-fired plant. Figure P4.12 of Chapter 4 shows the block diagram of two generators interconnected by a tie line (transmission line).

Data for generation set (steam turbine and generator) #1:

Frequency change ($\Delta\omega_1$) per change in generator output power ($\Delta P_1$) having the droop characteristic

of $R_1 = \dfrac{\Delta\omega_1}{\Delta P_1}$ pu $= 0.01$ pu,

Load change ($\Delta P_{L1}$) per frequency change ($\Delta\omega_1$)

yielding $D_1 = \dfrac{\Delta P_{L1}}{\Delta\omega_1} = 0.8$ pu,

Step load change $\Delta P_{L1}(s) = \dfrac{\Delta P_{L1}}{s}$ pu $= \dfrac{0.2}{s}$ pu,

Angular momentum of steam turbine and generator set $M_1 = 4.5$,

Base apparent power $S_{base} = 500$ MVA,

Governor time constant $T_{G1} = 0.01$ s,

Valve changing (charging) time constant $T_{CH1} = 0.5$ s,

(Load reference set point)$_1 = 0.8$ pu.

Data for generation set (wind turbine and generator) #2:

Frequency change ($\Delta\omega_2$) per change in generator output power ($\Delta P_2$) having the droop characteristic

of $R_2 = \dfrac{\Delta\omega_2}{\Delta P_2}$ pu $= 0.5$ pu,

Load change ($\Delta P_{L2}$) per frequency change ($\Delta\omega_2$)

yielding $D_2 = \dfrac{\Delta P_{L2}}{\Delta\omega_2} = 1.0$ pu,

Step load change $\Delta P_{L2}(S) = \dfrac{\Delta P_{L2}}{s}$ pu $= \dfrac{-0.2}{s}$ pu,

Angular momentum of wind turbine and generator set $M_2 = 6$,

Base apparent power $S_{base} = 500$ MVA,

Governor time constant $T_{G2} = 0.01$ s,

Valve (blade control) changing (charging) time constant $T_{CH2} = 0.7$ s,

(Load reference set point)$_2 = 0.8$ pu.

Data for tie line: $T = \dfrac{377}{X_{tie}}$ with $X_{tie} = 0.2$ pu.

a) List the ordinary differential equations and the algebraic equations of the block diagram of Figure P4.12 of Chapter 4.

b) Use either Mathematica or MATLAB to establish transient and steady-state conditions by imposing a step function for load reference set

point $(s)_1 = \dfrac{0.8}{s}$ pu, load reference set point $(s)_2 = \dfrac{0.8}{s}$ pu, and run the program with no load changes $\Delta P_{L1} = 0, \Delta P_{L2} = 0$ for 5 s. After 5 s impose

step load change $\Delta P_{L1}(s) = \dfrac{\Delta P_{L1}}{s}$ pu $= \dfrac{0.2}{s}$ pu, and

after 7 s impose step load change $\Delta P_{L2}(s) = \dfrac{\Delta P_{L2}}{s}$ pu $= \dfrac{-0.2}{s}$ pu to find the transient and steady-state responses $\Delta\omega_1(t)$ and $\Delta\omega_2(t)$.

c) Repeat part b for $R_1 = 0.01$ pu, $R_2 = 0.5$ pu, $T_{CH1} = 0.1$ s, and $T_{CH2} = 0.1$ s.

d) Repeat part b for $R_1 = 0.01$ pu, $R_2 = 0.01$ pu, $T_{CH1} = 0.1$ s, and $T_{CH2} = 0.1$ s.

## Problem 8.7: Load Sharing Control of Wind and Solar Power Plants

This problem addresses the frequency and load-sharing control of an interconnected power system broken into two areas, each having one 5 MW wind-power plant and one 5 MW photovoltaic plant. Figure P4.12 of Chapter 4 shows the block diagram of two generators interconnected by a tie line (transmission line).

Data for generation set (wind turbine and generator) #1:

Frequency change $(\Delta\omega_1)$ per change in generator output power $(\Delta P_1)$ having the droop characteristic

of $R_1 = \dfrac{\Delta\omega_1}{\Delta P_1}\,\text{pu} = 0.5\,\text{pu}$

Load change $(\Delta P_{L1})$ per frequency change $(\Delta\omega_1)$

yielding $D_1 = \dfrac{\Delta P_{L1}}{\Delta\omega_1} = 0.8\,\text{pu}$,

Step load change $\Delta P_{L1}(s) = \dfrac{\Delta P_{L1}}{s}\,\text{pu} = \dfrac{0.2}{s}\,\text{pu}$,

Angular momentum of wind turbine and generator set $M_1 = 4.5$,

Base apparent power $S_{\text{base}} = 5$ MVA,

Governor time constant $T_{G1} = 0.01$ s,

Valve (blade control) changing (charging) time constant $T_{CH1} = 0.1$ s,

(Load reference set point)$_1 = 0.8$ pu.

Data for generation set (photovoltaic array and inverter) #2:

Frequency change $(\Delta\omega_2)$ per change in inverter output power $(\Delta P_2)$ having the droop characteristic

of $R_2 = \dfrac{\Delta\omega_2}{\Delta P_2}\,\text{pu} = 0.5\,\text{pu}$,

Load change $(\Delta P_{L2})$ per frequency change $(\Delta\omega_2)$

$D_2 = \dfrac{\Delta P_{L2}}{\Delta\omega_2} = 1.0\,\text{pu}$,

Step load change $\Delta P_{L2}(s) = \dfrac{\Delta P_{L2}}{s}\,\text{pu} = \dfrac{-0.2}{s}\,\text{pu}$,

Equivalent angular momentum $M_2 = 6$,

Base apparent power $S_{\text{base}} = 5$ MVA,

Governor time constant $T_{G2} = 0.02$ s,

Equivalent valve changing (charging) time constant $T_{CH2} = 0.1$ s,

(Load reference set point)$_2 = 0.8$ pu.

Data for tie line: $T = \dfrac{377}{X_{\text{tie}}}$ with $X_{\text{tie}} = 0.2$ pu.

a) List the ordinary differential equations and the algebraic equations of the block diagram of Fig. P4.12 of Chapter 4.

b) Use either Mathematica or MATLAB to establish transient and steady-state conditions by imposing a step function for load reference set point $(s)_1 = \dfrac{0.8}{s}\,\text{pu}$, load reference set point $(s)_2 = \dfrac{0.8}{s}\,\text{pu}$, and run the program with no load changes $\Delta P_{L1} = 0$, $\Delta P_{L2} = 0$ for 5 s. After 5 s impose step load change $\Delta P_{L1}(s) = \dfrac{\Delta P_{L1}}{s}\,\text{pu} = \dfrac{0.2}{s}\,\text{pu}$, and after 7 s impose step load change $\Delta P_{L2}(s) = \dfrac{\Delta P_{L2}}{s}\,\text{pu} = \dfrac{-0.2}{s}\,\text{pu}$ to find the transient and steady-state responses $\Delta\omega_1(t)$ and $\Delta\omega_2(t)$.

c) Repeat part b for $R_1 = 0.1$ pu and $R_2 = 0.05$ pu.

## Problem 8.8: Calculation of Uncertainty of Losses $P_{\text{Loss}} = P_{\text{in}} - P_{\text{out}}$ Based on the Conventional Approach

Even if current and voltage sensors (e.g., CTs, PTs) as well as current meters and voltmeters with error limits of 0.1% are employed, the conventional approach – where the losses are the difference between the input and output powers – always leads to relatively large uncertainties (>15%) in the loss measurements of highly efficient transformers.

The uncertainty of power loss measurement will be demonstrated for the 25 kVA, 240 V/7200 V single-phase transformer of Fig. E8.8.1, where the low-voltage winding is connected as primary, and rated currents are therefore $I_{2\text{rat}} = 3.472$ A, $I_{1\text{rat}} = 104.167$ A [72]. The instruments and their error limits are listed in Table E8.8.1, where errors of all current and potential transformers (e.g., CTs, PTs) are referred to the meter sides. In Table E8.8.1, error limits for all ammeters and voltmeters are given based on the full-scale errors, and those for all current and voltage transformers are derived based on the measured values of voltages and currents (see Appendix A6.1).

Because the measurement error of an ammeter or a voltmeter is a random error, the uncertainty of the instrument is $\varepsilon/\sqrt{3}$, if the probability distribution of the true value within the error limits $\pm\varepsilon$ (as shown in Table E8.8.1) is assumed to be uniform. However, the measurement error of an instrument transformer is a systematic error, and therefore its uncertainty equals the error limit (see Appendix A6.1).

For the conventional approach, power loss is computed by

$$p_{\text{loss}} = v_1 i_1 - v_2 i_2,$$

where $v_1$ is measured by $PT_1$ and voltmeter $V_1$ (see Fig. E8.8.1). The type B variance of measured $v_1$ is $u_{v1}^2 = \varepsilon_{PT1}^2 + \varepsilon_{V1}^2/3 = 0.24^2 + 0.3^2/3^2$.

Similarly, the variances of measured $i_1$, $v_2$, and $i_2$ referred to the primary sides of the instrument transformers are, with $k_{CT1} = 100$ A/5 A = 20, $k_{PT2} = 7200$ V/240 V = 30,

$$u_{i1}^2 = k_{CT1}^2(\varepsilon_{CT1}^2 + \varepsilon_{A1}^2/3),$$
$$u_{v2}^2 = k_{PT2}^2(\varepsilon_{PT2}^2 + \varepsilon_{V2}^2/3),$$
$$u_{i2}^2 = \varepsilon_{CT2}^2 + \varepsilon_{A2}^2/3.$$

The variance of the measured power loss is

$$u_{loss}^2 = \left(\frac{\partial p_{loss}}{\partial v_1}\right)^2 u_{v1}^2 + \left(\frac{\partial p_{loss}}{\partial i_1}\right)^2 u_{i1}^2 + \left(\frac{\partial p_{loss}}{\partial v_2}\right)^2 u_{v2}^2 +$$
$$\left(\frac{\partial p_{loss}}{\partial i_2}\right)^2 u_{i2}^2$$
$$= (i_1 u_{v1})^2 + (v_1 u_{i1})^2 + (i_2 u_{v2})^2 + (v_2 u_{i2})^2.$$

Answer: The uncertainty of power loss measurement is 61.4 W, resulting in the relative uncertainty of 15.7%.

## Problem 8.9: Calculation of Uncertainty of Losses Based on the New Approach

The determination of the losses from voltage and current differences greatly reduces the relative uncertainty of the power loss measurement. The copper and iron-core losses are measured based on $p_{cu} = i_2'(v_1 - v_2') = i_2' v_0$ and $p_{fe} = v_1(i_1 - i_2') = v_1 i_0$, respectively, as shown in Fig. E8.9.1.

The series voltage drop and exciting current at rated operation referred to the low voltage side of the 25 kVA, 240 V/7200 V single-phase transformer are $(v_1 - v_2') = 4.86$ V and $(i_1 - i_2') = 0.71$ A, respectively [70, 71]. The instruments and their error limits are listed in Table E8.9.1.

In Fig. E8.9.1, $v_1$ is measured by $PT_1$ and voltmeter $V_1$. Therefore, the relative uncertainty of the $v_1$ measurement is

$$\frac{u_{v1}}{v_1} = \frac{1}{v_1}\sqrt{\varepsilon_{PT1}^2 + \varepsilon_{V1}^2/3}.$$

Similarly, the relative uncertainties of the $(i_1 - i_2')$, $(v_1 - v_2')$, and $i_2$ measurements are, with $k_{CT1} = 100$ A/5 A = 20,

$$\frac{u_{i0}}{i_0} = \frac{k_{CT1}}{(i_1 - i_2')}\sqrt{\varepsilon_{CT1}^2 + \varepsilon_{CT2}^2 + \varepsilon_{A1}^2/3},$$

$$\frac{u_{v0}}{v_0} = \frac{1}{(v_1 - v_2')}\sqrt{\varepsilon_{PT1}^2 + \varepsilon_{PT2}^2 + \varepsilon_{V2}^2/3},$$

$$\frac{u_{i2}}{i_2} = \frac{1}{i_2}\sqrt{\varepsilon_{CT2}^2 + \varepsilon_{A2}^2/3}.$$

The relative uncertainty of iron-core loss measurement is

$$\frac{u_{Pfe}}{P_{fe}} = \sqrt{(u_{v1}/v_1)^2 + (u_{i0}/i_0)^2}.$$

The relative uncertainty of copper loss measurement is

$$\frac{u_{Pcu}}{P_{cu}} = \sqrt{(u_{v0}/v_0)^2 + (u_{i2}/i_2)^2}.$$

Answer: The uncertainty of power loss measurement is

$$u_{loss} = \sqrt{(0.176 P_{fe})^2 + (0.071 P_{cu})^2},$$

and the relative uncertainty is $(25.8/390)100\% = 6.6\%$.

The uncertainty of power loss measurements can be reduced further by using three-winding current and potential transformers as shown in Fig. E8.9.2a,b for the measurement of exciting current and series voltage drop. The input and output currents pass though the primary and secondary windings of the current transformer, and the current in the tertiary winding represents the exciting current of the power transformer. Similarly, the input and output voltages are applied to the two primary windings of the potential transformers and the secondary winding measures the voltage drop of the power transformer. For this 25 kVA single-phase transformer, the turns ratio for both the $CT$ and the $PT$ is $N_{p1} : N_{p2} : N_s = 1 : 30 : 1$. The power loss uncertainty by using three-winding current and potential transformers can be reduced to as small a value as to that of the back-to-back method, which will be described next.

## Problem 8.10: Calculation of Uncertainty of Losses $P_{Loss} = P_{in} - P_{out}$ Based on Back-to-back Approach

If two transformers of the same type and manufacturing batch are available, the copper and iron-core losses can be measured via the back-to-back method. The testing circuit for this method is shown in Fig. E8.10.1. The series voltage drop $(v_1 - v_2')$ and exciting current $(i_1 - i_2')$ at rated operation referred to the low-voltage side of the two 25 kVA, 240 V/7200 V single-phase transformers are therefore 9.72 V and 1.41 A, respectively (see Table 2 of [70], where the data were measured via the back-to-back approach, and therefore the exciting current and series voltage drop are those of two transformers in series). The instruments and their error limits are listed in Table E8.10.1.

The relative uncertainties of $(i_1 - i_2')$, $v_1$, $(v_1 - v_2')$, and $i_2$ measurements are, for $k_{CT2} = 5$ A/5 A = 1,

$$\frac{u_{i0}}{i_0} = \frac{1}{(i_1 - i_2')}\sqrt{\varepsilon_{CT1}^2 + \varepsilon_{AI}^2/3},$$

$$\frac{u_{v1}}{v_1} = \frac{1}{v_1}\sqrt{\varepsilon_{PT1}^2 + \varepsilon_{V1}^2/3},$$

$$\frac{u_{v0}}{v_0} = \frac{1}{(v_1 - v_2')}\sqrt{\varepsilon_{PT2}^2 + \varepsilon_{V2}^2/3},$$

$$\frac{u_{i2}}{i_2} = \frac{k_{CT2}}{i_2}\sqrt{\varepsilon_{CT2}^2 + \varepsilon_{A2}^2/3}.$$

The relative uncertainty of iron-core loss measurement is

$$\frac{u_{Pfe}}{P_{fe}} = \sqrt{(u_{v1}/v_1)^2 + (u_{i0}/i_0)^2}.$$

The relative uncertainty of copper loss measurement is

$$\frac{u_{Pcu}}{P_{cu}} = \sqrt{(u_{v0}/v_0)^2 + (u_{i2}/i_2)^2}.$$

Answer: The uncertainty of power loss measurement is $u_{loss} = \sqrt{(0.00178 P_{fe})^2 + (0.00163 P_{cu})^2}$, and the relative uncertainty is $(0.54/390)100\% = 0.14\%$.

## Problem 8.11: Fault-current Limiter

a) Design a fault-current limiter (FCL) based on the inductive current limiter of Fig. E8.12.2b employing a current sensor (e.g., low-inductive shunt with optical separation of potentials via optocoupler, Hall sensor) and a semiconductor switch (e.g., thyristor, triac, MOSFET, IGBT).

b) Use the FCL designed in part a and analyze the single-phase circuit of Fig. E8.12.1 using PSpice for the parameters as given in Application Example 8.12 for a three-phase fault occurring at location #1 corresponding to a distance $\ell = 0$ km and a three-phase fault at location #2 at a distance of either $\ell = 1$ km, 4 km, or 8 km from the FCL connected in series with the circuit breaker (CB). In particular $L_{FCL} = 2.6$ mH, $C_{FCL} = 1$ nF, $C_{stray} = 100$ nF, $V_{L-N} = 159$ kV, $L_{tr} = 0.8$ mH/km, and $C_{tr} = 15$ nF/km.

## Problem 8.12: Batteries Used for Peak-power Tracking Connected to a Small Residential Photovoltaic Power Plant

Figure P8.12.1 shows the circuit diagram of a battery directly connected to a solar generator (array), and Fig. P8.12.2 depicts the 12 V lead–acid battery and solar power plant characteristics. The operating points (intersections of battery and solar panel characteristics) can be made rather close to the maximum power points of the solar power plant.

Note that a 12 V lead–acid battery has a rated voltage of $V_{BAT} = E_0 = 12.6$ V if fully charged and a voltage of $V_{BAT} = E_0 = 11.7$ V if almost empty, whereby the voltage drops across the electrochemical capacitance $C_B$ and the nonlinear capacitance $C_b = f(V_{BAT\_over})$ of the battery are zero. A voltage of $V_{BAT} = V_{BAT\_oc} + V_{BAT\_over} = 14$ V if fully charged provided the voltage drops across the electrochemical capacitance $C_B$ and the nonlinear capacitance $C_b = f(V_{BAT\_over})$ of battery are included.

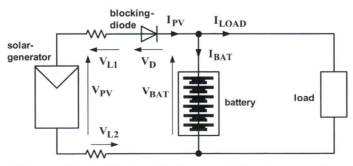

**FIGURE P8.12.1** Circuit diagram of lead–acid battery directly connected to solar power plant.

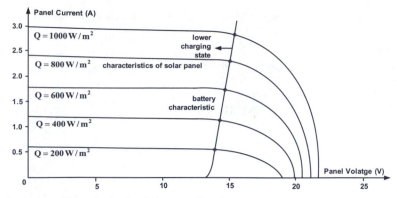

**FIGURE P8.12.2** Operating points of a lead–acid battery directly connected to a solar plant. Q is the insolation.

**TABLE P8.13.1** Interruptions or Transients during February 2006

| Number of interruptions or transients | Interruption or transient duration | Voltage (% of rated value) | Total number of customers affected | Cause of interruption or transient |
|---|---|---|---|---|
| 110 | 2 s | 90 | 2,000 | Switching transients |
| 60 | 1 s | 130 | 3,000 | Capacitor switching |
| 90 | 0.5 s | 40 | 15,000 | Short-circuits due to snow and ice |

Design a peak-power tracking system for a 360 V lead–acid battery and an appropriate photovoltaic system.

### Problem 8.13: Computer Tolerance Curves CBEMA, ITIC, and Semiconductor Manufacturing Ride-Through Standard SEMI F47

a) Determine whether the voltage sags or swells of Table P8.13.1 are within the CBEMA and ITIC voltage tolerance regions. That is, can the computing equipment ride-through and withstand the either low or high voltages of short duration? In this problem February has been chosen because there are line switching, capacitor switching, and snow and ice induced faults as is indicated in Table P8.13.1. The total number of customers is 40,000.

b) Determine whether the voltage sags of Table P8.13.1 lead to semiconductor plant outages according to the equipment ride-through standard SEMI F47.

### Problem 8.14: Emergency Standby Power Supply Based on Diesel-Generator Set

An emergency standby generating set for a commercial building is to be sized. The building has an air-conditioning unit (10 tons, SEER (seasonal energy efficiency ratio) = 14) and an air handler (consisting of two variable-speed drives for the indoor air circulation fan and the outdoor condenser), two high-efficiency refrigerator/freezers (18 cubic feet each), 100 lightbulbs, three fans, two microwave ovens, three drip coffee makers, one clothes washer (not including energy to heat hot water), one clothes dryer, one range (with oven), radio, TV and VCR, three computers, three printers, and one 3 hp motor for a water pump.

Find the wattage of the required emergency standby diesel-generator set, provided the typical wattage values of Table P8.14.1 are used.

### Problem 8.15: Emergency Power Supply Based on Fuel Cells

The transient performance of an inverter – feeding into three-phase power system – supplied by a fuel cell is investigated in this problem. Replace the DC source (with the voltage $V_Q = 390$ V) of Fig. E1.5.1 of Application Example 1.5 of Chapter 1 by the equivalent circuit of the fuel cell as described in Fig. 2 of the paper by Wang, Nehrir, and Shaw [176]. You may assume that in Fig. 2 of that paper the voltage changes $E = 390$ V $\pm 30$ V, where the superimposed rectangular voltage $\pm 30$ V has a period of $T_{\pm 30 \, V} = 1$ s. The remaining parameters of the fuel cell equivalent circuit can be extrapolated from Table III of the paper by Wang et al. for $E = 390$ V.

## Problem 8.16: Operation of Induction Generator with Injected Voltage in Wound-rotor Circuit as Used for State-of-the-art, Variable-speed Wind-power Plants [177]

Induction machines with wound-rotor circuits are used in wind-power plants as generators as shown in Fig. P8.16. To minimize the rating of the solid-state rectifier and the PWM inverter the three-phase stator winding is directly connected to the power system and the rectifier/inverter combination is supplying or extracting power into or from the rotor circuit of the induction generator. A $P_{out} = 5.7$ MW, $V_{L-L} = 4$ kV, $f = 60$ Hz, 8-pole Y-connected wound-rotor induction motor has these parameters: $R_S = 0.02$ Ω, $X_S = 0.3$ Ω, $R'_r = 0.05$ Ω, $X'_r = 0.4$ Ω, $X_m \to \infty$, and a stator-to-rotor turns ratio $a_{T1} = 10$. The rated slip of this machine when operation as a motor is $s_{rat} = 0.02$.

When the induction machine is operated as a motor, determine the following:

a) phase voltage $|\tilde{V}_{S\_ph\_rat}|$
b) synchronous mechanical speed $n_{ms\_rat}$
c) synchronous angular velocity $\omega_{ms-rat}$
d) speed of the shaft $n_{m\_rat}$

When the machine is operated as a generator with injected voltage $\tilde{V}_r = |\tilde{V}_r| \angle \Phi_r$ in the rotor circuit:

e) Derive a complex, second-order equation for the rotor voltage $\tilde{V}_r = |\tilde{V}_r| \angle \Phi_r$, provided the induction generator operates at $n_{m\_new} = 600$ rpm, rated (negative) torque $T_{rat}$, and a leading power factor of $\cos(\varepsilon) = 0.8$, that is, the generator current $\tilde{I}'_r$ leads the generator phase voltage $\tilde{V}_{S\_ph\_rat} = |\tilde{V}_{S\_ph\_rat}| \angle 0°$ by $\varepsilon = 36.87°$. Solve this complex, second-order equation for $|\tilde{V}_r|$ and $\Phi_r$.
f) Determine the ratio of the mechanical gear provided the rated speed of the wind turbine is $n_{m\_wind\ turbine} = 20$ rpm.

**TABLE P8.14.1** Typical Wattage of Appliances

| Appliance | Value (W) |
|---|---|
| Clothes dryer | 4800 |
| Clothes washer (not including energy to heat hot water) | 500 |
| Drip coffee maker | 1500 |
| Fan (ceiling) | 100–200 |
| Microwave oven | 1450 |
| Range (with conventional oven) | 3200 |
| Range (with self-cleaning oven) | 4000 |
| High efficiency refrigerator/freezer (18 cubic feet) | 300 |
| Radio | 60 |
| VCR (including operation of TV) | 100 |
| Lightbulbs | 30–200 |
| Computer | 200 |
| Printer | 100 |
| Compressor of air conditioning unit (8 tons, SEER = 14) | 6000 |
| Indoor air handler and outdoor condenser of air conditioning unit | 2000 |

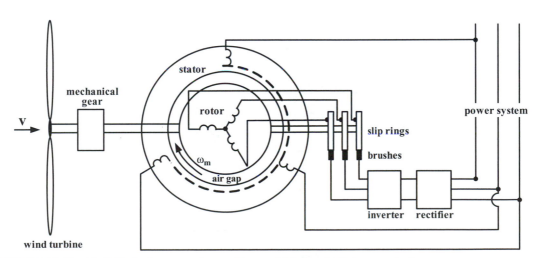

**FIGURE P8.16** Doubly fed induction generator used in a variable-speed wind-power plant.

## Problem 8.17: Design of a Flywheel Storage Power Plant

Design a flywheel storage system which can provide for about 6 minutes power of 100 MW, that is, energy of 10 MWh. The flywheel power plant consists (see Fig. P8.17.1) of a flywheel, mechanical gear, synchronous machine, inverter/rectifier set, and a step-up transformer. The individual components of this plant must be designed as follows:

a) Possible flywheel configurations. For the flywheels (made from steel) based on a spherical ($R_{1o} = 1.5$ m), cylindrical ($R_{1o} = 1.5$ m, $h = 0.9$ m), and wheel-type (with four spokes) configurations as shown in Fig. P8.17.2 ($h = 1.8$ m, $R_{1o} = 1.5$ m, $R_{1i} = 1.3$ m, $R_{2o} = 0.50$ m, $R_{2i} = 0.10$ m, $b = 0.2$ m), compute the ratio of inertia ($J$) to the weight ($W$) of the flywheel, that is, $J/W$. You may assume that the flywheel has magnetic bearings; see http://www.skf.com/portal/skf_rev/home.

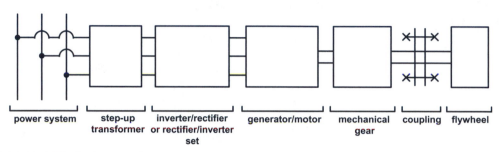

| power system | step-up transformer | inverter/rectifier or rectifier/inverter set | generator/motor | mechanical gear | coupling | flywheel |

**FIGURE P8.17.1** Flywheel-power plant.

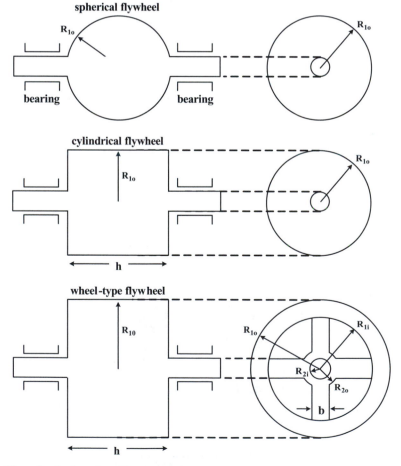

**FIGURE P8.17.2** Three flywheel configurations.

The axial moment of inertia of a rotating sphere is

$$J_{\text{sphere}} = \frac{8}{15} \cdot \pi \cdot \gamma \cdot R_{1o}^5 \, [\text{kgm}^2], \qquad \text{(P17-1)}$$

where the specific mass of iron is

$$\gamma_{\text{steel}} = 7.86 \, [\text{kg/dm}^3]. \qquad \text{(P17-2)}$$

The axial moment of inertia of a rotating cylinder is

$$J_{\text{cylinder}} = \frac{1}{2} \cdot \pi \cdot \gamma \cdot h \cdot R_{1o}^4 \, [\text{kgm}^2]. \qquad \text{(P17-3)}$$

The axial moment of inertia of a rotating wheel-type configuration with 4 spokes is

$$J_{\text{wheel}} = \frac{1}{2} \cdot \pi \cdot \gamma \cdot h \cdot (R_{1o}^4 - R_{1i}^4) + \frac{4}{3} \cdot h \cdot b \cdot \gamma \cdot (R_{1i}^3 - R_{2o}^3) + \frac{1}{2} \cdot \pi \cdot \gamma \cdot h \cdot (R_{2o}^4 - R_{2i}^4) \, [\text{kgm}^2]. \qquad \text{(P17-4)}$$

b) For the given values $h = 1.8$ m, $R_{1o} = 1.5$ m, $R_{1i} = 1.3$ m, $R_{2o} = 0.50$ m, $R_{2i} = 0.10$ m, $b = 0.2$ m, calculate for the wheel-type configuration the stored energy $E_{\text{stored rated}}$ provided the flywheel rotates at $n_{\text{fly-wheel rated}} = 15{,}000$ rpm.

c) Would a homopolar machine [178–182] be more suitable for this application than a synchronous machine?

## Problem 8.18: Steady-state Characteristics of a DC Series Motor Powered by Solar Cells [183]

The performance of a DC series motor supplied by a solar cell array (Fig. P8.18) is to be analyzed. Load matching (solar array, DC series motor, and mechanical load) plays an important role. In this case there is no need for a peak-power tracker because the load matching provides inherent peak-power tracking at steady-state operation.

The solar array is governed by the relation

$$V_g = -0.9 I_g + \frac{1}{0.0422} \ln\left(1 + \frac{I_{phg} - I_g}{0.0081}\right), \qquad \text{(P18-1)}$$

where $V_g$ is the output voltage and $I_g$ is the output current of the solar array. The current $I_{phg}$ is the photon current, which is proportional to the insolation $Q$. The value $I_{phg} = 13.6$ A corresponds to 100% of insolation, that is, $Q = 1000$ W/m$^2$.

The series DC motor has the rated parameters $V_{\text{rat}} = 120$ V, $I_{\text{rat}} = 9.2$ A, $n_{\text{rat}} = 1500$ rpm, $R_a = 1.5$ $\Omega$, $R_f = 0.7$ $\Omega$, $M_{af} = 0.0675$ H and can be described by the relation

$$\omega = \frac{V_g - (T_m/M_{af})^{1/2} R}{M_{af}(T_m/M_{af})^{1/2}}, \qquad \text{(P18-2)}$$

where $R = R_a + R_f$ and $T_m = M_{af} I_a^2$.

The mechanical load (water pump) can be described by the relation

$$T_m = T_{m1} = A + B\omega + T_{L1} = 4.2 + 0.002387\omega, \qquad \text{(P18-3)}$$

where $A = 0.2$ Nm, $B = 0.0002387$ Nm/(rad/s), and $T_{L1} = 4.0$ Nm.

a) Plot the solar array characteristic $I_g$ versus $V_g$ for the three insolation levels of $Q = 100\%, 80\%$, and 60% using Eq. P18-1 with $I_{phg}$ as parameter.

b) Based on Eqs. P18-2 and P18-3 plot the motor torque relation (Eq. P18-2) and the water pump torque relation (Eq. P18-3) in the angular velocity ($\omega$) versus torque ($T_m$) plane.

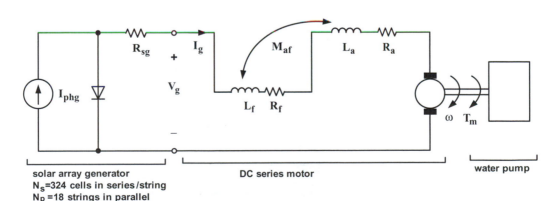

solar array generator
$N_S$=324 cells in series/string
$N_p$=18 strings in parallel

DC series motor

water pump

**FIGURE P8.18** Solar array powering a DC series motor that drives an (irrigation) water pump.

c) Using the relations $V_m = V_g = M_{af}I_a\omega + RI_a$ and $T_m = M_{af}I^2_a$, where $I_a = I_g$, plot the load line $T_{m1}$ in the $I_g$–$V_g$ plane. Does a good load matching exist?

d) Calculate the input power of the series DC motor $P_{in}$ from $V_{rat}$ and $I_{rat}$.

e) Calculate the power delivered to the pump by the solar array $P_{solar}$ at $Q = 100\%$, $80\%$, and $60\%$ insolation.

f) How does $P_{solar}$ compare with $P_{in}$ for the three different insolation levels?

## Problem 8.19: Design of a Compressed-Air Storage Facility [61, 62]

Design a compressed-air energy storage (CAES) plant for $P_{rated} = 100$ MW, $V_{L-L} = 13800$ V, and $\cos\Phi = 1$ that can deliver rated power for 2 hours. The overall block diagram of such a plant is shown in Fig. P8.19. It consists of a compressor, cooler 1, booster compressor, cooler 2, booster-compressor three-phase 2-pole induction motor with a $1:2$ mechanical gear ratio, underground air-storage

reservoir, combustor fired either by natural gas, oil, or coal, a modified (without compressor) gas turbine, and a three-phase 2-pole synchronous generator/motor.

Design data are as follows:

Air inlet temperature of compressor: $50°F \equiv 283.16°K$ at ambient pressure 1 atm $\equiv 14.696$ psi $\equiv$ $101.325$ kPa(scal).

Output pressure of compressor: 11 atm $\equiv 161$ psi $\equiv$ $1114.5$ kPa.

Output temperature of cooler 2: $120°F \equiv 322.05°K$ at $1000$ psi $\equiv 6895$ kPa.

Output temperature of combustor: $1500°F \equiv 1089°K$ at $650$ psi $\equiv 4482$ kPa.

Output temperature of gas turbine: $700°F \equiv 644°K$ at ambient pressure 1 atm $\equiv 14.696$ psi $\equiv$ $101.325$ kPa.

Generation operation: 2 h at 100 MW.

Recharging (loading) operation: 8 h at 40 MW.

Capacity of air storage reservoir: $3.5 \cdot 10^6$ ft$^3 \equiv$ $97.6 \cdot 10^3$ m$^3 \equiv$ (46 m $\times$ 46 m $\times$ 46 m).

Start-up time: 6 minutes.

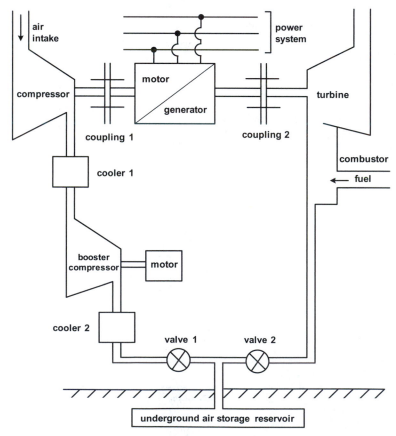

**FIGURE P8.19** Compressed-air storage (CAES) power plant.

Turbine and motor/generator speed: $n_{s\ mot/gen} = 3600$ rpm.

Compressor speed: $n_{s\ comp} = 7200$ rpm.

a) Calculate the Carnot efficiency of the gas turbine:

$$\eta_{carnot\ gas\ turbine} = \eta_{compressor\_turbine} \cdot \eta_{combustor\_turbine}.$$

b) Note that the compressor of a gas turbine has an efficiency of $\eta_{compressor\_turbine} = 0.50$. In this case the compressor is not needed because pressurized air is available from the underground reservoir. The absence of a compressor increases the overall efficiency of the CAES. If "free," that is, low-cost wind power is available for charging the reservoir, this large-scale energy storage method is suitable for most geographic locations relying on either natural or man-made reservoirs. Provided the motor/generator set has an efficiency of $\eta_{mot/gen} = 0.9$, the two air compressors (charging the reservoir) including the two coolers require an input power of 40 MW, and the combustor input fuel must deliver 5500 BTU/kWh, calculate the overall efficiency of the CAES:

$$\eta_{overall} = \frac{P_{out\_generator}}{P_{combustor} + P_{2\_compressors+2\_coolers}}.$$

c) The CAES plant delivers the energy $E = 200$ MWh per day for which customers pay $0.30/kWh due to peak-power generation. What is the payback period of this CAES plant if the construction price is $2,000 per installed power of 1 kW, the cost for pumping (recharging) is $0.03/kWh, and the interest rate is $3%?

**Hint:** For the calculation of the Carnot efficiency of the gas turbine and of the compressors you may use software available on the Internet: http://hyperphysics.phy-astr.gsu.edu/hbase/thermo/adiab.html#c3.

## Problem 8.20: Design of a 5 MW Variable-speed Wind-power Plant for an Altitude of 1600 m

Design a wind-power plant supplying the rated power $P_{out} = 5$ MW into the three-phase distribution system at a line-to-line voltage of $V_{L-L system} = 12.47$ kV. The wind-power plant consists of the following:

- One (Y-grounded/Δ) three-phase, step-up transformer, $N_{inverter}$ parallel-connected three-phase PWM inverters with an input voltage of $V_{DC} = 1200$ V, where each inverter delivers an output AC current $I_{phase\_inverter}$ at unity power factor $\cos \Phi = 1$ to the low-voltage winding of the transformer (e.g., the angle $\Phi$ between the line-to-neutral voltage of the inverter $V_{L-n\_inverter}$ and the phase current of the inverter $I_{phase\_inverter}$ is zero).
- $N_{rectifier} (=N_{inverter})$ three-phase rectifiers, each one equipped with one self-commutated switch and 6 diodes operating at a duty cycle of $\delta = 0.5$. Note that each rectifier feeds one inverter.
- One synchronous generator.
- One mechanical gear.
- One wind turbine.

a) Sketch the block diagram of the entire wind-power plant. At rated operation, the efficiencies of (one) gear box, (one) generator, (one) rectifier, (one) inverter, and (one) transformer are $\eta = 0.95$, each. Note there are $N_{rectifier}$ parallel rectifiers and $N_{inverter}$ parallel inverters, and each rectifier feeds one inverter. At an operation of less than rated load the efficiencies will be smaller. The parallel configuration of inverters and rectifiers permits an efficiency increase, because at light and medium loads only a few inverters and rectifiers must be operated, and some can be disconnected.

b) Determine the output power of each component, e.g., transformer, inverters, rectifiers, generator, gear box, wind turbine, at rated operation.

c) Determine for a modulation index of $m = 0.8$ the inverter output line-to-neutral voltage $V_{L-n inverter}$ such that the inverter can deliver at unity power factor an approximately sinusoidal current to the transformer. $N_{inverter}$ commercially available three-phase PWM inverters are connected in parallel and one inverter has the output power rating $P_{inverter} = 500$ kW. What is the number of inverters required? Determine $I_{phase\_inverter}$ of one inverter, the resulting output current of all $N_{inverter}$ inverters, that is, $\Sigma I_{phase\_inverter}$, and the transformation ratio $a = (N_s/N_p)$ of the (Y-grounded/Δ) step-up transformer. For your calculations you may assume (one) ideal transformer and an ideal power system, where all resistances and leakage inductances are neglected. However, the resistances of the transformer are taken into account in the efficiency calculation of the transformer. Why do we use a Y-grounded/Δ transformer configuration?

d) Each of the $N_{inverter}$ inverters is fed by one three-phase rectifier with one self-commuted PWM-operated switch (IGBT) and 6 diodes, and all $N_{rectifier}$ three-phase rectifiers are fed by (one) three-phase synchronous generator with $p = 8$ poles at $n_s = 900$ rpm, utilization factor $C = 3.5$ kWmin/m$^3$, $D_{rotor} = 1.0$ m. The input line-to-line voltage $V_{L-L\,rectifier} = 1.247$ kV of one three-phase rectifier results at a duty cycle of $\delta = 0.5$ in the DC output voltage (of one rectifier, including filter) $V_{DC} = 1200$ V. Determine the line-to-line output voltage $V_{L-L\,generator}$ of the nonsalient-pole (round-rotor) synchronous generator such that it matches the required input line-to-line voltage of the rectifier $V_{L-L\,rectifier}$; thus a transformer between synchronous generator and rectifier can be avoided.

Find all pertinent parameters of the synchronous generator including its synchronous reactance $X_s$ in per unit and in ohms.

e) Design the mechanical gear between wind turbine and synchronous generator so that the wind turbine can operate at $n_{turbine} = 10$ rpm.

f) Determine the geometric dimensions of the wind turbine – using the Lanchester–Betz equation – for the power efficiency coefficient $c_p = 0.4$ of the wind turbine, and the rated wind velocity of $v = 8$ m/s.

g) What are the advantages and disadvantages of this variable-speed wind-power plant design?

## 8.10 REFERENCES

1) Billinton, R.; and Allan, R.N.; *Reliability Evaluation of Power Systems*, Plenum Press, New York, 1996.

2) Pahwa, A.; Redmon, J.; and Clinard, K.; "Impact of automation on distribution reliability," *Proc. IEEE Power Engineering Society Summer Meeting*, Vancouver, Canada, 2001, Vol. 1, p. 736.

3) Goel, L.; and Billinton, R.; "Determination of reliability worth for distribution system planning," *IEEE Transactions on Power Delivery*, Vol. 9, Issue 3, 1994, pp. 1577–1583.

4) Goel, L.; "A comparison of distribution system reliability indices for different operating configurations," *Journal of Electric Machines and Power Systems*, 27, 1999, pp. 1029–1039.

5) Redmon, J.R.; "Traditional versus automated approaches to reliability," *Proceedings of the IEEE Power Engineering Society Summer Meeting*, Vancouver, Canada, 2001, Vol. 1, pp. 739–742.

6) Pahwa, A.; "Role of distribution automation in restoration of distribution systems after emergencies," *Proceedings of the IEEE Power Engineering Society Summer Meeting*, Vancouver, Canada, 2001, Vol. 1, pp. 737–738.

7) Rigler, D.M.; Hodgkins, W.R.; and Allan, R.N.; "Quantitative reliability analysis of distribution systems: automation," *Power Engineering Journal*, Vol. 13, Issue 4, 1999, pp. 201–204.

8) Allan, R.N.; Billinton, R.; Sjarief, I.; Goel, L.; and So, K.S.; "A reliability test system for educational purposes: basis distribution system data and results," *IEEE Transactions on Power Systems*, Vol. 6, Issue 2, 1991, pp. 813–820.

9) Conti, S; and Tina, G.; "Reliability assessment for distribution systems: automated vs. traditional configurations," *International Journal of Power and Energy Systems*, 2006, digital object identifier (DOI): 10.2316/Journal.203.2006.2.203-3407.

10) Kraft, L.A.; "Modeling lightning performance of transmission systems using PSpice," *IEEE Transactions on Power Systems*, Vol. 6, Issue 2, May 1991, pp. 543–549.

11) Jiang, X.; Shu, L.; Sima, W.; Xie, S.; Hu, J.; and Zhang, Z.; "Chinese transmission lines' icing characteristics and analysis of severe ice accidents," *International Journal of Offshore and Polar Engineering*, Vol. 14, No. 3, Sept. 2004 (ISSN 1053-5381).

12) Zhang, J.H.; and Hogg, B.W.; "Simulation of transmission line galloping using finite element method," *Proceedings of the IEE 2nd International Conference on Advances in Power System Control, Operation and Management*, Dec. 1993, Hong Kong, 5 pages.

13) Sullivan, C.R.; Petrenko, V.F.; McCurdy, J.D.; and Kozliouk, V.; "Breaking the ice [transmission line ice]," *IEEE Industry Applications Magazine*, Vol. 9, Issue 5, Sept.–Oct. 2003, pp. 49–54.

14) http://www.isope.org/publications/journals/ijope-14-3/ijope-14-3-p196-abst-JC-348-Jiang.pdf

15) http://tdworld.com/overhead_transmission/power_deutsche_bahn_installs/

16) McCurdy, J.D.; Sullivan, C.R.; and Petrenko, V.F.; "Using dielectric losses to de-ice power transmission lines with 100 kHz high-voltage excitation," *Conf. Rec. of IEEE Industry Applications Society Annual Meeting*, 2001, pp. 2515–2519.

17) Petrenko, V.F.; and Whitworth, R.W.; *Physics of Ice,* London: Oxford Univ. Press, 1999.

18) Petrenko, V.F.; and Sullivan, C.R.; *Methods and systems for removing ice from surfaces*, U.S. Patent Application PCT/US99/28330, 1999.

19) Kieny, C.; Le Roy, G.; and Sbai, A.; "Ferroresonance study using Galerkin method with pseudo-arclength continuation method," *IEEE Transactions on Power Delivery*, Vol. 4, Issue 1, Jan. 1989, pp. 1841–1847.

20) Grainger, J.J.; and Stevenson, Jr., W.D.; *Power System Analysis*, McGraw-Hill, 1994.

21) Boteler, D.H.; Shier, R.M.; Watanabe, T.; and Horita, R.E.; "Effects of geomagnetically induced currents in the BC Hydro 500 kV system," *IEEE Transactions on Power Delivery*, Vol. 4 Issue 1, Jan. 1989, pp. 818–823.

22) Agrawal, B.L.; and Farmer, R.G.; "Effective damping for SSR analysis of parallel turbine generators," *IEEE Transactions on Power Systems*, Vol. 3, Issue 4, Nov. 1988, pp. 1441–1448.

23) McConnell, B.W.; Barnes, P.R.; Tesche, T.M.; and Schafer, D.A.; "Impact of quasi-dc currents on three-phase distribution transformer installations," ORNL/Sub/89-SE912/1, *Report prepared by the Oak Ridge National Laboratory*, Oak Ridge, Tennessee 37831-6285, June 1992.

24) Amin, M.; "Security challenges for the electricity infrastructure," *Computer*, April 2002, Vol. 35, Issue 4, pp. 8–10, ISSN 0018-9162.

25) EPRI, *Electricity Infrastructure Security Assessment*, Vol. I–II, EPRI, Palo Alto, CA, Nov. and Dec. 2001.

26) Amin, M.; "EPRI/DOD Complex interactive networks/systems initiative: self-healing infrastructures," *Proc. 2nd DARPA-JFACC Symp.*

*Advances in Enterprise Control*, IEEE Computer Soc. Press, Los Alamitos, CA, 2000.

27) Silberman, S.; "The energy web," *Wired*, Vol. 9, No. 7, July 2001.

28) Jones, P.D.; "Recent warming in global temperature series," *Geophysical Research Letters*, Vol. 21, Issue 12, 15 June 1994.

29) Ellsaesser, H.W.; MacCracken, M.C.; Walton, J.J.; and Grotch, S.L.; "Global climatic trends as revealed by the recorded data," *Rev. Geophys. Space Phys.*, Vol. 24, Issue 4.

30) Koch, H.; and Hillers, T.; "Second-generation gas-insulated line," *Power Engineering Journal*, June 2002, pp. 111–116.

31) You, Y.; Fuchs, E.F.; Lin, D.; and Barnes, P.R.; "Reactive power demand of transformers with dc bias," *IEEE Industry Applications Society Magazine*, Vol. 2, No. 4, July/August 1996, pp. 45–52.

32) Fuchs, E.F.; "Are harmonic recommendations according to IEEE 519 and CEI/IEC 555 too restrictive?" *Proceedings of Annual Frontiers of Power Conference*, Oklahoma State University, Stillwater, Oklahoma, Oct. 28–29, 2002, pp. IV-1 to IV-18.

33) Fuller, J.F.; Fuchs, E.F.; and Roesler, D.J.; "Influence of harmonics on power system distribution protection," *IEEE Transactions on Power Delivery*, Vol. PWRD-3, No. 2, April 1988, pp. 546–554.

34) Dennhart, A.; *Grundzüge der Tonfrequenz-Rundsteuertechnik und ihre Anwendung*, Verlags- und Wirtschaftsgesellschaft der Elektrizitätswerke m.b.H. VWEW, 6000 Frankfurt (Main) 70- Stresemannalleee 23, 1971.

35) *Electrical Transmission and Distribution Reference Book*, Westinghouse Electric Corporation, East Pittsburgh, PA, 1964.

36) Xiao Hui; Zeng Xiangjun; Fan Shaosheng; Wu Xuebin; and Kuang Lang; "Medium-voltage power line carrier communication system," *Proceedings of the 2004 International Conference on Power SystemTechnology, PowerCon 2004*, Vol. 2, 21–24 Nov. 2004, pp. 1536–1539.

37) Franklin, G.A.; and Hsu, S.-M.; "Power line carrier interference caused by DC electric arc furnaces," *Proceedings of the IEEE Power Engineering Society General Meeting 2004*, Vol. 2, 13–17 July 2003.

38) Arnborg, S.; Andersson, G.; Hill, D.J.; and Hiskens, I.A.; "On undervoltage load shedding in power systems," *International Journal of Electrical Power and Energy Systems*, Vol. 19, No. 2, February 1997, pp. 141–149(9).

39) Feng, Z.; Ajjarapu, V.; and Maratukulam, D.J.; "A practical minimum load shedding strategy to mitigate voltage collapse," *IEEE Transactions on Power Systems*, Vol. 13, No. 4, pp. 1285–1290. Nov. 1998.

40) Ando, N.; Takashima, M.; Doi, Y.; Sugitate, Y.; and Menda, N.; "Automatic meter reading system adopting automatic routing technology," *IEEE/PES Transmission and Distribution Conference and Exhibition 2002: Asia Pacific*, Vol. 3, 6–10 Oct. 2002, pp. 2305–2309.

41) Thumaty, S.; Pham, A.; and van Wyk, H.; "Development of a low-IF receiver and a fixed wireless utility network," *Microwave Symposium Digest, 2002 IEEE MTT-S International*, Vol. 2, 2–7 June 2002, pp. 935–938.

42) Chih-Hung Wu; Shun-Chien Chang; and Yu-Wei Huang; "Design of a wireless ARM-based automatic meter reading and control system," *IEEE Power Engineering Society General Meeting, 2004*, Vol. 1, 6–10 June 2004, pp. 957–962.

43) Gaul, A.J.; Handschin, E.; and Hoffmann, W.; "Evolutionary strategies applied for an optimal management of electrical loads," *Proceedings of the International Conference on Intelligent Systems Applications to Power Systems, 1996, ISAP '96*, 28 Jan.–2 Feb. 1996, pp. 368–372.

44) Marti, J.R.; Castellanos, F.; and Santiago, N.; "Wide-band corona circuit model for transient simulations," *IEEE Transactions on Power Systems*, Vol. 10, Issue 2, May 1995, pp. 1003–1013.

45) Podporkin, G.V.; and Sivaev, A.D.; "Lightning impulse corona characteristics of conductors and bundles," *IEEE Transactions on Power Delivery*, Vol. 12, Issue 4, Oct. 1997, pp. 1842–1847.

46) *EHV Transmission Line Reference Book*, Edison Electric Institute, 750 Third Ave., New York, NY, 1968.

47) *Transmission Line Reference Book*, Electric Research Institute, Palo Alto, CA, 1982.

48) Filippopoulos, G.; and Tsanakas, D.; "Analytical calculation of the magnetic field produced by electric power lines," *IEEE Transactions on Power Delivery*, Vol. 20, No. 2, April 2005, pp. 1474–1482.

49) Leeper, E.A.; *Silencing the Fields: A Practical Guide to Reducing AC Magnetic Fields*, Symmetry Books, Boulder, CO, 2001.

50) Barnes, F.S.; "Interaction of DC and ELF electric fields with biological materials and systems," *Handbook on Biological Effects of Electromagnetic Fields*, C. Polk and E. Postow, Eds., pp. 103–147, CRC Press, Boca Raton, FL.

51) Barnes, F.S.; "Typical electric and magnetic field exposures at power-line frequencies and their coupling to biological systems," Chapter 3 of *Electromagnetic Fields*, published by American Chemical Society, 1995, pp. 37–55.

52) Publications of the Public Utilities Commission of the State of California.

53) Silva, J.M.; Fleischmann, H.H.; and Shih, C.H.; "Transmission line corona and X-rays," *IEEE Transactions on Power Delivery*, Vol. 19, No. 3, July 2004, pp. 1472–1482.

54) Anaya-Lara, O.; Hughes, F.M.; Jenkins, N.; and Strbac, G.; "Contribution of DFIG-based wind farms to power system short-term frequency regulation," *IEE Proceedings – Generation, Transmission and Distribution*, Vol. 153, Issue 2, 16 March 2006, pp. 164–170.

55) Ekanayake, J.; and Jenkins, N.; "Comparison of the response of doubly fed and fixed-speed induction generator wind turbines to changes in network frequency," *IEEE Transactions on Energy Conversion*, Vol. 19, Issue 4, Dec. 2004, pp. 800–802.

56) Masoum, M.A.S.; Dehbonei, H.; and Fuchs, E.F.; "Theoretical and experimental analyses of photovoltaic systems with voltage- and current-based maximum power point tracking," *IEEE Trans. on Energy Conversion*, Vol. 17, No. 4, December 2002, pp. 514–522.

57) Wood, A.J.; and Wollenberg, B.F.; *Power Generation Operation and Control*, John Wiley & Sons, Inc., 1984.

58) Mandelbaum, R.; "Reap the wild wind [offshore wind farm]," *IEEE Spectrum*, Vol. 39, Issue 10, Oct. 2002, pp. 34–39.

59) 90 MW Pumpspeicherwerk Glems, Technische Werke der Stadt Stuttgart A.G., 1964.

60) http://www.tva.gov/sites/raccoonmt.htm

61) Mattick, W.; Haddenhorst, H.G.; Weber, O.; and Stys, Z.S., "Huntorf- the world's first 290 MW gas turbine air storage peaking plant," *Proceedings of the American Power Conference*, Vol. 37, 1975, pp. 322–330.

62) Vosburgh, K.G., "Compressed air energy storage," *J. Energy*, Vol. 2, No. 2, March–April 1978, pp. 106–112.

63) Park, R.H.; "Fast turbine valving," *IEEE Transactions on Power Apparatus and Systems*, Vol. PAS-92, Issue 3, May 1973, pp. 1065–1073.

64) Cushing, E.W.; Drechsler, G.E.; Killgoar, W.P.; Marshall, H.G.; and Stewart, H.R.; "Fast valving as an aid to power system transient stability and prompt resynchronization and rapid reload after full load rejection," *IEEE Transactions on Power Apparatus and Systems*, Vol. PAS-91, Issue 4, July 1972, pp. 1624–1636.

65) Hsiao-Dong Chiang; Cheng-Shan Wang; and Flueck, A.J.; "Look-ahead voltage and load margin contingency selection functions for large-scale power systems," *IEEE Transactions on Power Systems*, Vol. 12, Issue 1, Feb. 1997, pp. 173–180.

66) Sharif, S.S.; and Taylor, J.H.; "Real-time load forecasting by artificial neural networks," *Proceedings of the IEEE Power Engineering Society Summer Meeting 2000*, Vol. 1, 16–20 July 2000, pp. 496–501.

67) Kandil, M.S.; El-Debeiky, S.M.; and Hasanien, N.E.; "Long-term load forecasting for fast developing utility using a knowledge-based expert system," *IEEE Transactions on Power Systems*, Vol. 17, Issue 2, May 2002, pp.491–496.

68) Nagasaka, K.; and Al Mamun, M.; "Long-term peak demand prediction of 9 Japanese power utilities using radial basis function networks," *Proceedings of the IEEE Power Engineering Society General Meeting 2004*, Vol. 1, 6–10 June 2004, pp. 315–322.

69) Dranetz – Basic Measuring Instruments (BMI), 1000 New Durham Rd, Edison, NJ 08818.

70) Lin, D.; Fuchs, E.F.; and Doyle, M.; "Computer-aided testing program of electrical apparatus supplying nonlinear loads," *IEEE Transactions on Power Systems*, Vol. 12, No. 1, Feb. 1997, pp. 11–21.

71) Fuchs, E.F.; and Fei, R.; "A new computer-aided method for the efficiency measurement of low-loss transformers and inductors under non-sinusoidal operation," *IEEE Transactions on Power Delivery*, January 1996, Vol. PWRD-11, No. 1, pp. 292–304.

72) Fuchs, E.F.; Yildirim, D.; and Batan, T.; "Innovative procedure for measurement of losses of transformers supplying nonsinusoidal loads," *IEE Proceedings – Generation, Transmission and Distribution*, Vol. 146, No. 6, November 1999, IEE Proceedings online no. 19990684.

73) Fuchs, E.F.; Lin, D.; and Martynaitis, J.; "Measurement of three-phase transformer derating and reactive power demand under nonlinear loading conditions," *Transactions on Power Delivery*, Vol. 21, Issue 2, April 2006, pp. 665–672.

74) Lin, D.; and Fuchs, E.F.; "Real-time monitoring of iron-core and copper losses of three-phase transformers under (non)sinusoidal operation," *IEEE Transactions on Power Delivery*, Vol. 21, No. 3, July 2006, pp. 1333–1341.

75) Stensland, T.; Fuchs, E.F.; Grady, W.M.; and Doyle, M.; "Modeling of magnetizing and core-loss currents in single-phase transformers with voltage harmonics

for use in power flow," *IEEE Transactions on Power Delivery*, Vol. 12, No. 2, April 1997, pp. 768–774.

76)  *IEEE Standard 519*, "IEEE recommended practices and requirements for harmonic control in electric power systems," *IEEE-519*, 1992.

77)  IEC 61000-3-2 (2001–10): *Consolidated Edition*, "Electromagnetic compatibility – Part 3-2: limits for harmonic current emissions."

78)  Heydt, G.T.; and Grady, W.M.; "User manual HARMFLOW, Version 3.1," *Electric Power Research Institute Publication, EL-4366-CMM*, Palo Alto, CA, 1985.

79)  Fuchs, E.F.; Yildirim, D.; and Grady, W.M.; "Measurement of eddy-current loss coefficient $P_{EC-R}$, derating of single-phase transformers, and comparison with K-factor approach," *IEEE Transactions on Power Delivery*, Vol. 15, No. 1, January 2000, pp. 148–154, and Vol. 15, No. 4, October 2000, pp. 1331–1333 (discussions/closure) and p. 1357.

80)  Fuchs, E.F.; and You, Y.; "Measurement of ($\lambda$-i) characteristics of asymmetric three-phase transformers and their applications," *IEEE Transactions on Power Delivery*, Vol. 17, No. 4, Oct. 2002, pp. 983–990.

81)  *Current and Voltage Transducer Catalog*, Third Edition, LEM USA, Inc., 6643 West Mill Road, Milwaukee, WI 53218.

82)  Rieman, S.; "An optically isolated data acquisition system," *Independent Study Report*, University of Colorado at Boulder, December 1997.

83)  Yildirim, D.; and Fuchs, E.F.; "Computer-aided measurements of inductor losses at high-frequencies (0 to 6 kHz)," *Proceedings of the 14th Annual Applied Power Electronics Conference and Exposition*, March 14–18, 1999, Dallas, TX.

84)  EMTP program, *Electric Power Research Institute*, Palo Alto, CA.

85)  Stöckl, M.; *Elektrische Messtechnik*, Reihe 1, Band IV, B.G. Teubner Verlagsgesellschaft, Stuttgart, 2. Auflage, 1961.

86)  Lin, D.; Batan, T.; Fuchs, E.F.; and Grady, W.M.; "Harmonic losses of single-phase induction motors under nonsinusoidal voltages," *IEEE Transactions on Energy Conversion*, Vol. 11, No. 2, June 1996, pp. 273–286.

87)  Lin, D.; and Fuchs, E.F.; "On-line measurement of transformer losses under nonsinusoidal operation," *Proceedings of the IASTED International Conference on Power and Energy Systems (PES 2007)*, Clearwater, Florida, January 2007, pp. 239–247.

88)  Fuchs, E.F.; Stensland, T.; Grady, W.M.; and Doyle, M.; "Measurement of harmonic losses of pole transformers and single-phase induction motors," *Proceedings of the IEEE-IAS Industry Applications Society Annual Meeting*, Denver, CO, October 2–7, 1994, Paper No. EM-03-162, pp. 128–134.

89)  American National Standards Institute, IEEE standard test code for liquid-immersed distribution, power, and regulating transformers and IEEE guide for short-circuit testing of distribution and power transformers, ANSI/IEEE C57.12.90-1987, Institute of Electrical and Electronics Engineers, Inc., New York, NY, 1988.

90)  Arri, E.; Locci, N.; and Mocci, F.; "Measurement of transformer power losses and efficiency in real working conditions," *IEEE Transactions on Instrumentation and Measurement*, Vol. 40, No. 2, April 1991, 384–387.

91)  Mocci, F.; "Un nuovo metodo per la determinazione della potenza assorbita dai doppi bipoli," *L'Energia Elettrica*, No. 7–8, 1989, 323–329.

92)  Current and voltage transducer catalog, Third edition, LEM USA Inc., Milwaukee, WI 53218.

93)  Fuchs, E.F.; "Transformers, liquid filled," Encyclopedia of Electrical and Electronics Engineering online, September 1, 2000, J. Webster ed., John Wiley & Sons, New York, NY 10158, paper No. 934C, 40 pages, http://www.interscience. wiley.com:38/eeee/32/6132/w.6132-toc.html

94)  Fuchs, E.F.; You, Y.; and Lin, D.; "Development and validation of GIC transformer models," Final Report for Contract 19X-SK205V, Prepared for Martin Marietta Energy Systems, Inc., Oak Ridge, TN, June 1996.

95)  *DAS-50, User's Guide*, Keithley MetraByte Corp., Taunton, MA 02780, Part # 24851.

96)  *uCDAS-16G Manual*, Keithley MetraByte Corp., Taunton, MA 02780, Part # 24845.

97)  *Requirements for Instrument Transformers* (IEEE Standard ANSI/IEEE C57.13- 1993).

98)  *Instrument Transformers – Part 8: Electronic Current Transformers* (IEC 60044-8-2002-07).

99)  *Messwandler für Elektrizitätszähler*, PTB-A-20.2, Physikalisch-Technische Bundesanstalt (PTB), Braunschweig.

100)  *AEG-Hilfsbuch*, 9th Edition, Allgemeine Elektrizitäts-Gesellschaft, Grunewald, Berlin, 5/97-5/104, 1965.

101)  Grady, W.M.; Fuchs, E.F.; Lin, D.; and Stensland, T.D.; "The potential effects of single-phase power electronic-based loads on power system distortion and losses, Volume 2: single-phase transformers and induction motors," *Technical Report* (Electric

Power Research Institute, Palo Alto, CA, 1000655, 2003).

102) ANSI/NCSL Z540-2-1997(R2002), American National Standard for Calibration – U.S. Guide to the Expression of Uncertainty in Measurement, American National Standard Institute, Nov. 12, 2002.

103) Adamo, F.; Attivissimo, F.; Cavone, G.; and Giaquinto, N.; "SCADA/HMI systems in advanced educational courses," *IEEE Transactions on Instrumentation and Measurement*, Vol. 56, Issue 1, Feb. 2007, pp. 4–10.

104) Barnes, F.S.; Fuchs, E.F.; Tietjen, J.S.; Silverstein, J.; Ko, H-Y.; Lockabaugh, T.; Brown, T.X.; and Sicker, D.C.; "A proposal for an integrated utilities engineering management M.E. program," *Proceedings of the IASTED International Conference on Power and Energy Systems (EuroPES 2005)*, Paper # 468-153, Benalmadena, Spain, June 15–17, 2005.

105) Stahlkopf, K.; "Dream jobs 2004, engineer in paradise," *IEEE Spectrum*, February 2004, p. 32.

106) U.S.–Canada Power System Outage Task Force, *Final Report on the August 14, 2003 Blackout in the United States and Canada: Causes and Recommendations*, April 2004, 228 pages.

107) Fairley, P.; "The unruly power grid," *IEEE Spectrum*, Vol. 41, Issue 8, Aug. 2004, pp. 22–27.

108) Hassan, M.; and Klein, J.; "Distributed photovoltaic systems: utility interface issues and their present status," *Report DOE/ET-20356-3*, Jet Propulsion Laboratory, California Institute of Technology, Pasadena, CA, September 15, 1981.

109) Yildirim, D.; Fuchs, E.F.; and Batan, T.; "Test results of a 20 kW, direct drive, variable-speed windpower plant," *Proceedings of the International Conference on Electrical Machines*, Istanbul, Turkey, September 2–4, 1998, pp. 2039–2044.

110) Fuchs, E.F.; Roesler, D.J.; and Masoum, M.A.S.; "Are harmonic recommendations according to IEEE and IEC too restrictive?" *IEEE Transactions on Power Delivery*," Vol. 19, No. 4, October 2004, pp. 1775–1786.

111) Ragaller, K.; *Current Interruption in High-Voltage Networks*, Plenum Press, New York, 1978.

112) Cha, Y.S.; Yang, Z.; Turner, R.; and Poeppel, B.; "Analysis of a passive superconducting fault current limiter," *IEEE Transactions on Applied Superconductivity*, Vol. 8, No. 1, March 1998, pp. 20–25.

113) Matsumura, T.; Uchii, T.; and Yokomizu, Y.; "Development of flux-lock-type fault current limiter with high $T_C$ superconducting element," *IEEE*

Transactions on Applied Superconductivity*, Vol. 7, No. 2, June 1997, pp. 1001–1004.

114) Hara, T.; Okuma, T.; Yamamoto, T.; Ito, D.; Takaki, K.; and Tsurunaga, K.; "Development of a new 6.6 kV/1000 A class superconducting fault current limiter for electric power systems," *IEEE Transactions on Power Delivery*, Vol. 8, No. 1, January 1993, pp. 182–191.

115) Ueda, T.; Morita, M.; Arita, H.; Kida, J.; Kurosawa, Y.; and Yamagiwa, T.; "Solid-state current limiter for power distribution system," *IEEE Transactions on Power Delivery*, Vol. 8, No. 4, October 1993, pp. 1796–1801.

116) Smith, R.K.; Slade, P.G.; Sarkozi, M.; Stacey, E.J.; Bonk, J.J.; and Mehta, H.; "Solid state distribution current limiter and circuit breaker: application requirements and control strategies," *IEEE Transactions on Power Delivery*, Vol. 8, No. 3, July 1993, pp. 1155–1164.

117) King, E.F.; Chikhani, A.Y.; Hackam, R.; and Salama, M.M.A.; "A microprocessor-controlled variable impedance adaptive fault current limiter," *IEEE Transactions on Power Delivery*, Vol. 5, No. 4, November 1990, pp. 1830–1838.

118) Sugimoto, S.; Kida, J.; Fukui, C.; and Yamagiwa, T.; "Principle and characteristics of a fault current limiter with series compensation," *IEEE Transactions on Power Delivery*, Vol. 11, No. 2, April 1996, pp. 842–847.

119) Skindhoj, J.; Glatz-Reichenbach, J.; and Strümpler, R.; "Repetitive current limiter based on polymer PTC resistor," *IEEE Transactions on Power Delivery*, Vol. 13, No. 2, April 1998, pp. 489–494.

120) Engelman, N.; Schreurs, E.; and Drugge, B.; "Field test results for a multi-shot 12.47 kV fault current limiter," *IEEE Transactions on Power Delivery*, Vol. 6, No. 3, July 1991, pp. 1081–1086.

121) Tosato, F.; and Quania, S.; "Reducing voltage sags through fault current limitation," *IEEE Transactions on Power Delivery*, Vol. 16, No. 1, January 2001, pp. 12–17.

122) Lee, S.; Lee, C.; Ko, T.K.; and Hyun, O.; "Stability analysis of a power system with fault current limiter installed," *IEEE Transactions on Applied Superconductivity*, Vol. 11, No. 1, March 2001, pp. 2098–2101.

123) Sjöström, M.; Cherkaoui, R.; and Dutoit, B.; "Enhancement of power system transient using superconducting fault current limiters," *IEEE Transactions on Applied Superconductivity*, Vol. 9, No. 2, June 1999, pp. 1328–1330.

124) Iioka, D.; Shimizu, H.; Yokomizu, Y.; and Kano, I.; "The effect of resistive and inductive fault current

limiter on shaft torsional torque of synchronous generator installed in consumer systems," *Proceedings of the IASTED International Conference, Power and Energy Systems*, November 19–22, 2001, Tampa, FL, pp. 329–333.

125) Grover, W.; Lecture notes, 2000.

126) Guyton, A.; *Textbook of Medical Physiology*, Eighth Edition, W.B. Saunders Co., 1991.

127) Fuchs, E.F.; You, Y.; and Roesler, D.J.; "Modeling, simulation and their validation of three-phase transformers with three legs under DC bias," *IEEE Transactions on Power Delivery*, Vol. 14, No. 2, April 1999, pp. 443–449.

128) *Military standards* MIL 454, MIL 461C, MIL 1275A, MILB 5087, MIL 220A, and MIL 810E.

129) Hughes, D.; *Grounding for the Control of EMI*, DWCI Virginia, 1983.

130) Vance, E.F.; "Electromagnetic interference control," *IEEE Transactions on EMC*, Vol. 22, EMC-11, pp. 319–328, Nov. 1980.

131) Masoum, M.A.S.; Ladjevardi, M.; Fuchs, E.F.; and Grady, W.M.; "Optimal placement and sizing of fixed and switched capacitor banks under nonsinusoidal operating conditions," *Proceedings of the Summer Power Meeting 2002*, Chicago, July 21–25, 2002, pp. 807–813.

132) Masoum, M.A.S.; Ladejvardi, M.; Fuchs, E.F.; and Grady, W.M.; "Application of local variations and maximum sensitivities selection for optimal placement of shunt capacitor banks under nonsinusoidal operating conditions," *International Journal of Electrical Power and Energy Systems*, Elsevier Science Ltd., Vol. 26, Issue 10, December 2004, pp. 761–769.

133) Masoum, M.A.S.; Jafarian, A.; Ladjevardi, M.; Fuchs, E.F.; and Grady, W.M.; "Fuzzy approach for optimal placement and sizing of capacitor banks in the presence of harmonics," *IEEE Transactions on Power Delivery*, Vol. 19, No. 2, April 2004, pp. 822–829.

134) Masoum, M.A.S.; Ladjevardi, M.; Jafarian, A.; and Fuchs, E.F.; "Optimal placement, replacement and sizing of capacitor banks in distorted distribution networks by genetic algorithms," *IEEE Transactions on Power Delivery*, Vol. 19, No. 4, October 2004, pp. 1794–1801.

135) Gupta, B.K.; et al.; "Turn insulation capability of large AC motors, part 1 – surge monitoring," *IEEE Transactions on Energy Conversion*, Vol. EC-2, No. 4, December 1987, pp. 658–665.

136) Gupta, B.K.; et al.; "Turn insulation capability of large ac motors, part 2 – surge strength," *IEEE Transactions on Energy Conversion*, Vol. EC-2, No. 4, December 1987, pp. 666–673.

137) Gupta, B.K.; et al.; "Turn insulation capability of large AC motors, part 3 – insulation coordination," *IEEE Transactions on Energy Conversion*, Vol. EC-2, No. 4, December 1987, pp. 674–679.

138) Dick, E.P.; et al.; "Practical calculation of switching surges at motor terminals," *IEEE Transactions on Energy Conversion*, Vol. 3, No. 4, December 1988, pp. 864–872.

139) Narang, A.; et al.; "Measurement and analysis of surge distribution in motor windings," *IEEE Transactions on Energy Conversion*, Vol. 4, No. 1, March 1989, pp. 126–134.

140) Acosta, J.; and Cornick, K.J.; "Field investigations into the factors governing the severity of pre-striking transients," *IEEE Transactions on Energy Conversion*, Vol. EC-2, No. 4, December 1987, pp. 638–645.

141) Guardado, J.L.; and Cornick, K.J.; "The effects of coil parameters on the distribution of steep-fronted surges in machine windings," *IEEE Transactions on Energy Conversion*, Vol. 7, No. 3, September 1992, pp. 552–556.

142) Guardado, J.L.; and Cornick, K.J.; "A computer model for calculating steep-fronted surge distribution in machine windings," *IEEE Transactions on Energy Conversion*, Vol. 4, No. 1, March 1989, pp. 95–101.

143) Oyegoke, B.S.; "Voltage distribution in the stator winding of an induction motor following a voltage surge," *Electrical Engineering 82 (2000)*, Springer-Verlag, 2000, pp. 199–205.

144) Oyegoke, B.S.; "A comparative analysis of methods for calculating the transient voltage distribution within the stator winding of an electric machine subjected to steep-fronted surge," *Electrical Engineering 82 (2000)*, pp. 173–182.

145) Oyegoke, B.S.; "Indirect boundary integral equation method as applied for wave propagation in the stator winding of an electric machine," *IEE Proc. Sci. Meas. Technol.*, 1997, 144, No. 5, pp. 223–227.

146) Oyegoke, B.S.; "Effect of the overhang capacitance on the fast-rising surge distribution in winding of a large motor," *IEE Proc. Sci. Meas. Technol.*, 1999, 146, No. 6, November 1999, pp. 299–303.

147) Oyegoke, B.S.; "A comparative analysis of methods for calculating the transient voltage distribution within the stator winding of an electric machine subjected to steep-fronted surge," *IEE 1997 Conference Publication No. 444.*

148) Kaufhold, M.; et al.; "Electrical stress and failure mechanism of the winding insulation in PWM-inverter-fed low-voltage induction motors," *IEEE Transactions on Industrial Electronics*, Vol. 47, No. 3, April 2000, pp. 396–402.

149) Agrawal, A. K.; et al.; "Application of modal analysis to the transient response of multi-conductor transmission lines with branches," *IEEE Transactions on Electromagnetic Compatibility*, Vol. EMC-21, No. 3, August 1979, pp. 256–262.

150) Wang, J.; McInerny, S.; and Haskew, T.; "Insulation fault detection in a PWM-controlled induction motor experimental design and preliminary results," *Proceedings of the 9th International Conference on Harmonics and Quality of Power*, Vol. 2, 1–4 Oct. 2000, pp. 487–492.

151) Bonnet, A.H.; "Analysis of the impact of pulse-width modulated inverter voltage waveforms on ac induction motors," *IEEE Transactions on Industry Applications*, Vol. 32, No. 2, March/April 1996, pp. 386–392.

152) Penrose, H.W.; "Electric motor repair for low voltage induction motors in PWM inverter duty environments," *Proceedings of Electrical Insulation and Electrical Manufacturing and Coil Winding Conference (Cat. No. 97CH36075)*, pp. 841–848.

153) EPRI Project RP2307.

154) Guide for testing turn-to-turn insulation on form-wound stator coils for AC rotating electric machines – for trial use, *IEEE Standard 522*.

155) Hirst, E.; "A good integrated resource plan: guidelines for electric utilities and regulators," ORNL/CON-354, Oak Ridge National Laboratory, Oak Ridge, TN.

156) Eto, J.; Vine, E.; Shown, L.; Sonnenblick, R.; and Payne, C.; "The cost and performance of utility commercial lighting programs," LBL-34967, Lawrence Berkley National Laboratory, Berkeley, CA, 1994.

157) Hadley, S.; and Hirst, E.; "Utility DSM programs from 1989 through 1998: continuation or crossroad," ORNL/CON-405, Oak Ridge National Laboratory, Oak Ridge, TN, 1995.

158) ERPI, "1992 Survey of utility demand-side management programs," prepared by Plexus Research Inc. and Scientific Communications Inc., Palo Alto, CA, 1993.

159) Eco Northwest; "Environmental externalities and electric utility regulation," Report prepared for National Association of Regulatory Utility Commissioners, Washington DC, 1993.

160) Leeb, S.B.; Shaw, S.; and Kirtley, J.L.; 1995, "Transient event detection in spectral envelope estimates for non-intrusive load monitoring," *IEEE Transactions on Power Delivery*, Vol. 10, No. 3, pp. 1200–1210.

161) Heydt, G.T.; Ayyanar, R.; and Thallam, R.; "Power acceptability," *IEEE Power Engineering Review*, September 2001, pp. 12–15.

162) Collins Jr., E.R.; and Morgan, R.L.; "Three phase sag generator for testing industrial equipment," *IEEE Transactions on Power Delivery*, Vol. 11, Jan. 1996, pp. 526–532.

163) Key, T.; "Predicting behavior of induction motors during service faults and interruptions," *IEEE Ind. Applicat. Mag.*, Vol. 1, No. 1, Jan./Feb. 1995, p. 7.

164) Waggoner, R.; "Understanding the CBEMA curve," *Elect. Construction and Maintenance*, Vol. 90, No. 10, Oct. 1991, pp. 55–57.

165) Turner, A.E.; and Collins, E.R.; "The performance of AC contactors during voltage sags," *Proc. IEEE Int. Conf. Harmonics and Quality of Power*, Las Vegas, NV, Oct. 1996, pp. 589–595.

166) McGranaghan, M.; and Roettger, W.; "Economic evaluation of power quality," *IEEE Power Engineering Review*, February 2002, pp. 8–12.

167) *IEEE Recommended Practice for Monitoring Electric Power Quality*, IEEE Standard 1159, 1995.

168) *Specification for Semiconductor Processing Equipment Voltage Sag Immunity*, SEMI F47-0200.

169) *IEEE Recommended Practice for Evaluating Electric Power System Compatibility with Electronic Process Equipment*, IEEE Standard 1346, 1998.

170) IEEE Std. 446-1980, IEEE Recommended Practice for Emergency and Standby Power Systems for Industrial and Commercial Applications (orange book), John Wiley & Sons, Inc.

171) http://www.upei.ca/~physics/p261/projects/flywheel1/flywheel1.htm

172) http://www.doc.ic.ac.uk/~matti/ise2grp/energystorage_report/node8.html

173) http://www.mpoweruk.com/supercaps.htm

174) http://electricitystorage.org/technologies_applications.htm

175) Masoum, M.S.; Dehbonei, H.; and Fuchs, E.F.; Closure of "Theoretical and experimental analyses of photovoltaic systems with voltage- and current-based maximum power tracking," *IEEE Transactions on Energy Conversion*, Vol. 19, No. 3, September 2004, pp. 652–653.

176) Wang, C.; Nehrir, M.H.; and Shaw, S.R.; "Dynamic models and model validation for PEM fuel cells using electrical circuits," *IEEE Transactions on Energy Conversion*, Vol. 20, No. 2, June 2005, pp. 442–451.

177) Dubey, G.K.; *Power Semiconductor Controlled Drives*, Prentice Hall, Englewood Cliffs, NJ, 1989.

178) Schenk, K.F.; Fuchs, E.F.; and Trutt, F.C.; "Load analysis of nonlinear homopolar inductor alternators," *IEEE Transactions on Power Apparatus and Systems*, March/April 1973, Vol. PAS-92, No. 2, pp. 442–448.

179) Siegl, M.; Fuchs, E.F.; and Erdelyi, E.A.; "Medium frequency homopolar inductor alternators with amortisseur bars connected to rectifiers, part 1: saturated position-dependent reactances," Proceedings of the Midwest Power Symposium, Cincinnati, Ohio, October 1973.

180) Siegl, M.; Fuchs, E.F.; and Erdelyi, E.A.; "Medium frequency homopolar inductor alternators with amortisseur bars connected to rectifiers, part 2: the effect of reduced subtransient reactance on rectifier operation," *Proceedings of the Midwest Power Symposium*, Cincinnati, Ohio, October 1973.

181) Fuchs, E.F.; and Frank, U.V.; "High-speed motors with reduced windage and eddy current losses, part I: mechanical design," *ETZ Archiv*, Vol. 5, January 1983, H.1, pp. 17–23.

182) Fuchs, E.F.; and Frank, U.V.; "High-speed motors with reduced windage and eddy current losses, part II: magnetic design," *ETZ Archiv*, Vol. 5, February 1983, H.2, pp. 55–62.

183) Appelbaum, J.; "Starting and steady-state characteristics of dc motors powered by solar cell generators," *IEEE Transactions on Energy Conversion*, Vol. EC-1, No. 1, March 1986, pp. 17–25.

## 8.11 ADDITIONAL BIBLIOGRAPHY

184) Hart, G.W.; "Non-intrusive appliance load monitoring," *Proceedings of IEEE, 80*, (12), 1992, pp. 1870–1891.

185) Krarti, M.; Kreider, J.; Cohen, D.; and Curtiss, P.; "Estimation of energy savings for building retrofits using neural networks," *ASME Transactions Journal of Solar Energy Engineering*, Vol. 120, No. 3. 1998.

186) Krarti, M.; *Energy Audit for Building Systems: An Engineering Approach*, CRC Press, Boca Raton, FL, 2000.

187) Fuchs, E.F.; and Fuchs, H.A.; "Distributed generation and frequency/load control of power systems," *Proceedings of the 40th Annual Frontiers of Power Conference*, Oklahoma State University, Stillwater, Oklahoma, October 29–30, 2007, pp. II-1 to II-9.

188) Fuchs, E.F.; and Fuchs, F.S.; "Intentional islanding – maintaining power systems operation during major emergencies," *Proceedings of the 10th IASTED International Conference on Power and Energy Systems (PES 2008)*, Baltimore, Maryland, April 16–18, 2008.

189) Fuchs, E.F.; and Fuchs, H.A.; "Power quality of electric machines and power systems," *Proceedings of the 8th IASTED International Conference on Power and Energy Systems (EuroPES 2008)*, Corfu, Greece 2008, June 16–18, 2008.

# The Roles of Filters in Power Systems

The implementation of harmonic filters has become an essential element of electric power networks. With the advancements in technology and significant improvements of power electronic devices, utilities are continually pressured to provide high-quality and reliable energy. Power electronic devices such as computers, printers, fax machines, fluorescent lighting, and most other office equipment generate harmonics. These types of devices are commonly classified as nonlinear loads. These loads (e.g., rectifiers, inverters) create harmonics by drawing current in short pulses rather than in a sinusoidal manner.

Harmonic currents and the associated harmonic powers must be supplied from the utility system. The major issues associated with the supply of harmonics to nonlinear loads are severe overheating: increased operating temperatures of generators and transformers degrade the insulation material of their windings. If this heating were continued to the point at which the insulation fails, a flashover would occur. This would partially or permanently damage the device and result in loss of generation or transmission that could cause blackouts. Impacts of harmonics on the insulation of transformers and electrical machines are discussed in Chapter 6.

One solution to this problem is to install a harmonic filter for each nonlinear load connected to the power system. There are different types of harmonic filters including passive, active, and hybrid configurations [1–8]. The installation of active filters proves indispensable for solving power quality problems in distribution networks such as the compensation of harmonic current and voltage, reactive power, voltage sags, voltage flicker, and negative-phase sequence currents. Ultimately, this would ensure a system with increased reliability and better power quality.

Filters can only compensate harmonic currents and/or harmonic voltages at the installed bus and do not consider the power quality of other buses. On the other hand, as the number of nonlinear loads increases, more harmonic filters are required. New generations of active filters are unified power quality conditioners (UPQCs) and active power line

conditioners (APLCs). The UPQCs are discussed in Chapter 11.

This chapter begins with a discussion of the different types of nonlinear loads. Section 9.2 presents a broad classification of filters as they are employed in supply systems, and thereafter the subclassification of filters according to their topology. The next three sections focus on different types of filter topologies including passive, active, and hybrid configurations. Block diagrams of active filters are reviewed in Section 9.6. Detailed discussion and classification of control methods for active and hybrid filters are presented in Section 9.7. The chapter ends with general comments and conclusions.

## 9.1 TYPES OF NONLINEAR LOADS

There are three main types of nonlinear loads in power systems: current-source loads, voltage-source loads, and their combinations.

The first category of nonlinear loads includes the current-source (current-fed or current-stiff) loads. Traditionally, most nonlinear loads have been represented as current sources because their current waveforms on the AC side are distorted. Examples of this category include the phase-controlled thyristor rectifier with a filter inductance on the DC side of the rectifier resulting in DC currents (Fig. 9.1); thyristor rectifiers convert an AC voltage source to a DC current source supplying current-source inverters (CSIs) and high-voltage DC (HVDC) systems, and diode rectifiers with sufficient filter inductance supply DC loads. In general, highly inductive loads are served by silicon-controlled rectifiers (SCRs) converting AC power to DC power. Of course, the inverse operation – where an inverter supplies AC power from a DC power source – results in similar current distortions (e.g., harmonics) on the AC side of the inverter. The transfer characteristics and resulting harmonic currents of current-source nonlinear loads are less dependent on the circuit parameters of the AC side than those on the DC side. Accordingly, passive shunt, active shunt, or hybrid (e.g., a combination of active shunt in series

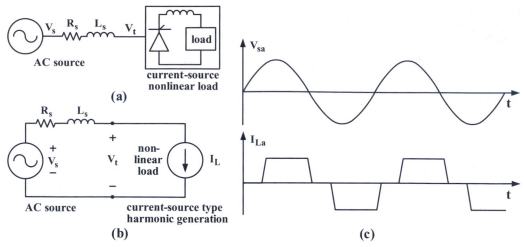

**FIGURE 9.1** Typical current-source nonlinear load: (a) thyristor rectifier for DC drives, (b) per-phase equivalent circuit of thyristor rectifier, (c) typical phase-voltage and line-current waveforms of thyristor rectifier.

with passive shunt) filters for harmonic compensation are commonly applied on the AC side of the converters serving nonlinear loads. The principle of the passive shunt filter is to provide a low-impedance shunt branch for the harmonic currents caused by the nonlinear load. These passive filters have been applied in Chapter 1. The principle of the active shunt filter is to inject into the power system harmonic currents with the same amplitudes and opposite phases of the harmonic currents due to nonlinear loads, thus eliminating harmonic current flowing into the AC source.

The second category of nonlinear loads comprises voltage-source (or voltage-fed or voltage-stiff) loads, such as diode rectifiers with a capacitive filter at the DC link feeding variable-frequency voltage-source inverter (VSI)-based AC motor drives, power supplies with front-end diode rectifier and capacitive filters installed in computers and other household appliances, battery chargers, etc. These voltage-stiff loads draw discontinuous and nonsinusoidal currents resulting in very high $THD_i$, low power factor, and distortions of the AC terminal voltage at the point of common coupling (PCC) defined in Chapter 1. These loads generate harmonic voltage sources, rather than harmonic current sources (Fig. 9.2). Their harmonic amplitudes are greatly affected by the AC side impedance and source voltage imbalance, whereas their rectified voltages are less dependent on the impedance of the AC system. Therefore, diode rectifiers behave like a voltage source, rather than a current source. Accordingly, passive series, active series, or hybrid (e.g., a combination of active

series with passive series) filters are relied on to compensate distortions due to nonlinear loads. Current-source and voltage-source nonlinear loads exhibit dual relations to one another with respect to circuits and properties, and can be effectively compensated by parallel and series filters discussed in Chapter 1, respectively [1].

The third category of nonlinear loads is characterized by the combination of current- and voltage-source loads. They are neither of the current-source nor of the voltage-source type and may contain loads of both kinds. For example, adjustable-speed drives behave as several types of nonlinear loads; variable-frequency VSI-fed AC motors perform as voltage-stiff loads, whereas CSI-fed AC motor drives act as current-stiff loads. For these loads a hybrid topology consisting of active series with passive shunt filter elements is appropriate, provided that the power supply is ideal and adjustable reactive power compensation is not required.

The assessment of harmonic injection of nonlinear loads may not be easy and depends on the following [4]:

- type and topology of nonlinear loads producing harmonics. A voltage-source converter with diode front end may inject an entirely different harmonic spectrum as compared with a current-source converter. Switching power supplies, pulse-width-modulation (PWM) drives, cycloconverters, and arc furnaces have their own specific harmonic spectra;
- interaction of nonlinear loads with the AC system impedance; and

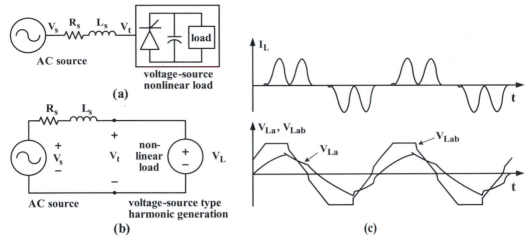

**FIGURE 9.2** Typical voltage-source nonlinear load: (a) diode rectifier for AC drives, electronic equipment, etc., (b) per-phase equivalent circuit of diode rectifier, (c) typical line-current and phase and line-to-line voltage waveforms of thyristor rectifier.

• the harmonic spectra may vary as a function of the nonlinear load. For example, the input AC waveform to a current-source converter and its ripple content depends on the firing angle $\alpha$ and the length of the commutation interval – due to the commutating inductance – as pointed out in Chapter 1. In addition, a small amount of noncharacteristic harmonics may be generated. Therefore, worst-case scenarios should be considered for the analyses of these load types.

## 9.2 CLASSIFICATION OF FILTERS EMPLOYED IN POWER SYSTEMS

During the past few decades the number and types of nonlinear loads have tremendously increased. This has motivated the utilities and consumers of electric power to implement different harmonic filters. Filters are capable of compensating harmonics of nonlinear loads through current-based compensation. They can also improve the quality of the AC supply, for example, compensating voltage harmonics, sags, swells, notches, spikes, flickers, and imbalances through voltage-based techniques. Filters are usually installed near or close to the points of distortion, for example, across nonlinear loads, to ensure that harmonic currents do not interact with the power system. They are designed to provide a bypass for the harmonic currents, to block them from entering the power system, or to compensate them by locally supplying harmonic currents and/or harmonic voltages.

Filters can be designed to trap harmonic currents and shunt them to ground through the use of capacitors, coils, and resistors. Due to the lower impedance of a filter element (at a certain frequency) in comparison to the impedance of the AC source, harmonic currents will circulate between the load and the filter and do not affect the entire system; this is called series resonance as defined in Chapter 1. A filter may contain several of these filter elements, each designed to compensate a particular harmonic frequency or an array of frequencies. Filters are often the most common solution approaches that are used to mitigate harmonics in power systems. Unlike other solutions, filters offer a simple and inexpensive alternative with high benefits.

Classifications of filters may be performed based on various criteria, including the following:

• number and type of elements (e.g., one, two, or more passive and/or active filters),
• topology (e.g., shunt-connected, series-connected, or a combination of the two),
• supply system (e.g., single-phase, three-phase three-wire, and three-phase four-wire),
• type of nonlinear load such as current-source and/or voltage-source loads,
• power rating (e.g., low, medium, and high power),
• compensated variable (e.g., harmonic current, harmonic voltage, reactive power, and phase balancing, as well as multiple compensation),

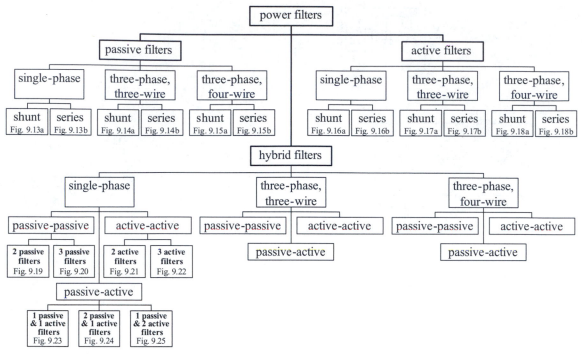

**FIGURE 9.3** Classification of filters used for power quality improvement [1–5].

- converter type (e.g., VSI and/or CSI to realize active elements of filter),
- control technique (e.g., open loop, constant capacitor voltage, constant inductor current, linear voltage control, or optimal control), and
- reference estimation technique (e.g., time and/or frequency current/voltage reference estimation).

Various topologies such as passive filter (PF), active filter (AF), and hybrid filter (HF) in shunt, series, and shunt/series for single-phase, three-phase three-wire, and three-phase four-wire systems have been installed using current-source inverters and voltage-source inverters for implementation of the control of active filters. Figure 9.3 shows the classification of filters as used in power systems based on the supply system with the topology as a subclassification [1–5]. Further classification is made on the basis of the numbers and types of elements employed in different topologies. According to Fig. 9.3, there are 156 valid configurations of passive, active, and hybrid filters. Each class of filters offers its own unique solution to improve the quality of electric power. The choice of filter depends on the nature of the power quality problem, the required level and speed of compensation, as well as the economic cost associated with its implementation.

## 9.3 PASSIVE FILTERS AS USED IN POWER SYSTEMS

If a nonlinear load is locally causing significant harmonic distortion, passive filters may be installed to prevent the harmonic currents from being injected into the system. Passive filters are inexpensive compared with most other mitigating devices. They are composed of only passive elements (inductances, capacitances, and resistances) tuned to the harmonic frequencies of the currents or voltages that must be attenuated. Passive filters have better performances when they are placed close to the harmonic-producing nonlinear loads. They create a sharp parallel resonance at a frequency below the notch (tuned) frequency as introduced in Chapter 1. The resonant frequency of the power system must be carefully placed far from any significant harmonic distortion caused by the nonlinear load. For this reason, passive filters should be tuned slightly lower than the harmonic to be attenuated. This will provide a margin of safety in case there is some change in the system parameters. Otherwise, variations in either filter capacitance and/or filter inductance (e.g., with temperature or failure) might shift resonance conditions such that harmonics cause problems in the power system. In practice, passive filters are added to the system starting with the lowest order of harmonics that must be filtered; for example, installing a seventh-harmonic filter usually requires that a fifth-harmonic filter also be included.

There are various types of passive filters for single-phase and three-phase power systems in shunt and series configurations. Shunt passive filters are the most common type of filters in use. They provide low-impedance paths for the flow of harmonic currents. A shunt-connected passive filter carries only a fraction of the total load current and will have a lower rating than a series-connected passive filter that must carry the full load current. Consequently, shunt filters are desirable due to their low cost and fine capability to supply reactive power at fundamental frequency. It is possible to use more than one passive filter in either shunt and/or series configuration. These topologies are classified as passive–passive single- and three-phase hybrid filters (Fig. 9.3) and will be discussed in Section 9.5. The structure chosen for implementation depends on the type of the dominant harmonic source (voltage source, current source) and the required compensation function (e.g., harmonic current or voltage, reactive power).

### 9.3.1 Filter Transfer Function

Transfer functions are extensively used for filter design and system modeling. Figure 9.4a shows the single-phase presentation of a balanced three-phase distribution system with a nonlinear load and a passive shunt filter installed at the PCC. At harmonic frequencies, the system can be approximated by the equivalent circuit of Fig. 9.4b. Three main types of transfer functions can be defined: impedance, current-divider, and voltage-divider transfer functions.

**Impedance Transfer Functions.** Impedance transfer functions are the most basic blocks for design and modeling of power systems. Two types of filter impedance transfer functions have been defined (Fig. 9.4):

- *Filter impedance transfer function* is the impedance frequency response of the filter expressed in the s domain as

$$H_f(s) = Z_f(s) = \frac{V_f(s)}{I_f(s)}, \qquad (9\text{-}1a)$$

where $Z_f(s)$, $V_f(s)$, and $I_f(s)$ are the filter impedance, complex voltage, and current, respectively, and $s = j\omega$ is the Laplace operator.

- *Filter system impedance transfer function* is defined after the filter is designed and connected to the network. It is the impedance frequency response of the system (with the filter installed) expressed in the s domain:

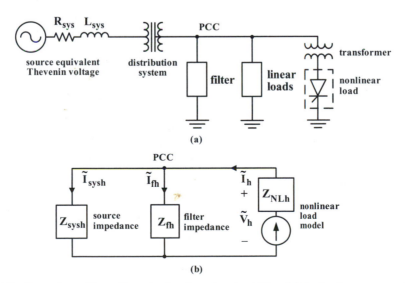

(a)

(b)

**FIGURE 9.4** Distribution system with nonlinear load and passive shunt filter; (a) single-phase representation, (b) equivalent circuit for harmonic (of order h) filter installation, see Chapter 1.

$$H_{fsys}(s) = Z_{fsys}(s) = \frac{V_{sys}(s)}{I_f(s) + I_{sys}(s)} = \frac{1}{\dfrac{1}{Z_f(s)} + \dfrac{1}{Z_{sys}(s)}},$$

$$(9\text{-}1b)$$

where $Z_{fsys}(s)$ is the impedance of the system after the connection of the filter and $V_{sys}(s)$ and $I_{sys}(s)$ are the system complex voltage and current, respectively.

**Current-Divider Transfer Functions.** Two types of current-divider transfer functions are derived for a filter (Fig. 9.4):

- $H_{cds}(s)$ is the ratio of system (sys) harmonic current to the injected harmonic nonlinear (NL) current:

$$H_{cds}(s) = \frac{I_{sys}(s)}{I_{NL}(s)} = \frac{Z_f(s)}{Z_f(s) + Z_{sys}(s)} = \rho_{sys}(s). \quad (9\text{-}2a)$$

- $H_{cdf}(s)$ is the ratio of filter (f) harmonic current to injected harmonic nonlinear current:

$$H_{cdf}(s) = \frac{I_f(s)}{I_{NL}(s)} = \frac{Z_{sys}(s)}{Z_f(s) + Z_{sys}(s)} = \rho_f(s). \quad (9\text{-}2b)$$

In Eqs. 9-2, $\rho_{sys}(s)$ and $\rho_f(s)$ are complex quantities that determine the distribution of the harmonic currents in the filter and the system and have been introduced in Chapter 1. A properly designed filter will have $\rho_f(s)$ close to unity, typically 0.995, and $\rho_{sys}(s)$ about 0.045 with the corresponding angles of about 2.6° and 81°, respectively. It is desirable that $\rho_{sys}(s) = \rho_{sys}(j\omega)$ be small at the various occurring harmonics.

According to Eqs. 9-2 the system impedance plays an important role in the filtering process. Without a filter, all harmonic currents pass through the system. For a large system impedance and a given harmonic frequency, the bypass via the filter is perfect and all current harmonics will flow through the filter impedance. Conversely, for a system of low impedance at a given harmonic frequency most current of harmonic frequency will flow into the system.

It is easy to show that

$$H_{cdf}(s) = \frac{Z_{sys}(s)}{Z_f(s)} H_{cds}(s) = \frac{1}{Z_f(s)} H_{fsys}(s), \quad (9\text{-}3a)$$

$$H_{cds}(s) = \frac{1}{Z_{sys}(s)} H_{fsys}(s). \quad (9\text{-}3b)$$

When designing a filter the impedance transfer functions (Eqs. 9-1) can be used to evaluate the overall system performance. After the filter is installed, computed and measured current-divider ratios (Eqs. 9–2) can be plotted to evaluate filter performance and determine whether harmonic current distortion limits comply with the IEEE-519 Standard [11], as discussed in Chapter 1 (Application Examples 1.8 and 1.9).

**Voltage-Divider Transfer Functions.** Similar transfer functions can be developed for systems with loads that are sensitive to harmonic voltages. Filters are used to provide frequency detuning or voltage distortion control. These transfer functions are based on a voltage division between an equivalent harmonic voltage source – representing nonlinear loads – and harmonic sensitive loads. Applications of these transfer functions are concerned with the voltage distortion amplification of shunt power factor capacitors due to parallel and series resonances.

### 9.3.2 Common Types of Passive Filters for Power Quality Improvement

The configuration of a filter depends on the frequency spectrum and nature of the distortion. Figure 9.5 depicts common types of passive filters.

The first-order damped and undamped high-pass filter (Fig. 9.5a with and without the resistor, respectively) is used to attenuate high-frequency current harmonics that cause telephone interference, reduce voltage notching caused by commutation of SCRs, and provide partial displacement power factor correction of the fundamental load current. The single-tuned low-pass (also called band-pass or second-order series resonant band-pass) filter (Fig. 9.5b) is most commonly applied for mitigation of a single dominant low-order harmonic. Fig. 9.5c represents a second-order damped high-pass filter. Most nonlinear loads and devices generate more than one harmonic, however, and one filter is not usually adequate to effectively compensate, mitigate, or reduce all harmonics within the power system. There are different approaches to overcome this problem.

One approach is to use two single-tuned filters with identical characteristics to form a double band-pass filter (Fig. 9.5d). In practice, there is more than one dominant low-order harmonic and combinations of three or more single-tuned filters are applied. Starting with the lowest harmonic order (e.g., 5th), each filter is tuned to one harmonic frequency. This approach is not practical when the number of domi-

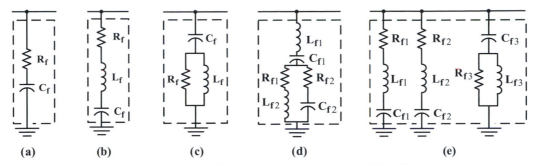

**FIGURE 9.5** Common types of passive filters; (a) first-order damped high-pass filter, (b) second-order series resonant band-pass filter, (c) second-order damped high-pass filter, (d) fourth-order double band-pass filter, (e) composite filter consisting of two band-pass filters and one high-pass filter.

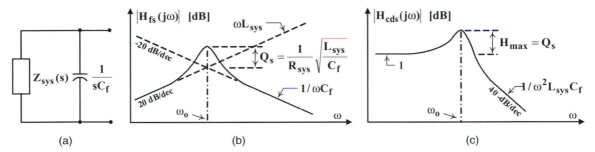

**FIGURE 9.6** First-order high-pass filter; (a) equivalent circuit, (b) filter system impedance transfer function, (c) current-divider transfer function.

nant low-order harmonics is large and/or higher order harmonics are present.

Another approach is to use a first-order high-pass filter (consisting of a resistor connected in series with a capacitor) or a second-order high-pass filter (consisting of a series capacitor and a parallel combination of inductor and resistor; Fig. 9.5c) and set the resonant frequency below the lowest order dominant harmonic frequency. However, the $Z - \omega$ plot of a second-order high-pass filter shows that the minimum impedance of this filter (e.g., in the pass-band region) is higher than that of a single-tuned band-pass filter. Therefore, the high-pass filter allows a percentage of all harmonics above its notch frequency to pass through and will result in large fundamental filter rating and high resistive losses. This filter is commonly applied for higher frequencies and notch reduction.

The third and most implemented approach is to install a composite filter consisting of two or more branches of band-pass filters tuned at low-order harmonic frequencies and a parallel branch of high-pass filter tuned at higher frequencies (called a passive filter element; Figs. 9.5e, 9.11, and 9.12). This con-

figuration is suitable for most residential, domestic, and industrial nonlinear loads such as arc furnaces. In addition, there are other configurations of higher order filters (e.g., third-order damped filter) for transmission systems.

This section derives the equations and general relations of basic filter types.

### 9.3.2.1 First-Order, High-Pass Filter

A first-order, high-pass filter consists of a shunt capacitor (Fig. 9.6a) and can be used to attenuate high-frequency harmonic current components causing telephone interference, to reduce voltage notching (e.g., caused by rectifier commutation), and to improve the displacement (fundamental) load power factor. Such filters are amply described in currently available circuit textbooks [12, 13].

If the system impedance is purely inductive $(Z_{sys}(s) = sL_{sys})$, filter impedance and filter system impedance transfer functions can be derived using Eqs. 9-1:

$$H_f(s) = Z_f(s) = \frac{V_f(s)}{I_f(s)} = \frac{1}{sC_f}, \qquad (9\text{-}4a)$$

$$H_{fsys}(s) = Z_{fsys}(s) = \frac{V_{sys}(s)}{I_f(s) + I_{sys}(s)} = \frac{1}{\frac{1}{Z_f(s)} + \frac{1}{Z_{sys}(s)}}$$

$$= \frac{1}{sC_f + \frac{1}{sL_{sys}}}.$$

(9-4b)

The order of this filter, that is, the highest exponent of the characteristic (denominator) polynomial of $H_f(s)$, is one.

The current-divider transfer function, Eq. 9-2a, can be derived as

$$H_{cds}(s) = H_{cds}(j\omega) = \frac{\frac{1}{j\omega C_f}}{j\omega L_{sys} + \frac{1}{j\omega C_f}}.$$

(9-5)

Simple algebra yields analytical (closed-form) expressions for asymptotes, local maxima, and minima of these transfer functions as is illustrated in Fig. 9.6b,c. The procedures for plotting $H_{cds}(j\omega)$ on a logarithm paper as shown in Fig. 9.6c are

- For low frequencies
  $(\omega \ll 1/\sqrt{L_{sys}C_f} \Rightarrow |1/j\omega C_f| \gg |j\omega L_{sys}|)$:

$$H_{cds}(j\omega) \approx \frac{\frac{1}{j\omega C_f}}{\frac{1}{j\omega C_f}} = 1.$$

Therefore, the roll-off of the low-frequency components is one or 0 dB per decade.

- For high frequencies
  $(\omega \gg 1/\sqrt{L_{sys}C_f} \Rightarrow |1/j\omega C_f| \ll |j\omega L_{sys}|)$:

$$H_{cds}(j\omega) \approx \frac{\frac{1}{j\omega C_f}}{j\omega L_{sys}} = \frac{-1}{\omega^2 L_{sys}C_f}.$$

Therefore, the roll-off of the high frequency components is $1/\omega^2$ or $-40$ dB per decade.

If system resistance is included ($Z_{sys}(s) = R_{sys} + sL_{sys}$), the maximum can be found as

$$H_{cds}(s = j\omega)|_{\omega = \frac{1}{\sqrt{L_{sys}C_f}}} = -j\frac{1}{R_{sys}}\sqrt{\frac{L_{sys}}{C_f}}$$

and

$$H_{max} = |H_{cds}(s = j\omega)|_{\omega = \frac{1}{\sqrt{L_{sys}C_f}}} \quad \text{with}$$

$$Q = \frac{1}{R_{sys}}\sqrt{\frac{L_{sys}}{C_f}}, \text{ therefore, } H_{max} = Q.$$

Similar procedures are relied on to plot $H_{fs}(j\omega)$ as shown in Fig. 9.6b.

### 9.3.2.2 First-Order Damped High-Pass Filter

Figure 9.7a shows the configuration of a first-order damped high-pass filter. A series connected resistance ($R_f$) is included to provide a damping characteristic.

Filter transfer functions are

$$H_f(s) = Z_f(s) = \frac{V_f(s)}{I_f(s)} = R_f + \frac{1}{sC_f}.$$

(9-6a)

$$H_{fsys}(s) = Z_{fsys}(s) = \frac{1}{\frac{1}{R_f + 1/sC_f} + \frac{1}{sL_{sys}}},$$

(9-6b)

$$H_{cds}(s) = H_{cds}(j\omega) = \frac{R_f + \frac{1}{j\omega C_f}}{R_f + \frac{1}{j\omega C_f} + j\omega L_{sys}}.$$

(9-6c)

As discussed in the previous section, $H_{cds}(s)$ can be determined analytically (Fig. 9.7c):

- For low frequencies
  $(\omega \ll 1/\sqrt{L_{sys}C_f} \Rightarrow |1/j\omega C_f| \gg |j\omega L_{sys}|)$:

$$H_{cds}(j\omega) \approx \frac{\frac{1}{j\omega C_f}}{\frac{1}{j\omega C_f}} = 1.$$

- For moderate high frequencies
  $(1/\sqrt{L_{sys}C_f} < \omega < 1/R_f C_f \Rightarrow |1/j\omega C_f| \ll |j\omega L_{sys}|)$:

$$H_{cds}(j\omega) \approx \frac{\frac{1}{j\omega C_f}}{j\omega L_{sys}} = \frac{-1}{\omega^2 L_{sys}C_f}$$

$\Rightarrow$ roll-off is $1/\omega^2$ or $-40$ dB per decade.

- For high frequencies ($\omega \gg 1/R_f C_f \Rightarrow R_f \gg |1/j\omega C_f|$ and $|j\omega L_{sys}| \gg |R_f + 1/j\omega C_f|$):

$$H_{cds}(j\omega) = \frac{R_f + \frac{1}{j\omega C_f}}{j\omega L_{sys} + \left(R_f + \frac{1}{j\omega C_f}\right)} \approx \frac{R_f}{j\omega L_{sys}}$$

$\Rightarrow$ roll-off is $1/\omega$ or $-20$ dB per decade.

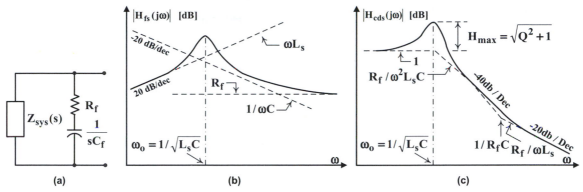

**FIGURE 9.7** First-order damped high-pass filter; (a) equivalent circuit, (b) filter impedance transfer function, (c) current-divider transfer function.

- If the system resistance is included $(Z_{sys}(s) = R_{sys} + sL_{sys})$, the maximum can be found as

$$H_{cds}(s = j\omega)\Big|_{\omega=\frac{1}{\sqrt{L_{sys}C_f}}} = 1 - j\frac{1}{R_{sys}}\sqrt{\frac{L_{sys}}{C_f}}$$

and

$$H_{max} = |H_{cds}(s = j\omega)|_{\omega=\frac{1}{\sqrt{L_{sys}C_f}}} = \sqrt{Q^2+1} \approx Q$$

with $Q = \frac{1}{R_{sys}}\sqrt{\frac{L_{sys}}{C_f}}$.

Similar procedures are employed to plot $H_{fs}(j\omega)$ as illustrated in Fig. 9.7b. Comparison of Figs. 9.6 and 9.7 indicates the following:

- the damping resistance significantly limits the high-frequency performance (e.g., roll-off of high-frequency components is only $1/\omega$ or $-20$ dB per decade). Therefore, the damped filter is less desirable for telephone interference reduction applications; and
- for parallel resonant conditions (e.g., cancellation of $L_{sys}$ by $C_f$), if the resonant frequency falls on or near a critical harmonic frequency, a high-pass damping resistance ($R_f$) can be used to control and reduce the amplification. However, this will increase the fundamental frequency power loss and reduces the effectiveness of the high-pass attenuation above the frequency $1/(R_fC_f)$.

### 9.3.2.3 Second-Order Band-Pass Filter

The second-order band-pass filter (also called single-tuned or series-resonant filter) is a series combination of capacitor ($C_f$), inductor ($L_f$), and a small damping resistor ($R_f$), as depicted in Fig. 9.8a. The damping resistance is usually due to the internal resistance of either the inductor and/or capacitor. This filter is tuned to attenuate one single low-order harmonic. Typically, combinations of band-pass filters are used because the power system contains a number of dominant low-order harmonics.

Assuming a pure inductive system ($Z_{sys}(s) = sL_{sys}$), filter transfer functions can be derived using Eqs. 9-1a and 9-2a:

$$H_f(s) = Z_f(s) = \frac{V_f(s)}{I_f(s)} = R_f + sL_f + \frac{1}{sC_f}$$
$$= \frac{A}{s}\left[1 + \frac{1}{Q}\left(\frac{s}{\omega_0}\right) + \left(\frac{s}{\omega_0}\right)^2\right], \quad (9\text{-}7)$$

$$H_{cds}(s) = H_{cds}(j\omega) = \frac{R_f + j\omega L_f + \frac{1}{j\omega C_f}}{R_f + j\omega L_f + \frac{1}{j\omega C_f} + j\omega L_{sys}}. \quad (9\text{-}8)$$

where A is the gain. The filter quality factor (Q) and the series-resonant frequency ($\omega_0$) are

$$Q = \frac{1}{R_f}\sqrt{\frac{L_f}{C_f}}, \quad (9\text{-}9a)$$

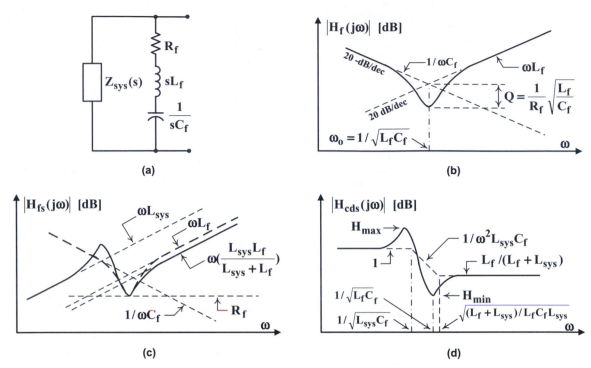

**FIGURE 9.8** Second-order band-pass filter; (a) equivalent circuit, (b) filter impedance transfer function, (c) filter system impedance transfer function with $L_{sys} > L_f$, (d) current-divider transfer function with $L_{sys} > L_f$.

$$\omega_0 = \frac{1}{\sqrt{L_f C_f}}. \qquad (9\text{-}9b)$$

For high-voltage applications, the filter current has a low rms magnitude and does not require large current carrying conductors; therefore, air-core inductors are regularly used and the filter quality factor is relatively large (e.g., $50 < Q < 150$). For low-voltage applications, gapped iron-core inductors and conductors with a large cross section are employed (e.g., $10 < Q < 50$).

The filter impedance transfer function ($H_f(j\omega)$) is presented graphically as shown in Fig. 9.8b. It is evaluated at low and high frequencies to determine its asymptotes:

- At low frequencies ($\omega \ll \omega_0 \Rightarrow |1/j\omega C_f| \gg |R + j\omega L_{sys}|$), the filter is dominantly capacitive and provides reactive power to the system. Therefore, $H_f(s) \approx 1/j\omega C_f$ and the roll-off of the low frequency components is –20 dB per decade.
- At high frequencies ($\omega \gg \omega_0 \Rightarrow |j\omega L_{sys}| \gg |R + 1/j\omega C_f|$), the filter is dominantly inductive and consumes reactive power with little influence on high-frequency distortions. Therefore, $H_f(s) \approx j\omega L_f$ and the roll-off of the high frequency components is 20 dB per decade.

- At the resonant frequency ($\omega = \omega_0$), the filter is entirely resistive, that is, $H_f(s) = R_f$ More attenuation is provided with lower filter resistances; however, there are practical limits to the value of $R_f$.

Similar procedures are applied to plot $H_{fs}(j\omega)$ and $H_{cds}(j\omega)$ as illustrated in Fig. 9.8c,d, respectively. Note that the minima of the transfer functions ($H_{min}$) are very close to the series resonant frequency; however, due to the interaction between the capacitive reactance and the filter inductive reactance, $H_{max}$ does not occur at the parallel resonant frequency formed by filter capacitance and system inductance. The derivatives of the transfer functions are usually evaluated numerically to determine frequencies where maxima occur.

The second-order band-pass (series resonant) filters are very popular and have many applications in power system design including transmission (e.g., connected in parallel at the AC terminals of HVDC converters), distribution (e.g., for detuning power factor capacitor banks to avoid parallel resonant frequencies), and utilization (e.g., across terminals of nonlinear loads).

### 9.3.2.4 Second-Order Damped Band-Pass Filter

A popular topology of the second-order damped filter is shown in Fig. 9.9a where a bypass resistance ($R_{bp}$) is included across the inductor to provide attenuation for harmonic frequency components over a wide frequency range. The filter impedance transfer function can be expressed in normalized form

$$H_f(s) = Z_f(s) = \frac{V_f(s)}{I_f(s)} = \frac{1}{\dfrac{1}{R_{bp}} + \dfrac{1}{R_f + sL_f}} + \frac{1}{sC_f}$$

$$= \frac{A}{s\left(1 + \dfrac{s}{\omega_p}\right)}\left[1 + \frac{1}{Q_{bp}}\left(\frac{s}{\omega_0}\right) + \left(\frac{s}{\omega_0}\right)^2\right],$$

(9-10a)

where the gain A, the series resonant angular frequency $\omega_0$, the quality factor $Q_{bp}$, and the pole frequency [13] for typical cases (whereby $R_f \ll R_{bp}$) are

$$A = \frac{1}{C_f}, \quad \omega_0 = \sqrt{\frac{R_f + R_{bp}}{R_{bp}L_fC_f}} \approx \frac{1}{\sqrt{L_fC_f}},$$

$$Q_{bp} = \frac{R_f + R_{bp}}{\omega_0(R_{bp}L_fC_f + L_f)}, \quad \omega_p = \frac{R_f + R_{bp}}{L_f} \approx \frac{R_{bp}}{L_f}.$$

(9-10b)

The series resistance $R_f$ is determined by the desired quality factor of the second-order band-pass filter and $R_{bp}$ is selected to achieve the required high-pass response and series resonant attenuation. Typical values of the bypass quality factor are $0.5 < Q_{bp} < 2.0$. However, there is a trade-off between larger values (e.g., more series resonant attenuation and less high-pass response) and smaller values (e.g., less series resonant attenuation and greater high-pass response) of the quality factor.

### 9.3.2.5 Composite Filter

Higher order filters are constructed by increasing the number of storage elements (e.g., capacitors, induc-

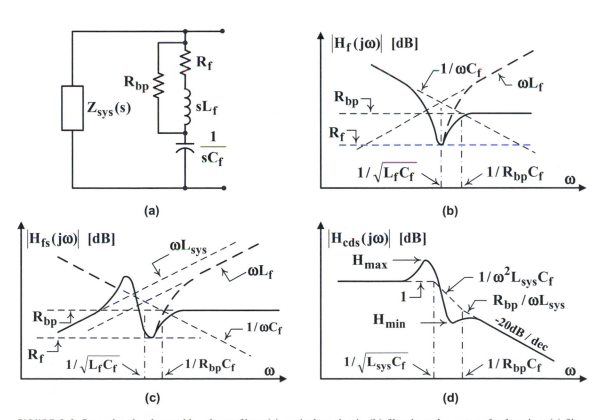

**FIGURE 9.9** Second-order damped band-pass filter; (a) equivalent circuit, (b) filter impedance transfer function, (c) filter system impedance transfer function, (d) current-divider transfer function.

tors). However, their application in power systems is limited due to economic and reliability factors.

A third-order filter may be constructed by adding a series capacitance $C_{f2}$ to the inductor bypass resistance (Fig. 9.10a) to limit $I_{bp}$ and reduce the corresponding fundamental frequency losses or by including $C_{f2}$ in series with $L_f$ (Fig. 9.10b) and sizing it to form a series resonant branch at the fundamental frequency to reduce $I_{bp}$ and to increase filter efficiency. If a second inductor $L_{f2}$ is added to the circuit of Fig. 9.10a, a fourth-order double band-pass filter will be obtained, as shown in Fig. 9.10c.

A common type of an nth-order composite filter for power quality improvement is shown in Fig. 9.11. Several band-pass filters are connected in parallel and individually tuned to selected harmonic frequencies to provide compensation over a wide frequency range. The last branch is a high-pass filter attenuating high-order harmonics, which usually are a result of fast switching actions. Composite filters are only applied when even-order harmonics are small since a parallel resonance will occur between any two adjacent band-pass filter branches and cause amplification of the distortion in that frequency range. For example, composite filter systems with shunt branches tuned at the 5th and 7th harmonics will have a resonant frequency about at the 6th harmonic. The impedance transfer function of the nth-order composite filter is

$$H_f(s) = Z_f(s) = \frac{V_f(s)}{I_f(s)}$$

$$= \frac{1}{\dfrac{1}{Z_f^{(5)}} + \dfrac{1}{Z_f^{(7)}} + \dfrac{1}{Z_f^{(11)}} + \dots + \dfrac{1}{Z_f^{(n)}}}. \quad (9\text{-}11)$$

It is time-consuming to derive the transfer functions of multiple-order filters in terms of factorized expressions of poles and zeros [13]. Therefore, numerical approaches are typically applied to plot the transfer function, and an iterative design procedure is used to optimize a filter configuration.

### 9.3.3 Classification of Passive Power Filters

Passive filters cause harmonic currents to flow at their resonant frequency. Due to this resonance, the harmonic currents are attenuated in the power system through the LC filter circuits tuned to the harmonic orders that require filtering. In practice, a single passive filter is not adequate since there is more than one harmonic to be attenuated; therefore, a shunt or series passive composite filter unit (Fig. 9.12) is used. Passive filter units are generally designed to remove a few low-order harmonics, for example, the 5th, 7th, and 11th. Normally passive filter units employ three (or more for high-power applications such as HVDC) tuned filters, the first two being for the lowest dominant harmonics followed by a high-pass filter. In a shunt passive filter unit, two lossless LC components are connected in series to sink harmonic currents and all three (or more) filters are connected in parallel (Fig. 9.12a). However, in a series passive filter unit, two lossless LC components are connected in parallel for creating a harmonic dam and all three (or more) filters are connected in series, as shown in Fig. 9.12b.

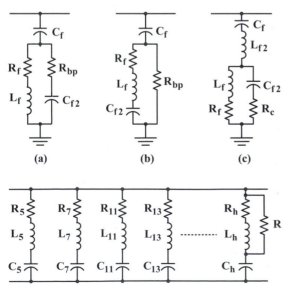

**(a)**    **(b)**    **(c)**

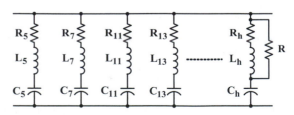

**FIGURE 9.10** Higher order filters; (a) and (b) third-order filters, (c) fourth-order filter.

**FIGURE 9.11** Nth-order composite filter.

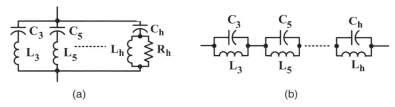

(a)             (b)

**FIGURE 9.12** Typical passive filter unit employing a number of tuned filter units, the first $(h-1)$ units are tuned to the lowest dominant harmonic frequencies followed by a high-pass element; (a) shunt passive filter block, (b) series passive filter block.

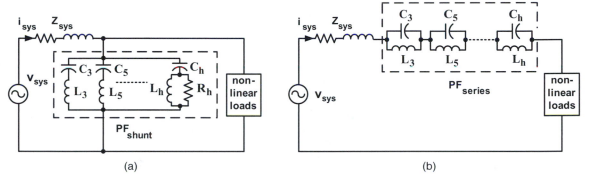

(a)             (b)

**FIGURE 9.13** Single-phase passive filter; (a) shunt configuration, (b) series configuration.

The inductors should have high quality factors to reduce system losses and must not saturate in the entire current operating region. Capacitor values are decided by the required reactive power of the system, and then inductor values are calculated by tuning them for the particular harmonic frequencies. Resistor values are calculated by the corresponding quality factors based on the desired sharpness of the characteristics. Similarly, the resistance of the high-pass filter is computed based on the sharpness or attenuation of the higher order harmonics on the one hand, and reducing the losses on the other hand to achieve an optimum value for the quality factor. Passive filter units with shunt and series configurations are extensively used in single-phase (Fig. 9.13), three-phase three-wire (Fig. 9.14), and three-phase four-wire (Fig. 9.15) systems to improve the quality of power and mitigate harmonic currents and voltages.

### 9.3.4 Potentials and Limitations of Passive Power Filters

Design, construction, and implementation of passive filters are straightforward and easier than their counterparts, for example, synchronous condensers and active filters. They have many advantages:

- they are robust, economical, and relatively inexpensive;

- they can be implemented in large sizes of MVArs and require low maintenance;
- they can provide a fast response time (e.g., one cycle or less with SVC), which is essential for some power quality problems such as flickering voltage dips due to arc-furnace loads;
- unlike rotating machines (e.g., synchronous motors or condensers) passive filters do not contribute to the short-circuit currents; and
- in addition to power quality improvement, well-designed passive filters can be configured to provide power factor correction, reactive power compensation, or voltage support (e.g., on critical buses in case of a source outage) and reduce the starting impact and associated voltage drops due to large loads (e.g., large induction motors).

These advantages have resulted in broad applications of passive filters in different sectors of power systems. However, power engineers have also encountered some limitations and shortcomings of passive filters, which have called for alternative solutions such as active and hybrid filter systems. Some limitations of passive filters are as follows:

- fixed compensation or harmonic mitigation, large size, and possible resonance with the supply system impedance at fundamental and/or other harmonic frequencies;

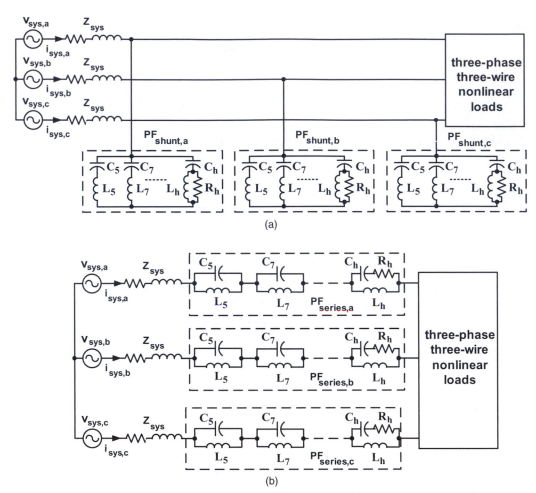

**FIGURE 9.14** Three-phase three-wire passive filter; (a) shunt configuration, (b) series configuration [4].

- compensation is limited to a few orders of harmonics;
- once installed, the tuned frequency and/or filter size cannot be changed easily;
- unsatisfactory performance (e.g., detuning) may occur due to variations of filter parameters (caused by aging, deterioration, and temperature effects) and nonlinear load characteristics;
- change in system operating conditions (e.g., inclusion of capacitors and/or other filters) has large influence on filter designs. Therefore, to be effective, filter impedance should be less than system impedance, which can become a problem for strong or stiff systems [2];
- resonance between system and filter (e.g., shifted resonance frequency) for single- and double-

tuned filters can cause an amplification of characteristics and noncharacteristic harmonic currents. Therefore, the designer has a limited choice in selecting tuned frequencies and ensuring adequate bandwidth between shifted frequencies and integer (even and odd) harmonics [2];
- special protective and monitoring devices might be required to control switching surges, although filter reactors will reduce the magnitude of the switching inrush current magnitude and its associated high frequency; and
- stepless control (of reactive power, power factor correction, and voltage support) is not possible since the filter can either be switched on or switched off only.

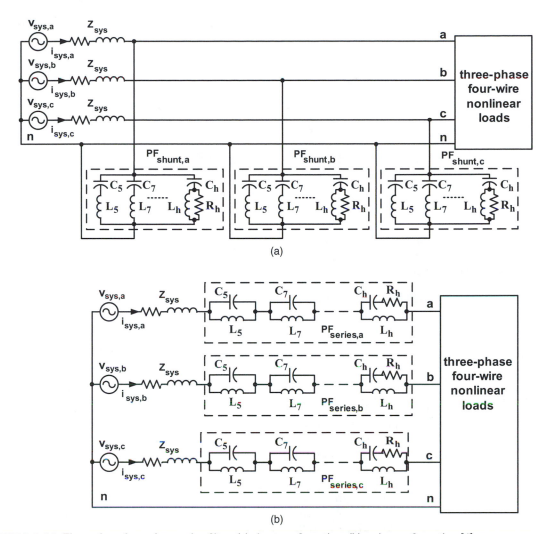

**FIGURE 9.15** Three-phase four-wire passive filter; (a) shunt configuration, (b) series configuration [4].

### 9.3.5 Application Example 9.1: Hybrid Passive Filter Design to Improve the Power Quality of the IEEE 30-Bus Distribution System Serving Adjustable-Speed Drives

Consider the IEEE 23 kV, 30-bus distribution system [9, 10] shown in Fig. E9.1.1. System transmission-line data, bus data, and capacitor data are provided in Tables E9.1.1 to E9.1.3, respectively. To investigate the impact of large AC drive systems on the power quality of the distribution system, a PWM adjustable-speed drive (675 kW, 439 kVAr) and a variable-frequency drive (350 kW, 175 kVAr) are connected to buses 15 and 18, respectively. Table E9.1.4 depicts the spectra of the current harmonics injected by typical AC drives into the system. Use harmonic current sources to model the two nonlinear loads and utilize the decoupled harmonic power flow algorithm (Chapter 7, Section 7.5.1) to simulate the system. Determine the total harmonic voltage distortion ($THD_v$) of the system and at the individual buses before and after the installation of the hybrid passive filter banks listed below. Plot the voltage and current waveforms at bus 15 before and after filtering.

a) Two filter banks are placed at the terminals of nonlinear loads (buses 15 and 18). Each bank consists of a number of shunt-connected series

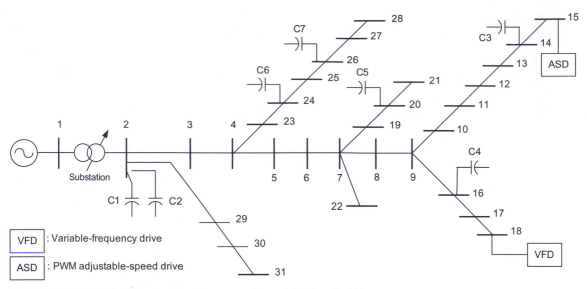

**FIGURE E9.1.1** IEEE 30-bus system with two nonlinear AC drives [9, 10].

**TABLE E9.1.1** Line Data for the IEEE 30-Bus System (Base kV = 23 kV, Base MVA = 100 MVA)

| From bus | To bus | R (ohm) | X (ohm) | R (pu) | X (pu) |
|---|---|---|---|---|---|
| 1 | 2 | 0.0021 | 0.0365 | 0.0004 | 0.0069 |
| 2 | 3 | 0.2788 | 0.0148 | 0.0527 | 0.0028 |
| 3 | 4 | 0.4438 | 0.4391 | 0.0839 | 0.0830 |
| 4 | 5 | 0.8639 | 0.7512 | 0.1633 | 0.1420 |
| 5 | 6 | 0.8639 | 0.7512 | 0.1633 | 0.1420 |
| 6 | 7 | 1.3738 | 0.7739 | 0.2597 | 0.1463 |
| 7 | 8 | 1.3738 | 0.7739 | 0.2597 | 0.1463 |
| 8 | 9 | 1.3738 | 0.7739 | 0.2597 | 0.1463 |
| 9 | 10 | 1.3738 | 0.7739 | 0.2597 | 0.1463 |
| 10 | 11 | 1.3738 | 0.7739 | 0.2597 | 0.1463 |
| 11 | 12 | 1.3738 | 0.7739 | 0.2597 | 0.1463 |
| 12 | 13 | 1.3738 | 0.7739 | 0.2597 | 0.1463 |
| 13 | 14 | 1.3738 | 0.7739 | 0.2597 | 0.1463 |
| 14 | 15 | 1.3738 | 0.7739 | 0.2597 | 0.1463 |
| 9 | 16 | 0.8639 | 0.7512 | 0.1633 | 0.1420 |
| 16 | 17 | 1.3738 | 0.7739 | 0.2597 | 0.1463 |
| 17 | 18 | 1.3738 | 0.7739 | 0.2597 | 0.1463 |
| 7 | 19 | 0.8639 | 0.7512 | 0.1633 | 0.1420 |
| 19 | 20 | 0.8639 | 0.7512 | 0.1633 | 0.1420 |
| 20 | 21 | 1.3738 | 0.7739 | 0.2597 | 0.1463 |
| 7 | 22 | 0.8639 | 0.7512 | 0.1633 | 0.1420 |
| 4 | 23 | 0.4438 | 0.4391 | 0.0839 | 0.0830 |
| 23 | 24 | 0.4438 | 0.4391 | 0.0839 | 0.0830 |
| 24 | 25 | 0.8639 | 0.7512 | 0.1633 | 0.1420 |
| 25 | 26 | 0.8639 | 0.7512 | 0.1633 | 0.1420 |
| 26 | 27 | 0.8639 | 0.7512 | 0.1633 | 0.1420 |
| 27 | 28 | 1.3738 | 0.7739 | 0.2597 | 0.1463 |
| 2 | 29 | 0.2788 | 0.0148 | 0.0527 | 0.0028 |
| 29 | 30 | 0.2788 | 0.0148 | 0.0527 | 0.0028 |
| 30 | 31 | 1.3738 | 0.7739 | 0.2597 | 0.1463 |

**TABLE E9.1.2** Bus Data for the IEEE 30-Bus System (Base kV = 23 kV, Base MVA = 100 MVA)

| Bus number | P (kW) | Q (kVAr) | P (pu) | Q (pu) |
|---|---|---|---|---|
| 1[a] | 0 | 0 | 0 | 0 |
| 2 | 52.2 | 17.2 | 0.522 | 0.172 |
| 3 | 0 | 0 | 0 | 0 |
| 4 | 0 | 0 | 0 | 0 |
| 5 | 93.8 | 30.8 | 0.938 | 0.308 |
| 6 | 0 | 0 | 0 | 0 |
| 7 | 0 | 0 | 0 | 0 |
| 8 | 0 | 0 | 0 | 0 |
| 9 | 0 | 0 | 0 | 0 |
| 10 | 18.9 | 6.2 | 0.189 | 0.062 |
| 11 | 0 | 0 | 0 | 0 |
| 12 | 33.6 | 11 | 0.336 | 0.11 |
| 13 | 65.8 | 21.6 | 0.658 | 0.216 |
| 14 | 78.4 | 25.8 | 0.784 | 0.258 |
| 15 | 73 | 24 | 0.73 | 0.24 |
| 16 | 47.8 | 15.7 | 0.478 | 0.157 |
| 17 | 55 | 18.1 | 0.55 | 0.181 |
| 18 | 47.8 | 15.7 | 0.478 | 0.157 |
| 19 | 43.2 | 14.2 | 0.432 | 0.142 |
| 20 | 67.3 | 22.1 | 0.673 | 0.221 |
| 21 | 49.6 | 16.3 | 0.496 | 0.163 |
| 22 | 20.7 | 6.8 | 0.207 | 0.068 |
| 23 | 52.2 | 17.2 | 0.522 | 0.172 |
| 24 | 192 | 63.1 | 1.92 | 0.631 |
| 25 | 0 | 0 | 0 | 0 |
| 26 | 111.7 | 36.7 | 1.117 | 0.367 |
| 27 | 55 | 18.1 | 0.55 | 0.181 |
| 28 | 79.3 | 26.1 | 0.793 | 0.261 |
| 29 | 88.3 | 29 | 0.883 | 0.29 |
| 30 | 0 | 0 | 0 | 0 |
| 31 | 88.4 | 29 | 0.884 | 0.29 |

[a]Swing bus.

**TABLE E9.1.3** Capacitor Data for the IEEE 30-Bus System

| Capacitor | C1 | C2 | C3 | C4 | C5 | C6 | C7 |
|---|---|---|---|---|---|---|---|
| Bus location | 2 | 2 | 14 | 16 | 20 | 24 | 26 |
| kVAr | 900 | 600 | 600 | 600 | 300 | 900 | 900 |

**TABLE E9.1.4** Typical Harmonic Spectrum of Variable-Frequency and PWM Adjustable-Speed Drives [14]

| | Variable-frequency drive | | PWM adjustable-speed drive | |
|---|---|---|---|---|
| Harmonic order h | Magnitude (%) | Phase angle (degree) | Magnitude (%) | Phase angle (degree) |
| 1 | 100 | 0 | 100 | 0 |
| 5 | 23.52 | 111 | 82.8 | −135 |
| 7 | 6.08 | 109 | 77.5 | 69 |
| 11 | 4.57 | −158 | 46.3 | −62 |
| 13 | 4.2 | −178 | 41.2 | 139 |
| 17 | 1.8 | −94 | 14.2 | 9 |
| 19 | 1.37 | −92 | 9.7 | −155 |
| 23 | 0.75 | −70 | 1.5 | −158 |
| 25 | 0.56 | −70 | 2.5 | 98 |
| 29 | 0.49 | −20 | 0 | 0 |
| 31 | 0.54 | 7 | 0 | 0 |

resonance passive filters tuned at the respective dominant harmonic frequencies. Assume $R_f = 100\ \Omega$ and $L_f = 100\ mH$ for all filters and compute the corresponding values of $C_f$.

b) Repeat part a with one filter bank at bus 15.
c) Repeat part b with one filter bank at bus 9.
d) Compare the results of parts a to c and make comments regarding the impact of numbers and locations of filters on the power quality of the system.

## 9.4 ACTIVE FILTERS

Active filters (AFs) are feasible alternatives to passive filters (PFs). For applications where the system configuration and/or the harmonic spectra of nonlinear loads (e.g., orders, magnitudes, and phase angles) change, active elements may be used instead of the passive components to provide dynamic compensation.

An active filter is implemented when the order numbers of harmonic currents are varying. This may be due to the nature of nonlinear loads injecting time-dependent harmonic spectra (e.g., variable-speed drives) or may be caused by a change in the system configuration. The structure of an active filter may be that of series or parallel architectures. The

proper structure for implementation depends on the types of harmonic sources in the power system and the effects that different filter solutions would cause to the overall system performance. Active filters rely on active power conditioning to compensate undesirable harmonic currents replacing a portion of the distorted current wave stemming from the nonlinear load. This is achieved by producing harmonic components of equal amplitude but opposite phase angles, which cancel the injected harmonic components of the nonlinear loads. The main advantage of active filters over passive ones is their fine response to changing loads and harmonic variations. In addition, a single active filter can compensate more than one harmonic, and improve or mitigate other power quality problems such as flicker.

Active filters are expensive compared with their passive counterparts and are not feasible for small facilities. The main drawback of active filters is that their rating is sometimes very close to the load (up to 80% in some typical applications), and thus it becomes a costly option for power quality improvement in a number of situations. Moreover, a single active filter might not provide a complete solution in many practical applications due to the presence of both voltage and current quality problems. For such cases, a more complicated filter design consisting of two or three passive and/or active filters (called a hybrid filter) is recommended as discussed in Section 9.5.

### 9.4.1 Classification of Active Power Filters Based on Topology and Supply System

Active filters have become a mature technology for harmonic and reactive power compensation of single- and three-phase electric AC power networks with high penetration of nonlinear loads. This section classifies active filters based on the supply system taking into consideration their (shunt or/and series) topology. Therefore, there are six types of shunt and series connected active filters for single-phase two-wire, three-phase three-wire, and three-phase four-wire AC networks, as shown in Figs. 9.16, 9.17, and 9.18, respectively. Composite filters consisting of two or more active filters are realized as hybrid filters and are discussed in Section 9.5.

### 9.4.2 Classification of Active Power Filters Based on Power Rating

Active filters can also be classified according to their power rating and speed of response required for their application in a compensated system [5]:

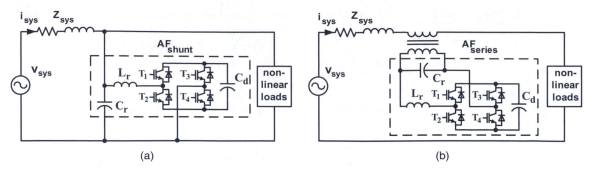

**FIGURE 9.16** Single-phase active filter (AF); (a) shunt configuration, (b) series configuration [4].

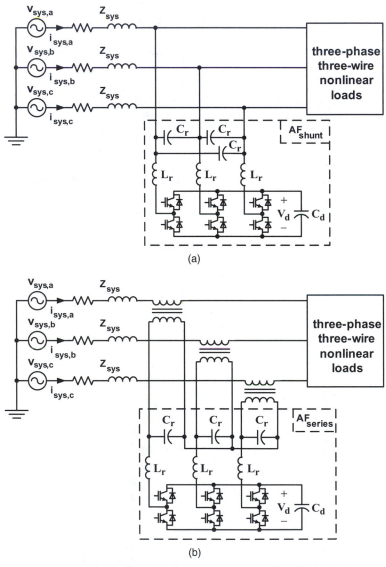

**FIGURE 9.17** Three-phase three-wire active filter; (a) shunt configuration, (b) series configuration.

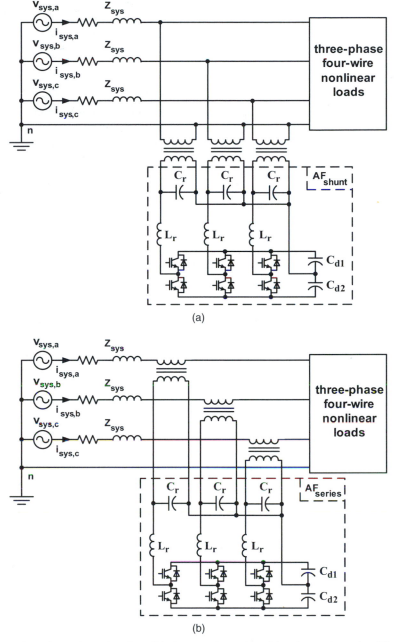

**FIGURE 9.18** Three-phase four-wire active filter; (a) shunt configuration, (b) series configuration [4].

- *Low-power active filters* (with power ratings below 100 kVA and response times of about 10 μs to 10 ms) are mainly used in residential areas, small to medium-sized factories, electric drives, commercial buildings, and hospitals. These applications require sophisticated dynamic filtering techniques and include single-phase and three-phase systems.
- Single-phase active filters are mainly available for low power ratings, and hence can be operated at relatively high frequencies leading to improved performance and lower prices. They have many retrofit applications such as commercial and educational buildings with many computer loads.
- For balanced three-phase low power applications a three-phase filter can be used. However, for unbalanced load currents or unsymmetrical supply voltages, especially in three-phase four-wire distribution systems, three single-phase filters or a hybrid filter are usually required.

- *Medium-power active filters* (with power ratings of about 100 kVA to 10 MVA and response times of about 100 ms to 1 s) are mainly associated with medium- to high-voltage distribution systems and high-power, high-voltage drive systems, where the effect of phase imbalance is negligible. Due to economic concerns and problems associated with high-voltage systems (isolation, series or parallel connections of switches, etc.), these filters are usually designed to perform harmonic cancellation, and reactive-power compensation is not included in their control algorithms. At high voltage, other approaches of reactive power compensation such as capacitive or inductive static compensators, synchronous condensers, controlled LC filters, tuneable harmonic filters, line-commutated thyristor converters, and VAr compensators should be considered.

- *High-power active filters* (with power ratings above 10 MVA and response times of tens of seconds) are mainly associated with power-transmission grids, ultrahigh-power DC drives, and HVDC. Due to the availability of high switching-frequency power devices with high-voltage and high-power ratings, these filters are not cost effective. It is possible to use series–parallel combinations of low-power switches; however, there are many implementation issues and cost considerations. Fortunately, there are not too many power quality problems associated with high-voltage power systems.

## 9.5 HYBRID POWER FILTERS

A major drawback of active filters is their high rating (e.g., up to 80% of the nonlinear load in some practical applications). In addition, a single active filter cannot offer a complete solution for the simultaneous compensation of both voltage and current power quality disturbances. Due to higher ratings and cost considerations, the acceptability of active filters has been limited in practical applications. In response to these factors, different structures of hybrid filters

have evolved as a cost-effective solution for the compensation of nonlinear loads. Hybrid filters are found to be more effective in providing complete compensation of various types of nonlinear loads.

### 9.5.1 Classification of Hybrid Filters

Hybrid filters combine a number of passive and/or active filters and their structure may be of series or parallel topology or a combination of the two. They can be installed in single-phase, three-phase three-wire, and three-phase four-wire distorted systems. The passive circuit performs basic filtering action at the dominant harmonic frequencies (e.g., 5th or 7th) whereas the active elements, through precise control, mitigate higher harmonics. This will effectively reduce the overall size and cost of active filtering.

In the literature, there are different classifications of active and hybrid filters based on power rating, supply system (e.g., number of wires and phases), topology (e.g., shunt and/or series connection), number of (passive and active) elements, speed of response, power circuit configuration, system parameter(s) to be compensated, control approach, and reference-signal estimation technique. In this book, classification of hybrid filters is based on the supply system with the topology as a further subclassification. If there are a maximum of three (passive and active) filters in each phase, then 156 types of hybrid filters are expected for single-phase two-wire systems, three-phase three-wire, and three-phase four-wire AC networks. Figures 9.19 to 9.25 show the 52 types of hybrid filter topologies for single-phase two-wire systems. These topologies can be easily extended to illustate the other 104 types of hybrid filters for three-phase systems.

Figures 9.19 and 9.20 depict hybrid filters consisting of two and three passive filters, respectively, whereas Figs. 9.21 and 9.22 show similar configurations for two and three active filters. There are 8 topologies of hybrid filters consisting of one passive and one active filter as shown in Fig. 9.23. There are many possible combinations if three filters are

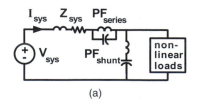
(a)

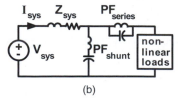
(b)

**FIGURE 9.19** Single-phase hybrid filter (including two passive filters) as a combination of (a) passive-series and passive-shunt filters, (b) passive-shunt and passive-series filters [4].

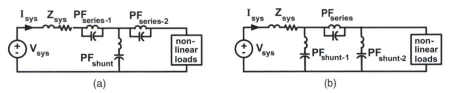

(a)                                                          (b)

**FIGURE 9.20** Single-phase hybrid filter (including three passive filters) as a combination of (a) passive-series, passive-series, and passive-shunt filters, (b) passive-shunt, passive-series, and passive-shunt filters [4].

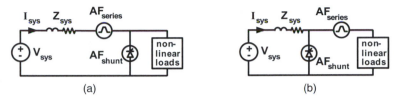

(a)                                                          (b)

**FIGURE 9.21** Single-phase hybrid filter (including two active filters) as a combination of (a) active-series and active-shunt filters, (b) active-shunt and active-series filters [4].

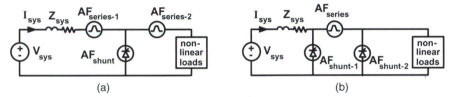

(a)                                                          (b)

**FIGURE 9.22** Single-phase hybrid filter (including three active filters) as a combination of (a) active-series, active-series, and active-shunt filters, (b) active-shunt, active-series, and active-shunt filters [4].

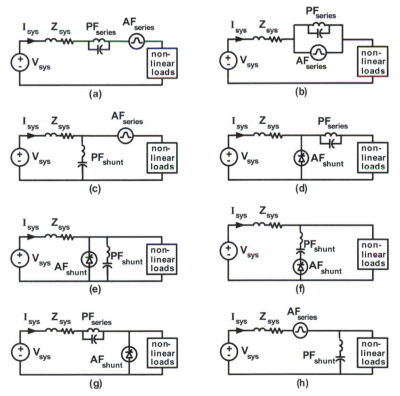

(a)                                                          (b)

(c)                                                          (d)

(e)                                                          (f)

(g)                                                          (h)

**FIGURE 9.23** Single-phase hybrid filter (including one passive and one active filter) as a combination of (a) series-connected passive-series and active-series filters, (b) parallel-connected passive-series and active-series filters, (c) passive-shunt and active-series filters, (d) active-shunt and passive-series filters, (e) active-shunt and passive-shunt filters, (f) series-connected passive-shunt and active-shunt filters, (g) passive-series and active-shunt filters, (h) active-series and passive-shunt filters [4].

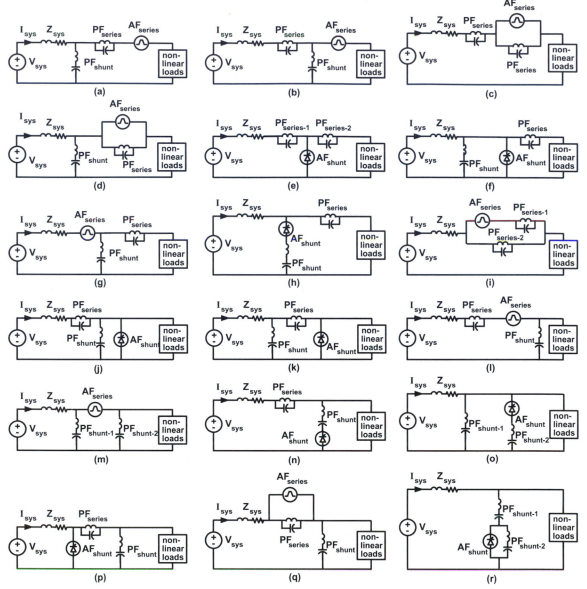

**FIGURE 9.24** Single-phase hybrid filter (including two passive and one active filter) as a combination of (a) passive-shunt, passive–series, and active-series filters, (b) passive-series, passive shunt, and active-series filters, (c) passive-series in series with parallel-connected active-series and passive-series filters, (d) passive-shunt and parallel-connected active-series and passive-series filters, (e) passive-series, active-shunt, and passive-series filters, (f) parallel-connected passive-shunt with active-shunt and passive-series filters, (g) active-series, passive-shunt, and passive-series filters, (h) series-connected passive-shunt with active-series and passive-series filters, (i) series-connected passive-series with active-series in parallel with passive-series filters, (j) passive-series and parallel-connected passive-shunt with active-shunt filters, (k) passive-shunt, passive-series, and active-shunt filters, (l) series-connected passive-series with active-series and passive-shunt filters, (m) passive-shunt, active-series, and passive-shunt filters, (n) passive-series and series-connected passive-shunt with active-shunt filters, (o) passive-shunt and series-connected active-series with passive-shunt filters, (p) combination of active-shunt, passive-series, and passive-shunt filters, (q) parallel-connected active-series with passive-series and passive-shunt filters, (r) passive-shunt and parallel-connected passive-shunt with active-series filters [4].

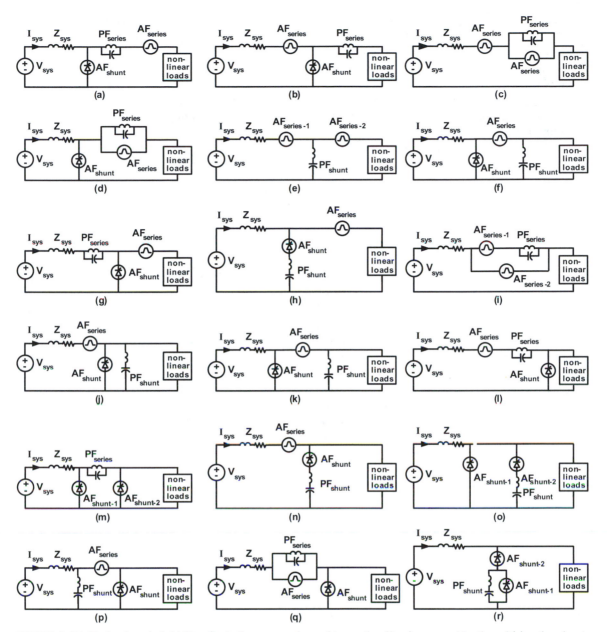

**FIGURE 9.25** Single-phase hybrid filter (including one passive and two active filters) as a combination of (a) active-shunt, passive-series, and active-series filters, (b) active-series, active-shunt, and passive-series filters, (c) active-series, parallel-connected passive-series with active-series filters, (d) active-shunt, parallel-connected passive–series with active-series filters, (e) active-series, passive-shunt, and active-series filters, (f) active-shunt, passive-shunt, and active-series filters, (g) passive-series, active-shunt, and active-series filters, (h) series-connected active-shunt with passive-shunt and actives-series filters, (i) active-series in parallel with passive-series and active-series filters, (j) active-series, active-shunt and passive-shunt filters, (k) active-shunt, active-series, and passive-shunt filters, (l) active-series, passive-series, and active-shunt filters, (m) active-shunt, passive-series, and active-shunt filters, (n) active-series, series-connected active-shunt, and passive-shunt filters, (o) active-shunt, series-connected active-shunt, and passive-shunt filters, (p) passive-shunt, active-series, and active-shunt filters, (q) parallel-connected passive-series with active-series and active-shunt filters, (r) active-shunt in series with parallel-connected active-shunt and passive-shunt filters [4].

combined. Figure 9.24 illustrates the 18 possible hybrid filters consisting of two passive filters and one active filter. There are also 18 possible hybrid filters consisting of one passive filter and two active filters as represented in Fig. 9.25.

The rating of active filters is reduced through augmenting them by passive filters to form hybrid filters. This reduces the overall cost and in many cases provides better compensation than when either passive or active filters alone are employed. However, a more efficient approach is to combine shunt and series active filters, which can provide both current and voltage compensation. This (active–active) hybrid filter is known as a unified power quality conditioner (UPQC) or universal active filter (Fig. 9.21). Therefore, the development in hybrid filter technology began from the arrangement of (two or three) passive filters (Figs. 9.19 and 9.20) and progressed to the more effective combination of a number of shunt and/or series active filters (Figs. 9.21 to 9.25), yielding a cost-effective solution and complete compensation.

Hybrid filters are usually considered a cost-effective option for power quality improvement, compensation of the poor power quality effects due to nonlinear loads, or to provide a sinusoidal AC supply to sensitive loads. There are a large number of low-power nonlinear loads in a single-phase power system, such as ovens, air conditioners, fluorescent lamps, TVs, computers, power supplies, printers, copiers, and battery chargers. Low-cost harmonic compensation of these residential nonlinear loads can be achieved using passive filters (Figs. 9.19 and 9.20). Compensation of single-phase high-power traction systems are effectively performed with hybrid filters (Figs. 9.21 and 9.22). Three-phase three-wire power systems are supplying a large number of nonlinear loads with moderate power

levels – such as adjustable-speed drives – up to large power levels associated with HVDC transmission systems. These loads can be compensated using either a group of passive filters (e.g., a passive filter unit as shown in Fig. 9.13a) or a combination of active and passive filters of different configurations (Figs. 9.23 to 9.25) depending on the properties of the AC system.

## 9.6 BLOCK DIAGRAM OF ACTIVE FILTERS

The generalized block diagram of shunt-connected active filters is shown in Fig. 9.26. It consists of the following basic elements [6]:

- **Sensors and transformers** (not shown) are used to measure waveforms and inject compensation signals. Nonsinusoidal voltage and current waveforms are sensed via potential transformers (PTs), current transformers (CTs), Hall-effect sensors, and isolation amplifiers (e.g., optocouplers). Connection transformers in shunt and series with the power system are employed to inject the compensation current ($I_{APF}$, see Fig. 9.26) and voltage ($V_{APF}$), respectively.
- **Distortion identifier** is a signal-processing function that takes the measured distorted waveform, $d(t)$ (e.g., line current or phase voltage) and generates a reference waveform, $r(t)$, to reduce the distortion.
- **Inverter** is a power converter (with the corresponding coupling inductance and transformer) that reproduces the reference waveform with appropriate amplitude for shunt ($I_{APF}$) and/or series ($V_{APF}$) active filtering.
- **Inverter controller** is usually a pulse-width modulator with local current control loop to ensure $I_{APF}$ (and/or $V_{APF}$) tracks $r(t)$.

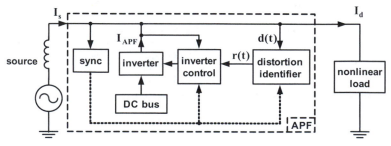

**FIGURE 9.26** Control diagram of shunt-connected active power filters (APFs) [6]. For series-connected active filters, the inverter injects the series voltage $V_{APF}$ via a series-connected transformer. For hybrid filters, two inverters are used for the simultaneous injection of $I_{APF}$ and $V_{APF}$.

- *Synchronizer* is a signal-processing block (based on phase-lock-loop techniques) to ensure compensation waveforms ($I_{APF}$ and/or $V_{APF}$) are correctly synchronized with the power system voltage. Certain control methods do not require this block.
- *DC bus* is an energy storage device that supplies the fluctuating instantaneous power demand of the inverter.

The control diagram of the series-connected active filter is identical to that given in Fig. 9.26; however, the inverter injects the series voltage $V_{APF}$ via a series-connected transformer. For the implementation of hybrid filters, the control diagram of Fig. 9.26 can be expanded to include both shunt ($I_{APF}$) and series ($V_{APF}$) compensation. Detailed implementation and control of a hybrid filter with active shunt and series converters (UPQC) is presented in Chapter 11.

The inverter block represents either a voltage-source inverter (VSI) or a current-source inverter (CSI). The VSI configuration includes an AC inductor $L_r$ along with optional small AC capacitor $C_r$ to form a ripple filter (for eliminating the switching ripple and improving the voltage profile) and a self-supporting DC bus with capacitor $C_d$. The CSI arrangement uses inductive energy storage at the DC link with current control and the shunt AC capacitors form a filter element. However, VSI structures have more advantages (lower losses, smaller size, less noise, etc.) and are usually preferred. Depending on the supply system, the inverter may be a single-phase two-arm bridge, a three-phase three-arm bridge and a three-phase four-arm bridge, or a mid point or three single-phase (e.g., for a four-wire system) bridge.

The inverter can be connected directly in series with the power system in single phase (to reduce the cost) or through connection transformers (usually with a higher number of turns on the inverter side) to act as a high active impedance blocking harmonic current and representing a low impedance for the fundamental current. In the same manner, the inverter may be connected in shunt either directly to the power system or through step-down transformers to act as an adjustable sink for harmonic currents.

The solid-state switching device is a MOSFET (metal oxide semiconductor field-effect transistor) for low power ratings, an IGBT (insulated-gate bipolar transistor) for medium power ratings, and a GTO (gate turn-off) thyristor for very high power ratings. These switching devices are manufactured in

modular form (with gating, protection, and interfacing elements) to reduce the overall size, cost, and weight.

An essential component of the active filter is the control unit, which consists of a microprocessor. The voltage and current signals are sensed using potential transformers (PTs), current transformers (CTs), Hall-effect sensors, and isolation amplifiers. The control approach (Section 9.7) is implemented online by the microprocessor after receiving input signals through analog-to-digital (A/D) converter channels, phase-lock loop (PLL), and synchronized interrupt signals. All control tasks are carried out concurrently in specially designed processors at a very low cost.

## 9.7 CONTROL OF FILTERS

The initial stage for any filtering process is the selection of the filter configuration. Depending on the nature of the distortion and system structure, as well as the required precision and speed of the compensation, an appropriate filter configuration (e.g., shunt, series, passive, active, or hybrid, as presented in Sections 9.2 to 9.5) is selected. Implementation of passive filters is relatively straightforward, as discussed in Section 9.3. This section presents a classification of control approaches for active and hybrid power filters and identifies their performance strengths.

The control configuration of Fig. 9.26 is open loop because load current $I_d$ (not supply current $I_s$) is measured, load distortion is identified, and a correction signal is fed-forward to compensate the distortion in the supply current. The open-loop approach has some limitations including inadequate correction (due to errors in calculations and/or inaccuracy in current injection), unstable compensation (e.g., injected current perturbs load and/or source currents such that the distortion is changed), and processing delay (e.g., slow transient response). In contrast, the closed-loop control approach (e.g., measuring the supply current $I_s$) identifies remaining distortion and updates the injected compensation signal.

The choice of control strategy strongly depends on the required compensation response. Active and hybrid filters have the capability to correct a wide range of power quality problems:

- *Harmonic distortion* is taken as a core function of all filter designs. Shunt and series filters are capable of compensating current and voltage distortion, respectively, whereas hybrid filters (with shunt and series components) can correct both current and voltage waveform distortions.

- **Reactive power control** at fundamental frequency is often included in the performance of active filters. It is possible to include harmonic reactive power control with more sophisticated control strategies.

- **Correction of negative- and zero-sequence components** at fundamental frequency due to imbalance and neutral current may be compensated with three-wire and four-wire three-phase hybrid active filters, respectively.

- **Flicker** (e.g., low irregular modulation of power flow) can also be compensated with active filters.

- **Voltage sag and voltage swell** are usually considered to be compensated by a dynamic voltage restorer (DVR); however, sophisticated hybrid active filters such as the UPQC (Fig. 9.21 and Chapter 11) can perform these types of compensations as well.

It is possible to consider and include all various types of power quality problems by adding the corresponding control functions to the basic active filter converter circuit; however, power rating and cost factors will become important and will raise challenging issues, especially in high-power applications.

There are three stages for the implementation of active (or hybrid) power filters (Fig. 9.26) [8]:

- **Stage One: Measurement of Distorted Waveform d(t).** For implementation and monitoring of the control algorithm, as well as recording various performance indices (e.g., either/both voltage or/and current THDs, power factor, active and reactive power), instantaneous distorted voltage and/or current waveforms need to be measured. This is done using voltage transformers (PTs), current transformers (CTs), or Hall-effect sensors including isolation amplifiers. Measured waveforms are filtered to avoid associated noise problems.

- **Stage Two: Derivation of Reference Signal r(t).** Measured distorted waveforms (consisting of fundamental component and distortion) are used to derive the reference signal (e.g., distortion components) for the inverter. This is done by the distortion identifier block. There are three main control approaches for the separation of the reference signal from the measured waveform, as discussed in Sections 9.7.1 to 9.7.3. The selected control method affects the rating and performance of the filter.

- **Stage Three: Generation of Compensation Signal $I_{APF}$ (or $V_{APF}$).** The inverter reproduces the refer-

ence waveform with appropriate amplitude for both/either shunt ($I_{APF}$) and/or series ($V_{APF}$) active filtering. These compensation signals are injected into the system to perform the required compensation task. The inverter control block uses reference-following techniques (discussed in Section 9.7.5) to generate the gating signals for the solid-state switches of the inverter.

Shunt active filters are often placed in low-voltage distribution networks, where there is usually a considerable amount of voltage distortion due to the propagation of current harmonics injected by nonlinear loads. Nonsinusoidal voltages at the terminals of an active filter will deteriorate its performance and require additional control functions.

Control objectives of filters are classified based on the supply current components to be compensated and by the response required to correct the distorted system voltage. An important question for the selection of filter control is, what is the most desirable response (of the combined filter and load) to the distortion of the supply voltage? Considering this question, there are three main control categories of compensation (Stage Two, as discussed in Section 9.7.1):

- **Waveform compensation** is based on achieving a sinusoidal supply current. The combination of filter and load is resistive at fundamental frequency and behaves like an open circuit for harmonic frequencies.

- **Instantaneous power compensation** relies on drawing a constant instantaneous three-phase power from the supply. In response to supply harmonic voltages, significant harmonic currents are drawn that result in nonsinusoidal current flow. The combination of filter and load generates a complicated and nonlinear response to the distorted excitation that cannot be described in terms of impedance.

- **Impedance synthesis** responds to voltage distortion with a resistive characteristic and draws harmonic currents in phase with the harmonic voltage excitation. The combination of filter and load draws power at all harmonic frequencies of the supply voltage and can be made to closely approximate a passive system.

There are a variety of reference-following techniques (Stage Three) that can be implemented to generate the compensation signal $I_{APF}$, including hysteresis-band, PWM, deadbeat, sliding, and fuzzy control, etc. A number of these control approaches are discussed in Section 9.7.5.

### 9.7.1 Derivation of Reference Signal using Waveform Compensation

The objective of waveform compensation is to achieve a sinusoidal supply current. There are many signal-processing techniques that can be used to decompose the distorted waveform into its fundamental (to be retained) and harmonic (to be cancelled) components. This is essentially a filtering task; however, in addition to the analytical approaches, there are also pattern-learning techniques that may be implemented.

#### 9.7.1.1 Waveform Compensation using Time-Domain Filtering

Distortion identification is a filtering task that can be performed by time-domain techniques or by Fourier-based frequency decomposition. The nonideal properties of real filters such as attenuation (e.g., preserving the identified components and heavily attenuating other components with a narrow bandwidth), phase distortion (e.g., preserving the phase of the identified components), and time response (e.g., considering the time-varying nature of the distortion signal and enforcing rapid filter response without large overshoot) must be recognized. The performance considerations of time-domain filters are linked, for example, smoothness of the magnitude response has to be traded off versus the phase response, and a well-damped transient response is in conflict with a narrow transition region. There are two general approaches to identify the distortion components in the time domain (Fig. 9.27a):

- **Direct distortion identification** applies a high-pass filter to the distorted signal $d(t)$ to determine the reference $r(t)$. Therefore, the identification process is subject to the transient response of the filter.
- **Indirect distortion identification** uses the filter to determine the fundamental component $f(t)$ and

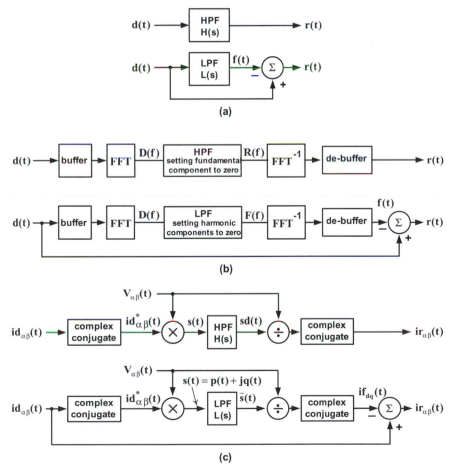

**(a)**

**(b)**

**(c)**

**FIGURE 9.27** Direct and indirect distortion identification techniques in (a) time domain, where $f(t)$ is the identified fundamental frequency signal that is allowed to remain, (b) frequency domain, where $D(f)$, $R(f)$, and $F(f)$ are the frequency-domain representations of distortion, reference, and fundamental component of the distorted signal, respectively [6], (c) instantaneous power domain.

subtracts it from the distorted signal $d(t)$ to form the reference $r(t)$.

Due to the inherent time lag of actual filters, the distortion term of the direct method will be out of date and the distortion cancellation will be in error. The disadvantage of the indirect method is the out-of-date fundamental term that requires additional exchange of real power through the inverter, resulting in DC-bus voltage disturbance and increased rating of the inverter. Most active filters use indirect distortion identification to achieve the best distortion cancellation during transients. Direct methods are preferred in applications where specific ranges of harmonics are to be compensated or where different groups of harmonics are to be treated differently.

To facilitate the separation of fundamental and harmonic components of the signal, active filter control algorithms rely on transformation techniques to transfer the signal from the conventional three-phase (abc) reference frame to an orthogonal two-phase representation. This will reduce the burden of intensive computations. Two well-known transfor-

mations, the stationary $\alpha\beta0$ reference and the rotating $dq0$ reference (also called Park transformation) frames, are widely employed (see Fig. 9.27). The spectra of a nonlinear load current (supplied by undistorted sinusoidal supply voltage) in abc, $\alpha\beta0$, and $dq0$ reference frames are shown in Fig. 9.28a,b,c, respectively. Analysis of these graphs indicates:

- The nonlinear load is assumed to be three-wire, three-phase and unbalanced that injects harmonics of the order $(6k \pm 1)$, where $k$ is any positive integer. This is clearly demonstrated by the frequency spectrum in the (abc) phase domain (Fig. 9.28a). Note that there is no distinction between harmonic components of the same order with different phase rotation.

- Both $\alpha\beta0$ and $dq0$ decompositions (Fig. 9.28b,c, respectively) transfer the zero-sequence component into a separate component set. Harmonic components with the same order and opposite phase rotation (e.g., $-h$ and $h$) are separated because positive and negative sets are shifted to orders $(h - 1)$ and $-(h - 1)$, respectively.

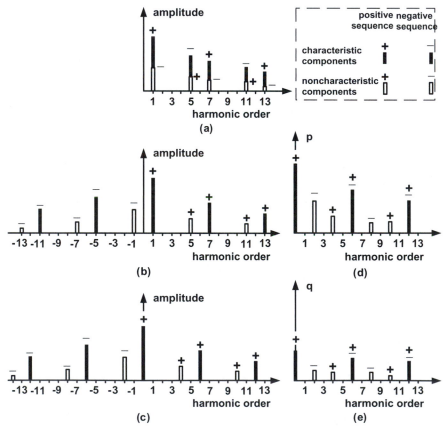

**FIGURE 9.28** Frequency spectrum of an unbalanced, three-wire three-phase nonlinear load (injecting current harmonics of order $h = (6k \pm 1)$: (a) current in phase domain, (b) current in $\alpha\beta0$ domain, (c) current in $dq0$ domain, (d) active power (p) in instantaneous power domain, (e) reactive power (q) in instantaneous power domain [6].

- The $\alpha\beta 0$ transformation (Fig. 9.28b) has no effect on the frequency spectra and does not separate sequence sets. The sequence can be determined from whether either the "$\alpha$" or the "$\beta$" component leads. The negative harmonic term of order $-1$ indicates the imbalance of the fundamental. Balanced nonlinear loads produce characteristic harmonics of order $-5, +7, -11, +13, \ldots, 6k+1$. Unbalanced nonlinear loads generate in addition noncharacteristic harmonic components of order $+5, -7, +11, -13, \ldots, 6k-1$.

- With the rotating $dq0$ transformation (Fig. 9.28c) all frequency components are shifted downward by one harmonic order. The positive-sequence fundamental becomes a DC term and the negative-sequence fundamental becomes a double-frequency term (order $-2$). The characteristic and noncharacteristic distortions are of orders $6k$ and $(6k-2)$, respectively.

- There is no difference between filtering in the abc or $\alpha\beta 0$ domains, but the $dq0$ domain is a better alternative because it offers frequency separation of fundamental, positive, and negative sequences, which makes the filtering task simpler. In the abc domain the fundamental may be separated using a low-pass filter that cuts off between orders 1 and 5 (or 1 and 3 if there is zero-sequence/four-wire harmonic distortion). In the $dq0$ domain, the filter cutoff should be between 0 and 6 for balanced conditions. For unbalanced conditions, a cutoff between 0 and 2 will allow the negative-sequence (unbalanced) fundamental to be corrected and a cutoff between 2 and 4 will not cancel the unbalanced fundamental but will cancel harmonic distortion. Zero-sequence distortion is conveniently placed in a separate component set that may be cancelled or retained as required.

### 9.7.1.2 Waveform Compensation Using Frequency-Domain Filtering

Filtering in the frequency domain requires discrete or fast Fourier transform (DFT or FFT) on a section of the signal that contains at least one cycle of the lowest frequency of interest and that has been sampled at over twice the highest harmonic frequency as prescribed by the Nyquist theorem. The advantage of filtering in the frequency domain is that completely abrupt cutoffs (with no transition band, low-pass band ripple, and phase distortion) can be obtained. The main disadvantage is that the filtering process is not very suitable for real-time filters and

sufficient time is required for sampling the signal and performing the transformation. In addition, the signals must be steady state and periodic because FFT implicitly assumes periodicity of the sampled waveform. If the FFT window is properly synchronized to the fundamental signal then the phase and magnitudes of the components can be accurately determined. If the sampling window does not cover an integer number of fundamental cycles, accuracy will be degraded (spectral leakage).

Distortion identification may also be performed with a direct or indirect approach using frequency-domain filtering (Fig. 9.27b). FFT is performed to transfer the distorted signal $d(t)$ to the frequency domain $D(f)$. In the direct method, the fundamental component is set to zero to form a cancellation reference. The filter may also cancel the reactive component if only the real (in-phase) component of the signals are set to zero. In the indirect method, all harmonic terms are set to zero; the fundamental is subtracted from the instantaneous signal to generate the reference signal. The filter may also cancel the reactive component if the imaginary (quadrature) fundamental component is set to zero. The final step is the application of the inverse FFT to generate $r(t)$.

A problem with frequency-domain filtering is that during transient conditions, the periodicity of the distortion is lost and the cancellation is not accurate. There is also time delay in filtering that can affect the distortion identification performance. For most applications, one complete cycle of data is stored in a buffer, processed during the next cycle, and de-buffered for use as a cancellation reference in the following cycle, that is, the cancellation reference is two cycles out of date. For applications requiring fast compensation, shorter update rates are possible by performing a FFT every half-cycle using one half-cycle of fresh data plus data from the previous half-cycle. This will reduce the filtering delay to one cycle.

### 9.7.1.3 Other Methods for Waveform Compensation

There are other waveform compensation schemes that attempt to overcome the disadvantages of time-domain techniques (e.g., compromises or imperfections in filter design) and the inherent time delay of frequency-domain approaches, such as heterodyne methods and neural networks.

The heterodyne methods of waveform compensation involve multiplying the distorted signal by a sinusoid [6]. If the sinusoid is of fundamental frequency then the fundamental frequency component of the distorted signal will be transformed to DC and double-frequency terms. The DC term can be separated with a low-pass filter. If the heterodyning sinusoid is in phase with the voltage, the fundamental active current is identified; otherwise, the fundamental reactive current can be identified using a quadrature sinusoid. As with the time-domain approaches, a low-order filter provides poor separation of the DC and double-frequency terms and a sharp (high-order) filter will have a long step response. Heterodyne methods of waveform compensation are popular in single-phase systems where $\alpha\beta0$ and $dq0$ transformations are not applicable.

Separation of a signal into the fundamental and distorted components can also be performed by using neural networks [6]. Both direct and indirect principles can be implemented. Neural networks can be used as harmonic identifiers to estimate the Fourier coefficients of the distorted signal. They can be trained to learn the characteristics of the load current (or that of the local grid system) to produce the cancellation reference. They may also be employed to provide fast-frequency decomposition to identify the fundamental component (e.g., for indirect algorithms) or low-order harmonics for a range of load current amplitudes.

### 9.7.2 Derivation of Compensated Signals Using Instantaneous Power Compensation

Another approach to control active filters is to cancel the fluctuating component of instantaneous power and perhaps compensate the fundamental component of the reactive power. A common approach is to start with the definitions of instantaneous active and reactive powers and express them in terms of average and alternating (fluctuating) components. This can be performed in either the abc, $\alpha\beta0$, or $dq0$ domain:

- In the $\alpha\beta0$ domain:

$$\begin{bmatrix} v_0 \\ v_\alpha \\ v_\beta \end{bmatrix} = \sqrt{\frac{2}{3}} \begin{bmatrix} 1/\sqrt{2} & 1/\sqrt{2} & 1/\sqrt{2} \\ 1 & -1/2 & -1/2 \\ 0 & \sqrt{3}/2 & -\sqrt{3}/2 \end{bmatrix} \begin{bmatrix} v_a \\ v_b \\ v_c \end{bmatrix}. \quad (9\text{-}12)$$

Without the zero-sequence component, the instantaneous power can be written as

$$\begin{aligned} s(t) &= p(t) + jq(t) \\ &= v_{\alpha\beta} i_{\alpha\beta}^* = (v_\alpha + jv_\beta)(i_\alpha - ji_\beta) \\ &= (v_\alpha i_\alpha + v_\beta i_\beta) + j(v_\beta i_\alpha - v_\alpha i_\beta), \quad (9\text{-}13a) \end{aligned}$$

where the superscript * denotes the complex conjugate. Therefore,

$$\begin{bmatrix} p \\ q \end{bmatrix} = \begin{bmatrix} (v_\alpha i_\alpha + v_\beta i_\beta) \\ (v_\beta i_\alpha - v_\alpha i_\beta) \end{bmatrix} = \begin{bmatrix} v_\alpha & v_\beta \\ v_\beta & -v_\alpha \end{bmatrix} \begin{bmatrix} i_\alpha \\ i_\beta \end{bmatrix}. \quad (9\text{-}13b)$$

- In the $dq0$ domain:

$$\begin{bmatrix} v_d \\ v_q \\ v_0 \end{bmatrix} = \sqrt{\frac{2}{3}} \begin{bmatrix} \cos\theta_d & \cos(\theta_d - 2\pi/3) \\ -\sin\theta_d & -\sin(\theta_d - 2\pi/3) \\ 1/\sqrt{2} & 1/\sqrt{2} \end{bmatrix}$$
$$\begin{matrix} \cos(\theta_d + 2\pi/3) \\ -\sin(\theta_d + 2\pi/3) \\ 1/\sqrt{2} \end{matrix} \begin{bmatrix} v_a \\ v_b \\ v_c \end{bmatrix}. \quad (9\text{-}14)$$

In this equation $\theta_d = \omega_d + \varphi$, where $\omega_d$ is the angular velocity of voltages and $\varphi$ is the initial voltage angle. With no zero-sequence components:

$$\begin{aligned} s(t) &= p(t) + jq(t) \\ &= v_{dq} i_{dq}^* = (v_d + jv_q)(i_d - ji_q) \\ &= (v_d i_d + v_q i_q) + j(v_q i_d - v_d i_q). \quad (9\text{-}15) \end{aligned}$$

In four-wire systems there is the additional term of zero-sequence instantaneous (real or reactive [6]) power; with Eqs. 9-13 and 9-15 the instantaneous power may be written as

$$p(t) = \bar{p} + \tilde{p}(t)$$
$$q(t) = \bar{q} + \tilde{q}. \quad (9\text{-}16)$$

Figure 9.28a,d,e shows the spectra of current, active, and reactive powers for the case of a sinusoidal supply voltage, which is harmonically distorted and generates unbalanced currents. Characteristic harmonics of order $h = 6k + 1$ (e.g., $-11, -5, 7, 13, \ldots$, where negative orders represent negative-sequence harmonics) produce power terms of order $6|k|$, whereas noncharacteristic harmonics of order $h = 6k - 1$ (e.g., $-13, -7, 5, 11, \ldots$) yield power terms of order $6|k| \pm 2$.

The inverse transform (e.g., conversion from power domain to the phase domain) is

$$i_{\alpha\beta}^* = i_\alpha - ji_\beta = \frac{s}{v_{\alpha\beta}} = \frac{(pv_\alpha + qv_\beta) + j(qv_\alpha - pv_\beta)}{v_\alpha^2 + v_\beta^2} \quad (9\text{-}17)$$

or

$$\begin{bmatrix} i_\alpha \\ i_\beta \end{bmatrix} = \frac{1}{v_\alpha^2 + v_\beta^2} \begin{bmatrix} v_\alpha & v_\beta \\ v_\beta & -v_\alpha \end{bmatrix} \begin{bmatrix} p \\ q \end{bmatrix}. \quad (9\text{-}18)$$

Figure 9.27c depicts direct and indirect distortion identification approaches based on the instantaneous power domain as given by Eq. 9-18. The concept is analogous to that of the frequency domain and similar comments apply.

Equation 9-18 assumes sinusoidal supply voltages and distorted currents. However, active filter control relying on the instantaneous power method causes harmonic currents to flow when the supply voltage is either distorted or unbalanced (see Application Examples 9.3 and 9.4). The following comments apply to the instantaneous power decomposition:

- Since p and q terms are readily available (e.g., by setting the $q$-term to zero in the indirect method and not removing it by filtering in the direct method), it is easy to include the compensation of fundamental reactive power in the active filter control algorithm and perform optimal power follow.
- Fundamental-frequency negative-sequence current appears as a second harmonic in the instantaneous power domain and may be easily filtered.
- The instantaneous power method applied in the $\alpha\beta 0$ domain has low processing burden because it does not require rotational transformation and explicit synchronization.
- The inverse transformation of Eq. 9-18 assumes sinusoidal supply voltages with current distortions. If both the supply voltage and the load current are distorted, active filter (with instantaneous power compensation) will result in harmonic current flow through the supply (Eq. 9-22). A similar scenario occurs when the supply contains negative-sequence voltages introducing imbalance in the currents and harmonic distortion [6]. Efforts such as modifying the inverse transformation and altering the definition of p and q have been made to overcome this problem.
- This section has focused on the control of shunt active filters. Similar discussions apply to the control of series active filters for which dual equations can be generated.

### 9.7.2.1 Application Example 9.2: Instantaneous Power for Sinusoidal Supply Voltages and Distorted Load Currents

Assuming a sinusoidal supply voltage and a distorted load current (with fundamental and a single harmonic component of order h), find the equations for instantaneous power as well as average and fluc-

tuating, pulsating, or oscillating real and reactive powers.

**Solution:**

$$p(t) = V_1 I_1 + V_1 I_h \cos[(1-h)\omega t]$$
$$\Rightarrow \begin{cases} \bar{p} = V_1 I_1 \\ \tilde{p} = V_1 I_h \cos[(1-h)\omega t] \end{cases}$$
$$q(t) = -V_1 I_h \sin[(1-h)\omega t] \Rightarrow \begin{cases} \bar{q} = 0 \\ \tilde{q} = -V_1 I_h \sin[(1-h)\omega t] \end{cases}$$
$$(9\text{-}19)$$

### 9.7.2.2 Application Example 9.3: Instantaneous Power Consumed by a Resistive Load Subjected to Distorted Supply Voltages

A three-phase distorted voltage waveform consisting of the fundamental and one harmonic component of order $h$ is connected across the terminals of a balanced three-phase resistor bank in ungrounded $Y$ configuration. Compute the components of voltages and currents in the $\alpha\beta 0$ domain and determine the average and the fluctuating active power consumed by the load.

**Solution:**

The load is balanced and has no zero-sequence terms due to the ungrounded Y. The $\alpha$ and $\beta$ components are

$$v_\alpha = V^{(1)} \cos(\omega t) + V^{(h)} \cos(h\omega t)$$
$$v_\beta = V^{(1)} \sin(\omega t) + V^{(h)} \sin(h\omega t)$$
$$\Rightarrow \begin{aligned} i_\alpha &= \frac{V^{(1)}}{R} \cos(\omega t) + \frac{V^{(h)}}{R} \cos(h\omega t) \\ i_\beta &= \frac{V^{(1)}}{R} \sin(\omega t) + \frac{V^{(h)}}{R} \sin(h\omega t) \end{aligned}$$
$$(9\text{-}20)$$

Applying Eqs. 9–13 and 9–16:

$$p(t) = (v_\alpha i_\alpha + v_\beta i_\beta) = \bar{p} + \tilde{p}(t)$$
$$\Rightarrow \begin{cases} \bar{p} = \dfrac{(V^{(1)})^2}{R} + \dfrac{(V^{(h)})^2}{R} \\ \tilde{p}(t) = 2\dfrac{V^{(1)}V^{(h)}}{R} \cos[(1-h)\omega t] \end{cases}$$
$$(9\text{-}21)$$

Note that an oscillatory power with the difference frequency (e.g., $(1-h)f_1$) is generated. Based on these equations, a nonsinusoidal current must be drawn from a distorted supply voltage to maintain a constant instantaneous power. Similarly, a nonsinusoidal current is required to draw constant instantaneous power from an unbalanced supply voltage (containing a fundamental-frequency negative-sequence component). Therefore, both/either distorted and/or unbalanced supply voltages will force the active filter (with instantaneous power based controller) to inject harmonic currents.

### 9.7.2.3 Application Example 9.4: Supply Current Distortion Caused by Active Filters with Instantaneous Power-Based Controllers

An active filter with an instantaneous power controller is applied to the system of Application Example 9.3. Compute $\alpha$ and $\beta$ components of the supply current after the filter has been installed and the nonsinusoidal load current (through the resistive load) has been compensated.

**Solution:**
Relying on Eqs. 9-18 and 9-21, the supply current corresponding to constant power is

$$i_\alpha = \frac{pv_\alpha}{v_\alpha^2 + v_\beta^2}$$
$$= \frac{p[V^{(1)}\cos(\omega t) + V^{(h)}\cos(h\omega t)]}{(V^{(1)})^2 + 2V^{(1)}V^{(h)}\cos[(1-h)\omega t] + (V^{(h)})^2}$$
$$i_\beta = \frac{-pv_\beta}{v_\alpha^2 + v_\beta^2}$$
$$= \frac{p[V^{(1)}\sin(\omega t) + V^{(h)}\sin(h\omega t)]}{(V^{(1)})^2 + 2V^{(1)}V^{(h)}\cos[(1-h)\omega t] + (V^{(h)})^2}.$$

$$(9\text{-}22)$$

Note the presence of different harmonics in the numerators and denominators which are not of the same order as that of the voltage excitation (e.g., order $h$). Therefore, a single voltage harmonic distortion gives rise to an infinite series of current harmonics when the active filter controller is based on instantaneous power compensation.

### 9.7.3 Derivation of Compensating Signals Using Impedance Synthesis

Impedance synthesis is based on the idea of modifying the inverse transform of the instantaneous power (Eq. 9-17) to prevent injection of higher frequency current harmonics. If Eq. 9-17 is modified using the root-mean-square (rms) value of the current, that is, replace $v_\alpha^2 + v_\beta^2$ with $(v_\alpha)_{rms} + (v_\beta)_{rms}$, the denominator will not be time varying and a modulation of the

current is avoided. There are two impedance synthesis approaches where the voltage and current are related by a chosen value of the impedance: impedance-based blocking and impedance-based compensation.

#### 9.7.3.1 Impedance-Based Blocking

Impedance-based blocking control presents resistive characteristics at all frequencies. A filter arrangement with a number of shunt and series (passive and/or active) branches is used to block or enable a signal path. An inverter acting as an impedance relating voltage to current by a simple transfer function can form one branch of the filter. This normally takes the form of a hybrid filter consisting of active and passive elements, although combinations of active units – without passive elements – are also possible.

There are many forms of hybrid filters (Section 9.5) that can be used for impedance-based blocking. For example, a hybrid combination consisting of a series active filter and a shunt passive filter can be used to perform impedance synthesis. The role of the series branch is to block harmonic currents by representing a high impedance. The instantaneous power theory can be utilized to separate the distortion terms of the current. The active filter then injects a voltage proportional to this current to synthesize a moderately high impedance. In practice, the filter employed to separate the oscillatory terms in the instantaneous power may not be perfect and might influence the impedance presented by the overall active filter [6].

#### 9.7.3.2 Impedance-Based Compensation

A difficulty faced in designing the control of active filters for general purposes (e.g., correcting a group of nonlinear loads) is that the system dynamics are unknown. Furthermore, the compensation current – in response to the detected distortion – might activate one of the system resonances or perturb the voltage and cause instability [6].

One solution is to use a shunt device that draws a current proportional to the voltage. This is done by utilizing a dissipative active filter that absorbs energy at harmonic frequencies; thus voltage harmonics are detected and damped. This is a passive controller because it presents a resistive characteristic. Therefore, the operation of the active filter is stable regardless of the system structure.

### 9.7.4 DC Bus Energy Balance

The DC buses of inverters in active power filter systems are not DC links but energy storage devices. If a voltage-source inverter is used, the link contains a capacitor; otherwise an inductor acts as the DC link for the current-source inverter.

In practice, the link voltage will not usually remain constant due to various effects:

- real power exchange between the active filter and the AC grid,
- converter (conduction and switching) losses,
- undesirable real power components in the compensation reference signal due to magnitude and phase errors of the distortion identifier or because of time delays in responding to a transient condition,
- the error between the identified correction current and the actual output current of the inverter that may lead to some power exchange, and
- distortion in the grid voltage that might cause some real power exchange.

To overcome these deficiencies, converter power losses are compensated by drawing a balancing power into the DC bus from the AC grid. This may be performed through a DC-bus voltage regulator (Fig. 9.29) as follows:

- The voltage error signal is generated based on a reference DC voltage;
- The error is passed through a controller;
- A PI compensator operates on the voltage error. It should have a low crossover frequency to avoid interaction with the compensation system;
- Controller output is taken as a demand for fundamental active current and is added to the compensation reference;
- The control algorithm can be implemented in any convenient domain (e.g., abc, $\alpha\beta0$, or $dq0$ domain); and
- The bandwidth of the controller must be carefully selected: if it is too high (comparable with the

fundamental frequency), the injected balancing current will contain components at harmonic frequencies that will add distortion to the system; if it is too low, the DC-link voltage may deviate from its nominal value.

### 9.7.5 Generation of Compensation Signal Using Reference-Following Techniques

The identified reference signal (with one of the techniques discussed in Sections 9.7.1 to 9.7.3) must be generated or reproduced by the inverter (Fig. 9.26) and injected into the system to eliminate the power quality problem. Generation of the compensation signal that accurately follows the identified reference may be difficult because of the following:

- the high rate of change and the wide bandwidth of the reference signal,
- the errors in the generated reference caused by the $di/dt$ limit of a voltage-source inverter,
- the voltage imposed on the inverter (e.g., difference between the available DC-bus voltage and the instantaneous voltage of the AC system),
- the impact of increasing the DC-bus voltage – to improve the reference following – on the ratings of the inverter, and
- the effect of the processing delay on the formation of the reference waveform.

Three approaches to force the inverter output current to follow a reference signal are widely implemented: PWM current regulation, hysteresis regulation, and deadbeat or predictive control.

***PWM Current Regulation.*** The current error, for example, the difference between identified (reference) and generated signals, is acted on by a PI controller and its output is passed through a PWM or space-voltage vector modulation to define the voltage to be produced by the inverter. The controller may be of the open- or closed-loop type and can be implemented in any convenient (abc, $\alpha\beta0$, or $dq0$) domain.

***Hysteresis Regulation.*** The main idea is to force the compensation signal to follow the reference signal within a given hysteresis band error (e.g., $i_{compensation} = i_{reference} \pm \Delta i_{hysteresis}/2$). The implementation of this controller is easy as it requires an op-amp operating in the hysteresis mode. Hysteresis regulation is unconditionally stable, does not require detailed plant data, and can achieve high rates of change in the controlled variable. Unfor-

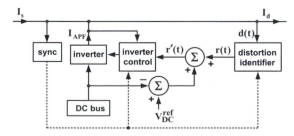

**FIGURE 9.29** A simple DC-bus voltage regulator [6].

tunately, there are several drawbacks such as the unpredictable switching frequency and the broadband switching noise injected by the nonconstant switching frequency. Consequently, different approaches have been considered to maintain constant (or near constant) switching frequency and to synchronize the switching of all three phases.

***Deadbeat or Predictive Control.*** Deadbeat or predictive control is attractive for following rapidly changing reference signals of active filters because it selects a voltage command for the inverter that will force the generated current to reach a target value by the end of the next sample period [6]. However, it requires knowledge of the system to accurately model the rate of change of current ($di/dt$) and to properly select the voltage command. The interface inductance of the inverter is a key parameter and the opposing AC system voltage must be measured or estimated. Deadbeat controllers may be applied when there is some parameter uncertainty (e.g., the coupling inductor). Adaptive filters can be employed to predict the current reference, and the deadbeat controller can use the predicted current at the end of the sample period as the target current [6].

### 9.7.6 Application Example 9.5: Hybrid of Passive and Active Power Filters for Harmonic Mitigation of Six-Pulse and Twelve-Pulse Rectifier Loads [16]

This example demonstrates the application of hybrid filters to compensate harmonics of a nonlinear load consisting of two six-pulse rectifiers. The two rectifiers are connected to the secondary of a ($Y/\Delta$, $Y$) transformer. The simultaneous operation of the two six-pulse rectifiers represents a twelve-pulse rectifier load. Two arrangements of hybrid filters are considered:

***Hybrid Filter #1*** is composed of a shunt-hybrid filter in parallel with a passive filter as shown in Fig. E9.5.1. The passive filter consists of a set of tuned branches. The hybrid power filter is a series combination of an active filter with a small rating (converter) and a passive branch. Due to the use of a parallel passive filter, the converter rating is decreased and the performance and resonance problems of the passive filter are expected to improve.
***Hybrid Filter #2*** is composed of a shunt-hybrid filter in parallel with two passive filters tuned at the 5th and 7th harmonics as shown in Fig. E9.5.2. The hybrid filter consists of an active filter with a small rating in series with two passive branches and is tuned at the dominant 11th and 13th harmonics.

a) Plot the load current waveforms and calculate $THD_i$ with twelve-pulse and six-pulse rectifier configurations.
b) Derive the equations for power system and passive and hybrid filter currents.
c) With a six-pulse rectifier load, plot the waveforms of the power system, load, passive filter, and hybrid filter branch currents before and after the installation of hybrid filters #1 and #2.
d) Plot the power system current under resonant conditions before and after the addition of hybrid filter #2.
e) Compare the performances of hybrid filters #1 and #2.

## 9.8. SUMMARY

With the increase of harmonic currents and voltages in present-day power systems, the installation of passive, active, and hybrid filters gains in importance. In the future, recommended practices such as IEEE-519 [11] will result in further increased use of filters within the distribution system. There are a great variety of filter configurations (Figs. 9.3 and 9.13 to 9.25). Classification of filters can be based on different criteria such as supply type, filter connection, number of filter elements, power rating, compensation type, speed of response, and control technique. In this chapter the classification of power filters is performed based on the type of supply (Fig. 9.3) and the control method (Section 9.7). Proper selection of filter configurations depends primarily on the properties of nonlinear loads, type of supply, and system rating:

- Passive filter blocks (Fig. 9.12 and 9.13) and a combination of passive filters (Figs. 9.19 and 9.20) are the most qualified candidates for single-phase systems – with high penetration of nonlinear loads – and low-power applications.
- Hybrid filters (Figs. 9.21 and 9.22) are considered as a cost-effective option for power quality improvement, compensation of the poor power quality effects of nonlinear loads, or to provide a sinusoidal AC supply to sensitive loads, especially for high-power applications. They are also good options for compensation of single-phase systems and high-power applications.
- A group of passive filters (Fig. 9.13a) or a combination of active and passive filters in different configurations (Figs. 9.23 to 9.25) should be used for the compensation of three-phase three-wire power systems that have a great amount of current or

voltage harmonics due to a large number of small- and moderate-rating nonlinear loads, as well as feeding terminals of HVDC transmission systems.

Selection of the control approach (Sections 9.7.1 to 9.7.3 and 9.7.5) greatly depends on the choice of the control objective, for example, the required waveform of the supply current after compensation. The selection of the domain (e.g., time or frequency) and the reference frame (e.g., $dq0$ or $\alpha\beta0$) will affect the computational burden required to implement appropriate filters. All active filters perform harmonic mitigation. However, depending on the selected control strategies, additional forms of compensation (e.g., fundamental reactive power, negative-sequence, zero-sequence, phase balancing, sag, swell, and flicker) can be accommodated within the apparent power rating of an active filter. There exist three different definitions for the control objective:

**First Definition:** Supply currents (after filtering) should contain only fundamental frequency components. This is the main task of active filter controllers using waveform compensation (Section 9.7.1).

**Second Definition:** Supply currents (after filtering) should be such that the instantaneous active power is constant while the instantaneous reactive power is zero. This is performed by active filter controllers that use instantaneous power compensation (Section 9.7.2).

**Third Definition:** Supply currents (after filtering) should be entirely active. This is the main function of active filter controllers using impedance-based compensation (Section 9.7.3).

If the supply voltage is not distorted (with nearly sinusoidal waveform), the following remarks should be considered for the selection of the control method:

- The three aforementioned objective definitions are equivalent and the choice of the control method should be based on the ease of their implementation and their instrumentation requirements.
- Given the harmonic limits provided by relevant guidelines, recommended practices, or standards such as IEEE-519 [11], waveform compensation (first definition) is the most important approach. The techniques for this approach separate the fundamental from the harmonic components. Time-domain filters have the advantage of continuity during transient conditions, whereas frequency-domain filters have a better selectivity under steady-state conditions. The synchronous $dq0$ reference frame can ease the filter implementation of three-phase systems at the expense of longer processing time.

- The instantaneous power methods (second definition) are beneficial in three-phase systems due to their ease of filter design without the need for $dq0$ transformation.
- The active current approaches (third definition) are relatively easy to implement because they require only the identification of the (average) real power. These approaches are implicit in controllers that use the energy balance at the DC bus to identify the desired current.

Further complications regarding the selection of the control method for active filters arise if the supply voltage (before filtering) is distorted. This is the case for most practical power quality problems, for example, when active filters are operating on weak feeders of the distribution network. If the supply voltage is distorted, the following guidelines should also be considered for the selection of the control method:

- Distorted supply voltage may increase the apparent power rating of the active power filter.
- Waveform compensation approaches only draw fundamental current and do not generate extra distortion; however, they will not provide damping of the existing distortion.
- Instantaneous active power methods respond to harmonic supply voltages by introducing additional harmonic currents at other frequencies (Eq. 9-22). In particular, a negative-sequence fundamental voltage will cause these controllers to inject harmonic currents.
- Controllers with impedance-based compensation – resulting in active supply current – represent resistive characteristics at all frequencies, and therefore convert nonlinear loads into linear loads. They inherently provide damping and reduce the propagation of harmonic voltages. However, the design of active filters will have to be coordinated with existing approaches for the damping of network resonances.

Active filters can be designed to perform phase balancing if the exchange of instantaneous power between phases is within their ratings.

- All three control approaches (waveform compensation, instantaneous power compensation, and impedance synthesis) can enforce phase balancing. Waveform compensation and impedance synthesis (active current) methods can be implemented on a phase-by-phase basis to avoid rebalancing.

- Instantaneous power methods (in their original form) have an unacceptable response under unbalanced operating conditions, and will inject additional harmonic currents at different frequencies. Modifications (e.g., redefining the inverse transformation) are necessary to control harmonic generation.

- Waveform compensation methods result in balanced currents despite unbalanced voltages.

- Impedance synthesis methods present balanced impedances to unbalanced voltages. As a result, they draw the least current from the smallest phase voltage. In contrast, the constant instantaneous power approaches draw the greatest current from the smallest phase voltage.

## 9.9 REFERENCES

1) Peng, F.Z.; "Harmonic sources and filtering approaches," *IEEE Industrial Application Magazine*, Vol. 7, pp. 18–25, 2001.

2) Das, J.C.; "Passive filters – potentialities and limitations," *IEEE Transactions on Power Industry Applications*, Vol. 40, no. 1, pp. 232–241, 2004.

3) Akagi, H.; "Active harmonic filters," *Proceedings of the IEEE*, Vol. 93, no. 12, pp. 2128–2141, 2005.

4) Singh, B.; Verma, V.; Chandra, A.; and Al-Hadded, K.; "Hybrid filters for power quality improvement," *IEE Proceedings – Generation, Transmission and Distribution*, Vol. 152, no. 3, pp. 365–378, 2005.

5) El-Habrouk, M.; Darwish, M.K.; and Mehta, P.; "Active power filters: a review," *IEE Proceedings – Electric Power Application*, Vol. 147, no. 5, pp. 403–413, 2000.

6) Green, T.C.; Marks, J.H.; "Control techniques for active power filters," *IEE Proceedings – Electric Power Application*, Vol. 152, no. 2, pp. 369–381, 2005.

7) Phipps, J.K.; "A transfer function approach to harmonic filter design," *IEEE Industrial Application Magazine*, pp. 68–82, 1997.

8) Singh, B.; Al-Haddad, K.; and Chandra, A.; "A review of active filters for power quality improvement," *IEEE Transactions on Industrial Electronics*, Vol. 46, no. 5, pp. 960–971, 1999.

9) Grainger, J.J.; and Civanlar, S.; "Volt/VAr control on distribution systems with lateral branches using shunt capacitors and voltage regulators. Part III: the numerical result," *IEEE Transactions on Power Apparatus and Systems*, Vol. 104, no. 11, pp. 3291–3297, 1985.

10) Hu, Z.; Wang, X.; Chen, H.; and Taylor, G.A.; "Volt/VAr control in distribution systems using a time-interval based approach," *IEE Proceedings – Generation, Transmission and Distribution*, Vol. 150, no. 5, pp. 548–554, 2003.

11) *IEEE Recommended Practice and Requirements for Harmonic Control in Electric Power Systems*, IEEE Standard 519, 1992.

12) Nilsson, J.W.; *Electric Circuits*, Addison-Wesley Publishing Company, Reading, Massachusetts, 1983.

13) Thomas, R.E.; and Rosa, A. J.; *The Analysis and Design of Linear Circuits*, 4th Edition, John Wiley & Sons, Inc., 2004.

14) Masoum, M.A.S.; Ulinuha, A.; Islam S.; and Tan, K.; "Hybrid passive filter design for distribution systems with adjustable speed drives," *The Seventh International Conference on Power Electronics and Drive Systems (PEDS-2007)*, November 27–30, 2007, Bangkok, Thailand.

15) Jou, H.L.; Wu, J.C.; and Wu, K.D.; "Parallel operation of passive power filter and hybrid power filter for harmonic suppression," *IEE Proc. on Generation, Transmission and Distribution*, Vol. 148, Issue 1, pp. 8–14, Jan 2001.

16) Gholamrezaei, M.; Masoum, M.A.S.; and Kalantar, M.; "Parallel combination of passive and hybrid power filters for harmonic mitigation," *18th International Power System Conference*, Vol. 1, pp. 7–15, PSC 2003.

# 10

# Optimal Placement and Sizing of Shunt Capacitor Banks in the Presence of Harmonics

At the distribution level about 13% of the total generated power consists of losses due to active and reactive current components. Research over the past several decades has shown that proper placement of shunt-capacitor banks can reduce the losses caused by reactive currents [1–34]. In addition to the reduction of energy and peak-power losses, effective capacitor installation can also release additional reactive power capacity within the distribution system and can improve the system voltage profile. Studies [35–46] indicate that the proper placement of shunt capacitors in the presence of voltage and current harmonics can also improve power quality and reduce the total voltage harmonic distortion ($THD_v$) of distribution systems. Thus, the problem of capacitor allocation involves the determination of the optimal locations, sizes, and number of capacitors to be installed within a distribution system such that maximum benefits are achieved, while all operational constraints (e.g., voltage profile, $THD_v$, and total number of capacitor banks at each bus and the entire system) are satisfied at different loading levels. There are a great many papers describing capacitor placement algorithms: the Capacitor Subcommittee of the IEEE Transmission and Distribution Committee has published 10 bibliographies on power capacitors from 1950 to 1980 [33]. Moreover, the VAr Management Working Group of the IEEE System Control Subcommittee has published another bibliography on reactive power and voltage control [34]. The total publication count listed in these bibliographies is over 400, and many of the papers are specific to the problem of optimal capacitor allocation [1–46]. Recent efforts mainly focus on the capacitor placement under nonsinusoidal operating conditions.

In power systems, the predominantly inductive nature of loads and distribution feeders as well as transformers and lines accounts for significant power losses due to lagging currents. The introduction of strategically sized and placed shunt capacitors within the distribution system helps to reduce the losses created by an inductive system, thus increasing the system capacity, reducing system losses, and improving the voltage profile. Due to the fact that there are many constraints and variables present, the process for calculating the placement and sizing of shunt capacitors involves the evaluation of an optimization function. Often this function will not only address technical aspects of a problem, but will evaluate its cost as well. Therefore, it becomes necessary to design an approach that will optimize the solution at minimal cost.

Analyses of this chapter are restricted to the application of shunt capacitor banks for reactive power compensation of distribution systems under sinusoidal as well as nonsinusoidal operating conditions. The capacitor placement problem of a distribution system consists of determining the optimal locations, sizes, and types of capacitor banks in order to achieve maximum annual benefits within the permissible region of constraints (e.g., bus voltage deviation, total harmonic distortion, number of capacitor banks at each bus and the entire system). Various techniques and algorithms have been proposed and implemented for solving this problem under normal sinusoidal conditions and in the presence of voltage and current harmonics. All such methods have their own advantages, disadvantages, and limitations as stated by their assumptions.

The goal of this chapter is to show the progression of research in optimal capacitor placement for sinusoidal operating conditions and to introduce a number of algorithms for solving this problem in the presence of harmonics. Section 10.1 presents the concept of reactive power compensation based on capacitors. Section 10.2 addresses common types of shunt capacitor banks as used in power systems. A survey of various optimization algorithms for the placement and sizing of shunt capacitor banks under sinusoidal operating conditions is presented in Section 10.3. The last part of the chapter includes analysis and comparison of various solution approaches for the optimal placement and sizing of fixed capacitor banks in the presence of voltage and current harmonics.

## 10.1 REACTIVE POWER COMPENSATION

Reactive power compensation is an important issue in electric power systems involving operational, economic, and quality of service aspects. Consumer loads (residential, commercial, industrial, service sector, etc.) impose active and reactive power demands. Active power is converted into useful energy, such as light, heat, and rotary motion. Reactive power must be compensated to guarantee an efficient delivery of active power to loads.

Capacitor banks are widely employed in distribution systems for reactive power compensation in order to achieve power and energy loss reduction, system capacity release, and acceptable voltage profiles. The extent of these benefits depends on the location, size, type, and number of shunt capacitors (sources of reactive power) and also on their control settings. Hence an optimal solution for placement and sizing of shunt capacitors in a distribution system is a very important aspect of power system analysis.

### 10.1.1 Benefits of Reactive Power Compensation

Shunt capacitors applied at the receiving end of a power system feeder supplying a load at lagging power factor have several benefits, which may be the reason for their extensive applications. This section highlights some advantages of reactive power compensation.

**Improved Voltage Profile.** Highly utilized feeders with high reactive power demands have poor voltage profiles and experience large voltage variations when loading levels change significantly. In a power system, it is desirable to regulate bus voltages within a narrow range (e.g., +5%, −8% of their nominal values) and to have a balanced load on all three phases (eliminating negative- and zero-sequence currents with undesirable consequences such as equipment heating and torque pulsations in generators and turbines). However, loads of the power system fluctuate and can result in voltage deviations beyond their acceptable limits. In view of the fact that the internal impedance of the AC system is mainly inductive, it is the reactive power change of the load that has the most adverse effect on voltage regulation.

Traditionally, line voltage drop is compensated with distributed capacitors along a feeder, and by switching them according to the reactive power demand a rather constant voltage profile is achieved. Furthermore, by keeping the power factor close to

unity along the section in question, the number of expensive voltage regulators can be minimized.

Impacts of active and reactive current variations on the terminal load voltage ($V_t$) are demonstrated in Fig. 10.1. An increase of $\Delta Q$ in the lagging VArs drawn by the load (indicated by quantities with primes in Fig. 10.1b) causes the reactive current component to increase to ($I_q + \Delta I_q$) while the active current component $I_p$ is assumed to be unchanged. This will result in a terminal voltage drop of $\Delta V_t$ and decreases the real power. Figure 10.1c shows the phasor diagram where the active current component increases to ($I_p + \Delta I_p$) with $\Delta I_p = \Delta I_q$ (of Fig. 10.1b) while $I_q$ is assumed to be unchanged. Comparison of phasor diagrams indicates that variations in reactive power (Fig. 10.1b) have more impact on voltage regulation as compared with active power variations (Fig. 10.1c).

**Reduced Power System Losses.** Correcting the power factor can substantially reduce network losses in between the point in the network where the correction is performed and the source of generation. This can result in an annual gross return of as much as 15% on the capacitor investment [48]. To maximize benefits, power factor correction should be implemented as close to the customer load as possible. However, due to economies of scale, installation of capacitors at the low-voltage (LV) level is very costly when compared with high-voltage (HV) options.

In most industrial plant power distribution systems, the $I^2R$ losses vary from 2.5 to 7.5% of the load kWh. This depends on the hours of full-load and no-load plant operation, wire size, and length of the main and branch feeder circuits. Capacitors are effective in reducing only that portion of the losses that is due to the reactive current. Losses are proportional to the current squared, and since current is reduced in direct proportion to power factor improvement, the losses are inversely proportional to the square of the power factor. Hence, as the power factor is increased with the addition of capacitor banks to the system, the amount of the losses is reduced. The connected capacitor banks have losses, but they are relatively small. Losses account for approximately one-third of one percent of the reactive power rating [48].

**Released Power System Capacity.** When capacitors are placed in a power system, they deliver reactive power and furnish magnetizing current for motors, transformers, and other electromagnetic devices. This increases the power factor and it means that for

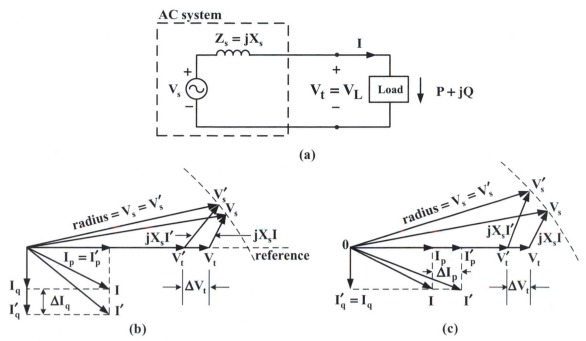

**FIGURE 10.1** Effect of active ($I_p$) and reactive ($I_q$) current on terminal voltage $V_t$; (a) Thevenin's equivalent circuit of the AC system with a purely inductive internal impedance, (b) change in $I_q$, (c) change in $I_p$ [47].

a lower current (or apparent power) level a higher real power utilization occurs. Therefore, capacitor banks can be used to reduce overloading or permit additional load to be added to the existing system. Release of system capacity by power factor improvement due to capacitors is becoming extremely important because of the associated economic and system benefits.

Reactive power compensation improves network utilization; that is, capacity is released and it allows more customers to be connected to what could have been a highly utilized feeder, without going to the costly extreme of adding an additional feeder.

**Increased Plant Ratings.** Another undesirable effect of a network with poor power factor is a possible reduction in plant ratings. This is often noticeable when transformer ratings are limited by overvoltage available from tapping. Correcting power factor to unity at the zone substation will maximize the real-power rating of transformers by providing access to thermal emergency ratings.

**Capital Deferment.** As ratings are based on apparent power rather than active power, reduced apparent power utilization reduces operating risks and can therefore defer expenditure related to the network expansion. This deferment can be applied from the

distribution feeder, back to the zone substation, subtransmission, terminal station, transmission network, and generation.

## 10.2 DRAWBACKS OF REACTIVE POWER COMPENSATION

Although capacitor banks can undoubtedly provide many benefits to a power system, there are several situations where they can cause deterioration of the power system. They may result in power quality problems, and in damage of connected equipment. In some situations, capacitor banks themselves may also become victims of unfortunate circumstances. Optimal placement and sizing of shunt capacitor banks as discussed in Sections 10.3 and 10.4 will prevent most of these problems. This section highlights some drawbacks of reactive power compensation.

**Resonance.** Resonance is the condition where the capacitive and inductive reactances of a system cancel each other, thus leaving the resistive elements in the system as the only form of impedance. The frequency at which this offsetting effect takes place is called the resonant frequency of the system. Resonance conditions will cause unacceptable high currents and overvoltages to occur with the potential to

**TABLE 10.1** Advantages and Disadvantages of Distributed (Pole-Mounted) Capacitors over Zone Substation Installations

| Component | Distributed capacitors | Zone substation capacitors |
|---|---|---|
| Initial capital outlay | Less expensive | More expensive |
| Reduction in losses | More effective | Less effective |
| Network-utilization improvement | More effective | Less effective |
| Voltage-profile improvement | Effective | Ineffective |
| Capital deferment | More effective | Less effective |
| Remote monitoring and control | More expensive | Less expensive |
| Maintenance and setting review | More expensive | Less expensive |
| Accommodation/space | Overhead network required | Space required at zone substation |
| Failure of a capacitor unit | Small impact on apparent power load | Large impact on apparent power load |

damage not only the installed capacitor but also other equipment operating on the system. In case of resonance large power losses could be experienced by the system.

**Harmonic Resonance.** If the resonant frequency happens to coincide with one generated by a harmonic source (e.g., nearby nonlinear load) then voltages and currents will increase disproportionately, causing damage to capacitors and other electrical equipment. Harmonic resonance may cause capacitor failure due to harmonic overvoltages and overcurrents.

**Magnification of Capacitor-Switching Transients.** Capacitor-switching transients typically occur when a large capacitor on the high-voltage side of the power system (usually at the utility side) is energized. This results in magnification of transients at low-voltage capacitors. The magnified transient at a low-voltage remote end can reach up to 400% of its rated values.

**Overvoltages.** The voltages of a power system are often maintained within a specified upper and lower limit of the rated value. Capacitors, especially switched types, can cause overvoltage problems by increasing the voltage beyond its maximum desired value. This causes complications in the power system and can damage and deteriorate the system.

## 10.2 COMMON TYPES OF DISTRIBUTION SHUNT CAPACITOR BANKS

There are many reasons and benefits for utilizing distribution capacitors; reactive power compensation remains one of the most effective solutions for permitting growth in a network. It has long been popular to have reactive compensation based at zone substations. However, there has been a growing trend to distribute reactive compensation (e.g., along high-voltage overhead feeders). Moving capacitor banks closer to the load is an effective way to reduce network losses, defer costly capital expenditure, and improve network utilization as well as voltage profiles of the distribution network.

Pole-mounted capacitor banks can be of either the switched or fixed type. They have the ability to be distributed along overhead feeders with ease and as such are a very popular choice for reactive compensation. Fixed capacitors are usually sized for minimum base-load conditions, whereas switched units are designed for loads above the minimum. Switched capacitor banks provide the capability and the flexibility to control the reactive power along the feeder as the load changes. Table 10.1 shows some of the different advantages of distributed versus zone substation capacitors.

There are several control strategies available for switched capacitors, the number of which is mainly constrained by the needs and the complexity of the capacitor bank. Some of the more well used and documented techniques are

- Voltage control: Capacitor switching relies on a known upper and lower voltage limit set by the utility. The bank is switched in and switched out at those voltage limits.
- Time control: It does not require any measured values, as capacitor switching is solely based on time. This approach is often used with well-known, time-dependent load conditions.
- Reactive-power control: Capacitor switching is performed when the measured reactive power flow of the line exceeds a preset level.

### 10.2.1 Open-Rack Shunt Capacitor Bank

The open-rack shunt capacitor bank [49] is one of the most common types of capacitor banks used by utilities. It is available with internally fused,

**TABLE 10.2** Typical Ratings and Locations of Distribution Shunt Capacitor Banks [49]

| Type of shunt capacitor bank | | Maximum power (MVAr) | Maximum voltage (V)] | Location |
|---|---|---|---|---|
| Open-rack | Internally fused | Up to 600 | Up to 700 | Outdoor |
| | Externally fused | 1.2–600 | 2.4–765 | Outdoor |
| | Fuseless | 1.2–600 | 35–765 | Outdoor |
| Pole-mounted | | 150–3600 | 2.4–36 | |
| Modular | | Up to 100 | 5–138 | |
| Enclosed fixed | | 15 | 24 | |
| Enclosed switched | | 1000 kVAr/step | 24 | |

externally fused, or fuseless capacitor units (Table 10.2). The major advantage with this type of capacitor bank is its compact design, small footprint, and ease of maintenance.

**Internally Fused Open-Rack Capacitor Bank.** These banks have internal fuses for the current-limiting action; one fuse is connected in series with each element within the capacitor. They are designed and coordinated to isolate internal faults at the capacitor-element level and allow continued operation of the remaining elements of that capacitor unit. A very small part of the capacitor is being disconnected (self-healing), and therefore the capacitor bank remains in service. The designs are customized per customer specifications and generally include racks, capacitors, insulators, switches, imbalance protection scheme, and current-limiting reactors. An internally fused open-rack capacitor bank provides the following main advantages:

- cost-effective,
- utilizes large kVAr size units,
- few live parts are exposed,
- well-known technology (over 50 years of experience), and
- decreased installation and maintenance time and cost with proven reliability in the field.

**Externally Fused Open-Rack Capacitor Bank.** These open-rack capacitor banks [49] are typically used at substations and utilize an external fuse (current limiting) as a means of unit protection. The designs are customized per customer specifications and generally include racks, capacitors, insulators, fuses, elevation structures, switches, imbalance protection scheme, and current-limiting reactors. An externally fused open-rack capacitor bank provides the following main advantages:

- visual indication of blown fuses,
- utilizes small kVAr size units, and
- well-known technology (over 70 years of experience).

**Fuseless Open-Rack Capacitor Bank.** These types of open-rack substation banks [49] utilize an imbalance protection scheme as the primary means of protection. There are no internal or external fuses in this design. The "string" configuration provides a very reliable and cost-effective design for banks rated 35 kV and above. A fuseless open-rack capacitor bank provides the following main advantages:

- cost-effective and compact size,
- utilizes large kVAr size units,
- few live parts are exposed,
- reduced installation and maintenance time as well as reduced cost, and
- proven reliability in the field.

### 10.2.2 Pole-Mounted Capacitor Bank

Pole-mounted capacitor banks [49] provide an economical way to apply capacitors to a distribution feeder. The aim is to provide voltage support, lower system losses, release system capacity, and eliminate power-factor penalties. They are normally factory prewired and assembled, ready for installation. Table 10.2 outlines some of the typical ratings of pole-mounted capacitor banks.

Pole-mounted capacitor banks are typically supplied with the following set of options:

- aluminum or galvanized steel rack,
- oil or vacuum switches,
- junction box,
- switching controller,
- wildlife guards,
- insulated conductor,
- control power transformer,
- distribution class arresters,
- fused cutout, and
- current-limiting reactors.

### 10.2.3 Modular Capacitor Bank

The modular capacitor bank [49] is preengineered with a power circuit breaker, protection, and control panel, all factory mounted and tested on a steel structure. A modular capacitor bank requires minimal field installation and commissioning work. Table 10.2 outlines some of the typical power and voltage ratings of a modular capacitor bank.

A modular capacitor bank provides the following main advantages:

- many standardized configurations with flexibility to fit customer needs,
- factory tested and assembled, reducing environmental project delays and enhancing quality,
- available as a stand-alone unit or integrated into preexisting substation,
- easily relocatable, and
- future expandability with lower costs than traditional banks.

### 10.2.4 Enclosed Fixed Capacitor Bank

The enclosed fixed bank [49] is a fully insulated and fixed reactive compensation system. Its compact design and its full insulation makes the enclosed fixed bank easy to place and it requires no fencing or extra protection. Table 10.2 outlines some of the typical power and voltage ratings of an enclosed fixed capacitor bank.

### 10.2.5 Enclosed Switched Capacitor Bank

The enclosed switched bank [49] is a reactive compensation system with modular, multistage switched capacitor steps that will automatically compensate the network to maintain a preset power-factor level. Table 10.2 outlines some of the typical power and voltage ratings of an enclosed switched capacitor bank.

The enclosed switched bank is available with the following options:

- inrush or detuning reactor,
- cooling or heating,
- imbalance protection,
- stainless steel,
- user defined step sizes,
- switched or fixed,
- indoor or outdoor, and
- automatic or manual.

### 10.3 CLASSIFICATION OF CAPACITOR ALLOCATION TECHNIQUES FOR SINUSOIDAL OPERATING CONDITIONS

The problem of capacitor allocation for loss reduction in electric distribution systems has been extensively researched and many documents have been published [1]. Solution techniques for the capacitor

allocation problem can be classified into following categories:

- analytical methods,
- numerical programming (including dynamic programming (DP), local variations (LV), mixed integer programming, and integer quadratic programming),
- heuristic approaches,
- artificial intelligence-based algorithms (including expert systems, simulated annealing, artificial neural networks, fuzzy set theory, and genetic algorithms),
- graph search algorithm,
- particle swarm algorithm,
- tabu search algorithm, and
- sequential quadratic programming.

This section introduces each method and presents their merits and shortcomings.

### 10.3.1 Analytical Methods

All early works on optimal capacitor allocation under sinusoidal operating conditions with linear loads used analytical methods. These algorithms were devised when computing resources were unavailable or expensive. These methods mainly rely on analytical approaches and calculus to determine the maximum of a capacitor savings function. The savings function has often been given by

$$S = K_E \Delta E + K_P \Delta P - K_C C, \qquad (10\text{-}1)$$

where $K_E \Delta E$ and $K_P \Delta P$ are the economic energy and peak-power loss reductions due to capacitor placement, respectively, and $K_C C$ is the capacitor installation cost.

Initial techniques (by Neagle and Samson [2], Cook [3], Schmill [4], Chang [5], and Bae [6]) achieved simple closed-form solutions to maximize some form of the cost function (Eq. 10-1) based on unrealistic assumptions such as constant feeder conductor sizes and uniform current loading. This early work established the famous *two-thirds rule* that advocates for maximum loss reduction, a capacitor rated at two-thirds of the peak-reactive load, and installed at a position two-thirds of the distance along the total feeder length. Despite these unrealistic assumptions on which the two-thirds rule relies, some utilities today still implement their capacitor placement program based on this rule and some capacitor manufacturers list this rule in their application guides.

Later techniques (Grainger [7] and Salama [8]) achieve more accurate results by formulating equivalent normalized feeders with sections of different conductor sizes and nonuniformly distributed loads, as well as the inclusion of switching capacitors and load variations.

**Drawbacks of Analytical Methods.** One drawback of all analytical methods is the assumption of continuous variables. The calculated capacitor sizes may not match the available standard sizes and the optimized placements may not coincide with the physical node locations in the distribution system. The results would need to be rounded up or down to the nearest practical value and may result in an overvoltage situation or a loss saving less than the calculated one.

Therefore, the earlier works provide a rough rule of thumb for capacitor placement on feeders with evenly distributed loads. More sophisticated analytical methods are suitable for distribution systems of considerable sizes, but they require more distribution system information and more time to implement.

### 10.3.2 Numerical Programming Methods

Various numerical programming methods were devised to solve the capacitor allocation problem as computing power became more readily available and computer memory less expensive. These are iterative optimization techniques used to maximize (or minimize) an objective function of decision variables. The values of the decision variables must also satisfy a set of constraints. For optimal capacitor allocation, the saving represents the objective function, and operational restrictions (e.g., locations, sizes, and numbers of capacitors, bus voltages, and currents) are the decision variables that must all satisfy operational constraints. Numerical programming methods allow the use of more elaborate cost functions for the optimal capacitor placement problem and the objective functions can consider all voltage and line loading constraints, discrete sizes of capacitors, and physical locations of nodes. Using numerical programming methods, the capacitor allocation problem can be formulated as follows:

$$\text{maximize} \quad S = K_L \Delta L - K_C C$$
$$\text{subject to} \quad \Delta V \le \Delta V_{max} \quad , \quad (10\text{-}2)$$

where $K_L \Delta L$ is the cost saving, which may include energy and peak-power loss reductions and released capacity, $K_C C$ is the installation costs of the capaci-

tors, and $\Delta V$ is the change in voltage due to capacitor installation, which must not exceed a maximum of $\Delta V_{max}$.

The first applications of dynamic programming (DP) to the capacitor placement problem (by Duran [9] and Fawzi *et al.* [10]) were simple algorithms considering only the energy-loss reduction, released apparent power, and the discrete nature of capacitor sizes. Later works include applications of local variations (by Ponnavaikko and Rao [11]), mixed integer programming and integer quadratic programming (by Baran and Wu [12]) to account for the effects of load growth, switched capacitors for varying load, and regulators in a distribution system.

**Drawbacks of Numerical Programming Methods.** The level of sophistication and the complexity of the numerical programming models increase in chronological order of their publication time. This progression concurs with the advancement in computing capability. At the present time, computing is relatively inexpensive and many general numerical optimization packages are available to implement any of the above algorithms. Some of the numerical programming methods have the advantage of considering feeder node locations and capacitor sizes as discrete variables, which is an advantage over analytical methods. However, the data preparation and interface development for numerical techniques may require more time than for analytical methods. The main drawback of all numerical programming approaches is the possibility of arriving at a local optimal solution. Therefore, one must determine the convexity of the capacitor placement problem to determine whether the result yielded by a numerical programming technique is either a local or global extreme.

### 10.3.3 Heuristic Methods

Heuristics are rules of thumb that are developed through intuition, experience, and judgment. They produce fast and practical strategies that reduce the exhaustive search space and can lead to a solution that is near optimal with a good level of confidence. In applying heuristic search strategies to the optimum placement and sizing of capacitors in a distribution system, often a small number of nodes, named sensitive nodes, are selected for installing capacitors that optimize the net savings while achieving a large overall loss reduction. This is to decrease the exhaustive search space, while keeping the end result (objective function) at an optimal or near optimal value. Heuristic approaches, due to the reduced computational time compared to other techniques, can be suitable for large distribution systems and also in on-line implementation.

Abdel-Salam *et al.* [13] proposed a heuristic technique to determine the sensitive nodes in sections of the distribution system that have the greatest effect on loss reduction and then maximized the power-loss reduction from capacitor compensation. Chis *et al.* [14] improved on the work of [13] by determining the sensitive nodes that have the greatest impact on loss reduction for the entire distribution system and the effect of load variations.

**Drawbacks of Heuristic Methods.** Heuristic methods are intuitive, easier to understand, and simpler to implement than other optimization techniques such as analytical and numerical programming approaches. However, the results produced by heuristic algorithms are not guaranteed to be optimal. Other disadvantages associated with the use of a heuristic search algorithm are as follows:

- The criterion of selecting the nodes with the highest impact on the losses in a feeder section is not totally representative for the state of the power losses of the whole system and may result in inaccurate results;
- Compensation is based on loss reduction in the system rather than net dollar savings;
- It requires a large number of power flow runs;
- It only considers fixed capacitors; and
- It does not have the flexibility to control the balance between the global and local exploration of the search space, and its results may not be globally optimal.

### 10.3.4 Artificial Intelligence–Based (AI-Based) Methods

The popularity of artificial intelligence (AI) has led many researchers to investigate its applications in power system engineering and electrical machines. In particular, genetic algorithms (GAs), simulated annealing (SA), expert systems (ESs), artificial neural networks (ANNs), and fuzzy set theory (FST) have been proposed and implemented for optimal placement and sizing of shunt capacitor banks.

All of the above AI methods can be implemented using commercially available AI development shells or be hardcoded using any programming language with relative ease.

**Drawbacks of AI-Based Methods.** The user may encounter convergence problems with GAs, SA, and

ANNs methods. For on-line applications, ANNs can only be used for a particular load pattern. ANNs would need to be retrained for every different set of load curves that are characteristic of the distribution system. However, the use of ESs is better suited for on-line and dynamic applications.

#### 10.3.4.1 Genetic Algorithms

The progressive introduction of genetic algorithms (GAs) as powerful tools allows complicated, multi-nodal, discrete optimization problems to be solved. Genetic algorithms use biological evolution to develop a series of search space points toward an optimal solution. Similar to the natural evolution of life, GAs tend to evolve a group of initial poorly generated solutions to a set of acceptable solutions through successive generations. This approach involves coding of the parameter set rather than working with the parameters themselves. GAs operate by selecting a population of the coded parameters with the highest fitness levels (e.g., parameters yielding the best results) and performing a combination of mating, crossover, and mutation operations on them to generate a better set of coded parameters.

Most traditional optimization methods move from one point in the decision hyperspace to another using some deterministic rule. The problem with this is that the solution is likely to get trapped at a local optimum. Genetic algorithms start with a diverse set (population) of potential solutions (hyperspace vectors). This allows for the exploration of many optima in parallel, lowering the probability of getting trapped at a local optimum. Genetic algorithms are simple to implement and are capable of locating the global optimal solution.

Boone and Chiang [15] devised a method based on GAs to determine optimal capacitor sizes and locations. The sizes and locations of capacitors are encoded into binary strings, and crossover is performed to generate a new population. However, their problem formulation only considered the costs of the capacitors and the reduction of peak-power losses. Sundhararajan and Pahwa [16] performed a similar work with the exception of including energy losses and implementing an elitist strategy whereby the coded strings chosen for the next generation do not undergo mutation or crossover procedures. Miu *et al.* [17] revisited the GA formulation to include additional features of capacitor replacement and control for unbalanced distribution systems.

There are five components that are required to implement a genetic algorithm: representation,

**FIGURE 10.2** Basic chromosome structure.

initialization, evaluation function, genetic operators, and genetic parameters.

**Genetic Representation.** Genetic algorithms are derived from a study of biological systems. In biological systems evolution takes place on organic components used to encode the structure of living beings. These organic components are known as chromosomes. A living being is only a decoded structure of the chromosomes. Natural selection is the link between chromosomes and the performance of their decoded structures. In genetic algorithms the design variables or features that characterize an individual are represented in an ordered list called a string. Each design variable corresponds to a gene and the string of genes corresponds to a chromosome. A group of chromosomes is called a population. The number of chromosomes in a population is usually selected to be between 30 and 300. Each chromosome is a string of binary codes (genes) and may contain substrings. The merit of a string is judged by the fitness function, which is derived from the objective function and is used in successive genetic operations. During each iterative procedure (referred to as generation), a new set of strings with improved performance is generated using three GA operators (namely, reproduction, crossover, and mutation).

**Structure of Chromosomes.** Figure 10.2 shows an example illustrating the basic structure of a chromosome. Bit strings consisting of 0's and 1's have been found to be most effective in representing a wide variety of information in the problem domain involving function optimization. Many other representation schemes such as ordered lists, embedded lists, and variable-element lists have been developed for specific industrial applications of genetic algorithms.

**Initialization.** Genetic algorithms operate with a set of strings instead of a single string. This set of strings is known as a population and is put through the process of evolution to produce new individual strings. To begin with, the initial population could be seeded with heuristically chosen strings or at random. In either case, the initial population should contain a wide variety of structures to ensure the complete problem space is covered.

**Evaluation (Fitness) Function.** The evaluation function is a procedure to determine the fitness of each chromosome in the population and is very much application orientated. Since genetic algorithms proceed in the direction of evolving the fittest chromosomes, and the fitness value is the only information available to the algorithm, the performance is highly sensitive to the fitness values. In the case of optimization routines, the fitness is the value of the objective function to be optimized. Genetic algorithms are basically unconstrained search procedures in the given problem domain. Any constraints associated with the problem can be incorporated into the objective function as penalty functions.

**Genetic Operators.** Genetic operators are the stochastic transition rules applied to each chromosome during each generation procedure to generate a new improved population from an old one. A genetic algorithm usually consists of reproduction, crossover, and mutation operators.

- *Reproduction* is a probabilistic process for selecting two parent strings from the population of strings on the basis of roulette-wheel mechanism, using their fitness values. This process is accomplished by calculating the fitness function for each chromosome in the population and normalizing their values. A roulette wheel is constructed where each section of the wheel represents a certain chromosome and its relative fitness value. This ensures that the expected number of times a string is selected is proportional to its fitness relative to the rest of the population. Therefore, strings with higher fitness values have a higher probability of contributing offspring and are simply copied on into the next generation.
- *Crossover* is the process of selecting a random position in the string and swapping the characters either left or right of this point with another similarly partitioned string. This random position is called the crossover point. For example, if two parent strings are $x_1 = 1010 : 11$ and $x_2 = 1001 : 00$ and the crossover point has been selected as shown by the colon, then if swapping is implemented to the right of this point, the resulting offspring structures would be $y_1 = 1010 : 00$ and $y_2 = 1001 : 11$ (e.g., the two digits "11" to the right of the colon in $x_1$ are swapped with the two digits "00" to the right of the colon in $x_2$). The probability of

crossover occurring for two parent chromosomes is usually set to a large value in range of 0.7 to 1.0.

- *Mutation* is the process of random modification of a string position by changing "0" to "1" or vice versa, with a small probability. It prevents complete loss of genetic material through reproduction and crossover by ensuring that the probability of searching any region in the problem space is never zero. The probability of mutation is usually assumed to be small (e.g., between 0.01 and 0.1).

**Genetic Parameters.** Genetic parameters are the entities that help to regulate and improve the performance of a genetic algorithm. Some of the parameters that can characterize the genetic space are as follows:

- *Population size* affects the efficiency and performance of the genetic algorithm. A smaller population would not cover the entire problem space, resulting in poor performance. A larger population would cover more space and prevent premature convergence to local solutions; however, it requires more evaluations per generation and may slow down the convergence rate.
- *Crossover rate* affects the rate at which the process of crossover is applied. In each new population, the number of strings that undergo the process of crossover can be depicted by a chosen probability (crossover rate). A higher crossover rate introduces new strings more quickly into the population; however, if the rate is too high, high-performance strings are eliminated faster than selection can produce improvements. A low crossover rate may cause stagnations due to the lower exploration rate, and convergence problems may occur.
- *Mutation rate* is the probability with which each bit position of each string in the new population undergoes a random change after the selection process. A low mutation rate helps to prevent any bit position from getting stuck to a single value, whereas a high mutation rate can result in essentially random search.

**Convergence Criterion.** The iterations (regenerations) of genetic algorithms are continued until all generated chromosomes become equal or the maximum number of iterations is achieved. Due to the randomness of the GA method, the solution tends to differ for each run, even with the same initial population. For this reason, it is suggested to perform

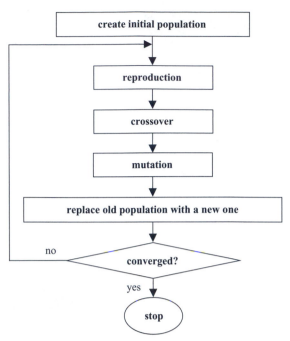

**FIGURE 10.3** Simplified flowchart of a typical genetic algorithm structure.

multiple runs and select the "most acceptable" solution (e.g., with most benefits, within the permissible region of constraints).

**Flowchart of Genetic Algorithm for Capacitor Placement and Sizing.** Figure 10.3 shows a simplified flowchart of an iterative genetic algorithm for optimal placement and sizing of capacitor banks in distribution system under sinusoidal operating conditions. An initial population is randomly selected, genetic operators are applied, and the old population is replaced with the new generation. This process is repeated until the algorithm has converged.

**Drawbacks of Genetic Algorithms.** There are disadvantages associated with the use of a genetic algorithm:

- It does not have flexibility to control the balance between global and local exploration of search space;
- It requires relatively large computational time and effort; and
- Convergence problems may occur, which may be overcome by solving the problem a number of times with different initial populations and selecting the best solution.

**TABLE 10.3** Yearly Costs of Fixed Capacitors [37, 45]

| $Q_p$ (kVAr) | 150 | 300 | 450 | 600 | 750 | 900 |
|---|---|---|---|---|---|---|
| $K_{cfp}$ ($/kVAr) | 0.500 | 0.350 | 0.253 | 0.220 | 0.276 | 0.183 |
| $Q_p$ (kVAr) | 1050 | 1200 | 1350 | 1500 | 1650 | 1800 |
| $K_{cfp}$ ($/kVAr) | 0.228 | 0.170 | 0.207 | 0.201 | 0.193 | 0.187 |
| $Q_p$ (kVAr) | 1950 | 2100 | 2250 | 2400 | 2550 | 2700 |
| $K_{cfp}$ ($/kVAr) | 0.211 | 0.176 | 0.197 | 0.170 | 0.189 | 0.187 |
| $Q_p$ (kVAr) | 2850 | 3000 | 3150 | 3300 | 3450 | 3600 |
| $K_{cfp}$ ($/kVAr) | 0.183 | 0.180 | 0.195 | 0.174 | 0.188 | 0.170 |

### 10.3.4.2 Expert Systems

Expert systems (ESs) or knowledge-based systems consist of a collection of rules, facts (knowledge), and an inference engine to perform logical reasoning. The ES concept has been proposed and implemented for many power system problems that require decision-making, empirical judgments, or heuristics such as fault diagnosis, planning, scheduling, and reactive power control.

Salama and colleagues [18] developed an ES containing technical literature expertise (TLE) and human expertise (HE) for reactive power control of a distribution system. The TLE includes the capacitor allocation method based on [8] for maximum savings from the reduction of peak-power and energy losses. The HE component of the knowledge base contains information to guide the user to perform reactive power control for the planning, operation, and expansion stages of distribution systems. Application of ESs for the capacitor placement problem is yet to be explored and requires more attention.

### 10.3.4.3 Simulated Annealing

Simulated annealing (SA) is an iterative optimization algorithm that is based on the annealing of solids. When a material is annealed, it is heated to a high temperature and slowly cooled according to a cooling schedule to reach a desired state. At the highest temperature, particles of the material are arranged in a random formation. As the material is cooled, the particles become organized into a lattice-like structure, which is a minimum energy state. For the capacitor allocation problem, a total cost function is formulated instead of a savings function. Analogous to reaching a minimum energy state of the annealing of solids, Ananthapadmanabha *et al.* [19] used SA to minimize a total cost function given by

$$T_{loss} = K_P P_{loss} + K_E E_{loss} + K_C C, \qquad (10\text{-}3)$$

where $C$ is the size of capacitor, $K_P P_{loss}$ is the cost of peak-power losses, $K_E E_{loss}$ is the cost of energy losses, and $K_C C$ is the capacitor installation costs. Also, $K_C = K_{cfp}$ (see Table 10.3) is the cost of kVAr.

### 10.3.4.4 Artificial Neural Networks

Application of artificial neural networks (ANNs) in power engineering is gaining interest. An ANN is the connection of artificial neurons that simulates the nervous system of a human brain. The potential advantage of the ANN is its ability to handle the nonlinear mapping of an input–output space. The difficulty in designing an ANN is to generate the training patterns. To overcome this problem, a reinforcement learning process is usually proposed. An ANN typically consists of three types of layers: an input layer, one or more hidden layers, and an output layer. The ANN accepts given input data and minimizes the difference between known output data and generated output information. The relationships between inputs and outputs are embedded as parameters in the hidden layer. Correct output patterns can be generated by the ANN provided that there are enough hidden layers and nodes to encode the input–output pattern, and enough known data to train the ANN. Once an ANN is trained, it can provide very fast results for any given set of inputs.

Santoso and Tan [20] used ANNs for the optimal control of switched capacitors. In their work, two neural networks are designed and trained. One network predicts the load profile from a set of previous load values obtained from direct measurement at various buses. A second network determines the optimal capacitor tap positions based on the load profile as predicted by the first network. The first network is trained with a set of prerecorded load profiles and the second network is trained to maximize the energy loss reduction for a given load condition. To reduce the complexity of training, the distribution system is partitioned into many smaller subsystems.

**Drawbacks of Artificial Neural Networks.** ANNs may not be suitable for a large distribution system since many smaller subsystems are required and the training time becomes excessive. However, once networks are trained, iterative calculations are no longer required and a fast solution for a given set of inputs can be provided. Gu and Rizy [21] followed the method of [20] and included an additional functionality of regulator control.

### 10.3.4.5 Fuzzy Set Theory

The concept of fuzzy set theory (FST) was introduced by Zadeh [22] in 1965 as a formal tool for dealing with uncertainty and soft modeling. A fuzzy variable is modeled by a membership function, which assigns a degree of membership to a set. Usually, this degree of membership varies from zero to one.

Chin [23] used FST for the capacitor allocation problem. Three membership functions (for power loss, bus voltage deviation, and harmonic distortion) are defined. A fuzzy decision variable (e.g., the intersection of the three membership functions for each node in the distribution system) is calculated to determine the most appropriate buses for capacitor placement. However, no mathematical optimization procedure is provided for capacitor sizing. Ng *et al.* [24–25] applied FST to the capacitor placement problem by using fuzzy approximate reasoning. Voltage and power loss indices of the distribution system buses are modeled by membership functions, and a fuzzy expert system (FES) containing a set of heuristic rules performs the inferencing to determine a capacitor placement suitability index of each node. Capacitors are placed at the nodes with the highest suitability.

**Fuzzy Logic and Fuzzy Inference System.** Fuzzy logic in terms of fuzzy sets is about mapping an input space to an output space. The main operation to achieve this end is through the use of if–then statements called rules. These rules are used to refer to variables and the adjectives that describe those variables. This is the main theory behind the operation of a fuzzy inference system. For the system to interpret these rules, they must first be defined in terms of the data they will control and the adjectives that will be used. A good example of this is a system defining how hot the water is coming out of a hot water system. We need to define the range of the expected water temperatures, as well as what is meant by the word "hot."

**Fuzzy Sets.** A fuzzy set is a set without a crisp, clearly defined boundary. It can contain elements with partial, full, or no membership. This differs quite dramatically from the notion of a classical set, which is defined by the inclusion or exclusion of certain elements.

Fuzzy logic is based on the fact that the truth of any statement becomes a matter of degree. The true power of using fuzzy logic is not that it is an extremely complex tool which can give a direct answer, rather that it can model human perception, language, and thinking, thus being able to use linguistic descriptors to answer a problem that would normally need or expect a yes–no answer with a not-quite-yes–no answer. This is done by expanding the familiar Boolean logic of 1 for yes and 0 for no. This means that if absolute truth is defined by 1 and that absolute falsehood is defined by 0, fuzzy logic will permit alternative logical answers between the range of [0 1] such as 0.35 or 0.6521 and by doing so be able to recreate such linguistic descriptors in relation to an element's relationship to a set as medium, high, low, very low, very high, and so on. Now with the possibility of a range of logical answers a more descriptive set can be created. A plot mapping an input space to an output space is considered to be a membership function.

**Membership Function.** A membership function is a curve that defines how each point in the input space is mapped to a membership value (or degree of membership) between 0 and 1. The main requirement for a membership function is that it varies between 1 and 0. The shape of membership functions are created from such requirements as simplicity, convenience, speed, and efficiency.

A fuzzy set is an extension of a classical set. If $X$ is the universe of discourse and its elements are denoted by $x$, then a fuzzy set $A$ in $X$ is defined as a set of ordered pairs

$$A = \{x, \mu_A(x)|xX\},$$

where $\mu_A(x)$ (called the membership function of $x$ in $A$) maps each element of $X$ to a membership value between 0 and 1.

Many types of membership functions have been proposed and implemented. They can be created from scratch or even tailored or trained to conform to a set of data. There are many types of membership functions such as triangular, trapezoidal, symmetric Gaussian (defines a $f(x;\sigma,c) = \exp(-(x - c)^2/2\sigma^2)$)) and sigmoidal. The possibility of creating member-

**TABLE 10.4** Comparison of Boolean and Fuzzy Logic Operations in Terms of Two Sets A and B

| Boolean operation | Fuzzy logic operation |
|---|---|
| A AND B | min(A,B) |
| A OR B | max(A,B) |
| NOT A | 1 − A |

ship functions based on already existing mathematical formulas and functions is endless, such as membership functions created from the difference and/or product of two sigmoidal functions.

**Fuzzy Logic Operators.** The logical operations associated with fuzzy logic can be described as a superset of standard Boolean logic. The operations in Boolean known as AND, OR, and NOT can all be repeated in fuzzy logic. Shown in Table 10.4 are these three integral Boolean operations with their related logical operations in fuzzy.

**Fuzzy If–Then Rules.** If–then rules are used to define the relationships between variables based on defined logic and membership functions. An if–then rule is created in the form

If $x$ is $A$ then $y$ is $B$.

$A$ and $B$ are linguistic values defined by fuzzy sets on the ranges (universes of discourse) $X$ and $Y$, respectively. The if-part of the rule is called the antecedent (or premise), whereas the then-part of the rule ($y$ is $B$) is called the consequent (or conclusion).

Interpreting if-then rules is a three-part process:

- Fuzzify inputs: Resolve all fuzzy statements in the antecedent to a degree of membership between 0 and 1. If there is only one part to the antecedent, this is the degree of support for the rule.
- Apply fuzzy operator to multiple part antecedents: If there are multiple parts to the antecedent, apply fuzzy logic operators and resolve the antecedent to a single number between 0 and 1. This is the degree of support for the rule.
- Apply implication method: Use the degree of support for the entire rule to shape the output fuzzy set. The consequent of a fuzzy rule assigns an entire fuzzy set to the output. This fuzzy set is represented by a membership function that is chosen to indicate the qualities of the consequent. If the antecedent is only partially true (i.e., is assigned a value less than 1), then the output

fuzzy set is truncated according to the implication method.

**Implication.** Implication is the method used to create a final consequent function or set. The if–then rule is often not sufficient enough and more rules may be necessary to create a satisfying result. Consider an extension to this rule shown below,

If $x$ is $A$ OR $y$ is $B$ then $z$ is $C$.

This *expanded if–then* statement now has two rules involved that are controlled by the logical operation OR. Implication becomes the means by which the consequent is created from this rule, which is the "$z$ is $C$" part.

Two main methods of implication are often used. These are entitled minimum (similar to the AND operation previously discussed) and the product methods. The minimum method will truncate the output fuzzy set, and the product method will scale the output fuzzy set deriving an aggregated output from a set of triggered rules.

**Fuzzy Inference.** The Mamdani implication methods [50] are usually used for inferencing. Two implication methods are relied on to determine the aggregated output from a set of triggered rules:

- *The Mamdani max–min implication method* of truncating the consequent membership function of each discarded rule (at the minimum membership value of all the antecedents) and determining the final aggregated membership function.
- *The Mamdani max–prod implication method* of scaling the consequent membership function of each discarded rule (to the minimum membership value of all the antecedents) and determining the final aggregated membership function.

**Defuzzification.** Defuzzification is the process of obtaining a single number from the output of the aggregated fuzzy set. It is used to transfer fuzzy inference results into a crisp output. In other words, defuzzification is realized by a decision-making algorithm that selects the best crisp value based on a fuzzy set. There are several forms of defuzzification including center of gravity (COG), mean of maximum (MOM), and center average methods. The COG method returns the value of the center of area under the curve and the MOM approach can be regarded as the point where balance is obtained on a curve.

**Fuzzy Expert System for Capacitor Placement and Sizing.** A fuzzy expert system based on approximate reasoning is defined and relied on to determine suitable candidate buses in a distribution system for capacitor placement. Voltages and power loss reduction indices of distribution system buses are modeled by fuzzy membership functions. A fuzzy expert system (FES) containing a set of heuristic rules is then employed to determine the capacitor placement suitability of each bus in the distribution system. Capacitors are then placed on the nodes with the highest suitability using the following steps:

- A load flow program calculates the power-loss reduction (PLR) of the system with respect to each bus when the reactive load current at that bus is compensated. The PLRs are then linearly normalized into a [0, 1] range with the highest loss reduction being 1 and the lowest being 0.
- The PLR indices along with the per-unit nodal voltages (before PLR compensation) are the two inputs that are used by the fuzzy expert system. The FES then determines the most suitable node for capacitor installation by fuzzy inferencing.
- A savings function relating to the annual cost of the capacitor (determined by size) and the savings achieved from the power-loss reduction is maximized to determine the optimal sizing of capacitor bank to be placed at that node.
- The above process will be repeated, finding another node for capacitor placement, until no more financial savings can be realized (Fig. 10.4).

**Drawbacks of Fuzzy Set Theory.** The fuzzy optimization approach is very simple and naturally fast as compared to other optimization methods. However, for some problems the procedure might get trapped in a local optimal point and fail to converge to the global (or near global) optimal solution.

### 10.3.5 Graph Search Algorithm

In a graph search algorithm a graph is commonly defined by a set of nodes connected by arcs. Each node corresponds to a possible combination of capacitor sizes and locations for a given number of capacitors to be placed within the system. A separate graph is used for each possible number of capacitors to be placed at a certain location. This makes it possible to search the nodes of a graph in an attempt to determine the optimal solution. Beginning with some minimum number of capacitors to be placed (user-defined parameter), a search is performed to locate

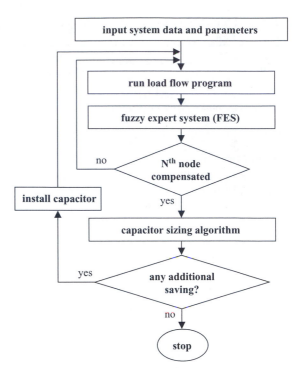

**FIGURE 10.4** Flowchart of an iterative fuzzy algorithm for optimal placement and sizing of capacitor banks in distribution system under sinusoidal operating conditions.

the node in the graph that produces maximum savings. The mathematical simplicity of this method makes it possible to include many features (e.g., capacitor location constraints) in the algorithm that would be rather difficult with other solution methods. Carlisle [26] developed graph search algorithms for the capacitor placement and sizing problem.

**Drawbacks of Graph Search Algorithm.** Many assumptions and constraints are required to obtain an optimal solution. This is not always possible for the operating conditions of actual distribution systems. Therefore, solutions based on graph search algorithms may not be very accurate.

### 10.3.6 Particle Swarm Algorithm

Similar to an evolutionary algorithm, the particle swarm optimization (PSO) technique conducts searches using a population of particles, corresponding to individuals. PSO systems combine two models:

- A social-only model suggesting that individuals ignore their own experience and adjust their behavior according to the successful beliefs of the individual in their neighborhood.

- A cognition-only model treating individuals as isolated beings.

A particle (representing a candidate solution to the optimal reactive power problem) changes its position in a multidimensional search space using these models. PSO uses payoff (performance index or objective function) information to guide the search in the problem space and can easily deal with nondifferentiable objective functions. Additionally, this property relieves PSO of many assumptions and approximations that are often required by traditional optimization models. Mantawy [27] proposed a particle swarm optimization algorithm for reactive power optimization problem.

**Drawbacks of Particle Swarm Algorithm.** The performance of particle swarm optimization is not competitive with respect to some problems. In addition, representation of the weights is difficult and the genetic operators have to be carefully selected or developed.

### 10.3.7 Tabu Search Algorithm

The tabu search strategy is a metaheuristic search method superimposed on another heuristic approach and is employed for solving combinatorial optimization problems. It has been applied for various problems and its capability to obtain high-quality solutions with reasonable computing times has been demonstrated. The approach is to avoid entrainment in cycles by penalizing moves that take the solution to points in the solution space previously visited (hence "tabu"). This ensures new regions of solution space will be investigated with the goal of avoiding local minima and ultimately capturing the optimal solution. The tabu search method is built on a descent mechanism of a search process that biases the search toward points with a lower objective function, while special features are added to avoid being trapped at a local minimum. To avoid retracing previous steps, recent moves (iterations) are recorded in one or more tabu lists (memories). The basic differences between various implementations of the tabu method are the size, variability, and adaptability of the tabu memory to a particular problem domain. Yang *et al.* [28], Mori [29], and Chang [30] applied the tabu search algorithm to the capacitor placement problem in radial distribution systems.

**Drawbacks of Tabu Search Algorithm.** The main disadvantage of tabu approach is its low solution speed. In addition, it uses certain control parameters that may be system dependent and difficult to determine.

### 10.3.8 Sequential Quadratic Programming

Sequential quadratic programming is a generalized Newton method for constrained optimization. This algorithm replaces the objective function with quadratic approximation and constraint functions with linear approximations to solve a quadratic programming problem at each iteration.

Abril [31] presents an application of the sequential quadratic programming algorithm to find the optimal sizing of shunt capacitors and/or passive filters for maximizing the net present value resulting from energy-loss reduction while taking investment cost into account and complying with set standards. The approach considers fixed and switched reactive power (VAr) compensators by considering a characteristic load variation curve for each linear and nonlinear load.

**Drawbacks of Sequential Quadratic Programming.** Large numbers of computations are involved in sequential quadratic programming. Many mathematical assumptions have to be made in order to simplify the problem; otherwise it is very difficult to calculate the variables.

### 10.3.9 Application Example 10.1: Fuzzy Capacitor Placement in an 11 kV, 34-Bus Distribution System with Lateral Branches under Sinusoidal Operating Conditions

Application of the fuzzy set theory for the capacitor allocation problem will be demonstrated for the 34-bus, three-phase, 11 kV radial distribution system of Fig. E10.1.1. The line and load data of this system are given in Tables E10.1.1 and E10.1.2, respectively [32]. One load level with fixed (not switched) capacitors under balanced operating conditions is assumed. The cost of fixed capacitor banks are listed in Table 10.3 where the size of the capacitor ($Q_P$ in kVAr) as a function of the yearly cost ($F_{cost}$ in \$/kVAr) is given by $K_{cfp}$ (e.g., $F_{cost} = K_{cfp}Q_p$).

### 10.3.10 Application Example 10.2: Genetically Optimized Placement of Capacitor Banks in an 11 kV, 34-Bus Distribution System with Lateral Branches under Sinusoidal Operating Conditions

Solve Application Example 10.1 using a genetic algorithm.

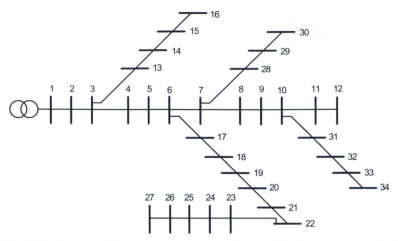

**FIGURE E10.1.1** Single-line diagram of the 11 kV, 34-bus, three-phase, radial distribution system [32].

**TABLE E10.1.1** Bus Data [32] of the 11 kV, 34-Bus, Three-Phase, Radial Distribution System (Fig. E10.1.1)

| Bus number | Load | | Generator | | | | Injected MVAr |
|---|---|---|---|---|---|---|---|
| | MW | MVAr | MW | MVAr | Qmin | Qmax | |
| 1 | 0.0000 | 0.0000 | 0 | 0 | 0 | 0 | 0 |
| 2 | 0.2300 | 0.1425 | 0 | 0 | 0 | 0 | 0 |
| 3 | 0.0000 | 0.0000 | 0 | 0 | 0 | 0 | 0 |
| 4 | 0.2300 | 0.1425 | 0 | 0 | 0 | 0 | 0 |
| 5 | 0.2300 | 0.1425 | 0 | 0 | 0 | 0 | 0 |
| 6 | 0.0000 | 0.0000 | 0 | 0 | 0 | 0 | 0 |
| 7 | 0.0000 | 0.0000 | 0 | 0 | 0 | 0 | 0 |
| 8 | 0.2300 | 0.1425 | 0 | 0 | 0 | 0 | 0 |
| 9 | 0.2300 | 0.1425 | 0 | 0 | 0 | 0 | 0 |
| 10 | 0.0000 | 0.0000 | 0 | 0 | 0 | 0 | 0 |
| 11 | 0.2300 | 0.1425 | 0 | 0 | 0 | 0 | 0 |
| 12 | 0.1370 | 0.0840 | 0 | 0 | 0 | 0 | 0 |
| 13 | 0.0720 | 0.0450 | 0 | 0 | 0 | 0 | 0 |
| 14 | 0.0720 | 0.0450 | 0 | 0 | 0 | 0 | 0 |
| 15 | 0.0720 | 0.0450 | 0 | 0 | 0 | 0 | 0 |
| 16 | 0.0135 | 0.0075 | 0 | 0 | 0 | 0 | 0 |
| 17 | 0.2300 | 0.1425 | 0 | 0 | 0 | 0 | 0 |
| 18 | 0.2300 | 0.1425 | 0 | 0 | 0 | 0 | 0 |
| 19 | 0.2300 | 0.1425 | 0 | 0 | 0 | 0 | 0 |
| 20 | 0.2300 | 0.1425 | 0 | 0 | 0 | 0 | 0 |
| 21 | 0.2300 | 0.1425 | 0 | 0 | 0 | 0 | 0 |
| 22 | 0.2300 | 0.1425 | 0 | 0 | 0 | 0 | 0 |
| 23 | 0.2300 | 0.1425 | 0 | 0 | 0 | 0 | 0 |
| 24 | 0.2300 | 0.1425 | 0 | 0 | 0 | 0 | 0 |
| 25 | 0.2300 | 0.1425 | 0 | 0 | 0 | 0 | 0 |
| 26 | 0.2300 | 0.1425 | 0 | 0 | 0 | 0 | 0 |
| 27 | 0.1370 | 0.0850 | 0 | 0 | 0 | 0 | 0 |
| 28 | 0.0750 | 0.0480 | 0 | 0 | 0 | 0 | 0 |
| 29 | 0.0750 | 0.0480 | 0 | 0 | 0 | 0 | 0 |
| 30 | 0.0750 | 0.0480 | 0 | 0 | 0 | 0 | 0 |
| 31 | 0.0570 | 0.0345 | 0 | 0 | 0 | 0 | 0 |
| 32 | 0.0570 | 0.0345 | 0 | 0 | 0 | 0 | 0 |
| 33 | 0.0570 | 0.0345 | 0 | 0 | 0 | 0 | 0 |
| 34 | 0.0570 | 0.0345 | 0 | 0 | 0 | 0 | 0 |

**TABLE E10.1.2** Line Data [32] of the 11 kV, 34-Bus, 3-Phase, Radial Distribution System (Fig. E10.1.1)

**TABLE E10.1.2** Line Data [32] of the 11 kV, 34-Bus, 3-Phase, Radial Distribution System (Fig. E10.1.1)

| Line | | | | |
|------|------|-----------------|-----------------|-------------|
| From bus | To bus | $R_{line}$ (pu) | $X_{line}$ (pu) | Tap setting |
| 1 | 2 | 0.0967 | 0.0397 | 1 |
| 2 | 3 | 0.0886 | 0.0364 | 1 |
| 3 | 4 | 0.1359 | 0.0377 | 1 |
| 4 | 5 | 0.1236 | 0.0343 | 1 |
| 5 | 6 | 0.1236 | 0.0343 | 1 |
| 6 | 7 | 0.2598 | 0.0446 | 1 |
| 7 | 8 | 0.1732 | 0.0298 | 1 |
| 8 | 9 | 0.2598 | 0.0446 | 1 |
| 9 | 10 | 0.1732 | 0.0298 | 1 |
| 10 | 11 | 0.1083 | 0.0186 | 1 |
| 11 | 12 | 0.0866 | 0.1488 | 1 |
| 3 | 13 | 0.1299 | 0.0223 | 1 |
| 13 | 14 | 0.1732 | 0.0298 | 1 |
| 14 | 15 | 0.0866 | 0.1488 | 1 |
| 15 | 16 | 0.0433 | 0.0074 | 1 |
| 6 | 17 | 0.1483 | 0.0412 | 1 |
| 17 | 18 | 0.1359 | 0.0377 | 1 |
| 18 | 19 | 0.1718 | 0.0391 | 1 |
| 19 | 20 | 0.1562 | 0.0355 | 1 |
| 20 | 21 | 0.1562 | 0.0355 | 1 |
| 21 | 22 | 0.2165 | 0.0372 | 1 |
| 22 | 23 | 0.2165 | 0.0372 | 1 |
| 23 | 24 | 0.2598 | 0.0446 | 1 |
| 24 | 25 | 0.1732 | 0.0298 | 1 |
| 25 | 26 | 0.1083 | 0.0186 | 1 |
| 26 | 27 | 0.0866 | 0.1488 | 1 |
| 7 | 28 | 0.1299 | 0.0223 | 1 |
| 28 | 29 | 0.1299 | 0.0223 | 1 |
| 29 | 30 | 0.1299 | 0.0223 | 1 |
| 10 | 31 | 0.1299 | 0.0223 | 1 |
| 31 | 32 | 0.1732 | 0.0298 | 1 |
| 32 | 33 | 0.1299 | 0.0223 | 1 |
| 33 | 34 | 0.0866 | 0.1488 | 1 |

## 10.4 OPTIMAL PLACEMENT AND SIZING OF SHUNT CAPACITOR BANKS IN THE PRESENCE OF HARMONICS

The problem of shunt capacitor allocation in distribution networks for reactive power compensation, voltage regulation, power factor correction, and power or energy loss reduction is discussed in the previous section. Optimal capacitor bank placement is a well-researched subject. However, limited attention is given to this problem in the presence of voltage and current harmonics.

With the increasing application of nonlinear loads and (power electronic and electromagnetic) devices causing voltage and current harmonics in distribution networks, special attention should be given to placement and sizing of capacitors in such systems with nonsinusoidal voltages and currents. If (shunt) capacitor banks are not properly sized and placed in such a power system, they may cause various operational and power quality problems:

- amplification and propagation of current and voltage harmonics,
- deterioration of power quality to unacceptable levels,
- increased losses at fundamental and harmonic frequencies,
- extensive reactive power demand and overvoltages at fundamental and harmonic frequencies,
- decreased capacities of distribution lines and transformers,
- poor power-factor conditions, and
- harmonic parallel resonances.

The problem of optimal capacitor placement and sizing under nonsinusoidal operating conditions is a newly researched area with limited publications that requires more attention. Early works by Baghzouz [35, 36], Gou [37], Hsu et al. [39], Wu [40], and Chin [43] capture local optimal points and ignore harmonic couplings. This will result in inaccurate solutions. More recent documents by Masoum et al. [41, 42, 44–46] achieve accurate (near) global optimal solutions by incorporating artificial intelligence–based (AI-based) techniques and relying on a harmonic power flow that takes into account harmonic couplings caused by nonlinear loads. Document [38] addresses the problem under unbalanced three-phase conditions.

Proposed solution techniques for the capacitor allocation problem in the presence of harmonics can be classified into following categories:

- exhaustive search [35],
- local variations [36],
- mixed integer-nonlinear programming [37, 38],
- heuristic methods [39],
- maximum sensitivities selection [40–42],
- fuzzy set theory [43, 44],
- genetic algorithms [45], and
- genetic algorithm with fuzzy logic (approximate reasoning) [46].

Most of these techniques are fast, but they suffer from the inability to escape local optimal solutions. Simulated annealing (SA), tabu search (TS), and genetic algorithms (GAs) are three near global optimization techniques that have demonstrated fine capabilities for capacitor placement, but the computational burden is heavy. This section introduces some of these methods and presents their merits and shortcomings.

### 10.4.1 Reformulation of the Capacitor Allocation Problem to Account for Harmonics

#### 10.4.1.1 System Model at Fundamental and Harmonic Frequencies

A harmonic load flow algorithm is required to model the distribution system with nonlinear loads at fundamental and harmonic frequencies. This book relies on the Newton-based harmonic load flow formulation and notations of Chapter 7 (Section 7.4). System solution is achieved by forcing total (fundamental and harmonic) mismatch active and reactive powers as well as mismatch active and reactive fundamental and harmonic currents to zero.

Define bus 1 to be the swing bus, buses 2 through $m - 1$ to be the linear ($PQ$ and $PV$) buses, and buses m through n as nonlinear buses ($n$ = total number of buses). We assume that nonlinear load models – representing the coupling between harmonic voltages and currents – are given either in the frequency domain (e.g., $\tilde{V}^{(h)}$ and $\tilde{I}^{(h)}$ characteristics) or in the time domain (e.g., $v(t)$ and $i(t)$ characteristics). These models are available for many nonlinear loads and systems such as power electronic devices, nonlinear transformers, discharge lighting, and EHV and HVDC networks [44]. In the above formulation of harmonic power flow, we assume that the capacitors are shunt capacitor banks with variable reactances, and capacitor placement is possible for MC number of buses (candidate buses). We also use the proposed formulation of [35] in the harmonic power flow for the admittance of linear loads at harmonic frequencies:

$$y_i^{(h)} = \frac{P_i^{(1)} - jQ_i^{(1)}/h}{\left|V_i^{(1)}\right|^2} \quad \text{for } i = 2, \ldots, m-1. \quad (10\text{-}4)$$

#### 10.4.1.2 Constraints

Voltage constraints will be taken into account by specifying upper (e.g., $V^{max} = 1.1$ pu) and lower (e.g., $V^{min} = 0.9$ pu) bounds of rms voltage ($n$ = total number of buses, $i$ = bus number, $h$ = harmonic order):

$$V^{min} \leq \sqrt{\sum_h \left(V_i^{(h)}\right)} \leq V^{max} \quad \text{for } i = 1, \ldots, n. \quad (10\text{-}5)$$

The distortion of voltage is considered by defining the maximum total harmonic distortion ($THD_v$) of voltages:

$$THD_{v,i} = \left(\left[\sqrt{\sum_{h \neq 1} (V_i^{(h)})^2}\right] \middle/ V_i^{(1)}\right) \cdot 100\% \leq THD_v^{max}$$
$$\text{for } i = 1, \ldots, n. \quad (10\text{-}6)$$

Bounds for Eq. 10-6 are specified by the IEEE-519 standard [51]. Let $u^{max}$ denote the maximum number of capacitor units allowed at each bus. The number of capacitors on bus $i$ is limited by

$$u = u_{fi} \leq u^{max} \quad \text{for } i \in MC, \quad (10\text{-}7)$$

where $u_{fi}$ denotes the number of fixed capacitor banks at bus $i$ and $MC$ is the set of possible buses for capacitor placement.

#### 10.4.1.3 Objective Function (Cost Index)

The objective function (*cost*) selected for capacitor placement in the presence of harmonics is [35–37, 44–46]

$$\min F = F_{loss} + F_{cap.} + F_{capacity}$$
$$= K_E T P_{loss}\left(V^{(1)}, \ldots, V^{(L)}, C\right) + \sum_{i \in MC} K_{cfp} C_{fi}$$
$$+ K_A P_{loss}\left(V^{(1)}, \ldots, V^{(L)}, C\right), \quad (10\text{-}8)$$

where

$F_{loss}$ = energy loss cost,

$F_{cap.}$ = cost of fixed capacitors,

$F_{capacity}$ = cost corresponding to losses (e.g., used capacity of the system),

$P_{loss}$ = total system losses,

$V^{(h)}$ = bus voltage vector at harmonic $h$,

$L$ = highest order of harmonics considered,

$C$ = size of connected capacitors,

$T$ = duration of load (hours/year),

$MC$ = set of possible shunt capacitor buses,

$K_{cfp}$ = cost per unit of fixed capacitance (Table 10.3 [37, 45]),

$K_E$ = cost per MWh (e.g., $K_E = 50$ \$/MWh [40, 45]), and

$K_A$ = saving per MW for reduction in losses (e.g., $K_A = 120,000$ \$/MW [40]).

Total losses can be computed using harmonic power flow output results

$$P_{loss} = \sum_{h=1}^{L}\left[\sum_{i=1}^{n}\sum_{j=1}^{n} V_i^{(h)} V_j^{(h)} y_{ij}^{(h)} \cos(\theta_i^{(h)} - \theta_j^{(h)} - \delta_{ij}^{(h)})\right], \quad (10\text{-}9)$$

where $V_i^{(h)}$ and $\theta_i^{(h)}$ are magnitude and phase of the hth harmonic voltage at bus $i$ and $y_{ij}^{(h)}$ and $\delta_{ij}^{(h)}$ are magnitude and phase of hth harmonic line admittance between buses $i$ and $j$, respectively.

## 10.4.2 Application of Maximum Sensitivities Selection (MSS) for the Capacitor Allocation Problem

This section outlines the application of maximum sensitivities selection (MSS) in conjunction with a Newton-based harmonic load flow program to generate the local solution of the capacitor placement and sizing problem in the presence of voltage and current harmonics. If the initial condition is near the global point, the MSS algorithm will generate a near global solution. Otherwise, the MSS algorithm has a high chance of getting stuck at a local point. This section introduces the application of MSS for the capacitor allocation problem as proposed by [41].

### 10.4.2.1 Sensitivity Functions for MSS

Sensitivities of the objective function (Eq. 10-8) to capacitance values at compensation buses can be computed using partial derivatives:

$$\frac{\partial P_{\text{loss},k}}{\partial Q} = \sum_{h=1}^{L} W_h \frac{\partial P_{\text{loss},k}^{(h)}}{\partial Q^{(h)}}$$
$$\frac{dF_{\text{cost}}}{dQ_p} = \frac{F_{\text{cost}}^{(p+1)} - F_{\text{cost}}^{(p)}}{Q_{(p+1)} - Q_p}, \tag{10-10}$$

where $W_h = 1/h$ is the weight function at harmonic $h$, and $P_{\text{loss},k}^{(h)}$ are the total system losses at harmonic $h$ for load level $k$ and $F_{\text{cost}}^{(p+1)}$ is the cost corresponding to capacitor of size $Q_{(p+1)}$. Subscript $p$ indicates the discrete step sizes of capacitor cost as listed in Table 10.3 (e.g., $Q_1 = 150$ kVAr, $Q_2 = 300$ kVAr, . . .). Note the application of linear interpolation in Eq. 10-10 due to the discrete nature of the capacitor cost (Table 10.3).

The partial derivatives of $P_{\text{loss},k}^{(h)}$ are computed as follows:

$$\begin{bmatrix} \dfrac{\partial P_{\text{loss},k}^{(h)}}{\partial P^{(h)}} \\[2mm] \dfrac{\partial P_{\text{loss},k}^{(h)}}{\partial Q^{(h)}} \\[2mm] \dfrac{\partial P_{\text{loss},k}^{(h)}}{\partial I_r^{(h)}} \\[2mm] \dfrac{\partial P_{\text{loss},k}^{(h)}}{\partial I_i^{(h)}} \end{bmatrix} = \begin{bmatrix} \dfrac{\partial P^{(h)}}{\partial \theta^{(h)}} & \dfrac{\partial Q^{(h)}}{\partial \theta^{(h)}} & \dfrac{\partial I_r^{(h)}}{\partial \theta^{(h)}} & \dfrac{\partial I_i^{(h)}}{\partial \theta^{(h)}} \\[2mm] \dfrac{\partial P^{(h)}}{\partial V^{(h)}} & \dfrac{\partial Q^{(h)}}{\partial V^{(h)}} & \dfrac{\partial I_r^{(h)}}{\partial V^{(h)}} & \dfrac{\partial I_i^{(h)}}{\partial V^{(h)}} \end{bmatrix}^{-1} \begin{bmatrix} \dfrac{\partial P_{\text{loss},k}^{(h)}}{\partial \theta^{(h)}} \\[2mm] \dfrac{\partial P_{\text{loss},k}^{(h)}}{\partial V^{(h)}} \end{bmatrix},$$
$$\tag{10-11}$$

where all right-hand side matrix entries are easily computed using the outputs of the harmonic power flow, as outlined in Chapter 7 (Section 7.4). The subscripts $r$ and $i$ denote real and imaginary components, respectively.

### 10.4.2.2 The MSS Algorithm

The shunt capacitor placement and sizing problem in the presence of linear and nonlinear loads is an optimization problem with discrete variables (e.g., discrete values of capacitors). This problem is solved using the harmonic power flow algorithm and the maximum sensitivities selection (MSS) method as follows (Fig. 10.5):

**Step 1:** Input system parameters (e.g., line and load specifications, system topology, and number of load levels). Select the first load level (e.g., $k = 1$) and set initial capacitance values at all possible compensation nodes to zero. For other load levels ($k \geq 2$), use the capacitor values, computed for the preceding load levels, as the initial capacitor values of the compensation buses.

**Step 2:** Calculate harmonic power flow at load level $k$ and compute the objective function (Eq. 10-8).

**Step 3:** Calculate sensitivities of the objective function with respect to capacitances at compensation buses (Eqs. 10-10 to 10-11).

**Step 4:** Select $SC$ candidate buses with maximum sensitivities.

**Step 5 (temporary change of system topology):** Add one unit of capacitor to one of the candidate buses. Calculate harmonic power flow and the related objective function. Remove the newly added capacitor unit from the corresponding candidate bus.

**Step 6:** Repeat Step 5 for all candidate buses.

**Step 7 (change of system topology):** Among all candidate buses, select the one with the minimum objective function value and place one unit of capacitor at this bus. Record the corresponding system topology, $F$, $THD_v$, and $V_{\text{rms}}$.

**Step 8 (check convergence):** If the reactive power of the total capacitance of shunt capacitors does not exceed the sum of reactive loads, the algorithm has not converged; go to Step 3.

**Step 9:** According to the calculated information (objective functions, $THD_v$, and $V_{\text{rms}}$ values), select the topology with the minimum objective function that satisfies the problem constraints (Eqs. 10-5 to 10-7). This is the optimal solution at load level value $k$. If such topology does not exist, that is, there is no solution at load level $k$, go to Step 11.

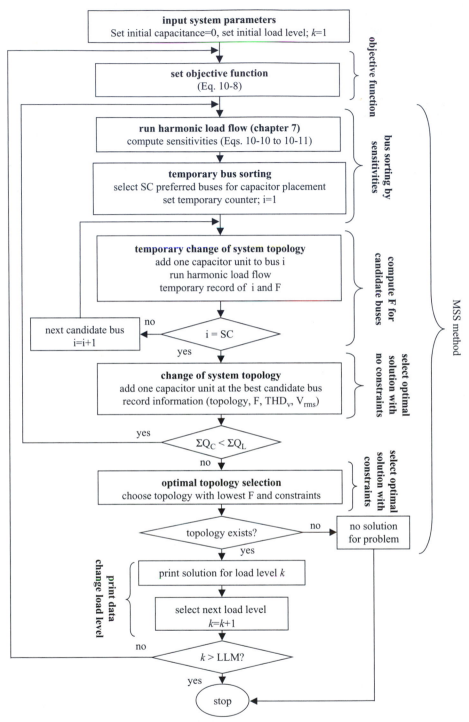

**FIGURE 10.5** Iterative algorithm for optimal placement and sizing of shunt capacitor banks under nonsinusoidal operating conditions using harmonic power flow and MSS method. SC refers to the preferred (candidate) buses for capacitor placement.

**Step 10:** Print the solution for this load level. If the final load level is not reached, select the next load level and go to Step 2.

**Step 11:** Stop.

### 10.4.2.3 Convergence of the MSS Algorithm

The burden of calculation in the MSS algorithm is much less than those of the analysis and dynamic programming (DP) methods. Assuming $N_c$ number of capacitor units and $K_c$ number of compensation buses, the average number of power flow calculations in DP algorithm is $K_c(K_c + 1)(N_c + 1)$. Total number of power flow calculations of the MSS algorithm is $N_c K_c$, which is much less than that of the DP method, especially when $N_c$ is large.

### 10.4.3 Application of Local Variation (LV) for the Capacitor Allocation Problem

The shunt capacitor placement and sizing problem in the presence of linear and nonlinear loads is solved using the harmonic power flow algorithm (Chapter 7, Section 7.4) and the local variations (LV) method [36]. The solution procedure as proposed by [42] is as follows (Fig. 10.6):

**Step 1:** Input system parameters (e.g., line and load specifications, system topology, and number of load levels). Select the first load level (e.g., $k = 1$).

**Step 2:** Select SC buses as optimal (or preferred) locations for capacitor bank placements. If such locations are not available, select all buses. Set iteration, capacitor location, and step interval numbers ($i = 1$, $j = 1$, $L_j = 1$). Set initial value for the optimal objective function ($F^* = 0$).

**Step 3:** Start with a random combination of capacitor sizes $Q^c = \{Q_1^c, Q_2^c, \ldots, Q_{SC}^c\}$.

**Step 4:** Set the objective function (Eq. 10-8).

**Step 5:** Perform harmonic power flow to compute fundamental and harmonic bus voltages.

**Step 6:** If power quality constraints (Eqs. 10-5 and 10-6) are satisfied go to Step 7, else:

**Step 6a:** If $i = 1$, go to Step 3 and repeat Steps 3–5 until a feasible combination is found.

**Step 6b:** If $i \neq 1$ go to Step 8.

**Step 7:** Compute objective function (Eq. 10-8). If $i = 1$, increase $i$, save the objective function value ($F^* = F$) and capacitor locations $Q^{C^*} = \{Q_1^C, Q_2^C, \ldots, Q_{SC}^C\}$. Else, compare $F$ with $F^*$.

**Step 7a:** If $F \geq F^*$, go to Step 8.

**Step 7b:** Else, set $F^* \geq F$ and $Q^{C^*} = \{Q_1^C, Q_2^C, \ldots, Q_{SC}^C\}$.

**Step 8:** Change capacitor size at capacitor location $j$ in small steps from $Q_o$ to $LQ_o$ while keeping the capacitance of other capacitor locations unchanged. Repeat Steps 5–7 for each step change ($Q_j = L_j Q_o$, $L_j = 1, \ldots, L$).

**Step 9:** Select a combination of capacitor sizes that corresponds to the lowest objective function ($Q^C = Q^{C^*}$). This will be the starting point of Step 10.

**Step 10:** Repeat Steps 8 and 9 for each capacitor location (e.g., $j \in MC$).

**Step 11:** If solution has not converged (e.g., $F^*$ is changing in two subsequent iterations), set $i = i + 1$, $j = 1$ and repeat Steps 8–10.

**Step 12:** This is the optimal solution at load level $k$. If the final load level is reached, go to Step 14.

**Step 13:** Select the next load level ($k = k + 1$), put $Q^C$ as system parameters and go to Step 2.

**Step 14:** Print the solution and stop.

### 10.4.4 A Hybrid MSS-LV Algorithm for the Capacitor Allocation Problem

Reference [42] presents a hybrid algorithm based on a combination of maximum sensitivities selection (to direct solution toward the permissible region by selecting candidate buses) and local variations (to improve accuracy). MSS is used as a fast technique to obtain the proper initial conditions for the LV method. Therefore, this algorithm takes advantage of MSS speed and LV accuracy. Unlike the LV scheme, however, its solution does not strongly depend on the initial conditions and is not limited to a local optimum. Therefore, solution accuracy is improved at the expense of more computing time.

If the initial conditions generated by the MSS method do not satisfy the power quality constraints (e.g., MSS algorithm fails to locate a solution), the proposed algorithm will apply these constraints in the second iteration of the LV algorithm.

### 10.4.5 Application Example 10.3: Optimal Placement and Sizing of Capacitor Banks in the Distorted 18-Bus IEEE Distribution System by MSS and MSS-LV Methods

The MSS and MSS-LV algorithms are applied to the 23 kV, 18-bus, distorted IEEE distribution system (Fig. E10.3.1). Specifications of this system are given in reference [42]. Simulation results are shown in Table E10.3.1 and Fig. E10.3.2 for the following three operating conditions:

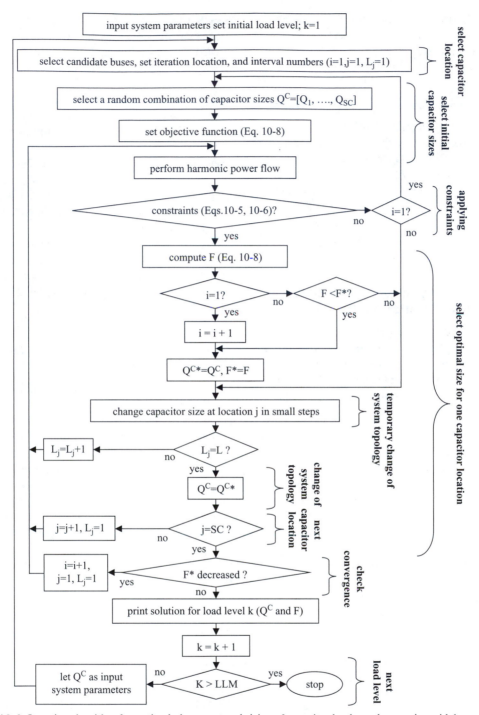

**FIGURE 10.6** Iterative algorithm for optimal placement and sizing of capacitor banks under nonsinusoidal operating conditions using harmonic power flow and the local variations (LV) method.

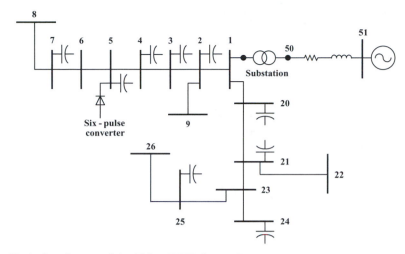

**FIGURE E10.3.1** Single-line diagram of the 18-bus IEEE distorted system.

**Case 1: High Harmonic Distortion.** The nonlinear load in Fig. E10.3.1 is a six-pulse rectifier with active and reactive powers of 0.3 pu (3 MW) and 0.226 pu (2.26 MVAr), respectively. Results based on harmonic power flow show for this system a maximum voltage $THD_v$ of 8.48%. The MSS and MSS-LV algorithms are applied to this system for optimal placement and sizing of capacitor banks.

Results of the MSS method show a yearly benefit of $20360 (last row of Table E10.3.1) and different locations and sizes of capacitor banks before (column 2 of Table E10.3.1) and after (column 3 of Table E10.3.1) optimization. The maximum voltage $THD_v$ is only reduced to 6.37% (row 19 of Table E10.3.1 and Fig. E10.3.2). Therefore, considerable yearly benefit is achieved but power quality constraints are not satisfied.

By applying the MSS-LV algorithm, we can decrease the level of voltage $THD_v$ to 4.72% (row 19 of Table E10.3.1). As expected (due to power quality improvement), yearly benefit is slightly decreased (to $18162). If in highly harmonic distorted systems the MSS-LV algorithm does not satisfy power quality constraints, the application of passive filters and active filters (Chapter 9) before capacitor placement is recommended.

**Case 2: Medium Harmonic Distortion.** The size of the nonlinear load is reduced such that its active and reactive powers are 0.12 pu and 0.166 pu, respectively. After optimal placement and sizing of capacitor banks by the MSS method, voltage $THD_v$ decreases from 5.18 to 5.00% and a yearly benefit of $11650 is achieved. Applying the hybrid MSS-LV algorithm causes minor changes (e.g., eliminating

one capacitor unit). Therefore, the performances of the two algorithms are similar for this case. This means that the output of MSS algorithm is near the system global solution.

**Case 3: Low Harmonic Distortion.** The size of the nonlinear load is further decreased such that its active and reactive powers are 0.02 pu and 0.046 pu, respectively. After optimal placement and sizing of capacitor banks by the MSS method, voltage $THD_v$ increases from 2.1 to 2.95% (which is lower than the desired limit of 3%), total capacitance of capacitor banks decreases from 1.005 to 0.63 pu, and a yearly benefit of $12608 results. By applying the MSS-LV algorithm, considerable benefit is achieved ($44787), while voltage $THD_v$ is at 2.87%.

### 10.4.6 Fuzzy Approach for the Optimal Placement and Sizing of Capacitor Banks in the Presence of Harmonics

All mathematical optimization methods require that the objective function be (pseudo-) convex and the constraints be (quasi-) convex. Only if convexity exists is there one optimal solution within the feasible or permissible multidimensional region. Experience teaches that the above-mentioned conditions of convexity are not necessarily met. The method of the steepest decent or ascent may lead to a conditional optimum of which several may exist due to the violation of convexity of the objective and constraint functions.

To avoid a trajectory leading to a conditional but not necessarily best conditional optimum, fuzzification of the objective and constraint functions has been introduced [44]. This process makes it possible

**TABLE E10.3.1** Simulation Results for the 18-Bus, Distorted IEEE Distribution System (Fig. E10.3.1) at Different Distortion Levels with per Unit VA = 10 MVA, per Unit V = 23 kV, and Swing Bus Voltage = 1.05 pu[a]

| | Case 1 (high distortion) | | | Case 2 (medium distortion) | | | Case 3 (low distortion) | | |
|---|---|---|---|---|---|---|---|---|---|
| | Before optimization | MSS optimization | MSS-LV optimization | Before optimization | MSS optimization | MSS-LV optimization | Before optimization | MSS optimization | MSS-LV optimization |
| Capacitor bank locations and sizes (kVAr) | | | | | Q1 = 0.090 | Q1 = 0.090 | | | |
| | Q2 = 0.105 | Q2 = 0.030 | Q2 = 0.120 | Q2 = 0.105 | Q2 = 0.120 | Q2 = 0.120 | Q2 = 0.105 | Q2 = 0.090 | Q2 = 0.090 |
| | Q3 = 0.060 | Q3 = 0.090 | Q3 = 0.060 | Q3 = 0.060 | Q3 = 0.060 | Q3 = 0.060 | Q3 = 0.060 | | Q3 = 0.120 |
| | Q4 = 0.060 | Q4 = 0.180 | Q4 = 0.180 | Q4 = 0.060 | Q4 = 0.060 | Q4 = 0.060 | Q4 = 0.060 | Q4 = 0.060 | Q4 = 0.030 |
| | Q5 = 0.180 | Q5 = 0.240 | Q5 = 0.240 | Q5 = 0.180 | Q5 = 0.180 | Q5 = 0.180 | Q5 = 0.180 | Q5 = 0.180 | Q5 = 0.180 |
| | Q7 = 0.060 | Q7 = 0.120 | Q7 = 0.120 | Q7 = 0.060 | Q7 = 0.060 | Q7 = 0.060 | Q7 = 0.060 | Q7 = 0.060 | Q7 = 0.060 |
| | | | | | Q8 = 0.060 | Q8 = 0.060 | | | |
| | | | | | Q9 = 0.030 | Q9 = 0.030 | | | |
| | Q20 = 0.06 | Q20 = 0.09 | Q20 = 0.09 | Q20 = 0.06 | Q20 = 0.06 | Q20 = 0.06 | Q20 = 0.06 | Q20 = 0.03 | Q20 = 0.09 |
| | Q21 = 0.12 | Q21 = 0.12 | Q21 = 0.12 | Q21 = 0.12 | Q21 = 0.12 | Q21 = 0.12 | Q21 = 0.12 | | |
| | Q24 = 0.15 | | | Q24 = 0.15 | Q24 = 0.15 | Q24 = 0.15 | Q24 = 0.15 | Q24 = 0.18 | Q24 = 0.06 |
| | Q25 = 0.09 | Q25 = 0.03 | | Q25 = 0.09 | | | Q25 = 0.09 | Q25 = 0.03 | Q25 = 0.09 |
| | Q50 = 0.12 | | | Q50 = 0.12 | Q50 = 0.03 | | Q50 = 0.12 | | Q50 = 0.09 |
| Total capacitor (pu) | Qt = 1.005 | Qt = 0.900 | Qt = 0.930 | Qt = 1.005 | Qt = 1.020 | Qt = 0.990 | Qt = 1.005 | Qt = 0.630 | Qt = 0.810 |
| Minimum voltage (pu) | 1.029 | 1.016 | 1.013 | 1.044 | 1.048 | 1.043 | 1.050 | 1.020 | 1.065 |
| Maximum voltage (pu) | 1.055 | 1.056 | 1.059 | 1.062 | 1.074 | 1.071 | 1.073 | 1.050 | 1.05 |
| Maximum THD$_v$ (%) | **8.48** | **6.37** | **4.72** | **5.18** | **5.00** | **5.00** | **2.10** | **2.95** | **2.87** |
| Losses (kW) | 282.93 | 246.43 | 250.37 | 213.69 | 192.81 | 192.84 | 189.91 | 167.32 | 109.64 |
| Capacitor cost ($) | 2432.70 | 2204.7 | 2206.8 | 2432.70 | 2535.3 | 2430.3 | 2432.70 | 1754.70 | 1830.6 |
| Total cost ($/year) | 157872.98 | 137512.00 | 139710.47 | 119237.90 | 107587.47 | 107604.72 | 105969.20 | 93361.15 | 61181.74 |
| Benefits ($/year) | | **20360** | **18162.51** | | **11650.43** | **11633.18** | | **12608** | **44787.46** |

[a]All capacitors were removed before the optimization process.

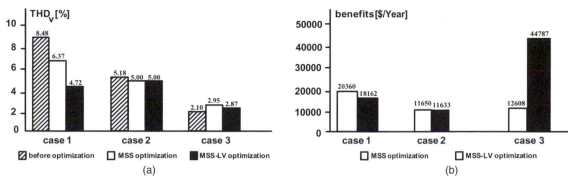

**FIGURE E10.3.2** Simulated results of Fig. E10.3.1: (a) maximum voltage $THD_v$ before and after optimal capacitor placement, (b) yearly benefit after optimal sizing and placement of capacitor banks.

that the solution trajectory – consisting of a band which is broad – avoids a conditional optimum, which exists due to the pseudo- or quasi-convex objective and constraint functions.

### 10.4.6.1 Sensitivity of Objective Function and $THD_v$

Sensitivities of constraints and objective function are used to limit the number of calculations while a fuzzy combination of constraints and objective function is constructed to guarantee the monotonous convergence of the solution.

**Sensitivity of Objective Function.** Sensitivities of objective function (Eq. 10-8) to capacitance values at compensation buses can be computed using Eqs. 10-10 and 10-11.

**Sensitivity of $THD_v$.** Sensitivities of $THD_v$ (Eq. 10-6) to capacitance values at compensation buses is computed using the partial derivatives

$$S_{THDv,j} = \frac{\partial THD_{v,j}}{\partial Q} = \sum_{h=1}^{L} W_h \frac{\partial THD_{v,j}}{\partial Q^{(h)}}, \quad (10\text{-}12)$$

where $W_h = 1/h$ is the proposed weighting function for harmonic order h. The partial derivatives of $THD_{v,j}$ are computed as follows:

$$\begin{bmatrix} \dfrac{\partial THD_{v,j}}{\partial P^{(h)}} \\[2mm] \dfrac{\partial THD_{v,j}}{\partial Q^{(h)}} \\[2mm] \dfrac{\partial THD_{v,j}}{\partial I_r^{(h)}} \\[2mm] \dfrac{\partial THD_{v,j}}{\partial I_i^{(h)}} \end{bmatrix} = \begin{bmatrix} \dfrac{\partial P^{(h)}}{\partial \theta^{(h)}} & \dfrac{\partial Q^{(h)}}{\partial \theta^{(h)}} & \dfrac{\partial I_r^{(h)}}{\partial \theta^{(h)}} & \dfrac{\partial I_i^{(h)}}{\partial \theta^{(h)}} \\[2mm] \dfrac{\partial P^{(h)}}{\partial V^{(h)}} & \dfrac{\partial Q^{(h)}}{\partial V^{(h)}} & \dfrac{\partial I_r^{(h)}}{\partial V^{(h)}} & \dfrac{\partial I_i^{(h)}}{\partial V^{(h)}} \end{bmatrix}^{-1} \begin{bmatrix} \dfrac{\partial THD_{v,j}}{\partial \theta^{(h)}} \\[2mm] \dfrac{\partial THD_{v,j}}{\partial V^{(h)}} \end{bmatrix},$$

$$(10\text{-}13)$$

where all right-hand-side matrix entries are easily computed using results from the application of a harmonic power flow program.

### 10.4.6.2 Fuzzy Implementation

Fuzzy set theory (FST) is relied on to determine the candidate buses and to select the most suitable bus for capacitor placement.

**Fuzzification of Sensitivities.** For the sensitivities of objective function (Eqs. 10-10 and 10-11) and constraints (Eqs. 10-12 and 10-13) with respect to the reactive power injection at each bus, the exponential fuzzifier functions $\mu_F^{sensitivity}$, $\mu_{THDv_j}^{sensitivity}$, $\mu_{v,j}^{sensitivity}$ are employed. Figure 10.7a shows the fuzzifier for the sensitivity of $THD_{v,j}$. $S_{THD}^{max}$ is a large number, that is, buses with higher sensitivities have higher values of membership.

**Fuzzification of Objective Function and Constraints.** For the objective function (Eq. 10-8) a membership function ($\mu_F$) is defined as depicted in Fig. 10.7b. $F_o$ is the initial value of objective function and $F_1$ is the value of objective function after compensating the total reactive (inductive) loads at every bus of the distribution system. High objective values are given a low membership function value, whereas low objective values are assigned a high membership function value.

For voltage constraints, it is suitable to use a membership function ($\mu_{v,j}$) with trapezoidal form, as shown in Fig. 10.7c. $V^{min}$ and $V^{max}$ correspond to Eq. 10-5 such that buses with voltages values between the two limit values $V_{min}$ and $V_{max}$ have full membership. $V_o^-$ and $V_o^+$ correspond to initial values of the bus effective voltage. Therefore, a bus with high rms voltage deviation (Eq. 10-5) is given a low value and a bus with low rms voltage deviation is given a high value.

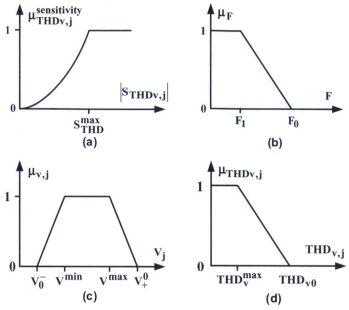

**FIGURE 10.7** Membership functions for (a) sensitivity of the voltage $THD_v$ (other sensitivity membership functions are similar), (b) objective function, (c) rms voltage, (d) voltage $THD_{v,j}$ at bus $j$.

For $THD_{v,j}$ a membership function ($\mu_{THD\,\theta,j}$) is defined as illustrated in Fig. 10.7d. $THD_v^{max}$ and $THD_{VO}$ are the limit and the initial value of $THD_{v,j}$ corresponding to Eq. 10-6, respectively. Buses are given a high value (e.g., $\mu_{THDJ,j} = 1$) unless their $THD_v$ are higher than the IEEE-519 limits.

**Fuzzy Combination of Membership Functions.** The algebraic product of the membership functions is used such as the t-norm for $THD_{v,j}$:

$$\mu_{THDv} = t_1(\mu_{THDv,1} \cdot \mu_{THDv,2} \cdots \mu_{THDv,n}) = \prod_{j=1}^{n} \mu_{THDv,j}.$$

(10-14)

The same formulations apply to t-norms of the bus voltages, sensitivity of $THD_{v,j}$ and sensitivity of voltage constraints, the combination of objective function and constraints, as well as their sensitivities to reactive power injection:

$$\mu_{total}^{sensitivity} = t_2(\mu_{THDv}^{sensitivity}, \mu_V^{sensitivity}, \mu_F^{sensitivity}), \quad (10\text{-}15a)$$

$$\mu_{total} = t_3(\mu_{THDv}, \mu_V, \mu_F). \quad (10\text{-}15b)$$

**Alpha-Cut Process.** In order to direct the algorithm toward the permissible region bordered by constraints, the $\alpha$-cut process is used at each iteration.

To do this, the values of $\alpha$ for F, $THD_v$, and $V$ are set to their corresponding membership values in the previous iteration (e.g., $\alpha_F^{\ell+1} = \alpha_F^\ell$, where $\ell$ is the iteration number).

**Analysis.** The initial starting point of Eq. 10-8 (e.g., the initial compensation buses) does not usually reside inside the permissible solution region. Therefore, some type of criteria (based on simultaneous improvements of F, $THD_v$, and $V$ in consecutive iterations) is required to direct the solution toward the permissible region and to select the most appropriate buses for capacitor placement. The section uses the fuzzy set theory to measure individual improvements, generate a single criterion to direct the solution, and guarantee the monotonous convergence of the solution at each iteration. Other approaches with similar characteristics (e.g., analytical hierarchy process) could also be used.

Note that the locations of candidate buses as indicated by the sensitivities of F, $THD_v$, and $V$ are usually different. A single criterion – a fuzzy combination – is employed to determine the optimal locations of candidate buses. The exponential sensitivity functions of Fig. 10.7a are solely used for ranking of candidate buses. Other methods could be employed for this ranking (e.g., weighted average, statistical methods). The selected method will not have any impact on the capacitor placement algorithm. However, the selected trapezoidal membership functions of Fig. 10.7b–d for $F$, $V$, and $THD_v$ have major impacts on the solution process (e.g., improving or deteriorating convergence).

The inclusion of sensitivity functions and fuzzification of objective and power quality constraints will automatically eliminate all solutions generating extreme values of voltages and/or currents and prevents fundamental and harmonic parallel resonances.

### 10.4.6.3 Solution Methodology

The shunt capacitor placement and sizing problem in the presence of linear and nonlinear loads is an optimization problem with discrete variables (e.g., discrete values of capacitors). This problem is solved using the harmonic power flow algorithm and the fuzzy set theory (Fig. 10.8) as follows:

**Step 1:** Input system parameters (e.g., line and load specifications and system topology). Initial capacitor values of all possible compensation buses are set to zero.

**Step 2:** Calculate harmonic power flow.

**Step 3:** Evaluate the objective function (Eq. 10-8), sensitivities of objective function, constraints, and their fuzzy combination (Eq. 10-15a) in order to determine the list of candidate buses.

**Step 4:** Sort candidate buses (CB) and from among them select several buses with maximum sensitivities and total number of installed capacitor banks less than $u^{max}$.

**Step 5:** Apply $\alpha$-cut process and compute the membership functions for the objective function

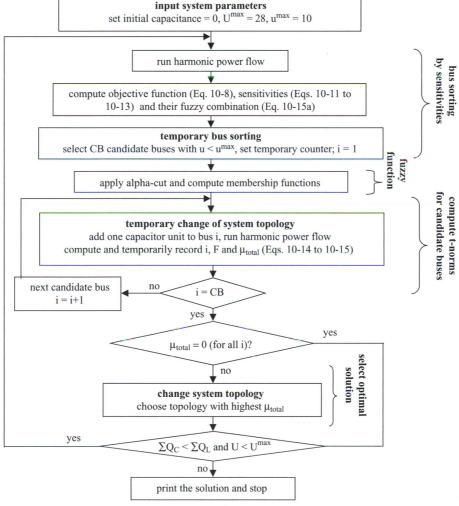

**FIGURE 10.8** Iterative fuzzy algorithm for optimal placement and sizing of capacitor banks in the presence of harmonics.

and constraints with respect to the present topology.

**Step 6:** Add one unit of capacitor to one of the candidate buses. Calculate harmonic power flow and record the values of related membership functions. Remove the newly added capacitor unit from the corresponding candidate bus.

**Step 7:** Repeat Step 6 for all other candidate buses.

**Step 8:** If all total membership functions are zero, go to Step 11.

**Step 9:** Among all candidate buses, select the one with maximum value of total membership function and place one unit of capacitor at this bus.

**Step 10:** If the reactive power of the total capacitance of shunt capacitors does not exceed the sum of reactive loads and the total number of installed capacitor banks are less than $U^{max}$, go to Step 2.

**Step 11:** Print the solution and stop.

## 10.4.7 Application Example 10.4: Optimal Placement and Sizing of Capacitor Banks in the Distorted 18-Bus IEEE Distribution System by Fuzzy Expert System

The fuzzy and MSS algorithms for capacitor placement and sizing are applied to the 23 kV, 18-bus distorted IEEE distribution system (Fig. E10.3.1). Simulation results are compared in Table E10.4.1 for the following operating conditions.

**Case 1: High Harmonic Distortion.** The nonlinear load in Fig. E10.3.1 is a six-pulse rectifier with active and reactive powers of 0.3 pu (3 MW) and 0.226 pu (2.26 MVAr), respectively. Results of harmonic power flow indicate a maximum voltage $THD_v$ of 8.48% for this system (column 2 of Table E10.4.1).

**TABLE E10.4.1** Simulation Results for the 18-Bus, Distorted IEEE Distribution System (Fig. E10.3.1) at Different Distortion Levels with per Unit S = 10 MVA, per Unit V = 23 kV, and Swing Bus Voltage = 1.05 pu[a]

| | Case 1 (high distortion) | | | Case 2 (low distortion) | | |
|---|---|---|---|---|---|---|
| | Before optimization | After MSS optimization | After fuzzy optimization | Before optimization | After MSS optimization | After fuzzy optimization |
| Capacitor bank locations and sizes (pu) | Q2 = 0.105 | Q2 = 0.030 | Q2 = 0.015 | Q2 = 0.105 | Q2 = 0.090 | |
| | Q3 = 0.060 | Q3 = 0.090 | Q3 = 0.030 | Q3 = 0.060 | | Q3 = 0.255 |
| | Q4 = 0.060 | Q4 = 0.180 | Q4 = 0.060 | Q4 = 0.060 | Q4 = 0.060 | Q4 = 0.030 |
| | Q5 = 0.180 | Q5 = 0.240 | Q5 = 0.240 | Q5 = 0.180 | Q5 = 0.180 | Q5 = 0.060 |
| | Q7 = 0.060 | Q7 = 0.120 | Q7 = 0.210 | Q7 = 0.060 | Q7 = 0.060 | Q7 = 0.030 |
| | Q20 = 0.060 | Q20 = 0.090 | Q20 = 0.090 | Q20 = 0.060 | Q20 = 0.030 | Q20 = 0.165 |
| | Q21 = 0.120 | Q21 = 0.120 | Q21 = 0.120 | Q21 = 0.120 | | Q21 = 0.015 |
| | Q24 = 0.150 | | | Q24 = 0.150 | Q24 = 0.180 | Q24 = 0.075 |
| | Q25 = 0.090 | Q25 = 0.030 | | Q25 = 0.090 | Q25 = 0.030 | Q25 = 0.045 |
| | Q50 = 0.120 | | | Q50 = 0.120 | | |
| Total capacitor (pu) | Qt = 1.005 | Qt = 0.900 | Qt = 0.765 | Qt = 1.005 | Qt = 0.630 | Qt = 0.675 |
| Minimum voltage (pu) | 1.029 | 1.016 | 0.998 | 1.050 | 1.020 | 1.021 |
| Maximum voltage (pu) | 1.055 | 1.056 | 1.050 | 1.073 | 1.050 | 1.054 |
| Max. THD_v (%) | 8.48 | 6.37 | 4.89 | 2.10 | 2.95 | 2.63 |
| Losses (kW) | 282.93 | 246.43 | 257.46 | 189.91 | 167.32 | 160.48 |
| Capacitor cost ($/year) | 1978.2 | 1692 | 1458.30 | 1978.20 | 1311.90 | 1538.25 |
| Total cost ($/year) | 159853.14 | 139199.94 | 145120.98 | 107947.98 | 94676.46 | 91086.09 |
| Benefits ($/year) | | 20653 | 14732 | | 13271 | 16861 |

[a]All capacitors were removed before the optimization process.

Application of the MSS method to Fig. E10.3.1 shows that optimal capacitor placement results in considerable yearly benefit (e.g., $20653 per year) but it does not limit $THD_v$ to the desired level of 5% (column 3 of Table E10.4.1). This is expected from the MSS method dealing with rich harmonic configurations, where capacitor placement is not the primary solution for harmonic mitigation.

The fuzzy algorithm of Fig. E10.3.1 is also applied to this system for optimal placement and sizing of capacitor banks. Results show a yearly benefit of $14732 per year (last row of Table E10.4.1) and different locations and sizes of capacitor banks before (column 2 of Table E10.4.1) and after (column 4 of Table E10.4.1) the fuzzy optimization. In addition, maximum voltage $THD_v$ is reduced to 4.89% (row 16 of Table E10.4.1) and total capacitance is decreased by 24%.

**Case 2: Low Harmonic Distortion.** The size of the nonlinear load is decreased such that its active and reactive powers are 0.02 pu and 0.046 pu, respectively. After optimal placement and sizing of capacitor banks with MSS method, $THD_v$ increases from 2.1 to 2.95% (which is lower than the desired limit of 5%), total capacitance of capacitor banks is decreased from 1.005 pu to 0.630 pu and a yearly benefit of $13271 is achieved (column 6 of Table E10.4.1).

Application of the fuzzy algorithm results in $THD_v$ increase from 2.1 to 2.63% (which is in the permitted region) and also total capacitance decreases from 1.005 to 0.675 pu (column 7 of Table E10.4.1). In addition, annual savings are increased to $16861 (e.g., 27% greater savings than that of the MSS method). These results demonstrate the fine capability of the fuzzy method for capacitor placement in the presence of harmonics.

### 10.4.8 Optimal Placement, Replacement, and Sizing of Capacitor Banks in Distorted Distribution Networks by Genetic Algorithms

Most traditional optimization methods move from one point in the decision hyperspace to another using some deterministic rule. The problem with this is that the solution is likely to get stuck at a local optimum. Genetic algorithms (GAs) start with a diverse set (population) of potential solutions (hyperspace vectors). This allows for exploration of many optima in parallel, lowering the probability of getting stuck at a local optimum. Even though GAs are probabilistic, they are not strictly random searches. The stochastic operators used on the popu-

lation direct the search toward regions of the hyperspace that are likely to have higher fitness values.

This section presents a genetic method to formulate the capacitor placement and sizing problem in the presence of voltage and current harmonics, taking into account fixed capacitors with a limited number of capacitor banks at each bus and the entire feeder [45]. Operational and power quality constraints include bounds for rms voltage, $THD_v$, and the number and size of installed capacitors.

#### 10.4.8.1 Genetic Algorithm

Genetic algorithms use the principles of natural evolution and population genetics to search and arrive at a near global solution. The required design variables are encoded into a binary string as a set of genes corresponding to chromosomes in biological systems. Unlike traditional optimization techniques that require one starting point, GAs use a set of points as initial conditions. Each point is called a chromosome. A group of chromosomes is called a population. The number of chromosomes in a population is usually selected to be from 30 to 300. Each chromosome is a string of binary codes (genes) and may contain substrings. The merit of a string is judged by the fitness function, which is derived from the objective function and is used in successive genetic operations. During each iterative procedure (referred to as a generation), a new set of strings with improved performance is generated relying on three GA operators (i.e., reproduction, crossover, and mutation).

**Structure of Chromosomes.** Different types of chromosomes for GAs have been proposed and implemented. This section employs the chromosome structure of Fig. 10.9 consisting of MC substrings of binary numbers, where MC denotes the number of possible compensation (candidate) buses for capacitor placement in the entire feeder. The binary numbers indicate the size of the installed capacitor at the bus under consideration.

**Fitness Functions.** The inverse algebraic product of penalty functions (Fig. 10.10) is used as the fitness

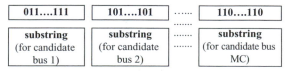

**FIGURE 10.9** A typical chromosome structure for the genetic algorithm.

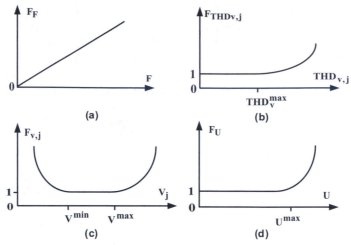

**FIGURE 10.10** Penalty functions of the GA used to compute fitness (Eq. 10-16): (a) for the objective function, (b) for $THD_{v,j}$, (c) for $v,i$, (d) for total number of capacitors of the entire feeder (U).

function combining objective and constraint functions:

$$F_{fitness} = 1/(F_F \cdot F_{THDv} \cdot F_V \cdot F_U)$$

$$F_{THDv} = \prod_{i=1}^{n} F_{THDv,i} \qquad (10\text{-}16)$$

$$F_v = \prod_{i=1}^{n} F_{v,i}$$

where penalty functions $F_F$, $F_{THDv,j}$, $F_{v,i}$, and $F_U$ are shown in Fig. 10.10.

**Genetic Operators.** Three genetic operators including reproduction (based on the roulette-wheel mechanism), crossover (with crossover probability of 0.6 to 1.0), and mutation (with mutation probability of 0.01 to 0.1) are established.

**Convergence Criterion.** The iterations (regenerations) of the genetic algorithm are continued until all generated chromosomes become equal or the maximum number of iterations ($N^{max} = 40$) is reached. Due to the randomness of the GA method, the solution tends to differ for each run, even with the same initial population. For this reason, it is suggested to perform multiple runs and select the "most acceptable" solution (e.g., with most benefits, within the permissible region bounded by constraints).

### 10.4.8.2 Solution Methodology

The shunt capacitor placement and sizing problem in the presence of linear and nonlinear loads is solved

based on the genetic algorithm of Fig. 10.11 as follows:

**Step 1:** Input system parameters (e.g., system topology, line and load specifications). Input the initial population with $N_{ch}$ chromosomes.

**Step 2:** Set initial counter and parameter values (e.g., $N_{ch} = N_{it} = N_R = 1$ and $F_{MIN} = $ a high number).

**Step 3 (Fitness Process):**

**Step 3a:** Run harmonic power flow for $N_{ch}$ chromosomes and save results.

**Step 3b:** Compute proposed penalty functions (Fig. 10.10) employing data generated by harmonic power flow program. Compute fitness functions (Eq. 10-16) for chromosome $N_{ch}$. Set $N_{ch} = N_{ch} + 1$.

**Step 3c:** If $N_{ch} \leq N_{chrom}$ go to Step 3a.

**Step 4 (Reproduction Process):**

**Step 4a:** Define total fitness as the sum of all fitness values for all chromosomes.

**Step 4b:** Select a percentage of "roulette wheel" for each chromosome that is equal to the ratio of its fitness value to the total fitness value.

**Step 4c:** Improve generation by rolling the roulette wheel $N_{ch}$ times. Select a new combination of chromosomes.

**Step 5 (Crossover Process):**

**Step 5a:** Select a random number ($R_1$) for mating two parent chromosomes.

**Step 5b:** If $R_1$ is between 0.6 and 1.0 then combine the two parents, generate two offspring, and go to Step 5d.

**Step 5c:** Else, transfer the chromosome with no crossover.

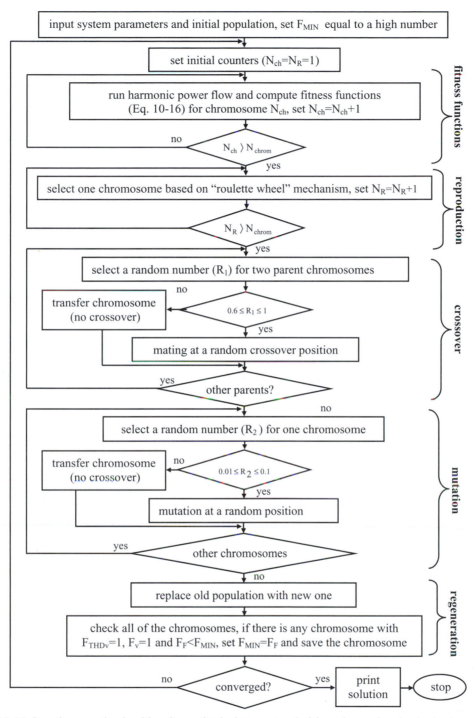

**FIGURE 10.11** Iterative genetic algorithm for optimal placement and sizing of capacitor banks in the presence of harmonics.

**Step 5d:** Repeat steps 5a to 5c for all chromosomes.

**Step 6 (Mutation Process):**

**Step 6a:** Select a random number ($R_2$) for mutation of one chromosome.

**Step 6b:** If $R_2$ is between 0.01 and 0.1 then apply the mutation process at a random position and go to Step 6d.

**Step 6c:** Else, transfer the chromosome with no mutation.

**Step 6d:** Repeat steps 6a to 6c for all chromosomes.

**Step 7 (Updating Populations):** Replace the old population with the improved population generated by Steps 2 to 6. Check all chromosomes; if there is any chromosome with $F_{THDv} = 1$, $F_v = 1$, and $F_F < F_{MIN}$, set $F_{MIN} = F_F$ and save it. Set $N_{it} = N_{it} + 1$.

**Step 8 (Convergence):** If all chromosomes are the same or the maximum number of iterations is reached ($N_{it} = N^{max}$), then print the solution and stop, else go to Step 2.

### 10.4.9 Application Example 10.5: Optimal Placement and Sizing of Capacitor Banks in the 6-Bus IEEE Distorted System

The MSS, MSS-LV, fuzzy, and genetic algorithms for optimal capacitor sizing and placement are applied to the 69 kV, 6-bus, distorted distribution system (Fig. E10.5.1). Specifications of this system are given in reference [42]. Simulation results are compared in Table E10.5.1.

The nonlinear load in Fig. E10.5.1 is a six-pulse rectifier with active and reactive powers of 0.2 pu (2 MW) and 0.25 pu (2.5 MVAr), respectively. Results of a harmonic power flow program indicate a maximum voltage $THD_v$ of 5.270% (column 2 of Table E10.5.1).

After optimal placement and sizing of capacitor banks with the MSS algorithm, the voltage $THD_v$ decreases to 5.007% by allocating 0.15 pu capacitor banks and a yearly benefit of $35968 is achieved (column 3 of Table E10.5.1). Applying the MSS-LV algorithm to the system yields a yearly benefit of $40098 and a voltage $THD_v$ of 4.970 (column 4 of Table E10.5.1). Optimization of the system based on the fuzzy set algorithm generates voltage $THD_v$ of 4.990% and a yearly benefit of $33758 (column 5 of Table E10.5.1). Application of the genetic algorithm

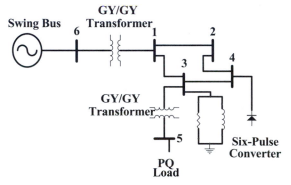

**FIGURE E10.5.1** Single-line diagram of the 6-bus [43] IEEE distorted system used for simulation and analysis. GY is a grounded transformer.

**TABLE E10.5.1** Simulation Results for the 6-Bus, Distorted IEEE System of Fig. E10.5.1 with Base Apparent Power S = VA = 10 MVA, Base Voltage V = 69 kV, and Swing Bus Voltage = 1.00 pu[a]

|  | Before optimization | After MSS optimization | After MSS-LV optimization | After fuzzy optimization | After GA optimization |
|---|---|---|---|---|---|
| Capacitor bank locations and sizes (pu) | Q1 = 0.000 | Q1 = 0.000 | Q1 = 0.405 | Q1 = 0.000 | Q1 = 0.120 |
|  | Q2 = 0.000 | Q2 = 0.000 | Q2 = 0.000 | Q2 = 0.000 | Q2 = 0.000 |
|  | Q3 = 0.000 | Q3 = 0.000 | Q3 = 0.315 | Q3 = 0.000 | Q3 = 0.45 |
|  | Q4 = 0.000 | Q4 = 0.000 | Q4 = 0.000 | Q4 = 0.000 | Q4 = 0.00 |
|  | Q5 = 0.000 | Q5 = 0.150 | Q5 = 0.150 | Q5 = 0.135 | Q5 = 0.405 |
|  | Q6 = 0.000 | Q6 = 0.000 | Q6 = 0.000 | Q6 = 0.000 | Q6 = 0.000 |
| Total capacitor (pu) | Qt = 0.000 | Qt = 0.150 | Qt = 0.870 | Qt = 0.135 | Qt = 0.405 |
| Minimum voltage (pu) | 0.935 | 0.965 | 1.000 | 0.962 | 0.996 |
| Maximum voltage (pu) | 1.005 | 1.013 | 1.070 | 1.012 | 1.044 |
| Maximum $THD_v$ (%) | 5.270 | 5.007 | 4.970 | 4.990 | 4.943 |
| Losses (kW) | 360 | 295 | 285.2 | 299 | 276.7 |
| Capacitor cost ($/year) | 0.00 | 301.50 | 1640.7 | 279.45 | 725.85 |
| Total cost ($/year) | 200880.0 | 164911.5 | 160782.0 | 167121.45 | 155101.85 |
| Computing time (sec)[b] | – | 6 | 115 | 10 | 639 |
| Benefits ($/year) |  | 35968.5 | 40098.0 | 33758.55 | 45778.15 |

[a]All capacitors were removed before the optimization process.
[b]Using a Pentium III (550 MHz) personal computer.

(with $N_{ch} = 100$) results in about the same voltage $THD_v$ (e.g., 4.943%), and the annual savings are increased to $45778 (column 6 of Table E10.5.1).

## 10.4.10 Application Example 10.6: Optimal Placement and Sizing of Capacitor Banks in the 18-Bus IEEE Distorted System

The MSS, MSS-LV, fuzzy, and genetic algorithms for capacitor placement and sizing are applied to the 23 kV, 18-bus distorted IEEE distribution system (Fig. E10.3.1).

The nonlinear load in Fig. E10.3.1 is a six-pulse rectifier with active and reactive powers of 0.3 pu (3 MW) and 0.226 pu (2.26 MVAr), respectively. Results of a harmonic power flow program show a maximum voltage $THD_v$ of 8.486% for this system (column 2 of Table E10.6.1). Application of MSS method to Fig. E10.3.1 shows that optimal capacitor placement results in considerable yearly benefit (e.g., $20653 per year), but it does not limit voltage $THD_v$ to the desired level of 5% (column 3 of Table E10.6.1). Applying the MSS-LV algorithm yields a yearly benefit of $17939 and voltage $THD_v$ is reduced to 4.72% (column 4 of Table E10.6.1). Application of the fuzzy algorithm results in an acceptable voltage $THD_v$ level and benefits (4.899% and $14732, column 5 of Table E10.6.1).

The genetic algorithm (GA) of Fig. 10.11 is also applied to this system for optimal placement and sizing of capacitor banks ($N_{ch} = 250$). Results reveal a yearly benefit of $18949 per year (last row of Table E10.6.1); maximum voltage $THD_v$ is limited to 4.880% (column 6 of Table E10.6.1), and the total allocated capacitance is decreased by 16%.

Greater power quality control mitigation is achieved with the GA at the expense of lower benefits compared with the MSS solution. The GA yields greater yearly benefits while the power quality conditions are the same than those of the MSS-LV solution. Compared with the fuzzy solution, the same power quality conditions result (rows 12–15 of Table E10.6.1), but greater benefits (e.g., 28% greater savings) are achieved with GA.

The results of Tables E10.5.1 and E10.6.1 indicate that the genetic algorithm captures more advantageous solutions than the MSS, MSS-LV, and fuzzy algorithms. This is expected because the genetic algorithm has the capability of computing the near global solution.

**Computing Time.** Unfortunately, the application of GAs often looks like a never-ending process. In fact, only a compromise between population size, mutation rate, and so on can lead to an adequate

**TABLE E10.6.1** Simulation Results of MSS, MSS-LV, Fuzzy, and Genetic Algorithms for the 18-Bus, Distorted IEEE Distribution System (Fig. E10.3.1) with per Unit VA = 10 MVA, per Unit V = 23 kV, and Swing Bus Voltage = 1.05 pu[a]

|  | Before optimization | After MSS optimization | After MSS-LV optimization | After fuzzy optimization | After GA optimization |
|---|---|---|---|---|---|
| Capacitor bank locations and sizes (pu) | Q2 = 0.105 | Q2 = 0.030 | Q2 = 0.120 | Q2 = 0.015 | Q2 = 0.030 |
|  | Q3 = 0.060 | Q3 = 0.090 | Q3 = 0.060 | Q3 = 0.030 | Q3 = 0.000 |
|  | Q4 = 0.060 | Q4 = 0.180 | Q4 = 0.180 | Q4 = 0.060 | Q4 = 0.165 |
|  | Q5 = 0.180 | Q5 = 0.240 | Q5 = 0.240 | Q5 = 0.240 | Q5 = 0.330 |
|  | Q7 = 0.060 | Q7 = 0.120 | Q7 = 0.120 | Q7 = 0.210 | Q7 = 0.105 |
|  | Q20 = 0.060 | Q20 = 0.090 | Q20 = 0.090 | Q20 = 0.090 | Q20 = 0.060 |
|  | Q21 = 0.120 | Q21 = 0.120 | Q21 = 0.120 | Q21 = 0.120 | Q21 = 0.090 |
|  | Q24 = 0.150 | Q24 = 0.000 | Q24 = 0.000 | Q24 = 0.000 | Q24 = 0.015 |
|  | Q25 = 0.090 | Q25 = 0.030 | Q25 = 0.000 | Q25 = 0.000 | Q25 = 0.015 |
|  | Q50 = 0.120 | Q50 = 0.000 | Q50 = 0.000 | Q50 = 0.000 | Q50 = 0.030 |
| Total capacitor (pu) | Qt = 1.005 | Qt = 0.900 | Qt = 0.930 | Qt = 0.765 | Qt = 0.840 |
| Minimum voltage (pu) | 1.029 | 1.016 | 1.013 | 0.998 | 1.003 |
| Maximum voltage (pu) | 1.055 | 1.056 | 1.059 | 1.050 | 1.050 |
| Maximum THD$_v$ (%) | 8.486 | 6.370 | 4.720 | 4.899 | 4.883 |
| Losses (kW) | 282.93 | 246.43 | 250.37 | 257.46 | 249.31 |
| Capacitor cost ($/year) | 1978.20 | 1692.00 | 2206.80 | 1458.30 | 1788.75 |
| Total cost ($/year) | 159853.14 | 139199.94 | 141913.26 | 145120.98 | 140903.73 |
| Computing time (sec)[b] | – | 96 | 141 | 48 | 2780 |
| Benefits ($/year) |  | 20653.6 | 17939.88 | 14732.16 | 18949.41 |

[a]All capacitors were removed before the optimization process.
[b]Using a Pentium III (550 MHz) personal computer.

algorithm that finds good solutions in a relatively short time. However, capacitor bank placement is usually considered a planning problem and its computing time is not of great concern. The computational burden associated with the genetic algorithm is included and compared with those of the MSS, MSS-LV, and fuzzy methods as shown in row 15 of Table E10.5.1 and row 19 of Table E10.6.1. As expected, the computing times of the genetic algorithm are much longer.

## 10.4.11 Genetically Optimized Fuzzy Placement and Sizing of Capacitor Banks in Distorted Distribution Networks

This section presents a genetic algorithm in conjunction with fuzzy logic (GA-FL) to formulate the capacitor placement and sizing problem in the presence of voltage and current harmonics, taking into account fixed capacitors with a limited number of capacitor banks at each bus [46]. Operational and power quality constraints include bounds for rms voltage, $THD_v$, and the number as well as the size of installed capacitors. In this hybrid global optimization method, fuzzy approximate reasoning is introduced to determine the suitability (fitness function) of each GA chromosome for capacitor placement. This will improve the evolution process of the GA, lowering the probability of getting stuck at local optima. Fuzzy logic has shown good results for capacitor bank allocation when combined with genetic algorithms under sinusoidal operating conditions [1].

### 10.4.11.1 Solution Method

Suitability of $THD_v$ and voltage are defined for each chromosome based on fuzzy approximate reasoning. These variables are used with a cost index to determine the suitability (fitness function) of each GA chromosome for capacitor placement.

Fuzzy set theory (FST) contains a set of rules that are developed from qualitative descriptions. When suitability of $THD_v$, voltage, and cost of the distribution system are studied, an engineer can choose the most appropriate locations and sizes of shunt capacitor banks. These rules are defined to determine the suitability of each chromosome ($S_{chrom}$) and may be executed employing fuzzy inferencing. $S_{chrom}$ serves as the fitness function in the GA to further improve the evolution process, reducing the probability of converging to a local optimum.

**Structure of Chromosomes.** The chromosome structure of Fig. 10.9 is used.

**Fuzzy Fitness (Suitability) Functions.** The suitability level of each chromosome ($S_{chrom}$) is computed by defining fuzzy membership functions for suitability of total harmonic distortion ($\mu_{S_{THD}}$), voltage ($\mu_{S_V}$), and cost ($\mu_{cost}$). A fuzzy expert system (FES) uses the number of buses with high $THD_v$ and voltage values and the average of the $THD_v$ and voltage deviation (at these buses) to determine the suitability of $THD_v$ and voltage.

A concern for the development of fuzzy expert systems is the assignment of appropriate membership functions, which could be performed based on intuition, rank ordering, or probabilistic methods. However, the choice of membership degree in the interval [0, 1] does not matter, as it is the order of magnitude that is important [1].

**Suitability of $THD_v$ ($S_{THD}$).** For a given chromosome, define the average unacceptable value of $THD_v$ as

$$THD_{avg}^{high} = \Sigma_{j=1}^{N_{THD}} THD_j / N_{THD}, \qquad (10\text{-}17)$$

where $N_{THD}$ is the number of buses with high total voltage harmonic distortions (e.g., $THD_j > THD_v^{max}$). The following steps are performed to compute $S_{THD}$:

- Fuzzification of $THD_{avg}^{high}$ and $N_{THD}$ (based on membership functions of Fig. 10.12a and 10.12b, respectively) to compute the corresponding membership values ($\mu_{THD}$ and $\mu_{N_{THD}}$).
- Fuzzy inferencing and defuzzification based on the decision matrix of Table 10.5, the membership function of Fig. 10.13, *Mamdani max-prod* implication, and center average methods [50]:

$$S_{THD} = \Sigma \bar{y} \mu_{N_{THD}} \mu_{THD} / \Sigma \mu_{N_{THD}} \mu_{THD}. \qquad (10\text{-}18)$$

**Suitability of Voltage ($S_V$).** For a given chromosome, define the average unacceptable value of voltage deviation as

$$\Delta V_{avg} = \frac{\sum_{j=1}^{N_{\Delta|V|}} \Delta V_j}{N_{\Delta|V|}}, \quad \begin{cases} \Delta V_j = |V| - V^{max} \ (\text{if } |V| > V^{max}) \\ \Delta V_j = V^{min} - |V| \ (\text{if } |V| < V^{min}) \end{cases}$$
$$(10\text{-}19)$$

where $N_{\Delta|V|}$ is the number of buses with unacceptable (high) values of voltage. $V^{max} = 1.1$ pu and $V^{min} = 0.9$ pu

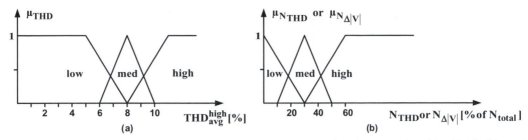

**FIGURE 10.12** Membership functions for (a) the average of unacceptable (high) $THD_v$ values (Eq. 10-17), (b) number of buses with unacceptable $THD_v$ or voltage, where "avg" means average.

**TABLE 10.5** Decision Matrix for Determining Suitability of $THD_v$ (and V)

|  |  | $THD_v$ (or $\Delta|V|$) | | |
|---|---|---|---|---|
| AND | | Low | Medium | High |
| $N_{THD}$ (or $N_{\Delta|V|}$) | Low | *High* | *Medium* | *Medium* |
| | Medium | *Medium* | *Low* | *Low* |
| | High | *Medium* | *Low* | *Low* |

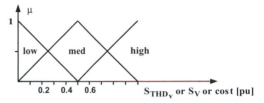

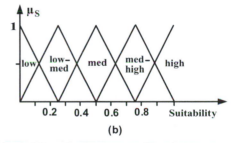

**FIGURE 10.13** Membership functions for suitability of $THD_v$, V, and "cost."

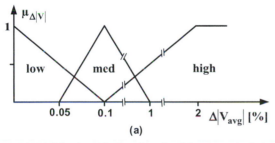

**FIGURE 10.14** Membership functions for (a) voltage deviation, (b) suitability of a chromosome ($S_{chrom}$).

are upper and lower bounds of rms voltage, respectively. The following steps are performed to compute $S_V$:

- Fuzzification of $\Delta V_{ave}$ and $N_{\Delta|V|}$ (based on membership functions of Fig. 10.14a and 10.14b, respectively) to compute the corresponding membership values ($\mu_{\Delta|V|}$ and $N_{N_{\Delta|V|}}$).
- Fuzzy inferencing and defuzzification (using the decision matrix of Table 10.5, Fig. 10.13, and Eq. 10-18) to determine the suitability of voltage ($S_V$).

***Cost Index (cost).*** Compute the cost for all chromosomes (Eq. 10-8) and linearly normalize them into [0, 1] range with the largest cost having a value of one and the smallest one having a value of zero.

***Fitness (Suitability) of a Chromosome ($S_{chrom}$).*** Suitability level for a given chromosome is computed from $S_{THD}$, $S_V$, and cost using the following steps:

- Fuzzification of $S_{THD}$, $S_V$, and cost (based on Fig. 10.13) to compute the corresponding membership values ($\mu_{S_{THD}}$, $\mu_{S_V}$, and $\mu_{cost}$).
- Fuzzy inferencing and defuzzification (using the decision matrix of Table 10.6, Fig. 10.14b, and Eq. 10-18) to determine the suitability of the chromosome ($S_{chrom}$).

**Genetic Operators.** Three genetic operators including reproduction (based on the "roulette-wheel" mechanism), crossover (with crossover probability of 0.6 to 1.0), and mutation (with mutation probability of 0.01 to 0.1) are used.

**TABLE 10.6** Decision Matrix for Determining Suitability of a Chromosome ($S_{chrom}$)

| $S_v$ | | $S_{THD}$ Low ("cost") | $S_{THD}$ Medium ("cost") | $S_{THD}$ High ("cost") |
|---|---|---|---|---|
| Low | | Low / Low-medium | Medium / Low | High / Low |
| Medium | | Medium / Low-medium | Medium / Medium | High / Low-medium |
| High | | Medium / Low-medium | Medium / Medium-high | Medium / Medium |

Figure 10.15 illustrates the block diagram of the fuzzy expert system for the determination of the suitability of a chromosome for capacitor placement.

### 10.4.11.2 Solution Methodology

The shunt capacitor placement and sizing problem in the presence of linear and nonlinear loads is solved using the GA-FL algorithm of Fig. 10.16 as follows:

**Step 1:** Input system parameters and the initial population.

**Step 2: Fuzzy Fitness or Suitability.** Perform the following steps for all chromosomes (Fig. 10.15):

**Step 2a:** Run the Newton-based harmonic power flow program.

**Step 2b:** Compute suitability of $THD_v$ and voltage (Fig. 10.13, Table 10.5, and Eq. 10-18).

**Step 2c:** Compute cost index.

**Step 2d:** Compute suitability, $S_{chrom}$ (Fig. 10.14b, Table 10.6, and Eq. 10-18).

**Step 3: Reproduction Process.** Select a new combination of chromosomes by rolling the roulette wheel.

**Step 4: Crossover Process.** Select a random number for mating two parent chromosomes. If it is between 0.6 and 1.0 then combine the two parents, and generate two offspring. Else, transfer the chromosome with no crossover.

**Step 5: Mutation Process.** Select a random number for mutation of one chromosome. If it is between 0.01 and 0.1 then apply the mutation process at a random position. Else, transfer the chromosome with no mutation.

**Step 6: Updating Populations.** Replace the old population with the improved population generated by Steps 2 to 5. Check all chromosomes and save the one with minimum cost, satisfying all constraints.

**Step 7: Convergence.** If all chromosomes are the same or the maximum number of iterations is achieved ($N_{max} = 40$), then print the solution and stop, else go to Step 2.

### 10.4.12 Application Example 10.7: Genetically Optimized Fuzzy Placement and Sizing of Capacitor Banks in the 18-Bus IEEE Distorted System

The hybrid GA-FL method (Fig. 10.16) and the capacitor placement techniques of prior sections (MSS, MSS-LV, fuzzy, and genetic) are applied to the 23 kV, 18-bus distorted IEEE distribution system (Fig. E10.3.1). The six-pulse rectifier (with P = 0.3 pu = 3 MW and Q = 0.226 pu = 2.26 MVAr)

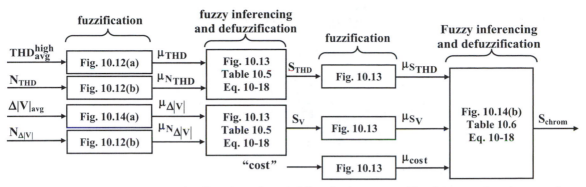

**FIGURE 10.15** Fuzzy expert system (FES) to determine suitability of a chromosome ($S_{chrom}$) for capacitor placement from total harmonic distortion ($THD_v$), voltage deviation ($\Delta|V|_{avg}$), "cost," number of buses with unacceptable $THD_v$ ($N_{THD}$) and unacceptable $|V|$ ($N_{\Delta|V|}$) [46].

causes a maximum voltage $THD_v$ of 8.486% (column 2 of Table E10.7.1).

Simulation results for the MSS, MSS-LV, fuzzy, and genetic algorithms are listed in columns 3 to 6 of Table E10.7.1, respectively. Among these algorithms, the GA approach results in a near global solution with the best yearly benefit of $18949 per year (column 6 of Table E10.7.1).

The GA-FL approach of Fig. 10.16 is also applied to this system for optimal placement and sizing of capacitor banks. Results show a yearly benefit of $19550 per year (last row of Table E10.7.1) and maximum voltage $THD_v$ is limited to 4.982% (column 7 of Table E10.7.1). Compared to the GA approach, total yearly benefit is increased by $600 (e.g., 3.2%), and the total allocated capacitance is decreased by 0.015 pu.

**Analysis and Convergence Criterion of the GA-FL Algorithm.** The iterations of the GA-FL algorithm are continued until all generated chromosomes become equal or the maximum number of iterations is reached.

At each iteration, a fuzzy expert system (Fig. 10.15) is used to compute the fitness function of each chromosome and to select the parent population. As a result, new generations have more variety of chromosomes and higher probability of capturing the global solution. As demonstrated in Fig. E10.7.1, the relatively narrow search band of the conventional GA does not allow a large variation of chromosomes between iterations and the solution (e.g., benefits) is trapped at a near global point after the 14th iteration. However, the larger search space of the GA-FL approach (allowing large variations of benefits in consecutive iterations) has resulted in a better near global solution at the 35th iteration.

The inclusion of objective function and power quality constraints will automatically eliminate all solutions generating extreme values for voltages and/or currents and prevents fundamental and harmonic parallel resonances.

### 10.4.13 Application Example 10.8: Genetically Optimized Fuzzy Placement and Sizing of Capacitor Banks in the 123-Bus IEEE System with 20 Nonlinear Loads

The GA (Fig. 10.11) and GA-FL (Fig. 10.16) methods are applied to the 4.16 kV, 123-bus IEEE radial distribution network of Fig. E10.8.1. Specifications of this system are given in reference [51]. To introduce harmonic distortion, 20 nonlinear loads (Table E10.8.1) with different harmonic spectra (Table E10.8.2) are included. These loads are modeled as harmonic current sources. The coupling between harmonics is not considered and the decoupled harmonic power flow algorithm of Chapter 7 (Section 7.5.1) is used.

The high penetration of these nonlinear loads causes a maximum $THD_v$ of 8.03% (column 2, Table E10.8.3). The results of the GA-FL algorithm show a 5.5% increase in total annual benefit compared with those generated by the GA method (last row of Table E10.8.3), whereas the maximum $THD_v$ is significantly limited to 2.76%.

**Analysis and Convergence Criterion.** The generations (iterations) of the GA-FL algorithm are continued until all chromosomes of the population become equal or the maximum number of generations is achieved. The initial conditions for Eq. 10-8 (e.g., the initial capacitor values) do not usually reside inside the permissible solution region. In this section, a GA-FL approach is used to minimize the objective

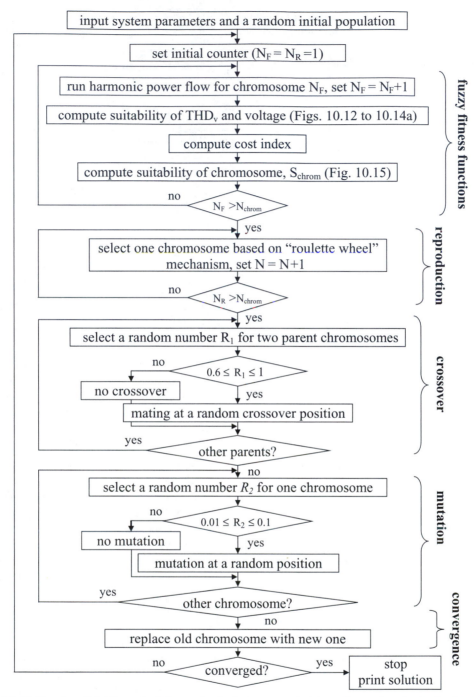

**FIGURE 10.16** Iterative GA-FL method for optimal placement and sizing of capacitor banks in the presence of harmonics.

**TABLE E10.7.1** Simulation Results of MSS, MSS-LV, Fuzzy, Genetic and GA-FL Algorithms for the 18-Bus, Distorted IEEE Distribution System of Fig. E10.3.1 with per Unit VA = 10 MVA, per Unit V = 23 kV, and Swing Bus Voltage = 1.05 pu[a]

| | | Local optimization | | | (Near) global optimization | |
|---|---|---|---|---|---|---|
| | Before optimization | After MSS optimization | After MSS-LV optimization | After fuzzy optimization | After GA optimization | After GA-FL optimization |
| Capacitor bank | Q2 = 0.105 | Q2 = 0.030 | Q2 = 0.120 | Q2 = 0.015 | Q2 = 0.030 | Q2 = 0.030 |
| locations and sizes | Q3 = 0.060 | Q3 = 0.090 | Q3 = 0.060 | Q3 = 0.030 | Q3 = 0.000 | Q3 = 0.000 |
| (pu) | Q4 = 0.060 | Q4 = 0.180 | Q4 = 0.180 | Q4 = 0.060 | Q4 = 0.165 | Q4 = 0.195 |
| | Q5 = 0.180 | Q5 = 0.240 | Q5 = 0.240 | Q5 = 0.240 | Q5 = 0.330 | Q5 = 0.300 |
| | Q7 = 0.060 | Q7 = 0.120 | Q7 = 0.120 | Q7 = 0.210 | Q7 = 0.105 | Q7 = 0.105 |
| | Q20 = 0.060 | Q20 = 0.090 | Q20 = 0.090 | Q20 = 0.090 | Q20 = 0.060 | Q20 = 0.090 |
| | Q21 = 0.120 | Q21 = 0.120 | Q21 = 0.120 | Q21 = 0.120 | Q21 = 0.090 | Q21 = 0.075 |
| | Q24 = 0.150 | Q24 = 0.000 | Q24 = 0.000 | Q24 = 0.000 | Q24 = 0.015 | Q24 = 0.015 |
| | Q25 = 0.090 | Q25 = 0.030 | Q25 = 0.000 | Q25 = 0.000 | Q25 = 0.015 | Q25 = 0.015 |
| | Q50 = 0.120 | Q50 = 0.000 | Q50 = 0.000 | Q50 = 0.000 | Q50 = 0.030 | Q50 = 0.000 |
| Total capacitor (pu) | Qt = 1.005 | Qt = 0.900 | Qt = 0.930 | Qt = 0.765 | Qt = 0.840 | Qt = 0.825 |
| Minimum voltage (pu) | 1.029 | 1.016 | 1.013 | 0.998 | 1.003 | 1.005 |
| Maximum voltage (pu) | 1.055 | 1.056 | 1.059 | 1.050 | 1.050 | 1.050 |
| Maximum THD$_v$ (%) | 8.486 | 6.370 | 4.720 | 4.899 | 4.883 | 4.982 |
| Losses (kW) | 282.93 | 246.43 | 250.37 | 257.46 | 249.31 | 248.18 |
| Capacitor cost ($/year) | 1978.20 | 1692.00 | 2206.80 | 1458.30 | 1788.75 | 1817.55 |
| Total cost ($/year) | 159853.14 | 139199.94 | 141913.26 | 145120.98 | 140903.73 | 140302.8 |
| Benefits ($/year) | | 20653.6 | 17939.88 | 14732.16 | 18949.41 | 19550.34 |

[a]All capacitors were removed before the optimization process.

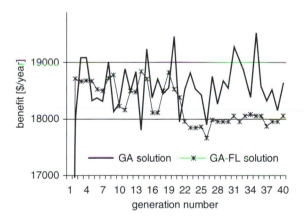

**FIGURE E10.7.1** Comparison of the solution progress for the GA-FL and the GA methods (Table E10.7.1) for the 18-bus, distorted IEEE distribution system (Fig. E10.3.1) using the same initial population.

function while directing the constraints toward the permissible region.

To select the parent population at each generation, a fuzzy expert system (Fig. 10.15) is used to compute the fitness function of each chromosome considering the uncertainty of decision making based on the objective function and constraints. As a result, new generations have a higher probability of capturing the global solution.

As demonstrated in Fig. E10.8.2, the relatively narrow search band of the conventional GA does not allow a large variation of chromosomes between generations, and thus it is more difficult to escape from local optima. For the 18-bus and 123-bus systems, the solution is trapped at a near global point after the 14th and 22nd iterations, respectively. In the GA-FL approach, there is more variety in the generated parent population as the chromosome evaluation is based on fuzzy logic. The larger search space (allowing large variations of benefits in consecutive iterations) has resulted in a better near global solution at the 35th and 29th iterations for the 18-bus and 123-bus systems, respectively.

The inclusion of objective function and power quality constraints will automatically eliminate all solutions that generate extreme values for voltages and/or currents and prevents fundamental and harmonic parallel resonances.

## 10.5 SUMMARY

Related to reactive power compensation is the placement and sizing of capacitor banks so that the transmission efficiency can be increased at minimum cost. First filter designs are addressed for sinusoidal conditions. Various optimization methods are described such as analytical methods, numerical programming,

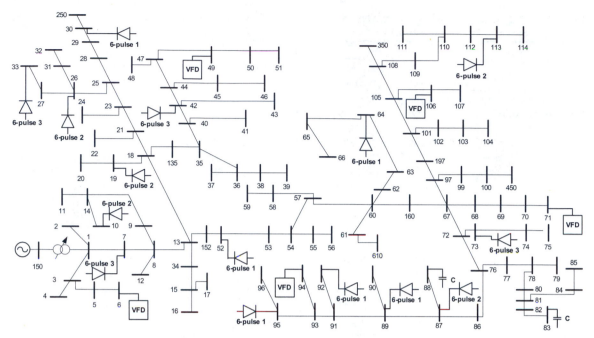

**FIGURE E10.8.1** The IEEE 123-bus distribution system [51] with 20 nonlinear loads.

**TABLE E10.8.1** Nonlinear Load Data for the IEEE 123-Bus System (Fig. E10.8.1)

| Nonlinear bus | Nonlinear load type (Table E10.8.2) | kW | kVAr |
|---|---|---|---|
| 6 | Six-pulse variable-frequency drive | 15 | 10 |
| 7 | Six-pulse 3 | 15 | 10 |
| 10 | Six-pulse 2 | 24 | 3 |
| 19 | Six-pulse 2 | 24 | 3 |
| 26 | Six-pulse 2 | 15 | 10 |
| 30 | Six-pulse 1 | 24 | 3 |
| 33 | Six-pulse 3 | 24 | 3 |
| 42 | Six-pulse 3 | 15 | 10 |
| 49 | Six-pulse variable-frequency drive | 60 | 45 |
| 52 | Six-pulse 1 | 24 | 3 |
| 64 | Six-pulse 1 | 24 | 3 |
| 71 | Six-pulse variable-frequency drive | 15 | 10 |
| 73 | Six-pulse 3 | 24 | 3 |
| 87 | Six-pulse 2 | 60 | 45 |
| 89 | Six-pulse 1 | 24 | 3 |
| 92 | Six-pulse 1 | 24 | 3 |
| 94 | Six-pulse variable-frequency drive | 15 | 10 |
| 95 | Six-pulse 1 | 24 | 3 |
| 106 | Six-pulse variable-frequency drive | 15 | 10 |
| 113 | Six-pulse 2 | 24 | 3 |

**TABLE E10.8.2** Harmonic Spectra of Nonlinear Loads for the IEEE 123-Bus System (Fig. E10.8.1)

| | Nonlinear load | | | | | | | |
|---|---|---|---|---|---|---|---|---|
| | Six-pulse 1 | | Six-pulse 2 | | Six-pulse 3 | | Six-pulse VFD | |
| Order | Magnitude (%) | Phase (deg) | Magnitude (%) | Phase (deg) | Magnitude (%) | Phase (deg) | Magnitude (%) | Phase (deg) |
| 1 | 100 | 0 | 100 | 0 | 100 | 0 | 100 | 0 |
| 5 | 20 | 0 | 19.1 | 0 | 20 | 0 | 23.52 | 111 |
| 7 | 14.3 | 0 | 13.1 | 0 | 14.3 | 0 | 6.08 | 109 |
| 11 | 9.1 | 0 | 7.2 | 0 | 9.1 | 0 | 4.57 | −158 |
| 13 | 7.7 | 0 | 5.6 | 0 | 0 | 0 | 4.2 | −178 |
| 17 | 5.9 | 0 | 3.3 | 0 | 0 | 0 | 1.8 | −94 |
| 19 | 5.3 | 0 | 2.4 | 0 | 0 | 0 | 1.37 | −92 |
| 23 | 4.3 | 0 | 1.2 | 0 | 0 | 0 | 0.75 | −70 |
| 25 | 4 | 0 | 0.8 | 0 | 0 | 0 | 0.56 | −70 |
| 29 | 3.4 | 0 | 0.2 | 0 | 0 | 0 | 0.49 | −20 |
| 31 | 3.2 | 0 | 0.2 | 0 | 0 | 0 | 0.54 | 7 |
| 35 | 2.8 | 0 | 0.4 | 0 | 0 | 0 | 0 | 0 |
| 37 | 2.7 | 0 | 0.5 | 0 | 0 | 0 | 0 | 0 |

**TABLE E10.8.3** Simulation Results of GA (Fig. 10.11) and GA-FL (Fig. 10.16) Algorithms for the 123-Bus, Distorted IEEE Distribution System (Fig. E10.8.1) Using Decoupled Harmonic Power Flow (Chapter 7, Section 7.5.1)

| | Before optimization | GA optimization | GA-FL optimization |
|---|---|---|---|
| Capacitor bank locations and sizes | Q83 = 200 | Q26 = 150 | Q42 = 300 |
| (pu) | Q88 = 50 | Q42 = 300 | Q52 = 300 |
| | | Q72 = 150 | Q72 = 300 |
| | | Q76 = 300 | Q105 = 150 |
| Total capacitor (kW) | Qt = 350 | Qt = 900 | Qt = 1050 |
| Minimum voltage (pu) | 0.874 | 0.904 | 0.915 |
| Maximum voltage (pu) | 0.980 | 0.980 | 0.980 |
| Maximum THD$_v$ (%) | 8.0319 | 3.1018 | 2.7603 |
| Losses (kW) | 94.96 | 82.021 | 81.28 |
| Capacitor cost ($/year) | 125 | 360 | 390 |
| Total cost ($/year) | 53114.5 | 46127.72 | 45744.24 |
| Benefits ($/year) | | 6986.8 | 7370.3 |

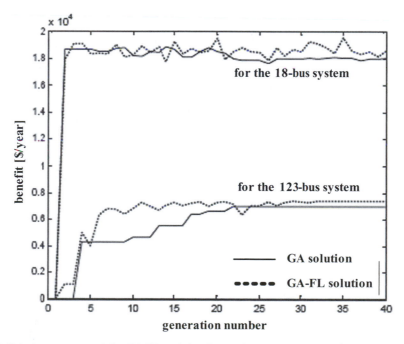

**FIGURE E10.8.2** Solution progress of the GA-FL and the GA methods for the 18-bus (Fig. E10.3.1) and 123-bus (Fig. E10.8.1) distorted IEEE distribution systems (Table E10.8.3).

heuristic approaches, and artificial intelligence (AI-based) methods. The latter ones comprise genetic algorithms, expert systems, simulated annealing, artificial neural networks, and fuzzy set theory. Furthermore, graph search, particle swarm, tabu search, and sequential quadratic programming algorithms are explained. Thereafter, the optimal placement and sizing of capacitor banks is studied under the influence of harmonics relying on some of the same optimization methods as used for sinusoidal conditions. Thus, the nonsinusoidal case is an extension of the sinusoidal approach. It is recommended that the reader first applies some of the optimization procedures to the sinusoidal case and then progresses to nonsinusoidal conditions.

Eight application examples highlight the advantages and disadvantages of the various optimization methods under given constraints (e.g., $THD_v$, $V_{rms}$) and objective functions (e.g., cost). Unfortunately, there is not one single method that incorporates all the advantages and no disadvantages. It appears that the more advanced methods such as fuzzy set theory, genetic algorithms, artificial neural networks, and particle swarm methods are more able to identify the global optimum than the straightforward search methods where the gradient (e.g., steepest decent) is employed guiding the optimization process. This is so because the more advanced methods proceed not along a single trajectory but search for the global optimum within a certain region in a simultaneous manner.

## 10.6 REFERENCES

1) Ng, H.N.; Salama, M.M.A.; and Chikhani, A.Y.; "Classification of capacitor allocation techniques," *IEEE Transactions on Power Delivery,* Vol. 15, No. 1, pp. 387–392, January 2000.

2) Neagle, N.M.; and Samson, D.R.; "Loss reduction from capacitors installed on primary feeders," *AIEE Transactions*, Vol. 75, pp. 950–959, Oct. 1956.

3) Cook, R.F.; "Optimizing the application of shunt capacitors for reactive-voltampere control and loss reduction," *AIEE Transactions*, Vol. 80, pp. 430–444, Aug. 1961.

4) Schmill, J.V.; "Optimum size and location of shunt capacitors on distribution feeders," *IEEE Transactions on Power Apparatus and Systems*, Vol. 84, No. 9, pp. 825–832, Sept. 1965.

5) Chang, N.E.; "Locating shunt capacitors on primary feeder for voltage control and loss reduction," *IEEE Transactions on Power Apparatus and Systems*, Vol. 88, No. 10, pp. 1574–1577, Oct. 1969.

6) Bae, Y.G.; "Analytical method of capacitor allocation on distribution primary feeders," *IEEE Transactions on Power Apparatus and Systems*, Vol. 97, No. 11, pp. 1232–1238, July/Aug. 1978.

7) Lee, S.H.; and Grainger, J.J.; "Optimum placement of fixed and switched capacitors on primary distribution feeders," *IEEE Transactions on Power Apparatus and Systems*, Vol. 100, No. 1, pp. 345–352, Jan. 1981.

8) Salama, M.M.A.; Chikhani, A.Y.; and Hackam, R.; "Control of reactive power in distribution systems with an end-load and fixed load conditions," *IEEE Transactions on Power Apparatus and Systems*, Vol. 104, No. 10, pp. 2779–2788, Oct. 1985.

9) Duran, H.; "Optimum number, location, and size of shunt capacitors in radial distribution feeders, a dynamic programming approach," *IEEE Transactions on Power Apparatus and Systems*, Vol. 87, No. 9, pp. 1769–1774, Sept. 1968.

10) Fawzi, T.H.; El-Sobki, S.M.; and Abdel-Halim, M.A.; "New approach for the application of shunt capacitors to the primary distribution feeders," *IEEE Transactions on Power Apparatus and Systems*, Vol. 102, No. 1, pp. 10–13, Jan. 1983.

11) Ponnavaikko, M.; and Prakasa Rao, K.S.; "Optimal choice of fixed and switched shunt capacitors on radial distributors by the method of local variations," *IEEE Transactions on Power Apparatus and Systems*, Vol. 102, No. 6, pp. 1607–1615, June 1983.

12) Baran, M.E.; and Wu, F.F.; "Optimal sizing of capacitors placed on a radial distribution system," *IEEE Transactions on Power Delivery*, Vol. 4, No. 1, pp. 735–743, Jan. 1989.

13) Abdel-Salam, T.S.; Chikhani, A.Y.; and Hackam, R.; "A new technique for loss reduction using compensating capacitors applied to distribution systems with varying load condition," *IEEE Transactions on Power Delivery*, Vol. 9, No. 2, pp. 819–827, Apr. 1994.

14) Chis, M.; Salama, M.M.A.; and Jayaram, S.; "Capacitor placement in distribution systems using heuristic search strategies," *IEE Proceedings on Generation, Transmission and Distribution*, Vol. 144, No. 2, pp. 225–230, May 1997.

15) Boone, G.; and Chiang, H.D.; "Optimal capacitor placement in distribution systems by genetic algorithm," *Electrical Power and Energy Systems*, Vol. 15, No. 3, pp. 155–162, 1993.

16) Sundhararajan, S.; and Pahwa, A.; "Optimal selection of capacitors for radial distribution systems using a genetic algorithm," *IEEE Transactions on Power Systems*, Vol. 9, No. 3, pp. 1499–1507, Aug. 1994.

17) Miu, K.N.; Chiang, H.D.; and Darling, G.; "Capacitor placement, replacement and control in large-scale distribution systems by a GA-based two-stage algorithm," *IEEE Transactions on Power Systems*, Vol. 12, No. 3, pp. 1160–1166, Aug. 1997.

18) Laframboise, J.R.P.R.; Ferland, G.; Chikhani, A.Y.; and Salama, M.M.A.; "An expert system for reactive power control of a distribution system, part 2: system implementation," *IEEE Transactions on Power Systems*, Vol. 10, No. 3, pp. 1433–1441, Aug. 1995.

19) Ananthapadmanabha, T.; Kulkarni, A.D.; Gopala Rao, A.S.; and Rao, K.R.; "Knowledge-based expert system for optimal reactive power control in distribution system," *Electrical Power and Energy Systems*, Vol. 18, No. 1, pp. 27–31, 1996.

20) Santoso, N.I.; and Tan, O.T.; "Neural-net based real-time control of capacitors installed on distribution systems," *IEEE Transactions on Power Delivery*, Vol. 5, No. 1, pp. 266–272, Jan. 1990.

21) Gu, Z.; and Rizy, D.T.; "Neural networks for combined control of capacitor banks and voltage regulators in distribution systems," *IEEE Transactions on Power Delivery*, Vol. 11, No. 4, pp. 1921–1928, Oct. 1996.

22) Zadeh, L.A.; "Fuzzy sets," *Information and Control*, Vol. 8, pp. 338–353, 1965.

23) Chin, H.C.; "Optimal shunt capacitor allocation by fuzzy dynamic programming," *Electric Power Systems Research*, Vol. 35, pp. 133–139, 1995.

24) Ng, H.N.; Salama, M.M.A.; and Chikhani, A.Y.; "Capacitor allocation by approximate reasoning: fuzzy capacitor placement," *IEEE Transactions on Power Delivery*, Vol. 15, Issue 1, Jan. 2000, pp. 393–398

25) Ng, H.N.; and Salama, M.M.A.; "Fuzzy optimal capacitor sizing and placement," *Proceedings of*

*Canadian Conference on Electrical and Computer Engineering*, Vol. 2, pp. 680–683, 1995.

26) Carlisle, J.C.; and El-Keib, A.A.; "A graph search algorithm for optimal placement of fixed and switched capacitors on radial distribution systems," *IEEE Transactions on Power Delivery*, Vol. 15, No. 1, pp. 423–428, January 2000.

27) Mantawy, A.H.; and Al-Ghamdi, M.S.; "A new reactive power optimization algorithm," *IEEE Conference*, Bologna, Italy, June 2003.

28) Yang, H.T.; Huang, Y.C.; and Huang, C.L.; "Solution to capacitor placement in a radial distribution system using tabu search method," *Proceedings of International Conference on Energy Management and Power Delivery*, EMPD 95, Vol. 1, pp. 388–393, 1995.

29) Mori, H.; and Ogita, Y.; "Parallel tabu search for capacitor placement in radial distribution systems," *Proceedings of the IEEE Power Engineering Society Winter Meeting 2000*, Vol. 4, 23–27 Jan. 2000, pp. 2334–2339.

30) Chang, C.S.; and Lern, L.P.; "Application of tabu search strategy in solving non-differentiable savings function for the calculation of optimum savings due to shunt capacitor installation in a radial distribution system," *Proceedings of the Power Engineering Society Winter Meeting 2000*, Vol. 4, 23–27 Jan. 2000 pp. 2323–2328.

31) Abril, I.P.; and Quintero, J.A.G.; "VAR compensation by sequential quadratic programming," *IEEE Transactions on Power Systems*, Vol. 18, No. 1, pp. 36–41, February 2003.

32) Ng, H.N.; Salama, M.M.A.; and Chikhani, A.Y.; "Capacitor allocation by approximate reasoning: fuzzy capacitor placement," *IEEE Transactions on Power Delivery*, Vol. 15, No. 1, pp. 393–398, January 2000.

33) Capacitor Subcommittee of the IEEE Transmission and Distribution Committee, "Bibliography on power capacitors 1975–1980," *IEEE Transactions on Power Apparatus and Systems*, Vol. 102, No. 7, pp. 2331–2334, July 1983.

34) IEEE VAR Management Working Group of the IEEE System Control Subcommittee, "Bibliography on reactive power and voltage control," *IEEE Transactions on Power Systems*, Vol. 2, No. 2, pp. 361–370, May 1997.

35) Baghzouz, Y.; "Effects of nonlinear loads on optimal capacitor placement in radial feeders," *IEEE Transactions on Power Delivery*, Vol. PD-6, No. 1, pp. 245–251, January 1991.

36) Baghzouz, Y.; and Ertem, S.; "Shunt capacitor sizing for radial distribution feeders with distorted substation voltage," *IEEE Transactions on Power Delivery*, Vol. PD-5, No. 2, pp. 650–657, 1990.

37) Gou, B.; and Abur, A.; "Optimal capacitor placement for improving power quality," *Proceedings of the IEEE Power Engineering Society Summer Meeting 1999*, Vol. 1, 18–22 July 1999, pp. 488–492.

38) Carpinelli, G.; Varilone, P.; Vito, V.D.; and Abur, A.; "Capacitor placement in three-phase distribution systems with nonlinear and unbalanced loads," *IEE Proceedings on Generation, Transmission and Distribution*, Vol. 152, No. 1, pp. 47–51, 2005.

39) Hsu, C.T.; Yan, Y.H.; Chen, C.S.; and Her, S.L.; "Optimal reactive power planning for distribution systems with nonlinear loads," *IEEE Region 10 International Conference on Computer, Communication, Control and Power Engineering TENCON '93*, pp. 330–333, 1993.

40) Wu, Z.Q.; and Lo, K.L.; "Optimal choice of fixed and switched capacitors in radial distributions with distorted substation voltage," *IEE Proceedings on Generation, Transmission and Distribution*, Vol. 142, No. 1, pp. 24–28, January 1995.

41) Masoum, M.A.S.; Ladjevardi, M.; Fuchs, E.F.; and Grady, W.M.; "Optimal sizing and placement of fixed and switched capacitor banks under non-sinusoidal operating conditions," *Proceedings of the IEEE Summer Power Meeting 2002*, pp. 807–813, July 2002.

42) Masoum, M.A.S.; Ladjevardi, M.; Fuchs, E.F.; and Grady, W.M.; "Application of local variations and maximum sensitivities selections for optimal placement of shunt capacitor banks under nonsinusoidal operating conditions," *34th Annual North American Power Symposium*, pp. 507–515, NAPS-2002.

43) Chin, H.C.; "Optimal shunt capacitor allocation by fuzzy dynamic programming," *Electric Power Systems Research*, Vol. 35, pp. 133–139, 1995.

44) Masoum, M.A.S.; Jafarian, A.; Ladjevardi, M.; Fuchs, E.F.; and Grady, W.M.; "Fuzzy approach for optimal placement and sizing of capacitor banks in the presence of harmonics," *IEEE Transactions on Power Delivery*, Vol. 19, No. 2, April 2004.

45) Masoum, M.A.S.; Ladjevardi, M.; Jafarian, M.; and Fuchs, E.F.; "Optimal placement, replacement and sizing of capacitor banks in distorted distribution networks by genetic algorithms," *IEEE Transactions on Power Delivery*, Vol. 19, No. 4, October 2004.

46) Ladjevardi, M.; and Masoum, M.A.S.; "Genetically optimized fuzzy placement and sizing of capacitor

banks in distorted distribution networks," *IEEE Transactions on Power Delivery*, Accepted for Publication, Paper Number TPWRD-00741-2005. R1.

47) Mohan, N.; Undeland, T.M.; and Robbins, W.P.; *Power Electronics, Converters, Applications and Design*, John Wiley & Sons, USA, 2003.

48) Saric, A.T.; Calovic, M.S.; and Djukanovic, M.B.; "Fuzzy optimization of capacitors in distribution systems," *IEE Proceedings on Generation, Transmission and Distribution*, Vol. 144, No. 5, pp. 415–422, September 1997.

49) http://www.abb.com, Retrieved: Feb. 2005.

50) Mamdani, E.H.; and Asilian, S.; "An experiment in linguistic synthesis with a fuzzy logic controller," *International Journal of Man–Machine Studies*, Vol. 7, pp. 1–13, 1975.

51) IEEE recommended practices and requirements for harmonic control in electric power systems, *IEEE Standard 519-1992*, New York, NY, 1993.

# 11

# Unified Power Quality Conditioner (UPQC)

One of the main responsibilities of a utility system is to supply electric power in the form of sinusoidal voltages and currents with appropriate magnitudes and frequency for the customers at the points of common coupling (PCC). Design and analysis of system equipment and user apparatus are based on sinusoidal current and voltage waveforms with nominal or rated magnitudes and frequency. Although the generated voltages of synchronous machines in power plants are nearly sinusoidal, some undesired and/or unpredictable conditions – such as lightning and short-circuit faults – and nonlinear loads cause steady-state and/or transient voltage and current disturbances. For example, electric arc furnaces cause voltage fluctuations, power electronic converters generate current harmonics and distort voltage waveforms, and short-circuit faults result in voltage sags and swells. On the other hand, most customer loads such as computers, microcontrollers, programmable logic controllers (PLCs), and hospital equipment are sensitive and unprotected to power quality disturbances and their proper operation depends on the quality of the voltage that is delivered to them.

To optimize the performance and stability of power systems and to overcome power quality problems, many techniques and devices have been proposed and implemented. Conventional devices used for voltage regulation, power factor correction, and power quality improvement include shunt or series power capacitors, phase shifters, and passive or active filters (Chapter 9). Advanced technical solutions are flexible AC transmission system (FACTS) devices [1, 2] to improve voltage regulation, as well as steady-state and instantaneous active and reactive power control at fundamental frequency. Other approaches are based on custom power devices [3, 4] that offer a variety of options including series and shunt active (P) and reactive (Q) power compensation to improve power quality, voltage regulation, as well as steady-state and dynamic performances at fundamental and harmonic frequencies.

The concept of hybrid active filtering with shunt and series branches was introduced by Akagi and colleagues [5–7]. The most complete configuration of hybrid filters is the unified power quality conditioner (UPQC), which is also known as the universal active filter [8–13]. UPQC is a multifunction power conditioner that can be used to compensate various voltage disturbances of the power supply, to correct voltage fluctuation, and to prevent the harmonic load current from entering the power system. It is a custom power device designed to mitigate the disturbances that affect the performance of sensitive and/or critical loads. UPQC has shunt and series compensation capabilities for (voltage and current) harmonics, reactive power, voltage disturbances (including sags, swells, flicker, etc.), and power-flow control. Normally, a UPQC consists of two voltage-source converters with a common DC link designed in single-phase, three-phase three-wire, or three-phase four-wire configurations. One converter is connected in series through a transformer between the source and the critical load at the PCC and operates as a voltage-source inverter (VSI). The other converter is connected in shunt at the PCC through a transformer and operates as a current-source inverter (CSI). The active series converter compensates for voltage supply disturbances (e.g., including harmonics, imbalances, negative and zero-sequence components, sags, swells, and flickers), performs harmonic isolation, and damps harmonic oscillations. The active shunt converter compensates for load current waveform distortions (e.g., caused by harmonics, imbalances, and neutral current) and reactive power, and performs the DC link voltage regulation.

UPQC is specifically designed to protect the critical load at the point of installation connected to distorted distribution systems by correcting any of the following shortcomings:

- harmonic distortions (of source voltage and load current) at the utility–consumer PCC,
- voltage disturbances (sags, swells, flickers, imbalances, and instability),
- voltage regulation,
- reactive power flow at fundamental and harmonic frequencies,

**443**

- neutral and negative-sequence currents, and
- harmonic isolation (e.g., between a subtransmission and a distribution system).

Therefore, UPQC has the capability of improving power quality at the point of installation of power distribution systems. This includes compensation of voltage-based disturbances across the sensitive load (such as voltage harmonics, voltage imbalances, voltage flickers, sags, and swells) and current-based distortions of the system (such as current harmonics, reactive power, and neutral current). However, UPQC is not capable of improving the power quality of the entire system.

This chapter starts with a section about compensation devices. Section 11.2 introduces the UQPC including its structure and operation. Section 11.3 discusses the control system of the UPQC. The following two sections provide detailed analyses and discussions of two main approaches for implementing UPQC control systems including the Park ($dq0$) transformation and the instantaneous real and reactive power approach. Section 11.6 explores UPQC performance through six application examples.

## 11.1 COMPENSATION DEVICES AT FUNDAMENTAL AND HARMONIC FREQUENCIES

Over the past several decades the transmission and distribution of electric power has become extremely complicated and subject to very tight operational and power quality specifications: the loads are increasing, power transfer grows, and the power system becomes increasingly more difficult to operate and less secure with respect to major disturbances and outages. The full potential of transmission interconnections may not be utilized due to

- large power flow with inadequate control pooling,
- excessive reactive power in various part of the power system, and
- large dynamic power swings between different parts of the system.

The main problems with interconnected transmission lines are

- at no-load or light-load condition the voltage increases at receiving ends resulting in reduced stability,
- at loaded conditions the voltage drop increases resulting in the decrease of transmittable power,
- voltage collapse and voltage instability,

- power and power angle variation due to small disturbances, and
- transient instability due to sudden and severe power variations or oscillations.

As the length of a line increases these problems become more pronounced. The ultimate solution is the instantaneous control of the bus voltage (V), the active power (P), the reactive power (Q), and the total harmonic voltage distortion ($THD_v$).

### 11.1.1 Conventional Compensation Devices

The conventional approaches for voltage regulation and power compensation at fundamental frequency are relatively simple and inexpensive. They include

- shunt compensators (capacitors, reactors),
- series compensators (series capacitors), and
- phase shifters.

These devices are mechanically controlled at low speed and there are wear-and-tear problems resulting in greater operating margins and redundancies. If conventional compensation devices are employed, the system is not fully controlled and optimized from a dynamic and steady-state point of view.

Conventional devices used for power quality improvement are passive and active power filters, as discussed in Chapter 9. Passive filters are simple and inexpensive. However, they have fixed (tuned) frequencies and will not perform properly under changing system configurations and/or variable (nonlinear) load conditions. Active filters eliminate harmonics at the point of installation without considering the power quality status of the entire system. In addition, they do not solve the stability problem, and are not generally capable of optimizing the performance (e.g., voltage regulation, reactive power compensation) of power systems.

### 11.1.2 Flexible AC Transmission Systems (FACTS)

In order to meet stringent specifications and the established contractual obligations set out in supply agreements, various technologies have been used. Among the newly developed technical solutions are FACTS, which were introduced by GyuGyi [1, 2]. They offer a variety of options including series and shunt compensation of power at fundamental frequency to achieve instantaneous and individual or simultaneous control of V, P, and Q. FACTS devices are designed to improve the reliability and the quality of power transmission. However, they do not consider the quality of power that is delivered to customers. This section briefly introduces different types of FACTS controllers [2].

**Shunt FACTS Controllers.** Three types of shunt FACTS controllers are employed to inject compensation current into the system. An injected current in phase quadrature with the phase voltage results in Q control; other phase relations provide simultaneous control of P and Q.

- *Variable impedance shunt FACTS controllers* are also called static VAr generators (SVGs) or static VAr compensators (SVCs). They are thyristor-controlled reactors and/or thyristor-switched capacitors and consist of the following types:
  - Thyristor-controlled reactors (TCRs) and thyristor-switched reactors (TSRs),
  - Thyristor-switched capacitor (TSC),
  - Fixed capacitor, thyristor-controlled reactor (FC-TCR), and
  - Thyristor-switched capacitor, thyristor-controlled reactor (TSC-TCR).
- *Switching converter SVG shunt FACTS controllers* are switching converter based controllable synchronous voltage sources and consist of the following type:
  - Static synchronous compensator (STATCOM).
- *Hybrid SVG and switching converter shunt FACTS controllers* are switching type SVGs with fixed (or controllable) capacitors and/or reactors. They include:
  - STATCOM with fixed capacitors (STATCOM & FC),
  - STATCOM with fixed reactors (STATCOM & FR), and
  - STATCOM with TSC and TCR units (STATCOM & TSC-TCR).

**Series FACTS Controllers.** There are two basic types of series compensation:

- *Variable impedance series FACTS controllers* use thyristor-switches to provide series variable reactive impedances. They include:
  - Gate turn-off (GTO) thyristor-controlled series capacitor (GCSC),
  - Thyristor-switched series capacitor (TSSC), and
  - Thyristor-controlled series capacitor (TCSC).
- *Switching converter series FACTS controllers* use switching converters to provide variable series voltage sources. They include
  - Static synchronous series compensator (SSSC).

**Combined Shunt–Series FACTS Controllers.** These controllers may consist of a number of shunt, a number of series, or a combination of shunt and series FACTS devices. Three popular combined shunt–series FACTS controllers are

- Unified power flow controller (UPFC) has one converter in series and the other in shunt with the transmission line,
- Interline power flow controller (IPFC) has both converters in series with usually a different line, and
- Back-to-back STATCOM has both converters in shunt, each connected to a different power system.

**FACTS Controllers with Storage.** Any of the converter-based FACTS devices can generally accommodate storage such as capacitors, batteries, and superconducting coils. These controllers are more effective for controlling system dynamics due to their ability to dynamically provide/extract real power P to/from a system.

**Development of FACTS Controllers.** The first generation of FACTS (e.g., SVC, TSSC, TCSC, thyristor controlled phase shifters, etc) use thyristor controlled switches as control devices. The second generation (e.g., STATCOM, SSSC, UPFC, IPFC, etc) are converter-based devices and use switching static converters as fast controllable voltage and current sources.

The application of FACTS devices is increasing as their advantages become more evident to power system engineers and researchers. FACTS controllers are primarily designed to operate at fundamental power frequency under sinusoidal conditions and do not necessarily improve the quality of electric power. In fact, most FACTS controllers use switching devices that inject current and voltage harmonics into power systems. Corrective approaches (e.g., sinusoidal PWM techniques) have been proposed to minimize their limitations.

### 11.1.3 Custom Power Devices

The introduction of power electronic loads has raised much concern about power quality problems caused by harmonics, distortions, interruptions, and surges. Equipment such as large industrial drives (e.g., cycloconverters) generate significantly high voltage and current (inter-, sub-) harmonics and create extensive voltage fluctuation. This is more evident in power systems that make use of long transmission lines with limited reactive capacities.

The application of harmonic filters and SVCs to radial transmission systems can offer partial solution to high THD levels and voltage fluctuations. Yet, the lack of dynamic capabilities of these devices limits them to bulk correction. In addition, they might be effective in one application but fail to correct other power quality issues.

The concept of custom power was introduced by Hingorani [3] as the solution to V, P, and Q compensation and power quality problems at the expense of high cost and network complexity. As FACTS controllers improve the reliability and quality of power transmission – by simultaneously enhancing both power transfer capacity and stability – custom power devices enhance the quality and reliability of power delivered to the customer. With a custom power device, a customer (e.g., a sensitive load) will be able to receive a prespecified quality of electric power with a combination of specifications including but not limited to

- magnitude and duration of over- and under-voltages with specified limits,
- low harmonic distortion in the supply, load voltages, and currents,
- small phase imbalance,
- low flicker in the supply voltage,
- control of power interruptions, and
- control of supply voltage frequency within specified limits.

Custom power devices are classified based on their power electronic controllers, which can be either of the network-reconfiguration type or of the compensation type. The network-reconfiguration devices – also called switchgear – include the solid-state and/or static versions of current limiting, current breaking, and current transferring components. The compensation-type custom power devices either compensate a load (e.g., correct its power factor, imbalance) or improve the quality of the supply voltage (e.g., eliminate its harmonics). They are either connected in shunt or in series or a combination of both. Custom power devices are classified as follows [4]:

- **Network-reconfiguration custom power devices** include
  - Solid-state current limiter (SSCL),
  - Solid-state breaker (SSB), and
  - Solid-state transfer switch (SSTS).
- **Compensation-custom power devices** include
  - Distribution STATCOM (DSTATCOM),
  - Dynamic voltage restorer/regulator (DVR), and
  - Unified power quality conditioner (UPQC).

Custom power devices are designed to improve the quality of power at their point of installation of the power distribution systems. They are not primarily designed to improve the power quality of the entire system.

### 11.1.4 Active Power Line Conditioner (APLC)

The APLC is a converter-based compensation device designed to improve the power quality of the entire distribution system by injecting corrective harmonic currents at selected (sensitive) buses. It is usually necessary to use more than one APLC unit to improve the power quality of the entire distribution system. Therefore, APLC units can be considered as a group of shunt active filters; their placement, sizing, and compensation levels (e.g., orders, magnitudes, and phases of injected current harmonics) are optimally designed to improve the power quality of the entire distribution system. The number of required APLC units depends on the severity of distortion, the nature of the distribution system, and types of nonlinear loads, as well as the required quality of electric power (e.g., as specified by IEEE-519). At present, the design of APLCs does not consider transient distortions and stability issues.

### 11.1.5 Remark Regarding Compensation Devices

In the absence of saturation and nonlinearity, power system voltage and current waveforms are sinusoidal and FACTS controllers are employed to solve the associated steady-state, transient, and stability problems. Custom power devices are designed to improve the steady-state and transient problems of distorted distribution systems at the point of installation. Power filters and APLCs mitigate power quality problems at a given location and throughout the power system, respectively. However, at present, none of them are designed to simultaneously consider stability and power quality issues. This is highly related to the fact that in conventional centralized power systems, power quality problems are limited to the low- and medium-voltage distribution systems, whereas stability issues are mainly occurring in high-voltage transmission systems. With the introduction of distributed generation (DG) and the fast growing rate of renewable energy sources in today's power system, there will be an overlap between stability and power quality issues. More attention and further research is required to analyze and examine stability issues in decentralized distorted networks, as well as isolated and DG power systems.

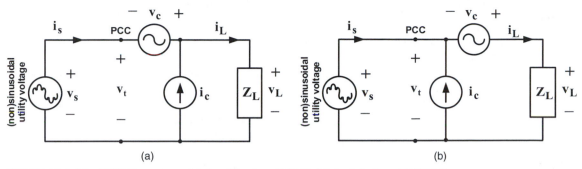

FIGURE 11.1  Ideal UPQC structure; (a) the right-shunt UPQC, (b) the left-shunt UPQC [4].

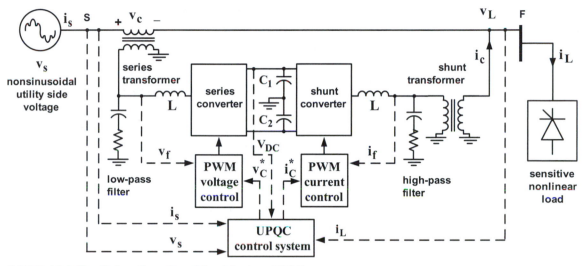

FIGURE 11.2  Detailed configuration of the right-shunt UPQC.

## 11.2 UNIFIED POWER QUALITY CONDITIONER (UPQC)

Unified power quality conditioners are viable compensation devices that are used to ensure that delivered power meets all required standards and specifications at the point of installation. The ideal UPQC can be represented as the combination of a voltage-source converter (injecting series voltage $v_c$), a current-source converter (injecting shunt current $i_c$), and a common DC link (connected to a DC capacitor). There are two possible ways of connecting the unit to the terminal voltage ($v_t$) at PCC:

- Right-shunt UPQC (Fig. 11.1a), where the shunt compensator ($i_c$) is placed at the right side of the series compensator ($v_c$).
- Left-shunt UPQC (Fig. 11.1b), where the shunt compensator ($i_c$) is placed at the left side of $v_c$.

These two structures have similar features; however, the overall characteristics of the right-shunt UPQC are superior (e.g., operation at zero

power injection/absorption mode, achieving unity power factor at load terminals, and full reactive power compensation). In this chapter, a right-shunt UPQC configuration is assumed and analyzed. References [4] and [9] present detailed analysis of the left-shunt structure.

### 11.2.1 UPQC Structure

The detailed structure of the right-shunt UPQC is shown in Fig. 11.2. Assuming a three-phase, four-wire configuration with an unbalanced and distorted system voltage ($v_s$) and nonsinusoidal load current ($i_L$), the UPQC shall perform the following functions:

- convert the feeder (system) current $i_s$ to balanced sinusoids through the shunt compensator,
- convert the load voltage $v_L$ to balanced sinusoids through the series compensator,
- ensure zero real power injection (and/or absorption) by the compensators, and

- supply reactive power to the load (Q compensation).

UPQC employs two converters that are connected to a common DC link with an energy storage capacitor. The main components of a UPQC are series and shunt power converters, DC capacitors, low-pass and high-pass passive filters, and series and shunt transformers:

- **Series converter** is a voltage-source converter connected in series with the AC line and acts as a voltage source to mitigate voltage distortions. It eliminates supply voltage flickers or imbalance from the load terminal voltage and forces the shunt branch to absorb current harmonics generated by the nonlinear load. Control of the series converter output voltage is usually performed using sinusoidal pulse-width modulation (SPWM). The gate pulses are generated by the comparison of a fundamental voltage reference signal with a high-frequency triangular waveform.
- **Shunt converter** is a voltage-source converter connected in shunt with the same AC line and acts as a current source to cancel current distortions, to compensate reactive current of the load, and to improve the power factor. It also performs the DC-link voltage regulation, resulting in a significant reduction of the DC capacitor rating. The output current of the shunt converter is adjusted (e.g., using a dynamic hysteresis band) by controlling the status of semiconductor switches such that output current follows the reference signal and remains in a predetermined hysteresis band.
- **Midpoint-to-ground DC capacitor bank** is divided into two groups, which are connected in series. The neutrals of the secondary transformers are directly connected to the DC link midpoint. As the connection of both three-phase transformers is $Y/Y_0$, the zero-sequence voltage appears in the primary winding of the series-connected transformer in order to compensate for the zero-sequence voltage of the supply system. No zero-sequence current flows in the primary side of both transformers. It ensures the system current to be balanced even when the voltage disturbance occurs.
- **Low-pass filter** is used to attenuate high-frequency components at the output of the series converter that are generated by high-frequency switching.
- **High-pass filter** is installed at the output of shunt converter to absorb current switching ripples.
- **Series and shunt transformers** are implemented to inject the compensation voltages and currents, and

for the purpose of electrical isolation of UPQC converters.

The UPQC is capable of steady-state and dynamic series and/or shunt active and reactive power compensations at fundamental and harmonic frequencies. However, the UPQC is only concerned about the quality of the load voltage and the line current at the point of its installation, and it does not improve the power quality of the entire system.

### 11.2.2 Operation of the UPQC

The shunt converter forces the feeder (system) current to become a balanced (harmonic-free) sinusoidal waveform, while the series converter ensures a balanced, sinusoidal, and regulated load voltage.

#### 11.2.2.1 Operation of the UPQC with Unbalanced and Distorted System Voltage and Load Current

In most cases the three-phase system voltages and load currents are unbalanced and contain higher order frequency components (harmonics), and a UPQC unit is used to compensate for both types of distortions. The UPQC takes the following two steps to compensate for the imbalance and the distortion of system voltage and line current [4].

**Step 1: Obtaining Balanced Voltage at Load Terminals.** The imbalance or distortion of a three-phase system may consist of positive, negative, and zero-sequence fundamental and harmonic components. The system (utility) voltage at point $S$ (Fig. 11.2) can be expressed as

$$v_S(t) = v_{S+}(t) + v_{S-}(t) + v_{S0}(t) + \sum v_{Sh}(t), \quad (11\text{-}1)$$

where subscripts +, −, and 0 refer to the positive-, negative-, and zero-sequence fundamental components, respectively; $v_{s+}(t) = V_{s+} \sin(\omega t + \phi_+)$, $v_{s-}(t) = V_{s-} \sin(\omega t + \phi_-)$, and $v_{s0}(t) = V_{s0} \sin(\omega t + \phi_0)$ are positive-, negative-, and zero-sequence fundamental frequency components of the system voltage; $\sum V_{sh}(t) = \sum V_{sh} \sin(h\omega t + \delta_h)$ represents the harmonics in the voltage, $h$ is harmonic order; and $\phi_+$, $\phi_-$, $\phi_0$, and $\phi_h$ are the corresponding voltage phase angles.

Usually, the voltage at the load at PCC is expected to be sinusoidal with a fixed amplitude $V_L$:

$$v_L(t) = V_L \sin(\omega t + \phi_+). \quad (11\text{-}2)$$

Hence the series converter will need to compensate for the following components of voltage:

$$v_c(t) = v_L(t) - v_S(t) = (V_L - V_{s+})\sin(\omega t + \phi_+) - v_{s-}(t)$$
$$- v_{s0}(t) - \sum v_{sh}(t).$$

$$(11\text{-}3)$$

The control system should automatically control the series converter so that its generated voltage at its output terminals is $v_c(t)$ (in Fig. 11.2) and matched with Eq. 11-3.

**Step 2: Obtaining Balanced System Current Through the Feeder.** The distorted nonlinear load current (Fig. 11.2) can be expressed as

$$i_L(t) = i_{L+}(t) + i_{L-}(t) + i_{L0}(t) + \sum i_{Lh}(t), \quad (11\text{-}4)$$

where $i_{L+}(t) = I_{L+}\sin(\omega t + \delta_+)$, $i_{L-}(t) = I_{L-}\sin(\omega t + \delta_-)$, and $i_{L0}(t) = I_{L0}\sin(\omega t + \delta_0)$ are the fundamental frequency positive-, negative- and zero-sequence components of the load current and $\sum i_{Lh}(t) = \sum I_{Lh}\sin(h\omega t + \delta_h)$ represents the harmonics of the load current. $\delta_+$, $\delta_-$, $\delta_0$, and $\delta_h$ are the corresponding current phase angles.

The shunt converter is supposed to provide compensation of the load harmonic currents to reduce voltage distortion. It should act as a controlled current source and its output current must consist of harmonic, negative-, and zero-sequence currents to cancel the load current distortions. It is usually desired to have a certain phase angle (displacement power factor angle), $\theta_L$, between the positive sequences of voltage and current at the load terminals:

$$\theta_L = (\delta_+ - \phi_+) \Rightarrow \delta_+ = (\phi_+ + \theta_L). \quad (11\text{-}5)$$

Substituting Eq. 11-5 into Eq. 11-4 and simplifying yields

$$i_L(t) = I_{L+}\sin(\omega t + \phi_+)\cos\theta_L + [I_{L+}\cos(\omega t + \phi_+)$$
$$\sin\theta_L + i_{L-}(t) + i_{Lo}(t) + \sum i_{Lh}(t)].$$

$$(11\text{-}6)$$

It is clear that the output current of the shunt converter should be controlled and must assume a wave shape as specified by the 2nd, 3rd, 4th, and 5th terms of Eq. 11-6. That is,

$$i_C(t) = I_{L+}\cos(\omega t + \phi_+)\sin\theta_L + i_{L-}(t) + i_{L0}(t)$$
$$+ \sum i_{Lh}(t).$$

$$(11\text{-}7)$$

This will ensure that the system (feeder) current has a sinusoidal waveform:

$$i_s(t) = i_L(t) - i_c(t) = (I_{L+}\cos\theta_L)\sin(\omega t + \phi_+). \quad (11\text{-}8)$$

Equations 11-3 and 11-7 demonstrate the principles of an ideal UPQC. If these equations are dynamically implemented by the UPQC controller, terminal load voltage and system current will be sinusoidal. In addition, system voltage and current will be in phase if $\theta_L = 0$ (e.g., unity power factor), and neither positive nor negative reactive power will be supplied by the source.

### 11.2.2.2 Operation of UPQC with Unbalanced System Voltages and Load Currents

Assuming that the system voltages and load currents are unbalanced but not distorted (harmonic-free), one can generate UPQC equations as shown below. The UPQC takes the following four steps to compensate the imbalance of the system voltages and line currents [4].

**Step 1: Obtaining Balanced Voltages at Load Terminals.** From the structure of the ideal right-hand UPQC (Fig. 11.2) follows

$$v_L(t) = v_s(t) + v_c(t). \quad (11\text{-}9)$$

To obtain balanced voltages at the load terminals, $v_c(t)$ must cancel the imbalance of the system voltages; therefore,

$$V_{c0} = -V_{s0}, \ V_{c-} = -V_{s-}, \quad (11\text{-}10)$$

which implies that $V_L$ is of positive sequence. Note that the positive-sequence magnitude of the load voltage ($V_L$) should be set to the desired regulated voltage, whereas its phase angle depends on the load displacement power factor.

One defines an injected positive-sequence voltage ($|V_{c+}|$) such that the series compensator does not require any positive-sequence (real) power. Because $i_s$ flows through the series compensator, $v_{c+}$ must have a phase difference of 90° with $i_{s+}$. This implies

$$V_{L+} = V_{s+} + V_{c+}(a + jb), \quad (11\text{-}11)$$

where $(a + jb)$ is the unit vector that is perpendicular to $I_{s+}$. Assuming $V_{L+} = V_L\angle 0°$, Eq. 11-11 results in the following second-order equation:

$$|V_{c+}|^2 - 2a|V_L||V_{c+}| + |V_L|^2 - |V_{s+}|^2 = 0. \quad (11\text{-}12)$$

If the desired regulated voltage level $V_L$ is achievable, this quadratic equation will have two real solutions for $|V_{c+}|$. The minimum solution should be selected because it will result in less cost (e.g., smaller rating) of the UPQC.

**Step 2: Obtaining Balanced System Currents Through the Feeder.** From the structure of the ideal right-hand UPQC (Fig. 11.2) one gets

$$i_L(t) = i_s(t) + i_c(t). \qquad (11\text{-}13)$$

To obtain balanced system currents, the injected current $i_c(t)$ must become zero and one gets for the sequence components of the load current

$$I_{c0} = I_{L0}, \ I_{c+} = 0, \ I_{c-} = I_{L-}. \qquad (11\text{-}14)$$

This will ensure a positive-sequence current (equal to the positive-sequence load current) only passing through the system.

**Step 3: Ensuring Zero Real Power Injection/Absorption by the Compensators.** Equations 11-9 to 11-14 guarantee that imbalances in both the system voltages and the load currents are cancelled by the UPQC. Moreover,

- The series compensator (converter) will consume zero average (real) power because its voltage and current are in quadrature (e.g., $\tilde{V}_{c+}$ is in quadrature with $\tilde{I}_{s+}$).
- The shunt compensator (converter) will also inject zero average (real) power since its positive-sequence injected current is zero ($I_{c+} = 0$) and the load voltage is of positive sequence only.

Therefore, the UPQC does neither inject nor absorb real power.

**Step 4: Reactive Power Injection by the Compensators.** The UPQC can supply part of the load reactive power through its shunt compensator (converter). To do this Eq. 11-14 has to be modified while keeping in mind that the zero and negative-sequence components of the load current must also be canceled by the shunt converter. The new conditions for the injected shunt currents are now

$$I_{c0} = I_{L0}, \ I_{c+} \neq 0, \ I_{c-} = I_{L-}. \qquad (11\text{-}15)$$

We can divide the average component of the load reactive power between the shunt converter, the series converter, and the system. Due to the fact that the average values of the active and the reactive powers of the shunt converter depend only on the

positive-sequence current and its real power consumption (or absorption) must be zero, one gets

$$|V_{L+}| \, |I_{c+}| \cos \theta = 0 \Rightarrow \theta = \pm 90°$$
$$\text{because } (|V_{L+}| \neq 0 \text{ and } |I_{c+}| \neq 0), \qquad (11\text{-}16a)$$

$$|V_{L+}| \, |I_{c+}| \sin \theta = \beta(Q_{L\text{-}avg}), \qquad (11\text{-}16b)$$

where $\theta$ is the angle between $V_{L+}$ and $I_{c+}$, scalar $\beta$ indicates how much reactive power must be supplied by the shunt converter, and $Q_{L\text{-}avg}$ is the average of the load reactive power.

For inductive loads (e.g., negative $Q_{L\text{-}avg}$), compensator (converter) current $I_{c+}$ must lag $V_L$ by 90° (to satisfy Eq. 11-16a) and one can write

$$|I_{c+}| = \beta \frac{|Q_{L\text{-}avg}|}{|V_L|}. \qquad (11\text{-}17)$$

The cancellation of the supply and load current imbalances using an UPQC is addressed in [4, pp. 383–385] and the reactive power compensation with a UPQC is discussed in [4, pp. 386–388].

## 11.3 THE UPQC CONTROL SYSTEM

It is possible to distinguish between disturbances related to the quality of the system (supply) voltage and those related to the quality of the load current. The UPQC corrects both problems at the point of installation (e.g., at the terminals of a sensitive or critical load). It combines a shunt active filter together with a series active filter in a back-to-back configuration to simultaneously improve the power quality of the supply current through the power system and the load voltage at PCC. The main functions of the UPQC are summarized in Table 11.1.

The main components of a UPQC are the series and shunt converters representing active filters, high

**TABLE 11.1** Main Functions of UPQC Controller

| Series converter | Shunt converter |
|---|---|
| • Compensation of voltage harmonics (including negative- and zero-sequence components at fundamental frequency) | • Compensation of current harmonics (including negative- and zero-sequence components at fundamental frequency) |
| • Mitigation of harmonic currents through the power line (harmonic isolation) | • Compensation of the reactive power at the load |
| • Improving system stability (damping) | • Regulation of the DC capacitor voltage |
| • Controlling active and reactive powers | • Regulation of the load voltage |

and low-pass passive filters, DC capacitors, and series and shunt transformers, as shown in Fig. 11.2. The shunt converter is a STATCOM that injects nonsinusoidal corrective current to improve the power quality, regulate load voltage, and control the load power factor. It also compensates for the load reactive power demand, corrects supply current harmonics, and regulates the DC-link voltage. The series converter represents a dynamic voltage regulator (DVR) that introduces corrective harmonic voltages to ensure sinusoidal supply current and constant voltage conditions under steady-state and transient operating conditions (e.g., during voltage sags and swells).

The power electronic converters are frequently voltage-source inverters (VSIs) with semiconductor switches. The output current of the shunt converter is adjusted using an appropriate control scheme (e.g., dynamic hysteresis band). The states of semiconductor switches are controlled so that the output current tracks the reference signal and remains in a predetermined hysteresis band. The installed passive high-pass filter – at the output of shunt converter – absorbs the current switching ripples.

Control of the series converter output voltage is performed using PWM techniques. The gate pulses are generated by the comparison of a fundamental-voltage reference signal with a high-frequency triangular waveform. The low-pass filter attenuates high-frequency components at the output of the series converter, generated by high-frequency switching.

For the compensation of the zero-sequence components, the split-capacitor scheme is usually used on the DC side of converters and the capacitor's midpoint is connected to the neutral. In order to inject the generated voltages and currents, and for the purpose of electrical isolation of UPQC converters, series and shunt transformers may also be included in the UPQC configuration.

### 11.3.1 Pattern of Reference Signals

Provided unbalanced and distorted source voltages and nonlinear load currents exist, then – based on Eqs. 11-1 to 11-17 – the output voltage of the series converter ($v_c(t)$, Fig. 11.2) and the output current of shunt converter ($i_c(t)$, Fig. 11.2) must obey the following equations:

$$\begin{cases} v_c(t) = (V_L - V_{s+})\sin(\omega t + \phi_+) - v_{s-}(t) - v_{s0}(t) \\ \qquad - \sum v_{sh}(t) \\ i_c(t) = I_{L+}\cos(\omega t + \delta_+)\sin\theta_L + i_{L-}(t) \\ \qquad + i_{L0}(t) + \sum i_{Lh}(t), \end{cases}$$
(11-18)

so that the current drawn from the source and the voltage appearing across the load will be sinusoidal.

Generation, tracking, and observation of these reference signals are performed by the UPQC control system. There are two main classes of UPQC control: the first one is based on the Park transformation [15], and the second one on the instantaneous real and imaginary power theory [9, 13, 14]. These will be introduced in the following sections. There are other control techniques for active and hybrid (e.g., UPQC) filters as discussed in Section 9.7 of Chapter 9.

## 11.4 UPQC CONTROL USING THE PARK (*DQ*0) TRANSFORMATION

It is clear that the UPQC should first separate the fundamental positive-sequence component from other components (e.g., derivation of reference signal using either waveform compensation, instantaneous power compensation, or impedance synthesis, as discussed in Sections 9.7.1 to 9.7.3, respectively), and then control the outputs of series and shunt converters according to Eq. 11-18 (e.g., generation of compensation signal using reference-following techniques as discussed in Section 9.7.5). The conventional approach to perform the first task for the UPQC relies on the synchronous *dq*0 reference frame.

### 11.4.1 General Theory of the Park (*dq*0) Transformation

In the study of power systems, mathematical transformations are often used to decouple variables, to facilitate the solution of time-varying equations, and to refer variables to a common reference frame. A widely used transformation is the Park (*dq*0) transformation [15]. It is can be applied to any arbitrary (sinusoidal or distorted) three-phase, time-dependent signals.

The Park (*dq*0) transformation is defined as

$$\begin{bmatrix} x_d \\ x_q \\ x_0 \end{bmatrix} = \sqrt{\frac{2}{3}} \begin{bmatrix} \cos\theta_d & \cos(\theta_d - 2\pi/3) & \cos(\theta_d + 2\pi/3) \\ -\sin\theta_d & -\sin(\theta_d - 2\pi/3) & -\sin(\theta_d + 2\pi/3) \\ 1/\sqrt{2} & 1/\sqrt{2} & 1/\sqrt{2} \end{bmatrix} \begin{bmatrix} x_a \\ x_b \\ x_c \end{bmatrix}.$$
(11-19)

In this equation, $\theta_d = \omega_d + \varphi$, where $\omega_d$ is the angular velocity of signals to be transformed and $\varphi$ is the initial angle.

As discussed in Chapter 9 (Section 9.7), the main advantage of the $dq0$ transformation is that the fundamental (positive-sequence) component of waveform will be transferred to a constant DC term under steady-state conditions. In addition, a harmonic component with frequency $\omega_h$ will appear as a sinusoidal signal with frequency of $\omega_h - \omega_d$ in the $dq0$ domain.

### 11.4.2 Control of Series Converter Based on the $dq0$ Transformation

The system (supply) voltage may contain negative- and zero-sequence components as well as harmonic distortions that need to be eliminated by the series compensator. The control of the series compensator based on the $dq0$ transformation is shown in Fig. 11.3.

The system voltages are first detected and then transformed into the synchronous $dq0$ reference frame. If the system voltages are unbalanced and contain harmonics, then the Park transformation (Eq. 11-19) results in

$$
\begin{bmatrix} v_d \\ v_q \\ v_0 \end{bmatrix} = \sqrt{\frac{2}{3}} \begin{bmatrix} \cos\theta_d & \cos(\theta_d - 2\pi/3) & \cos(\theta_d + 2\pi/3) \\ -\sin\theta_d & -\sin(\theta_d - 2\pi/3) & -\sin(\theta_d + 2\pi/3) \\ 1/\sqrt{2} & 1/\sqrt{2} & 1/\sqrt{2} \end{bmatrix}
$$

$$
\begin{bmatrix} v_a \\ v_b \\ v_c \end{bmatrix}
$$

$$
= \sqrt{\frac{3}{2}} \left\{ \begin{bmatrix} V_{sp}\cos\phi_p \\ V_{sp}\sin\phi_p \\ 0 \end{bmatrix} + \begin{bmatrix} V_{sn}\cos(2\omega t + \phi_n) \\ -V_{sn}\sin(2\omega t + \phi_n) \\ 0 \end{bmatrix} \right.
$$

$$
+ \begin{bmatrix} 0 \\ 0 \\ V_{s0}\cos(2\omega t + \phi_0) \end{bmatrix}
$$

$$
\left. + \begin{bmatrix} \sum V_{sh}\cos[(h-1)(2\omega t + \phi_h)] \\ \sum V_{sh}\sin[(h-1)(2\omega t + \phi_h)] \\ 0 \end{bmatrix} \right\}
$$

$$
= \begin{bmatrix} V_{dp} \\ V_{qp} \\ 0 \end{bmatrix} + \begin{bmatrix} V_{dn} \\ V_{qn} \\ 0 \end{bmatrix} + \begin{bmatrix} 0 \\ 0 \\ V_{00} \end{bmatrix} + \begin{bmatrix} V_{dh} \\ V_{qh} \\ 0 \end{bmatrix}
$$

$$(11-20)$$

where $\phi_p$ is the phase difference between the positive-sequence component and the reference voltage. Based on Eq. 11-20, the fundamental positive-sequence component of the voltage is represented by DC terms (e.g., $V_{sp}\cos\phi_p$ and $V_{sp}\sin\phi_p$) in the $dq0$ reference frame.

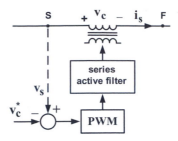

**FIGURE 11.3** Control block of the UPQC series compensator using Park (dq0) transformation [15].

The load bus voltage should be maintained sinusoidal with constant amplitude even if the source voltage is distorted. Therefore, the expected load voltage in the $dq0$ domain will only consist of one value:

$$
V_{dq0}^{expected} = \begin{bmatrix} V_{dF} \\ V_{qF} \\ V_{0F} \end{bmatrix} = [T_{dq0}(\theta_d)]V_{abc}^{expected} = \begin{bmatrix} V_m \\ 0 \\ 0 \end{bmatrix}, \quad (11-21)
$$

where

$$
V_{abc}^{expected} = \begin{bmatrix} V_S\cos(\omega t) \\ V_S\cos(\omega t + 120) \\ V_S\cos(\omega t - 120) \end{bmatrix}
$$

are sinusoids and the compensation reference voltage is

$$
V_{hdq0}^{ref} = V_{dq0}^{expected} - V_{sdq0}, \quad (11-22)
$$

where $V_{sdq0}$ is the distorted source voltage in the $dq0$ frame (Eq. 11-20). Therefore, $V_{dp}$ of Eq. 11-20 should be maintained at $V_m$ and all other terms must be eliminated by the compensation voltage.

The compensation reference voltage of Eq. 11-22 is then inversely transformed into the abc reference frame. Comparing the compensation reference voltage with a triangular wave, the output voltage of the series converter can be obtained by a PWM voltage control unit.

### 11.4.3 Control of Shunt Converter Relying on the $dq0$ Transformation

Similar transformation as employed for voltages can be applied to current waveforms as well:

$$
\begin{bmatrix} i_d \\ i_q \\ i_0 \end{bmatrix} = \sqrt{\frac{2}{3}} \begin{bmatrix} \cos\theta_d & \cos(\theta_d - 2\pi/3) & \cos(\theta_d + 2\pi/3) \\ -\sin\theta_d & -\sin(\theta_d - 2\pi/3) & -\sin(\theta_d + 2\pi/3) \\ 1/\sqrt{2} & 1/\sqrt{2} & 1/\sqrt{2} \end{bmatrix}
$$

$$
\begin{bmatrix} i_a \\ i_b \\ i_c \end{bmatrix}
$$

$$
= \sqrt{\frac{3}{2}} \left\{ \begin{bmatrix} I_{LP}\cos\theta_p \\ I_{Lsp}\sin\theta_p \\ 0 \end{bmatrix} + \begin{bmatrix} I_{Ln}\cos(2\omega t + \theta_n) \\ -I_{Ln}\sin(2\omega t + \theta_n) \\ 0 \end{bmatrix} \right.
$$

$$
\left. + \begin{bmatrix} \sum I_{Lh}\cos[(h-1)(2\omega t + \theta_h)] \\ \sum I_{Lh}\sin[(h-1)(2\omega t + \theta_h)] \\ 0 \end{bmatrix} \right\}. \tag{11-23}
$$

Unlike the voltage waveforms, which are maintained at their rated values, the load currents will change with the load characteristics at the PCC and it will not be possible to have a fixed expected reference current value. To solve this problem, different techniques are proposed and implemented to extract the fundamental positive-sequence component (which becomes a DC term in the $dq0$ reference frame) and the corresponding harmonic components, as discussed in Sections 9.7.1 to 9.7.3.

Therefore, the expected current in the $dq0$ reference frame has only one value

$$
i_{dq0}^{\text{expected}} = \begin{bmatrix} I_{Lp}\cos\theta_p \\ 0 \\ 0 \end{bmatrix}. \tag{11-24}
$$

The compensation reference current is

$$
i_{\text{hdq0}}^{\text{ref}} = i_{dq0}^{\text{expected}} - i_{\text{sdq0}}, \tag{11-25}
$$

where $i_{\text{sdq0}}$ is the distorted load current in the $dq0$ frame (Eq. 11-23).

The shunt converter operates as a controlled current source and different techniques (e.g., hysteresis, predictive, and suboscillation control) are employed to ensure that its output current follows the corresponding reference signal $i_{\text{hdq0}}^{\text{ref}}$.

### 11.4.4 Control of DC Link Voltage using the $dq0$ Transformation

In the UPQC configuration, the DC link voltage is usually divided into two capacitor units $C_1$ and $C_2$ connected in series. The voltages of these capacitors are such that $V_{C1} = V_{C2}$ under balanced operation. Normally, the DC link voltage is maintained at a desired/nominal level for all operating conditions.

## 11.5 UPQC CONTROL BASED ON THE INSTANTANEOUS REAL AND IMAGINARY POWER THEORY

In this section, the UPQC control technique is based on the $\alpha\beta0$ transformation and the instantaneous active and reactive power theory is introduced [9, 13, 14].

### 11.5.1 Theory of Instantaneous Real and Imaginary Power

The $\alpha\beta0$ transformation applied to abc three-phase voltages in a four-wire system is

$$
\begin{bmatrix} v_0 \\ v_\alpha \\ v_\beta \end{bmatrix} = \sqrt{\frac{2}{3}} \begin{bmatrix} 1/\sqrt{2} & 1/\sqrt{2} & 1/\sqrt{2} \\ 1 & -1/2 & -1/2 \\ 0 & \sqrt{3/2} & -\sqrt{3/2} \end{bmatrix} \begin{bmatrix} v_a \\ v_b \\ v_c \end{bmatrix}. \tag{11-26}
$$

The inverse $\alpha\beta0$ transformation is

$$
\begin{bmatrix} v_a \\ v_b \\ v_c \end{bmatrix} = \sqrt{\frac{2}{3}} \begin{bmatrix} 1/\sqrt{2} & 1 & 0 \\ 1/\sqrt{2} & -1/2 & \sqrt{3/2} \\ 1/\sqrt{2} & -1/2 & -\sqrt{3/2} \end{bmatrix} \begin{bmatrix} v_0 \\ v_\alpha \\ v_\beta \end{bmatrix}. \tag{11-27}
$$

Identical relations hold for the currents.

The instantaneous three-phase active power is given by [14]

$$
\begin{aligned} p_{\text{three-phase}}(t) &= v_a i_a + v_b i_b + v_c i_c = p_a(t) + p_b(t) + p_c(t) \\ &= v_\alpha i_\alpha + v_\beta i_\beta + v_0 i_0 = p_\alpha(t) + p_\beta(t) + p_0(t) \\ &= p(t) + p_0(t), \end{aligned} \tag{11-28}
$$

where $p(t) = p_\alpha(t) + p_\beta(t) = v_\alpha i_\alpha + v_\beta i_\beta$ corresponds to the instantaneous three-phase real power and $p_0(t) = v_0 i_0$ is the instantaneous zero-sequence power. Note the separation of the zero-sequence component in the $\alpha\beta0$ system.

The instantaneous three-phase reactive power is defined as

$$
\begin{aligned} q_{\text{three-phase}}(t) &= q_\alpha(t) + q_\beta(t) = v_\alpha i_\beta - v_\beta i_\alpha \\ &= -[(v_a - v_b)i_c + (v_b - v_c)i_a + (v_c - v_a)i_b]/\sqrt{3}. \end{aligned} \tag{11-29}
$$

Therefore, the instantaneous real, reactive, and zero-sequence powers are calculated from

$$
\begin{bmatrix} p_0 \\ p \\ q \end{bmatrix} = \begin{bmatrix} v_0 & 0 & 0 \\ 0 & v_\alpha & v_\beta \\ 0 & -v_\beta & v_\alpha \end{bmatrix} \begin{bmatrix} i_0 \\ i_\alpha \\ i_\beta \end{bmatrix}. \tag{11-30}
$$

Here, $p$ and $q$ (in the $\alpha\beta0$ space) are not expressed in the conventional watts and VArs.

From Eq. 11-30, the $\alpha$ and $\beta$ components of currents may be calculated as

$$\begin{bmatrix} i_\alpha \\ i_\beta \end{bmatrix} = \frac{1}{\Delta} \begin{bmatrix} v_\alpha & -v_\beta \\ v_\beta & v_\alpha \end{bmatrix} \begin{bmatrix} p \\ q \end{bmatrix} \quad \text{with} \quad \Delta = v_\alpha^2 + v_\beta^2. \quad (11\text{-}31)$$

These are divided into two types of currents

$$\begin{bmatrix} i_\alpha \\ i_\beta \end{bmatrix} = \frac{1}{\Delta} \left\{ \begin{bmatrix} v_\alpha & -v_\beta \\ v_\beta & v_\alpha \end{bmatrix} \begin{bmatrix} p \\ 0 \end{bmatrix} + \begin{bmatrix} v_\alpha & -v_\beta \\ v_\beta & v_\alpha \end{bmatrix} \begin{bmatrix} 0 \\ q \end{bmatrix} \right\}$$
$$= \begin{bmatrix} i_{\alpha p} \\ i_{\beta p} \end{bmatrix} + \begin{bmatrix} i_{\alpha q} \\ i_{\beta q} \end{bmatrix}, \quad (11\text{-}32)$$

where

$i_{\alpha p} = v_\alpha p/\Delta$ is the $\alpha$-axis instantaneous active current,

$i_{\alpha p} = -v_\alpha q/\Delta$ is the $\alpha$-axis instantaneous reactive current,

$i_{\beta p} = v_\beta p/\Delta$ is the $\beta$-axis instantaneous active current, and

$i_{\beta q} = v_\beta q/\Delta$ is the $\beta$-axis instantaneous reactive current.

Now, the power components of the phases $\alpha$ and $\beta$ can be separated as

$$\begin{bmatrix} p_\alpha \\ p_\beta \end{bmatrix} = \begin{bmatrix} v_\alpha i_\alpha \\ v_\beta i_\beta \end{bmatrix} = \begin{bmatrix} v_\alpha i_{\alpha p} \\ v_\beta i_{\beta p} \end{bmatrix} + \begin{bmatrix} v_\alpha i_{\alpha q} \\ v_\beta i_{\beta q} \end{bmatrix} = \begin{bmatrix} p_{\alpha p} \\ p_{\beta p} \end{bmatrix} + \begin{bmatrix} p_{\alpha q} \\ p_{\beta q} \end{bmatrix}, \quad (11\text{-}33)$$

where

$$p_{\alpha p} = v_\alpha i_{\alpha p} = v_\alpha^2 p/\Delta,$$

$$p_{\alpha p} = v_\alpha i_{\alpha p} = -v_\alpha v_\beta q/\Delta,$$

$$p_{\beta p} = v_\beta i_{\beta p} = v_\beta^2 p/\Delta,$$

$$p_{\beta q} = v_\beta i_{\alpha q} = v_\alpha v_\beta q/\Delta,$$

because $p_{\alpha q} + p_{\beta q} = 0$, and thus $p_{\text{three-phase}} = p_\alpha + p_\beta + p_o = p_{\alpha p} + p_{\beta p} + p_0$.

All these instantaneous quantities are valid under steady-state, transient, and nonsinusoidal conditions of voltage and/or current waveforms. Furthermore,

- The conventional concept of reactive power corresponds to the peak value of $p(t)$ that has a zero average value. However, the new concept of instantaneous reactive power in the $\alpha\beta0$ space cor-

responds to the parts of $p(t)$ that are dependent on the instantaneous imaginary power $q$, which independently exist in each phase but vanish when added (e.g., $p_{\alpha q} + p_{\beta q} = 0$). If these powers ($p_{\alpha q}$ and $p_{\beta q}$) are transformed back to the original abc system, they add up to zero.

- The instantaneous real power ($p$) represents the net energy per second being transferred from source to load and vice versa at any time. It depends only on the voltages and currents in phases $\alpha$ and $\beta$. It does not have any zero-sequence power components.

- The imaginary power ($q$) does not contribute to the total instantaneous flow of energy. It relates to a block of energy exchanged between the phases. For example, consider a balanced three-phase capacitor bank connected to a balanced three-phase voltage source. The energy stored in each capacitor changes with time $W(t) = Cv^2(t)/2$. However, $v_a^2(t) + v_b^2(t) + v_c^2(t)$ is constant. Therefore,

$$\begin{cases} p_{\text{three-phase}} = p + p_0 = 0 \\ q = \bar{q} = -3V_+I_+ \sin(\phi_+ - \delta_+) \\ \qquad = 3V_+I_+ = -Q_{\text{three-phase}}. \end{cases}$$

### 11.5.1.1 Application Example 11.1: The $\alpha\beta0$ Transformation for Three-Phase Sinusoidal System Supplying a Linear Load

Assume sinusoidal system voltages and sinusoidal load currents [14]:

$$\begin{cases} v_{as}(t) = \sqrt{2}V_S \sin \omega t \\ v_{bs}(t) = \sqrt{2}V_S \sin(\omega t + 120), \quad \text{(E11.1-1a)} \\ v_{CS}(t) = \sqrt{2}V_S \sin(\omega t - 120) \end{cases}$$

$$\begin{cases} i_{aL}(t) = \sqrt{2}I_L \sin(\omega t - \phi) \\ i_{bL}(t) = \sqrt{2}I_L \sin(\omega t + 120 - \phi). \quad \text{(E11.1-1b)} \\ i_{cL}(t) = \sqrt{2}I_L \sin(\omega t - 120 - \phi) \end{cases}$$

Then

$$\begin{cases} v_\alpha = \sqrt{3}V_S \sin \omega t \\ v_\beta = \sqrt{3}V_S \cos \omega t \quad \text{and} \\ v_0 = 0 \end{cases} \begin{cases} i_\alpha = \sqrt{3}I_L \sin(\omega t - \phi) \\ i_\beta = \sqrt{3}I_L \cos(\omega t - \phi). \\ i_0 = 0 \end{cases}$$
$$\text{(E11.1-2)}$$

Therefore, from Eq. 11-30

$$\begin{aligned} p &= v_\alpha i_\alpha + v_\beta i_\beta = 3V_s I_L \cos \phi = P_{3\text{-phase}} \\ q &= v_\alpha i_\beta - v_\beta i_\alpha = 3V_s I_L \sin \phi = Q_{3\text{-phase}} \end{aligned} \quad \text{(E11.1-3)}$$

Equation E11.1-3 is equivalent to the traditional active and reactive power concept.

### 11.5.1.2 Application Example 11.2: The $\alpha\beta0$ Transformation for Three-Phase Sinusoidal System Supplying a Nonlinear Load

Assume sinusoidal system voltages (Eq. E11.1-1a) and nonsinusoidal load currents [14]:

$$\begin{cases} i_{aL}(t) = \sum_{h=1}^{\infty} \sqrt{2} I_{hL} \sin(h\omega t - \phi_h) \\ i_{bL}(t) = \sum_{h=1}^{\infty} \sqrt{2} I_{hL} \sin[h(\omega t + 120) - \phi_h] \\ i_{CL}(t) = \sum_{h=1}^{\infty} \sqrt{2} I_{hL} \sin[h(\omega t - 120) - \phi_h]. \end{cases} \quad (E11.2-1)$$

Then

$$\begin{cases} i_{\alpha} = \sum_{h=1}^{\infty} \frac{2}{\sqrt{3}} I_{hL} \sin(h\omega t - \phi_h)[1 - \cos(h120)] \\ i_{\beta} = \sum_{h=1}^{\infty} 2h I_L \cos(h\omega t - \phi_h) \sin(h120) \\ i_0 = \frac{1}{\sqrt{3}}(i_a + i_b + i_c) = \sum_{h=1}^{\infty} \sqrt{6} I_{3hL} \sin(3h\omega t - \phi_{3h}). \end{cases}$$

$$(E11.2-2)$$

It is interesting to note that only the triplen harmonics (e.g., $h = 3k$) are present in $i_0$. The power components are

$$\begin{cases} p(t) = v_{\alpha} i_{\alpha} + v_{\beta} i_{\beta} = \bar{p} + \tilde{p} \\ \quad = 3V_S I_{1L} \cos\phi_1 - 3V_S I_{2L} \cos(3\omega t - \phi_2) + \\ \quad \quad 3V_S I_{4L} \cos(3\omega t + \phi_4) - 3V_S I_{5L} \cos(6\omega t - \phi_5) \\ \quad \quad + 3V_S I_{7L} \cos(6\omega t + \phi_7) - \cdots \\ q(t) = v_{\alpha} i_{\beta} - v_{\beta} i_{\alpha} = \bar{q} + \tilde{q} \\ \quad = 3V_S I_{1L} \sin\phi_1 - 3V_S I_{2L} \sin(3\omega t - \phi_2) + \\ \quad \quad 3V_S I_{4L} \sin(3\omega t + \phi_4) - 3V_S I_{5L} \sin(6\omega t - \phi_5) \\ \quad \quad + 3V_S I_{7L} \sin(6\omega t + \phi_7) - \cdots \\ p_0 = v_0 i_0 \end{cases}$$

$$(E11.2-3)$$

where the bar (–) and the tilde (~) correspond to the average value and the alternating components, respectively. In this example, the following relationships exist between the conventional and the instantaneous power theories:

- The average values of power ($\bar{p} = P_{\text{there-phas}}$ and $\bar{q} = Q_{\text{there-phas}}$) correspond to the conventional active and reactive powers, respectively.
- The alternating components of the real power ($\tilde{p}$) represent the energy per second that is being transferred from source to load or vice versa at any time. This is the energy being stored (or released)

at the three-phase (abc) or two-phase ($\alpha\beta$) load or source.
- The alternating (fluctuating) component of the imaginary power ($\tilde{q}$) is responsible for the harmonic reactive power (due to the nonlinear nature of the load) in each phase but will vanish when added.
- The instantaneous energy transferred does not depend on the imaginary power ($q = \bar{q} + \tilde{q}$), although reactive currents exist in each phase and flow in the conductor.
- The power H including harmonics can be expressed as $H = \sqrt{\bar{P}^2 + \bar{Q}_2}$ where $\bar{P}$ and $\bar{Q}$ are the root-mean-square values of $\bar{p}$ and $\bar{q}$, respectively.

### 11.5.1.3 Application Example 11.3: The $\alpha\beta0$ Transformation for Unbalanced Three-Phase, Four-Wire System Supplying a Linear Load

Unbalanced three-phase system voltages may be expressed in terms of their symmetrical components using [14]

$$\begin{bmatrix} V_0 \\ V_+ \\ V_- \end{bmatrix} = \frac{1}{3} \begin{bmatrix} 1 & 1 & 1 \\ 1 & \alpha & \alpha^2 \\ 1 & \alpha^2 & \alpha \end{bmatrix} \begin{bmatrix} V_a \\ V_b \\ V_c \end{bmatrix}, \\ \begin{bmatrix} V_a \\ V_b \\ V_c \end{bmatrix} = \frac{1}{3} \begin{bmatrix} 1 & 1 & 1 \\ 1 & \alpha^2 & \alpha \\ 1 & \alpha & \alpha^2 \end{bmatrix} \begin{bmatrix} V_0 \\ V_+ \\ V_- \end{bmatrix}, \quad (E11.3-1)$$

where $\alpha = e^{j2\pi/3}$ and subscripts +, –, and 0 correspond to the positive-, negative-, and zero-sequence components, respectively. Applying these relations to the unbalanced system voltages, we have

$$\begin{cases} v_a = \sqrt{2} V_{S0} \sin(\omega t + \phi_0) + \sqrt{2} V_{S+} \sin(\omega t + \phi_+) + \\ \quad \sqrt{2} V_{S-} \sin(\omega t + \phi_-) \\ v_b = \sqrt{2} V_{S0} \sin(\omega t + \phi_0) + \sqrt{2} V_{S+} \sin(\omega t - 120° + \phi_+) + \\ \quad \sqrt{2} V_{S-} \sin(\omega t + 120° + \phi_-) \\ v_c = \sqrt{2} V_{S0} \sin(\omega t + \phi_0) + \sqrt{2} V_{S+} \sin(\omega t + 120° + \phi_+) + \\ \quad \sqrt{2} V_{S-} \sin(\omega t - 120° + \phi_-) \end{cases}$$

$$(E11.3-2)$$

Expressing these instantaneous voltages in terms of the $\alpha\beta0$-coordinate system yields

$$\begin{cases} v_{\alpha} = \sqrt{3} V_{S+} \sin(\omega t + \phi_+) + \sqrt{3} V_{S-} \sin(\omega t + \phi_-) \\ v_{\beta} = -\sqrt{3} V_{S+} \sin(\omega t + \phi_+) + \sqrt{3} V_{S-} \sin(\omega t + \phi_-). \\ v_0 = \sqrt{6} V_{So} \sin(\omega t + \phi_0) \end{cases}$$

$$(E11.3-3)$$

Similarly, the load currents may be expresses in terms of their instantaneous $\alpha\beta0$ components

$$\begin{cases} i_\alpha = \sqrt{3}I_{L+}\sin(\omega t + \delta_+) + \sqrt{3}I_{L-}\sin(\omega t + \delta_-) \\ i_\beta = -\sqrt{3}I_{L+}\sin(\omega t + \delta_+) + \sqrt{3}I_{L-}\sin(\omega t + \delta_-). \\ i_0 = \sqrt{6}I_{L0}\sin(\omega t + \delta_0) \end{cases}$$

(E11.3-4)

For this case, the instantaneous active, reactive, and zero-sequence powers are

$$\begin{cases} p(t) = \bar{p} + \tilde{p} \\ q(t) = \bar{q} + \tilde{q} \\ p_0(t) = \bar{p}_0 + \tilde{p}_0 \end{cases}$$

(E11.3-5)

where

- instantaneous average real power is

$$\bar{p} = 3V_+I_{L+}\cos(\phi_+ - \delta_+) + 3V_-I_{L-}\cos(\phi_- - \delta_-)$$

(E11.3-6a)

- instantaneous average imaginary power is

$$\bar{q} = -3V_+I_{L+}\sin(\phi_+ - \delta_+) + 3V_-I_{L-}\sin(\phi_- - \delta_-)$$

(E11.3-6b)

- instantaneous alternating real power is

$$\tilde{p} = -3V_+I_{L-}\cos(2\omega t + \phi_+ + \delta_-) \\ - 3V_-I_{L+}\cos(2\omega t + \phi_- + \delta_+)$$

(E11.3-6c)

- instantaneous alternating imaginary power is

$$\tilde{q} = 3V_+I_{L-}\sin(2\omega t + \phi_+ + \delta_-) - 3V_-I_{L+}\cos(2\omega t + \phi_- + \delta_+)$$

(E11.3-6d)

- instantaneous average zero-sequence power is

$$\bar{p}_0 = 3V_0I_{L0}\cos(\phi_0 - \delta_0)$$  (E11.3-6e)

- instantaneous alternating zero-sequence power is

$$\tilde{p}_0 = -3V_0I_{L0}\cos(2\omega t + \phi_0 + \delta_0)$$  (E11.3-6f)

### 11.5.2 UPQC Control System Based on Instantaneous Real and Imaginary Powers

Operation of the UPQC isolates the utility voltage from current quality problems of the load and, at the same time, isolates load from voltage quality problems of the utility. The most important subject in the UPQC control is the generation of reference compensating currents of the shunt converter (e.g., signal $i_C^* \equiv [i_{ca}^*, i_{cb}^*, i_{cc}^*]$ Fig. 11.2) and the reference compensation voltages of the series converter (e.g., signal $v_C^* \equiv [v_{ca}^*, v_{cb}^*, v_{cc}^*]$ Fig. 11.2). PWM control blocks will use these references to generate corresponding switching signals of the shunt and series converters.

In this section, the concept of instantaneous real and imaginary power will be relied on to generate reference signals $i_C^*$ and $v_C^*$. The three-phase voltage and current signals are transformed from the abc reference frame to the $\alpha\beta0$ frame (Eq. 11-26) and the instantaneous real, imaginary, and zero-sequence powers are calculated (Eq. 11-30). The decomposition of voltage and current harmonic components into symmetrical components shows that the average of real ($\bar{p}$) and imaginary ($\bar{q}$) powers consists of interacting components of voltage and current that have the same frequency and sequence order (Eq. E11.3-6). Therefore, with the calculation of instantaneous powers and the derivation of their oscillation parts (e.g., $\tilde{p}$ and $\tilde{q}$), the voltage and current reference signals can be obtained and applied to the PWM controls of power electronic converters (Fig. 11.2). Before analyzing the control systems of the shunt and series converters, the circuits of the phase-lock loop (PLL) and the positive-sequence voltage detector (PSVD) will be introduced.

### 11.5.2.1 Phase-Lock Loop (PLL) Circuit

The phase-lock loop (PLL) circuit (Fig. 11.4) is one of the components of the positive-sequence voltage

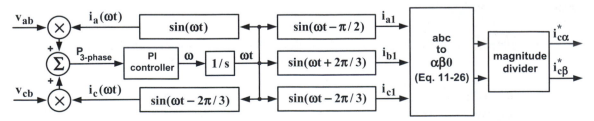

**FIGURE 11.4** Synchronizing circuit of the phase-lock loop (PLL) [9].

detector (Fig. 11.5). It detects the fundamental angular frequency ($\omega_1$) and generates synchronous sinusoidal signals that correspond to positive-sequence auxiliary currents under sinusoidal as well as highly distorted and unbalanced system voltages. The inputs of the PLL are $v_{ab} = v_a - v_b$ and $v_{cb} = (v_c - v_b)$.

Operation of PLL is based on the instantaneous active three-phase power theory ($p_{\text{three–phase}} = \bar{p} + \tilde{p} = v_{ab}i_a + v_{cb}i_c$):

- The current feedback signals $i_a(\omega t) = \sin(\omega t)$ and $i_c(\omega t) = \sin(\omega t - 2\pi/3)$ are generated by sine generator circuits through integration of $\omega$ at the output of a PI-controller. These currents are 120 degrees out of phase and represent a feedback from a positive-sequence component at angular frequency $\omega$.

- The PLL circuit can reach a stable point of operation only if the input of the PI-controller has a zero average value ($\bar{p} = 0$) with minimized low-frequency oscillations (e.g., small $\tilde{p}$). Because $i_a(\omega t)$ and $i_c(\omega t)$ contain only positive-sequence components and have unity magnitudes, the components of power (Eqs. E11.3-6a and E11.3-6c) simplify to

$$\bar{p} = 3V_+(1)\cos(\phi_+ - \delta_+) = 0$$
$$\tilde{p} = -3V_- I_+ \cos(2\omega t + \delta_+) \approx 0 \qquad (11\text{-}34)$$

where $\phi_+$ and $\delta_+$ are the initial phase angles of the fundamental positive-sequence voltage and current $i_a(\omega t)$, respectively. According to these equations $\phi_+ - \delta_+ = \pi/2$. This means that the auxiliary currents $i_a(\omega t) = \sin(\omega t)$ and $i_c(\omega t) = \sin(\omega t - 2\pi/3)$ become orthogonal to the fundamental positive-sequence component of the measured voltages $v_a$ and $v_c$, respectively. Therefore, $i_{a1}(\omega t) = \sin(\omega t - \pi/2)$ is in phase with the fundamental positive-sequence component of $v_a$. Similar relations hold for currents $i_{b1}$ and $i_{c1}$.

- Finally, signals $i_{a1}$, $i_{b1}$, and $i_{c1}$ are transformed to $i''_\alpha$ and $i''_\beta$ (with unity magnitudes) to the $\alpha\beta0$ domain by using an $\alpha\beta0$ block and magnitude divider blocks (which divide each signal by its magnitude).

### 11.5.2.2 Positive-Sequence Voltage Detector (PSVD)

The positive-sequence voltage detector (PSVD) circuit (Fig. 11.5) is one of the components of the shunt converter controller (Fig. 11.6). It extracts the

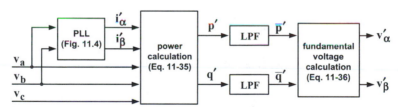

**FIGURE 11.5** Circuit of the positive-sequence voltage detector (PSVD) [9]; prime (') indicates instantaneous voltages in the $\alpha\beta0$ reference frame.

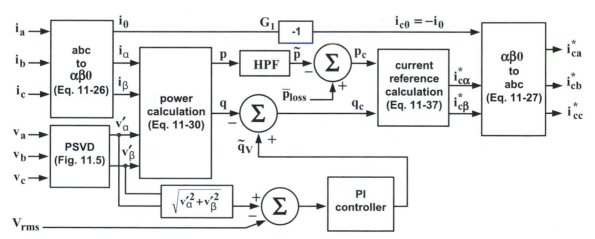

**FIGURE 11.6** Block diagram of the UPQC shunt converter control system based on instantaneous power theory [9, 13].

instantaneous $\alpha$ and $\beta$ values ($v'_\alpha$ and $v'_\beta$) of the fundamental positive-sequence components of the three-phase voltages $v_a$, $v_b$, and $v_c$ (Fig. 11.5).

The operation of the positive-sequence voltage detector is as follows:

- A phase-lock loop (PLL) circuit (Fig. 11.4) produces $i'_\alpha = \sin(\omega_1 t)$ and $i'_\beta = \cos(\omega_1 t)$ that correspond to an auxiliary fundamental positive-sequence current transformed into the $\alpha\beta0$ reference frame.
- Because $i'_\alpha$ and $i'_\beta$ contain only the positive-sequence component, the average values of real power and imaginary powers comprise only the fundamental positive-sequence components of the voltages, and Eq. E11.3-6 simplifies to

$$\bar{p} = 3V_+I'_+ \cos(\phi_+ - \delta_+)$$
$$\bar{q} = -3V_+I'_+ \sin(\phi_+ - \delta_+)$$

(11-35)

- The average values of instantaneous powers $\bar{p}'$ and $\bar{q}'$ are extracted using low-pass filters (LPFs). Consequently, the instantaneous values of the positive-sequence voltage can be calculated as

$$\begin{bmatrix} v'_\alpha \\ v'_\beta \end{bmatrix} = \frac{1}{i'^2_\alpha + i'^2_\beta} \begin{bmatrix} i'_\alpha & i'_\beta \\ i'_\beta & -i'_\alpha \end{bmatrix} \begin{bmatrix} \bar{p}' \\ \bar{q}' \end{bmatrix}.$$

(11-36)

### 11.5.2.3 Control of Shunt Converter using Instantaneous Power Theory

Inputs of the shunt converter control system (Fig. 11.6) are three-phase load voltages ($v_a$, $v_b$, $v_c$) and the currents ($i_a$, $i_b$, $i_c$) and its outputs are the compensating reference currents $i^*_{ca}$, $i^*_{cb}$, and $i^*_{cc}$ (e.g., signal $i^*_C$ in Fig. 11.2).

The shunt controller operates as follows:

- A positive-sequence voltage detector (PSVD, Fig. 11.5) is used to generate the instantaneous voltages $v'_\alpha$ and $v'_\beta$. Primes indicate instantaneous voltages in the $\alpha\beta0$ reference frame, which only corresponds to the positive-sequence voltage in the abc reference frame.
- Three-phase load currents are transformed to the $\alpha\beta0$ space.
- A power calculation block (Eq. 11-30) is relied on for the generation of the instantaneous real and reactive powers. Due to the presence of the PSVD, the average values of the calculated powers (Eqs. E11.3-6a,b) contain only the fundamental positive-sequence component of the voltage (e.g., $\bar{p} = 3V_+I_{L+} \cos(\phi_+ - \delta_+)$).

- The shunt converter is supposed to supply the oscillating instantaneous power $\tilde{p}$ and the instantaneous reactive power $q$ to the load while the load receives $\bar{p}$ from the source. Therefore, a high-pass passive filter is used to extract $\tilde{p}$ from the total power ($p = \bar{p} + \tilde{p}$). The oscillating part of the real power ($\tilde{p}$) consists of harmonic and negative-sequence components of the load current. Furthermore, the UPQC must also compensate for losses $\bar{p}_{loss}$ that will neutralize DC voltage variations caused by the series converter. Therefore, the amount of necessary instantaneous real power compensation is $p_c = -\tilde{p} + \bar{p}_{loss}$.
- In addition, the shunt converter will also compensate some reactive power ($\bar{q}_v$) required for the voltage regulation of the series converter. Therefore the amount of necessary instantaneous reactive power is $q_c = -q + \bar{q}_v = -(\bar{q} + \tilde{q}) + \bar{q}_v$. The imaginary power $\bar{q}_v$ is used for regulating the magnitude of the voltage ($|v|$), which is changed by the compensation voltage $v_C$ of the series converter. As shown in Fig. 11.6, the rms voltage value ($v_{rms}$) can be controlled to match $\bar{q}_v$ with the desired value $\sqrt{v'^2_\alpha + v'^2_\beta}$ through a PI controller.
- A current reference calculation block is employed to generate the reference compensation current in the $\alpha\beta0$ reference frame:

$$\begin{bmatrix} i^*_{c\alpha} \\ i^*_{c\beta} \end{bmatrix} = \frac{1}{v'^2_\alpha + v'^2_\beta} \begin{bmatrix} v'_\alpha & v'_\beta \\ v'_\beta & v'_\alpha \end{bmatrix} \begin{bmatrix} -\tilde{p} + \bar{p}_{loss} \\ -q + \bar{q}_V \end{bmatrix}.$$

(11-37)

- To compensate for zero-sequence power, the zero-sequence current that is generated by the shunt converter must be equal to that of the load but negative. The sign is inverted by a gain block ($G_1$).
- Finally, the $\alpha\beta0$ transformation block (based on Eq. 11-27) is relied on to generate the reference compensation currents ($i^*_{ca}$, $i^*_{cb}$, and $i^*_{cc}$) in the abc space.

### 11.5.2.4 Control of DC Voltage using Instantaneous Power Theory

Figure 11.7 illustrates the block diagram of the DC voltage control system. There are two reasons for the variation of the DC capacitor average voltages.

- The first reason is the real power losses of power electronic converters and the power injected into the network by the series branch. When a voltage sag or voltage swell appears, the series branch of the UPQC injects a fundamental component voltage into the network. The required (observed) real power to be injected is generated by the converter via the DC link, causing variation of the DC-link average voltage.

- Another factor that causes a variation of the average DC capacitor voltages is the zero-sequence component of the current, which is injected by the shunt converter into the system. These currents have no effect on the DC-link voltage but make the two capacitor voltages asymmetric.

The DC voltage controller performs the following two functions (Fig. 11.7):

- Regulating the DC-link average voltage by adding two capacitor voltages, comparing the sum with the reference DC voltage ($V_{ref}$), and applying the average voltage error to a PI controller. The output of this controller ($\bar{p}_{loss}$) is used by the shunt converter control system to absorb this power from the network and regulate the DC-link voltage. Note that the deviation of voltages is filtered by a low-pass passive filter and the PI controller matches the desired $\bar{p}_{loss}$ that neutralizes the DC bus voltage variations. The low-pass filter makes the voltage regulator insensitive to fundamental frequency voltage variations, which appear when the shunt active filter compensates the fundamental zero-sequence current of the load.
- The correction of the difference between two capacitor average voltages is performed by calculating the difference voltage signal, passing it through a low-pass filter, applying it to a limiter, and adding it (e.g., signal $\varepsilon$ in Fig. 11.7) with the upper and the lower hysteresis band limits. This process absorbs a zero-sequence current by the shunt converter and corrects the difference of the capacitor average voltages. Similarly, another low-pass filter is applied to filter the signal $V_{C2} - V_{C1}$. The filtered voltage difference produces $\varepsilon$ according to the following limit function:

$$\begin{cases} \varepsilon = -1 & \Leftrightarrow \Delta V < -0.05 V_{ref} \\ \varepsilon = \dfrac{\Delta V}{0.05 V_{ref}} & \Leftrightarrow -0.05 V_{ref} \leq \Delta V \leq 0.05 V_{ref}, \\ \varepsilon = 1 & \Leftrightarrow \Delta V \geq 0.05 V_{ref} \end{cases} \quad (11\text{-}38)$$

where $V_{ref}$ is the DC bus voltage reference.

- Signal $\varepsilon$ and the reference compensation currents ($i_C^* \equiv [i_{ca}^*, i_{cb}^*, i_{cc}^*]$, Fig. 11.6) are employed to generate gating signals for the shunt converter switches (Fig. 11.2). For example, $\varepsilon$ can act as a dynamic offset level that is added to the hysteresis-band limits in the PWM current control of Fig. 11.2:

$$\begin{cases} \text{upper hysteresis band limit} = i_C^* + \Delta(1+\varepsilon) \\ \text{lower hysteresis band limit} = i_C^* - \Delta(1-\varepsilon) \end{cases},$$
$$(11\text{-}39)$$

where $i_C^* \equiv [i_{ca}^*, i_{cb}^*, i_{cc}^*]$ are the instantaneous current references (computed by the shunt converter control system) and $\Delta$ is a fixed semi-bandwidth of the hysteresis control. Therefore, signal $\varepsilon$ shifts the hysteresis band to change the switching times of the shunt converter such that

$$\begin{cases} \varepsilon > 0 & \text{increases } V_{C1} \text{ and decreases } V_{C2} \\ \varepsilon < 0 & \text{increases } V_{C2} \text{ and decreases } V_{C1} \end{cases}.$$

### 11.5.2.5 Control of Series Converter using Instantaneous Power Theory

Inputs of the series converter control system (Fig. 11.8) are the three-phase supply voltages ($v_{sa}, v_{sb}, v_{sc}$), load voltages ($v_a, v_b, v_c$), and the line currents ($i_a, i_b, i_c$). Its outputs are the reference-compensating voltages $v_{ca}^*, v_{cb}^*,$ and $v_{cc}^*$ (e.g., signal $v_C^*$ in Fig. 11.2). The compensating voltages consist of two components:

- One is the voltage-conditioning component $v_s' - v_s$ (e.g., the difference between the supply voltage $v_s \equiv [v_{sa}, v_{sb}, v_{sc}]$ and its fundamental positive-sequence component $v_s' \equiv [v_{sa}', v_{sb}', v_{sc}']$) that com-

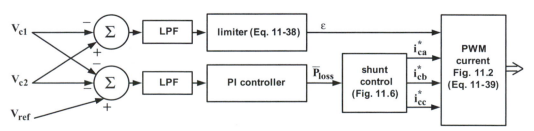

**FIGURE 11.7** Block diagram of the DC voltage control system based on instantaneous power theory [9, 13].

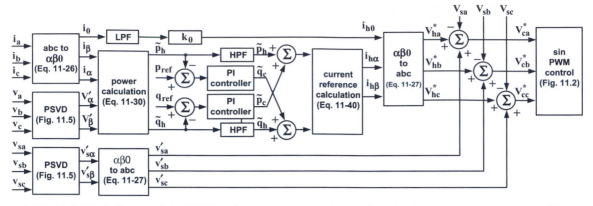

**FIGURE 11.8** Block diagram of the UPQC series converter control system based on instantaneous power theory [9].

pensates harmonics, fundamental negative-, and zero-sequence components of voltage, and regulates the fundamental component magnitude.

- The other voltage component ($v_h \equiv [v_{ha}, v_{hb}, v_{hc}]$) performs harmonic isolation and zero-sequence current damping, as well as damping of harmonic oscillations caused by the series resonance due to the series inductances and the shunt capacitor banks. $v_h$ can also compensate for the oscillating instantaneous active and reactive powers ($\tilde{p}_h$ and $\tilde{q}_h$).

The series controller operates as follows (Fig. 11.8):

- PSVD, $\alpha\beta 0$, and power calculation blocks are relied on to compute oscillating instantaneous powers $\tilde{p}_h$ and $\tilde{q}_h$. This procedure is similar to the shunt converter control system.
- $\tilde{p}_h$ and $\tilde{q}_h$ are compared with their reference values ($p_{\mathrm{ref}}$ and $q_{\mathrm{ref}}$) to generate the desired compensation powers. The reference active power ($p_{\mathrm{ref}}$) forces the converter to inject a fundamental voltage component orthogonal to the load voltage. This will cause change in the power angle $\delta$ (phase angle displacement between the voltages at both ends of transmission line) and mostly affects the active power flow. The reference reactive power ($q_{\mathrm{ref}}$) will cause a compensation voltage in phase with the load voltage. This will change the magnitudes of the voltages and strongly affects the reactive power flow of the system.
- A current reference calculation block is used to generate the reference compensation currents in the $\alpha\beta 0$ space:

$$\begin{bmatrix} i_{h\alpha} \\ i_{c\beta} \end{bmatrix} = \frac{1}{v_\alpha'^2 + v_\beta'^2} \begin{bmatrix} v_\alpha' & -v_\beta' \\ v_\beta' & v_\alpha' \end{bmatrix} \begin{bmatrix} \tilde{p}_h + \tilde{p}_C \\ \tilde{q}_h + \tilde{q}_C \end{bmatrix}. \quad (11\text{-}40)$$

- Signal $i_{h0}$, which provides harmonic isolation and damping for zero-sequence currents flowing through the series converter, is separated from the signal $i_0$ by a low-pass filter.
- The $\alpha\beta 0$ transformation block is employed to generate the reference compensation voltages for harmonic isolation and harmonic oscillation damping ($v_h \equiv [v_{ha}, v_{hb}, v_{hc}]$). These voltages are added with $v_s' \equiv [v_{sa}', v_{sb}', v_{sc}']$ and subtracted from the supply voltages $v_s \equiv [v_{sa}, v_{sb}, v_{sc}]$ to form the reference compensation voltage for the series converter:

$$\begin{cases} v_{ca}^* = v_{ha} + (v_{sa}' - v_{sa}) \\ v_{cb}^* = v_{hb} + (v_{sb}' - v_{sb}). \\ v_{cc}^* = v_{hc} + (v_{sc}' - v_{sc}) \end{cases} \quad (11\text{-}41)$$

- The reference compensation voltages and a sinusoidal PWM control approach are relied on to produce the gating signals of the series converter (Fig. 11.2).

The UPQC series control system (Fig. 11.8) may be simplified by eliminating the reference compensation currents as shown in Fig. 11.9. Voltage regulation is controlled by $V_{\mathrm{ref}}$. The converter is designed to behave like a resistor at harmonic frequencies to perform damping of harmonic oscillations. To do this, harmonic components of the line current are detected and multiplied by a damping coefficient.

## 11.6 PERFORMANCE OF THE UPQC

To demonstrate the performance of the UPQC in the presence of several current and voltage disturbance sources, the model distribution system of Fig. 11.10 consisting of seven (linear and nonlinear) buses is considered. System specifications are listed in Table 11.2.

The loads of this network (Fig. 11.10) are

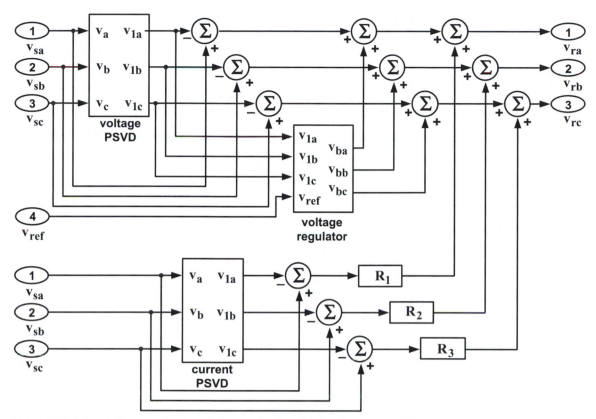

**FIGURE 11.9** Block diagram of a simplified controller for UPQC series converter [13].

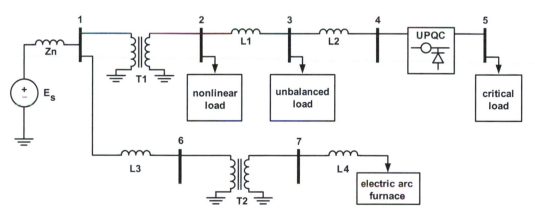

**FIGURE 11.10** Model distribution network used to analyze the performance of the UPQC [13].

**TABLE 11.2** Specifications of the Model Distribution System (Fig. 11.10)

| | |
|---|---|
| Medium voltage network | $V_L = 20\,kV$, $Z_n = (25 + j2.5)\,\Omega$, $R_g = 10\,\Omega$ |
| Distribution transformer ($T_1$) | $S = 2\,MVA$, $V$: 20 kV/400 V, $Z = (0.002 + j0.1)$ pu |
| Arc furnace transformer ($T_2$) | $S = 5\,MVA$, $V$: 20 kV/700 V, $Z = (0.01 + j0.1)$ pu |
| Nonlinear load | Thyristor-controlled rectifier, $\alpha = 0°$, $I_{DC} = 400\,A$ |
| Unbalanced linear load | $Z_a = (0.08 + j0.06)\,\Omega$, $Z_b = (1.0 + j0.7)\,\Omega$, $Z_c = (0.14 + j0.12)\,\Omega$ |
| Transmission lines | $Z_{L1} = (0.01 + j0.003)\,\Omega$, $Z_{L2} = (0.03 + j0.006)\,\Omega$, $Z_{L3} = (0.1 + j0.3)\,\Omega$, $Z_{L4} = (0.01 + j0.08)\,\Omega$ |

**Load 1 (Nonlinear Load at Bus 2):** a three-phase thyristor-controlled rectifier providing 700 A of DC current. This nonlinear load produces harmonic currents and causes network voltage distortion.

**Load 2 (Unbalanced Load at Bus 3):** an unbalanced three-phase linear load consists of three different linear impedances. This load draws unbalanced currents and causes a voltage imbalance condition.

**Load 3 (Critical Load at Bus 5):** a three-phase nonlinear load that is considered to be the critical load to be protected by the UPQC. This load consists of three power electronic converters:

- A three-phase thyristor-controlled rectifier with a firing angle of 30° and a DC current of 10 A.
- A single-phase thyristor-controlled rectifier placed in phase A, with a firing angle of 45° and a DC current of 10 A.
- A single-phase uncontrolled (diode) rectifier placed in phase B, with a DC current of 5 A.

**Load 4 (Electric Arc Furnace Load at Bus 7):** an electric arc furnace produces voltage fluctuation (flicker). Due to its nonlinear volt-ampere characteristic, this load acts like a time-varying nonlinear resistor. This nonlinearity is considered for the positive half-cycle as follows [16]:

$$v_{arc} = v_t + c/(i_{arc} + d),  \qquad (11\text{-}42)$$

where $v_{arc}$ and $i_{arc}$ are the voltage and current of the arc furnace, $c$ and $d$ are constant coefficients, and $v_t$ is the threshold voltage:

$$v_t = v_e + kL.  \qquad (11\text{-}43)$$

Here $v_e$ is the voltage drop across the furnace electrodes and $kL$ corresponds to the variable voltage drop proportional to the arc length ($L$). There are different models for the arc length time variation [13]. In Fig. 11.10, the sinusoidal time variation law is used and arc length is considered as a sinusoidal function of time with a fixed frequency:

$$L = L_1 + L_2 \sin \omega t.  \qquad (11\text{-}44)$$

Since human eyes are most sensitive to flickers with a frequency of about 8–10 Hz, the frequency is assumed to be $f = 10$ Hz.

Simulation of the model distribution network and UPQC (with the control systems of Figs. 11.6, 11.7,

and 11.9) is performed using the MATLAB/Simulink software package. The specifications of UPQC are listed in Table 11.3. UPQC performance is examined in the following six application examples [13].

### 11.6.1 Application Example 11.4: Dynamic Behavior of UPQC for Current Compensation

In this example, only the critical load at bus 5 (load 3) of Fig. 11.10 is activated. In order to investigate the dynamic behavior of the UPQC, the single-phase and the three-phase controlled rectifiers are operating continuously in the network, while the single-phase diode rectifier is connected and then disconnected at $t = 0.06$ s and $t = 0.12$ s, respectively. The gate pulses of the UPQC converter are activated at $t = 0.04$ s. The results including load 3 current, shunt converter current, line current, and DC capacitor voltages are shown in Fig. E11.4.1. As expected sinusoidal balanced currents are achieved and the neutral current is reduced to zero, as soon as UPQC is activated. Current $THD_i$ of phase A is reduced from 19.2 to 0.6%. This figure shows the fine dynamic performance of the UPQC for nonlinear and unbalanced load current variations. During the time when the single-phase diode rectifier is activated ($t = 0.06$ s to $t = 0.12$ s), the UPQC distributes its load demand between the three phases and, as a result, the line currents increase. Compensation of the oscillatory parts of real power and the zero-sequence current has caused some variation of the DC capacitor voltages, as shown in Fig. E11.4.1d.

### 11.6.2 Application Example 11.5: UPQC Compensation of Voltage Harmonics

The nonlinear load at bus 2 (load 1) of Fig. 11.10 is connected to the network. Therefore, harmonic currents are injected and voltage distortion is generated (Fig. E11.5.1). As expected, UPQC compensates for voltage harmonics and the load voltage is corrected to a sinusoidal waveform. Voltage $THD_v$ is reduced from 9.1 to 0.1%.

**TABLE 11.3** Specifications of the UPQC Components in Fig. 11.10

| | |
|---|---|
| Shunt transformer | S = 4.6 kVA, V: 230/100 V, Z = (0.01 + j0.05) pu |
| Series transformer | S = 2 kVA, V: 95/190 V, Z = (0.01 + j0.05) pu |
| DC capacitors | C = 2000 μF, V = 300 V |
| Low-pass filter | L = 2 mH, C = 5 μF, R = 4 Ω |
| High-pass filter | C = 15 μF, R = 5 Ω |
| Commutation inductance | L = 3 mH |

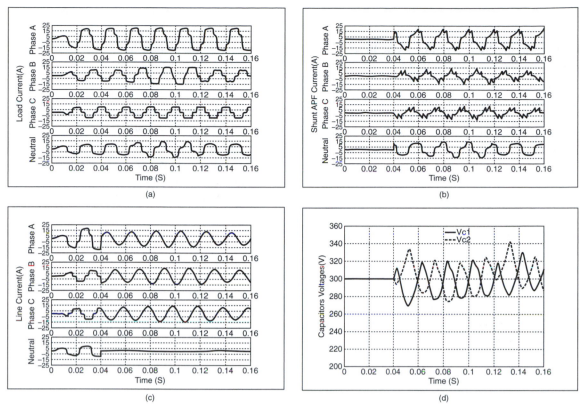

**FIGURE E11.4.1** UPQC performance including compensation of reactive power, harmonic currents, and current imbalance: (a) load current, (b) shunt converter current, (c) line current, (d) DC capacitor voltages [13].

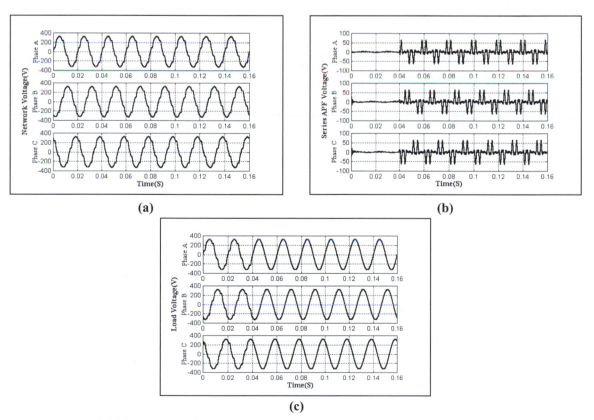

**FIGURE E11.5.1** UPQC performance including compensation of voltage harmonics: (a) network voltage, (b) series converter voltage, (c) load voltage [13].

### 11.6.3 Application Example 11.6: UPQC Compensation of Voltage Imbalance

The linear load at bus 3 (load 2) of Fig. 11.10 is connected to the network and causes voltage imbalance (Fig. E11.6.1). Operation of the UPQC and compensation of negative- and zero-sequence currents results in the balanced voltage conditions at bus 4. The ratios of negative- and zero-sequence voltages to the positive-sequence voltage are reduced from 24 and 8.4% to about 0.

### 11.6.4 Application Example 11.7: Dynamic Performance of UPQC for Sudden Voltage Variation

With loads 2 and 3 connected (e.g., imbalance and harmonic distortion) in Fig. 11.10, a single-phase short-circuit with a fault resistance of 45 Ω is imposed at phase B of bus 6 at $t = 0.06$ s and is cleared at $t = 0.12$ s. (Fig. E11.7.1). Such an asymmetrical fault causes a voltage sag in phase $B$ and voltage swells in

phases $A$ and $C$. However, the fast response of the UPQC (e.g., a delay of only 1 ms) does not allow any disturbances in the voltage of bus 5.

### 11.6.5 Application Example 11.8: Damping of Harmonic Oscillations Using a UPQC

In this example, a 1 kVAr three-phase capacitor bank is installed at bus 5 of Fig. 11.10 for reactive power compensation. Loads 1, 2, and 3 are activated. As a result, harmonic distortion and voltage imbalance are imposed on the network. Shunt and series branches of the UPQC start to operate at $t = 0.04$ s and $t = 0.08$ s, respectively. Figure E11.8.1 shows the voltage of bus 5 and the current of line 2. Although the harmonic components of load 3 are compensated by the shunt converter, the line current includes a high-frequency component before the operation of series branch occurs. This is so because of a series resonance between the series inductances and the shunt capacitor bank due to voltage distortion at bus 2. This condition occurs frequently in distribution

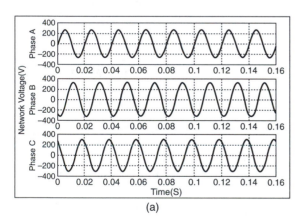

(a)

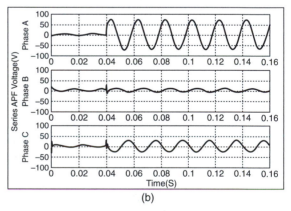

(b)

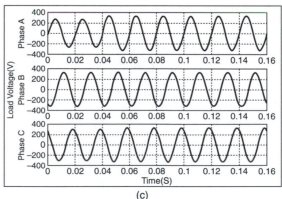

(c)

**FIGURE E11.6.1** UPQC performance including compensation of voltage imbalance: (a) network voltage, (b) series converter voltage, (c) load voltage [13].

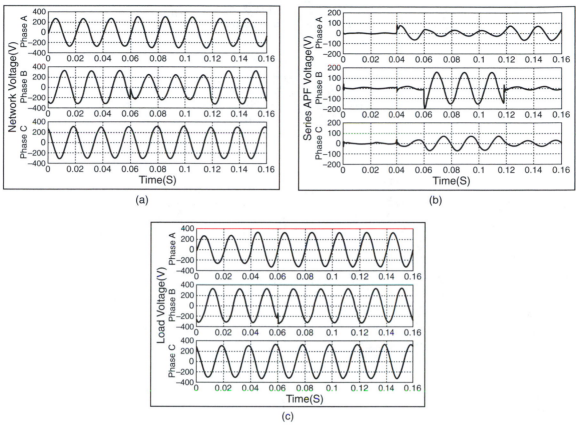

(a)

(b)

(c)

**FIGURE E11.7.1** UPQC performance including compensation of voltage sag and swell: (a) network voltage, (b) series converter voltage, (c) load voltage [13].

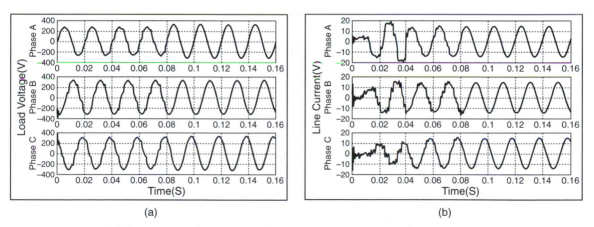

(a)

(b)

**FIGURE E11.8.1** UPQC performance including damping of harmonic resonance: (a) load voltage, (b) line current [13].

networks where significant harmonic distortion exists. Because the capacitor bank currents are not recorded by shunt branch current sensors, no compensation is performed. After the operation of the series converter has occurred at $t = 0.08$ s, resonant currents are damped significantly because the series branch acts like a resistor and increases the impedance of the resonance path.

### 11.6.6 Application Example 11.9: UPQC Compensation of Flicker

In the presence of all three loads of Fig. 11.10, the electric arc furnace (load 4) is connected to the network and causes fluctuation of the voltage with a magnitude change of 20% at a frequency of 10 Hz (Fig. E11.9.1). At $t = 0.04$ s, the series branch of the

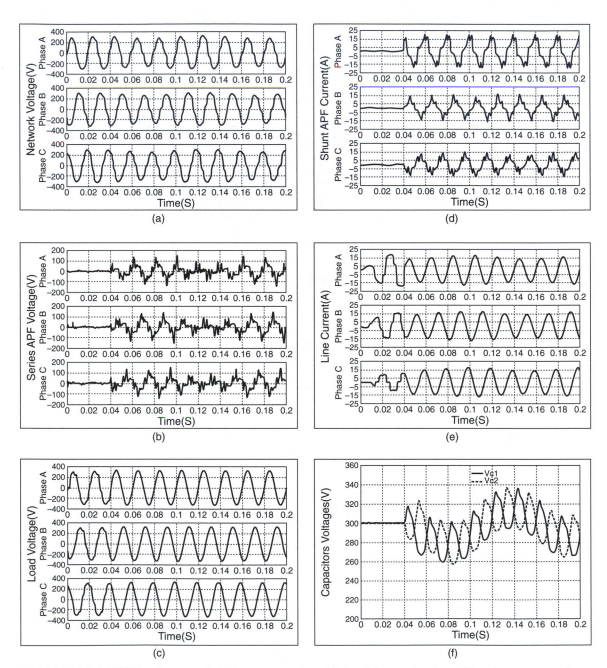

**FIGURE E11.9.1** UPQC performance including compensation of flicker: (a) network voltage, (b) series converter voltage, (c) load voltage, (d) shunt converter current, (e) line current, (f) DC capacitor voltages [13].

UPQC starts to operate, injecting voltage harmonics of high order (to correct voltage harmonic distortion) and low order (to compensate for voltage fluctuation) into the network. As a result, the voltage of bus 5 is restored to a sinusoid. Figure E11.9.1d,e,f illustrates shunt converter current, line 2 current, and DC capacitor voltages, respectively. By compensating harmonics, and negative- and zero-sequence components of the nonlinear load currents, balanced three-phase sinusoidal line currents are achieved. Some small fluctuations appear in the line currents because the UPQC requires a variable power to compensate voltage fluctuations. This power is provided by the DC capacitors and, as a result, the DC-link voltage includes low-frequency oscillations (Fig. E11.9.1f). This slow variation is sensed by the DC voltage control system, and for its compensation the UPQC absorbs an oscillatory power from the network. Therefore, both the shunt branch current and the line current will have low frequency variations.

Other application examples discussed in [4, pp. 117–129] and [3, pp. 129–130] include

- operation of solid-state current limiters (SSCLs),
- load compensation (with stiff source) using a STATCOM,
- load compensation (with nonstiff source) using a STATCOM,
- voltage regulation using a STATCOM, and
- protecting sensitive loads using a DVR.

## 11.7 SUMMARY

The unified power quality conditioner (UPQC) is an advanced hybrid filter that consists of a series active power filter for compensation of voltage disturbances and a shunt active power filter for eliminating current distortions. This custom power device is mainly employed to protect a critical nonlinear load by improving the quality of voltage across it, and to improve the waveform of the supply current. Different control approaches may be implemented to simultaneously perform these tasks. This chapter introduces two commonly relied on UPQC control approaches based on the $dq0$ transformation and the instantaneous real and reactive powers. Detailed UPQC block diagrams are provided for these approaches and application examples are used to demonstrate their effectiveness. Different control approaches for hybrid filters (which are part of UPQCs) are also presented in Chapter 9. Using an appropriate control algorithm, a UPQC can improve the utility current and protect the critical load at the point of installation in distorted distribution systems by taking a number of actions including reducing harmonic distortions, voltage disturbance compensation, voltage regulation, reactive (harmonic) power flow, harmonic isolation, and neutral and negative-sequence current compensations.

## 11.8 REFERENCES

1) GyuGyi, L.; "A unified flow control concept for flexible AC transmission systems," *Proc. Inst. Elect. Eng.*, Vol. 139, Part C, No. 4, pp. 323–331, 1992.

2) Hingorani, N.G.; and GyuGyi, L.; *Understanding FACTS – Concepts and Technology of Flexible Transmission Systems*, IEEE Press, 2000.

3) Hingorani, N.G.; "Introducing custom power," *IEEE Spectrum*, Vol. 32, No. 6, pp. 41–48, June 1995.

4) Ghosh, A.; and Ledwich, G.; *Power Quality Enhancement Using Custom Power Devices*, Kluwer Academic Publishers, Boston, 2002.

5) Peng, F.Z.; Akagi, H.; and Nabae, A.; "A new approach to harmonic compensation in power system – a combined system of shunt passive and series active filters," *IEEE Transactions on Industry Applications*, Vol. 26, No. 6, pp. 983–990, 1990.

6) Akagi, H.; "New trends in active filters for power conditioning," *IEEE Transactions on Industry Applications*, Vol. 32, No. 6, pp. 1312–1322, 1996.

7) Fujita, H.; and Akagi, H.; "The unified power quality conditioner: the integration of series and shunt–active power filters," *IEEE Transactions on Power Electronics*, Vol. 13, No. 2, pp. 315–322, March 1998.

8) Kamran, F.; and Habetler, T.G.; "Combined deadbeat control of a series-parallel converter combination used as universal power filter," *Proceedings of the IEEE Power Electronic Specialist Conference (PESC)*, pp. 196–201, 1995, *IEEE Transactions on Power Electronics*, Vol. 13, No. 1, pp. 160–168, Jan. 1998.

9) Aredes, M.; Heumann, K.; and Watanabe, E.H.; "An universal active power line conditioner," *IEEE Transactions on Power Delivery*, Vol. 13, No. 2, pp. 545–551, April 1998.

10) Tolbert, L.M.; and Peng, F.Z.; "A multilevel converter-based universal power conditioner," *IEEE Transactions on Industrial Applications*, Vol. 36, No. 6, pp. 596–603, March/April 2000.

11) Graovac, D.; Katic, V.; and Rufer, A.; "Power quality compensation using universal power quality conditioning system," *IEEE Power Engineering Review*, pp. 58–60, Dec. 2000.

12) Ghosh, A.; and Ledwich, G.; "A unified power quality conditioner (UPQC) for simultaneous voltage and current compensation," *Electric Power Systems Research*, Vol. 59, pp. 55–63, 2001.

13) Mosavi, S.A.; and Masoum, M.A.S.; "Mitigation of temporary and sustained power quality disturbances by UPQC," *International Journal of Engineering*, Iran, Vol. 15, No. 3, pp. 109–120, 2004.

14) Watanabe, E.H.; Stephan, R.M.; and Aredes, M.; "New concepts of instantaneous active and reactive powers in electrical systems with generic loads," *IEEE Transactions on Power Delivery*, Vol. 8, No. 2, pp. 697–703, April 1993.

15) Vilathgamuwa, M.; Zhang, Y.H.; and Choi, S.S.; "Modeling, analysis and control of unified power quality conditioner," *Proceedings of the 8th International Conference on Harmonics and Quality of Power*, Vol. 2, pp. 1035–1040, 14–16 Oct, 1998.

16) Montanari, G.C.; Loggini, M.; Cavallini, A.; Pitti L.; and Zaninelli, D.; "Arc furnace model for the study of flicker compensation in electrical networks," *IEEE Transactions on Power Delivery*, Vol. 9, pp. 2026–2036, 1994.

# Sampling Techniques

There are two sampling techniques:

- series (sequential) sampling with interpolation, and
- parallel sampling with holding network [79, Chapter 2].

**Series Sampling with Interpolation.** The derating measurements as described in Section 2.8 are based on sequential sampling utilizing the soft- and hardware specified by the UCDAS-16G Manual of the Keithley Corporation [80, Chapter 2]. The software program is written in Quick BASIC language. Sampling programs for two and five channels have been used [81, Chapter 2]. Some of the frequently asked questions concerning sequential sampling are addressed as follows.

## 1.1 WHAT CRITERION IS USED TO SELECT THE SAMPLING RATE (SEE LINE 500 OF TWO-CHANNEL PROGRAM [81, CHAPTER 2])?

The A/D converter UCDAS-16G has a maximum throughput (sampling) rate of 70 kHz in the direct memory access (DMA) mode, where the gain is 1.0.

**In the Two-Channel Program [81, Chapter 2]**

select
$$D\%(0) = 4$$
$$D\%(1) = 58$$

from this follows the SAMPLE RATE
$$= 10,000,000/[D\%(0) \cdot D\%(1)]$$
$$= 10,000,000/[4 \cdot 58] = 43.1 \text{ kHz} \approx 43 \text{ kHz}$$

**In the Five-Channel Program [81, Chapter 2]**

select
$$D\%(0) = 5$$
$$D\%(1) = 37$$

from this follows the SAMPLE RATE
$$= 10,000,000/[D\%(0) \cdot D\%(1)]$$
$$= 10,000,000/[5 \cdot 37] \approx 54 \text{ kHz}$$

## 1.2 WHAT CRITERION IS USED TO SELECT THE TOTAL NUMBER OF CONVERSIONS (LINE 850 OF THE TWO-CHANNEL PROGRAM [81, CHAPTER 2])?

**In the Two-Channel Program.** One period $T = 16.66$ ms of fundamental frequency ($f = 60$ Hz) signal is approximated by 360 equidistant intervals or $360 + 1 = 361$ points as indicated in Fig. A1.1. For two channels and sequential sampling $361 \cdot 2 = 722$ sample points are required. Due to interpolation procedure #1 at the end of the sampled period (discussed later), one chooses a somewhat larger number $722 + 11 = 733$.

**In the Five-Channel Program.** One period $T = 16.66$ ms of fundamental frequency ($f = 60$ Hz) signal is approximated by 180 equidistant intervals or $180 + 1 = 181$ points as explained in Fig. A1.2. For five channels and sequential sampling $181 \cdot 5 = 905$ sample points are required. Due to interpolation one chooses a somewhat larger number $905 + 9 = 914$.

## 1.3 WHY ARE THE TWO-CHANNEL PROGRAM DIMENSION AND THE ARRAY FOR THE CHANNEL NUMBER NOT USED FOR THE FIVE-CHANNEL PROGRAM [81, CHAPTER 2]?

There are two ways to store the data:

- For the two-channel program, the array $DT\%(N)$ comprises the sampled points of both (2) channels (e.g., in an interleaved manner). Therefore, $CH\%(N)$ is needed to separate the sample points for each channel, and $DT\%(N)$ results in $DA(N + 10)$ (increased due to interpolation).
- For the five-channel program, the array $DA(368)$ contains the sample points of channel #1, $DB(368)$ contains the sample points of channel #2, etc. That is, the sampled data are separated for each channel and no channel number $CH\%(N)$ is required.

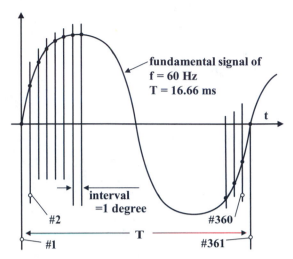

**FIGURE A1.1** Approximation of one 60 Hz period by 360 equidistant intervals.

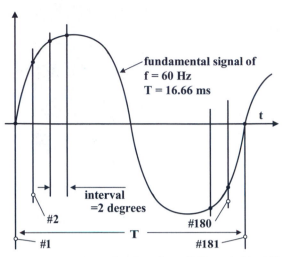

**FIGURE A1.2** Approximation of one 60 Hz period by 180 equidistant intervals.

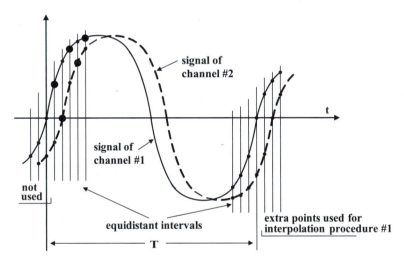

**FIGURE A1.3** Sampling of two signals in two-channel configuration.

### 1.4 WHAT IS THE CRITERION FOR SELECTING THE MULTIPLYING FACTOR IN STEP 9 (0.0048 82812 ≈ 0.004883) FOR THE TWO- AND FIVE-CHANNEL CONFIGURATIONS?

The UCDAS-16G is a 12 bit converter, that is, ±10 V is divided into $2^{12} = 4096$ steps. For a voltage peak-to-peak value of $10\ V - (-10\ V) = 20\ V$ one step corresponds to $20\ V/4096 = 0.004882812\ V \approx 0.004883\ V$.

### 1.5 WHY IS IN STEP 9 OF THE TWO-CHANNEL PROGRAM (LINE 1254) THE ARRAY DA(N + 10) OR DA(733), AND IN THE FIVE-CHANNEL PROGRAM (LINE 1233) THE ARRAY IS DA(368)?

For the two-channel program the sampled data of both channels are stored in $DA(733)$ as indicated in Fig. A1.3.

For the five-channel program the sampled data are stored separately for each channel, that is, for channel

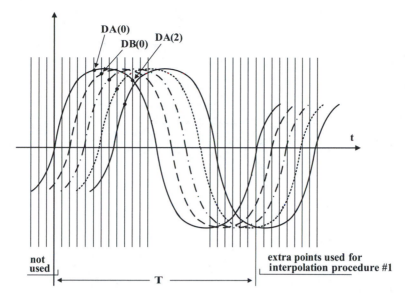

**FIGURE A1.4** Sampling of five signals in five-channel configuration.

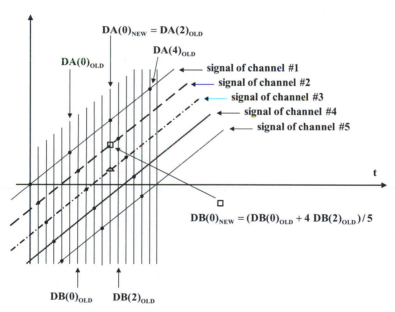

**FIGURE A1.5** Interpolation for five-channel configuration.

#1 in *DA*, for channel #2 in *DB*, etc. explained in Fig. A1.4.

For $I = 0$ (line 1235):

```
DA(0)  =  DT%(0+5)  ·  0.004883
DB(0)  =  DT%(0+6)  ·  0.004883
           ·              ·
           ·              ·
           ·              ·
DE(0)  =  DT%(0+9)  ·  0.004883
```

For $I = 2$ (line 1235):

```
DA(2)  =  DT%(5+5)  ·  0.004883
DB(2)  =  DT%(5+6)  ·  0.004883
           ·              ·
           ·              ·
           ·              ·
DE(2)  =  DT%(5+9)  ·  0.004883
```

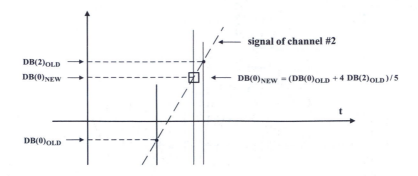

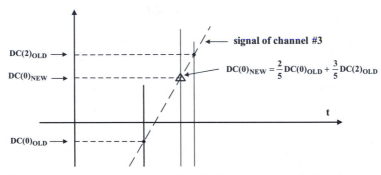

**FIGURE A1.6** Detailed example for interpolation (see Fig. A1.5) as performed by five-channel program: (a) signal for second channel, (b) signal for third channel.

For I = 4:

```
DA(4)  =  DT%(10+5)  ·  0.004883
DB(4)  =  DT%(10+6)  ·  0.004883
         .              .
         .              .
         .              .
DE(4)  =  DT%(10+9)  ·  0.004883
         .              .
         .              .
         .              .
```

One concludes that although there are only 181 sampled points for each channel of the five-channel program, the Fourier program [81, Chapter 2] requires $181 \cdot 2 = 362$ or $362 + 6 = 368$ dimensions for array $DA$, that is, $DA(368)$. The points between sampled points are obtained by interpolation procedure #2. Lines 1241 to 1247 of the five-channel program perform this interpolation. The sequential sampling does not provide at a given instant of time values for all five channels; therefore, at each instant of time a channel signal datum is generated by interpolation procedure #2 for each and every sampled function. The interpolation procedure is explained in Figs. A1.5 and A1.6.

For $J = 0$:

$$DA(0)_{NEW} = DA(2)_{OLD} \quad \text{and} \quad DB(0)_{NEW} = \{DB(0)_{OLD} + DB(2)_{OLD} \cdot 4\}/5, \text{ and}$$

$$DB(0)_{NEW} = \frac{DB(0)_{OLD}}{5} + DB(2)_{OLD} \cdot \frac{4}{5},$$

where (1/5) is the weighting function for $DB(0)_{OLD}$ and (4/5) is the weighting function for $DB(2)_{OLD}$.

Statements (line 1248 to line 1257) restore the data because FORTRAN cannot handle $DA(0)$, that is, $DA(0)$ becomes $DA(1)$, etc.

# Program List for Fourier Analysis [81, Chapter 2]

This appendix contains the program list to compute the Fourier coefficient of a periodic nonsinusoidal function. An output example is given for a rectangular wave.

## A2.1 FOURIER ANALYSIS PROGRAM LIST

Following is the program list (in FORTRAN) to compute the Fourier terms (DC offset, even and odd harmonics) of a periodic nonsinusoidal function:

```
C       PROGRAM ANAFOR
C       test, square wave
        COMMON NW,DEG(361),RAD(361),
          XV(361),YVALUE(361)
        COMMON
        NPTS,NPS,INT,INC,NH,GA,
          PI,WW,OM,NC,RF,ER,EL
        COMMON ERR,ITMAX
        DIMENSION Y(30,365),SCALE(30),
          YMAX(30),X(365)
        DIMENSION
        CURVT(100),
          CURVC(100),CURVS(100),CURVP(100)
        DIMENSION AMP(30,100),
          ANG(30,100)
C       READ DATA FROM DATA FILE.
C
        OPEN(5,FILE='INDUCTOR.
          DAT',STATUS='OLD',
          ACCESS='SEQUENTIAL')
C       REWIND 5
C*******************
        NSETS=1
        NPTS=42
C***************
        SCALE(1)=1.0
C       CLOSE(UNIT=5)
C       OPEN(3,FILE='FODA.DAT',
          STATUS='OLD',
          ACCESS='SEQUENTIAL')
C       REWIND 3
C       DO 5 J=1,NSETS
C 5     READ(3,132)
          (Y(J,I),I=1,NPTS+4)
  132   FORMAT(E11.4,2X,E11.4,2X,
          E11.4,2X,E11.4,2X,E11.4)
C       Input of data
        Do 1320 J=1,20
 1320   Y(1,J)=1.0
        Do 1321 J=21,41
 1321   Y(1,J)=-1.0
        Y(1,42)=1.0
C       ECHO THE DATA ON THE SCREEN.
        DO 10 J=1,NSETS
   10   WRITE(0,1000)
          J,SCALE(J),(Y(J,I),I=1,NPTS)
 1000   FORMAT(//' SET NO.',I4,' SCALE
          FACTOR:'
        *,F16.10/ 'Y-COORDINATES
          BEFORE SCALING:'(/5E11.4))
C       SUBTRACT D.C. COMPONENT AND
          MULTIPLY THE SCALE FACTORS
        DO 30 J=1,NSETS
        YMAX(J)=0.
        DC=0.0
        DO 37 K=1,NPTS-1
   37   DC=DC+Y(J,K)
        DC=DC/(NPTS-1)
        WRITE(0,*) 'DC=', DC
        DO 20 I=1,NPTS
        Y(J,I)=SCALE(J)*(Y(J,I)-DC)
   20   IF (ABS(Y(J,I)) .GT. YMAX(J))
          YMAX(J)=ABS(Y(J,I))
   30   CONTINUE
C       WRITE THE SCALED DATA
        DO 40 J=1,NSETS
   40   WRITE(0,1010)
          J,YMAX(J),(Y(J,I),I=1,NPTS)
 1010   FORMAT(// ,SET NO.',I4,' PEAK
          ABSOLUTE VALUE:',
        *F16.10/' Y-COORDINATES AFTER
          SCALING:'(/5E11.4))
        DO 50 I=1,NPTS
   50   X(I)=FLOAT(I-1)
        NSPACE=NPTS-1
        DEGMULT = 360.0/NSPACE
```

```
        DO 19 J=1,NPTS
        X(J) = X(J)*DEGMULT
  19    IF (J.EQ.NPTS) X(J) =
        X(J)+0.001
C       OPEN(10,FILE='FODA01.
        DAT',STATUS='OLD')
        WRITE (0,*) 'THE NUMBER OF
        KNOWN POINT (NPTS) IS
        ',NPTS
        WRITE (0,*) 'THE NUMBER OF
        DATA SETS (NSETS) IS'
        ,NSETS
        WRITE (0,*) 'THE KNOWN POINTS
        OF X AND F(X) ARE : '
C       THIS PART IS BASED ON THE
        ALGORITHMS IN PAGE 119
C       OF VANDERGRAFT.
        OM = 314.1592654
        GA = 0.017453292
        PI=3.141593
        INTF=42
        INT=42
        INC=5
        NW=1
        NH=29
        RADIN = PI/20.5
        DEGIN = 180./20.5
        T1 = -RADIN
        T2 = -DEGIN
        DO 22 I=1,INT
        T1 = T1+RADIN
        RAD(I) = T1
        T2 = T2+DEGIN
        DEG(I) = T2
  22    CONTINUE
        Write(0,*) (RAD(I), I=1,NPTS)
        Write(0,*) (DEG(I), I=1,NPTS)
        DO 1 J=1,NH+1
        CURVC(J) =0.0
        CURVS(J) =0.0
        CURVT(J) =0.0
        CURVP(J) =0.0
  1     CONTINUE
        DO 17 I=1,361
        XV(I)=FLOAT(I)-1.0
  17    CONTINUE
        DO 12 J=1,NSETS
        NC=J
        DO 13 I=1,361
        YVALUE(I) = Y(J,I)
  13    CONTINUE
        YMAX(J)=0.0
        DO 137 I=1,361
        IF(YVALUE(I) .GE. YMAX(J))
        YMAX(J)=YVALUE(I)
  137   CONTINUE
        WRITE(0,*)'J=',J
        WRITE(0,*)'YMAX(J)=',YMAX(J)
        WRITE(0,1013)
        J,YMAX(J),(YVALUE(I),I=1,NPTS)
  1013  FORMAT(// 'SET NO.',I4,' PEAK
        ABSOLUTE VALUE:',
       *F16.10/' Y-COORDINATES AFTER
        SCALING:'(/5E11.4))
        CALL HARM
        (YVALUE,CURVC,CURVS,CURVT,CURVP)
        DO 133 I=1,NH
  133   AMP(J,I)=CURVT(I)
        DO 144 I=1,NH
  144   ANG(J,I)=CURVP(I)
        DO 14 K=1,NH
        CURVP(K) = CURVP(K)/GA
  14    CONTINUE
        WRITE(0,*)
        '********MAIN**************'
        WRITE(0,*) 'CURVT(',NH,J,')=:'
        WRITE(0,100)(CURVT(K),K=1,NH)
        WRITE(0,*) 'THE COSINE BASED
        PHASOR ANGLES ARE:'
        WRITE(0,100)(-CURVT(K)-90.,
        K=1,NH)
        WRITE(0,*)'THE MAGNITUDE OF
        THE HARMONICS AS A',
       *'PERCENTAGE OF THE
        FUNDAMENTAL:'
        IF (CURVT(1) .EQ.0.00000) THEN
        WRITE(0,100) (0.0, K=1,NH)
        ELSE
        WRITE(0,100)(100.*CURVT(K)/
        CURVT(1),K=1,NH)
        END IF
  12    CONTINUE
  100   FORMAT (6(1X,1PE11.4))
        ENDFILE(10)
        REWIND 10
        CLOSE(10)
        THDI=0.0
        Do 1111 K=2,NH
  1111  THDI=THDI+(CURVT(K))*(CURVT(K))
        THDIFINAL=(SQRT(THDI))/
        CURVT(1)
        WRITE(0,*) THDIFINAL
        RMSIWITHOUT=0
        Do 1112 K=1,NH
  1112  RMSIWITHOUT=RMSIWITHOUT
        +(CURVT(K))*(CURVT(K))
```

```
       RMSIWITHOUTF=(SQRT
         (RMSIWITHOUT))/10.
       WRITE(0,*) RMSIWITHOUTF
       RMSIWITH=(DC)*(DC)
       Do 1113 K=1,NH
 1113  RMSIWITH=RMSIWITH+(CURVT(K))*
         (CURVT(K))
       RMSIWITHF=(SQRT(RMSIWITH))/10.
       WRITE(0,*) RMSIWITHF
       STOP
       END
       SUBROUTINE HARM (VALUE,COMPC,
         COMPS,COMPT,COMPP)
       COMMON NW,DEG(361),RAD(361),
         XV(361),YVALUE(361)
       COMMON NPTS,NPS,INT,INC,NH,GA,
         PI,WW,OM,NC,RF,ER,EL
       COMMON ERR,ITMAX
       DIMENSION VALUE(361),
         COMPC(100),COMPS(100),
         COMPT(100)
       DIMENSION COMPP(100),
         VALUEC(361)
       CHECK=1.0
       WRITE(0,1013) (VALUE(I),
         I=1,NPTS)
 1013  FORMAT(// ' Y-COORDINATES
         AFTER SCALING:'(/5E11.4))
       P=SQRT(0.6)
       A1=5./9.
       A2=8./9.
       FK1=0.5*P*(1.+P)
       FK2=(1.+P)*(1.-P)
       FK3=0.5*P*(P-1.)
       DO 44 J=1,NH
       DL1=0.0
       DL2=0.0
       H=FLOAT(J)
       DO 33 I=1,INT-2,2
       T1=H*(RAD(I))
       T2=H*(RAD(I+1))
       T3=H*(RAD(I+2))
       X1=FK1*T1+FK2*T2+FK3*T3
       X2=T2
       X3=FK3*T1+FK2*T2+FK1*T3
       C1=VALUE(I)
       C2=VALUE(I+1)
       C3=VALUE(I+2)
       C4=FK1*C1+FK2*C2+FK3*C3
       C6=FK3*C1+FK2*C2+FK1*C3
       D1=C4*COS(X1)
       D2=C2*COS(X2)
       D3=C6*COS(X3)
       D4=C4*SIN(X1)
       D5=C2*SIN(X2)
       D6=C6*SIN(X3)
       DL1=DL1+A1*(D1+D3)+A2*D2
       DL2=DL2+A1*(D4+D6)+A2*D5
   33  CONTINUE
       COMPC(J)=DL1*2./41.
       COMPS(J)=DL2*2./41.
       COMPT(J)=SQRT(COMPC(J)**2
         +COMPS(J)**2)
       If (COMPS(J).EQ.0.00000) THEN
       COMPP(J)=0.0
       ELSE

         COMPP(J)=-ATAN(COMPC(J)/COMPS(J))
       END IF
       IF (COMPS(J) .LT. 0.0)
         COMPP(J)= COMPP(J)+PI
   44  CONTINUE
       WRITE(0,*)
         '*******SUBROUTINE*********'
       WRITE(0,*) 'COMPT(',NH,J,')=:'
       WRITE(0,100)(COMPT(K),K=1,NH)
       WRITE(0,*) 'THE COSINE BASED
         PHASOR ANGLES ARE:'
       WRITE(0,100)(-COMPT(K)-90.,
         K=1,NH)
       WRITE(0,*)'THE MAGNITUDE OF
         THE HARMONICS AS A',
       *'PERCENTAGE OF THE
         FUNDAMENTAL:'
  100  FORMAT (6(1X,1PE11.4))
       NWW=1*NW
       IF ((CHECK.EQ.1) .AND. (NC
         .EQ. NWW)) THEN
       EMDIF = 0
       DO 51 I=1,INT
       VALUEC(I)=0.
       DO 52 J=1,NH,1
       H=FLOAT(J)
       VALUEC(I)=VALUEC(I)
         +COMPC(J)*COS(H*RAD(I))+
       *COMPS(J)*SIN(H*RAD(I))
   52  CONTINUE
       DIF=ABS(VALUE(I)-VALUEC(I))
       IF (DIF.GT.EMDIF) THEN
       EMDIF=DIF
       DIFN=FLOAT(I)
       ENDIF
   51  CONTINUE
       WRITE(0,*)
         '***CHECKING FOURIER
         ANALYSIS*'
```

```
      WRITE(0,*)
      WRITE(0,*) 'VALUE(361) AT
        NC=',NC
      WRITE(0,1001) (VALUE(I),
        I=1,42)
      WRITE(0,1001) (VALUEC(I),
        I=1,42)
 1001 FORMAT (10(1X,1PE11.4))
      WRITE(0,*) 'THE MAXIMUM
        DIFFERENCE IS :',EMDIF,'AT
        I=', DIFN
      ENDIF
      RETURN
      END
```

## A2.2 OUTPUT OF THE FOURIER ANALYSIS PROGRAM

The output of the Fourier program for a square wave is as follows:

```
 SET NO.   1 SCALE FACTOR:
  1.0000000000
Y-COORDINATES BEFORE SCALING:

  0.1000E+01  0.1000E+01  0.1000E+01
  0.1000E+01  0.1000E+01

  0.1000E+01  0.1000E+01  0.1000E+01
  0.1000E+01  0.1000E+01

  0.1000E+01  0.1000E+01  0.1000E+01
  0.1000E+01  0.1000E+01

  0.1000E+01  0.1000E+01  0.1000E+01
  0.1000E+01  0.1000E+01

-0.1000E+01-0.1000E+01-0.1000E
  +01-0.1000E+01-0.1000E+01

-0.1000E+01-0.1000E+01-0.1000E
  +01-0.1000E+01-0.1000E+01

-0.1000E+01-0.1000E+01-0.1000E
  +01-0.1000E+01-0.1000E+01

-0.1000E+01-0.1000E+01-0.1000E
  +01-0.1000E+01-0.1000E+01

-0.1000E+01  0.1000E+01
  DC=   -2.43902E-02

 SET NO.   1 PEAK ABSOLUTE VALUE:
  1.0243902206
 Y-COORDINATES AFTER SCALING:

  0.1024E+01  0.1024E+01  0.1024E+01
  0.1024E+01  0.1024E+01
```

```
  0.1024E+01  0.1024E+01  0.1024E+01
  0.1024E+01  0.1024E+01

  0.1024E+01  0.1024E+01  0.1024E+01
  0.1024E+01  0.1024E+01

0.1024E+01  0.1024E+01  0.1024E+01
  0.1024E+01  0.1024E+01

-0.9756E+00-0.9756E+00-0.9756E
  +00-0.9756E+00-0.9756E+00

-0.9756E+00-0.9756E+00-0.9756E
  +00-0.9756E+00-0.9756E+00

-0.9756E+00-0.9756E+00-0.9756E
  +00-0.9756E+00-0.9756E+00

-0.9756E+00-0.9756E+00-0.9756E
  +00-0.9756E+00-0.9756E+00

-0.9756E+00  0.1024E+01

THE NUMBER OF KNOWN POINT (NPTS)
 IS    42
THE NUMBER OF DATA SETS (NSETS)
 IS  1
THE KNOWN POINTS OF X AND F(X) ARE :
   0.    0.153248    0.306497
 0.459745    0.612994    0.766242
 0.919491    1.07$
        1.22599    1.37924    1.53248
 1.68573    1.83898    1.99223
 2.14548
        2.29873    2.45198    2.60522
 2.75847    2.91172    3.06497
 3.21822
        3.37147    3.52472    3.67796
 3.83121    3.98446    4.13771
 4.29096
        4.44421    4.59745    4.75070
 4.90395    5.05720    5.21045
 5.36370
        5.51694    5.67019    5.82344
 5.97669    6.12994    6.28319
   0.      8.78049    17.5610
 26.3415    35.1220    43.9024
 52.6829    61.4$
        70.2439    79.0244    87.8049
 96.5854    105.366    114.146
 122.927
        131.707    140.488    149.268
 158.049    166.829    175.610
 184.390
        193.171    201.951    210.732
 219.512    228.293    237.073
 245.854
```

```
      254.634        263.415        272.195
   280.976        289.756        298.537
   307.317
      316.098        324.878        333.659
   342.439        351.220        360.000
J=   1
YMAX(J)=        1.02439
```

SET NO.   1 PEAK ABSOLUTE VALUE:
 1.0243902206
 Y-COORDINATES AFTER SCALING:

 0.1024E+01 0.1024E+01 0.1024E+01
  0.1024E+01 0.1024E+01

 0.1024E+01 0.1024E+01 0.1024E+01
  0.1024E+01 0.1024E+01

 0.1024E+01 0.1024E+01 0.1024E+01
  0.1024E+01 0.1024E+01

0.1024E+01 0.1024E+01 0.1024E+01
  0.1024E+01 0.1024E+01

-0.9756E+00-0.9756E+00-0.9756E
  +00-0.9756E+00-0.9756E+00

-0.9756E+00-0.9756E+00-0.9756E
  +00-0.9756E+00-0.9756E+00

-0.9756E+00-0.9756E+00-0.9756E
  +00-0.9756E+00-0.9756E+00

-0.9756E+00-0.9756E+00-0.9756E
  +00-0.9756E+00-0.9756E+00

-0.9756E+00 0.1024E+01

 Y-COORDINATES AFTER SCALING:

0.1024E+01 0.1024E+01 0.1024E+01
  0.1024E+01 0.1024E+01

 0.1024E+01 0.1024E+01 0.1024E+01
  0.1024E+01 0.1024E+01

 0.1024E+01 0.1024E+01 0.1024E+01
  0.1024E+01 0.1024E+01

0.1024E+01 0.1024E+01 0.1024E+01
  0.1024E+01 0.1024E+01

-0.9756E+00-0.9756E+00-0.9756E
  +00-0.9756E+00-0.9756E+00

-0.9756E+00-0.9756E+00-0.9756E
  +00-0.9756E+00-0.9756E+00

-0.9756E+00-0.9756E+00-0.9756E
  +00-0.9756E+00-0.9756E+00

-0.9756E+00-0.9756E+00-0.9756E
  +00-0.9756E+00-0.9756E+00

-0.9756E+00 0.1024E+01
 *************SUBROUTINE**********
 COMPT(   29   30)=:
  1.2717E+00   3.4488E-02   4.1969E-01
  3.6450E-02   2.4608E-01   3.8742E-02
  1.6870E-01   4.0489E-02   1.2299E-01
  4.1067E-02   9.1722E-02   4.0183E-02
  6.8664E-02   3.7860E-02   5.1330E-02
  3.4436E-02   3.8801E-02   3.0685E-02
  3.1712E-02   3.8456E-02   3.7949E-02
  2.5107E-02   2.4685E-02   2.1601E-02
  2.5122E-02   2.0924E-02   2.7380E-02
  2.2261E-02   3.0684E-02
 THE COSINE BASED PHASOR ANGLES ARE:
 -9.1272E+01 -9.0034E+01 -9.0420E+01
  -9.0036E+01 -9.0246E+01 -9.0039E+01
  -9.0169E+01 -9.0040E+01 -9.0123E+01
  -9.0041E+01 -9.0092E+01 -9.0040E+01
  -9.0069E+01 -9.0038E+01 -9.0051E+01
  -9.0034E+01 -9.0039E+01 -9.0031E+01
  -9.0032E+01 -9.0038E+01 -9.0038E+01
  -9.0025E+01 -9.0025E+01 -9.0022E+01
  -9.0025E+01 -9.0021E+01 -9.0027E+01
  -9.0022E+01 -9.0031E+01
 THE MAGNITUDE OF THE HARMONICS AS
 APERCENTAGE OF THE FUNDAMENTAL:
 *** CHECKING FOURIER ANALYSIS*

VALUE(361) AT NC=   1
  1.0244E+00   1.0244E+00   1.0244E+00
  1.0244E+00   1.0244E+00   1.0244E+00
  1.024$
  1.0244E+00   1.0244E+00   1.0244E+00
  1.0244E+00   1.0244E+00   1.0244E+00
  1.024$
  -9.7561E-01  -9.7561E-01  -9.7561E-01
  -9.7561E-01  -9.7561E-01  -9.7561E-01
  -9.756$
  -9.7561E-01  -9.7561E-01  -9.7561E-01
  -9.7561E-01  -9.7561E-01  -9.7561E-01
  -9.756$
  -9.7561E-01   1.0244E+00
   5.0335E-01   1.1156E+00   9.8548E-01
   1.0407E+00   9.6388E-01   1.0831E+00
   9.720$
   9.5716E-01   1.0451E+00   9.9427E-01
   1.0748E+00   9.3542E-01   1.0637E+00
   1.022$
  -8.8539E-01  -9.6836E-01  -9.0364E-01
  -1.0717E+00  -9.4528E-01  -9.9542E-01
  -9.457$
```
```

```
-9.5329E-01  -1.0331E+00  -9.2584E-01
-1.0258E+00  -9.5068E-01  -1.0254E+00
-9.310$
-5.0985E-01   5.0335E-01
```
THE MAXIMUM DIFFERENCE IS :
```
 0.521040AT I=      1.00000
```
**********MAIN******************

```
 CURVT(  29   1)=:
  1.2717E+00   3.4488E-02   4.1969E-01
  3.6450E-02   2.4608E-01   3.8742E-02
  1.6870E-01   4.0489E-02   1.2299E-01
  4.1067E-02   9.1722E-02   4.0183E-02
  6.8664E-02   3.7860E-02   5.1330E-02
  3.4436E-02   3.8801E-02   3.0685E-02
  3.1712E-02   3.8456E-02   3.7949E-02
  2.5107E-02   2.4685E-02   2.1601E-02
  2.5122E-02   2.0924E-02   2.7380E-02
  2.2261E-02   3.0684E-02
```
THE COSINE BASED PHASOR ANGLES ARE:
```
 -9.1272E+01  -9.0034E+01  -9.0420E+01
 -9.0036E+01  -9.0246E+01  -9.0039E+01
 -9.0169E+01  -9.0040E+01  -9.0123E+01
 -9.0041E+01  -9.0092E+01  -9.0040E+01
-9.0069E+01  -9.0038E+01  -9.0051E+01
 -9.0034E+01  -9.0039E+01  -9.0031E+01
 -9.0032E+01  -9.0038E+01  -9.0038E+01
 -9.0025E+01  -9.0025E+01  -9.0022E+01
 -9.0025E+01  -9.0021E+01  -9.0027E+01
 -9.0022E+01  -9.0031E+01
```
THE MAGNITUDE OF THE HARMONICS AS
APERCENTAGE OF THE FUNDAMENTAL:
```
 1.0000E+02   2.7118E+00   3.3001E+01
 2.8661E+00   1.9350E+01   3.0463E+00
 1.3265E+01   3.1837E+00   9.6712E+00
 3.2292E+00   7.2123E+00   3.1597E+00
 5.3992E+00   2.9770E+00   4.0362E+00
 2.7078E+00   3.0510E+00   2.4128E+00
 2.4936E+00   3.0239E+00   2.9840E+00
 1.9742E+00   1.9410E+00   1.6985E+00
 1.9754E+00   1.6453E+00   2.1530E+00
 1.7504E+00   2.4127E+00
    0.444086
    0.139151
    0.139172
```

# Program List for Propagation of a Surge through a Distribution Feeder with an Insulator Flashover (Application Example 2.9)

```
lightning performance of a
  transmission line
*lightning current source
i1 1 0 exp(0 15.4k 0 1.0u 2.0111u
  5u)
*span#1
t1 1 0 3 0 z0=300 f=60 NL=100u
*insulator #1
d1 13 3 DMOD
L1 13 53 20u
vs1 23 53 dc 0
w1 23 33 vs1 wmod1
rvs1 33 43 20
vd1 0 43 3700k
rft1 23 0 20
*span#2
t2 3 0 5 0 z0=300 f=60 NL=200u
*insulator #2
d2 15 5 dmod
L2 15 55 20u
vs2 25 55 dc 0
w2 25 35 vs2 wmod1
rvs2 35 45 20
vd2 0 45 3700k
rft2 25 0 20
*span#3
t3 5 0 7 0 z0=300 f=60 NL=200u
*z
RL3 7 0 300
*span#4
t4 1 0 9 0 z0=300 f=60 NL=500u
*z
RL4 9 0 300
.model dmod d(is=1e-14)
.model wmod1 iswitch(ron=1 roff=1E6
  ion=1 ioff=1.01`)
.tran 0.1u 20u 0 0.05u
.options ITL4=200 ITL5=0
  LIMPTS=100000 RELTOL=0.001
.probe
*.print tran v(3) v(5) i(vs1) i(vs2)
.end
```

# 4

# Program List for Lightning Arrester Operation
## (Application Example 2.10)

```
lightning performance of a
  transmission line, lightning
  arrester operation
*lightning current source
i1 1 0 exp(0 7.44k 0 1.0u 2.011u 5u)
*span#1
t1 1 0 3 0 z0=300 f=60 NL=100u
*insulator #1
d1 13 3 DMOD
L1 13 53 20u
vs1 23 53 dc 0
w1 23 33 vs1 wmod1
rvs1 33 43 20
vd1 0 43 3700k
rft1 23 0 20
*span#2
t2 3 0 5 0 z0=300 f=60 NL=200u
*insulator #2
d2 15 5 dmod
L2 15 55 20u
vs2 25 55 dc 0
w2 25 35 vs2 wmod1
rvs2 35 45 20
vd2 0 45 600k
rft2 25 0 20
*span#3
t3 5 0 7 0 z0=300 f=60 NL=200u
*z
RL3 7 0 300
*span#4
t4 1 0 9 0 z0=300 f=60 NL=500u
*z
RL4 9 0 300
.model dmod d(is=1e-14)
.model wmod1 iswitch(ron=1 roff=1E6
  ion=1 ioff=1.01)
.tran 0.1u 20u 0 0.05u
.options ITL4=25 ITL5=0 LIMPTS=100000
  RELTOL=0.001
.probe
*.print tran v(3) v(5) i(vs1) i(vs2)
.end
```

## 5
APPENDIX

# Equipment for Tests

## A5.1 THE 9 kVA THREE-PHASE TRANSFORMER BANK

A *Y–Y* connected three-phase transformer bank with 416/416 V of rated line-to-line voltages consists of three single-phase transformers. The type numbers of the three single-phase transformers are J7065, J7610, and J7065 (bank #1), and the nameplate data are

| Powerformer | Dry-type transformer |
|---|---|
| CAT. NO.: 211-101 | 3 kVA, single-phase, 60 Hz |
| High voltage: 240/480 V | Low voltage: 120/240 V |

## A5.2 THE 4.5 kVA THREE-PHASE TRANSFORMER BANK #1

Single-phase transformers with the nameplate data listed in Section A5.1 are used in transformer bank #1 and are connected in such a way that the rated line-to-line voltages are 240/240 V. Only the two high-voltage windings of each single-phase transformer are employed as primary and secondary, and the rated apparent power of the three-phase transformer bank is therefore 4.5 kVA.

## A5.3 THE 4.5 kVA THREE-PHASE TRANSFORMER BANK #2

Single-phase transformers with the nameplate data listed in Section A5.1 having the type numbers J7605, J7606, and J7605 are used in transformer bank #2, and are connected in such a way that the rated line-to-line voltages are 240/240 V. Only the two high-voltage windings of each single-phase transformer are employed as primary and secondary, and the rated apparent power of the three-phase transformer bank is therefore 4.5 kVA.

## A5.4 THE 15 kVA THREE-PHASE TRANSFORMER BANK

The 15 kVA, 240/240, Δ–Δ connected bank consists of three 5 kVA transformers. The original rated apparent power of the single-phase units is 10 kVA. The two high-voltage windings are used as primary and secondary, and the rated power of the bank is therefore 15 kVA. The nameplate data are

| Westinghouse | Type: EP transformer |
|---|---|
| Frame NO.: 179 | 10 kVA, single-phase, 60 Hz |
| High voltage: 240/480 V | Low voltage: 120/240 V |

## A5.5 THREE-PHASE DIODE BRIDGE

$I_{max} = 100$ A, $V_{max} = 1200$ V.

## A5.6 HALF-CONTROLLED THREE-PHASE SIX-STEP INVERTER

General Electric AC adjustable frequency drive:

Model NO.: 6VGAW2007CI

| Input: | Output: |
|---|---|
| Volts: 230 | HP: 7.5 |
| Hertz: 50/60 | kVA: 8.8 |
| Amps.: 22 | Volts: variable |
| Phase 3 | Hertz: 6–60 |
| | Amps. AC: variable |
| | Phase 3 |

## A5.7 CONTROLLED THREE-PHASE RESONANT RECTIFIER [12, CHAPTER 2]

University of Colorado:

ID NO.: 5484

| Input: | Output: |
|---|---|
| Volts: 340–600 AC | 20 kW |
| Hertz: 12–60 | Volts: 380 DC |
| Amps.: 45 | Amps.: 60 DC |
| Phase 3 | |

## A5.8 CONTROLLED THREE-PHASE PWM INVERTER [12, CHAPTER 2]

University of Colorado:

ID NO.: 5485

| Input: | Output: |
|---|---|
| Volts: 360 DC | 20 kW |
| Amps.: 70 DC | Volts: 240 AC |
| | Amps.: 80 AC |
| | Power factor: 0.7 (lead, generator notation) to 1.0 pu |
| | Hertz: 50/60 |

ISBN 978-0-12-369536-9

# Measurement Error of Powers

## A6.1 MEASUREMENT ERROR OF POWERS

Assume the input power of a load with power factor (lagging') between 0.6 and 1.0 is measured with a $CT$, an ammeter, a $PT$, and a voltmeter, as shown in Fig. A6.1. The ammeter and the voltmeter have an error of $\varepsilon_A$ and $\varepsilon_V$, respectively. The $CT$ has a ratio correction factor $RCF_{CT}$ and a phase angle shift $\gamma$ measured in minutes. The $PT$ has a ratio correction factor $RCF_{PT}$ and a phase angle shift $\beta$ measured in minutes [97–100, Chapter 8].

The input power is expressed as

$$P = V\,I \cos \Omega.$$

The measurement error of the input power is given by

$$\frac{\Delta P}{P} = \frac{\Delta V}{V} + \frac{\Delta I}{I} - \frac{\sin\varphi \cdot \Delta\varphi}{\cos\varphi},$$

$$= \left(RCF_{PT} - 1 + \frac{\varepsilon_V}{V}\right) + \left(RCF_{CT} - 1 + \frac{\varepsilon_A}{I}\right)$$
$$- \frac{\sin\varphi}{\cos\varphi}\frac{\pi(\beta-\gamma)}{180\times60}.$$

For power factors ranging from 0.6 to 1.0 (lagging, based on consumer notation), the worst case in power measurement error is when $\cos\varphi = 0.6$, therefore,

$$\frac{\Delta P}{P} \approx \left(RCF_{PT} - 1 + \frac{\varepsilon_V}{V}\right) + \left(RCF_{CT} - 1 + \frac{\varepsilon_A}{I}\right) - \frac{(\beta-\gamma)}{2600},$$

$$= \left(TCF_{PT} - 1 + \frac{\varepsilon_V}{V}\right) + \left(TCF_{CT} - 1 + \frac{\varepsilon_A}{I}\right)$$
$$= \left(\frac{\varepsilon_{PT} + \varepsilon_V}{V}\right) + \left(\frac{\varepsilon_{CT} + \varepsilon_A}{I}\right),$$

where

$$TCF_{CT} = RCF_{CT} + \gamma/2600,$$

$$TCF_{PT} = RCF_{PT} - \beta/2600,$$

$$\varepsilon_{CT} = (TCF_{CT} - 1) \cdot I,$$

$$\varepsilon_{PT} = (TCF_{PT} - 1) \cdot V.$$

According to American National Standard [97, Chapter 8], transformer correction factors $TCF_{CT}$ and $TCF_{PT}$ ($\varepsilon_{CT}$ and $\varepsilon_{PT}$) can directly be obtained based on the accuracy class from the device nameplates. Normally, the maximum absolute errors for ammeters and voltmeters are based on the full-scale errors. For instrument transformers, the maximum absolute errors ($\varepsilon_{CT}$ and $\varepsilon_{PT}$) are proportional to the actually measured values of currents and voltages, as derived above.

Note that the maximum errors of an instrument transformer derived above are systematic errors that can be reduced by ratio and angle corrections. If the measured results are not corrected, the overall uncertainty of measurement is derived from the systematic uncertainty and random uncertainty:

$$u = \sqrt{u_s^2 + u_r^2}.$$

The random uncertainty is normally much smaller than the systematic uncertainty and can be ignored; therefore,

$$u = u_s = \varepsilon_{CT} \text{ or } \varepsilon_{PT}.$$

## A6.2 NAMEPLATE DATA OF MEASURED TRANSFORMERS

### 3 kVA single-phase transformers

| Powerformer | Dry-type transformer |
|---|---|
| CAT. NO.: 211-101 | 3 kVA, single-phase, 60 Hz |
| High voltage: 240/480 V | Low voltage: 120/240 V |

**4.5 kVA three-phase transformer bank.** Three single-phase transformers with the nameplate data just listed for the 3 kVA single-phase transformers having the type numbers J7065, J7610, and J7065 are used in the three-phase transformer bank and are connected in $\Delta$–$Y_0$ with rated phase voltages of

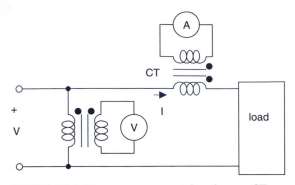

**FIGURE A6.1** Power measurement based on a *CT*, an ammeter, a *PT*, and a voltmeter.

240/240 V. Only the two high-voltage windings of each single-phase transformer are employed as primary and secondary, and the rated apparent power of the three-phase transformer bank is therefore 4.5 kVA.

### 25 kVA single-phase transformer [101, Chapter 8]

| ABB | MFG Date: July 93 |
|---|---|
| Serial #: 93A312106 | 25 kVA, 60 Hz |
| High voltage: 7200 V | Low voltage: 120/240 V |

# Solutions to Examples from Chapters

## CHAPTER 1 SOLUTIONS

### 1.4.18 Application Example 1.1: Calculation of Input/Output Currents and Voltages of a Three-Phase Thyristor Rectifier: Solution (see also Figure E1.1.3)

```
*thyristor rectifier
Va 1 2 SIN(0V   339.41V 60 Hz 0 0)
Vb 2 3 SIN(0V   339.41V 60 Hz 5.556m
 0)
Vc 3 3a SIN(0V 339.41V 60 Hz 11.111m
 0)
R3 3a 1 0.01
Rarb 14 0 .01
*5 deg delay = .231
*50 deg delay = 2.31
*150 deg delay = 6.94
*Vfet1 sp1 A1 pulse(0 15 .231m     1u
 1u 9m 16.667ms)
*Vfet2 sp2 A2 pulse(0 15 3.01m     1u
 1u 9m 16.667ms)
*Vfet3 sp3 A3 pulse(0 15 5.79m     1u
 1u 9m 16.667ms)
*Vfet4 sp4 A4 pulse(0 15 8.56m    1u
 1u 9m 16.667ms)
*Vfet5 sp5 A5 pulse(0 15 11.32m    1u
 1u 9m 16.667ms)
*Vfet6 sp6 A6 pulse(0 15 14.12m    1u
 1u 9m 16.667ms)
Vfet1 sp1 A1 pulse(0 15 2.31m     1u
 1u 9m 16.667ms)
Vfet2 sp2 A2 pulse(0 15 5.089m    1u
 1u 9m 16.667ms)
Vfet3 sp3 A3 pulse(0 15 7.869m    1u
 1u 9m 16.667ms)
Vfet4 sp4 A4 pulse(0 15 10.639m   1u
 1u 9m 16.667ms)
Vfet5 sp5 A5 pulse(0 15 13.399m   1u
 1u 9m 16.667ms)
Vfet6 sp6 A6 pulse(0 15 16.199m   1u
 1u 9m 16.667ms)
*Vfet1 sp1 A1 pulse(0 15 6.94m    1u
 1u 9m 16.667ms)
```

```
*Vfet2 sp2 A2 pulse(0 15 9.72m     1u
 1u 9m 16.667ms)
*Vfet3 sp3 A3 pulse(0 15 12.50m    1u
 1u 9m 16.667ms)
*Vfet4 sp4 A4 pulse(0 15 15.27m    1u
 1u 9m 16.667ms)
*Vfet5 sp5 A5 pulse(0 15 18.04m    1u
 1u 9m 16.667ms)
*Vfet6 sp6 A6 pulse(0 15 20.83m    1u
 1u 9m 16.667ms)
RAB 7 8 10Meg
M1 7   sp1 A1 A1 SMM
M2 14 sp2 A2 A2 SMM
M3 8   sp3 A3 A3 SMM
M4 14 sp4 A4 A4 SMM
M5 9   sp5 A5 A5 SMM
M6 14 sp6 A6 A6 SMM
La 1   4 300uH
Ra 4   7 0.05
Lb 2   5 300uH
Rb 5   8 0.05
Lc 3   6 300uH
Rc 6   9 0.05
d1 A1 13 D1N4001
d2 A2 12 D1N4001
L2 12   9 5nH
d3 A3 13 D1N4001
d4 A4 10 D1N4001
L4 10   7 5nH
d5 A5 13 D1N4001
d6 A6 11 D1N4001
L6 11   8 5nH
Lf 13 15 5mH
Cf 15 14 500u
R1 15 14 10
.model D1N4001 D(IS=1E-12)
.model SMM nmos(level=3 gamma=0
 delta=0 eta=0 theta=0
+ kappa=0 vmax=0 xj=0 tox=100n
 uo=600 phi=0.6 rs=42.69M kp=20.87u
+ l=2u w=2.9 vto=3.487 rd=0.19
 cbd=200n pb=0.8 mj=0.5 cgso=3.5n
+ cgdo=100p rg=1.2 is=10f)
```

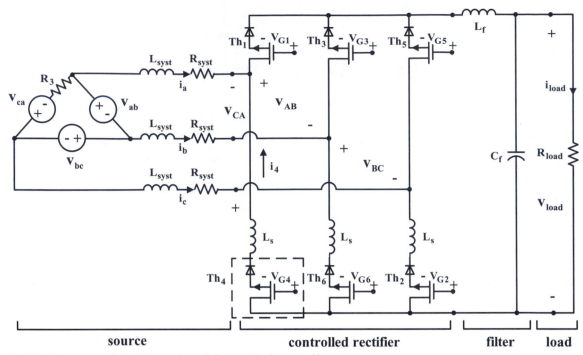

**FIGURE E1.1.1** Controlled three-phase, full-wave thyristor rectifier.

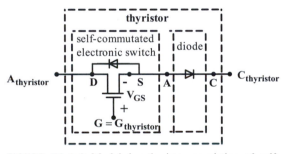

**FIGURE E1.1.2** Model for thyristor consisting of self-commuted switch and diode.

```
.tran .01u 60m 30m uic
.options abstol=1m chgtol=1m
  reltol=10m vntol=10m gmin=1m
.four 60 I(ra) V(RAB)
.probe
.end
```

### 1.4.19 Application Example 1.2: Calculation of Input/Output Currents and Voltages of a Three-Phase Rectifier with One Self-Commutated Electronic Switch: Solution (see also Figure E1.2.2)

```
*rectifier with one self-commutated
  switch
```

```
Va 1 2 SIN(0V 339.41V 60 Hz 0 0)
Vb 2 3 SIN(0V 339.41V 60 Hz 5.56m 0)
Vc 3 3a SIN(0V 339.41V 60 Hz 11.11m
  0)
R3 3a 1 0.1
Rarb 14 0 .1
RAB 7 8 10Meg
Vfet sp 15 pulse(0 15 0 1u 1u 833u
  1.67m)
M1 13 sp 15 15 SMM
La 1  4 10mH
Ra 4  7 0.05
Lb 2  5 10mH
Rb 5  8 0.05
Lc 3  6 10mH
Rc 6  9 0.05
d1 7 13 D1N4001
d2 14 12 D1N4001
L2 12 9 5nH
d3 8 13 D1N4001
d4 14 10 D1N4001
L4 10 7 5nH
d5 9 13 D1N4001
d6 14 11 D1N4001
L6 11 8 5nH
Df  14 15 D1N4001
Lf 15 16 5mH
Cf 16  14 500u
R1 16 14 10
```

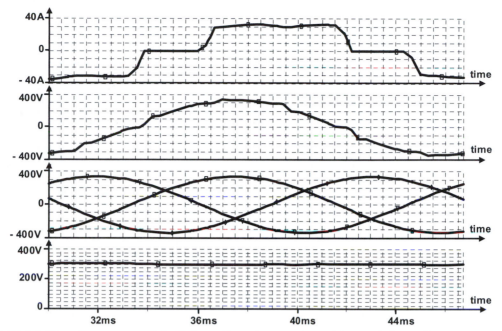

**FIGURE E1.1.3** Plots of input current $i_a$ (top), line-to-line voltage $v_{AB}$ (second from top), line-to-line voltages $v_{ab}$, $v_{bc}$, $v_{ca}$, and output voltage $v_{load}$ for a firing angle of $\alpha = 50°$.

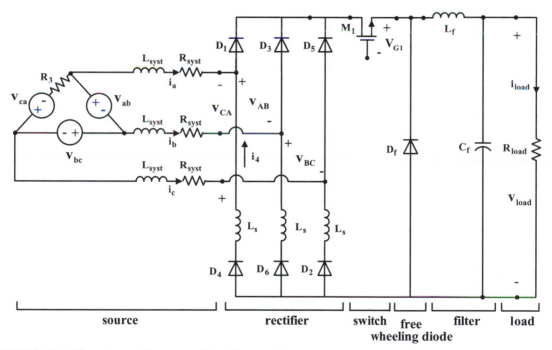

**FIGURE E1.2.1** Three-phase, full-wave rectifier with one self-commutated switch.

```
.model D1N4001 D(IS=1E-12)
.model SMM nmos(level=3 gamma=0
  delta=0 eta=0 theta=0
+ kappa=0 vmax=0 xj=0 tox=100n
  uo=600 phi=0.6 rs=42.69M kp=20.87u
+ l=2u w=2.9 vto=3.487 rd=0.19
  cbd=200n pb=0.8 mj=0.5 cgso=3.5n
```
```
+ cgdo=100p rg=1.2 is=10f)
.tran .010u 60m 30m uic
.options abstol=1m chgtol=1m
  reltol=10m vntol=10m gmin=1m
.four 60 I(ra) V(RAB)
.probe
.end
```

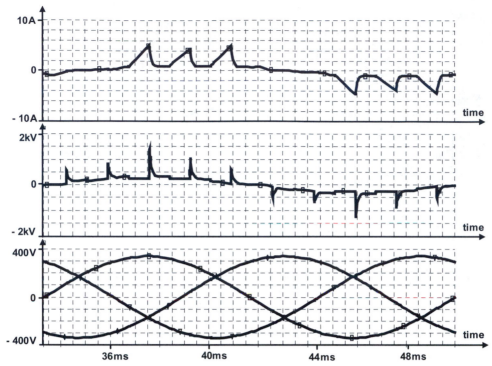

**FIGURE E1.2.2** Input line current $i_a$ (top), line-to-line voltage $v_{AB}$ with (extreme) commutating notches (second from top), line-to-line voltages $v_{ab}$, $v_{bc}$, $v_{ca}$.

### 1.4.20 Application Example 1.3: Calculation of Input Currents of a Brushless DC Motor in Full-on Mode (Three-Phase Permanent-Magnet Motor Fed by a Six-Step Inverter): Solution (see also Figures E1.3.3a to E1.3.3d)

```
*brushless DC motor drive
Va   1a st SIN(0 160V 1500 Hz   0 0
  60)
Vb   3a st SIN(0 160V 1500 Hz   0 0
  180)
Vc   5a st SIN(0 160V 1500 Hz   0 0
  300)
Vdc1   7 0 DC 150V
Vdc2   8 0 DC -150
Rshunt st 0 10Meg
*5 deg delay = .231
*50 deg delay = 2.31
*150 deg delay = 6.94
Vfet1 sp1 2 pulse(0 15 0
  1u 1u 222.22us 666.66us)
Vfet2 sp2 6 pulse(0 15 222.22us 1u
  1u 222.22us 666.66us)
Vfet3 sp3 4 pulse(0 15 444.44us 1u
  1u 222.22us 666.66us)
Vfet4 sp4 8 pulse(0 15 333.33us 1u
  1u 222.22us 666.66us)
```

```
Vfet5 sp5 8 pulse(0 15 555.51us 1u
  1u 222.22us 666.66us)
Vfet6 sp6 8 pulse(0 15 111.11us 1u
  1u 222.22us 666.66us)
Mau 7 sp1 2 2 SMM
Mbu 7 sp2 6 6 SMM
Mcu 7 sp3 4 4 SMM
Mal 2 sp4 8 8 SMM
Mbl 6 sp5 8 8 SMM
Mcl 4 sp6 8 8 SMM
L1   1a 1 50u
R1   1 2 0.5
L3   3a 3 50u
R3   3 4 0.5
L5   5a 5 50u
R5   5 6 0.5
C1   7 0 100u
C2   0 8 100u
Dau 2 7 D1N4001
Dbu 6 7 D1N4001
Dal 8 2 D1N4001
Dbl 8 6 D1N4001
Dcu 4 7 D1N4001
Dcl 8 4 D1N4001
.model D1N4001 D(IS=1E-12)
.model SMM nmos(level=3 gamma=0
  delta=0 eta=0 theta=0
```

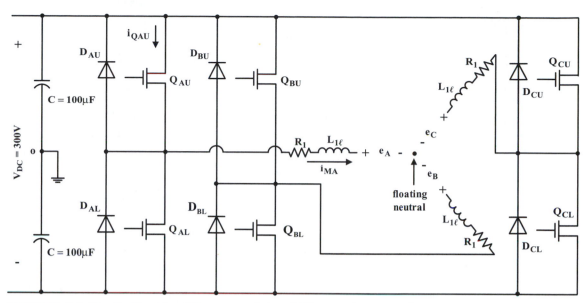

**FIGURE E1.3.1** Circuit of brushless DC motor consisting of DC source, inverter, and permanent-magnet machine.

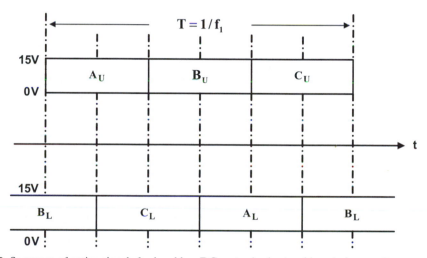

**FIGURE E1.3.2** Sequence of gating signals for brushless DC motor in six-step (six-pulse) operation.

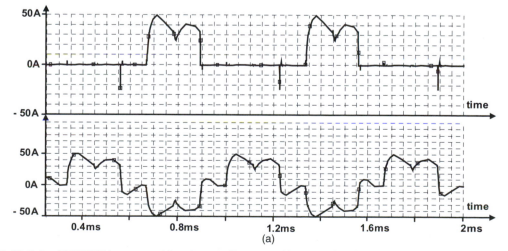

**FIGURE E1.3.3a** MOSFET (upper graph) and motor (lower graph) current for phase angle $\theta = 0°$.

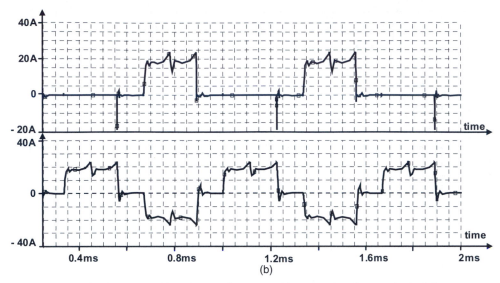

**FIGURE E1.3.3b** MOSFET (upper graph) and motor (lower graph) current for phase angle $\theta = 30°$.

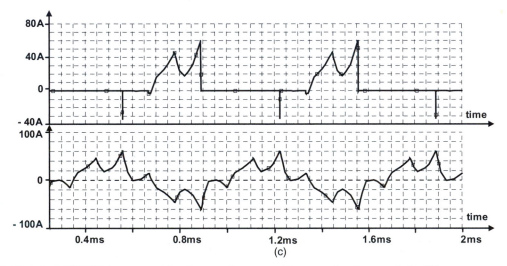

**FIGURE E1.3.3c** MOSFET (upper graph) and motor (lower graph) current for phase angle $\theta = 60°$.

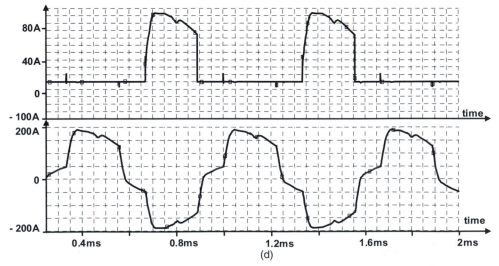

**FIGURE E1.3.3d** MOSFET (upper graph) and motor (lower graph) current for phase angle $\theta = -60°$.

```
+  kappa=0 vmax=0 xj=0 tox=100n
   uo=600 phi=0.6 rs=42.69m kp=20.87u
+  l=2u w=2.9 vto=3.487 rd=0.19
   cbd=200n pb=0.8 mj=0.5 cgso=3.5n
+  cgdo=100p rg=1.2 is=10f)
.tran .05u 2m uic
.options abstol=1m chgtol=1m
   reltol=10m vntol=10m
.four 1500 I(L1)
.probe
.end
```

b) The reversed phase sequence will result in nonperiodic waveshapes, and thus this solution is not meaningful.

### 1.4.21 Application Example 1.4: Calculation of the Efficiency of a Polymer Electrolyte Membrane (PEM) Fuel Cell Used as an Energy Source in a Variable-Speed Drive: Solution

a) The power efficiency is defined by

$\eta_{power}$ = (rated electrical output power)/(hydrogen input power) = $P_{out}/P_{in}$.

**Method #1.** The PEM fuel cell generates 0.87 liters or 0.87 kg of water per hour by converting hydrogen and oxygen according to the relation

$$2H_2 + O_2 = 2H_2O.$$

The atomic weights are

$$1H \triangleq 1.0080,$$
$$1O \triangleq 16.0000,$$
$$1H_2O \triangleq 18.0160,$$

The required mass of hydrogen to generate 0.87 liters of water per hour is $\frac{2}{18.0160} 0.87$ kg = 0.09658 kg. The energy input per hour in form of hydrogen is (0.09658 kg)(28 kWh/kg) = 2704.26 Wh.

Output (electrical) power of PEM fuel cell
$P_{out}$ = 1200 W.
Input (hydrogen) power (or energy per hour) of PEM fuel cell $P_{in}$ = 2704.26 W.
Power (energy per hour) efficiency
$\eta_{power} = P_{out}/P_{in}$ = 0.4437 pu ≈ 44.4%.

**Method #2.** Hydrogen input power of PEM fuel cell is

$P_{in}$ = (mass density of hydrogen) (hydrogen consumption during one hour) (energy density of hydrogen).

With (hydrogen consumption during one hour) = 18.5 SLPM·60 minutes/hour = (18.5)(60) liters/hour = 1110 liters/hour, one obtains the input power in form of hydrogen:

$$P_{in} = (0.0899 \cdot 10^{-3} \text{ kg/liter}) (1110 \text{ liters/hour})$$
$$(28 \cdot 10^{+3} \text{ W/kg}) = 2794 \text{ W}.$$
Power (energy per hour) efficiency
$\eta_{power} = P_{out}/P_{in}$ = 0.4295 pu ≈ 43%.

**Conclusion.** Both methods generate about the same efficiency for a PEM fuel cell.

b) The specific power density of this PEM fuel cell expressed in W/kg is defined as

$$\frac{P_{out}}{mass} = 1200 \text{ W}/13 \text{ kg} = 92.3 \text{ W/kg}.$$

c) The specific power density of a PEM fuel cell is 92.3 W/kg as compared with that of a lead-acid battery of about 150 W/kg [66 in Chapter 1].

### 1.4.22 Application Example 1.5: Calculation of the Currents of a Wind-Power Plant PWM Inverter Feeding Power into the Power System: Solution

See Fig. E1.5.3

### 1.8.3.1 Application Example 1.6: Interharmonic Reduction by Dedicated Transformer: Solution

Computations are shown in Figs. E1.6.5 and E1.6.6 for the above three cases. As illustrated, for $V_{load1} = V_{load2}$ the 1128th interharmonic amplitude is $V_{1128} = \frac{(122.64 - 120)}{120}(100\%) = 2.2\%$, whereas for $V_{load3}$ this interharmonic is only 0.006%. This demonstrates the effectiveness of a dedicated line or transformer, other than an isolation transformer with a turns ratio 1 : 1.

### 1.8.6.1 Application Example 1.7: Hand Calculation of Harmonics Produced by Twelve-Pulse Converters: Solution

**Calculating Current Harmonics at PCC #1.**
At PCC #1:

• $S_{ph} = \frac{|\tilde{V}||\tilde{I}|}{3} = \frac{50 \text{ MVA}}{3} = 16.67 \text{ MVA}$

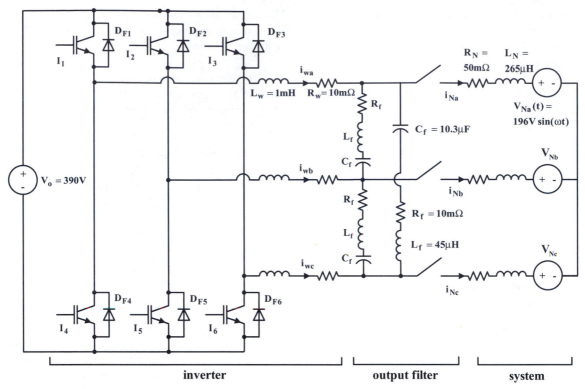

**FIGURE E1.5.1** Current-controlled PWM inverter feeding into utility system.

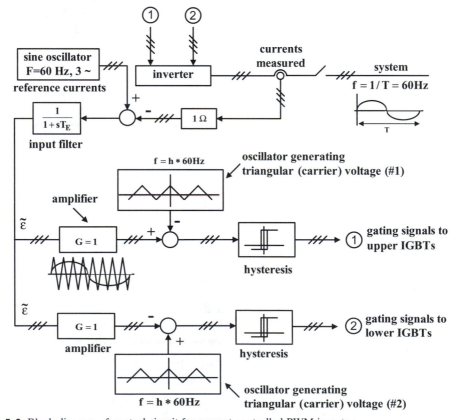

**FIGURE E1.5.2** Block diagram of control circuit for current-controlled PWM inverter.

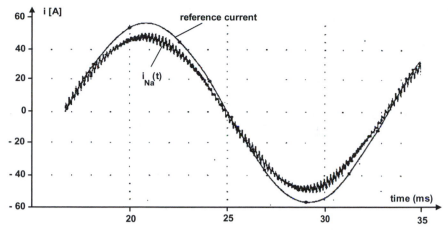

**FIGURE E1.5.3** Phase reference current and phase current $i_{Na}(t)$ delivered by inverter to power system at about unity power factor.

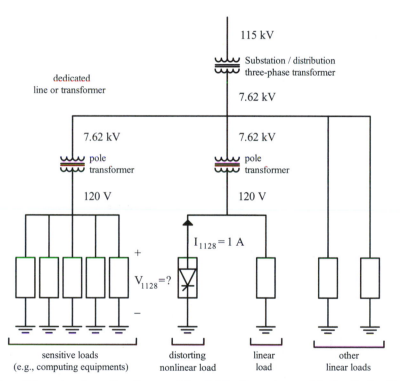

**FIGURE E1.6.1** Overall (per phase) one-line diagram of the distribution system used in Application Example 1.6.

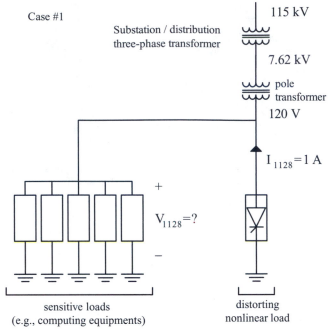

**FIGURE E1.6.2** Case #1 of Application Example 1.6: distorting nonlinear load and sensitive loads are fed from same pole transformer.

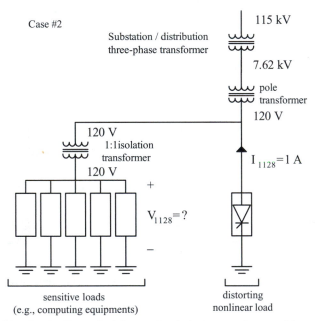

**FIGURE E1.6.3** Case #2 of Application Example 1.6: use of an isolation transformer with a turns ratio 1 : 1 between distorting (nonlinear) load and sensitive loads.

- $V_{ph} = \dfrac{115\,\text{kV}}{\sqrt{3}} = 66.4\,\text{kV}$

- $I_{ph} = \dfrac{S_{ph}}{V_{ph}} = \dfrac{50\,\text{MVA}}{3(66.4\,\text{kV})}$

**TABLE E1.7.1** Harmonic Current ($I_h$) Generated by Six-Pulse and Twelve-Pulse Converters [64] Based on $X_t^h = 0.12$ pu and $\alpha = 30°$

| Harmonic order ($h$) | $I_h$ for 6-pulse converter (pu) | $I_h$ for 12-pulse converter (pu) |
|---|---|---|
| 1 | 1.000 | 1.000 |
| 5 | 0.192 | **0.0192** |
| 7 | 0.132 | **0.0132** |
| 11 | 0.073 | 0.073 |
| 13 | 0.057 | 0.057 |
| 17 | 0.035 | **0.0035** |
| 19 | 0.027 | **0.0027** |
| 23 | 0.020 | 0.020 |
| 25 | 0.016 | 0.016 |
| 29 | 0.014 | **0.0014** |
| 31 | 0.012 | **0.0012** |
| 35 | 0.011 | 0.011 |
| 37 | 0.010 | 0.010 |
| 41 | 0.009 | **0.0009** |
| 43 | 0.008 | **0.0008** |
| 47 | 0.008 | 0.008 |
| 49 | 0.007 | 0.007 |

- $I_L = I_{ph} = 251$ A.

- At a short-circuit ratio of $R_{SC} = \dfrac{I_{SC}}{I_L} = 40$ (at PCC #1)

- The system's impedance is $Z_{sys}$ (10 MVA base) $= 0.5\% = 0.005$ pu,

- $I_{SC}^{\text{PCC#1}} = R_{SC} I_L = 40\,(251\text{ A}) = 10000$ A.

According to Table 1.7, for PCCs from 69 to 138 kV, the current harmonic limits should be divided by two. Therefore, for $I_{SC}/I_L = (20$ to $50)$:

$$I_{\text{hlimit}} = \frac{7.0}{2} = 3.5\% \quad \text{for } 5 \le h < 11;$$

$$I_{\text{hlimit}} = \frac{3.5}{2} \cong 1.8\% \quad \text{for } 11 \le h < 17;$$

$$I_{\text{hlimit}} = \frac{2.5}{2} \cong 1.3\% \quad \text{for } 17 \le h < 23.$$

The actually occurring harmonic currents are for $I_{spc} = (251/2)$ A $= 125.5$ A for 12-pulse converter (SPC means static power converter):

$$I_5^{\text{actual}} = I_{spc}\,(0.0192) = 125.5(0.0192) = 2.41 \text{ A}$$
$$I_7^{\text{actual}} = 125.5(0.0132) = 1.65 \text{ A}$$
$$I_{11}^{\text{actual}} = 9.15 \text{ A}$$
$$I_{13}^{\text{actual}} = 7.15 \text{ A}$$

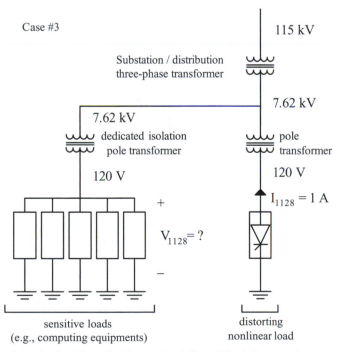

**FIGURE E1.6.4** Case #3 of Application Example 1.6: use of a dedicated (isolation transformer with turns ratio 7620 : 120) pole transformer between distorting nonlinear load and sensitive loads.

Analysis of dedicated isolated transformer

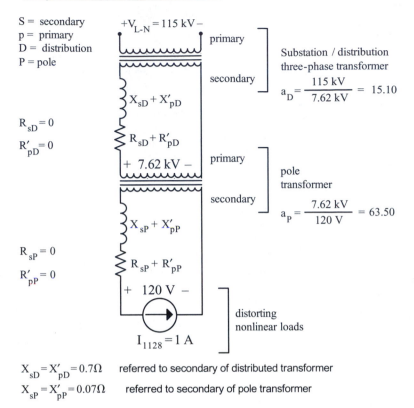

$$X_{sD} = X'_{pD} = 0.7\Omega \quad \text{referred to secondary of distributed transformer}$$
$$X_{sP} = X'_{pP} = 0.07\Omega \quad \text{referred to secondary of pole transformer}$$

**FIGURE E1.6.5** Equivalent circuit of distribution and pole transformers.

Analysis of dedicated isolated transformer) continued (:

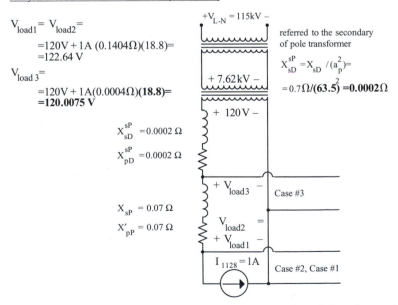

$V_{load1} = V_{load2} =$

$\quad = 120V + 1A\,(0.1404\Omega)(18.8) =$
$\quad = 122.64\ V$

$V_{load3} =$

$\quad = 120V + 1A(0.0004\Omega)(18.8) =$
$\quad = \mathbf{120.0075\ V}$

referred to the secondary
of pole transformer

$X_{sD}^{sP} = X_{sD}\,/(a_p^2) =$
$\quad = 0.7\Omega/(63.5)^2 = \mathbf{0.0002\Omega}$

$X_{sD}^{sP} = 0.0002\ \Omega$

$X_{pD}^{sP} = 0.0002\ \Omega$

$X_{sP} = 0.07\ \Omega$

$X'_{pP} = 0.07\ \Omega$

**FIGURE E1.6.6** Equivalent circuit referring all lumped reactances to the secondary of the pole transformer.

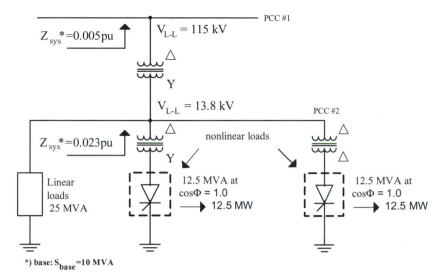

**FIGURE E1.7.1** One-line diagram of a large industrial plant fed from transmission voltage [64 in Chapter 1].

with $I_L = 251$ A

$$I_5^{actual\%} = \frac{I_5^{actual}}{I_L} 100\%$$

$\left. \begin{array}{l} I_5^{actual\%} = 0.96\% \\ I_7^{actual\%} = 0.66\% \end{array} \right\}$ allowed limits are 3.5%

$\left. \begin{array}{l} I_{11}^{actual\%} = 3.65\% \\ I_{13}^{actual\%} = 2.85\% \end{array} \right\}$ allowed limits are 1.8%.

As can be seen, $I_{11}^{actual}$ and $I_{13}^{actual}$ are too large!

**Calculating Voltage Harmonics at PCC #1.** Calculation of short-circuit (apparent) power $S_{SC}$:

- $I_{SC} = I_L \cdot R_{SC} = 251(40) = 10040$ A

- $\dfrac{I_{SC}}{I_L} = R_{SC}$

- $S_{SC} = 3 \cdot V_{ph} \cdot I_{SC} = 3\left(\dfrac{115kV}{\sqrt{3}}\right)(10.04 kA)$

  $= 2000 MVA$

Checking:

$$S_{SC} = MVA_{SC} = \frac{S_{base}[MVA]}{Z_{syst}[\text{in per unit at base MVA}]}$$

$$= \frac{10 MVA}{0.005 pu} = 2,000 MVA.$$

Calculation of voltage harmonics (at PCC #1) for the current harmonics:

- $V_h = \dfrac{I_h}{I_{base}}(h) \cdot Z_{sys} \cdot 100\%$
- $S_{base} = 10$ MVA (all three phases)

- $I_{base} = \dfrac{S_{basephase}}{V_{phase}}$

- $I_{base} = \dfrac{10 MVA/3}{115 kV/\sqrt{3}} = 50.2$ A

Harmonic (actually occurring) voltages:

$\left. \begin{array}{l} V_5 = \dfrac{2.41}{50.2}(5)(0.005)(100\%) \\ V_5 = 0.120\% \\ V_7 = \dfrac{1.65}{50.2}(7)(0.005)(100\%) \\ V_7 = 0.115\% \\ V_{11} = 1.0\% \\ V_{13} = 0.93\% \end{array} \right\}$ allowed limit is 1.5% for individual harmonic voltage (Table 1.8).

Also:

- $THD_V^{limit} \le 2.5\%$

- $THD_V^{actual} = \dfrac{\sqrt{\sum_{h=5,7,...} V_h^2}}{V_{phase}}(100\%) = 1.64\%$

**Calculating Current Harmonics at PCC #2.**
At PCC #2:

- $S_{ph} = 16.67$ MVA

- $V_{ph} = \dfrac{13.8 kV}{\sqrt{3}} = 7.968 kV$

- $I_{ph} = \dfrac{16.67 MVA}{7.968 kV} = 2092$ A

- At a short-circuit ratio of $R_{SC} = \dfrac{I_{SC}}{I_L} = 8.7$ (at PCC #2)

- The system's impedance is $Z_{sys}$ (10 MVA base) = 2.3% = 0.023 pu,

- $I_{SC}^{PCC\#2} = R_{SC} \times I_L = 8.7(2092\ A) = 18.2\ kA$.

    Calculation of short-circuit (apparent) power $S_{SC}$:

- $S_{phase} = \dfrac{50\,MVA}{3} = 16.67\,MVA$

- $V_{phase} = \dfrac{13.8\,kV}{\sqrt{3}} = 7.97\,kV$

- $I_{phaseL} = \dfrac{50\,MVA}{3} \times \dfrac{\sqrt{3}}{13.8\,kV} = 2.092\,kA$

- $I_{SCphase} = R_{SC} \times I_{phaseL} = 8.7(2.092) = 18.2\ kA$

- $I_{SPCphase} = \dfrac{I_{phaseL}}{2} = 1.046\,kA$

- $R_{SC} = \dfrac{I_{SCphase}}{I_{phaseL}} = \dfrac{18.2}{2.092} = 8.7$

- $Z_{sys} = 2.3\% = 0.023$ pu.

Actually occurring harmonic currents at PCC #2 for 12-pulse converter:

$$I_5^{actual} = 1046(0.0192) = 20.08\ A$$
$$I_7^{actual} = 1046(0.0132) = 13.81\ A$$
$$I_{11}^{actual} = 1046(0.073) = 76.35\ A$$
$$I_{13}^{actual} = 1046(0.057) = 59.62\ A$$

in percent

$$I_5^{actual\%} = \dfrac{I_5^{actual}}{I_{phaseL}}(100\%)$$

$$I_5^{actual\%} = \dfrac{20.08}{2092}(100\%) = 0.96\%$$

$$I_7^{actual\%} = 0.66\%$$

$$\left.\begin{array}{l} I_{11}^{actual\%} = 3.65\% \\ I_{13}^{actual\%} = 2.85\% \end{array}\right\} \begin{array}{l}\text{above limits} \\ \text{(see analysis for PCC \#1)}\end{array}$$

**Calculating Voltage Harmonics at PCC #2.**

$$V_h = \dfrac{I_h}{I_{base}}(h) \cdot Z_{syst} \cdot 100\%.$$

where

$$I_{base} = \dfrac{10\,MVA/3}{13.8\,kV/\sqrt{3}}$$

$I_{base} = 418.4\ A$

$$\left.\begin{array}{l} V_5 = \dfrac{20.08}{418.4}(5)(0.023)100\% \\ V_5 = 0.552\% \\ \\ V_7 = \dfrac{13.81}{418.4}(7)(0.023)100\% \\ V_7 = 0.531\% \end{array}\right\} \text{below limit of 3\%.}$$

$$\left.\begin{array}{l} V_{11} = 4.61\% \\ V_{13} = 4.26\% \end{array}\right\} \text{above limit of 3\%.}$$

Note that on the 13.8 kV bus, the current and voltage distortions are greater than recommended by IEEE-519 (Table 1.8). A properly sized harmonic filter applied on the 13.8 kV bus would reduce the current distortion and the voltage distortion to within the current limits and the voltage limits on the 13.8 kV bus.

### 1.8.6.2 Application Example 1.8: Filter Design to Meet IEEE-519 Requirements: Solution

**System Analysis.** At harmonic frequencies, the circuit of Fig. E1.8.1 can be approximated by the equivalent circuit shown by Fig. E1.8.2. This circuit should be analyzed at each frequency of interest by calculating series and parallel resonances.

For series resonance $\tilde{I}_{fh}$ is large whereas for parallel resonance $\tilde{I}_{fh}$ and $\tilde{I}_{sysh}$ are large. The major impedance elements in the circuit respond differently as frequency changes. The impedance of the transmission line $Z_{lineh}$ is a complex relationship between the inductive and capacitive reactances. Using the fundamental frequency resistance R and inductance of the transmission line, however, gives acceptable results. For most industrial systems $Z_{th}$ and $Z_{lineh}$ can be approximated by the short-circuit impedance if low-frequency phenomena are considered.

**The Impedance versus Frequency Characteristic of a Transformer.** The impedance versus frequency characteristic of a transformer depends on its design, size, voltage, etc. Its load loss, $I^2R$, will constitute 75 to 85% of the total transformer loss and about 75% of this is not frequency dependent (skin effect). The remainder varies with the square of the frequency. The no-load loss (core loss) constitutes between 15 and 25% of the total loss and, depending on flux density, the loss varies as $f^{3/2}$ to $f^3$. From this, with the reactance increasing directly with frequency (inductance L is assumed to be constant), it can be seen that the harmonic ($X_h/R_h$) ratios will be less than the fundamental ($h = 1$) frequency ($X_1/R_1$) ratio, that is,

$$\left(\dfrac{X_h}{R_h}\right) < \left(\dfrac{X_1}{R_1}\right).$$

If the fundamental frequency ratio ($X_1/R_1$) is used, there will be less damping of the high-frequency current than in actuality.

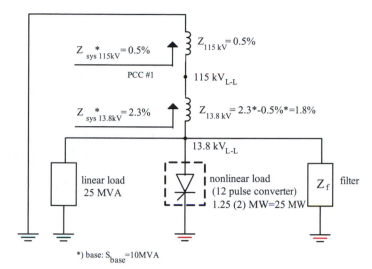

**FIGURE E1.8.1** One-line diagram of a large industrial plant fed from transmission voltage (Fig. E1.7.1) with a passive filter placed at PCC #2.

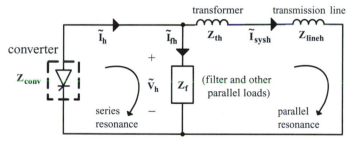

**FIGURE E1.8.2** Equivalent circuit of Fig. E1.8.1 at harmonic frequencies.

**Adjacent Capacitor Banks.** If there are large capacitor banks or filters connected to the utility system, it is necessary to consider their effect.

**Converter as a Harmonic Generator.** The converter is usually considered to be a generator of harmonic currents $i_h$, and is considered to be a constant-current source. Thus $Z_{conv}$ is very large and can be ignored. If the converter is a constant-voltage source $Z_{conv}$ should be included.

**Circuit Analysis.** Using Ohm's and Kirchhoff's laws (Fig. E1.8.2):

$$Z_{sysh} = Z_{th} + Z_{lineh}$$
$$\tilde{I}_h = \tilde{I}_{sysh} + \tilde{I}_{fh}$$
$$\tilde{I}_{fh} Z_{fh} = \tilde{I}_{sysh} Z_{sysh},$$

or

$$\tilde{I}_{fh} = \tilde{I}_{sysh}\left(\frac{Z_{sysh}}{Z_{fh}}\right)$$

$$\tilde{I}_{sysh} = \tilde{I}_h - \tilde{I}_h$$

$$\tilde{I}_{sysh} = \tilde{I}_h - \tilde{I}_{sysh}\left(\frac{Z_{sysh}}{Z_{fh}}\right),$$

or

$$\tilde{I}_{sysh}\left(1 + \frac{Z_{sysh}}{Z_{fh}}\right) = \tilde{I}_{fh}$$

$$\tilde{I}_{sysh}\left(\frac{Z_{fh} + Z_{sysh}}{Z_{fh}}\right) = \tilde{I}_h,$$

or

$$\tilde{I}_{sysh} = \left(\frac{Z_{fh}}{Z_{fh} + Z_{sysh}}\right)\tilde{I}_h.$$

Correspondingly,

$$\tilde{I}_{fh} = \left(\frac{Z_{sysh}}{Z_{fh} + Z_{sysh}}\right)\tilde{I}_h.$$

Define

$$\rho_{sysh} = \frac{Z_{fh}}{Z_{fh} + Z_{sysh}},$$

$$\rho_{fh} = \frac{Z_{sysh}}{Z_{fh} + Z_{sysh}}.$$

Then

$$\tilde{I}_{sysh} = \rho_{sysh}\tilde{I}_h,$$
$$\tilde{I}_{fh} = \rho_{fh}\tilde{I}_h.$$

Because of

$$\tilde{I}_h = \tilde{I}_{sysh} + \tilde{I}_{fh}$$
$$\rho_{sysh} + \rho_{fh} = 1.$$

Note that $\rho_{sysh}$ and $\rho_{fh}$ are complex quantities. It is desirable that $\rho_{sysh}$ be small at the various occurring harmonics. Typical values for a series tuned filter are (at the tuned frequency $hf_1$)

$$\rho_{sysh} = 0.045 \angle -80.6°, \text{ (for series tuned filters)}$$
$$\rho_{fh} = 0.994 \angle +2.6°, \text{ (for series tuned filters)}.$$

Parallel resonances occur between $Z_{fh}$ and $Z_{sysh}$ if $\rho_{sysh}$ and $\rho_{fh}$ are large at the tuned frequency ($hf_1$). Typical values are

$$\rho_{sysh} = 16.67 \angle -92.9°,$$
$$\rho_{fh} = 16.75 \angle +83.6°.$$

The approximate 180° phase difference emphasizes why a parallel resonance cannot be tolerated at a frequency near a harmonic current generated by the converter: a current of the resonance frequency will excite the circuit and a 16.67 pu current will oscillate between the two energy storage units, the system impedance $Z_{sysh}$ and that of the filter capacitors $Z_{fh}$.

A plot of $\rho_{sysh}$ versus $h$ is a useful display of filter performance. Frequently a plot of $\log(\rho_{sysh})$ is more convenient. The harmonic voltage $\tilde{V}_h$ is

$$\tilde{V}_h = Z_{sysh}\tilde{I}_{sysh} = Z_{fh}\tilde{I}_{fh}$$
$$\tilde{V}_h = \rho_{sysh}(Z_{fh} + Z_{sysh})\tilde{I}_{fh}$$
$$\tilde{V}_h = \rho_{sysh}(Z_{fh} + Z_{sysh})\rho_{fh}\tilde{I}_h$$

$$\tilde{V}_h = \rho_{sysh}(Z_{fh} + Z_{sysh})\frac{Z_{sysh}}{(Z_{fh} + Z_{sysh})}\tilde{I}_h$$

$$\tilde{V}_h = \rho_{sysh} \cdot Z_{sysh} \cdot \tilde{I}_h,$$

or

$$\tilde{V}_h = \frac{Z_{fh}}{(Z_{fh} + Z_{sysh})}(Z_{fh} + Z_{sysh})\frac{Z_{sysh}}{(Z_{fh} + Z_{sysh})}\tilde{I}_h$$

$$\frac{Z_{sysh}}{(Z_{fh} + Z_{sysh})} = \rho_{fh}$$

$$\tilde{V}_h = \rho_{fh} \cdot Z_{fh}\tilde{I}_h.$$

### 1.8.6.3 Application Example 1.9: Several Users on a Single Distribution Feeder

**Calculation of Harmonic Current for User #1 (Case A, No Filter).** For user #1:

- $S_{SC} = 350 \text{ MVA}$
- $S_{SCphase} = \dfrac{S_{SC}}{3} = 116.68 \text{ MVA}$
- $S_{SCphase} = I_{SCphase}V_{phase}$
- $I_{SCphase} = \dfrac{S_{SCphase}}{V_{phase}} = \dfrac{116.67 \text{ MVA}}{7.959 \text{ kV}}$
- $I_{SCphase} = 14.65 \text{ kA}$
- $S_{load} = 2.5 \text{ MVA}$
- $S_{loadphase} = \dfrac{S_{load}}{3} = 0.833 \text{ MVA}$
- $I_{loadphase} = \dfrac{S_{loadphase}}{V_{phase}}$
- $I_{loadphase} = \dfrac{0.833 \text{ MVA}}{7.959 \text{ kV}} = 104.7 \text{ A}.$

For this user, there is a 25% static power converter (SPC) load:

- $I_{loadSPC} = \dfrac{104.7 \text{ A}}{4} = 26.2 \text{ A}$
- $R_{SC} = \dfrac{I_{SCphase}}{I_{loadphase}} = \dfrac{14.65 \text{ kA}}{104.7 \text{ A}} = 140.$

Therefore, harmonic currents for the six-pulse static power converter of user #1 are

- $I_{5[A]} = 26.2(0.192) = 5.03 \text{ A} \rightarrow$
  $$I_{5[\%]} = \frac{I_{5[A]}}{I_{loadphase}}(100\%) = \frac{5.03 \text{ A}}{104.7 \text{ A}}(100\%) = 4.8\%$$
- $I_{7[A]} = 26.2(0.132) = 3.46 \text{ A} \rightarrow$
  $$I_{7[\%]} = \frac{I_{7[A]}}{I_{loadphase}}(100\%) = \frac{3.46}{104.7} = 3.3\%$$
- $I_{11[A]} = 1.91 \text{ A} \rightarrow I_{11[\%]} = 1.82\%$
- $I_{13[A]} = 1.49 \text{ A} \rightarrow I_{13[\%]} = 1.42\%$

**Calculation of Harmonic Current for User #2 (Case A, No Filter).** For user #2:

- $S_{SC} = 300 \text{ MVA}$
- $S_{SCphase} = \dfrac{S_{SC}}{3} = 100 \text{ MVA}$

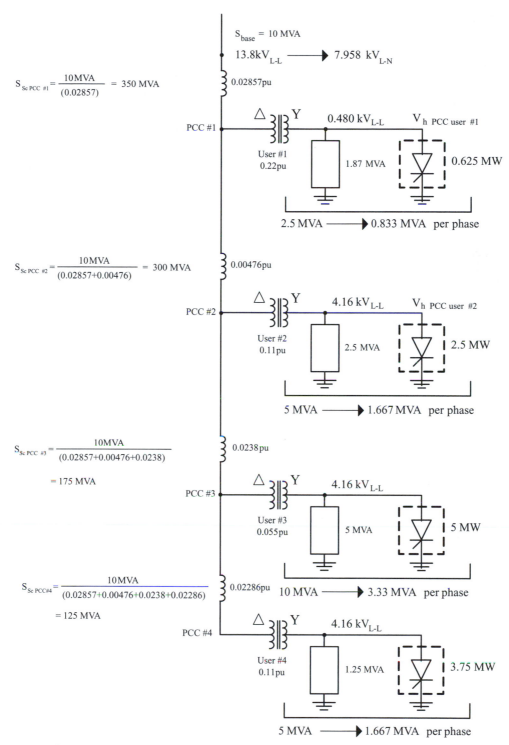

$$S_{Sc\ PCC\ \#1} = \frac{10\,\text{MVA}}{(0.02857)} = 350\ \text{MVA}$$

$S_{base} = 10$ MVA

$13.8\,\text{kV}_{L-L} \longrightarrow 7.958\ \text{kV}_{L-N}$

0.02857 pu

PCC #1

Y

$0.480\ \text{kV}_{L-L}$  $V_{h\ PCC\ user\ \#1}$

User #1
0.22 pu

1.87 MVA

0.625 MW

2.5 MVA $\longrightarrow$ 0.833 MVA  per phase

$$S_{Sc\ PCC\ \#2} = \frac{10\,\text{MVA}}{(0.02857+0.00476)} = 300\ \text{MVA}$$

0.00476 pu

PCC #2

Y

$4.16\ \text{kV}_{L-L}$  $V_{h\ PCC\ user\ \#2}$

User #2
0.11 pu

2.5 MVA

2.5 MW

5 MVA $\longrightarrow$ 1.667 MVA  per phase

$$S_{Sc\ PCC\ \#3} = \frac{10\,\text{MVA}}{(0.02857+0.00476+0.0238)}$$
$$= 175\ \text{MVA}$$

0.0238 pu

PCC #3

Y

$4.16\ \text{kV}_{L-L}$

User #3
0.055 pu

5 MVA

5 MW

10 MVA $\longrightarrow$ 3.33 MVA  per phase

$$S_{Sc\ PCC\#4} = \frac{10\,\text{MVA}}{(0.02857+0.00476+0.0238+0.02286)}$$
$$= 125\ \text{MVA}$$

0.02286 pu

PCC #4

Y

$4.16\ \text{kV}_{L-L}$

User #4
0.11 pu

1.25 MVA

3.75 MW

5 MVA $\longrightarrow$ 1.667 MVA  per phase

**FIGURE E1.9.1** Overall one-line diagram of the distribution system feeder containing four users with six-pulse converters [64].

- $V_{phase} = 7.9585 \text{ kV}$
- $I_{SCphase} = \dfrac{S_{SCphase}}{V_{phase}} = \dfrac{100\,\text{MVA}}{7.9585\,\text{kV}} = 12.56\,\text{kA}$
- $S_{load} = 5 \text{ MVA}$
- $S_{loadphase} = 1.667 \text{ MVA}$
- $I_{loadphase} = \dfrac{S_{loadphase}}{V_{phase}}$
- $I_{loadphase} = 209.5 \text{ A}.$

The 50% static power converter (SPC) load of this user yields:

- $I_{loadSPC} = 104.73 \text{ A}$
- $R_{SC} = \dfrac{I_{SCphase}}{I_{loadphase}} = \dfrac{12560\,\text{A}}{209.5\,\text{A}} = 59.95$
- $R_{SC} \approx 60.$

Therefore, harmonic currents for the six-pulse static power converter of user #2 are

- $I_{5[A]} = 104.73(0.192) = 20.11\text{A} \rightarrow$
  $I_{5[\%]} = \dfrac{I_{5[A]}}{I_{loadphase}}(100\%) = \dfrac{20.11A}{209.5A}(100\%) = 9.6\%,$
- $I_{7[A]} = 104.73(0.132) = 13.82\text{A} \rightarrow$
  $I_{7[\%]} = \dfrac{I_{7[A]}}{I_{loadphase}}(100\%) = \dfrac{13.82A}{209.5A}(100\%) = 6.6\%,$
- $I_{11[A]} = 7.64 \text{ A} \rightarrow I_{11[\%]} = 3.65\%$
- $I_{13[A]} = 5.96 \text{ A} \rightarrow I_{13[\%]} = 2.9\%$

**Calculation of Harmonic Voltages $V_h$ (Case A, No Filter).** The harmonic equivalent circuit of Fig. E1.9.1 in pu and ohms is shown by Fig. E1.9.2a and E1.9.2b, respectively. Using the harmonic equivalent circuits of Fig. E.1.9.2:

- $S_{base} = 10 \text{ MVA}$
- $S_{basephase} = \dfrac{10}{3}\text{MVA}$
- $I_{base} = \dfrac{S_{basephase}}{V_{phase}} = \dfrac{10\,\text{MVA}}{3(7.9585\,\text{kV})}$

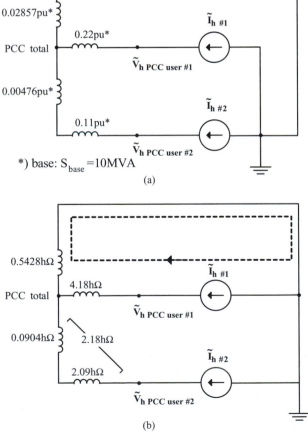

(a)

(b)

**FIGURE E1.9.2** Harmonic equivalent circuit of Fig. E1.9.1 with impedances expressed in (a) pu without harmonic order h, (b) ohms.

- $I_{base} = 418.83$ A
- $Z_{base} = \dfrac{V_{phase}}{I_{base}} = \dfrac{V_{base}}{I_{base}}$

Note that $V_{base} = V_{phase}$; therefore,

$$Z_{base} = \frac{7.9585}{418.83} = 19\,\Omega.$$

Voltage harmonics are computed using the total harmonic currents. For example, for the fifth harmonic:

$$\left.\begin{array}{l} I_{5\#1} = 5.03\,\text{A} \\ I_{5\#2} = 20.11\,\text{A} \end{array}\right\} I_{5\#1} + I_{5\#2} = 25.14\,\text{A}.$$

The fifth voltage harmonic amplitudes for users 1 and 2 are

- $V_{5PCCuser\#1} = 4.18(5)(5.03) + 0.5428(5)(25.14)$
  $V_{5PCCuser\#1} = 173.38$ V
  $V_{5PCCuser\#1} = 2.18\%$,
- $V_{5PCCuser\#2} = 2.18(5)(20.11) + 0.5428(5)(25.14)$
  $V_{5PCCuser\#2} = 287.46$ V
  $V_{5PCCuser\#2} = 3.61\%$.

The total fifth harmonic voltage is

$$V_{5PCCtot} = 0.5428(5)(25.14) = 68.26\,\text{V}$$

$$V_{5PCCtot} = \frac{68.26}{7.96\,\text{kV}}(100\%) = 0.86\%.$$

## CHAPTER 2 SOLUTIONS

### 2.2.4.3 Application Example 2.3: Application of the Direct-Loss Measurement Technique to a Single-Phase Transformer: Solution

Transformer losses are

$$P_{loss} = P_1 - P_2 = \frac{1}{T}\int_0^T v_1 i_1\,dt - \frac{1}{T}\int_0^T v_2' i_2'\,dt. \qquad \text{(E2.3-1)}$$

Rearrangement of terms results in

$$P_{loss} = \frac{1}{T}\int_0^T i_1(v_1 - v_2')\,dt + \frac{1}{T}\int_0^T v_2'(i_1 - i_2')\,dt, \qquad \text{(E2.3-2)}$$

$$P_{series} = \frac{1}{T}\int_0^T i_1(v_1 - v_2')\,dt, \qquad \text{(E2.3-3)}$$

$$P_{shunt} = \frac{1}{T}\int_0^T v_2'(i_1 - i_2')\,dt. \qquad \text{(E2.3-4)}$$

Note that $P_{loss}$ consists of the sum (not the difference) of two components where $(v_1 - v_2')$ and $(i_1 - i_2')$ can be accurately measured and calibrated. Thus the difference of $P_{in}$ and $P_{out}$ is not computed as for the indirect, conventional method of Section 2.2.4.1. This method can be applied to polyphase systems as discussed in [8, 9, in Chapter 2]. The two available sampling methods – series and parallel sampling – are addressed in Appendix 1.

### 2.3.3.1 Application Example 2.4: Sensitivity of K- and $F_{HL}$-Factors and Derating of 25 kVA Single-Phase Pole Transformer with Respect to the Number and Order of Harmonics: Solution

a) The Fourier analysis program is listed in Appendix A2.1.
b) Output of the Fourier program for a square wave is listed in Appendix A2.2.
c) Taking into account the Nyquist criterion one can compute based on 42 sampled points per period the harmonics up to the 27th order. These harmonics are listed in Table E2.4.1 and the resulting Fourier approximation is shown in Fig. E2.4.1.
d) Taking into account the harmonics up to the 27th order (Table E2.4.2) one obtains for the K-factor: $K_{27} = 9.86$, and for the $F_{HL}$-factor: $F_{HL27} = 8.22$. Thereby it is assumed that the 25 kVA single-phase pole transformer [11] is loaded only by non-linear loads having square wave currents. Using these factors and knowing the $P_{EC-R}$, one can compute the derating of a transformer. The $P_{EC-R}$ can be obtained from the manufacturer or it can be measured [11, in Chapter 2]. For $P_{EC-R} = 0.0186$ pu one obtains using the K-factor, Eq. 2-25 the derating:

$$I_{max27K\text{-}factor}(pu) = 0.94,$$

and the derating based on the $F_{HL}$-factor, with Eqs. 2-24 and Eq. 2-26:

$$I_{max27FHL\text{-}factor}(pu) = \sqrt{\frac{(1 + P_{EC-R})}{1 + F_{HL}P_{EC-R}}} = 0.94.$$

**Conclusion.** The calculation of the derating of transformers with the K- and $F_{HL}$-factors leads to the same result provided the same number and order of harmonics are considered.

To show that the K- and $F_{HL}$-factors and, therefore the derating calculation, critically depend on the number and order of harmonics (used in their calculation), in the following the number of harmonics is

**TABLE E2.4.1** Spectrum of Square Wave

| $I_{DC}$ (pu) | $I_1$ (pu) | $I_2$ (pu) | $I_3$ (pu) | $I_4$ (pu) | $I_5$ (pu) | $I_6$ (pu) |
|---|---|---|---|---|---|---|
| 0 | 1.27 | 0.04 | 0.42 | 0.04 | 0.25 | 0.04 |
| $I_7$ (pu) | $I_8$ (pu) | $I_9$ (A) | $I_{10}$ (pu) | $I_{11}$ (pu) | $I_{12}$ (pu) | $I_{13}$ (pu) |
| 0.17 | 0.04 | 0.12 | 0.04 | 0.09 | 0.04 | 0.07 |
| $I_{14}$ (pu) | $I_{15}$ (pu) | $I_{16}$ (pu) | $I_{17}$ (pu) | $I_{18}$ (pu) | $I_{19}$ (pu) | $I_{20}$ (pu) |
| 0.04 | 0.05 | 0.03 | 0.04 | 0.03 | 0.03 | 0.04 |
| $I_{21}$ (pu) | $I_{22}$ (pu) | $I_{23}$ (pu) | $I_{24}$ (pu) | $I_{25}$ (pu) | $I_{26}$ (pu) | $I_{27}$ (pu) |
| 0.04 | 0.03 | 0.03 | 0.02 | 0.03 | 0.02 | 0.03 |

reduced from 27 to 13. Thus one obtains $K_{13} = 6.04$ and $F_{HL13} = 5.08$, resulting in

$$I_{max13} \,(pu) = 0.97.$$

**Conclusion**

$K_{27} > K_{13}$ and $F_{HL27} > F_{HL13}$, therefore $I_{max13}$ (pu) $> I_{max27}$ (pu).

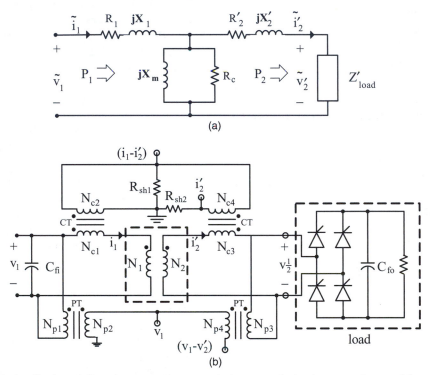

**FIGURE E2.3.1** Application of direct-loss measurement technique to a single-phase transformer; (a) equivalent circuit, (b) measuring circuit.

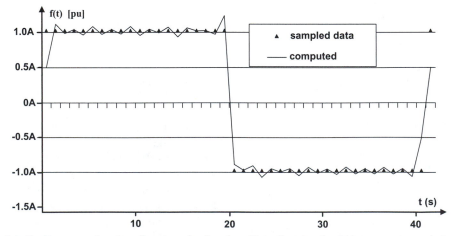

**FIGURE E2.4.1** Fourier approximation of rectangular function. Note: the triangles (▲) represent sampled values and the continuously drawn function represents the Fourier approximation.

**TABLE E2.4.2** Parameters for the Calculation of the K- and $F_{HL}$-Factors (Note the Values for the $I_h$ (pu) Are Truncated)

| h | $I_h$ (pu) | $I_h^2$ | $h^2 I_h^2$ |
|---|---|---|---|
| 1 | 1.0 | 1.0 | 1.0 |
| 2 | 0.03 | 0.001 | 0.004 |
| 3 | 0.33 | 0.109 | 0.984 |
| 4 | 0.03 | 0.001 | 0.016 |
| 5 | 0.20 | 0.040 | 0.969 |
| 6 | 0.03 | 0.001 | 0.036 |
| 7 | 0.13 | 0.018 | 0.878 |
| 8 | 0.03 | 0.001 | 0.063 |
| 9 | 0.09 | 0.009 | 0.723 |
| 10 | 0.03 | 0.001 | 0.099 |
| 11 | 0.07 | 0.005 | 0.608 |
| 12 | 0.03 | 0.001 | 0.143 |
| 13 | 0.06 | 0.003 | 0.513 |
| 14 | 0.03 | 0.001 | 0.194 |
| 15 | 0.04 | 0.002 | 0.349 |
| 16 | 0.02 | 0.0006 | 0.143 |
| 17 | 0.03 | 0.001 | 0.287 |
| 18 | 0.02 | 0.0006 | 0.181 |
| 19 | 0.02 | 0.0006 | 0.201 |
| 20 | 0.03 | 0.001 | 0.397 |
| 21 | 0.03 | 0.001 | 0.438 |
| 22 | 0.02 | 0.0006 | 0.270 |
| 23 | 0.02 | 0.0006 | 0.295 |
| 24 | 0.02 | 0.0002 | 0.143 |
| 25 | 0.02 | 0.0006 | 0.348 |
| 26 | 0.02 | 0.0002 | 0.168 |
| 27 | 0.02 | 0.0006 | 0.407 |
| Σ | | **1.2006** | **9.860** |

### 2.3.3.2 Application Example 2.5: K- and $F_{HL}$-Factors and Their Application to Derating of 25 kVA Single-Phase Pole Transformer Loaded by Variable-Speed Drives: Solution

Taking into account the harmonics up to the 27th order (Tables E2.5.1 and E2.5.2) one obtains for the K-factor: $K_{27} = 50.52$, and for the $F_{HL}$-factor: $F_{HL27} = 19.42$. Using these factors and knowing the value of $P_{EC-R}$ one can compute the derating of a transformer. The $P_{EC-R}$ can be obtained from the manufacturer or it can be measured [11]. For $P_{EC-R} = 0.0186$ pu one obtains the derating using the K-factor, Eq. 2-25:

$$I_{max27K\text{-factor}}(pu) = 0.87,$$

and the derating based on the $F_{HL}$-factor, with Eq. 2-24 and Eq. 2-26:

$$I_{max27FHL\text{-factor}}(pu) = \sqrt{\frac{(1 + P_{EC-R})}{1 + F_{HL} P_{EC-R}}} = 0.87.$$

Note that the influence of the DC component of the current has been disregarded. As is shown in [9], this

**TABLE E2.5.1** Current Spectrum of the Air Handler of a Variable-Speed Drive with (Medium or) 50% of Rated Output Power

| $I_{DC}$ (A) | $I_1$ (A) | $I_2$ (A) | $I_3$ (A) | $I_4$ (A) | $I_5$ (A) | $I_6$ (A) |
|---|---|---|---|---|---|---|
| −0.82 | 10.24 | 0.66 | 8.41 | 1.44 | 7.27 | 1.29 |
| $I_7$ (A) | $I_8$ (A) | $I_9$ (A) | $I_{10}$ (A) | $I_{11}$ (A) | $I_{12}$ (A) | $I_{13}$ (A) |
| 4.90 | 0.97 | 3.02 | 0.64 | 1.28 | 0.48 | 0.57 |
| $I_{14}$ (A) | $I_{15}$ (A) | $I_{16}$ (A) | $I_{17}$ (A) | $I_{18}$ (A) | $I_{19}$ (A) | $I_{20}$ (A) |
| 0.53 | 0.73 | 0.57 | 0.62 | 0.31 | 0.32 | 0.14 |
| $I_{21}$ (A) | $I_{22}$ (A) | $I_{23}$ (A) | $I_{24}$ (A) | $I_{25}$ (A) | $I_{26}$ (A) | $I_{27}$ (A) |
| 0.18 | 0.29 | 0.23 | 0.39 | 0.34 | 0.39 | 0.28 |

**TABLE E2.5.2** Parameters for the Calculation of the K- and $F_{HL}$-Factors (Note the Values for the $I_h$ (pu) Are Rounded)

| h | $I_h$ (pu) | $I_h^2$ | $h^2 I_h^2$ |
|---|---|---|---|
| 1 | 1.024 | 1.0486 | 1.0486 |
| 2 | 0.066 | 0.0044 | 0.0176 |
| 3 | 0.841 | 0.7073 | 6.3657 |
| 4 | 0.144 | 0.0207 | 0.3312 |
| 5 | 0.727 | 0.5285 | 13.2125 |
| 6 | 0.129 | 0.01664 | 0.5990 |
| 7 | 0.490 | 0.2401 | 11.7649 |
| 8 | 0.097 | 0.0094 | 0.6016 |
| 9 | 0.302 | 0.0910 | 7.3710 |
| 10 | 0.064 | 0.0041 | 0.4100 |
| 11 | 0.128 | 0.0164 | 1.9844 |
| 12 | 0.048 | 0.0023 | 0.3312 |
| 13 | 0.057 | 0.0033 | 0.5577 |
| 14 | 0.053 | 0.00281 | 0.5508 |
| 15 | 0.073 | 0.0053 | 1.1925 |
| 16 | 0.057 | 0.0033 | 0.8448 |
| 17 | 0.062 | 0.0038 | 1.0982 |
| 18 | 0.031 | 0.0010 | 0.3240 |
| 19 | 0.032 | 0.0010 | 0.3610 |
| 20 | 0.014 | 0.0002 | 0.0800 |
| 21 | 0.018 | 0.0003 | 0.1323 |
| 22 | 0.029 | 0.0008 | 0.3872 |
| 23 | 0.023 | 0.0005 | 0.2645 |
| 24 | 0.039 | 0.0015 | 0.8640 |
| 25 | 0.034 | 0.0012 | 0.7500 |
| 26 | 0.039 | 0.0015 | 1.0140 |
| 27 | 0.028 | 0.0007 | 0.5103 |
| Σ | | **2.7271** | **52.9690** |

DC component contributes to additional AC losses and an additional temperature rise further increasing the derating.

Measured current wave shape of the air handler operating at variable speed with (medium or) 50% of rated output power with a DC component of $I_{DC} = -0.82$ A is shown in Fig. E2.5.1. The corresponding Fourier approximation (based on the Fourier coefficients of Table E2.5.1) without the DC offset is compared with measured current waveform in Fig. E2.5.2. In Table E2.5.2 the fundamental

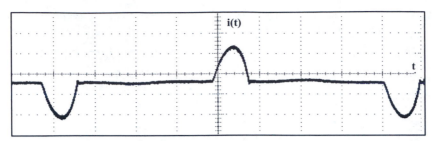

**FIGURE E2.5.1** Measured 60 Hz current wave shape of the air handler operating at variable speed with (medium or) 50% of rated output power.

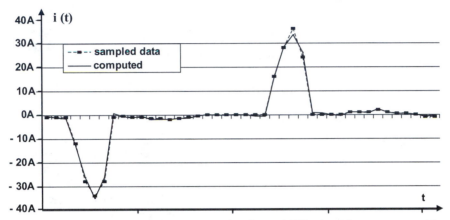

**FIGURE E2.5.2** Fourier analysis of current of air handler at (medium or) 50% of rated output power without DC component of $I_{DC} = -0.82$ A (see Fig. E2.5.1). The rectangles (■) represent measured (sampled) values and the continuous line is the Fourier approximation based on the Fourier coefficients of Table E2.5.2. All ordinate values must be divided by 10 to obtain pu values.

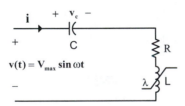

**FIGURE E2.7.1** Ferroresonant nonlinear circuit.

current is not normalized to 1.00 pu; this does not affect the calculation of the K- and $F_{HL}$-factors.

### 2.5.3.1 Application Example 2.7: Calculation of Ferroresonant Currents within Transformers: Solution

a) To show that the equations are true use the circuit of Fig. E2.7.1:

$i(t) = a\lambda^3$, $v(t) = V_{max}\sin(\omega t)$

By inspection of the circuit of Figure E2.7.1:

$$V_{max}\sin(\omega t) = v_c + aR\lambda^3 + \frac{d\lambda}{dt}$$

$$i_C(t) = i(t) = a\lambda^3 = C\frac{dv_c}{dt}, \text{ so } \frac{dv_c}{dt} = \frac{a\lambda^3}{C}$$

Taking the derivative of the circuit equation:

$$\omega V_{max}\cos(\omega t) = \frac{a\lambda^3}{C} + 3aR\lambda^2\frac{d\lambda}{dt} + \frac{d^2\lambda}{dt^2}$$

For $k = 3aR$, $b = \omega V_{max}$, and $C = a$,

$$\frac{d^2\lambda}{dt^2} + k\lambda^2\frac{d\lambda}{dt} + \lambda^3 = b\cos(\omega t)$$

b) To compute numerical values of $\lambda$, $d\lambda/dt$, and $i$ as a function of time for $0 < t < 0.5$ s ($f = 60$ Hz, $R = 0.1$ Ω, $C = a = 100$ $\mu$F, $1000\sqrt{2}$ V $< V_{max} < 20,000\sqrt{2}$ V, all initial conditions = 0) use the following MATLAB codes:

```
%     Matlab code
%
%     getting started
clear all;
close all;
nfig = 0;
%
%     define integration variables and
initial conditions
to = 0;    tf = 0.5;     % time in
seconds
```

```
xo = 0;    vo = 0;    % initial
conditions
zo = [xo vo]';    % initial
conditions in vector form

%Vmax =
[1000*sqrt(2):100:20000*sqrt(2)];
%Vmax = 20000*sqrt(2);
%b = 2*pi*60*Vmax;
%w = 2*pi*60;
%k = 3*100e-6*0.1;

%    Evaluate z numerically
[t,z] = ode23('Hw5eqnfile',[to
tf],zo);
%x = dsolve('D2x+k*x^2*Dx+x^3=b*cos(w
*t)', 'x(0)=0');
%y = rk2('D2y+k*y^2*Dy+y^3=
b*cos(w*t)', [0 0.5], [0 0 0])
%dxdt = [x(2);
b*cos(w*t)-Dx(2)-k*x(1)^2*x(2)-x(1)^3];
%
%    Evaluate x(t) analytically
%       te = linspace(to,tf,200);
%       xe = 0.7737*exp(-3*te).
*cos(2*te-1.1903)-0.3095*cos(5*te-0.3805);
%
%    Plot profiles
        nfig = nfig+1;
        figure(nfig)
        plot(t,z(:,1)),grid
        xlabel('Time
(s)'),ylabel('Inductor Flux Linkage')
        nfig = nfig+1;
        figure(nfig)
        plot(t,z(:,2)),grid
        xlabel('Time
(s)'),ylabel('Voltage')
        nfig = nfig+1;
        figure(nfig)
        plot(t,100e-6*z(:,1).^3),grid
        xlabel('Time
(s)'),ylabel('Current')
%    Plot dx/dt versus x(t)
        nfig = nfig+1;
        figure(nfig)

plot(z(:,2),z(:,1),zo(1),zo(2),'rx'),grid

xlabel('Voltage'),ylabel('Inductor
Flux Linkage')
%    Plot dx/dt versus i(t)
        nfig = nfig+1;
        figure(nfig)
```

```
plot(100e-6*z(:,1).^3,z(:,2),zo(1),zo(2),'rx'),grid

xlabel('Current'),ylabel('Voltage')
```

c) Plots of $\lambda$ versus $d\lambda/dt$ for the values as given in part $b$ are shown in Figs. E2.7.2 and E2.7.3.

d) Plots of $\lambda$ and $d\lambda/dt$ as a function of time for the values given in part $b$ are shown in Figs. E2.7.4 through E2.7.7.

e) Plots of $i$ as a function of time for the values as given in part $b$ are shown in Figs. E2.7.8 and E2.7.9.

f) Plots of voltage versus $d\lambda \neq dt$ $i$ for the values as given in part $b$ are shown in Figs. E2.7.10 and E2.7.11.

### 2.6.1 Application Example 2.8: Calculation of Magnetic Field Strength $\vec{H}$

Magnetic field strength $\vec{H}$ of a wire carrying 100 A at a distance of $R = 1$ $m$ (Fig. E2.8.1) is

$$|\vec{H}| = \frac{I}{2\pi R} = \frac{100\,\text{A}}{2\pi\,1\,\text{m}} = 15.92\,\text{A/m}$$

or

$$|\vec{B}| = \mu_0 |\vec{H}| = 20\mu\text{T}.$$

### 2.7.1.2 Application Example 2.9: Propagation of a Surge through a Distribution Feeder with an Insulator Flashover: Solution

a) The normalized length of a distance traveled by a wave (e.g., span) NL is a function of the frequency $f$, the span length $l$, and the speed of light $c = 2.99792 \cdot 10^8$ m/s [58]: $NL = \frac{l}{c} f$ or the length of a span is $l = c \frac{NL}{f}$. The time to travel the distance $NL \cdot \lambda$, where $\lambda$ is the wavelength, is

$$TD = \frac{NL}{f} = \frac{l}{c} = NL \cdot \lambda .$$

Length of span #1: $l_1 = 2500$ m, length of span #2: $l_2 = 500$ m, length of span #3: $l_3 = 1000$ m, length of span #4: $l_4 = 1000$ m.

b) The PSpice program is listed in Appendix 3. One obtains the results of Figs. E2.9.2 and E2.9.3.

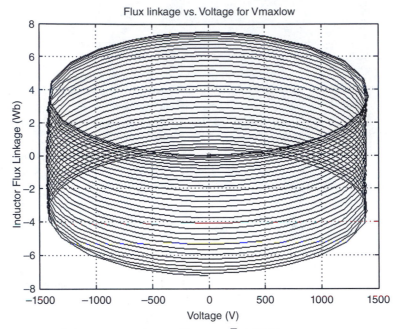

**FIGURE E2.7.2** Inductor flux linkage versus voltage at $V_{max\_low} = \sqrt{2} \cdot 1000$ V.

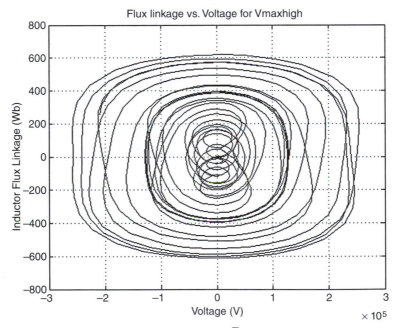

**FIGURE E2.7.3** Inductor flux linkage versus voltage at $V_{max\_high} = \sqrt{2} \cdot 20000$ V.

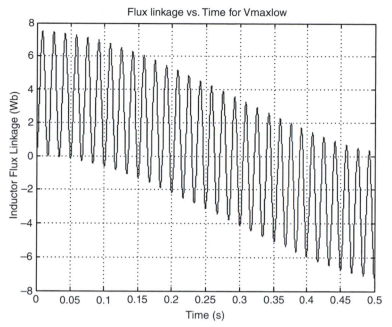

**FIGURE E2.7.4** Inductor flux linkage versus time at $V_{max\_low} = \sqrt{2} \cdot 1000$ V.

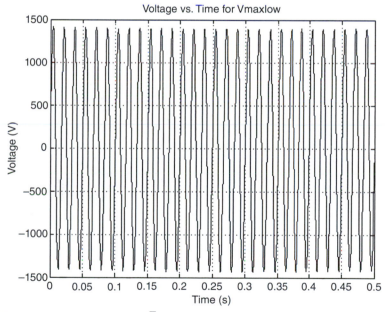

**FIGURE E2.7.5** Voltage versus time at $V_{max\_low} = \sqrt{2} \cdot 1000$ V.

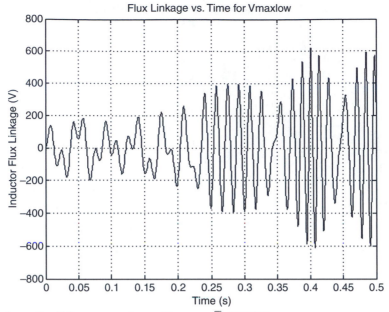

**FIGURE E2.7.6** Inductor flux linkage versus time at $V_{max\_high} = \sqrt{2} \cdot 20000$ V.

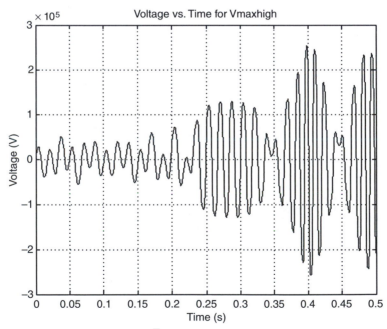

**FIGURE E2.7.7** Voltage versus time at $V_{max\_high} = \sqrt{2} \cdot 20000$ V.

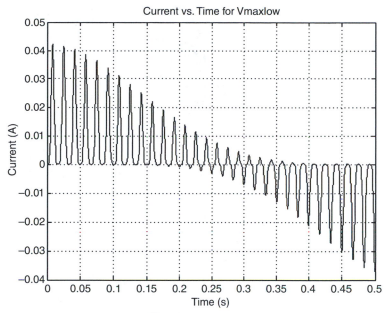

**FIGURE E2.7.8** Current versus time at $V_{max\_low} = \sqrt{2} \cdot 1000$ V.

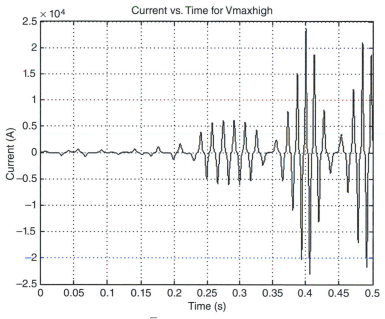

**FIGURE E2.7.9** Current versus time at $V_{max\_high} = \sqrt{2} \cdot 20000$ V.

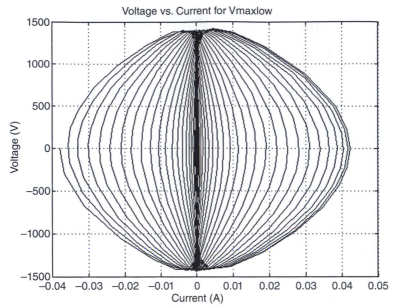

**FIGURE E2.7.10** Voltage versus current at $V_{max\_low} = \sqrt{2} \cdot 1000$ V.

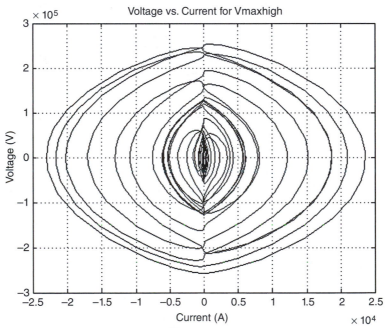

**FIGURE E2.7.11** Voltage versus current at $V_{max\_high} = \sqrt{2} \cdot 20000$ V.

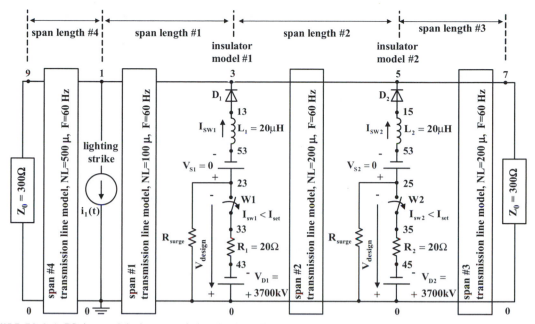

**FIGURE E2.9.1** PSpice model of a transmission line including insulators.

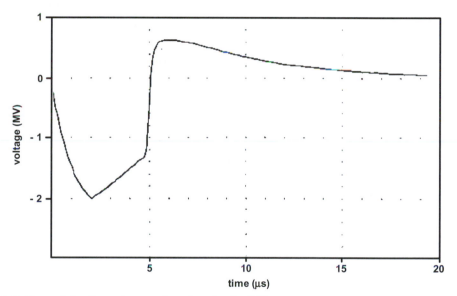

**FIGURE E2.9.2** Transmission line voltage at the point of strike V(1).

### 2.7.1.3 Application Example 2.10: Lightning Arrester Operation: Solution

a) Calculations are performed using PSpice, and the plots of the transmission line voltage (in MV) as a function of time (in $\mu$s) at the point of strike, at

the insulator, and at the lightning arrester are shown in Figs. E2.10.2 and E2.10.3. The PSpice program is listed in Appendix 4.

b) One obtains the results of Figs. E2.10.4 and E2.10.5.

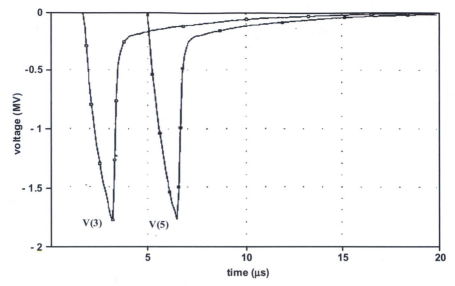

**FIGURE E2.9.3** Transmission line voltage at the first insulator V(3), and second insulator V(5).

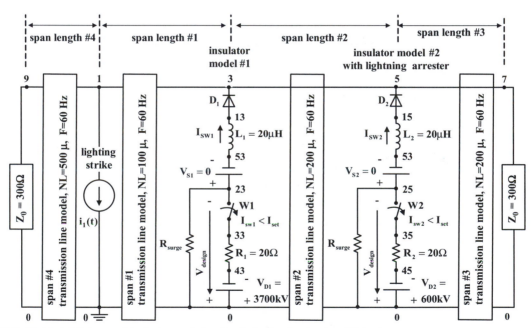

**FIGURE E2.10.1** PSpice model of a transmission line including insulator and lightning arrester.

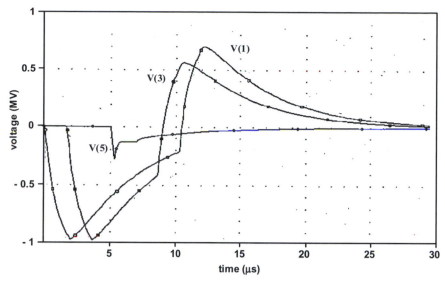

**FIGURE E2.10.2** Transmission line voltage at the point of strike V(1), at the first insulator V(3), and at the lightning arrester V(5).

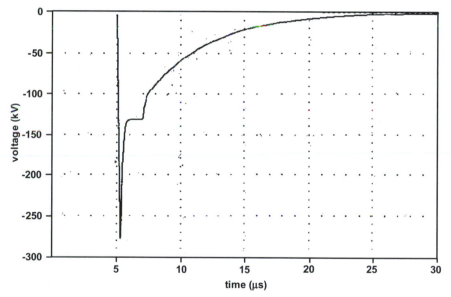

**FIGURE E2.10.3** Transmission line voltage at the lighning arrester (magnified) V(5).

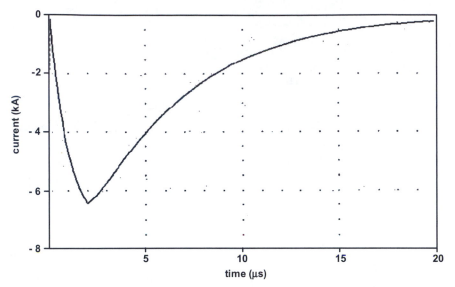

**FIGURE E2.10.4** Transmission line current at the point of strike $I(i_1(t))$.

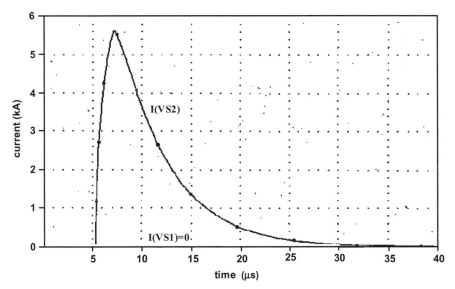

**FIGURE E2.10.5** Transmission line current at the insulator $I(VS1) = 0$ and the lightning arrester $I(VS2)$.

## CHAPTER 3 SOLUTIONS

### 3.1.1 Application Example 3.1: Steady-State Operation of Induction Motor at Undervoltage: Solution

a) Based on the small-slip approximation $s \propto \frac{1}{V^2}$, the new slip at the low voltage is

$$s_{\text{low}} = 0.03(480/430)^2 = 0.0374.$$

b) With the synchronous speed $n_s = \frac{120f}{p} = \frac{120(60)}{6} = 1200\,\text{rpm}$ rpm one calculates the shaft speed at low voltage as

$$n_{m\_\text{low}} = (1 - s_{\text{low}})n_s = (1 - 0.0374)(1200)$$
$$= 1155.14 \text{ rpm}.$$

c) The output power at low voltage is $P_{\text{out}} = \omega_m \cdot T_m$. For constant mechanical torque one gets

$$P_{\text{out\_low}} = \frac{(1 - 0.0374)}{(1 - 0.03)}(100\,\text{hp}) = 99.23\,\text{hp}.$$

d) Due to the relation $T = \frac{P_g}{\omega_{s1}}$ (where $P_g$ is the air-gap power), the torque is proportional to $P_g$, and therefore, for constant torque operation the air-gap power is constant. Thus the rotor loss is

$$P_{\text{cur\_low}} = s_{\text{low}}P_g = \frac{0.0374}{0.03}P_{\text{cur\_rat}} = 1.247\,P_{\text{cur\_rat}}.$$

One concludes that a decrease of the terminal voltage increases the rotor loss (temperature) of an induction motor.

### 3.1.2 Application Example 3.2: Steady-State Operation of Induction Motor at Overvoltage: Solution

a) Based on the small-slip approximation $s \propto \frac{1}{V^2}$, the new slip at the low voltage is

$$s_{\text{high}} = 0.03(480/510)^2 = 0.02657.$$

b) With the synchronous speed $n_s = \frac{120f}{p} = \frac{120(60)}{6} = 1200\,\text{rpm}$ rpm one calculates the shaft speed at low voltage as
$n_{m\_\text{high}} = (1 - s_{\text{high}})n = (1 - 0.02657)(1200) = 1168.11$ rpm.

c) The output power at low voltage is $P_{\text{out}} = \omega_m \cdot T_m$. For constant mechanical torque one gets

$$P_{\text{out\_high}} = \frac{(1 - 0.02657)}{(1 - 0.03)}(100\,\text{hp}) = 100.35\,\text{hp}.$$

d) Due to the relation $T = \frac{P_g}{\omega_{s1}}$, where $P_g$ is the air-gap power, the torque is proportional to $P_g$, and therefore, for constant torque operation the air-gap power is constant. Thus the rotor loss is

$$P_{\text{cur\_high}} = s_{\text{high}}P_g = \frac{0.02657}{0.03}P_{\text{cur\_rat}} = 0.886\,P_{\text{cur\_rat}}.$$

One concludes that an increase of the terminal voltage decreases the rotor loss (temperature) of an induction motor.

### 3.1.3 Application Example 3.3: Steady-State Operation of Induction Motor at Undervoltage and Under-Frequency: Solution

The small-slip approximation results in the torque equation

$$T = \frac{q_1 \cdot V_s^2 \cdot s}{\omega_{s1} \cdot R_r'}.$$

a) Applying the relation to rated conditions (1) one gets

$$T^{(1)} = \frac{q_1 \cdot V_s^{2(1)} \cdot s^{(1)}}{\omega_{s1}^{(1)} \cdot R_r'}.$$

Applying the relation to low-voltage and low-frequency conditions (2) one gets

$$T^{(2)} = \frac{q_1 \cdot V_s^{2(2)} \cdot s^{(2)}}{\omega_{s1}^{(2)} \cdot R_r'}.$$

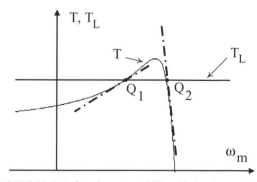

**FIGURE E3.4.1** Steady-state stability of induction machine with constant load torque.

With $T^{(1)} = T^{(2)}$ and the relations

$$\omega_{s1}^{(1)} = \frac{2\pi f^{(1)}}{p/2} \quad \text{and} \quad \omega_{s1}^{(2)} = \frac{2\pi f^{(2)}}{p/2}$$

one obtains

$$s^{(2)} = s^{(1)} \frac{(V_s^{(1)})^2}{(V_s^{(2)})^2} \frac{f^{(2)}}{f^{(1)}} = 0.0367.$$

b) The synchronous speed $n_s^{(2)}$ at $f^{(2)} = 59$ Hz
   is $n_s^{(2)} = \dfrac{120(59)}{6} = 1180\,\text{rpm}$, or the corresponding shaft speed $n_m^{(2)} = n_s^{(2)}(1 - s^{(2)})$
   $= 1137$ rpm.

c) The torque-output power relations are

$$T^{(1)} = \frac{P_{out}^{(1)}}{\omega_m^{(1)}} \quad \text{and} \quad T^{(2)} = \frac{P_{out}^{(2)}}{\omega_m^{(2)}}.$$

With $T^{(1)} = T^{(2)}$ one gets

$$\frac{P_{out}^{(1)}}{\omega_m^{(1)}} = \frac{P_{out}^{(2)}}{\omega_m^{(2)}} \quad \text{or} \quad \frac{P_{out}^{(1)}}{\omega_{s1}^{(1)}(1-s^{(1)})} = \frac{P_{out}^{(2)}}{\omega_{s1}^{(2)}(1-s^{(2)})} \quad \text{or}$$

$$\frac{P_{out}^{(1)}}{\dfrac{120 f^{(1)}}{p}(1-s^{(1)})} = \frac{P_{out}^{(2)}}{\dfrac{120 f^{(2)}}{p}(1-s^{(2)})}.$$

For the reduced output power it follows

$$P_{out}^{(2)} = \frac{(1-s^{(2)})}{(1-s^{(1)})} \frac{f^{(2)}}{f^{(1)}} P_{out}^{(1)} = 97.65\,\text{hp}.$$

d) The copper losses of the rotor are $P_{cur} = q_1(I_r')^2 R_r'$
   and the output power is $P_{out} = q_1(I_r')^2 R_r' \dfrac{(1-s)}{s}$.

The application of the relations to the two conditions (1) and (2) yields

$$P_{out}^{(1)} = q_1(I_r'^{(1)})^2 R_r' \frac{(1-s^{(1)})}{s^{(1)}} \quad \text{and}$$

$$P_{out}^{(2)} = q_1(I_r'^{(2)})^2 R_r' \frac{(1-s^{(2)})}{s^{(2)}} \quad \text{or}$$

$$\frac{(I_r'^{(2)})^2}{(I_r'^{(1)})^2} = \frac{P_{out}^{(2)}}{P_{out}^{(1)}} \frac{s^{(2)}}{s^{(1)}} \frac{(1-s^{(1)})}{(1-s^{(2)})} = 1.20.$$

The rotor loss increase becomes now

$$\frac{P_{cur}^{(2)}}{P_{cur}^{(1)}} = \frac{(I_r'^{(2)})^2}{(I_r'^{(1)})^2} = 1.20.$$

One concludes that the simultaneous reduction of voltage and frequency increases the rotor losses (temperature) in a similar manner as discussed in Application Example 3.1. It is advisable not to lower the power system voltage beyond the levels as specified in standards.

### 3.3.1 Application Example 3.4: Unstable and Stable Steady-State Operation of Induction Machines: Solution

The stability conditions at the two points where the two curves intersect are

$$\mathbf{Q_1:} \quad \left.\begin{array}{l} \dfrac{\Delta T_L}{\Delta \omega_m} = 0 \\[2mm] \dfrac{\Delta T}{\Delta \omega_m} > \text{positive.} \end{array}\right\} 0 < \text{positive} \ggg \textbf{unstable.}$$

$$\mathbf{Q_2:} \quad \left.\begin{array}{l} \dfrac{\Delta T_L}{\Delta \omega_m} = 0 \\[2mm] \dfrac{\Delta T}{\Delta \omega_m} = \text{negative.} \end{array}\right\} 0 > \text{negative} \ggg \textbf{stable.}$$

Therefore, operation at $Q_1$ is unstable and will result in rapid speed reduction. This will stop the operation of the induction machine, while operating point $Q_2$ is stable.

### 3.3.2 Application Example 3.5: Stable Steady-State Operation of Induction Machines: Solution

The stability conditions at the point where the two curves intersect is

$$\mathbf{Q_1:} \quad \left.\begin{array}{l} \dfrac{\Delta T_L}{\Delta \omega_m} = \text{positive} \\[2mm] \dfrac{\Delta T}{\Delta \omega_m} = \text{negative} \end{array}\right\} \ggg \textbf{stable.}$$

Therefore, an induction machine connected to a mechanical load with parabolic torque-speed characteristics as shown in Fig. E3.5.1 will not experience any steady-state stability problems.

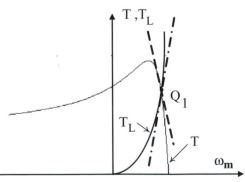

**FIGURE E3.5.1** Steady-state stability of induction machine with nonlinear load torque.

**TABLE E3.6.1** Main-Phase (m) Input Impedance at Single-Phase No-Load Operation as a Function of Input Voltage for R333MC

| $V_m$ (V) | $R_m$ (Ω) | $X_m$ (Ω) | $Z_m$ (Ω) | $L_m$ (mH) |
|---|---|---|---|---|
| 250.3 | 9.65 | 46.6 | 47.6 | 126 |
| **229.9** | **11.2** | **56.9** | **58.0** | **154** |
| 210.3 | 13.2 | 67.8 | 69.1 | 183 |
| 189.8 | 15.1 | 77.3 | 78.7 | 209 |
| 170.6 | 17.0 | 82.4 | 84.1 | 223 |
| 149.5 | 18.6 | 85.4 | 87.4 | 232 |
| 129.8 | 20.5 | 86.3 | 88.7 | 235 |
| 109.2 | 22.6 | 86.7 | 89.5 | 238 |
| 89.9 | 22.5 | 87.8 | 90.6 | 240 |
| 69.4 | 26.3 | 87.0 | 90.9 | 241 |

**TABLE E3.6.2** Auxiliary-Phase (a) Input Impedance at Single-Phase No-Load Operation as a Function of Input Voltage for R333MC

| $V_a$ (V) | $R_a$ (Ω) | $X_a$ (Ω) | $Z_a$ (Ω) | $L_a$ (mH) |
|---|---|---|---|---|
| **298.88** | **30.6** | **94.7** | **99.5** | **264** |
| 279.9 | 33.6 | 106.0 | 111.2 | 295 |
| 259.3 | 36.8 | 117.9 | 123.5 | 328 |
| 239.8 | 39.1 | 126.2 | 132.1 | 351 |
| 219.4 | 41.1 | 132.5 | 138.7 | 368 |
| 199.0 | 43.8 | 135.6 | 142.5 | 378 |
| 179.3 | 45.6 | 137.4 | 144.7 | 384 |
| 160.0 | 46.7 | 138.6 | 146.3 | 388 |
| 139.1 | 49.9 | 139.0 | 147.6 | 392 |
| 120.8 | 50.3 | 140.2 | 148.9 | 395 |
| 100.1 | 54.3 | 139.3 | 149.5 | 397 |

### 3.7.2 Application Example 3.6: Measurement of Harmonics within Yoke (Back Iron) and Tooth Flux Densities of Single-Phase Induction Machines

#### 3.7.2.1 Data of Motor R333MC: Solution

**Stator Parameters**
- 60 Hz AC main-phase resistance: 4.53 Ω,
- main-phase leakage reactance: 5.02 Ω,
- auxiliary-phase resistance: 17.77 Ω,
- auxiliary-phase leakage reactance: 6.70 Ω,
- turns ratio between auxiliary- and main-phase windings: 1.34, and
- angle enclosed by the axes of auxiliary- and main-phase windings: 90.8°.

**Rotor Parameters**
- DC resistance: 4.687 Ω,
- 60 Hz AC resistance: 4.703 Ω,
- 120 Hz AC resistance: 4.751 Ω,
- linear leakage reactance: 5.635 Ω,
- skin-effect coefficient: $k_r = 0.0572$, and
- nonlinear rotor leakage $(\lambda - i)$ characteristic (referred to stator) is shown in Fig. E3.6.1.

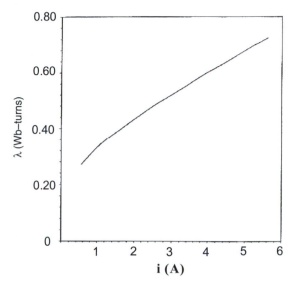

**FIGURE E3.6.1** Nonlinear rotor leakage $(\lambda - i)$ characteristic of R333MC.

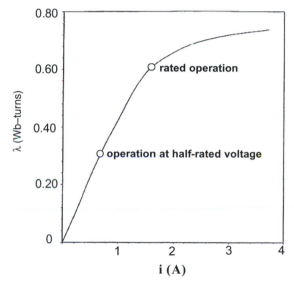

**FIGURE E3.6.2** Nonlinear magnetizing $(\lambda - i)$ characteristic of R333MC.

**Magnetizing Parameters**
- unsaturated magnetizing reactance: 166.1 Ω,
- resistance corresponding to iron-core loss: 580 Ω, and
- nonlinear magnetizing $(\lambda - i)$ characteristic is depicted in Fig. E3.6.2.

**No-Load Input Impedance.** The main- and auxiliary-phase input impedances are measured and recorded in Tables E3.6.1 and E3.6.2, respectively, at single-phase no-load operation as a function of the rms value of the input voltage. Note that in Table E3.6.1 the parameters $Z_m$ and $L_m$ are referred to the main-phase $(m)$ winding, whereas in Table E3.6.2 $Z_a$ and $L_a$ are referred to the auxiliary-phase $(a)$ winding.

**Geometric Data**

- length of iron core: 80 mm,
- width of stator tooth: 3.5 mm,
- height of main-phase yoke: 14.85 mm,
- height of auxiliary-phase yoke: 11.5 mm, and
- iron-core stacking factor: 0.95.

**Measured Waveforms of Induced Voltages and Flux Densities.** Figures E3.6.3a and E3.6.3b show the induced voltage waveforms within search coils wound on the teeth located at the axes of the main-phase and auxiliary-phase windings, respectively.

Figures E3.6.4a and E3.6.4b depict the flux density waveforms associated with the axis of the main-phase winding of all three machines as listed in Table 3.3. Figures E3.6.4c and E3.6.4d depict the flux density waveforms associated with the axis of the auxiliary-phase winding for all three machines.

Figures E3.6.5a, E3.6.5b, E3.6.5c, and E3.6.5d show that the flux densities are approximately a linear function of the input voltage. In Figs. E3.6.6a and E3.6.6b comparison of the flux densities in the teeth and yoke (back iron) are presented for the methods discussed in the prior subsections of Section 3.7. The no-load currents of R333MC and 80664346 are depicted in Figs. E3.6.7a and E3.6.7b, respectively.

### 3.7.2.2 Discussion of Results and Conclusions

The permanent-split capacitor (PSC) motors of the air conditioners 80664346 and R333MC have the same output power and about the same overall

dimensions (length and outer diameter). However, their stator and rotor windings are different. Measurements indicate that the 80664346 motor [36] generates a nearly sinusoidal flux density in the stator teeth (Fig. E3.6.4a), whereas that of the R333MC motor exhibits nonsinusoidal waveform although its $(B_{mt})_{max}$ is smaller. In addition, the no-load current of the 80664346 (Fig. E3.6.7b) is about half that of R333MC (Fig. E3.6.7a), although the latter's maximum flux density in stator teeth and back iron (yoke) is less (Figs. E3.6.4a and E3.6.4c). This indicates that the proper designs of the stator and rotor windings are important for the waveform of the stator tooth flux density and its associated iron-core loss, and for the efficiency optimization of PSC motors.

The yoke flux densities (Figs. E3.6.4b and E3.6.4d) are mostly sinusoidal for all machines tested [36 in Chapter 3]. Although 80664346 has a larger flux density than R333MC, the no-load losses of 80664346 are smaller as can be seen from their stator, rotor, and magnetizing parameters. References [30], [37], and Appendix of [38] address how time-dependent waveforms of the magnetizing current depend on the spatial distribution/pitch of a winding. Figures E3.6.5a through E3.6.5d show that in these three PSC motors saturation in stator teeth and yokes is small as compared to transformers, and therefore the principle of superposition with respect to flux densities and losses can be applied. Figures E3.6.6a and E3.6.6b demonstrate that both the computer sam-

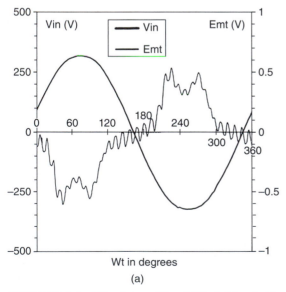

(a)

**FIGURE E3.6.3a** Waveforms of input ($V_{in}$, left-hand side ordinate) and main-phase winding tooth induced voltages ($e_{mt} \equiv E_{mt}$, right-hand side ordinate) of R333MC.

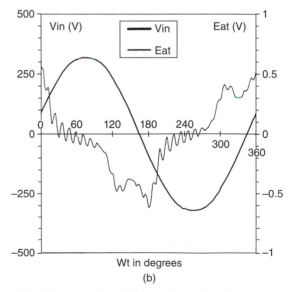

(b)

**FIGURE E3.6.3b** Waveforms of input ($V_{in}$, left-hand side ordinate) and auxiliary-phase winding tooth induced voltages ($e_{at} \equiv E_{at}$, right-hand side ordinate) of R333MC.

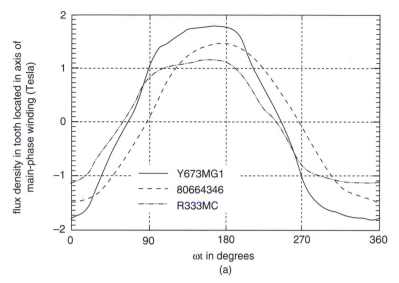

**FIGURE E3.6.4a** Flux density waveforms of tooth ($B_{mt}$) located in axis of main-phase winding at rated voltage of R333MC, 80664346, and Y673MG1.

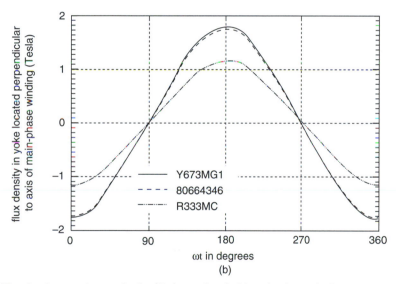

**FIGURE E3.6.4b** Flux density waveforms of yoke ($B_{my}$) associated with main-phase winding at rated voltage of R333MC, 80664346, and Y673MG1.

pling and the oscilloscope methods lead to similar results with respect to maximum flux densities. One concludes that although a single-phase motor (e.g., 80664346) has relatively large flux densities in stator yoke and teeth the no-load current can be relatively small, although it is quite nonsinusoidal. One of the reasons for this is that the power factor at no load is relatively large. The best design of a PSC motor is thus not necessarily based on sinusoidal current wave shapes. Further work should investigate the following:

- the losses generated by alternating flux densities as they occur in the stator teeth, and rotating flux densities as they occur in the stator yoke, and
- the interrelation between time and space harmonics and their effects on loss generation.

### 3.8.3 Application Example 3.7: Computation of Forward-Rotating Subharmonic Torque in a Voltage-Source-Fed Induction Motor: Solution

Parameters of Eqs. 3-47 to 3-54 are listed as a function of the fundamental ($h = 1$) and

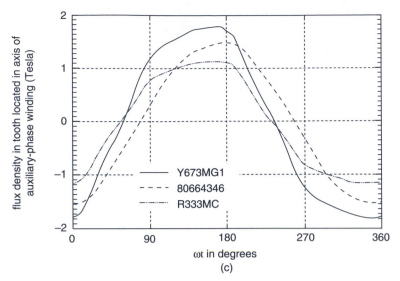

**FIGURE E3.6.4c** Flux density waveforms of tooth ($B_{at}$) located in axis of auxiliary-phase winding at rated voltage of R333MC, 80664346, and Y673MG1.

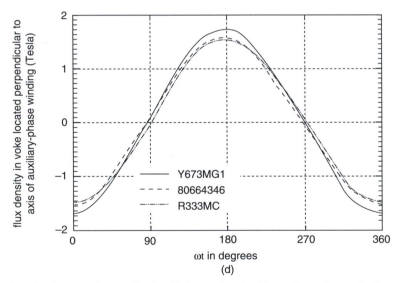

**FIGURE E3.6.4d** Flux density waveforms of yoke ($B_{ay}$) associated with auxiliary-phase winding at rated voltage of R333MC, 80664346, and Y673MG1.

subharmonic $h = 0.1$ in Table E3.7.1. Fundamental torque values $T_{e1}$ are listed in Table E3.7.2 as a function of the fundamental slip $s_1$.

The plot of the positively rotating fundamental torque characteristic $T_{e1} = f(s_1)$ is shown in Fig. E3.7.1 and the positively rotating torque $Te_{0.1} = f(s_{0.1})$ is depicted in Fig. E3.7.2, where the 6 Hz voltage component has an amplitude of 5%, that is, $V_{6Hz} = 0.05 \cdot V_{60Hz} = 0.05(240 \text{ V}) = 12 \text{ V}$.

Subharmonic torque values $T_{e0.1}$ are listed in Table E3.7.3 as a function of the subharmonic slip $s_{0.1}$.

Using Eqs. 3-4d and 3-25 one can compute the fundamental torque $T_{e1}$ and the subharmonic torque $T_{e0.1}$ as a function of the fundamental slip $s_1$ (Table E3.7.4). For example a subharmonic ($h = 0.1$) slip of $s_{0.1} = 1$ corresponds to a fundamental slip of (see Eqs. 3-31b, 3-32, and Table E3.7.3)

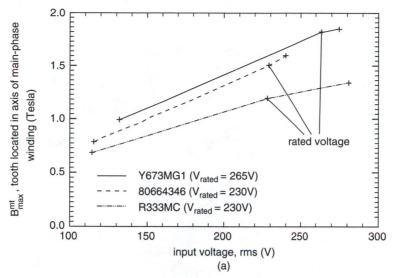

**FIGURE E3.6.5a** Maximum flux density in tooth located in axis of main-phase winding as a function of input voltage of R333MC, 80664346, and Y673MG1.

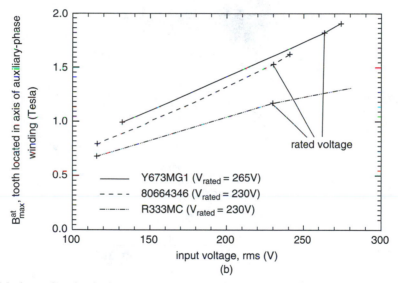

**FIGURE E3.6.5b** Maximum flux density in tooth located in axis of auxiliary-phase winding as a function of input voltage of R333MC, 80664346, and Y673MG1.

$$s_1 = (0.1)(s_{0.1}) + 0.9 = 1.0. \qquad \text{(E3.7-2)}$$

The plot of fundamental and subharmonic torque values $(T_{e1} + T_{e0.1})$ as a function of fundamental slip $(s_1)$ is shown in Fig. E3.7.3.

### 3.8.4 Application Example 3.8: Rationale for Limiting Harmonic Torques in an Induction Machine: Solution

For subharmonic slip $s_{0.1} = 1.0$ one obtains with Eq. E3.7-2 the fundamental slip $s_1 = 1.0$. Correspond-

ingly for subharmonic slip $s_{0.1} = 0.82$ one obtains $s_1 = 0.982$, and for subharmonic slip $s_{0.1} = -0.82$ one obtains $s_1 = 0.818$ (see Table E3.7.3).

One notes that the subharmonic torque at 6 Hz ($h = 0.1$) is relatively large in the generator region ($T_{e0.1} = -4.61$ Nm). If the worst case is considered as defined by Eq. 3-52 the corresponding low-frequency harmonic torque is even larger. This is the reason why the low-frequency harmonic voltages should be limited to less than a few percent (e.g., 2%) of rated voltage.

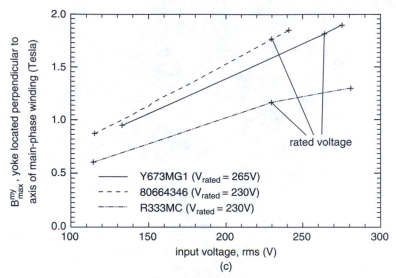

**FIGURE E3.6.5c** Maximum flux density in yoke located in axis of main-phase winding as a function of input voltage of R333MC, 80664346, and Y673MG1.

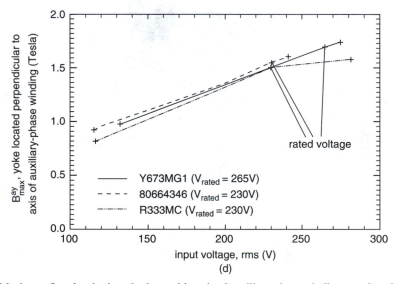

**FIGURE E3.6.5d** Maximum flux density in yoke located in axis of auxiliary-phase winding as a function of input voltage of R333MC, 80664346, and Y673MG1.

### 3.8.5 Application Example 3.9: Computation of Forward-Rotating Subharmonic Torque in Current-Source-Fed Induction Motor: Solution

The parameters of Eqs. 3-53 and 3-54 and the corresponding fundamental and subharmonic torques are listed in Table E3.9.1 through E3.9.3 as a function of the fundamental $h = 1$ and the subharmonic $h = 0.1$.

One notes that the low-frequency harmonic torques at 6 Hz and 5% harmonic current amplitudes are very small (about 0.0063 Nm).

### 3.9.1 Application Example 3.10: Computation of Rotating MMF with Time and Space Harmonics: Solution

a) For $A_1 = 1$ pu, $A_H = 0.05$ pu, $I_{m1} = 1$ pu, $I_{mh} = 0.05$ pu, $H = 5$, and $h = 5$ one obtains the rotating mmf as

$$F_{total} = 1.5\cos(\theta - \omega_1 t) + 0.075\cos(\theta - 5\omega_1 t)$$
$$+ 0.075\cos(5\theta + \omega_1 t) + 0.00375\cos(5\theta - 5\omega_1 t)$$

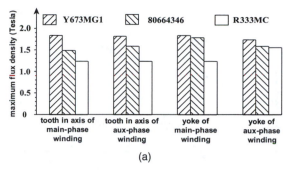

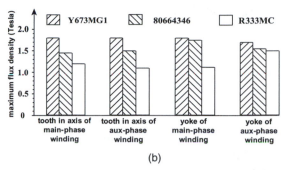

(a)

(b)

**FIGURE E3.6.6a** Comparison of maximum flux densities in teeth located in axis of main- and auxiliary-phase windings, and yokes located in axis of main- and auxiliary-phase windings for Y673MG1, 80664346, and R333MC (computer-sampling method).

**FIGURE E3.6.6b** Comparison of maximum flux densities in teeth located in axis of main- and auxiliary-phase windings, and yokes located in axis of main- and auxiliary-phase windings for Y673MG1, 80664346, and R333MC (oscilloscope method).

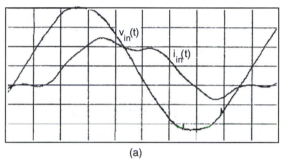

(a)

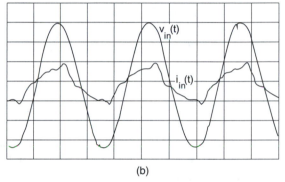

(b)

**FIGURE E3.6.7a** Oscillogram of total input current waveform of R333MC, 2 A/div for current $i_{in}(t)$ and input AC voltage $v_{in}(t)$ with 100 V/div at minimum load at $f = 60$ Hz.

**FIGURE E3.6.7b** Oscillogram of total input current waveform of 80664346, 1 A/div for current $i_{in}(t)$ and input AC voltage $v_{in}(t)$ of 230.5 $V_{rms}$ at minimum load at $f = 60$ Hz.

**TABLE E3.7.1** Maximum Torques for Fundamental (h = 1) and Subharmonic (h = 0.1) Voltages, Where the Fundamental Voltage is 100% and the Subharmonic Voltage Is 5% of the Fundamental Voltage

| $h$ | $V_{shTH}$ (V) | $\pm S_{hmax}$ (−) | $\omega_{msh}$ (rad/s) | $T_{eh\ max}^{mot}$ (Nm) | $T_{eh\ max}^{gen}$ (Nm) |
|---|---|---|---|---|---|
| 1 | 223.34 | 0.312 | 188.49 | 17.98 | −39.89 |
| 0.1 | 9.62 | 0.820 | 18.849 | 0.696 | −4.61 |

**TABLE E3.7.2** Fundamental Torque $T_{e1}$ as a Function of Fundamental Slip $s_1$

| $S_1$ (−) | 2.0 | 1.0 | 0.9 | 0.8 | 0.7 |
|---|---|---|---|---|---|
| $T_{e1}$ (Nm) | 6.78 | 11.61 | 12.44 | 13.37 | 14.40 |
| $S_1$ (−) | 0.6 | 0.5 | 0.4 | 0.3 | 0.2 |
| $T_{e1}$ (Nm) | 15.5 | 16.63 | 17.59 | 17.97 | 16.76 |
| $S_1$ (−) | 0.1 | 0.0344 | 0 | −0.1 | −0.2 |
| $T_{e1}$ (Nm) | 11.8 | 4.99 | 0 | −18.46 | −34.32 |

**TABLE E3.7.3** Subharmonic Torque $T_{e0.1}$ as a Function of Subharmonic Slip $s_{0.1}$

| $S_{0.1}$ (−) | 1.0 | 0.9 | 0.82 | 0.8 | 0.7 | 0.6 |
|---|---|---|---|---|---|---|
| $S_1$ (−) | 1.0 | 0.99 | 0.982 | 0.98 | 0.97 | 0.96 |
| $T_{e0.1}$ (Nm) | 0.687 | 0.693 | 0.696 | 0.695 | 0.69 | 0.676 |
| $S_{0.1}$ (−) | 0.5 | 0.4 | 0.3 | 0.2 | 0.1 | 0 |
| $S_1$ (−) | 0.95 | 0.94 | 0.93 | 0.92 | 0.91 | 0.9 |
| $T_{e0.1}$ (Nm) | 0.648 | 0.602 | 0.528 | 0.415 | 0.247 | 0 |
| $S_{0.1}$ (−) | −0.1 | −0.2 | −0.3 | −0.4 | −0.5 | −0.6 |
| $S_1$ (−) | 0.89 | 0.88 | 0.87 | 0.86 | 0.85 | 0.84 |
| $T_{e0.1}$ (Nm) | −0.353 | −0.842 | −1.488 | −2.28 | −3.12 | −3.88 |
| $S_{0.1}$ (−) | −0.7 | −0.8 | −0.82 | −0.9 | −1 | −1.1 |
| $S_1$ (−) | 0.83 | 0.82 | 0.818 | 0.81 | 0.8 | 0.79 |
| $T_{e0.1}$ (Nm) | −4.40 | −4.60 | −4.61 | −4.53 | −4.29 | −3.95 |
| $S_{0.1}$ (−) | −1.2 | −1.3 | −1.4 | −1.5 | −1.6 | −1.7 |
| $S_1$ (−) | 0.78 | 0.77 | 0.76 | 0.75 | 0.74 | 0.73 |
| $T_{e0.1}$ (Nm) | −3.60 | −3.26 | −2.96 | −2.68 | −2.45 | −2.24 |
| $S_{0.1}$ (−) | −2.0 | −3.0 | −4.0 | −5.0 | −6.0 | −7.0 |
| $S_1$ (−) | 0.70 | 0.60 | 0.50 | 0.40 | 0.30 | 0.20 |
| $T_{e0.1}$ (Nm) | −1.76 | −0.98 | −0.67 | −0.51 | −0.40 | −0.34 |
| $S_{0.1}$ (−) | −8.0 | −9.0 | −10.0 | −11.0 | −12.0 | −13.0 |
| $S_1$ (−) | 0.1 | 0 | −0.1 | −0.20 | −0.30 | −0.40 |
| $T_{e0.1}$ (Nm) | −0.29 | −0.25 | −0.22 | −0.20 | −0.18 | −0.17 |

with the angular velocities

$$\frac{d\theta_{h=1}}{dt} = \omega_1, \quad \frac{d\theta_{h=5}}{dt} = 5\omega_1, \quad \frac{d\theta_{H=5}}{dt} = -\omega_1/5,$$

$$\frac{d\theta_{H=5,h=5}}{dt} = \omega_1,$$

$$(E3.10\text{-}1)$$

b) For $A_1 = 1$ pu, $A_H = 0.05$ pu, $I_{m1} = 1$ pu, $I_{mh} = 0.05$ pu, $H = 13$, and $h = 0.5$ one obtains the rotating mmf as

$$F_{\text{total}} = 1.5\cos(\theta - \omega_1 t) + 0.075\cos(\theta - 0.5\omega_1 t)$$
$$+ 0.075\cos(13\theta - \omega_1 t)$$
$$+ 0.0035\cos(13\theta - 0.5\omega_1 t)$$

with the angular velocities

$$\frac{d\theta_{h=1}}{dt} = \omega_1, \quad \frac{d\theta_{h=0.5}}{dt} = 0.5\omega_1, \quad \frac{d\theta_{H=13}}{dt} = \omega_1/13,$$

$$\frac{d\theta_{H=13,h=0.5}}{dt} = \omega_1/26.$$

$$(E3.10\text{-}2)$$

**TABLE E3.7.4** Fundamental and Subharmonic Torque $(T_{e1} + T_{e0.1})$ as a Function of Fundamental Slip $s_1$

| $S_1$ (–) | 1.0 | 0.99 | 0.982 | 0.98 | 0.97 | 0.96 |
|---|---|---|---|---|---|---|
| $T_{e1}$ (Nm) | 11.61 | 11.69 | 11.75 | 11.77 | 11.85 | 11.93 |
| $T_{e1} + T_{e0.1}$ (Nm) | 12.30 | 12.38 | 12.45 | 12.46 | 12.54 | 12.60 |
| $S_1$ (–) | 0.95 | 0.94 | 0.93 | 0.92 | 0.91 | 0.9 |
| $T_{e1}$ (Nm) | 12.01 | 12.10 | 12.18 | 12.27 | 12.35 | 12.44 |
| $T_{e1} + T_{e0.1}$ (Nm) | 12.66 | 12.70 | 12.71 | 12.68 | 12.60 | 12.44 |
| $S_1$ (–) | 0.89 | 0.88 | 0.87 | 0.86 | 0.85 | 0.83 |
| $T_{e1}$ (Nm) | 12.53 | 12.62 | 12.71 | 12.80 | 12.89 | 13.08 |
| $T_{e1} + T_{e0.1}$ (Nm) | 12.18 | 11.78 | 11.22 | 10.52 | 9.77 | 8.68 |
| $S_1$ (–) | 0.82 | 0.818 | 0.81 | 0.8 | 0.79 | 0.78 |
| $T_{e1}$ (Nm) | 13.18 | 13.19 | 13.27 | 13.37 | 13.47 | 13.57 |
| $T_{e1} + T_{e0.1}$ (Nm) | 8.58 | 8.58 | 8.74 | 9.08 | 9.52 | 9.97 |
| $S_1$ (–) | 0.77 | 0.76 | 0.75 | 0.74 | 0.73 | 0.70 |
| $T_{e1}$ (Nm) | 13.67 | 13.77 | 13.87 | 13.97 | 14.08 | 14.40 |
| $T_{e1} + T_{e0.1}$ (Nm) | 10.41 | 10.81 | 11.19 | 11.52 | 11.84 | 12.64 |
| $S_1$ (–) | 0.60 | 0.50 | 0.40 | 0.30 | 0.20 | 0.10 |
| $T_{e1}$ (Nm) | 15.50 | 16.63 | 17.59 | 17.97 | 16.76 | 11.80 |
| $T_{e1} + T_{e0.1}$ (Nm) | 14.52 | 15.96 | 17.08 | 17.57 | 16.42 | 11.51 |
| $S_1$ (–) | 0 | −0.10 | −0.20 | −0.30 | −0.40 | — |
| $T_{e1}$ (Nm) | 0 | −18.46 | −34.32 | −39.84 | −38.01 | — |
| $T_{e1} + T_{e0.1}$ (Nm) | −0.25 | −18.68 | −34.52 | −40.02 | −38.19 | — |

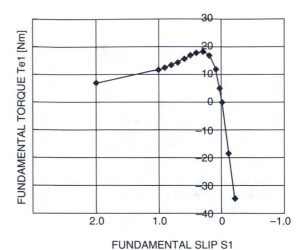

**FIGURE E3.7.1** Fundamental torque $T_{e1}$ as a function of fundamental slip $s_1$.

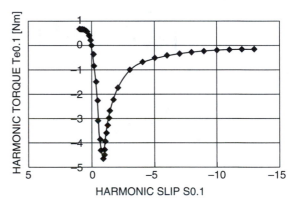

**FIGURE E3.7.2** Subharmonic torque $T_{e0.1}$ as a function of subharmonic slip $s_{0.1}$.

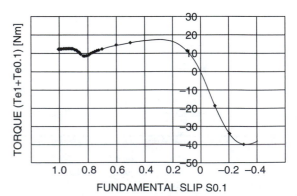

**FIGURE E3.7.3** Fundamental and subharmonic torque $(T_{e1} + T_{e0.1})$ as a function of fundamental slip $s_1$.

**TABLE E3.9.1** Maximum Torques for Fundamental and Subharmonic Currents, Where the Fundamental Current is 1.73 A and the Subharmonic Current is 5% of the Fundamental Current

| $h$ | $|\tilde{V}_{sh}|$ (V)] | $\pm S_{h\,max}$ (−) | $\omega_{msh}$ (rad/s) | $T_{eh\,max}^{mot}$ (Nm) | $T_{eh\,max}^{gen}$ (Nm) |
|---|---|---|---|---|---|
| 1 | 190.3 | 0.044 | 188.49 | 2.53 | −2.35 |
| 0.1 | 0.952 | 0.439 | 18.849 | 0.0063 | −0.0063 |

**TABLE E3.9.2** Torque as a Function of Slip $s_1$ for Fundamental ($h = 1$) Currents

| $S_1$ (−) | 1.0 | 0.9 | 0.8 | 0.7 | 0.6 |
|---|---|---|---|---|---|
| $T_{e1}$(Nm) | 0.221 | 0.246 | 0.276 | 0.316 | 0.368 |
| $S_1$ (−) | 0.5 | 0.4 | 0.3 | 0.2 | 0.1 |
| $T_{e1}$(Nm) | 0.440 | 0.548 | 0.724 | 1.058 | 1.860 |

**TABLE E3.9.3** Torque as a Function of Slip $s_{0.1}$ for Subharmonic ($h = 0.1$) Currents of 5% of the Fundamental Current

| $S_{0.1}$ (−) | 1.0 | 0.9 | 0.8 | 0.7 | 0.6 |
|---|---|---|---|---|---|
| $T_{e0.1}$ (Nm) | 0.0047 | 0.0050 | 0.0053 | 0.0057 | 0.0060 |
| $S_{0.1}$ (−) | 0.5 | 0.4 | 0.3 | 0.2 | 0.1 |
| $T_{e0.1}$ (Nm) | 0.0063 | 0.0063 | 0.0059 | 0.0048 | 0.0027 |

### 3.9.2 Application Example 3.11: Computation of Rotating MMF with Even Space Harmonics: Solution

$$F_{total} = 1.5\cos(\theta - \omega_1 t) + 0.075\cos(\theta - 0.1\omega_1 t)$$

with the angular velocities

$$\frac{d\theta_{h=1}}{dt} = \omega_1, \quad \frac{d\theta_{h=0.1}}{dt} = 0.1\omega_1. \quad (E3.11\text{-}1)$$

### 3.9.3 Application Example 3.12: Computation of Rotating MMF with Noninteger Space Harmonics: Solution

$$\begin{aligned}
F_{total} &= 1.5\cos(\theta - \omega_1 t) + 0.75\cos(\theta - 0.1\omega_1 t) \\
&\quad + 0.025\{\cos(13.5\theta - \omega_1 t) \\
&\quad + \cos(13.5\theta - 60° - \omega_1 t) \\
&\quad + \cos(13.5\theta + 60° - \omega_1 t)\} \\
&\quad + 0.025\{\cos(13.5\theta + \omega_1 t) \\
&\quad + \cos(13.5\theta + 60° + \omega_1 t) \\
&\quad + \cos(13.5\theta - 60° + \omega_1 t)\} \\
&\quad + 0.00125\{\cos(13.5\theta - 0.1\omega_1 t) \\
&\quad + \cos(13.5\theta - 60° - 0.1\omega_1 t) \\
&\quad + \cos(13.5\theta - 60° - 0.1\omega_1 t)\} \\
&\quad + 0.00125\{\cos(13.5\theta - 0.1\omega_1 t) \\
&\quad + \cos(13.5\theta + 60° + 0.1\omega_1 t) \\
&\quad + \cos(13.5\theta - 60° + 0.1\omega_1 t)\},
\end{aligned}$$

with the angular velocities

$$\frac{d\theta_{h=1}}{dt} = \omega_1, \quad \frac{d\theta_{h=0.1}}{dt} = 0.1\omega_1, \quad \frac{d\theta_{H=13.5}}{dt} = \pm\omega_1/13.5,$$

$$\frac{d\theta_{H=13.5,\,h=0.1}}{dt} = \pm\omega_1/13.5.$$

$$(E3.12\text{-}1)$$

### 3.11.1 Application Example 3.13: Calculation of Winding Stress Due to PMW Voltage-Source Inverters: Solution #1 (based on Mathematica)

a) Solution based on simplified equivalent circuit is as follows.

```
Hval=10;
Lval=0;
period=200*^-6;
duty=.5;
Vs[t_]:=If[Mod[t,period]<duty*period,Hval,Lval];
    Plot[Vs[t],{t,0,.001},
    PlotRange→All,AxesLabel→
    {"t","Vs(t)"}]
```

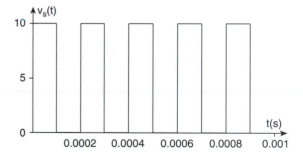

```
R1=25*^-6;R2=25*^-6;R3=25*^-
6;R4=25*^-6;R5=25*^-6;R6=25*^-6;
R7=25*^-6;R8=25*^-6;
L1=1*^-3;L3=1*^-3;L5=1*^-3;L7=1*^-3;
L2=10*^-3;L4=10*^-3;L6=10*^-3;L8=10*^-3;
L15=5*^-6;L37=5*^-6;
L26=10*^-6;L48=10*^-6;
C1=.7*^-12;C3=.7*^-12;C5=.7*^-12;C7=.7*^-12;
C2=7*^-12;C4=7*^-12;C6=7*^-12;C8=7*^-12;
C15=.35*^-12;C37=.35*^-12;
C26=3.5*^-12;C48=3.5*^-12;
G1=1/5000;G3=1/5000;G5=1/5000;G7=1/5000;
G2=1/500;G4=1/500;G6=1/500;G8=1/500;
    Z=1*^-6;
```

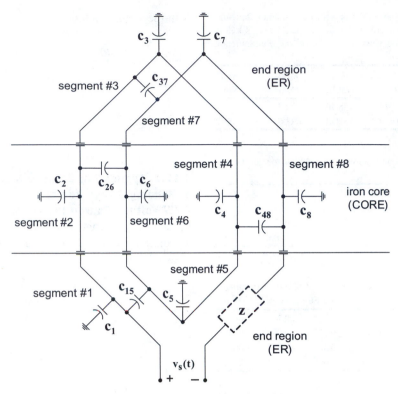

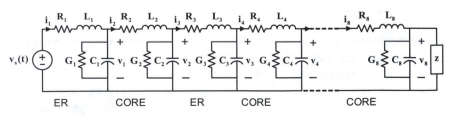

**FIGURE E3.13.1** Winding consisting of two turns, and each turn consists of four segments.

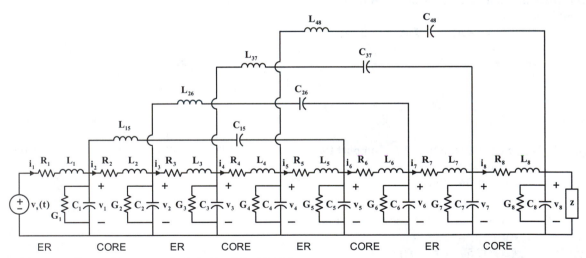

**FIGURE E3.13.2** Detailed equivalent circuit of Fig. E3.13.1.

**FIGURE E3.13.3** Simplified equivalent circuit of Fig. E3.13.1.

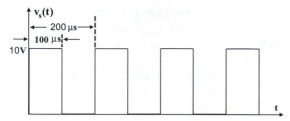

**FIGURE E3.13.4** PWM voltage source $v_s(t)$.

```
I1d=I1'[t]==-R1/L1*I1[t]-V1[t]/L1+Vs[t]/L1;
ic1=I1[0]==0;
V1d=V1'[t]==
I1[t]/C1-G1/C1*V1[t]-I2[t]/C1;
ic2=V1[0]==0;
I2d=I2'[t]==-R2/L2*I2[t]-V2[t]/L2+V1[t]/L2;
ic3=I2[0]==0;
V2d=V2'[t]==I2[t]/C2-G2/C2*V2[t]-I3[t]/C2;
ic4=V2[0]==0;
I3d=I3'[t]==-R3/L3*I3[t]-V3[t]/L3+V2[t]/L3;
ic5=I3[0]==0;
V3d=V3'[t]==I3[t]/C3-G3/C3*V3[t]-I4[t]/C3;
ic6=V3[0]==0;
I4d=I4'[t]==-R4/L4*I4[t]-V4[t]/L4+V3[t]/L4;
ic7=I4[0]==0;
V4d=V4'[t]==I4[t]/C4-G4/C4*V4[t]-I5[t]/C4;
ic8=V4[0]==0;
I5d=I5'[t]==-R5/L5*I5[t]-V5[t]/L5+V4[t]/L5;
ic9=I5[0]==0;
V5d=V5'[t]==I5[t]/C5-G5/C5*V5[t]-I6[t]/C5;
ic10=V5[0]==0;
I6d=I6'[t]==-R6/L6*I6[t]-V6[t]/L6+V5[t]/L6;
ic11=I6[0]==0;
V6d=V6'[t]==I6[t]/C6-G6/C6*V6[t]-I7[t]/C6;
ic12=V6[0]==0;
I7d=I7'[t]==-R7/L7*I7[t]-V7[t]/L7+V6[t]/L7;
ic13=I7[0]==0;
V7d=V7'[t]==I7[t]/C7-G7/C7*V7[t]-I8[t]/C7;
ic14=V7[0]==0;
I8d=I8'[t]==-R8/L8*I8[t]-V8[t]/L8+V7[t]/L8;
ic15=I8[0]==0;
V8d=V8'[t]==I8[t]/C8-V8[t]*(1/(Z*C8)+G8/C8);
ic16=V8[0]==0;
sol1=NDSolve[{I1d,V1d,I
2d,V2d,I3d,V3d,I4d,V4d,
I5d,V5d,I6d,V6d,I7d,V7d
,I8d,V8d,ic1,ic2,ic3,
ic4,ic5,ic6,ic7,ic8,
ic9,ic10,ic11,ic12,
ic13,ic14,ic15,ic16},
{I1[t],V1[t],I2[t],
```

```
V2[t],I3[t],V3[t],
I4[t],V4[t],I5[t],
V5[t],I6[t],V6[t],
I7[t],V7[t],I8[t],
V8[t]},{t,0,.0021},
MaxSteps→100000]
Plot[Vs[t]-Evaluate[V1[t]/.sol1],{t,0,
.002},AxesLabel→{"Time", "Vs - V1"}]
Plot[Evaluate[V1[t]/.sol1] -
Evaluate[V5[t]/.sol1],{t,0,.002},Axes
Label→{"Time", "V1 - V5"}]
Plot[Evaluate[V2[t]/.sol1] -
Evaluate[V6[t]/.sol1],{t,0,.002},Axes
Label→{"Time", "V2 - V6"}]
Plot[Evaluate[V3[t]/.sol1] -
Evaluate[V7[t]/.sol1],{t,0,.002},Axes
Label→{"Time", "V3 - V7"}]
     Plot[Evaluate[V4[t]/.sol1] -
     Evaluate[V8[t]/.sol1],{t,0,
     .002},AxesLabel→{"Time",
     "V4 - V8"}]
```

b) The detailed equivalent circuit (Fig. E3.13.2) generates about the same result as the simplified equivalent circuit (Fig. E3.13.3) and the results are therefore not given.

### Solution #2 (Based on MATLAB)

a) Graphs shown below apply to a short-circuited load, for example, $Z = 0$.

b) The state variable equations from above are still valid for the detailed equivalent circuit. The following four equations need to be added (assuming that the initial current through all the branches is zero):

$$\frac{di_{15}}{dt} = \frac{v_1}{L_{15}} - \frac{i_{15}}{L_{15}C_{15}} - \frac{v_5}{L_{15}}, \quad \text{(E3.13-9)}$$

$$\frac{di_{26}}{dt} = \frac{v_2}{L_{26}} - \frac{i_{26}}{L_{26}C_{26}} - \frac{v_6}{L_{26}}, \quad \text{(E3.13-10)}$$

$$\frac{di_{37}}{dt} = \frac{v_3}{L_{37}} - \frac{i_{37}}{L_{37}C_{37}} - \frac{v_7}{L_{37}}, \quad \text{(E3.13-11)}$$

$$\frac{di_{48}}{dt} = \frac{v_4}{L_{48}} - \frac{i_{48}}{L_{48}C_{48}} - \frac{v_8}{L_{48}}. \quad \text{(E3.13-12)}$$

For an open-circuited winding (e.g., $Z = 1\ M\Omega$) the results are about the same as above.

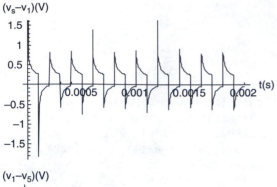

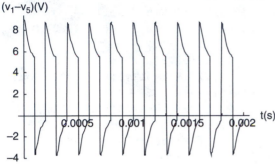

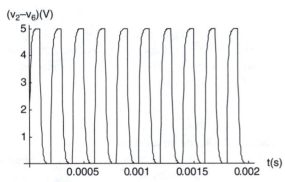

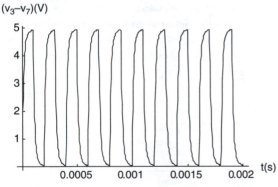

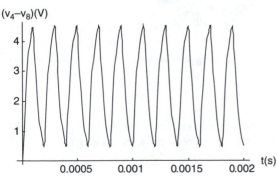

## 3.11.2 Application Example 3.14: Calculation of Winding Stress Due to PWM Current-Source Inverters: Solution

To demonstrate the sensitivity of the induced voltage stress as a function of the rise time of the current ripple, a rise time of less that 5 $\mu$s has been assumed first and the result is plotted in Fig. E3.14.5a,b. The voltage across capacitance C15 (that is, the voltage between the first turn and the second turn at the end region (ER)) is very large.

Figure E3.14.6 illustrates the voltage stress $(v_1 - v_5)$ for the load $Z = 1$ M$\Omega$ at a rise time of 5 $\mu$s. That is the considered winding is open-circuited.

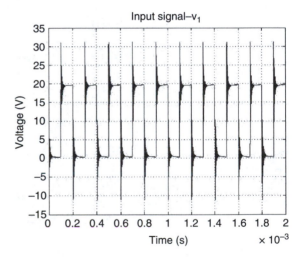

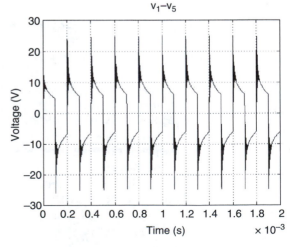

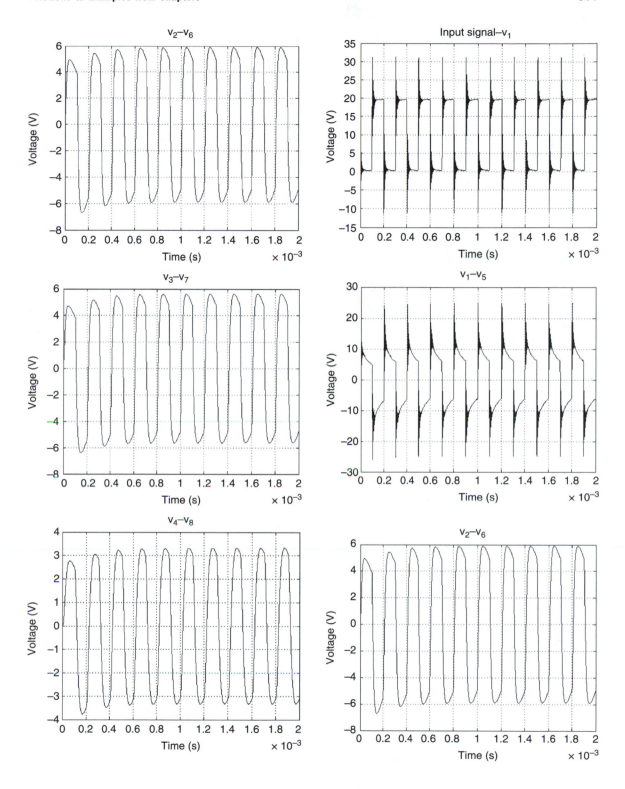

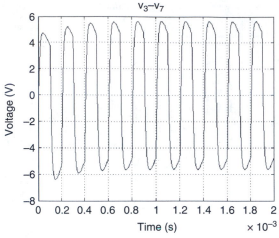

$v_3 - v_7$

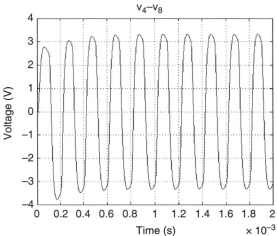

$v_4 - v_8$

a) Solution for simplified equivalent circuit for a short-circuited winding $Z = 1\ \mu\Omega$ is as follows.

```
amp = 5;
period = 200*^-6;
duty=0.5;
trise = 5*^-6;
Is[t_]:=If[Mod[t,period]>duty*period,
0,If[Mod[t,period]<trise,amp/(trise)
*Mod[t,period],amp-amp/(period*
duty-trise)*(Mod[t,period]-trise)]]
Plot[Is[t],{t,0,.001},PlotRange→All,
AxesLabel→{"t","Is(t)"}]
R1=25*^-6;R2=25*^-6;R3=25*^-6;
R4=25*^-6;R5=25*^-6;R6=25*^-6;R7=25*^-6;R8=25*^-6;
L1=1*^-3;L3=1*^-3;L5=1*^-3;L7=1*^-3;
L2=10*^-3;L4=10*^-3;L6=10*^-3;L8=10*^-3;
L15=5*^-6;L37=5*^-6;
L26=10*^-6;L48=10*^-6;
```

```
C1=.7*^-12;C3=.7*^-12;C5=.7*^-12;C7=.7*^-12;
C2=7*^-12;C4=7*^-12;C6=7*^-12;C8=7*^-12;
C15=.35*^-12;C37=.35*^-12;
C26=3.5*^-12;C48=3.5*^-12;
G1=1/5000;G3=1/5000;G5=1/5000;G7=1/5000;
G2=1/500;G4=1/500;G6=1/500;G8=1/500;
Z=1*^-6;
V1d=V1'[t]==Is[t]/C1-G1/C1*V1[t]-I2[t]/C1;
ic2=V1[0]==0;
I2d=I2'[t]==-R2/L2*I2[t]-V2[t]/L2+V1[t]/L2;
ic3=I2[0]==0;
V2d=V2'[t]==I2[t]/C2-G2/C2*V2[t]-I3[t]/C2;
ic4=V2[0]==0;
I3d=I3'[t]==-R3/L3*I3[t]-V3[t]/L3+V2[t]/L3;
ic5=I3[0]==0;
V3d=V3'[t]==I3[t]/C3-G3/C3*V3[t]-I4[t]/C3;
ic6=V3[0]==0;
I4d=I4'[t]==-R4/L4*I4[t]-V4[t]/L4+V3[t]/L4;
ic7=I4[0]==0;
V4d=V4'[t]==I4[t]/C4-G4/C4*V4[t]-I5[t]/C4;
ic8=V4[0]==0;
I5d=I5'[t]==-R5/L5*I5[t]-V5[t]/L5+V4[t]/L5;
ic9=I5[0]==0;
V5d=V5'[t]==I5[t]/C5-G5/C5*V5[t]-I6[t]/C5;
ic10=V5[0]==0;
I6d=I6'[t]==-R6/L6*I6[t]-V6[t]/L6+V5[t]/L6;
ic11=I6[0]==0;
V6d=V6'[t]==I6[t]/C6-G6/C6*V6[t]-I7[t]/C6;
ic12=V6[0]==0;
I7d=I7'[t]==-R7/L7*I7[t]-V7[t]/L7+V6[t]/L7;
ic13=I7[0]==0;
V7d=V7'[t]==I7[t]/C7-G7/C7*V7[t]-I8[t]/C7;
ic14=V7[0]==0;
I8d=I8'[t]==-R8/L8*I8[t]-V8[t]/L8+V7[t]/L8;
ic15=I8[0]==0;
V8d=V8'[t]==I8[t]/C8-V8[t]*(1/(Z*C8)+G8/C8);
ic16=V8[0]==0;

sol1=NDSolve[{V1d,I2d,V2d,I3d,V3d,
I4d,V4d,I5d,V5d,I6d,V6d,I7d,V7d,I8d,
V8d,ic2,ic3,ic4,ic5,ic6,ic7,ic8,ic9,
ic10,ic11,ic12,ic13,ic14,ic15,ic16},
{V1[t],I2[t],V2[t],I3[t],V3[t],I4[t],
V4[t],I5[t],V5[t],I6[t],V6[t],I7[t],
V7[t],I8[t],V8[t]},{t,0,.002},
MaxSteps→100000]

Plot[Evaluate[V1[t]/.sol1] -
Evaluate[V5[t]/.sol1],{t,0,.002},Axes
Label→{"Time", "V1 - V5"}]
Plot[Evaluate[V2[t]/.sol1] -
Evaluate[V6[t]/.sol1],{t,0,.002},Axes
Label→{"Time", "V2 - V6"}]
```

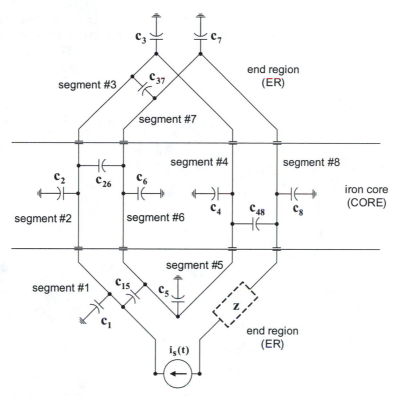

**FIGURE E3.14.1** Winding consisting of two turns, and each turn consists of four segments.

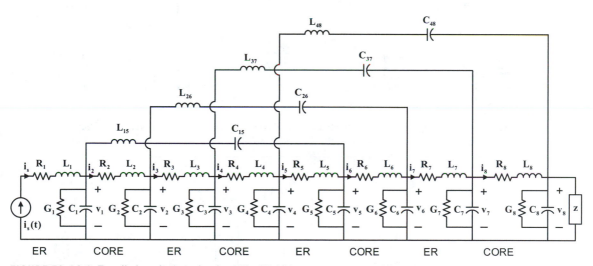

**FIGURE E3.14.2** Detailed equivalent circuit of Fig. E3.14.1.

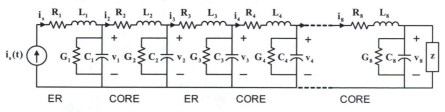

**FIGURE E3.14.3** Simplified equivalent circuit of Fig. E3.14.1.

```
Plot[Evaluate[V3[t]/.sol1] -
Evaluate[V7[t]/.sol1],{t,0,.002},Axes
Label→{"Time", "V3 - V7"}]
Plot[Evaluate[V4[t]/.sol1] -
Evaluate[V8[t]/.sol1],{t,0,.002},Axes
Label→{"Time", "V4 - V8"}]
```

b) The detailed equivalent circuit (Fig. E3.14.2) generates about the same result as the simplified equivalent circuit (Fig. E3.14.3) and the results are therefore not given.

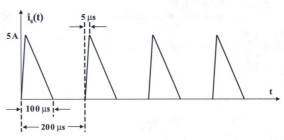

**FIGURE E3.14.4** PWM current source $i_s(t)$.

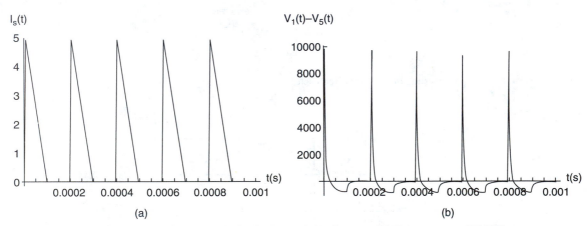

**FIGURE E3.14.5** (a) Current ripple of 5 A of induction motor fed by inverter. (b) Voltage stress of 10 kV between segment #1 (voltage $v_1(t)$) and segment #5 (voltage $v_5(t)$) of induction motor winding; the time t is measured in seconds.

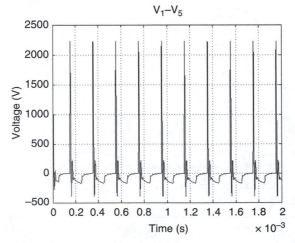

**FIGURE E3.14.6** Voltage ($v_1 - v_5$) for $Z = 1\,M\Omega$ at a rise time of 5 $\mu$s.

$I_s(t)$

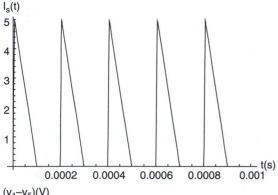

$(v_1-v_5)(V)$

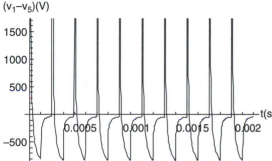

$(v_2-v_6)(V)$

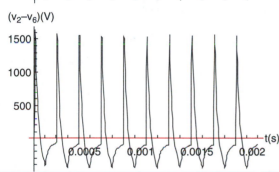

$(v_3-v_7)(V)$

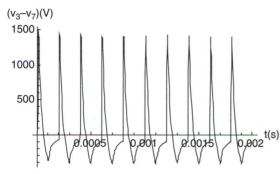

$(v_4-v_8)(V)$

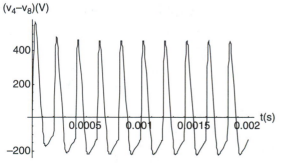

## CHAPTER 4 SOLUTIONS

### 4.2.3 Application Example 4.1: Steady-State Analysis of a Nonsalient-Pole (Round-Rotor) Synchronous Machine: Solution

a) The per-phase equivalent circuit of a nonsalient-pole synchronous machine based on generator notation is shown in Fig. 4.15a. The parameters of the machine are in ohms: $R_a = 0.05 \cdot Z_{base} = (0.05)(17.14\ \Omega) = 0.857\ \Omega$, $X_S = 2 \cdot Z_{base} = 2(17.14\ \Omega) = 34.28\ \Omega$.

b) $S_{rated} = 3 \cdot V_{a\_rated} \cdot I_{a\_rated} = 3(24\ kV)(1.4\ kA)$
$= 100.8\ MVA$. $\qquad$ (E4.1-1)

c) $P_{rated} = S_{rated} \cdot \cos \varphi = 80.64\ MW$. $\qquad$ (E4.1-2)

d) The phasor diagram is given in Fig. E4.1.1.

$$\varphi = \cos^{-1}(0.8) = 0.6435\ rad$$
$$\text{or } 36.87° \text{ lagging} \qquad (E4.1-3)$$

$$|\tilde{V}_{a-rated}| = 1\ pu = 24kV = 3\ inches$$

$$|\tilde{I}_{a-rated}| = 1\ pu = 1.4\ kA = 2.5\ inches$$

$$|\tilde{I}_{a-rated}| \cdot R_a = 1.4kA(0.05)(17.141\ \Omega)$$
$$= 1199.8\ V = 0.15\ inches$$

$$|\tilde{I}_{a-rated}| \cdot X_s = 1.4\ kA(2)(17.141\ \Omega)$$
$$= 47992\ V = 6\ inches.$$

e) For the induced voltage one obtains

$$E_a = \sqrt{\begin{array}{c}(|\tilde{V}_a| + |\tilde{V}_R|\cos\varphi + |\tilde{V}_X|\sin\varphi)^2 \\ + (|\tilde{V}_X|\cos\varphi - |\tilde{V}_R|\sin\varphi)^2\end{array}} = 2.735\,pu$$
$$(E4.1-4)$$

where

$|\tilde{V}_a| = 1\ pu$, $|\tilde{V}_X| = 2\ pu$, $|\tilde{V}_R| = 0.05\ pu$, $\sin \varphi = 0.6$, and $\cos \varphi = 0.8$.

The torque angle $\delta$ is

$$\delta = \cos^{-1}\{(|\tilde{V}_a| + |\tilde{V}_R|\cos \varphi + |\tilde{V}_X|\sin \varphi)/E_a\}$$
$$= 0.611\,rad \text{ or } 35°.$$

f) Rated speed is

$$n_S = 120f/p = 120(60)/4 = 1800\ rpm. \quad (E4.1-5)$$

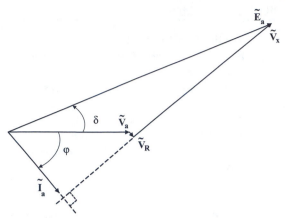

**FIGURE E4.1.1** Per-phase phasor diagram of nonsalient-pole (round-rotor) synchronous motor for lagging (overexcited) operation (drawn to-scale).

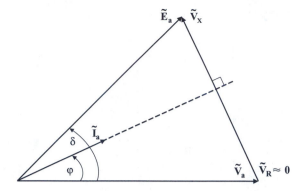

**FIGURE E4.1.2** The per-phase phasor diagram of a nonsalient-pole synchronous motor for leading (underexcited) operation (drawn to-scale).

g) Angular velocity is

$$\omega_S = 2\,\pi n_S/60 = 188.7 \text{ rad/s}. \quad \text{(E4.1-6)}$$

h) Approximate shaft power is

$$P_{shaft} \approx 3(E_a \cdot V_a \cdot \sin \delta)/X_S = 3[(2.735)(24k)(24k)$$
$$\sin(0.611 \text{ rad})]/[(2)(17.14)] = 79.1 \text{ MW}.$$
$$\text{(E4.1-7)}$$

i) The shaft torque $T_{shaft}$ is

$$T_{shaft} = P_{shaft}/\omega_S = 419.14 \text{ kNm}. \quad \text{(E4.1-8)}$$

j) The total ohmic loss of the motor is

$$P_{loss} = 3 \cdot R_a |\tilde{I}_a|^2 = 3(0.05)(17.14)(1.4k)^2 = 5.04 \text{ MW}.$$
$$\text{(E4.1-9)}$$

k) Repeat the above analysis for underexcited (leading current based on generator reference system) displacement power factor $\cos \varphi = 0.8$ leading and $|\tilde{I}_a| = 0.5$ pu.

a) see Fig. 4.15a
b) $S_{rated} = 3 \cdot V_{a\_rated} \cdot I_{a\_rated} = 3(24\text{kV})(0.7\text{kA})$
$\qquad = 50.4 \text{ MVA}.$
c) $P_{rated} = S_{rated} \cdot \cos \varphi = 40.32 \text{ MW}.$
d) The phasor diagram is given in Fig. E4.1.2.

$$\varphi = \cos^{-1}(0.8) = 0.6435 \text{ rad or } 36.87° \text{ leading}$$
$$\text{(underexcited)}$$

$$|\tilde{V}_{a-rated}| = 1 \text{ pu} = 24\text{kV} = 3 \text{ inches}$$

$$|\tilde{I}_{a-rated}| = 0.5 \text{ pu} = 700\text{kA} = 1.25 \text{ inches}$$

$$|\tilde{I}_{a-rated}| \cdot R_a = 700\text{A}(0.05)(17.141 \text{ }\Omega) = 599.9 \text{ V}$$
$$= 0.075 \text{ inches}$$

$$|\tilde{I}_{a-rated}| \cdot X_S = 700\text{A}(2)(17.14 \text{ }\Omega) = 23.996\text{kV}$$
$$= 3 \text{ inches}$$

e) For the induced voltage one obtains

$$E_a = \sqrt{\begin{array}{l}(|\tilde{V}_a| + |\tilde{V}_R|\cos\varphi - |\tilde{V}_X|\sin\varphi)^2 \\ + (|\tilde{V}_X|\cos\varphi + |\tilde{V}_R|\sin\varphi)^2\end{array}}$$
$$= 0.917\,\text{pu or } 22.004 \text{ kV},$$

where

$$|\tilde{V}_a| = 1 \text{ pu}, |\tilde{V}_X| = 1 \text{ pu}, |\tilde{V}_R| = 0.025 \text{ pu}, \sin \varphi = 0.6,$$
$$\cos \varphi = 0.8.$$

The torque angle $\delta$ is

$$\delta = \cos^{-1}\{(|\tilde{V}_a| + |\tilde{V}_R|\cos \varphi - |\tilde{V}_X|\sin \varphi)/E_a\} = 1.095 \text{ rad}$$
$$\text{or } 62.74°.$$

f) Rated speed is

$$n_S = 120f/p = 120 \cdot 60/4 = 1800 \text{ rpm}.$$

g) Angular velocity is

$$\omega_S = 2\pi n_S/60 = 188.7 \text{ rad/s}.$$

h) Approximate shaft power is

$$P_{shaft} \approx 3(E_a \cdot V_a \cdot \sin \delta)/X_S = 3[(0.917)(24k)(24k)$$
$$\sin(1.095 \text{ rad})]/[(2)(17.14)] = 41.09 \text{ MW}.$$

i) The shaft torque $T_{shaft}$ is

$$T_{shaft} = P_{shaft}/\omega_S = 217.75 \text{ kNm}.$$

j) The total ohmic loss of the motor is

$$P_{loss} = 3 \cdot R_a |\tilde{I}_a|^2 = 3(0.05)(17.14)(0.7k)^2 = 1.26 \text{ MW}.$$

## 4.2.4 Application Example 4.2: Calculation of the Synchronous Reactance $X_S$ of a Cylindrical-Rotor (Round-Rotor, Nonsalient-Pole) Synchronous Machine: Solution

a) Calculation of the distribution (breadth) factor of the stator winding $k_{st\ b} = k_{st\ d}$ is as follows. The breadth factor is defined for one phase belt as

$$k_{stb} = \frac{\text{geometric sum} \sum\limits_{i=1}^{8} F_i}{\text{algebraic sum} \sum\limits_{i=1}^{8} F_i}. \qquad \text{(E4-2.1)}$$

It is convenient to evaluate Eq. E4.2-1 graphically, as shown in Fig. E4.2.4. Note that one slot pitch corresponds to $360°/48 = 7.5°$. Thus, one gets from Fig. E4.2.4

$$k_{stb} = 0.96. \qquad \text{(E4-2.2)}$$

b) Calculation of the pitch factor of the stator winding $k_{Stp1}$ is as follows. One stator turn spans

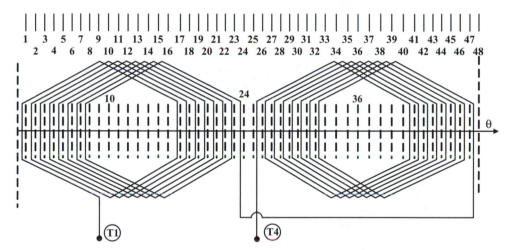

**FIGURE E4.2.1** Stator winding of two-pole, three-phase synchronous machine.

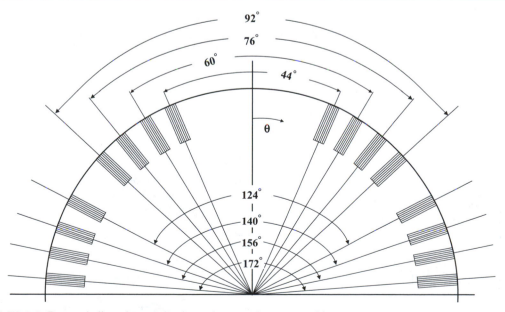

**FIGURE E4.2.2** Rotor winding of two-pole, three-phase synchronous machine.

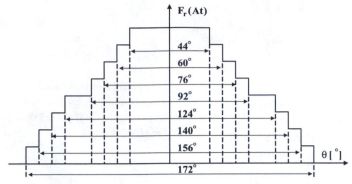

**FIGURE E4.2.3** Rotor mmf $F_r$ of two-pole, three-phase synchronous machine.

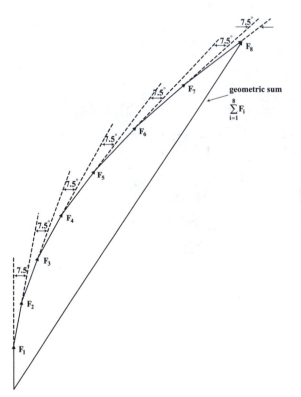

**FIGURE E4.2.4** Graphical determination of the stator breadth or distribution factor $k_{stp1}$.

16 slots out of 24 slots, that is, it has a pitch of (2/3); therefore, one obtains for the fundamental $n = 1$

$$k_{stp1} = \left|\sin\left(\frac{n\beta}{2}\right)\right| = \left|\sin\left(\frac{(1)(120)}{2}\right)\right| = 0.866 \quad (E4\text{-}2.3)$$

or for the 5th harmonic $n = 5$

$$k_{stp5} = \left|\sin\left(\frac{(5)(120)}{2}\right)\right| = 0.866.$$

The 5th harmonic is not attenuated; therefore, the choice of a 2/3 winding pitch is not recommended.

The total fundamental winding factor of the stator winding is now

$$k_{st} = k_{stb} \cdot k_{stp1} = 0.96(0.866) = 0.831.$$

c) Calculate the pitch factor of the rotor (r) winding $k_{r\ p}$ as follows. The fundamental ($n = 1$) pitch-factor of the rotor winding is correspondingly

$$k_{rp} = \frac{\left|\sum_{i=1}^{8}\sin\left(\frac{n\beta_i}{2}\right)\right|}{8} \quad (E4\text{-}2.4)$$

$$k_{rp} = \left|\sin\left(\frac{44°}{2}\right)\right| + \left|\sin\left(\frac{60°}{2}\right)\right| + \left|\sin\left(\frac{76°}{2}\right)\right| + \left|\sin\left(\frac{92°}{2}\right)\right|$$
$$+ \left|\sin\left(\frac{124°}{2}\right)\right| + \left|\sin\left(\frac{140°}{2}\right)\right| + \left|\sin\left(\frac{156°}{2}\right)\right|$$
$$+ \left|\sin\left(\frac{172°}{2}\right)\right|$$

$$k_{rp} = 0.751.$$

d) Determination of the stator flux $\Phi_{Stmax}$ is as follows. The induced voltage in the stator winding per phase is

$$E_{ph} = 4.44 \cdot f \cdot N_{stph} \cdot k_{stb} \cdot K_{stp1} \cdot \Phi_{stmax} \quad (E4\text{-}2.5)$$

or one obtains for the stator flux

$$\Phi_{st\,max} = \frac{E_{L-L}}{\sqrt{3} \cdot 4.44 \cdot f \cdot N_{stph} \cdot k_{stb} \cdot k_{stp1}} \quad (E4\text{-}2.6)$$

with $E_{L-L} = V_{L-L} = 13,800$ V at no load, one gets the flux for no-load condition:

$$\Phi_{st\,max}^{no-load} = \frac{13800}{\sqrt{3}(4.44)(60)(16)(0.96)(0.866)} = 2.25 \frac{Wb}{pole}.$$

(E4-2.7)

e) Compute the area (area$_p$) of one pole:

$$area_p = \frac{\Phi_{st\,max}^{no-load}}{B_{max}} = 2.25\,m^2.$$

(E4-2.8)

f) Compute the field (f) current $I_{fo}$ required at no load:

$$I_{fo} = \frac{B_{st\,max} \cdot p \cdot g \cdot \pi}{4 \cdot \mu_0 \cdot k_{rp} \cdot N_{rtot}} = \frac{1(2)(0.0127)(\pi)}{4(4\pi \times 10^{-7})(0.751)(288)}$$

$$= 73.4\,A.$$

(E4-2.9)

g) Determine the synchronous (S) reactance (per phase value) $X_S$ in ohms. The self-inductance of phase a is

$$L_{aa} = \frac{\lambda_{aa}}{i_a} = \frac{N_{st} \cdot \phi_{st\,max}}{i_a} = \frac{N_{st} \cdot B_{st\,max} \cdot area_p}{i_a}$$

$$= \frac{\mu_0 \cdot 4 \cdot k_{stb} \cdot k_{stp1} \cdot (Nstph)^2 \cdot area_p}{p \cdot g \cdot \pi}$$

(E4-2.10)

$$L_{aa} = \frac{4\pi \times 10^{-7}(4)(0.96)(0.866)(16)^2(2.25)}{2(0.0127)(\pi)}$$

$$= 30.16\,mH.$$

According to Concordia [7 in Chapter 4]

$$L_d = L_{aa} + L_{ab} + 1.5 L_{aa2}. \quad (E4-2.11)$$

Neglecting $L_{aa2}$ and $L_{aa} \approx L_{ab}$ one gets for the synchronous inductance $L_d$

$$L_d \approx 2 \cdot L_{aa} = 60.33\,mH$$

The synchronous reactance of the round-rotor machine is

$$X_S = \omega L_d = \omega L_S = 22.74\,\Omega. \quad (E4-2.12)$$

The base voltage is

$$V_{base} = \frac{V_{L-L}}{\sqrt{3}} = V_{phase} = 7,967\,V. \quad (E4-2.13)$$

The base apparent power is per phase

$$S_{base} = S_{phase} = \frac{S_{rat}}{3} = 5.33\,MVA \quad (E4-2.14)$$

and the base current is

$$I_{base} = I_{phase} = \frac{S_{base}}{V_{base}} = 669\,A. \quad (E4-2.15)$$

h) Find the base impedance $Z_{base}$ expressed in ohms. The base impedance becomes

$$Z_{base} = \frac{V_{base}}{I_{base}} = 11.9\,\Omega. \quad (E4-2.16)$$

i) Express the synchronous reactance $X_S$ in per unit (pu). The per-unit synchronous reactance is now

$$X_S^{pu} = \frac{X_S}{Z_{base}} = \frac{22.74}{11.9} = 1.91\,pu. \quad (E4-2.17)$$

### 4.2.5 Application Example 4.3: dq0 Modeling of a Salient-Pole Synchronous Machine: Solution

a) Balanced Three-Phase Short-Circuit. If the terminals of a synchronous generator are short circuited at rated no-load excitation one obtains the balanced short-circuit currents in per unit [31 in Chapter 4]:

$$i_a(\varphi) = \frac{1}{X_d}\cos(\varphi+\alpha) + \frac{(X_d - X_d')}{X_d X_d'}e^{-\varphi/T_d'}\cos(\varphi+\alpha)$$

$$-\frac{1}{2}(\frac{1}{X_d'}+\frac{1}{X_q})e^{-\varphi/T_a}\cos\alpha - \frac{1}{2}(\frac{1}{X_d'}+\frac{1}{X_q})e^{-\varphi/T_a}$$

$$\cos(2\varphi+\alpha)$$

(E4.3-1)

$$i_b(\varphi) = \frac{1}{X_d}\cos(\varphi+\alpha-120°) + \frac{(X_d - X_d')}{X_d X_d'}e^{-\varphi/T_d'}$$

$$\cos(\varphi+\alpha-120°) + \frac{1}{4}(\frac{1}{X_d'}+\frac{1}{X_q})e^{-\varphi/T_a}\cos\alpha$$

$$+\frac{1}{2}(\frac{1}{X_d'}-\frac{1}{X_q})e^{-\varphi/T_a}\cos(2\varphi+\alpha+\delta)$$

(E4.3-2)

$$i_c(\varphi) = \frac{1}{X_d}\cos(\varphi+\alpha+120°) + \frac{(X_d - X_d')}{X_d X_d'}e^{-\varphi/T_d'}$$

$$\cos(\varphi+\alpha+120°) + \frac{1}{4}(\frac{1}{X_d'}+\frac{1}{X_q})e^{-\varphi/T_a}$$

$$\cos\alpha + \frac{1}{2}(\frac{1}{X_d'}-\frac{1}{X_q})e^{-\varphi/T_a}\cos(2\varphi+\alpha-\delta),$$

(E4.3-3)

where $\tan\delta=\sqrt{3}$, $\varphi=\omega t$.

The field current is

$$i_f(\varphi)=1+\frac{(X_d-X'_d)}{X'_d}(e^{-\varphi/T'_{do}}-\cos\varphi\,e^{-\varphi/T_a}) \quad \text{(E4.3-4)}$$

and the torque is

$$T_{3-phase}=\frac{(X'_d-X_q)}{2X'_dX_q}\cdot\sin(2\varphi)+\frac{1}{X'_d}\sin\varphi. \quad \text{(E4.3-5)}$$

The maximum torque occurs at the angle $\varphi_{max}$ which can be obtained from

$$\cos\varphi_{max}=\frac{1}{4}\left(\frac{X_q}{(X_q-X'_d)}\pm\sqrt{\left(\frac{X_q}{(X'_d-X_q)}\right)^2+8}\right). \quad \text{(E4.3-6)}$$

Figure E4.3.1 illustrates the torque of Eqs. E4.3-5 and E4.3-6.

b) Line-to-Line Short-Circuit. The short-circuit of the two terminals b and c of a synchronous generator results in the line-to-line short-circuit current [31 in Chapter 4]

$$i_b=\frac{\sqrt{3}(\cos\varphi-\cos\varphi_o)}{(X_q+X'_d)-(X_q-X'_d)\cos 2\varphi} \quad \text{(E4.3-7)}$$

or for $\varphi_o=0$

$$i_b=\frac{\sqrt{3}(\cos\varphi-1)}{(X_q+X'_d)-(X_q-X'_d)\cos 2\varphi} \quad \text{(E4.3-8)}$$

and the torque is

$$T_{L-L}=-\frac{2}{\sqrt{3}}(\sin\varphi)i_b+\frac{2}{3}\sin 2\varphi(X'_d-X_q)i_b^2 \quad \text{(E4.3-9)}$$

or

$$T_{L-L}=\frac{-2\sin(\cos\varphi-1)}{[(X_q+X'_d)-(X_q-X'_d)\cos 2\varphi]}+\frac{2\sin 2\varphi(X'_d-X_q)(\cos\varphi-1)^2}{[(X_q+X'_d)-(X_q-X'_d)\cos 2\varphi]^2}. \quad \text{(E4.3-10)}$$

Figure E4.3.2 illustrates the torques of Eqs. E4.3-5, E4.3-6 and E4.3-10.

The open-circuit voltage induced in phase a in case of a short-circuit of line b with line c is

$$e_a=-\cos\varphi-b\left(\frac{[-\sin\varphi\sin 2\varphi+(\cos\varphi-\cos\varphi_o)2\cos 2\varphi]}{[a-b\cos 2\varphi]}-\frac{(\cos\varphi-\cos\varphi_o)(\sin 2\varphi)^2 2b}{[a-b\cos 2\varphi]^2}\right), \quad \text{(E4.3-11)}$$

where

$$a=(X_q+X'_d) \text{ and } b=(X_q-X'_d). \quad \text{(E4.3-12a,b)}$$

The maximum value of $e_a$ occurs for $\varphi_o=0$

Figure E4.3.3 illustrates the induced voltage $e_a$ of Eqs. E4.3-11 and E4.3-12a,b.

c) Line-to-Neutral Short-Circuit. The current when phase a of a synchronous machine at no load is short-circuited is for $\theta=90°+\varphi$ [31 in Chapter 4]

$$i_a=\frac{3(\cos\theta-\cos\theta_o)}{(X_q-X'_d)}\times\frac{1}{\frac{(X'_d+X_q+X_o)}{(X_q-X'_d)}-\cos 2\theta} \quad \text{(E4.3-13)}$$

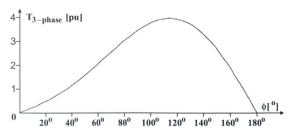

**FIGURE E4.3.1** First half-cycle of torque described by Eqs. E4.3-5 and E4.3-6 (for the parameters $X'_d=0.2909$ pu and $X_q=0.8$ pu).

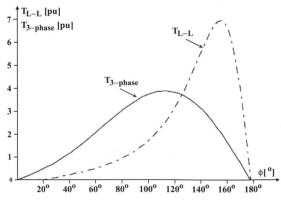

**FIGURE E4.3.2** First half-cycle of torque described by Eqs. E4.3-5, E4.3-6, and E4.3-10 (for the parameters $X'_d=0.2909$ pu and $X_q=0.8$ pu).

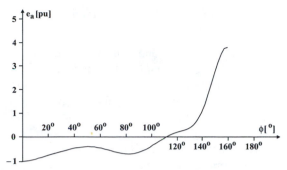

**FIGURE E4.3.3** Induced voltage $e_a$ during first half-cycle at an initial condition $\varphi_o = 0$ described by Eqs. E4.3-11 and E4.3-12a,b (for the parameters $X'_d = 0.2909$ pu and $X_q = 0.8$ pu).

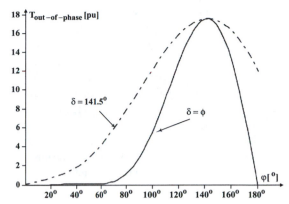

**FIGURE E4.3.4** Out-of-phase synchronizing torque when $\delta = 141.5°$ (worst case) and when $\varphi = \delta$ in Eq. E4.3-18 (for the parameters $X'_d = 0.2909$ pu and $X_q = 0.80$ pu).

or in terms of an infinite series

$$i_a = \frac{3(\cos\theta - \cos\theta_o)}{(X_q - X'_d)} \cdot \frac{2a}{1-a^2}(1 + 2a\cos 2\theta + 2a^2\cos 4\theta + \ldots),$$

(E4.3-14)

where

$$a = \frac{\sqrt{X_q + \dfrac{X_o}{2}} - \sqrt{X'_d + \dfrac{X_o}{2}}}{\sqrt{X_q + \dfrac{X_o}{2}} + \sqrt{X'_d + \dfrac{X_o}{2}}}.$$

(E4.3-15)

Note that $i_a$ must be initially zero, that is, $\varphi = \varphi_o = 0$ or $\theta_o = 90°$.

The fundamental component becomes now

$$i_{a1} = \frac{3\cos\theta}{X'_d + \sqrt{X_q + \dfrac{X_o}{2}} \cdot \sqrt{X'_d + \dfrac{X_o}{2}} + \dfrac{X_o}{2}}$$

(E4.3-16)

or in terms of symmetrical components

$$i_{a1} = \frac{3\cos\theta}{X'_d + X_2 + X_o}.$$

(E4.3-17)

d) Out-of-Phase Synchronization. When synchronizing a generator (either machine or inverter) with the power system the following conditions must be met:

• Phase sequence must be the same;
• The phase angle between the two voltage systems must be sufficiently small; and
• The slip must be small.

Synchronizing out of phase is an important issue because it generates the highest torques and forces (in particular within the end winding of a machine) when paralleled with the power system [31 in Chapter 4]. The torque during out-of-phase synchronization is as a function of time $\varphi = \omega t$ and the angle $\delta$ between the power system's voltage and that of the synchronous machine voltage at the time of paralleling:

$$T_{\text{out-of-phase}} = \frac{1}{X_q}[\sin\delta(1 - \cos\varphi) + (1 - \cos\delta)\sin\varphi]$$
$$\left\{1 + \frac{(X_q - X'_d)}{X'_d}[(1 - \cos\delta)(1 - \cos\varphi)\right.$$
$$\left. - \sin\delta\sin\varphi]\right\}.$$

(E4.3-18)

Note this expression is symmetrical in $\delta$ and $\varphi$, that is, $\delta$ and $\varphi$ can be interchanged without altering the relation. Introducing $\delta = \varphi$ in Eq. E4.3-18, one obtains for the torque

$$T_{\text{out-of-phase}} = \frac{2\sin\delta(1 - \cos\delta)}{X'_d} - \sin\delta\left[\frac{X_q - X'_d}{X_q X'_d}\right]$$
$$[2 + 2\cos\delta - 8(\cos\delta)^2 + 4(\cos\delta)^3].$$

(E4.3-19)

The corresponding field current is

$$i_f(\varphi) = \frac{X_d - X'_d}{X'_d}[1 - \cos\delta + (\cos\delta - 1)\cos\varphi - \sin\delta\sin\varphi]$$
$$+ 1.$$

(E4.3-20)

Figure E4.3.4 illustrates the very large torque if the generator is out-of-phase synchronized with the worst case angle of $\delta = 141.5°$ and when $\varphi = \delta$ in Eq. E4.3-18.

Experimental results [32–34] during out-of-phase synchronization show that for small synchronous machines below 100 kVA a voltage difference of $\Delta V \le \pm 10\%$, a phase angle difference of $\Delta \delta \le \pm 5°$, and a slip of $s \le 1\%$ can be permitted. For large machines these conditions must be more stringent; that is, the voltage difference $\Delta V \le \pm 1\%$, the phase angle difference $\Delta \delta \le \pm 0.5°$, and the slip $s \le 0.1\%$. During the past first analog [32–34 in Chapter 4] and then digital paralleling devices have been employed to connect synchronous generators to the grid. Although these devices are very reliable, once in a while a faulty paralleling occurs.

Figure E4.3.5 depicts the various parameters when out-of-phase synchronization occurs with an angle of $\delta = 255°$ – which is not the worst-case condition (see

Fig. E4.3.4). In this analysis subtransient reactances are taken into account.

- Stator currents $i_a$, $i_b$, and $i_c$ are amplitude modulated and reach maximum values of 4 pu;
- Zero-sequence component current ($i_o = i_a + i_b + i_c$) flowing in the grounded neutral is relatively small and about 0.5 pu;
- Electrical generator ($G$) torque $T_G$ assumes a maximum value of about 4 pu;
- Rotor (torque) angle exceeds 90°;
- Real $P_G$ and reactive $Q_G$ power oscillations occur;
- Slip $s$ reaches values of more than 3 pu; and
- Mechanical torque $T_M$ between generator and turbine is below 1 pu.

Not shown in this graph are the extremely large winding forces in the end region of the stator winding,

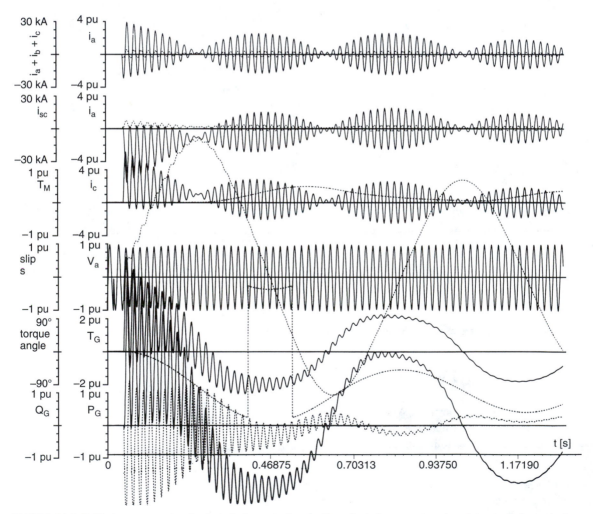

**FIGURE E4.3.5** Various parameters for out-of-phase synchronization of a turbogenerator when the out-of-phase synchronizing angle is $\delta = 255°$.

and the excessive heating of the amortisseur bars and the solid-rotor pole face. The latter issue will be addressed in a later example. In Fig. E4.3.5 the left scale pertains to the dotted signals and the right scale belongs to the full-line signals.

Figure E4.3.6a illustrates the conditions under which a 30 kVA inverter of a wind-power plant was paralleled with the power system [29, 30 in Chapter 4]. The switching frequency and therefore the fundamental (nominally 60 Hz) frequency was phase-locked with the power system's frequency and thus the slip is by definition $s = 0$. However, the voltage differences and the phase differences must be within certain limits. Experiments showed it was advisable to choose the inverter rms value of the line-to-line voltage always somewhat larger (e.g., $V_{L\text{-Linverter}} = 260\ V_{rms}$) than that of the power system's line-to-line voltage (e.g., $V_{L\text{-Lpower system}} = 240\ V_{rms}$). Figure

E4.3.6b depicts the transient current flowing from inverter to power system when inverter is paralleled at an inverter voltage of $V_{L\text{-Linverter}} = 260\ V_{rms}$ at an out-of-phase voltage angel of $\delta = 24°$, where the inverter voltage leads that of the power system. As can be seen, the transient current is more than 200 A.

Figure E4.3.7 illustrates the stator flux distribution when a large synchronous generator is paralleled out-of-phase [21 in Chapter 4]: the stator flux leaves the iron core (stator back iron) and currents are induced in the key bars of the iron-core suspension as well as in the aluminum bars within the dovetails of the iron core. The flux density within the space between the iron core and the frame is 17 mT at a terminal voltage of 1.23 pu. This flux density is sufficient to heat up the solid iron key bars and melt the aluminum bars within the dovetails of the iron core.

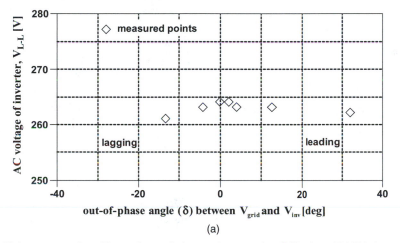

(a)

**FIGURE E4.3.6a** Voltage rms values $V_{L\text{-}L}$ and out-of-phase voltage angles $\delta$ (°) when 30 kVA inverter is out-of-phase synchronized with power system.

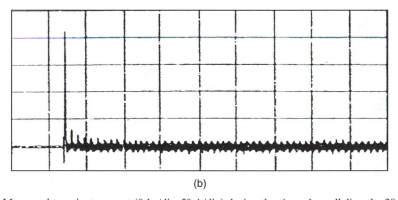

(b)

**FIGURE E4.3.6b** Measured transient current (0.1 s/div, 50 A/div) during the time of paralleling the 30 kVA inverter with the power system at $V_{L\text{-}L\ power\ system} = 240\ V_{rms}$ and out-of-phase synchronizing angle is $\delta = 24°$ (inverter voltage leads the power system voltage).

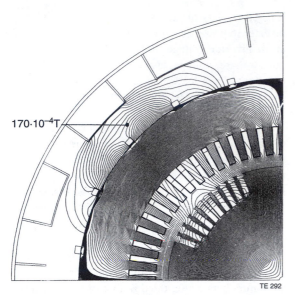

$170 \cdot 10^{-4} T$

TE 292

**FIGURE E4.3.7** Flux distribution when a large generator is out-of-phase synchronized with power system [21 in Chapter 4].

e) Synchronizing and Damping Torques. When a synchronous generator is paralleled with the power system one obtains for the angle perturbation $\Delta \sin \omega t$ from the steady-state angle $\delta_0$ the synchronizing $T_S$ and damping $T_D$ torques in addition to the steady-state torque $T_0$ [31 in Chapter 4]:

$$T = T_0 + T_S(\Delta \sin \omega t) + T_D(\omega \cdot \Delta \cos \omega t), \quad (E4.3\text{-}21)$$

where

$$T_0 = \frac{V_{l-n} \cdot V_{l-n0}}{X_d} \sin \delta_0 + V_{l-n}^2 \left[ \frac{X_d - X_q}{2 X_q X_d} \right] \sin 2\delta_0, \quad (E4.3\text{-}22)$$

$$T_S = \frac{V_{l-n} \cdot V_{l-n0}}{X_d} \cos \delta_0 + V_{l-n}^2 \left( \frac{1}{X_q} - \frac{1}{X_d} \right) \cos 2\delta_0$$
$$+ V_{l-n}^2 \frac{(X_d - X_d')}{X_d X_d'} \cdot \frac{(\omega T_d')^2}{1 + (\omega T_d')^2} (\sin \delta_0)^2, \quad (E4.3\text{-}23)$$

$$T_D = V_{l-n}^2 (\sin \delta_0)^2 \left[ \frac{(X_d - X_d')}{X_d X_d'} \right] \frac{T_d'}{1 + (\omega T_d')^2}, \quad (E4.3\text{-}24)$$

where $T_{d0}' = \dfrac{X_F}{R_F} = \dfrac{X_d}{X_d'} = T_d'.$

f) Torques during Reclosing. In order to minimize power quality problems after short-circuits have occurred, it is common practice to reclose the switches that have been opened due to the short-circuit. Short-circuits can be interrupted within four to six 60 Hz cycles. In the hope that the short-circuit has been cleared, the utilities reclose the previously opened switches after nine to fifteen 60 Hz cycles have elapsed. If the short-circuit still persists, then the fault will be cleared again after additional four to six 60 Hz cycles. At the most three to five reclosing operations are performed. The timing of the reclosing is important: if the timing is unfavorable the mechanical torques of the turbine stages become excessive, but under favorable conditions these mechanical torques can be minimized. Only an accurate dynamic calculation can approximately predetermine the most favorable reclosing times. Transient electromagnetic and shaft torques of synchronous generators can become as high as 5–6 during reclosing [35–39 in Chapter 4]. The magnitudes of the torques depend on the reclosing time. If the reclosing occurs at a time when the electrical torque due to reclosing adds to the transient shaft torque, the resulting mechanical shaft becomes large as compared to the case where the electrical torque due to the reclosing opposes the transient shaft torque; in the latter case the resulting shaft torque will be relatively small. This amplification effect of the mechanical transient torque has its analogy in the "swing effect" one encounters on a playground.

Joyce *et al.* [35, 36 in Chapter 4] have recommended introducing a measure for the theoretical lifetime of a turbine shaft: the occurrence of many high torsional torques reduces the lifetime of a shaft more quickly than the occurrence of low torsional torques. Figure E4.3.8 illustrates a common connection of a power plant to the high-voltage system. One generator feeds two transformers, which are connected to high-voltage feeders each. This is to minimize the outage of a power plant due to external faults such as short-circuits within the power system. The short-circuit is caused by making the impedances $Z_{Fa}$, $Z_{Fb}$, and $Z_{Fc}$ zero. Breaker #2 interrupts the short-circuit and recloses after a preprogrammed time.

Figures E4.3.9a and E4.3.9b depict the various parameters when reclosing with different reclosing times but with the same short-circuit duration (104 ms) occurs. In these graphs the left scale pertains to the dotted signals and the right scale belongs to the full-line signals.

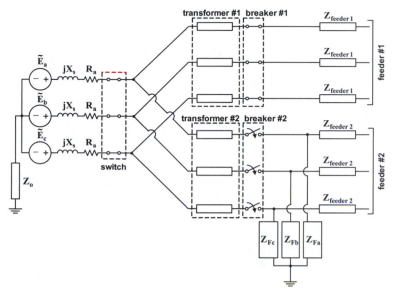

**FIGURE E4.3.8** Large synchronous generator feeding two transmission lines via two transformers.

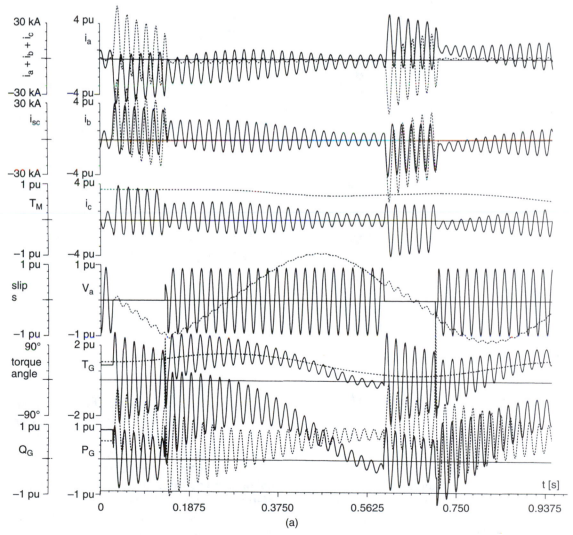

**FIGURE E4.3.9a** Three-phase short-circuit lasting 104 ms followed by clearing of short-circuit for 450 ms and thereafter followed by reclosing whereby the short-circuit has not cleared and lasts 104 ms, thereafter again clearing of short-circuit.

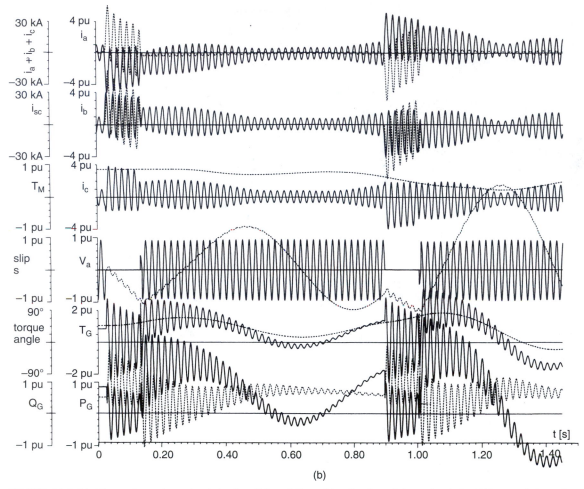

**FIGURE E4.3.9b** Three-phase short-circuit lasting 104 ms followed by clearing of short-circuit for 745 ms and thereafter followed by reclosing whereby the short-circuit has not cleared and lasts 104 ms, thereafter again clearing of short-circuit.

Summarizing one can state the following:

- During the three-phase short-circuit the stator currents $i_a$, $i_b$, and $i_c$ are amplitude modulated and exceed maximum values of 4 pu;
- Zero-sequence component current $(i_o = i_a + i_b + i_c)$ flowing in the grounded neutral is relatively large and exceeds 4 pu;
- Electrical torque $T_G$ assumes a maximum value of about 3 pu;
- Rotor (torque) angle is below 90°;
- Very large (about 2 pu) real $P_G$ and reactive $Q_G$ power swings occur;
- Maximum slip reaches values of more than 1 pu;
- Mechanical torque $T_M$ between generator and turbine is below 1 pu, but it has an alternating component causing mechanical stress; and
- In both cases the stability of the synchronous generators is maintained due to the torque angle $\delta$ being less than 90°.

Not shown is the graphs are the extremely large winding forces in the end region of the stator winding, and the excessive heating of the amortisseur bars residing on the rotor. As can be expected, the electrical and mechanical stresses depend on the reclosing times, which are 450 ms and 745 ms for Figs. E4.3.9a and E4.3.9b, respectively.

### 4.2.6 Application Example 4.4: Calculation of the Amortisseur (Damper Winding) Bar Losses of a Synchronous Machine during Balanced Three-Phase Short-Circuit, Line-to-Line Short-Circuit, Out-of-Phase Synchronization, and Unbalanced Load Based on the Natural abc Reference System [40]: Solution

Following the notation of [7 in Chapter 4] one obtains the voltage equations:

$$e_a = p\Psi_a - r \cdot i_a \qquad \text{(E4.4-25a)}$$

$$e_b = p\Psi_b - r \cdot i_b \qquad \text{(E4.4-25b)}$$

$$e_c = p\Psi_c - r \cdot i_c \qquad \text{(E4.4-25c)}$$

$$e_{fd} = p\Psi_{fd} + r_{fd} \cdot i_{fd} \qquad \text{(E4.4-25d)}$$

$$0 = p\Psi_{kd} - r_{kd} \cdot i_{kd} + v_{kd} \qquad \text{(E4.4-25e)}$$

$$0 = p\Psi_{kq} - r_{kq} \cdot i_{kq} + v_{kq} \qquad \text{(E4.4-25f)}$$

for $k = 1, 2, 3, \ldots, 8$ damper windings in d- or q-axes.

The flux linkage equations are

$$\Psi_a = -L_{aa}i_a - L_{ab}i_b - L_{ac}i_c + L_{afd}i_{fd} + L_{akd}i_{kd} + L_{akq}i_{kq} \qquad \text{(E4.4-26a)}$$

$$\Psi_b = -L_{ba}i_a - L_{bb}i_b - L_{bc}i_c + L_{bfd}i_{fd} + L_{bkd}i_{kd} + L_{bkq}i_{kq} \qquad \text{(E4.4-26b)}$$

$$\Psi_c = -L_{ca}i_a - L_{cb}i_b - L_{cc}i_c + L_{cfd}i_{fd} + L_{ckd}i_{kd} + L_{ckq}i_{kq} \qquad \text{(E4.4-26c)}$$

$$\Psi_{fd} = -L_{fda}i_a - L_{fdb}i_b - L_{fdc}i_c + L_{fdfd}i_{fd} + L_{fdkd}i_{kd} \qquad \text{(E4.4-26d)}$$

$$\Psi_{kd} = -L_{kda}i_a - L_{kdb}i_b - L_{kdc}i_c + L_{kdfd}i_{fd} + L_{kd1d}i_{1d} + L_{kd2d}i_{2d} + \ldots + L_{kdld}i_{ld} \qquad \text{(E4.4-26e)}$$

$$\Psi_{kq} = -L_{kqa}i_a - L_{kqb}i_b - L_{kqc}i_c + L_{kk1q}i_{1q} + L_{kq2q}i_{2q} + \ldots + L_{kqlq}i_{lq} \qquad \text{(E4.4-26f)}$$

for $k = 1, 2, \ldots, 8$ and $l = 1, 2, \ldots, 8$.

Following the notation of Fig. E4.4.2, the self-inductance of the kth damper winding is

$$L_{kd}^R = 2 \cdot I_R \cdot A_{Rkd}/I_{kd} \qquad \text{(E4.4-27)}$$

and the leakage inductance with respect to stator winding is

$$L_{kdl}^R = 2 \cdot I_R \cdot (A_{Rkd} - A_{Skd})/I_{kd} \qquad \text{(E4.4-28)}$$

or referred to the stator reference frame, where $m_1$ and $m_2$ are the number of stator and rotor phases, respectively:

$$L_{kd} = m_1(2 \cdot l_R \cdot A_{Rkd}/I_{kd})M_{kda}^2/m_2 \qquad \text{(E4.4-29)}$$

$$L_{kdl} = m_1\{2 \cdot l_R \cdot (A_{Rkd} - A_{Skd})/I_{kd}\}M_{kda}^2/m_2. \qquad \text{(E4.4-30)}$$

The mutual inductances between the kth and the lth d-axis windings are

$$L_{kdld} = m_1 \cdot L_{kdld}^{kd} \cdot M_{kdld}^2 \cdot M_{lda}^2/m_2) \qquad \text{(E4.4-31a)}$$

$$L_{ldkd} = m_1 \cdot L_{ldkd}^{ld} \cdot M_{ldkd}^2 \cdot M_{kda}^2/m_2. \qquad \text{(E4-4.31b)}$$

Correspondingly the ohmic resistance is

$$r_{kd} = m_1 \cdot R_{kd} \cdot M_{kda}^2/m_2. \qquad \text{(E4.4-32)}$$

The same procedure applies to the q-axis parameters, as indicated in Fig. E4.4.3. In these equations $l_R$ is the rotor length. $M_{kda}$, $M_{kdld}$, $M_{lda}$, and $M_{ldkd}$ are transformation ratios. The calculation of the stator inductances (Fig. E4.4.4) is performed in the same manner as for the rotor inductances.

If Eqs. E4.4-26 to E4.4-32 are introduced in Eq. E4.4-25 one obtains

$$[L] \cdot \left[\frac{di}{dt}\right] = [V], \qquad \text{(E4.4-33)}$$

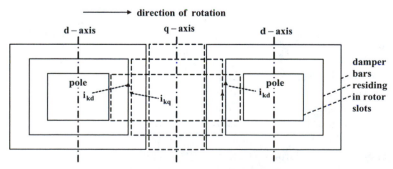

**FIGURE E4.4.1** Decomposition of amortisseur winding of a two-pole generator [40 in Chapter 4]. In this graph only six damper windings are shown: three in the d-axis and three in the q-axis. Figures E4.4.2 to E4.4.4 and E4.4.11 to E4.4.13 provide details about the location of the 16 damper bars within the rotor slots.

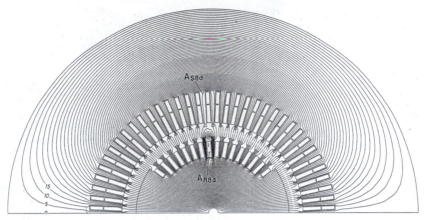

**FIGURE E4.4.2** Calculation of d-axis rotor (leakage, mutual, and self-) inductances, $I = I_{kd} = 500$ kA, where $k = 8$. All vector potentials must be multiplied by $10^{-2}$ Wb/m [40 in Chapter 4].

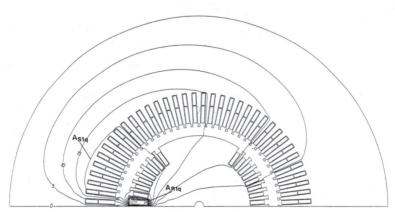

**FIGURE E4.4.3** Calculation of q-axis rotor (leakage, mutual, and self-) inductances, $I = I_{kq} = 500$ kA, where $k = 1$. All vector potentials must be multiplied by $10^{-2}$ Wb/m [40 in Chapter 4].

where $[L]$ is an inductance matrix with time-varying coefficients, $\left[\dfrac{di}{dt}\right]$ is the derivative of the solution vector $[i]$ and $[V]$ represents the forcing sources. Using Gaussian elimination one gets

$$\left[\frac{di}{dt}\right] = [L]^{-1} \cdot [V]. \qquad (E4.4\text{-}34)$$

Subsequently a Runge-Kutta integration procedure of the fourth order provides the first starting steps for the Adams-Moulton method, which is a predictor-corrector integration procedure, and one obtains the solution vector $[i]$. A line-to-line short-circuit including the (generator) transformer leakage impedance is performed. The line-to-neutral voltages in per unit (referred to the rated voltage of the generator) are shown in Fig. E4.4.5. The stator currents are depicted in Fig. E4.4.6, and field and damper bar currents are shown in Figs. E4.4.7 and E4.4.8a,b. The resulting electrical and mechanical torques are depicted in Figs. E4.4.9 and E4.4.10.

The integral $\int_1^{t_2} (i_{k\text{bar}})^2 dt$ is a measure for the energy absorbed by the damper bar k due to the current $i_{k\text{bar}}$. Therefore, the energy dissipated in the damper bar k during time $(t_2 - t_1)$ is

$$E_{k\,\text{bar}} = \left\{ \int_1^{t_2} (i_{k\,\text{bar}})^2 dt \right\} R_{k\,\text{bar}}. \qquad (E4.4\text{-}35)$$

It is well known that the bar (e.g., #16) of the amortisseur (see Fig. E4.4.11) located before the rotor pole as viewed in the direction of rotation incurs greater loss than the bar (e.g., #1) of the amortisseur located behind the rotor pole as viewed in the direction of rotation. Figure E4.4.4.12 shows the

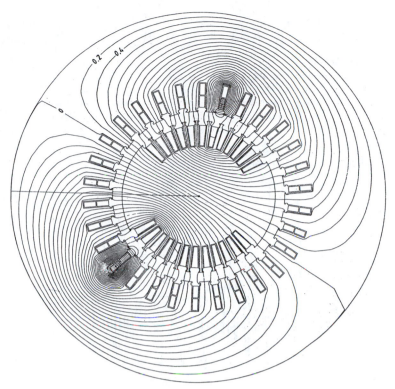

**FIGURE E4.4.4** Calculation of stator (leakage, mutual, and self-) inductances. One stator winding turn is excited with the current $I = 50$ A; all vector potentials must be multiplied by $10^{-3}$ Wb/m [40 in Chapter 4].

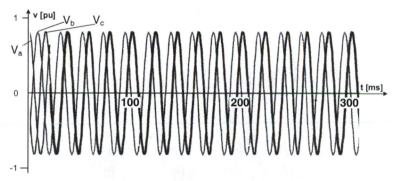

**FIGURE E4.4.5** Line-to-neutral voltages $v_a$, $v_b$, and $v_c$ as a function of time. The line-to-line short-circuit occurs at 25 ms at no load and $v_c = v_a$. The voltages are referred to rated line-to-neutral voltage.

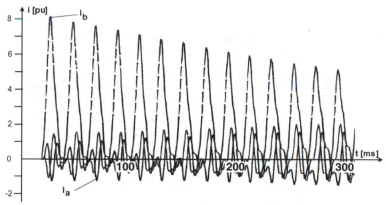

**FIGURE E4.4.6** Phase currents $i_a$, $i_b$, and $i_c$ as a function of time. The line-to-line short-circuit occurs at 25 ms at no load and $v_c = v_a$. The currents are referred to the rated stator current.

**FIGURE E4.4.7** Field currents $i_f$, slip s, and change in rotor angle $\delta$ as a function of time. The line-to-line short-circuit occurs at 25 ms at no load and $v_c = v_a$. The field current is referred to the rated stator current.

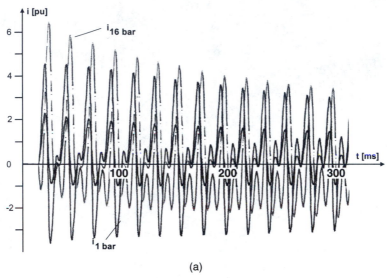

(a)

**FIGURE E4.4.8a** Rotor bar currents $i_{1bar}$, $i_{16bar}$ (see Fig. E4.4.11) as a function of time. The line to-line short-circuit occurs at 25 ms at no load and $v_c = v_a$. The currents are referred to the rated stator current.

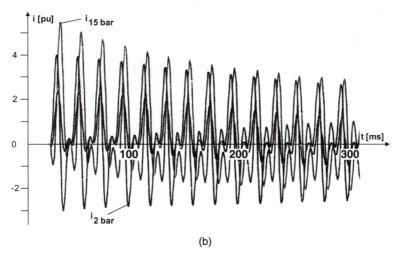

(b)

**FIGURE E4.4.8b** Rotor bar currents $i_{2bar}$, $i_{15bar}$ (see Fig. E4.4.11) as a function of time. The line-to-line short-circuit occurs at 25 ms at no load and $v_c = v_a$. The currents are referred to the rated stator current.

**TABLE E4.4.1** Dissipated Energies in Amortisseur #1 and #16 Bars for Different Fault Conditions

| Fault condition | Energy dissipated in per unit[a] in bar #1 during first 1.25 s | Energy dissipated in per unit[a] in bar #16 during first 1.25 s |
|---|---|---|
| Line-to-line short-circuit | 0.95 | 1.05 |
| Balanced three-phase short-circuit | 0.92 | 1.12 |
| Out-of-phase synchronization with a 120° leading rotor angle | 2.36 | 2.43 |
| Out-of-phase synchronization with a 120° lagging rotor angle | 1.25 | 1.57 |
| Unbalanced stator currents resulting in an inverse stator current of $|\tilde{I}_2| = 8.7\%$. | 0.049 | 0.053 |

[a]Energy dissipated in a bar is referred to the heat energy required to melt one bar if there is no cooling.

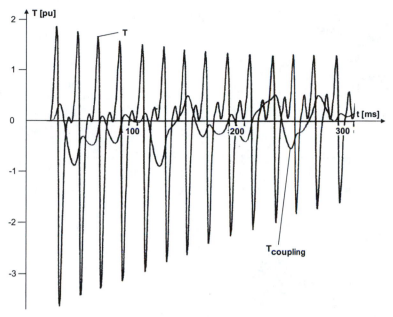

**FIGURE E4.4.9** Electrical generator torque T and mechanical torque $T_{coupling}$ at coupling between generator and steam turbine stage #1 as a function of time. The line-to-line short-circuit occurs at 25 ms at no load and $v_c = v_a$. The torques are referred to the rated torque of generator.

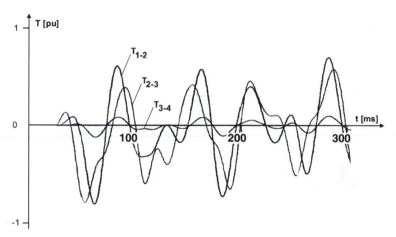

**FIGURE E4.4.10** Mechanical torques at the couplings 1–2 (stages 1 and 2), 2–3 (stages 2 and 3), and 3–4 (generator and exciter) as a function of time. The line-to-line short-circuit occurs at 25 ms at no load and $v_c = v_a$. The torques are referred to the rated torque of generator.

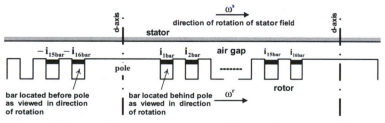

**FIGURE E4.4.11** Definition of the bars of the amortisseur located before (#16) and after (#1) the rotor pole as viewed in direction of rotation.

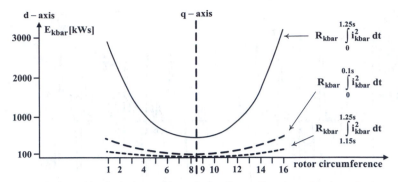

**FIGURE E4.4.12** Energy dissipated in the k damper bars during line-to-line short-circuit at the secondary terminals of a transformer fed by a generator [40 in Chapter 4].

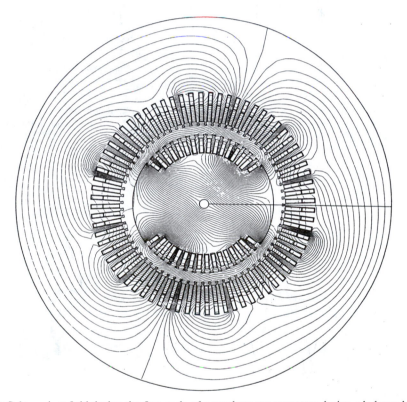

**FIGURE E4.4.13** Subtransient field during the first cycle of a synchronous generator during a balanced three-phase short circuit.

values $E_{kbar}$ for a sudden line-to-line short-circuit at the terminals including transformer impedance for various time intervals ($t_2 - t_1$). Figure E4.4.13 depicts the subtransient field of a generator during the first cycle of a three-phase short-circuit.

Table E4.4.1 summarizes the dissipated energies in the 1st and 16th amortisseur bars during various fault conditions of a generator including the (generator) transformer leakage impedance:

1) line-to-line short-circuit,
2) balanced three-phase short-circuit,

3) out-of-phase synchronization with a 120° leading rotor angle,
4) out-of-phase synchronization with a 120° lagging rotor angle, and
5) unbalanced stator currents resulting in an inverse stator current of $|\tilde{I}_2| = 8.7\%$.

**Discussion of Results**
- Most transients decay within 1 to 2 s.
- The 2nd harmonic in the stator currents of Fig. E4.4.6 are confirmed by measurements.
- The thermal stress of the amortisseur bars due to a balanced short-circuit can be tolerated, provided

there are no subsequent short-circuits within a brief time frame.

- An imbalance due to an inverse current of $|\tilde{I}_2| = 8.7\%$ is acceptable—even during steady-state operation.
- There is a great difference when synchronizing with either a leading rotor angle or a lagging rotor angle. Out-of-phase synchronization with a leading rotor angle generates more losses within the generator than in the case of synchronization with a lagging rotor angle. In the first case the machine acts as a generator and in the latter it acts as a motor during the synchronization process. Note that an out-of-phase angle in the neighborhood of 140° leads to the largest currents and torques (see Fig. E4.3.4).
- To mitigate or reduce the thermal stress of damper bars #1 and #16 manufacturers of synchronous generators embed damper bars also on the surface of the rotor poles. In this manner the thermal stress of all amortisseur bars will be made more equal.
- Not neglecting saturation and the generation of eddy currents within the solid rotor poles will in addition mitigate the thermal stress of the entire amortisseur because the inductive coupling between stator and rotor damper windings will be reduced.
- A similar analysis applies to the induced currents of an induction machine. Under normal operation the currents induced in the rotor bars of an induction machine are uniform and have the same rated magnitude $I_{bar\_rated}$ in each and every bar. However, if a bar is broken (that is, no current can flow) then the neighboring bars will have currents that can be significantly larger than the rated magnitude. Such an unequal current distribution may result in a thermal imbalance. As a consequence due to this imbalance and the very small air gap of induction machines, the rotor may become eccentric and rub against the stator core, and this may lead to failure of the induction machine. See Chapter 3 for more details.

The abc-reference frame employed in this example has been subsequently used in references [41–43]. The above results are confirmed in these references. However, the out-of-phase synchronizing angle of 180° as used in [43] does not yield the largest currents, losses, and torques (see Fig. E4.3.4). These references teach that the large losses in the damper bars next to the poles (e.g., bar #1 and bar #16) can be reduced by the deployment of amortis-

seur bars, which are embedded within the pole faces.

## 4.2.9 Application Example 4.7: Design of a Low-Speed 20 kW Permanent-Magnet Generator for a Wind-Power Plant: Solution

a) Synchronous speed

$$n_s = \frac{120f}{p} \quad \text{or} \quad f = \frac{n_s p}{120} = \frac{60(12)}{120} = 6\,\text{Hz}. \quad \text{(E4.7-1)}$$

b) Ampere's law

$$H_m l_m + H_g l_g = 0. \quad \text{(E4.7-2)}$$

c) Continuity of flux

$$\Phi = B_m A_m = B_g A_g, \quad \text{(E4.7-3)}$$

d) Constituent relation

$$B_g = \mu_0 H_g. \quad \text{(E4.7-4)}$$

From Eq. E4.7-2

$$H_m l_m = -H_g l_g. \quad \text{(E4.7-5)}$$

With Eq. E4.7-4

$$H_m l_m = -B_g l_g / \mu_0, \quad \text{(E4.7-6)}$$

and with Eq. E4.7-3

$$B_g = B_m \left( \frac{A_m}{A_g} \right), \quad \text{(E4.7-7)}$$

the load-line equation becomes

$$H_m = -\left( \frac{A_m}{A_g} \right)\left( \frac{l_g}{l_m} \right)\frac{B_m}{\mu_0}. \quad \text{(E4.7-8)}$$

Introducing the given numerical values into load-line equation gives

$$H_m = -91.7\,B_m \left[ \frac{kA}{m} \right]. \quad \text{(E4.7-9)}$$

This equation is plotted in Fig. E4.7.4. Note that the magnet material is not operated in the point of the maximum energy product.

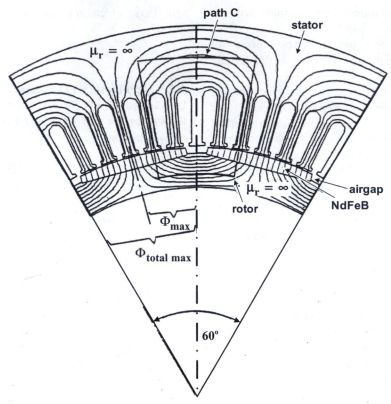

**FIGURE E4.7.1** Cross section of one period (two pole pitches) of permanent-magnet machine with no-load field.

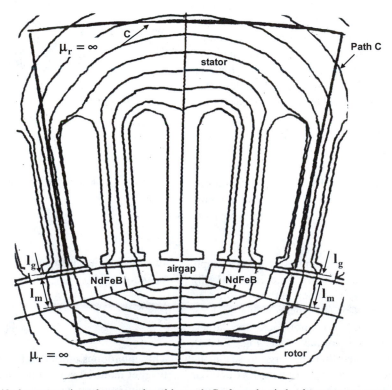

**FIGURE E4.7.2** Magnified cross section of area enclosed by path C of a pole pitch of permanent-magnet machine (see Fig. E4.7.1).

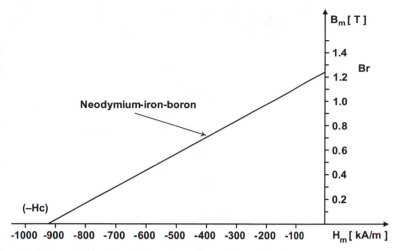

**FIGURE E4.7.3** $B_m$–$H_m$ characteristic of neodymium–iron–boron (NdFeB) permanent-magnet material.

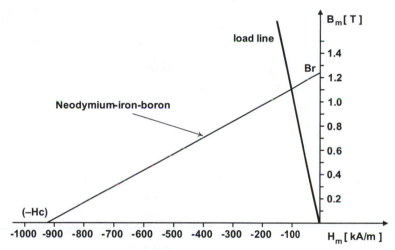

**FIGURE E4.7.4** Magnet (NdFeB) characteristic and load line.

e) Recoil permeability of NdFeB (slope of the permanent-magnet characteristic)

$$\mu_R = \frac{(y_2 - y_1)}{(x_2 - x_1)} = \frac{(B_r - 0)}{(0 - (-H_c))} = \frac{1.25}{950(10^3)} \quad \text{(E4.7-10)}$$
$$= 1.316 \times 10^{-6} \,(\text{H/m})$$

or

$$\mu_R = 1.047 \,\mu_0. \quad \text{(E4.7-11)}$$

f) Induced voltage

$$\Phi_{\max} = B_m A_m, \quad \text{(E4.7-12)}$$

$$\Phi_{\text{totalmax}} = 2\Phi_{\max} = 2B_m A_m, \quad \text{(E4.7-13)}$$

The flux is defined as

$$\Phi(t) = \Phi_{\text{totalmax}} \cos \omega t. \quad \text{(E4.7-14)}$$

Therefore, the induced voltage is

$$E_{\max} = N(d\Phi/dt) = \omega N_{ph} k_w k_s \Phi_{\text{totalmax}}. \quad \text{(E4.7-15)}$$

From load line and magnet characteristic one obtains the operating point:

$$B_{m0} = 1.1 \text{ T}, \quad \text{(E4.7-16a)}$$

$$H_{m0} = -100\text{k A/m}. \quad \text{(E4.7-16b)}$$

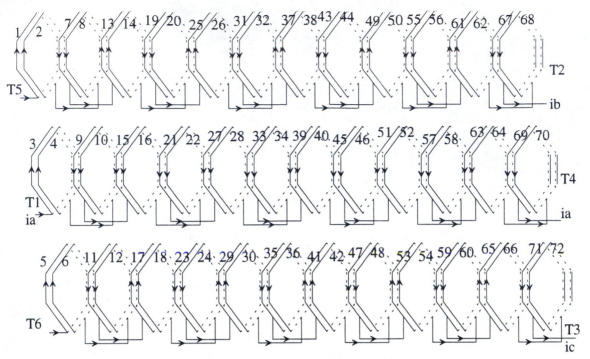

**FIGURE E4.7.5** Stator winding of permanent-magnet generator. Input terminals are $T_1$, $T_2$, and $T_3$. In Y-configuration connect $T_4$, $T_5$, and $T_6$.

Therefore,

$$\Phi = \Phi_{totalmax} = (2)(1.1)(11668 \times 10^{-6})\text{Wb} = 0.0257 \text{ Wb},$$
$$\text{(E4.7-17)}$$

and the induced voltage is now

$$e(t) = (360)\,(0.0257)\,(2\pi)\,(6)\,(0.8)\,(0.94)\sin \omega t$$
$$= E_{max} \sin \omega t,$$
$$\text{(E4.7-18a)}$$

or

$$E_{max} = 262.3 \text{ V}, \qquad \text{(E4.7-18b)}$$

$$E_{rms} = E_{max}/\sqrt{2} = 185.48 \text{ V}, \qquad \text{(E4.7-18c)}$$

and the line-to-line terminal voltage is approximately

$$V_{L\text{-}L} = \sqrt{3}\ V_{L\text{-}N} = \sqrt{3}(1.05)E_{rms} = 337.3 \text{ V}. \qquad \text{(E4.7-19)}$$

g) Determination of wire diameter of the stator winding. Figure E4.7.5 depicts the stator winding housed in 72 slots. It is a full-pitch, double-layer winding. Using a copper-fill factor of $k_{cu} = 0.62$ one obtains for a stator slot cross section of

$A_{st\_slot} = 368$ mm$^2$ and $(360/12) = 30$ turns per stator pole and per phase with the copper cross section of one stator slot $A_{st\_slot\_cu} = A_{st\_slot} \cdot k_{cu} = 368 \cdot 0.62$ mm$^2 = 30\pi D^2/4$ or a wire diameter of $D = 3.1$ mm. For a rated current of $I_{rat} = 31$ A one obtains a current density of $J = 31/7.6 = 4.07$ A/mm$^2$. This current density is acceptable for a ventilated machine.

### 4.2.10 Application Example 4.8: Design of a 10 kW Wind-Power Plant Based on a Synchronous Machine: Solution

a) $P_{out\_transformer} = 10$ kW, $P_{out\_inverter} = 10.53$ kW, $P_{out\_rectifier} = 11.08$ kW, $P_{out\_generator} = 11.67$ kW, $P_{out\_gear} = 12.28$ kW, $P_{out\_wind\ turbine} = 12.92$ kW.
$$\text{(E4.8-1)}$$

b) $$\left(\frac{N_Y}{N_\Delta}\right) = \frac{1}{89.9}. \qquad \text{(E4.8-2)}$$

c) Note $\tilde{V}_{L\text{-}L\ system} = \tilde{V}_{L\text{-}L\Delta}$ and $\tilde{V}_{L\text{-}L\ inverter} = \tilde{V}_{L\text{-}LY}$; therefore,

$$\tilde{V}_{L\text{-}L\Delta} = \tilde{V}_{L\text{-}LY}\angle + 30°. \qquad \text{(E4.8-3)}$$

d) $$\tilde{I}_{L\ system} = \tilde{I}_{L\ inverter}\angle + 30°. \qquad \text{(E4.8-4)}$$

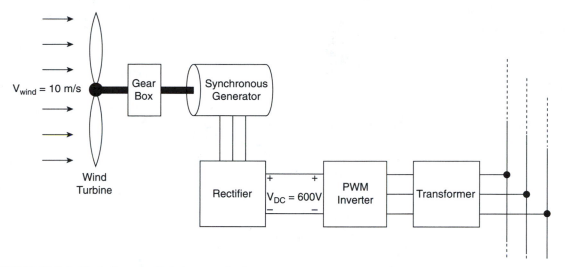

**FIGURE E4.8.1** Block diagram of wind-power plant.

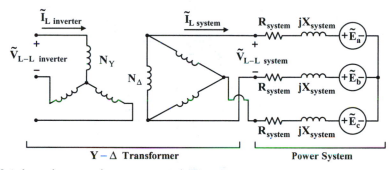

**FIGURE E4.8.2** Y–Δ three-phase transformer connected to power system.

One concludes that the line-to-line voltages of the secondary Δ as well as the line currents fed into the power system are shifted by +30 degrees.

e) The computer program for the inverter is given in Chapter 1, resulting in an inverter output current of $I_{\text{out\_inverter}} = 25.33$ A and an inverter input current of $I_{\text{in\_inverter}} = 18.47$ A. The inverter output phase voltage and output line current are shown in Fig. E4.8.3.

f) Computer program for rectifier is given in Chapter 1 resulting in a rectifier output current of $I_{\text{out\_rectifier}} = 18.47$ A and an rectifier input current of $I_{\text{in\_rectifier}} = 14.04$ A. The output voltage of the rectifier is depicted in Fig. E4.8.4.

g) Design of synchronous generator

Stator frequency $f = \dfrac{n_s \cdot p}{120} = \dfrac{(3600)(2)}{120} = 60\,\text{Hz}$,

$$\text{(E4.8-5)}$$

phase voltage $V_{L-N} = 480\ \text{V}/1.732 = 277.14\ \text{V}$,

$$\text{(E4.8-6a)}$$

base voltage $V_{\text{base}} = V_{L-N} = 277.14\ \text{V}$,　　　(E4.8-6b)

base (rated) current $I_{\text{base}} = \dfrac{P_{\text{out\_generator}}}{3 \cdot V_{L-N}} = \dfrac{11670}{3(277.14)}$
$= 14.04\ \text{A}$,

$$\text{(E4.8-7)}$$

base impedance $Z_{\text{base}} = \dfrac{V_{\text{base}}}{I_{\text{base}}} = \dfrac{277.14\ \text{V}}{14.04\ \text{A}} = 19.74\,\Omega$,

$$\text{(E4.8-8)}$$

generator loss $P_{\text{loss}} = P_{\text{in\_generator}} - P_{\text{out\_generator}}$
$= (12280 - 11670)\ \text{W} = 610\ \text{W}$,

$$\text{(E4.8-9)}$$

stator resistance $R_a = \dfrac{P_{\text{loss}}}{3 \cdot I_{\text{base}}^2} = 1.03\,\Omega$,

$$\text{(E4.8-10)}$$

ideal core length $l_i = \dfrac{P_{\text{out\_generator}}}{D_{\text{rotor}}^2 \cdot C \cdot n_s} = \dfrac{11670}{(0.2^2)(1.3)(3600)}$
$= 0.0623\,\text{m},$

(E4.8-11)

actual iron core length (with interlamination

insulation) $l_{\text{act}} = \dfrac{l_i}{k_{fe}} = \dfrac{0.0623}{0.95}\,\text{m} = 0.0656\,\text{m},$

(E4.8-12)

stator synchronous reactance $X_S = X_S[\text{pu}] \cdot Z_{\text{base}}$
$= (1.5)(19.74) = 29.61\,\Omega,$

(E4.8-13)

with $X_d \approx X_S = X_{aa} + X_{ab} + \dfrac{3}{2} X_{aa2}$ one gets approximately $X_S = 2 \cdot (2\pi f) L_{aa}$ or for the self-inductance $L_{aa}$ of one phase

$$L_{aa} = \frac{X_S}{4\pi f} = 0.0393\,\text{H} \qquad \text{(E4.8-14)}$$

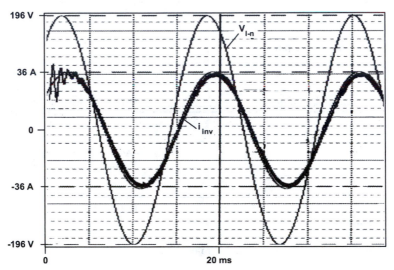

**FIGURE E4.8.3** Inverter output phase voltage and line current.

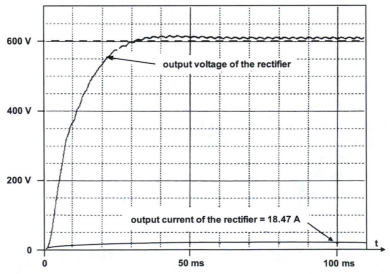

**FIGURE E4.8.4** Rectifier output voltage and output current.

and the leakage inductance of one stator (s) phase

$$L_{s\ell} = 0.1\left(\frac{29.61}{2\pi 60}\right) = 0.0079\,\text{H}. \quad \text{(E4.8-15)}$$

The area per pole is

$$\text{area}_p = \frac{2\pi \cdot D_{\text{rotor}}}{2} \cdot \frac{l_i}{2} = \frac{2\pi(0.2)(0.0623)}{4} = 0.0196\,\text{m}^2$$

$$\text{(E4.8-16)}$$

the maximum flux linked with one pole is

$$\Phi_{s\_\max} = B_{\max} \cdot \text{area}_p = 0.7(0.0196) = 0.0137\,\text{Wb}.$$

$$\text{(E4.8-17)}$$

The induced voltage in one phase of the stator winding becomes (generator operation)

$$E_{ph} = 1.05 \cdot V_{L-N} = 4.44 \cdot f \cdot N_s \cdot k_s \cdot \Phi_{s\_\max} = 291.0\,\text{V}$$

$$\text{(E4.8-18)}$$

or the number of stator turns per phase is

$$N_s = \frac{E_{ph}}{4.44 \cdot f \cdot k_s \cdot \Phi_{s\_\max}} = \frac{291}{4.44(60)(0.8)(0.0137)}$$
$$= 100\,\text{turns}.$$

$$\text{(E4.8-19)}$$

The one-sided air-gap length can be obtained from

$$L_{aa} = \frac{\lambda_{aa}}{i_a} = \frac{N_s \cdot \Phi_{s\_\max}}{i_{\max}}, \quad \text{(E4.8-20)}$$

where

$$\Phi_{s\_\max} = \frac{\mu_o \cdot \dfrac{4}{\pi} \cdot k_s \cdot N_s \cdot i_{\max} \cdot \text{area}_p}{p \cdot g}. \quad \text{(E4.8-21)}$$

Therefore,

$$L_{aa} = \frac{\mu_o \cdot \dfrac{4}{\pi} \cdot k_s \cdot N_s^2 \cdot \text{area}_p}{p \cdot g}$$

or solved for the air-gap length

$$g = \frac{\mu_o \cdot \dfrac{4}{\pi} \cdot k_s \cdot N_s^2 \cdot \text{area}_p}{p \cdot L_{aa}}$$

$$= \frac{(4\pi \times 10^{-7})\left(\dfrac{4}{\pi}\right)(0.8)(10000)(0.0196)}{2(0.0393)}$$

$$\text{(E4.8-22)}$$

$$= 0.0032\,\text{m} = 3.2\,\text{mm}.$$

The torque of the generator is

$$(-T) = \frac{P_{\text{out\_generator}}}{\omega_s} = \frac{11670}{377} = 30.94\,\text{Nm}. \quad \text{(E4.8-23)}$$

For a torque angle of $\delta = 30°$ one gets with

$$T = -\frac{p \cdot \pi \cdot D_{\text{rotor}} \cdot l_{\text{act}} \cdot B_{\max} \cdot F_r}{4} \cdot \sin\delta, \quad \text{(E4.8-24)}$$

or the rotor mmf

$$F_r = \frac{(-T)\cdot 4}{p \cdot \pi \cdot D_{\text{rotor}} \cdot l_{\text{act}} \cdot B_{\max} \cdot \sin\delta}$$

$$= \frac{(30.94)(4)}{(2)(\pi)(0.2)(0.0656)(0.7)(0.5)} = 4289.6\,\text{At}.$$

$$\text{(E4.8-25)}$$

The ampere turns for the rotor are $F_r = \dfrac{4}{\pi} \cdot k_r \cdot \dfrac{N_{r(\text{tot})}}{p} \cdot I_r$

or solved for the ampere turns

$$N_{r(\text{tot})} \cdot I_r = \frac{F_r \cdot p \cdot \pi}{4 \cdot k_r} = \frac{(4289.6)(2)(\pi)}{(4)(0.8)} = 8422.4\,\text{At}.$$

$$\text{(E4.8-26)}$$

Selecting 16 rotor slots with 20 turns each, one obtains the field (rotor) current:

$$I_r = 8422.4\ \text{Aturns}/(16 \cdot 20\text{turns}) = 26.3\,\text{A}. \quad \text{(E4.8-27)}$$

h) Gear ratio. The gear ratio is

$$\text{gear ratio} = \frac{n_{\text{generator}}}{n_{\text{wind turbine}}} = \frac{3600}{30} = 120. \quad \text{(E4.8-28)}$$

i) Radius of (three) wind turbine blades. The limiting efficiency for wind turbines is according to Lanchester–Betz [45 in Chapter 4] $\eta_{\text{LanchesterBetz}} = (16/27) = 0.59$. The wind turbine design depends on the density of the air and the ambient tempera-

**TABLE E4.8.1** Air Density as a Function of Altitude

| Altitude (m) | −500 | 0 | 500 | 1000 | 1500 | 2000 | 2500 | 3000 |
|---|---|---|---|---|---|---|---|---|
| Air (mass) density ρ (kg/m³) | 1.2854 | 1.2255 | 1.1677 | 1.112 | 1.0583 | 1.0067 | 0.957 | 0.9092 |
| Altitude (m) | 3500 | 4000 | 4500 | | | | | |
| Air (mass) density ρ (kg/m³) | 0.8623 | 0.8191 | 0.7768 | | | | | |

ture. Table E4.8.1 provides the air density as a function of altitude.

The wind turbine has to provide a rated output power of $P_{\text{out\_wind turbine}} = 12.92$ kW at a rated wind velocity of $v = 10$ m/s. The efficiency (coefficient of performance) is $c_p = 0.3$, where $c_p$ is the ratio between the power delivered by the rotor of the wind turbine to the power in the wind with velocity v striking the area A swept by the rotor.

According to [45 in Chapter 4] the output power of the wind turbine is

$$P_{\text{wind\_turbine}} = \frac{1}{2} \cdot c_p \cdot v^3 \cdot A \cdot \rho. \quad \text{(E4.8-29)}$$

The area swept by the rotor is

$$A = \frac{2 \cdot P_{\text{wind-turbine}}}{c_p \cdot v^3 \cdot \rho} = \frac{2(12920)}{0.3(1000)(1.05)} = 82\,\text{m}^2, \quad \text{(E4.8-30)}$$

or the rotor blade radius is

$$R_{\text{blade}} = \sqrt{\frac{A}{\pi}} = \sqrt{\frac{82}{\pi}} = 5.1\,\text{m}. \quad \text{(E4.8-31)}$$

The height of the tower can be assumed to be about 2 times the blade radius, that is, $H_{\text{tower}} = 10$ m.

## CHAPTER 5 SOLUTIONS

### 5.1.2.1 Application Example 5.1: Computation of Displacement Power Factor (DPF) and Total Power Factor (TPF): Solution

a) PSpice code

```
*Computation of DPF and TPF
Van 2p 1p SIN(0V 489.3V 60Hz 0 0 0)
Vbn 3p 1p SIN(0V 489.3V 60Hz 0 0
-120)
Vcn 4p 1p SIN(0V 489.3V 60Hz 0 0
-240)
Rgnd 1p 0 1u
Rsysa 2p 5p 0.05
Rsysb 3p 6p 0.05
Rsysc 4p 7p 0.05
Lsysa 5p 1 265u
Lsysb 6p 2 265u
Lsysc 7p 3 265u
D1a 5 1 D1N4001
D1b 5 6 D1N4001
M1 1 7 5 5 SMM
R1 6 23 1u
V1 7 5 PULSE(0 2.7778ms 0 0
10.4167ms 16.66667ms)
D2a 8 24 D1N4001
D2b 8 9 D1N4001
M2 24 10 8 8 SMM
R2 9 3 1u
V2 10 8 PULSE(0 5.5556ms 0 0
10.4167ms 16.66667ms)
D3a 11 2 D1N4001
D3b 11 12 D1N4001
M3 2 13 11 11 SMM
R3 12 23 1u
V3 13 11 PULSE(0 8.3333ms 0 0
10.4167ms 16.66667ms)
D4a 14 24 D1N4001
D4b 14 15 D1N4001
M4 24 16 14 14 SMM
R4 15 1 1u
V4 16 14 PULSE(0 11.1111ms 0 0
10.4167ms 16.66667ms)
D5a 17 3 D1N4001
D5b 17 18 D1N4001
M5 3 19 17 17 SMM
R5 18 23 1u
V5 19 17 PULSE(0 13.8888ms 0 0
10.4167ms 16.66667ms)
D6a 20 24 D1N4001
D6b 20 21 D1N4001
M6 24 22 20 20 SMM
R5 21 2 1u
V6 22 20 PULSE(0 0ms 0 0 10.4167ms
16.66667ms)
Cf 23 24 500u
Rload 23 24 10
```

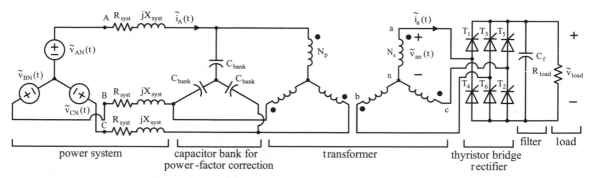

FIGURE E5.1.1 Connection of a wye/wye three-phase transformer with a thyristor rectifier, filter, load, and a bank of power-factor correction capacitors.

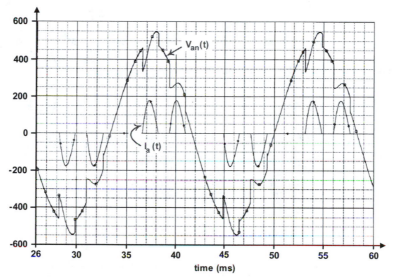

FIGURE E5.1.2 Phase voltage $v_{an}(t)$ and phase current for $i_a(t)$ for $\alpha = 10°$.

```
Rgnd2 24 0 1Meg
.model SMM NMOS(Level=3 Gamma=0
Delta=0 Eta=0 Theta=0 Kappa=0 Vmax=0
Xj=0
+   Tox=100n Uo=600 Phi=0.6 Rs=42.69m
Kp=20.87u L=2u W=2.9 Vto=3.487
+   Rd=0.19 Cbd=200n Pb=0.8 Mj=0.5
Cgso=3.5n Cgdo=100p Rg=1.2 Is=10f)
.model D1N4001 D(Is=1e-12)
.options list node opts
.options abstol=0.01 chgtol=0.01
reltol=0.1 vntol=100m trtol=20
.options itl1=500 itl2=500 itl4=100
itl5=0
.tran 0.1u 60ms 26ms 1u UIC
.four 60Hz V(1) I(Lsysa)
.probe
.end
```

b) to d) Figures E5.1.2 to E5.1.4 show the line to neutral (phase) voltage $v_{an}(t)$ and the phase current $i_a(t)$ for $\alpha = 10°$, 40°, and 80°, respectively. Table E5.1.1 lists the Fourier coefficients (magnitude and phase angles) for the phase voltage $v_{an}(t)$ and phase current $i_a(t)$ as a function of $\alpha$.

e) Table E5.1.2 gives a summary of the DPF and TPF and their ratio (TPF/DPF) as a function of the firing angle $\alpha$, Fig. E5.1.5 shows the DPF and TPF as a function of $\alpha$, and Fig. E5.1.6 illustrates the ratio (TPF/DPF). Figure E5.1.5 shows that the displacement power factor (DPF) as a function of the firing angle $\alpha$ obeys the relation $\cos \varphi = 0.955$ $\cos \alpha$ [14, in Chapter 5]. Figure E5.1.6 illustrates the ratio (TPF/DPF) as a function of the firing angle $\alpha$.

f) By introducing the capacitor bank into the above PSpice program and performing a series of runs for capacitor bank capacitance values $C_{bank}$ from 400 to 10 $\mu F$ one finds that for $C_{bank} = 115\ \mu F$ a DPF value of 0.9178 lagging is obtained. Figure E5.1.7 illustrates the voltage $v_{an}(t)$ and $i_a(t)$ for this

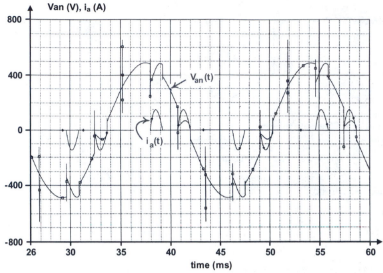

**FIGURE E5.1.3**  Phase voltage $v_{an}(t)$ and phase current for $i_a(t)$ for $\alpha = 40°$.

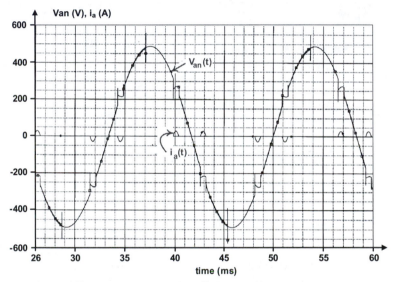

**FIGURE E5.1.4**  Phase voltage $v_{an}(t)$ and phase current for $i_a(t)$ for $\alpha = 80°$.

**TABLE E5.1.1**  Fourier Components of $v_{an}(t)$ and $i_a(t)$ for the Firing Angles $\alpha = 10°, 40°$, and $80°$

| $\alpha = 10°$ | $v_{anh}(V)$ | $\angle v_{anh}(°)$ | $i_{ah}(A)$ | $\angle i_{ah}(°)$ |
|---|---|---|---|---|
| $h = 1$ | 482.5 | −144.6 | 75.9 | −170.5 |
| $h = 5$ | 30.52 | −47.91 | 60.76 | 47.8 |
| $h = 7$ | 33.62 | −153.5 | 47.93 | −112.4 |
| $\alpha = 40°$ | $v_{anh}(V)$ | $\angle v_{anh}(°)$ | $i_{ah}(A)$ | $\angle i_{ah}(°)$ |
| $h = 1$ | 483.8 | −144.1 | 49.86 | 163.1 |
| $h = 5$ | 21.92 | 179.9 | 43.64 | −84.45 |
| $h = 7$ | 26.63 | −32.15 | 37.98 | 61.99 |
| $\alpha = 80°$ | $v_{anh}(V)$ | $\angle v_{anh}(°)$ | $i_{ah}(A)$ | $\angle i_{ah}(°)$ |
| $h = 1$ | 488.8 | −144 | 4.666 | 129.8 |
| $h = 5$ | 2.274 | 12.87 | 4.537 | 109.0 |
| $h = 7$ | 3.097 | 94.55 | 4.403 | −171.4 |

**TABLE E5.1.2**  Displacement Power Factor (DPF), Total Power Factor (TPF) and Their Ratio TPF/DPF

| $\alpha(°)$ | DPF (pu) | TPF (pu) | TPF/DPF (pu) |
|---|---|---|---|
| 0 | 0.9537 | 0.6831 | 0.7162 |
| 10 | 0.8996 | 0.6214 | 0.6908 |
| 20 | 0.8211 | 0.5518 | 0.6720 |
| 30 | 0.7230 | 0.4749 | 0.6569 |
| 40 | 0.6046 | 0.3892 | 0.6437 |
| 50 | 0.4726 | 0.2982 | 0.6311 |
| 60 | 0.3338 | 0.2064 | 0.6183 |
| 70 | 0.1942 | 0.1173 | 0.6038 |
| 80 | 0.0663 | 0.0388 | 0.5857 |

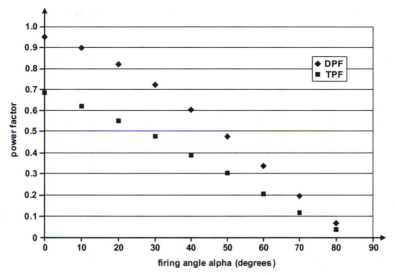

**FIGURE E5.1.5** DPF and TPF as a function of the firing angle $\alpha$.

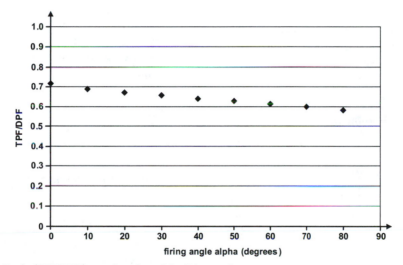

**FIGURE E5.1.6** Ratio (TPF/DPF) as a function of the firing angle $\alpha$.

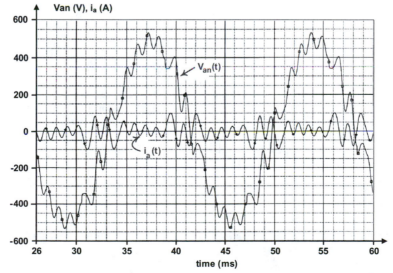

**FIGURE E5.1.7** Voltage $v_{an}(t)$ and current $i_a(t)$ for a capacitance value of $C_{bank} = 115\ \mu\text{F}$.

case. These wave shapes demonstrate that a resonance condition exists: the capacitor bank with the capacitance $C_{bank} = 115\ \mu F$ is able to increase the DPF but is unable to reduce the harmonic components of voltages and currents, indeed the harmonic components are increased due to the resonance condition.

g) As mentioned above the displacement power factor can be increased but the harmonics are increased due to resonance. The solution to this power factor correction problem is the replacement of the capacitor bank (with the capacitance value $C_{bank}$) by a passive filter $Z_f$ consisting of a series connection of an inductor $L_f$ and a capacitor $C_f$, which can be tuned to an appropriate frequency.

### 5.3.1 Application Example 5.2: Design of a Tuned Harmonic Filter: Solution

The following steps will be followed to design the filter:

- Calculation of the required reactive power $Q_{required}$ for improving DPF to the desired value:

$$Q_{required} = P(\tan \cos^{-1} 0.42 - \tan \cos^{-1} 0.97)$$
$$= 1180.38\ kVAr/phase. \qquad (E5.2\text{-}1)$$

- According to IEEE-519 [8], $THD_i$ should drop from 30.47% to 4.0%. We employ about 30% of the required $Q_{required}$ for filtering (tuned) and the remaining for (untuned) DPFC:

$$Q_{filter} = 0.30(1180.38) \approx 354\ kVar. \qquad (E5.2\text{-}2)$$

- Assuming star-connected (filter) capacitor banks one obtains

$$\text{capacitor/phase} = C = 1/(2\pi \cdot f \cdot X_C^{(1)})$$
$$= 1/(2\pi \cdot 50 \cdot 354 \cdot 1000) \approx 9\ \mu F/phase.$$

From this follows for the fundamental (1) reactance and the 5th harmonic (5) reactance $X_C^{(1)} = 354.11\ \Omega/phase$ and $X_C^{(5)} = 354.11/5 = 69.16\ \Omega/phase$, respectively. To obtain resonance at the 5th harmonic we require

$$X_L^{(5)} = X_C^{(5)} = 70.8\ \Omega/phase \Rightarrow X_L^{(1)} = 70.8/5$$
$$= 14.16\ \Omega/phase \Rightarrow L = 45\ mH. \qquad (E5.2\text{-}3)$$

- The current rating of the capacitor at $f = 50$ Hz is

$$\begin{cases} I_C^{(1)} = V^{(1)}/X_C^{(1)} = 6.35k/354.11 \approx 19\ A \\ I_C^{(5)} \approx 40\ A\ (given) \end{cases} \Rightarrow I_{rms} \approx 44\ A.$$
$$(E5.2\text{-}4)$$

- The reactive power rating of the reactor of the tuned filter is

$$\begin{cases} V_L = V_L^{(1)} + V_L^{(5)} = I^{(1)} \cdot X_L^{(1)} \\ \quad + I^{(5)} \cdot X_L^{(5)} \approx 3.1 kV \\ I_{rms} = 44\ A \end{cases} \Rightarrow Q_{rating}^{reactor} = 137\ kVAr.$$
$$(E5.2\text{-}5)$$

- The reactive power rating of the selected capacitor is

$$\begin{cases} V_C = V_C^{(1)} + V_C^{(5)} = I^{(1)} \cdot X_C^{(1)} \\ \quad + I^{(5)} \cdot X_C^{(5)} \approx 9.5 kV \\ I_{rms} = 44\ A \end{cases} \Rightarrow Q_{rating}^{capacitor} = 418 kVAr$$
$$(E5.2\text{-}6)$$

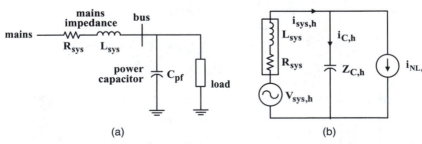

(a)  (b)

**FIGURE E5.3.1** A simplified industrial power system with capacitor and nonlinear load; (a) configuration, (b) the harmonic equivalent circuit [2 in Chapter 5] with $Z_{sys,h} = R_{sys} + jhL_{sys}$.

## 5.4.3 Application Example 5.3: Harmonic Resonance in a Distorted Industrial Power System with Nonlinear Loads: Solution

**i)** To observe the effect of the nonlinear load, the harmonic voltage source (representing the distorted utility voltage) is assumed to be short-circuited. The harmonic current injected into the capacitor due to the nonlinear load is

$$\tilde{I}_{C,h}^{\text{due }NL} = \frac{Z_{sys,h}}{Z_{sys,h} + Z_{C,h}} \tilde{I}_{NL,h}. \qquad (E5.3\text{-}3)$$

A parallel resonance occurs when the denominator approaches zero. The frequency of the parallel resonance is

$$f_{\text{parallel}} = \frac{1}{2\pi\sqrt{L_{sys}C}} = \frac{1}{2\pi\sqrt{0.1m \cdot 1300\mu}} \approx 420\,\text{Hz}.$$

$$(E5.3\text{-}4)$$

This condition will result in a large harmonic current being injected into the capacitor bank. The magnitude of the injected harmonic current can be several times the rated capacitor current, and may result in extensive harmonic voltages across the capacitor. Therefore, parallel resonance may result in the destruction of the capacitor due to overvoltage or over-current conditions.

**ii)** To investigate the effect of the distorted utility voltage on the capacitor, the harmonic current source (approximating the nonlinear load) is assumed to be ideal, that is, its impedance approaches infinity and behaves like an open circuit. The harmonic current injected into the capacitor due to the distorted utility source is

$$\tilde{I}_{C,h}^{\text{due }sys} = \frac{1}{Z_{sys,h} + Z_{C,h}} \tilde{V}_{sys,h}. \qquad (E5.3\text{-}5)$$

A series resonance occurs when the denominator approaches zero. Therefore, for this system the frequencies of the series and parallel resonances are identical:

$$f_{\text{series}} = f_{\text{parallel}} = \frac{1}{2\pi\sqrt{L_{sys}C}} \approx 420\,\text{Hz}. \qquad (E5.3\text{-}6)$$

The harmonic current injection into the power capacitor will be increased if the utility voltage contains harmonic components near resonance frequencies.

**iii)** The total harmonic current injected into the power capacitor is

$$\tilde{I}_{C,h} = \tilde{I}_{C,h}^{\text{due }NL} + \tilde{I}_{C,h}^{\text{due }sys} = \frac{Z_{sys,h}}{Z_{sys,h} + Z_{C,h}} \tilde{I}_{NL,h}$$

$$+ \frac{1}{Z_{sys,h} + Z_{C,h}} \tilde{V}_{sys,h}.$$

$$(E5.3\text{-}7)$$

Therefore, the 5th harmonic component of the injected current into the capacitor is

$$\tilde{I}_{C,5} = \tilde{I}_{C,5}^{\text{due }NL} + \tilde{I}_{C,5}^{\text{due }sys}$$

$$\tilde{I}_{C,5}^{\text{due }NL} = \frac{Z_{sys,5}}{Z_{sys,5} + Z_{C,5}} \tilde{I}_{NL,5}$$

$$= \frac{[0.06 + j5(2\pi \cdot 60 \cdot 0.11 \cdot 10^{-3})]\tilde{I}_{NL,5}}{[0.06 + j5(2\pi \cdot 60 \cdot 0.11 \cdot 10^{-3})]}$$

$$+ [j/5(2\pi \cdot 60 \cdot 1300 \cdot 10^{-6})]$$

$$= \frac{0.06 + j0.2074}{0.06 + j0.2074 - j0.408} \tilde{I}_{NL,5}$$

$$= \frac{0.06 + j0.2074}{0.06 - j0.2} \tilde{I}_{NL,5} = \frac{0.2159\angle73.87°}{0.2\angle-73.3°} \tilde{I}_{NL,5}$$

$$= (1.08\angle147.17°)\tilde{I}_{NL,5}$$

$$= 1.08\angle147.17°)(\sqrt{2}(0.3)\angle0°)$$

$$= \sqrt{2}(0.324)\angle147.17°\,\text{A}$$

$$\tilde{I}_{C,5}^{\text{due }sys} = \frac{1}{Z_{sys,5} + Z_{C,5}} \tilde{V}_{sys,5} = \frac{1}{0.2\angle-73.3°} \tilde{V}_{sys,5}$$

$$= (5\angle73.3°)\tilde{V}_{sys,5} = (5\angle73.3°)(\sqrt{2}(3)\angle0°)$$

$$= \sqrt{2}(15)\angle73.3°\,\text{A}$$

$$\Rightarrow\Rightarrow \tilde{I}_{C,5} = \tilde{I}_{C,5}^{\text{due }NL} + \tilde{I}_{C,5}^{\text{due }sys}$$

$$= (\sqrt{2}(0.324)\angle147.17°) + (\sqrt{2}(15)\angle73.3°$$

$$= \sqrt{2}(4.04 + j14.54) = \sqrt{2}(15.1)\angle74.47°\,\text{A}.$$

$$(E5.3\text{-}8)$$

The 7th harmonic component of the injected current into the power capacitor is

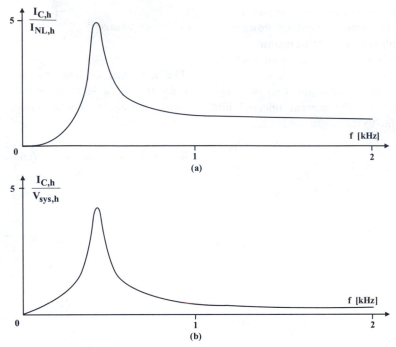

**FIGURE E5.3.2** Frequency response of capacitor current (a) divided by the nonlinear load harmonic current, (b) divided by the utility harmonic voltage [2 in Chapter 5].

$$\tilde{I}_{C,7} = \tilde{I}_{C,7}^{due\, NL} + \tilde{I}_{C,7}^{due\, sys}$$

$$\tilde{I}_{C,7}^{due\, NL} = \frac{Z_{sys,7}}{Z_{sys,7} + Z_{C,7}} \tilde{I}_{NL,7}$$

$$= \frac{[0.06 + j7(2\pi \cdot 60 \cdot 0.11 \cdot 10^{-3})] \tilde{I}_{NL,7}}{0.06 + j7(2\pi \cdot 60 \cdot 0.11 \cdot 10^{-3})}$$

$$\quad + j/7(2\pi \cdot 60 \cdot 1300 \cdot 10^{-6})$$

$$= \frac{0.06 + j7 \cdot 0.04147}{0.06 + j7 \cdot 0.04141 - j2.040/7} \tilde{I}_{NL,7}$$

$$= \frac{0.06 + j0.29029}{0.06 - j0.0011} \tilde{I}_{NL,7} = \frac{0.2964 \angle 78.32°}{0.06 \angle -1.05°} \tilde{I}_{NL,7}$$

$$= (4.964 \angle 79.37°) \tilde{I}_{NL,7}$$

$$= (4.964 \angle 79.37°)(\sqrt{2}(0.2) \angle 0°)$$

$$= \sqrt{2}(0.993) \angle 79.37° \text{ A}$$

$$\tilde{I}_{C,7}^{due\, sys} = \frac{1}{Z_{sys,7} + Z_{C,7}} \tilde{V}_{sys,7} = \frac{1}{0.06 \angle -1.05°} \tilde{V}_{sys,7}$$

$$= (16.7 \angle 1.05°) \tilde{V}_{sys,7}$$

$$= (16.7 \angle 1.05°)(\sqrt{2}(2) \angle 0°)$$

$$= \sqrt{2}(33.4) \angle 1.05° \text{ A}$$

$$\Rightarrow \Rightarrow \tilde{I}_{C,7} = \tilde{I}_{C,7}^{due\, NL} + \tilde{I}_{C,7}^{due\, sys}$$

$$= (\sqrt{2}(0.993) \angle 79.37°) + (\sqrt{2}(33.4) \angle 1.05°)$$

$$= \sqrt{2}(33.58 + 1.59)$$

$$= \sqrt{2}(33.62) \angle 2.7° \text{ A.}$$

$$(E5.3-9)$$

**iv)** As expected, the 7th harmonic current of the nonlinear load and the distorted utility voltage are magnified (it is a relatively weak system; $R_{sys} = 0.06\ \Omega$ and $X_{sys} = 2\pi(60)(0.11 \times 10^{-3})$ $= 0.042\ \Omega$) due to the parallel and series and resonances at $f = 420$ Hz.

**v)** The frequency responses of the capacitor current divided by the load harmonic current and divided by the utility harmonic voltage are shown in Fig. E5.3.2.

### 5.4.4 Application Example 5.4: Parallel Resonance Caused by Capacitors: Solution

The equivalent circuit (neglecting resistances) is shown in Fig E5.4.1b. The resonance frequency is with $S_{sc} = MVA_{sc}$ and $Q_{cap} = MVAr_{cap}$

$$h = \sqrt{\frac{MVA_{SC}}{MVAr_{cap}}} = \sqrt{\frac{72.6}{0.6}} = 11. \quad (E5.4-2)$$

This is a characteristic frequency of the six-pulse nonlinear load. Figure E5.4.1c shows a plot of the equivalent system impedance magnitude as seen from bus 1. This plot is known as a frequency scan.

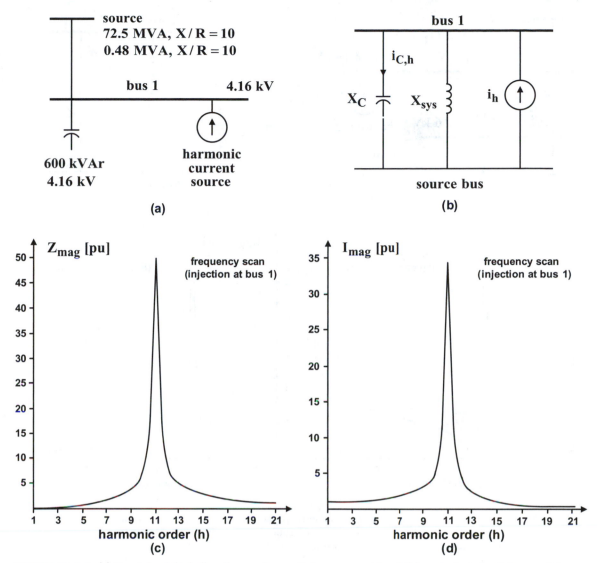

**FIGURE E5.4.1** (a) Example of single-line diagram for parallel resonance at bus 1, (b) equivalent circuit for parallel resonance, (c) frequency scan of bus 1 equivalent impedance, (d) current amplification of bus 1 parallel resonant circuit [7 in Chapter 5]. The harmonic-current source is a six-pulse converter.

A plot of the current amplification is given in Fig. E5.4.1d. Note that 1 A of the 11th harmonic current injection by the nonlinear load (into the system) will result in about 33 A of harmonic current flowing into the source and the capacitor.

### 5.4.5 Application Example 5.5: Series Resonance Caused by Capacitors: Solution

The series resonance frequency is

$$h = \sqrt{\frac{X_C}{X_L}} = \sqrt{\frac{X_C/h^2}{X_L}}\sqrt{\frac{28.843}{1.154}} = 5. \quad \text{(E5.5-1)}$$

This circuit produces a low-frequency path for the 5th harmonic current. A plot of the magnitude of the system impedance – as seen from bus 1 – is shown in Fig. E5.5.1c. The value of $Z = j0.049pu$ at $h = 5$ and the phase angle (crossing the zero axis at $h = 5$) are not shown. A plot of the current amplification through the series-resonant circuit is given in Fig. E5.5.1d, pertaining to the tuned point, $h = 5$, and $I_{pu} = 0.99$. The current amplification should be near unity at the tuned frequency. Note that this circuit also has a parallel resonance at $h = 4.6$, where the amplification factor (AF) is $I_{pu} = 5.9$. This is the frequency at which the series-resonant circuit has a capacitive reactance equal in magnitude to the inductive reactance of the source.

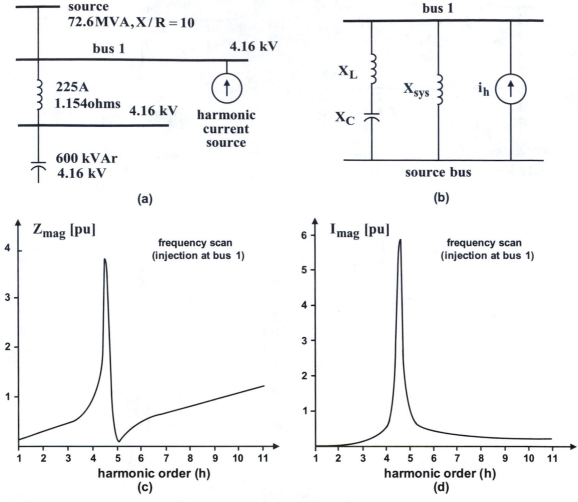

**FIGURE E5.5.1** (a) Example of single-line diagram for series resonance at bus 1 of power system, (b) equivalent circuit for series resonance-tuned filter, (c) frequency scan of bus 1 equivalent impedance, (d) current amplification across the tuning reactor [7 in Chapter 5]. The harmonic-current source is a six-pulse converter.

### 5.4.6 Application Example 5.6: Protecting Capacitors by Virtual Harmonic Resistors: Solution

**i)** The virtual resistor is only activated at harmonic frequencies; therefore, the 5th harmonic component of the injected current in the capacitor is

$$\tilde{I}_{C,5} = \tilde{I}_{C,5}^{due\,NL} + \tilde{I}_{C,5}^{due\,sys}$$

$$\tilde{I}_{C,5}^{due\,NL} = \frac{Z_{sys,5}}{Z_{sys,5} + Z_{C,5} + R_{vir}} \tilde{I}_{NL,5}$$

$$= \frac{0.06 + j0.2074}{0.06 + j0.2074 - j0.408 + 0.5} \tilde{I}_{NL,5}$$

$$= \frac{0.06 + j0.2074}{0.56 - j0.2} \tilde{I}_{NL,5} = \frac{0.2159\angle 73.87°}{0.595\angle -19.65°} \tilde{I}_{NL,5}$$

$$= (0.363\angle 93.52°)\tilde{I}_{NL,5}$$

$$= (0.363\angle 93.52°)(\sqrt{2}(0.3)\angle 0°)$$

$$= \sqrt{2}(0.109)\angle 93.52° \text{ A}$$

$$\tilde{I}_{C,5}^{due\,sys} = \frac{1}{Z_{sys,5} + Z_{C,5} + R_{vir}} \tilde{V}_{sys,5}$$

$$= \frac{1}{0.595\angle -19.65°} \tilde{V}_{sys,5} = (1.68\angle 19.65°)\tilde{V}_{sys,5}$$

$$= (1.68\angle 19.65°)(\sqrt{2}(3)\angle 0°)$$

$$= \sqrt{2}(5.4)\angle 19.65° \text{ A}$$

$$\Rightarrow\Rightarrow \tilde{I}_{C,5} = \tilde{I}_{C,5}^{due\,NL} + \tilde{I}_{C,5}^{due\,sys}$$

$$= (\sqrt{2}(0.109)\angle 93.52°) + (\sqrt{2}(5.4)\angle 19.65°)$$

$$= \sqrt{2}(5.09 + j1.93) = \sqrt{2}(5.44)\angle 20.77° \text{ A}.$$
$$(E5.6-1)$$

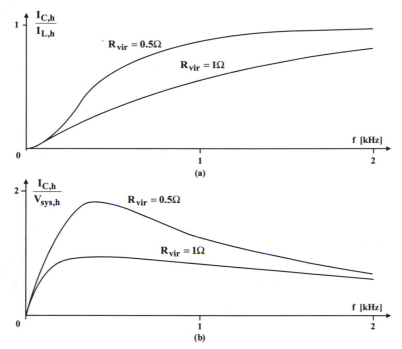

**FIGURE E5.6.1** Frequency response of capacitor bank as a function of (a) nonlinear load harmonic current, (b) utility harmonic voltage [2 in Chapter 5].

Note that the magnitude of the 5th harmonic capacitor current is decreased by 64% (e.g., from $\sqrt{2} \cdot 15.1$A, without the virtual resistor, to $\sqrt{2} \cdot 5.44$A).

**ii)** Similarly, the 7th harmonic component of the injected current in the capacitor is

$$\tilde{I}_{C,7} = \tilde{I}_{C,7}^{\text{due }NL} + \tilde{I}_{C,7}^{\text{due }sys}$$

$$\tilde{I}_{C,7}^{\text{due }NL} = \frac{Z_{sys,7}}{Z_{sys,7} + Z_{C,7} + R_{vir}} \tilde{I}_{NL,7}$$

$$= \frac{0.06 + j0.29029}{0.06 - j0.0011 + 0.5} \tilde{I}_{NL,7}$$

$$= \frac{0.2964 \angle 78.32°}{0.56 \angle -0.11°} \tilde{I}_{NL,7}$$

$$= (0.53 \angle 78.43°) \tilde{I}_{NL,7}$$

$$= (0.53 \angle 78.43°)(\sqrt{2}(0.2) \angle 0°)$$

$$= \sqrt{2}(0.106) \angle 78.43° \text{ A}$$

$$\tilde{I}_{C,7}^{\text{due }sys} = \frac{1}{Z_{sys,7} + Z_{C,7} + R_{vir}} \tilde{V}_{sys,7}$$

$$= \frac{1}{0.56 \angle -0.11°} \tilde{V}_{sys,5} = (1.79 \angle 0.11°) \tilde{V}_{sys,7}$$

$$= (1.79 \angle 0.11°)(\sqrt{2}(2) \angle 0°)$$

$$= \sqrt{2}(3.58) \angle 0.11° \text{ A}$$

$$\Rightarrow\Rightarrow \tilde{I}_{C,7} = \tilde{I}_{C,7}^{\text{due }NL} + \tilde{I}_{C,7}^{\text{due }sys}$$

$$= (\sqrt{2}(0.106) \angle 78.43°) + (\sqrt{2}(3.58) \angle 0.11°)$$

$$= \sqrt{2}(3.60 + 0.11) = \sqrt{2}(3.60) \angle 1.75° \text{ A}.$$

$$(E5.6-2)$$

Note that the magnitude of the 7th harmonic capacitor current is decreased by about 90% (e.g., from $\sqrt{2} \cdot 33.62$A, without the virtual resistor, to $\sqrt{2} \cdot 3.60$A).

**iii)** The improved frequency responses of power capacitor current for virtual harmonic resistors of $R_{vir} = 0.5$ Ω and 1 Ω as a function of load harmonic current and utility harmonic voltage are shown in Fig. E5.6.1.

### 5.5.1 Application Example 5.7: Frequency and Capacitance Scanning [9]: Solution

The frequency scan can be done by calculating and inverting the admittance matrix of the network. Magnitude and phase angle plots of the driving-point impedance at bus 3 (Fig. E5.7.2a) indicate a resonance at the 71th harmonic, where the magnitude rises to about 1729 Ω. This is also evident from the zero crossing of the phase plot.

A capacitance scan is performed in a similar manner by computing the driving-point impedance at bus 3 for different capacitance values of the capacitor. The result is shown in Fig. E5.7.2b for several values of h. Note that resonances at the 5th and the 7th harmonics occur for a capacitance in the range of $0.1/\omega_o$ F. These resonances should be avoided in the presence of six-pulse rectifiers.

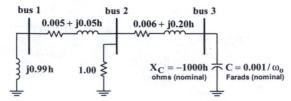

**FIGURE E5.7.1** The three-bus example system [9 in Chapter 5].

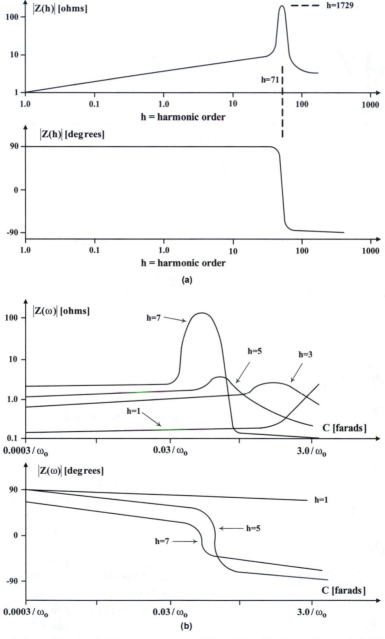

**FIGURE E5.7.2** Simulation results for the three bus system of Fig. E5.7.1; (a) frequency scan as seen at bus 3, (b) capacitance scan for impedance at bus 3 [9].

### 5.6.5 Application Example 5.8: Harmonic Limits for Capacitors when Used in a Three-Phase System: Solution

The feasible operating region is shown in Fig. E5.8.1.

### 5.7.1 Application Example 5.9: Harmonic Losses of Capacitors: Solution

***Single-Phase System.*** The application of Eq. 5-19 for $h = 1$ yields the fundamental loss for the single-phase system at $R_{s1} = 0.01 \, \Omega$, $P_1 = (100 \times 10^{-6})^2 (0.01)[(377)(1)]^2 (460)^2 = 3005$ mW.

For $h = h_{max} = 19$, Eq. 5-19 results for the single-phase system and $R_{s1} = 0.01 \, \Omega$ in the loss $P_{19} = (100 \times 10^{-6})^2 (0.01)[(377)(19)]^2 (4.6)^2 = 109$ mW.

a) The total harmonic losses including the fundamental loss is for the single-phase system at $R_{sh} = 0.01 \, \Omega$, $P_{total\_single-phase} = 3.608$ W.
b) The total harmonic losses including the fundamental loss is for the single-phase system at $R_{sh} = 0.01(h) \, \Omega$, $P_{total\_single-phase} = 10.035$ W.
c) The total harmonic losses including the fundamental loss is for the single-phase system at $R_{sh} = 0.01(h^2) \, \Omega$, $P_{total\_single-phase} = 107.62$ W.

Fundamental and harmonic losses for the single-phase system are listed in columns 3–5 of Table E5.9.1.

***Three-Phase System.*** The application of Eq. 5-19 for $h = 1$ yields the fundamental loss for the three-phase system provided the capacitor bank is in $\Delta$-connection at $R_{s1} = 0.01 \, \Omega$, $P_1 = (100 \times 10^{-6})^2 (0.01)[(377)(1)]^2 (460)^2 = 3005$ mW.

For $h = h_{max} = 19$ Eq. 5-19 results for the three-phase system at $R_{s1} = 0.01 \, \Omega$, $P_{19} = (100 \times 10^{-6})^2 (0.01)[(377)(19)]^2 (2.3)^2 = 21.7$ mW.

a) The total harmonic losses including the fundamental loss is for the three-phase system at $R_{sh} = 0.01 \, \Omega$, $P_{total\_three-phase} = 3.636$ W.
b) The total harmonic losses including the fundamental loss is for the three-phase system at $R_{sh} = 0.01(h) \, \Omega$, $P_{total\_three-phase} = 8.571$ W.
c) The total harmonic losses including the fundamental loss is for the three-phase system at $R_{sh} = 0.01(h^2) \, \Omega$, $P_{total\_three-phase} = 62.54$ W.

Fundamental and harmonic losses for the single-phase system are listed in columns 7-9 of Table E5.9.1.

**Conclusion.** The results of this application example stress that the harmonic losses of capacitors are indeed one of the reasons why the higher order harmonic voltage amplitudes must be limited as recommended in Table E5.9.1.

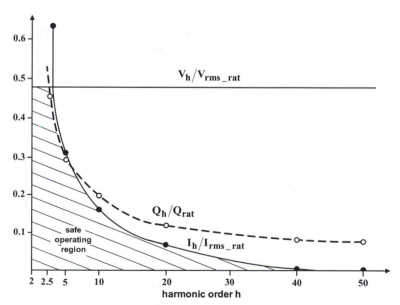

**FIGURE E5.8.1** Feasible operating region of capacitors with voltage, current, and reactive power harmonics (only one harmonic is present at any time).

**TABLE E5.9.1** Possible Voltage Spectra and Associated Capacitor Losses

| $h$ | $\left(\dfrac{V_h}{V_{60Hz}}\right)_{1\Phi}$ (%) | $P_h$ for part $a$ (mW) | $P_h$ for part $b$ (mW) | $P_h$ for part $c$ (mW) | $\left(\dfrac{V_h}{V_{60Hz}}\right)_{3\Phi}$ (%) | $P_h$ for part $a$ (mW) | $P_h$ for part $b$ (mW) | $P_h$ for part $c$ (mW) |
|---|---|---|---|---|---|---|---|---|
| 1 | 100 | 3005 | 3005 | 3005 | 100 | 3005 | 3005 | 3005 |
| 2 | 0.5 | 0.3 | 0.6 | 1.2 | 0.5 | 0.3 | 0.6 | 1.2 |
| 3 | 4.0 | 40 | 120 | 360 | $2.0^a$ | 10.8 | 32.4 | 97.2 |
| 4 | 0.3 | 0.4 | 1.6 | 6.4 | 0.5 | 1.2 | 4.8 | 19.2 |
| 5 | 3.0 | 67.7 | 339 | 1695 | 5.0 | 188 | 940 | 4700 |
| 6 | 0.2 | 0.43 | 2.6 | 15.6 | 0.2 | 0.43 | 2.6 | 15.6 |
| 7 | 2.0 | 58.9 | 412 | 2880 | 3.5 | 181 | 1267 | 8869 |
| 8 | 0.2 | 0.77 | 6.16 | 49.3 | 0.2 | 0.77 | 6.16 | 49.3 |
| 9 | 1.0 | 24.4 | 220 | 1980 | 0.3 | 2.19 | 19.71 | 177.4 |
| 10 | 0.1 | 0.3 | 3 | 30 | 0.1 | 0.3 | 3 | 30 |
| 11 | 1.5 | 81.9 | 900 | 9900 | 1.5 | 81.9 | 900 | 9900 |
| 12 | 0.1 | 0.433 | 5.2 | 62.4 | 0.1 | 0.433 | 5.2 | 62.4 |
| 13 | 1.5 | 114 | 1480 | 19240 | 1.0 | 114 | 1482 | 19270 |
| 14 | 0.1 | 0.59 | 8.26 | 115.6 | 0.05 | 0.15 | 2.58 | 28.81 |
| 15 | 0.5 | 16.9 | 254 | 3810 | 0.1 | 0.68 | 10.2 | 153.0 |
| 16 | 0.05 | 0.192 | 3.07 | 49.1 | 0.05 | 0.192 | 3.07 | 49.1 |
| 17 | 1.0 | 87 | 1480 | 25140 | 0.5 | 22 | 374 | 6358 |
| 18 | 0.05 | 0.244 | 4.4 | 79.2 | 0.01 | 0.0097 | 0.175 | 3.143 |
| 19 | 1.0 | 109 | 2070 | 39200 | 0.5 | 27.1 | 513 | 9750 |
| | | | | All higher voltage harmonics < 0.5% | | | | |

$^a$Under certain conditions (e.g., DC bias of transformers as discussed in Chapter 2, and the harmonic generation of synchronous generators as outlined in Chapter 4) triplen harmonics are not of the zero-sequence type and can therefore exist in a three-phase system.

# CHAPTER 6 SOLUTIONS

## 6.5.1 Application Example 6.1: Temperature Rise of a Single-Phase Transformer Due to Single Harmonic Voltage: Solution

The rated temperature rise is $\Delta T_{rated} = T_{rated} - T_{amb} = T_{rise\ rat} = 60°C$. For the harmonic with order $h = 3$ and amplitude $V_3 = 0.10$ pu $\equiv 10\%$, one obtains (through linear extrapolation) the weighted-harmonic factor by replacing $\dfrac{V_{ph}}{V_{p1}}$ by 10:

$$\sum_{h=3}\frac{1}{(3)^{0.9}}(10)^{1.75} = 20.98,$$

resulting with Fig. 6.14 in the additional temperature increase due to the 3rd harmonic $\Delta T_{h=3} = 3.3\%$ or $\Delta T_{h=3} = 1.98°C$.

## 6.5.2 Application Example 6.2: Temperature Rise of a Single-Phase Induction Motor Due to Single Harmonic Voltage: Solution

The rated temperature rise is $\Delta T_{rated} = T_{rated} - T_{amb} = T_{rise\ rat} = 60°C$. For the harmonic with order $h = 3$ and amplitude $V_3 = 0.10$ pu $\equiv 10\%$ one obtains the weighted-harmonic factor by replacing $\dfrac{V_{ph}}{V_{p1}}$ by 10:

$$\sum_{h=3}\frac{1}{(3)^{0.85}}(10)^{1.40} = 9.88,$$

resulting with Fig. 6.14 in the additional temperature increase due to the 3rd harmonic $\Delta T_{h=3} = 11\%$ or $\Delta T_{h=3} = 6.6°C$.

## 6.8.1 Application Example 6.3: Aging of a Single-Phase Induction Motor with $E = 0.74$ eV Due to a Single Harmonic Voltage: Solution

The slope $(E/K)$ is $E/K = 8.58 \times 10^{3}°$Kelvin. Thus one obtains from $t_1 = 40e^{-8.58\times10^3 \frac{\Delta T_h}{373(373+\Delta T_h)}}$,

$$t_1 = 40\ e$$

with $\Delta T_{h=3} = 6.6°C$, a reduced lifetime of $t_1|_{\Delta T_{h=3}=6.6°C} = 26.81$ years.

## 6.8.2 Application Example 6.4: Aging of a Single-Phase Induction Motor with $E = 0.51$ eV Due to a Single Harmonic Voltage: Solution

The slope $(E/K)$ is $E/K = 5.91 \times 10^3$ °Kelvin. Thus one obtains from

$$t_1 = 40e^{-5.91\times10^3 \frac{\Delta T_h}{373(373+\Delta T_h)}},$$

with $\Delta T_h = 6.6°C$, a reduced lifetime of $t_1|_{\Delta T_{h=3}=6.6°C} = 30.37$ years.

From Application Examples 6.3 and 6.4 one concludes that lower values of the activation energy E result in lower decreases in lifetime.

### 6.10.1 Application Example 6.5: Estimation of Lifetime Reduction for Given Single-Phase and Three-Phase Voltage Spectra with High-Harmonic Penetration with $E = 1.1$ eV: Solution

Calculation of weighted harmonic factor for single-phase spectrum of Table E6.5.1 based on the values for $k_{avg} = 0.85$ and $l_{avg} = 1.4$ for single-phase induction motors:

$$\frac{1}{2^{0.85}}(2.5)^{1.4} = 2.000, \frac{1}{3^{0.85}}(5.71)^{1.4} = 4.506,$$

$$\frac{1}{4^{0.85}}(1.6)^{1.4} = 0.594,\ldots,\frac{1}{17^{0.85}}(0.23)^{1.4} = 0.0115,$$

$$\frac{1}{18^{0.85}}(0.22)^{1.4} = 0.0103, \frac{1}{19^{0.85}}(1.03)^{1.4} = 0.0853.$$

Summing all contributions results in the weighted harmonic-voltage factor for single-phase induction motors:

**TABLE E6.5.1** Possible Voltage Spectra with High-Harmonic Penetration

| $h$ | $\left(\frac{V_h}{V_{60Hz}}\right)_{1\Phi}$ (%) | $\left(\frac{V_h}{V_{60Hz}}\right)_{3\Phi}$ (%) |
|---|---|---|
| 1 | 100 | 100 |
| 2 | 2.5 | 0.5 |
| 3 | 5.71 | 1.0 |
| 4 | 1.6 | 0.5 |
| 5 | 1.25 | 7.0 |
| 6 | 0.88 | 0.2 |
| 7 | 1.25 | 5.0 |
| 8 | 0.62 | 0.2 |
| 9 | 0.96 | 0.3 |
| 10 | 0.66 | 0.1 |
| 11 | 0.30 | 2.5 |
| 12 | 0.18 | 0.1 |
| 13 | 0.57 | 2.0 |
| 14 | 0.10 | 0.05 |
| 15 | 0.10 | 0.1 |
| 16 | 0.13 | 0.05 |
| 17 | 0.23 | 1.5 |
| 18 | 0.22 | 0.01 |
| 19 | 1.03 | 1.0 |
| | All higher harmonics < 0.2% | |

$$\sum_{h=2}^{h\max} \frac{1}{h^k}\left(\frac{V_{ph}}{V_{p1}}\right)^l \approx 8.5.$$

Calculation of weighted harmonic factor for three-phase spectrum of Table E6.5.1 using the values for $k_{avg} = 0.95$ and $l_{avg} = 1.6$ for three-phase induction motors:

$$\frac{1}{2^{0.95}}(0.5)^{1.6} = 0.1707, \frac{1}{3^{0.95}}(1.0)^{1.6} = 0.3522,$$

$$\frac{1}{4^{0.95}}(0.5)^{1.6} = 0.0884,\ldots,\frac{1}{17^{0.95}}(1.5)^{1.6} = 0.1297,$$

$$\frac{1}{18^{0.95}}(0.01)^{1.6} = 0.00004, \frac{1}{19^{0.95}}(1.0)^{1.6} = 0.0609.$$

Summing all contributions results in the weighted harmonic-voltage factor for three-phase induction motors:

$$\sum_{h=2}^{h\max} \frac{1}{h^k}\left(\frac{V_{ph}}{V_{p1}}\right)^l \approx 8.6.$$

Calculation of weighted harmonic factor for single-phase spectrum of Table E6.5.1 based on the values for $k_{avg} = 0.90$ and $l_{avg} = 1.75$ for single-phase transformers:

$$\frac{1}{2^{0.9}}(2.5)^{1.75} = 2.664, \frac{1}{3^{0.9}}(5.71)^{1.75} = 7.847,$$

$$\frac{1}{4^{0.9}}(1.6)^{1.75} = 0.6536,\ldots,\frac{1}{17^{0.9}}(0.23)^{1.75} = 0.00597,$$

$$\frac{1}{18^{0.9}}(0.22)^{1.75} = 0.00524, \frac{1}{19^{0.9}}(1.03)^{1.75} = 0.744.$$

Summing all contributions results in the weighted harmonic-voltage factor for single-phase transformers

$$\sum_{h=2}^{h\max} \frac{1}{h^k}\left(\frac{V_{ph}}{V_{p1}}\right)^l \approx 12.3.$$

Calculation of weighted harmonic factor for three-phase spectrum of Table E6.5.1 based on the values for $k_{avg} = 0.90$ and $l_{avg} = 1.75$ for three-phase transformers:

$$\frac{1}{2^{0.9}}(0.5)^{1.75} = 0.1593, \frac{1}{3^{0.9}}(1)^{1.75} = 0.37203,$$

$$\frac{1}{4^{0.9}}(0.5)^{1.75} = 0.0854,\ldots,\frac{1}{17^{0.9}}(1.5)^{1.75} = 0.15876,$$

$$\frac{1}{18^{0.9}}(0.01)^{1.75} = 0.0000234, \frac{1}{19^{0.9}}(1)^{1.75} = 0.07065.$$

Summing all above contributions results in the weighted harmonic-voltage factor for three-phase transformers:

$$\sum_{h=2}^{h\max} \frac{1}{h^k}\left(\frac{V_{ph}}{V_{p1}}\right)^l \approx 11.8.$$

Calculation of weighted harmonic factor for single-phase spectrum of Table E6.5.1 based on the values for $k_{avg} = 1.0$ and $l_{avg} = 2.0$ for universal motors:

$$\frac{1}{2}(2.5)^2 = 3.125, \frac{1}{3}(5.71)^2 = 10.868, \frac{1}{4}(1.6)^2 = 0.640,$$

$$\ldots, \frac{1}{17}(0.23)^2 = 0.00311, \frac{1}{18}(0.22)^2 = 0.0027,$$

$$\frac{1}{19}(1.03)^2 = 0.0558.$$

Summing all contributions results in the weighted harmonic-voltage factor for universal motors:

$$\sum_{h=2}^{h\max} \frac{1}{h^k}\left(\frac{V_{ph}}{V_{p1}}\right)^l \approx 15.6.$$

Based on Fig. 6.14, $T_2 = 85°C$, $T_{amb} = 23°C$, $E = 1.1$ eV, and rated lifetime of $t_2 = 40$ years the above harmonic factors result in the additional temperature rises and lifetime reductions of Table E6.5.2.

## 6.10.2 Application Example 6.6: Estimation of Lifetime Reduction for Given Single-Phase and Three-Phase Voltage Spectra with Moderate-Harmonic Penetration with $E = 1.1$ eV: Solution

Calculation of weighted harmonic factor for single-phase spectrum of Table E6.6.1 based on the values for $k_{avg} = 0.85$ and $l_{avg} = 1.4$ for single-phase induction motors:

$$\frac{1}{2^{0.85}}(0.5)^{1.4} = 0.2103, \frac{1}{3^{0.85}}(3)^{1.4} = 1.8297,$$

$$\frac{1}{4^{0.85}}(0.3)^{1.4} = 0.05703, \ldots, \frac{1}{17^{0.85}}(0.5)^{1.4} = 0.0341,$$

$$\frac{1}{18^{0.85}}(0.05)^{1.4} = 0.0013, \frac{1}{19^{0.85}}(0.4)^{1.4} = 0.0227.$$

Summing all contributions results in the weighted harmonic-voltage factor for single-phase induction motors:

$$\sum_{h=2}^{h\max} \frac{1}{h^k}\left(\frac{V_{ph}}{V_{p1}}\right)^l \approx 3.4.$$

Calculation of weighted harmonic factor for three-phase spectrum of Table E6.6.1 using the values for $k_{avg} = 0.95$ and $l_{avg} = 1.6$ for three-phase induction motors:

$$\frac{1}{2^{0.95}}(0.5)^{1.6} = 0.1707, \frac{1}{3^{0.95}}(0.5)^{1.6} = 0.1162,$$

$$\frac{1}{4^{0.95}}(0.5)^{1.6} = 0.0884, \ldots, \frac{1}{17^{0.95}}(0.3)^{1.6} = 0.00986,$$

$$\frac{1}{18^{0.95}}(0.01)^{1.6} = 0.00004, \frac{1}{19^{0.95}}(0.2)^{1.6} = 0.00464.$$

Summing all contributions results in the weighted harmonic-voltage factor for three-phase induction motors:

$$\sum_{h=2}^{h\max} \frac{1}{h^k}\left(\frac{V_{ph}}{V_{p1}}\right)^l \approx 2.6.$$

Calculation of weighted harmonic factor for single-phase spectrum of Table E6.6.1 based on the values for $k_{avg} = 0.90$ and $l_{avg} = 1.75$ for single-phase transformers:

$$\frac{1}{2^{0.90}}(0.5)^{1.75} = 0.1593, \frac{1}{3^{0.90}}(3)^{1.75} = 2.544,$$

$$\frac{1}{4^{0.90}}(0.3)^{1.75} = 0.0349, \ldots, \frac{1}{17^{0.90}}(0.5)^{1.75} = 0.0232,$$

$$\frac{1}{18^{0.90}}(0.05)^{1.75} = 0.000392, \frac{1}{19^{0.90}}(0.4)^{1.75} = 0.01421.$$

Summing all contributions results in the weighted harmonic-voltage factor for single-phase transformers:

**TABLE E6.5.2** Additional Temperature Rise and Associated Lifetime Reduction of Induction Motors, Transformers, and Universal Motors Due to the Harmonic Spectra of Table E6.5.1

|  | Three-phase induction motors | Single-phase transformers | Three-phase transformers | Universal motors |
|---|---|---|---|---|
| $\Delta T_h$ (%) | 9.2 | 4.5 | 1.9 | 1.8 | 2.4 |
| $\Delta T_h$ (°C) | 5.7 | 2.8 | 1.2 | 1.1 | 1.5 |
| Lifetime reduction (%) | 43 | 24 | 11 | 10 | 14 |

$$\sum_{h=2}^{h\max} \frac{1}{h^k}\left(\frac{V_{ph}}{V_{p1}}\right)^l \approx 4.1.$$

Calculation of weighted harmonic factor for three-phase spectrum of Table E6.6.1 based on the values for $k_{avg} = 0.90$ and $l_{avg} = 1.75$ for three-phase transformers:

$$\frac{1}{2^{0.90}}(0.5)^{1.75} = 0.1593, \frac{1}{3^{0.90}}(0.5)^{1.75} = 0.1106,$$

$$\frac{1}{4^{0.90}}(0.5)^{1.75} = 0.0854, \ldots, \frac{1}{17^{0.90}}(0.3)^{1.75} = 0.00949,$$

$$\frac{1}{18^{0.90}}(0.01)^{1.75} = 0.0000234, \frac{1}{19^{0.90}}(0.2)^{1.75} = 0.00423.$$

Summing all contributions results in the weighted harmonic-voltage factor for three-phase transformers:

$$\sum_{h=2}^{h\max} \frac{1}{h^k}\left(\frac{V_{ph}}{V_{p1}}\right)^l \approx 3.1.$$

**TABLE E6.6.1** Possible Voltage Spectra with Moderate-Harmonic Penetration

| $h$ | $\left(\dfrac{V_h}{V_{60Hz}}\right)_{1\Phi}$ (%) | $\left(\dfrac{V_h}{V_{60Hz}}\right)_{3\Phi}$ (%) |
|---|---|---|
| 1 | 100 | 100 |
| 2 | 0.5 | 0.5 |
| 3 | 3.0 | 0.5 |
| 4 | 0.3 | 0.5 |
| 5 | 2.0 | 3.0 |
| 6 | 0.2 | 0.2 |
| 7 | 1.0 | 2.5 |
| 8 | 0.2 | 0.2 |
| 9 | 0.75 | 0.3 |
| 10 | 0.1 | 0.1 |
| 11 | 1.0 | 1.0 |
| 12 | 0.1 | 0.1 |
| 13 | 0.9 | 0.85 |
| 14 | 0.1 | 0.05 |
| 15 | 0.3 | 0.1 |
| 16 | 0.05 | 0.05 |
| 17 | 0.5 | 0.3 |
| 18 | 0.05 | 0.01 |
| 19 | 0.4 | 0.2 |
| All higher harmonics < 0.2% | | |

Calculation of weighted harmonic factor for single-phase spectrum of Table E6.6.1 based on the values for $k_{avg} = 1.0$ and $l_{avg} = 2.0$ for universal motors:

$$\frac{1}{2}(0.5)^2 = 0.125, \frac{1}{3}(3)^2 = 3.0, \frac{1}{4}(0.3)^2 = 0.0225, \ldots,$$

$$\frac{1}{17}(0.5)^2 = 0.0147, \frac{1}{18}(0.05)^2 = 0.000139,$$

$$\frac{1}{19}(0.4)^2 = 0.00842.$$

Summing all contributions results in the weighted harmonic-voltage factor for universal motors:

$$\sum_{h=2}^{h\max} \frac{1}{h^k}\left(\frac{V_{ph}}{V_{p1}}\right)^l \approx 4.4.$$

Based on Fig. 6.14 the above harmonic factors, $T_2 = 85°C$, $T_{amb} = 23°C$, $E = 1.1$ eV, and rated lifetime of $t_2 = 40$ years result in the additional temperature rises and lifetime reductions of Table E6.6.2.

### 6.13.1 Application Example 6.7: Temperature Increase of Rotating Machine with a Step Load: Solution

With Eq. 6-39 one obtains

$$0.95 \cdot \theta_f = \theta_f - (\theta_f - \theta_{amb})e^{\left(\frac{-t_{95\%}}{\tau_\theta}\right)}$$

$$= 0.95 \cdot 120 = 120 - (120 - 40)e^{\left(\frac{-t_{95\%}}{\tau_\theta}\right)}$$

or

$$t_{95\%} = 2.59 \cdot \tau_\theta = 7.77 \text{ h}.$$

### 6.14.6 Application Example 6.8: Steady State with Superimposed Periodic Intermittent Operation with Irregular Load Steps: Solution

The induction motor is operated at 60 Hz (natural torque-speed characteristic) [46 in Chapter 6].

a)

$n_S = 1800$ rpm, $\Omega_{ms} = 188.49$ rad/s, $\omega_m = 180.1$ 1rad/s, $s_{rat} = 0.044$, $|\tilde{I}_s(t)| = 49.82$A, $|\tilde{I}'_r(t)| = 47.66$ A, $|\tilde{E}| = 221.75$ V, and $T_{rat} = 164.31$ Nm    (E6.8-1)

**TABLE E6.6.2** Additional Temperature Rise and Associated Lifetime Reduction of Induction Motors, Transformers, and Universal Motors Due to the Harmonic Spectra of Table E.6.6.1

| | Single-phase induction motors | Three-phase induction motors | Single-phase transformers | Three-phase transformers | Universal motors |
|---|---|---|---|---|---|
| $\Delta T_h$ (%) | 3.7 | 1.3 | 0.6 | 0.4 | 0.7 |
| $\Delta T_h$ (°C) | 2.3 | 0.81 | 0.37 | 0.25 | 0.43 |
| Lifetime reduction (%) | 20 | 7.7 | 3.6 | 2.5 | 4.2 |

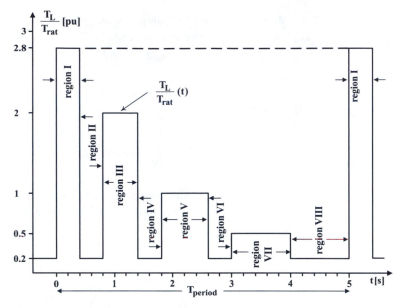

**FIGURE E6.8.1** Steady-state/intermittent load cycle.

b) Due to constant flux control, the $\Omega_m$–$T$ or the $s$–$T$ characteristic consists of a family of parallel lines as indicated in Fig. E6.8.2.

The maximum torque $T_{max}$ and the slip $s_m$ where the maximum torque occurs are [46]

$$T_{max} = 391.32 \text{ Nm}, \ s_m = 0.20. \qquad \text{(E6.8-2)}$$

From the relation (see Fig. E6.8.2)

$$\frac{T_{rat}}{T_{max\_fict}} = \frac{s_{rat}}{1} \qquad \text{(E6.8-3)}$$

one obtains for the maximum fictitious torque

$$T_{max\_fict} = T_{rat}/s_{rat} = 3734.3 \text{ Nm.} \qquad \text{(E6.8-4)}$$

c) Equation of motion (or fundamental torque equation) is

$$T - T_L - J\frac{d\omega_m}{dt} = 0, \qquad \text{(E6.8-5)}$$

where $T_L$ and $T$ are the load and the motor torques, respectively. $J$ is the total axial moment of inertia $J = J_m + J_{load}$. Dividing this equation by $T_{max\_fict}$ one gets

$$\frac{T}{T_{max\_fict}} - \frac{T_L}{T_{max\_fict}} - \frac{J}{T_{max\_fict}} \cdot \frac{d\omega_m}{dt} = 0 \qquad \text{(E6.8-6)}$$

or with $\dfrac{T}{T_{max\_fict}} = s$ follows

$$s - \frac{T_L}{T_{max\_fict}} - \frac{J}{T_{max\_fict}} \cdot \frac{d\omega_m}{dt} = 0. \qquad \text{(E6.8-7)}$$

with $\omega_m = \omega_{ms}(1 - s)$ the differential equation becomes

$$\tau_m \frac{ds}{dt} + s = \frac{T_L}{T_{max\_fict}}, \qquad \text{(E6.8-8)}$$

where

$$\tau_m = (J \cdot \omega_{ms})/T_{max\_fict} = (5 \cdot 188.49)/3734.3 = 0.252 \text{ s} \qquad \text{(E6.8-9)}$$

is the mechanical time constant.

The general solution of this differential equation is

$$s(t) = \text{const} \cdot e^{\left(-\frac{t}{\tau_m}\right)} + \frac{T_L}{T_{max\_fict}}. \qquad \text{(E6.8-10)}$$

For the initial condition where at $t = 0$ one assigns $s = s(0)$

$$s(t) = \left(s(0) - \frac{T_L(t)}{T_{max\_fict}}\right) \cdot e^{\left(-\frac{t}{\tau_m}\right)} + \frac{T_L(t)}{T_{max\_fict}}. \qquad \text{(E6.8-11)}$$

d) Application of s(t) to eight different regions of Fig. E6.8.1.

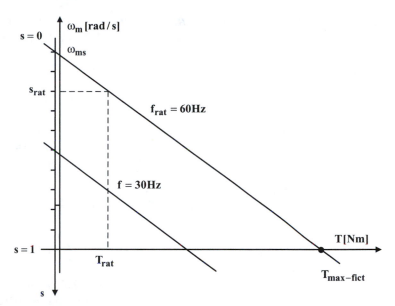

**FIGURE E6.8.2** $\Omega_m$–$T$ or $s$–$T$ characteristics for $f_{rat} = 60$ Hz (natural characteristic) and for $f = 30$ Hz.

**Region I.** It is assumed that the motor at time $t = 0$ s (at the start of the load cycle) has the steady-state slip of $s(0)$. This slip is obtained from

$$\frac{T(0)}{T_{rat}} = \frac{s(0)}{s_{rat}} \qquad \text{(E6.8-12)}$$

and

$$T(0) = 0.2 \cdot T_{rat} = 32.86 \text{ Nm},$$
$$s(0) = 32.86 \cdot 0.044/(164.31) = 0.0088.$$

Therefore, the slip at the start of region $I$ is

$$s(t^{(I)} = 0) = 0.0088.$$

It is best to introduce for each region a new coordinate system for the time axis. Thus for region $I$ $0 \leq t^{(I)} \leq 0.4$ s one obtains from Fig. E6.8.1 the load torque

$$T_L^{(I)} = 2.8 \times T_{rat} = 460.07 \text{ Nm}.$$

With Eq. E6.8-11

$$s(t^{(I)}) = \left( s(t^{(I)} = 0) - \frac{T_L^{(I)}(t^{(I)})}{T_{max\_fict}} \right) \cdot e^{\left( -\frac{t^{(I)}}{\tau_m} \right)} + \frac{T_L^{(I)}(t^{(I)})}{T_{max\_fict}}$$

$$= \left( 0.0088 - \frac{460.07}{3734.3} \right) e^{\left( -\frac{t^{(I)}}{\tau_m} \right)} + \frac{460.07}{3734.3}$$

$$s(t^{(I)}) = -0.1144 e^{\left( -\frac{t^{(I)}}{\tau_m} \right)} + 0.1232$$

$$s(t^{(I)} = 0.4 \text{ s}) = 0.0998 = s(t^{(II)} = 0) \text{ and}$$
$$s(t^{(I)} = 0.2 \text{ s}) = 0.07147.$$

**Region II.** For region II $(0 \leq t^{(II)} \leq 0.4$ s$)$ one obtains from Fig. E6.8.1 the load torque

$$T_L^{(II)} = 0.2 \cdot T_{rat} = 32.86 \text{ Nm}.$$

With Eq. E6.8-11

$$s(t^{(II)}) = \left( s(t^{(II)} = 0) - \frac{T_L^{(II)}(t^{(II)})}{T_{max\_fict}} \right) \cdot e^{\left( -\frac{t^{(II)}}{\tau_m} \right)} + \frac{T_L^{(II)}(t^{(II)})}{T_{max\_fict}}$$

$$= \left( 0.0998 - \frac{32.86}{3734.3} \right) e^{\left( -\frac{t^{(II)}}{\tau_m} \right)} + \frac{32.86}{3734.3}$$

$$s(t^{(II)}) = 0.091 e^{\left( -\frac{t^{(II)}}{\tau_m} \right)} + 0.0088$$

$$s(t^{(II)} = 0.4 \text{ s}) = 0.0274 = s(t^{(III)} = 0) \text{ and}$$
$$s(t^{(II)} = 0.2 \text{ s}) = 0.04994.$$

**Region III.** For region III $(0 \leq t^{(III)} \leq 0.6$ s$)$ one obtains from Fig. E6.8.1 the load torque

$$T_L^{(III)} = 2 \cdot T_{rat} = 328.62 \text{ Nm}.$$

With Eq. E6.8-11

$$s(t^{(III)}) = \left( s(t^{(III)} = 0) - \frac{T_L^{(III)}(t^{(III)})}{T_{max\_fict}} \right) \cdot e^{\left( -\frac{t^{(III)}}{\tau_m} \right)} + \frac{T_L^{(III)}(t^{(III)})}{T_{max\_fict}}$$

$$= \left( 0.0274 - \frac{328.62}{3734.3} \right) e^{\left( -\frac{t^{(III)}}{\tau_m} \right)} + \frac{328.62}{3734.3}$$

$$s(t^{(III)}) = -0.0606 e^{\left( -\frac{t^{(III)}}{\tau_m} \right)} + 0.088$$

$$s(t^{(III)} = 0.6 \text{ s}) = 0.0824 = s(t^{(IV)} = 0) \text{ and}$$
$$s(t^{(III)} = 0.3 \text{ s}) = 0.06957.$$

***Region IV.*** For region IV $(0 \le t^{(IV)} \le 0.4\text{ s})$ one obtains from Fig. E6.8.1 the load torque

$$T_L^{(IV)} = 0.2 \cdot T_{rat} = 32.86 \text{ Nm.}$$

With Eq. E6.8-11

$$s(t^{(IV)}) = \left(s(t^{(IV)} = 0) - \frac{T_L^{(IV)}(t^{(IV)})}{T_{max\_fict}}\right) \cdot e^{\left(\frac{-t^{(IV)}}{\tau_m}\right)} + \frac{T_L^{(IV)}(t^{(IV)})}{T_{max\_fict}}$$

$$= \left(0.0824 - \frac{32.86}{3734.3}\right)e^{\left(\frac{-t^{(IV)}}{\tau_m}\right)} + \frac{32.86}{3734.3},$$

$$s(t^{(IV)}) = 0.0736e^{\left(\frac{-t^{(IV)}}{\tau_m}\right)} + 0.0088,$$

$$s(t^{(IV)} = 0.4\text{ s}) = 0.02385 = s(t^{(V)} = 0) \text{ and}$$
$$s(t^{(IV)} = 0.2\text{ s}) = 0.04208.$$

***Region V.*** For region V $(0 \le t^{(V)} \le 0.8\text{ s})$ one obtains from Fig. E6.8.1 the load torque

$$T_L^{(V)} = T_{rat} = 164.3 \text{ Nm.}$$

With Eq. E6.8-11

$$s(t^{(V)}) = \left(s(t^{(V)} = 0) - \frac{T_L^{(V)}(t^{(V)})}{T_{max\_fict}}\right) \cdot e^{\left(\frac{-t^{(V)}}{\tau_m}\right)} + \frac{T_L^{(V)}(t^{(V)})}{T_{max\_fict}}$$

$$= \left(0.02385 - \frac{164.31}{3734.3}\right)e^{\left(\frac{-t^{(V)}}{\tau_m}\right)} + \frac{164.31}{3734.3}$$

$$s(t^{(V)}) = -0.0202e^{\left(\frac{-t^{(V)}}{\tau_m}\right)} + 0.044$$

$$s(t^{(v)} = 0.8\text{ s}) = 0.0432 = s(t^{(VI)} = 0) \text{ and}$$
$$s(t^{(V)} = 0.4\text{ s}) = 0.03987.$$

***Region VI.*** For region VI $(0 \le t^{(VI)} \le 0.4\text{ s})$ one obtains from Fig. E6.8.1 the load torque

$$T_L^{(IV)} = 0.2 \cdot T_{rat} = 32.86 \text{ Nm.}$$

With Eq. E6.8-11

$$s(t^{(VI)}) = \left(s(t^{(VI)} = 0) - \frac{T_L^{(VI)}(t^{(VI)})}{T_{max\_fict}}\right) \cdot e^{\left(\frac{-t^{(VI)}}{\tau_m}\right)} + \frac{T_L^{(VI)}(t^{(VI)})}{T_{max\_fict}}$$

$$= \left(0.0432 - \frac{32.86}{3734.3}\right)e^{\left(\frac{-t^{(VI)}}{\tau_m}\right)} + \frac{32.86}{3734.3}$$

$$s(t^{(VI)}) = 0.0344e^{\left(\frac{-t^{(VI)}}{\tau_m}\right)} + 0.0088$$

$$s(t^{(VI)} = 0.4\text{ s}) = 0.01583 = s(t^{(VII)} = 0) \text{ and}$$
$$s(t^{(VI)} = 0.2\text{ s}) = 0.0244.$$

***Region VII.*** For region VII $(0 \le t^{(VII)} \le 1\text{ s})$ one obtains from Fig. E6.8.1 the load torque

$$T_L^{(VII)} = 0.5 \cdot T_{rat} = 82.155 \text{ Nm.}$$

With Eq. E6.8-11

$$s(t^{(VII)}) = \left(s(t^{(VII)} = 0) - \frac{T_L^{(VII)}(t^{(VII)})}{T_{max\_fict}}\right) \cdot e^{\left(\frac{-t^{(VII)}}{\tau_m}\right)} + \frac{T_L^{(VII)}(t^{(VII)})}{T_{max\_fict}}$$

$$= \left(0.01583 - \frac{82.155}{3734.3}\right)e^{\left(\frac{-t^{(VII)}}{\tau_m}\right)} + \frac{82.155}{3734.3}$$

$$s(t^{(VII)}) = -0.00617e^{\left(\frac{-t^{(VII)}}{\tau_m}\right)} + 0.022$$

$$s(t^{(VII)} = 1\text{ s}) = 0.0219 = s(t^{(VIII)} = 0) \quad \text{and}$$
$$s(t^{(VII)} = 0.5\text{ s}) = 0.02115.$$

***Region VIII.*** For region VIII $(0 \le t^{(VIII)} \le 1\text{ s})$ one obtains from Fig. E6.8.1 the load torque

$$T_L^{(IV)} = 0.2 \cdot T_{rat} = 32.86 \text{ Nm.}$$

With Eq. E6.8-11

$$s(t^{(VIII)}) = \left(s(t^{(VIII)} = 0) - \frac{T_L^{(VIII)}(t^{(VIII)})}{T_{max\_fict}}\right) \cdot e^{\left(\frac{-t^{(VIII)}}{\tau_m}\right)}$$

$$+ \frac{T_L^{(VIII)}(t^{(VIII)})}{T_{max\_fict}} = \left(0.0219 - \frac{32.86}{3734.3}\right)e^{\left(\frac{-t^{(VIII)}}{\tau_m}\right)}$$

$$+ \frac{32.86}{3734.3}$$

$$s(t^{(VIII)}) = 0.0131e^{\left(\frac{-t^{(VIII)}}{\tau_m}\right)} + 0.0088$$

$s(t^{(VIII)} = 1\text{ s}) = 0.0090 \approx s(t^{(I)} = 0) = 0.0088$. That is, the periodicity of the slip s(t) is about satisfied.
$s(t^{(VIII)} = 0.5\text{ s}) = 0.0106$.
The plot of $s(t)$ is given in Fig. E6.8.3.

e) Calculation of the torque $T(t)$ and its plot from $t = 0$ to $t = 5$ s. From

$$\frac{T}{T_{max\_fict}} = \frac{s}{1}$$

follows

$$T = s \cdot T_{max\_fict}. \qquad \text{(E6.8-13)}$$

### Region I

$$T(t^{(I)}) = s(t^{(I)}) \cdot T_{max\_fict} \qquad (E6.8\text{-}14)$$

$T(t^{(I)} = 0) = 0.0088 \cdot 3734.3 \; 32.86 \; \text{Nm} \equiv 0.2 \; pu*,$
$T(t^{(I)} = 0.2 \text{ s}) = 266.9 \; \text{Nm} \equiv 1.62 \; pu,$
$T(t^{(I)} = 0.4 \text{ s}) = 372.7 \; \text{Nm} \equiv 2.27 \; pu.$

### Region II

$T(t^{(II)} = 0 \text{ s}) = 372.7 \; \text{Nm} \equiv 2.27 \; pu,$
$T(t^{(II)} = 0.2 \text{ s}) = 186.5 \; \text{Nm} \equiv 1.14 \; pu,$
$T(t^{(II)} = 0.4 \text{ s}) = 102.3 \; \text{Nm} \equiv 0.62 \; pu.$

### Region III

$T(t^{(III)} = 0 \text{ s}) = 102.3 \; \text{Nm} \equiv 0.62 \; pu,$
$T(t^{(III)} = 0.3 \text{ s}) = 259.8 \; \text{Nm} \equiv 1.58 \; pu,$
$T(t^{(III)} = 0.6 \text{ s}) = 307.7 \; \text{Nm} \equiv 1.87 \; pu.$

### Region IV

$T(t^{(IV)} = 0 \text{ s}) = 307.7 \; \text{Nm} \equiv 1.87 \; pu,$
$T(t^{(IV)} = 0.2 \text{ s}) = 157.1 \; \text{Nm} \equiv 0.96 \; pu,$
$T(t^{(IV)} = 0.4 \text{ s}) = 89.1 \; \text{Nm} \equiv 0.54 \; pu.$

### Region V

$T(t^{(V)} = 0 \text{ s}) = 89.1 \; \text{Nm} \equiv 0.54,$
$T(t^{(V)} = 0.4 \text{ s}) = 148.9 \; \text{Nm} \equiv 0.91 \; pu,$
$T(t^{(V)} = 0.8 \text{ s}) = 161.3 \; \text{Nm} \equiv 0.98 \; pu.$

### Region VI

$T(t^{(VI)} = 0 \text{ s}) = 161.3 \; \text{Nm} \equiv 0.982 \; pu,$
$T(t^{(VI)} = 0.2 \text{ s}) = 9.91 \; \text{Nm} \equiv 0.55 \; pu,$
$T(t^{(VI)} = 0.4 \text{ s}) = 59.1 \; \text{Nm} \equiv 0.36 \; pu.$

### Region VII

$T(t^{(VII)} = 0 \text{ s}) = 59.1 \; \text{Nm} \equiv 0.36 \; pu,$
$T(t^{(VII)} = 0.5 \text{ s}) = 79 \; \text{Nm} \equiv 0.48 \; pu,$
$T(t^{(VII)} = 1 \text{ s}) = 81.8 \; \text{Nm} \equiv 0.50 \; pu.$

### Region VIII

$T(t^{(VIII)} = 0 \text{ s}) = 81.8 \; \text{Nm} \equiv 0.50 \; pu,$
$T(t^{(VIII)} = 0.5 \text{ s}) = 39.6 \; \text{Nm} \equiv 0.24 \; pu,$
$T(t^{(VIII)} = 1 \text{ s}) = 33.6 \; \text{Nm} \equiv 0.024 \; pu.$

(*) referred to rated torque $T_{rat} = 164.31$ Nm. Note that the motor torque is at any time less than $T_{max} = 391.32$ Nm. The motor torque $T(t)$ is proportional to the slip $s(t)$ and is depicted in Fig. E6.8.4.

f) Point-wise calculation of the stator current $|\tilde{I}_s(t)|$ for the eight different regions. For the calculation of the stator current $|\tilde{I}_s(t)|$ the equivalent circuit of Fig. E6.8.5 is employed.

For $s(t^{(I)} = 0) = 0.0088$ one gets

$$\frac{R'_r}{s} = \frac{0.2}{0.0088} = 22.73 \,\Omega.$$

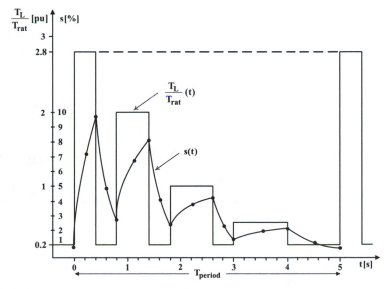

**FIGURE E6.8.3** Slip $s(t)$.

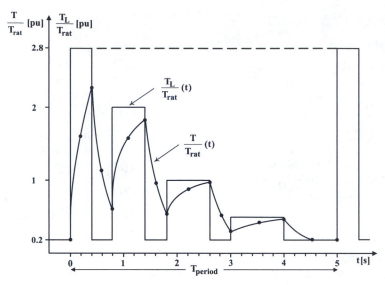

**FIGURE E6.8.4** Motor torque $T(t)$.

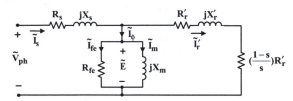

**FIGURE E6.8.5** Equivalent circuit of induction motor.

With $\tilde{E} = 221.75\ V\angle 0°$ the rotor current becomes

$$\tilde{I}'_r = \frac{\tilde{E}}{\left(\dfrac{R'_r}{s} + jX'_r\right)} = \frac{221.75\,\mathrm{V}\angle 0°}{(22.73 + j)} = (9.737 - j0.428)\,A.$$

The magnetizing current is

$$\tilde{I}_m = \frac{\tilde{E}}{jX_m} = \frac{221.75\angle°}{j30\Omega} = -j7.392\,A$$

and the stator current

$$\tilde{I}_s = \tilde{I}_m + \tilde{I}'_r = (9.737 - j7.82)\Omega$$

where $\tilde{E}$ is the reference phasor.

   Table E6.8.1 summarizes the point-wise calculation of the rms values of $|\tilde{I}_s(t)|$ over a load cycle.

g) Figure E6.8.6 shows the plot of $|\tilde{I}_s(t)|$ as a function of time and Fig. E6.8.7 depicts $|\tilde{I}_s(t)|^2$.

h) The rms current value of the stator current $I_{s\_rms\_actual}$ for the load cycle of Fig. E6.8.1 can be calculated from Fig. E6.8.7 using

$$I_{s\_rms\_actual} = \sqrt{\frac{1}{T}\int_t^{t+T} |\tilde{I}_s(t)|^2 \, \mathrm{d}t}.$$

The evaluation of this integral can be done mathematically or graphically. The graphic evaluation leads to $I_{s\_rms\_actual} = 47.7$A.

i) Comparison of rated $I_{s\_rms\_rat}$ and actual current $I_{s\_rms\_actual}$. Is the motor over- or underdesigned?

The losses of an induction motor are about proportional to the square of the stator current $I_{s\_rms}$. The ratio $\left(\dfrac{I_{s\_rms\_actual}}{I_{s\_rms\_rat}}\right)^2 = \left(\dfrac{47.4}{49.82}\right)^2 = 0.91$ indicates that the motor rating can be reduced by 9%, that is, $P_{out\_reduced} = 0.91 \cdot P_{out\_rat} = 0.91 \cdot 29.594\ \mathrm{kW} = 26.93$ kW or 36.1 hp.

   Let us assume that the stator current is larger than the rated current, say, $I_{s\_rms\_high} = 52$A; then the motor will experience a lifetime reduction – if the service factor of the motor is 1.0 – based on the ratio $\left(\dfrac{I_{s\_rms\_high}}{I_{s\_rms\_rat}}\right)^2 = \left(\dfrac{52}{49.82}\right)^2 = 1.09$, that is, the losses are about $\Delta P_{loss} = 9\%$ larger than the rated losses. This leads for a steady-state rated temperature of the hottest spot of $T_2 = 85°C \equiv 273 + 85 = 358$ °Kelvin at an ambient temperature of $T_{amb} = 40°C$ to an additional temperature rise of $\Delta T = 9\%$ or $\Delta T = 4.05°C$. With an activation energy of $E = 0.74$ eV and a rated lifetime of 40 years the reduced lifetime of the three-phase induction motor is with $\Delta T = 4.05°C$

**583**

**TABLE E6.8.1** Point-Wise Calculation of the rms Values of $|\tilde{I}_s(t)|$ Measured in Amperes

| Region I | $|\tilde{I}_s|$ | Region II | $|\tilde{I}_s|$ | Region III | $|\tilde{I}_s|$ |
|---|---|---|---|---|---|
| $s(t^{(I)} = 0) = 0.0088$ | 12.49 | $s(t^{(II)} = 0) = 0.0998$ | 102.69 | $s(t^{(III)} = 0) = 0.0274$ | 31.95 |
| $s(t^{(I)} = 0.2 \text{ s}) = 0.07147$ | 77.44 | $s(t^{(II)} = 0.2 \text{ s}) = 0.04994$ | 56.04 | $s(t^{(III)} = 0.3 \text{ s}) = 0.06957$ | 75.56 |
| $s(s^{(I)} = 0.4 \text{ s}) = 0.0998$ | 102.69 | $s(t^{(II)} = 0.4 \text{ s}) = 0.0274$ | 31.95 | $s(t^{(III)} = 0.6 \text{ s}) = 0.0824$ | 101.74 |
| Region IV | $|\tilde{I}_s|$ | Region V | $|\tilde{I}_s|$ | Region VI | $|\tilde{I}_s|$ |
| $s(t^{(IV)} = 0) = 0.082$ | 101.74 | $s(t^{(V)} = 0) = 0.02385$ | 28.11 | $s(t^{VI}) = 0) = 0.0432$ | 48.92 |
| $s(t^{(IV)} = 0.2 \text{ s}) = 0.04208$ | 47.76 | $s(t^{(V)} = 0.4 \text{ s}) = 0.03987$ | 45.38 | $s(t^{(VI)} = 0.2 \text{ s}) = 0.0244$ | 28.706 |
| $s(t^{(IV)} = 0.4 \text{ s}) = 0.02385$ | 28.11 | $s(t^{(V)} = 0.8 \text{ s}) = 0.0432$ | 48.92 | $s(t^{(VI)} = 0.4 \text{ s}) = 0.01583$ | 19.53 |
| Region VII | $|\tilde{I}_s|$ | Region VIII | $|\tilde{I}_s|$ | | |
| $s(t^{(VII)} = 0) = 0.01583$ | 19.53 | $s(t^{(VIII)} = 0) = 0.0219$ | 26 | | |
| $s(t^{(VIII)} = 0.5 \text{ s}) = 0.02115$ | 25.19 | $s(t^{(VIII)} = 0.5 \text{ s}) = 0.0106$ | 14.2 | | |
| $s(t^{(VII)} = 1 \text{ s}) = 0.0219$ | 26 | $s(t^{(VIII)} = 1 \text{ s}) = 0.009$ | 12.68 | | |

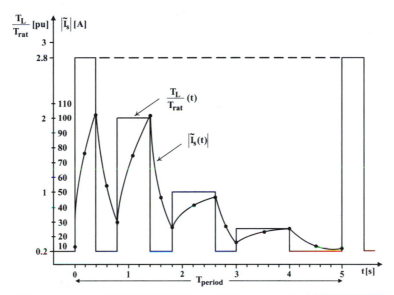

**FIGURE E6.8.6** $|\tilde{I}_s(t)|$.

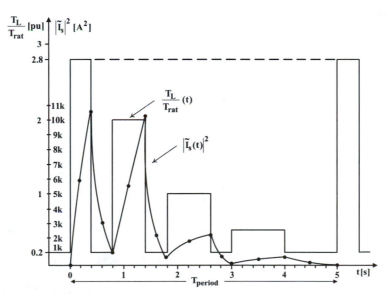

**FIGURE E6.8.7** $|\tilde{I}_s(t)|^2$.

$$t_1 = 40e^{-8.58 \times 10^3 \left( \frac{\Delta T}{358(358 + \Delta T)} \right)} = 30.6 \, \text{years.}$$

Most induction motors are designed with a service factor of 1.1–1.15, that is, the motor can be overloaded by 10 to 15 %. If this motor were designed with a service factor of 1.10 then no lifetime reduction would occur for a stator current of $I_{s\_rms\_high} = 52 \text{A}$.

### 6.14.8 Application Example 6.9: Reduction of Harmonic Torques of a Piston-Compressor Drive with Synchronous Motor as Prime Mover: Solution

a) $n_{ms} = (120 \cdot f)/p = 257.14 \, \text{rpm}, \, \omega_{ms} = (2\pi \cdot n_{ms})/60 = 26.927 \, \text{rad/s}$,

$$T_{rat} = \frac{P_{rat}}{\omega_{ms}} = 278.53 \, \text{kNm}, \qquad (\text{E6.9-5})$$

$$\left| \tilde{I}_{a\_rat} \right| = \frac{P_{rat}}{\sqrt{3} \cdot V_{L-L\,rat} \cos\phi_{rat} \eta_{rat}} = 827.6 \, \text{A.} \quad (\text{E6.9-6})$$

The average load torque $T_{Lavg} = 265.53$ kNm leads with

$$T_{Lavg} = \frac{3 |\tilde{E}_a| |\tilde{V}_{a\_rat}|}{X_s \omega_{ms}} \sin\delta \quad \text{and} \quad T_{max} = \frac{3 |\tilde{E}_a| |\tilde{V}_{a\_rat}|}{X_s \omega_{ms}}$$

to the ratio

$$\frac{T_{Lavg}}{T_{max}} = \sin\delta \quad \text{or} \quad \delta = \sin^{-1}\left( \frac{T_{Lavg}}{2.3 \cdot T_{rat}} \right) \qquad (\text{E6.9-7})$$
$$= 0.427 \, \text{rad or } 24.48°.$$

Figure E6.9.1 defines the torque angle $\delta$.

From the phasor diagram (overexcited, consumer notation) one obtains

$$\left| \tilde{E}_a \right|^2 = \left( |\tilde{V}_{a\_rat}| + |\tilde{I}_a| X_s \sin\phi \right)^2 + \left( |\tilde{I}_a| X_s \cos\phi \right)^2$$

or

$$\left| \tilde{E}_a \right|^2 = |\tilde{V}_{a\_rat}|^2 + 2 |V_{a\_rat}| |I_a| X_s \sin\phi + \left( |\tilde{I}_a| X_s \right)^2$$
$$(\text{E6.9-8})$$

and with

$$P = \frac{3 |\tilde{E}_a| |\tilde{V}_{a\_rat}| \sin\delta}{X_s} \qquad (\text{E6.9-9})$$

follows the second-order equation for $X_s$

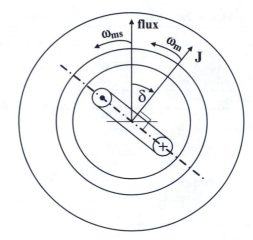

**FIGURE E6.9.1** Definition of torque angle $\delta$ without induction motor damping.

$$\left[ \left( \frac{P}{3 |\tilde{V}_{a\_rat}| \sin\delta} \right)^2 - |\tilde{I}_a|^2 \right] X_s^2$$
$$- \left( 2 |\tilde{V}_{a\_rat}| |\tilde{I}_a| \sin\phi \right) X_s - |\tilde{V}_{a\_rat}|^2 = 0. \qquad (\text{E6.9-10})$$

For the given operating point one gets

$$P = T_{Lavg} \cdot \omega_s = 7.14 \, \text{MW}, \, |\tilde{V}_{a\_rat}| = 3.464 \, \text{kV},$$
$$\sin\delta = \frac{T_{lavg}}{T_{max}} = \frac{T_{Lavg}}{2.3 \cdot T_{rat}} = 0.4145. \, \cos\Phi = 0.9,$$
$$\sin\Phi = 0.4359,$$

$$|\tilde{I}_a| = \frac{T_{Lavg} \cdot \omega_s}{3 \cdot |\tilde{V}_{a\_rat}| \cdot \cos\phi \cdot \eta_{rat}} = 789 \, \text{A}, \qquad (\text{E6.9-11})$$

and therefore

$$X_S = 3.07 \, \Omega. \qquad (\text{E6.9-12})$$

With the base impedance of

$$Z_{base} = \frac{|\tilde{V}_{a\_rat}|}{|\tilde{I}_{a\_rat}|} = 4.18 \Omega \qquad (\text{E6.9-13})$$

the per-unit synchronous reactance is

$$X_{s\_pu} = 0.74 \, \text{pu} \qquad (\text{E6.9-14})$$

b) Nonlinear differential equation for the motor torque.

The torque equation is

$$T - T_L - J \frac{d\omega_m}{dt} = 0, \qquad (\text{E6.9-15})$$

where

$$T = T_{max} \cdot \sin \delta, \delta = \frac{p}{2}\alpha, T_L = T_{Lavg} + T_{L1}, \alpha = \omega_{ms}t - \omega_m t,$$

$$\frac{d\alpha}{dt} = \omega_{ms} - \omega_m, \frac{d\delta}{dt} = \frac{p}{2} \cdot \frac{d\alpha}{dt}, \frac{d\delta}{dt} = \frac{p}{2}(\omega_{ms} - \omega_m),$$

$$\omega_m = \omega_{ms} - \frac{1}{p/2} \cdot \frac{d\delta}{dt}, \frac{d\omega_m}{dt} = -\frac{1}{p/2}\frac{d^2\delta}{dt^2}.$$

(E6.9-16)

The nonlinear differential equation becomes

$$T_{max} \sin \delta - T_{Lavg} - T_{L1} + \frac{J}{p/2}\frac{d^2\delta}{dt^2} = 0. \quad (E6.9-17)$$

Linearization of the nonlinear differential equation is based on Fig. E6.9.2.
The motor torque is

$$T = T_{max}\sin(\delta_0 + \delta_1) = T_{max}\sin\delta_0 + T_{max}\delta_1\cos\delta_0$$

(E6.9-18)

or with the synchronizing torque $S = T_{max} \cos \delta_0$, the linearized differential equation becomes

$$\delta_1 + \frac{J}{(p/2)S}\frac{d^2\delta_1}{dt^2} = \frac{T_{L1}}{S}. \quad (E6.9-19)$$

The eigen-angular frequency of the undamped machine can be defined by

$$\omega_{eigen} = \sqrt{\frac{(p/2)S}{J}}. \quad (E6.9-20)$$

c) Repetition of part b, but not neglecting damping provided by induction motor action as is indicated in Fig. E6.9.3. The small-slip approximation of an induction motor is depicted in Fig. E6.9.4.

From Fig. E6.9.4 one obtains for the motor torque developed by induction-motor action

$$T_{induction} = \frac{T_{rat} \cdot s}{s_{rat}}, s = 1 - \frac{n_m}{n_{ms}} = 1 - \frac{\omega_m}{\omega_{ms}},$$

$$T_{induction} = \frac{T_{rat}}{s_{rat}}\left(1 - \frac{\omega_m}{\omega_{ms}}\right), \quad (E6.9-21a)$$

or

$$T_{induction} = \frac{T_{rat}}{S_{rat}}\frac{1}{(p/2) \cdot \omega_{ms}} \cdot \frac{d\delta}{dt}. \quad (E6.9-21b)$$

With the power system's frequency $\omega_s = 2\pi f$ and $\omega_s = (p/2\omega_{ms})$ one obtains the total motor torque

$$T = T_{max}\sin\delta + \frac{T_{rat}}{s_{rat}}\frac{1}{\omega_s}\cdot\frac{d\delta}{dt} \quad (E6.9-21c)$$

or the nonlinear differential equation including the damping due to induction-motor action

$$\frac{J}{p/2}\frac{d^2\delta}{dt^2} + \frac{T_{rat}}{s_{rat}\cdot\omega_s}\frac{d\delta}{dt} + T_{max}\sin\delta = T_{Lavg} + T_{L1}.$$

(E6.9-22)

Linearization leads to the differential equation

$$\frac{J}{p/2}\frac{d^2\delta_1}{dt^2} + D\frac{d\delta_1}{dt} + S\delta_1 = T_{L1}, \quad (E6.9-23)$$

where the damping factor is

$$D = \frac{T_{rat}}{s_{rat}\cdot\omega_s}. \quad (E6.9-24)$$

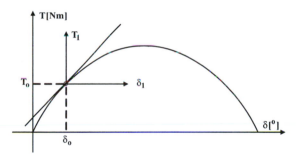

**FIGURE E6.9.2** Linearization of sine function for a given operating point $\delta = \delta_0 + \delta_1$, where $\delta_1 \ll \delta_0$.

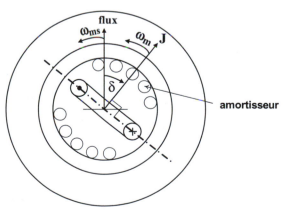

**FIGURE E6.9.3** Definition of torque angle $\delta$ with induction motor damping.

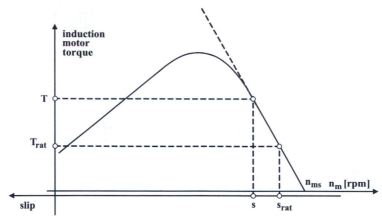

**FIGURE E6.9.4** Small-slip approximation of induction motor characteristic.

d) Solution of the linearized differential equations of part b – without damping – and determining (by introducing phasors indicated by "~") the axial moment of inertia J required so that the first harmonic motor torque will be reduced to ±4% of the average torque (Eq. E6.9-2).

Neglecting damping one obtains for the kth harmonic the differential equation

$$\delta_1 + \frac{1}{\omega_{eigen}^2} \frac{d^2\delta_1}{dt^2} = \frac{T_{Lk}}{S}. \qquad (E6.9\text{-}25)$$

Replacing $\frac{d}{dt}$ by $j\omega_k$ for sine/cosine variations of harmonic torques one gets

$$\tilde{\delta}_1 + \frac{1}{\omega_{eigen}^2}(j\omega_k)^2 \tilde{\delta}_1^2 = \frac{\tilde{T}_{Lk}}{S}. \qquad (E6.9\text{-}26)$$

With

$$\tilde{T}_k = T_{max}\cos\delta_0\, \tilde{\delta}_1 = S \cdot \tilde{\delta}_1 \qquad (E6.9\text{-}27)$$

follows

$$\frac{\tilde{T}_k}{\tilde{T}_{Lk}} = \frac{1}{\left(1 - \dfrac{\omega_k^2}{\omega_{eigen}^2}\right)}, \qquad (E6.9\text{-}28)$$

or magnitudes only where "∧" indicates amplitude

$$\frac{|\tilde{T}_k|}{|\tilde{T}_{Lk}|} = \frac{\hat{T}_k}{\hat{T}_{Lk}} = \frac{1}{\left[1 - \left(\dfrac{\omega_k}{\omega_{eigen}}\right)^2\right]}, \qquad (E6.9\text{-}29)$$

where $\tan(\varepsilon) = 0$ for $\omega_k < \omega_{eigen}$ and $\tan(\varepsilon) = -180°$ for $\omega_k > \omega_{eigen}$.

For the first harmonic torque one can write

$$T_{L1} = 31.04 \text{ kNm } \sin(\omega_{ms} - \pi),\ \hat{T}_1 = \pm 0.04T_{rat} = \pm 0.04T_{Lavg} \qquad (E6.9\text{-}30)$$

$$\frac{\hat{T}_1}{\hat{T}_{L1}} = \frac{1}{\left(1 - \dfrac{\omega_1^2}{\omega_{eigen}^2}\right)}. \qquad (E6.9\text{-}31)$$

The required inertia becomes for the worst case (– sign):

$$J = \frac{S(p/2)}{\omega_1^2}\left(1 - \frac{\hat{T}_{L1}}{\hat{T}_1}\right) = \frac{2.3 \cdot T_{rat} \cdot \cos\delta_o \cdot (p/2)^3}{\omega^2}$$
$$\left(1 - \frac{\hat{T}_{L1}}{\hat{T}_1}\right) = 11.256 \cdot 10^3 \left(1 - \frac{31.04}{\pm 0.04 \cdot 265.53}\right)$$
$$= 44 \cdot 10^3 \text{ kgm}^2.$$

$$(E6.9\text{-}32)$$

Solution of the linearized differential equations of part c – with damping – and determining (by introducing phasors) the axial moment of inertia J required so that the first harmonic motor torque will be reduced to ±4% of the average torque (Eq. E6.9-2).

Not neglecting damping one obtains for the kth harmonic the differential equation

$$\frac{J}{p/2}\frac{d^2\delta_1}{dt^2} + D\frac{d\delta_1}{dt} + S\delta_1 = T_{Lk} \qquad (E6.9\text{-}33)$$

or for sine/cosine variations of harmonic load torques

$$(j\omega_k)^2 \frac{J}{(p/2)} \cdot \tilde{\delta}_1 + j\omega_k D \tilde{\delta}_1 + S \tilde{\delta}_1 = \tilde{T}_{Lk} \quad \text{(E6.9-34)}$$

with

$$S \tilde{\delta}_1 = \tilde{T}_k \quad \text{(E6.9-35)}$$

$$\frac{\tilde{T}_k}{\tilde{T}_{Lk}} = \frac{1}{(j\omega_k)^2 \cdot \frac{J}{(p/2) \cdot S} + j\omega_k \frac{D}{S} + 1} \quad \text{(E6.9-36)}$$

or the magnitudes only

$$\frac{|\tilde{T}_k|}{|\tilde{T}_{Lk}|} = \frac{\hat{T}_k}{\hat{T}_{Lk}} = \frac{1}{\sqrt{\left(1 - \omega_k^2 \frac{J}{(p/2) \cdot S}\right)^2 + \left(\omega_k \frac{D}{S}\right)^2}}$$

$$\text{(E6.9-37)}$$

with

$$\tan(\varepsilon) = \frac{\omega_k \frac{D}{S}}{1 - \omega_k^2 \frac{J}{(p/2) \cdot S}}. \quad \text{(E6.9-38)}$$

The inertia is now

$$J = \frac{(p/2) \cdot S}{\omega_k^2} \left[1 \mp \sqrt{\left(\frac{\hat{T}_{Lk}}{\hat{T}_k}\right)^2 - \left(\omega_k \frac{D}{S}\right)^2}\right] \quad \text{(E6.9-39)}$$

with

$$\omega_k = \omega_{ms} = \frac{\omega_s}{(p/2)}, \hat{T}_{Lk} = \hat{T}_{L1} = 31.04 \,\text{kNm},$$

$$\hat{T}_k = \hat{T}_1 = 0.04 \cdot 265.53 \,\text{kNm} = 10.62 \,\text{kNm},$$

one obtains with D and S defined above

$$J = 43.66 \cdot 10^3 \,\text{kgm}^2 \approx 44 \cdot 10^3 \,\text{kgm}^2, \quad \text{(E6.9-40)}$$

***Conclusion.*** Damping does not significantly influence the calculation of the required axial moment of inertia!

e) Comparison of the eigen frequency $f_{\text{eigen}}$ with the pulsation frequencies occurring at the eccentric shaft.

The angular eigen (resonance) frequency of the undamped oscillation is

$$\omega_{\text{eigen}} = \sqrt{\frac{S(p/2)}{J}} = 13.62 \,\text{rad/s}$$

or the eigen frequency is

$$f_{\text{eigen}} = \frac{\omega_{\text{eigen}}}{2\pi} = 2.17 \frac{1}{s}.$$

The first harmonic pulsation (given by the load) frequency of the eccentric shaft is

$$\omega_1 = \omega_{ms} = \frac{\omega_s}{(p/2)} = \frac{2\pi 60}{14} = 26.93 \frac{1}{s}. \quad \text{(E6.9-41)}$$

The ratio between first harmonic pulsation frequency and resonance frequency is

$$\frac{\omega_1}{\omega_{\text{eigen}}} = \frac{f_1}{f_{\text{eigen}}} = 1.98 \approx 2.0. \quad \text{(E6.9-42)}$$

The fourth harmonic pulsation (given by the load) frequency of the eccentric shaft is

$$\omega_4 = 4 \cdot \omega_{ms} = 4 \cdot \omega_1. \quad \text{(E6.9-43)}$$

The ratio between fourth harmonic pulsation frequency and resonance frequency is

$$\frac{\omega_4}{\omega_{\text{eigen}}} = \frac{f_4}{f_{\text{eigen}}} = 4 \cdot 1.98 \approx 8.0. \quad \text{(E6.9-44)}$$

Note that from a dynamic performance point of view it is desirable if

$$\frac{\omega_k}{\omega_{\text{eigen}}} \geq 3 - 4. \quad \text{(E6.9-45)}$$

As can be seen the ratio $\omega_1/\omega_{\text{eigen}} = 1.98 \approx 2.0$ does not satisfy this condition and therefore the calculated axial moment of inertia of $44 \cdot 10^3 \,\text{kgm}^2$ is not large enough. It is recommended to select a desirable axial moment of inertia of $J_{\text{desirable}}$ that is larger than $44 \times 10^3 \,\text{kgm}^2$ and results in $\omega_1/\omega_{\text{eigen}} \approx 3.0$.

f) List the total motor torque T(t) as a function of time (neglecting damping) for one revolution of the eccentric shaft and compare it with the load torque $T_L(t)$. Why are they different?

The total motor torque is now

$$\vec{T} = \vec{T}_{\text{avg}} + \vec{T} + \vec{T} \quad \text{(E6.9-46)}$$

where

$$\bar{T}_{avg} = 265.53 \text{kNm} \qquad \text{(E6.9-47)}$$

and the motor torque for the kth harmonic is

$$\vec{T}_k = \frac{\vec{T}_{Lk}}{1 - \left(\dfrac{\omega_k^2}{\omega_{eigen}^2}\right)}. \qquad \text{(E6.9-48)}$$

For the first harmonic load torque one obtains

$$T_{L1}(t) = 31.04 \text{ kNm} \sin(\omega_{ms}t - \pi)$$
$$= -31.04 \text{ kNm} \sin(\omega_{ms}t) \qquad \text{(E6.9-49)}$$

and the first harmonic motor torque becomes

$$T_1(t) = \frac{-31.04 \text{ kNm}}{1 - \left(\dfrac{\omega_1^2}{\omega_{eigen}^2}\right)} \sin\omega_{ms}t = 10.35 \text{kNm} \sin(\omega_{ms}t).$$

$$\text{(E6.9-50)}$$

For the fourth harmonic motor torque one gets

$$T_4(t) = \frac{50.65 \text{kNm}}{1 - \left(\dfrac{\omega_1^2}{\omega_{eigen}^2}\right)} \sin\left(\omega_{ms}t - \frac{\pi}{6}\right)$$

$$\text{(E6.9-51)}$$

$$= -0.804 \text{kNm} \sin\left(\omega_{ms}t - \frac{\pi}{6}\right).$$

***Conclusions.*** The total load torque is

$$T_L(t) = 265.53 \text{ kNm} + 31.04 \text{ kNm} \sin(\omega_{ms}t - \pi) +$$
$$50.6 \text{ kNm} \sin(4\omega_{ms} - \pi/6) \qquad \text{(E6.9-52)}$$

and the total motor torque becomes

$$T(t) = 265.53 \text{kNm} + 10.35 \text{kNm} \sin(\omega_{ms}t)$$
$$-0.804 \text{kNm} \sin\left(\omega_{ms}t - \frac{\pi}{6}\right). \qquad \text{(E6.9-53)}$$

The difference in the harmonic torques is provided by the flywheel with inertia of $J = 44 \cdot 10^3 \text{ kgm}^2$.

### 6.14.10 Application Example 6.10: Temperature-Rise Equations for a Totally Enclosed Fan-Cooled 100 hp Motor: Solution

From Fig. E6.10.1 one notes that the stator ohmic loss generates a temperature drop across the slot insulation $(\Delta T_i)$. The temperature drop across the stator back iron or yoke $(\Delta T_{y1})$ is produced by the sum of the stator and rotor ohmic losses, stator teeth

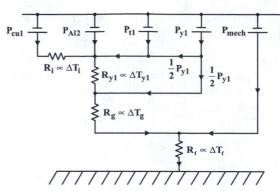

**FIGURE E6.10.1** Thermal network of a totally enclosed fan-cooled induction motor

loss, and half of the stator back iron loss. The calculation of $\Delta T_{y1}$ is based on half of the stator back iron loss because this loss is assumed to be generated uniformly throughout the stator back iron; therefore, the stator back iron loss generated radially through the stator back iron varies from zero to $P_{y1}$ from the inner to the outer surfaces of the stator back iron, respectively. The temperature drop across the small air gap between stator back iron and the frame $(\Delta T_g)$ is produced by the sum of the ohmic and the iron-core losses of the motor. Finally, the total losses cause a temperature drop from the motor frame to the moving air $(\Delta T_f)$.

The temperature difference between the stator winding and the iron core (across the stator winding insulation) is

$$\Delta T_i = P_{cu1} \frac{\beta_i}{\lambda_i \cdot S_i}, \qquad \text{(E6.10-2)}$$

where

$\beta_i$ is the thickness of stator slot insulation
$\lambda_i$ is the thermal conductivity (W/m°C) of stator slot insulation
$S_i$ is the total surface area of (all) stator slots

The temperature drop across the stator back iron can be expressed as

$$\Delta T_{y1} = \left(P_{cu1} + \frac{1}{2}P_{y1} + P_{t1} + P_{sp1} + P_{Al2} + P_{sp2}\right) \frac{h_{y1}}{\lambda_c \cdot S_{y1}},$$

$$\text{(E6.10-3a)}$$

where

$h_{y1}$ is the radial height of stator back iron

$\lambda_c$ is the thermal conductivity (W/m°C) of the stator laminations

$S_{y1}$ is the cross-sectional area of the stator core at the radial middle of the back iron

$P_{sp1}$ is the surface and pulsation loss caused by stator teeth (note that these losses are very small and have been neglected in Fig. E6.10.1)

$P_{sp2}$ is the surface and pulsation loss caused by rotor teeth (note that these losses are very small and have been neglected in Fig. E6.10.1)

The temperature difference across the small air gap between the stator back iron and the frame can be expressed as

$$\Delta T_\delta = \left(P_{cu1} + P_{y1} + P_{t1} + P_{sp1} + P_{Al2} + P_{sp2}\right)\frac{\beta_\delta}{\lambda_\delta \cdot S_\delta},$$

$$(E6.10\text{-}3b)$$

where

$\beta_\delta$ is the air gap length between stator back iron and frame

$\lambda_\delta$ is the thermal conductivity (W/m°C) of the airgap

$S_\delta$ is the outside surface of stator core

$P_{sp1}$ is the surface and pulsation loss caused by stator teeth (note that these losses are very small and have been neglected in Fig. E6.10.1)

$P_{sp2}$ is the surface and pulsation loss caused by rotor teeth (note that these losses are very small and have been neglected in Fig. E6.10.1)

The heat of the motor is dissipated to the air at ambient temperature through the end bells of the motor frame to the stationary air, and through the cylindrical frame surface containing the cooling ribs (which increase the frame surface) to the moving air. The temperature difference between the outer surface of the frame and the ambient temperature is

$$\Delta T_f = \frac{\sum P}{\alpha_o S_o + \alpha_v S_v} \qquad (E6.10\text{-}4)$$

where

$\Sigma P$ represents total losses of the motor

$\alpha_o$ is the surface thermal coefficient (W/m²°C) for dissipating the heat from the surface of the frame to the stationary air (near end bells)

$\alpha_v$ is the surface thermal coefficient (W/m²°C) for dissipating the heat from the surface of the frame to the moving air (near cylindrical frame)

$S_o$ is the surface area in contact with stationary air

$S_v$ is the surface area in contact with moving air

The relation between $\alpha_o$ and $\alpha_v$ is given by

$$\alpha_v = \alpha_o(1 + k_v \sqrt{v}) \qquad (E6.10\text{-}5)$$

where

$k_v$ is a coefficient, and $v$ is the air velocity (m/s)

### 6.14.11 Application Example 6.11: Temperature-Rise Equations for a Drip-Proof 5 hp Motor: Solution

From the thermal network of Fig. E6.11.1 one obtains for the temperature drop across the stator winding insulation

$$\Delta T_i = P_{em}\frac{\beta_i}{\lambda_i \cdot S_i}, \qquad (E6.11\text{-}1)$$

and the temperature drop across the stator back iron is

$$\Delta T_{y1} = \left(P_{em} + \frac{1}{2}P_{y1} + P_{t1} + P_{sp1} + P_b + P_{sp2}\right)\frac{h_{y1}}{\lambda_c \cdot S_{y1}},$$

$$(E6.11\text{-}2)$$

where $P_b$ are the ohmic losses in the rotor bars.

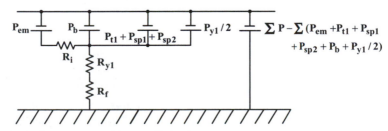

**FIGURE E6.11.1** Thermal network of a drip-proof induction motor.

Finally, the temperature drop from the motor frame to the ambient temperature is

$$\Delta T_f = \frac{\left(P_{em} + P_b + P_{t1} + P_{sp1} + P_{sp2} + \frac{1}{2}P_{y1}\right)}{\alpha_o \pi (D + 2h_1 + 2h_{y1} + 2h_f)(L + L_e)} \quad \text{(E6.11-3)}$$

where

$h_1$ is the radial height of stator tooth

$h_f$ is the radial thickness of stator frame

$h_{y1}$ is the radial height of stator back iron

$L$ is the embedded (cylindrical) length of stator winding

$L_e$ is the length of the two stator end windings

The temperature rise of the stator winding is therefore

$$\Delta T = \Delta T_i + \Delta T_{y1} + \Delta T_f. \quad \text{(E6.11-4)}$$

## CHAPTER 7 SOLUTIONS

### 7.2.3.1 Application Example 7.7: Jacobian Matrices: Solution

Given

$$\begin{bmatrix} 1 & 0 \\ l_{21} & 1 \end{bmatrix}\begin{bmatrix} u_{11} & u_{12} \\ 0 & u_{22} \end{bmatrix} = \begin{bmatrix} J_{11} & J_{12} \\ J_{21} & J_{22} \end{bmatrix}. \quad \text{(E7.7-1)}$$

Find $u_{11}, u_{12}, u_{22}$ and $l_{21}$.

$$J_{11} = u_{11} \qquad \text{or} \qquad u_{11} = J_{11}$$
$$J_{12} = u_{12} \qquad \text{or} \qquad u_{12} = J_{12}$$
$$J_{21} = l_{21}u_{11} \qquad \text{or} \qquad l_{21} = J_{21}/u_{11}$$
$$J_{22} = l_{21}u_{12} + u_{22} \quad \text{or} \quad u_{22} = J_{22} - l_{21}u_{12}$$

In general:

$$u_{11} = J_{11}$$
$$u_{12} = J_{12}$$
$$\cdots \cdots$$
$$u_{1c} = J_{1c}$$
$$\cdots \cdots$$
$$u_{1n} = J_{1n}$$

Since the first row of the table of factors is row 1 of $\bar{U}$, the first row of the table of factors is simply row 1 of $\bar{J}$.

In row 2 of $\bar{J}$:

$$l_{21}u_{11} = J_{21}$$
$$l_{21}u_{12} + u_{22} = J_{22}$$
$$l_{21}u_{13} + u_{23} = J_{23}$$
$$\cdots \cdots$$

These equations yield $l_{21}, u_{22}, u_{23}, u_{24}, \ldots$

$$l_{21} = J_{21}/u_{11}$$
$$u_{22} = J_{22} - l_{21}u_{12}$$
$$u_{23} = J_{23} - l_{21}u_{13}$$
$$\cdots \cdots$$

Thus in the expansion of row 2 of $\bar{J}$, one $l$ and $(n-1)$ $u$'s are formed. The generalization for the expansion of row r of $\bar{J}$ is straightforward:

$$l_{rc} = \frac{J_{rc} - \sum_{q=1}^{c-1} l_{rq}u_{qc}}{u_{cc}}, \quad \text{for } c = 1, 2, \ldots, (r-1) \quad \text{(E7.7-2)}$$

$$u_{rc} = J_{rc} - \sum_{q=1}^{r-1} l_{rq}u_{qc}, \quad \text{for } c = r, r+1, \ldots, n. \quad \text{(E7.7-3)}$$

The rules for the formation of the table of factors *in situ* are as follows:

1) Row 1 of $\bar{J}$ is not modified.
2) Row 2 of $\bar{J}$ is modified to yield one $l$ and $(n-1)$ $u$'s. These entries are found using the above two equations for $l_{rc}$ and $u_{rc}$. For these equations, when the summation upper index is zero, there are no terms in the sum (e.g., ignore the summation term). When the upper and lower indices of summation are the same, there is a single term in the sum.
3) Rule 2 is repeated for each row and $\bar{J}$ is replaced by the table of factors. In row r, there will be $(r-1)$ $l$-terms calculated and $(n-r+1)$ $u$-terms. The above equations for $l_{rc}$ and $u_{rc}$ are used.

### 7.3.5 Application Example 7.8: Computation of Fundamental Admittance Matrix: Solution

For this example $y_{14} = -\tilde{y}_{14}$, where $\tilde{y}_{14}$ is the admittance between bus 1 and bus 4 and $y_{14}$ is an entry of the bus admittance matrix $Y_{bus}$. Note the minus sign.

$$y_{14} = -\tilde{y}_{14} = \frac{-1}{0.01 + j0.02} = \frac{-1}{0.02236\angle 1.1071\text{rad}}$$
$$= \frac{1\angle \pm 3.1415\text{rad}}{0.02236\angle 1.1071\text{rad}} = 44.72\text{pu}\angle 2.034\text{rad}$$

$$y_{12} = -\tilde{y}_{12} = \frac{-1}{0.01 + j0.01} = \frac{-1}{0.01414\angle 0.7854}$$
$$= \frac{1\angle \pm 3.1415\text{rad}}{0.01414\angle 0.7854} = 70.72\text{pu}\angle 2.356\text{rad}$$

$$y_{11} = -\{y_{14} + y_{12}\} = -\{44.72\cos(2.034) +$$
$$70.72\cos(2.356) + j[44.72\sin(2.034) +$$
$$70.72\sin(2.356)]\} = -\{-19.98 - 49.997 + j[40.008 +$$
$$50.02]\} = (69.98 - j90.03) = 114.03\text{pu}\angle -0.910\text{rad}$$

$y_{13} = -\tilde{y}_{13} = 0$ (because there is no direct line between buses 1 and 3)

$$y_{23} = -\tilde{y}_{23} = \frac{-1}{0.02 + j0.08} = \frac{1\angle \pm 3.1415\,\text{rad}}{0.082462\angle 1.3258\,\text{rad}}$$
$$= 12.127\,\text{pu}\angle 1.816\,\text{rad}$$

$$y_{34} = -\tilde{y}_{34} = \frac{-1}{0.01 + j0.02} = \frac{1\angle \pm 3.1415\,\text{rad}}{0.02236\angle 1.1071\,\text{rad}}$$
$$= 44.72\,\text{pu}\angle 2.034\,\text{rad}$$

$$y_{22} = -(y_{12} + y_{23}) = -\{70.72\cos(2.356) + 12.127\cos(1.816) + j(70.72\sin(2.356) + 12.127\sin(1.816)]\} = -\{-49.997 - 2.994 +$$

$$j(50.102 + 11.764)\} = -\{-52.941 + j61.862\} = 81.42\,\text{pu}\angle -0.863\,\text{rad}$$

$$y_{33} = -(y_{23} + y_{34}) = -\{12.127\cos(1.816) + 44.72\cos(2.034) + j[12.127\sin(1.8161) + 44.72\sin(2.034)]\} = -\{-2.944 - 19.982 + j(11.764 + 40.008)\} = -\{-22.93 + j51.772\} = 56.62\,\text{pu}\angle -1.154$$

$$y_{44} = -(y_{14} + y_{34}) = -\{44.72\cos(2.034) + 44.72\cos(2.034) + j(44.72\sin(2.034) + 44.72\sin(2.034)]\} = -\{-19.982 - 19.982 + j(40.008 + 40.008)\} = \{-39.964 + j80.016\} = 89.44\,\text{pu}\angle -1.1076\,\text{rad}.$$

The bus admittance matrix is now

$$\bar{Y}_{\text{bus}} = \begin{bmatrix} 114.03\,\text{pu}\angle -0.910\,\text{rad} & 70.72\,\text{pu}\angle 2.356\,\text{rad} & 0 & 44.72\,\text{pu}\angle 2.034\,\text{rad} \\ 70.72\,\text{pu}\angle 2.356\,\text{rad} & 81.42\,\text{pu}\angle -0.863\,\text{rad} & 12.127\,\text{pu}\angle 1.816\,\text{rad} & 0 \\ 0 & 12.127\,\text{pu}\angle 1.816\,\text{rad} & 56.62\,\text{pu}\angle -1.154\,\text{rad} & 44.72\,\text{pu}\angle 2.034\,\text{rad} \\ 44.72\,\text{pu}\angle 2.034\,\text{rad} & 0 & 44.72\,\text{pu}\angle 2.034\,\text{rad} & 89.44\,\text{pu}\angle -1.1076\,\text{rad} \end{bmatrix}.$$

Note that $\bar{Y}_{\text{bus}}$ is singular; that is, $\bar{Y}_{\text{bus}}^{-1}$ cannot be found, because there is no admittance connected to the ground bus. This is unimportant for the Newton-Raphson approach because for forming of the Jacobian the entries of $\bar{Y}_{\text{bus}}$ are used and $\bar{Y}_{\text{bus}}$ is not inverted.

### 7.3.6 Application Example 7.9: Evaluation of Fundamental Mismatch Vector: Solution

$$\Delta\bar{W}^0 = (P_2 + F_{r,2},\ Q_2 + F_{i,2},\ P_3 + F_{r,3},\ Q_3 + F_{i,3},\ P_4 + F_{r,4},\ Q_4 + F_{i,4})^t$$

$$\Delta P_2 = P_2 + F_{r,2} = P_2 + \sum_{j=1}^{4} y_{2j} V_j V_2 \cos(-\theta_{2j} - \delta_j + \delta_2)$$
$$= P_2 + y_{21} V_1 V_2 \cos(-\theta_{21} - \delta_1 + \delta_2)$$
$$+ y_{22} V_2 V_2 \cos(-\theta_{22} - \delta_2 + \delta_2)$$
$$+ y_{23} V_3 V_2 \cos(-\theta_{23} - \delta_3 + \delta_2)$$
$$+ y_{24} V_4 V_2 \cos(-\theta_{24} - \delta_4 + \delta_2)$$

$$\Delta P_2 = P_2 + F_{r,2} = 0.10 + \{70.72 \cdot 1 \cdot 1 \cdot \cos(-2.356 - 0 + 0) + 81.42 \cdot 1 \cdot 1 \cdot \cos(0.863 - 0 + 0) + 12.127 \cdot 1 \cdot 1 \cdot \cos(-1.816 - 0 + 0) + 0\} = 0.10 - 49.996 + 52.936 - 2.942 = 0.0978$$

$$\Delta Q_2 = Q_2 + F_{i,2} = Q_2 + \sum_{j=1}^{4} y_{2j} V_j V_2 \sin(-\theta_{2j} - \delta_j + \delta_2)$$
$$= Q_2 + y_{21} V_1 V_2 \sin(-\theta_{21} - \delta_1 + \delta_2)$$
$$+ y_{22} V_2 V_2 \sin(-\theta_{22} - \delta_2 + \delta_2)$$
$$+ y_{23} V_3 V_2 \sin(-\theta_{23} - \delta_3 + \delta_2)$$
$$+ y_{24} V_4 V_2 \sin(-\theta_{24} - \delta_4 + \delta_2)$$

$$\Delta Q_2 = Q_2 + F_{i,2} = 0.10 + \{70.72 \cdot 1 \cdot 1 \cdot \sin(-2.356 - 0 + 0) + 81.42 \cdot 1 \cdot 1 \cdot \sin(0.863 - 0 + 0) + 12.127 \cdot 1 \cdot 1 \cdot \sin(-1.816 - 0 + 0) + 0\} = 0.10 - 50.016 + 61.86 - 11.57 = 0.374$$

$$\Delta P_3 = P_3 + F_{r,3} = P_3 + y_{31} V_1 V_3 \cos(-\theta_{31} - \delta_1 + \delta_3) + y_{32} V_2 V_3 \cos(-\theta_{32} - \delta_2 + \delta_3) + y_{33} V_3 V_3 \cos(-\theta_{33} - \delta_3 + \delta_3) + y_{34} V_4 V_3 \cos(-\theta_{34} - \delta_4 + \delta_3)$$

$$\Delta P_3 = P_3 + F_{r,3} = 0.0 + 0.0 + \{12.127 \cdot 1 \cdot 1 \cdot \cos(-1.816 - 0 + 0) + 56.62 \cdot 1 \cdot 1 \cdot \cos(1.154 - 0 + 0) + 44.72 \cdot 1 \cdot 1 \cdot \cos(-2.034 - 0 + 0)\} = 0.0 + 0.0 - 2.944 + 22.92 - 19.98 = -0.0056$$

$$\Delta Q_3 = Q_3 + F_{i,3} = Q_3 + y_{31} V_1 V_3 \sin(-\theta_{31} - \delta_1 + \delta_3) + y_{32} V_2 V_3 \sin(-\theta_{32} - \delta_2 + \delta_3) + y_{33} V_3 V_3 \sin(-\theta_{33} - \delta_3 + \delta_3) + y_{34} V_4 V_3 \sin(-\theta_{34} - \delta_4 + \delta_3)$$

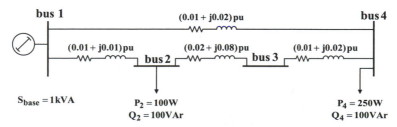

**FIGURE E7.8.1** Four-bus power systems with per unit (pu) line impedances.

$$\Delta Q_3 = Q_3 + F_{i,3} = 0.0 + 0.0 + \{12.127 \cdot 1 \cdot 1 \cdot$$
$$\sin(-1.816 - 0 + 0) + 56.62 \cdot 1 \cdot 1 \cdot \sin(1.154 - 0 + 0)$$
$$+ 44.72 \cdot 1 \cdot 1 \cdot \sin(-2.034 - 0 + 0)\} = 0.0 + 0.0 -$$
$$11.764 + 51.77 - 40.008 = -0.00166$$

$$\Delta P_4 = P_4 + F_{r,4} = P_4 + y_{41}V_1V_4\cos(-\theta_{41} - \delta_1 + \delta_4) +$$
$$y_{42}V_2V_4\cos(-\theta_{42} - \delta_2 + \delta_4) + y_{43}V_3V_4\cos(-\theta_{43} - \delta_3 + \delta_4)$$
$$+ y_{44}V_4V_4\cos(-\theta_{44} - \delta_4 + \delta_4)$$

$$\Delta P_4 = P_4 + F_{r,4} = 0.25 + \{44.72 \cdot 1 \cdot 1 \cdot \cos(-2.034 -$$
$$0 + 0) + 0 + 44.72 \cdot 1 \cdot 1 \cdot \cos(-2.034 - 0 + 0) +$$
$$89.44 \cdot 1 \cdot 1 \cdot \cos(1.1076 - 0 + 0)\} = 0.25 -$$
$$19.98 + 0 - 19.98 + 39.96 = 0.2527$$

$$\Delta Q_4 = Q_4 + F_{i,4} = Q_4 + y_{41}V_1V_4\sin(-\theta_{41} - \delta_1 + \delta_4) +$$
$$y_{42}V_2V_4\sin(-\theta_{42} - \delta_2 + \delta_4) + y_{43}V_3V_4\sin(-\theta_{43} -$$
$$\delta_3 + \delta_4) + y_{44}V_4V_4\sin(-\theta_{44} - \delta_4 + \delta_4)$$

$$\Delta Q_4 = Q_4 + F_{i,4} = 0.1 + \{44.72 \cdot 1 \cdot 1 \cdot \sin(-2.034 -$$
$$0 + 0) + 0 + 44.72 \cdot 1 \cdot 1 \cdot \sin(-2.034 - 0 + 0) +$$
$$89.44 \cdot 1 \cdot 1 \cdot \sin(1.1076 - 0 + 0) + 0\} = 0.1 -$$
$$40.0076 + 0 - 40.0076 + 80.016 = 0.1004.$$

The mismatch power vector is now

$$\Delta \overline{W}^0 = (0.0978, 0.374, -0.0056,$$
$$-0.00166, 0.2527, 0.1004)^t.$$

### 7.3.7 Application Example 7.10: Evaluation of Fundamental Jacobian Matrix: Solution

$$J_{11} = \frac{\partial \Delta P_2}{\partial \delta_2} = -\sum_{\substack{j=1 \\ \neq 2}}^{4} y_{2j}V_2V_j \sin(\delta_2 - \delta_j - \theta_{2j})$$
$$= -y_{21}V_2V_1 \sin(\delta_2 - \delta_1 - \theta_{21})$$
$$- y_{23}V_2V_3 \sin(\delta_2 - \delta_3 - \theta_{23})$$
$$- y_{24}V_2V_4 \sin(\delta_2 - \delta_4 - \theta_{24})$$

$$J_{11} = \frac{\partial \Delta P_2}{\partial \delta_2} = -70.72 \cdot 1 \cdot 1 \cdot \sin(0 - 0 - 2.356)$$
$$- 12.127 \cdot 1 \cdot 1 \cdot \sin(0 - 0 - 1.816) - 0$$
$$= 61.68$$

$$J_{12} = \frac{\partial \Delta P_2}{\partial V_2} = \sum_{\substack{j=1 \\ \neq 2}}^{4} y_{2j}V_j \cos(\delta_2 - \delta_j - \theta_{2j})$$
$$+ 2y_{22}V_2 \cos(-\theta_{22})$$
$$= y_{21}V_1 \cos(\delta_2 - \delta_1 - \theta_{21}) + y_{23}V_3 \cos(\delta_2 - \delta_3 - \theta_{23})$$
$$+ y_{24}V_4 \cos(\delta_2 - \delta_4 - \theta_{24}) + 2y_{22}V_2 \cos(-\theta_{22})$$

$$J_{12} = \frac{\partial \Delta P_2}{\partial V_2} = 70.72 \cdot 1 \cdot \cos(0 - 0 - 2.356)$$
$$+ 12.127 \cdot 1 \cdot \cos(0 - 0 - 1.816) + 0$$
$$+ 2 \cdot 1 \cdot 81.42 \cos(0.863)$$
$$= -49.996 - 2.944 + 0 + 105.872 = 52.932$$

$$J_{13} = \frac{\partial \Delta P_2}{\partial \delta_3} = y_{23}V_1V_3 \sin(\delta_2 - \delta_3 - \theta_{23})$$
$$= 12.127 \cdot 1 \cdot 1 \cdot \sin(0 - 0 - 1.816) = -11.764$$

$$J_{14} = \frac{\partial \Delta P_2}{\partial V_3} = y_{23}V_2 \cos(\delta_2 - \delta_3 - \theta_{23})$$
$$= 12.127 \cdot 1 \cdot \cos(0 - 0 - 1.816) = -2.944$$

$$J_{15} = \frac{\partial \Delta P_2}{\partial \delta_4} = y_{24}V_1V_5 \sin(\delta_2 - \delta_5 - \theta_{25}) = 0$$

$$J_{16} = \frac{\partial \Delta P_2}{\partial V_4} = y_{24}V_2 \cos(\delta_2 - \delta_4 - \theta_{24}) = 0$$

$$J_{21} = \frac{\partial \Delta Q_2}{\partial \delta_2} = \sum_{\substack{j=1 \\ \neq 2}}^{4} y_{2j}V_2V_j \cos(\delta_2 - \delta_j - \theta_{2j})$$
$$= y_{21}V_2V_1 \cos(\delta_2 - \delta_1 - \theta_{21})$$
$$+ y_{23}V_2V_3 \cos(\delta_2 - \delta_3 - \theta_{23})$$
$$+ y_{24}V_2V_4 \cos(\delta_2 - \delta_4 - \theta_{24})$$

$$J_{21} = \frac{\partial \Delta Q_2}{\partial \delta_2} = 70.72 \cdot 1 \cdot 1 \cdot \cos(0 - 0 - 2.3560)$$
$$+ 12.127 \cdot 1 \cdot 1 \cdot \cos(0 - 0 - 1.816) + 0$$
$$= -49.996 - 2.944 + 0 = -52.94$$

$$J_{22} = \frac{\partial \Delta Q_2}{\partial V_2} = \sum_{\substack{j=1 \\ \neq 2}}^{4} y_{2j}V_j \sin(\delta_2 - \delta_j - \theta_{2j})$$
$$+ 2V_2y_{22} \sin(-\theta_{22})$$
$$= y_{21}V_1 \sin(\delta_2 - \delta_1 - \theta_{21}) + y_{23}V_3 \sin(\delta_2 - \delta_3 - \theta_{23})$$
$$+ y_{24}V_4 \sin(\delta_2 - \delta_4 - \theta_{24}) + 2V_2y_{22} \sin(-\theta_{22})$$

$$J_{22} = \frac{\partial \Delta Q_2}{\partial V_2} = 70.72 \cdot 1 \cdot \sin(0 - 0 - 2.356)$$
$$+ 12.127 \cdot 1 \cdot \sin(0 - 0 - 1.816)$$
$$+ 0 + 2 \cdot 1 \cdot 81.42 \sin(0.863)$$
$$= -50.016 - 11.764 + 0 + 123.725 = 61.945$$

$$J_{23} = \frac{\partial \Delta Q_2}{\partial \delta_3} = -y_{23}V_2V_3 \cos(\delta_2 - \delta_3 - \theta_{23})$$
$$= -12.127 \cdot 1 \cdot 1 \cdot \cos(0 - 0 - 1.816)$$
$$= 2.944$$

$$J_{24} = \frac{\partial \Delta Q_2}{\partial V_3} = y_{23}V_2 \sin(\delta_2 - \delta_3 - \theta_{23})$$
$$= 12.127 \cdot 1 \cdot \sin(0 - 0 - 1.816) = -11.764$$

$$J_{25} = \frac{\partial \Delta Q_2}{\partial \delta_4} = -y_{24}V_2V_4 \cos(\delta_2 - \delta_4 - \theta_{24}) = 0$$

$$J_{26} = \frac{\partial \Delta Q_2}{\partial V_4} = y_{24}V_2 \sin(\delta_2 - \delta_4 - \theta_{24}) = 0$$

$$J_{31} = \frac{\partial \Delta P_3}{\partial \delta_2} = y_{32} V_3 V_2 \sin(\delta_3 - \delta_2 - \theta_{32})$$
$$= 12.127 \cdot 1 \cdot 1 \cdot \sin(0 - 0 - 1.816) = -11.764$$

$$J_{32} = \frac{\partial \Delta P_3}{\partial V_2} = y_{32} V_3 \cos(\delta_3 - \delta_2 - \theta_{32})$$
$$= 12.127 \cdot 1 \cdot \cos(0 - 0 - 1.816) = -2.944$$

$$J_{33} = \frac{\partial \Delta P_3}{\partial \delta_3} = -\sum_{\substack{j=1 \\ \neq 3}}^{4} y_{3j} V_3 V_j \sin(\delta_3 - \delta_j - \theta_{3j})$$

$$J_{33} = \frac{\partial \Delta P_3}{\partial \delta_3} = -y_{31} V_3 V_1 \sin(\delta_3 - \delta_1 - \theta_{31})$$
$$- y_{32} V_3 V_2 \sin(\delta_3 - \delta_2 - \theta_{32})$$
$$- y_{34} V_3 V_4 \sin(\delta_3 - \delta_4 - \theta_{34})$$
$$= 0 - 12.127 \cdot 1 \cdot 1 \cdot \sin(0 - 0 - 1.816)$$
$$- 44.72 \cdot 1 \cdot 1 \cdot \sin(0 - 0 - 2.034) = 51.772$$

$$J_{34} = \frac{\partial \Delta P_3}{\partial V_3} = \sum_{\substack{j=1 \\ \neq 3}}^{4} y_{3j} V_j \cos(\delta_3 - \delta_j - \theta_{3j})$$
$$+ 2 V_3 y_{33} \cos(-\theta_{33})$$
$$= y_{31} V_1 \cos(\delta_3 - \delta_1 - \theta_{31}) + y_{32} V_2 \cos(\delta_3 - \delta_2 - \theta_{32})$$
$$+ y_{34} V_4 \cos(\delta_3 - \delta_4 - \theta_{34}) + 2 V_3 y_{33} \cos(-\theta_{33})$$
$$= 0 + 12.127 \cdot 1 \cdot \cos(0 - 0 - 1.816) + 44.72 \cdot 1 \cdot$$
$$\cos(0 - 0 - 2.034) + 2 \cdot 1 \cdot 56.62 \cos(1.154)$$
$$= 0 - 2.944 - 19.982 + 45.843 = 22.917$$

$$J_{35} = \frac{\partial \Delta P_3}{\partial \delta_4} = y_{34} V_3 V_4 \sin(\delta_3 - \delta_4 - \theta_{34})$$
$$= 44.72 \cdot 1 \cdot 1 \cdot \sin(0 - 0 - 2.034) = -40.0076$$

$$J_{36} = \frac{\partial \Delta P_3}{\partial V_4} = y_{34} V_3 V_4 \cos(\delta_3 - \delta_4 - \theta_{34})$$
$$= 44.72 \cdot 1 \cdot 1 \cdot \cos(0 - 0 - 2.034) = -19.98$$

$$J_{41} = \frac{\partial \Delta Q_3}{\partial \delta_2} = -y_{32} V_3 V_2 \cos(\delta_3 - \delta_2 - \theta_{32})$$
$$= -12.127 \cdot 1 \cdot 1 \cdot \cos(0 - 0 - 1.816) = 2.944$$

$$J_{42} = \frac{\partial \Delta Q_3}{\partial V_2} = y_{32} V_3 \sin(\delta_3 - \delta_2 - \theta_{32})$$
$$= 12.127 \cdot 1 \cdot \sin(0 - 0 - 1.816) = -11.764$$

$$J_{43} = \frac{\partial \Delta Q_3}{\partial \delta_3} = \sum_{\substack{j=1 \\ \neq 3}}^{4} y_{3j} V_3 V_j \cos(\delta_3 - \delta_j - \theta_{3j})$$

$$J_{43} = \frac{\partial \Delta Q_3}{\partial \delta_3} = y_{31} V_3 V_1 \cos(\delta_3 - \delta_1 - \theta_{31})$$
$$+ y_{32} V_3 V_2 \cos(\delta_3 - \delta_2 - \theta_{32})$$
$$+ y_{34} V_3 V_4 \cos(\delta_3 - \delta_4 - \theta_{34})$$

$$J_{43} = \frac{\partial \Delta Q_3}{\partial \delta_3} = 0 + 12.127 \cdot 1 \cdot 1 \cdot \cos(0 - 0 - 1.816)$$
$$+ 44.72 \cdot 1 \cdot 1 \cdot \cos(0 - 0 - 2.034)$$
$$= -2.944 - 19.982 = -22.926$$

$$J_{44} = \frac{\partial \Delta Q_3}{\partial V_3} = \sum_{\substack{j=1 \\ \neq 3}}^{4} y_{3j} V_j \sin(\delta_3 - \delta_j - \theta_{3j})$$
$$+ 2 V_3 y_{33} \sin(-\theta_{33})$$
$$= y_{31} V_1 \sin(\delta_3 - \delta_1 - \theta_{31}) + y_{32} V_2 \sin(\delta_3 - \delta_2 - \theta_{32})$$
$$+ y_{34} V_4 \sin(\delta_3 - \delta_4 - \theta_{34}) + 2 V_3 y_{33} \sin(-\theta_{33})$$

$$J_{44} = \frac{\partial \Delta Q_3}{\partial V_3} = 0 + 12.127 \cdot 1 \cdot \sin(0 - 0 - 1.816)$$
$$+ 44.72 \cdot 1 \cdot \sin(0 - 0 - 2.034)$$
$$+ 2 \cdot 1 \cdot 56.62 \sin(1.154)$$
$$= -11.764 - 40.0076 + 103.546 = 51.773$$

$$J_{45} = \frac{\partial \Delta Q_3}{\partial \delta_4} = -y_{34} V_3 V_4 \cos(\delta_3 - \delta_4 - \theta_{34})$$
$$= -44.72 \cdot 1 \cdot 1 \cdot \cos(0 - 0 - 2.034) = 19.982$$

$$J_{46} = \frac{\partial \Delta Q_3}{\partial V_4} = y_{34} V_3 \sin(\delta_3 - \delta_4 - \theta_{34})$$
$$= 44.72 \cdot 1 \cdot \sin(0 - 0 - 2.034) = -40.0076$$

$$J_{51} = \frac{\partial \Delta P_4}{\partial \delta_2} = y_{42} V_4 V_2 \sin(\delta_4 - \delta_2 - \theta_{42}) = 0$$

$$J_{52} = \frac{\partial \Delta P_4}{\partial V_2} = y_{42} V_4 \cos(\delta_4 - \delta_2 - \theta_{42}) = 0$$

$$J_{53} = \frac{\partial \Delta P_4}{\partial \delta_3} = y_{43} V_4 V_3 \sin(\delta_4 - \delta_2 - \theta_{43})$$
$$= 44.72 \cdot 1 \cdot 1 \cdot \sin(0 - 0 - 2.034) = -40.0076$$

$$J_{54} = \frac{\partial \Delta P_4}{\partial V_3} = y_{43} V_4 \cos(\delta_4 - \delta_2 - \theta_{43})$$
$$= 44.72 \cdot 1 \cdot \cos(0 - 0 - 2.034) = -19.982$$

$$J_{55} = \frac{\partial \Delta P_4}{\partial \delta_4} = -\sum_{\substack{j=1 \\ \neq 4}}^{4} y_{4j} V_4 V_j \sin(\delta_4 - \delta_j - \theta_{4j})$$
$$= -y_{41} V_4 V_1 \sin(\delta_4 - \delta_1 - \theta_{41})$$
$$- y_{42} V_4 V_2 \sin(\delta_4 - \delta_2 - \theta_{42})$$
$$- y_{43} V_4 V_3 \sin(\delta_4 - \delta_3 - \theta_{43})$$

$$J_{55} = \frac{\partial \Delta P_4}{\partial \delta_4} = -44.72 \cdot 1 \cdot 1 \cdot \sin(0 - 0 - 2.034) - 0$$
$$- 44.72 \cdot 1 \cdot 1 \cdot \sin(0 - 0 - 2.034)$$
$$= 40.0076 - 0 + 40.0076 = 80.015$$

$$J_{56} = \frac{\partial \Delta P_4}{\partial V_4} = -\sum_{\substack{j=1 \\ \neq 4}}^{4} y_{4j} V_j \cos(\delta_4 - \delta_j - \theta_{4j})$$
$$+ 2V_4 y_{44} \cos(-\theta_{44})$$

$$J_{56} = \frac{\partial \Delta P_4}{\partial V_4} = y_{41} V_1 \cos(\delta_4 - \delta_1 - \theta_{41})$$
$$+ y_{42} V_2 \cos(\delta_4 - \delta_2 - \theta_{42})$$
$$+ y_{43} V_3 \cos(\delta_4 - \delta_3 - \theta_{43})$$
$$+ 2V_4 y_{44} \cos(-\theta_{44})$$

$$J_{56} = \frac{\partial \Delta P_4}{\partial V_4} = 44.72 \cdot 1 \cdot \cos(0 - 0 - 2.034) + 0$$
$$+ 44.72 \cdot 1 \cdot \cos(0 - 0 - 2.034)$$
$$+ 2 \cdot 1 \cdot 89.44 \cos(1.1076)$$

$$J_{56} = \frac{\partial \Delta P_4}{\partial V_4} = -19.982 - 19.982 + 79.925 = 39.961$$

$$J_{61} = \frac{\partial \Delta Q_4}{\partial \delta_2} = -y_{42} V_4 V_2 \cos(\delta_4 - \delta_2 - \theta_{42}) = 0$$

$$J_{62} = \frac{\partial \Delta Q_4}{\partial V_2} = y_{42} V_4 \sin(\delta_4 - \delta_2 - \theta_{42}) = 0$$

$$J_{63} = \frac{\partial \Delta Q_4}{\partial \delta_3} = -y_{43} V_4 V_3 \cos(\delta_4 - \delta_3 - \theta_{43})$$
$$= -44.72 \cdot 1 \cdot 1 \cdot \cos(0 - 0 - 2.034) = 19.982$$

$$J_{64} = \frac{\partial \Delta Q_4}{\partial V_3} = y_{43} V_4 \sin(\delta_4 - \delta_3 - \theta_{43})$$
$$= 44.72 \cdot 1 \cdot \sin(0 - 0 - 2.034) = -40.0076$$

$$J_{65} = \frac{\partial \Delta Q_4}{\partial \delta_4} = \sum_{\substack{j=1 \\ \neq 4}}^{4} y_{4j} V_4 V_j \cos(\delta_4 - \delta_j - \theta_{4j})$$

$$J_{65} = \frac{\partial \Delta Q_4}{\partial \delta_4} = y_{41} V_4 V_1 \cos(\delta_4 - \delta_1 - \theta_{41})$$
$$+ y_{42} V_4 V_2 \cos(\delta_4 - \delta_2 - \theta_{42})$$
$$+ y_{43} V_4 V_3 \cos(\delta_4 - \delta_3 - \theta_{43})$$

$$J_{65} = \frac{\partial \Delta Q_4}{\partial \delta_4} = 44.72 \cdot 1 \cdot 1 \cdot \cos(0 - 0 - 2.034) + 0$$
$$+ 44.72 \cdot 1 \cdot 1 \cdot \cos(0 - 0 - 2.034)$$
$$= -19.982 - 19.982 = -39.963$$

$$J_{66} = \frac{\partial \Delta Q_4}{\partial V_4} = \sum_{\substack{j=1 \\ \neq 4}}^{4} y_{4j} V_j \sin(\delta_4 - \delta_j - \theta_{4j})$$
$$+ 2V_4 y_{44} \sin(-\theta_{44})$$

$$J_{66} = \frac{\partial \Delta Q_4}{\partial V_4} = y_{41} V_1 \sin(\delta_4 - \delta_1 - \theta_{41})$$
$$+ y_{42} V_2 \sin(\delta_4 - \delta_2 - \theta_{42}) + y_{43} V_3 \sin(\delta_4 - \delta_3 - \theta_{43})$$
$$+ 2V_4 y_{44} \sin(-\theta_{44})$$

$$J_{66} = \frac{\partial \Delta Q_4}{\partial V_4} = 44.72 \cdot 1 \cdot \sin(0 - 0 - 2.034)$$
$$+ 0 + 44.72 \cdot 1 \cdot \sin(0 - 0 - 2.034)$$
$$+ 2 \cdot 1 \cdot 89.44 \sin(1.1076)$$

$$J_{66} = \frac{\partial \Delta Q_4}{\partial V_4} = -40.0076 - 40.0076 + 160.031 = 80.$$

The Jacobian matrix is

$$\bar{J} = \begin{bmatrix} J_{11} & J_{12} & J_{13} & J_{14} & J_{15} & J_{16} \\ J_{21} & J_{22} & J_{23} & J_{24} & J_{25} & J_{26} \\ J_{31} & J_{32} & J_{33} & J_{34} & J_{35} & J_{36} \\ J_{41} & J_{42} & J_{43} & J_{44} & J_{45} & J_{46} \\ J_{51} & J_{52} & J_{53} & J_{54} & J_{55} & J_{56} \\ J_{61} & J_{62} & J_{63} & J_{64} & J_{65} & J_{66} \end{bmatrix}$$

$$\bar{J}^0 = \begin{bmatrix} 61.68 & 52.932 & -11.764 & -2.944 & 0 & 0 \\ -52.94 & 61.945 & 2.944 & -11.764 & 0 & 0 \\ -11.764 & -2.944 & 51.772 & 22.917 & -40.0076 & -19.98 \\ 2.944 & -11.764 & -22.926 & 51.773 & 19.982 & -40.0076 \\ 0 & 0 & -40.0076 & -19.982 & 80.015 & 39.961 \\ 0 & 0 & 19.982 & -40.0076 & -39.963 & 80 \end{bmatrix}.$$

## 7.3.8 Application Example 7.11: Calculation of the Inverse of Jacobian Matrix: Solution

Calculation of the inverse of $\bar{J}^0$, that is. $(\bar{J}^0)^{-1}$.

$$(\bar{J}^0)^{-1} = \frac{\text{Adj}(\bar{J}^0)}{\det(\bar{J}^0)},$$

where $\text{Adj}(\bar{J}^0)$ is the adjoint of matrix $\bar{J}^0$ and $\det(\bar{J}^0) = |\bar{J}^0|$ is the determinant of $\bar{J}^0$.

The adjoint of $\bar{J}^0$ is obtained by first taking the transpose of $\bar{J}^0$, that is, $(\bar{J}^0)^t$, and then replacing each element in the transpose with its cofactor:

$$[(\bar{J}^0)^t]_{ij} = [\bar{J}^0]_{ji}.$$

The transpose of $\bar{J}^0$ is

$$
(\bar{J}^0)^t = \begin{bmatrix}
J_{11} & J_{21} & J_{31} & J_{41} & J_{51} & J_{61} \\
J_{12} & J_{22} & J_{32} & J_{42} & J_{52} & J_{62} \\
J_{13} & J_{23} & J_{33} & J_{43} & J_{53} & J_{63} \\
J_{14} & J_{24} & J_{34} & J_{44} & J_{54} & J_{64} \\
J_{15} & J_{25} & J_{35} & J_{45} & J_{55} & J_{65} \\
J_{16} & J_{26} & J_{36} & J_{46} & J_{56} & J_{66}
\end{bmatrix}
$$

$$
(\bar{J}^0)^t = \begin{bmatrix}
61.68 & -52.94 & -11.764 & 2.944 & 0 & 0 \\
52.932 & 61.945 & -2.944 & -11.764 & 0 & 0 \\
-11.764 & 2.944 & 51.772 & -22.926 & -40.0076 & -19.982 \\
-2.944 & -11.764 & 22.917 & 51.773 & -19.982 & -40.0076 \\
0 & 0 & -40.0076 & 19.982 & 80.015 & -39.963 \\
0 & 0 & -19.98 & -40.0076 & 39.961 & 80
\end{bmatrix}
$$

$$
Adj(\bar{J}^0) = \begin{bmatrix}
+Adj(\bar{J}^0)_{11} - Adj(\bar{J}^0)_{12} + Adj(\bar{J}^0)_{13} - Adj(\bar{J}^0)_{14} + Adj(\bar{J}^0)_{15} - Adj(\bar{J}^0)_{16} \\
-Adj(\bar{J}^0)_{21} + Adj(\bar{J}^0)_{22} - Adj(\bar{J}^0)_{23} + Adj(\bar{J}^0)_{24} - Adj(\bar{J}^0)_{25} + Adj(\bar{J}^0)_{26} \\
+Adj(\bar{J}^0)_{31} - Adj(\bar{J}^0)_{32} + Adj(\bar{J}^0)_{33} - Adj(\bar{J}^0)_{34} + Adj(\bar{J}^0)_{35} - Adj(\bar{J}^0)_{36} \\
-Adj(\bar{J}^0)_{41} + Adj(\bar{J}^0)_{42} - Adj(\bar{J}^0)_{43} + Adj(\bar{J}^0)_{44} - Adj(\bar{J}^0)_{45} + Adj(\bar{J}^0)_{46} \\
+Adj(\bar{J}^0)_{51} - Adj(\bar{J}^0)_{52} + Adj(\bar{J}^0)_{53} - Adj(\bar{J}^0)_{54} + Adj(\bar{J}^0)_{55} - Adj(\bar{J}^0)_{56} \\
-Adj(\bar{J}^0)_{61} + Adj(\bar{J}^0)_{62} - Adj(\bar{J}^0)_{63} + Adj(\bar{J}^0)_{64} - Adj(\bar{J}^0)65 + Adj(\bar{J}^0)_{66}
\end{bmatrix}
$$

or

$$
Adj(\bar{J}^0) =
\begin{bmatrix}
+\begin{vmatrix} J_{22}J_{32}J_{42}J_{52}J_{62} \\ J_{23}J_{33}J_{43}J_{53}J_{63} \\ J_{24}J_{34}J_{44}J_{54}J_{64} \\ J_{25}J_{35}J_{45}J_{55}J_{65} \\ J_{26}J_{36}J_{46}J_{56}J_{66} \end{vmatrix}
-\begin{vmatrix} J_{12}J_{32}J_{42}J_{52}J_{62} \\ J_{13}J_{33}J_{43}J_{53}J_{63} \\ J_{14}J_{34}J_{44}J_{54}J_{64} \\ J_{15}J_{35}J_{45}J_{55}J_{65} \\ J_{16}J_{36}J_{46}J_{56}J_{66} \end{vmatrix}
+\begin{vmatrix} J_{12}J_{22}J_{42}J_{52}J_{62} \\ J_{13}J_{23}J_{43}J_{53}J_{63} \\ J_{14}J_{24}J_{44}J_{54}J_{64} \\ J_{15}J_{25}J_{45}J_{55}J_{65} \\ J_{16}J_{26}J_{46}J_{56}J_{66} \end{vmatrix}
-\begin{vmatrix} J_{12}J_{22}J_{32}J_{52}J_{62} \\ J_{13}J_{23}J_{33}J_{53}J_{63} \\ J_{14}J_{24}J_{34}J_{54}J_{64} \\ J_{15}J_{25}J_{35}J_{55}J_{65} \\ J_{16}J_{26}J_{36}J_{56}J_{66} \end{vmatrix}
+\begin{vmatrix} J_{12}J_{22}J_{32}J_{42}J_{62} \\ J_{13}J_{23}J_{33}J_{43}J_{63} \\ J_{14}J_{24}J_{34}J_{44}J_{64} \\ J_{15}J_{25}J_{35}J_{45}J_{65} \\ J_{16}J_{26}J_{36}J_{46}J_{66} \end{vmatrix}
-\begin{vmatrix} J_{12}J_{22}J_{32}J_{42}J_{52} \\ J_{13}J_{23}J_{33}J_{43}J_{53} \\ J_{14}J_{24}J_{34}J_{44}J_{54} \\ J_{15}J_{25}J_{35}J_{45}J_{55} \\ J_{16}J_{26}J_{36}J_{46}J_{56} \end{vmatrix} \\[3em]
-\begin{vmatrix} J_{21}J_{31}J_{41}J_{51}J_{61} \\ J_{23}J_{33}J_{43}J_{53}J_{63} \\ J_{24}J_{34}J_{44}J_{54}J_{64} \\ J_{25}J_{35}J_{45}J_{55}J_{65} \\ J_{26}J_{36}J_{46}J_{56}J_{66} \end{vmatrix}
+\begin{vmatrix} J_{11}J_{31}J_{41}J_{51}J_{61} \\ J_{13}J_{33}J_{43}J_{53}J_{63} \\ J_{14}J_{34}J_{44}J_{54}J_{64} \\ J_{15}J_{35}J_{45}J_{55}J_{65} \\ J_{16}J_{36}J_{46}J_{56}J_{66} \end{vmatrix}
-\begin{vmatrix} J_{11}J_{21}J_{41}J_{51}J_{61} \\ J_{13}J_{23}J_{43}J_{53}J_{63} \\ J_{14}J_{24}J_{44}J_{54}J_{64} \\ J_{15}J_{25}J_{45}J_{55}J_{65} \\ J_{16}J_{26}J_{46}J_{56}J_{66} \end{vmatrix}
+\begin{vmatrix} J_{11}J_{21}J_{31}J_{51}J_{61} \\ J_{13}J_{23}J_{33}J_{53}J_{63} \\ J_{14}J_{24}J_{34}J_{54}J_{64} \\ J_{15}J_{25}J_{35}J_{55}J_{65} \\ J_{16}J_{26}J_{36}J_{56}J_{66} \end{vmatrix}
-\begin{vmatrix} J_{11}J_{21}J_{31}J_{41}J_{61} \\ J_{13}J_{23}J_{33}J_{43}J_{63} \\ J_{14}J_{24}J_{34}J_{44}J_{64} \\ J_{15}J_{25}J_{35}J_{45}J_{65} \\ J_{16}J_{26}J_{36}J_{46}J_{66} \end{vmatrix}
+\begin{vmatrix} J_{11}J_{21}J_{31}J_{41}J_{51} \\ J_{13}J_{23}J_{33}J_{43}J_{53} \\ J_{14}J_{24}J_{34}J_{44}J_{54} \\ J_{15}J_{25}J_{35}J_{45}J_{55} \\ J_{16}J_{26}J_{36}J_{46}J_{56} \end{vmatrix} \\[3em]
+\begin{vmatrix} J_{21}J_{31}J_{41}J_{51}J_{61} \\ J_{22}J_{32}J_{42}J_{52}J_{62} \\ J_{24}J_{34}J_{44}J_{54}J_{64} \\ J_{25}J_{35}J_{45}J_{55}J_{65} \\ J_{26}J_{36}J_{46}J_{56}J_{66} \end{vmatrix}
\quad -Adj(\bar{J}^0)_{32} \quad +Adj(\bar{J}^0)_{33} \quad -Adj(\bar{J}^0)_{34} \quad +Adj(\bar{J}^0)_{35} \quad -Adj(\bar{J}^0)_{36} \\[3em]
-\begin{vmatrix} J_{21}J_{31}J_{41}J_{51}J_{61} \\ J_{22}J_{32}J_{42}J_{52}J_{62} \\ J_{23}J_{33}J_{43}J_{53}J_{63} \\ J_{25}J_{35}J_{45}J_{55}J_{65} \\ J_{26}J_{36}J_{46}J_{56}J_{66} \end{vmatrix}
\quad +Adj(\bar{J}^0)_{42} \quad -Adj(\bar{J}^0)_{43} \quad +Adj(\bar{J}^0)_{44} \quad -Adj(\bar{J}^0)_{45} \quad +Ad(\bar{J}^0)j_{46} \\[3em]
+\begin{vmatrix} J_{21}J_{31}J_{41}J_{51}J_{61} \\ J_{22}J_{32}J_{42}J_{52}J_{62} \\ J_{23}J_{33}J_{43}J_{53}J_{63} \\ J_{24}J_{34}J_{44}J_{54}J_{64} \\ J_{26}J_{36}J_{46}J_{56}J_{66} \end{vmatrix}
\quad -Adj(\bar{J}^0)_{52} \quad +Adj(\bar{J}^0)_{53} \quad -Ad(\bar{J}^0)j_{54} \quad +Adj(\bar{J}^0)_{55} \quad -Adj(\bar{J}^0)_{56} \\[3em]
-\begin{vmatrix} J_{21}J_{31}J_{41}J_{51}J_{61} \\ J_{22}J_{32}J_{42}J_{52}J_{62} \\ J_{23}J_{33}J_{43}J_{53}J_{63} \\ J_{24}J_{34}J_{44}J_{54}J_{64} \\ J_{25}J_{35}J_{45}J_{55}J_{65} \end{vmatrix}
\quad +Adj(\bar{J}^0)_{62} \quad -Adj(\bar{J}^0)_{63} \quad +Ad(\bar{J}^0)j_{64} \quad -Adj(\bar{J}^0)_{65} \quad +Ad(\bar{J}^0)j_{66}
\end{bmatrix}
$$

As an example, the calculation of the first entry of the adjonint $\text{Adj}(\bar{J}^0)_{11}$ will be outlined:

$$\text{Adj}(\bar{J}^0)_{11} = \begin{bmatrix} J_{22} & J_{32} & J_{42} & J_{52} & J_{62} \\ J_{23} & J_{33} & J_{43} & J_{53} & J_{63} \\ J_{24} & J_{34} & J_{44} & J_{54} & J_{64} \\ J_{25} & J_{35} & J_{45} & J_{55} & J_{65} \\ J_{26} & J_{36} & J_{46} & J_{56} & J_{66} \end{bmatrix} = \begin{bmatrix} 61.945 & -2.944 & -11.764 & 0 & 0 \\ 2.944 & 51.772 & -22.926 & -40.0076 & 19.982 \\ -11.764 & 22.917 & 51.773 & -19.982 & -40.0076 \\ 0 & -40.0076 & 19.982 & 80.015 & -39.963 \\ 0 & -19.98 & -40.0076 & 39.961 & 80.0 \end{bmatrix}$$

$$\text{Adj}(\bar{J}^0)_{11} = 61.945 \begin{bmatrix} 51.772 & -22.926 & -40.0076 & 19.982 \\ 22.917 & 51.772 & -19.982 & -40.0076 \\ -40.0076 & 19.982 & 80.015 & -39.963 \\ -19.98 & -40.0076 & 39.961 & 80 \end{bmatrix}$$

$$+ 2.944 \begin{bmatrix} 2.944 & -22.917 & -40.0076 & 19.98 \\ -11.764 & 51.773 & -19.982 & -40.0076 \\ 0 & 19.982 & 80.015 & -39.963 \\ 0 & -40.0076 & 39.961 & 80 \end{bmatrix}$$

$$- 11.764 \begin{vmatrix} 2.944 & 51.772 & -40.0076 & 19.98 \\ -11.764 & 22.917 & -19.982 & -40.0076 \\ 0 & -40.0076 & 80.015 & -39.963 \\ 0 & -19.982 & 39.961 & 80 \end{vmatrix}.$$

The determinant is

$$\det(\bar{J}) = J_{11} \begin{bmatrix} J_{22}J_{23}J_{24}J_{25}J_{26} \\ J_{32}J_{33}J_{34}J_{35}J_{36} \\ J_{42}J_{43}J_{44}J_{45}J_{46} \\ J_{52}J_{53}J_{54}J_{55}J_{56} \\ J_{62}J_{63}J_{64}J_{65}J_{66} \end{bmatrix} - J_{12} \begin{bmatrix} J_{21}J_{23}J_{24}J_{25}J_{26} \\ J_{31}J_{33}J_{34}J_{35}J_{36} \\ J_{41}J_{43}J_{44}J_{45}J_{46} \\ J_{51}J_{53}J_{54}J_{55}J_{56} \\ J_{61}J_{63}J_{64}J_{65}J_{66} \end{bmatrix}$$

$$+ J_{13} \begin{bmatrix} J_{21}J_{22}J_{24}J_{25}J_{26} \\ J_{31}J_{32}J_{34}J_{35}J_{36} \\ J_{41}J_{43}J_{44}J_{45}J_{46} \\ J_{51}J_{52}J_{54}J_{55}J_{56} \\ J_{61}J_{62}J_{64}J_{65}J_{66} \end{bmatrix} - J_{14} \begin{bmatrix} J_{21}J_{22}J_{23}J_{25}J_{26} \\ J_{31}J_{32}J_{33}J_{35}J_{36} \\ J_{41}J_{42}J_{43}J_{45}J_{46} \\ J_{51}J_{52}J_{53}J_{55}J_{56} \\ J_{61}J_{62}J_{63}J_{65}J_{66} \end{bmatrix}$$

$$+ J_{15} \begin{bmatrix} J_{21}J_{22}J_{23}J_{24}J_{26} \\ J_{31}J_{32}J_{33}J_{34}J_{36} \\ J_{41}J_{42}J_{43}J_{44}J_{46} \\ J_{51}J_{52}J_{53}J_{54}J_{56} \\ J_{61}J_{62}J_{63}J_{64}J_{66} \end{bmatrix} - J_{16} \begin{bmatrix} J_{21}J_{22}J_{23}J_{24}J_{25} \\ J_{31}J_{32}J_{33}J_{34}J_{35} \\ J_{41}J_{42}J_{43}J_{44}J_{45} \\ J_{51}J_{52}J_{53}J_{54}J_{55} \\ J_{61}J_{62}J_{63}J_{64}J_{65} \end{bmatrix}.$$

The $(\bar{J}_{11}^0)^{-1}$ entry is now

$$(J_{11}^0)^{-1} = \frac{\text{Adj}(\bar{J}^0)}{\det(\bar{J}^0)}$$

The cofactors of $A^t$ are

$$A_{11}^t = (-1)^{(1+1)} \begin{vmatrix} -4 & 1 \\ 2 & 1 \end{vmatrix} = (-4-2) = -6,$$

### 7.3.9 Application Example 7.12: Inversion of a 3 × 3 Matrix: Solution

Transpose of $\bar{A}$ is

$$A_{12}^t = (-1)^{(1+2)} \begin{vmatrix} 3 & 1 \\ -1 & 1 \end{vmatrix} = -(3+1) = -4,$$

$$A_{13}^t = (-1)^{(1+3)} \begin{vmatrix} 3 & -4 \\ -1 & 2 \end{vmatrix} = (6-4) = 2,$$

$$\bar{A}^t = \begin{bmatrix} 1 & 2 & 3 \\ 3 & -4 & 1 \\ -1 & 2 & 1 \end{bmatrix}.$$

$$A_{21}^t = (-1)^{(2+1)} \begin{vmatrix} 2 & 3 \\ 2 & 1 \end{vmatrix} = (-2+6) = 4,$$

$$A_{22}^t = (-1)^{(2+2)} \begin{vmatrix} 1 & 3 \\ -1 & 1 \end{vmatrix} = (1+3) = 4,$$

$$A_{23}^t = (-1)^{(2+3)} \begin{vmatrix} 1 & 2 \\ -1 & 2 \end{vmatrix} = -(2+2) = -4,$$

$$A_{31}^t = (-1)^{(3+1)} \begin{vmatrix} 2 & 3 \\ -4 & 1 \end{vmatrix} = (2+24) = 14,$$

$$A_{32}^t = (-1)^{(3+2)} \begin{vmatrix} 1 & 3 \\ 3 & 1 \end{vmatrix} = -(1-9) = 8,$$

$$A_{33}^t = (-1)^{(3+3)} \begin{vmatrix} 1 & 2 \\ 3 & -4 \end{vmatrix} = (-4-6) = -10.$$

The determinant is

$$|\bar{A}| = 1 \cdot \begin{vmatrix} -4 & 2 \\ 1 & 1 \end{vmatrix} - 3 \cdot \begin{vmatrix} 2 & 2 \\ 3 & 1 \end{vmatrix} - 1 \cdot \begin{vmatrix} 2 & -4 \\ 3 & 1 \end{vmatrix}$$
$$= -4 - 2 - 6 + 18 - 2 - 12 = -26 + 18 = -8.$$

The inverse of $\bar{A}$ is now

$$(\bar{A})^{-1} = \frac{\text{Adj}(\bar{A})}{|\bar{A}|} = -\frac{1}{8} \begin{bmatrix} -6 & -4 & 2 \\ 4 & 4 & -4 \\ 14 & 8 & -10 \end{bmatrix}.$$

### 7.3.10 Application Example 7.13: Computation of the Correction Voltage Vector: Solution

Based on the results of Application Examples 7.9 to 7.11, the correction voltage vector $\Delta \bar{x}^0$ can be computed from

$$\Delta \bar{x}^0 = (\bar{J}^0)^{-1} \Delta \bar{W}(\bar{x}^0) = (2.68E-4, \ 2.39E-3,$$
$$2.79E-3, \ 3.57E-3, \ 3.39E-3, \ 4.03E-3)^t.$$

The updated mismatch power vector is

$$\Delta \bar{W}^1 = (1.41E-4, \ 3.01E-4, \ -9.63E-6, \ 7.62E-5,$$
$$1.20E-3, \ 5.96E-4)^t.$$

Since the components of $\Delta \bar{W}^1$ are larger than the specified convergence tolerance (e.g., 0.0001), the Newton–Raphson solution has not converged. However, it will converge in two iterations:

Fundamental power flow iteration summary (Fig. E7.8.1):

| Iteration | Absolute real power mismatch | | | Absolute reactive power mismatch | | |
|---|---|---|---|---|---|---|
| | Average (%) | Worst (%) | Worst bus | Average (%) | Worst (%) | Worst bus |
| 0 | 11.67 | 25 | 4 | 6.67 | 10.00 | 4 |
| 1 | 0.05 | 0.12 | 4 | 0.03 | 0.06 | 4 |
| 2 | 0.00 | 0.00 | 4 | 0.00 | 0.00 | 4 |

Fundamental power flow output solution (Fig. E7.8.1):

| From bus | Bus voltage | | Bus generation (G) and load (L) powers | | | | | Line power* | |
|---|---|---|---|---|---|---|---|---|---|
| | $|V|$ (%) | $\delta$ (°) | $P_G$ (%) | $Q_G$ (%) | $P_L$ (%) | $Q_L$ (%) | To bus | $P_{line}$ (%) | $Q_{line}$ (%) |
| 1 | 100.0 | 0.00 | 35.09 | 20.15 | 0.00 | 0.00 | 2 | 13.35 | 10.69 |
| | | | | | | | 4 | 21.73 | 9.47 |
| 2 | 99.76 | −0.02 | 0.00 | 0.00 | 10.00 | 10.00 | 1 | −13.32 | −10.66 |
| | | | | | | | 3 | 3.32 | 0.66 |
| 3 | 99.64 | −0.16 | 0.00 | 0.00 | 0.00 | 0.00 | 2 | −3.32 | −0.65 |
| | | | | | | | 4 | 3.32 | 0.65 |
| 4 | 99.59 | −0.20 | 0.00 | 0.00 | 25.00 | 10.00 | 3 | −3.32 | −0.65 |
| | | | | | | | 1 | −21.68 | −9.35 |

*Active and reactive powers in the line between "from bus" and "to bus."

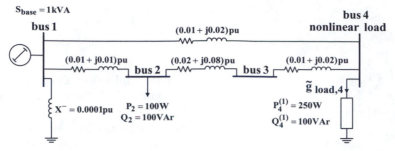

**FIGURE E7.14.1** Four-bus power system with one nonlinear load bus, where $n = m = 4$, that is, there is one nonlinear bus (bus 4), 2 linear buses (buses 2 and 3) and the swing bus (bus 1).

### 7.4.7 Application Example 7.14: Computation of Harmonic Admittance Matrix: Solution

The fundamental bus admittance matrix ($X^-$ does not exist for fundamental quantities) is computed in Application Example 7.8. The 5th harmonic bus admittance matrix takes into account $X^-$ and its entries are calculated as follows:

$$y_{11}^{(5)} = \frac{1}{0.01 + j(0.02 \cdot 5)} + \frac{1}{0.01 + j(0.01 \cdot 5)} + \frac{1}{j(0.0001 \cdot 5)}$$
$$= 2029.137 \text{pu} \angle -1.5684 \text{rad}$$

$$y_{12}^{(5)} = \frac{-1}{0.01 + j(0.01 \cdot 5)} = 19.612 \text{pu} \angle 1.7681 \text{rad}$$

$$y_{13}^{(5)} = 0$$

$$y_{14}^{(5)} = \frac{-1}{0.01 + j(0.02 \cdot 5)} = 9.9503 \text{pu} \angle 1.6704 \text{rad}$$

$$y_{21}^{(5)} = y_{12}^{(5)}$$

$$y_{22}^{(5)} = \frac{1}{0.01 + j(0.01 \cdot 5)} + \frac{1}{0.02 + j(0.08 \cdot 5)}$$
$$= 22.084 \text{pu} \angle -1.3900 \text{rad}$$

$$y_{23}^{(5)} = \frac{-1}{0.02 + j(0.08 \cdot 5)} = 2.4969 \text{pu} \angle 1.62066 \text{rad}$$

$$y_{24}^{(5)} = 0$$

$$y_{31}^{(5)} = y_{13}^{(5)}$$

$$y_{32}^{(5)} = y_{23}^{(5)}$$

$$y_{33}^{(5)} = \frac{1}{0.02 + j(0.08 \cdot 5)} + \frac{1}{0.01 + j(0.02 \cdot 5)}$$
$$= 12.445 \text{pu} \angle -1.4811 \text{rad}$$

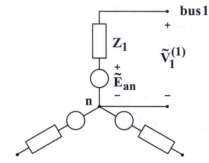

**FIGURE E7.14.2** Representation of swing bus (bus 1) for fundamental (h = 1) frequency.

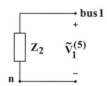

**FIGURE E7.14.3** Equivalent circuit of swing bus for the 5th harmonic (h = 5).

$$y_{34}^{(5)} = \frac{-1}{0.01 + j(0.02 \cdot 5)} = 9.9504 \text{pu} \angle 1.6704 \text{rad}$$

$$y_{41}^{(5)} = y_{14}^{(5)}$$

$$y_{42}^{(5)} = y_{24}^{(5)}$$

$$y_{43}^{(5)} = y_{34}^{(5)}$$

$$y_{44}^{(5)} = \frac{1}{0.01 + j(0.02 \cdot 5)} + \frac{1}{0.01 + j(0.02 \cdot 5)}$$
$$= 19.901 \text{pu} \angle -1.4711 \text{rad}.$$

Therefore,

$$\bar{Y}_{bus}^{(5)} = \begin{bmatrix} 2029.137\text{pu}\angle - 1.5684\text{rad} & 19.61\text{pu}\angle 1.7681\text{rad} & 0 & 9.9503\text{pu}\angle 1.6704\text{rad} \\ 19.61\text{pu}\angle 1.7681\text{rad} & 22.0847\text{pu}\angle - 1.3900\text{rad} & 2.4969\text{pu}\angle 1.62066\text{rad} & 0 \\ 0 & 2.4969\text{pu}\angle 1.62066\text{rad} & 12.445\text{pu}\angle - 1.4811\text{rad} & 9.9504\text{pu}\angle 1.6704\text{rad} \\ 9.9503\text{pu}\angle 1.6704\text{rad} & 0 & 9.9504\text{pu}\angle 1.6704\text{rad} & 19.901\text{pu}\angle - 1.4711\text{rad} \end{bmatrix}.$$

Note that: $-y_{ij}\angle\theta_{ij} = y_{ij}\angle(\theta_{ij} + \pi)$.

### 7.4.8 Application Example 7.15: Computation of Nonlinear Load Harmonic Currents: Solution

The Initial Step of Harmonic Load Flow is as follows:

The real and imaginary nonlinear device currents $G_{r,4}^{(1)}$ and $G_{i,4}^{(1)}$ – referred to the swing bus – may be computed as follows, where the following powers are referred to the nonlinear bus 4:

$$\begin{cases} P_4^{(1)} = V_4^{(1)} I_4^{(1)} \cos(\delta_4^{(1)} - \gamma_4^{(1)}) \\ Q_4^{(1)} = V_4^{(1)} I_4^{(1)} \sin(\delta_4^{(1)} - \gamma_4^{(1)}) \end{cases}$$

$I_4^{(1)}$ and $\gamma_4^{(1)} = \delta_4^{(1)} - \tan^{-1}(Q_4^{(1)}/P_4^{(1)})$ are the magnitude and phase angle of the fundamental nonlinear device current, respectively. The diagram of Fig. E7.15.1 demonstrates the phasor relationships.

The real part of the fundamental of the nonlinear device current at bus 4 (referred to the swing bus) is

$$G_{r,4}^{(1)} = I_4^{(1)}\cos\gamma_4^{(1)}$$

where

$$I_4^{(1)} = \frac{P_4^{(1)}}{V_4^{(1)}\cos(\delta_4^{(1)} - \gamma_4^{(1)})}.$$

Thus

$$G_{r,4}^{(1)} = \frac{P_4^{(1)}\cos\gamma_4^{(1)}}{V_4^{(1)}\cos(\delta_4^{(1)} - \gamma_4^{(1)})}.$$

Correspondingly one obtains for the imaginary part of the fundamental of the nonlinear device current at bus 4 (referred to the swing bus):

$$G_{i,4}^{(1)} = \frac{P_4^{(1)}\sin\gamma_4^{(1)}}{V_4^{(1)}\cos(\delta_4^{(1)} - \gamma_4^{(1)})}.$$

The real and imaginary harmonic (5th) nonlinear device currents are given as (referred to bus 4)

$$\begin{cases} g_{r,4}^{(5)} = 0.3(V_4^{(1)})^3 \cos(3\delta_4^{(1)}) + 0.3(V_4^{(5)})^2 \cos(3\delta_4^{(5)}) \\ g_{i,4}^{(5)} = 0.3(V_4^{(1)})^3 \sin(3\delta_4^{(1)}) + 0.3(V_4^{(5)})^2 \sin(3\delta_4^{(5)}) \end{cases}$$

where $\delta_4^{(5)}$ is the angle between $\tilde{V}_4^{(5)}$ and $\tilde{V}_1^{(1)}$, the swing bus voltage, $I_4^{(5)}$ is the fifth harmonic current magnitude

$$I_4^{(5)} = \sqrt{(g_{r,4}^{(5)})^2 + (g_{i,4}^{(5)})^2}$$

and $\gamma_4^{(5)}$ is the phase angle of $\tilde{I}_4^{(5)}$ with respect to the swing bus voltage $\tilde{V}_1^{(1)}$.

Assuming bus 1 to be the swing bus ($\delta_1^{(1)} = 0$ radians, $|\tilde{V}_1^{(1)}| = 1.00$ pu), the solution procedure is as follows.

***Step #1 of Harmonic Load Flow.*** Using the solution vector $\bar{x}$ of the fundamental power flow analysis (see Application Example 7.13) and assuming harmonic voltage magnitudes and phase angles of 0.1 pu and 0 radians, respectively, the bus vector $\bar{U}^0$ will be

$$\bar{U}^0 = (\delta_2^{(1)}, V_2^{(1)}, \delta_3^{(1)}, V_3^{(1)}, \delta_4^{(1)}, V_4^{(1)}, \delta_1^{(5)}, V_1^{(5)}, \delta_2^{(5)}, V_2^{(5)},$$
$$\delta_3^{(5)}, V_3^{(5)}, \delta_4^{(5)}, V_4^{(5)}, \alpha_4, \beta_4)^t$$

$$\bar{U}^0 = (-2.671 \cdot 10^{-4}, 0.9976, -2.810 \cdot 10^{-3}, 0.9964,$$
$$-3.414 \cdot 10^{-3}, 0.9960, 0, 0.1, 0, 0.1, 0, 0.1, 0, 0.1, 0, 0)^t$$

Note that the nonlinear device variables ($\alpha_4$ and $\beta_4$) are assumed to be zero because there are no device variables defined in this example for the nonlinear load at bus 4.

***Step #2 of Harmonic Load Flow.*** Use $\bar{U}^0$ to compute the nonlinear device currents. With $P_4^{(1)} = 250$ W corresponding to 0.25 pu, $Q_4^{(1)} = 100$VAr corresponding to 0.100 pu, $V_4^{(1)} = 0.9960$ pu, and $\delta_4^{(1)} = -3.414 \cdot 10^{-3}$ radians, one gets

$$\varepsilon_4^{(1)} = \tan^{-1}\left(\frac{Q_4^{(1)}}{P_4^{(1)}}\right) = \tan^{-1}\left(\frac{0.1}{0.25}\right) = 0.3805\text{rad}$$

$$\gamma_4^{(1)} = \delta_4^{(1)} - \varepsilon_4^{(1)} = (-0.003414 - 0.3805) = -0.38391 \text{ rad}$$

$$G_{r,4}^{(1)} = \frac{P_4^{(1)}\cos(\gamma_4^{(1)})}{V_4^{(1)}\cos(\delta_4^{(1)} - \gamma_4^{(1)})}$$
$$= \frac{0.25 \cdot \cos(-0.38391)}{0.9960 \cdot \cos(-0.003414 - 0.38391)} = 0.2514\text{pu}$$

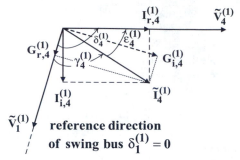

**FIGURE E7.15.1** Phasor diagram for fundamental quantities, where $\varepsilon_4^{(1)} = \tan^{-1}\ (Q_4^{(1)}/P_4^{(1)}) = (\delta_4^{(1)} - \gamma_4^{(1)})$ or $\delta_4^{(1)} = \varepsilon_4^{(1)} + \gamma_4^{(1)}$.

$$G_{i,4}^{(1)} = \frac{P_4^{(1)}\sin(\gamma_4^{(1)})}{V_4^{(1)}\cos(\delta_4^{(1)} - \gamma_4^{(1)})}$$

$$= \frac{0.25 \cdot \sin(-0.38391)}{0.9960 \cdot \cos(-0.003414 - 0.38391)} = -0.1015\text{pu}.$$

The fundamental magnitude of the nonlinear current is

$$I_4^{(1)} = \sqrt{(G_{r,4}^{(1)})^2 + (G_{i,4}^{(1)})^2} = \sqrt{(0.2514)^2 + (-0.1015)^2}$$
$$= 0.2711\text{pu} \quad \text{with} \quad \gamma_4^{(1)} = -0.38391\text{rad}.$$

The fifth harmonic current components of the nonlinear load current at bus 4 are (referred to bus 4)

$$\begin{cases} g_{r,4}^{(5)} = 0.3(V_4^{(1)})^3\cos(3\delta_4^{(1)}) + 0.3(V_4^{(5)})^2\cos(3\delta_4^{(5)}) \\ g_{i,4}^{(5)} = 0.3(V_4^{(1)})^3\sin(3\delta_4^{(1)}) + 0.3(V_4^{(5)})^2\sin(3\delta_4^{(5)}) \end{cases}.$$

With $V_4^{(1)} = 0.9960\text{pu}$, $\delta_4^{(1)} = -0.003414\text{rad}$, $V_4^{(5)} = 0.1\text{pu}$, $\delta_4^{(5)} = 0\text{rad}$, one obtains

$$\begin{cases} g_{r,4}^{(5)} = 0.3(0.9960)^3\cos(-3 \cdot 0.003414) \\ \qquad + 0.3(0.1)^2\cos(3 \cdot 0) = 0.2994\text{pu} \\ g_{i,4}^{(5)} = 0.3(0.9960)^3\sin(-3 \cdot 0.003414) \\ \qquad + 0.3(0.1)^2\sin(3 \cdot 0) = -0.003036\text{pu} \end{cases}$$

$$I_4^{(5)} = \sqrt{(g_{r,4}^{(5)})^2 + (g_{i,4}^{(5)})^2} = \sqrt{(0.2994)^2 + (-0.003036)^2}$$
$$= 0.2994\text{pu}$$

$$\varepsilon_4^{(5)} = \tan^{-1}\left(\frac{|g_{i,4}^{(5)}|}{|g_{r,4}^{(5)}|}\right) = 0.01014\text{rad}.$$

Therefore, the fifth harmonic currents at bus 4 referred to the swing bus are $G_{r,4}^{(5)} = I_4^{(5)}\cos\gamma_4^{(5)}$ and $G_{i,4}^{(5)} = I_4^{(5)}\sin\gamma_4^{(5)}$.

The phasor diagram for the 5th harmonic is shown in Fig. E7.15.2, where the angles are

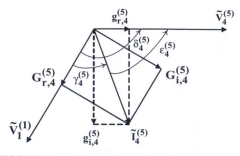

**FIGURE E7.15.2** Phasor diagram for 5th harmonic.

$$\delta_4^{(5)} = \varepsilon_4^{(5)} + \gamma_4^{(5)}$$

or

$$\gamma_4^{(5)} = \delta_4^{(5)} - \varepsilon_4^{(5)} = 0 - \varepsilon_4^{(5)} = -0.01014\text{rad}$$

resulting in $G_{r,4}^{(5)} = 0.2994\text{pu}$ and $G_{i,4}^{(5)} = -0.003036\text{pu}$.

These nonlinear currents referred to the swing bus are about the same as those referred to bus 4 because the angle $\gamma_4^{(5)}$ is very small for this first iteration.

Note that

$$\begin{cases} P_4^{(1)} = I_4^{(1)}V_4^{(1)}\cos(\delta_4^{(1)} - \gamma_4^{(1)}) \\ \quad = 0.2711 \cdot 0.9960 \cdot \cos(-0.003414 + 0.38391) \\ \quad = 0.250\text{pu} \\ Q_4^{(1)} = I_4^{(1)}V_4^{(1)}\sin(\delta_4^{(1)} - \gamma_4^{(1)}) \\ \quad = 0.2711 \cdot 0.9960 \cdot \sin(-0.003414 + 0.38391) \\ \quad = 0.100\text{pu} \end{cases}$$

are as expected. Correspondingly,

$$\begin{cases} P_4^{(5)} = I_4^{(5)}V_4^{(5)}\cos(\delta_4^{(5)} - \gamma_4^{(5)}) \\ \quad = 0.2994 \cdot 0.1 \cdot \cos(0 + 0.01014) = 0.02994\text{pu} \\ Q_4^{(5)} = I_4^{(5)}V_4^{(5)}\sin(\delta_4^{(5)} - \gamma_4^{(5)}) \\ \quad = 0.2994 \cdot 0.1 \cdot \sin(0 + 0.01014) = 0.000304\text{pu}. \end{cases}$$

Therefore, the total injected real and reactive powers at the nonlinear bus 4 are

$$\begin{cases} P_4^t = P_4^{(1)} + P_4^{(5)} = 0.250\text{pu} + 0.02994\text{pu} \\ \quad = 0.27994\text{pu} \\ Q_4^t = Q_4^{(1)} + Q_4^{(5)} = 0.100\text{pu} + 0.000304\text{pu} \\ \quad = 0.100304\text{pu}. \end{cases}$$

### 7.4.9 Application Example 7.16: Evaluation of Harmonic Mismatch Vector: Solution

*Step #3 of Harmonic Load Flow.* Evaluation of the mismatch vector

$$\Delta \overline{M}^0 = (\Delta \overline{W}, \Delta \overline{I}^{(5)}, \Delta \overline{I}^{(1)})^t = (P_2^{(1)} + F_{r,2}^{(1)}, Q_2^{(1)} + F_{i,2}^{(1)},$$
$$P_3^{(1)} + F_{r,3}^{(1)}, Q_3^{(1)} + F_{i,3}^{(1)}, P_4^t + F_{r,4}^{(1)} + F_{r,4}^{(5)}, Q_4^t + F_{i,4}^{(1)} + F_{i,4}^{(5)},$$
$$I_{r,1}^{(5)}, I_{i,1}^{(5)}, I_{r,2}^{(5)}, I_{i,2}^{(5)}, I_{r,3}^{(5)}, I_{i,3}^{(5)}, I_{r,4}^{(5)} + G_{r,4}^{(5)}, I_{i,4}^{(5)} + G_{i,4}^{(5)},$$
$$I_{r,4}^{(1)} + G_{r,4}^{(1)}, I_{i,4}^{(1)} + G_{i,4}^{(1)}).$$

To compute these mismatches use equations Eqs. 7-80 and 7-81

$$\Delta P_i = P_i + F_{r,i} = P_i + \sum_{j=1}^{n} y_{ij} |\tilde{V}_j| |\tilde{V}_i| \cos(-\theta_{ij} - \delta_j + \delta_i)$$

$$\Delta Q_i = Q_i + F_{i,i} = Q_i + \sum_{j=1}^{n} y_{ij} |\tilde{V}_j| |\tilde{V}_i| \sin(-\theta_{ij} - \delta_j + \delta_i)$$

and Eqs. 7-119a,b

$$\Delta I_{r,i}^{(h)} = I_{r,i}^{(h)} + G_{r,i}^{(h)} = \sum_{j=1}^{n} y_{ij}^{(h)} |\tilde{V}_j^{(h)}| \cos(\theta_{ij}^{(h)} + \delta_j^{(h)}) + G_{r,i}^{(h)}$$

$$\Delta I_{i,i}^{(h)} = I_{i,i}^{(h)} + G_{i,i}^{(h)} = \sum_{j=1}^{n} y_{ij}^{(h)} |\tilde{V}_j^{(h)}| \sin(\theta_{ij}^{(h)} + \delta_j^{(h)}) + G_{i,i}^{(h)}.$$

Hence,

$$\Delta P_2^{(1)} = P_2^{(1)} + F_{r,2}^{(1)} = P_2 + y_{21} V_1 V_2 \cos(-\theta_{21} - \delta_1 + \delta_2) +$$
$$y_{22} V_2 V_2 \cos(-\theta_{22} - \delta_2 + \delta_2) + y_{23} V_3 V_2 \cos(-\theta_{23} - \delta_3 + \delta_2)$$
$$+ y_{24} V_4 V_2 \cos(-\theta_{24} - \delta_4 + \delta_2) = -0.0170 \, \text{pu}$$

$$\Delta Q_2^{(1)} = Q_2^{(1)} + F_{i,2}^{(1)} = Q_2 + y_{21} V_1 V_2 \sin(-\theta_{21} - \delta_1 + \delta_2) +$$
$$y_{22} V_2 V_2 \sin(-\theta_{22} - \delta_2 + \delta_2) + y_{23} V_3 V_2 \sin(-\theta_{23} - \delta_3 + \delta_2)$$
$$+ y_{24} V_4 V_2 \sin(-\theta_{24} - \delta_4 + \delta_2) = 0.083 \, \text{pu}$$

$$\Delta P_3^{(1)} = P_3^{(1)} + F_{r,3}^{(1)} = P_3 + y_{31} V_1 V_3 \cos(-\theta_{31} - \delta_1 + \delta_3) +$$
$$y_{32} V_2 V_3 \cos(-\theta_{32} - \delta_2 + \delta_3) + y_{33} V_3 V_3 \cos(-\theta_{33} - \delta_3$$
$$+ \delta_3) + y_{34} V_4 V_3 \cos(-\theta_{34} - \delta_4 + \delta_3) = -0.00325 \, \text{pu}$$

$$\Delta Q_3^{(1)} = Q_3^{(1)} + F_{i,3}^{(1)} = Q_3 + y_{31} V_1 V_3 \sin(-\theta_{31} - \delta_1 + \delta_3) +$$
$$y_{32} V_2 V_3 \sin(-\theta_{32} - \delta_2 + \delta_3) + y_{33} V_3 V_3 \sin(\theta_{33} - \delta_3 + \delta_3)$$
$$+ y_{34} V_4 V_3 \sin(-\theta_{34} - \delta_4 + \delta_3) = 0.00192 \, \text{pu}$$

$$\Delta P_4^t = P_4^t + F_{r,4}^{(1)} + F_{r,4}^{(5)} = P_4^t$$
$$+ \sum_{j=1}^{4} y_{4j} V_j^{(1)} V_4^{(1)} \cos(-\theta_{4j} - \delta_j^{(1)} + \delta_4^{(1)})$$
$$+ \sum_{j=1}^{4} y_{4j}^{(5)} V_j^{(5)} V_4^{(5)} \cos(-\theta_{4j}^{(5)} - \delta_j^{(5)} + \delta_4^{(5)}) = P_4^t$$
$$+ y_{41} V_1^{(1)} V_4^{(1)} \cos(-\theta_{41} - \delta_1^{(1)} + \delta_4^{(1)})$$
$$+ y_{43} V_3^{(1)} V_4^{(1)} \cos(-\theta_{43} - \delta_3^{(1)} + \delta_4^{(1)})$$
$$+ y_{44} V_4^{(1)} V_4^{(1)} \cos(-\theta_{44} - \delta_4^{(1)} + \delta_4^{(1)})$$
$$+ y_{41}^{(5)} V_1^{(5)} V_4^{(5)} \cos(-\theta_{41}^{(5)} - \delta_1^{(5)} + \delta_4^{(5)})$$
$$+ y_{43}^{(5)} V_3^{(5)} V_4^{(5)} \cos(-\theta_{43}^{(5)} - \delta_3^{(5)} + \delta_4^{(5)})$$
$$+ y_{44}^{(5)} V_4^{(5)} V_4^{(5)} \cos(-\theta_{44}^{(5)} - \delta_4^{(5)} + \delta_4^{(5)}) = 0.037 \, \text{pu}$$

$$\Delta Q_4^t = Q_4^t + F_{i,4}^{(1)} + F_{i,4}^{(5)} =$$
$$Q_4^t + \sum_{j=1}^{4} y_{4j} V_j^{(1)} V_4^{(1)} \sin(-\theta_{4j} - \delta_j^{(1)} + \delta_4^{(1)})$$
$$+ \sum_{j=1}^{4} y_{4j}^{(5)} V_j^{(5)} V_4^{(5)} \sin(-\theta_{4j}^{(5)} - \delta_j^{(5)} + \delta_4^{(5)}) =$$
$$Q_4^t + y_{41} V_1^{(1)} V_4^{(1)} \sin(-\theta_{41} - \delta_1^{(1)} + \delta_4^{(1)})$$
$$+ y_{43} V_3^{(1)} V_4^{(1)} \sin(-\theta_{43} - \delta_3^{(1)} + \delta_4^{(1)})$$
$$+ y_{44} V_4^{(1)} V_4^{(1)} \sin(-\theta_{44} - \delta_4^{(1)} + \delta_4^{(1)})$$
$$+ y_{41}^{(5)} V_1^{(5)} V_4^{(5)} \sin(-\theta_{41}^{(5)} - \delta_1^{(5)} + \delta_4^{(5)})$$
$$+ y_{43}^{(5)} V_3^{(5)} V_4^{(5)} \sin(-\theta_{43}^{(5)} - \delta_3^{(5)} + \delta_4^{(5)})$$
$$+ y_{44}^{(5)} V_4^{(5)} V_4^{(5)} \sin(-\theta_{44}^{(5)} - \delta_4^{(5)} + \delta_4^{(5)}) = -0.188 \, \text{pu}$$

$$I_{r,1}^{(5)} = \sum_{j=1}^{4} y_{1j}^{(5)} |\tilde{V}_j^{(5)}| \cos(\theta_{1j}^{(5)} + \delta_j^{(5)})$$
$$= y_{11}^{(5)} |\tilde{V}_1^{(5)}| \cos(\theta_{11}^{(5)} + \delta_1^{(5)})$$
$$+ y_{12}^{(5)} |\tilde{V}_2^{(5)}| \cos(\theta_{12}^{(5)} + \delta_2^{(5)})$$
$$+ y_{13}^{(5)} |\tilde{V}_3^{(5)}| \cos(\theta_{13}^{(5)} + \delta_3^{(5)})$$
$$+ y_{14}^{(5)} |\tilde{V}_4^{(5)}| \cos(\theta_{14}^{(5)} + \delta_4^{(5)}) = 0.201 \, \text{pu}$$

$$I_{i,1}^{(5)} = \sum_{j=1}^{4} y_{1j}^{(5)} |\tilde{V}_j^{(5)}| \sin(\theta_{1j}^{(5)} + \delta_j^{(5)})$$
$$= y_{11}^{(5)} |\tilde{V}_1^{(5)}| \sin(\theta_{11}^{(5)} + \delta_1^{(5)})$$
$$+ y_{12}^{(5)} |\tilde{V}_2^{(5)}| \sin(\theta_{12}^{(5)} + \delta_2^{(5)})$$
$$+ y_{13}^{(5)} |\tilde{V}_3^{(5)}| \sin(\theta_{13}^{(5)} + \delta_3^{(5)})$$
$$+ y_{14}^{(5)} |\tilde{V}_4^{(5)}| \sin(\theta_{14}^{(5)} + \delta_4^{(5)}) = -201.98 \, \text{pu}$$

$$I_{r,2}^{(5)} = \sum_{j=1}^{4} y_{2j}^{(5)} |\tilde{V}_j^{(5)}| \cos(\theta_{2j}^{(5)} + \delta_j^{(5)})$$
$$= y_{21}^{(5)} |\tilde{V}_1^{(5)}| \cos(\theta_{21}^{(5)} + \delta_1^{(5)})$$
$$+ y_{22}^{(5)} |\tilde{V}_2^{(5)}| \cos(\theta_{22}^{(5)} + \delta_2^{(5)})$$
$$+ y_{23}^{(5)} |\tilde{V}_3^{(5)}| \cos(\theta_{23}^{(5)} + \delta_3^{(5)})$$
$$+ y_{24}^{(5)} |\tilde{V}_4^{(5)}| \cos(\theta_{24}^{(5)} + \delta_4^{(5)}) = 0.000255 \, \text{pu}$$

$$I_{i,2}^{(5)} = \sum_{j=1}^{4} y_{2j}^{(5)} |\tilde{V}_j^{(5)}| \sin(\theta_{2j}^{(5)} + \delta_j^{(5)})$$
$$= y_{21}^{(5)} |\tilde{V}_1^{(5)}| \sin(\theta_{21}^{(5)} + \delta_1^{(5)})$$
$$+ y_{22}^{(5)} |\tilde{V}_2^{(5)}| \sin(\theta_{22}^{(5)} + \delta_2^{(5)}) + y_{23}^{(5)} |\tilde{V}_3^{(5)}| \sin(\theta_{23}^{(5)} + \delta_3^{(5)})$$
$$+ y_{24}^{(5)} |\tilde{V}_4^{(5)}| \sin(\theta_{24}^{(5)} + \delta_4^{(5)}) = -0.000620 \, \text{pu}$$

$$I_{r,3}^{(5)} = \sum_{j=1}^{4} y_{3j}^{(5)} |\tilde{V}_j^{(5)}| \cos(\theta_{3j}^{(5)} + \delta_j^{(5)})$$
$$= y_{31}^{(5)} |\tilde{V}_1^{(5)}| \cos(\theta_{31}^{(5)} + \delta_1^{(5)})$$
$$+ y_{32}^{(5)} |\tilde{V}_2^{(5)}| \cos(\theta_{32}^{(5)} + \delta_2^{(5)})$$
$$+ y_{33}^{(5)} |\tilde{V}_3^{(5)}| \cos(\theta_{33}^{(5)} + \delta_3^{(5)})$$
$$+ y_{34}^{(5)} |\tilde{V}_4^{(5)}| \cos(\theta_{34}^{(5)} + \delta_4^{(5)}) = 0.000104 \, \text{pu}$$

$$I_{i,3}^{(5)} = \sum_{j=1}^{4} y_{3j}^{(5)} |\tilde{V}_j^{(5)}| \sin(\theta_{3j}^{(5)} + \delta_j^{(5)})$$

$$= y_{31}^{(5)} |\tilde{V}_1^{(5)}| \sin(\theta_{31}^{(5)} + \delta_1^{(5)})$$
$$+ y_{32}^{(5)} |\tilde{V}_2^{(5)}| \sin(\theta_{32}^{(5)} + \delta_2^{(5)})$$
$$+ y_{33}^{(5)} |\tilde{V}_3^{(5)}| \sin(\theta_{33}^{(5)} + \delta_3^{(5)})$$
$$+ y_{34}^{(5)} |\tilde{V}_4^{(5)}| \sin(\theta_{34}^{(5)} + \delta_4^{(5)}) = 0.0000082\,\text{pu}$$

$$I_{r,4}^{(5)} + G_{r,4}^{(5)} = \sum_{j=1}^{4} y_{4j}^{(5)} |\tilde{V}_j^{(5)}| \cos(\theta_{4j}^{(5)} + \delta_j^{(5)}) + G_{r,4}^{(5)}$$

$$= y_{41}^{(5)} |\tilde{V}_1^{(5)}| \cos(\theta_{41}^{(5)} + \delta_1^{(5)}) + y_{43}^{(5)} |\tilde{V}_3^{(5)}| \cos(\theta_{43}^{(5)} + \delta_3^{(5)})$$
$$+ y_{44}^{(5)} |\tilde{V}_4^{(5)}| \cos(\theta_{44}^{(5)} + \delta_4^{(5)}) + G_{r,4}^{(5)} = 0.2996\,\text{pu}$$

$$I_{i,4}^{(5)} + G_{i,4}^{(5)} = \sum_{j=1}^{4} y_{4j}^{(5)} |\tilde{V}_j^{(5)}| \sin(\theta_{4j}^{(5)} + \delta_j^{(5)}) + G_{i,4}^{(5)}$$

$$= y_{41}^{(5)} |\tilde{V}_1^{(5)}| \sin(\theta_{41}^{(5)} + \delta_1^{(5)}) + y_{43}^{(5)} |\tilde{V}_3^{(5)}| \sin(\theta_{43}^{(5)} + \delta_3^{(5)})$$
$$+ y_{44}^{(5)} |\tilde{V}_4^{(5)}| \sin(\theta_{44}^{(5)} + \delta_4^{(5)}) + G_{i,4}^{(5)} = -0.00325\,\text{pu}$$

$$I_{r,4}^{(1)} + G_{r,4}^{(1)} = \sum_{j=1}^{4} y_{4j}^{(1)} |\tilde{V}_j^{(1)}| \cos(\theta_{4j}^{(1)} + \delta_j^{(1)}) + G_{r,4}^{(1)}$$

$$= y_{41} |\tilde{V}_1^{(1)}| \cos(\theta_{41}^{(1)} + \delta_1^{(1)}) + y_{43} |\tilde{V}_3^{(1)}| \cos(\theta_{43}^{(1)} + \delta_3^{(1)})$$
$$+ y_{44} |\tilde{V}_4^{(1)}| \cos(\theta_{44}^{(1)} + \delta_4^{(1)}) + G_{r,4}^{(1)} = 0.0027\,\text{pu}$$

$$I_{i,4}^{(1)} + G_{i,4}^{(1)} = \sum_{j=1}^{4} y_{4j}^{(5)} |\tilde{V}_j^{(1)}| \sin(\theta_{4j}^{(1)} + \delta_j^{(1)}) + G_{i,4}^{(1)}$$

$$= y_{41} |\tilde{V}_1^{(1)}| \sin(\theta_{41}^{(1)} + \delta_1^{(1)}) + y_{43} |\tilde{V}_3^{(1)}| \sin(\theta_{43}^{(1)} + \delta_3^{(1)})$$
$$+ y_{44} |\tilde{V}_4^{(1)}| \sin(\theta_{44}^{(1)} + \delta_4^{(1)}) + G_{i,4}^{(1)} = -0.0054\,\text{pu}.$$

The mismatch vector $\Delta \bar{M}^0$ is now

$$\Delta \bar{M}^0 = (\Delta \bar{W}, \Delta \bar{I}^{(5)}, \Delta \bar{I}^{(1)})^t = (P_2^{(1)} + F_{r,2}^{(1)}, Q_2^{(1)} + F_{i,2}^{(1)},$$
$$P_3^{(1)} + F_{r,3}^{(1)}, Q_3^{(1)} + F_{i,3}^{(1)}, P_4 + F_{r,4}^{(1)} + F_{r,4}^{(5)}, Q_4^t + F_{i,4}^{(1)} +$$
$$F_{i,4}^{(5)}, I_{r,1}^{(5)}, I_{i,1}^{(5)}, I_{r,2}^{(5)}, I_{i,2}^{(5)}, I_{r,3}^{(5)}, I_{i,3}^{(5)}, I_{r,4}^{(5)} + G_{r,4}^{(5)}, I_{i,4}^{(5)} + G_{i,4}^{(5)},$$
$$I_{r,4}^{(1)} + G_{r,4}^{(1)}, I_{i,4}^{(1)} + G_{i,4}^{(1)})^t$$

or

$$\Delta \bar{M}^0 = (-0.0170, 0.083, -0.00325, 0.00192, 0.037,$$
$$-0.188, 0.201, -201.98, 0.000255, -0.000620,$$
$$0.000104, 0.0000082, 0.2996, -0.00325, 0.0027,$$
$$-0.0054).$$

Note that for a mismatch residual of 0.0001 pu the solution has not converged yet.

### 7.4.10 Application Example 7.17: Evaluation of Fundamental and Harmonic Jacobian Submatrices: Solution

*Step #4 of Harmonic Load Flow.* The matrix equation shown below makes it easy to identify the entries of the Jacobian matrix:

$$
\begin{array}{c}
\begin{array}{cccccccccccccccc}
\delta_2^{(1)} & V_2^{(1)} & \delta_3^{(1)} & V_3^{(1)} & \delta_4^{(1)} & V_4^{(1)} & \delta_1^{(5)} & V_1^{(5)} & \delta_2^{(5)} & V_2^{(5)} & \delta_3^{(5)} & V_3^{(5)} & \delta_4^{(5)} & V_4^{(5)} & \alpha & \beta \\
\downarrow & \downarrow & \downarrow & \downarrow & \downarrow & \downarrow & \downarrow & \downarrow & \downarrow & \downarrow & \downarrow & \downarrow & \downarrow & \downarrow & \downarrow & \downarrow
\end{array}
\end{array}
$$

$$
\bar{J}^0 =
\begin{vmatrix}
J^0_{1,1} & J^0_{1,2} & J^0_{1,3} & J^0_{1,4} & J^0_{1,5} & J^0_{1,6} & J^0_{1,7} & J^0_{1,8} & J^0_{1,9} & J^0_{1,10} & J^0_{1,11} & J^0_{1,12} & J^0_{1,13} & J^0_{1,14} & J^0_{1,15} & J^0_{1,16} \\
J^0_{2,1} & J^0_{2,2} & J^0_{2,3} & J^0_{2,4} & J^0_{2,5} & J^0_{2,6} & J^0_{2,7} & J^0_{2,8} & J^0_{2,9} & J^0_{2,10} & J^0_{2,11} & J^0_{2,12} & J^0_{2,13} & J^0_{2,14} & J^0_{2,15} & J^0_{2,16} \\
J^0_{3,1} & J^0_{3,2} & J^0_{3,3} & J^0_{3,4} & J^0_{3,5} & J^0_{3,6} & J^0_{3,7} & J^0_{3,8} & J^0_{3,9} & J^0_{3,10} & J^0_{3,11} & J^0_{3,12} & J^0_{3,13} & J^0_{3,14} & J^0_{3,15} & J^0_{3,16} \\
J^0_{4,1} & J^0_{4,2} & J^0_{4,3} & J^0_{4,4} & J^0_{4,5} & J^0_{4,6} & J^0_{4,7} & J^0_{4,8} & J^0_{4,9} & J^0_{4,10} & J^0_{4,11} & J^0_{4,12} & J^0_{4,13} & J^0_{4,14} & J^0_{4,15} & J^0_{4,16} \\
J^0_{5,1} & J^0_{5,2} & J^0_{5,3} & J^0_{5,4} & J^0_{5,5} & J^0_{5,6} & J^0_{5,7} & J^0_{5,8} & J^0_{5,9} & J^0_{5,10} & J^0_{5,11} & J^0_{5,12} & J^0_{5,13} & J^0_{5,14} & J^0_{5,15} & J^0_{5,16} \\
J^0_{6,1} & J^0_{6,2} & J^0_{6,3} & J^0_{6,4} & J^0_{6,5} & J^0_{6,6} & J^0_{6,7} & J^0_{6,8} & J^0_{6,9} & J^0_{6,10} & J^0_{6,11} & J^0_{6,12} & J^0_{6,13} & J^0_{6,14} & J^0_{6,15} & J^0_{6,16} \\
J^0_{7,1} & J^0_{7,2} & J^0_{7,3} & J^0_{7,4} & J^0_{7,5} & J^0_{7,6} & J^0_{7,7} & J^0_{7,8} & J^0_{7,9} & J^0_{7,10} & J^0_{7,11} & J^0_{7,12} & J^0_{7,13} & J^0_{7,14} & J^0_{7,15} & J^0_{7,16} \\
J^0_{8,1} & J^0_{8,2} & J^0_{8,3} & J^0_{8,4} & J^0_{8,5} & J^0_{8,6} & J^0_{8,7} & J^0_{8,8} & J^0_{8,9} & J^0_{8,10} & J^0_{8,11} & J^0_{8,12} & J^0_{8,13} & J^0_{8,14} & J^0_{8,15} & J^0_{8,16} \\
J^0_{9,1} & J^0_{9,2} & J^0_{9,3} & J^0_{9,4} & J^0_{9,5} & J^0_{9,6} & J^0_{9,7} & J^0_{9,8} & J^0_{9,9} & J^0_{9,10} & J^0_{9,11} & J^0_{9,12} & J^0_{9,13} & J^0_{9,14} & J^0_{9,15} & J^0_{9,16} \\
J^0_{10,1} & J^0_{10,2} & J^0_{10,3} & J^0_{10,4} & J^0_{10,5} & J^0_{10,6} & J^0_{10,7} & J^0_{10,8} & J^0_{10,9} & J^0_{10,10} & J^0_{10,11} & J^0_{10,12} & J^0_{10,13} & J^0_{10,14} & J^0_{10,15} & J^0_{10,16} \\
J^0_{11,1} & J^0_{11,2} & J^0_{11,3} & J^0_{11,4} & J^0_{11,5} & J^0_{11,6} & J^0_{11,7} & J^0_{11,8} & J^0_{11,9} & J^0_{11,10} & J^0_{11,11} & J^0_{11,12} & J^0_{11,13} & J^0_{11,14} & J^0_{11,15} & J^0_{11,16} \\
J^0_{12,1} & J^0_{12,2} & J^0_{12,3} & J^0_{12,4} & J^0_{12,5} & J^0_{12,6} & J^0_{12,7} & J^0_{12,8} & J^0_{12,9} & J^0_{12,10} & J^0_{12,11} & J^0_{12,12} & J^0_{12,13} & J^0_{12,14} & J^0_{12,15} & J^0_{12,16} \\
J^0_{13,1} & J^0_{13,2} & J^0_{13,3} & J^0_{13,4} & J^0_{13,5} & J^0_{13,6} & J^0_{13,7} & J^0_{13,8} & J^0_{13,9} & J^0_{13,10} & J^0_{13,11} & J^0_{13,12} & J^0_{13,13} & J^0_{13,14} & J^0_{13,15} & J^0_{13,16} \\
J^0_{14,1} & J^0_{14,2} & J^0_{14,3} & J^0_{14,4} & J^0_{14,5} & J^0_{14,6} & J^0_{14,7} & J^0_{14,8} & J^0_{14,9} & J^0_{14,10} & J^0_{14,11} & J^0_{14,12} & J^0_{14,13} & J^0_{14,14} & J^0_{14,15} & J^0_{14,16} \\
J^0_{15,1} & J^0_{15,2} & J^0_{15,3} & J^0_{15,4} & J^0_{15,5} & J^0_{15,6} & J^0_{15,7} & J^0_{15,8} & J^0_{15,9} & J^0_{15,10} & J^0_{15,11} & J^0_{15,12} & J^0_{15,13} & J^0_{15,14} & J^0_{15,15} & J^0_{15,16} \\
J^0_{16,1} & J^0_{16,2} & J^0_{16,3} & J^0_{16,4} & J^0_{16,5} & J^0_{16,6} & J^0_{16,7} & J^0_{16,8} & J^0_{16,9} & J^0_{16,10} & J^0_{16,11} & J^0_{16,12} & J^0_{16,13} & J^0_{16,14} & J^0_{16,15} & J^0_{16,16}
\end{vmatrix}
\begin{array}{l}
\leftarrow \Delta P_2^{(1)} \\
\leftarrow \Delta Q_2^{(1)} \\
\leftarrow \Delta P_3^{(1)} \\
\leftarrow \Delta Q_3^{(1)} \\
\leftarrow \Delta P_4^t \\
\leftarrow \Delta Q_4^t \\
\leftarrow \Delta I_{r,1}^{(5)} \\
\leftarrow \Delta I_{i,1}^{(5)} \\
\leftarrow \Delta I_{r,2}^{(5)} \\
\leftarrow \Delta I_{i,2}^{(5)} \\
\leftarrow \Delta I_{r,3}^{(5)} \\
\leftarrow \Delta I_{i,3}^{(5)} \\
\leftarrow \Delta I_{r,4}^{(5)} \\
\leftarrow \Delta I_{i,4}^{(5)} \\
\leftarrow \Delta I_{r,4}^{(1)} \\
\leftarrow \Delta I_{i,4}^{(1)}
\end{array}
$$

The top row, which is not part of the Jacobian, lists the variables:

$$
\delta_2^{(1)}\ V_2^{(1)}\ \delta_3^{(1)}\ V_3^{(1)}\ \delta_4^{(1)}\ V_4^{(1)}\ \delta_1^{(5)}\ V_1^{(5)}\ \delta_2^{(5)}\ V_2^{(5)}\ \delta_3^{(5)}\ V_3^{(5)}\ \delta_4^{(5)} \\
V_4^{(5)}\ \alpha\ \beta
$$

and the right-hand side column, which is not part of the Jacobian, lists the mismatch quantities that must be satisfied:

$$
[\Delta P_2^{(1)}\ \Delta Q_2^{(1)}\ \Delta P_3^{(1)}\ \Delta Q_3^{(1)}\ \Delta P_4^t\ \Delta Q_4^t\ I_{r,1}^{(5)}\ \Delta I_{i,1}^{(5)}\ \Delta I_{r,2}^{(5)}\ \Delta I_{i,2}^{(5)} \\
\Delta I_{r,3}^{(5)}\ \Delta I_{i,3}^{(5)}\ \Delta I_{r,4}^{(5)}\ \Delta I_{i,4}^{(5)}\ \Delta I_{r,4}^{(1)}\ \Delta I_{i,4}^{(1)}]^t.
$$

These two identifications help in defining the partial derivatives of the Jacobian. For example,

$$
J^0_{3,5} = \frac{\partial(\Delta P_3^{(1)})}{\partial(\delta_4^{(1)})}, \quad J^0_{3,10} = \frac{\partial(\Delta P_3^{(1)})}{\partial(V_2^{(5)})},
$$

$$
J^0_{8,5} = \frac{\partial(\Delta I_{i,1}^{(5)})}{\partial(\delta_4^{(1)})}, \quad J^0_{8,10} = \frac{\partial(\Delta I_{i,i}^{(5)})}{\partial(V_2^{(5)})},
$$

$$
J^0_{9,11} = \frac{\partial(\Delta I_{r,2}^{(5)})}{\partial(\delta_3^{(5)})}, \ \text{ and } \ J^0_{12,16} = \frac{\partial(\Delta I_{i,3}^{(5)})}{\partial(\beta)}.
$$

In the solution below submatrices are defined, which is an alternative way of defining the Jacobian entries.

***Step #4 of Harmonic Load Flow.*** Evaluate the Jacobian matrix $\bar{J}$ and compute $\Delta \bar{U}^0$. The Jacobian matrix has the following form ($\bar{0}_{n\times m}$ denotes an $n$ by $m$ zero submatrix with zero entries):

$$
\bar{J} =
\begin{vmatrix}
\bar{J}^{(1)} & \bar{J}^{(5)} & \begin{matrix}\bar{0}_{6,2} \\ \bar{0}_{6,2}\end{matrix} \\
\bar{G}^{(5,1)} & \bar{Y}^{(5,5)} + \bar{G}^{(5,5)} & \bar{H}^{(5)} \\
\bar{Y}^{(1,1)} + \bar{G}^{(1,1)} & \bar{G}^{(1,5)} & \bar{H}^{(1)}
\end{vmatrix}.
$$

The Jacobian submatrix for the fundamental is

$$
\bar{J}^{(1)} =
\begin{vmatrix}
61.68 & 52.932 & -11.764 & -2.944 & 0 & 0 \\
-52.94 & 61.945 & 2.944 & -11.764 & 0 & 0 \\
-11.764 & -2.944 & 51.772 & 22.917 & -40.0076 & -19.98 \\
2.944 & -11.764 & -22.926 & 51.773 & 19.982 & -40.0076 \\
0 & 0 & -40.0076 & -19.982 & 80.015 & 39.961 \\
0 & 0 & 19.982 & -40.0076 & -39.963 & 80
\end{vmatrix}.
$$

The Jacobian submatrix for the 5$^{th}$ harmonic is

$$\bar{J}^{(5)} = \begin{vmatrix} & & & \bar{0}_{4\times8} & & & & \\ -0.099 & -0.099 & 0 & 0 & -0.099 & -0.09 & 0.198 & 0.198 \\ 0.010 & -0.099 & 0 & 0 & 0.010 & -0.99 & -0.020 & 1.98 \end{vmatrix}.$$

Also

$$\bar{Y}^{(1,1)} = \begin{vmatrix} 0 & 0 & -39.889 & -19.89 & 79.81 & 39.73 \\ 0 & 0 & -19.81 & 40.06 & 39.56 & -80.14 \end{vmatrix}$$

$$\bar{Y}^{(5,5)} = \begin{vmatrix} 202.91 & 4.84 & -1.92 & -3.85 & 0 & 0 & -0.99 & -0.99 \\ 0.48 & -2029.1 & -0.39 & 19.23 & 0 & 0 & -0.10 & 9.90 \\ -1.92 & -3.846 & 2.17 & 3.97 & -0.25 & -0.13 & 0 & 0 \\ -0.39 & 19.23 & 0.40 & -21.73 & -0.01 & 2.45 & 0 & 0 \\ 0 & 0 & -0.25 & -0.13 & 1.24 & 1.12 & -0.99 & -0.99 \\ 0 & 0 & -0.01 & 2.49 & 0.11 & -12.40 & -0.10 & 9.90 \\ -0.99 & -0.99 & 0 & 0 & -0.99 & -0.99 & 1.98 & 1.98 \\ -0.01 & 9.90 & 0 & 0 & -0.01 & 9.90 & 0.20 & -19.80 \end{vmatrix}.$$

In this example, $\bar{H}^{(1)}$ and $\bar{H}^{(5)}$ are zero submatrices since there are no nonlinear device variables defined for the system of Fig. E7.14.1. However, setting $\bar{H}^{(1)}$ and $\bar{H}^{(5)}$ to zero introduces two columns of zeros in the Jacobian matrix, which makes it singular. In order to avoid this situation, $\bar{H}^{(1)}$ and $\bar{H}^{(5)}$ are set to random matrices. Note that the solution does not depend on the values of these submatrices.

In order to compute $\bar{G}^{(1,1)}$ and $\bar{G}^{(1,5)}$, we need to differentiate $G_{r,4}^{(1)}$ and $G_{i,4}^{(1)}$ as follows:

$$\frac{\partial G_{r,4}^{(1)}}{\partial \delta_4^{(1)}} = -\frac{P_4^{(1)} \sin\gamma_4^{(1)}}{V_4^{(1)} \cos(\delta_4^{(1)} - \gamma_4^{(1)})} = 0.1016$$

$$\frac{\partial G_{r,4}^{(1)}}{\partial V_4^{(1)}} = -\frac{P_4^{(1)} \cos\gamma_4^{(1)}}{(V_4^{(1)})^2 \cos(\delta_4^{(1)} - \gamma_4^{(1)})} = -0.252$$

$$\frac{\partial G_{i,4}^{(1)}}{\partial \delta_4^{(1)}} = -\frac{P_4^{(1)} \cos\gamma_4^{(1)}}{V_4^{(1)} \cos(\delta_4^{(1)} - \gamma_4^{(1)})} = -0.250$$

$$\frac{\partial G_{i,4}^{(1)}}{\partial V_4^{(1)}} = -\frac{P_4^{(1)} \sin\gamma_4^{(1)}}{(V_4^{(1)})^2 \cos(\delta_4^{(1)} - \gamma_4^{(1)})} = -0.102.$$

The partial derivatives of $G_{r,4}^{(1)}$ and $G_{i,4}^{(1)}$ with respect to $V_4^{(5)}$ and $\delta_4^{(5)}$ are zero. Therefore, $\bar{G}^{(1,5)}$ is a $2 \times 8$ submatrix of zeros and

$$\bar{G}^{(1,1)} = \begin{vmatrix} & 0.102 & -0.252 \\ \bar{0}_{2\times6} & & \\ & 0.250 & 0.102 \end{vmatrix}.$$

In order to compute $\bar{G}^{(5,1)}$ and $\bar{G}^{(5,5)}$, we need to differentiate $G_{r,4}^{(5)}$ and $G_{i,4}^{(5)}$ as follows:

$$\frac{\partial G_{r,4}^{(5)}}{\partial \delta_4^{(1)}} = -0.9(V_4^{(1)})^3 \sin(3\delta_4^{(1)}) = 0.00896$$

$$\frac{\partial G_{r,4}^{(5)}}{\partial V_4^{(1)}} = 0.9(V_4^{(1)})^2 \cos(3\delta_4^{(1)}) = 0.893$$

$$\frac{\partial G_{i,4}^{(5)}}{\partial \delta_4^{(1)}} = 0.9(V_4^{(1)})^3 \cos(3\delta_4^{(1)}) = 0.8894$$

$$\frac{\partial G_{i,4}^{(5)}}{\partial V_4^{(1)}} = 0.9(V_4^{(1)})^2 \sin(3\delta_4^{(1)}) = -0.009.$$

Also

$$\frac{\partial G_{r,4}^{(5)}}{\partial \delta_4^{(5)}} = -0.9(V_4^{(5)})^2 \sin(3\delta_4^{(5)}) = 0.00$$

$$\frac{\partial G_{r,4}^{(5)}}{\partial V_4^{(5)}} = 0.6V_4^{(5)} \cos(3\delta_4^{(5)}) = 0.06$$

$$\frac{\partial G_{i,4}^{(5)}}{\partial \delta_4^{(5)}} = 0.9(V_4^{(5)})^2 \cos(3\delta_4^{(5)}) = 0.006$$

$$\frac{\partial G_{i,4}^{(5)}}{\partial V_4^{(5)}} = 0.6V_4^{(5)} \sin(3\delta_4^{(5)}) = 0.00.$$

Therefore,

$$\bar{G}^{(5,1)} = \begin{vmatrix} & \bar{0}_{6\times 8} & \\ & & 0.00896 & 0.893 \\ \bar{0}_{2\times 6} & & & \\ & & 0.8894 & -0.009 \end{vmatrix},$$

$$\bar{G}^{(5,5)} = \begin{vmatrix} & \bar{0}_{6\times 8} & \\ & & 0 & 0.06 \\ \bar{0}_{2\times 6} & & & \\ & & 0.006 & 0 \end{vmatrix}.$$

The Jacobian matrix is

$$
\begin{vmatrix}
-61.57 & 52.71 & -11.70 & -2.90 & 0 & 0 & 0 & 0 & 0 & 0 & 0 & 0 & 0 & 0 & 0 & 0 \\
-52.79 & 61.52 & 2.89 & -11.74 & 0 & 0 & 0 & 0 & 0 & 0 & 0 & 0 & 0 & 0 & 0 & 0 \\
-11.69 & -2.96 & 51.37 & 22.86 & -39.71 & -19.90 & 0 & 0 & 0 & 0 & 0 & 0 & 0 & 0 & 0 & 0 \\
2.95 & -11.72 & -22.77 & 51.58 & 19.82 & -39.87 & 0 & 0 & 0 & 0 & 0 & 0 & 0 & 0 & 0 & 0 \\
0 & 0 & -39.66 & -19.94 & 79.45 & 39.59 & -0.099 & -0.099 & 0 & 0 & -0.099 & -0.099 & 0.198 & 0.198 & 0 & 0 \\
0 & 0 & 19.86 & -39.83 & -39.93 & 79.58 & 0.010 & -0.99 & 0 & 0 & 0.010 & -0.99 & -0.020 & 1.98 & 0 & 0 \\
0 & 0 & 0 & 0 & 0 & 0 & 202.9 & 4.84 & -1.92 & -3.85 & 0 & 0 & -0.99 & -0.99 & 0 & 0 \\
0 & 0 & 0 & 0 & 0 & 0 & 0.48 & -2029 & -0.39 & 19.23 & 0 & 0 & -0.10 & 9.90 & 0 & 0 \\
0 & 0 & 0 & 0 & 0 & 0 & -1.92 & -3.846 & 2.17 & 3.97 & -0.25 & -0.13 & 0 & 0 & 0 & 0 \\
0 & 0 & 0 & 0 & 0 & 0 & -0.39 & 19.23 & 0.40 & -21.73 & -0.01 & 2.45 & 0 & 0 & 0 & 0 \\
0 & 0 & 0 & 0 & 0 & 0 & 0 & 0 & -0.25 & -0.13 & 1.24 & 1.12 & -0.99 & -0.99 & 0 & 0 \\
0 & 0 & 0 & 0 & 0 & 0 & 0 & 0 & -0.01 & 2.49 & 0.11 & -12.40 & -0.10 & 9.90 & 0 & 0 \\
0 & 0 & 0 & 0 & 0.009 & 0.893 & -0.99 & -0.99 & 0 & 0 & -0.99 & -0.99 & 1.98 & 2.04 & 0 & 0 \\
0 & 0 & 0 & 0 & 0.889 & -0.009 & -0.01 & 9.90 & 0 & 0 & -0.01 & 9.90 & 0.201 & -19.80 & 0 & 0 \\
0 & 0 & -39.89 & -19.89 & 79.91 & 39.48 & 0 & 0 & 0 & 0 & 0 & 0 & 0 & 0 & 1 & 2 \\
0 & 0 & -19.81 & 40.06 & 39.81 & -80.04 & 0 & 0 & 0 & 0 & 0 & 0 & 0 & 0 & 3 & 4
\end{vmatrix}.
$$

## 7.4.11 Application Example 7.18: Computation of the Correction Bus Vector and Convergence of Harmonic Power Flow: Solution

***Step #5 of Harmonic Load Flow.*** Evaluate $\Delta \bar{U}^0$ and update $\Delta \bar{U}^0$.

The correction bus vector $\Delta \bar{U}^0$ is computed as follows:

$$\Delta \bar{U}^0 = (\bar{J}^0)^{-1} \begin{bmatrix} 3.43472\,E-6 \\ -1.33365\,E-6 \\ -7.62939\,E-6 \\ -3.81470\,E-6 \\ 2.99421\,E-2 \\ 2.97265\,E-4 \\ 4.217029\,E-6 \\ -200.0 \\ -4.09782\,E-8 \\ 2.384186\,E-7 \\ -1.93715\,E-7 \\ -5.960464\,E-8 \\ 0.29935 \\ -3.03515\,E-3 \\ -1.78814\,E-7 \\ 8.86619\,E-7 \end{bmatrix} = \begin{bmatrix} 1.20815\,E-6 \\ -8.59742\,E-7 \\ 9.552948\,E-6 \\ 1.54111\,E-6 \\ 1.184274\,E-5 \\ 2.3572\,E-6 \\ 1.46588\,E-3 \\ 0.10000228 \\ 2.38763\,E-2 \\ 0.1005407 \\ 0.2042002 \\ 0.1021379 \\ 0.249194 \\ 0.1027635 \\ -- \\ -- \end{bmatrix}.$$

The updated bus vector is

$$\overline{U}^1 = \overline{U}^0 - \Delta\overline{U}^0 = (-2.68332E - 4, 0.99759,$$
$$-2.80029E - 3, 0.99640, -3.4257E - 3, 0.99594,$$
$$-1.465844E - 3, -2.28E - 6, -2.38763E - 2,$$
$$-5.407E - 4, -0.2042, -2.138E - 3, -0.24919,$$
$$-2.7635E - 3, 0, 0)^t.$$

Note that all harmonic voltage magnitudes are negative. To prevent negative magnitudes, the corresponding phase angles are shifted by $\pi$ radians; therefore,

$$\overline{U}^1 = (-2.68332E - 4, 0.99759, -2.80029E - 3,$$
$$0.99640, -3.4257E - 3, 0.99594, 3.14013, 2.28E - 6,$$
$$3.11772, 5.407E - 4, 2.93738, 2.138E - 3, 2.89239,$$
$$2.7635E - 3, 0, 0)^t.$$

The new (updated) nonlinear device currents are $G_{r,4}^{(1)} = 0.256629$, $G_{i,4}^{(1)} = -0.1011142$, $G_{r,4}^{(5)} = 0.2965726$ and $G_{i,4}^{(5)} = -2.59613E - 3$. The new (updated) mismatch vector is

$$\Delta\overline{W}^1 = (5.342E - 6, -3.800E - 7, 5.722E - 6,$$
$$3.815E - 6, -1.403E - 3, 4.332E - 4, -2.269E - 3,$$
$$-3.300E - 2, -2.677E - 3, 6.475E - 3, -1.021E - 3,$$
$$-2.105E - 3, 0.3023, 3.015E - 2, -6.264E - 4,$$
$$-1.473E - 4)^t.$$

This mismatch vector is not small enough and the solution has not converged. However, the iterative procedure will converge in eight iterations. The final solution is (all magnitudes are in percentage of the base values and all angles are in degrees):

Fundamental power flow output solution (Fig. E7.14.1):

| | Bus voltage | | Bus generation (G) and load (L) powers | | | | | Line power* | |
|---|---|---|---|---|---|---|---|---|---|
| | $|V|$ | $\delta$ | $P_G$ | $Q_G$ | $P_L$ | $Q_L$ | | $P_{line}$ | $Q_{line}$ |
| From Bus | (%) | (°) | (%) | (%) | (%) | (%) | To Bus | (%) | (%) |
| 1 | 100.0 | 0.00 | 35.16 | 20.90 | 0.00 | 0.00 | 2 | 13.35 | 10.81 |
| | | | | | | | 4 | 21.81 | 10.09 |
| 2 | 99.76 | −0.01 | 0.00 | 0.00 | 10.00 | 10.00 | 1 | −13.33 | −10.78 |
| | | | | | | | 3 | 3.32 | 0.78 |
| 3 | 99.64 | −0.16 | 0.00 | 0.00 | 0.00 | 0.00 | 2 | −3.32 | −0.77 |
| | | | | | | | 4 | 3.32 | 0.77 |
| 4 | 99.58 | −0.19 | 0.00 | 0.00 | 25.00 | 10.00 | 3 | −3.32 | −0.77 |
| | | | | | | | 1 | −21.75 | −9.98 |

*a*Fundamental active and reactive powers in the line between "from bus" and "to bus."

Harmonic power flow output solution for harmonic number 5, frequency = 300 Hz (Fig. E7.14.1):

| | Bus voltage | | | Line power*a* | | Line current*a* | |
|---|---|---|---|---|---|---|---|
| | $|V|$ | $\delta$ | | $P_{line}$ | $Q_{line}$ | Magnitude | Angle |
| From bus | (%) | (°) | To bus | (%) | (%) | (%) | (°) |
| 1 | 0.0148 | −90.57 | 2 | 0.000015 | −0.000675 | 4.563896 | −1.88 |
| | | | 4 | −0.000015 | −0.003707 | 25.042787 | −0.33 |
| | | | Neutral shunt | 0.000000 | 0.004382 | 29.605302 | 179.43 |
| 2 | 0.2472 | −102.44 | 1 | 0.002067 | 0.011090 | 4.563896 | 178.12 |
| | | | 3 | −0.002067 | −0.011090 | 4.563895 | −1.88 |
| 3 | 2.0731 | −95.66 | 2 | 0.006233 | 0.094407 | 4.563895 | 178.12 |
| | | | 4 | −0.006233 | −0.094406 | 4.563865 | −1.88 |
| 4 | 2.5315 | −96.01 | 3 | 0.008315 | 0.115235 | 4.563865 | 178.12 |
| | | | 1 | 0.062730 | 0.630848 | 25.042787 | 179.67 |
| | | | Neutral nonlinear | −0.071046 | −0.746085 | 29.605308 | −0.57 |

*a*Harmonic power/current in the line between "from bus" and "to bus."

Total current/power:

| From bus | To bus | Line current | | | | Line power | | | |
|----------|--------|--------------------------|-----------------|----------------|----------------------|--------|--------|--------|--------|
| | | Fundamental value (%) | rms value (%) | Peak value (%) | THD$_i$ (%) | P (%) | Q (%) | D (%) | S (%) |
| 1 | 2 | 17.18 | 17.78 | 19.26 | 0.9665 | 13.35 | 10.81 | 4.57 | 17.78 |
| 1 | 4 | 24.03 | 34.71 | 46.94 | 0.6923 | 21.81 | 10.09 | 25.04 | 34.71 |
| 2 | 1 | 17.18 | 17.78 | 19.26 | 0.9665 | −13.32 | −10.77 | 4.59 | 17.73 |
| 2 | 3 | 3.42 | 5.70 | 7.90 | 0.6000 | 3.32 | 0.77 | 4.56 | 5.69 |
| 3 | 2 | 3.42 | 5.70 | 7.90 | 0.6000 | −3.32 | −0.68 | 4.57 | 5.69 |
| 3 | 4 | 3.42 | 5.71 | 7.90 | 0.6000 | 3.32 | 0.68 | 4.57 | 5.69 |
| 4 | 3 | 3.42 | 5.71 | 7.90 | 0.6000 | −3.31 | −0.65 | 4.57 | 5.69 |
| 4 | 1 | 24.03 | 34.71 | 46.94 | 0.6923 | −21.69 | −9.35 | 25.25 | 34.57 |
| 4 | Neutral nonlinear | 27.04 | 40.09 | 54.76 | 0.6744 | 24.93 | 9.25 | 29.80 | 39.94 |

Bus voltage summary:

| Bus number | Fundamental value (%) | rms value (%) | Peak value (%) | THD$_v$ (%) |
|------------|-----------------------|---------------|----------------|-------------|
| 1 | 100.00 | 100.00 | 100.00 | 1.0000 |
| 2 | 99.76 | 99.76 | 99.71 | 1.0000 |
| 3 | 99.63 | 99.65 | 99.97 | 0.9998 |
| 4 | 99.58 | 99.61 | 100.11 | 0.9997 |

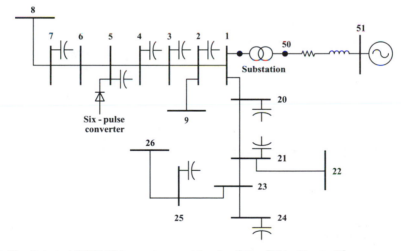

**FIGURE E7.18.1** The distorted IEEE 18-bus system used for simulation [29 in Chapter 7].

### 7.5.7 Application Example 7.19: Accuracy of Decoupled Harmonic Power Flow (DHPF): Solution

The Newton-based HPF (e.g., using an accurate converter model that includes harmonic couplings [5, 6 in Chapter 7]) and the DHPF (modeling the con-verter with harmonic current sources of Table E7.18.1) are applied to the 18 bus system. Simulation results are compared in Table E7.18.2. Small levels of the maximum and average deviations (last two rows of Table E7.18.2) show the acceptable accuracy of the decoupled approach.

**TABLE E7.18.1** Magnitude of Harmonic Currents (as a Percentage of the Fundamental) Used to Model Six-Pulse Converters[a]

| Harmonic order | Magnitude (%) | Harmonic order | Magnitude (%) | Harmonic order | Magnitude (%) |
|---|---|---|---|---|---|
| 1 | 100 | 19 | 2.4 | 37 | 0.5 |
| 5 | 19.1 | 23 | 1.2 | 41 | 0.5 |
| 7 | 13.1 | 25 | 0.8 | 43 | 0.5 |
| 11 | 7.2 | 29 | 0.2 | 47 | 0.4 |
| 13 | 5.6 | 31 | 0.2 | 49 | 0.4 |
| 17 | 3.3 | 35 | 0.4 | | |

[a]All harmonic phase angles are assumed to be zero.

**TABLE E7.18.2** Comparison of Simulation Results for the 18 Bus System (Fig. E7.18.1) using Coupled (Section 7.4) and Decoupled (Section 7.5.1) Harmonic Power Flow

| Bus | Fundamental voltage (pu) Decoupled HPF | Coupled HPF | rms voltage (pu) Decoupled HPF | Coupled HPF | Distortion factor D (%) Decoupled HPF | Coupled HPF | THD of voltage (%) Decoupled HPF | Coupled HPF |
|---|---|---|---|---|---|---|---|---|
| 1 | 1.0545 | 1.0545 | 1.0549 | 1.0548 | 0.996 | 0.9997 | 2.7159 | 2.3828 |
| 2 | 1.0511 | 1.0510 | 1.0516 | 1.0516 | 0.9995 | 0.9995 | 3.1353 | 3.1942 |
| 3 | 1.0456 | 1.0455 | 1.0463 | 1.0465 | 0.9993 | 0.9991 | 3.6591 | 4.2593 |
| 4 | 1.0425 | 1.0424 | 1.0433 | 1.0436 | 0.9992 | 0.9989 | 3.9386 | 4.7737 |
| 5 | 1.0359 | 1.0358 | 1.0370 | 1.0377 | 0.9988 | 0.9981 | 4.8692 | 6.1886 |
| 6 | 1.0348 | 1.0347 | 1.0360 | 1.0367 | 0.9988 | 0.9981 | 4.8692 | 6.1886 |
| 7 | 1.0326 | 1.0325 | 1.0339 | 1.0347 | 0.9987 | 0.9979 | 5.1238 | 6.5404 |
| 8 | 1.0268 | 1.0267 | 1.0281 | 1.0289 | 0.9987 | 0.9978 | 5.1528 | 6.5775 |
| 9 | 1.0496 | 1.0496 | 1.0501 | 1.0501 | 0.9995 | 0.9995 | 3.1397 | 3.1985 |
| 20 | 1.0505 | 1.0504 | 1.0509 | 1.0509 | 0.9996 | 0.9996 | 2.9329 | 2.9248 |
| 21 | 1.0496 | 1.0495 | 1.0505 | 1.0506 | 0.9991 | 0.9990 | 4.1782 | 4.4609 |
| 22 | 1.0479 | 1.0479 | 1.0488 | 1.0489 | 0.9991 | 0.9990 | 4.1847 | 4.4682 |
| 23 | 1.0451 | 1.0451 | 1.0474 | 1.0477 | 0.9978 | 0.9974 | 6.6676 | 7.1620 |
| 24 | 1.0485 | 1.0485 | 1.0519 | 1.0523 | 0.9968 | 0.9964 | 7.9886 | 8.4865 |
| 25 | 1.0419 | 1.0418 | 1.0449 | 1.0452 | 0.9971 | 0.9967 | 7.6002 | 8.1187 |
| 26 | 1.0415 | 1.0414 | 1.0445 | 1.0449 | 0.9971 | 0.9967 | 7.6029 | 8.1220 |
| 50 | 1.0501 | 1.0501 | 1.0501 | 1.0501 | 1.0000 | 1.0000 | 0.2482 | 0.1217 |
| 51 | 1.0500 | 1.0500 | 1.0500 | 1.0500 | 1.0000 | 1.0000 | 0.2482 | 0.0035 |
| MD[a] | 0.0097 | – | 0.778 | – | 0.0009 | – | 1.42473 | – |
| AD[b] | 0.0059 | – | 0.0278 | – | 0.003 | – | 0.57242 | – |

[a]Maximum deviation with respect to the coupled HPF (%).
[b]Average deviation with respect to the couple HPF (%).

## CHAPTER 8 SOLUTIONS

### 8.1.1 Application Example 8.1: Calculation of Reliability Indices: Solution

The reliability indices are

SAIDI = (sum of the number of customer interruption durations)/(total number of affected customers) = 0.003 h/customer, where interruptions with 0 V are taken into account,

SAIFI = (sum of the number of customer interruptions)/(total number of affected customers) = 0.004 interruptions/customer, where interruptions with 0 V are taken into account,

$$SARFI_{\%V} = \Sigma N_i / N_T = 67.5\%,$$

where $SARFI_{\%V}$ is the number of specified short-duration rms variations per system customer (60 s aggregation). The notation "%V" refers to the bus voltage percent deviation that is counted as an event $SARFI_{\%V} = \Sigma N_i / N_T$, where %V is the rms voltage threshold to define an event, $N_i$ is the number of customers that see the event, and $N_T$ is the total number of customers in the study. In this case %V = ±10%. For the SARFI index, the voltage threshold allows assessment of compatibility for voltage-sensitive devices. Typical annual value for SARFI is about 15% for a 70% voltage (%V) event.

**TABLE P8.1.1** Electricity interruptions/transients

| Number of interruptions/ transients | Number of interruptions/ transients duration | voltage in % of rated value | total number of customers affected | cause of interruption/ transient |
|---|---|---|---|---|
| 10 | 4.5 h | 0 | 1000 | overload |
| 15 | 3 h | 0 | 1500 | repair |
| 60 | 1 min | 90 | 500 | switching transients |
| 90 | 1 s | 130 | 500 | capacitor switching |
| 110 | 0.1 s | 40 | 15000 | short-circuits |
| 95 | 0.1 s | 150 | 4000 | lightning |
| 130 | 0.01 s | 50 | 3000 | short-circuits |

### 8.2.5.1 Application Example 8.5: Frequency Control of an Interconnected Power System Broken into Two Areas: The First One with a 300 MW Coal-Fired Plant and the Other One with a 5 MW Wind-Power Plant: Solution

a) List of ordinary differential equations

$$\varepsilon_{11} = loadref_1(s) - \frac{\Delta\omega_1}{R_1} \quad \text{(E8.5-1)}$$

$$\Delta P_{valve1} + T_{G1}\frac{d(\Delta P_{valve1})}{dt} = \varepsilon_{11} \quad \text{(E8.5-2)}$$

$$\Delta P_{mech1} + T_{CH1}\frac{d(\Delta P_{mech1})}{dt} = \Delta P_{valve1} \quad \text{(E8.5-3)}$$

$$\varepsilon_{21} = \Delta P_{mech1} - \Delta P_{L1} - \Delta P_{tie} \quad \text{(E8.5-4)}$$

$$\Delta\omega_1 D_1 + M_1\frac{d(\Delta\omega_1)}{dt} = \varepsilon_{21} \quad \text{(E8.5-5)}$$

$$\frac{\Delta P_{tie}}{\varepsilon_3} = T/s \quad \text{(E8.5-6)}$$

$$s\Delta P_{tie} = T\varepsilon_3 \quad \text{(E8.5-7)}$$

$$\frac{1}{T}\frac{d(\Delta P_{tie})}{dt} = T\varepsilon_3 \quad \text{(E8.5-8)}$$

$$\varepsilon_{12} = loadref_2(s) - \frac{\Delta\omega_2}{R_2} \quad \text{(E8.5-9)}$$

$$\Delta P_{valve2} + T_{G2}\frac{d(\Delta P_{valve2})}{dt} = \varepsilon_{12} \quad \text{(E8.5-10)}$$

$$\Delta P_{mech2} + T_{CH2}\frac{d(\Delta P_{mech2})}{dt} = \Delta P_{valve2} \quad \text{(E8.5-11)}$$

$$\varepsilon_{22} = \Delta P_{mech2} - \Delta P_{L2} + \Delta P_{tie} \quad \text{(E8.5-12)}$$

$$\Delta\omega_2 D_2 + M_2\frac{d(\Delta\omega_2)}{dt} = \varepsilon_{22}. \quad \text{(E8.5-13)}$$

There are 11 independent equations and 11 unknowns. The latter are $\varepsilon_{11}$, $\Delta P_{valve1}$, $\Delta P_{mech1}$, $\varepsilon_{21}$, $\Delta P_{tie}$, $\Delta\omega_1$, $\varepsilon_{12}$, $\Delta P_{valve2}$, $\Delta P_{mech2}$, $\varepsilon_{22}$, $\Delta\omega_2$.

b)

**Mathematica Program**

```
Remove[Dw1, R1, Loadref1, E11,
DPvalve1, DPmech1, DPl1, E12, Tg1,
Tch1, M1, D1, Dw2, R2, Loadref2,
E22, DPvalve2, DPmech2, DPl2, E21,
Tch2, M2, D2, Sbase, eq1, eq2, eq3,
ic1, ic2, ic3, DPtie, E3, Tie, Xtie,
ic4, ic5, ic6, ic7];
Sbase=500000000;
M1=4.5;
Tg1=0.01;
Tch1=0.5;
R1=0.01; D1=0.8; M2=6;
Tg2=0.02;
Tch2=0.75;
R2=0.02;
D2=1;
Xtie=0.2;
Tie=377/Xtie;
Loadref1[t_]:=If[t<0,0,0.8];
Loadref2[t_]:=If[t<0,0,0.8];
DPl1[t_]:=If[t<5,0,0.2];
DPl2[t_]:=If[t<5,0,-0.2];
ic1=Dw1[0]= =0;
ic2=DPmech1[0]= =0;
ic3=DPvalve1[0]= =0;
ic5=Dw2[0]= =0;
ic6=DPmech2[0]= =0;
ic7=DPvalve2[0]= =0;
ic4=DPtie[0]= =0;
E11[t_]:=Loadref1[t]-(Dw1[t]/R1);
E12[t_]:=DPmech1[t]-(DPl1[t]-DPtie[t]);
E3[t_]:=Dw1[t]-Dw2[t];
E22[t_]:=Loadref2[t]-(Dw2[t]/R2);
E21[t_]:=DPmech2[t]-(DPl2[t]-DPtie[t]);
eq1=Dw1'[t]=
=(1/M1)*(E12[t]-D1*Dw1[t]);
eq2=DPmech1'[t]=
=(1/Tch1)*(DPvalve1[t]-DPmech1[t]);
```

```
eq3=DPvalve1'[t]=
=(1/Tg1)*(E11[t]-DPvalve1[t]);
eq5=Dw2'[t]=
=(1/M2)*(E21[t]-D2*Dw2[t]);
eq6=DPmech2'[t]= =(1/Tch2)*(DPvalve2
[t]-DPmech2[t]);
eq7=
DPvalve2'[t]=
=(1/Tg2)*(E22[t]-DPvalve2[t]);
eq4=DPtie'[t]= =Tie*E3[t];
sol=NDSolve[{eq1,eq2,eq3,
eq4,eq5,eq6,eq7,ic1,ic2,ic3,ic4,ic5,ic6,ic7},Dw1[t],
DPmech1[t],DPvalve1[t],DPtie[t],
Dw2[t],DPmech2[t], Dpvalve2[t]},
{t,0,100},MaxSteps->
1000000];
Plot[Dw1[t]/sol,{t,0,100}, PlotRange-
>All, AxesLabel->{,,t","Dw1(t)"}];
Plot[Dw2[t]/sol,{t,0,100}, PlotRange-
>All, AxesLabel->{,,t","Dw2(t)"}];
```

b) The angular frequency deviations $\Delta\omega_1(t)$ and $\Delta\omega_2(t)$ based on Mathematica for droop characteristics with $R_1 = 0.01$ pu, $R_2 = 0.02$ pu are shown in Figs. E8.5.1 and E8.5.2. Note that there is instability, and both plots result in the same frequency response.

c) The angular frequency deviations $\Delta\omega_1(t)$ and $\Delta\omega_2(t)$ based on Mathematica for droop characteristics with $R_1 = 0.01$, $R_2 = 0.5$ are shown in Figs. E8.5.3 and E8.5.4. Note that there is stability, and both plots result in the same frequency

response.

**MATLAB Program**
```
%defining the input
Sbase=500e6;
M1=4.5;
Tg1=0.01;
Tch1=0.5;
R1=0.01;
D1=0.8;
M2=6;
Tg2=0.02;
Tch2=0.75;
R2=0.02;
D2=1;
Xtie=0.2;
Tie=377/Xtie;
%time span
T1=0:0.001:20;
T=T1.';
%solving system of equations
opts=odeset('AbsTol',1e-11,'RelTol',1e-7;
[t0,x0]=ode45)@F2,T,[0;0;0;0;0;0;0],o
pts,M1,Tg1,Tch1,R1,D1,M2,Tg2,Tch2,R2,
D2,Tie);
%plot results
figure;
Plot(t0,x0(:;4));
Grid on, zoom on;
Title(,Transient response of \Delta\
omega1(t) vs t'));
xlabel('t');
```

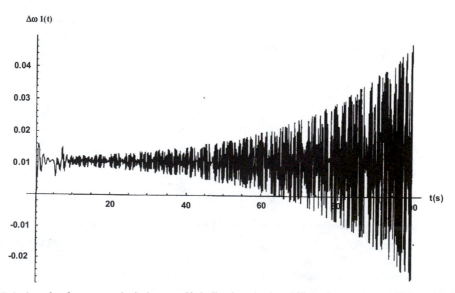

**FIGURE E8.5.1** Angular frequency deviation $\Delta\omega_1(t)$ indicating the instability of two power plants with similar droop characteristics.

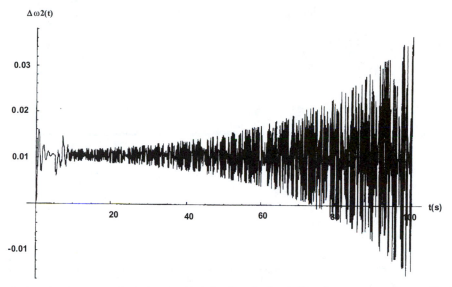

**FIGURE E8.5.2** Angular frequency deviation $\Delta\omega_2(t)$ indicating the instability of two power plants with similar droop characteristics.

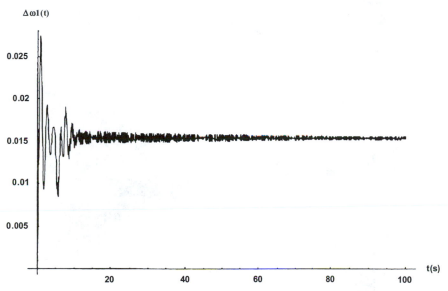

**FIGURE E8.5.3** Angular frequency deviation $\Delta\omega_1(t)$ indicating the stability of two power plants with dissimilar droop characteristics.

```
ylabel('\Delta\omega1(t)');
figure;
Plot(t0,x0(:,1));
Grid on, zoom on;
Title(,Transient response of \Delta\
omega2(t) vs t'));
xlabel('t');
ylabel('\Delta\omega2(t)');
where the function F2 is:
function xprime=F2(t,x,M1,Tg1,Tch1,
R1,D1,M2,Tg2,Tch2,R2,D2,Tie);
if t<5;
```

```
DPl1=0;
elseDPl1=0.2;
end
if t<7;
DPl2=0;
elseDPl2=-0.2;
end
if t<0;
load1=0;
else load1=0.8;
end
if t<0;
```

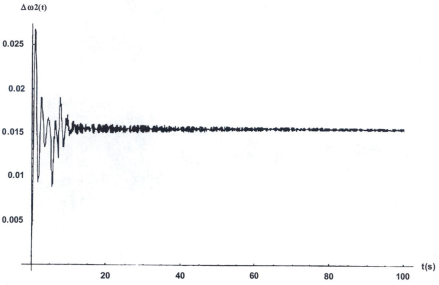

**FIGURE E8.5.4** Angular frequency deviation $\Delta\omega_2(t)$ indicating the stability of two power plants with dissimilar droop characteristics.

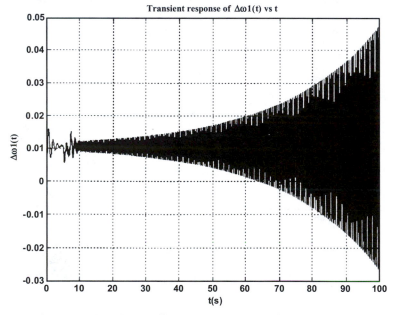

**FIGURE E8.5.5** Angular frequency deviation $\Delta\omega_1(t)$ indicating the instability of two power plants with similar droop characteristics.

```
load2=0;
else load2=0.8;
end
xprime=zeros(7,1);
xprime(1)=(1/M1)*(x(2)-DPl1-x(7)-(D1*x(1)));
xprime(2)=(1/Tch1)*(x(3)-x(2);
xprime(3)=(1/Tg1)*(load1-(x(1))*(1/R1)-x(3)));
xprime(4)=(1/M2)*(x(5)-DPl2-x(7)-(D2*x(4)));
xprime(5)=(1/Tch2)*(x(6)-x(5);
```

```
xprime(6)=(1/Tg2)*(load2-(x(4))*(1/R2)-x(6)));
xprime(7)=(Tie)*(x(12)-x(4));
```

The angular frequency deviations $\Delta\omega_1(t)$ and $\Delta\omega_2(t)$ based on MATLAB for droop characteristics with $R_1 = 0.01$ pu, $R_2 = 0.02$ pu are shown in Figs. E8.5.5 and E8.5.6 confirming the results of Figs. E8.5.1 and E8.5.2.

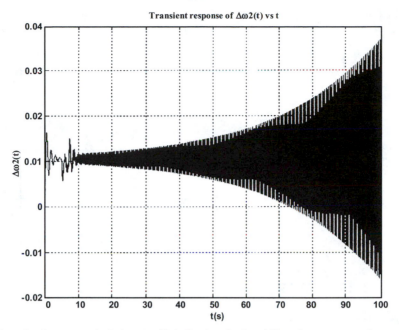

**FIGURE E8.5.6** Angular frequency deviation $\Delta\omega_2(t)$ indicating the instability of two power plants with similar droop characteristics.

The angular frequency deviations $\Delta\omega_1(t)$ and $\Delta\omega_2(t)$ based on MATLAB for droop characteristics with $R_1 = 0.01$ pu, $R_2 = 0.5$ pu are shown in Figs. E8.5.7 and E8.5.8 confirming the results of Figs. E8.5.3 and E8.5.4.

### 8.2.5.2 Application Example 8.6: Frequency Control of an Interconnected Power System Broken into Two Areas: The First One with a 5 MW Wind-Power Plant and the Other One with a 5 MW Photovoltaic Plant: Solution

The angular frequency deviations $\Delta\omega_1(t)$ and $\Delta\omega_2(t)$ for droop characteristics with $R_1 = 0.5$ pu, $R_2 = 0.5$ pu are much larger than those in Application Example 8.5, that is, there is poor frequency control. See Problems 8.5 through 8.7.

### 8.3.4 Application Example 8.9: New Approach $p_{cu} = i'_2(v_1 - v'_2)$ and $p_{fe} = v_1(i_1 - i'_2)$: Solution

The exciting current is measured with the error of (taking into account the maximum error limits of the instruments)

$$\frac{\Delta(i_1 - i'_2)}{(i_1 - i'_2)} = \frac{\varepsilon_{CT1} + \varepsilon_{CT2} + \varepsilon_{A1}}{(i_1 - i'_2)/20}$$
$$= \frac{(\pm 5.2 \pm 3.47 \pm 0.1)\,\text{mA}}{35.5\,\text{mA}} = \pm 0.247.$$

The input voltage is measured with the error of

$$\frac{\Delta v_1}{v_1} = \frac{\varepsilon_{PT1} + \varepsilon_{V1}}{v_1} = \frac{\pm 0.24 \pm 0.3}{240} = \pm 0.0023$$

and the iron-core loss is measured with the error of

$$\frac{\Delta P_{fe}}{P_{fe}} = \frac{\Delta v_1}{v_1} + \frac{\Delta(i_1 - i'_2)}{(i_1 - i'_2)} = \pm 0.249.$$

The series voltage drop is measured with the error of

$$\frac{\Delta(v_1 - v'_2)}{(v_1 - v'_2)} = \frac{\varepsilon_{PT1} + \varepsilon_{PT2} + \varepsilon_{V2}}{(v_1 - v'_2)} = \frac{\pm 0.24 \pm 0.24 \pm 0.01}{4.86}$$
$$= \pm 0.1008.$$

The output current is measured with the error of

$$\frac{\Delta i_2}{i_2} = \frac{\varepsilon_{CT2} + \varepsilon_{A2}}{i_2} = \frac{\pm 0.00347 \pm 0.005}{3.472} = \pm 0.00244$$

and the copper loss is measured with the error of

$$\frac{\Delta P_{cu}}{P_{cu}} = \frac{\Delta i_2}{i_2} + \frac{\Delta(v_1 - v'_2)}{(v_1 - v'_2)} = \pm 0.1032.$$

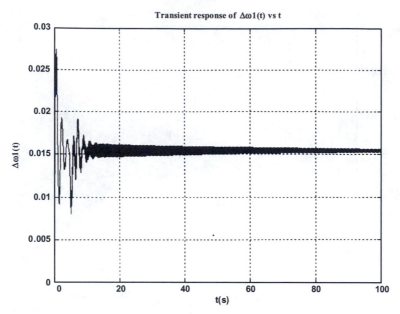

**FIGURE E8.5.7** Angular frequency deviation $\Delta\omega_1(t)$ indicating the stability of two power plants with dissimilar droop characteristics.

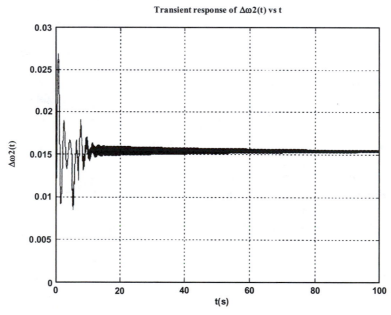

**FIGURE E8.5.8** Angular frequency deviation $\Delta\omega_2(t)$ indicating the stability of two power plants with dissimilar droop characteristics.

Therefore, the total loss is measured with the error of

$$\frac{\Delta P_{\text{Loss}}}{P_{\text{Loss}}} = \frac{\Delta P_{\text{cu}} + \Delta P_{\text{Fe}}}{P_{\text{Loss}}} = \frac{\pm 0.1032 \times 325 \pm 0.249 \times 65}{390}$$

$$= \pm 0.128 = \pm 12.8\%.$$

This maximum error can be reduced further by using three-winding current and potential transform-

ers as shown in Fig. E8.9.2a,b for the measurement of exciting currents and voltage drops. The input and output currents pass though the primary and secondary windings of the current transformer, and the current in the tertiary winding represents the exciting current of the power transformer. Similarly, the input and output voltages are applied to the two primary windings of the potential transformers, and the secondary winding measures the voltage drop of

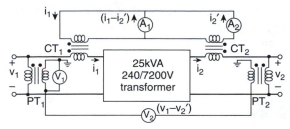

**FIGURE E8.9.1** Loss measurements of single-phase transformers from voltage and current differences.

**TABLE E8.9.1** Instruments and Their Error Limits of Fig. E8.9.1

| Instrument | Rating | Error limits |
|---|---|---|
| $PT_1$ | 240 V/240 V | $\varepsilon_{PT1} = \pm0.24$ V |
| $PT_2$ | 7200 V/240 V | $\varepsilon_{PT2} = \pm0.24$ V |
| $CT_1$ | 100 A/5 A | $\varepsilon_{CT1} = \pm5.2$ mA |
| $CT_2$ | 5 A/5 A | $\varepsilon_{CT2} = \pm3.47$ mA |
| $V_1$ | 300 V | $\varepsilon_{V1} = \pm0.3$ V |
| $V_2$ | 10 V | $\varepsilon_{V2} = \pm0.01$ V |
| $A_1$ | 100 mA | $\varepsilon_{A1} = \pm0.1$ mA |
| $A_2$ | 5 A | $\varepsilon_{A2} = \pm5$ mA |

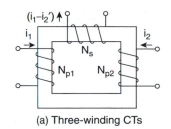

(a) Three-winding CTs

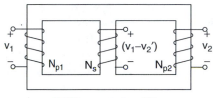

(b) Three-winding PTs

**FIGURE E8.9.2** (a) Three-winding current and (b) potential transformers as applied to new method of loss measurement.

**TABLE E8.10.1** Instruments and Their Error Limits of Fig. E8.10.1

| Instrument | Rating | Error limits |
|---|---|---|
| $PT_1$ | 240 V/240 V | $\varepsilon_{PT1} = \pm0.24$ V |
| $PT_2$ | 12 V/12 V | $\varepsilon_{PT2} = \pm9.7$ mV |
| $CT_1$ | 2 A/2 A | $\varepsilon_{CT1} = \pm1.4$ mA |
| $CT_2$ | 100 A/5 A | $\varepsilon_{CT2} = \pm5.2$ mA |
| $V_1$ | 300 V | $\varepsilon_{V1} = \pm0.3$ V |
| $V_2$ | 10 V | $\varepsilon_{V2} = \pm0.01$ V |
| $A_1$ | 2 A | $\varepsilon_{A1} = \pm2$ mA |
| $A_2$ | 5 A | $\varepsilon_{A2} = \pm5$ mA |

the power transformer. For this 25 kVA single-phase transformer, the turns ratio for both the CT and the PT is $N_{p1}:N_{p2}:N_s = 1:30:1$. The maximum error by using three-winding current and potential transformers can be reduced to as small a value as to that of the back-to-back method, which will be described below.

### 8.3.5 Application Example 8.10: Back-to-Back Approach of Two Transformers Simulated with CTs and PTs: Solution

The exciting current is measured with the error of

$$\frac{\Delta(i_1 - i_2')}{(i_1 - i_2')} = \frac{\varepsilon_{CT1} + \varepsilon_{A1}}{(i_1 - i_2')} = \frac{\pm0.0014 \pm 0.002}{1.41} = \pm0.0024.$$

The input voltage is measured with the error of

$$\frac{\Delta v_1}{v_1} = \frac{\varepsilon_{PT1} + \varepsilon_{V1}}{v_1} = \frac{\pm0.24 \pm 0.3}{240} = \pm0.0023$$

and the iron-core loss is measured with the error of

$$\frac{\Delta P_{fe}}{P_{fe}} = \frac{\Delta v_1}{v_1} + \frac{\Delta(i_1 - i_2')}{(i_1 - i_2')} = \pm0.0047.$$

The series voltage drop is measured with the error of

$$\frac{\Delta(v_1 - v_2')}{(v_1 - v_2')} = \frac{\varepsilon_{PT2} + \varepsilon_{V2}}{(v_1 - v_2')} = \frac{\pm0.0097 \pm 0.01}{9.72} = \pm0.002.$$

The output current is measured with the error of

$$\frac{\Delta i_2}{i_2} = \frac{\pm\varepsilon_{CT2} \pm \varepsilon_{A2}}{i_2/20} = \frac{\pm0.0052 \pm 0.005}{5.2} = \pm0.002$$

and the copper loss is measured with the error of

$$\frac{\Delta P_{cu}}{P_{cu}} = \frac{\Delta i_2}{i_2} + \frac{\Delta(v_1 - v_2')}{(v_1 - v_2')} = \pm0.004.$$

The total loss is measured with the error of

$$\frac{\Delta P_{\text{Loss}}}{P_{\text{Loss}}} = \frac{\Delta P_{\text{cu}} + \Delta P_{\text{Fe}}}{P_{\text{Loss}}} = \frac{\pm 0.004 \times 650 \pm 0.0047 \times 130}{780}$$
$$= \pm 0.0041 = \pm 0.41\%.$$

This error analysis shows that the maximum errors of the measurement approach of this section range from 0.5 to 15% mainly depending on the accuracy of the voltage and current sensors (e.g., PTs, CTs). The back-to-back method generates extremely small maximum errors because voltage and current differences are directly measured by one PT and one CT, respectively. In Fig. E8.9.1, the errors of voltage and current differences are large because they are measured by two PTs and two CTs. If three-winding PT and CT (see Fig. E8.9.2a,b) are employed to measure the voltage and current differences, the maximum errors will be as small as those of the back-to-back method – even though instruments with maximum errors of 0.5% are used the maximum errors in loss measurements will not exceed 2.5%. Hall sensors with maximum errors of 0.5% [92 in Chapter 8] are suitable for low-efficient transformers only.

### 8.4.7 Application Example 8.13: Ride-Through Capability of Computers and Semiconductor Manufacturing Equipment: Solution

a) Ride-through capability based on the CBEMA or ITIC voltage-tolerance curves:

1min at 90% of rated voltage: ride-through is possible,

1 s at 130% of rated voltage: ride-through is not possible,

0.1 s at 40% of rated voltage: ride-through is not possible,

0.1 s at 150% of rated voltage: ride-through is not possible,

0.01 s at 50% of rated voltage: ride-through is possible.

b) Ride-through capability based on SEMI F47 standard:

1 min at 90% of rated voltage: ride-through is possible (borderline),

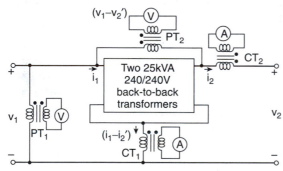

**FIGURE E8.10.1** Instrumentation for single-phase transformer employing back-to-back method of loss measurement.

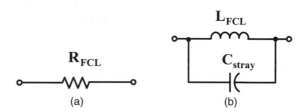

**FIGURES E8.12.2** Model for resistive (a) and inductive (b) fault current limiter.

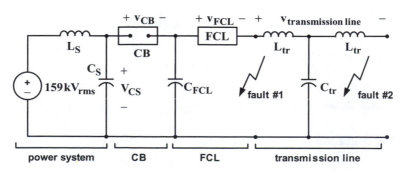

**FIGURE E8.12.1** Single-phase equivalent circuit of a circuit breaker (CB) connected in series with a fault-current limiter (FCL) with two three-phase to ground faults at different locations.

1 s at 130% of rated voltage: ride-through is possible,

0.1 s at 40% of rated voltage: ride-through is not possible,

0.1 s at 150% of rated voltage: ride-through is possible,

0.01 s at 50% of rated voltage: ride-through is possible.

## CHAPTER 9 SOLUTIONS

### 9.3.5 Application Example 9.1: Hybrid Passive Filter Design to Improve the Power Quality of the IEEE 30-Bus Distribution System Serving Adjustable-Speed Drives: Solution [14 in Chapter 9]

*a) Case A. Harmonic Compensation with Two Hybrid Passive Filter Banks at Buses 15 and 18.* Harmonic analysis before the installation of passive filters indicates a very high distortion of both voltage and current waveforms (Fig. E9.1.3 and Table E9.1.5) due to the two AC drives with relatively high power ratings. According to the IEEE-519 standard [11 in Chapter 9], the individual harmonic current injections at bus 15 (for 5th, 7th, 11th, 13th, 17th, and 19th harmonics) and bus 18 (for 5th, 7th, and 11th harmonics) are higher than the allowable level of 4%, and thus the electric utility has the right to disconnect (reject) these nonlinear loads. In addition, prop-

agation of the injected harmonic currents has resulted in unacceptable voltage THD$_v$ levels (e.g., larger than 5%) at buses 12 to 15. Buses 11 and 16 to 18 are also experiencing relatively high THD$_v$ of more than 4% (Fig. E9.1.6).

The common procedure is to require the owners of nonlinear loads (at buses 15 and 18) to install passive filter banks at the terminals of the AC drives. Figure E9.1.3 and Table E9.1.5 list the simulation results after the installation of the following two filter banks (Fig. E9.1.2), tuned at the dominant frequencies:

- *Filter Bank 1:* Passive filter (Fig. E9.1.2) consisting of five shunt branches tuned to the 5th, 7th, 11th, 13th, and 17th harmonic frequencies and installed at bus 15 with (Eq. 9-9b with $f_1 = 50$ Hz)

$R_f = 100\,\Omega,\ L_f^{(h)} = 100\,\text{mH}$ (for all branches),

$\omega_n^{(h)} = 2\pi h f = 1/\sqrt{L_f^{(h)}C_f^{(h)}} \Rightarrow C_f^{(5)} \Rightarrow 4.053\,\mu\text{F},$

$C_f^{(7)} = 2.068\,\mu\text{F}, C_f^{(11)} = 0.837\,\mu\text{F}, C_f^{(13)} = 0.680\,\mu\text{F},$

$C_f^{(17)} = 0.351\,\mu\text{F}.$

- *Filter Bank 2:* Passive filters (Fig. E9.1.2) with only two shunt branches tuned to the 5th and 7th harmonic frequencies and installed at bus 18 with (Eq. 9-9b with $f_1 = 50$ Hz)

$R_f = 100\,\Omega,\ L_f^{(h)} = 100\,\text{mH}$ (at all branches),

$\omega_n^{(h)} = 2\pi h f = 1/\sqrt{L_f^{(h)}C_f^{(h)}} \Rightarrow C_f^{(5)} \Rightarrow 4.053\,\mu\text{F},$

$C_f^{(7)} = 2.068\,\mu\text{F}.$

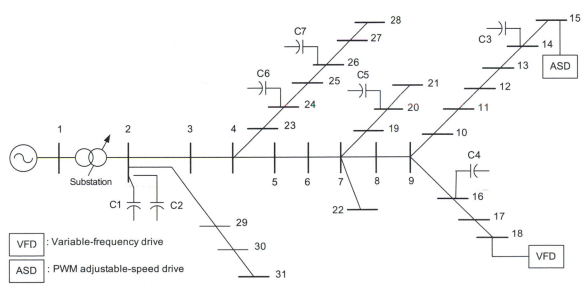

**FIGURE E9.1.1** IEEE 30-bus system with two nonlinear AC drives [9, 10 in Chapter 9].

**TABLE E9.1.1** Line Data for the IEEE 30-Bus System (Base kV = 23 kV, Base MVA = 100 MVA)

| From bus | To bus | R (ohm) | X (ohm) | R (pu) | X (pu) |
|---|---|---|---|---|---|
| 1 | 2 | 0.0021 | 0.0365 | 0.0004 | 0.0069 |
| 2 | 3 | 0.2788 | 0.0148 | 0.0527 | 0.0028 |
| 3 | 4 | 0.4438 | 0.4391 | 0.0839 | 0.0830 |
| 4 | 5 | 0.8639 | 0.7512 | 0.1633 | 0.1420 |
| 5 | 6 | 0.8639 | 0.7512 | 0.1633 | 0.1420 |
| 6 | 7 | 1.3738 | 0.7739 | 0.2597 | 0.1463 |
| 7 | 8 | 1.3738 | 0.7739 | 0.2597 | 0.1463 |
| 8 | 9 | 1.3738 | 0.7739 | 0.2597 | 0.1463 |
| 9 | 10 | 1.3738 | 0.7739 | 0.2597 | 0.1463 |
| 10 | 11 | 1.3738 | 0.7739 | 0.2597 | 0.1463 |
| 11 | 12 | 1.3738 | 0.7739 | 0.2597 | 0.1463 |
| 12 | 13 | 1.3738 | 0.7739 | 0.2597 | 0.1463 |
| 13 | 14 | 1.3738 | 0.7739 | 0.2597 | 0.1463 |
| 14 | 15 | 1.3738 | 0.7739 | 0.2597 | 0.1463 |
| 9 | 16 | 0.8639 | 0.7512 | 0.1633 | 0.1420 |
| 16 | 17 | 1.3738 | 0.7739 | 0.2597 | 0.1463 |
| 17 | 18 | 1.3738 | 0.7739 | 0.2597 | 0.1463 |
| 7 | 19 | 0.8639 | 0.7512 | 0.1633 | 0.1420 |
| 19 | 20 | 0.8639 | 0.7512 | 0.1633 | 0.1420 |
| 20 | 21 | 1.3738 | 0.7739 | 0.2597 | 0.1463 |
| 7 | 22 | 0.8639 | 0.7512 | 0.1633 | 0.1420 |
| 4 | 23 | 0.4438 | 0.4391 | 0.0839 | 0.0830 |
| 23 | 24 | 0.4438 | 0.4391 | 0.0839 | 0.0830 |
| 24 | 25 | 0.8639 | 0.7512 | 0.1633 | 0.1420 |
| 25 | 26 | 0.8639 | 0.7512 | 0.1633 | 0.1420 |
| 26 | 27 | 0.8639 | 0.7512 | 0.1633 | 0.1420 |
| 27 | 28 | 1.3738 | 0.7739 | 0.2597 | 0.1463 |
| 2 | 29 | 0.2788 | 0.0148 | 0.0527 | 0.0028 |
| 29 | 30 | 0.2788 | 0.0148 | 0.0527 | 0.0028 |
| 30 | 31 | 1.3738 | 0.7739 | 0.2597 | 0.1463 |

**TABLE E9.1.2** Bus Data for the IEEE 30-Bus System (Base kV = 23 kV, Base MVA = 100 MVA)

| Bus number | P (kW) | Q (kVAr) | P (pu) | Q (pu) |
|---|---|---|---|---|
| 1[a] | 0 | 0 | 0 | 0 |
| 2 | 52.2 | 17.2 | 0.522 | 0.172 |
| 3 | 0 | 0 | 0 | 0 |
| 4 | 0 | 0 | 0 | 0 |
| 5 | 93.8 | 30.8 | 0.938 | 0.308 |
| 6 | 0 | 0 | 0 | 0 |
| 7 | 0 | 0 | 0 | 0 |
| 8 | 0 | 0 | 0 | 0 |
| 9 | 0 | 0 | 0 | 0 |
| 10 | 18.9 | 6.2 | 0.189 | 0.062 |
| 11 | 0 | 0 | 0 | 0 |
| 12 | 33.6 | 11 | 0.336 | 0.11 |
| 13 | 65.8 | 21.6 | 0.658 | 0.216 |
| 14 | 78.4 | 25.8 | 0.784 | 0.258 |
| 15 | 73 | 24 | 0.73 | 0.24 |
| 16 | 47.8 | 15.7 | 0.478 | 0.157 |
| 17 | 55 | 18.1 | 0.55 | 0.181 |
| 18 | 47.8 | 15.7 | 0.478 | 0.157 |
| 19 | 43.2 | 14.2 | 0.432 | 0.142 |
| 20 | 67.3 | 22.1 | 0.673 | 0.221 |
| 21 | 49.6 | 16.3 | 0.496 | 0.163 |
| 22 | 20.7 | 6.8 | 0.207 | 0.068 |
| 23 | 52.2 | 17.2 | 0.522 | 0.172 |
| 24 | 192 | 63.1 | 1.92 | 0.631 |
| 25 | 0 | 0 | 0 | 0 |
| 26 | 111.7 | 36.7 | 1.117 | 0.367 |
| 27 | 55 | 18.1 | 0.55 | 0.181 |
| 28 | 79.3 | 26.1 | 0.793 | 0.261 |
| 29 | 88.3 | 29 | 0.883 | 0.29 |
| 30 | 0 | 0 | 0 | 0 |
| 31 | 88.4 | 29 | 0.884 | 0.29 |

[a]Swing bus.

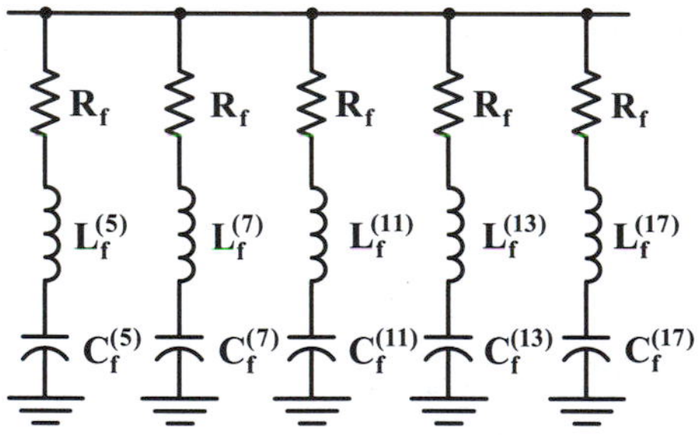

**FIGURE E9.1.2** Passive filter employing five series-resonance filter branches tuned at 5th, 7th, 11th, 13th, and 17th harmonics.

**TABLE E9.1.3** Capacitor Data for the IEEE 30-Bus System

| Capacitor | C1 | C2 | C3 | C4 | C5 | C6 | C7 |
|---|---|---|---|---|---|---|---|
| Bus location | 2 | 2 | 14 | 16 | 20 | 24 | 26 |
| kVAr | 900 | 600 | 600 | 600 | 300 | 900 | 900 |

According to Table E9.1.5 the performance of the above filters is satisfactory: the maximum level of THD$_v$ has dropped from 6.51 (at bus 15) to 2.17% (at bus 18) while the average system THD$_v$ is within the permissible limits of the IEEE-519 standard [11].

**b)** *Case B. Harmonic Compensation with One Passive Filter Bank at Bus 15.* This case examines the possibility of meeting IEEE-519 [11 in Chapter 9] limits with only one filter bank. Simulation results before and after the installation of Filter Bank 1 at bus 15 are shown in Fig. E9.1.4 and Table E9.1.5. Therefore, by performing harmonic load flow analysis, it may be possible to eliminate unnecessary filter banks, and therefore reduce the overall cost of harmonic compensation.

**TABLE E9.1.4** Typical Harmonic Spectrum of Variable-Frequency and PWM Adjustable-Speed Drives [14]

| Harmonic order h | Variable-frequency drive | | PWM adjustable-speed drive | |
|---|---|---|---|---|
| | Magnitude (%) | Phase angle (degree) | Magnitude (%) | Phase angle (degree) |
| 1 | 100 | 0 | 100 | 0 |
| 5 | 23.52 | 111 | 82.8 | −135 |
| 7 | 6.08 | 109 | 77.5 | 69 |
| 11 | 4.57 | −158 | 46.3 | −62 |
| 13 | 4.2 | −178 | 41.2 | 139 |
| 17 | 1.8 | −94 | 14.2 | 9 |
| 19 | 1.37 | −92 | 9.7 | −155 |
| 23 | 0.75 | −70 | 1.5 | −158 |
| 25 | 0.56 | −70 | 2.5 | 98 |
| 29 | 0.49 | −20 | 0 | 0 |
| 31 | 0.54 | 7 | 0 | 0 |

*c) Case C. Harmonic Compensation with One Passive Filter Bank at Bus 9.* To show the impact of filter location on the overall power quality of the distribution system, Filter Bank 1 is installed between the two drive systems at bus 9. The resulting voltage and current waveforms are depicted in Fig. E9.1.5 and the corresponding THD$_v$ levels are listed in Table E9.1.5. It is clear that improper placement of filters will not only deteriorate their performances, but might also favor harmonic propagation and cause additional power quality problems in other sectors of the system (e.g., in buses 10–15 and 16–18 for Case C).

*d) Comparison of Results.* A comparison of the THD$_v$ levels as well as the rms voltage values before and after the placement of filters (Cases A, B, and C) is listed in Table E9.1.5 and shown in Fig. 9.1.6.

The decoupled harmonic power flow algorithm (Chapter 7, Section 7.5.1) is used to simulate the distribution system. Nonlinear AC drives are modeled with harmonic current sources. This approach is very practical and convenient for the analysis of large distorted industrial systems with inadequate information about the nonlinear loads (e.g., parameters and ratings of AC drives). Simulation results of this application example indicate that the common approach of placing filter banks at the terminals of each AC drive system is not always the most economical solution. It might be possible to limit the overall system distortion, as well as the individual bus THD$_v$ levels, with fewer filters. To do this, system conditions before and after compensa-

tion need to be carefully studied, which requires fast algorithms, as presented in this example.

### 9.7.6 Application Example 9.5: Hybrid of Passive and Active Power Filters for Harmonic Mitigation of Six-Pulse and Twelve-Pulse Rectifier Loads [16 in Chapter 9]: Solution [16]

a) Load-Current Waveforms without Filtering

Figure E9.5.3 illustrates simulated power system current waveforms for a twelve-pulse rectifier load, indicating a THD$_i$ of 9.36%. Provided one of the rectifiers (Fig. E9.5.2) is out of service (e.g., six-pulse rectifier configuration), the power system current THD$_i$ increases to 26.28%.

b) Current Equations with Hybrid Filter #2

The converter of hybrid filter #2 operates at harmonic frequencies like a resistor in series with the power system impedance. Therefore, the effective harmonic impedance of the power system is increased and the resonance of the passive power filter is suppressed. In addition, the filter performance is improved by imposing a voltage harmonic waveform at its terminals with the following amplitude:

$$v_{a1}(t) = K_1 \, i_{sh}(t), \tag{E9.5-1}$$

where $i_{sh}(t)$ is the harmonic content of the power system current, and $K_1$ is the converter gain. The power converter consumes fundamental power. This power is stored in the DC capacitor to regulate its voltage and compensate the inverter losses [15, 16 in Chapter 9]. Therefore, DC capacitor voltage regulation is performed without any external power supply. The output voltage can be written as

$$v_a(t) = v_{a1}(t) + v_{a2}(t). \tag{E9.5-2}$$

Hybrid filter #2 is controlled using the synchronous reference frame (e.g., dq0 transformation), which changes the conventional three abc rotating phasors into direct (d), quadrature (q), and zero (0) sequence components. The fundamental components are transformed into a DC term and harmonic components are shifted by $\omega_s$ in the dq0 reference frame (Eq. 9-14).

The block diagram of the control scheme is shown in Fig. E9.5.4 where the control block is divided into two parts:

- The performance improvement block mitigates the resonance problem of the passive power filter and the overall performance of the filtering system.

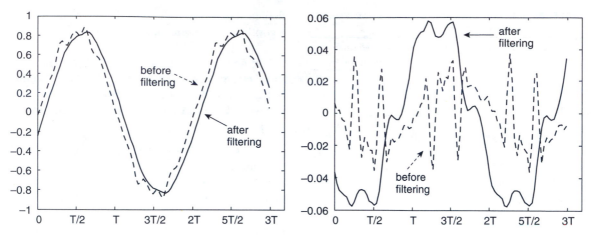

**FIGURE E9.1.3** Voltage and current waveforms at bus 15 before and after the installation of two filter banks at buses 15 and 18 (Case A).

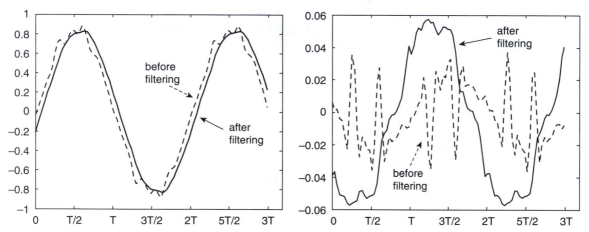

**FIGURE E9.1.4** Voltage and current waveforms at bus 15 before and after the installation of one filter bank at bus 15 (Case B).

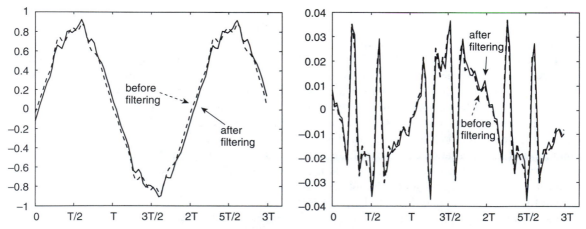

**FIGURE E9.1.5** Voltage and current waveforms at bus 15 before and after the installation of one filter bank at bus 9 (Case C).

**TABLE E9.1.5** Summary of Simulation Results for the IEEE 30-Bus System before and after the Installation of Filter(s)

| | Before filtering | | After filtering (Case A) | |
|---|---|---|---|---|
| | Minimum | Maximum | Minimum | Maximum |
| rms voltage | 0.8382 pu at bus 15 | 1.0003 pu at bus 1 | 0.8295 pu at bus 15 | 1.0003 pu at bus 2 |
| $THD_v$ | 1.8492% at bus 5 | 6.5089% at bus 15 | 1.0548% at bus 11 | 2.1691% at bus 18 |
| System $THD_v$ | 3.1820% | | 1.4818% | |

| | After filtering (Case B) | | After filtering (Case C) | |
|---|---|---|---|---|
| | Minimum | Maximum | Minimum | Maximum |
| rms voltage | 0.8337 pu at bus 5 | 1.0001 pu at bus 2 | 0.8469 pu at bus 15 | 1.0002 pu at bus 2 |
| $THD_v$ | 1.2520% at bus 22 | 1.8629% at bus 18 | 1.0938% at bus 6 | 5.8453% at bus 15 |
| System $THD_v$ | 1.4950% | | 1.8707% | |

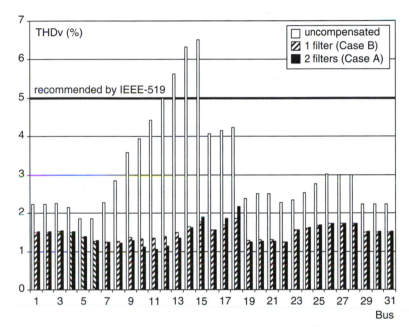

**FIGURE E9.1.6** Comparison of $THD_v$ levels of the IEEE 30-bus system before and after the installation of one (Case B) and two (Case A) filter banks [14 in Chapter 9].

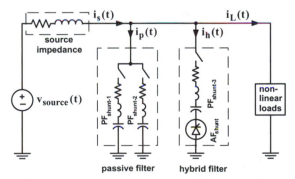

**FIGURE E9.5.1** Hybrid filter #1, a shunt hybrid power filter (series combination of a small rating active filter and a tuned passive branch) in parallel with tuned passive filters [15 in Chapter 9].

It contains high-pass filters in the dq0 frame to filter the fundamental component of the power system current. The output is passed through an amplifier and $v_{a1}(t)$ is obtained.

• The DC voltage regulator block maintains the DC bus voltage at a constant level and reduces possible voltage fluctuations due to the harmonic real power injection (for filter performance improvement) and the power loss of the converter.

A low-pass filter with a cutoff at line frequency is also employed to extract the fundamental component of the DC-link voltage. The DC capacitor voltage is compared with its reference value, and the output of the comparator serves as input to the PI

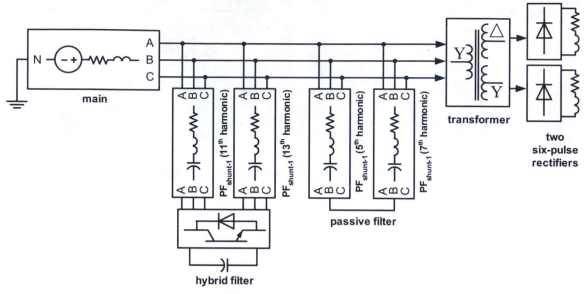

**FIGURE E9.5.2** Hybrid filter #2, a shunt hybrid power filter (a small rating active filter in series with two passive branches) in parallel with two passive filters [16 in Chapter 9].

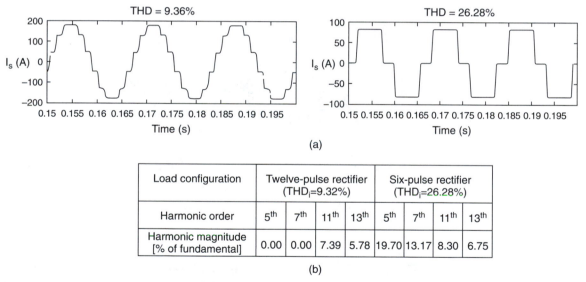

| Load configuration | Twelve-pulse rectifier ($\text{THD}_i$=9.32%) | | | | Six-pulse rectifier ($\text{THD}_i$=26.28%) | | | |
|---|---|---|---|---|---|---|---|---|
| Harmonic order | 5th | 7th | 11th | 13th | 5th | 7th | 11th | 13th |
| Harmonic magnitude [% of fundamental] | 0.00 | 0.00 | 7.39 | 5.78 | 19.70 | 13.17 | 8.30 | 6.75 |

(b)

**FIGURE E9.5.3** Load characteristics for a twelve-pulse rectifier and a six-pulse rectifier: (a) simulated power system current waveforms, (b) harmonic spectra.

controller. Owing to the use of the PI controller, the DC voltage will be maintained at its desired reference value at steady-state conditions. Finally, the desired voltage $v_a(t)$ (e.g., the summation of $v_{a1}(t)$ and $v_{a2}(t)$) serves as input to the PWM generator to produce the switching gate signals of the converter/inverter. With this arrangement, the converter will generate the desired output voltage.

Figure E9.5.5 illustrates the single-phase equivalent circuit of hybrid filter #2. The converter is modeled by the controllable voltage source $v_{ah}(t)$, and the nonlinear load (e.g., twelve- or six-pulse rectifier) is represented as the current source $i_{Lh}(t)$. $Z_{sh}$, $Z_{ph}$, and $Z_{hh}$ are the source, passive branch, and hybrid branch impedances, respectively. Harmonic currents of power system, passive filter, and hybrid

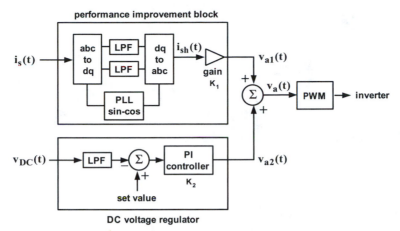

**FIGURE E9.5.4** Block diagram of the controller for hybrid filter #2. Details of the phase-lock-loop (PLL) block are provided in Fig. 11.4 (Chapter 11).

filter are determined by the following equations [14 in Chapter 9]:

$$i_{sh}(t) = \frac{1}{Z_{sh}}\left(\frac{1}{Z_{eq}} + \frac{1}{Z_{sh}Z_{hh}/K_1}\right)^{-1} i_{Lh}(t), \quad \text{(E9.5-3)}$$

$$Z_{eq} = \frac{Z_{ph}Z_{hh}Z_{sh}}{Z_{ph}Z_{hh} + Z_{sh}Z_{hh} + Z_{sh}Z_{ph}}, \quad \text{(E9.5-4)}$$

and

$$i_{ph}(t) = \frac{1}{Z_{ph}}\left(\frac{1}{Z_{eq}} + \frac{1}{Z_{sh}Z_{hh}/K_1}\right)^{-1} i_{Lh}(t), \quad \text{(E9.5-5)}$$

$$i_{hh}(t) = \frac{1}{Z_{hh}}\left(\frac{1}{Z_{hh}} + \frac{1}{Z_{sh}Z_{hh}/K_1}\right)^{-1} i_{Lh}(t). \quad \text{(E9.5-6)}$$

The power system impedance $Z_{sh}$ is usually inductive. Therefore, to avoid parallel resonance and to damp undesired resonant magnitudes, $Z_{hh}$ should be capacitive and the tuned frequency of the hybrid power filter must be larger than the tuned frequency of the passive filter.

In addition, $K_1$ acts as a resistor to damp parallel resonance between power system and passive filters. Figure E9.5.6 shows filtering characteristics for three selected cases (e.g., $K_1 = 0$, $K_1 = 1$, and $K_1 = 2$). It is found [14] that if $K_1 \neq 0$, the filtering characteristics are improved for all harmonic frequencies and no parallel resonance is encountered (Figs. E9.5.7 through E9.5.10).

**c) Current Waveforms with Hybrid Filters #1 and #2**

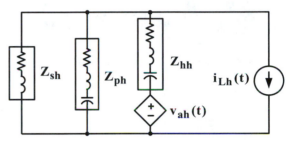

**FIGURE E9.5.5** Single-phase equivalent circuit of hybrid filter #2 at harmonic order h.

Hybrid filter #2 consists of a power converter in series with two passive filter branches, tuned at the dominant low-order harmonics (11th, 13th). Although a twelve-pulse rectifier does not produce significant 5th and 7th harmonic currents, a filter tuned at 11th and 13th harmonics cannot completely compensate the distortions of the power system current, because of series and parallel resonances between the line impedance and the passive filter branches that are likely to occur close to the 5th and 7th harmonic frequencies. For this reason, 5th and 7th LC tuned filters are installed to suppress 5th and 7th harmonic frequencies, respectively. In addition, this configuration will have more satisfactory results under the worst operating condition, that is, when only one rectifier is connected to the power system.

Simulated current waveforms for the worst case when only one six-pulse rectifier is activated and the passive branches of the hybrid power filter are tuned at the 5th and 7th harmonics (Table E9.5.1) are shown in Fig. E9.5.7. The rms and THD$_i$ values of

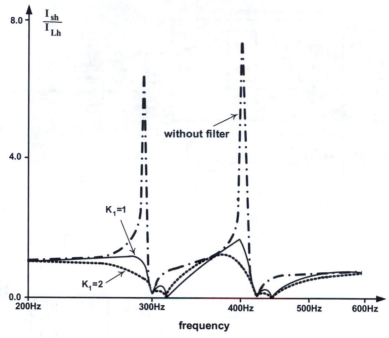

**FIGURE E9.5.6** System frequency responses of hybrid filter #2 for different values of $K_1$.

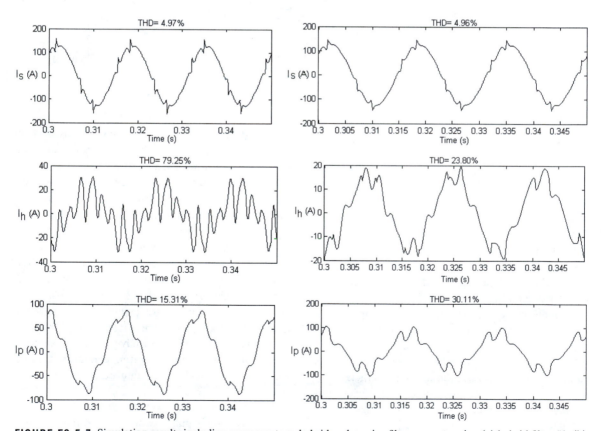

**FIGURE E9.5.7** Simulation results including power system, hybrid and passive filter currents using (a) hybrid filter #1, (b) hybrid filter #2.

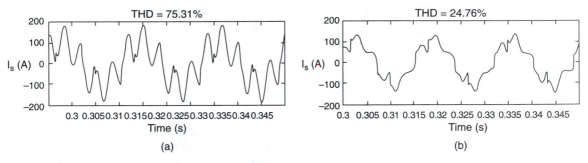

**FIGURE E9.5.8** Power system current under resonant condition (a) with no filter, (b) with hybrid filter #2 installed.

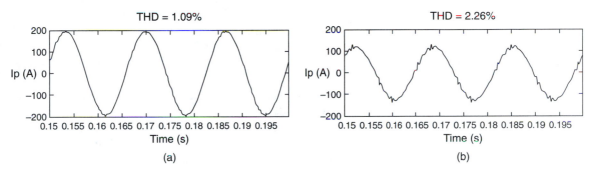

**FIGURE E9.5.9** Power system current waveform with hybrid filter #2 installed: (a) twelve-pulse rectifier load, (b) six-pulse rectifier load.

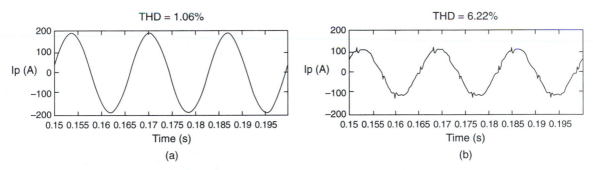

**FIGURE E9.5.10** Power system current waveform under resonance conditions with hybrid filter #2 installed: (a) twelve-pulse rectifier load, (b) six-pulse rectifier load.

**TABLE E9.5.1** Selected Parameters for Hybrid Filter #2 ($C_{DC\text{-}link}$ = 2500 μF, $V_{DC}$ = 80 V)

|  | Power system | Passive filter | | Hybrid filter | |
|---|---|---|---|---|---|
|  |  | 5th | 7th | 5th | 7th |
| L (mH) | 0.05 | 1.13 | 0.575 | 5.1 | 2.6 |
| C (μF) | – | 250 | 250 | 50 | 50 |
| R (Ω) | 0.003 | 0.01 | 0.01 | 0.02 | 0.02 |

**TABLE E9.5.2** RMS and $THD_i$ Values of Currents for Hybrid Filters #1 and #2

|  |  | RMS (A) | $THD_i$ (%) |  |  | RMS (A) | $THD_i$ (%) |
|---|---|---|---|---|---|---|---|
| Hybrid | $i_s$ | 91.36 | 4.97 | Hybrid | $i_s$ | 90.19 | 4.96 |
| filter #1 | $i_h$ | 17.74 | 79.25 | filter #2 | $i_h$ | 11.12 | 23.80 |
|  | $i_p$ | 55.20 | 15.31 |  | $i_p$ | 55.65 | 30.11 |

**TABLE E9.5.3** Selected Parameters of Hybrid Filter #2 for Resonant Compensation

| Filter parameter | 11th harmonic | 13th harmonic |
|---|---|---|
| L (mH) | 1.85 | 1.34 |
| C (μF) | 30 | 30 |
| R (Ω) | 0.02 | 0.02 |

these currents are compared in Table E9.5.2. The power system current for both hybrid filters are about the same. However, the magnitude of the hybrid filter current and the corresponding $THD_i$ is less when hybrid filter #2 is implemented, which results in a smaller inverter rating.

d) Resonance Conditions with Hybrid Filter #2

The performance of hybrid filter #2 is simulated and analyzed in the presence of resonance. To do so, the inductance and capacitance of the passive power filter (that were tuned for the 5th harmonic) are now modified (e.g., 125 $\mu F$, 2.2 mH) to produce a resonant condition. Figure E9.5.8 depicts waveforms of the power system current before and after the installation of hybrid filter #2 with the passive branches tuned at 5th and 7th harmonic frequencies, respectively. Note that the value of current $THD_i$ is decreased from 75.31 to 24.76%.

e) Comparison of Hybrid Filters #1 and #2

Although with hybrid filter #1 the $THD_i$ value of the power system current is limited, the individual magnitudes of 11th and 13th harmonic currents (e.g., 3.6%, 3.11%, respectively) are higher than the harmonic limits of IEEE-519 [11 in Chapter 9]. In addition, system $THD_i$ under resonant condition is much higher than the above-mentioned limits.

After the installation of hybrid filter #2 (with proper selection of filter parameters, Table E9.5.3), system power quality is considerably improved:

- With both six-pulse rectifiers activated, the $THD_i$ of the power system current is reduced from 9.36 (Fig. E9.5.3) to 1.09% (Fig. E9.5.9a).
- Under the worst operating conditions (e.g., with one six-pulse rectifier in service), the $THD_i$ of the power system current is reduced from 26.28 (Fig. E9.5.3) to 2.26% (Fig. E9.5.9b).
- Individual magnitudes of the 11th and 13th harmonic currents are within the allowable limits of IEEE-519 [11 in Chapter 9].
- Under resonant condition, the $THD_i$ of the power system currents (Fig. E9.5.10) are improved sig-

nificantly (e.g., from 6.38% and 75.31% to 1.06% and 6.22% for twelve-pulse and six-pulse rectifier loads, respectively).

## CHAPTER 10 SOLUTIONS

### 10.3.9 Application Example 10.1: Fuzzy Capacitor Placement in an 11 kV, 34-Bus Distribution System with Lateral Branches under Sinusoidal Operating Conditions: Solution

The following savings function is used to maximize the economic savings associated with capacitor placement (after a suitable bus is found):

$$S = K_P \Delta L_P + K_E \Delta L_E - K_C C \quad \text{(E10-1)}$$

where $\Delta L_P$ and $\Delta L_E$ are the loss reductions in peak demand and energy, respectively, due to capacitor installation. $C$ is the size of capacitor in kVAr, and $K_P = 120$, $K_E = 50$, $K_C = K_{cfp}$ (see Table 10.3) are the costs of peak demand, energy, and capacitors per kW, kWh, and kVAr, respectively. The savings function is based on annual savings and thus the cost of energy per hour is multiplied with the hours in a year, namely, 8760 hours.

The fuzzy algorithm consists of the following steps (Fig. 10.14):

**Step 1. Input System Parameters:** Input system parameters (Fig. E10.1.1, Tables E10.1.1 and E10.1.2). Run load flow and compute the savings function.

**Step 2. Fuzzification:** Fuzzy rules and membership functions (for power loss index (PLI), voltage (pu), and capacitor suitability) are shown in Table E10.1.3 and Fig. E10.1.2, respectively.

**Step 3. Inferencing:** For inferencing, the *Mamdani max-prod method* [50 in Chapter 10] is used to determine the capacitor suitability membership function $\mu_S(i)$. Therefore, the capacitor placement suitability membership function for bus $i$ becomes:

$$\mu_S = \max[\mu_P(i) \cdot \mu_V(i)], \quad \text{(E10-2)}$$

where, $\mu_p(i)$ and $\mu_V(i)$ are the membership functions of the power loss index and voltage level of bus $i$, respectively.

**Step 4. Defuzzification:** For defuzzification, the *center of gravity (COG)* method is used. There-

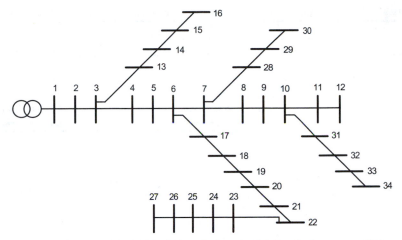

**FIGURE E10.1.1** Single-line diagram of the 11 kV, 34-bus, 3-phase, radial distribution system [32].

**TABLE E10.1.1** Bus Data [32] of the 11 kV, 34-Bus, 3-Phase, Radial Distribution System (Fig. E10.1.1)

| Bus number | Load | | Generator | | | | Injected |
|---|---|---|---|---|---|---|---|
| | MW | MVAr | MW | MVAr | $Q_{min}$ | $Q_{max}$ | MVAr |
| 1 | 0.0000 | 0.0000 | 0 | 0 | 0 | 0 | 0 |
| 2 | 0.2300 | 0.1425 | 0 | 0 | 0 | 0 | 0 |
| 3 | 0.0000 | 0.0000 | 0 | 0 | 0 | 0 | 0 |
| 4 | 0.2300 | 0.1425 | 0 | 0 | 0 | 0 | 0 |
| 5 | 0.2300 | 0.1425 | 0 | 0 | 0 | 0 | 0 |
| 6 | 0.0000 | 0.0000 | 0 | 0 | 0 | 0 | 0 |
| 7 | 0.0000 | 0.0000 | 0 | 0 | 0 | 0 | 0 |
| 8 | 0.2300 | 0.1425 | 0 | 0 | 0 | 0 | 0 |
| 9 | 0.2300 | 0.1425 | 0 | 0 | 0 | 0 | 0 |
| 10 | 0.0000 | 0.0000 | 0 | 0 | 0 | 0 | 0 |
| 11 | 0.2300 | 0.1425 | 0 | 0 | 0 | 0 | 0 |
| 12 | 0.1370 | 0.0840 | 0 | 0 | 0 | 0 | 0 |
| 13 | 0.0720 | 0.0450 | 0 | 0 | 0 | 0 | 0 |
| 14 | 0.0720 | 0.0450 | 0 | 0 | 0 | 0 | 0 |
| 15 | 0.0720 | 0.0450 | 0 | 0 | 0 | 0 | 0 |
| 16 | 0.0135 | 0.0075 | 0 | 0 | 0 | 0 | 0 |
| 17 | 0.2300 | 0.1425 | 0 | 0 | 0 | 0 | 0 |
| 18 | 0.2300 | 0.1425 | 0 | 0 | 0 | 0 | 0 |
| 19 | 0.2300 | 0.1425 | 0 | 0 | 0 | 0 | 0 |
| 20 | 0.2300 | 0.1425 | 0 | 0 | 0 | 0 | 0 |
| 21 | 0.2300 | 0.1425 | 0 | 0 | 0 | 0 | 0 |
| 22 | 0.2300 | 0.1425 | 0 | 0 | 0 | 0 | 0 |
| 23 | 0.2300 | 0.1425 | 0 | 0 | 0 | 0 | 0 |
| 24 | 0.2300 | 0.1425 | 0 | 0 | 0 | 0 | 0 |
| 25 | 0.2300 | 0.1425 | 0 | 0 | 0 | 0 | 0 |
| 26 | 0.2300 | 0.1425 | 0 | 0 | 0 | 0 | 0 |
| 27 | 0.1370 | 0.0850 | 0 | 0 | 0 | 0 | 0 |
| 28 | 0.0750 | 0.0480 | 0 | 0 | 0 | 0 | 0 |
| 29 | 0.0750 | 0.0480 | 0 | 0 | 0 | 0 | 0 |
| 30 | 0.0750 | 0.0480 | 0 | 0 | 0 | 0 | 0 |
| 31 | 0.0570 | 0.0345 | 0 | 0 | 0 | 0 | 0 |
| 32 | 0.0570 | 0.0345 | 0 | 0 | 0 | 0 | 0 |
| 33 | 0.0570 | 0.0345 | 0 | 0 | 0 | 0 | 0 |
| 34 | 0.0570 | 0.0345 | 0 | 0 | 0 | 0 | 0 |

**TABLE E10.1.2** Line Data [32] of the 11 kV, 34-Bus, 3-Phase, Radial Distribution System (Fig. E10.1.1)

| Line | | | | |
|---|---|---|---|---|
| From bus | To bus | $R_{line}$ (pu) | $X_{line}$ (pu) | Tap setting |
| 1 | 2 | 0.0967 | 0.0397 | 1 |
| 2 | 3 | 0.0886 | 0.0364 | 1 |
| 3 | 4 | 0.1359 | 0.0377 | 1 |
| 4 | 5 | 0.1236 | 0.0343 | 1 |
| 5 | 6 | 0.1236 | 0.0343 | 1 |
| 6 | 7 | 0.2598 | 0.0446 | 1 |
| 7 | 8 | 0.1732 | 0.0298 | 1 |
| 8 | 9 | 0.2598 | 0.0446 | 1 |
| 9 | 10 | 0.1732 | 0.0298 | 1 |
| 10 | 11 | 0.1083 | 0.0186 | 1 |
| 11 | 12 | 0.0866 | 0.1488 | 1 |
| 3 | 13 | 0.1299 | 0.0223 | 1 |
| 13 | 14 | 0.1732 | 0.0298 | 1 |
| 14 | 15 | 0.0866 | 0.1488 | 1 |
| 15 | 16 | 0.0433 | 0.0074 | 1 |
| 6 | 17 | 0.1483 | 0.0412 | 1 |
| 17 | 18 | 0.1359 | 0.0377 | 1 |
| 18 | 19 | 0.1718 | 0.0391 | 1 |
| 19 | 20 | 0.1562 | 0.0355 | 1 |
| 20 | 21 | 0.1562 | 0.0355 | 1 |
| 21 | 22 | 0.2165 | 0.0372 | 1 |
| 22 | 23 | 0.2165 | 0.0372 | 1 |
| 23 | 24 | 0.2598 | 0.0446 | 1 |
| 24 | 25 | 0.1732 | 0.0298 | 1 |
| 25 | 26 | 0.1083 | 0.0186 | 1 |
| 26 | 27 | 0.0866 | 0.1488 | 1 |
| 7 | 28 | 0.1299 | 0.0223 | 1 |
| 28 | 29 | 0.1299 | 0.0223 | 1 |
| 29 | 30 | 0.1299 | 0.0223 | 1 |
| 10 | 31 | 0.1299 | 0.0223 | 1 |
| 31 | 32 | 0.1732 | 0.0298 | 1 |
| 32 | 33 | 0.1299 | 0.0223 | 1 |
| 33 | 34 | 0.0866 | 0.1488 | 1 |

**TABLE E10.1.3** Decision Matrix for Application Example 10.1, Determining Suitable Capacitor Locations

| AND | | Voltage | | | | |
|---|---|---|---|---|---|---|
| | | Low | Low-normal | Normal | High-normal | High |
| Power loss | Low | *Low–medium* | *Low–medium* | *Low* | *Low* | *Low* |
| index (PLI) | Low–medium | *Medium* | *Low–medium* | *Low–medium* | *Low* | *Low* |
| | Medium | *High–medium* | *Medium* | *Low–medium* | *Low* | *Low* |
| | High–medium | *High–medium* | *High–medium* | *Medium* | *Low–medium* | *Low* |
| | High | High | *High–medium* | *Medium* | *Low–medium* | *Low–medium* |

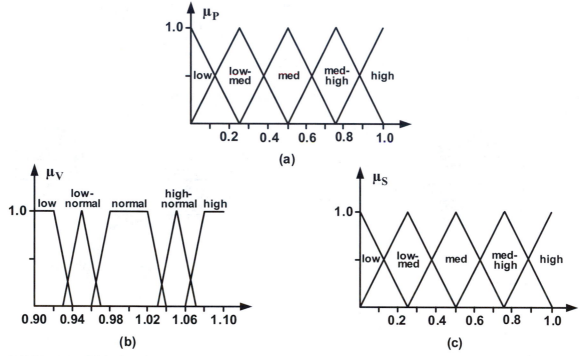

**FIGURE E10.1.2** Membership functions of Application Example 10.1 defined for (a) power loss index, (b) voltage (pu), and (c) capacitor suitability.

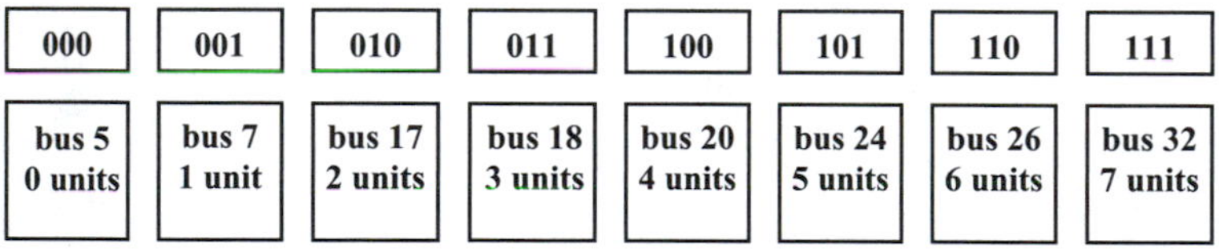

| 000 | 001 | 010 | 011 | 100 | 101 | 110 | 111 |
|---|---|---|---|---|---|---|---|
| bus 5 0 units | bus 7 1 unit | bus 17 2 units | bus 18 3 units | bus 20 4 units | bus 24 5 units | bus 26 6 units | bus 32 7 units |

**FIGURE E10.2.1** Selected chromosome structure for Application Example 10.2.

**TABLE E10.2.1** Binary Representation of Capacitor Bank Sizes for Application Example 10.2

| Binary representation | Number of 250 kVAr units | Capacitor bank size (kVAr) |
|---|---|---|
| 000 | 0 | 0 |
| 001 | 1 | 250 |
| 010 | 2 | 500 |
| 011 | 3 | 750 |
| 100 | 4 | 1000 |
| 101 | 5 | 1250 |
| 110 | 6 | 1500 |
| 111 | 7 | 1750 |

fore, the final capacitor placement suitability index is determined by the following formula:

$$S = \frac{\int \mu_S(z) \cdot z \, dz}{\int \mu_S(z) \cdot dz}, \quad (E10\text{-}3)$$

where $S$ is the suitability index and $\mu_S(z)$ is the capacitor placement suitability membership function.

**Step 5. Final Solution:** The final solution of the fuzzy placement and sizing of capacitor banks for the distribution system of Fig. E10.1.1 is shown in Table E10.2.2.

**TABLE E10.2.2** Simulation Results for Placement and Sizing of Capacitor Banks in the 34-Bus System (Fig. E10.1.1) under Sinusoidal Operating Conditions using Different Optimization Methods[a]

| | | After optimization | | | | |
|---|---|---|---|---|---|---|
| | Before optimization | Heuristic algorithm [1] | Expert systems [1] | Fuzzy [32] | Fuzzy (Section 10.3.10) | Genetic (Section 10.3.11) |
| Capacitor bank locations and sizes (kVAr) | No compensation | Q12 = 400 Q20 = 70 Q21 = 250 Q26 = 700 | Q5 = 750 Q22 = 900 | Q7 = 450 Q17 = 750 Q24 = 1500 | Q11 = 750 Q18 = 450 Q24 = 1350 | Q17 = 250 Q20 = 500 Q24 = 1000 Q32 = 500 |
| Total capacitance (kVAr) | $Q_t = 0$ | $Q_t = 1420$ | $Q_t = 1650$ | $Q_t = 2700$ | $Q_t = 2550$ | $Q_t = 2250$ |
| Minimum voltage (pu) | 0.942 | 0.948 | 0.947 | 0.951 | 0.951 | 0.950 |
| Maximum voltage (pu) | 1.000 | 1.000 | 1.000 | 1.000 | 1.000 | 1.000 |
| Average voltage (pu) | 0.966 | 0.969 | 0.968 | 0.971 | 0.971 | 0.971 |
| Total losses (kW) | 221.68 | 169.136 | 173.579 | 168.990 | 164.158 | 162.034 |
| Power loss reduction (kW) | 0 | 52.544 | 48.101 | 52.69 | 57.52 | 59.646 |
| Capacitor cost ($/year) | 0 | 710.00 | 825.00 | 1350.00 | 1275.00 | 1125.00 |
| Total cost ($/year) | 84859.10 | 65455.26 | 67271.04 | 66039.37 | 64114.68 | 63151.61 |
| Benefits ($/year) | | 19403.84 | 17588.06 | 18819.73 | 20744.42 | 21707.49 |

[a]All capacitors were removed before the optimization process.

### 10.3.10 Application Example 10.2: Genetically Optimized Placement of Capacitor Banks in an 11 kV, 34-Bus Distribution System with Lateral Branches under Sinusoidal Operating Conditions: Solution

The selected chromosome structure consists of N substrings (genes) of binary numbers (Fig. E10.2.1). N denotes the number of possible compensation (candidate) buses for shunt capacitor placement within the entire feeder. The binary numbers indicate the size of the installed capacitors at the bus under consideration. Candidate locations for capacitor placement are assumed to be buses 5, 7, 17, 18, 20, 24, 26, and 32. The 3-bit binary numbers (Fig. E10.2.1) indicate the size of the installed capacitor in terms of units. Only 250 kVAr banks are assumed to be available (Table E10.2.1).

The selected fitness function (incorporating energy and power losses in the system as well as the cost of the installed capacitor banks) is

$$\text{fitness} = K_e \sum_{i=1}^{n} T_i P_i + K_p p_O + \frac{K_c}{L} K_c \sum_{j=1}^{M} C_j, \quad \text{(E10-4)}$$

where $K_e = 0.03\$/kWh$, $K_p = 120\$/kW/year$, and $K_c = 5\$/kVAr$ are constants for energy, peak power, and capacitor costs, respectively. $P_i$ is the power loss at load level $i$ with a time duration $T_i$. $P_o$ represents the peak power loss of the system. $C_j$ represents the

size of the capacitor units to be installed at each of the M = 8 candidate buses. The constant denoted by L (= 10 years) represents the life of the capacitor. Therefore, a chromosome that consists of a large number of capacitor banks and a high energy and power loss will result in an undesirable high fitness value.

The genetic algorithm consists of the following steps (Fig. 10.3):

**Step 1. Input System Parameters:** Input system parameters (Fig. E10.1.1, Tables E10.1.1 and E10.1.2) and an initial random population with $N_{chrom}$ chromosomes.

**Step 2. Fitness Process:**

Step 2A: Run load flow algorithm for chromosome $N_{ch}$, save value of total active power loss and the computed fitness function (FF).

Step 2B: Calculate 1/FF for chromosome $N_{ch}$.

Step 2C: If $N_{ch} \le N_{chrom}$ go to Step 2A.

**Step 3. Reproduction Process:**

Step 3A: Define total fitness as a sum of values obtained in Step 2B for all chromosomes.

Step 3B: Select a percentage of the "roulette wheel" for each chromosome that is equal to the ratio of its fitness value to the total fitness value.

Step 3C: Normalize the values on the "roulette wheel" to be between zero and one.

Step 3D: Choose random numbers ($R_1$) between 0.0 and 1.0 to select an individual parent chromosome $N_{parent}$.

Step 3E: If $N_{parent} \leq N_{chrom}$ go to Step 3D.

**Step 4. Crossover Process:**

Step 4A: Select a random number ($R_2$) for mating two parent chromosomes.

Step 4B: If $R_2$ is between 0.6 and 1.0 then combine the two parents to generate two offsprings and go to step 4C. Else, transfer the chromosome with no crossover.

Step 4C: Repeat steps 4A to 4B for all chromosomes.

**Step 5. Mutation Process:**

Step 5A: Select a random number ($R_3$) for mutation of one chromosome.

Step 5B: If $R_3$ is between 0.01 and 0.1 then apply the mutation process at a random position in the string and go to Step 5C. Else, transfer the chromosome with no mutation.

Step 5C: Repeat Steps 5A to 5B for all chromosomes.

**Step 6. Updating Population:** Replace the old population with the new improved population generated by Steps 2 to 5.

**Step 7. Convergence:** If all chromosomes are the same or the maximum number of iterations have been reached ($N_{iterations} = N^{max}$) print the solution and stop, else go to Step 2.

**Step 8. Final Solution:** The final solution of the genetic placement and sizing of capacitor banks for the distribution system of Fig. E10.1.1 is shown in Table E10.2.2.

# Index